Applets for Data Analysis and Probability

Go to: highschool.bfwpub.com/spa3e and click "Go to Student Site"

Four new applets that progress from basic data analysis to statistical inference

- **One Quantitative Variable**
- **Two Quantitative Variables**
- **One Categorical Variable**
- **Two Categorical Variables**

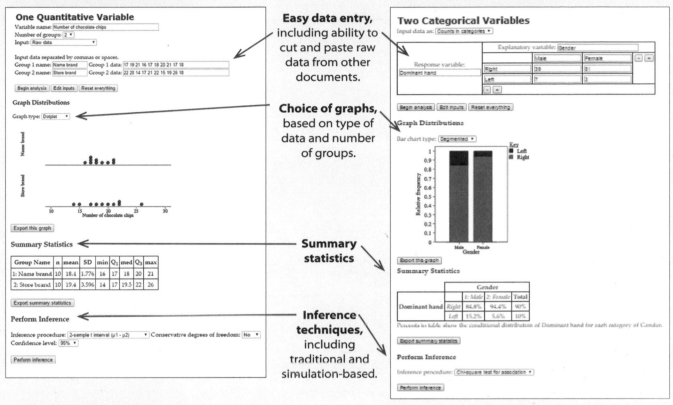

Easy data entry, including ability to cut and paste raw data from other documents.

Choice of graphs, based on type of data and number of groups.

Summary statistics

Inference techniques, including traditional and simulation-based.

A single probability applet that includes

- **Normal distributions**
- **Binomial distributions**
- **Discrete probability distributions**
- **Counting methods**

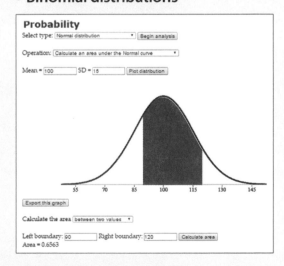

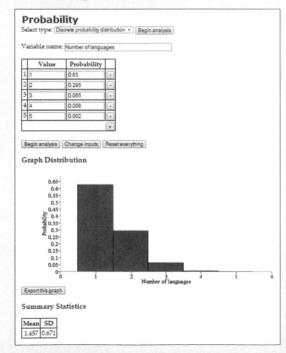

Statistics and Probability with Applications
Third Edition

DAREN STARNES
The Lawrenceville School

JOSH TABOR
Canyon del Oro High School

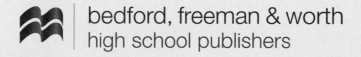
bedford, freeman & worth
high school publishers

VICE PRESIDENT, SOCIAL SCIENCE: Charles Linsmeier

SENIOR PUBLISHER, HIGH SCHOOL: Ann Heath

DEVELOPMENTAL EDITOR: Don Gecewicz

SENIOR MARKETING MANAGER: Lisa Erdely

MARKETING ASSISTANT: Peter Marucci

ASSOCIATE MEDIA EDITOR: Kimberly Morte

DIRECTOR OF DIGITAL PRODUCTION: Keri deManigold

SENIOR MEDIA PRODUCER: Alison Lorber

ASSOCIATE MEDIA PRODUCER: Hanna Squire

EDITORIAL ASSISTANT: Corrina Santos

DIRECTOR OF DESIGN, CONTENT MANAGEMENT: Diana Blume

COVER AND INTERIOR DESIGN: Marsha Cohen

DIRECTOR, CONTENT MANAGEMENT ENHANCEMENT: Tracey Kuehn

MANAGING EDITOR: Lisa Kinne

SENIOR PROJECT EDITOR: Vivien Weiss

PRODUCTION MANAGER: Susan Wein

ILLUSTRATIONS: codeMantra

SENIOR PHOTO EDITOR: Robin Fadool

PHOTO RESEARCH ASSISTANT: Candice Cheesman

ART MANAGER: Matthew McAdams

COMPOSITION: codeMantra

PRINTING AND BINDING: LSC Communications

COVER PHOTO CREDIT: Dorien Brouwers/Animal Photography/Alamy

Photo credits for design elements used throughout the book:

PEACOCK HEAD: Marcus Lindstrom/iStock/Getty Images

PEACOCK FEATHERS: Michael Fitzsimmons/iStock/Getty Images

Library of Congress Control Number: 2016946696
ISBN-13: 978-1-4641-2216-3
ISBN-10: 1-4641-2216-4

W. H. Freeman and Company
Bedford, Freeman & Worth
One New York Plaza
Suite 4500
New York, NY 10004-1562
http://www.highschool.bfwpub.com/catalog

BRIEF CONTENTS

About the Authors / xv
Acknowledgments / xviii
To the Student / xxi

Chapter 1   ANALYZING ONE-VARIABLE DATA   **2**

Lesson 1.1   Statistics: The Science and Art of Data / 4
Lesson 1.2   Displaying Categorical Data / 11
Lesson 1.3   Displaying Quantitative Data: Dotplots / 21
Lesson 1.4   Displaying Quantitative Data: Stemplots / 30
Lesson 1.5   Displaying Quantitative Data: Histograms / 38
Lesson 1.6   Measuring Center / 49
Lesson 1.7   Measuring Variability / 58
Lesson 1.8   Summarizing Quantitative Data: Boxplots and Outliers / 67
Lesson 1.9   Describing Location in a Distribution / 77

Chapter 2   ANALYZING TWO-VARIABLE DATA   **94**

Lesson 2.1   Relationships Between Two Categorical Variables / 96
Lesson 2.2   Relationships Between Two Quantitative Variables / 105
Lesson 2.3   Correlation / 113
Lesson 2.4   Calculating the Correlation / 121
Lesson 2.5   Regression Lines / 130
Lesson 2.6   The Least-Squares Regression Line / 137
Lesson 2.7   Assessing a Regression Model / 149
Lesson 2.8   Fitting Models to Curved Relationships / 160

Chapter 3   COLLECTING DATA   **180**

Lesson 3.1   Introduction to Data Collection / 182
Lesson 3.2   Sampling: Good and Bad / 187
Lesson 3.3   Simple Random Samples / 193
Lesson 3.4   Estimating a Margin of Error / 204
Lesson 3.5   Sampling and Surveys / 214
Lesson 3.6   Observational Studies and Experiments / 220
Lesson 3.7   How to Experiment Well / 228
Lesson 3.8   Inference for Experiments / 235
Lesson 3.9   Using Studies Wisely / 245

Chapter 4   PROBABILITY   **260**

Lesson 4.1   Randomness, Probability, and Simulation / 262
Lesson 4.2   Basic Probability Rules / 270
Lesson 4.3   Two-Way Tables and Venn Diagrams / 277
Lesson 4.4   Conditional Probability and Independence / 286
Lesson 4.5   The General Multiplication Rule and Tree Diagrams / 294
Lesson 4.6   The Multiplication Rule for Independent Events / 302
Lesson 4.7   The Multiplication Counting Principle and Permutations / 309
Lesson 4.8   Combinations and Probability / 317

Chapter 5   RANDOM VARIABLES   **330**

Lesson 5.1   Two Types of Random Variables / 332
Lesson 5.2   Analyzing Discrete Random Variables / 338

Lesson 5.3   Binomial Random Variables / 348
Lesson 5.4   Analyzing Binomial Random Variables / 356
Lesson 5.5   Continuous Random Variables / 365
Lesson 5.6   The Standard Normal Distribution / 375
Lesson 5.7   Normal Distribution Calculations / 384

Chapter 6   **SAMPLING DISTRIBUTIONS**   **398**

Lesson 6.1   What Is a Sampling Distribution? / 400
Lesson 6.2   Sampling Distributions: Center and Variability / 409
Lesson 6.3   The Sampling Distribution of a Sample Count
              (The Normal Approximation to the Binomial) / 417
Lesson 6.4   The Sampling Distribution of a Sample Proportion / 424
Lesson 6.5   The Sampling Distribution of a Sample Mean / 432
Lesson 6.6   The Central Limit Theorem / 439

Chapter 7   **ESTIMATING A PARAMETER**   **450**

Lesson 7.1   The Idea of a Confidence Interval / 452
Lesson 7.2   What Affects the Margin of Error? / 458
Lesson 7.3   Estimating a Proportion / 464
Lesson 7.4   Confidence Intervals for a Proportion / 470
Lesson 7.5   Estimating a Mean / 476
Lesson 7.6   Confidence Intervals for a Mean / 484

Chapter 8   **TESTING A CLAIM**   **498**

Lesson 8.1   The Idea of a Significance Test / 500
Lesson 8.2   Significance Tests and Decision Making / 508
Lesson 8.3   Testing a Claim about a Proportion / 515
Lesson 8.4   Significance Tests for a Proportion / 521
Lesson 8.5   Testing a Claim about a Mean / 529
Lesson 8.6   Significance Tests for a Mean / 536

Chapter 9   **COMPARING TWO POPULATIONS OR TREATMENTS**   **550**

Lesson 9.1   Estimating a Difference Between Two Proportions / 552
Lesson 9.2   Testing a Claim about a Difference Between Two Proportions / 561
Lesson 9.3   Estimating a Difference Between Two Means / 571
Lesson 9.4   Testing a Claim about a Difference Between Two Means / 581
Lesson 9.5   Analyzing Paired Data: Estimating a Mean Difference / 591
Lesson 9.6   Testing a Claim about a Mean Difference / 601

Chapter 10   **INFERENCE FOR DISTRIBUTIONS AND RELATIONSHIPS**   **618**

Lesson 10.1   Testing the Distribution of a Categorical Variable / 620
Lesson 10.2   Chi-Square Tests for Goodness of Fit / 627
Lesson 10.3   Testing the Relationship Between Two Categorical Variables / 636
Lesson 10.4   Chi-Square Tests for Association / 645
Lesson 10.5   Testing the Relationship Between Two Quantitative Variables / 655
Lesson 10.6   Inference for the Slope of a Least-Squares Regression Line / 667

Notes and Data Sources   N-1
Solutions   S-1
Glossary/Glosario   G-0
Index   I-1

Table A   **Standard Normal Probabilities / T-1**
Table B   ***t* Distribution Critical Values / T-3**
Table C   **Chi-Square Distribution Critical Values / T-4**
Table D   **Random Digits / T-5**

CONTENTS

About the Authors / xv
Acknowledgments / xviii
To the Student / xxi

| Chapter 1 | ANALYZING ONE-VARIABLE DATA | 2 |

Lesson 1.1   STATISTICS: THE SCIENCE AND ART OF DATA / 4
Classifying Data 5
Summarizing Data 7
Lesson App 1.1   What are my classmates like? 8
Lesson Exercises 9

Lesson 1.2   DISPLAYING CATEGORICAL DATA / 11
Bar Charts and Pie Charts 12
Comparing Distributions with Bar Charts 14
Graphs: Good and Bad 14
Lesson App 1.2   Which cell phone speaks to you? 16
Lesson Exercises 17

Lesson 1.3   DISPLAYING QUANTITATIVE DATA: DOTPLOTS / 21
Making and Interpreting Dotplots 21
Describing Shape 23
Describing and Comparing Distributions 24
Lesson App 1.3   How can we check the health of a stream? 26
Lesson Exercises 27

Lesson 1.4   DISPLAYING QUANTITATIVE DATA: STEMPLOTS / 30
Making and Interpreting Stemplots 30
Comparing Distributions with Stemplots 33
Lesson App 1.4   How many shoes are too many shoes? 34
Lesson Exercises 35

Lesson 1.5   DISPLAYING QUANTITATIVE DATA: HISTOGRAMS / 38
Making and Interpreting Histograms 38
Comparing Distributions with Histograms 42
Lesson App 1.5   How old are U.S. presidents? 42
Lesson Exercises 44

Lesson 1.6   MEASURING CENTER / 49
The Median 49
The Mean 51
Comparing the Mean and the Median 53
Lesson App 1.6   Is the pace of life slower in smaller cities? 55
Lesson Exercises 55

Lesson 1.7   MEASURING VARIABILITY / 58
The Range 59
The Interquartile Range 60
The Standard Deviation 61
Lesson App 1.7   Have we found the beef? 64
Lesson Exercises 65

Lesson 1.8   **SUMMARIZING QUANTITATIVE DATA: BOXPLOTS AND OUTLIERS / 67**

Identifying Outliers 68
Making and Interpreting Boxplots 69
Comparing Distributions with Boxplots 72

Lesson App 1.8   Which is best at reducing stress? 73
Lesson Exercises 74

Lesson 1.9   **DESCRIBING LOCATION IN A DISTRIBUTION / 77**

Finding and Interpreting Percentiles 78
Cumulative Relative Frequency Graphs 79
Finding and Interpreting Standardized Scores (*z*-Scores) 81

Lesson App 1.9   Which states are rich? 82
Lesson Exercises 83

Chapter 1   **MAIN POINTS / 86**
Chapter 1   **REVIEW EXERCISES / 88**
Chapter 1   **PRACTICE TEST / 90**

Chapter 2   **ANALYZING TWO-VARIABLE DATA**   **94**

Lesson 2.1   **RELATIONSHIPS BETWEEN TWO CATEGORICAL VARIABLES / 96**

Displaying Relationships Between Two Categorical Variables 96
Association 99

Lesson App 2.1   Which finger is longer? 100
Lesson Exercises 101

Lesson 2.2   **RELATIONSHIPS BETWEEN TWO QUANTITATIVE VARIABLES / 105**

Distinguishing Explanatory and Response Variables 105
Making a Scatterplot 106
Describing a Scatterplot 107

Lesson App 2.2   More sugar, more calories? 108
Lesson Exercises 110

Lesson 2.3   **CORRELATION / 113**

Estimating and Interpreting the Correlation 114
Correlation and Cause and Effect 116

Lesson App 2.3   If I eat more chocolate, will I win a Nobel Prize? 117
Lesson Exercises 118

Lesson 2.4   **CALCULATING THE CORRELATION / 121**

Calculating the Correlation 121
Properties of the Correlation 123
Outliers and Correlation 124

Lesson App 2.4   Flying dinosaur or early bird? 126
Lesson Exercises 127

Lesson 2.5   **REGRESSION LINES / 130**

Making Predictions 131
Residuals 132
Interpreting a Regression Line 133

Lesson App 2.5   Do cut flowers benefit from sugar in the water? 134
Lesson Exercises 134

Lesson 2.6   **THE LEAST-SQUARES REGRESSION LINE / 137**

Calculating the Equation of the Least-Squares Regression Line 138
Calculating the Equation of the Least-Squares Regression Line Using Summary Statistics 141
Outliers and the Least-Squares Regression Line 142

Lesson App 2.6   Did the Broncos buck the trend? 143
Lesson Exercises 144

Lesson 2.7   **ASSESSING A REGRESSION MODEL / 149**

Residual Plots 149
Standard Deviation of the Residuals 151
The Coefficient of Determination r^2 152

Lesson App 2.7   Do higher priced tablets have better battery life? 154
Lesson Exercises 156

Lesson 2.8   **FITTING MODELS TO CURVED RELATIONSHIPS / 160**

Quadratic Models 160
Exponential Models 164
Choosing a Model 167

Lesson App 2.8   How does life insurance work? 168
Lesson Exercises 169

Chapter 2 **MAIN POINTS / 174**
Chapter 2 **REVIEW EXERCISES / 176**
Chapter 2 **PRACTICE TEST / 178**

Chapter 3   **COLLECTING DATA**   **180**

Lesson 3.1   **INTRODUCTION TO DATA COLLECTION / 182**

Asking Statistical Questions 182
Populations and Samples 183
Observational Studies and Experiments 184

Lesson App 3.1   Do you have dinner plans? 185
Lesson Exercises 185

Lesson 3.2   **SAMPLING: GOOD AND BAD / 187**

How to Sample Poorly: Convenience Samples 188
How to Sample Poorly: Voluntary Response Samples 189
How to Sample Well: Random Samples 190

Lesson App 3.2   Still on the phone? 191
Lesson Exercises 191

Lesson 3.3   **SIMPLE RANDOM SAMPLES / 193**

Choosing a Simple Random Sample 194
Sampling Variability 196
Inference for Sampling 197

Lesson App 3.3   Do you tweet? 199
Lesson Exercises 200

Lesson 3.4   **ESTIMATING A MARGIN OF ERROR / 204**

Margin of Error: Estimating a Population Proportion 204
Margin of Error: Estimating a Population Mean 206

Lesson App 3.4   Can you roll your tongue? 209
Lesson Exercises 210

Lesson 3.5   **SAMPLING AND SURVEYS / 214**

Undercoverage 214
Nonresponse 215
Other Sources of Bias 216
How to Survey Well 217

Lesson App 3.5   Who did you say is calling? *Literary Digest?* 217
Lesson Exercises 218

Lesson 3.6   **OBSERVATIONAL STUDIES AND EXPERIMENTS / 220**

Confounding 221
Experiments: Comparison 222
Experiments: The Placebo Effect 223

Lesson App 3.6   What happens when physicians study themselves? 225
Lesson Exercises 225

Lesson 3.7   **HOW TO EXPERIMENT WELL / 228**

Experiments: Random Assignment 229
Experiments: Other Sources of Variability 231

Lesson App 3.7   Multitasking? Or multiple distractions? 232
Lesson Exercises 232

Lesson 3.8   **INFERENCE FOR EXPERIMENTS / 235**

Completely Randomized Designs 235
Statistical Significance 237
Determining Statistical Significance with Simulation 238

Lesson App 3.8   Does fish oil affect blood pressure? 240
Lesson Exercises 241

Lesson 3.9   **USING STUDIES WISELY / 245**

The Scope of Inference 245
Data Ethics 247

Lesson App 3.9   Is foster care better for children than an orphanage? 249
Lesson Exercises 249

Chapter 3   **MAIN POINTS / 253**
Chapter 3   **REVIEW EXERCISES / 255**
Chapter 3   **PRACTICE TEST / 257**

Chapter 4   **PROBABILITY**   **260**

Lesson 4.1   **RANDOMNESS, PROBABILITY, AND SIMULATION / 262**

The Idea of Probability 262
Myths about Randomness 264
Simulation 265

Lesson App 4.1   Will the train arrive on time? 266
Lesson Exercises 267

Lesson 4.2   **BASIC PROBABILITY RULES / 270**

Probability Models 270
Basic Probability Rules 271

Lesson App 4.2   How prevalent is high cholesterol? 274
Lesson Exercises 274

Lesson 4.3   **TWO-WAY TABLES AND VENN DIAGRAMS / 277**

Two-Way Tables and the General Addition Rule 277
Venn Diagrams and Probability 279

Lesson App 4.3   Who owns a home? 282
Lesson Exercises 283

Lesson 4.4   **CONDITIONAL PROBABILITY AND INDEPENDENCE / 286**

What Is Conditional Probability? 286
Conditional Probability and Independence 289

Lesson App 4.4   Who earns A's in college? 291
Lesson Exercises 291

Lesson 4.5   **THE GENERAL MULTIPLICATION RULE AND TREE DIAGRAMS / 294**

The General Multiplication Rule 294
Tree Diagrams and Conditional Probability 296

Lesson App 4.5   Not milk? 299
Lesson Exercises 300

Lesson 4.6   **THE MULTIPLICATION RULE FOR INDEPENDENT EVENTS / 302**
Calculating Probabilities with the Multiplication Rule for Independent Events 303
Use the Multiplication Rule for Independent Events Wisely 305
Lesson App 4.6   How should we interpret genetic screening? 306
Lesson Exercises 307

Lesson 4.7   **THE MULTIPLICATION COUNTING PRINCIPLE AND PERMUTATIONS / 309**
The Multiplication Counting Principle 309
Permutations 311
Lesson App 4.7   Do you scream for ice cream? 313
Lesson Exercises 315

Lesson 4.8   **COMBINATIONS AND PROBABILITY / 317**
Combinations 317
Counting and Probability 319
Lesson App 4.8   How many ways can you set up an iPod play list? 321
Lesson Exercises 322

Chapter 4   **MAIN POINTS / 325**
Chapter 4   **REVIEW EXERCISES / 327**
Chapter 4   **PRACTICE TEST / 328**

Chapter 5   **RANDOM VARIABLES**   **330**

Lesson 5.1   **TWO TYPES OF RANDOM VARIABLES / 332**
Discrete Random Variables 332
Finding Probabilities for Discrete Random Variables 334
Continuous Random Variables 335
Lesson App 5.1   Making the grade? 336
Lesson Exercises 336

Lesson 5.2   **ANALYZING DISCRETE RANDOM VARIABLES / 338**
Displaying Discrete Probability Distributions: Histograms and Shape 338
Measuring Center: The Mean (Expected Value) of a Discrete Random Variable 339
Measuring Variability: The Standard Deviation of a Discrete Random Variable 341
Lesson App 5.2   How much do college grades vary? 343
Lesson Exercises 345

Lesson 5.3   **BINOMIAL RANDOM VARIABLES / 348**
Binomial Settings 348
Calculating Binomial Probabilities 350
Binomial Distributions and Shape 351
Lesson App 5.3   Is the train binomial? 352
Lesson Exercises 354

Lesson 5.4   **ANALYZING BINOMIAL RANDOM VARIABLES / 356**
The Mean and Standard Deviation of a Binomial Distribution 356
Cumulative Binomial Probabilities 358
Lesson App 5.4   Free lunch? 361
Lesson Exercises 362

Lesson 5.5   **CONTINUOUS RANDOM VARIABLES / 365**
Finding Probabilities for Continuous Random Variables 366
The Mean and Median of a Continuous Random Variable 368
Normal Distributions 369
Lesson App 5.5   Still waiting for the server? 371
Lesson Exercises 372

Lesson 5.6　　**THE STANDARD NORMAL DISTRIBUTION / 375**
The 68–95–99.7 Rule 375
The Standard Normal Distribution 378
Lesson App 5.6　What's a good batting average? 381
Lesson Exercises 382

Lesson 5.7　　**NORMAL DISTRIBUTION CALCULATIONS / 384**
Calculating Probabilities 384
Finding Values from Probabilities 387
Lesson App 5.7　What cholesterol levels are unhealthy for teen boys? 389
Lesson Exercises 390

Chapter 5　**MAIN POINTS / 393**
Chapter 5　**REVIEW EXERCISES / 395**
Chapter 5　**PRACTICE TEST / 396**

Chapter 6　　**SAMPLING DISTRIBUTIONS**　　　　　　　　　　　　**398**

Lesson 6.1　　**WHAT IS A SAMPLING DISTRIBUTION? / 400**
Parameters and Statistics 401
Sampling Distributions 402
Using Sampling Distributions to Evaluate Claims 403
Lesson App 6.1　How cold is it inside the cabin? 404
Lesson Exercises 405

Lesson 6.2　　**SAMPLING DISTRIBUTIONS: CENTER AND VARIABILITY / 409**
Unbiased Estimators 409
Sampling Variability 411
Putting It All Together: Center and Variability 412
Lesson App 6.2　How many tanks does the enemy have? 413
Lesson Exercises 414

Lesson 6.3　　**THE SAMPLING DISTRIBUTION OF A SAMPLE COUNT (THE NORMAL APPROXIMATION TO THE BINOMIAL) / 417**
Center and Variability 418
Shape 419
Finding Probabilities Involving X 421
Lesson App 6.3　How can we check for bias in a survey? 422
Lesson Exercises 422

Lesson 6.4　　**THE SAMPLING DISTRIBUTION OF A SAMPLE PROPORTION / 424**
Center and Variability 425
Shape 427
Finding Probabilities Involving $\hat{p}$ 428
Lesson App 6.4　What's that spot on my potato chip? 429
Lesson Exercises 430

Lesson 6.5　　**THE SAMPLING DISTRIBUTION OF A SAMPLE MEAN / 432**
Center and Variability 433
Shape 434
Finding Probabilities Involving $\bar{x}$ 435
Lesson App 6.5　Are college women taller? 436
Lesson Exercises 437

Lesson 6.6　　**THE CENTRAL LIMIT THEOREM / 439**
The Central Limit Theorem 440
Probabilities Involving $\bar{x}$ 441
Lesson App 6.6　Keeping things cool with statistics? 442
Lesson Exercises 443

Chapter 6   **MAIN POINTS / 445**
Chapter 6   **REVIEW EXERCISES / 447**
Chapter 6   **PRACTICE TEST / 448**

Chapter 7	ESTIMATING A PARAMETER	450

Lesson 7.1   **THE IDEA OF A CONFIDENCE INTERVAL / 452**

Interpreting Confidence Intervals 453
Building a Confidence Interval 454
Using Confidence Intervals to Make Decisions 455

Lesson App 7.1   Do you approve of the president's job performance? 455
Lesson Exercises 456

Lesson 7.2   **WHAT AFFECTS THE MARGIN OF ERROR? / 458**

Interpreting Confidence Level 459
Factors That Affect the Margin of Error 459
What the Margin of Error Doesn't Account For 461

Lesson App 7.2   Do you like my photos? 461
Lesson Exercises 462

Lesson 7.3   **ESTIMATING A PROPORTION / 464**

Conditions for Estimating p 464
Critical Values 466
Calculating a Confidence Interval for p 467

Lesson App 7.3   Do you know your government? 468
Lesson Exercises 469

Lesson 7.4   **CONFIDENCE INTERVALS FOR A PROPORTION / 470**

Putting It All Together: The Four-Step Process 471
Determining the Sample Size 472

Lesson App 7.4   TV in bed? 473
Lesson Exercises 474

Lesson 7.5   **ESTIMATING A MEAN / 476**

Conditions for Estimating μ 477
The Problem of Unknown σ 477
t^* Critical Values 479
Calculating a Confidence Interval for μ 480

Lesson App 7.5   What does an Oreo weigh? 481
Lesson Exercises 482

Lesson 7.6   **CONFIDENCE INTERVALS FOR A MEAN / 484**

The Normal/Large Sample Condition 484
Putting It All Together: The Four-Step Process 485

Lesson App 7.6   How tense are the video screens? 487
Lesson Exercises 488

Chapter 7   **MAIN POINTS / 492**
Chapter 7   **REVIEW EXERCISES / 494**
Chapter 7   **PRACTICE TEST / 495**

Chapter 8	TESTING A CLAIM	498

Lesson 8.1   **THE IDEA OF A SIGNIFICANCE TEST / 500**

Stating Hypotheses 501
Interpreting *P*-Values 503
Making Conclusions 504

Lesson App 8.1   Do people kiss the "right" way? 505
Lesson Exercises 505

Lesson 8.2   **SIGNIFICANCE TESTS AND DECISION MAKING / 508**
Determining Statistical Significance 508
Type I and Type II Errors 509

Lesson App 8.2   Are these potatoes keepers? 512
Lesson Exercises 512

Lesson 8.3   **TESTING A CLAIM ABOUT A PROPORTION / 515**
Conditions for Testing a Claim about p 515
Calculations: Standardized Test Statistic and P-Value 516

Lesson App 8.3   Is it better to be last? 519
Lesson Exercises 519

Lesson 8.4   **SIGNIFICANCE TESTS FOR A PROPORTION / 521**
Putting It All Together: The Four-Step Process 521
Two-Sided Tests 523

Lesson App 8.4   Who feels job stress? 525
Lesson Exercises 527

Lesson 8.5   **TESTING A CLAIM ABOUT A MEAN / 529**
Conditions for Testing a Claim about μ 529
Calculating the Standardized Test Statistic 530
Finding P-Values 531

Lesson App 8.5   Who needs an aspirin? 534
Lesson Exercises 534

Lesson 8.6   **SIGNIFICANCE TESTS FOR A MEAN / 536**
Putting It All Together: Testing a Claim about a Population Mean 536

Lesson App 8.6   Do our employees have high blood pressure? 541
Lesson Exercises 541

Chapter 8   **MAIN POINTS / 545**
Chapter 8   **REVIEW EXERCISES / 547**
Chapter 8   **PRACTICE TEST / 548**

Chapter 9   COMPARING TWO POPULATIONS OR TREATMENTS   550

Lesson 9.1   **ESTIMATING A DIFFERENCE BETWEEN TWO PROPORTIONS / 552**
The Sampling Distribution of a Difference Between Two Proportions 552
Conditions for Estimating $p_1 - p_2$ 554
Constructing and Interpreting a Confidence Interval for $p_1 - p_2$ 555

Lesson App 9.1   Who likes rap music more? 557
Lesson Exercises 558

Lesson 9.2   **TESTING A CLAIM ABOUT A DIFFERENCE BETWEEN TWO PROPORTIONS / 561**
Stating Hypotheses and Checking Conditions for a Test about $p_1 - p_2$ 561
Calculations: Standardized Test Statistic and P-Value 562
Performing a Test about $p_1 - p_2$ 564

Lesson App 9.2   Does taking aspirin help prevent heart attacks? 566
Lesson Exercises 567

Lesson 9.3   **ESTIMATING A DIFFERENCE BETWEEN TWO MEANS / 571**
The Sampling Distribution of a Difference Between Two Means 571
Conditions for Estimating $\mu_1 - \mu_2$ 573
Constructing and Interpreting a Confidence Interval for $\mu_1 - \mu_2$ 574

Lesson App 9.3   Do bigger apartments cost more money? 576
Lesson Exercises 578

Lesson 9.4   **TESTING A CLAIM ABOUT A DIFFERENCE BETWEEN TWO MEANS / 581**

Stating Hypotheses and Checking Conditions for a Test about $\mu_1 - \mu_2$ 582
Calculations: Standardized Test Statistic and *P*-Value 583
Performing a Test about $\mu_1 - \mu_2$ 585

Lesson App 9.4   Is name-brand popcorn better than store-brand
popcorn? 587
Lesson Exercises 587

Lesson 9.5   **ANALYZING PAIRED DATA: ESTIMATING A MEAN DIFFERENCE / 591**

Analyzing Paired Data 592
Constructing and Interpreting a Confidence Interval for μ_{diff} 596

Lesson App 9.5   Is caffeine dependence real? 598
Lesson Exercises 599

Lesson 9.6   **TESTING A CLAIM ABOUT A MEAN DIFFERENCE / 601**

Performing a Test about μ_{diff} 602
Paired Data or Two Samples? 605

Lesson App 9.6   Does generic ice cream melt faster? 607
Lesson Exercises 607

Chapter 9   **MAIN POINTS / 611**
Chapter 9   **REVIEW EXERCISES / 613**
Chapter 9   **PRACTICE TEST / 615**

Chapter 10   INFERENCE FOR DISTRIBUTIONS AND RELATIONSHIPS   618

Lesson 10.1   **TESTING THE DISTRIBUTION OF A CATEGORICAL VARIABLE / 620**

Stating Hypotheses 621
Calculating Expected Counts 622
The Chi-Square Test Statistic 623

Lesson App 10.1   Are fruit flies predictable? 625
Lesson Exercises 626

Lesson 10.2   **CHI-SQUARE TESTS FOR GOODNESS OF FIT / 627**

Conditions for a Chi-Square Test for Goodness of Fit 628
Calculating *P*-values 628
The Chi-Square Test for Goodness of Fit 631

Lesson App 10.2   Is this die fair? 632
Lesson Exercises 634

Lesson 10.3   **TESTING THE RELATIONSHIP BETWEEN TWO CATEGORICAL
VARIABLES / 636**

Stating Hypotheses 637
Calculating Expected Counts 638
The Chi-Square Test Statistic 640

Lesson App 10.3   Is there an association between gender
and superpower preference? 642
Lesson Exercises 642

Lesson 10.4   **CHI-SQUARE TESTS FOR ASSOCIATION / 645**

Conditions for a Chi-Square Test for Association 645
Calculating *P*-Values 646
Performing a Chi-Square Test for Association 648

Lesson App 10.4   Should angry people go to the sauna? 650
Lesson Exercises 652

Lesson 10.5   **TESTING THE RELATIONSHIP BETWEEN TWO QUANTITATIVE VARIABLES / 655**

Stating Hypotheses 656
Checking Conditions for a Test about the Slope 657
Calculating the Test Statistic and P-value 659

Lesson App 10.5   Do beavers benefit beetles? 661
Lesson Exercises 662

Lesson 10.6   **INFERENCE FOR THE SLOPE OF A LEAST-SQUARES REGRESSION LINE / 667**
Calculating the Test Statistic and P-Value Using Technology 667
Performing a t Test for the Slope of a Least-Squares Regression Line 670
Confidence Intervals for the Slope of a Least-Squares Regression Line 671

Lesson App 10.6   How fit can you get? 674
Lesson Exercises 675

Chapter 10   **MAIN POINTS / 679**
Chapter 10   **REVIEW EXERCISES / 680**
Chapter 10   **PRACTICE TEST / 682**

EXTRA   LESSONS

These eleven extra lessons are available online at www.highschool.bfwpub.com/spa3e.

Lesson 1.9A   **TRANSFORMING DATA**

Lesson 2.2A   **TIMEPLOTS**

Lesson 3.5A   **OTHER RANDOM SAMPLING METHODS**

Lesson 3.8A   **BLOCKING**

Lesson 5.7A   **ASSESSING NORMALITY**

Lesson 5.7B   **TRANSFORMING RANDOM VARIABLES**

Lesson 5.7C   **COMBINING RANDOM VARIABLES**

Lesson 8.6A   **POWER OF A TEST**

Lesson 9.6A   **TESTING A CLAIM ABOUT PAIRED DATA: NONPARAMETRIC TESTS**

Lesson 10.6A   **MULTIPLE REGRESSION, PART 1**

Lesson 10.6B   **MULTIPLE REGRESSION, PART 2**

NOTES AND DATA SOURCES / N-1

SOLUTIONS / S-1

GLOSSARY/GLOSARIO / G-0

INDEX / I-1

TABLES

Table A   STANDARD NORMAL PROBABILITIES / T-1

Table B t DISTRIBUTION CRITICAL VALUES / T-3

Table C   CHI-SQUARE DISTRIBUTION CRITICAL VALUES / T-4

Table D   RANDOM DIGITS / T-5

Ann Heath

DAREN S. STARNES is Mathematics Department Chair and holds the Robert S. and Christina Seix Dow Distinguished Master Teacher Chair in Mathematics at The Lawrenceville School near Princeton, New Jersey. He earned his MA in mathematics from the University of Michigan and his BS in mathematics from the University of North Carolina at Charlotte. Daren is also an alumnus of the North Carolina School of Science and Mathematics. Daren has led numerous one-day and weeklong AP® Statistics institutes for new and experienced teachers, and he has been a Reader, Table Leader, and Question Leader for the AP® Statistics exam since 1998. Daren is a frequent speaker on the subject of statistics education at local, state, regional, national, and international conferences. From 2004 to 2009, Daren served on the ASA/NCTM Joint Committee on the Curriculum in Statistics and Probability (which he chaired in 2009). While on the committee, he edited the Guidelines for Assessment and Instruction in Statistics Education (GAISE) pre-K–12 report and coauthored (with Roxy Peck) *Making Sense of Statistical Studies*, a capstone module in statistical thinking for high school students. He currently serves as head judge for the American Statistical Association's Project Competition. Daren is lead author of *The Practice of Statistics*, Fifth Edition, the best-selling textbook for AP® Statistics.

JOSH TABOR has enjoyed teaching general and AP® Statistics to high school students for more than 21 years—most recently at his alma mater, Canyon del Oro High School in Oro Valley, Arizona. He received a BS in mathematics from Biola University, in La Mirada, California. In recognition of his outstanding work as an educator, Josh was named one of the five finalists for Arizona Teacher of the Year in 2011. He is a past member of the AP® Statistics Development Committee (2005–2009), as well as an experienced Table Leader and Question Leader at the AP® Statistics Reading. In 2013, Josh was named to the SAT® Mathematics Development Committee. Each year, Josh leads many workshops and frequently speaks at local, national, and international conferences. In addition to teaching and speaking, Josh has authored articles in *The Mathematics Teacher, STATS Magazine*, and *The Journal of Statistics Education*. Combining his love of statistics and sports, Josh teamed with Christine Franklin to write *Statistical Reasoning in Sports*, an innovative textbook for on-level statistics courses. He is also the coauthor of *The Practice of Statistics*, Fifth Edition, and the companion *Annotated Teacher's Edition*, which is the market-leading program for AP® Statistics.

Jonathan Osters, Lindsey Gallas, Luke Wilcox, Michael Legacy, Doug Tyson

Ann Heath

Ann Cannon—Cornell College, Mount Vernon, Iowa
Contributor and Accuracy Checker

Ann is the Watson M. Davis Professor of Mathematics and Statistics at Cornell College in Mount Vernon, Iowa. Ann has taught statistics at the college level for 25 years and recently added introductory courses in epidemiology and data science to her repertoire. She is very active in the Statistics Education Section of the American Statistical Association, currently serving as the secretary/treasurer. Ann has served as Reader, Table Leader, Question Leader, and Assistant Chief Reader for the AP® Statistics exam for the past 17 years. She is coauthor of *STAT2: Building Models for a World of Data* (W. H. Freeman).

Doug Tyson—Central York High School, York, PA
Content Advisor, Teacher's Edition, Teacher's Resource Materials

Doug has taught mathematics and statistics to high school and undergraduate students for over two decades. He is an AP® Statistics teacher and has served as an AP® Reader and Table Leader. Doug began the on-level statistics course in his district in 2007 and also teaches a Statistical Reasoning in Sports course. He is the coauthor of a curriculum module for the College Board, conducts student review sessions nationally, and gives workshops on teaching statistics. Doug also serves on the ASA/NCTM Joint Committee on Curriculum in Statistics and Probability.

Tim Brown—The Lawrenceville School, Lawrenceville, NJ
Content Advisor, Exercise Author, Test Bank, Pilot Tester

Tim piloted an AP® Statistics course the year before the first exam was administered, and he has been an AP® Reader and Table Leader many times since 1997. He has taught mathematics and statistics at The Lawrenceville School since 1982 and currently holds the Bruce McClellan Distinguished Teaching Chair. Besides teaching AP® Statistics, he has for many years taught an on-level statistics course and a mathematical statistics course for students who have completed calculus.

Luke Wilcox—East Kentwood High School, Kentwood, MI
Content Advisor, Solution Manual, Pilot Tester

Luke has been a math teacher for 15 years and is currently teaching introductory statistics and AP® Statistics. Luke recently received the Presidential Award for Excellence in Math and Science Teaching and was also a finalist for Michigan Teacher of the Year 2015–2016. He facilitates professional development for teachers in curriculum, instruction, assessment, and strategies for motivating students. Lindsey Gallas and Luke are the co-bloggers at TheStatsMedic (thestatsmedic.com), a site dedicated to improving statistics education, which includes activities and lessons for this textbook.

Jason Molesky—Lakeville Area Public Schools, Lakeville, MN

Media Coordinator, Worked Example and Lesson Videos, PPT Lectures

Jason has been teaching since 1998, including over 10 years in AP® Statistics. He has served as an AP® Statistics Reader and Table Leader since 2006 and established the FRAPPY process for developing understanding and preparing for the AP® exam. He provides professional development to teachers nationwide and maintains the StatsMonkey website, a clearinghouse for AP® Statistics resources. Jason is currently his district's Executive Director of Technology & Data Services, overseeing research and evaluation, assessment, and digital learning efforts. He tweets at @StatsMonkey.

Lindsey Gallas—East Kentwood High School, Kentwood, MI

Chapter Tests and Quizzes, Pilot Tester

Lindsey currently teaches introductory statistics and algebra. Her teaching centers on creating contexts inside and outside the classroom that spark students' curiosity. Before teaching at the high school level, Lindsey taught 7th- and 8th-grade math for five years. She, together with Luke Wilcox, has created TheStatsMedic (thestatsmedic.com), a site that illustrates how to teach high school statistics effectively—which includes daily lesson planning for this textbook.

Monica DeBold—Harrison High School, Harrison, NY

Worked Example, Chapter Review Exercise, and Lesson Videos

Monica has taught for nine years at both the high school and college levels. She is experienced in probability and statistics, as well as AP® Statistics and International Baccalaureate math courses. Monica has served as a mentor teacher in her home district and, more recently, as an AP® Statistics Reader. Monica believes successful teaching involves giving students a view of the big picture and enlivening the classroom with real-world articles and applications. She enjoys writing new curricula and vertically aligning courses. She tweets at @DeBoldMath.

Paul Rodriguez—Troy High School, Fullerton, CA

Test Bank

Paul Rodriguez has taught high school mathematics for 22 years, including 19 years as an AP® Statistics teacher. Paul has participated in the AP® Statistics Reading since 2004 as a Reader, Table Leader, and Rubric team member. Since 2011, he has served on the AP® Statistics Test Development Committee, with the responsibility of writing the AP® exam, and he is currently the co-chair of the committee. Paul has also conducted several 2-day workshops and summer institutes sponsored by the College Board®.

Bob Amar—The Lovett School, Atlanta, GA

Applets

Bob has taught math for 10 years, including five years of statistics and three years of AP® Statistics. He currently serves as the Upper School Mathematics Department Chair at The Lovett School, an independent college preparatory school in Atlanta. In his spare time, Bob develops statistical software for teachers and students and is an active accompanist, arranger, and musical coach. For this edition of the textbook, he developed the five wonderful applets for analyzing data, which reside at highschool.bfwpub.com/spa3e.

Michael Legacy—Greenhill School, Dallas, TX

Content Advisor, Accuracy Checker

Michael is a past member of the AP® Statistics Development Committee and has served as Table Leader at the AP® Statistics Reading and for the Alternate and International Exams. Michael also leads Mock Readings and Student Prep Sessions for the National Math and Science Initiative. He is a lead teacher at many AP® Summer Institutes. Michael is the author of the 2007 College Board AP® Statistics Teacher's Guide and was named the Texas 2009–2010 AP® Math/Science Teacher of the Year by the Siemens Corporation.

Jonathan Osters—The Blake School, Minneapolis, MN

Worked Example, Chapter Review Exercise, and Lesson Videos

Jonathan has taught high school mathematics for 11 years. He teaches AP® Statistics, Probability & Statistics, and Geometry at The Blake School. His goal is to get students to enjoy thinking mathematically about the world around them. Jonathan enjoys connecting with other math professionals, having spoken at both state-level and national conferences, and has been a Reader for the AP® Statistics exam for eight years. He writes a blog about teaching at experiencefirstmath.org and tweets at @callmejosters.

Nathan Kidwell—Dubuque, IA

Test Bank

After teaching high school mathematics and statistics courses for 12 years, most recently at Dubuque Community School District in Dubuque, Iowa, Nathan is embarking on a new venture as an international school teacher. He has served as an AP® Statistics Reader since 2012 and has been active in the statistics community, including service as a Judge for the ASA/NCTM National Statistical Project competition (2011–2016). Nathan was the recipient of the Holly High School Excellence in Best Practice Award in 2009.

ACKNOWLEDGMENTS

What does it take to reimagine a book after two editions? *Remarkable teamwork!* It has been the key to the rebirth of *Statistics and Probability with Applications,* Third Edition (SPA 3e). We are indebted to each and every member of the team for the time, energy, and passion that they have invested in making our collective vision for SPA 3e a reality over the past 24 months.

To our team captain, Ann Heath, we offer our utmost gratitude. Managing a revision project of this scope is a Herculean task! Ann has a knack for troubleshooting thorny issues with an uncanny blend of forthrightness and finesse. She assembled an all-star cast to collaborate on SPA and trusted each of us to deliver an excellent finished product. We hope you'll agree that the results speak for themselves. Thank you, Ann, for your unwavering support, patience, and friendship throughout the production of this edition.

It was a pleasure to work again with our Developmental Editor, Don Gecewicz. His keen mind and sharp eye were evident in the many probing questions and insightful suggestions offered at all stages of the project. Thanks, Don, for your willingness to push the boundaries of our thinking.

Working behind the scenes, Corrina Santos busily prepared the art and manuscript chapters for production and helped to assemble the glossary and format key resources. Kim Morte took on the sometimes tedious task of building our LaunchPad and overseeing the creation of our student and teacher e-Books. For the countless hours that Corrina and Kim invested sweating the details, we offer our sincere appreciation.

We are deeply grateful to Marsha Cohen and to Diana Blume for their aesthetic contributions to the eye-catching design of SPA 3e. Our heartfelt thanks also go to Vivien Weiss and Susan Wein for their skillful oversight of the production process. Patti Brecht did a superb job copyediting a very complex manuscript. Thank you to Assunta Petrone and the folks at codeMantra who did an excellent job laying out the pages in a thoughtful and efficient manner.

A special thank you goes to our good friends on the high school sales and marketing staff at Bedford, Freeman, and Worth (BFW) Publishers. We feel blessed to have such enthusiastic professionals on our extended team. In particular, we want to thank our chief cheerleader, Lisa Erdely, along with her inner circle of Pete Marucci and Nicole Desiato, for their willingness to promote *Statistics and Probability with Applications* at every opportunity.

We cannot say enough about the members of our Content Advisory Board and Supplements Team. This is a remarkable group of statistics educators! We'll start with Ann Cannon, who reviewed the statistical content of every chapter. Ann also checked the solutions to every exercise in the book. What a task! More than that, Ann offered us sage advice about virtually everything between the covers of this book. We are so grateful to Ann for all that she has done to enhance the quality of *Statistics and Probability with Applications.*

Tim Brown, Daren's colleague at Lawrenceville, stayed busy creating many new high-quality exercises for SPA 3e. In addition to each of the Chapter Review and Chapter Practice Test exercises, Tim wrote the Recycle and Review exercises and filled in many additional exercises in each lesson. In his down time, he provided excellent feedback on each of the lessons and was always available for a quick conversation in the office.

Doug Tyson and Luke Wilcox, along with Lindsey Gallas, have made an impact on the SPA 3e project from start to finish. Doug and Luke read early drafts of each lesson, offering wise advice from their years of experience working with high school students and teachers. The importance of Doug's work in creating the new *Teacher's Edition* and corresponding *Teacher's Resource Materials* cannot be overstated. These are incredibly useful resources that will make teaching this general statistics course a pleasurable and successful experience for both new and experienced teachers. Thank you, Doug.

As final drafts were produced, Luke wrote solutions to every exercise in the book for the *Teacher's Edition* and *Solutions Manual,* which is an enormous undertaking. Luke took great care in ensuring the accuracy of every solution, including adherence to the language and approach of the book. Luke and Lindsey piloted the entire book with over 150 students and blogged about their experience on www.thestatsmedic.com. After piloting the book, Lindsey created the prepared Tests and Quizzes that appear in the Teacher's Resource Materials. We owe each of these fine educators a debt of gratitude and can't wait to work with them again in the future.

Jason Molesky, aka "Stats Monkey," created the Lecture PowerPoint presentations along with many of the student and teacher facing videos for the book. In short, he is our Media Coordinator extraordinaire. We feel incredibly fortunate to have such a creative, energetic, and deeply thoughtful person at the helm as the media side of our project has exploded in many new directions at once. Jason is supported by a talented media team. Working with Jason, Jonathan Osters and Monica DeBold created videos for each of the 210 worked examples in the book, more than 100 chapter review exercises, and 32 lesson videos. We are sure that teachers and students will find these videos to be extremely helpful throughout the course.

Bob Amar did an amazing job creating the applets that are a key feature of *Statistics and Probability with Applications*. These applets are both powerful and easy to use, due largely to Bob's experience as a statistics teacher and computer programmer. We are excited that students will be able to use all kinds of devices—not just a graphing calculator—to analyze data and perform inference via traditional methods or simulation.

We offer our continued thanks to Michael Legacy for his contributions to the previous edition: composing the quizzes and tests, as well as authoring several chapters in the *Teacher's Resource Binder*. Much of Michael's work has found a new home in SPA 3e.

Although Dan Yates and David Moore both retired several years ago, their influence lives on in SPA 3e. They both had a dramatic impact on our thinking about how statistics is best taught and learned through their pioneering work as textbook authors. Without their early efforts, *Statistics and Probability with Applications* would not exist!

Thanks to all of the teachers we have met at workshops, conferences, and the AP® Reading over the years who took the time to "talk stats" with us. We carefully considered your thoughts as we were developing the plan for SPA 3e.

—*Daren Starnes and Josh Tabor*

A final note from Daren: It has been a privilege for me to work so closely with Josh Tabor on SPA 3e for the past two years. He is a gifted statistics educator; a successful author in his own right; and a caring parent, colleague, and friend. Josh's influence is present on virtually every page of the book. His talent for keeping the big picture in mind while ironing out the details is impressive! I feel incredibly fortunate that Josh agreed to tackle this mammoth project with me.

The most vital member of the SPA 3e team for me is my wonderful wife, Judy. She has read page proofs, typed in data sets, and endured countless statistical conversations in the car, in airports, on planes, in restaurants, and on our frequent strolls. If writing is a labor of love, then I am truly blessed to share my labor with the person I love more than anyone else in the world. Judy, thank you so much for making the seemingly impossible become reality, time and time again. And to our three sons, Simon, Nick, and Ben—thanks for the inspiration, love, and unwavering support throughout the lengthy duration of this project. I owe you, your spouses, and the grandkids some "work-free" vacation time.

A final note from Josh: When Daren asked me to join the SPA author team, I was excited to work with him on another book. No one I know works harder and holds himself to a higher standard than does Daren. His wealth of experience and creative vision for this book made him an excellent writing partner. And his thoughtful feedback and encouragement made a huge difference from beginning to end. Every time I was feeling stressed and overworked, I felt better knowing I have a colleague who understands the challenges of writing a book essentially from scratch. For all of these things—and many others—thanks!

I especially want to thank the two most important people in my life. To my wife Anne, your patience while I spent countless hours working on this project is greatly appreciated. I couldn't have survived without your consistent support and encouragement. To my daughter Jordan, I look forward to being home more often and spending less time on my computer when I am there. We have a lot of fun and games to catch up on. I love you both very much.

Pilot Testers of the Third Edition

We are indebted to a group of seasoned statistics teachers who class-tested all or part of the manuscripts of this new edition. Your willingness to work with draft materials and offer feedback is greatly appreciated.

- Luke Wilcox, East Kentwood High School, Kentwood, MI
- Lindsey Gallas, East Kentwood High School, Kentwood, MI
- Matt Powell, Canyon del Oro High School, Oro Valley, AZ
- Vicki Greenberg, The Weber School, Atlanta, GA

And the following teachers at The Lawrenceville School, Lawrenceville, NJ:

- Tim Brown
- Courtney Doyle
- Frank Fernandez
- Hardy Gieske
- Charise Hall
- Melissa Clore
- George Negroponte
- Stephen Wallis

Looking for the videos, applets, and other helpful resources or digital options?

Go to

highschool.bfwpub.com/spa3e

**and select "Go to Student Site"
from the menu**

Students

Go to Student Site

Get Help

View Purchasing Options

How to get the most from this program

Are you taking statistics because you want to be better prepared for social science or life science courses in college? Do you plan to open a restaurant or shop? Or do you want to become a more knowledge-able participant in the world at large? Whatever your reason, learning how to use the *Statistics and Probability with Applications* program effectively will help you achieve success in this course and in life.

CHECK OUT THE SIMPLE STRUCTURE. Each chapter is divided into 6 to 9 short lessons.

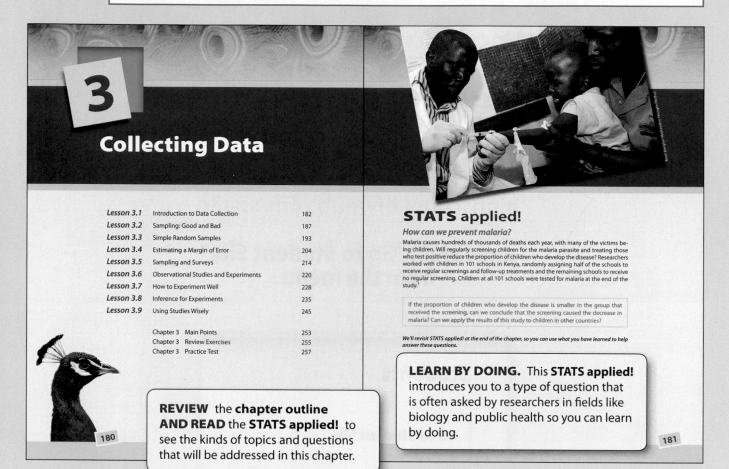

3 Collecting Data

Lesson 3.1 Introduction to Data Collection 182
Lesson 3.2 Sampling: Good and Bad 187
Lesson 3.3 Simple Random Samples 193
Lesson 3.4 Estimating a Margin of Error 204
Lesson 3.5 Sampling and Surveys 214
Lesson 3.6 Observational Studies and Experiments 220
Lesson 3.7 How to Experiment Well 228
Lesson 3.8 Inference for Experiments 235
Lesson 3.9 Using Studies Wisely 245

Chapter 3 Main Points 253
Chapter 3 Review Exercises 255
Chapter 3 Practice Test 257

STATS applied!

How can we prevent malaria?

Malaria causes hundreds of thousands of deaths each year, with many of the victims be-ing children. Will regularly screening children for the malaria parasite and treating those who test positive reduce the proportion of children who develop the disease? Researchers worked with children in 101 schools in Kenya, randomly assigning half of the schools to receive regular screenings and follow-up treatments and the remaining schools to receive no regular screening. Children at all 101 schools were tested for malaria at the end of the study.[1]

If the proportion of children who develop the disease is smaller in the group that received the screening, can we conclude that the screening caused the decrease in malaria? Can we apply the results of this study to children in other countries?

We'll revisit STATS applied! at the end of the chapter, so you can use what you have learned to help answer these questions.

LEARN BY DOING. This **STATS applied!** introduces you to a type of question that is often asked by researchers in fields like biology and public health so you can learn by doing.

REVIEW the **chapter outline AND READ** the **STATS applied!** to see the kinds of topics and questions that will be addressed in this chapter.

180
181

252 CHAPTER 3 • Collecting Data

Chapter 3 **RESOLVED**

STATS applied!

How can we prevent malaria?

In the STATS applied! on page 181, we described an experiment that investigated the effect of regular screening for malaria. Researchers worked with children in 101 schools in Kenya, randomly assigning half of the schools to receive regular screenings and follow-up treatments and the remaining schools to receive no regular screening.

1. Why was it necessary to include a group of schools that didn't receive the screening? Does excluding some schools from screening raise any ethical concerns?
2. Describe how you could randomly assign the 101 schools into groups of 51 and 50.
3. What is the purpose of random assignment in this experiment?
4. If the researchers found convincing evidence that the proportion with malaria is lower for children in schools that are regularly screened, would it be reasonable to say that screening caused the reduction in malaria? Would these results apply to all schools in Africa? Explain.
5. Unfortunately, the results of the study were not statistically significant. Explain what this means in the context of this study.

SOLVE A REAL PROBLEM. Once you have learned more statistical techniques, you have a chance to **RESOLVE the STATS applied!** by thinking through a series of related questions. These wrap-ups appear at the end of every chapter.

Easy-to-read lessons display all information clearly

CHECK THE SCOPE. Take note of the **Learning Targets** at the beginning of each lesson (there are usually only three of them—short and sweet). They are revisited at the end of the lesson and closely connected to supporting examples and exercises.

Lesson 1.4

Displaying Quantitative Data: Stemplots

LEARNING TARGETS

- Make stemplots of quantitative data.
- Interpret stemplots.
- Compare distributions of quantitative data with stemplots.

Another simple type of graph for displaying quantitative data is a **stemplot** (also called a *stem-and-leaf plot*).

DEFINITION Stemplot

A **stemplot** shows each data value separated into two parts: a *stem*, which consists of all but the final digit, and a *leaf*, the final digit. The stems are ordered from least to greatest and arranged in a vertical column. The leaves are arranged in increasing order out from the appropriate stems.

WATCH FOR DEFINITIONS. Be sure that you learn the vocabulary as you study each lesson. You can also look up key terms in the English and Spanish **Glossary/Glosario** at the end of the book.

Figure 1.4 shows a stemplot of data on the caffeine content per 8-ounce serving for several popular soft drinks. You'll learn how to make and interpret stemplots in this lesson.

```
1 | 556
2 | 033344
2 | 55667778888899
3 | 113
3 | 55567778
4 | 33
4 | 77
```

KEY: 2 | 8 means the soft drink contains 28 mg of caffeine per 8-ounce serving.

FIGURE 1.4 Stemplot showing the caffeine content (in milligrams or mg per 8-ounce serving) of various soft drinks.

Making and Interpreting Stemplots

It is fairly easy to make a stemplot by hand for small sets of quantitative data. Stemplots give us a quick picture of a distribution that includes the actual numerical values in the graph.

How to Make a Stemplot

1. **Make stems.** Separate each observation into a stem, consistin[g] and a leaf, the final digit. Write the stems in a vertical column [...] Draw a vertical line at the right of this column. Do not skip any [...] data value for a particular stem.

2. **Add leaves.** Write each leaf in the row to the right of its stem. [...]

3. **Order leaves.** Arrange the leaves in increasing order out from [...]

4. **Add a key.** Provide a key that explains in context what the ste[m]

30

FOLLOW THESE STEPS. Consult the **"How to" boxes** for step-by-step instructions about how to create graphs, perform calculations, or use a statistical process successfully.

How to Make a Segmented Bar Chart

1. **Identify the variables.** Determine which variable is the explanatory variable and which is the response variable.

2. **Draw and label the axes.** Put the name of the explanatory variable under the horizontal axis. To the left of the vertical axis, indicate if the graph shows the percent (or proportion) of individuals in each category of the response variable.

3. **Scale the axes.** Write the names of the categories of the explanatory variable at equally spaced intervals under the horizontal axis. On the vertical axis, start at 0% (or 0) and place tick marks at equal intervals until you reach 100% (or 1).

4. **Draw "100%" bars** above each of the category names for the explanatory variable on the horizontal axis so that each bar ends at the top of the graph. Make the bars equal in width and leave gaps between them.

5. **Segment each of the bars.** For each category of the explanatory variable, calculate the relative frequency for each category of the response variable. Then divide the corresponding bar so that the area of each segment corresponds to the proportion of individuals in each category of the response variable.

6. **Include a key** that identifies the different categories of the response variable.

WATCH FOR SUMMARY BOXES. On-the-spot summaries give a quick recap of the key things to remember about important statistical techniques.

Choosing a Measure of Center

- If a distribution of quantitative data is roughly symmetric and has no outliers, use the mean to measure center.
- If the distribution is strongly skewed or has outliers, use the median to measure center.

Worked examples help you understand concepts and techniques

EXAMPLE

Who survived on the Titanic?

Making a segmented bar chart

In 1912 the luxury liner *Titanic*, on its first voyage across the Atlantic, struck an iceberg and sank. Some passengers got off the ship in lifeboats, but many died. The two-way table gives information about adult passengers who survived and who died, by class of travel.

PROBLEM: Make a segmented bar chart to display the relationship between survival status and class of travel for passengers on the *Titanic*.

SOLUTION:

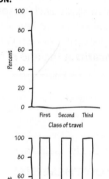

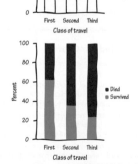

		Class of travel			
		First	Second	Third	Total
Survival status	Survived	197	94	151	442
	Died	122	167	476	765
	Total	319	261	627	1207

1. **Identify the variables.** Use class of travel for the explanatory variable because class might help predict whether or not a passenger survived.

2. **Draw and label the axes.** Label the horizontal axis with the explanatory variable "Class of travel" and the vertical axis with "Percent."

3. **Scale the axes.** Scale the horizontal axis with the values of the explanatory variable: "First," "Second," and "Third." Scale the vertical axis from 0 to 100% in increments of 20%.

4. **Draw "100%" bars** for each category of the explanatory variable.

5. **Segment each of the bars.**
• In first class, 197/319 = 62% survived and 122/319 = 38% died.
• In second class, 94/261 = 36% survived and 167/261 = 64% died.
• In third class, 151/627 = 24% survived and 476/627 = 76% died.

6. **Include a key** to identify the categories of the response variable.

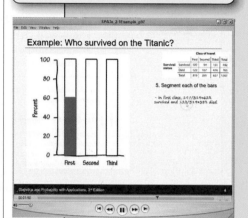

EXAMPLE

Is there an association between class and survival on the Titanic?

Checking for association

PROBLEM: Use the graph to determine if there is an association between survival status and class of travel for passengers on the *Titanic*. Explain your reasoning. If there is an association, briefly describe it.

SOLUTION:
There is an association between survival status and class of travel because the survival rates were not the same for the different classes of travel. Passengers in first class were the most likely to survive, and passengers in third class were the least likely to survive.

FOR PRACTICE   TRY EXERCISE 9.

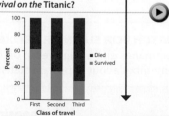

Knowing a passenger's class of travel helps us predict whether or not the passenger survived—higher-class passengers were more likely to survive. This means that there is an association between class of travel and survival status.

Apply what you've learned
(because that's how you retain concepts)

LESSON APP 1.4

How many shoes are too many shoes?

How many pairs of shoes does a typical teenager own? To find out, a group of statistics students surveyed separate random samples of 20 female students and 20 male students from their large high school. Then they recorded the number of pairs of shoes that each person owned. Here are the data.

Females	50	26	26	31	57	19	24	22	23	38
	13	50	13	34	23	30	49	13	15	51
Males	14	7	6	5	12	38	8	7	10	10
	10	11	4	5	22	7	5	10	35	7

1. Make a stemplot of the female data. Do not split stems.

2. Describe the shape of the distribution.

3. Explain why we should split stems for the male data.

4. The back-to-back stemplot with split stems displays the data for both genders. Write a few sentences comparing the male and female distributions.

> **KEEPING IT REAL.** Apply what you've learned in the lesson to a real-world situation by answering the questions in the **Lesson App.**

LESSON APP 1.3

How can we check the health of a stream?

Nitrates are organic compounds that are a main ingredient in fertilizers. When those fertilizers run off into streams, the nitrates can have a toxic effect on fish. An ecologist studying nitrate pollution in two streams measures nitrate concentrations at 42 places on Stony Brook and 42 places on Mill Brook. The parallel dotplots display the data.

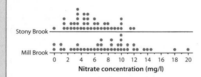

Nitrate concentration (mg/l)

1. Explain what the dot above 12 in the Stony Brook graph represents.

2. What percent of the nitrate concentration measurements for each stream exceeded 10 milligrams per liter (mg/l)?

3. Compare the centers of these two distributions.

4. Is the variability in nitrate concentrations for the two streams similar or different? Justify your answer.

TECH CORNER

Using an Applet To Make

You can use the *One Quantitative Variable* applet at highschool.bfwpub.com/spa3e to make a stemplot. For the electoral vote data on page 31:

1. Enter Electoral votes as the Variable name.

2. Select 1 as the number of groups and Raw data as the input method.

3. Enter the data. Be sure to separate the data values with commas or spaces as you type them.

4. Click Begin analysis.

5. Change the Graph type to a stemplot.

6. Split stems to get a better picture of the distribution.

Graph Distributions

Graph type: Stemplot ▾ Split stems: Yes, into two ▾

TECH CORNER

Making a Scatterplot with Technology

You can use a graphing calculator or an applet to make a scatterplot. We'll illustrate using the Faster and farther data from the example on page 106.

Applet

Launch the *Two Quantitative Variables* applet at highschool.bfwpub.com/spa3e.

1. Enter the name of the explanatory variable (Dash time) and the values of the explanatory variable in the first row of boxes. Then, enter the name of the response variable (Long-jump distance) and the values of the response variable in the second row of boxes.

Two Quantitative Variables

Variable	Name	Observations (separated by commas or spaces) *Keep individuals in the same order.*
Explanatory	Dash time (sec)	5.41 5.05 7.01 7.17 6.73 5.68 5.78 6.31 6.44 6.50 6.80 7.25
Response	Long-jump distan	171 184 90 65 78 130 173 143 92 139 120 110

Begin analysis   Edit inputs   Reset everything

2. Press the "Begin analysis" button to see the scatterplot.

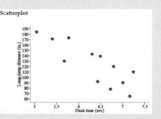

Scatterplot

TI-83/84

1. Enter the Sprint times in L_1 and the Long-jump distances in L_2.

- Press STAT and choose Edit…
- Type the values into L_1 and L_2.

> **TECH MAKES IT EASY.** Follow the steps in the **Tech Corners** to use the applets or your TI-83/84 graphing calculator to perform simulations and generate graphs and summary statistics. You can find the applets online at highschool.bfwpub.com/spa3e.

Review and practice to bring it home

Lesson 1.4

WHAT DID YOU LEARN?

LEARNING TARGET	EXAMPLES	EXERCISES
Make stemplots of quantitative data.	p. 31	1–4
Interpret stemplots.	p. 32	5–8
Compare distributions of quantitative data with stemplots.	p. 33	9–12

Exercises   Lesson 1.4

Mastering Concepts and Skills

1. **Science gets your heart beating!** Here are the resting heart rates of 26 ninth-grade biology students. Make a stemplot of these data. Do not split stems.
   pg 31

   61 78 77 81 48 75 70 77 70 76 86 55 65
   60 63 79 62 71 72 74 74 64 66 71 66 68

2. **Hot enough for you?** Here are the high temperature readings in degrees Fahrenheit for Phoenix, Arizona, for each day in July 2013. Make a stemplot of these data. Do not split stems.

   111 107 115 108 106 109 111 113 104 103 97
   99 104 110 109 100 105 107 102 101 84 93
   106 109

   study on salmon
   pH of 25 salmon
   a stemplot of these

   6.26 6.24 6.37 6.32
   6.32 6.32 6.48

   ner legumes are a
   wing data give the
   varieties of beans,
   ed beans.[25] Make a
   t stems.

   8.2 9.1 9.0 9.0
   7.0 7.5 13.5 8.3

5. **Science gets your heart beating!** Here is a stemplot using split stems for the heart-rate data from Exercise 1.
   pg 32

   ```
   4 | 8
   5 |
   5 | 5
   6 | 01234
   6 | 5668
   7 | 0011244
   7 | 567789
   8 | 1
   8 | 6
   ```

   (a) What percent of these ninth-grade biology students have resting heart rates below 70 beats per minute?
   (b) Describe the shape of the distribution.
   (c) Which value appears to be an outlier? Give the stemplot a key using this value.

6. **Hot enough for you?** Here is a stemplot using split stems for the daily high temperature in Phoenix data from Exercise 2.

   ```
   8 | 4
   8 |
   9 | 3
   9 | 799
   10 | 011223444
   10 | 556667788999
   11 | 0113
   11 | 5
   ```

   (a) What percent of days in this month were hotter than 100 degrees Fahrenheit (°F)?
   (b) Describe the shape of the distribution.
   (c) Which value appears to be an outlier? Give the stemplot a key using this value.

7. **Where are the older folks?** Following is a stemplot of the percents of residents aged 65 and older in the 50 states and the District of Columbia.[26]

Applying the Concepts

13. **Chasing food dollars** A marketing consultant observed 50 consecutive shoppers at a supermarket to find out how much each shopper spent in the store. Here are the data (in dollars), arranged in increasing order:

    3.11 8.88 9.26 10.81 12.69 13.78 15.23 15.62 17.00 17.39
    18.36 18.43 19.27 19.50 19.54 20.16 20.59 22.22 23.04 24.47
    24.58 25.13 26.24 26.26 27.65 28.06 28.08 28.38 32.03 34.98
    36.37 38.64 39.16 41.02 42.97 44.08 44.67 45.40 46.69 48.65
    50.39 52.75 54.80 59.07 61.22 70.32 82.70 85.76 86.37 93.34

    (a) Round each amount to the nearest dollar. Then make a stemplot using tens of dollars as the stems and dollars as the leaves.
    (b) Make another stemplot of the data by splitting stems. Which graph shows the shape of the distribution better?
    (c) Write a few sentences describing the amount of money spent by shoppers at this supermarket.

14. **What does ERA mean?** One way to measure the effectiveness of baseball pitchers is to use their *earned run average*, which measures how many earned runs opposing teams score, on average, every nine innings pitched. The overall earned run average for all pitchers in the major leagues in 2013 was 3.86. Here are the earned run averages for all 25 players who pitched for the Boston Red Sox in 2013.

    3.75 3.52 4.57 4.32 1.74 1.09 3.16 1.81 2.64 3.77 4.04 4.97 4.86
    5.34 4.88 4.62 3.86 3.52 5.56 3.60 8.60 5.40 6.35 9.82 9.00

    (a) Truncate the hundredths place of each data value and make a stemplot, using the ones digit as the stem and the tenths digit as the leaf.
    (b) Make another stemplot of the data by splitting stems. Which graph shows the shape of the distribution better?
    (c) Write a few sentences describing the distribution of earned run averages of Boston's pitchers in 2013.

Extending the Concepts

Sometimes, the variability in a data set is so small that splitting stems in two doesn't produce a stemplot that shows the shape of the distribution well. We can often solve this problem by splitting the stem into five parts, each consisting of two leaf values: 0 and 1, 2 and 3, 4 and 5, and so on.

Exercises 15 and 16 refer to the following setting. Here are the weights, in ounces, of 36 navel oranges selected from a large shipment to a grocery store.

5.7 5.4 5.8 5.3 4.6 4.9 5.6 5.3 5.5 5.5 5.4 5.8
5.3 5.5 5.5 5.4 5.8 5.9 5.4 5.1 5.0 5.6 5.3 5.4
5.0 5.3 5.1 5.2 5.7 5.6 5.8 4.5 5.2 5.4 5.7 5.6

15. **Weights of oranges** Make a stemplot of the data by splitting stems into two parts. Explain why this graph does not display the distribution of orange weights effectively.

16. **Splitting the oranges** Make a stemplot of the data by splitting stems into five parts. Describe the shape of the distribution.

Recycle and Review

17. **More gas guzzlers (1.2)** The EPA-estimated highway fuel efficiency for four different sedans is given in the bar chart.[28] Explain how this graph is misleading.

18. **Comparing tuition (1.3)** The dotplot shows 2014 out-of-state tuition for the 40 largest colleges and universities in North Carolina.[29] Describe the overall pattern of the distribution and identify any clear departures from the pattern.

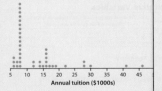

REVIEW YOUR KNOWLEDGE. See how well you have learned the contents of the chapter. Use the chapter **Main Points** as a quick review and work the **chapter review exercises**. If you have trouble, watch the **video clip** to get help step by step or check the answer in the **Solutions Appendix.**

Main Points — Chapter 1

- **Statistics** is the science and art of collecting, analyzing, and drawing conclusions from data.

Organizing Data

- A data set contains information about a number of **individuals**. Individuals may be people, animals, or things.

- A **variable** describes some characteristic of an individual, such as a person's height, gender, or salary.

- A **categorical variable** assigns a label th[at] each individual into one of several grou[ps] as male or female.

- A **quantitative variable** has numerical values that measure some characteristic of each individual, such as height in centimeters or salary in dollars.

- The **distribution** of a variable describes what values the variable takes and how often it takes them. You can use a **frequency table** or a **relative frequency table** to quickly summarize the distribution of a variable.

Displaying Data

Categorical Variables

---

88   CHAPTER 1 • Analyzing One-Variable Data

Chapter 1   Review Exercises

1. **U.S. presidents** (1.1) Here is some information about the first 10 U.S. presidents. Identify the individuals and variables in this data set. Classify each variable as categorical or quantitative.

Name	Political party	Age at inaug.	Age at death	State of birth
G. Washington	Federalist	57	67	Virginia
J. Adams	Federalist	61	90	Massachusetts
T. Jefferson	Democratic-Republican	57	83	Virginia
J. Madison	Democratic-Republican	57	85	Virginia
J. Monroe	Democratic-Republican	58	73	Virginia
J. Q. Adams	Democratic-Republican	57	80	Massachusetts
A. Jackson	Democrat	61	78	South Carolina
M. Van Buren	Democrat	54	79	New York
W. H. Harrison	Whig	68	68	Virginia
J. Tyler	Whig	51	71	Virginia

3. **Music preferences** (1.2) Australian high school students were asked to choose their favorite type of music from a list. The bar chart compares the preferences for samples of 250 students in surveys in 2012 and 2013. Write a few sentences comparing the distribution of music preferences for the two years.

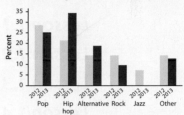

4. **Binge-watching** (1.2) Do you "binge-watch" television series by viewing multiple episodes of a series at one sitting? A survey of 800 people who "binge-watch" were asked how many episodes is too many to watch in one viewing session. The results are displayed in the bar chart.[64] Explain how this graph is misleading.

Exercise: U.S. Presidents

Here is some information about the first 10 U.S. presidents. Identify the individuals and variables in this data set. Classify each variable as categorical or quantitative.

Individuals: The first ten U.S. Presidents.

Categorical variables: Political party, state of birth.

Statistics and Probability with Applications, 3rd Edition

---

90   CHAPTER 1 • Analyzing One-Variable Data

Chapter 1   Practice Test

Section I: Multiple Choice *Select the best answer for each question.*

1. You record the age, marital status, and earned income of a sample of 1463 women. The number and type of variables you have recorded is
   (a) 3 quantitative, 0 categorical.
   (b) 3 quantitative, 1 categorical.
   (c) 2 quantitative, 1 categorical.
   (d) 2 quantitative, 2 categorical.

2. Here is a pie chart of how a randomly selected group of people described the cost of their health insurance. Which of the following bar charts is equivalent to the pie chart?

3. The accompanying bar chart summarizes responses of dog owners to the question "Where in the car do you let your dog ride?" Which of the following statements is true?

TEST YOURSELF. Finally, take the **Practice Test** to gauge your mastery of the material. Check your answers against those in the **Solutions appendix**. All of the Practice Test solutions are given there.

1

Analyzing One-Variable Data

Lesson 1.1 Statistics: The Science and Art of Data   4

Lesson 1.2 Displaying Categorical Data   11

Lesson 1.3 Displaying Quantitative Data: Dotplots   21

Lesson 1.4 Displaying Quantitative Data: Stemplots   30

Lesson 1.5 Displaying Quantitative Data: Histograms   38

Lesson 1.6 Measuring Center   49

Lesson 1.7 Measuring Variability   58

Lesson 1.8 Summarizing Quantitative Data: Boxplots and Outliers   67

Lesson 1.9 Describing Location in a Distribution   77

Chapter 1   Main Points   86
Chapter 1   Review Exercises   88
Chapter 1   Practice Test   90

STATS applied!

Does hand sanitizer work?

Is soap better than hand sanitizer for getting rid of unwanted bacteria? Daniel and Kate designed an experiment to find out. Using 30 identical petri dishes, they randomly assigned 10 students to press one hand in a dish after washing with soap, 10 students to press one hand in a dish after using hand sanitizer, and 10 students to press one hand in a dish after using nothing. After three days of incubation, they counted the number of bacteria colonies on each petri dish.[1]

> Which petri dishes had the most bacteria colonies? What conclusion did Daniel and Kate make based on the data?

We'll revisit STATS applied! at the end of the chapter, so you can use what you have learned to help answer these questions.

Lesson 1.1

Statistics: The Science and Art of Data

LEARNING TARGETS

- Identify the individuals and variables in a data set, then classify the variables as categorical or quantitative.
- Summarize the distribution of a variable with a frequency table or a relative frequency table.

We live in a world of *data*. Every day, the media report poll results, outcomes of medical studies, and analyses of data on everything from gasoline prices to standardized test scores to consumption of bottled water to new technology. The data are trying to tell us a story. To understand what the data are saying, you need to learn more about **statistics.**

DEFINITION Statistics

Statistics is the science and art of collecting, analyzing, and drawing conclusions from data.

A solid understanding of statistics will help you make good decisions based on data in your daily life.

ACTIVITY

The "1 in 6 wins" game

This activity will give you a "taste" of what statistics is about: drawing conclusions from data.

As a special promotion for its 20-ounce bottles of soda, a soft-drink company printed a message on the inside of each bottle cap. Some of the caps said, "Please try again!" while others said, "You're a winner!" The company advertised the promotion with the slogan "1 in 6 wins a prize." The prize is a free 20-ounce bottle of soda.

Jorge's statistics class wonders if the company's claim holds true at a nearby convenience store. To find out, all 30 students in the class go to the store, and each buys one 20-ounce bottle of the soda. Two of them get caps that say, "You're a winner!" Does this result give convincing evidence that the company's 1-in-6 claim is false? You and your classmates will perform a simulation to help answer this question.

For now, let's assume that the company is telling the truth and that every 20-ounce bottle of soda it fills has a 1-in-6 chance of getting a cap that says, "You're

a winner!" We can model the status of an individual bottle with a six-sided die: Let 1 through 5 represent "Please try again!" and 6 represent "You're a winner!"

1. Roll your die 30 times to imitate the process of the students in Jorge's statistics class buying their sodas. How many of them won a prize?

2. Your teacher will draw and label axes for a class dotplot. Plot the number of prize winners you got in Step 1 on the graph.

3. Have some students repeat Steps 1 and 2 until you have a total of at least 40 repetitions of the simulation for your class.

4. Discuss the results with your classmates. What percent of the time did Jorge's statistics class get two or fewer prizes, just by chance? Does it seem plausible (believable) that the company is telling the truth but that the class just got unlucky? Explain.

The previous activity outlines the steps in the **statistical problem-solving process.** You'll learn more about the details of this process in future lessons.

> #### DEFINITION Statistical problem-solving process[2]
>
> - **Ask Questions:** Clarify the research problem and ask one or more valid statistics questions.
> - **Collect Data:** Design and carry out an appropriate plan to collect the data.
> - **Analyze Data:** Use appropriate graphical and numerical methods to analyze the data.
> - **Interpret Results:** Draw conclusions based on the data analysis. Be sure to answer the research question(s)!

Classifying Data

The table displays data on several roller coasters that have opened since April 2014.[3]

Roller coaster	Type	Height (ft)	Design	Speed (mph)	Duration (s)
Wildfire	Wood	187	Sit down	70.2	120
Skyline	Steel	131.3	Inverted	50	90
Goliath	Wood	165	Sit down	72	105
Helix	Steel	134.5	Sit down	62.1	130
Banshee	Steel	167	Inverted	68	160
Black Hole	Steel	22.7	Sit down	25.5	75

replace

Most data tables follow this format—each row describes an **individual** and each column holds the values of a **variable.** (Sometimes the individuals in a data set are called *cases* or *observational units*.)

> #### DEFINITION Individual, Variable
>
> An **individual** is a person, animal, or thing described in a set of data.
>
> A **variable** is any attribute that can take different values for different individuals.

For the roller coaster data set, the *individuals* are the 6 roller coasters. The five *variables* recorded for each coaster are: type, height (in feet), design, speed (in miles per hour), and duration (in seconds). Type and design are **categorical variables.** Height, speed, and duration are **quantitative variables.**

> #### DEFINITION Categorical variable, Quantitative variable
>
> A **categorical variable** assigns labels that place individuals into particular groups.
>
> A **quantitative variable** takes number values for which it makes sense to find an average.

Not every variable that takes number values is quantitative. Zip code is one example. Although zip codes are numbers, it doesn't make sense to talk about the average zip code. In fact, zip codes place individuals (people or dwellings) into categories based on location.

CAUTION

EXAMPLE

So you want to be happy?

Individuals and variables

PROBLEM: The American Statistical Association sponsors a Web-based project that collects data about primary and secondary school students using surveys. We used the site's "Random Sampler" to choose 40 U.S. high school students who completed the survey in a recent year.[4] The table displays data for the first 10 students chosen. The rightmost column gives students' answers to the question:

Which would you prefer to be? Select one.

| _____ Rich | _____ Happy | _____ Famous | _____ Healthy |

State	Grade level	Gender	Age	Birth month	Height (cm)	Arm span (cm)	Preferred status
SC	12	Male	17	January	177	161	Famous
UT	9	Female	14	March	162	153	Healthy
NM	12	Female	17	August	164	167	Healthy
CA	12	Female	17	April	153	154	Famous
GA	12	Female	17	June	172	169	Happy
MI	11	Male	17	March	170	173	Famous
IN	12	Female	18	January	168	163	Happy
CO	9	Female	14	June	152	160	Happy
NJ	10	Female	16	November	165	174	Famous
CO	9	Male	15	January	190	177	Rich
...							

Identify the individuals and variables in this data set. Classify each variable as categorical or quantitative.

SOLUTION:

Individuals: 40 U.S. high school students who completed an online survey.

Variables:

- Categorical: State where student lives, grade level, gender, birth month, preferred status
- Quantitative: Age (years), height (centimeters), arm span (centimeters)

> Grade level is a categorical variable even though it takes number values. The numbers place the students into categories: 9 = freshman, 10 = sophomore, 11 = junior, and 12 = senior.

FOR PRACTICE   TRY EXERCISE 1.

The proper method of data analysis depends on whether a variable is categorical or quantitative. For that reason, it is important to distinguish between these two types of variables. To make life simpler, we sometimes refer to "categorical data" or "quantitative data" instead of identifying the variable as categorical or quantitative.

Summarizing Data

A variable generally takes values that vary from one individual to another. That's why we call it a variable! The **distribution** of a variable describes the pattern of variation of these values.

> **DEFINITION** Distribution
>
> The **distribution** of a variable tells us what values the variable takes and how often it takes these values.

We can summarize a variable's distribution with a **frequency table** or a **relative frequency table**.

> **DEFINITION** Frequency table, Relative frequency table
>
> A **frequency table** shows the number of individuals having each data value.
>
> A **relative frequency table** shows the proportion or percent of individuals having each data value.

Some people use the terms "frequency distribution" and "relative frequency distribution" instead. To make either kind of table, start by tallying the number of times that the variable takes each value.

EXAMPLE

Would you rather be happy or rich?

Frequency and relative frequency tables

PROBLEM: Here are the data on preferred status for all 40 students in the sample from the previous example:

Famous	Healthy	Healthy	Famous	Happy	Famous	Happy	Happy	Famous
Rich	Happy	Happy	Rich	Happy	Happy	Happy	Rich	Happy
Famous	Healthy	Rich	Happy	Happy	Rich	Happy	Happy	Rich
Healthy	Happy	Happy	Rich	Happy	Happy	Rich	Happy	Famous
Famous	Happy	Happy	Happy					

Summarize the distribution of preferred status with a frequency table and a relative frequency table.

SOLUTION:

Preferred status	Tally
Famous	JHT ||
Happy	JHT JHT JHT JHT	
Healthy	||||
Rich	JHT |||

> Start by tallying the number of students in each preferred status category.

Frequency table

Preferred status	Frequency
Famous	7
Happy	21
Healthy	4
Rich	8
Total	40

> The frequency table shows the *number* of students who chose each status.

Relative frequency table

Preferred status	Relative frequency
Famous	7/40 = 0.175 or 17.5%
Happy	21/40 = 0.525 or 52.5%
Healthy	4/40 = 0.100 or 10.0%
Rich	8/40 = 0.200 or 20.0%
Total	40/40 = 1.000 or 100%

> The relative frequency table shows the *proportion* or *percent* of students who chose each status.

FOR PRACTICE   TRY EXERCISE 5.

The same process can be used to summarize the distribution of a quantitative variable. Of course, it would be hard to make a frequency table or a relative frequency table for quantitative data that take many different values, like the ages of people attending a high school band concert. We'll look at a better option for quantitative variables with many possible values in Lesson 1.5.

LESSON APP 1.1

What are my classmates like?

On the first day of a statistics course, the instructor gave all 40 students in the class a survey. The table shows data from the first 10 students on the class roster.

Gender	Class	GPA	Pulse rate	Dominant hand	Children in family	Homework last night (min)	Sleep (h)	Have a smartphone?
F	Fr	3.22	72	R	3	0–14	10	Y
F	Fr	2.3	110	L	3	0–14	8	N
M	Ju	3.8	60	L	6	15–29	7	Y
M	So	3.1	72	R	2	15–29	7.5	Y
F	So	4.0	51	R	1	45–59	7	Y
F	So	3.4	68	R	4	0–14	8.5	Y
F	So	3.0	80	R	3	30–44	7	Y
M	So	3.5	59	R	2	30–44	7	Y
M	Fr	3.9	65	R	2	15–29	6	Y
M	Sr	3.5	104	R	2	0–14	7	N
. . .								

1. Identify the individuals and variables in this data set. Classify each variable as categorical or quantitative.

2. Here are the ages of the 40 students in the class:

17 16 17 17 17 16 18 14 16 15
16 16 17 18 17 16 17 16 15 14
17 14 14 17 17 17 16 15 17 17
17 18 18 14 15 18 17 17 17 16

Summarize the distribution of age with a frequency table and a relative frequency table.

Lesson 1.1

WHAT DID YOU LEARN?		
LEARNING TARGET	EXAMPLES	EXERCISES
Identify the individuals and variables in a data set, then classify the variables as categorical or quantitative.	p. 6	1–4
Summarize the distribution of a variable with a frequency table or a relative frequency table.	p. 7	5–8

Exercises   Lesson 1.1

The solutions to all exercises numbered in red are found in the Solutions Appendix, starting on page S-1.

Mastering Concepts and Skills

1. Box-office smash According to the Internet Movie Database, *Avatar* is tops based on box-office receipts worldwide. The table displays data on several popular movies.[5] Identify the individuals and variables in this data set. Classify each variable as categorical or quantitative.

pg 6

Movie	Year	Rating	Time (min)	Genre	Box office ($)
Avatar	2009	PG-13	162	Action	2,783,918,982
Titanic	1997	PG-13	194	Drama	2,207,615,668
Star Wars: The Force Awakens	2015	PG-13	136	Adventure	2,040,375,795
Jurassic World	2015	PG-13	124	Action	1,669,164,161
Marvel's The Avengers	2012	PG-13	142	Action	1,519,479,547
Furious 7	2015	PG-13	137	Action	1,516,246,709
The Avengers: Age of Ultron	2015	PG-13	141	Action	1,404,705,868
Harry Potter and the Deathly Hallows: Part 2	2011	PG-13	130	Fantasy	1,328,111,219
Frozen	2013	PG	108	Animation	1,254,512,386
Iron Man 3	2013	PG-13	129	Action	1,172,805,920

2. Tournament time A high school's lacrosse team is planning to go to Buffalo for a three-day tournament. The tournament's sponsor provides a list of available hotels, along with some information about each hotel. The following table displays data about hotel options. Identify the individuals and variables in this data set. Classify each variable as categorical or quantitative.

Hotel	Pool	Exercise room?	Internet ($/day)	Restaurants	Distance to site (mi)	Room service?	Room rate ($/day)
Comfort Inn	Out	Y	0	1	8.2	Y	149
Fairfield Inn & Suites	In	Y	0	1	8.3	N	119
Baymont Inn & Suites	Out	Y	0	1	3.7	Y	60
Chase Suite Hotel	Out	N	15	0	1.5	N	139
Courtyard	In	Y	0	1	0.2	Dinner	114
Hilton	In	Y	10	2	0.1	Y	156
Marriott	In	Y	9.95	2	0.0	Y	145

3. Portraits in data The table displays data on 10 randomly selected U.S. residents from a recent census. Identify the individuals and variables in this data set. Classify each variable as categorical or quantitative.

State	Number of family members	Age	Gender	Marital status	Yearly income	Travel time to work (min)
Kentucky	2	61	Female	Married	$31,000	20
Florida	6	27	Female	Married	$31,300	20
Wisconsin	2	27	Male	Married	$40,000	5
California	4	33	Female	Married	$36,000	10
Michigan	3	49	Female	Married	$25,100	25
Virginia	3	26	Female	Married	$35,000	15
Pennsylvania	4	44	Male	Married	$73,000	10
Virginia	4	22	Male	Never married/single	$13,000	0
California	1	30	Male	Never married/single	$50,000	15
New York	4	34	Female	Separated	$40,000	40

4.   **Who buys cars?** A new-car dealer keeps records on car buyers for future marketing purposes. The table gives information on the last 4 buyers. Identify the individuals and variables in this data set. Classify each variable as categorical or quantitative.

Buyer's name	Zip code	Gender	Buyer's distance from dealer (mi)	Car model	Engine type (cylinders)	Price
P. Smith	27514	M	13	Fiesta	4	$26,375
K. Ewing	27510	M	10	Mustang	8	$39,500
L. Shipman	27516	F	2	Fusion	4	$38,400
S. Reice	27243	F	4	F-150	6	$56,000

5.   **Choose your power** The online survey (page 6) also asked which superpower students would choose to have—fly, freeze time, invisibility, super strength, or telepathy (ability to read minds). Here are the responses from the 40 students in the sample. Summarize the distribution of superpower preference with a frequency table and a relative frequency table.

pg 7

Fly	Freeze time	Telepathy	Fly	Telepathy
Super strength	Telepathy	Telepathy	Fly	Super strength
Invisibility	Freeze time	Fly	Telepathy	Freeze time
Telepathy	Super strength	Fly	Freeze time	Telepathy
Freeze time	Freeze time	Freeze time	Fly	Fly
Fly	Freeze time	Invisibility	Fly	Invisibility
Telepathy	Telepathy	Fly	Telepathy	Fly
Fly	Telepathy	Telepathy	Fly	Fly

6.   **Birth months** Here are the reported birth months for the 40 students in the online sample. Summarize the distribution of birth month with a frequency table and a relative frequency table.

January	March	August	April	June
March	January	June	November	January
July	December	April	April	January
December	May	December	December	December
June	August	March	January	July
April	July	April	June	May
January	August	April	October	January
December	March	February	July	June

7.   **Get some sleep** The online survey also asked how much sleep students got on a typical school night. Here are the responses from the 40 students in the sample (in hours). Summarize the distribution of sleep amount with a frequency table and a relative frequency table.

9	8	6	7.5	7	8	4	7	7	8
8	8	6	7	8	8	7	7	6	8
9	7	6	5	7	8	8.5	7	9	6
6	6.5	8	9	5	8	7	7	7	7

8.   **Crowded house?** The online survey also asked how many people lived in the student's home. Here are the responses from the 40 students in the sample. Summarize the distribution of household size with a frequency table and a relative frequency table.

3	5	3	2	4	6	4	4	3	5
4	4	2	2	4	4	3	4	3	3
5	3	5	5	4	4	4	5	3	3
3	4	3	3	4	3	2	6	2	4

Applying the Concepts

9.   **Where did you go?** June and Barry are interested in where students at their school travel for spring break. So they survey 100 classmates who took a trip during spring break this year. Then they make a spreadsheet that includes the state or country visited, how many nights they spent there, mode of transportation to get to the destination, distance from home, and average cost per night for each student's trip. Identify the individuals in this data set. Classify each variable as categorical or quantitative.

10.   **Protecting history** How can we help wood surfaces resist weathering, especially when restoring historic wooden buildings? Researchers prepared wooden panels and then exposed them to the weather. Here are some of the variables recorded: type of wood (yellow poplar, pine, cedar); type of water repellent (solvent-based, water-based); paint thickness (in millimeters); paint color (white, gray, light blue); weathering time (in months). Identify the individuals in this data set. Classify each variable as categorical or quantitative.

11.   **Numerical but not quantitative** Give two examples of variables that take numerical values but are categorical.

12.   **Quantigorical?** In most data sets, age is classified as a quantitative variable. Explain how age could be classified as a categorical variable.

13.   **Car stats** Popular magazines rank car models based on their overall quality. Describe two categorical variables and two quantitative variables that might be considered in determining the rankings.

14.   **Social media** You are preparing to study the social media habits of high school students. Describe two categorical variables and two quantitative variables that you might record for each student.

Lesson 1.2

Displaying Categorical Data

- Make and interpret bar charts of categorical data.
- Interpret pie charts.
- Identify what makes some graphs of categorical data deceptive.

A frequency table or relative frequency table summarizes a variable's distribution with numbers. For instance, the Current Population Survey conducted by the U.S. Census Bureau collected data on the highest educational level achieved by U.S. 25- to 34-year-olds in 2014. The relative frequency table summarizes the data.[6] To display the distribution more clearly, use a graph.

Level of education	Percent
Less than high school	13.2
High school graduate	22.6
Some college	28.7
Bachelor's degree	24.9
Advanced degree	10.6

You can make a **bar chart** or a **pie chart** for categorical data. (Bar charts are sometimes called *bar graphs*. Pie charts are sometimes referred to as *circle graphs*.) We'll discuss graphs for quantitative data in the next few lessons.

DEFINITION Bar chart, Pie chart

A **bar chart** shows each category as a bar. The heights of the bars show the category frequencies or relative frequencies.

A **pie chart** shows each category as a slice of the "pie." The areas of the slices are proportional to the category frequencies or relative frequencies.

Figure 1.1 shows a bar chart and a pie chart of the data on the educational achievement of U.S. 25- to 34-year-olds in 2014. You can see that the most common level of education for this age group was "some college."

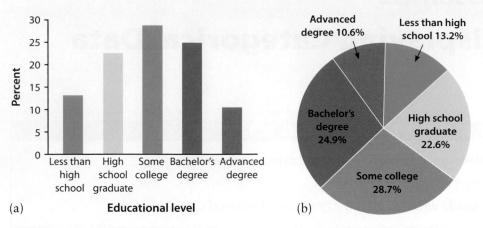

FIGURE 1.1 (a) Bar chart and (b) pie chart of the distribution of educational level attained in 2014 by people aged 25 to 34 in the United States.

Bar Charts and Pie Charts

It is fairly easy to make a bar chart by hand. Here's how you do it.

How to Make a Bar Chart

1. **Draw and label the axes.** Put the name of the categorical variable under the horizontal axis. To the left of the vertical axis, indicate whether the graph shows the frequency (count) or relative frequency (percent or proportion) of individuals in each category.

2. **"Scale" the axes.** Write the names of the categories at equally spaced intervals under the horizontal axis. On the vertical axis, start at 0 and place tick marks at equal intervals until you exceed the largest frequency or relative frequency in any category.

3. **Draw bars** above the category names. Make the bars equal in width and leave gaps between them. Be sure that the height of each bar corresponds to the frequency or relative frequency of individuals in that category.

Making a graph is not an end in itself. The purpose of a graph is to help us understand the data. When you look at a graph, always ask, "What do I see?"

EXAMPLE

Would students rather be happy or rich?

Making and interpreting a bar chart

PROBLEM: Here is a frequency table of the preferred status data for the 40 students in the "So you want to be happy?" example from Lesson 1.1. Make a bar chart to display the data. Describe what you see.

Frequency table

Preferred status	Frequency
Famous	7
Happy	21
Healthy	4
Rich	8
Total	40

SOLUTION:

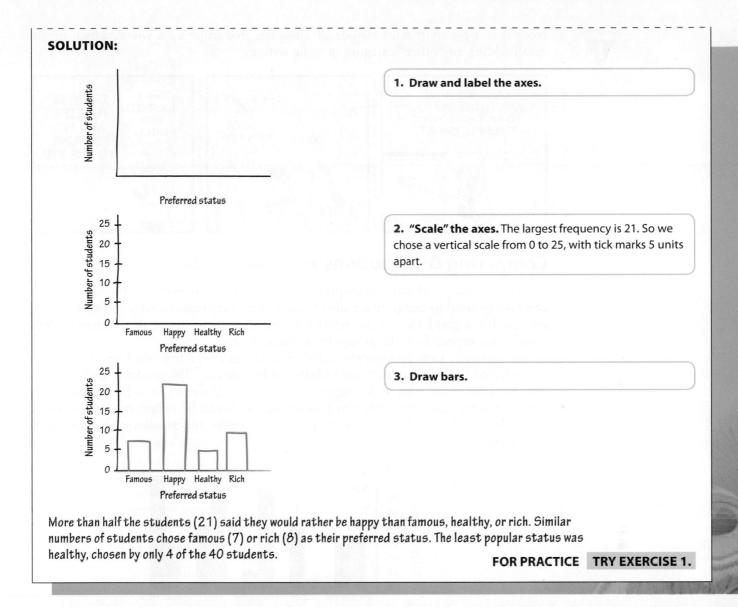

1. **Draw and label the axes.**

2. **"Scale" the axes.** The largest frequency is 21. So we chose a vertical scale from 0 to 25, with tick marks 5 units apart.

3. **Draw bars.**

More than half the students (21) said they would rather be happy than famous, healthy, or rich. Similar numbers of students chose famous (7) or rich (8) as their preferred status. The least popular status was healthy, chosen by only 4 of the 40 students.

FOR PRACTICE   TRY EXERCISE 1.

You can use a pie chart when you want to emphasize each category's relation to the whole. Pie charts are challenging to make by hand, but technology will do the job for you.

EXAMPLE

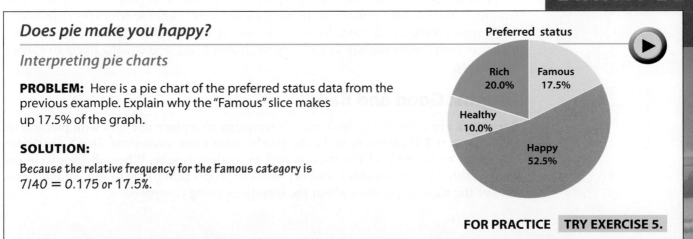

Does pie make you happy?

Interpreting pie charts

PROBLEM: Here is a pie chart of the preferred status data from the previous example. Explain why the "Famous" slice makes up 17.5% of the graph.

SOLUTION:
Because the relative frequency for the Famous category is $7/40 = 0.175$ or 17.5%.

FOR PRACTICE   TRY EXERCISE 5.

CAUTION

Note that a pie chart must include all categories that make up a whole, which might mean adding an "other" category in some settings.

Comparing Distributions with Bar Charts

Bar charts and pie charts can display a distribution of categorical data. A bar chart can also be used to compare the distribution of a categorical variable in two or more groups. It's a good idea to use relative frequencies (percents or proportions) when comparing, especially if the groups have different sizes.

For instance, a random sample of 200 children ages 9–17 from the United Kingdom was selected from an international website's online survey.[7] The gender of each student was recorded along with which superpower they would most like to have: invisibility, super strength, telepathy (ability to read minds), ability to fly, or ability to freeze time. The *side-by-side bar chart* shows the percents of males and females who chose each superpower.

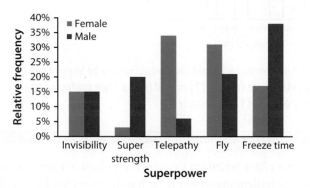

How do the distributions of superpower preference compare for the males and females in the sample? Females were much more likely to choose telepathy than males, while males were much more likely to choose super strength or freeze time than females. Females were slightly more likely to choose flying and equally likely to choose invisibility.

Graphs: Good and Bad

Bar charts are a bit dull to look at. It is tempting to replace the bars with pictures or to use special 3-D effects to make the graphs seem more interesting. Don't do it! Our eyes react to the area of the bars as well as to their height. When all bars have the same width, the area (width × height) varies in proportion to the height, and our eyes receive the right impression about the quantities being compared.

EXAMPLE

Who wants to party?

Beware the pictograph!

PROBLEM: The students in Mr. Tyson's high-school statistics class were recently asked if they would prefer a pasta party, a pizza party, or a donut party. Here are the data on the responses of that class about what type of party they would prefer.

Preferred party	Frequency
Donut	5
Pasta	18
Pizza	7
Total	30

(a) Here's a clever graph of the data that uses pictures instead of the more traditional bars. How is this graph misleading?

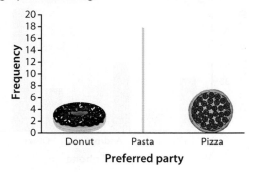

Preferred party

(b) Here is a bar chart of the data. Why could this graph be considered deceptive?

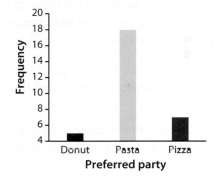

Preferred party

SOLUTION:

(a) The pictograph makes it seem like the number of students who preferred a pizza party (or a donut party) is much larger than the number of students who preferred a pasta party, which isn't the case. The area of the pasta bar is much smaller than the area of the donut and pizza images.

> Although the heights of the pictures are accurate, our eyes respond to the area of the pictures.

(b) By starting the vertical scale at 4 instead of 0, it looks like the number of students who preferred a pizza party is 3 times larger than the number of students who preferred a donut party, which isn't the case. In addition, it looks like the number of students who preferred a pasta party is more than 4 times larger than the number of students who preferred a pizza party, which also is not true.

> By starting the vertical scale at a number other than zero, we get a distorted impression of the relative numbers of students in the three categories.

FOR PRACTICE   TRY EXERCISE 9.

There are two important lessons to be learned from this example: (1) beware the pictograph, and (2) watch those scales.

CAUTION

LESSON APP 1.2

Which cell phone speaks to you?

The Pew Research Center asked a random sample of 2024 adult cell-phone owners from the United States which type of cell phone they own: iPhone, Android, or other (including non-smartphones). The frequency table displays the results.[8]

Type of cell phone	Frequency
iPhone	467
Android	503
Other	1,054
Total	**2,024**

1. Make a bar chart to display the distribution of phone ownership among all 2024 people in the sample. Describe what you see.

   The side-by-side bar chart displays the distribution of phone ownership for each of three age groups.

2. Write a few sentences comparing the distributions of phone ownership for the three age groups.

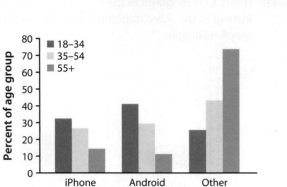

TECH CORNER

Making Bar Charts and Pie Charts with an Applet

You can use the *One Categorical Variable* applet at **highschool.bfwpub.com/spa3e** to make a bar chart or a pie chart. For the preferred-status data (page 12):

Preferred status	Frequency
Famous	7
Happy	21
Healthy	4
Rich	8
Total	40

1. Enter Preferred status in the Variable name box.

2. Select Single in the Groups menu and choose to input data as Counts in categories.

3. Type the category names and frequencies shown. Click on the + button to add rows to the frequency table.

One Categorical Variable
Variable name: Preferred status
Groups: Single ▼
Input data as: Counts in categories ▼

	Category Name	Frequency	
1	Famous	7	-
2	Happy	21	-
3	Healthy	4	-
4	Rich	8	-
			+

[Begin analysis] [Edit inputs] [Reset everything]

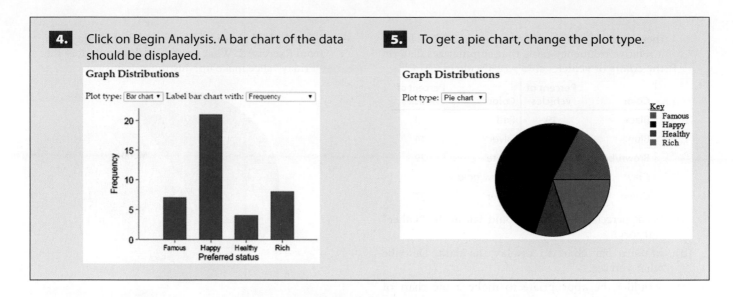

4. Click on Begin Analysis. A bar chart of the data should be displayed.

Graph Distributions

Plot type: Bar chart ▼ Label bar chart with: Frequency ▼

5. To get a pie chart, change the plot type.

Graph Distributions

Plot type: Pie chart ▼

Key
■ Famous
■ Happy
■ Healthy
■ Rich

Lesson 1.2

WHAT DID YOU LEARN?

LEARNING TARGET	EXAMPLES	EXERCISES
Make and interpret bar charts of categorical data.	p. 12	1–4
Interpret pie charts.	p. 13	5–8
Identify what makes some graphs of categorical data deceptive.	p. 15	9–12

Exercises

Lesson 1.2

Mastering Concepts and Skills

1. **Radio frequencies?** Arbitron, the rating service for radio audiences, places U.S. radio stations into categories that describe the kinds of programs they broadcast. The frequency table summarizes the distribution of station formats in a recent year.[9] Make a bar chart to display the data. Describe what you see.

pg **12**

Format	Count of stations
Adult contemporary	2,536
All sports	1,274
Contemporary hits	1,012
Country	2,893
News/talk/information	4,077
Oldies	831
Religious	3,884
Rock	1,636
Spanish language	878
Variety	1,579
Other formats	4,852

2. **What day were you born?** The frequency table summarizes the distribution of day of the week for all babies born in a single week in the United States. Make a bar chart to display the data. Describe what you see.

Day	Births
Sunday	7,374
Monday	11,704
Tuesday	13,169
Wednesday	13,038
Thursday	13,013
Friday	12,664
Saturday	8,459

3. **Cool colors** Popularity of colors for cars and light trucks changes over time. Silver passed green in 2000 to become the most popular color worldwide, then gave way to shades of white in 2007. Here is a relative frequency table that summarizes data on the colors of vehicles sold worldwide in 2014.[10]

Color	Percent of vehicles	Color	Percent of vehicles
Black	19	Red	9
Blue	6	Silver	14
Brown/beige	5	White	29
Gray	12	Yellow/gold	3
Green	1	Other	??

(a) What percent of vehicles would fall in the "Other" category?

(b) Make a bar chart to display the data. Describe what you see.

(c) Would it be appropriate to make a pie chart of these data? Explain.

4. **Slicing up spam** E-mail spam is the curse of the Internet. Here is a relative frequency table that summarizes data on the most common types of spam.[11]

Type of spam	Percent	Type of spam	Percent
Adult	19	Leisure	6
Financial	20	Products	25
Health	7	Scams	9
Internet	7	Other	??

(a) What percent of spam would fall in the "Other" category?

(b) Make a bar chart to display the data. Describe what you see.

(c) Would it be appropriate to make a pie chart of these data? Explain.

5. **Radio country** Here is a pie chart of the radio station format data from Exercise 1. What percent of the graph does the "Country" slice make up? Justify your answer.

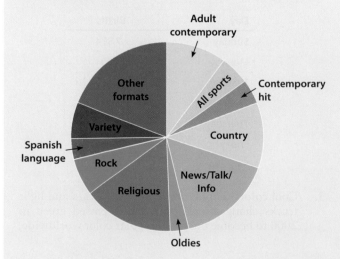

6. **Friday's child** Here is a pie chart of the birthday data from Exercise 2. What percent of the graph does the "Friday" slice make up? Justify your answer.

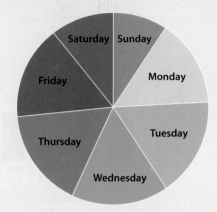

7. **What is your major?** About 3 million first-year students enroll in U.S. colleges and universities each year. The pie chart displays data on the percent of first-year students who plan to major in several disciplines.[12] About what percent of first-year students plan to major in business? In education?

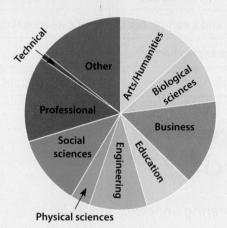

8. **Family origins** Here is a pie chart of Census Bureau data to show the countries from which the more than 14 million Asians in the United States in 2010 descend.[13] About what percent of Asians were of Chinese origin? Korean?

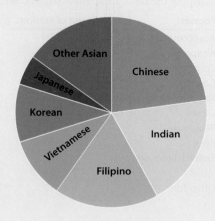

9. **Game on!** Students in a high school statistics class were given data about the favorite sport to play for a group of 35 girls. They produced the following pictograph. Explain how this graph is misleading.

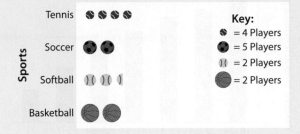

10. **Social media** The Pew Research Center surveyed a random sample of U.S. teens and adults about their use of social media in 2013. The following pictograph displays some results. Explain how this graph is misleading.

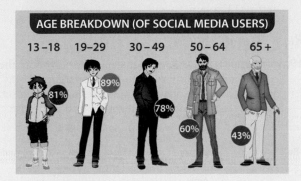

11. **Support the court?** A news network reported the results of a survey about a controversial court decision. The network initially posted on its website a bar chart of the data similar to the one that follows. Explain how this graph is misleading. (*Note:* When notified about the misleading nature of its graph, the network posted a corrected version.)

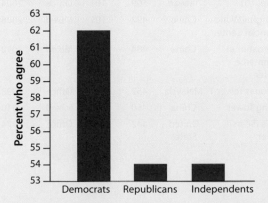

12. **Your favorite subject?** The bar chart shows the distribution of favorite subject for a sample of 1000 high school juniors. Explain how this graph is misleading.

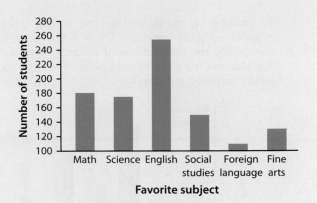

Applying the Concepts

13. **Frequent superpower?** The online survey from Lesson 1.1 (page 6) asked which superpower high school students would choose to have—fly, freeze time, invisibility, super strength, or telepathy. Here are the responses from the 40 students in the sample. Make a *relative frequency* bar chart for these data. Describe what you see.

Fly	Freeze time	Telepathy	Fly	Telepathy
Super strength	Telepathy	Telepathy	Fly	Super strength
Invisibility	Freeze time	Fly	Telepathy	Freeze time
Telepathy	Super strength	Fly	Freeze time	Telepathy
Freeze time	Freeze time	Freeze time	Fly	Fly
Fly	Freeze time	Invisibility	Fly	Invisibility
Telepathy	Telepathy	Fly	Telepathy	Fly
Fly	Telepathy	Telepathy	Fly	Fly

14. **Birth months** Here are the reported birth months for the 40 students in the online sample from Lesson 1.1 (page 6). Make a *relative frequency* bar chart for these data. Describe what you see.

January	March	August	April	June
March	January	June	November	January
July	December	April	April	January
December	May	December	December	December
June	August	March	January	July
April	July	April	June	May
January	August	April	October	January
December	March	February	July	June

15. **Far from home** A survey asked first-year college students, "How many miles is this college from your permanent home?" Students had to choose from the following options: 5 or fewer, 6 to 10, 11 to 50, 51 to 100, 101 to 500, or more than 500. The bar chart on the following page shows the percentage

of students at public and private 4-year colleges who chose each option.[14] Write a few sentences comparing the distributions of distance from home for students from private and public 4-year colleges who completed the survey.

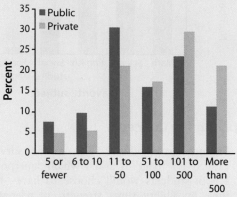

Distance from home (mi)

16. **Vehicle colors—U.S. versus Europe** Favorite vehicle colors may differ among countries. The bar chart displays data on the most popular car colors in a recent year for the United States and Europe. Write a few sentences comparing the distributions.

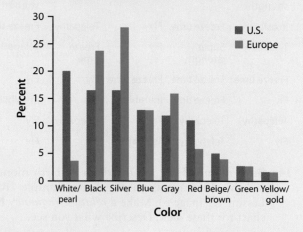

Color

Extending the Concepts

17. **Pareto charts** It is often more revealing to arrange the bars in a bar chart from tallest to shortest, moving from left to right. Some people refer to this type of bar chart as a *Pareto chart*, named after Italian economist Vilfredo Pareto. Make a Pareto chart for the data in Exercise 3. How is this graph more revealing than one with the bars ordered alphabetically?

18. **Who goes to movies?** The bar chart displays data on the percent of people in several age groups who attended a movie in the past 12 months:[15]

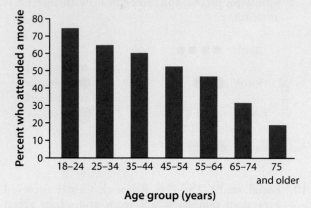

Age group (years)

(a) Describe what the graph reveals about movie attendance in the different age groups.

(b) Would it be appropriate to make a pie chart in this setting? Explain.

Recycle and Review

19. **Skyscrapers (1.1)** Here is some information about the tallest buildings in the world (completed by 2014).[16] Identify the individuals and variables in this data set. Classify each variable as categorical or quantitative.

Building	Country	Height (m)	Floors	Use	Year completed
Burj Khalifa	United Arab Emirates	828	163	Mixed	2010
Shanghai Tower	China	632	121	Mixed	2014
Makkah Royal Clock Tower Hotel	Saudi Arabia	601	120	Hotel	2012
One World Trade Center	United States	541	104	Office	2013
Taipei 101	Taiwan	509	101	Office	2004
Shanghai World Financial Center	China	492	101	Mixed	2008
International Commerce Center	China	484	118	Mixed	2010
Petronas Tower 1	Malaysia	452	88	Office	1998
Zifeng Tower	China	450	89	Mixed	2010
Willis (Sears) Tower	United States	442	108	Office	1974

Lesson 1.3
Displaying Quantitative Data: Dotplots

You can use a bar chart or pie chart to display categorical data. A **dotplot** is the simplest graph for displaying quantitative data.

DEFINITION Dotplot

A **dotplot** shows each data value as a dot above its location on a number line.

Figure 1.2 shows a dotplot of the number of siblings reported by each student in a statistics class. You'll learn how to make and interpret dotplots in this lesson.

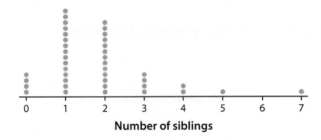

Number of siblings

FIGURE 1.2 Dotplot of data on the number of siblings reported by students in a statistics class.

Making and Interpreting Dotplots

For small sets of quantitative data, it is fairly easy to make a dotplot by hand.

How to Make a Dotplot

1. **Draw and label the axis.** Draw a horizontal axis and put the name of the quantitative variable underneath.
2. **Scale the axis.** Find the smallest and largest values in the data set. Start the horizontal axis at a number equal to or less than the smallest value and place tick marks at equal intervals until you equal or exceed the largest value.
3. **Plot the values.** Mark a dot above the location on the horizontal axis corresponding to each data value. Try to make all the dots the same size and space them out equally as you stack them.

Remember what we said in Lesson 1.2: Making a graph is not an end in itself. When you look at a graph, always ask, "What do I see?"

EXAMPLE

Which cars guzzle gas?

Making a dotplot

PROBLEM: The Environmental Protection Agency (EPA) is in charge of determining and reporting fuel economy ratings for cars. Think of those large window stickers on a new car. Here are the EPA estimates of highway gas mileage in miles per gallon (mpg) for a sample of 21 model year 2014 midsize cars.[17]

© D. Hurst/Alamy Stock Photo

Model	mpg	Model	mpg	Model	mpg
Acura RLX	31	Dodge Avenger	30	Mazda 6	40
Audi A8	28	Ford Fusion	31	Mercedes-Benz E350	30
BMW 550i	25	Hyundai Elantra	38	Nissan Maxima	26
Buick Lacrosse	28	Jaguar XF	30	Subaru Legacy	32
Cadillac CTS	27	Kia Optima	31	Toyota Prius	48
Chevrolet Malibu	30	Lexus ES 350	31	Volkswagen Passat	34
Chrysler 200	30	Lincoln MKZ	31	Volvo S80	25

(a) Make a dotplot of these data.

(b) Explain what the dot above 38 represents.

(c) What percent of the car models in the sample get more than 35 mpg on the highway?

SOLUTION:

(a)

Highway gas mileage (mpg)

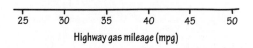

Highway gas mileage (mpg)

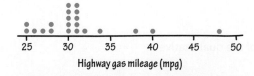

Highway gas mileage (mpg)

1. Draw and label the axis.

2. Scale the axis. The smallest value is 25 and the largest value is 48. So we chose a scale from 25 to 50 with tick marks 5 units apart.

3. Plot the values.

(b) The dot above 38 represents the 2014 Hyundai Elantra, which gets 38 mpg on the highway.

(c) 3/21 ≈ 0.143 or about 14.3% of the car models in the sample get more than 35 mpg on the highway.

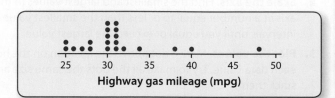

Highway gas mileage (mpg)

FOR PRACTICE   TRY EXERCISE 1.

Describing Shape

When you describe the shape of a dotplot or other graph of quantitative data, focus on the main features. Look for major peaks, not for minor ups and downs in the graph. Look for clusters of values and obvious gaps. Decide if the distribution is roughly **symmetric** or clearly **skewed.**

> **DEFINITION** Symmetric and skewed distributions
>
> A distribution is roughly **symmetric** if the right side of the graph (containing the half of the observations with larger values) is approximately a mirror image of the left side.
>
>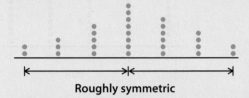
>
> **Roughly symmetric**
>
> A distribution is **skewed to the right** if the right side of the graph is much longer than the left side.
>
>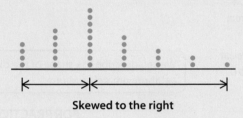
>
> **Skewed to the right**
>
> A distribution is **skewed to the left** if the left side of the graph is much longer than the right side.
>
>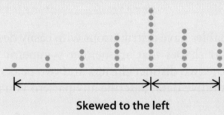
>
> **Skewed to the left**

For ease, we sometimes say "left-skewed" instead of "skewed to the left" and "right-skewed" instead of "skewed to the right." The direction of skewness is toward the long tail, not the direction where most observations are clustered. The drawing below is a cute but corny way to help you keep this straight. To avoid danger, Mr. Starnes skis on the gentler slope—in the direction of the skewness.

CAUTION

Skewed to the left!

EXAMPLE

What do distributions show?

Describing shape

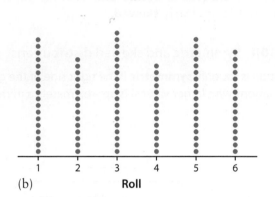

PROBLEM: The dotplots below display two different sets of quantitative data. Graph (a) shows the EPA highway gas mileage ratings for a sample of 21 model year 2014 midsize cars. Graph (b) shows the results of 100 rolls of a 6-sided die. Describe the shape of each distribution.

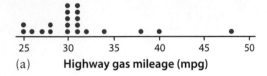

(a)   **Highway gas mileage (mpg)**

(b)   **Roll**

SOLUTION:

(a) The dotplot is right-skewed with a single peak near 30 to 31 mpg, one main cluster of dots between 25 and 34 mpg, a small gap from 34 to 38 mpg, and a large gap from 40 to 48 mpg.

(b) The distribution of die rolls is roughly symmetric with no clear peak. It has about the same height for all values from 1 to 6.

> We can describe the shape of the distribution of die rolls as "approximately uniform."

FOR PRACTICE   TRY EXERCISE 7.

Some quantitative variables have distributions with easily described shapes. But many distributions have irregular shapes that are neither symmetric nor skewed. Some data show other patterns, like the two distinct clusters and two peaks in Figure 1.3. When you examine a graph of quantitative data, describe any pattern you see as clearly as you can.

FIGURE 1.3 Dotplot displaying the duration, in minutes, of 220 eruptions of the Old Faithful geyser. This distribution has two main clusters of data and two clear peaks—one near 2 minutes and the other near 4.5 minutes. We could describe this graph as *bimodal* because it has two peaks.

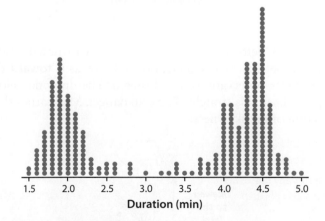

Duration (min)

Describing and Comparing Distributions

Here is a general strategy for describing a distribution of quantitative data.

How to Describe the Distribution of a Quantitative Variable

In any graph, look for the *overall pattern* and for clear *departures* from that pattern.

* You can describe the overall pattern of a distribution by its **shape**, **center**, and **variability**.

* An important kind of departure is an **outlier**, a value that falls outside the overall pattern.

We will discuss more formal ways to measure center and variability and to identify outliers in future lessons. For now, just use what you have learned about these ideas in previous math courses when describing or comparing graphs of quantitative data.

E X A M P L E

Anybody home?

Comparing distributions with dotplots

PROBLEM: How do the numbers of people living in households in the United Kingdom (U.K.) and South Africa compare? To help answer this question, we used Census At School's "Random Data Selector" to choose 50 students from each country. Here are parallel dotplots of the household sizes reported by the survey respondents. Compare the distributions of household size for these two countries.

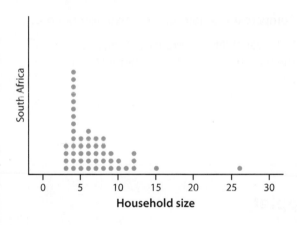

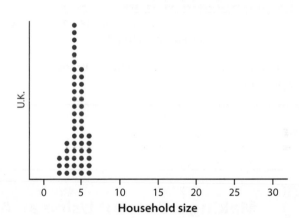

SOLUTION:

Shape: The South Africa distribution is skewed to the right and single-peaked, while the U.K. distribution is roughly symmetric and single-peaked.

Center: Household sizes for the South African students tend to be larger (center ≈ 6) than for the U.K. students (center ≈ 4).

Variability: The household sizes for the South African students vary more (from 3 to 26 people) than for the U.K. students (from 2 to 6 people).

Outliers: There aren't any obvious outliers in the U.K. dotplot. The two large values in the South Africa dotplot—students living in households with 15 and 26 people—appear to be outliers.

FOR PRACTICE **TRY EXERCISE 11.**

When comparing distributions of quantitative data, it's not enough just to list values for the center and variability of each distribution. You have to explicitly *compare* these values, using expressions like "greater than," "less than," or "about the same as."

LESSON APP 1.3

How can we check the health of a stream?

Nitrates are organic compounds that are a main ingredient in fertilizers. When those fertilizers run off into streams, the nitrates can have a toxic effect on fish. An ecologist studying nitrate pollution in two streams measures nitrate concentrations at 42 places on Stony Brook and 42 places on Mill Brook. The parallel dotplots display the data.

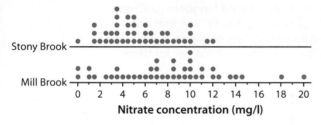

1. Explain what the dot above 12 in the Stony Brook graph represents.

2. What percent of the nitrate concentration measurements for each stream exceeded 10 milligrams per liter (mg/l)?

3. Compare the centers of these two distributions.

4. Is the variability in nitrate concentrations for the two streams similar or different? Justify your answer.

Goodluz/Shutterstock

TECH CORNER

Making a Dotplot Using an Applet

You can use the *One Quantitative Variable* applet at **highschool.bfwpub.com/spa3e** to make a dotplot. For the highway gas mileage data on page 22:

1. Enter Highway gas mileage as the Variable name.

2. Select 1 as the number of groups and Raw data as the input method.

3. Enter the data. Be sure to separate the data values with commas or spaces as you type them.

4. Click Begin analysis. A dotplot of the data should appear.

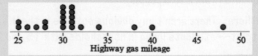

One Quantitative Variable

Variable name: Highway gas mileage
Number of groups: 1 ▾
Input: Raw data   ▾

Input data separated by commas or spaces.
Group 1 data: 31 28 25 28 27 30 30 30 31 38 30 31 31 31 40

Begin analysis   Edit inputs   Reset everything

The applet can also be used to make parallel dotplots for comparing the distribution of a variable in two or more groups, like the graph shown in the Household Size example on page 25.

Lesson 1.3

WHAT DID YOU LEARN?

LEARNING TARGET	EXAMPLES	EXERCISES
Make and interpret dotplots of quantitative data.	p. 22	1–6
Describe the shape of a distribution.	p. 24	7–10
Compare distributions of quantitative data with dotplots.	p. 25	11–14

Exercises   Lesson 1.3

Mastering Concepts and Skills

1. **Magic words** Here are data on the lengths of the
pg 22 first 25 words on a randomly selected page from
Harry Potter and the Prisoner of Azkaban.

    2   3   4   10   2   11   2   8   4   3   7   2   7

    5   3   6   4   4   2   5   8   2   3   4   4

(a) Make a dotplot of these data.

(b) Explain what the dot above 6 represents.

(c) What percent of the words have more than 4 letters?

2. **Frozen pizza** Here are the number of calories per
serving for 16 brands of frozen cheese pizza.[18]

    340   340   310   320   310   360   350   330

    260   380   340   320   310   360   350   330

(a) Make a dotplot of these data.

(b) Explain what the dot above 260 represents.

(c) What percent of the frozen pizzas have fewer than
330 calories?

3. **How fuel efficient?** Here are the EPA estimates
of city gas mileage in miles per gallon (mpg) for
the sample of 21 model year 2014 midsize cars.[19]
Make a dotplot of these data.

Model	mpg	Model	mpg
Acura RLX	20	Kia Optima	20
Audi A8	18	Lexus ES 350	21
BMW 550i	17	Lincoln MKZ	22
Buick Lacrosse	18	Mazda 6	28
Cadillac CTS	18	Mercedes-Benz E350	21
Chevrolet Malibu	21	Nissan Maxima	19
Chrysler 200	21	Subaru Legacy	24
Dodge Avenger	21	Toyota Prius	51
Ford Fusion	22	Volkswagen Passat	24
Hyundai Elantra	28	Volvo S80	18
Jaguar XF	19		

4. **Gooooaaal!** How good was the 2012 U.S. wom-
en's soccer team? With players like Abby Wambach, Megan Rapinoe, and Carli Lloyd, the team
put on an impressive showing en route to winning the gold medal at the 2012 Summer Olympics in London. Here are data on the number of
goals scored by the team in games played in the
12 months prior to the 2012 Olympics.[20] Make a
dotplot of these data.

    1   3   1   14   13   4   3   4   2   5   2   0   4

    1   3   4   3   4   2   4   3   1   2   4   2

5. **Visualizing fuel efficiency** The dotplot shows the
difference (Highway – City) in EPA mileage ratings
for each of the 21 model year 2014 midsize cars
from Exercise 3.

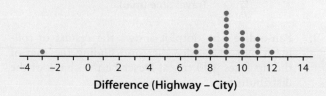

Difference (Highway – City)

(a) The dot above –3 is for the Toyota Prius. Explain
what this dot represents.

(b) What percent of these car models get fuel efficiency of at least 10 mpg more on the highway
than in the city?

6. **Look at that gooooaaal!** The dotplot shows
the difference in the number of goals scored in
each game (U.S. women's team – Opponent) in
Exercise 4.

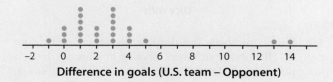

Difference in goals (U.S. team – Opponent)

(a) Explain what the dot above −1 represents.

(b) What percent of its games did the 2012 U.S. women's team win?

7. **Looking at the old** How old is the oldest person pg 24 you know? Prudential Insurance Company asked 400 people to place a blue sticker on a huge wall next to the age of the oldest person they have ever known. An image of the graph is shown here. Describe the shape of the distribution.

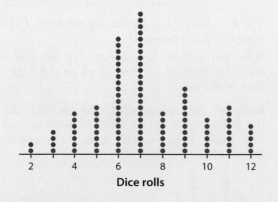

8. **Off to school** The dotplot displays data on the travel time to school (in minutes) reported by 50 Canadian students. Describe the shape of the distribution.

9. **Pair-a-dice** The dotplot shows the results of rolling a pair of 6-sided dice and finding the sum of the up faces 100 times. Describe the shape of the distribution.

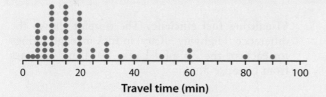

10. **Phone numbers** The dotplot displays the last digit of 100 phone numbers chosen at random from a phone book. Describe the shape of the distribution.

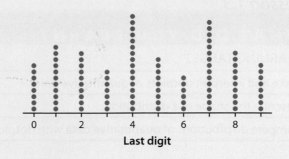

11. **Making money** The following parallel dotplots pg 25 show the total family income of randomly chosen individuals from Indiana (38 individuals) and New Jersey (44 individuals). Compare the distributions of total family incomes in these two samples.

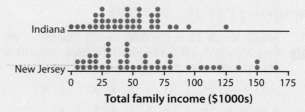

12. **Movie lengths** The following parallel dotplots display the lengths of the best 15 movies in each of three decades according to the Internet Movie Database (www.imdb.com). Compare the distributions of movie lengths for these three time periods.

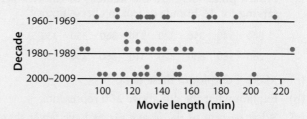

13. **Sugar high(er)?** Researchers collected data on 76 brands of cereal at a local supermarket.[21] For each brand, the values of several variables were recorded, including sugar (grams per serving), calories per serving, and the shelf in the store on which the cereal was located (1 = bottom, 2 = middle, 3 = top). Here are parallel dotplots of the data on sugar content by shelf.

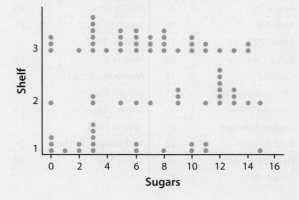

(a) Critics claim that supermarkets tend to put sugary kids' cereals on lower shelves, where the kids can see them. Do the data from this study support this claim? Justify your answer.

(b) Is the variability in sugar content of the cereals on the three shelves similar or different? Justify your answer.

14. **Enhancing creativity** Do external rewards—things like money, praise, fame, and grades—promote creativity? Researcher Teresa Amabile recruited 47 experienced creative writers who were college students and divided them at random into two groups. The students in one group were given a list of statements about external reasons (E) for writing, such as public recognition, making money, or pleasing their parents. Students in the other group were given a list of statements about internal reasons (I) for writing, such as expressing yourself and enjoying playing with words. Both groups were then instructed to write a poem about laughter. Each student's poem was rated separately by 12 different poets using a creativity scale.[22] These ratings were averaged to obtain an overall creativity score for each poem. Parallel dotplots of the two groups' creativity scores are shown here.

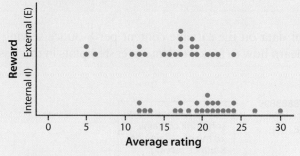

(a) What do you conclude about whether external rewards promote creativity? Justify your answer.

(b) Is the variability in creativity scores similar or different for the two groups? Justify your answer.

Applying the Concepts

15. **Bad dotplot** Janie asked 10 friends how many photos they posted on Instagram yesterday. Then she made the following dotplot to display the data. What's wrong with Janie's dotplot?

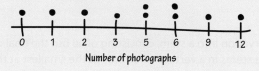

16. **Another bad dotplot** Herschel asked the students in his English class how many siblings they have. Then he made the following dotplot to display the data. What's wrong with Herschel's dotplot?

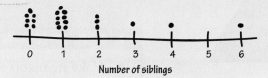

Recycle and Review

17. **Brands that sell (1.1)** The brands of the last 45 digital single-lens reflex (SLR) cameras sold on a popular Internet auction site are listed here. Summarize the distribution of camera brands with a frequency table and a relative frequency table.

Canon	Sony	Canon	Nikon	Fujifilm
Nikon	Canon	Sony	Canon	Canon
Nikon	Canon	Nikon	Canon	Canon
Canon	Nikon	Fujifilm	Canon	Nikon
Nikon	Canon	Canon	Canon	Canon
Olympus	Canon	Canon	Canon	Nikon
Olympus	Sony	Canon	Canon	Sony
Canon	Nikon	Sony	Canon	Fujifilm
Nikon	Canon	Nikon	Canon	Sony

18. **Divorce American Style (1.2)** The bar chart compares the marital status of U.S. adult residents (18 years old or older) in 1980 and 2010.[23] Write a few sentences comparing the distributions of marital status for these two years.

Lesson 1.4

Displaying Quantitative Data: Stemplots

- Make stemplots of quantitative data.
- Interpret stemplots.
- Compare distributions of quantitative data with stemplots.

Another simple type of graph for displaying quantitative data is a **stemplot** (also called a *stem-and-leaf plot*).

> **DEFINITION** Stemplot
>
> A **stemplot** shows each data value separated into two parts: a *stem*, which consists of all but the final digit, and a *leaf*, the final digit. The stems are ordered from least to greatest and arranged in a vertical column. The leaves are arranged in increasing order out from the appropriate stems.

Figure 1.4 shows a stemplot of data on the caffeine content per 8-ounce serving for several popular soft drinks. You'll learn how to make and interpret stemplots in this lesson.

```
1 | 556
2 | 033344
2 | 55667778888899
3 | 113
3 | 55567778
4 | 33
4 | 77
```

KEY: 2 | 8 means the soft drink contains 28 mg of caffeine per 8-ounce serving.

FIGURE 1.4 Stemplot showing the caffeine content (in milligrams or mg per 8-ounce serving) of various soft drinks.

Making and Interpreting Stemplots

It is fairly easy to make a stemplot by hand for small sets of quantitative data. Stemplots give us a quick picture of a distribution that includes the actual numerical values in the graph.

> ### How to Make a Stemplot
>
> 1. **Make stems.** Separate each observation into a stem, consisting of all but the final digit, and a leaf, the final digit. Write the stems in a vertical column with the smallest at the top. Draw a vertical line at the right of this column. Do not skip any stems, even if there is no data value for a particular stem.
> 2. **Add leaves.** Write each leaf in the row to the right of its stem.
> 3. **Order leaves.** Arrange the leaves in increasing order out from the stem.
> 4. **Add a key.** Provide a key that explains in context what the stems and leaves represent.

EXAMPLE

What does the Electoral College do?

Making a stemplot

PROBLEM: To become president of the United States, a candidate does not have to receive a majority of the popular vote. The candidate does, however, have to win a majority of the 538 electoral votes that are cast in the Electoral College. The table shows the number of electoral votes (EV) in 2016 for each of the 50 states and the District of Columbia. Make a stemplot of these data.

State	EV	State	EV	State	EV	State	EV
Alabama	9	Illinois	20	Montana	3	Rhode Island	4
Alaska	3	Indiana	11	Nebraska	5	South Carolina	9
Arizona	11	Iowa	6	Nevada	6	South Dakota	3
Arkansas	6	Kansas	6	New Hampshire	4	Tennessee	11
California	55	Kentucky	8	New Jersey	14	Texas	38
Colorado	9	Louisiana	8	New Mexico	5	Utah	6
Connecticut	7	Maine	4	New York	29	Vermont	3
Delaware	3	Maryland	10	North Carolina	15	Virginia	13
District of Columbia	3	Massachusetts	11	North Dakota	3	Washington	12
Florida	29	Michigan	16	Ohio	18	West Virginia	5
Georgia	16	Minnesota	10	Oklahoma	7	Wisconsin	10
Hawaii	4	Mississippi	6	Oregon	7	Wyoming	3
Idaho	4	Missouri	10	Pennsylvania	20		

SOLUTION:

```
0 |
1 |
2 |
3 |
4 |
5 |
```

1. Make stems. The smallest number of electoral votes for any state is 3 and the largest is 55. Thinking of single-digit numbers like 3 as 03, we use the first digit of each value as the stem. So we need stems from 0 to 5 (even though there is no value with a stem of 4).

```
0 | 9 3 6 9 7 3 3 4 4 6 6 8 8 4 6 3 5 6 4 5 3 7 7 4 9 3 6 3 5 3
1 | 1 6 1 0 1 6 0 0 4 5 8 3 2 1 0
2 | 9 0 9 0
3 | 8
4 |
5 | 5
```

Be sure to include this stem even though it contains no data.

2. Add leaves. For Alabama's 9 electoral votes, we place a 9 on the 0 stem. Then we place a 3 on the 0 stem for Alaska's 3 electoral votes. Next we place a 1 on the 1 stem for Arizona's 11 electoral votes. And so on.

```
0 | 3 3 3 3 3 3 3 3 4 4 4 4 4 5 5 5 6 6 6 6 6 6 7 7 7 8 8 9 9 9
1 | 0 0 0 0 1 1 1 1 2 3 4 5 6 6 8
2 | 0 0 9 9
3 | 8
4 |
5 | 5
```

KEY: 1|5 is a state with 15 electoral votes.

3. Order leaves.

4. Add a key. Explain in context what the stems and leaves represent.

FOR PRACTICE   TRY EXERCISE 1.

We can get a better picture of the electoral vote data by *splitting stems*. In Figure 1.5(a), the values from 0 to 9 are placed on the "0" stem. Figure 1.5(b) shows another stemplot of the same data. This time, values having leaves 0 through 4 are placed on one stem, while values ending in 5 through 9 are placed on another stem. Now we can see the shape of the distribution more clearly.

FIGURE 1.5 Two stemplots showing the electoral vote data. The graph in (b) improves on the graph in (a) by splitting stems.

(a)
```
0 | 333333333444444555566666677788999
1 | 000011112345668
2 | 0099
3 | 8
4 |
5 | 5
```
KEY: 1|5 is a state with 15 electoral votes.

(b)
```
0 | 3333333344444
0 | 55566666677788999
1 | 00001111234
1 | 5668
2 | 00
2 | 99
3 |
3 | 8
4 |
4 |
5 |
5 | 5
```

Here are a few tips to consider before making a stemplot:

- There is no magic number of stems to use. Too few or too many stems will make it difficult to see the distribution's shape. Five stems is a good minimum.
- If you split stems, be sure that each stem is assigned an equal number of possible leaf digits.
- When the data have too many digits, you can get more flexibility by rounding or truncating the data. See Exercises 13 and 14.

EXAMPLE

Who has the most votes?

Interpreting a stemplot

PROBLEM: Here once again is the stemplot with split stems for the electoral vote data. Use the stemplot and the data table from the previous example to answer these questions.

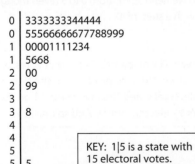

```
0 | 3333333344444
0 | 55566666677788999
1 | 00001111234
1 | 5668
2 | 00
2 | 99
3 |
3 | 8
4 |
4 |
5 |
5 | 5
```
KEY: 1|5 is a state with 15 electoral votes.

(a) What percent of states get 10 or more electoral votes?

(b) Describe the shape of the distribution.

(c) Which two values appear to be outliers? What states are these?

FOR PRACTICE TRY EXERCISE 5.

SOLUTION:

(a) 21/51 ≈ 0.412. About 41.2% of states get 10 or more electoral votes.

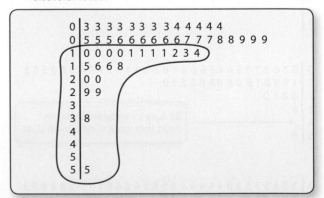

(b) The stemplot is skewed to the right with a single peak on the 05–09 stem and two clear gaps—one from 29 to 38 and the other from 38 to 55.

Note that skewed right always means skewed toward the larger values, regardless of the orientation of the graph.

(c) Apparent outliers: 38 (Texas) and 55 (California).

Comparing Distributions with Stemplots

You can use a *back-to-back stemplot* with common stems to compare the distribution of a quantitative variable in two groups. The leaves on each side are placed in order leading out from the common stem.

EXAMPLE

How can we distinguish oaks?

Comparing distributions with stemplots

PROBLEM: Of the many species of oak trees in the United States, 28 grow on the Atlantic Coast and 11 grow in California. How does the distribution of acorn sizes compare for oak trees in these two regions? Here are data on the average volumes of acorns (in cubic centimeters) for these 39 oak species:[24]

Atlantic	0.3	0.4	0.6	0.8	0.9	0.9	1.1	1.1	1.1	1.1	1.1	1.2	1.4	1.6
	1.8	1.8	1.8	2.0	2.5	3.0	3.4	3.6	3.6	4.8	6.8	8.1	9.1	10.5
California	0.4	1.0	1.6	2.0	2.6	4.1	5.5	5.9	6.0	7.1	17.1			

(a) Make a back-to-back stemplot for these data.

(b) Which of the two regions' oak tree species has the larger acorns?

(c) Are the shapes of the acorn size distributions similar or different in the two regions? Justify your answer.

SOLUTION:

(a)
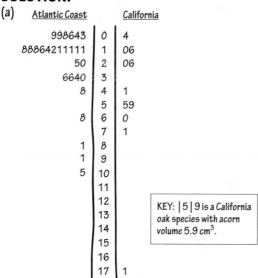

KEY: | 5 | 9 is a California oak species with acorn volume 5.9 cm³.

1. Make stems. The smallest acorn volume for any oak species is 0.3 cubic centimeter (cm³) and the largest volume is 17.1 cm³. We use the whole number part of each value as the stem and the decimal part as the leaf. So we need stems from 0 to 17.

2. Add leaves.

3. Order leaves. Put the Atlantic Coast oak species on the left side of the stem and the California oak species on the right side of the stem. The data values for both groups have been sorted in increasing order, so you can add the leaves on each stem from lowest to highest as you go. For the first Atlantic oak acorn, place a 3 to the left of the 0 stem. For the first California oak acorn volume, place a 4 on the right side of the 0 stem, and so on.

4. Add a key. Explain in context what the stems and leaves represent.

(b) California oak tree species tend to have larger acorn volumes (center ≈ 4.1 cm³) than Atlantic Coast oak tree species (center ≈ 1.7 cm³).

(c) The distributions of acorn size in the two regions are quite different. For Atlantic Coast oak tree species, the distribution is skewed to the right and single-peaked. For California oak tree species, the distribution has two distinct clusters (0.4 to 2.6 cm³ and 4.1 to 7.1 cm³), a large gap from 7.1 cm³ to 17.1 cm³, and no clear peak.

FOR PRACTICE   TRY EXERCISE 9.

LESSON APP 1.4

How many shoes are too many shoes?

How many pairs of shoes does a typical teenager own? To find out, a group of statistics students surveyed separate random samples of 20 female students and 20 male students from their large high school. Then they recorded the number of pairs of shoes that each person owned. Here are the data.

MissKadri/Getty Images

Females	50	26	26	31	57	19	24	22	23	38
	13	50	13	34	23	30	49	13	15	51
Males	14	7	6	5	12	38	8	7	10	10
	10	11	4	5	22	7	5	10	35	7

1. Make a stemplot of the female data. Do not split stems.

2. Describe the shape of the distribution.

3. Explain why we should split stems for the male data.

4. The back-to-back stemplot with split stems displays the data for both genders. Write a few sentences comparing the male and female distributions.

Females		Males
	0	4
	0	555677778
333	1	0000124
95	1	
4332	2	2
66	2	
410	3	
8	3	58
	4	
9	4	
100	5	
7	5	

KEY: 2|2 represents a male student with 22 pairs of shoes.

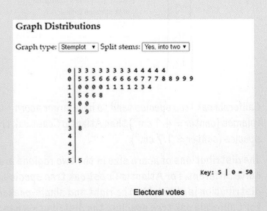

TECH CORNER

Using an Applet To Make a Stemplot

You can use the *One Quantitative Variable* applet at highschool.bfwpub.com/spa3e to make a stemplot. For the electoral vote data on page 31:

1. Enter Electoral votes as the Variable name.

2. Select 1 as the number of groups and Raw data as the input method.

3. Enter the data. Be sure to separate the data values with commas or spaces as you type them.

4. Click Begin analysis.

5. Change the Graph type to a stemplot.

6. Split stems to get a better picture of the distribution.

Graph Distributions

Graph type: Stemplot ▾ Split stems: Yes, into two ▾

```
0 | 3 3 3 3 3 3 3 4 4 4 4 4
0 | 5 5 5 6 6 6 6 6 7 7 7 8 8 9 9 9
1 | 0 0 0 0 1 1 1 1 2 3 4
1 | 5 6 6 8
2 | 0 0
2 | 9 9
3 |
3 | 8
4 |
4 |
5 |
5 | 5
```
Key: 5 | 0 = 50

Electoral votes

The applet can also make a back-to-back stemplot of the male and female shoe data, like the one shown in Lesson App 1.4.

Lesson 1.4

WHAT DID YOU LEARN?

LEARNING TARGET	EXAMPLES	EXERCISES
Make stemplots of quantitative data.	p. 31	1–4
Interpret stemplots.	p. 32	5–8
Compare distributions of quantitative data with stemplots.	p. 33	9–12

Exercises   Lesson 1.4

Mastering Concepts and Skills

1. **Science gets your heart beating!** Here are the resting heart rates of 26 ninth-grade biology students. Make a stemplot of these data. Do not split stems.

pg 31

61	78	77	81	48	75	70	77	70	76	86	55	65
60	63	79	62	71	72	74	74	64	66	71	66	68

2. **Hot enough for you?** Here are the high temperature readings in degrees Fahrenheit for Phoenix, Arizona, for each day in July 2013. Make a stemplot of these data. Do not split stems.

111	107	115	108	106	109	111	113	104	103	97
99	104	110	109	100	105	107	102	101	84	93
101	105	99	102	104	108	106	106	109		

3. **Something fishy** As part of a study on salmon health, researchers measured the pH of 25 salmon fillets. Here are the data. Make a stemplot of these data using split stems.

6.34 6.39 6.53 6.36 6.39 6.25 6.45 6.38 6.33 6.26 6.24 6.37 6.32

6.31 6.48 6.26 6.42 6.43 6.36 6.44 6.22 6.52 6.32 6.32 6.48

4. **Eat your beans!** Beans and other legumes are a great source of protein. The following data give the protein content of 31 different varieties of beans, in grams per 100 grams of cooked beans.[25] Make a stemplot of these data using split stems.

7.5	8.2	8.9	9.3	7.1	8.3	8.7	9.5	8.2	9.1	9.0	9.0
9.7	9.2	8.9	8.1	9.0	7.8	8.0	7.8	7.0	7.5	13.5	8.3
6.8	16.6	10.6	8.3	7.6	7.7	8.1					

5. **Science gets your heart beating!** Here is a stemplot using split stems for the heart-rate data from Exercise 1.

pg 32

```
4 | 8
5 |
5 | 5
6 | 01234
6 | 5668
7 | 0011244
7 | 567789
8 | 1
8 | 6
```

(a) What percent of these ninth-grade biology students have resting heart rates below 70 beats per minute?

(b) Describe the shape of the distribution.

(c) Which value appears to be an outlier? Give the stemplot a key using this value.

6. **Hot enough for you?** Here is a stemplot using split stems for the daily high temperature in Phoenix data from Exercise 2.

```
8  | 4
8  |
9  | 3
9  | 799
10 | 011223444
10 | 556667788999
11 | 0113
11 | 5
```

(a) What percent of days in this month were hotter than 100 degrees Fahrenheit (°F)?

(b) Describe the shape of the distribution.

(c) Which value appears to be an outlier? Give the stemplot a key using this value.

7. **Where are the older folks?** Following is a stemplot of the percents of residents aged 65 and older in the 50 states and the District of Columbia.[26]

```
 7 | 7
 8 |
 9 | 0
10 | 379
11 | 44
12 | 02333445899
13 | 02234555557788889
14 | 012334445689
15 | 49
16 | 0
17 | 3
```

KEY: 12|2 represents a state in which 12.2% of residents are 65 and older.

(a) What percent of states have more than 15% of residents aged 65 and older?

(b) Describe the shape of the distribution.

(c) Which value appears to be an outlier? Can you guess what state this is?

8. **South Carolina counties** Here is a stemplot of the areas of the 46 counties in South Carolina. Note that the data have been rounded to the nearest 10 square miles (mi²).

```
 3 | 9999
 4 | 0116689
 5 | 01115566778
 6 | 47899
 7 | 01245579
 8 | 0011
 9 | 13
10 | 8
11 | 233
12 | 2
```

KEY: 6 | 4 represents a county with an area of 640 square miles (rounded to the nearest 10 mi²)

(a) What percent of South Carolina counties have areas of less than 500 mi²?

(b) Describe the shape of the distribution.

(c) What is the area of the largest South Carolina county?

9. **Basketball scores** Here are the numbers of points scored by teams in the California Division I high school basketball playoffs in a single day's games.[27]

pg 33

71	38	52	47	55	53	76	65	77	63	65	63	68
54	64	62	87	47	64	56	78	64	58	51	91	74
71	41	67	62	106	46							

On the same day, the final scores of games in Division V were as follows:

98	45	67	44	74	60	96	54	92	72	93	46
98	67	62	37	37	36	69	44	86	66	66	58

(a) Make a back-to-back stemplot to compare the points scored by the 32 teams in the Division I playoffs and the 24 teams in the Division V playoffs.

(b) In which of the two divisions did teams score more points in their playoff games? Justify your answer.

(c) Are the shapes of the distributions of points scored similar or different in the two divisions? Justify your answer.

10. **Who's taller?** Who is taller, males or females? A sample of 14-year-olds from the United Kingdom was randomly selected. Here are the heights of the students (in centimeters).

Female	160	169	152	167	164	163	160	163	169	157	158
	153	161	165	165	159	168	153	166	158	158	166
Male	154	157	187	163	167	159	169	162	176	177	151
	175	174	165	165	183	180					

(a) Make a back-to-back stemplot for these data.

(b) Who tends to be taller in the United Kingdom: 14-year-old females or 14-year-old males? Justify your answer.

(c) Are the shapes of the male and female distributions of height similar or different? Justify your answer.

11. **Who hits the books more?** Researchers asked the students in a large first-year college class how many minutes they studied on a typical weeknight. The back-to-back stemplot displays the responses from random samples of 30 women and 30 men from the class, rounded to the nearest 10 minutes. Write a few sentences comparing the male and female distributions.

```
         Women |   | Men
               | 0 | 03333
            96 | 0 | 56668999
      22222222 | 1 | 02222222
888888888875555 | 1 | 558
          4440 | 2 | 00344
               | 2 |
               | 3 | 0
             6 | 3 |
```

KEY: 2 | 3 = 230 min

12. **Fill 'er up** The back-to-back stemplot displays the prices for regular gasoline at stations in Reading, Pennsylvania, and Yakima, Washington, in spring 2014. Write a few sentences comparing the two distributions.

```
        Reading |    | Yakima
              4 | 36 |
            996 | 36 | 7999
            422 | 37 | 333
     999999776655 | 37 | 5589999
              1 | 38 | 33
                | 38 | 55799
                | 39 |
                | 39 | 5
```

KEY: 36 | 7 = $3.67

Applying the Concepts

13. **Chasing food dollars** A marketing consultant observed 50 consecutive shoppers at a supermarket to find out how much each shopper spent in the store. Here are the data (in dollars), arranged in increasing order:

    3.11 8.88 9.26 10.81 12.69 13.78 15.23 15.62 17.00 17.39

    18.36 18.43 19.27 19.50 19.54 20.16 20.59 22.22 23.04 24.47

    24.58 25.13 26.24 26.26 27.65 28.06 28.08 28.38 32.03 34.98

    36.37 38.64 39.16 41.02 42.97 44.08 44.67 45.40 46.69 48.65

    50.39 52.75 54.80 59.07 61.22 70.32 82.70 85.76 86.37 93.34

    (a) Round each amount to the nearest dollar. Then make a stemplot using tens of dollars as the stems and dollars as the leaves.

    (b) Make another stemplot of the data by splitting stems. Which graph shows the shape of the distribution better?

    (c) Write a few sentences describing the amount of money spent by shoppers at this supermarket.

14. **What does ERA mean?** One way to measure the effectiveness of baseball pitchers is to use their *earned run average,* which measures how many earned runs opposing teams score, on average, every nine innings pitched. The overall earned run average for all pitchers in the major leagues in 2013 was 3.86. Here are the earned run averages for all 25 players who pitched for the Boston Red Sox in 2013.

    3.75 3.52 4.57 4.32 1.74 1.09 3.16 1.81 2.64 3.77 4.04 4.97 4.86

    5.34 4.88 4.62 3.86 3.52 5.56 3.60 8.60 5.40 6.35 9.82 9.00

    (a) Truncate the hundredths place of each data value and make a stemplot, using the ones digit as the stem and the tenths digit as the leaf.

    (b) Make another stemplot of the data by splitting stems. Which graph shows the shape of the distribution better?

    (c) Write a few sentences describing the distribution of earned run averages of Boston's pitchers in 2013.

Extending the Concepts

Sometimes, the variability in a data set is so small that splitting stems in two doesn't produce a stemplot that shows the shape of the distribution well. We can often solve this problem by splitting the stem into five parts, each consisting of two leaf values: 0 and 1, 2 and 3, 4 and 5, and so on.

Exercises 15 and 16 refer to the following setting. Here are the weights, in ounces, of 36 navel oranges selected from a large shipment to a grocery store.

5.7	5.4	5.8	5.3	4.6	4.9	5.6	5.3	5.5	5.5	5.4	5.8
5.3	5.5	5.5	5.4	5.8	5.9	5.4	5.1	5.0	5.5	5.7	4.9
5.0	5.3	5.1	5.2	5.7	5.6	5.8	4.5	5.2	5.4	5.7	5.6

15. **Weights of oranges** Make a stemplot of the data by splitting stems into two parts. Explain why this graph does not display the distribution of orange weights effectively.

16. **Splitting the oranges** Make a stemplot of the data by splitting stems into five parts. Describe the shape of the distribution.

Recycle and Review

17. **More gas guzzlers** (1.2) The EPA-estimated highway fuel efficiency for four different sedans is given in the bar chart.[28] Explain how this graph is misleading.

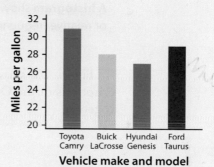

18. **Comparing tuition** (1.3) The dotplot shows the 2014 out-of-state tuition for the 40 largest colleges and universities in North Carolina.[29] Describe the overall pattern of the distribution and identify any clear departures from the pattern.

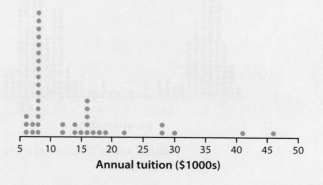

Lesson 1.5

Displaying Quantitative Data: Histograms

LEARNING TARGETS

● Make histograms of quantitative data.

● Interpret histograms.

● Compare distributions of quantitative data with histograms.

You can use a dotplot or stemplot to display quantitative data. Both graphs show every individual data value. For large data sets, this can make it difficult to see the overall pattern in the graph. We often get a cleaner picture of the distribution by grouping nearby values together. Doing so allows us to make a new type of graph: a **histogram.**

DEFINITION Histogram

A **histogram** shows each interval as a bar. The heights of the bars show the frequencies or relative frequencies of values in each interval.

Figure 1.6 shows a dotplot and a histogram of the durations (in minutes) of 220 eruptions of the Old Faithful geyser. Notice how the histogram groups nearby values together.

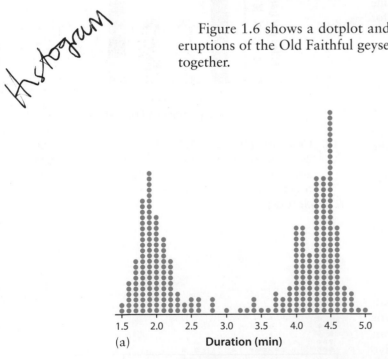

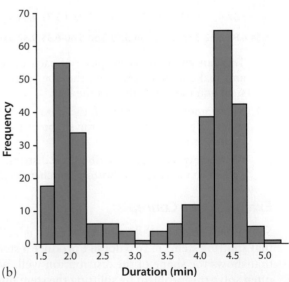

FIGURE 1.6 (a) Dotplot and (b) histogram of the duration (in minutes) of 220 eruptions of the Old Faithful geyser.

Making and Interpreting Histograms

It is fairly easy to make a histogram by hand. Here's how you do it.

How to Make a Histogram

1. **Choose equal-width intervals** that span the data. Five intervals is a good minimum.

2. **Make a table** that shows the frequency (count) or relative frequency (percent or proportion) of data values in each interval.

3. **Draw and label the axes.** Put the name of the quantitative variable under the horizontal axis. To the left of the vertical axis, indicate if the graph shows the frequency (count) or relative frequency (percent or proportion) of data values in each interval.

4. **Scale the axes.** Place equally spaced tick marks at the smallest value in each interval along the horizontal axis. On the vertical axis, start at 0 and place equally spaced tick marks until you exceed the largest frequency or relative frequency in any interval.

5. **Draw bars** above the intervals. Make the bars equal in width and leave no gaps between them. Be sure that the height of each bar corresponds to the frequency or relative frequency of data values in that interval. An interval with no data values will appear as a bar of height 0 on the graph.

It is possible to choose intervals of different widths when making a histogram. Such graphs are beyond the scope of this book.

E X A M P L E

Which states do immigrants choose?

Making a histogram

PROBLEM: How does the percent of foreign-born residents in your state compare to the rest of the country? The table presents the data for all 50 states in 2011.[30] Make a frequency histogram to display the data.

State	Percent	State	Percent	State	Percent
Alabama	3.4	Louisiana	3.8	Ohio	3.9
Alaska	6.2	Maine	3.3	Oklahoma	5.5
Arizona	13.4	Maryland	13.7	Oregon	9.5
Arkansas	4.3	Massachusetts	14.9	Pennsylvania	5.9
California	27.1	Michigan	6.1	Rhode Island	13.5
Colorado	9.7	Minnesota	7.4	South Carolina	4.7
Connecticut	13.3	Mississippi	2.3	South Dakota	2.9
Delaware	8.6	Missouri	4.1	Tennessee	4.7
Florida	19.4	Montana	2.0	Texas	16.5
Georgia	9.6	Nebraska	6.2	Utah	8.4
Hawaii	18.2	Nevada	19.2	Vermont	3.9
Idaho	5.9	New Hampshire	5.3	Virginia	11.1
Illinois	13.9	New Jersey	21.3	Washington	13.4
Indiana	4.6	New Mexico	10.2	West Virginia	1.3
Iowa	4.3	New York	22.2	Wisconsin	4.8
Kansas	6.7	North Carolina	7.3	Wyoming	2.9
Kentucky	3.3	North Dakota	2.4		

SOLUTION:

The data vary from 1.3% to 27.1%. We choose intervals of width 5, beginning at 0:

0 to <5   5 to <10   10 to <15   15 to <20   20 to <25   25 to <30

Interval	Frequency
0 to <5	19
5 to <10	15
10 to <15	9
15 to < 20	4
20 to <25	2
25 to <30	1

1. Choose equal-width intervals that span the data. Note that this choice results in more than the minimum of five intervals. Also notice that we follow the convention of including the left endpoint of an interval and excluding the right endpoint when making histograms.

2. Make a table. The table shows the count of data values in each interval. This type of table is sometimes called a *grouped frequency table.*

Percent foreign-born residents

3. Draw and label the axes. We label the horizontal axis with the name of the quantitative variable, "Percent foreign-born residents." On the vertical axis, we put "Number of states" for a frequency histogram.

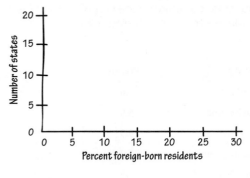

Percent foreign-born residents

4. Scale the axes. The scale on the horizontal axis runs from 0 to 30 with tick marks every 5 units to match the intervals we chose earlier. Because the highest frequency in an interval is 19, we scale the vertical axis from 0 to 20 with tick marks every 5 units.

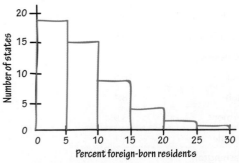

Percent foreign-born residents

5. Draw bars. The completed graph is shown at left.

FOR PRACTICE TRY EXERCISE 1.

CAUTION

Figure 1.7 shows two different histograms of the foreign-resident data. The one on the left (a) uses the intervals of width 5 from the previous example. The one on the right (b) uses intervals half as wide: 0 to <2.5, 2.5 to <5, and so on. The choice of intervals in a histogram can affect the appearance of a distribution. Histograms with more intervals show more detail but may have a less clear overall pattern.

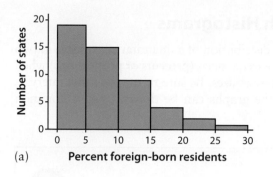

(a)   Percent foreign-born residents

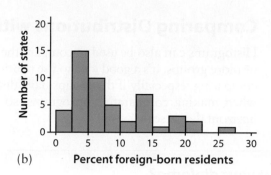

(b)   Percent foreign-born residents

FIGURE 1.7 (a) Frequency histogram of the percent of foreign-born residents in the 50 states with intervals of width 5, from the previous example. (b) Frequency histogram of the data with intervals of width 2.5.

EXAMPLE

Where do immigrants live?

Interpreting a histogram

PROBLEM: Use the data table from the previous example and the histograms in the figure above to answer these questions.

(a) What percent of states have less than 10% foreign-born residents?

(b) Describe the shapes of the two graphs.

(c) Which state is the possible outlier in the right-hand graph?

SOLUTION:

(a) $(19 + 15)/50 = 34/50$ or 68% of states have less than 10% foreign-born residents.

(b) The histogram with intervals of width 5 is skewed to the right with a single peak in the 0% to <5% interval. The histogram with intervals of width 2.5 is skewed to the right with two clear peaks—one in the 2.5% to <5% interval and the other in the 12.5% to <15% interval—and a gap from 22.5% to 25%.

(c) California, with 27.1% foreign-born residents.

FOR PRACTICE   TRY EXERCISE 5.

THINK ABOUT IT What are we actually doing when we make a histogram? The dotplot in part (a) shows the foreign-born resident data. We grouped the data values into intervals of width 5, beginning with 0 to <5, as indicated by the dashed lines. Then we counted the number of values in each interval. The dotplot in part (b) shows the results of that process. Finally, we drew bars of the appropriate height for each interval to get the completed histogram shown.

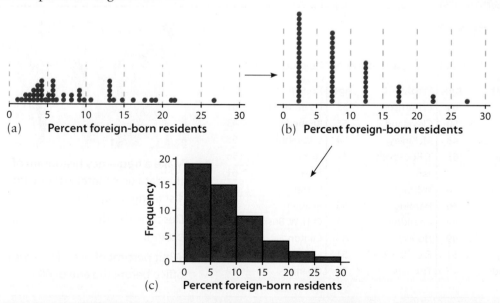

Comparing Distributions with Histograms

Histograms can also be used to compare the distribution of a quantitative variable in two or more groups. It's a good idea to use relative frequencies (percents or proportions) when comparing, especially if the groups have different sizes. Be sure to use the same intervals when making comparative histograms, so the graphs can be drawn using a common horizontal axis scale.

EXAMPLE

Why should you get your diploma?

Comparing distributions with histograms

PROBLEM: Is it true that students who graduate from high school earn more money than students who do not graduate from high school? To find out, we took a random sample of 371 U.S. residents aged 18 and older from a recent census. The educational level and total personal income of each person were recorded. Following are relative frequency histograms of the data for the 57 non-graduates (No) and the 314 graduates (Yes). Compare the distributions.

SOLUTION:

Shape: Both distributions are skewed to the right and single-peaked.

Center: The center of the distribution is larger for graduates, indicating that graduates typically have higher incomes than non-graduates in this sample.

Variability: The incomes for graduates vary a lot more than the incomes for non-graduates.

Outliers: There are some possible high outliers in the graduate distribution.

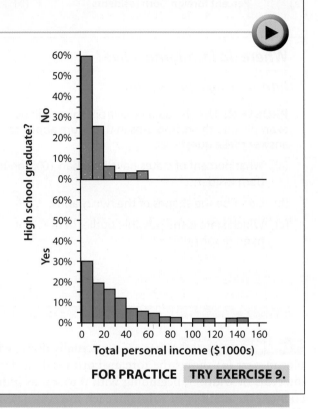

FOR PRACTICE   TRY EXERCISE 9.

LESSON APP 1.5

How old are U.S. presidents?

The table gives the ages of the first 44 U.S. presidents when they took office.

President	Age	President	Age	President	Age	President	Age
Washington	57	Taylor	64	B. Harrison	55	Eisenhower	61
J. Adams	61	Fillmore	50	Cleveland	55	Kennedy	43
Jefferson	57	Pierce	48	McKinley	54	L. B. Johnson	55
Madison	57	Buchanan	65	T. Roosevelt	42	Nixon	56
Monroe	58	Lincoln	52	Taft	51	Ford	61
J. Q. Adams	57	A. Johnson	56	Wilson	56	Carter	52
Jackson	61	Grant	46	Harding	55	Reagan	69
Van Buren	54	Hayes	54	Coolidge	51	G. H. W. Bush	64
W. H. Harrison	68	Garfield	49	Hoover	54	Clinton	46
Tyler	51	Arthur	51	F. D. Roosevelt	51	G. W. Bush	54
Polk	49	Cleveland	47	Truman	60	Obama	47

National Park Service

1. Make a frequency histogram of the data using intervals of width 4 starting at age 40.

2. Describe the shape of the distribution.

3. What percent of presidents took office before the age of 60?

TECH CORNER

Making a Histogram with an Applet or a Graphing Calculator

You can use an applet or a graphing calculator to make a histogram. The technology's default choice of intervals is a good starting point, but you should adjust the intervals to fit with common sense. For the foreign-born resident data (page 39), use either of the following.

Applet

1. Launch the *One Quantitative Variable* applet at highschool.bfwpub.com/spa3e.

2. Enter Percent foreign-born as the Variable name.

3. Select 1 as the number of groups and Raw data as the input method.

4. Enter the data from the example on page 39. Be sure to separate the data values with commas or spaces as you type them.

5. Click Begin analysis.

6. Change the Graph type to a histogram. You can adjust the intervals if desired.

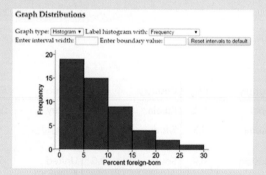

The applet can also be used to make parallel histograms for comparing the distribution of a variable in two or more groups, like the graph shown in the value of a high school diploma example on page 42.

TI-83/84

1. Enter the percent of foreign-born residents for each state in your statistics list editor.

● Press [STAT] and choose Edit….

● Type the values into list L_1.

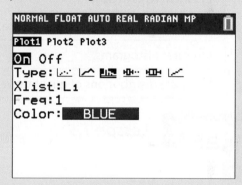

2. Set up a histogram in the statistics plots menu.

● Press [2nd] [Y=] (STAT PLOT).

● Press [ENTER] or [1] to go into Plot1.

● Adjust the settings as shown.

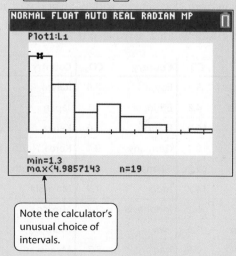

3. Use ZoomStat to let the calculator choose intervals and make a histogram.

● Press [ZOOM] and choose 9: ZoomStat.

● Press [TRACE] and [◄] [►] to examine the intervals.

Note the calculator's unusual choice of intervals.

4. Adjust the intervals to match those from the example and then graph the histogram.

● Press WINDOW. Enter the values shown.

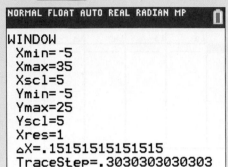

● Press GRAPH.
● Press TRACE and ◄ ► to examine the intervals.

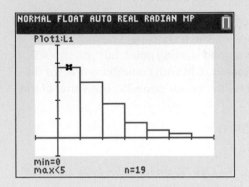

Lesson 1.5

WHAT DID YOU LEARN?

LEARNING TARGET	EXAMPLES	EXERCISES
Make histograms of quantitative data.	p. 39	1–4
Interpret histograms.	p. 41	5–8
Compare distributions of quantitative data with histograms.	p. 42	9–12

Exercises   *Lesson 1.5*

Mastering Concepts and Skills

1. **Measuring carbon dioxide** Burning fuels in power plants and motor vehicles emits carbon dioxide (CO_2), which may contribute to global warming. The table displays CO_2 emissions in metric tons per person from 48 countries with populations of at least 20 million.[31] Make a histogram of the data using intervals of width 2, starting at 0.

pg 39

Country	CO_2	Country	CO_2	Country	CO_2
Algeria	3.3	Egypt	2.6	Italy	6.7
Argentina	4.5	Ethiopia	0.1	Japan	9.2
Australia	16.9	France	5.6	Kenya	0.3
Bangladesh	0.4	Germany	9.1	Korea, North	11.5
Brazil	2.2	Ghana	0.4	Korea, South	2.9
Canada	14.7	India	1.7	Malaysia	7.7
China	6.2	Indonesia	1.8	Mexico	3.8
Colombia	1.6	Iran	7.7	Morocco	1.6
Congo	0.5	Iraq	3.7	Myanmar	0.2

Country	CO_2	Country	CO_2	Country	CO_2
Nepal	0.1	Russia	12.2	Turkey	4.1
Nigeria	0.5	Saudi Arabia	17.0	Ukraine	6.6
Pakistan	0.9	South Africa	9.0	United Kingdom	7.9
Peru	2.0	Spain	5.8	United States	17.6
Philippines	0.9	Sudan	0.3	Uzbekistan	3.7
Poland	8.3	Tanzania	0.2	Venezuela	6.9
Romania	3.9	Thailand	4.4	Vietnam	1.7

2. **Off to work I go** How long do people travel each day to get to work? The following table gives the average travel times to work (in minutes) for workers in each state and the District of Columbia who are at least 16 years old and don't work at home.[32] Make a histogram of the travel times using intervals of width 2 minutes, starting at 14 minutes.

State	Travel time to work (min)	State	Travel time to work (min)	State	Travel time to work (min)
AL	23.6	LA	25.1	OK	20.0
AK	17.7	ME	22.3	OR	21.8
AZ	25.0	MD	30.6	PA	25.0
AR	20.7	MA	26.6	RI	22.3
CA	26.8	MI	23.4	SC	22.9
CO	23.9	MN	22.0	SD	15.9
CT	24.1	MS	24.0	TN	23.5
DE	23.6	MO	22.9	TX	24.6
DC	29.2	MT	17.6	UT	20.8
FL	25.9	NE	17.7	VT	21.2
GA	27.3	NV	24.2	VA	26.9
HI	25.5	NH	24.6	WA	25.2
ID	20.1	NJ	29.1	WV	25.6
IL	27.9	NM	20.9	WI	20.8
IN	22.3	NY	30.9	WY	17.9
IA	18.2	NC	23.4		
KS	18.5	ND	15.5		
KY	22.4	OH	22.1		

3. **A bell curve?** The IQ scores of 60 randomly selected fifth-grade students from one school are shown here.[33]

145	139	126	122	125	130	96	110	118	118
101	142	134	124	112	109	134	113	81	113
123	94	100	136	109	131	117	110	127	124
106	124	115	133	116	102	127	117	109	137
117	90	103	114	139	101	122	105	97	89
102	108	110	128	114	112	114	102	82	101

(a) Make a histogram that displays the distribution of IQ scores effectively.

(b) Many people believe that the distribution of IQ scores follows a "bell curve," like the one shown at top right. Does the graph you drew in part (a) support this belief? Explain.

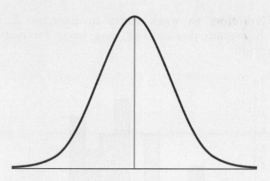

4. **Slow country tunes** Here are the lengths, in minutes, of the 50 most popular mp3 downloads of songs by country artist Dierks Bentley.

4.2	4.0	3.9	3.8	3.7
4.7	3.4	4.0	4.4	5.0
4.6	3.7	4.6	4.4	4.1
3.0	3.2	4.7	3.5	3.7
4.3	3.7	4.8	4.4	4.2
4.7	6.2	4.0	7.0	3.9
3.4	3.4	2.9	3.3	4.0
4.2	3.2	3.4	3.7	3.5
3.4	3.7	3.9	3.7	3.8
3.1	3.7	3.6	4.5	3.7

(a) Make a histogram that displays the distribution of song lengths effectively.

(b) Describe what you see.

5. **Carbon dioxide emissions** Refer to Exercise 1. pg 41 The histogram displays the data using intervals of width 1.

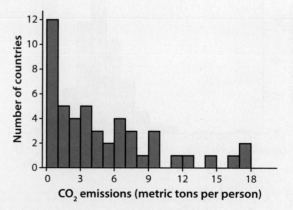

(a) In what percent of countries did CO_2 emissions exceed 10 metric tons per person?

(b) Describe the shape of the distribution.

(c) Which countries are possible outliers?

6. **Traveling to work** Refer to Exercise 2. The histogram displays the data using intervals of width 1.

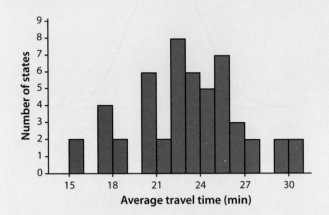

(a) In what percent of states is the average travel time at least 20 min?

(b) Describe the shape of the distribution.

(c) Which two states are possible outliers with average travel times of less than 16 min?

7. **Looking at returns on stocks** The return on a stock is the change in its market price plus any dividend payments made. Total return is usually expressed as a percent of the beginning price. The figure shows a histogram of the distribution of the monthly returns for all common stocks listed on U.S. markets over a 273-month period.[34]

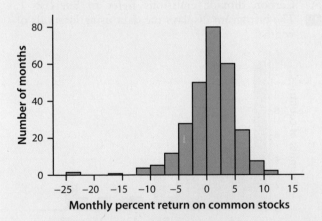

(a) A return less than zero means that stocks lost value in that month. About what percent of all months had returns less than zero?

(b) Describe the shape of the distribution.

(c) Identify the interval(s) that include(s) any possible outliers.

8. **Healthy cereal?** Researchers collected data on calories per serving for 77 brands of breakfast cereal. The following histogram displays the data.[35]

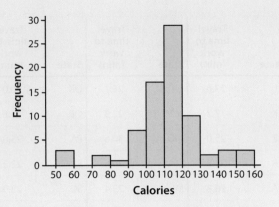

(a) About what percent of the cereal brands have 130 or more calories per serving?

(b) Describe the shape of the distribution.

(c) Identify the interval(s) that include(s) any possible outliers.

9. **Households and income** Rich and poor households differ in ways that go beyond income. Here are histograms that display the distributions of household size (number of people) for low-income and high-income households.[36] Low-income households had annual incomes less than $15,000, and high-income households had annual incomes of at least $100,000. Compare the distributions.

pg 42

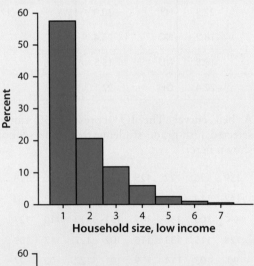

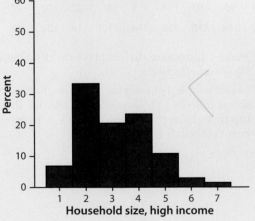

10. **The statistics of writing style** Numerical data can distinguish different types of writing and, sometimes, even individual authors. Here are histograms that display the distribution of word length in Shakespeare's plays and in articles from *Popular Science* magazine. Compare the distributions.

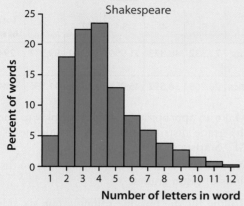

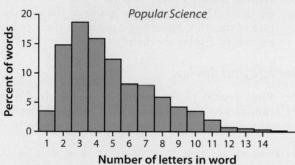

11. **The shape of populations** The following histograms show the distribution of age for 2015 in Vietnam and Australia from the U.S. Census Bureau's international database.

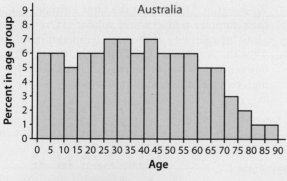

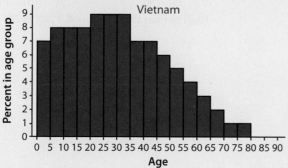

(a) The total population of Australia at this time was 22,751,014. Vietnam's population was 94,348,835. Why did we use percents rather than counts on the vertical axis of these graphs?

(b) What important differences do you see between the age distributions?

12. **Who makes more: men or women?** A manufacturing company is reviewing the salaries of its full-time employees below the executive level at a large plant. The following histograms display the distribution of salary for male and female employees:

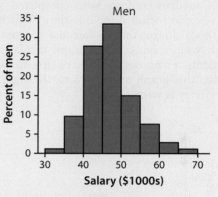

(a) There were 756 female employees and 2451 male employees at the plant. Why did we use percents rather than counts on the vertical axis of these graphs?

(b) Do men or women tend to earn higher salaries at this plant? Justify your answer.

Applying the Concepts

13. **Off to school** A random sample of 50 Canadian students was selected to complete an online survey in a recent year. The dotplot displays data on the travel time to school (in minutes) reported by each student. Make a histogram of the data. Describe what you see.

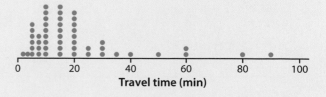

14. **Unprovoked gator!** The dotplot shows the total number of unprovoked attacks by wild alligators on people in Florida in each year from 1971 to 2013. Make a histogram of these data. Describe what you see.

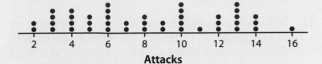

Attacks

15. **Communicate, eh?** We chose a random sample of 50 Canadian students who completed an online survey that included the question "Which of these methods do you most often use to communicate with your friends?" The graph displays data on students' responses. Jerry says that he would describe this graph as skewed to the right. Explain why Jerry is wrong.

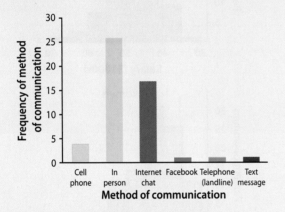

Method of communication

16. **The Brooklyn Half** The histogram shows the distribution of age for runners in the Brooklyn, New York, half-marathon in 2013.[37] Explain what is wrong with this histogram.

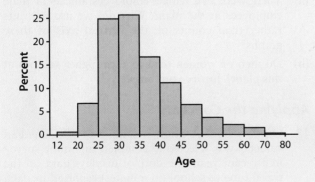

Age

Extending the Concepts

17. **Do the math** The table gives the distribution of grades earned by students taking the AP® Calculus AB and AP® Statistics exams in 2014.[38]

	Grade					
	5	4	3	2	1	Total no. of exams
Calculus AB	72,332	48,873	51,950	31,340	89,577	294,072
Statistics	26,265	38,512	45,052	32,748	41,596	184,173

(a) Make an appropriate graphical display to compare the grade distributions for AP® Calculus AB and AP® Statistics.

(b) Write a few sentences comparing the two distributions of exam grades.

18. **Rolling the die** Imagine rolling a fair, six-sided die 60 times. Draw a plausible graph of the distribution of die rolls. Should you use a bar chart or histogram to display the data?

Recycle and Review

19. **Runs Scored (1.4)** Listed here are the number of runs scored by players on the Chicago White Sox who played regularly during the 2013 season.[39]

24 41 46 68 46 43 84 57 60 38 14 15 19 8 7 4 6 7 2

(a) Make a stemplot of these data.

(b) Describe the shape of the distribution of runs scored.

20. **Hot enough for you? (1.3)** St. Louis, Missouri, and Washington, D.C., are at the same latitude, but are their summer temperatures similar? Here are dotplots for each city of the high temperature on July 4 for the years from 1980 through 2013.[40] Write a few sentences comparing the distributions of July 4 high temperature in these two cities.

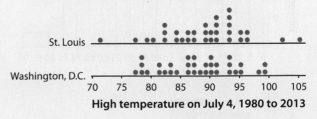

High temperature on July 4, 1980 to 2013

Lesson 1.6
Measuring Center

How long do people typically spend traveling to work? The answer may depend on where they live. Here are the travel times in minutes of 20 randomly chosen workers in New York state, along with a dotplot of the data:[41]

10 30 5 25 40 20 10 15 30 20 15 20 85 15 65 15 60 60 40 45

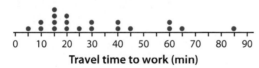

Travel time to work (min)

The distribution is right-skewed and single-peaked. The longest travel time (85 min) appears to be an outlier.

How should we describe where this (or some other) distribution of quantitative data is centered? The two most common ways to measure center are the *median* and the *mean*.

The Median

We could report the value in the "middle" of a distribution as its center. That's the idea of the **median.**

> **DEFINITION** The Median
>
> The **median** is the midpoint of a distribution, the number such that about half the observations are smaller and about half are larger. To find the median, arrange the data values from smallest to largest.
>
> - If the number *n* of data values is odd, the median is the middle value in the ordered list.
> - If the number *n* of data values is even, the median is the average of the two middle values in the ordered list.

The median is easy to find by hand for small sets of data.

EXAMPLE

How unhealthy are fast-food sandwiches?

Finding and interpreting the median when n *is odd*

PROBLEM: Here are data on the amount of fat (in grams) in 9 different fish and chicken sandwiches on McDonald's menu in 2014.[42] Find the median. Interpret this value in context.

Sandwich	Fat (g)
Filet-O-Fish®	19
McChicken®	16
Premium Crispy Chicken Classic Sandwich	22
Premium Crispy Chicken Club Sandwich	33
Premium Crispy Chicken Ranch Sandwich	27
Premium Grilled Chicken Classic Sandwich	9
Premium Grilled Chicken Club Sandwich	20
Premium Grilled Chicken Ranch Sandwich	14
Southern Style Crispy Chicken Sandwich	19

SOLUTION:

9   14   16   19   ⟨19⟩   20   22   27   33

The median is 19. About half of McDonald's fish and chicken sandwiches have more than 19 g of fat and about half have less.

> Sort the data values from smallest to largest. Because there are $n = 9$ data values (an odd number), the median is the middle value in the ordered list.

FOR PRACTICE   TRY EXERCISE 1.

A dotplot of the fat content data from the example is shown here. You can confirm that the median is 19 by "counting inward" from the minimum and maximum values.

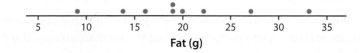

Fat (g)

EXAMPLE

Where's the beef?

Finding and interpreting the median when n *is even*

PROBLEM: Here are data on the amount of fat (in grams) in 12 different beef sandwiches from McDonald's 2014 menu, along with a dotplot. Find and interpret the median.

Sandwich	Fat (g)
Big Mac®	27
Cheeseburger	11
Daily Double	22
Double Cheeseburger	21
Bacon Clubhouse Burger	40
Hamburger	8
McDouble®	17
BBQ Ranch Burger	15
Quarter Pounder® Bacon and Cheese	29
Quarter Pounder® Bacon Habanero Ranch	31
Quarter Pounder® Deluxe	27
Quarter Pounder® with Cheese	26

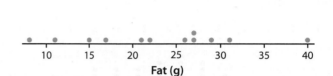

Fat (g)

SOLUTION:

8 11 15 17 21 (22 26) 27 27 29 31 40

The median is $\dfrac{22 + 26}{2} = 24$. About half of McDonald's beef sandwiches have more than 24 g of fat and about half have less.

> Sort the data values from smallest to largest. Because there are $n = 12$ data values (an even number), the median is the average of the middle two values in the ordered list.

FOR PRACTICE **TRY EXERCISE 3.**

The Mean

The most common measure of center is the **mean.**

> ### DEFINITION The mean $\bar{x}$
>
> The **mean** $\bar{x}$ (pronounced "x-bar") of a quantitative data set is the average of all n data values. To find the mean, add all the values and divide by n. That is,
>
> $$\bar{x} = \frac{\text{sum of data values}}{\text{number of data values}}$$

If the values in a data set are given by $x_1, x_2, \ldots, x_n$, we can rewrite the formula for calculating the mean as

$$\bar{x} = \frac{x_1 + x_2 + \cdots + x_n}{n} = \frac{\sum x_i}{n}$$

The Σ (capital Greek letter sigma) in the formula is short for "add them all up." The subscripts on the observations x_i are just a way of keeping the n data values distinct. They do not necessarily indicate order or any other special facts about the data.

Actually, the notation $\bar{x}$ refers to the mean of a *sample*. Most of the time, the data we'll encounter can be thought of as a sample from some larger population. When we need to refer to a *population* mean, we'll use the symbol μ (Greek letter mu, pronounced "mew").

EXAMPLE

How high is the score?

Finding the mean

PROBLEM: How many runs does the Lawrence High School baseball team usually score? Here are data on the number of runs scored by the team in all 21 games played during a recent season, along with a dotplot.

0 1 1 1 2 2 2 2 3 3 3 4 4 5 5 6 6 7 8 10 12

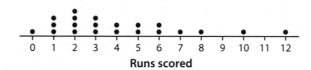

Runs scored

(a) Calculate the mean number of runs scored per game by the team during this season. Show your work.

(b) The dotplot suggests that the game in which the team scored 12 runs is a possible outlier. Calculate the mean number of runs scored per game by the team in the other 20 games this season. What do you notice?

SOLUTION:

(a) $\bar{x} = \dfrac{0 + 1 + 1 + \cdots + 10 + 12}{21} = \dfrac{87}{21} \approx 4.14$ runs per game

(b) $\bar{x} = \dfrac{0 + 1 + 1 + \cdots + 10}{20} = \dfrac{75}{20} \approx 3.75$ runs per game

You can calculate the mean with the formula:
$$\bar{x} = \frac{x_1 + x_2 + \cdots + x_n}{n} = \frac{\sum x_i}{n}$$

The team's 12-run game increased the mean number of runs scored per game by 0.39 runs.

FOR PRACTICE TRY EXERCISE 5.

CAUTION

The previous example illustrates an important weakness of the mean as a measure of center: The mean is not **resistant**.

DEFINITION Resistant

A measure of center (or variability) is **resistant** if it isn't influenced by unusually large or unusually small values in a distribution.

The median *is* a resistant measure of center. In the preceding example, the median number of runs scored by the Lawrence High baseball team in its 21-game season is 3. If we remove the possible outlier game in which the team scored 12 runs, the median number of runs scored in the remaining 20 games is still 3.

ACTIVITY

Mean as a "balance point"

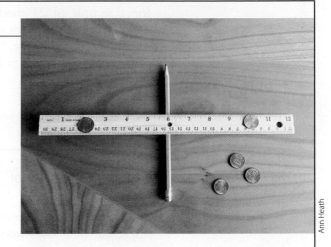

In this activity, you will investigate an important property of the mean.

1. Stack 5 pennies above the 6-inch mark on a ruler. Place a pencil under the ruler to make a "seesaw" on a desk or table. Move the pencil until the ruler balances. What is the relationship between the location of the pencil and the mean of the five data values: 6, 6, 6, 6, 6?

2. Move one penny off the stack to the 8-inch mark on your ruler. Now move one other penny so that the ruler balances again without moving the pencil. Where did you put the other penny? What is the mean of the five data values represented by the pennies now?

3. Move one more penny off the stack to the 2-inch mark on your ruler. Now move both remaining pennies from the 6-inch mark so that the ruler

still balances with the pencil in the same location. Is the mean of the data values still 6?

4. Discuss with your classmates: Why is the mean called the "balance point" of a distribution?

The activity gives a physical interpretation of the mean as the balance point of a distribution. For the data on runs scored in each game by the Lawrence High baseball team in a recent season, the dotplot balances at $\overline{x} = 4.14$ runs.

Comparing the Mean and the Median

Which measure—the mean or the median—should we report as the center of a distribution? That depends on both the shape of the distribution and whether there are any outliers.

Shape: Figure 1.8 shows the mean and median for dotplots with three different shapes. Notice how these two measures of center compare in each case. The mean is pulled in the direction of the long tail in a skewed distribution.

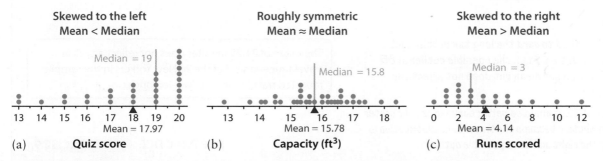

FIGURE 1.8 Dotplots that show the relationship between the mean and median in distributions with different shapes: (a) Scores on an easy statistics quiz, (b) usable capacity in a sample of 36 side-by-side refrigerators, and (c) runs scored by the Lawrence High School baseball team in 21 games played.

Outliers: We noted earlier that the mean is sensitive to extreme values. The left dotplot below shows the EPA estimates of city gas mileage in miles per gallon (mpg) for a sample of 21 model year 2014 midsize cars.[43] If we remove the clear outlier —the Toyota Prius, with its 51 mpg in the city—the mean falls to $\bar{x} = 21.1$ mpg, but the median stays at 21 mpg. See the right dotplot below. The median is a *resistant* measure of center, but the mean is not.

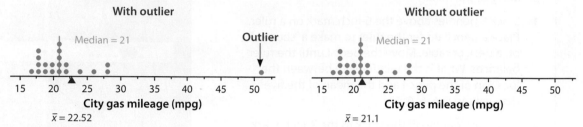

You can compare how the mean and median behave by using the *Mean and Median* applet at the book's website, highschool.bfwpub.com/spa3e.

Choosing a Measure of Center

- If a distribution of quantitative data is roughly symmetric and has no outliers, use the mean to measure center.
- If the distribution is strongly skewed or has outliers, use the median to measure center.

EXAMPLE

How bad is the traffic?

Comparing the mean and median

PROBLEM: At the beginning of the lesson, we presented data on the travel times in minutes of 20 randomly chosen New York workers. Here is a dotplot of the data.

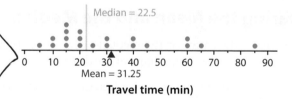

(a) Explain why the mean is so much larger than the median.

(b) Which measure of center better describes a typical travel time to work? Explain.

SOLUTION:

(a) The mean is pulled toward the long tail in this right-skewed distribution. Also, the possible outlier of 85 minutes inflates the mean but does not affect the median as much.

(b) The median of 22.5 minutes better summarizes the center of the distribution because the travel-time distribution is skewed to the right and has a possible outlier.

> The mean of 31.25 minutes does not reflect a typical travel time—only 7 of the 20 New Yorkers in the sample reported travel times this long or longer.

FOR PRACTICE   TRY EXERCISE 9.

LESSON APP 1.6

Is the pace of life slower in smaller cities?

Does it take less time to get to work in smaller cities? Here are the travel times in minutes for 15 workers in North Carolina, chosen at random by the Census Bureau, along with a dotplot of the data.[44]

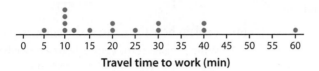

30 20 10 40 25 20 10 60 15 40 5 30 12 10 10

Travel time to work (min)

1. Find the median. Interpret this value in context.

2. Calculate the mean travel time. Show your work.

3. Which measure of center—the median or the mean— better describes a typical travel time to work for this sample of workers in North Carolina? Justify your answer.

Lesson 1.6

WHAT DID YOU LEARN?

LEARNING TARGET	EXAMPLES	EXERCISES
Find and interpret the median of a distribution of quantitative data.	pp. 50, 51	1–4
Calculate the mean of a distribution of quantitative data.	p. 52	5–8
Compare the mean and median of a distribution, and choose the more appropriate measure of center in a given setting.	p. 54	9–12

Exercises   *Lesson 1.6*

Mastering Concepts and Skills

1. **Quiz grades** Joey's first 13 quiz grades in a marking
pg **50** period are listed here. Find and interpret the median.

82 93 77 79 90 82 85 85 95 73 79 83 89

2. **Big boys** The roster of the Dallas Cowboys professional football team in a recent season included 7 defensive linemen. Their weights (in pounds) were 321, 285, 300, 285, 286, 293, and 298. Find and interpret the median.

3. **Large fries** Ryan and Brent were curious about
pg **51** the amount of french fries they would get in a large order from their favorite fast-food restaurant, Burger King. They went to several different Burger King restaurants over a series of days and ordered a total of 14 large fries. The weight of each order (in grams) is shown here. Find and interpret the median.

165 163 160 159 166 152 166 168 173 171 168 167 170 170

4. **Carrots** The weights (in grams) of 12 carrots in a single bag from a local grocery store are listed here. Find and interpret the median.

    44 56 48 41 66 55 42 33 51 44 61 65

5. **Skipped quiz** Refer to Exercise 1.
(a) Calculate Joey's mean quiz grade. Show your work.
(b) Joey has an unexcused absence for the 14th quiz, **pg 52** and he receives a score of zero. Recalculate the mean and median. Explain why the mean and median are so different now.

6. **Big outlier** Refer to Exercise 2.
(a) Calculate the mean weight of the 7 defensive linemen. Show your work.
(b) The defensive lineman that weighed 321 pounds may be an outlier. How did this player affect the mean? Justify your answer with an appropriate calculation.

7. **Mean fries** Refer to Exercise 3.
(a) Calculate the mean weight for the 14 orders of large fries. Show your work.
(b) Ryan and Brent noticed a shortage of fries in the order that weighed 152 grams. How did this order affect the mean? Justify your answer with an appropriate calculation.

8. **One more carrot** Refer to Exercise 4.
(a) Calculate the mean weight of the carrots. Show your work.
(b) The 13th carrot in the bag weighed 93 grams. Recalculate the mean and median. Explain why the mean and median are so different now.

9. **Birthrates in Africa** One of the important factors in **pg 54** determining population growth rates is the birthrate per 1000 individuals in a population. The dotplot shows the birthrates per 1000 individuals for 54 African nations:

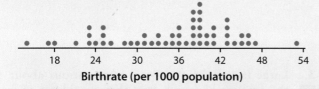

Birthrate (per 1000 population)

(a) Explain how the mean and median would compare.
(b) Which measure of center better describes a typical birthrate? Explain.

10. **Electing the president** To become president of the United States, a candidate does not have to receive a majority of the popular vote. The candidate does have to win a majority of the 538 electoral votes that are cast in the Electoral College. Here is a stemplot of the number of electoral votes in 2016 for each of the 50 states and the District of Columbia.

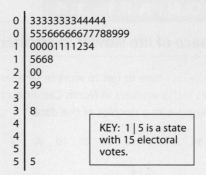

```
0 | 3333333344444
0 | 55566666677788999
1 | 00001111234
1 | 5668
2 | 00
2 | 99
3 |
3 | 8
4 |
4 |
5 |
5 | 5
```

KEY: 1 | 5 is a state with 15 electoral votes.

(a) Explain how the mean and median would compare.
(b) Which measure of center better describes a typical number of electoral votes? Explain.

11. **Smart kids** The histogram displays the IQ scores of 60 randomly selected fifth-grade students from one school. Which measure of center is the more appropriate choice in this setting? Explain.

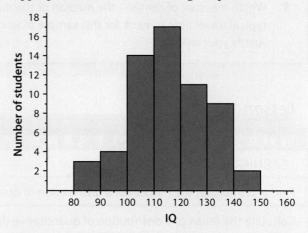

12. **Lightning** The histogram displays data from a study of lightning storms in Colorado.[45] It shows the distribution of time after midnight (in hours) until the first lightning flash for that day occurred. Which measure of center is the more appropriate choice in this setting? Explain.

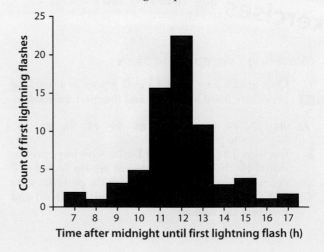

Time after midnight until first lightning flash (h)

Applying the Concepts

13. Do adolescent girls eat fruit? We all know that fruit is good for us. Following is a histogram of the number of servings of fruit per day claimed by 74 seventeen-year-old girls in a study in Pennsylvania.[46] Find the mean and median. Show your method clearly.

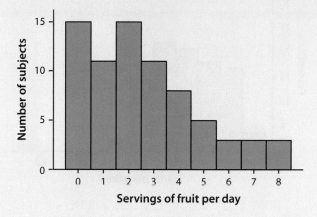

14. Shakespeare The following histogram shows the distribution of lengths of words used in Shakespeare's plays.[47] Find the mean and median. Show your method clearly.

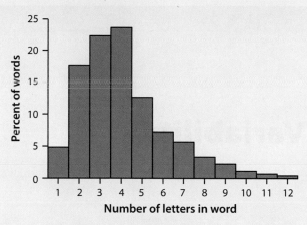

15. How much for that house? The mean and median selling prices of existing single-family homes sold in the United States in May 2014 were $260,700 and $213,600.[48]

(a) Which of these numbers is the mean and which is the median? Explain your reasoning.

(b) Write a sentence to describe how an unethical politician could use these statistics to argue that May 2014 home prices were too high.

16. How mean is this salary? Last year a small accounting firm paid each of its five clerks $44,000, two junior accountants $100,000 each, and the firm's owner $540,000. Write a sentence to describe how an unethical recruiter could use statistics to mislead prospective employees.

17. Baseball salaries, means and medians Suppose that a Major League Baseball team's mean yearly salary for its players is $2.3 million and that the team has 25 players on its active roster.

(a) What is the team's total annual payroll?

(b) If you knew only the median salary, would you be able to answer this question? Why or why not?

18. Mean or median? You are planning a party for 30 guests and want to know how many cans of soda to buy. Earl, the soda elf, offers to tell you either the mean number of cans guests will drink or the median number of cans. Which measure of center should you ask for? Why?

Extending the Concepts

Another measure of center for a quantitative data set is the *trimmed mean*. To calculate the trimmed mean, order the data set from lowest to highest, remove the same number of data values from each end, and calculate the mean of the remaining values. For a data set with 10 values, for example, we can calculate the 10% trimmed mean by removing the maximum and minimum value. Why? Because that's one value trimmed from each "end" of the data set out of 10 values, and 1/10 = 0.10 or 10%.

19. Shoes How many pairs of shoes does a typical teenage boy own? To find out, a group of statistics students surveyed a random sample of 20 male students from their large high school. Then they recorded the number of pairs of shoes that each boy owned. Here are the data, along with a dotplot.

14 7 6 5 12 38 8 7 10 10 10 11 4 5 22 7 5 10 35 7

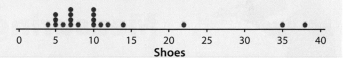

(a) Calculate the mean of the distribution.

(b) Calculate the 10% trimmed mean.

(c) Why is the trimmed mean a better summary of the center of this distribution than the mean?

Recycle and Review

20. File sizes (1.3) How much disk space does your music use? Here are the file sizes (in megabytes) for 18 randomly selected files on Gabriel's mp3 player.

| 2.4 | 2.7 | 1.6 | 1.3 | 6.2 | 1.3 | 5.6 | 1.1 | 2.2 |
| 1.9 | 2.1 | 4.4 | 4.7 | 3.0 | 1.9 | 2.5 | 7.5 | 5.0 |

(a) Make a dotplot to display the data.

(b) Explain what the dot above the number 5 on your dotplot represents.

(c) What percent of the files are larger than 2 megabytes?

(d) Use the dotplot to describe the shape of the distribution of file sizes.

21. He shoots, he scores! (1.5) Lebron James and Kevin Durant were two of the most prolific scorers in the National Basketball Association in 2013–2014. The following histograms display the distribution of points per game for all the regular season games in which each of them played. Compare the distributions.

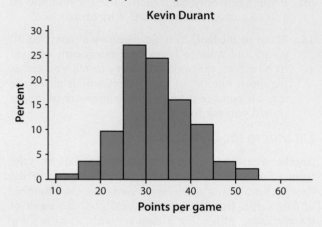

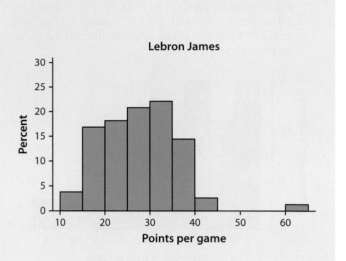

Lesson 1.7
Measuring Variability

LEARNING TARGETS

- Find the range of a distribution of quantitative data.
- Find and interpret the interquartile range.
- Calculate and interpret the standard deviation.

Being able to describe the shape and center of a distribution is a great start. However, two distributions can have the same shape and center, but still look quite different.

The parallel dotplots show the scores of two bowlers (Earl and Kelly) in their 100 most recent games. Both distributions are symmetric and single-peaked, with centers around 150. But the variability of these two distributions is quite different. It appears that Kelly is a much more consistent bowler than Earl. Her distribution of scores is much less variable than his.

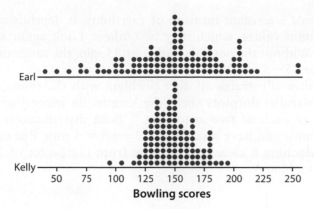

There are several ways to measure the variability of a distribution. The three most common are the *range, interquartile range,* and *standard deviation.*

The Range

The simplest measure of variability is the **range.**

> **DEFINITION Range**
>
> The **range** of a distribution is the distance between the minimum value and the maximum value. That is,
>
> $$range = maximum - minimum$$

Note that the range of a data set is a single number. In everyday language, people sometimes say things like, "The data values range from 5 to 85." Be sure to use the term *range* correctly, now that you know its statistical definition.

EXAMPLE

New Yorkers rushing to work?

Finding the range

PROBLEM: How long do people typically spend traveling to work? Here are the travel times in minutes of 20 randomly chosen New York workers, along with a dotplot.[49] Find the range of the distribution.

10   30   5   25   40   20   10   15   30   20   15   20   85   15   65   15   60   60   40   45

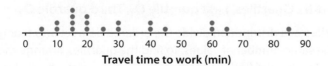

Travel time to work (min)

SOLUTION:

Range = Maximum − Minimum = 85 − 5 = 80 min

FOR PRACTICE   TRY EXERCISE 1.

The range is *not* a resistant measure of variability. It depends on only the maximum and minimum values, which may be outliers. Look again at the New York travel-time data. Without the possible outlier at 85 min, the range of the distribution would decrease to $65 - 5 = 60$ min.

The graph below illustrates another problem with the range as a measure of variability. The parallel dotplots show the lengths (in inches) of a sample of 11 nails produced by each of two machines.[50] Both distributions are centered at 70 millimeters (mm) and have a range of $72 - 68 = 4$ mm. But the lengths of the nails made by Machine B clearly vary more from the center of 70 mm than the nails made by Machine A.

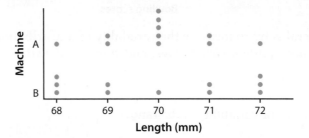

The Interquartile Range

We can avoid the impact of extreme values on our measure of variability by focusing on the middle of the distribution. Here's the idea. Order the data values from smallest to largest. Then find the **quartiles,** the values that divide the distribution into four groups of roughly equal size. The **first quartile Q_1** lies one-quarter of the way up the list. The second quartile is the median, which is halfway up the list. The **third quartile Q_3** lies three-quarters of the way up the list. The first and third quartiles mark out the middle half of the distribution.

For example, here are the amounts collected each hour by a charity at a local store: $19, $26, $25, $37, $31, $28, $22, $22, $29, $34, $39, and $31. The dotplot below displays the data. Because there are 12 data values, the quartiles divide the distribution into 4 groups of 3 values.

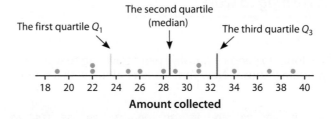

> **DEFINITION** Quartiles, First quartile Q_1, Third quartile Q_3
>
> The **quartiles** of a distribution divide the ordered data set into four groups having roughly the same number of values. To find the quartiles, arrange the data values from smallest to largest and find the median.
>
> The **first quartile Q_1** is the median of the data values that are to the left of the median in the ordered list.
>
> The **third quartile Q_3** is the median of the data values that are to the right of the median in the ordered list.

The **interquartile range (*IQR*)** measures the variability in the middle half of the distribution.

> ▌▌▌ **DEFINITION** Interquartile range (*IQR*)
>
> The **interquartile range (*IQR*)** is the distance between the first and third quartiles of a distribution. In symbols,
>
> $$IQR = Q_3 - Q_1$$

EXAMPLE

A healthy range to choose from?

Finding and interpreting the IQR

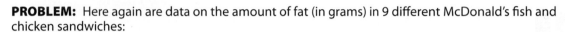

PROBLEM: Here again are data on the amount of fat (in grams) in 9 different McDonald's fish and chicken sandwiches:

| 19 | 16 | 22 | 33 | 27 | 9 | 20 | 14 | 19 |

Find the interquartile range. Interpret this value in context.

SOLUTION:

9   14   16   19   ⟨19⟩   20   22   27   33
                        Median

> Sort the data values from smallest to largest and find the median.

⌐9 _ _ 14 ⎸16 _ _ 19⌐   ⟨19⟩   20   22   27   33
  $Q_1 = 15$ Median

> Find the first quartile Q_1, which is the median of the data values to the left of the median in the ordered list.

9   14 ⎸16   19   ⟨19⟩ ⌐20 _ _ 22 ⎸27 _ _ 33⌐
  $Q_1 = 15$ Median $Q_3 = 24.5$

> Find the third quartile Q_3, which is the median of the data values to the right of the median in the ordered list.

The interquartile range is

$IQR = Q_3 - Q_1 - 24.5 - 15 = 9.5 \, g$

Interpretation: The range of the middle half of fat content values for these McDonald's fish and chicken sandwiches is 9.5 grams.

FOR PRACTICE   TRY EXERCISE 5.

The quartiles and the interquartile range are *resistant* because they are not affected by a few extreme values. For the fat content data, Q_3 would still be 24.5 and the *IQR* would still be 9.5 if the maximum were 53 rather than 33.

The Standard Deviation

When we use the median to measure the center of a distribution, the interquartile range (*IQR*) is our corresponding measure of variability. If we summarize the center of a distribution with the mean, then we should use the **standard deviation** to describe the variation of data values around the mean.

> ▌▌▌ **DEFINITION** Standard deviation
>
> The **standard deviation** measures the typical distance of the values in a distribution from the mean. To find the standard deviation s_x of a quantitative data set with *n* values:
>
> 1. Find the mean of the distribution.
> 2. Calculate the *deviation* of each value from the mean: deviation = value − mean.
> 3. Square each of the deviations.
> 4. Add all the squared deviations, divide by $n - 1$, and take the square root.

If the values in a data set are given by $x_1, x_2, \ldots, x_n$, we can rewrite the formula for calculating the standard deviation as

$$s_x = \sqrt{\frac{(x_1 - \bar{x})^2 + (x_2 - \bar{x})^2 + \cdots + (x_n - \bar{x})^2}{n - 1}} = \sqrt{\frac{\sum (x_i - \bar{x})^2}{n - 1}}$$

Actually, the notation s_x refers to the standard deviation of a *sample*. Most of the time, the data we'll encounter can be thought of as a sample from some larger population. When we need to refer to a *population* standard deviation, we'll use the symbol σ (Greek lowercase sigma). This standard deviation is calculated by dividing the sum of squared deviations by n instead of $n - 1$ before taking the square root.

EXAMPLE

How many pets?

Calculating and interpreting standard deviation

PROBLEM: Nine children were asked how many pets they had. Here are their responses, arranged from lowest to highest, along with a dotplot of the data.[51]

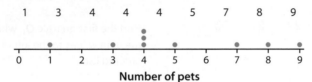

Calculate the standard deviation. Interpret this value in context.

SOLUTION:

$$\bar{x} = \frac{\sum x_i}{n} = \frac{1 + 3 + 4 + 4 + 4 + 5 + 7 + 8 + 9}{9} = \frac{45}{9} = 5 \text{ pets}$$

> **1.** Find the mean of the distribution.

Value x_i	Deviation from mean $x_i - \bar{x}$	Squared deviation $(x_i - \bar{x})^2$
1	$1 - 5 = -4$	$(-4)^2 = 16$
3	$3 - 5 = -2$	$(-2)^2 = 4$
4	$4 - 5 = -1$	$(-1)^2 = 1$
4	$4 - 5 = -1$	$(-1)^2 = 1$
4	$4 - 5 = -1$	$(-1)^2 = 1$
5	$5 - 5 = 0$	$0^2 = 0$
7	$7 - 5 = 2$	$2^2 = 4$
8	$8 - 5 = 3$	$3^2 = 9$
9	$9 - 5 = 4$	$4^2 = 16$
		Sum = 52

> **2.** Calculate the *deviation* of each value from the mean: deviation = value − mean

> **3.** Square each of the deviations.

> **4.** Add all the squared deviations, divide by $n - 1$, and take the square root to return to the original units (pets).

$$s_x = \sqrt{\frac{\sum (x_i - \bar{x})^2}{n - 1}} = \sqrt{\frac{52}{9 - 1}} \approx 2.55 \text{ pets}$$

Interpretation: The number of pets these children have typically varies by about 2.55 pets from the mean of 5 pets.

FOR PRACTICE   TRY EXERCISE 9.

THINK ABOUT IT **Why is the standard deviation calculated in such a complex way?** Add up the deviations from the mean in the previous example. You should get a sum of 0. Why? Because the mean is the balance point of the distribution. We square the deviations to avoid the positive and negative deviations balancing each other out and adding to 0. It might seem strange to "average" the squared deviations by dividing by $n - 1$. We'll explain the reason for doing this in Chapter 6. It's easier to understand why we take the square root: to return to the original units (pets).

More important than the details of calculating s_x are the properties of the standard deviation as a measure of variability:

- s_x **is always greater than or equal to 0.** $s_x = 0$ only when there is no variability, that is, when all values in a distribution are the same.

- **Larger values of s_x indicate greater variation** from the mean of a distribution. For instance, Earl's bowling scores have a standard deviation of about 40, while Kelly's scores have a standard deviation of about 20.

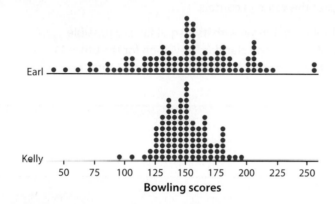

- s_x **is not resistant.** The use of squared deviations makes s_x even more sensitive than $\bar{x}$ to extreme values in a distribution. For example, the standard deviation of the travel times for the 20 New York workers is 21.88 min. If we omit the maximum value of 85 min, the standard deviation drops to 18.34 min.

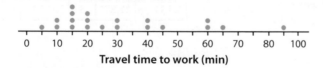

- s_x **measures variation about the mean.** It should be used only when the mean is chosen as the measure of center.

Choosing Measures of Center and Variability

The median and *IQR* are usually better than the mean and standard deviation for describing a skewed distribution or a distribution with outliers. Use $\bar{x}$ and s_x for roughly symmetric distributions that don't have outliers.

LESSON APP 1.7

Have we found the beef?

Here are data on the amount of fat (in grams) in 12 different McDonald's beef sandwiches, along with a dotplot. The mean fat content for these sandwiches is $\bar{x} = 22.833$ grams.

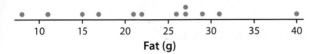

27   11   22   21   40   8   17   15   29   31   27   26

John E. Kelly/Getty Images

1. Find the range of the distribution.

2. Find the interquartile range. Interpret this value in context.

3. Calculate the standard deviation. Interpret this value in context.

4. The dotplot suggests that the Bacon Clubhouse Burger, with its 40 g of fat, is a possible outlier. Recalculate the range, interquartile range, and standard deviation for the other 11 sandwiches. Compare these values with the ones you obtained in Questions 1 through 3. Explain why each result makes sense.

TECH CORNER

Computing Numerical Summaries with Technology

You can use an applet or a graphing calculator to calculate measures of center and variability. That will free you up to concentrate on choosing the right numerical summaries and interpreting your results. For the New York travel-time data (page 59), use either of the following.

Applet

1. Go to highschool.bfwpub.com/spa3e and launch the *One Quantitative Variable* applet.

2. Enter Travel time to work (in minutes) as the Variable name.

3. Select 1 as the number of groups and Raw data as the input method.

4. Input the data. Be sure to separate the data values with commas or spaces as you type them.

5. Click Begin analysis to display the summary statistics.

Summary Statistics

n	mean	SD	min	Q_1	med	Q_3	max
20	31.25	21.877	5	20	22.5	42.5	85

TI-83/84

1. Type the values into list L_1.

2. Calculate numerical summaries using one-variable statistics.

• Press STAT ▶ (CALC) and choose 1–Var Stats.

OS 2.55 or later: In the dialog box, press 2nd 1 (L1) and ENTER to specify L1 as the List. Leave FreqList blank. Arrow down to Calculate and press ENTER.

Older OS: Press 2nd 1 (L1) and ENTER.

• Press ▼ to see the rest of the one-variable statistics.

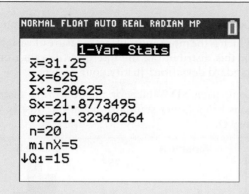

Lesson 1.7

WHAT DID YOU LEARN?

LEARNING TARGET	EXAMPLES	EXERCISES
Find the range of a distribution of quantitative data.	p. 59	1–4
Find and interpret the interquartile range.	p. 61	5–8
Calculate and interpret the standard deviation.	p. 62	9–12

Exercises   Lesson 1.7

Mastering Concepts and Skills

1. **Teens and shoes** How many pairs of shoes does a typical teenage boy own? To find out, a group of statistics students surveyed a random sample of 20 male students from their large high school. Then they recorded the number of pairs of shoes that each boy owned. Here are the data, along with a dotplot. Find the range of the distribution.

pg 59

14 7 6 5 12 38 8 7 10 10 10 11 4 5 22 7 5 10 35 7

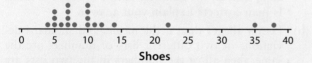

Shoes

2. **Traveling Tarheels!** Here are the travel times in minutes for 15 workers in North Carolina, chosen at random by the Census Bureau, along with a dotplot of the data. Find the range of the distribution.

30 20 10 40 25 20 10 60 15 40 5 30 12 10 10

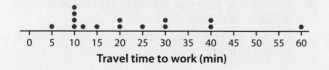

Travel time to work (min)

3. **Heavy Cowboys** The roster of the Dallas Cowboys professional football team in a recent season included 7 defensive linemen. Their weights (in pounds) were 321, 285, 300, 285, 286, 293, and 298. Find the range of the distribution.

4. **Pizza and calories** Here are data on the number of calories per serving for 16 brands of frozen cheese pizza.[52] Find the range of the distribution.

| 340 | 340 | 310 | 320 | 310 | 360 | 350 | 330 |
| 260 | 380 | 340 | 320 | 310 | 360 | 350 | 330 |

5. **Shoes and teens** Refer to Exercise 1. Find the interquartile range. Interpret this value in context.

pg 61

6. **Tarheels** Refer to Exercise 2. Find the interquartile range. Interpret this value in context.

7. **Cowboys** Refer to Exercise 3. Find and interpret the interquartile range.

8. **Frozen pizza** Refer to Exercise 4. Find and interpret the interquartile range.

9. **Well rested?** The first four students to arrive for a first-period statistics class were asked how much sleep (to the nearest hour) they got last night. Their responses were 7, 7, 8, and 10. Calculate the standard deviation. Interpret this value in context.

pg 62

10. **The rate of metabolism** A person's metabolic rate is the rate at which the body consumes energy. Metabolic rate is important in studies of weight gain, dieting, and exercise. Here are the metabolic rates of 7 men who took part in a study of dieting. (The units are calories per 24 hours. These are the same calories used to describe the energy content of foods.) Calculate the standard deviation. Interpret this value in context.

    1792   1666   1362   1614   1460   1867   1439

11. **Phosphate in blood** The level of various substances in the blood influences our health. Here are measurements of the level of phosphate in the blood of a patient, in milligrams of phosphate per deciliter of blood, made on 6 consecutive visits to a clinic. Calculate and interpret the standard deviation.

    5.6   5.2   4.6   4.9   5.7   6.4

12. **Foot lengths** Here are the foot lengths (in centimeters) for a random sample of seven 14-year-olds from the United Kingdom. Calculate and interpret the standard deviation.

    25   22   20   25   24   24   28

Applying the Concepts

13. **Varying fuel efficiency** The dotplot shows the difference (Highway − City) in EPA mileage ratings for each of 21 model year 2014 midsize cars.

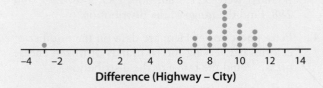

(a) Find the interquartile range of the distribution. Interpret this value in context.

(b) Calculate and interpret the standard deviation.

(c) Which is the more appropriate measure of variability for this distribution: the interquartile range or the standard deviation? Justify your answer.

14. **Another serving of carrots** The dotplot shows the weights (to the nearest gram) of 12 carrots in a single bag from a local grocery store.

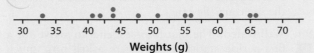

(a) Find the interquartile range of the distribution. Interpret this value in context.

(b) Calculate and interpret the standard deviation.

(c) Which is the more appropriate measure of variability for this distribution: the interquartile range or the standard deviation? Justify your answer.

15. **Comparing SD** Which of the following distributions has a larger standard deviation? Justify your answer.

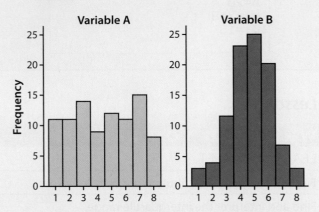

16. **Comparing SD** The parallel dotplots show the lengths (in millimeters) of a sample of 11 nails produced by each of two machines.[53] As mentioned on page 60, both distributions have a range of 4 mm. Which distribution has the larger standard deviation? Justify your answer.

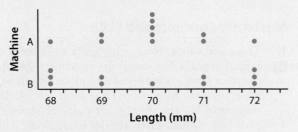

17. **Properties of the standard deviation**

(a) Juan says that, if the standard deviation of a list is zero, then all the numbers on the list are the same. Is Juan correct? Explain your answer.

(b) Letitia alleges that, if the means and standard deviations of two different lists of numbers are the same, then all of the numbers in the two lists are the same. Is Letitia correct? Explain your answer.

18. **SD contest** This is a standard deviation contest. You must choose four numbers from the whole numbers 0 to 10, with repeats allowed.

(a) Choose four numbers that have the smallest possible standard deviation.

(b) Choose four numbers that have the largest possible standard deviation.

(c) Is more than one choice possible in either part (a) or (b)? Explain.

Extending the Concepts

19. **Estimating SD** The dotplot shows the number of shots per game taken by NHL player Sidney Crosby in his 81 regular season games in a recent season.[54] Is the standard deviation of this distribution closest to 2, 5, or 10? Explain.

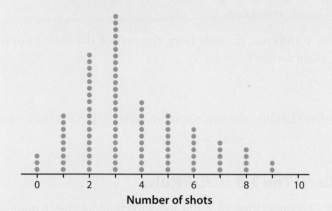

Number of shots

20. **Will Joey pass?** Joey's first 14 quiz grades in a marking period had a mean of 85 and a standard deviation of 8.

(a) Suppose Joey makes an 85 on the next quiz. Would the standard deviation of his 15 quiz scores be greater than, equal to, or less than 8? Justify your answer.

(b) Suppose instead that Joey has an unexcused absence and makes a 0 on the next quiz. Would the standard deviation of his 15 quiz scores be greater than, equal to, or less than 8? Justify your answer.

Recycle and Review

21. **Hurricanes (1.5, 1.6)** The histogram shows the distribution of the number of Atlantic hurricanes in every year from 1851 through 2012.[55]

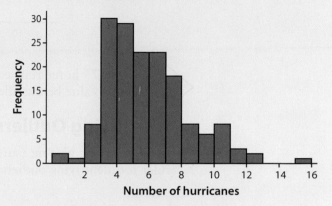

Number of hurricanes

(a) Describe the shape of the distribution.

(b) Which would be a better measure of the typical number of hurricanes per year: the mean or the median? Justify your answer.

22. **Salty nuggets (1.3, 1.4)** The sodium content, in milligrams per 3-oz serving, for 22 brands of breaded chicken nuggets and tenders are given here.[56]

| 340 | 360 | 310 | 370 | 300 | 310 | 210 | 230 | 240 | 480 | 330 |
| 240 | 450 | 180 | 270 | 240 | 420 | 330 | 560 | 440 | 350 | 210 |

(a) Make a stemplot of these data using split stems.

(b) Make a dotplot of these data.

(c) Describe any features of the distribution that are better illustrated by one graph than by the other.

Lesson 1.8

Summarizing Quantitative Data: Boxplots and Outliers

LEARNING TARGETS

- Use the 1.5 × *IQR* rule to identify outliers.
- Make and interpret boxplots of quantitative data.
- Compare distributions of quantitative data with boxplots.

Barry Bonds set the major league record by hitting 73 home runs in a single season in 2001. On August 7, 2007, Bonds hit his 756th career home run, which broke Hank Aaron's longstanding record of 755. By the end of the 2007 season when Bonds retired, he had increased the total to 762. The dotplot shows the number of home runs that Bonds hit in each of his 21 complete seasons:

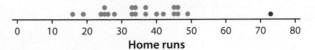

Home runs

Bonds's 73 home run season stands out (in red) from the rest of the distribution. Should this value be classified as an *outlier*?

Identifying Outliers

Besides serving as a measure of variability, the interquartile range (*IQR*) is used as a ruler for identifying outliers.

How to Identify Outliers: The 1.5 × *IQR* Rule

Call an observation an outlier if it falls more than 1.5 × *IQR* above the third quartile or more than 1.5 × *IQR* below the first quartile. That is,

Low Outliers < $Q_1 - 1.5 \times IQR$ High Outliers > $Q_3 + 1.5 \times IQR$

EXAMPLE

Home run king?

Identifying outliers

PROBLEM: Here are data on the number of home runs that Bonds hit in each of his 21 complete seasons. Identify any outliers in the distribution. Show your work.

16 25 24 19 33 25 34 46 37 33 42 40 37 34 49 73 46 45 45 26 28

SOLUTION:

16 19 24 25 25 | 26 28 33 33 34 (34) 37 37 40 42 45 | 45 46 46 49 73

$Q_1 = 25.5$ Median $Q_3 = 45$

$IQR = Q_3 - Q_1 = 45 - 25.5 = 19.5$

Outliers < $Q_1 - 1.5 \times IQR = 25.5 - 1.5 \times 19.5 = -3.75$

Outliers > $Q_3 + 1.5 \times IQR = 45 + 1.5 \times 19.5 = 74.25$

> Find the interquartile range (*IQR*). Use the method of Lesson 1.7.

> Calculate the upper and lower cutoff values for outliers.

Because there are no data values less than -3.75 or greater than 74.25, this distribution has no outliers.

> Barry Bonds's record-setting year with 73 home runs is not quite large enough to be classified as an outlier by the 1.5 × *IQR* rule.

FOR PRACTICE   TRY EXERCISE 1.

It is important to identify outliers in a distribution for several reasons:

1. **They might be inaccurate data values.** Maybe someone recorded a value as 10.1 instead of 101. Perhaps a measuring device broke down. Or maybe someone gave a silly response, like the student in a class survey who claimed to study 30,000 minutes per night!

2. **They can indicate a remarkable occurrence.** For example, in a graph of career golf earnings, Tiger Woods is likely to be an outlier.

3. **They can heavily influence the values of some summary statistics,** like the mean, range, and standard deviation.

Making and Interpreting Boxplots

You can use a dotplot, stemplot, or histogram to display the distribution of a quantitative variable. Another graphical option for quantitative data is a **boxplot** (sometimes called a *box-and-whisker* plot). A boxplot summarizes a distribution by displaying the location of 5 important values within the distribution, known as its **five-number summary.**

DEFINITION Five-number summary, Boxplot

The **five-number summary** of a distribution of quantitative data consists of the minimum, the first quartile Q_1, the median, the third quartile Q_3, and the maximum.

A **boxplot** is a visual representation of the five-number summary.

How to Make a Boxplot

1. **Find the five-number summary** for the distribution.
2. **Draw and label the axis.** Draw a horizontal axis and put the name of the quantitative variable underneath.
3. **Scale the axis.** Look at the smallest and largest values in the data set. Start the horizontal axis at a number equal to or below the smallest value and place tick marks at equal intervals until you equal or exceed the largest value.
4. **Draw a box** that spans from the first quartile (Q_1) to the third quartile (Q_3).
5. **Mark the median** with a vertical line segment that's the same height as the box.
6. **Identify outliers** using the $1.5 \times IQR$ rule.
7. **Draw whiskers**—lines that extend from the ends of the box to the smallest and largest data values that are *not* outliers. Mark any outliers with a special symbol such as an asterisk (*).

The top dotplot in the following figure shows Barry Bonds's home run data. We have marked the first quartile, the median, and the third quartile with blue lines. The process of testing for outliers with the $1.5 \times IQR$ rule is shown in red. Because there are no outliers, we draw the whiskers to the maximum and minimum data values, as shown in the finished boxplot at the bottom of the figure.

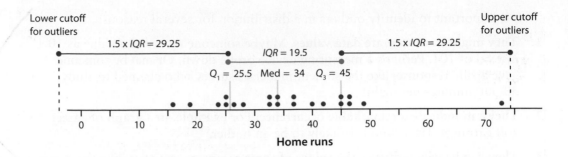

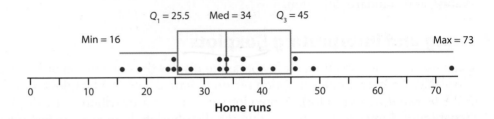

We see from the boxplot that the distance from the minimum to the median is much smaller than the distance from the median to the maximum. That is, the right side of the boxplot is more stretched out than the left side. So Barry Bonds's home run distribution is skewed to the right.

EXAMPLE

How big are the large fries?

Making and interpreting a boxplot

PROBLEM: Ryan and Brent were curious about the amount of french fries they would get in a large order from their favorite fast-food restaurant, Burger King. They went to several different Burger King locations over a series of days and ordered a total of 14 large fries. The weight of each order (in grams) is shown here.

165   163   160   159   166   152   166   168   173   171   168   167   170   170

(a) Make a boxplot to display the data.

(b) According to a nutrition website, Burger King's large fries weigh 160 grams, on average. Ryan and Brent suspect that their local Burger King restaurants may be skimping on fries. Does the graph in part (a) support their suspicion? Explain.

SOLUTION:

(a)

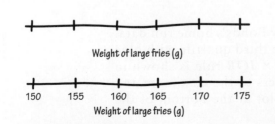

1. Find the five-number summary.

2. Draw and label the axis.

3. Scale the axis.

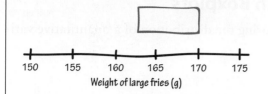

4. Draw a box that spans from the first quartile (Q_1) to the third quartile (Q_3).

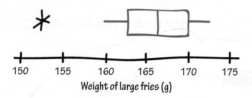

5. Mark the median with a vertical line segment that's the same height as the box.

$IQR = Q_3 - Q_1 = 170 - 163 = 7$

Low Outliers $< Q_1 - 1.5 \times IQR = 163 - 1.5 \times 7 = 152.5$

High Outliers $> Q_3 + 1.5 \times IQR = 170 + 1.5 \times 7 = 180.5$

The order of large fries that weighed 152 grams is an outlier.

6. Identify outliers.

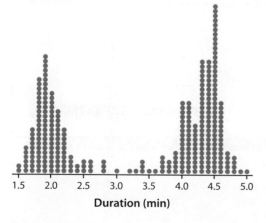

7. Draw whiskers—lines that extend from the ends of the box to the smallest and largest data values that are *not* outliers. Mark any outliers with an asterisk.

(b) No. From the boxplot, $Q_1 = 163$, so at least 75% of the large fries that Ryan and Brent bought from local Burger King restaurants weighed 163 grams or more. Only the outlier (152 grams) and one other order (159 grams) of large fries weighed less than 160 grams.

FOR PRACTICE   TRY EXERCISE 5.

Boxplots provide a quick summary of the center and variability of a distribution. The median is displayed as a vertical line in the central box, the interquartile range is the length of the box, and the range is the length of the entire plot, including outliers.

Boxplots do not display each individual value in a distribution. And boxplots don't show gaps, clusters, or peaks. For instance, the dotplot below displays the duration, in minutes, of 220 eruptions of the Old Faithful geyser. The distribution of eruption durations is clearly bimodal (two-peaked). But a boxplot of the data hides this important information about the shape of the distribution.

CAUTION

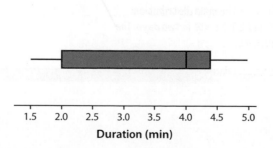

Comparing Distributions with Boxplots

Boxplots are especially effective for comparing the distribution of a quantitative variable in two or more groups.

EXAMPLE

Who is doing the texting?

Comparing distributions with boxplots

PROBLEM: For their final project, a group of statistics students wanted to compare the texting habits of males and females. They asked a random sample of students from their school to record the number of text messages sent and received over a two-day period. Here are their data.

Males	127	44	28	83	0	6	78	6	5	213	73	20	214	28	11
Females	112	203	102	54	379	305	179	24	127	65	41	27	298	6	130	0

Parallel boxplots of the data and numerical summaries are shown here. Compare the texting distributions for males and females.

	$\bar{x}$	s_x	Min	Q_1	Med	Q_3	Max	IQR
Male	62.4	71.4	0	6	28	83	214	77
Female	128.3	116.0	0	34	107	191	379	157

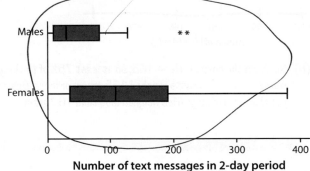

Number of text messages in 2-day period

SOLUTION:

Shape: Both distributions are strongly right-skewed.

Center: The females in the sample typically texted much more over a two-day period (median = 107) than the males did (median = 28). In fact, the median for the females is above the third quartile for the males. This indicates that at least 75% of the males texted less than the "typical" (median) female.

Variability: There is much more variation in texting among the females than the males. The IQR for females (157) is about twice the IQR for males (77).

Outliers: There are two outliers in the male distribution: students who reported 213 and 214 texts in two days. The female distribution has no outliers.

> Remember to compare shape, center, variability, and outliers!

> Due to the strong skewness and outliers, use the median and *IQR* instead of the mean and standard deviation when comparing center and variability.

FOR PRACTICE   TRY EXERCISE 9.

LESSON APP 1.8

Which is best at reducing stress?

If you are a dog lover, having your dog with you may reduce your stress level. Does having a friend with you reduce stress? To examine the effect of pets and friends in stressful situations, researchers recruited 45 women who said they were dog lovers. Fifteen women were assigned at random to each of three groups: to do a stressful task (1) alone, (2) with a good friend present, or (3) with their dogs present. The stressful task was to count backward by 13s or 17s. The woman's average heart rate during the task was one measure of the effect of stress. The following table shows the data.[57]

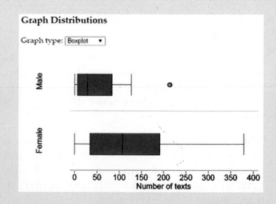

Winnie Au/Getty Images

Alone	62.6	70.9	73.3	75.5	77.8	80.4	84.5	84.7
	84.9	87.2	87.4	87.8	90.0	91.8	99.0	
Friend	76.9	80.3	81.6	83.4	87.0	88.0	89.8	91.4
	92.5	97.0	98.2	99.7	100.9	101.1	102.2	
Pet	58.7	64.2	65.4	68.9	69.2	69.2	69.5	70.2
	70.1	72.3	76.0	79.7	85.0	86.4	97.5	

1. Identify any outliers in the three groups. Show your work.

2. Make parallel boxplots to compare the heart rates of the women in the three groups.

3. Based on the data, does it appear that the presence of a pet or friend reduces heart rate during a stressful task? Justify your answer.

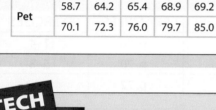

Making Boxplots with Technology

You can use an applet or a graphing calculator to make a boxplot. Let's use technology to make parallel boxplots of the male and female texting data.

Applet

1. Go to highschool.bfwpub.com/spa3e and launch the *One Quantitative Variable* applet.

2. Enter Number of texts as the Variable name.

3. Select 2 as the number of groups and Raw data as the input method.

4. Name Group 1 "Male" and Group 2 "Female." Then enter the data for each group. Be sure to separate the data values with spaces or commas as you type them.

5. Click Begin analysis. Parallel dotplots of the data should appear. Change the graph type to boxplot.

Graph Distributions

Graph type: [Boxplot ▼]

[Parallel boxplots of Male and Female number of texts, horizontal axis labeled "Number of texts" ranging from 0 to 400]

TI-83/84

1. Enter the texting data for males in list L_1 and for females in list L_2.

2. Set up two statistics plots: Plot1 to show a boxplot of the male data and Plot2 to show a boxplot of the female data. The setup for Plot1 is shown. When you define Plot2, be sure to change L1 to L2.

Note: The calculator offers two types of boxplots: one that shows outliers and one that doesn't. We'll always use the type that identifies outliers.

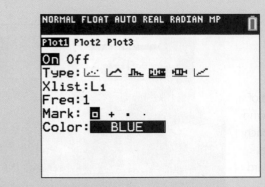

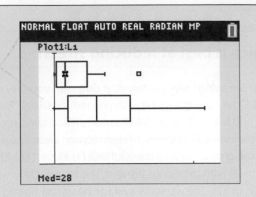

3. Press ZOOM and select ZoomStat to display the parallel boxplots. Then press TRACE to view the five-number summary.

Lesson 1.8

WHAT DID YOU LEARN?

LEARNING TARGET	EXAMPLES	EXERCISES
Use the $1.5 \times IQR$ rule to identify outliers.	p. 68	1–4
Make and interpret boxplots of quantitative data.	p. 70	5–8
Compare distributions of quantitative data with boxplots.	p. 72	9–12

Exercises — *Lesson 1.8*

Mastering Concepts and Skills

1. **Outlier Cowboys** The roster of the Dallas Cowboys professional football team in a recent season included 7 defensive linemen. Their weights (in pounds) were 321, 285, 300, 285, 286, 293, and 298. Identify any outliers in the distribution. Show your work.

pg 68

2. **Musical megabytes** How much disk space does your music use? Here are the file sizes (in megabytes) for 18 randomly selected files on Gabriel's mp3 player. Identify any outliers in the distribution.

2.4	2.7	1.6	1.3	6.2	1.3	5.6	1.1	2.2
1.9	2.1	4.4	4.7	3.0	1.9	2.5	7.5	5.0

3. **Pizza with outliers** The dotplot shows the number of calories per serving for 16 brands of frozen cheese pizza.[58] Identify any outliers in the distribution. Show your work.

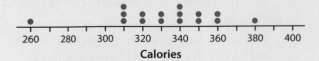

4. **Electoral College outliers** To become president of the United States, a candidate does not have to receive a majority of the popular vote. The candidate does, however, have to win a majority of the 538 electoral votes that are cast in the Electoral College. Here is a stemplot of the number of electoral votes in 2016 for each of the 50 states and the District of Columbia. Identify any outliers in the distribution. Show your work.

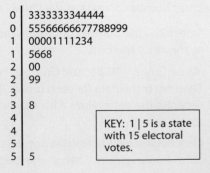

0	3333333344444
0	55566666677788999
1	00001111234
1	5668
2	00
2	99
3	
3	8
4	
4	
5	
5	5

KEY: 1 | 5 is a state with 15 electoral votes.

5. **No need to call** According to a study by Nielsen Mobile, "Teenagers ages 13 to 17 are by far the most prolific texters, sending or receiving 1742 messages a month." Mr. Williams, a high school

pg 70

statistics teacher, was skeptical about this claim. So he collected data from his first-period statistics class on the number of text messages and calls they had sent or received in the past 24 hours. Here are the data:

0   7   1   29   25   8   5   1   25   98   9   0   26
8   118   72   0   92   52   14   3   3   44   5   42

(a) Make a boxplot to display the data.

(b) Explain how the graph in part (a) gives evidence to contradict the claim in the article.

6. **Acing the first test** Here are the scores of Mrs. Liao's students on their first statistics test:

93 93 87.5 91 94.5 72 96 95 93.5 93.5 73 82 45 88 80 86
85.5 87.5 81 78 86 89 92 91 98 85 82.5 88 94.5 43

(a) Make a boxplot to display the data.

(b) How did the students do on Mrs. Liao's first test? Use the graph from part (a) to help justify your answer.

7. **Boxed Cowboys** Refer to Exercise 1.

(a) Make a boxplot to display the data.

(b) Which measure of variability—the *IQR* or standard deviation—would you report for these data? Use the graph from part (a) to help justify your choice.

8. **Variable memory** Refer to Exercise 2.

(a) Make a boxplot to display the data.

(b) Which measure of variability—the *IQR* or standard deviation—would you report for these data? Use the graph from part (a) to help justify your choice.

9. **Fat sandwiches, skinny sandwiches** The following
pg 72 boxplots summarize data on the amount of fat (in grams) in 12 McDonald's beef sandwiches and 9 McDonald's chicken or fish sandwiches. Compare the distributions of fat content for the two types of sandwiches.

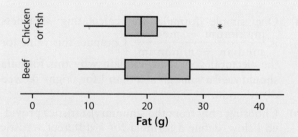

10. **Get to work!** The following boxplots summarize data on the travel times to work for 20 randomly chosen New Yorkers and 15 randomly chosen North Carolinians. Compare the distributions of travel time for the workers in these two states.

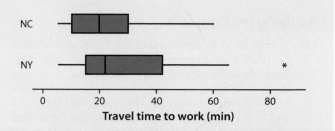

11. **Energetic refrigerators** In its May 2010 edition, *Consumer Reports* magazine rated different types of refrigerators, including those with bottom freezers, those with top freezers, and those with side freezers. One of the variables they measured was annual energy cost (in dollars). The following boxplots show the energy cost distributions for each of these types.

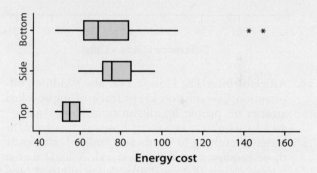

(a) What percentage of bottom freezers cost more than $60 per year to operate? What about side freezers and top freezers?

(b) Compare the energy cost distributions for the three types of refrigerators.

12. **Income in New England** The following boxplots show the total income of 40 randomly chosen households each from Connecticut, Maine, and Massachusetts, based on U.S. Census data from the American Community Survey for 2012.

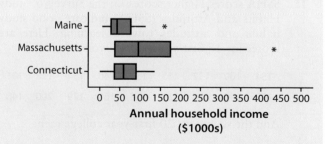

(a) Approximately what percentage of households in the Maine sample had annual incomes below $50,000? What about households in Massachusetts and Connecticut?

(b) Compare the distributions of annual incomes in the three states.

Applying the Concepts

13. **Text or talk?** In a September 28, 2008, article titled, "Letting Our Fingers Do the Talking," the *New York Times* reported that Americans now send more text messages than they make phone calls. Mr. Williams was curious about whether this claim was valid for high school students. So he collected data from his first-period statistics class on the number of text messages and calls they had sent or received in the past 24 hours. A boxplot of the difference (Texts – Calls) in the number of texts and calls for each student is shown here. Do these data support the claim in the article about texting versus calling? Justify your answer.

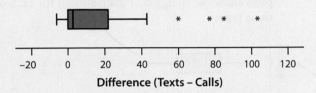

Difference (Texts – Calls)

14. **Alligator bites** The Florida Fish and Wildlife Conservation Commission keeps track of unprovoked attacks on people by alligators, defining "major" attacks as those requiring hospital treatment or (rarely) resulting in death and "minor" attacks as those requiring, at most, first aid. A local tourist bureau claims that most attacks are minor. A boxplot of the difference (Major – Minor) in reported number of attacks for each year from 1971 through 2013 is given here.[59] Do these data support the tourist bureau's claim? Justify your answer.

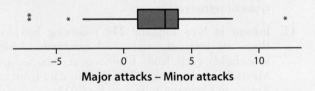

Major attacks – Minor attacks

15. **SSHA scores** Higher scores on the Survey of Study Habits and Attitudes (SSHA) indicate good study habits and attitudes toward learning. Here are scores for 18 first-year college women.

154	109	137	115	152	140	154	178	101
103	126	126	137	165	165	129	200	148

And the scores for 20 first-year college men:

108	140	114	91	180	115	126	92	169	146
109	132	75	88	113	151	70	115	187	104

(a) Make parallel boxplots to compare the distributions.

(b) Do these data support the belief that men and women differ in their study habits and attitudes toward learning? Give appropriate evidence to support your answer.

16. **Well connected** Who has more contacts—males or females? The data show the number of contacts that a sample of high school students had in their cell phones.

Male	124	41	29	27	44	87	85	260	290	31
	168	169	167	214	135	114	105	103	96	144
Female	30	83	116	22	173	155	134	180	124	33
	213	218	183	110						

(a) Make parallel boxplots to compare the distributions.

(b) Based on your graphs in part (a), which gender tends to have more contacts in their cell phones? Give appropriate evidence to support your answer.

Extending the Concepts

17. **Measuring skewness** Here is a boxplot of the number of electoral votes in 2016 for each of the 50 states and the District of Columbia, along with summary statistics. You can see that the distribution is skewed to the right with 3 high outliers. How might we compute a numerical measure of skewness?

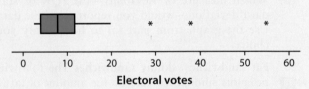

Electoral votes

Variable	n	Mean	Minimum	Q_1	Median	Q_3	Maximum
Electoral votes	51	10.55	3.00	4.00	8.00	12.00	55.00

(a) One simple formula for calculating skewness is $\dfrac{\text{maximum} - \text{median}}{\text{median} - \text{minimum}}$. Compute this value for the electoral vote data. Explain why this formula should yield a value greater than 1 for a right-skewed distribution.

(b) Choosing only from the summary statistics provided below, define a formula for a different statistic that measures skewness. Compute the value of this statistic for the electoral vote data. What values of the statistic might indicate that a distribution is skewed to the right? Explain.

Recycle and Review

18. Best in Show (1.1, 1.2) Here is a frequency table showing the breed group of the dog that won Best in Show at the Westminster Kennel Club dog show for 96 years.[60] Make a relative frequency bar chart for these data. Describe what you see.

Breed group	Frequency	Breed group	Frequency
Terrier	38	Toy	11
Sporting	18	Non-sporting	9
Working	15	Hound	5

19. Heartbeats (1.7) Here are the resting heart rates of 26 ninth-grade biology students. Decide on an appropriate measure of variability for these data and calculate it. Justify your choice.

48 55 60 61 62 63 64 65 66 66 68 70 70
71 71 72 74 74 75 76 77 77 78 79 81 86

Lesson 1.9

Describing Location in a Distribution

LEARNING TARGETS

- Find and interpret a percentile in a distribution of quantitative data.
- Estimate percentiles and individual values using a cumulative relative frequency graph.
- Find and interpret a standardized score (*z*-score) in a distribution of quantitative data.

Here are the scores of all 25 students in Mr. Pryor's statistics class on their first test:

79 81 80 77 73 83 74 93 78 80 75 67 73

77 83 **86** 90 79 85 83 89 84 82 77 72

The bold score is Jenny's 86. How did she perform on this test relative to her classmates?

The following dotplot displays the class's test scores, with Jenny's score marked in red. The distribution is roughly symmetric with no apparent outliers. From the dotplot, we can see that Jenny's score is above the mean (balance point) of the distribution. We can also see that Jenny did better on the test than most other students in the class.

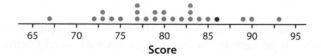

Score

Finding and Interpreting Percentiles

One way to describe Jenny's location in the distribution of test scores is to calculate her **percentile.**

DEFINITION Percentile

An individual's **percentile** is the percent of values in a distribution that are less than the individual's data value.

Using the dotplot, we see that Jenny's 86 places her fourth from the top of the class. Because 21 of the 25 observations (84%) are below her score, Jenny is at the 84th percentile in the class's test score distribution.

Be careful with your language when describing percentiles. Percentiles are specific locations in a distribution, so an observation isn't "in" the 84th percentile. Rather, it is "at" the 84th percentile.

EXAMPLE

What are the results of the first test?

Finding and interpreting percentiles

PROBLEM: Refer to the dotplot of scores on Mr. Pryor's first statistics test.

(a) Find the percentile for Norman, who scored 72.

(b) Maria's test score is at the 48th percentile of the distribution. Interpret this value in context. What score did Maria earn?

Score

SOLUTION:

(a) $1/25 = 0.04$, so Norman scored at the 4th percentile on this test.

> Only 1 of the 25 scores in the class is below Norman's 72.

(b) $(0.48)(25) = 12$, so Maria's score was higher than 12 of the 25 students in the class. Maria earned an 80 on the test.

> One other student in the class scored an 80 on the test. This student's score is also at the 48th percentile because 12 of the 25 students in the class earned lower scores.

FOR PRACTICE   TRY EXERCISE 1.

The median of a distribution is roughly the 50th percentile. For instance, 80 is the median score on Mr. Pryor's first test. As you saw in part (b) of the example, Maria's score of 80 put her at the 48th percentile of the distribution. The first quartile Q_1 is roughly the 25th percentile of a distribution because it separates the lowest 25% of values from the upper 75%. Likewise, the third quartile Q_3 is roughly the 75th percentile.

A high percentile is not always a good thing. For example, a man whose blood pressure is at the 90th percentile for his age group may need treatment for his high blood pressure!

Cumulative Relative Frequency Graphs

There are some interesting graphs that can be made with percentiles. One of the most common starts with a frequency table for a quantitative variable. For instance, this frequency table summarizes the ages of the first 44 U.S. presidents when they took office.

Age	Frequency
40–44	2
45–49	7
50–54	13
55–59	12
60–64	7
65–69	3

Let's expand this table to include columns for relative frequency, cumulative frequency, and cumulative relative frequency.

- To get the values in the relative frequency column, divide the count in each interval by 44, the total number of presidents. Multiply by 100 to convert to a percent.

- To fill in the *cumulative frequency* column, add the counts in the frequency column for the current interval and all intervals with smaller values of the variable.

- For the *cumulative relative frequency* column, divide the entries in the cumulative frequency column by 44, the total number of individuals. Multiply by 100 to convert to a percent.

Here is the original frequency table with the relative frequency, cumulative frequency, and cumulative relative frequency columns added.

Age	Frequency	Relative frequency	Cumulative frequency	Cumulative relative frequency
40–44	2	2/44 = 0.045, or 4.5%	2	2/44 = 0.045, or 4.5%
45–49	7	7/44 = 0.159, or 15.9%	9	9/44 = 0.205, or 20.5%
50–54	13	13/44 = 0.295, or 29.5%	22	22/44 = 0.500, or 50.0%
55–59	12	12/44 = 0.273, or 27.3%	34	34/44 = 0.773, or 77.3%
60–64	7	7/44 = 0.159, or 15.9%	41	41/44 = 0.932, or 93.2%
65–69	3	3/44 = 0.068, or 6.8%	44	44/44 = 1.000, or 100%

Now we can make a **cumulative relative frequency graph.**

DEFINITION Cumulative relative frequency graph

A **cumulative relative frequency graph** plots a point corresponding to the cumulative relative frequency in each interval at the smallest value of the *next* interval, starting with a point at a height of 0% at the smallest value of the first interval. Consecutive points are then connected with a line segment to form the graph.

Figure 1.9 on the next page shows the completed cumulative relative frequency graph for the presidential age at inauguration data. Notice the following:

- The leftmost point is plotted at a height of 0% at Age = 40, the smallest value in the first interval. This point tells us that none of the first 44 U.S. presidents took office before age 40.

- The next point to the right is plotted at a height of 4.5% at Age = 45. This point tells us that 4.5% of presidents (i.e., two of them) were inaugurated before they were 45 years old.
- The graph grows very gradually at first because few presidents were inaugurated when they were in their 40s. Then the graph gets very steep beginning at age 50 because most U.S. presidents were in their 50s when they were inaugurated. The rapid growth in the graph slows at age 60.
- The rightmost point on the graph is plotted above age 70 and has cumulative relative frequency 100%. That's because 100% of U.S. presidents took office before age 70.

FIGURE 1.9 Cumulative relative frequency graph for the ages of U.S. presidents when they took office.

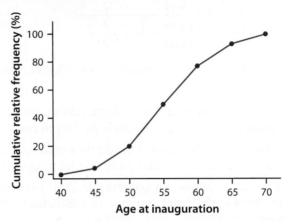

A cumulative relative frequency graph can be used to describe the position of an individual within a distribution or to locate a specified percentile of the distribution.

EXAMPLE

Is the president old?

Interpreting cumulative relative frequency graphs

PROBLEM: Use the graph in Figure 1.9 to help you answer each question:

(a) Was Barack Obama, who was first inaugurated at age 47, unusually young?

(b) Estimate and interpret the 65th percentile of the distribution.

SOLUTION:

(a)

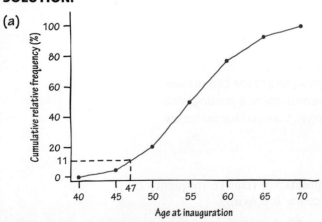

To find President Obama's location in the distribution, draw a vertical line up from his age (47) on the horizontal axis until it meets the graph. Then draw a horizontal line from this point to the vertical axis.

Barack Obama's inauguration age places him at about the 11th percentile. About 11% of all U.S. presidents first took office at a younger age than Obama did. So Obama was fairly young, but not unusually young, when he took office.

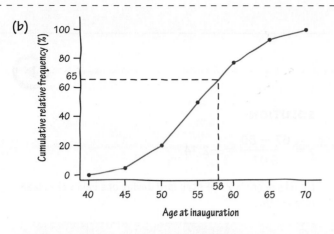

The 65th percentile of the distribution is the age with cumulative relative frequency 65%. To find this value, draw a horizontal line across from the vertical axis at a height of 65% until it meets the graph. Then draw a vertical line from this point down to the horizontal axis.

The 65th percentile is about 58. About 65% of all U.S. presidents were younger than 58 when they took office.

FOR PRACTICE   TRY EXERCISE 5.

Finding and Interpreting Standardized Scores (z-Scores)

A percentile is one way to describe the location of an individual in a distribution of quantitative data. Another way is to give the **standardized score** (*z*-score) for the observed value.

> **DEFINITION   Standardized score (z-score)**
>
> The **standardized score (z-score)** for an individual value in a distribution tells us how many standard deviations from the mean the value falls, and in what direction. To find the standardized score (*z*-score), compute
>
> $$z = \frac{\text{value} - \text{mean}}{\text{standard deviation}}$$

Values larger than the mean have positive *z*-scores. Values smaller than the mean have negative *z*-scores.

Let's return to the data from Mr. Pryor's first statistics test. The following dotplot displays the data, with Jenny's score marked in red. The table provides numerical summaries for these data.

Summary Statistics

n	mean	SD	min	Q_1	Med	Q_3	max
25	80	6.07	67	76	80	83.5	93

Where does Jenny's 86 fall within the distribution? Her standardized score (*z*-score) is

$$z = \frac{\text{value} - \text{mean}}{\text{standard deviation}} = \frac{86 - 80}{6.07} = 0.99$$

That is, Jenny's test score is 0.99 standard deviations above the mean score of the class.

EXAMPLE

How well did Lionel do?

Finding and interpreting z-scores

PROBLEM: Find the standardized score (z-score) for Lionel, who earned a 67 on Mr. Pryor's first test. Interpret this value in context.

FOR PRACTICE   TRY EXERCISE 9.

SOLUTION:

$$z = \frac{67 - 80}{6.07} = -2.14$$

Lionel's score is 2.14 standard deviations below the class mean of 80.

We often standardize observed values to express them on a common scale. For example, we might compare the heights of two children of different ages or genders by calculating their z-scores.

- At age 7, Jordan is 51 in. tall. Her height puts her at a z-score of 1. That is, Jordan is 1 standard deviation above the mean height of 7-year-old girls.
- Zayne's height at age 9 is 54 in. His corresponding z-score is 0.5. In other words, Zayne is one-half standard deviation above the mean height of 9-year-old boys.

So Jordan is taller relative to girls her age than Zayne is relative to boys his age. The standardized heights tell us where each child stands (pun intended!) in the distribution for his or her age group.

LESSON APP 1.9

Which states are rich?

The following cumulative relative frequency graph and the numerical summaries describe the distribution of median household incomes in the 50 states in a recent year.[61]

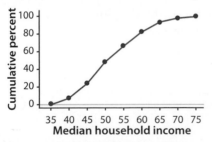

Use the information provided above and in this table to help you answer the following questions.

Median household income	
n	50
Mean	$51,742.44
SD	$8,210.642

© Andre Jenny/Alamy Stock Photo

1. At what percentile is North Dakota, with a median household income of $55,766?

2. Estimate and interpret the first quartile Q_1 of the distribution.

3. Find and interpret the standardized score (z-score) for New Jersey, with a median household income of $66,692.

Lesson 1.9

WHAT DID YOU LEARN?

LEARNING TARGET	EXAMPLES	EXERCISES
Find and interpret a percentile in a distribution of quantitative data.	p. 78	1–4
Estimate percentiles and individual values using a cumulative relative frequency graph.	p. 80	5–8
Find and interpret a standardized score (z-score) in a distribution of quantitative data.	p. 82	9–12

Exercises   Lesson 1.9

Mastering Concepts and Skills

1. Play ball! The dotplot shows the number of wins
pg **78** for each of the 30 Major League Baseball teams in the 2014 season:

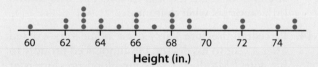

Wins

(a) Find the percentile for the Boston Red Sox, who won 71 games.

(b) The New York Yankees' number of wins is at the 60th percentile of the distribution. Interpret this value in context. How many games did New York win?

2. Stand up tall The dotplot shows the heights of the 25 students in Mrs. Navard's statistics class.

Height (in.)

(a) Find the percentile for Lynette, the student who is 65 in. tall.

(b) Asher's height is at the 88th percentile of the distribution. Interpret this value in context. How tall is Asher?

3. A boy and his shoes How many pairs of shoes does a typical teenage boy own? To find out, a group of statistics students surveyed a random sample of 20 male students from their large high school. Then they recorded the number of pairs of shoes that each boy owned. Here are the data.

| 14 | 7 | 6 | 5 | 12 | 38 | 8 | 7 | 10 | 10 |
| 10 | 11 | 4 | 5 | 22 | 7 | 5 | 10 | 35 | 7 |

(a) Martin is the student who reported owning 22 pairs of shoes. Find Martin's percentile.

(b) Luis is at the first quartile Q_1 of the distribution. How many pairs of shoes does Luis own?

4. Unlocked for sale The "sold" listings on a popular auction website included 21 sales of used "unlocked" phones of one popular model. Here are the sales prices.

450	415	495	300	325	430	370
400	325	400	235	330	304	415
355	405	449	355	425	299	345

(a) Find the percentile of the phone that sold for $325.

(b) What was the sales price of the phone that was at the third quartile Q_3?

5. Supermarket sweep The following figure is a cu-
pg **80** mulative relative frequency graph of the amount spent by a sample of 50 grocery shoppers at a store.

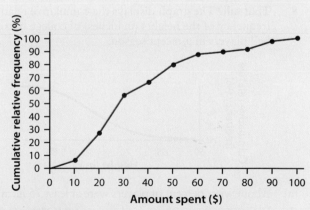

(a) What is the percentile for the shopper who spent $19.50?

(b) Estimate and interpret the 80th percentile of the distribution.

6. Light life The following graph is a cumulative relative frequency graph showing the lifetimes (in hours) of 200 lamps.[62]

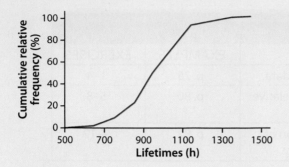

(a) What is the percentile for a lamp that lasted 900 hours?

(b) Estimate and interpret the 60th percentile of this distribution.

7. **Call me maybe?** The graph displays the cumulative relative frequency of the lengths of phone calls made from the math department office at Gabalot High last month.

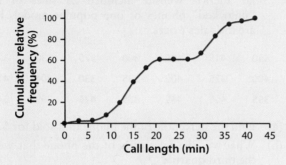

(a) About what percent of calls lasted 30 minutes or more?

(b) Estimate the interquartile range (*IQR*) of the distribution.

8. **That tall?** The graph displays the cumulative relative frequency of the heights (in inches) of college basketball players in a recent season.

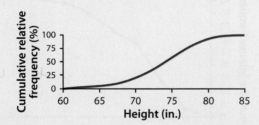

(a) About what percent of players were at least 75 in. tall?

(b) Estimate the interquartile range (*IQR*) of the distribution.

9. **The Nationals play** During the 2014 season, the mean number of wins for Major League Baseball teams was 81 with a standard deviation of 9.6 wins. Find the standardized score (*z*-score) for the Washington Nationals, who won 96 games. Interpret this value in context.

pg 82

10. **Stand tall** The heights of the 25 students in Mrs. Navard's statistics class have a mean of 67 in. and a standard deviation of 4.29 in. Find the standardized score (*z*-score) for Boris, a member of the class who is 76 in. tall. Interpret this value in context.

11. **Where are the old folks?** Based on data from the 2010 U.S. Census, the percent of residents aged 65 or older in the 50 states and the District of Columbia has mean 13.26% and standard deviation 1.67%.

(a) Find and interpret the standardized score (*z*-score) for the state of Colorado, which had 9.7% of its residents age 65 or older.

(b) The standardized score for Florida is $z = 2.60$. Find the percent of the state's residents that were 65 or older.

12. **Meaning of the Dow** The Dow Jones Industrial Average (DJIA) is a commonly used index of the overall strength of the U.S. stock market. In 2013 the mean daily change in the DJIA for the 252 days that the stock markets were open was 13.59 points with a standard deviation of 94.05 points.

(a) Find and interpret the standardized score (*z*-score) for the change in the DJIA on May 7, 2013, which was 87.31 points.

(b) The standardized score for May 1, 2013, was $z = -1.62$. Find the change in the DJIA for that date.

Applying the Concepts

13. **Setting speed limits** According to the *Los Angeles Times,* speed limits on California highways are set at the 85th percentile of vehicle speeds on those stretches of road. Explain to someone who knows little statistics what that means.

14. **Percentile pressure** Larry came home very excited after a visit to his doctor. He announced proudly to his wife, "My doctor says my blood pressure is at the 90th percentile among men like me. That means I'm better off than about 90% of similar men." How should his wife, who is a statistician, respond to Larry's statement?

15. **Big or little?** Mrs. Munson is interested to know how her son's height and weight compare with those of other boys his age. She uses an online calculator to determine that her son is at the 48th percentile for weight and the 76th percentile for height. Explain to Mrs. Munson what these values mean.

16. **Run faster** Peter is a star runner on the track team. In the league championship meet, Peter records a time that would fall at the 80th percentile of all his race times that season. But his performance places him at the 50th percentile in the league championship meet. Explain how this is possible. (Remember that lower times are better in this case!)

17. **SAT versus ACT** During her senior year, Courtney took both the SAT and ACT. She scored 680 on the SAT math test and 27 on the ACT math test. Scores on the math section of the SAT vary from 200 to 800, with a mean of 514 and standard deviation of 117. Scores on the math section of the ACT vary from 1 to 36, with a mean of 21.0 and a standard deviation of 5.3. Calculate Courtney's standardized score on each test. Which of her two test scores was better? Explain.

18. **Generational GPA** Rebecca and her father both graduated from the same high school. When her father looked at Rebecca's transcript, he noticed that her high school GPA (4.2) was higher than his high school GPA (3.9). After letting Rebecca gloat for a minute, he pointed out that there were no weighted grades when he went to school. To settle their argument, they called the registrar at the school and got information about the distribution of GPA in each of their graduation years. When the father graduated, the mean GPA was 2.8 with a standard deviation of 0.6. When Rebecca graduated, the mean GPA was 3.2 with a standard deviation of 0.7. Calculate Rebecca's and her father's standardized GPA. Who had the better GPA? Explain.

Extending the concepts

19. **Medical exam results** People with low bone density have a high risk of broken bones. Currently, the most common method for testing bone density is dual-energy X-ray absorptiometry (DEXA). A patient who undergoes a DEXA test usually gets bone density results in grams per square centimeter (g/cm^2) and in standardized units.

    Judy, who is 25 years old, has her bone density measured using DEXA. Her results indicate a bone density in the hip of 948 g/cm^2 and a standardized

score of $z = -1.45$. In the population of 25-year-old women like Judy, the mean bone density in the hip is 956 g/cm^2.[63]

(a) Judy has not taken a statistics class in a few years. Explain in simple language what the standardized score tells her about her bone density.

(b) Use the information provided to calculate the standard deviation of bone density in the population of 25-year-old women.

Recycle and Review

20. **Birthrates in Africa (1.6, 1.7, 1.8)** One of the important factors in determining population growth rates is the birthrate per 1000 individuals in a population. Here are a dotplot and five-number summary for the birthrates per 1000 individuals in 54 African nations.

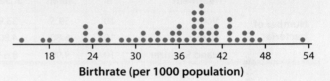

Birthrate (per 1000 population)

Minimum	Q_1	Median	Q_3	Maximum
14	29	37.5	41	53

(a) Construct a boxplot for these data.

(b) Suppose the maximum value of 53 was in error and should have been 45. For each statistic, indicate whether this correction would result in an increase, a decrease, or no change. Justify your answer in each case.

- Mean
- Median
- Standard deviation
- Interquartile range

Chapter 1   RESOLVED

STATS applied!

Does hand sanitizer work?

In the STATS applied! on page 3, recall Daniel and Kate's experiment. Using 30 identical petri dishes, they randomly assigned 10 students to press one hand in a dish after washing with soap, 10 students to press one hand in a dish after using hand sanitizer, and 10 students to press one hand in a dish after using nothing. After three days of incubation, the number of bacteria colonies on each petri dish was counted. Here are the data from Daniel and Kate's experiment.

None	108	97	92	81	57	49	41	38	29	3
Soap	18	10	10	6	6	5	4	4	4	1
Hand sanitizer	27	23	14	8	7	6	5	4	3	2

Parallel boxplots and numerical summaries of the data are shown here. You will now use what you have learned in this chapter to analyze the data.

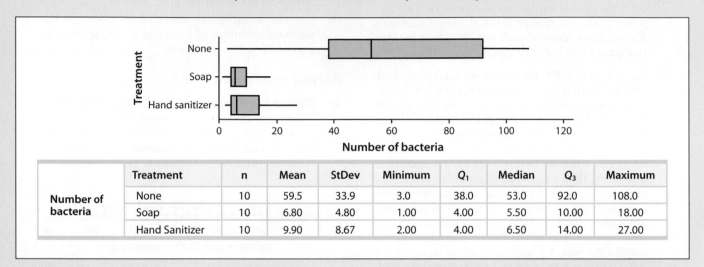

	Treatment	n	Mean	StDev	Minimum	Q_1	Median	Q_3	Maximum
Number of bacteria	None	10	59.5	33.9	3.0	38.0	53.0	92.0	108.0
	Soap	10	6.80	4.80	1.00	4.00	5.50	10.00	18.00
	Hand Sanitizer	10	9.90	8.67	2.00	4.00	6.50	14.00	27.00

1. Compare the shapes of the three distributions.

2. Explain why the mean is much greater than the median for the hand sanitizer group.

3. Give a possible reason why the *variability* in number of bacteria is so much larger for the group that didn't wash their hands at all.

4. Based on the data, does it appear that hand sanitizer kills bacteria? Better than soap and water? Give appropriate evidence to support your answers.

Main Points

Chapter 1

■ **Statistics** is the science and art of collecting, analyzing, and drawing conclusions from data.

Organizing Data

■ A data set contains information about a number of **individuals.** Individuals may be people, animals, or things.

■ A **variable** describes some characteristic of an individual, such as a person's height, gender, or salary.

■ A **categorical variable** assigns a label that places each individual into one of several groups, such as male or female.

■ A **quantitative variable** has numerical values that measure some characteristic of each individual, such as height in centimeters or salary in dollars.

■ The **distribution** of a variable describes what values the variable takes and how often it takes them. You can use a **frequency table** or a **relative frequency table** to quickly summarize the distribution of a variable.

Displaying Data

Categorical Variables

■ **Bar charts** and **pie charts** can be used to display the distribution of a categorical variable. You can also use a bar chart to compare the distribution of a categorical variable in two or more groups.

- Use counts or percents to describe the distribution of a categorical variable.

- Beware of graphs that mislead the eye. Look at the scales to see if they have been distorted to create a particular impression. Avoid making graphs that replace the bars of a bar chart with pictures whose height and width both change.

Quantitative Variables

- You can use a **dotplot, stemplot,** or **histogram** to show the distribution of a quantitative variable. A dotplot displays individual values on a number line. Stemplots separate each observation into a stem and a one-digit leaf. Histograms plot the counts (frequencies) or percents (relative frequencies) of values in equal-width intervals.

- Histograms are for quantitative data; bar charts are for categorical data. Be sure to use relative frequencies when comparing data sets of different sizes.

- When examining any graph of quantitative data, look for an *overall pattern* and for clear *departures* from that pattern. **Shape, center,** and **variability** describe the overall pattern of the distribution of a quantitative variable. **Outliers** are observations that lie outside the overall pattern of a distribution. Always look for outliers and try to explain them.

- Some distributions have simple shapes, such as **symmetric, skewed to the left,** or **skewed to the right.** The number of peaks is another aspect of overall shape.

- When comparing distributions of quantitative data, be sure to compare shape, center, variability, and outliers.

Numerical Summaries for Quantitative Data

- The **median** and the **mean** $\bar{x}$ measure the center of a distribution in different ways. The median is the midpoint of the distribution, the number such that about half the observations are smaller and half are larger. The mean is the average of the observations:

$$\bar{x} = \frac{\sum x_i}{n}$$

- The simplest measure of variability for a distribution of quantitative data is the **range,** which is the distance from the maximum value to the minimum value.

- When you use the median to describe the center of a distribution, measure its variability using the **interquartile range.** The **first quartile** Q_1 has about one-fourth of the observations below it, and the **third quartile** Q_3 has about three-fourths of the observations below it. The interquartile range (IQR) measures variability in the middle half of the distribution and is found by $IQR = Q_3 - Q_1$.

- When you use the mean to describe the center of a distribution, measure its variability using the **standard deviation.** The standard deviation s_x gives the typical distance of the values in a distribution from the mean. In symbols $s_x = \sqrt{\dfrac{\sum (x_i - \bar{x})^2}{n - 1}}$. The standard deviation s_x is zero when there is no variability and gets larger as variability from the mean increases.

- The median is a **resistant** measure of center because it tends not to be affected by extreme observations. The mean is not resistant. Among measures of variability, the IQR is resistant, but the standard deviation and range are not.

- The mean and standard deviation are good descriptions for roughly symmetric distributions with no outliers. The median and IQR are a better description for skewed distributions or distributions with outliers.

- According to the **1.5 × *IQR* rule,** an observation is an outlier if it is smaller than $Q_1 - 1.5 \times IQR$ or larger than $Q_3 + 1.5 \times IQR$.

- **Boxplots** are based on the **five-number summary** of a distribution, consisting of the minimum, Q_1, the median, Q_3, and the maximum. The box shows the variability in the middle half of the distribution. The median is marked within the box. Lines extend from the box to the smallest and the largest observations that are not outliers. Outliers are plotted with special symbols. Boxplots are especially useful for comparing distributions.

Describing Location in a Distribution

- Two ways of describing an individual's location in a distribution are **percentiles** and **z-scores.** An individual's percentile is the percent of values in a distribution that are less than the individual's data value.

- To standardize any data value, subtract the mean of the distribution and then divide the difference by the standard deviation. The resulting z-score

$$z = \frac{\text{value} - \text{mean}}{\text{standard deviation}}$$

measures how many standard deviations the data value lies above or below the mean of the distribution. We can also use percentiles and z-scores to compare the location of individuals in different distributions.

- A **cumulative relative frequency graph** allows us to examine location within a distribution. The completed graph allows you to estimate the percentile for an individual value, and vice versa.

Chapter 1 Review Exercises

1. U.S. presidents (1.1) Here is some information about the first 10 U.S. presidents. Identify the individuals and variables in this data set. Classify each variable as categorical or quantitative.

Name	Political party	Age at inaug.	Age at death	State of birth
G. Washington	Federalist	57	67	Virginia
J. Adams	Federalist	61	90	Massachusetts
T. Jefferson	Democratic-Republican	57	83	Virginia
J. Madison	Democratic-Republican	57	85	Virginia
J. Monroe	Democratic-Republican	58	73	Virginia
J. Q. Adams	Democratic-Republican	57	80	Massachusetts
A. Jackson	Democrat	61	78	South Carolina
M. Van Buren	Democrat	54	79	New York
W. H. Harrison	Whig	68	68	Virginia
J. Tyler	Whig	51	71	Virginia

2. Disc dogs (1.1, 1.2) Here is a list of the breeds of dogs that won the World Canine Disc Championships from 1975 through 2016.

Whippet	Labrador retriever	Australian shepherd
Whippet	Mixed breed	Border collie
Whippet	Australian shepherd	Mixed breed
Mixed breed	Border collie	Australian shepherd
Mixed breed	Australian shepherd	Australian shepherd
Other purebred	Mixed breed	Australian shepherd
Labrador retriever	Mixed breed	Border collie
Mixed breed	Mixed breed	Border collie
Mixed breed	Border collie	Australian shepherd
Border collie	Border collie	Border collie
Mixed breed	Australian shepherd	Border collie
Mixed breed	Border collie	Other purebred
Labrador retriever		Border collie
		Border collie

(a) Summarize the distribution of dog breed with a frequency table.

(b) Make a bar chart to display the data. Describe what you see.

3. Music preferences (1.2) Australian high school students were asked to choose their favorite type of music from a list. The bar chart compares the preferences for samples of 250 students in surveys in 2012 and 2013. Write a few sentences comparing the distribution of music preferences for the two years.

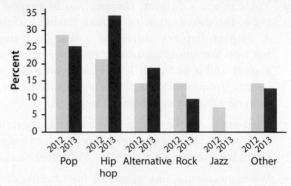

4. Binge-watching (1.2) Do you "binge-watch" television series by viewing multiple episodes of a series at one sitting? A survey of 800 people who "binge-watch" were asked how many episodes is too many to watch in one viewing session. The results are displayed in the bar chart.[64] Explain how this graph is misleading.

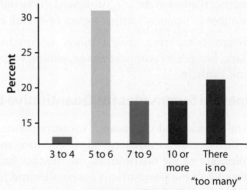

5. Comparing family income (1.3) The dotplots show the total family income of 40 randomly chosen individuals each from Connecticut, Indiana, and Maine, based on U.S. Census data. Write a few sentences to compare the distributions of income in these three states.

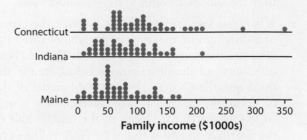

Exercises 6 and 7 refer to the following setting.

6. **Words and music (1.5, 1.6)** For a project in their statistics class, Alex and Tempe studied the impact of different types of background music on students' ability to remember words from a list they were allowed to study for 5 min. Here is a list of how many words one group of students who listened to Beethoven's Fifth Symphony were able to remember.

11	12	23	15	14	15	14	15
10	14	15	9	11	13	25	11
13	13	12	20	17	23	11	12
12	11	20	20	12	12	19	13
15	10	14	11	7	17	13	18

(a) Make a histogram that effectively displays the distribution of words recalled.

(b) We used technology to compute the mean and median of this distribution. One is 13 and the other is 14.3. Based on the histogram, explain how you know which is which without doing any calculations.

7. **Words and music II (1.7, 1.8)**

(a) Find the range of the data in Exercise 6.

(b) Find the interquartile range of these data.

(c) Identify any outliers in this distribution.

(d) The standard deviation is 4.05. Interpret this value in context.

8. **Comparing reaction time (1.8)** Catherine and Ana suspect that athletes (i.e., students who have been on at least one varsity team) typically have a faster reaction time than other students. To test this theory, they gave an online reflex test to 33 varsity athletes at their school and 30 other students. The following parallel boxplots display the reaction times (in milliseconds) for the two groups of students. What do the data suggest about Catherine and Ana's suspicion? Explain.

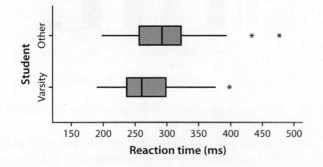

9. **Your Iowa home (1.9)** The following dotplot gives the sale prices for 40 houses in Ames, Iowa, sold during a recent month. The mean sale price was $203,388 with a standard deviation of $87,609.

(a) Find the percentile of the house that is indicated on the dotplot.

(b) Calculate and interpret the standardized score (*z*-score) for the house indicated by the red dot, which sold for $234,000.

10. **Algebra II exam times (1.9)** Here is a cumulative relative frequency graph for the length of time a group of 62 students spent on a no-time-limit final exam in Algebra II.

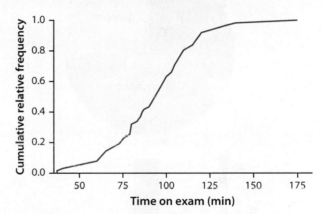

(a) Ace took 120 min to finish the exam. Estimate his percentile.

(b) Estimate the median and interquartile range for the amount of time these students spent on the exam.

Chapter 1 Practice Test

Section I: Multiple Choice *Select the best answer for each question.*

1. You record the age, marital status, and earned income of a sample of 1463 women. The number and type of variables you have recorded is

   (a) 3 quantitative, 0 categorical.

   (b) 3 quantitative, 1 categorical.

   (c) 2 quantitative, 1 categorical.

   (d) 2 quantitative, 2 categorical.

2. Here is a pie chart of how a randomly selected group of people described the cost of their health insurance. Which of the following bar charts is equivalent to the pie chart?

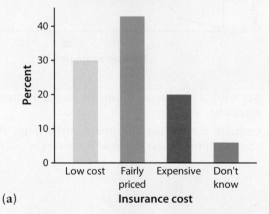

(a)

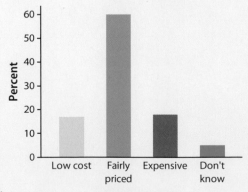

(b)

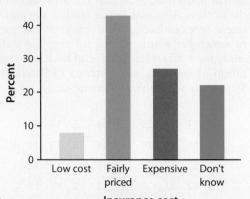

(c)

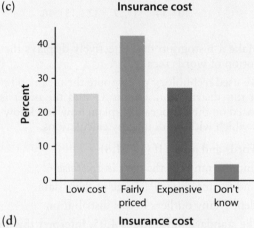

(d)

3. The accompanying bar chart summarizes responses of dog owners to the question "Where in the car do you let your dog ride?" Which of the following statements is true?

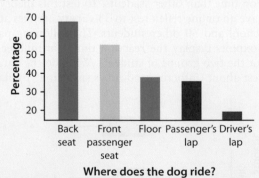

Where does the dog ride?

(a) A majority of owners do not allow their pets to ride in the front passenger seat.

(b) Roughly twice as many pets are allowed to sit in the front passenger seat as in the passenger's lap.

(c) The vertical scale of this graph exaggerates the difference between the percentage who let their dogs ride in the driver's lap versus a passenger's lap.

(d) These data could also be presented in a pie chart.

Questions 4 and 5 refer to the following setting. Forty students took a statistics exam having a maximum of 50 points. The distribution of score is given in the stem-and-leaf plot.

```
0 | 28
1 | 2245
2 | 01333358889
3 | 001356679
4 | 22444466788      KEY: 2 | 3 = 23 points
5 | 000
```

4. Which of the following are the correct median and interquartile range of this distribution?

(a) Median = 31; Interquartile range = 21

(b) Median = 32; Interquartile range = 21

(c) Median = 32; Interquartile range = 23 to 44

(d) Median = 33; Interquartile range = 21

5. Tess's score was 25. What was her percentile?

(a) 12th

(b) 30th

(c) 35th

(d) 70th

Questions 6 and 7 refer to the following setting. In a statistics class with 136 students, the professor records how much money (in dollars) each student has in his or her possession during the first class of the semester. The histogram shows the data that were collected.

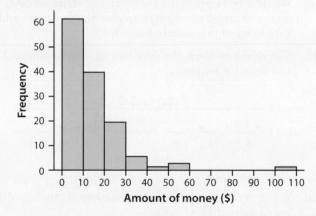

6. The percentage of students with less than $10 in their possession is closest to

(a) 30%.

(b) 35%.

(c) 45%.

(d) 60%.

7. Which of the following statements about this distribution can be concluded from the graph?

(a) The distribution is left-skewed.

(b) The median is between $10 and $20.

(c) The standard deviation of the distribution is more than $60.

(d) The mean is less than the median.

8. The mean speed of vehicles in the "cars only" lanes of the New Jersey turnpike is 68 mph. The mean speed of vehicles in the "any vehicle" lanes is 64 mph. What must be true about the mean speed of all vehicles on the turnpike, assuming these are the only types of lanes?

(a) It could be any number from 64 to 68 mph.

(b) It must be larger than the median speed.

(c) It must be larger than 66 mph.

(d) It must be 66 mph.

9. Which of the following distributions has the smallest standard deviation?

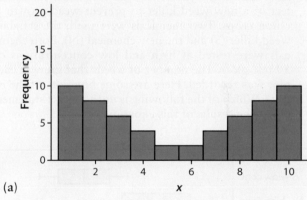

(a)

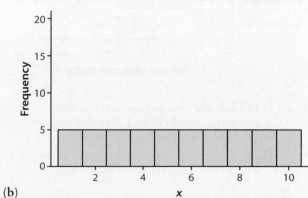

(b)

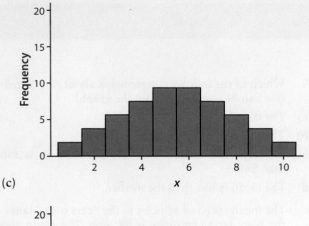

(c)

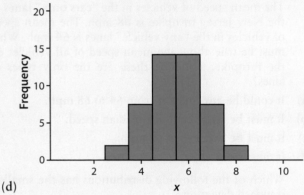

(d)

10. An experiment was conducted to investigate the effect of a new weed killer to prevent weed growth in onion crops. Two chemicals were used, the standard weed killer (S) and the new chemical (N). Both chemicals were tested at high and low concentrations on 50 test plots. The number of weeds that grew in each plot was recorded. Here are some boxplots of the results. Which of the following is *not* a correct statement about the results of this experiment?

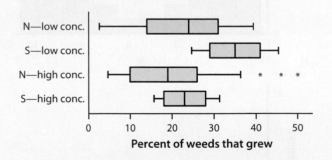

(a) At both high and low concentrations, the new chemical gives better weed control than the standard weed killer.

(b) For both chemicals, fewer weeds grew at higher concentrations than at lower concentrations.

(c) The results for the standard weed killer are less variable than those for the new chemical.

(d) The high concentration of the standard weed killer was more effective than the low concentration of the new chemical.

11. Until the scale was changed in 1995, SAT scores were based on a scale set many years ago. For Math scores, the mean under the old scale in the 1990s was 470 and the standard deviation was 110. In 2013 the mean was 515 and the standard deviation was 116. Gina took the SAT in 1994 and scored 500 on the Math test. Her cousin Colleen took the SAT in 2013 and scored 530 on the Math test. Who did better relative to their peers, and how can you tell?

(a) Colleen—she scored 30 points higher than Gina.

(b) Colleen—her standardized score is higher than Gina's.

(c) Gina—her standardized score is higher than Colleen's.

(d) Gina—the standard deviation was bigger in 2013.

Section II: Free Response

12. You are interested in how much time students spend on the Internet each day. Here are data on the time spent on the Internet (in minutes) for a particular day reported by a random sample of 30 students at a large high school.

7	20	24	25	25	28	28	30	32	35
42	43	44	45	46	47	48	48	50	51
72	75	77	78	79	83	87	88	135	151

(a) Construct a histogram of these data.

(b) Are there any outliers? Justify your answer.

(c) Would it be better to use the mean and standard deviation or the median and *IQR* to describe the center and variability of this distribution? Why?

13. The dotplots show the lifetimes of several Brand X and Brand Y batteries.

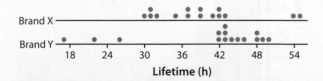

(a) What percentage of all the batteries (both brands) lasted more than 42 hours?

(b) Give a reason someone might prefer a Brand X battery.

(c) Give a reason someone might prefer a Brand Y battery.

14. During the early part of a Major League Baseball season, many fans and players noticed that the number of home runs being hit seemed unusually large. Here are the data on the number of home runs hit by American League and National League teams in the early part of that season:

American League	35	40	43	49	51	54	57	58	58	64	68	68	75	77
National League	29	31	42	46	47	48	48	53	55	55	55	63	63	67

Make parallel boxplots to compare the distributions of home runs for the two leagues. Describe what you see.

15. Mrs. Causey asked her students how much time they had spent using a computer during the previous week. The following figure shows a cumulative relative frequency graph of her students' responses.

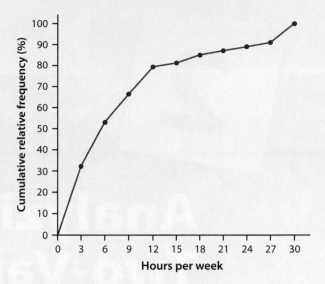

(a) At what percentile does a student who used her computer for 7 hours last week fall?

(b) Estimate the interquartile range (*IQR*) from the graph.

2

Analyzing Two-Variable Data

Lesson 2.1   Relationships Between Two Categorical Variables   96

Lesson 2.2   Relationships Between Two Quantitative Variables   105

Lesson 2.3   Correlation   113

Lesson 2.4   Calculating the Correlation   121

Lesson 2.5   Regression Lines   130

Lesson 2.6   The Least-Squares Regression Line   137

Lesson 2.7   Assessing a Regression Model   149

Lesson 2.8   Fitting Models to Curved Relationships   160

Chapter 2   Main Points   174

Chapter 2   Review Exercises   176

Chapter 2   Practice Test   178

94

© Frank Lukassek/Corbis

STATS applied!

When will the cherry trees blossom?

The Teague family is planning a trip to Japan to see the blossoming of the cherry trees. They can only afford to go for a week, but they aren't certain when the cherry trees will blossom. To help guide their decision, they gather some data. Here is a dotplot showing the day in April that the first blossom appeared on cherry trees over a 24-year period.

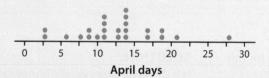

April days

About half the time, the first blossom occurs between April 10 and April 15. But the first blossom has occurred as early as April 3 and as late as April 28. With only a week to visit, when should the Teague family plan to arrive if they want to have the best chance to see the first blossom?

We'll revisit STATS applied! at the end of the chapter, so you can use what you have learned to help answer these questions. For now, keep this in mind: To understand one variable (like the date of first blossom), you often have to look at how it is related to other variables.

Lesson 2.1

Relationships Between Two Categorical Variables

LEARNING TARGETS

- Distinguish between explanatory and response variables for categorical data.
- Make a segmented bar chart to display the relationship between two categorical variables.
- Determine if there is an association between two categorical variables and describe the association if it exists.

In Chapter 1, you learned how to display the distribution of a single categorical variable using a pie chart or a bar chart. In this lesson, you will learn how to display and describe the relationship between two categorical variables. The first step is to determine which variable is the **response variable** and which variable is the **explanatory variable.**

||| **DEFINITION** Response variable, Explanatory variable

A **response variable** measures an outcome of a study.

An **explanatory variable** may help predict or explain changes in a response variable.

EXAMPLE

Tax reform and heart disease?

Identifying explanatory variables

PROBLEM: Identify the explanatory variable for the following relationships. Explain your reasoning.

(a) Opinion about tax reform and political party membership

(b) Anger level (low, moderate, or high) and whether or not a person has coronary heart disease

SOLUTION:

(a) The explanatory variable is political party membership because knowing what political party a person belongs to will help predict the person's opinion about tax reform.

(b) The explanatory variable is anger level because this might help explain whether or not a person develops coronary heart disease.

FOR PRACTICE   TRY EXERCISE 1.

In some relationships, there isn't a clear explanatory or response variable. For example, in the relationship between eye color and hair color, either variable could be used to predict or explain the other.

Displaying Relationships Between Two Categorical Variables

After identifying the explanatory variable, the next step is to display the distribution of the response variable for each category of the explanatory variable. This can be done with the side-by-side bar charts from Lesson 1.2 or with **segmented bar charts.**

| | **DEFINITION** Segmented bar chart |

A **segmented bar chart** displays the possible values of a categorical variable as segments of a rectangle, with the area of each segment proportional to the percent of individuals in the corresponding category.

In Lesson 1.2, we presented data from a random sample of 200 students from the United Kingdom. The gender and preferred superpower were recorded for each student in the sample. Figure 2.1 shows the side-by-side bar chart and corresponding segmented bar chart that display the relationship between gender and preferred superpower. Segmented bar charts are sometimes called "stacked" bar charts because they are the result of stacking the bars from a side-by-side bar chart.

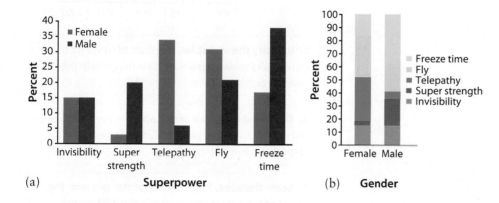

(a)   (b)

FIGURE 2.1 Side-by-side bar chart (a) and segmented bar chart (b) showing the relationship between gender and superpower preference for a sample of 200 children ages 9–17 from the U.K.

In both types of graphs, we calculate the relative frequencies separately *for each gender* because gender is the explanatory variable. That is, we calculate the percent *of the females* who prefer each superpower. Then, we calculate the percent *of the males* who prefer each superpower.

How to Make a Segmented Bar Chart

1. **Identify the variables.** Determine which variable is the explanatory variable and which is the response variable.

2. **Draw and label the axes.** Put the name of the explanatory variable under the horizontal axis. To the left of the vertical axis, indicate if the graph shows the percent (or proportion) of individuals in each category of the response variable.

3. **Scale the axes.** Write the names of the categories of the explanatory variable at equally spaced intervals under the horizontal axis. On the vertical axis, start at 0% (or 0) and place tick marks at equal intervals until you reach 100% (or 1).

4. **Draw "100%" bars** above each of the category names for the explanatory variable on the horizontal axis so that each bar ends at the top of the graph. Make the bars equal in width and leave gaps between them.

5. **Segment each of the bars.** For each category of the explanatory variable, calculate the relative frequency for each category of the response variable. Then divide the corresponding bar so that the area of each segment corresponds to the proportion of individuals in each category of the response variable.

6. **Include a key** that identifies the different categories of the response variable.

EXAMPLE

Who survived on the *Titanic*?

Making a segmented bar chart

In 1912 the luxury liner *Titanic,* on its first voyage across the Atlantic, struck an iceberg and sank. Some passengers got off the ship in lifeboats, but many died. The two-way table gives information about adult passengers who survived and who died, by class of travel.

PROBLEM: Make a segmented bar chart to display the relationship between survival status and class of travel for passengers on the *Titanic*.

SOLUTION:

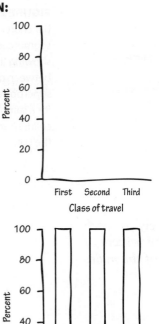

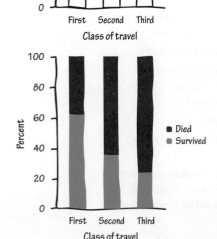

		First	Second	Third	Total
Survival status	Survived	197	94	151	442
	Died	122	167	476	765
	Total	319	261	627	1207

Class of travel (column header spanning First/Second/Third/Total)

1. Identify the variables. Use class of travel for the explanatory variable because class might help predict whether or not a passenger survived.

2. Draw and label the axes. Label the horizontal axis with the explanatory variable "Class of travel" and the vertical axis with "Percent."

3. Scale the axes. Scale the horizontal axis with the values of the explanatory variable: "First," "Second," and "Third." Scale the vertical axis from 0 to 100% in increments of 20%.

4. Draw "100%" bars for each category of the explanatory variable.

5. Segment each of the bars.
- In first class, 197/319 = 62% survived and 122/319 = 38% died.
- In second class, 94/261 = 36% survived and 167/261 = 64% died.
- In third class, 151/627 = 24% survived and 476/627 = 76% died.

6. Include a key to identify the categories of the response variable.

FOR PRACTICE   TRY EXERCISE 5.

Association

Once we make a graph, the final step is to determine whether there is an **association** between the two categorical variables.

DEFINITION Association

There is an **association** between two variables if knowing the value of one variable helps us predict the value of the other. If knowing the value of one variable does not help us predict the value of the other, then there is no association between the variables.

There appears to be an association between gender and superpower preference because knowing a person's gender helps us to predict which superpower he or she would prefer. For example, females are much more likely to choose telepathy or ability to fly than males. Males are much more likely to choose the ability to freeze time or super strength. We can "see" the association in the segmented bar chart because the segments for each superpower aren't the same size for the female graph and the male graph in Figure 2.2. In other words, the graphs look different.

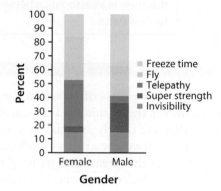

FIGURE 2.2 Segmented bar chart showing the relationship between gender and superpower preference for a sample of 200 children ages 9–17 from the U.K.

THINK ABOUT IT **What would the segmented bar chart look like *if there were no association between gender and superpower preference?*** If there were no association, then knowing a person's gender wouldn't help predict which superpower he or she prefers. In other words, the distribution of superpower preferences would be the same for both genders. You can "see" this in the segmented bar chart in Figure 2.3 because the sizes of the segments for each superpower are the same for *both* males and females.

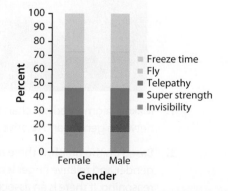

FIGURE 2.3 Segmented bar chart showing no association between gender and superpower preference.

EXAMPLE

Is there an association between class and survival on the Titanic?

Checking for association

PROBLEM: Use the graph to determine if there is an association between survival status and class of travel for passengers on the *Titanic*. Explain your reasoning. If there is an association, briefly describe it.

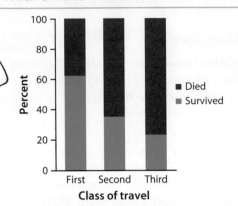

SOLUTION:

There is an association between survival status and class of travel because the survival rates were not the same for the different classes of travel. Passengers in first class were the most likely to survive, and passengers in third class were the least likely to survive.

Knowing a passenger's class of travel helps us predict whether or not the passenger survived—higher-class passengers were more likely to survive. This means that there is an association between class of travel and survival status.

FOR PRACTICE   TRY EXERCISE 9.

It may be true that being in a higher class of travel on the *Titanic* increased a passenger's chance of survival. However, there isn't always a cause-and-effect relationship between two variables even if they are clearly associated. For example, a recent study proclaimed that people who are overweight are less likely to die within a few years than people of normal weight. Does this mean that gaining weight will *cause* you to live longer? Not at all. The study included smokers, who tend to be thinner and also much more likely to die in a given period than non-smokers. Smokers increased the death rate for the normal-weight category, making it look like being overweight is better.[1] Don't confuse association with cause and effect!

CAUTION
⚠️

LESSON APP 2.1

Which finger is longer?

Is there a relationship between gender and relative finger length? To investigate, a random sample of 452 U.S. high school students was selected.[2] The two-way table shows the gender of each student and which finger was longer on his or her left hand (index finger or ring finger).

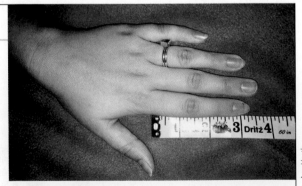

		Gender		
		Female	Male	Total
Longer finger	Index finger	78	45	123
	Ring finger	82	152	234
	Same length	52	43	95
	Total	212	240	452

1. Make a segmented bar chart to show the relationship between gender and relative finger length.

2. Based on the graph, is there an association between gender and relative finger length? Explain your reasoning. If there is an association, briefly describe it.

Analyzing Two Categorical Variables with Technology

We can use the *Two Categorical Variables* applet at **highschool.bfwpub.com/spa3e** to investigate the relationship between two categorical variables. We'll illustrate using the *Titanic* data.

1. On the top of the table, enter the name of the explanatory variable (Class of travel), along with the category names for this variable (First, Second, and Third). Press the "+" button in the upper right to get a third category for the explanatory variable.

2. On the left side of the table, enter the name of the response variable (Survival status), along with the category names for this variable (Survived and Died).

3. Enter the counts for each cell, as shown in the screen shot.

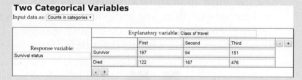

Two Categorical Variables
Input data as: Counts in categories ▾

Response variable: Survival status		Explanatory variable: Class of travel				
		First	Second	Third		
	Survivor	197	94	151	−	+
	Died	122	167	476		

4. Press the "Begin analysis" button. This will generate a segmented bar chart and a table showing the percent who survived and died within each class of travel. *Note:* To have the segmented bar chart display percents instead of proportions, click on the link to adjust preferences.

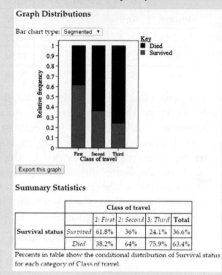

Graph Distributions

Bar chart type: Segmented ▾

Export this graph

Summary Statistics

Survival status		Class of travel			
		1: First	2: Second	3: Third	Total
Survival status	Survived	61.8%	36%	24.1%	36.6%
	Died	38.2%	64%	75.9%	63.4%

Percents in table show the conditional distribution of Survival status for each category of Class of travel.

Lesson 2.1

WHAT DID YOU LEARN?

LEARNING TARGET	EXAMPLES	EXERCISES
Distinguish between explanatory and response variables for categorical data.	p. 96	1–4
Make a segmented bar chart to display the relationship between two categorical variables.	p. 98	5–8
Determine if there is an association between two categorical variables and describe the association if it exists.	p. 100	9–12

 **Lesson 2.1**

Mastering Concepts and Skills

For Exercises 1–4: Identify the explanatory variable for the following relationships. Explain your reasoning.

1. Dorm life Residential status (on- or off-campus) and year in college.
pg **96**

2. Don't smoke Whether or not a person develops lung cancer and smoking status.

3. Everyone likes sports Gender and favorite sport.

4. Are you graduating? Participation in extracurricular activities during high school and graduation status.

5. **Cell-phone preferences** The Pew Research Center asked a random sample of 2024 adult cell-phone owners from the United States their age and which type of cell phone they own: iPhone, Android, or other (including non-smart phones). Use the information in the two-way table to make a segmented bar chart to show the relationship between age and type of cell phone.[3]

		Age			
		18–34	35–54	55+	Total
Type of cell phone	iPhone	169	171	127	467
	Android	214	189	100	503
	Other	134	277	643	1054
	Total	517	637	870	2024

6. **Warm cozy nests** How is the hatching of water python eggs influenced by the temperature of the snake's nest? Researchers assigned newly laid eggs to one of three water temperatures: hot, neutral, or cold. Hot duplicates the extra warmth provided by the mother python, and cold duplicates the absence of the mother. Use the information in the two-way table to make a segmented bar chart to show the relationship between nest temperature and hatching success.[4]

		Nest temperature			
		Cold	Neutral	Hot	Total
Hatching status	Hatched	16	38	75	129
	Didn't hatch	11	18	29	58
	Total	27	56	104	187

7. **Get rich quick** A survey of 4826 randomly selected young adults (aged 19 to 25) asked, "What do you think the chances are you will have much more than a middle-class income at age 30?" The gender of each respondent was also recorded. Use the information in the two-way table to make a segmented bar chart to show the relationship between gender and opinion about becoming rich.[5]

		Gender		
		Female	Male	Total
Opinion	Almost no chance	96	98	194
	Some chance but probably not	426	286	712
	A 50-50 chance	696	720	1416
	A good chance	663	758	1421
	Almost certain	486	597	1083
	Total	2367	2459	4826

8. **Growing into algebra?** At Springfield High School, there are freshman, sophomores, and juniors taking Algebra II. Ms. Fernandez, the chairman of the Mathematics Department, wonders if there is a relationship between year in school and performance in the course. Use the information in the two-way table to make a segmented bar chart to show the relationship between year in school and grade in Algebra II.

		Year in school			
		Freshman	Sophomore	Junior	Total
Algebra II grade	A	17	31	2	50
	B	14	34	10	58
	C or lower	3	27	7	37
	Total	34	92	19	145

9. **Still on a land line?** Telephone surveys made through random-digit dialing used to exclude cell-phone numbers. If the opinions of people who have only cell phones differ from those of people who have landline service, the poll results may not represent all adults. The Pew Research Center interviewed a sample of cell-only and landline telephone users who were younger than 30 years old. Here's what Pew found about how these people describe their political-party affiliation. Use the graph to determine if there is an association between type of phone use and political affiliation for the members of the sample. Explain your reasoning. If there is an association, briefly describe it.[6]

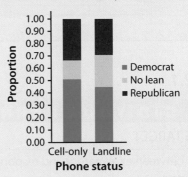

10. **Who buys Toyotas?** The segmented bar chart summarizes the distribution of sales of Toyotas in different regions of the world. Use the graph to determine if there is an association between region and year. Explain your reasoning. If there is an association, briefly describe it.[7]

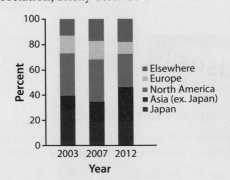

11. **Color my world** Favorite vehicle colors may differ among countries. The side-by-side bar chart shows data on the most popular colors of cars in a recent year for the United States and Europe. Based on the graph, is there an association between color preference and where you live? Explain your reasoning. If there is an association, briefly describe it.

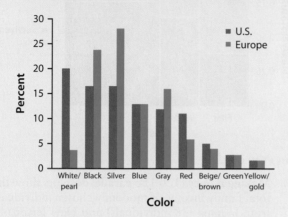

12. **How do you get to school?** Random samples of high school students in Australia and New Zealand were asked how they typically get to school. The side-by-side bar chart summarizes the data.[8] Based on the graph, is there an association between country and travel method for the members of the sample? Explain your reasoning. If there is an association, briefly describe it.

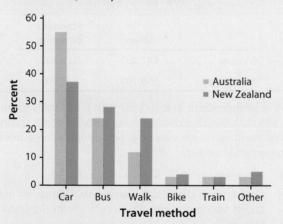

Applying the Concepts

13. **Green snowmobiles?** Yellowstone National Park surveyed a random sample of 1526 winter visitors to the park. The survey asked each person if he or she owned, rented, or had never used a snowmobile. Respondents were also asked whether or not they belonged to an environmental organization (like the Sierra Club). The two-way table summarizes the survey responses.

		Environmental club		
		No	Yes	Total
Snowmobile use	Never used	445	212	657
	Snowmobile renter	497	77	574
	Snowmobile owner	279	16	295
	Total	1221	305	1526

(a) Make a segmented bar chart to show the relationship between environmental club membership and snowmobile use for the members of the sample.

(b) Based on the graph, is there an association between these variables? Explain your reasoning. If there is an association, briefly describe it.

14. **Regional parties** The two-way table gives the party affiliation of all members of the House of Representatives from eight different regions in the United States in July 2014 (there were 3 vacant seats at that time).[9]

		Party affiliation		
		Democrat	Republican	Total
Region	New England	21	0	21
	Mid-Atlantic	39	26	65
	Great Lakes	26	39	65
	Midwest	9	20	29
	South Atlantic	24	50	74
	South Central	18	58	76
	Mountain	13	18	31
	Pacific	50	21	71
	Total	200	232	432

(a) Make a segmented bar chart to show the relationship between U.S. region and party affiliation for members of the House of Representatives.

(b) Based on the graph, is there an association between these variables? Explain your reasoning. If there is an association, briefly describe it.

15. **More snowmobiles** Refer to Exercise 13. Draw a segmented bar chart that would show no association between environmental club membership and snowmobile use for the members of the sample.

16. **More regional parties** Refer to Exercise 14. Draw a segmented bar chart that would show no association between U.S. region and party affiliation for members of the House of Representatives.

Extending the Concepts

17. Who owns a home? Using data from the 2010 census, a random sample of 348 U.S. residents aged 18 and older was selected. Among the variables recorded were gender (male or female), housing status (rent or own), and marital status (married or not married). The first two-way table summarizes the relationship between gender and housing status.

		Gender		
		Male	Female	Total
Housing status	Own	132	122	254
	Rent	50	44	94
	Total	182	166	348

(a) Make a graph to display the association between gender and housing status. Describe what you see.

(b) The second two-way table summarizes the relationship between marital status and housing status. For the members of the sample, is the relationship between marital status and housing status stronger or weaker than the relationship between gender and housing status that you described in part (a)? Justify your choice using the data provided in both two-way tables.

		Marital status		
		Married	Not married	Total
Housing status	Own	172	82	254
	Rent	40	54	94
	Total	212	136	348

18. Reversing the variables In the *Titanic* example from this lesson, we learned that there was an association between the class of travel (explanatory variable) and the survival status (response variable) for passengers on the *Titanic*. Would there still be an association if the variables were reversed and survival status was the explanatory variable? Make a segmented bar chart using survival status as the explanatory variable and discuss if there is still an association.

19. Mosaic plots A mosaic plot resembles a segmented bar chart except that the width of the bars corresponds to the proportion of individuals in each category of the explanatory variable. Here is an example of the mosaic plot for the *Titanic* data on page 98. Following this example, make a mosaic plot for the cell-phone data in Exercise 5.

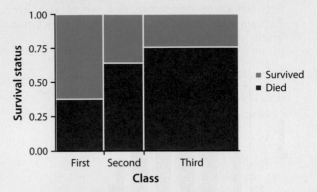

Recycle and Review

20. Making money (1.9) The parallel dotplots show the total family income of randomly chosen individuals from Indiana (38 individuals) and New Jersey (44 individuals). Means and standard deviations are given below the dotplots.

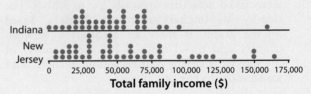

	Mean	Standard deviation
Indiana	$47,400	$29,400
New Jersey	$58,100	$41,900

Consider individuals in each state with total family incomes of $95,000. Which individual has a higher income, relative to others in his or her state? Use percentiles and z-scores to support your answer.

21. More money (1.7) Refer to Exercise 20.

(a) Find and interpret the interquartile range for total family income of the 38 individuals in the Indiana sample.

(b) Interpret the standard deviation for the total family income of the 38 individuals in the Indiana sample.

Lesson 2.2

Relationships Between Two Quantitative Variables

LEARNING TARGETS

- Distinguish between explanatory and response variables for quantitative data.
- Make a scatterplot to display the relationship between two quantitative variables.
- Describe the direction, form, and strength of a relationship displayed in a scatterplot, and identify outliers.

In Chapter 1, you learned how to display the distribution of a single quantitative variable using dotplots, stemplots, histograms, and boxplots. Although there are many ways to display the distribution of a single quantitative variable, a **scatterplot** is the best way to display the relationship between two quantitative variables.

DEFINITION Scatterplot

A **scatterplot** shows the relationship between two quantitative variables measured on the same individuals. The values of one variable appear on the horizontal axis, and the values of the other variable appear on the vertical axis. Each individual in the data set appears as a point in the graph.

Figure 2.4 is a scatterplot showing the relationship between the weight (in carats) and price (in $1000s) for 95 different round, clear, internally flawless, and excellently cut diamonds.[10]

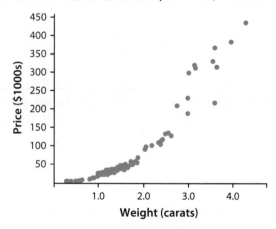

FIGURE 2.4 Scatterplot showing the relationship between carat weight and price for a sample of diamonds.

Distinguishing Explanatory and Response Variables

It is fairly easy to make a scatterplot by hand. The first step is to determine which variable is the response variable and which variable is the explanatory variable. As in Lesson 2.1, a response variable measures an outcome of a study and an explanatory variable may help predict or explain changes in the response variable. In the scatterplot above, we used carat weight as the explanatory variable because the size of a diamond helps explain how expensive it is.

However, in some cases, there isn't a clear explanatory or response variable. For example, in the relationship between SAT math and SAT critical reading scores, either variable could be used to predict or explain the other.

EXAMPLE

Life span and heights?

Identifying explanatory variables

PROBLEM: Identify the explanatory variable for the following relationships if possible. Explain your reasoning.

(a) The average income and life expectancy in a sample of countries

(b) The heights of husbands and the heights of wives in a sample of married couples

SOLUTION:

(a) The explanatory variable is the average income because the amount of money people have to spend on health care helps explain the average life span of people in a country.

(b) The explanatory variable could be either the heights of husbands or the heights of wives because either variable could be used to explain or predict the other.

FOR PRACTICE   TRY EXERCISE 1.

Making a Scatterplot

After identifying the explanatory and response variables, it is easy to complete the scatterplot.

How to Make a Scatterplot

1. **Label the axes.** The explanatory variable is plotted on the horizontal axis and the response variable is plotted on the vertical axis. If there is no explanatory variable, either variable can go on the horizontal axis.

2. **Scale the axes.** Put the name of the explanatory variable under the horizontal axis and place equally spaced tick marks along the axis beginning at a "friendly" number just below the smallest value of the explanatory variable and continuing until you exceed the largest value. Do the same for the response variable along the vertical axis.

3. **Plot individual data values.** For each individual, plot a point directly above that individual's value for the explanatory variable and directly to the right of that individual's value for the response variable.

EXAMPLE

Faster and farther?

Making a scatterplot

PROBLEM: Here are the 40-yard-dash times (in seconds) and long-jump distances (in inches) for a small class of 12 students. Make a scatterplot to display the relationship between 40-yard-dash times and long-jump distances.

Dash time (sec)	5.41	5.05	7.01	7.17	6.73	5.68	5.78	6.31	6.44	6.50	6.80	7.25
Long-jump distance (in.)	171	184	90	65	78	130	173	143	92	139	120	110

SOLUTION:

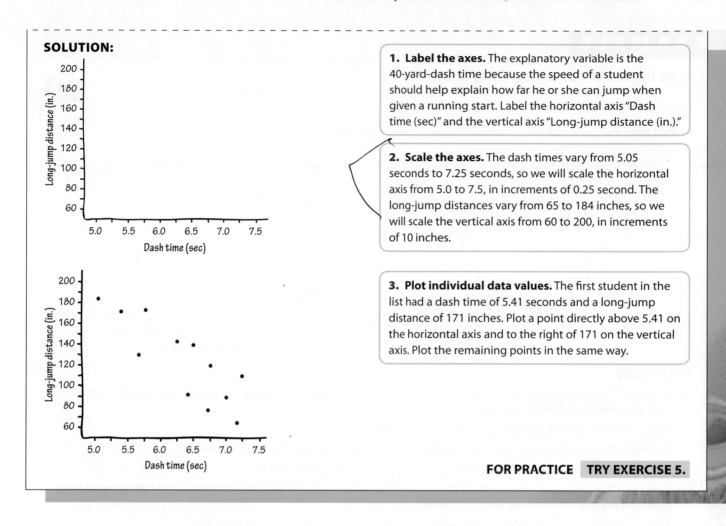

1. Label the axes. The explanatory variable is the 40-yard-dash time because the speed of a student should help explain how far he or she can jump when given a running start. Label the horizontal axis "Dash time (sec)" and the vertical axis "Long-jump distance (in.)."

2. Scale the axes. The dash times vary from 5.05 seconds to 7.25 seconds, so we will scale the horizontal axis from 5.0 to 7.5, in increments of 0.25 second. The long-jump distances vary from 65 to 184 inches, so we will scale the vertical axis from 60 to 200, in increments of 10 inches.

3. Plot individual data values. The first student in the list had a dash time of 5.41 seconds and a long-jump distance of 171 inches. Plot a point directly above 5.41 on the horizontal axis and to the right of 171 on the vertical axis. Plot the remaining points in the same way.

FOR PRACTICE   TRY EXERCISE 5.

Describing a Scatterplot

To describe the relationship shown in a scatterplot, follow the strategy from Chapter 1: Look for the *overall pattern* and for clear *departures* from that pattern.

How to Describe a Scatterplot

To describe a scatterplot, make sure to address the following four characteristics in the context of the data:

- **Direction:** A scatterplot can show a positive association, negative association, or no association. In a *positive association,* larger values of the explanatory variable tend to be paired with larger values of the response variable, and smaller values tend to be paired with smaller values. In a *negative association,* larger values of the explanatory variable tend to be paired with smaller values of the response variable and vice versa.

- **Form:** A scatterplot can show a linear form or a nonlinear form. The form is linear if the overall pattern follows a straight line. Otherwise, the form is nonlinear.

- **Strength:** A scatterplot can show a weak, moderate, or strong association. An association is strong if the points don't deviate much from the form identified. An association is weak if the points deviate quite a bit from the form identified.

- **Outliers:** Individual points that fall outside the overall pattern of the relationship.

EXAMPLE

How big a diamond, how long a jump?

Describing a scatterplot

PROBLEM: Describe the relationships shown in the scatterplots.

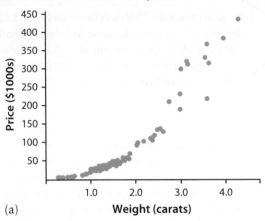
(a)

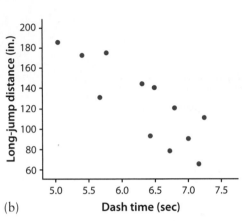
(b)

SOLUTION:

(a) Direction: There is a positive association between weight in carats and price of diamonds. Heavier diamonds tend to cost more and lighter diamonds tend to cost less.

Form: There is a curved (nonlinear) pattern in the scatterplot.

Strength: Because the points do not vary much from the curved pattern, the association is strong.

Outliers: There don't appear to be any individuals who depart from the curved pattern.

(b) Direction: There is a negative association between dash time and long-jump distance. Students with longer dash times tended to have shorter distances in the long jump and vice versa.

Form: There is a linear pattern in the scatterplot. That is, the overall pattern follows a straight line.

Strength: Because the points vary somewhat from the linear pattern, the association has moderate strength.

Outliers: There don't appear to be any individuals who depart from the linear pattern.

FOR PRACTICE   TRY EXERCISE 9.

Even when there is a clear relationship between two variables in a scatterplot, *the direction of the association only describes the overall trend*—not the relationship for each pair of points. For example, in some pairs of diamonds, the heavier diamond costs less than the lighter one.

LESSON APP 2.2

More sugar, more calories?

Is there a relationship between the amount of sugar (in grams) and the number of calories in movie-theater candy? Here are the data from a sample of 12 types of candy.[11]

Matthew Staver/Bloomberg/Getty Images

Name	Sugar (g)	Calories	Name	Sugar (g)	Calories
Butterfinger Minis	45	450	Reese's Pieces	61	580
Junior Mints	107	570	Skittles	87	450
M&M'S®	62	480	Sour Patch Kids	92	490
Milk Duds	44	370	SweeTarts	136	680
Peanut M&M'S	79	790	Twizzlers	59	460
Raisinets	60	420	Whoppers	48	350

1. Make a scatterplot to display the relationship between amount of sugar and the number of calories in movie-theater candy.

2. Describe the relationship shown in the scatterplot.

Making a Scatterplot with Technology

You can use a graphing calculator or an applet to make a scatterplot. We'll illustrate using the Faster and farther data from the example on page 106.

Applet

Launch the *Two Quantitative Variables* applet at **highschool.bfwpub.com/spa3e**.

1. Enter the name of the explanatory variable (Dash time) and the values of the explanatory variable in the first row of boxes. Then, enter the name of the response variable (Long-jump distance) and the values of the response variable in the second row of boxes.

Two Quantitative Variables

Variable	Name	Observations (separated by commas or spaces) *Keep individuals in the same order.*
Explanatory	Dash time (sec)	5.41 5.05 7.01 7.17 6.73 5.68 5.78 6.31 6.44 6.50 6.80 7.25
Response	Long-jump distan	171 184 90 65 78 130 173 143 92 139 120 110

Begin analysis Edit inputs Reset everything

2. Press the "Begin analysis" button to see the scatterplot.

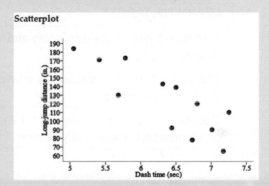

Scatterplot

TI-83/84

1. Enter the Sprint times in L₁ and the Long-jump distances in L₂.

- Press STAT and choose Edit...
- Type the values into L₁ and L₂.

2. Set up a scatterplot in the statistics plots menu.

- Press 2nd Y= (STAT PLOT).
- Press ENTER or 1 to go into Plot1.
- Adjust the settings as shown.

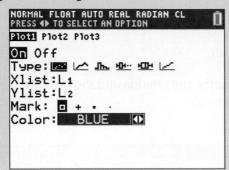

3. Use ZoomStat to let the calculator choose an appropriate window.

- Press ZOOM and choose 9: ZoomStat.

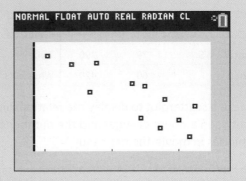

Lesson 2.2

WHAT DID YOU LEARN?

LEARNING TARGET	EXAMPLES	EXERCISES
Distinguish between explanatory and response variables for quantitative data.	p. 106	1–4
Make a scatterplot to display the relationship between two quantitative variables.	p. 106	5–8
Describe the direction, form, and strength of a relationship displayed in a scatterplot, and identify outliers.	p. 108	9–12

Exercises

Lesson 2.2

Mastering Concepts and Skills

For Exercises 1–4: Identify the explanatory variable for the following relationships if possible. Explain your reasoning.

1. **Studying stats?** The amount of time studying for a
pg 106 statistics exam and the grade on the exam.

2. **This is corny** The yield of corn in bushels per acre and the amount of rain in the growing season.

3. **Vitruvian man** The arm span and the height of high school seniors.

4. **Mixed communication** The number of texts and the number of calls for cell-phone users.

5. **Fast drive to an empty tank?** How does the fuel
pg 106 consumption of a car change as its speed increases? Here are data for a British Ford Escort.

Speed (km/h)	Fuel used (L/100 km)	Speed (km/h)	Fuel used (L/100 km)
10	21.00	90	7.57
20	13.00	100	8.27
30	10.00	110	9.03
40	8.00	120	9.87
50	7.00	130	10.79
60	5.90	140	11.77
70	6.30	150	12.83
80	6.95		

Speed is measured in kilometers per hour, and fuel consumption is measured in liters of gasoline used per 100 kilometers traveled. Make a scatterplot by hand to show the relationship between speed and fuel used.[12]

6. **How much to carry?** Ninth-grade students at the Webb Schools go on a backpacking trip each fall. Students are divided into hiking groups of size 8 by selecting names from a hat. Before leaving, students and their backpacks are weighed. The data here are from a group of hikers that one of the authors accompanied. Make a scatterplot by hand that shows how backpack weight relates to body weight.

Body weight (lb)	120	187	109	103	131	165	158	116
Backpack weight (lb)	26	30	26	24	29	35	31	28

7. **Coaster crazy** Many people like to ride roller coasters. Amusement parks try to increase attendance by building exciting new coasters. The table displays data on several roller coasters that were opened in a recent year.[13] Make a scatterplot to show the relationship between height and maximum speed.

Roller coaster	Height (ft)	Maximum speed (mph)
Apocalypse	100	55
Bullet	196	83
Corkscrew	70	55
Flying Turns	50	24
Goliath	192	66
Hidden Anaconda	152	65
Iron Shark	100	52
Stinger	131	50
Wild Eagle	210	61

8. **How's the weather up there?** The table presents data on the elevation (in feet) and average January temperature (in degrees Fahrenheit) for 10 cities and towns in Colorado.[14] Make a scatterplot to show the relationship between elevation and average temperature in January.

City	Elevation (ft)	Average January temperature (°F)
Limon	5452	27
Denver	5232	31
Golden	6408	29
Flagler	5002	29
Eagle	6595	21
Vail	8220	18
Glenwood Springs	7183	25
Rifle	5386	26
Grand Junction	4591	29
Dillon	9049	16

9. **State SAT stats** The scatterplot shows the percent of high school graduates in each state who took the SAT and the state's mean SAT Math score in a recent year. Describe the relationship.

pg 108

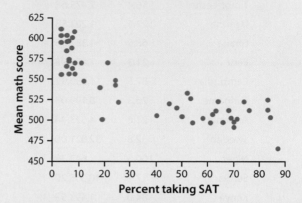

10. **IQ and GPA, connected?** The scatterplot shows the school grade point average (GPA) on a 12-point scale and the IQ score for all 78 seventh-grade students in a rural midwestern school. Describe the relationship.

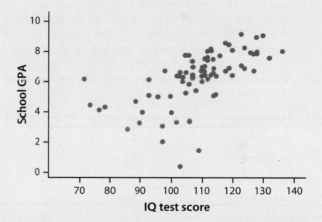

11. **Crazy coasters** Refer to Exercise 7. Describe the relationship between the height and maximum speed of the roller coasters.

12. **Is it colder up there?** Refer to Exercise 8. Describe the relationship between elevation and average January temperature for these Colorado towns.

Applying the Concepts

13. **Income and mortality** What does a country's income per person (measured in adjusted gross domestic product per person in dollars) tell us about the mortality rate for children under 5 years of age (per 1000 live births)? Here are the data for a random sample of 14 countries in a recent year.[15] Make a scatterplot of the relationship between income and child mortality. Describe what you see.

Country	Mortality rate	Income per person
Switzerland	4.4	38,003.90
Timor-Leste	56.4	2,475.68
Uganda	127.5	1,202.53
Ghana	68.5	1,382.95
Peru	21.3	7,858.97
Cambodia	87.5	1,830.97
Suriname	26.3	8,199.03
Armenia	21.6	4,523.44
Sweden	2.8	32,021.00
Niger	160.3	643.39
Serbia	7.1	10,005.20
Kenya	84.0	1,493.53
Fiji	17.6	4,016.20
Grenada	14.5	8,826.90

14. **High-protein beans** Beans and other legumes are an excellent source of protein. The table gives data on the total protein and carbohydrate content of a one-half cup portion of cooked beans for 12 different varieties.[16] Make a scatterplot of the relationship between protein and carbohydrate content for these bean varieties, using carbohydrate content as the explanatory variable. Describe what you see.

Bean variety	Carbohydrates (g)	Protein (g)
Soybeans	8.5	14.0
Garbanzos	22.5	7.0
Black beans	20.0	8.0
Adzuki beans	28.5	8.5
Cranberry beans	21.5	8.0
Great Northern beans	18.5	7.0
Kidney beans	20.0	7.5
Navy beans	23.5	7.5
Pinto beans	22.5	7.5
White beans	22.5	8.5
Lima beans	19.5	7.0
Mung beans	19.0	7.0

15. **Different forms** Draw two scatterplots with strong positive associations but different forms.

16. **Different strengths** Draw two scatterplots with positive linear associations but different strengths.

17. **Burning oil** Explain what it means if there is a negative association between the amount of fuel oil that a furnace uses and the outside temperature.

18. **Melting snow** Explain what it means if there is a positive association between the amount of snow in the Colorado mountains during winter and the volume of water in area rivers during the following summer.

Extending the Concepts

19. **Lagging or not?** For a long time, the South has lagged behind the rest of the United States in the performance of its schools. Efforts to improve education have reduced the gap. Does the South stand out if we look at state average SAT Math scores? The figure enhances the scatterplot in Exercise 9, but plots 12 southern states in red.

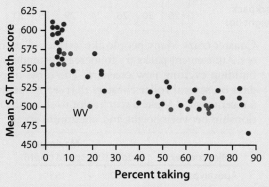

(a) What does the graph suggest about the southern states?

(b) The point for West Virginia is labeled in the graph. Explain how this state is an outlier.

20. **Lean body mass and energy** We have data on the lean body mass and resting metabolic rate for 12 women and 7 men who are subjects in a study of dieting. Lean body mass, given in kilograms, is a person's weight leaving out all fat. Metabolic rate is measured in calories burned per 24 hours. The researchers believe that lean body mass is an important influence on metabolic rate. Here is a scatterplot of the data for all 19 subjects, with separate symbols for males and females. Does the same overall pattern hold for both women and men? What difference between the sexes do you see in the graph?

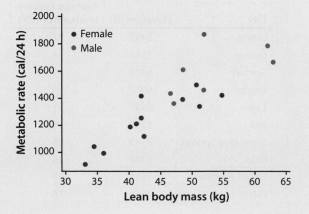

Recycle and Review

21. Two-income couples (2.1) Has the number of two-income couples changed over time? The segmented bar chart shows the employment arrangements of married couples with children under 18 in 1960 and in 2011.[17] Use the graph to determine if there is an association between year and employment arrangement. Explain your reasoning. If there is an association, briefly describe it.

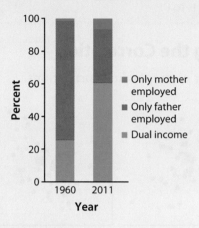

22. Are education and income related? (1.8) The boxplots compare the distributions of income for full-time workers between the ages of 25 and 64 with five levels of education, after eliminating the top 5% and bottom 5% of incomes in each category. Write a brief description of how the distribution of income changes with the highest level of education reached.[18]

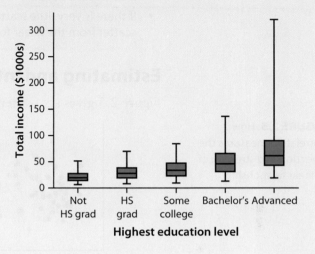

Lesson 2.3

Correlation

LEARNING TARGETS

- Estimate the correlation between two quantitative variables from a scatterplot.
- Interpret the correlation.
- Distinguish correlation from causation.

In the previous lesson, we used direction, form, and strength to describe the association between two quantitative variables. To quantify the strength of a *linear* relationship between two quantitative variables, we use the **correlation r.** You will learn how to calculate the correlation in the next lesson. In this lesson, we will focus on how to understand and interpret the correlation.

> **IIII DEFINITION** Correlation *r*
>
> The **correlation *r*** is a measure of the strength and direction of a linear relationship between two quantitative variables.
>
> - The correlation *r* is a value between −1 and 1 (−1 ≤ *r* ≤ 1).
> - If the relationship is negative, then *r* < 0. If the relationship is positive, then *r* > 0.
> - If *r* = 1 or *r* = −1, then there is a perfect linear relationship. In other words, all of the points will be exactly on a line.
> - If there is very little scatter from the linear form, then *r* is close to 1 or −1. The more scatter from the linear form, the closer *r* is to 0.

Estimating and Interpreting the Correlation

Figure 2.5 gives six scatterplots and their corresponding correlations.

FIGURE 2.5 How correlation measures the direction and strength of a linear relationship.

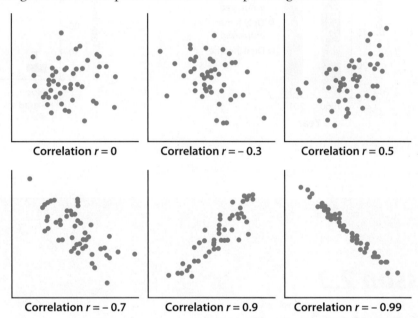

ACTIVITY

Guess the correlation

In this activity, we will have a class competition to see who can best guess the correlation.

1. Load the *Guess the Correlation* applet at www.rossmanchance.com/applets.

Correlation Guessing Game

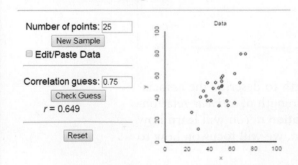

2. The teacher will press the "New Sample" button to see a "random" scatterplot. As a class, try to guess the correlation. Enter the guess in the Correlation Guess box and press Enter to see how the class did. Repeat several times to see more examples.

For the competition, there will be two rounds.

3. Starting on one side of the classroom and moving in order to the other side, the teacher will give each student *one* new sample and have him or her guess the correlation. The teacher will then record how far off the guess was from the true correlation.

4. Once every student has made an attempt, the teacher will give each student a second sample. This time, the students will go in reverse order so that the student who went first in Round 1 will go last in Round 2. The student who has the closest guess in either round wins the prize!

EXAMPLE

Saving manatees, choosing a seat?

Estimating correlation

Carol Grant/Getty Images

PROBLEM: For each of the following relationships, is $r > 0$ or $r < 0$? Closer to $r = 0$ or $r = \pm 1$? Explain your reasoning.

(a) Manatees are large, gentle, slow-moving creatures found along the coast of Florida. Many manatees are injured or killed by boats. Here is a scatterplot showing the relationship between the number of boats registered in Florida (in thousands) and the number of manatees killed by boats for the years 1977 to 2013.[19]

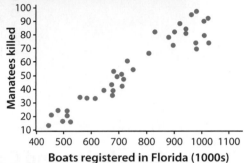

(b) To see if seat location affects test scores, a statistics teacher randomly assigned students to seat locations in his classroom for a particular chapter and recorded the test score for each student at the end of the chapter. The explanatory variable in this experiment is which row the student was assigned, where Row 1 is closest to the front and Row 7 is the farthest away. Here is a scatterplot showing the relationship between row and test score.

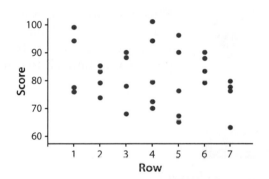

SOLUTION:

(a) Because the relationship between boats and manatees killed is positive, $r > 0$. Also, r is closer to 1 than 0 because the relationship is strong—there isn't much scatter from the linear pattern.

(b) Because the relationship between row and score is negative, $r < 0$. Also, r is closer to 0 than -1 because the relationship is weak—there is a lot of scatter from the linear pattern.

FOR PRACTICE   TRY EXERCISE 1.

EXAMPLE

Faster and farther?

Interpreting the correlation

PROBLEM: The scatterplot shows the relationship between 40-yard-dash times and long-jump distances from the example in Lesson 2.2. The correlation is $r = -0.838$. Interpret this value in context.

SOLUTION:

The correlation of -0.838 indicates that the linear relationship between the long-jump distance and dash time for these students is strong and negative.

FOR PRACTICE   TRY EXERCISE 5.

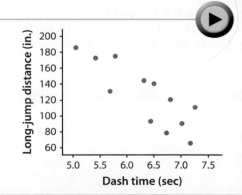

A correlation close to 1 or –1 doesn't necessarily mean an association is linear. For example, the scatterplot in Figure 2.6 is clearly nonlinear, yet the correlation is $r = 0.93$. Correlation *alone* doesn't provide any information about form. To determine the form of an association, you must look at a scatterplot.

FIGURE 2.6 Scatterplot showing the relationship between weight in carats and price for a sample of diamonds.

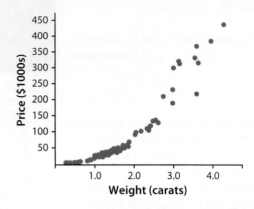

Correlation and Cause and Effect

While the correlation is a good way to measure the strength and direction of a linear relationship, it has limitations. Most importantly, correlation doesn't imply causation. In many cases, two variables might have a strong correlation, but changes in one variable are very unlikely to cause changes in the other variable.

EXAMPLE

Tangled on the slopes?

Correlation and causation

PROBLEM: For the years 2000–2009, the correlation between total revenue generated by skiing facilities in the United States and the number of people who died by becoming tangled in their bedsheets is $r = 0.97$.[20] Does the strong correlation between these two variables suggest that an increase in skiing revenue causes more people to die by becoming tangled in their bedsheets? Explain.

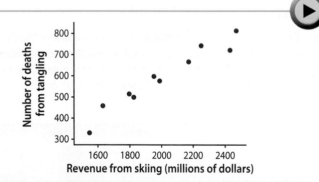

SOLUTION:

Probably not. Although there is a strong, positive correlation, an increase in skiing revenue is not likely to cause more people to get tangled in their bedsheets. It is likely that both of these variables are becoming larger over time because of population growth, which explains the association.

FOR PRACTICE **TRY EXERCISE 9.**

LESSON APP 2.3

If I eat more chocolate, will I win a Nobel Prize?

Most people love chocolate for its great taste. But does it also make you smarter? A scatterplot like this one recently appeared in the *New England Journal of Medicine*.[21] The explanatory variable is the chocolate consumption per person for a sample of countries. The response variable is the number of Nobel Prizes per 10 million residents of that country.

1. Interpret the correlation of $r = 0.791$.

2. If people in the United States started eating more chocolate, can we expect more Nobel Prizes to be awarded to residents of the United States? Explain.

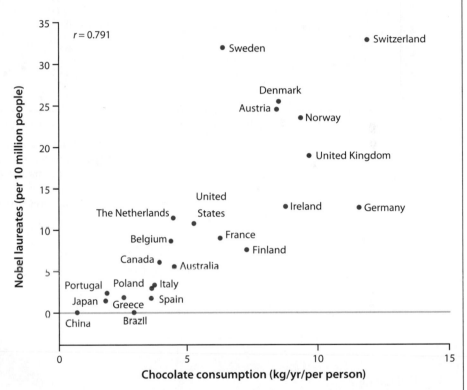

Lesson 2.3

WHAT DID YOU LEARN?

LEARNING TARGET	EXAMPLES	EXERCISES
Estimate the correlation between two quantitative variables from a scatterplot.	p. 115	1–4
Interpret the correlation.	p. 115	5–8
Distinguish correlation from causation.	p. 116	9–12

Exercises *Lesson 2.3*

Mastering Concepts and Skills

For Exercises 1–4: In each of the following relationships, is r > 0 or r < 0?
Closer to r = 0 or r = ± 1? Explain your reasoning.

1. Estimating *r*

pg 115

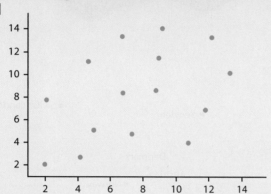

2. Estimating *r*

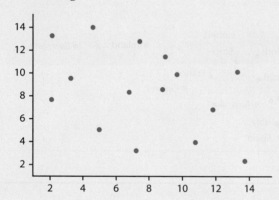

3. Estimating *r*

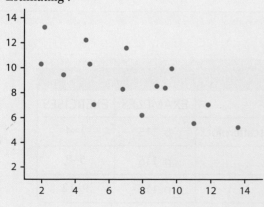

4. Estimating *r*

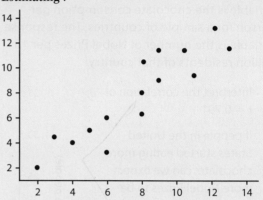

5. Points and turnovers There is a linear relationship
pg 115 between the number of turnovers and the number
of points scored for players in the 2013–2014 NBA
season.[22] The correlation for these data is $r = 0.92$.
Interpret this value.

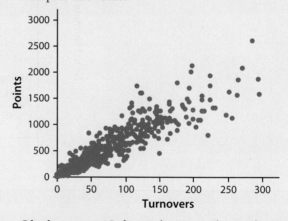

6. Oh, that smarts! Infants who cry easily may be more
easily stimulated than others. This may be a sign of
higher IQ. Child development researchers explored
the relationship between the crying of infants 4 to
10 days old and their IQ test scores at age 3. A snap
of a rubber band on the sole of the foot caused the
infants to cry. The researchers recorded the crying
and measured its intensity by the number of peaks
in the most active 20 seconds. The relationship
between these variables is linear, with a correlation
of $r = 0.45$.[23] Interpret this value.

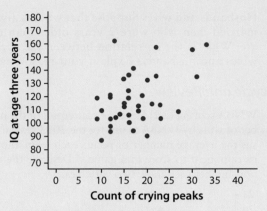

7. **It's warmer down here** There is a roughly linear relationship between the latitude of major American cities and the average temperature in July. The correlation for the 12 cities with the largest populations is $r = -0.55$.[24] Interpret this value.

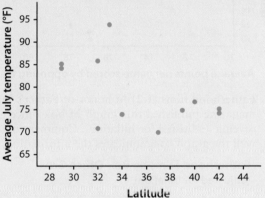

8. **A kettle of hawks** One of nature's patterns connects the percent of adult birds in a colony that return from the previous year and the number of new adults that join the colony. Research on 13 colonies of sparrowhawks shows a linear relationship between these two variables, with a correlation of $r = -0.75$.[25] Interpret this value.

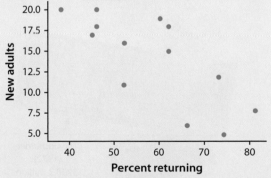

9. **GPA linked to pencils?** For their final statistics project, Jordynn and Angie surveyed a random sample of 50 students at their school. Each student in the survey reported his or her GPA and the number of pencils in his or her possession at the moment. The association between number of pencils and GPA

pg **116**

is linear, with a correlation of $r = -0.23$.[26] Does the negative association suggest that carrying fewer pencils will increase a student's GPA? Explain.

10. **Long courtship, long marriage?** A psychologist who specializes in divorce counseling finds that among his patients there is a linear relationship between the number of years of courtship before marriage and the number of years the couple stays married. The correlation is $r = 0.76$. Does this suggest that waiting a few years before getting married will make your marriage last longer? Explain.

11. **More points, more turnovers?** Refer to Exercise 5. Does the fact that $r = 0.92$ suggest that an increase in turnovers will cause NBA players to score more points? Explain.

12. **Crying for higher IQ?** Refer to Exercise 6. Does the fact that $r = 0.45$ suggest that making an infant cry will increase his or her IQ later in life? Explain.

Applying the Concepts

13. **Tipping at a buffet** Do customers who stay longer at buffets give larger tips? Charlotte, a statistics student who worked at an Asian buffet, decided to investigate this question for her second-semester project. While working as a hostess, she obtained a random sample of receipts, which included the length of time (in minutes) the party was in the restaurant and the amount of the tip (in dollars). Here is a scatterplot of these data.[27]

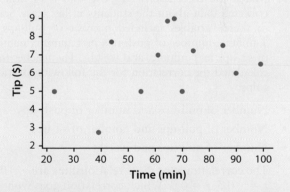

(a) Interpret the value $r = 0.36$.

(b) Does increasing the amount of time spent in the restaurant cause an increase in the amount of the tip? Explain.

14. **More math, more income?** Is there a relationship between a country's per capita income and its students' achievement in mathematics? Here is a scatterplot of per-capita gross domestic product versus average score on an international mathematics achievement test for 26 countries in Europe and Asia.[28]

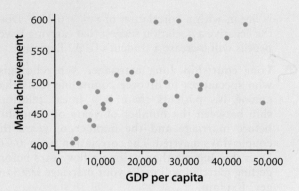

(a) Interpret the value $r = 0.59$.

(b) Will increasing per-capita income in a country lead to higher student achievement in mathematics? Explain.

15. **Correlation and form** Sketch a scatterplot where the association is nonlinear, but the correlation is close to $r = -1$.

16. **Correlation and form** Sketch a scatterplot where the association is linear, but the correlation is close to $r = 0$.

Extending the Concepts

17. **Rank the correlations** Consider each of the following relationships: the heights of fathers and the heights of their adult sons, the heights of husbands and the heights of their wives, and the heights of women at age 4 and their heights at age 18. Rank the correlations between these pairs of variables from largest to smallest. Explain your reasoning.

18. **Match the correlations** Suppose that a PE teacher collected data about the students in her class. Some of these variables included number of pull-ups in 1 minute, number of push-ups in 1 minute, number of sit-ups in 1 minute, and weight. The teacher then calculated the correlation for the following relationships:

- Number of pull-ups and number of push-ups
- Number of pull-ups and number of sit-ups
- Number of pull-ups and weight

The correlations for these relationships are $r = 0.9$, $r = -0.5$, $r = 0.3$. Which correlation goes with which relationship? Explain.

19. **Teachers and researchers** A college newspaper interviews a psychologist about student ratings of the teaching of faculty members. The psychologist says, "The evidence indicates that the correlation between the research productivity and teaching rating of faculty members is close to zero." The paper reports this as "Professor McDaniel said that good researchers tend to be poor teachers, and vice versa." Explain why the paper's report is wrong. Write a statement in plain language (don't use the word "correlation") to explain the psychologist's meaning.

20. **Husbands and wives** Suppose that women always married men who were 2 years older than they are. What is the correlation between the ages of wives and husbands? Explain your reasoning.

Recycle and Review

21. **WNBA scoring (2.2)** Here is a scatterplot of the number of wins by WNBA teams for the 2013 season versus the average number of points each team allowed its opponent to score in a game.[29] Describe the relationship between these two variables.

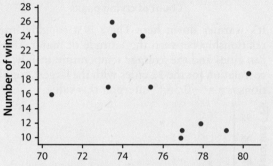

22. **Fatherhood films (1.2)** In honor of Father's Day, a magazine published this graph of box-office gross revenue (adjusted for inflation). Comment on how well the graph communicates the information.

Fatherhood Films by Box-Office Gross (adjusted for Inflation)

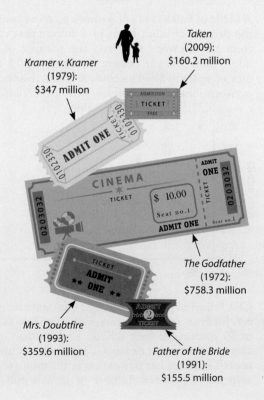

Lesson 2.4

Calculating the Correlation

In Lesson 2.3, you learned that the correlation r measures the strength and direction of the linear relationship between two quantitative variables. In this lesson, you will learn how to calculate the correlation and explore its properties.

Calculating the Correlation

How to Calculate the Correlation r

1. Find the mean $\bar{x}$ and the standard deviation s_x of the explanatory variable. Calculate the z-score for the value of the explanatory variable for each individual.

2. Find the mean y and the standard deviation s_y of the response variable. Calculate the z-score for the value of the response variable for each individual.

3. For each individual, multiply the z-score for the explanatory variable and the z-score for the response variable.

4. Add the z-score products and divide the sum by $n - 1$.

Here is the formula:

$$ r = \frac{\sum \left(\dfrac{x_i - \bar{x}}{s_x} \right)\left(\dfrac{y_i - \bar{y}}{s_y} \right)}{n - 1} = \frac{\sum z_x z_y}{n - 1} $$

To calculate the correlation, both variables must be quantitative. If one or both of the variables are categorical, we can consider the *association* between the two variables, but not the correlation.

EXAMPLE

What's the correlation between foot length and height?

Calculating the correlation

PROBLEM: The table shows the foot length (in centimeters) and the height (in centimeters) for a random sample of six high school seniors. Calculate the correlation for these data.

Foot length (cm)	Height (cm)	Foot length (cm)	Height (cm)
23	167	28	163
32	188	28	185
22	150	23	155

Eric Audras/© PhotoAlto sas/Alamy

SOLUTION:

x	y	$z_x = \dfrac{x - \bar{x}}{s_x}$	$z_y = \dfrac{y - \bar{y}}{s_y}$	$z_x z_y$
23	167	$\dfrac{23 - 26}{3.95} = -0.76$	$\dfrac{167 - 168}{15.54} = -0.06$	$(-0.76)(-0.06) = 0.0456$
32	188	$\dfrac{32 - 26}{3.95} = 1.52$	$\dfrac{188 - 168}{15.54} = 1.29$	$(1.52)(1.29) = 1.9608$
22	150	$\dfrac{22 - 26}{3.95} = -1.01$	$\dfrac{150 - 168}{15.54} = -1.16$	$(-1.01)(-1.16) = 1.1716$
28	163	$\dfrac{28 - 26}{3.95} = 0.51$	$\dfrac{163 - 168}{15.54} = -0.32$	$(0.51)(-0.32) = -0.1632$
28	185	$\dfrac{28 - 26}{3.95} = 0.51$	$\dfrac{185 - 168}{15.54} = 1.09$	$(0.51)(1.09) = 0.5559$
23	155	$\dfrac{23 - 26}{3.95} = -0.76$	$\dfrac{155 - 168}{15.54} = -0.84$	$(-0.76)(-0.84) = 0.6384$

$$r = \frac{0.0456 + 1.9608 + 1.1716 + (-0.1632) + 0.5559 + 0.6384}{6 - 1} = 0.84$$

1. The mean and standard deviation for x = foot length are $\bar{x} = 26$ and $s_x = 3.95$. The z-scores for foot length are in the third column.

2. The mean and standard deviation for y = height are $\bar{y} = 168$ and $s_y = 15.54$. The z-scores for height are in the fourth column.

3. The products of the z-scores are shown in the last column.

4. Add the z-score products and divide the sum by $n - 1$.

FOR PRACTICE   TRY EXERCISE 1.

THINK ABOUT IT   What does the correlation measure? The graphs provide more detail. The first is a scatterplot of the foot length and height data with two lines added—a vertical line at the group's mean foot length and a horizontal line at the mean height. Most of the points fall in the upper-right or lower-left "quadrants" of the graph. This confirms the positive association between the variables.

Students with above-average foot lengths tend to have above-average heights, and students with below-average foot lengths tend to have below-average heights.

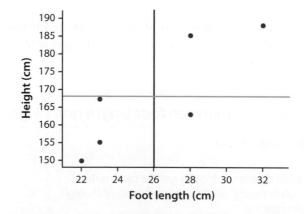

The second graph is a scatterplot of the standardized scores. To get this graph, we transformed both the x- and the y-values by subtracting their mean and dividing by their standard deviation.

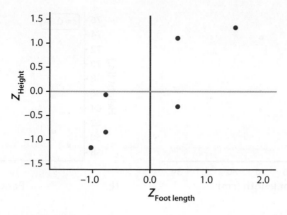

Standardizing the values of a variable converts the mean to 0 and the standard deviation to 1. That's why the vertical and horizontal lines in this graph are both at 0.

Notice that almost all of the products of the standardized values will be positive (positive × positive in Quadrant I and negative × negative in Quadrant III). This isn't surprising, considering the strong positive association between the variables.

Properties of the Correlation

1. *Correlation makes no distinction between explanatory and response variables.* It makes no difference which variable you call x and which you call y in calculating the correlation. As you can see in the formula, reversing the roles of x and y would only change the order of the multiplication, not the product:

$$r = \frac{\sum \left(\frac{x_i - \bar{x}}{s_x}\right)\left(\frac{y_i - \bar{y}}{s_y}\right)}{n - 1} = \frac{\sum z_x z_y}{n - 1}$$

Likewise, the scatterplots in Figure 2.7 show the same direction and strength, even though the variables are reversed in the second scatterplot.

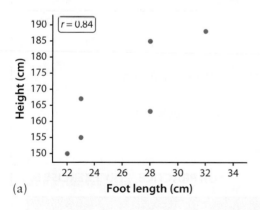

(a)

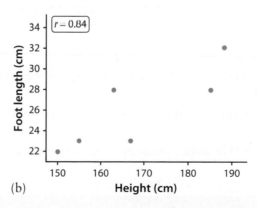

(b)

FIGURE 2.7 Scatterplots showing the association between (a) $x =$ foot length and $y =$ height, and (b) $x =$ height and $y =$ foot length. The correlation is the same for both associations.

2. Because r uses the standardized values of the observations, r *does not change when we change the units of measurement of x, y, or both.* Measuring foot length and height in inches rather than centimeters does not change the correlation between foot length and height. Figure 2.8 on the next page gives two scatterplots of the same six students. The graph on the left (a) uses centimeters for both measurements, and the graph on the right (b) uses inches for both measurements. The strength and direction are identical—only the scales on the axes have changed.

FIGURE 2.8 Scatterplots showing the association between (a) x = foot length in centimeters and y = height in centimeters, and (b) x = foot length in inches and y = height in inches. The correlation is the same for both associations.

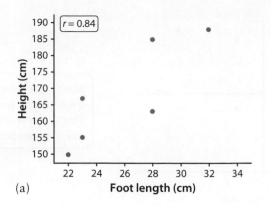

(a)

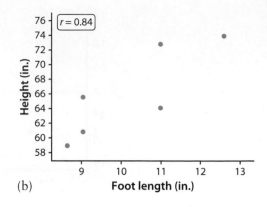

(b)

3. *The correlation r has no units of measurement* because we are using standardized values in the calculation and standardized values have no units.

EXAMPLE

Dash and jump, or jump and dash?

Properties of correlation

PROBLEM: The scatterplot shows the relationship between 40-yard-dash times and long-jump distances from Lesson 2.2. The correlation is $r = -0.838$.

(a) What would happen to the correlation if long-jump distance was plotted on the horizontal axis and dash time was plotted on the vertical axis? Explain.

(b) What would happen to the correlation if long-jump distance was measured in feet instead of inches? Explain.

(c) Sabrina claims that the correlation between long-jump distance and dash time is $r = -0.838$ inches per second. Is this correct?

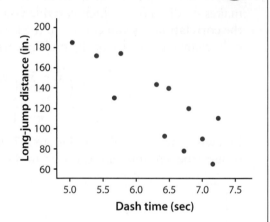

SOLUTION:

(a) The correlation would still be $r = -0.838$ because the correlation makes no distinction between explanatory and response variables.

(b) The correlation would still be $r = -0.838$ because the correlation doesn't change when we change the units of either variable.

(c) No. The correlation doesn't have units, so including "inches per second" is incorrect.

FOR PRACTICE   TRY EXERCISE 5.

Outliers and Correlation

While the correlation is a good way to measure the strength and direction of a linear relationship, it has limitations. In Lesson 2.3, we learned that correlation doesn't imply a cause-and-effect relationship and that correlation alone doesn't reveal the form of the relationship. The following activity illustrates another limitation.

ACTIVITY

Outliers and correlation

In this activity, you will investigate outliers and their influence on the correlation *r*.

1. Launch the *Correlation and Regression* applet at highschool.bfwpub.com/spa3e.

2. Click on the graphing area to add 10 points in the lower left corner so that the correlation is about *r* = 0.40. Also, check the box to Show the Mean X & Y lines as in the screen shot.

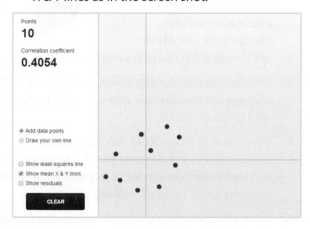

3. If you were to add a point at the right edge of the graphing area, but in the same linear pattern as the rest of the points, what do you think will happen to the correlation? Add the point to see if you were correct.

4. Click on the point you just added, and drag it up and down along the right edge of the graphing area. What happens to the correlation?

5. Now, move this point so that it is on the vertical $\bar{x}$ line. Drag the point up and down on the $\bar{x}$ line. What happens to the correlation? Do outliers in the vertical direction have as much influence on the correlation as outliers in the horizontal direction?

6. Briefly summarize how outliers influence the value of the correlation.

EXAMPLE

Netbooks: More charge for the price?

Outliers and correlation

PROBLEM: The scatterplot shows the relationship between cost (in dollars) and battery life (in hours) for a sample of netbooks.[30] How do the two points in the lower-right corner of the graph affect the correlation? Explain.

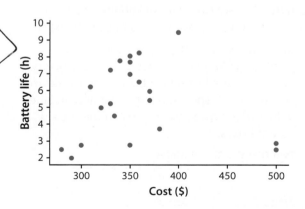

SOLUTION:

Most of the points show a positive association. However, because the two points in the lower right are separated from the rest of the points in the *x* direction, they are very influential and will make the correlation closer to 0 and possibly negative. Without these two points, the correlation is clearly positive.

FOR PRACTICE   TRY EXERCISE 9.

The formula for correlation involves the mean and standard deviation of both variables. Because the mean and standard deviation aren't resistant to outliers, it isn't surprising that the correlation isn't resistant to outliers either. For example, consider a scatterplot where the majority of points form a positive linear association. An outlier that is in the same pattern as the rest of the points will make the correlation closer to 1. An outlier that is not in the pattern of the rest of the points will make the correlation closer to 0—or possibly negative.

LESSON APP 2.4

Flying dinosaur or early bird?

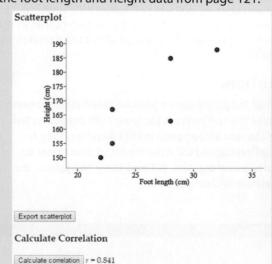

Archaeopteryx is an extinct beast having feathers like a bird, but teeth and a long bony tail like a reptile. Because the known specimens differ greatly in size, some scientists think they are different species rather than individuals from the same species. However, if the specimens belong to the same species and differ in size because some are younger than others, there should be a positive linear relationship between the lengths of a pair of bones from all individuals. An outlier from this relationship would suggest a different species. Here are data on the lengths (in centimeters) of the femur (a leg bone) and the humerus (a bone in the upper arm) for five specimens that preserve both bones.[31]

Length of femur (cm)	38	56	59	64	74
Length of humerus (cm)	41	63	70	72	84

1. Make a scatterplot using length of femur as the explanatory variable. Do you think that all five specimens come from the same species? Explain.

2. Find the correlation r step by step, using the formula on page 121. Explain how your value for r matches your graph in part (a).

3. Suppose that a new fossil was discovered. If the femur is 70 centimeters and the humerus is 40 centimeters, do you think this specimen came from the same species? Explain.

4. What effect will the new fossil have on the correlation? Explain.

TECH CORNER

Calculating the Correlation Using an Applet

You can use the *Two Quantitative Variables* applet at highschool.bfwpub.com/spa3e to calculate the correlation between two quantitative variables. Here is an example using the foot length and height data from page 121.

1. Enter the name of the explanatory variable (foot length) and the values of the explanatory variable in the first row of boxes. Then, enter the name of the response variable (height) and the values of the response variable in the second row of boxes.

Two Quantitative Variables

Variable	Name	Observations (separated by commas or spaces) *Keep individuals in the same order.*
Explanatory	Foot length (cm)	23, 32, 22, 28, 28, 23
Response	Height (cm)	167, 188, 150, 163, 185, 155

Begin analysis   Edit inputs   Reset everything

2. Press the "Begin analysis" button. This will display the scatterplot of height versus foot length. Then, press the "Calculate correlation" button under the scatterplot.

Scatterplot

Export scatterplot

Calculate Correlation

Calculate correlation $r = 0.841$

Lesson 2.4

WHAT DID YOU LEARN?

LEARNING TARGET	EXAMPLES	EXERCISES
Calculate the correlation between two quantitative variables.	p. 121	1–4
Apply the properties of the correlation.	p. 124	5–8
Describe how outliers influence the correlation.	p. 125	9–12

Exercises   Lesson 2.4

Mastering Concepts and Skills

1. **Go Spurs!** Here are the heights in inches and number of rebounds in the 2014 playoffs for 7 members of the NBA champion San Antonio Spurs.[32] Calculate the correlation.

pg 121

Name	Height (in.)	Rebounds
Tim Duncan	83	211
Kawhi Leonard	79	155
Tony Parker	74	45
Boris Diaw	80	111
Manu Ginobili	78	75
Danny Green	78	69
Tiago Splitter	83	140

2. **Crime on campus** Here are the number of enrolled students (in 1000s) and the number of burglaries for 8 public universities in one U.S. state.[33] Calculate the correlation.

Enrollment (1000s)	Burglaries in 1 year
5	3
36	23
22	9
29	34
18	20
11	2
8	9
31	42

3. **Conserving energy** Joan is concerned about how much energy she uses to heat her home, so she keeps a record of the natural gas her furnace consumes for each month from October to May. Because the months are not equally long, she divides each month's consumption by the number of days in the month to get the average number of cubic feet of gas used per day. She wants to see if there is a relationship between gas consumption and the average temperature (in degrees Fahrenheit) for each month. The data are given in the table. Calculate the correlation.

Month	Average temperature (°F)	Average fuel consumption (ft³/day)
October	49.4	520
November	38.2	610
December	27.2	870
January	28.6	850
February	29.5	880
March	46.4	490
April	49.7	450
May	57.1	250

4. **Flying from Philly** How strong is the relationship between air fare and distance traveled? The table gives the distance (in miles) from Philadelphia to 6 cities and the cost of the cheapest flight on a popular discount airline (in dollars).[34] Calculate the correlation.

Airport	Distance from Philadelphia (mi)	Cheapest fare ($)
Raleigh-Durham	340	131
Orlando	860	120
St. Louis	810	139
Denver	1550	180
Houston	1330	174
Phoenix	2070	188

5. **Metabolic rate** Metabolic rate, the rate at which the body consumes energy, is important in studies of weight gain, dieting, and exercise. We have data on the lean body mass and resting metabolic rate for 12 women who are subjects in a study of dieting. Lean body mass, given in kilograms, is a person's weight leaving out all fat. Metabolic rate is measured in calories burned per 24 hours. The scatterplot shows the relationship between metabolic rate and lean body mass. The correlation is $r = 0.88$.

pg 124

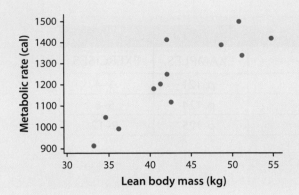

(a) What would happen to the correlation if metabolic rate was plotted on the horizontal axis and lean body mass was plotted on the vertical axis? Explain.

(b) What would happen to the correlation if lean body mass was measured in pounds instead of kilograms? Explain.

(c) Howard claims that the correlation between metabolic rate and lean body mass is $r = 0.88$ cal/kg. Is this correct? Explain.

6. **Cricket chirps** In a famous study published in 1948, communications engineer George Washington Pierce investigated the relationship between air temperature and the rate at which crickets chirp. The scatterplot displays his data on air temperature (in degrees Fahrenheit) and the chirp rate of the striped ground cricket (in chirps per minute). The correlation is $r = 0.835$.[35]

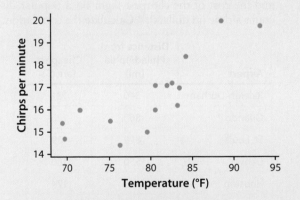

(a) What would happen to the correlation if chirp rate was plotted on the horizontal axis and air temperature was plotted on the vertical axis? Explain.

(b) What would happen to the correlation if air temperature was measured in degrees Celsius instead of degrees Fahrenheit? Explain.

(c) Mariko claims that the correlation between chirp rate and temperature is $r = 0.835$ chirps per minute per degree Fahrenheit. Is this correct? Explain.

7. **Price and mileage** Cars lose value as they pile up the road miles. The scatterplot shows the relationship between asking price (in thousands of dollars) and odometer mileage (in thousands of miles) for nineteen 2012 Ford Fusions for sale at www.carmax.com one day in August. The correlation is $r = -0.800$.

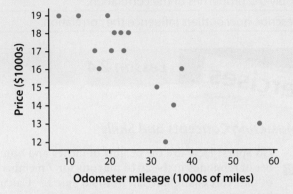

(a) What would happen to correlation if the odometer reading was plotted on the vertical axis and price was plotted on the horizontal axis? Explain.

(b) What would happen to the correlation if the distance each car had been driven was measured in kilometers rather than miles? Explain.

8. **Nutritious grains** The scatterplot displays the protein and carbohydrate content of different types of grains and flours. Both nutrients are measured in grams per cup. The correlation is $r = 0.545$.[36]

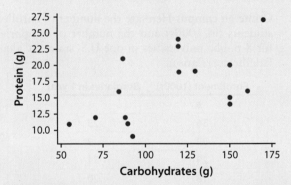

(a) What would happen to the correlation if carbohydrates was plotted on the vertical axis and protein was plotted on the horizontal axis? Explain.

(b) What would happen to the correlation if carbohydrates was measured in ounces per cup instead of grams per cup? Explain.

9. **Avoiding crashes** Should travelers avoid flying airlines that have had crashes in the past? Writer Nate Silver asked this question in July 2014 and found some interesting results.[37] Here is a scatterplot showing the adjusted number of incidents (incidents per 1 trillion seat kilometers) for 56 airlines in two periods: 1985–1999 and 2000–2014.

pg 125

How does the point highlighted in red (Aeroflot) affect the correlation? Explain.

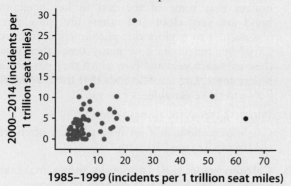

10. **Still avoiding crashes** Should travelers avoid flying airlines that have had crashes in the past? Writer Nate Silver asked this question in July 2014 and found some interesting results.[38] Here is a scatterplot showing the adjusted number of fatalities (deaths per 1 trillion seat kilometers) for 56 airlines in two periods: 1985–1999 and 2000–2014. How does the point highlighted in red (Kenya Airways) affect the correlation? Explain.

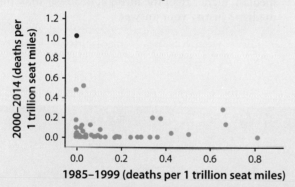

11. **Buying a stand mixer** The following scatterplot shows the weight (in pounds) and cost (in dollars) for a sample of 11 stand mixers.[39] The point highlighted in red is a stand mixer from Walmart. How does this point affect the correlation? Explain.

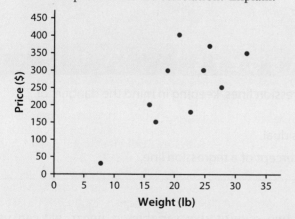

12. **Brainy or brawny?** The scatterplot shows the average brain weight in grams versus average body weight in kilograms for 96 species of mammals.[40] There are many small mammals whose points overlap at the lower left. How does the point for the elephant affect the correlation? Explain.

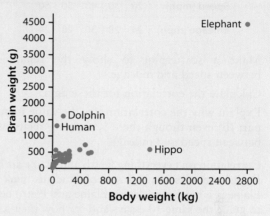

Applying the Concepts

13. **Spot the mistake** Each of the following statements contains an error. Explain what is wrong in each case.

(a) There is a high correlation between the gender of American workers and their income.

(b) We found a high correlation ($r = 1.09$) between SAT scores and GPA for a sample of high school students.

14. **Find the mistake** Each of the following statements contains an error. Explain what is wrong in each case.

(a) The correlation between the distance traveled by a hiker and the time spent hiking is $r = 0.9$ m/sec.

(b) There is a correlation of 0.54 between the position a football player plays and his weight.

15. **Netbooks charging** In the netbooks example on page 125, the correlation between battery life and cost was $r = -0.07$. If the price of each netbook was reduced by \$50, how would the correlation change? Explain.

16. **Back to the track** In the long-jump example on page 124, the correlation between dash time and long-jump distance was $r = -0.838$. After gathering the data, the students realized that they should have been measuring jump distances from the front of the takeoff board instead of the back. How will subtracting the width of the board (20 cm) affect the correlation? Explain.

Extending the Concepts

17. **How do speed and mileage relate?** The gas mileage of an automobile first increases and then decreases as the speed increases. Suppose that this relationship is very regular, as shown by the following data on speed (miles per hour) and mileage (miles per gallon).

Speed (mph)	20	30	40	50	60
Mileage (mpg)	24	28	30	28	24

(a) Make a scatterplot to show the relationship between speed and mileage.

(b) Calculate the correlation for these data.

(c) Explain why the correlation has the value found in part (b) even though there is a strong relationship between speed and mileage.

18. **Correlation isn't everything** Pietro and Elaine are both English teachers at a high school. Students think that Elaine is a harder grader, so Elaine and Pietro decide to grade the same 10 essays and see how their scores compare. The correlation was $r = 0.98$, but Elaine's scores were always lower than Pietro's. Draw a scatterplot that illustrates this situation. *Hint:* Include the line $y = x$ on your scatterplot.

Recycle and Review

19. **Wash your car!** (2.3) One weekend, a statistician notices that some of the cars in her neighborhood are very clean and others are quite dirty. She decides to explore this phenomenon and asks 15 of her neighbors how many times they wash their cars each year and how much they paid in car repair costs last year. She finds that the correlation between these variables is $r = -0.71$.

(a) Interpret this correlation in context.

(b) Can she conclude that washing your car often will reduce repair costs? Explain.

20. **Calculus grades** (1.6) The dotplot shows final exam scores for Mr. Miller's 25 calculus students.

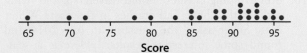

(a) Find the median exam score.

(b) Without doing any calculations, would you estimate that the mean is about the same as the median, higher than the median, or lower than the median? Justify your answer.

Lesson 2.5

Regression Lines

LEARNING TARGETS

- Make predictions using regression lines, keeping in mind the dangers of extrapolation.
- Calculate and interpret a residual.
- Interpret the slope and *y* intercept of a regression line.

When the relationship between two quantitative variables is linear, we can use a **regression line** to model the relationship and make predictions.

DEFINITION Regression line

A **regression line** is a line that describes how a response variable y changes as an explanatory variable x changes. Regression lines are expressed in the form $\hat{y} = a + bx$, where $\hat{y}$ (pronounced "y hat") is the predicted value of y for a given value of x.

We could also express the regression line in the form $y = mx + b$ like we do in algebra. However, statisticians prefer this reordered format because it works better when they use several explanatory variables. Just remember, the slope is always the coefficient of the x variable.

Making Predictions

The most common use of a regression line is to make predictions.

EXAMPLE

How much is that truck worth?

Making predictions

PROBLEM: Everyone knows that cars and trucks lose value the more miles they are driven. Can we predict the price of a Ford F-150 if we know how many miles it has on the odometer? A random sample of 16 Ford F-150 SuperCrew 4 × 4s was selected from among those listed for sale at autotrader.com. The number of miles driven and price (in dollars) were recorded for each of the trucks.[41] Here is a scatterplot of the data, along with the regression line $\hat{y} = 38{,}257 - 0.1629x$, where $x =$ miles driven and $y =$ price. Predict the cost of a Ford F-150 that has been driven 100,000 miles.

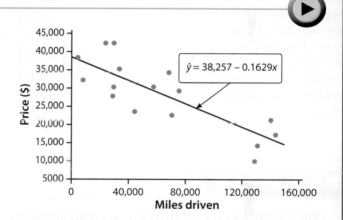

$\hat{y} = 38{,}257 - 0.1629x$

SOLUTION:

$\hat{y} = 38{,}257 - 0.1629(100{,}000)$

$\hat{y} = \$21{,}967$

FOR PRACTICE **TRY EXERCISE 1.**

To predict the cost of a Ford F-150 that has been driven 100,000 miles, substitute $x = 100{,}000$ into the equation and simplify.

Can we predict the price of a Ford F-150 with 300,000 miles driven? We can certainly substitute 300,000 into the equation of the line. The prediction is

$$\hat{y} = 38{,}257 - 0.1629(300{,}000) = -\$10{,}613$$

That is, we predict that we would need to pay someone else $10,613 just to take the truck off our hands! This prediction is an **extrapolation.**

DEFINITION Extrapolation

Extrapolation is the use of a regression line for prediction far outside the interval of x values used to obtain the line. Such predictions are often not accurate.

Residuals

Even when we are not extrapolating, our predictions are seldom perfect. For a specific point, the difference between the actual value of y and the predicted value of y is called a **residual**.

|||| DEFINITION Residual

A **residual** is the difference between an actual value of y and the value of y predicted by the regression line. That is,

$$\text{residual} = \text{actual } y - \text{predicted } y = y - \hat{y}$$

The scatterplot in Figure 2.9 shows the residual for the F-150 with 70,583 miles and an actual price of \$21,994. The predicted price is $\hat{y} = 38,257 - 0.1629(70,583) = \$26,759$, so the residual is $21,994 - 26,759 = -\$4765$. The negative value means that the price of this truck was \$4765 less than predicted, based on the number of miles it had been driven. It also means that the point is below the regression line on the scatterplot.

FIGURE 2.9 Scatterplot showing the relationship between miles driven and price for a sample of Ford F-150s, along with the regression line. The residual for one truck is illustrated with a vertical line segment.

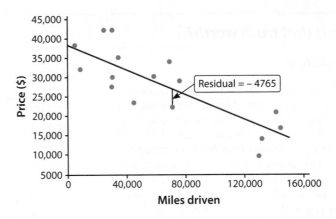

EXAMPLE

Does tapping a can prevent a soda shower?

Calculating and interpreting a residual

PROBLEM: Don't you hate it when you open a can of soda and some of the contents spray out of the can? Two students wanted to investigate if tapping on a can of soda would reduce the amount of soda expelled after the can has been shaken. For their experiment, they vigorously shook 40 cans of soda and randomly assigned each can to be tapped for 0 seconds, 4 seconds, 8 seconds, or 12 seconds. Then, after opening the can and cleaning up the mess, the students measured the amount of soda left in each can (in milliliters, or ml).[42] Here is the scatterplot, along with the regression line $\hat{y} = 248.6 + 2.63x$, where x = tapping time (in seconds) and y = amount of soda remaining (in milliliters). *Note:* Fewer than 40 points appear in the scatterplot because some cans had the same x and y values.

Davies and Starr/Getty Images

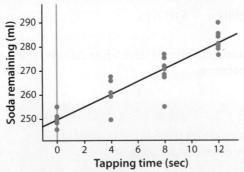

(a) Calculate the residual for the can that was tapped for 4 seconds and had 260 milliliters of soda remaining.

(b) Interpret the residual from part (a).

SOLUTION:

(a) $\hat{y} = 248.6 + 2.63(4) = 259.12$

   residual $= y - \hat{y} = 260 - 259.12 = 0.88 \, ml$

(b) This can had 0.88 milliliter more soda remaining than expected, based on the regression line using $x =$ tapping time.

> Find the predicted value of y using the equation of the regression line. Then subtract the predicted value of y from the actual value of y to find the residual.

FOR PRACTICE   TRY EXERCISE 5.

Interpreting a Regression Line

In the regression line $\hat{y} = a + bx$, a is the y **intercept** and b is the **slope**.

> ### DEFINITION y intercept, Slope
>
> The **y intercept** a is the predicted value of y when $x = 0$. The **slope** b of a regression line describes the predicted change in the y variable for each 1-unit increase in the x variable.

In the Ford F-150 example on page 131, the equation of the regression line is $\hat{y} = 38{,}257 - 0.1629x$. The slope is the coefficient of x, $b = -0.1629$. This means that the *predicted* price of a Ford F-150 goes down by 0.1629 dollars for each additional mile that the truck is driven. The y intercept is $a = 38{,}257$. This means that the *predicted* price of a truck that has been driven 0 miles is $38,257.

It is very important to include the word "predicted" (or its equivalent) in the interpretation of the slope and y intercept. Otherwise, it may seem that our predictions will be exactly correct.

CAUTION

EXAMPLE

More taps, more soda?

Interpreting a regression line

PROBLEM: In the preceding example about tapping on cans, the equation of the regression line is $\hat{y} = 248.6 + 2.63x$, where $x =$ tapping time (in seconds) and $y =$ amount of soda remaining (in milliliters).

(a) Interpret the slope of the regression line.

(b) Does the value of the y intercept have meaning in this context? If so, interpret the y intercept. If not, explain why.

SOLUTION:

(a) The predicted amount of soda remaining in a can increases by 2.63 milliliters for each additional second that the can is tapped.

(b) The y intercept has meaning because it is possible to tap a can for 0 seconds. If a can is tapped for 0 seconds, the predicted amount of soda remaining is 248.6 milliliters.

FOR PRACTICE   TRY EXERCISE 9.

In some contexts, the y intercept doesn't have meaning because a value of $x = 0$ doesn't make sense. For example, in a scatterplot relating $x =$ height and $y =$ weight for a sample of students, it wouldn't make sense to predict the weight for a student with height $= 0$!

LESSON APP 2.5

Do cut flowers benefit from sugar in the water?

Does adding sugar to the water in a vase help flowers stay fresh? To find out, two statistics students went to a flower shop and randomly selected 12 carnations. When they got home, the students prepared 12 identical vases with exactly the same amount of water in each vase. They put 1 tablespoon of sugar in 3 vases, 2 tablespoons of sugar in 3 vases, and 3 tablespoons of sugar in 3 vases. In the remaining 3 vases, they added no sugar. After the vases were prepared, the students randomly assigned 1 carnation to each vase and observed how many hours each flower continued to look fresh. Here is a scatterplot along with the regression line $\hat{y} = 180.8 + 15.8x$, where $x =$ amount of sugar (in tablespoons) and $y =$ hours of freshness.

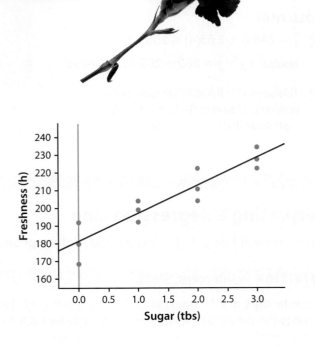

1. Calculate and interpret the residual for the flower that had 2 tablespoons of sugar and looked fresh for 204 hours.

2. Interpret the slope and y intercept of the regression line.

3. Would you be willing to use the regression line to predict the hours of freshness for a flower that received 10 tablespoons of sugar? Explain.

Lesson 2.5

WHAT DID YOU LEARN?

LEARNING TARGET	EXAMPLES	EXERCISES
Make predictions using regression lines, keeping in mind the dangers of extrapolation.	p. 131	1–4
Calculate and interpret a residual.	p. 132	5–8
Interpret the slope and y intercept of a regression line.	p. 133	9–12

Exercises   Lesson 2.5

Mastering Concepts and Skills

1. **Old Faithful** One of the major attractions in Yellowstone National Park is the Old Faithful geyser. The scatterplot shows the relationship between $x =$ the duration of the previous eruption (in minutes) and $y =$ the interval of time between eruptions (in minutes). The equation of the regression line relating these variables is $\hat{y} = 33.35 + 13.29x$. Predict the interval of time between eruptions if the previous eruption lasted 4 minutes.

pg 131

imagenavi/Getty Images

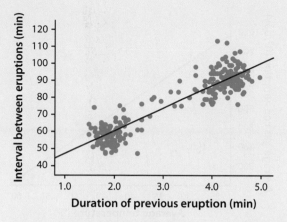

2. **More crickets chirping** The scatterplot shows the relationship between temperature in degrees Fahrenheit (x) and chirps per minute (y) for the striped ground cricket. The equation of the regression line relating these variables is $\hat{y} = -0.31 + 0.212x$. Predict the cricket chirp rate when the temperature is 82°F.

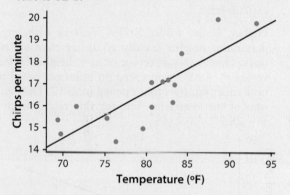

3. **Jumping predictions** The scatterplot shows the relationship between x = 40-yard-dash time (in seconds) and y = long-jump distance (in inches) for a sample of students. The equation of the regression line relating these variables is $\hat{y} = 415 - 45.7x$.

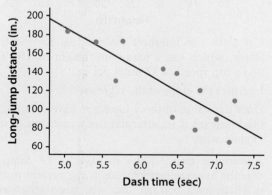

(a) Predict the long-jump distance for a student who had a 40-yard-dash time of 6 seconds.

(b) Would you be willing to use the regression line to predict the long-jump distance for a student whose 40-yard sprint time is 3 seconds? Explain.

4. **Pizza-powered athletes** The scatterplot shows the relationship between the number of slices of pizza eaten by each member of a football team (x) and the number of laps around the block the player could run immediately afterward (y). The equation of the regression line shown on the graph is $\hat{y} = 10 - 0.67x$.

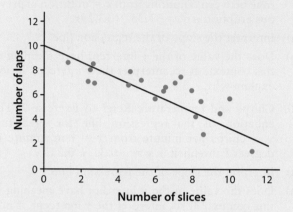

(a) Predict the number of laps a player who ate 7 slices can run.

(b) Would you be willing to use the regression line to predict the number of laps a player who ate 15 slices of pizza could run? Explain.

5. **A faithful residual** Refer to Exercise 1. The equation of the regression line for predicting y = time between eruptions from x = duration of previous eruption is $\hat{y} = 33.35 + 13.29x$.

pg **132**

(a) In one cycle, it took 62 minutes between eruptions, and the duration of the previous eruption was 2 minutes. Calculate the residual for this cycle.

(b) Interpret the residual for this cycle.

6. **Crickets aren't perfect** Refer to Exercise 2. The equation of the regression line for predicting y = chirps per minute from x = temperature in degrees Fahrenheit is $\hat{y} = -0.31 + 0.212x$.

(a) One observation in these data measured 16.2 chirps per minute at 83.3°F. Calculate the residual for this observation.

(b) Interpret the residual for this observation.

7. **Better than expected?** Refer to Exercise 3. The equation of the regression line for predicting y = long-jump distance from x = 40-yard-dash time is $\hat{y} = 415 - 45.7x$.

(a) One student had a 40-yard-dash time of 7.17 seconds and a long-jump distance of 65 inches. Calculate the residual for this student.

(b) Interpret the residual for this student.

8. **Leftover pizza** Refer to Exercise 4. The equation of the regression line for predicting y = the number of laps around the block a player could run from x = the number of slices of pizza eaten is $\hat{y} = 10 - 0.67x$.

(a) One player ate 8.5 pieces of pizza and ran 2.75 laps. Calculate the residual for this player.

(b) Interpret the residual for this player.

9. **Interpreting Old Faithful** Refer to Exercise 1. The equation of the regression line for predicting $y =$ time between eruptions from $x =$ duration of previous eruption is $\hat{y} = 33.35 + 13.29x$.

pg [133]

(a) Interpret the slope of the regression line.

(b) Does the value of the y intercept have meaning in this context? If so, interpret the y intercept. If not, explain why.

10. **Chirps and temperature** Refer to Exercise 2. The equation of the regression line for predicting $y =$ chirps per minute from $x =$ temperature in degrees Fahrenheit is $\hat{y} = -0.31 + 0.212x$.

(a) Interpret the slope of the regression line.

(b) Does the value of the y intercept have meaning in this context? If so, interpret the y intercept. If not, explain why.

11. **Stay on track** Refer to Exercise 3. The equation of the regression line for predicting $y =$ long-jump distance (in.) from $x =$ 40-yard-dash time (sec) is $\hat{y} = 415 - 45.7x$.

(a) Interpret the slope of the regression line.

(b) Does the value of the y intercept have meaning in this context? If so, interpret the y intercept. If not, explain why.

12. **More pizza, anyone?** Refer to Exercise 4. The equation of the regression line for predicting $y =$ the number of laps around the block a player could run from $x =$ the number of slices of pizza eaten is $\hat{y} = 10 - 0.67x$.

(a) Interpret the slope of the regression line.

(b) Does the value of the y intercept have meaning in this context? If so, interpret the y intercept. If not, explain why.

Applying the Concepts

13. **Predicting gas consumption** Joan is concerned about how much energy she uses to heat her home, so she keeps a record of the natural gas her furnace consumes for each month from October to May. Because the months are not equally long, she divides each month's consumption by the number of days in the month. She wants to see if there is a relationship between $x =$ average temperature (in degrees Fahrenheit) and $y =$ average gas consumption (in cubic feet per day) for each month. Here is a scatterplot along with the regression line $\hat{y} = 1425 - 19.87x$.

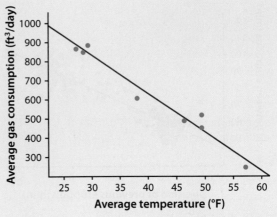

(a) Calculate and interpret the residual for March, when the average temperature was 49.4°F and Joan used an average of 520 cubic feet per day.

(b) Interpret the slope of the regression line.

(c) Does the value of the y intercept have meaning in this context? If so, interpret the y intercept. If not, explain why.

14. **Higher, faster, wilder** Roller coasters with larger maximum heights usually go faster than shorter ones. Here is a scatterplot of $x =$ height (in feet) versus $y =$ maximum speed (in miles per hour) for nine roller coasters that opened in 2012. The equation of the regression line for this relationship is $\hat{y} = 28.17 + 0.2143x$.

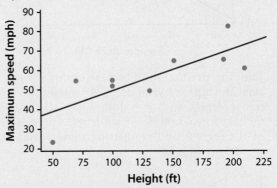

(a) Calculate and interpret the residual for the Iron Shark, which has a maximum height of 100 feet and a top speed of 52 miles per hour.

(b) Interpret the slope of the regression line.

(c) Does the value of the y intercept have meaning in this context? If so, interpret the y intercept. If not, explain why.

15. **A warming trend?** Refer to Exercise 13. Suppose that the average temperature in the current month is 10 degrees warmer than the previous month. About how much less gas should Joan expect to use this month than the previous month? Explain.

16. **An even higher coaster** Refer to Exercise 14. Suppose a park owner plans to increase the height of his roller coaster by 20 feet during the off-season. About how much faster can he expect the top speed to be? Explain.

Extending the Concepts

17. **Multiple regression** It is possible to use two or more explanatory variables at the same time to predict a response variable. Here is a multiple regression model for predicting y = long-jump distance (in inches) using x_1 = 40-yard-dash time (in seconds) and x_2 = gender (1 = male, 0 = female) for a sample of students: $\hat{y} = 293.56 - 31.05x_1 + 42.02x_2$.

(a) Predict the long-jump distance for a male student who had a dash time of 5.41 seconds.

(b) The student in part (a) had a long jump distance of 171 inches. Calculate and interpret the residual for this student.

(c) According to the model, about how much farther do males jump than females, after accounting for dash time? Explain.

Recycle and Review

18. **Explanatory and response variables (2.2)** Identify the explanatory variable for the following relationships if possible. Explain your reasoning.

(a) For many families, you have data on parents' income and the years of education their eldest child completes.

(b) Because elderly people may have difficulty standing to have their heights measured, a study looked at predicting overall height from height to the knee.

(c) Systolic blood pressure tends to be positively correlated with age in people over 40 years old.

19. **Salty dogs (2.3, 2.4)** Are hot dogs that are high in calories also high in salt? The scatterplot shows the relationship between calories and salt content (measured as milligrams of sodium) in 17 brands of meat hot dogs.[43]

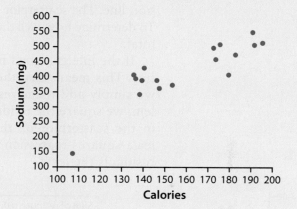

(a) The correlation for these data is $r = 0.87$. Explain what this value means.

(b) What effect does the hot-dog brand with the lowest calorie content have on the correlation? Justify your answer.

Lesson 2.6

The Least-Squares Regression Line

LEARNING TARGETS

- Calculate the equation of the least-squares regression line using technology.
- Calculate the equation of the least-squares regression line using summary statistics.
- Describe how outliers affect the least-squares regression line.

A good regression line makes the residuals as small as possible so that the predicted values are close to the actual values. For this reason, statisticians prefer using the **least-squares regression line.**

DEFINITION Least-squares regression line

The **least-squares regression line** is the line that makes the sum of the squared residuals as small as possible.

The scatterplots in Figure 2.10 show the relationship between the price of a Ford F-150 and the number of miles it has been driven, along with the least-squares regression line. The scatterplot on the left (a) shows the residuals as vertical line segments. To determine how well the line works, we could add the residuals and hope for a small total.

If the line is a good fit, some of the residuals will be positive and others negative. This means that the positive and negative residuals cancel each other out if we simply add them, resulting in a sum of approximately 0. To avoid this problem, we square the residuals to make them positive before adding them, as shown in the scatterplot on the right (b). Of all the possible lines we could use, the least-squares regression line is the one that makes the sum of the *squares* of the residuals the *least*.

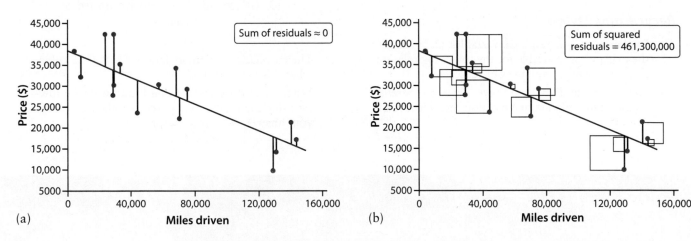

FIGURE 2.10 Scatterplots of the Ford F-150 data with the regression line added. (a) The residuals will add to approximately 0 when using a good regression line. (b) A good regression line should make the sum of squared residuals as small as possible.

Calculating the Equation of the Least-Squares Regression Line

Technology makes it easy to calculate the equation of the least-squares regression line.

TECH CORNER

Calculating the Equation of the Least-Squares Regression Line

You can use an applet or a graphing calculator to calculate the equation of the least-squares regression line. Here are the Ford F-150 data.

Miles driven	70,583	129,484	29,932	29,953	24,495	75,678
Price ($)	21,994	9,500	29,875	41,995	41,995	28,986
Miles driven	8,359	4,447	34,077	58,023	44,447	68,474
Price ($)	31,891	37,991	34,995	29,988	22,896	33,961
Miles driven	144,162	140,776	29,397	131,385		
Price ($)	16,883	20,897	27,495	13,997		

Applet

Launch the *Two Quantitative Variables* applet at **highschool.bfwpub.com/spa3e**.

1. Enter the name of the explanatory variable (miles driven) and the values of the explanatory variable in the first row of boxes. Then, enter the name of the response variable (price) and the values of the response variable in the second row of boxes.

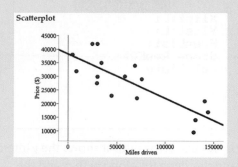

2. Press the "Begin analysis" button. This will display the scatterplot of price versus miles driven. Then, press the "Calculate least-squares regression line" button under Regression Models. The graph of the least-squares regression line appears on the scatterplot and a residual plot is created. The equation and other summary statistics are displayed in the Regression Models section. *Note:* We will learn about residual plots and the additional summary statistics in the next lesson.

Scatterplot

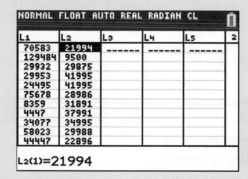

Regression Models

Calculate least-squares regression line

Equation	n	s	r^2
$\hat{y} = 38257.135 - 0.163x$	16	5740.131	0.664

TI-83/84

1. Enter the miles driven in L_1 and the prices in L_2.

- Press [STAT] and choose Edit…
- Type the values into L_1 and L_2.

2. Calculate the equation of the least-squares regression line.

- Press [STAT], arrow over to the CALC menu, and choose LinReg(a + bx).

• Adjust the settings as shown. **Older OS:** Complete the command LinReg(a+bx) L₁, L₂.

• Press Calculate. The output shows the y intercept (*a*) and the slope (*b*), along with r^2 and the correlation *r*. We will learn what r^2 measures in Lesson 2.7. *Note:* If *r* and r^2 do not show up in the output, you need to turn on the Stats Diagnostics. Find this option by pressing MODE or choosing DiagnosticOn in the Catalog (2nd 0).

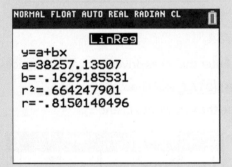

3. *Optional:* Graph the least-squares regression line on a scatterplot.

• Set up a scatterplot as in Lesson 2.2 (page 109).

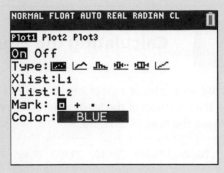

• Press Y= and enter the equation of the least-squares regression line for Y₁.

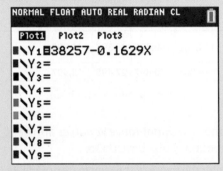

• Press ZOOM and choose ZoomStat.

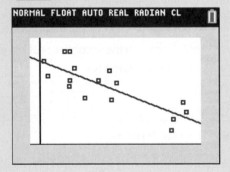

EXAMPLE

A foot taller?

Calculating the least-squares regression line using technology

PROBLEM: The following table shows the foot length (in centimeters) and the height (in centimeters) for a random sample of six high school seniors. Use technology to calculate the least-squares regression line for predicting height from foot length.

Foot length (cm)	Height (cm)	Foot length (cm)	Height (cm)
23	167	28	163
32	188	28	185
22	150	23	155

SOLUTION:

Using the *Two Quantitative Variables* applet, enter "Foot length" for the explanatory variable and "Height" for the response variable. The equation of the least-squares regression line is $\hat{y} = 82 + 3.308x$.

FOR PRACTICE   TRY EXERCISE 1.

Calculating the Equation of the Least-Squares Regression Line Using Summary Statistics

It is also possible to calculate the equation of the least-squares regression line using the means and standard deviations of each variable, along with their correlation.

How to Calculate the Least-Squares Regression Line Using Summary Statistics

If $\bar{x}$ and s_x are the mean and standard deviation of the explanatory variable, $\bar{y}$ and s_y are the mean and standard deviation of the response variable, and r is the correlation between the explanatory and response variables, the slope of the least-squares regression line is

$$\text{slope} = b = r\frac{s_y}{s_x}$$

Because the equation of the least-squares regression line always includes the point $(\bar{x}, \bar{y})$, the y intercept is

$$y \text{ intercept} = a = \bar{y} - b\bar{x}$$

EXAMPLE

Average taps, average soda?

Calculating the least-squares regression line using summary statistics

PROBLEM: In Lesson 2.5, we used tapping time to predict the amount of soda remaining in a vigorously shaken can. For these cans, the mean tapping time was 6 seconds, with a standard deviation of 4.53 seconds. The mean soda remaining was 264.45 milliliters, with a standard deviation of 12.92 milliliters. The correlation between tapping time and soda remaining was $r = 0.924$. Calculate the equation of the least-squares regression line for predicting the amount of soda remaining from tapping time.

SOLUTION:

$$\text{slope} = b = r\frac{s_y}{s_x} = 0.924\left(\frac{12.92}{4.53}\right) = 2.64$$

Letting x = tapping time and y = soda remaining, calculate the slope using $r = 0.924$, $s_y = 12.92$, and $s_x = 4.53$.

$$y \text{ intercept} = a = \bar{y} - b\bar{x} = 264.45 - 2.64(6) = 248.61$$

Then, find the intercept using $\bar{y} = 264.45$, $\bar{x} = 6$, and $b = 2.64$.

The equation of the least-squares regression line is $\hat{y} = 248.61 + 2.64x$.

FOR PRACTICE   TRY EXERCISE 5.

The slope and intercept in this example vary slightly from the equation we used in Lesson 2.5. This is because we used rounded values for the correlation, means, and standard deviations.

Outliers and the Least-Squares Regression Line

ACTIVITY

Outliers and least-squares regression lines

In this activity, we will investigate the influence outliers have on the least-squares regression line.

1. Launch the *Correlation and Regression* applet at highschool.bfwpub.com/spa3e.

2. Click on the graphing area to add 10 points in the lower-left corner so that the correlation is about *r* = 0.40. Also, check the boxes to show the Least-Squares Line and the Mean X & Y lines as in the screen shot. Notice that the least-squares regression line goes though the point $(\bar{x}, \bar{y})$.

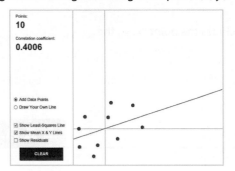

3. If you were to add a point on the least-squares regression line at the right edge of the graphing area, what do you think will happen to the equation of the least-squares regression line? Add the point to see if you were correct.

4. Click on the point you just added, and drag it up and down along the right edge of the graphing area. What happens to the equation of the least-squares regression line?

5. Now, move this point so that it is on the vertical $\bar{x}$ line. Drag the point up and down on the $\bar{x}$ line. What happens to the equation of the least-squares regression line? Do outliers in the vertical direction have as much influence on the least-squares regression line as outliers in the horizontal direction?

6. Briefly summarize how outliers influence the equation of the least-squares regression line. Does the least-squares regression line always go through the point $(\bar{x}, \bar{y})$?

As you learned in the activity, the equation of the least-squares regression line can be greatly influenced by outliers. However, some outliers are more influential than others!

EXAMPLE

Out of the mouths of babes?

Describing the effect of outliers

PROBLEM: Does the age at which a child begins to talk predict a later score on a test of mental ability? A study of the development of young children recorded the age in months at which each of 23 children spoke his or her first word and that child's Gesell Adaptive Score, the result of an aptitude test taken much later.[44] The scatterplot shows the relationship between Gesell score and age at first word, along with the least-squares regression line. Two outliers, Child 18 and Child 19, are identified on the scatterplot.

(a) Describe the effect Child 18 has on the equation of the least-squares regression line.

(b) Describe the effect Child 19 has on the equation of the least-squares regression line.

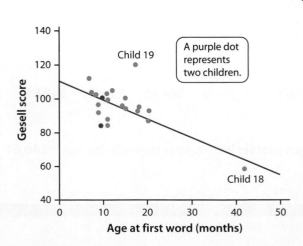

SOLUTION:

(a) Because the point for this child is below the line on the right, it is making the regression line steeper (more negative) and increasing the y intercept.

(b) Because this point is near $\bar{x}$ but above the rest of the points, it pulls the line up a little, which increases the y intercept but doesn't change the slope very much.

FOR PRACTICE **TRY EXERCISE 9.**

The formulas for the slope and y intercept of the least-squares regression line involve the mean and standard deviation of both variables. Because the mean and standard deviation are not resistant to outliers, it isn't surprising that the least-squares regression line is not resistant to outliers either.

Figure 2.11 shows the results of removing Child 18 and Child 19 on the correlation and the regression line. The graph adds two more regression lines, one calculated after leaving out Child 18 and the other after leaving out Child 19. You can see that removing the point for Child 18 changes the correlation and moves the line quite a bit. Because of Child 18's extreme position on the age scale, this point has a strong influence on both the correlation and the line. However, removing Child 19 has a small effect on the correlation and not much effect on the regression line.

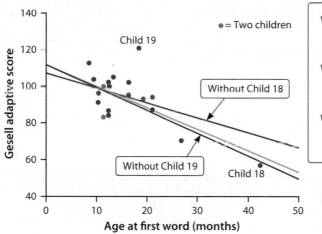

With all 19 children:
$r = -0.64$
$\hat{y} = 109.874 - 1.127x$

Without Child 19:
$r = -0.76$
$\hat{y} = 109.305 - 1.193x$

Without Child 18:
$r = -0.33$
$\hat{y} = 105.630 - 0.779x$

FIGURE 2.11 Three least-squares regression lines of Gesell score on age at first word. The green line is calculated from all the data. The dark blue line is calculated leaving out Child 18. Child 18 is an influential observation because leaving out this point moves the regression line quite a bit. The red line is calculated leaving out only Child 19.

LESSON APP 2.6

Did the Broncos buck the trend?

In 2013 the Denver Broncos football team set the NFL record for the most points scored in a season. On the next page is a scatterplot showing the relationship between passing yards and points scored for all 32 NFL teams in 2013, along with the least-squares regression line.[45]

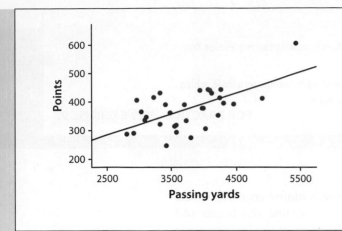

1. For passing yards, the mean is 3770 yards, with a standard deviation of 594 yards. For points scored, the mean is 375 points, with a standard deviation of 70 points. The correlation between passing yards and points scored is $r = 0.616$. Use this information to calculate the equation of the least-squares regression line.

2. The point for the Denver Broncos is highlighted in red. What effect does this point have on the equation of the least-squares regression line? Explain.

Lesson 2.6

WHAT DID YOU LEARN?

LEARNING TARGET	EXAMPLES	EXERCISES
Calculate the equation of the least-squares regression line using technology.	p. 140	1–4
Calculate the equation of the least-squares regression line using summary statistics.	p. 141	5–8
Describe how outliers affect the least-squares regression line.	p. 142	9–12

Exercises   Lesson 2.6

Mastering Concepts and Skills

1. **Candy calories** Here are data from a sample of 12  types of movie candy.[46] Use technology to calculate the equation of the least-squares regression line relating y = calories to x = amount of sugar (in grams).

Name	Sugar	Calories	Name	Sugar	Calories
Butterfinger Minis	45	450	Reese's Pieces	61	580
Junior Mints	107	570	Skittles	87	450
M&M'S®	62	480	Sour Patch Kids	92	490
Milk Duds	44	370	SweeTarts	136	680
Peanut M&M'S	79	790	Twizzlers	59	460
Raisinets	60	420	Whoppers	48	350

2. **Big TV prices** Here are prices and screen sizes (in inches, measured diagonally) for 7 different sizes of one brand of LED HD television.[47] Use technology to calculate the equation of the least-squares regression line relating y = price to x = screen size.

Screen size	Price	Screen size	Price
60	1000	42	430
55	800	39	400
50	700	32	300
47	600		

3. **Are beavers and beetles buddies?** Do beavers benefit beetles? Researchers laid out 23 circular plots, each 4 meters in diameter, in an area where beavers were cutting down cottonwood trees. In each plot, they counted the number of stumps from trees cut by beavers and the number of clusters of beetle larvae. Ecologists think

that the new sprouts from stumps are more tender than other cottonwood growth so that beetles prefer them. If so, more stumps should produce more beetle larvae.[48] Here are the data.

Stumps	2	2	1	3	3	4	3	1	2	5	1	3
Beetle larvae	10	30	12	24	36	40	43	11	27	56	18	40
Stumps	2	1	2	2	1	1	4	1	2	1	4	
Beetle larvae	25	8	21	14	16	6	54	9	13	14	50	

Use technology to calculate the equation of the least-squares regression line relating y = number of larvae to x = number of stumps.

4. **Mr. Boggs's soap** Mathematics and statistics teacher Rex Boggs, from Australia, weighed the bar of soap in his shower stall before showering in the morning. The data on day and soap weight (in grams) appear in the table. (Notice that he forgot to weigh the soap on some days!) Use technology to calculate the equation of the least-squares regression line relating y = weight of soap to x = number of days.

Day	Weight (g)	Day	Weight (g)	Day	Weight (g)
1	124	8	84	16	27
2	121	9	78	18	16
5	103	10	71	19	12
6	96	12	58	20	8
7	90	13	50	21	6

5. **Tall husbands, tall wives?** The mean height of married American women in their early 20s is 64.5 inches and the standard deviation is 2.5 inches. The husbands of these women have a mean height of 68.5 inches, with a standard deviation of 2.7 inches. The correlation between the heights of husbands and wives is about $r = 0.5$. Find the equation of the least-squares regression line for predicting a husband's height from his wife's height for married couples in their early 20s. Show your work.

pg 141

6. **Premier league soccer** In professional soccer, it's all about scoring goals. The number of games that a team wins is more strongly correlated with the number of goals the team scores than with the number of goals the team surrenders to its opponent. For the 2013–2014 Premier League in England, the mean number of wins per team was 15.1, with a standard deviation of 6.73. The mean number of goals scored by a team for the entire season was 52.6, with a standard deviation of 20.62. The correlation between these two variables was $r = 0.889$.[49] Find the equation of the least-squares regression line for predicting the number of wins from number of goals scored. Show your work.

7. **The January stock market** Some people think that the behavior of the stock market in January predicts its behavior for the rest of the year. Take the explanatory variable x to be the percent change in a stock market index in January and the response variable y to be the change in the index for the entire year. We expect a positive correlation between x and y because the change during January contributes to the full year's change. Calculation from data for an 18-year period gives

$$\bar{x} = 1.75\% \qquad s_x = 5.36\% \qquad \bar{y} = 9.07\%$$

$$s_y = 15.35\% \qquad r = 0.596$$

Find the equation of the least-squares line for predicting full-year change from January change. Show your work.

8. **Crossbills and food** An ecologist studying breeding habits of a bird called the common crossbill in different years finds that there is a linear relationship between the number of breeding pairs of crossbills and the abundance of spruce cones. In the table, statistics are given for 8 years of measurements, where x = average number of cones per tree and y = number of breeding pairs of crossbills in a certain forest.[50]

	Mean	Standard deviation
x = average number of cones/tree	23.0	16.2
y = number of crossbill pairs	18.0	15.1

The correlation between x and y is $r = 0.968$. Find the equation of the least-squares regression line for predicting the number of crossbill pairs from the average number of cones per tree.

9. **Buried change** Drilling down beneath a lake in Alaska yields chemical evidence of past changes in climate. Biological silicon, left by the skeletons of single-celled creatures called diatoms, is a measure of the abundance of life in the lake. A variable based on the ratio of certain isotopes gives an indirect measure of moisture, mostly from snow. As we drill down, we look further into the past. Here is a scatterplot of data from 2300 to 12,000 years ago. Describe the effect the point in the upper-right corner has on the equation of the least-squares regression line.

pg 142

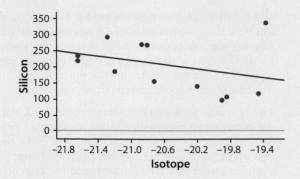

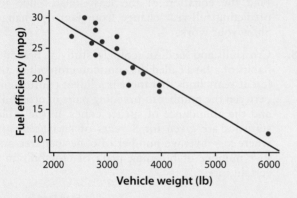

10. **Fuel efficiency in the city** Here is a scatterplot of the fuel efficiency in city driving (in miles per gallon) and weight (in pounds) of twenty 2014 sedans.[51] The dot in the lower-right corner represents the 2014 Bentley Mulsanne. (With a sticker price of $305,325, anyone buying it probably doesn't care about fuel efficiency!) Describe the effect this point has on the equation of the least-squares regression line.

11. **Reaction times and memory** Is there a relationship between a student's score in a memory game and his or her reaction time? A random sample of 14 high school students was selected. The scatterplot shows the relationship between scores in a memory game and reaction times (in seconds), along with the least-squares regression line. Two of the students are identified on the scatterplot as Student A and Student B.

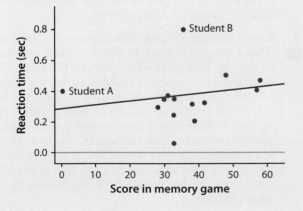

(a) Describe the effect Student A has on the equation of the least-squares regression line.

(b) Describe the effect Student B has on the equation of the least-squares regression line.

12. **Born to get old?** Is there a relationship between the gestational period (time from conception to birth) of an animal and its average life span? The figure shows a scatterplot of the gestational period and average life span for 43 species of animals.[52]

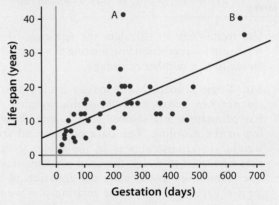

(a) Point A is the hippopotamus. Describe the effect this point has on the equation of the least-squares regression line.

(b) Point B is the Asian elephant. Describe the effect this point has on the equation of the least-squares regression line.

Applying the Concepts

13. **Measuring glucose** People with diabetes measure their fasting plasma glucose (FPG; measured in units of milligrams per milliliter) after fasting for at least 8 hours. Another measurement, made at regular medical checkups, is called HbA. This is roughly the percent of red blood cells that have a glucose molecule attached. It measures average exposure to glucose over a period of several months. The table gives data on both HbA and FPG for 18 diabetics five months after they had completed a diabetes education class.[53]

Subject	HbA (%)	FPG (mg/ml)	Subject	HbA (%)	FPG (mg/ml)
1	6.1	141	10	8.7	172
2	6.3	158	11	9.4	200
3	6.4	112	12	10.4	271
4	6.8	153	13	10.6	103
5	7.0	134	14	10.7	172
6	7.1	95	15	10.7	359
7	7.5	96	16	11.2	145
8	7.7	78	17	13.7	147
9	7.9	148	18	19.3	255

(a) Make a scatterplot to show the relationship between FPG and HbA, using HbA as the explanatory variable.

(b) Using technology, calculate the equation of the least-squares regression line relating y = FPG to x = HbA.

(c) What effect do you think Subject 18 has on the equation of the least-squares regression line? Calculate the equation of the least-squares regression without this subject to confirm your answer.

14. **The mighty bean** Beans and other legumes are an excellent source of protein. The table gives data on the total protein and carbohydrate content of a one-half-cup portion of cooked beans for 12 different varieties.[54]

Bean variety	Carbohydrates (g)	Protein (g)
Soybeans	8.5	14.0
Garbanzos	22.5	7.0
Black beans	20.0	8.0
Adzuki beans	28.5	8.5
Cranberry beans	21.5	8.0
Great Northern beans	18.5	7.0
Kidney beans	20.0	7.5
Navy beans	23.5	7.5
Pinto beans	22.5	7.5
White beans	22.5	8.5
Lima beans	19.5	7.0
Mung beans	19.0	7.0

(a) Make a scatterplot to show the relationship between protein and carbohydrate content for these bean varieties, using carbohydrate content as the explanatory variable. (You may have done this already in Lesson 2.2.)

(b) Using technology, calculate the equation of the least-squares regression line relating y = protein content to x = carbohydrate content.

(c) What effect do you think the observation for soybeans has on the equation of the least-squares regression line? Calculate the equation of the least-squares regression line without this variety to confirm your answer.

Extending the Concepts

15. **Reversing the variables** In Lesson 2.4, you learned that reversing the explanatory and response variables doesn't affect the correlation. Does reversing the variables affect the equation of the least-squares regression line?

(a) Using the movie candy data from Exercise 1, calculate the equation of the least-squares regression line relating y = sugar to x = calories.

(b) Are the slope and y intercept the same as in Exercise 1?

16. **Graph the data first** Here are 4 sets of data.[55]

Data Set A

x	10	8	13	9	11	14	6	4	12	7	5
y	8.04	6.95	7.58	8.81	8.33	9.96	7.24	4.26	10.84	4.82	5.68

Data Set B

x	10	8	13	9	11	14	6	4	12	7	5
y	9.14	8.14	8.74	8.77	9.26	8.10	6.13	3.10	9.13	7.26	4.74

Data Set C

x	10	8	13	9	11	14	6	4	12	7	5
y	7.46	6.77	12.74	7.11	7.81	8.84	6.08	5.39	8.15	6.42	5.73

Data Set D

x	8	8	8	8	8	8	8	8	8	8	19
y	6.58	5.76	7.71	8.84	8.47	7.04	5.25	5.56	7.91	6.89	12.50

(a) Using technology, calculate the equation of the least-squares regression line for each of the 4 data sets. What do you notice?

(b) Sketch scatterplots for each of the 4 data sets. Explain why knowing the equation of the least-squares regression line isn't enough to describe the relationship between two variables.

17. **Lean body mass and metabolic rate** The data and scatterplot show the lean body mass and metabolic rate for a sample of 5 adults. For each person, the lean body mass is the subject's total weight in kilograms less any weight due to fat. The metabolic rate is the number of calories burned in a 24-hour period.

Mass (kg)	33	43	40	55	49
Rate (cal/day)	1050	1120	1400	1500	1700

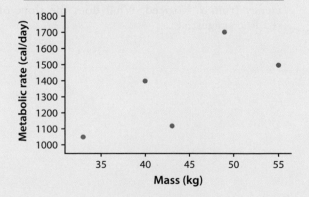

In this context, it makes sense to model the relationship between metabolic rate and body mass with a direct variation function in the form $y = kx$. After all, a person with no lean body mass should burn no calories, and functions in the form $y = kx$ always go through the point $(0, 0)$. But what value of k would be best?

Several different values of k, such as $k = 25$, $k = 26$, and so on, were used to predict the metabolic rate, and the sum of squared residuals (SSR) was calculated for each value of k. Here is a scatterplot showing the relationship between SSR and k. According to the scatterplot, what is the ideal value of k to use for predicting metabolic rate? Explain.

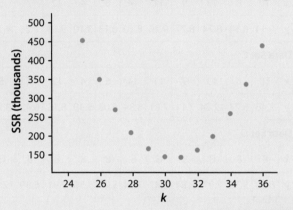

18. **Slow down to save** Fuel efficiency for many cars is related to how fast the car is driven. To explore this relationship, some researchers calculated the mean fuel efficiency for nine different car models at 13 different speeds.[56] Here are the results.

Speed (mph)	Mean fuel efficiency (mpg)	Speed (mph)	Mean fuel efficiency (mpg)	Speed (mph)	Mean fuel efficiency (mpg)
15	24.4	40	31.0	60	31.4
20	27.9	45	31.6	65	29.2
25	30.5	50	32.4	70	26.8
30	31.7	55	32.4	75	24.8
35	31.2				

(a) Use technology to calculate a least-squares regression line for predicting y = mean fuel efficiency from x = speed. What does the slope of the line suggest?

(b) Make a scatterplot of the relationship. What conclusion can you draw about the applicability of a least-squares regression line in this setting?

Recycle and Review

19. **Traffic accidents and marijuana (2.1, 2.3)** Researchers in New Zealand interviewed 907 drivers at age 21. The researchers had data on previous traffic accidents and asked the drivers about marijuana use. Here are data on whether these drivers had caused accidents at age 19, broken down by marijuana use at that age.[57]

		Frequency of marijuana use per year				
		Never	1–10 times	11–50 times	51+ times	Total
Caused accident	No	393	193	55	106	747
	Yes	59	36	15	50	160
		452	229	70	156	907

(a) Make a segmented bar chart to determine if there is an association between these variables.

(b) Explain why we can't conclude, using these data, that marijuana use *causes* accidents.

20. **Heavy diamonds (1.3, 1.5, 1.7)** Here are the weights (in milligrams) of 58 diamonds from a nodule carried up to the earth's surface in surrounding rock. These data represent a population of diamonds formed in a single event deep in the earth.[58] Make a histogram that shows the distribution of weights of these diamonds. Describe what you see. Give appropriate numerical measures of center and variability.

13.8	3.7	33.8	11.8	27	18.9	19.3	20.8	25.4	23.1	7.8
10.9	9	9	14.4	6.5	7.3	5.6	18.5	1.1	11.2	7
7.6	9	9.5	7.7	7.6	3.2	6.5	5.4	7.2	7.8	3.5
5.4	5.1	5.3	3.8	2.1	2.1	4.7	3.7	3.8	4.9	2.4
1.4	0.1	4.7	1.5	2	0.1	0.1	1.6	3.5	3.7	2.6
4	2.3	4.5								

Lesson 2.7

Assessing a Regression Model

Now that we have learned how to calculate a least-squares regression line, it is important to assess how well the line fits the data. We do this by asking two questions:

- Is a line the right model to use, or would a curve be better?
- If a line is the right model to use, how well does it make predictions?

Residual Plots

In Lesson 2.5, we learned how to calculate and interpret a residual. We can also use residuals to assess whether a regression model is appropriate by making a **residual plot.**

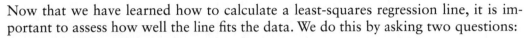

DEFINITION Residual plot

A **residual plot** is a scatterplot that plots the residuals on the vertical axis and the explanatory variable on the horizontal axis.

Figure 2.12(a) is a scatterplot showing the relationship between Super Bowl number and the cost of a 30-second commercial for the years 1967–2013, along with the least-squares regression line.[59] The resulting residual plot is shown in Figure 2.12(b).

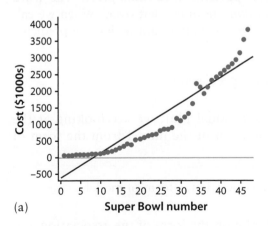

(a)

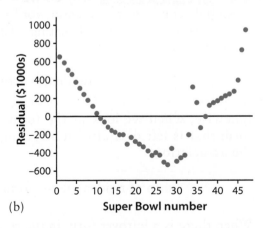

(b)

FIGURE 2.12 Shown here are the (a) scatterplot and (b) residual plot for the relationship between the cost of a Super Bowl ad and Super Bowl number.

The least-squares regression line clearly doesn't fit this association very well! In the early years, the actual cost of an ad is always greater than the line predicts, resulting in positive residuals. From Super Bowl 11 to Super Bowl 33, the actual cost is always less than the line predicts, resulting in negative residuals. After Super Bowl 33, the actual cost is almost always greater than the line predicts, again resulting in positive residuals. This positive-negative-positive pattern in the residual plot indicates that the linear form of our model doesn't match the form of the association. A curved model might be better in this case.

Figure 2.13 gives a scatterplot (a) showing the Ford F-150 data from Lesson 2.5, along with the corresponding residual plot (b). Looking at the scatterplot, the line seems to be a good fit for the association. You can "see" that the line is appropriate by the lack of a leftover pattern in the residual plot. In fact, the residuals look randomly scattered around the residual = 0 line.

FIGURE 2.13 Shown here are the (a) scatterplot and (b) residual plot for the relationship between price and miles driven for Ford F-150s.

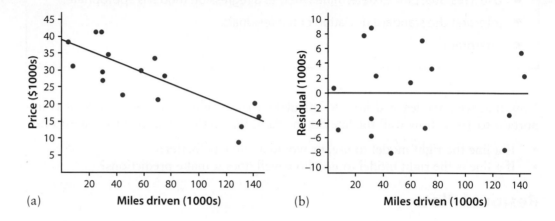

(a) (b)

Interpreting a Residual Plot

To determine whether the regression model is appropriate, look at the residual plot.

■ If there is no leftover pattern in the residual plot, the regression model is appropriate.

■ If there is a leftover pattern in the residual plot, the regression model is not appropriate.

THINK ABOUT IT **Why do we look for patterns in residual plots?** The word "residual" comes from the Latin word *residuum*, meaning "left over." When we calculate a residual, we are calculating what is left over after subtracting the predicted value from the actual value:

$$\text{residual} = \text{actual } y - \text{predicted } y$$

Likewise, when we look at the form of a residual plot, we are looking at the form that is left over after subtracting the form of the model from the form of the association:

$$\text{form of residual plot} = \text{form of association} - \text{form of model}$$

When there is a leftover form in the residual plot, the form of the association and form of the model are not the same. However, if the form of the association and form of the model are the same, the residual plot should have no form, other than random scatter.

EXAMPLE

Okay to tap on cans?

Interpreting a residual plot

PROBLEM: In Lesson 2.5, we used a least-squares regression line to model the relationship between the amount of soda remaining and the tapping time for cans of vigorously shaken soda. Here is the residual plot for that model. Use the residual plot to determine whether the regression model is appropriate.

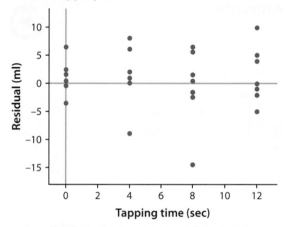

> The vertical stacks of dots in the residual plot are nothing to worry about, even though this might appear to be a "pattern." These stacks are present because only four different tapping times were used in the experiment. There would be a troublesome pattern if, for example, all the residuals in the first and last stack were positive and all the residuals in the middle two stacks were negative.

SOLUTION:

Because there is no leftover pattern in the residual plot, the least-squares regression line is an appropriate model for relating the amount of soda remaining to the tapping time.

FOR PRACTICE   TRY EXERCISE 1.

Standard Deviation of the Residuals

Once we have all the residuals, we can measure how well the line makes predictions with the **standard deviation of the residuals.**

DEFINITION Standard deviation of the residuals s
The **standard deviation of the residuals s** measures the size of a typical residual. That is, s measures the typical distance between the actual y values and the predicted y values.

To calculate the standard deviation of the residuals s, we square each of the residuals, add them, divide the sum by $n - 2$, and take the square root. We can write the formula for calculating the standard deviation of the residuals as

$$s = \sqrt{\frac{\text{sum of squared residuals}}{n - 2}} = \sqrt{\frac{\sum (y_i - \hat{y}_i)^2}{n - 2}}$$

In Lesson 2.5, we calculated a residual of -4765 dollars for the Ford F-150 with 70,583 miles and an actual price of \$21,994. The price of this truck was \$4765 less than predicted, based on the number of miles it had been driven. Here are the residuals for all 16 trucks in our sample using the equation $\hat{y} = 38{,}257 - 0.1629x$.

$$-4765 \quad -7664 \quad -3506 \quad 8617 \quad 7728 \quad 3057 \quad -5004 \quad 458$$
$$2289 \quad 1183 \quad -8121 \quad 6858 \quad 2110 \quad 5572 \quad -5973 \quad -2857$$

The standard deviation of the residuals is

$$s = \sqrt{\frac{(-4765)^2 + (-7664)^2 + \cdots + (-2857)^2}{16 - 2}} = \sqrt{\frac{461,264,136}{14}} = \$5740$$

EXAMPLE

How much does the price of a Ford F-150 vary from expected?

Interpreting s

PROBLEM: In Lesson 2.5, we used a least-squares regression line to model the relationship between the price of a Ford F-150 and the number of miles it had been driven. The standard deviation of the residuals for this model is $s = \$5740$. Interpret this value.

SOLUTION:

The actual price of a Ford F-150 is typically about $5740 away from its predicted price using the least-squares regression line with x = miles driven.

FOR PRACTICE TRY EXERCISE 5.

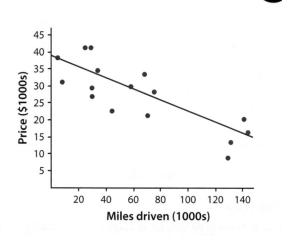

The Coefficient of Determination r^2

Besides the standard deviation of the residuals s, we can also use the **coefficient of determination r^2** to measure how well the regression line makes predictions.

> **DEFINITION Coefficient of determination r^2**
>
> The **coefficient of determination r^2** measures the percent reduction in the sum of squared residuals when using the least-squares regression line to make predictions, rather than the mean value of y. In other words, r^2 measures the percent of the variability in the response variable that is accounted for by the least-squares regression line.

Suppose that we wanted to predict the price of a particular Ford F-150, but we didn't know how many miles it had been driven. Our best guess would be the average cost of a Ford F-150, $\bar{y} = \$27,834$. Of course, this prediction is unlikely to be very good, as the prices vary quite a bit from the mean ($s_y = \$9570$). If we knew how many miles the truck had been driven, we could use the least-squares regression line to make a better prediction. How much better are predictions using the least-squares regression line with x = miles driven, rather than using only the average price? The answer is r^2.

The scatterplot in Figure 2.14(a) shows the squared residuals along with the sum of squared residuals (approximately 1,374,000,000) when using the average price as the predicted value. The scatterplot in Figure 2.14(b) shows the squared residuals along with the sum of squared residuals (approximately 461,300,000) when using the least-squares regression line with x = miles driven to predict the price.

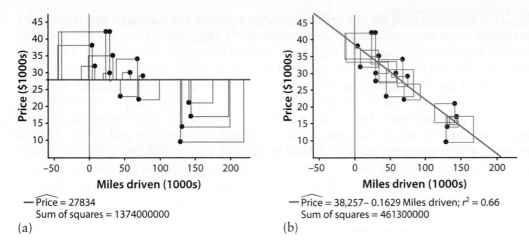

— $\widehat{Price}$ = 27834
Sum of squares = 1374000000
(a)

— $\widehat{Price}$ = 38,257 − 0.1629 Miles driven; r^2 = 0.66
Sum of squares = 461300000
(b)

FIGURE 2.14 (a) The sum of squared residuals is 1,374,000,000 if we use the mean price as our prediction for all 16 trucks. (b) The sum of squared residuals from the least-squares regression line is 461,300,000.

To find r^2, calculate the percent reduction in the sum of squared residuals:

$$r^2 = \frac{1,374,000,000 - 461,300,000}{1,374,000,000} = \frac{912,700,000}{1,374,000,000} = 0.66$$

The sum of squared residuals has been reduced by 66%. That is, 66% of the variability in the price of a Ford F-150 is accounted for by the least-squares regression line with x = miles driven. The remaining 34% is due to other factors, including age, color, condition, and other features of the truck.

The easiest way to calculate the value of r^2 is to square the value of the correlation r. The value of r^2 is also provided every time you calculate the equation of a least-squares regression line on your graphing calculator or use the *Two Quantitative Variables* applet at highschool.bfwpub.com/spa3e.

EXAMPLE

Does knowing tapping time help?

Calculating and interpreting r^2

PROBLEM: In Lesson 2.5, we used a least-squares regression line to model the relationship between the amount of soda remaining (in milliliters) and the tapping time (in seconds) for cans of vigorously shaken soda. Interpret the value $r^2 = 0.85$ for this model.

SOLUTION:

85% of the variability in the amount of soda remaining (milliliters) is accounted for by the least-squares regression line with x = tapping time (seconds).

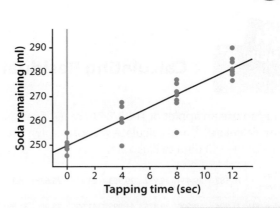

FOR PRACTICE **TRY EXERCISE 9.**

THINK ABOUT IT   What's the relationship between the standard deviation of the residuals s and the coefficient of determination r^2? They are both calculated from the sum of squared residuals. They also both attempt to answer the question "How well does the line fit the data?" The standard deviation of the residuals reports the size of a typical prediction error, in the same units as the response variable. In the truck example, $s = 5740$ *dollars*. Ideally, the value of s will be close to 0, indicating that our predictions are very close to the actual values. The value of r^2, however, does not have units and is usually expressed as a percentage between 0% and 100%, such as $r^2 = 66.4\%$. Ideally, the value of r^2 will be close to 100%, indicating that the model accounts for almost all of the variability in the response variable. Because these values assess how well the line fits the data in different ways, we recommend you follow the example of most statistical software and report them both.

LESSON APP 2.7

Do higher priced tablets have better battery life?

Can you predict the battery life of a tablet using the price? Using data from a sample of 15 tablets, the least-squares regression line $\hat{y} = 4.67 + 0.0068x$ was calculated using x = price (in dollars) and y = battery life (in hours).[60] A residual plot for this model is shown.

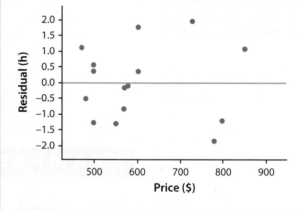

sturti/Getty Images

1. Use the residual plot to determine whether the regression model is appropriate.

2. Interpret the value $s = 1.21$ for this model.

3. Interpret the value $r^2 = 0.342$ for this model.

Calculating Residual Plots, s, and r^2

You can use an applet or TI-83/84 to make a residual plot, calculate r^2, and calculate s (applet only). Let's use the Ford F-150 data to illustrate.

Miles driven	70,583	129,484	29,932	29,953	24,495	75,678	8,359	4,447
Price ($)	21,994	9,500	29,875	41,995	41,995	28,986	31,891	37,991
Miles driven	34,077	58,023	44,447	68,474	144,162	140,776	29,397	131,385
Price ($)	34,995	29,988	22,896	33,961	16,883	20,897	27,495	13,997

Applet

You can use the *Two Quantitative Variables* applet at highschool.bfwpub.com/spa3e to make a residual plot, calculate s, and calculate r^2.

1.   Enter the name of the explanatory variable (miles driven) and the values of the explanatory variable in the first row of boxes. Then, enter the name of the response variable (price) and the values of the response variable in the second row of boxes.

Two Quantitative Variables

Variable	Name	Observations (separated by commas or spaces) *Keep individuals in the same order.*
Explanatory	Miles driven	70583 129484 29932 29953 24495 75678 8359 4447 34077 58023
Response	Price ($)	21994 9500 29875 41995 41995 28986 31891 37991 34995 29988

Begin analysis   Edit inputs   Reset everything

2. Press the "Begin analysis" button. This will display the scatterplot of price versus miles driven. Then, press the "Calculate least-squares regression line" button under Regression Models. The residual plot appears directly below the scatterplot. The equation of the least-squares regression line, *n, s*, and r^2 are displayed in the Regression Models section.

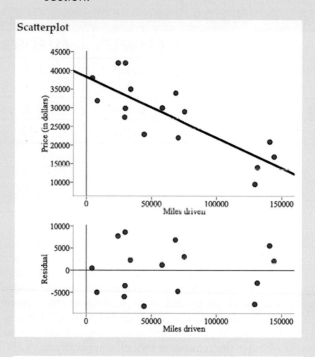

Scatterplot

Regression Models

Calculate least-squares regression line

Equation	n	s	r^2
$\hat{y} = 38257.135 - 0.163x$	16	5740.131	0.664

TI-83/84

1. Enter the miles driven in L_1 and the prices in L_2.

● Press STAT and choose Edit…

● Type the values into L_1 and L_2.

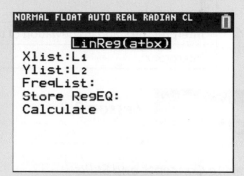

2. Calculate the equation of the least-squares regression line and r^2.

● Press STAT, arrow over to the CALC menu, and choose LinReg(a+bx).

● Adjust the settings as shown and press Calculate. In addition to the equation of the least-squares regression line, the calculator also displays the value of r^2. **Older OS:** Complete the command LinReg(a+bx) L_1, L_2.

3. To make a residual plot, set up a scatterplot in the statistics plots menu.

● Press 2nd Y= (STAT PLOT).

● Press ENTER or 1 to go into Plot1.

● Adjust the settings as shown on the next page. The RESID list is found in the List menu by pressing 2nd STAT. *Note:* You have to calculate the equation of the least-squares regression line using the calculator before making a residual plot. Otherwise, the RESID list will include the residuals from a different least-squares regression line.

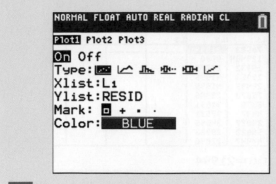

4. Use ZoomStat to let the calculator choose an appropriate window.

● Press [ZOOM] and choose 9:ZoomStat.

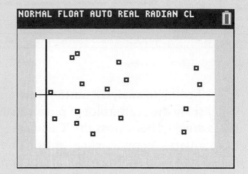

Lesson 2.7

WHAT DID YOU LEARN?

LEARNING TARGET	EXAMPLES	EXERCISES
Use a residual plot to determine whether a regression model is appropriate.	p. 151	1–4
Interpret the standard deviation of the residuals.	p. 152	5–8
Interpret r^2.	p. 153	9–12

Exercises   Lesson 2.7

Mastering Concepts and Skills

1. **Statistics is growing** More students are taking statistics in high school than ever before. One way to track this trend is to look at the number of AP® Statistics exams given each year. A linear model was used to predict y = number of AP® Statistics exams from x = year.[61] Here is the residual plot for that model. Use it to determine if the regression model is appropriate.

pg 151

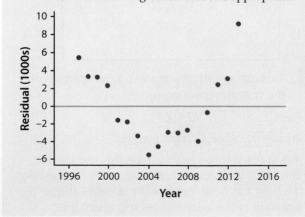

2. **Farther is better** In golf, being able to drive the ball farther is likely to lead to better (lower) scores. Using data from the Ladies Professional Golf Association (LPGA) in a recent year, a linear model was used to predict y = scoring average from x = average driving distance (in yards).[62] Here is the residual plot for that model. Use it to determine whether the regression model is appropriate.

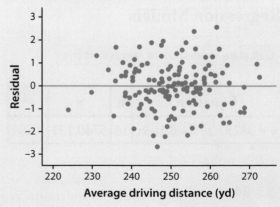

3. **Texting and GPA** Mac and Nick collected data from a random sample of students at their school to see if they could use a linear model to predict y = a student's grade point average from x = the number of text messages a student sent on the previous day. Here is the residual plot for that model. Use it to determine whether the regression model is appropriate.

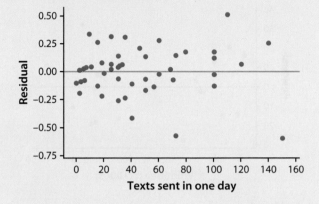

4. **Does the grill stay hot?** A curious statistics teacher wondered how fast his gas grill cools off once it has been turned off. He recorded the grill's thermometer reading at several times and used a linear model to predict y = grill temperature (in degrees Fahrenheit) from time (in minutes) elapsed since the grill was turned off. Here is the residual plot for that model. Use it to determine whether the regression model is appropriate.

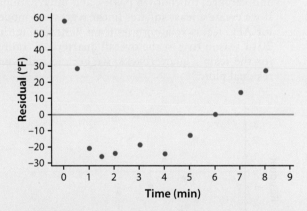

5. **Age versus height** A random sample of 195 students was selected from the United Kingdom. The scatterplot shows the relationship between y = height (in centimeters) and x = age (in years), along with the least-squares regression line. The standard deviation of the residuals for this model is s = 8.61. Interpret this value.[63]

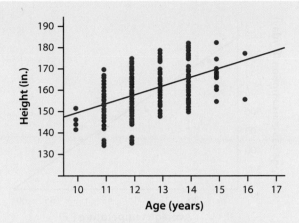

6. **More Mentos, more mess?** When Mentos are dropped into a newly opened bottle of Diet Coke, carbon dioxide is released from the Diet Coke very rapidly, causing the Diet Coke to be expelled from the bottle. Using 16-ounce (2-cup) bottles of Diet Coke, two statistics students dropped either 2, 3, 4, or 5 Mentos into a randomly selected bottle, waited for the fizzing to stop, and measured the number of cups remaining in the bottle.[64] Then, they subtracted this measurement from the original amount in the bottle to calculate the amount of Diet Coke expelled (in cups). The scatterplot shows the relationship between y = amount expelled (in cups) and x = number of Mentos, along with the least-squares regression line. The standard deviation of the residuals for this model is s = 0.067. Interpret this value.

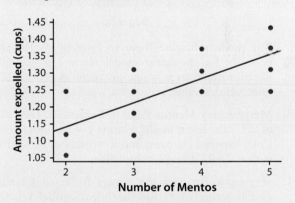

7. **Staying toasty all winter** Joan keeps a record of the natural gas that her furnace consumes for each month from October to May. The following scatterplot shows the relationship between x = average temperature (in degrees Fahrenheit) and y = average gas consumption (in cubic feet per day) for each month, along with the least-squares regression line. The standard deviation of the residuals for this model is s = 46.4. Interpret this value.

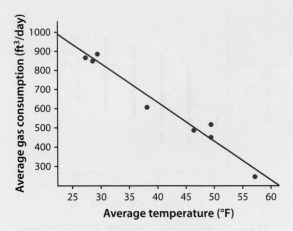

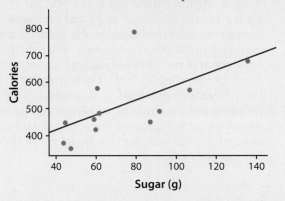

8. Residual candy Data were collected on a sample of 12 brands of movie candy. The scatterplot shows the relationship between x = amount of sugar (in grams) and y = number of calories, along with the least-squares regression line. The standard deviation of the residuals for this model is $s = 105.5$. Interpret this value.

9. Age predicts height? Refer to Exercise 5. The value of r^2 for the linear model relating y = height (incentimeters) to x = age (in years) is $r^2 = 0.274$. Interpret this value.

10. Messy, messy Mentos Refer to Exercise 6. The value of r^2 for the linear model relating y = amount of Diet Coke expelled (in cups) and x = number of Mentos is $r^2 = 0.602$. Interpret this value.

11. Staying warm over the winter Refer to Exercise 7. The value of r^2 for the linear model relating y = average gas consumption (in cubic feet per day) and x = average temperature is $r^2 = 0.966$. Interpret this value.

12. Candies, calories, and sugar Refer to Exercise 8. The value of r^2 for the linear model relating y = number of calories to x = amount of sugar (in grams) is $r^2 = 0.382$. Interpret this value.

Applying the Concepts

13. Wildebeest save the day Long-term records from the Serengeti National Park in Tanzania show interesting ecological relationships. When wildebeest are more abundant, they graze the grass more heavily, so there are fewer fires. Researchers collected data on one part of this cycle and computed a least-squares regression line relating y = the percent of the grass area burned to x = wildebeest abundance (in thousands of animals) in the same year. Here is a residual plot for this model.[65]

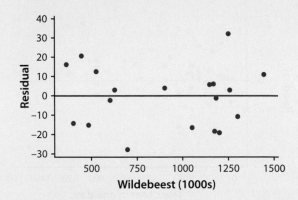

(a) Use the residual plot to determine whether the regression model is appropriate.

(b) Interpret the value $s = 15.99$ for this model.

(c) Interpret the value $r^2 = 0.646$ for this model.

14. Do quarterbacks matter? How important is the quarterback's skill to the success of an NFL football team? One way to measure a quarterback's effectiveness is with the "quarterback rating," a complex calculation based on passing accuracy and yardage, touchdown passes, and interceptions. If we create a least-squares linear regression model for AFC teams relating y = team wins during the 2013 season to x = the overall quarterback rating for the team's quarterbacks, we get the following residual plot.[66]

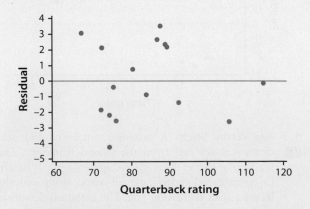

(a) Use the residual plot to determine whether the regression model is appropriate.

(b) Interpret the value $s = 2.473$ for this model.

(c) Interpret the value $r^2 = 0.457$ for this model.

15. **Statistics is still growing** Refer to Exercise 1. There is a positive association between x = year and y = number of AP® Statistics exams. Use this fact along with the residual plot from Exercise 1 to sketch a scatterplot of x = year versus y = number of AP® exams.

16. **Is the grill still hot?** Refer to Exercise 4. There is a negative association between x = time since the grill was turned off and y = temperature. Use this fact along with the residual plot from Exercise 4 to sketch a scatterplot of x = time versus y = temperature.

17. **Calories in sandwiches** The number of calories in a food item depends on many factors, including the amount of fat. The data in the table show the amount of fat (in grams) and the number of calories in 7 beef sandwiches at McDonalds.[67]

Sandwich	Fat (g)	Calories
Big Mac®	27	530
Quarter Pounder® with Cheese	26	520
Double Quarter Pounder® with Cheese	43	750
Hamburger	8	240
Cheeseburger	11	290
Double Cheeseburger	21	430
McDouble	17	380

(a) Sketch a scatterplot for these data and calculate the equation of the least-squares regression line using technology.

(b) Construct a residual plot for this model. Is a linear model appropriate? Explain.

(c) Using technology, calculate and interpret the standard deviation of the residuals for this model.

(d) Using technology, calculate and interpret the value of r^2 for this model.

18. **Stair climber** Alana's favorite exercise machine is a stair climber. On the "random" setting, the machine changes speeds at regular intervals, so the total number of simulated "floors" she climbs varies from session to session. She also exercises for different lengths of time each session. She decides to explore the relationship between the number of minutes she works out on the stair climber and the number of floors it tells her that she's climbed.

Alana records minutes of climbing time and number of floors climbed for 6 exercise sessions.

Minutes	Floors	Minutes	Floors
15	73	20	103
16	82	22	109
18	88	25	127

(a) Sketch a scatterplot for these data and calculate the equation of the least-squares regression line using technology.

(b) Construct a residual plot for this model. Is a linear model appropriate? Explain.

(c) Using technology, calculate and interpret the standard deviation of the residuals for this model.

(d) Using technology, calculate and interpret the value of r^2 for this model.

Extending the Concepts

19. **Age and metric height** Refer to Exercise 5. Suppose that the heights of students were recorded in meters instead of centimeters (1 meter = 100 centimeters). How would this change affect the values of s and r^2? Explain.

20. **Form and s** The value of s alone doesn't reveal information about the form of an association. Sketch a scatterplot showing a nonlinear association with a small value of s and a second scatterplot showing a linear association with a large value of s.

Recycle and Review

21. **How tall is tall?** (1.9) According to the National Center for Health Statistics, the distribution of heights for 15-year-old males has a mean of 170 centimeters and a standard deviation of 7.5 centimeters. Paul is 15 years old and 179 centimeters tall.

(a) Find the z-score corresponding to Paul's height. Explain what this value means.

(b) Paul's height puts him at the 85th percentile among 15-year-old males. Explain what this means to someone who knows no statistics.

22. **More stair climbing** (2.5) Refer to Exercise 18.

(a) Using the regression equation you calculated in Exercise 18, part (a), predict the number of simulated "floors" Alana would climb in 21 minutes.

(b) Would you be willing to use the regression line to predict the number of floors Alana would climb in 35 minutes? Explain.

Lesson 2.8

Fitting Models to Curved Relationships

LEARNING TARGETS

- Use technology to calculate quadratic models for curved relationships, then calculate and interpret residuals using the model.
- Use technology to calculate exponential models for curved relationships, then calculate and interpret residuals using the model.
- Use residual plots to determine the most appropriate model.

When the association between two quantitative variables is linear, we use a least-squares regression line to model the relationship. When an association is nonlinear, we can use technology to calculate other types of models.

Quadratic Models

In some cases, we can use a **quadratic model** to fit a nonlinear association between two quantitative variables.

DEFINITION Quadratic model

A **quadratic model** is a model in the form $\hat{y} = ax^2 + bx + c$. The graph of a quadratic model is a parabola.

The scatterplot in Figure 2.15 shows the relationship between passing yards and age for 32 quarterbacks in the 2010 NFL season.[68]

FIGURE 2.15 Scatterplot showing the relationship between passing yards and age for 32 quarterbacks in the 2010 NFL season.

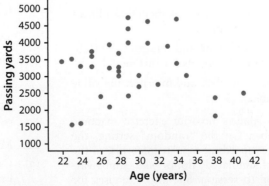

The association between passing yards and age is clearly nonlinear. Passing yards tend to be lower for younger and older quarterbacks and higher for "middle-aged" quarterbacks. The same scatterplot appears in Figure 2.16, along with the quadratic model $\hat{y} = -13.65x^2 + 835.5x - 9258$. The quadratic model fits the form of this association quite well.

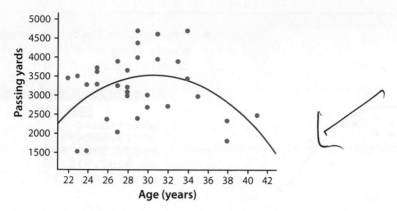

FIGURE 2.16 Scatterplot showing the relationship between passing yards and age for 32 quarter-backs in the 2010 NFL season, along with the quadratic model.

Brett Favre was 41 years old in 2010 and had 2509 passing yards. The quadratic model suggests that he would have $\hat{y} = -13.65(41)^2 + 835.5(41) - 9258 = 2052$ yards. This means that Favre had $2509 - 2052 = 457$ more passing yards than predicted, based on the quadratic model using $x =$ age. In other words, the residual for Brett Favre is 457 yards.

Calculating a Quadratic Model

You can use an applet or a graphing calculator to calculate the equation of a quadratic model. Here are the quarterback data.

Age	29	34	31	29	29	31	27	33
Passing yards	4710	4700	4620	4370	4002	3970	3922	3900
Age	25	28	25	23	22	34	25	24
Passing yards	3705	3653	3622	3512	3451	3377	3301	3291
Age	27	28	28	30	35	28	32	30
Passing yards	3274	3200	3116	3018	3001	3000	2734	2686
Age	41	29	26	38	27	38	24	23
Passing yards	2509	2387	2370	2365	2065	1823	1576	1558

Applet

Launch the *Two Quantitative Variables* applet at **highschool.bfwpub.com/spa3e**.

1. Enter "Age" for the explanatory variable and "Passing yards" for the response variable. Then, enter the data for both variables.

Two Quantitative Variables

Variable	Name	Observations (separated by commas or spaces) *Keep individuals in the same order.*
Explanatory	Age	29 34 31 29 29 31 27 33 25 28 25 23 22 34 25 24 27 28 28 30 35 28 32 30
Response	Passing yards	4710 4700 4620 4370 4002 3970 3922 3900 3705 3653 3622 3512

Begin analysis   Edit inputs   Reset everything

2. Press the "Begin analysis" button. This will display the scatterplot of passing yards versus age. In the Regression Models section, choose "Quadratic" from the drop down menu and press the "Calculate other regression model" button. The equation of the quadratic model, along with the values of *n*, *s*, and r^2, are shown in the output. In addition, the graph of the quadratic model has been added to the scatterplot and a residual plot has been created.

Regression Models

Calculate least-squares regression line

Calculate other regression model: Quadratic ▾   Calculate other regression model

Equation	n	s	r^2
$\hat{y} = -9257.916 + 835.5x - 13.65x^2$	32	802.124	0.176

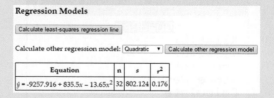

Scatterplot

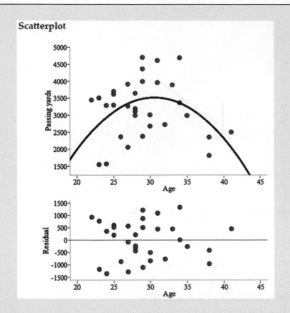

TI-83/84

1. Enter the ages in L_1 and the passing yards in L_2.

● Press STAT and choose Edit...

● Type the values into L_1 and L_2.

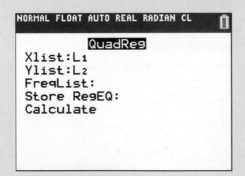

2. Calculate the equation of the quadratic model.

● Press STAT, arrow over to the CALC menu, and choose QuadReg.

● Adjust the settings as shown.

Older OS: Complete the command QuadReg L_1, L_2.

● Press Calculate. The output shows the values of a, b, and c, along with R^2. *Note:* Don't worry about the difference between R^2 and r^2.

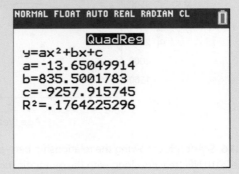

3. Display the quadratic model on a scatterplot.

● Press Y=, and enter the equation of the quadratic model in Y_1.

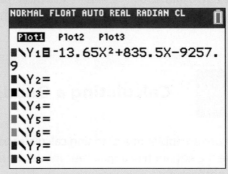

● Press 2nd Y= (STAT PLOT). Press ENTER or 1 to go into Plot1. Adjust the settings as shown.

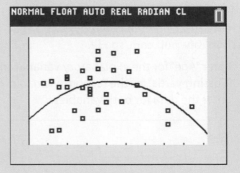

● Press ZOOM and choose 9:ZoomStat.

We interpret R^2 and s in the same way as for a linear model: 17.6% of the variability in passing yards is accounted for by the quadratic model using x = age. Also, when using the quadratic model with x = age, our predictions of passing yards are typically off by about 802 yards.

EXAMPLE

How does speed affect braking distance?

Calculating and using a quadratic model

How is the braking distance for a motorcycle related to the speed the motorcycle was going when the brake was applied? Statistics teacher Aaron Waggoner gathered data to answer this question. The table shows the speed (in miles per hour) and the distance needed to come to a complete stop when the brake was applied (in feet).

Speed (mph)	Distance (ft)	Speed (mph)	Distance (ft)
6	1.42	32	52.08
9	4.92	40	84.00
19	18.00	48	110.33
30	44.75		

PROBLEM:

(a) Calculate a quadratic model for these data using speed as the explanatory variable.

(b) Sketch the scatterplot, along with the quadratic model.

(c) Calculate and interpret the residual for the last observation in the table.

SOLUTION:

(a) Using the *Two Quantitative Variables* applet, the quadratic model is $\hat{y} = 0.043x^2 + 0.33x - 2.294$

(b) Here is the scatterplot, along with the quadratic model from the applet.

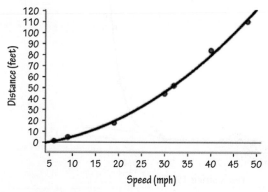

(c) $\hat{y} = 0.043\,(48)^2 + 0.33(48) - 2.294 = 112.6\,\text{feet}$

$y - \hat{y} = 110.33 - 112.6 = -2.27\,\text{feet}$

The braking distance for this trial was 2.27 feet less than expected, based on the quadratic model with x = speed.

> Calculate the predicted braking distance using the model. Then calculate the residual by subtracting the predicted braking distance from the actual braking distance.

FOR PRACTICE   TRY EXERCISE 1.

Exponential Models

We can also use an **exponential model** to fit a nonlinear association between two quantitative variables.

DEFINITION Exponential model

An **exponential model** is a model in the form $\hat{y} = a(b)^x$, where $b > 0$. If $b > 1$, the graph will show exponential growth. If $0 < b < 1$, the graph will show exponential decay.

Gordon Moore, one of the founders of Intel Corporation, predicted in 1965 that the number of transistors on an integrated circuit chip would double every 18 months. This is Moore's law, one way to measure the revolution in computing. Figure 2.17 is a scatterplot showing the number of transistors for Intel microprocessors and the number of years since 1970.[69]

FIGURE 2.17 Here is a scatterplot showing the number of transistors for Intel microprocessors and the number of years since 1970.

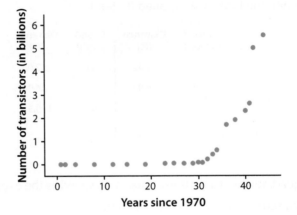

If Moore's prediction is correct, then an exponential model should be a good model for the data. Figure 2.18 is the same scatterplot, along with the exponential model $\hat{y} = 1277(1.431)^x$. The exponential model fits the form of this association fairly well.

FIGURE 2.18 Here is a scatterplot showing the number of transistors for Intel microprocessors and the number of years since 1970, along with the exponential model.

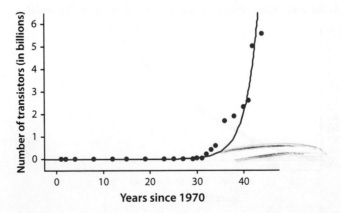

In 2014 there were 5,560,000,000 transistors on the Intel 18-core Xeon Haswell-E5 microprocessor. The exponential model suggests that it would have $\hat{y} = 1277(1.431)^{44}$ = 9,001,912,406 transistors. The residual for this value is 5,560,000,000 − 9,001,912,406 = −3,441,912,406 transistors. In other words, it has about 3.4 billion fewer transistors than expected, based on the exponential model.

Calculating an Exponential Model

You can use an applet or a graphing calculator to calculate the equation of an exponential model. Here are the Moore's law data.

Years since 1970	Number of transistors	Years since 1970	Number of transistors
1	2,300	30	42,000,000
2	3,500	31	45,000,000
4	4,500	32	220,000,000
8	29,000	33	410,000,000
12	134,000	34	592,000,000
15	275,000	36	1,700,000,000
19	1,180,235	38	1,900,000,000
23	3,100,000	40	2,300,000,000
25	5,500,000	41	2,600,000,000
27	7,500,000	42	5,000,000,000
29	9,500,000	44	5,560,000,000

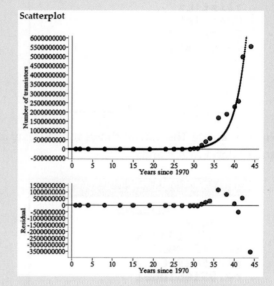

Applet

Launch the *Two Quantitative Variables* applet at **highschool.bfwpub.com/spa3e**.

1. Enter "Years since 1970" for the explanatory variable and "Number of transistors" for the response variable. Then, enter the data for each variable.

Two Quantitative Variables

Variable	Name	Observations (separated by commas or spaces) Keep individuals in the same order.
Explanatory	Years since 1970	1 2 4 8 12 15 19 23 25 27 29 30 31 32 33 34 36 38 40 41 42 44
Response	Number of transi	2300 3500 4500 29000 134000 275000 1180235 3100000 5500000

Begin analysis   Edit inputs   Reset everything

2. Press the "Begin analysis" button. This will display the scatterplot of number of transistors versus years since 1970. In the Regression Models section, choose "Exponential" from the drop down menu and press the "Calculate other regression model" button. The equation of the exponential model, along with the values of n, s, and R^2, are shown in the output. In addition, the graph of the exponential model has been added to the scatterplot and a residual plot has been created.

TI-83/84

1. Enter the years in L_1 and the numbers of transistors in L_2.

- Press $\boxed{\text{STAT}}$ and choose Edit…
- Type the values into L_1 and L_2.

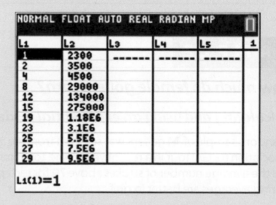

2. Calculate the equation of the exponential model.

- Press STAT, arrow over to the CALC menu, and choose ExpReg.
- Adjust the settings as shown.

Older OS: Complete the command ExpReg L₁, L₂.

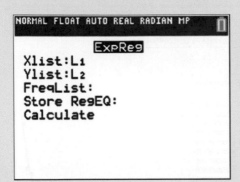

- Press Calculate. The output shows the values of *a* and *b*, along with r^2 and *r*.

3. Display the exponential model on a scatterplot.

- Press Y= and enter the equation of the exponential model in Y₁.

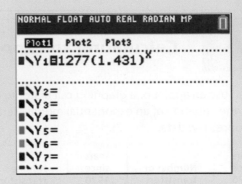

- Press 2nd Y= (STAT PLOT). Press ENTER or 1 to go into Plot1. Adjust the settings as shown.

- Press ZOOM and choose 9:ZoomStat.

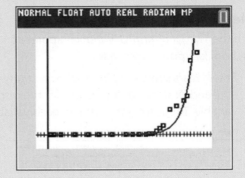

EXAMPLE

How much do female golfers earn?

Calculating and using an exponential model

A random sample of 14 golfers was selected from the 147 players on the Ladies Professional Golf Association (LPGA) tour in a recent year.[70] The total amount of money won during the year (in dollars) and the average number of strokes above 70 for each player in the sample were recorded. Lower scoring averages are better in golf.

Adjusted scoring average	Earnings ($)	Adjusted scoring average	Earnings ($)	Adjusted scoring average	Earnings ($)
4.88	22,927	4.10	36,403	1.02	448,048
3.57	44,332	2.78	121,347	2.98	70,790
1.86	459,449	1.38	636,734	4.50	17,944
1.87	104,603	2.79	212,352	0.16	1,626,297
3.41	122,540	2.08	149,872		

PROBLEM:

(a) Calculate an exponential model for these data using adjusted scoring average as the explanatory variable.

(b) Sketch the scatterplot, along with the exponential model.

(c) Calculate and interpret the residual for Lorena Ochoa, the last golfer in the list.

SOLUTION:

(a) Using the *Two Quantitative Variables* applet, the exponential model is $\hat{y} = 1,479,696(0.405)^x$.

(b) Here is the scatterplot, along with the exponential model from the applet.

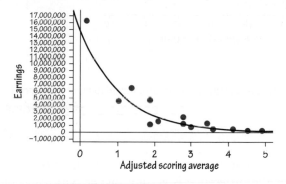

(c) $\hat{y} = 1,479,696(0.405)^{0.16} = \$1,280,458$

$y - \hat{y} = 1,626,297 - 1,280,458 = \$345,839$

The earnings for Lorena Ochoa were $345,839 greater than expected, based on the exponential model with x = adjusted scoring average.

> Calculate the predicted earnings using the model. Then calculate the residual by subtracting the predicted earnings from the actual earnings.

FOR PRACTICE   TRY EXERCISE 5.

Choosing a Model

In some cases, it is hard to tell if a quadratic model or an exponential model would be better to use. To decide, look at the residual plots for both models and choose the model with the residual plot that has the most random scatter.

EXAMPLE

How high can the cost of Super Bowl ads go?

Choosing a model

PROBLEM: In Lesson 2.7, we looked at the relationship between the cost of a 30-second commercial during the Super Bowl and the Super Bowl number. The scatterplot to the right clearly shows that the relationship is nonlinear. An exponential model and a quadratic model were calculated for the relationship between cost and Super Bowl number. Here are the residual plots for these models. Based on the residual plots, which model is more appropriate? Explain.

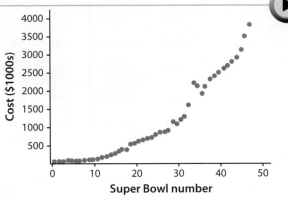

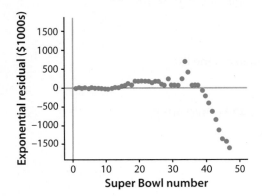
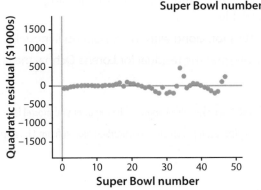

SOLUTION:

Because the residual plot for the quadratic model seems more randomly scattered, the quadratic model is more appropriate for these data than the exponential model.

FOR PRACTICE **TRY EXERCISE 9.**

For the most recent Super Bowls, predictions from the exponential model are consistently too high—and the size of the difference is getting bigger with each passing year. In Super Bowl 47, the prediction is off by more than $1.5 million! The residual plot for the quadratic model has some leftover patterns, but it is definitely more scattered than the exponential model—and the residuals are typically much smaller as well.

LESSON APP 2.8

How does life insurance work?

Many adults try to protect their families by buying life insurance. The policyholder makes regular payments (premiums) to the insurance company in return for the coverage. When the insured person dies, a payment is made to designated family members or other beneficiaries.

Age	Premium ($)
40	29
45	46
50	68
55	106
60	157
65	257

How do insurance companies decide how much to charge for life insurance? They rely on a staff of highly trained actuaries—people with expertise in probability, statistics, and advanced mathematics—to determine premiums. If someone wants to buy life insurance, the premium will depend on the type and amount of the policy as well as on personal characteristics like age, gender, and health status. The table shows monthly premiums (in dollars) for a 10-year term-life insurance policy worth $1,000,000 for people of various ages (in years).[71]

1. Calculate a quadratic model for these data.

2. Calculate an exponential model for these data.

3. Which model is more appropriate? Justify your answer.

4. Using your chosen model, calculate and interpret the residual for the 65-year-old.

Lesson 2.8

WHAT DID YOU LEARN?

LEARNING TARGET	EXAMPLES	EXERCISES
Use technology to calculate quadratic models for curved relationships, then calculate and interpret residuals using the model.	p. 163	1–4
Use technology to calculate exponential models for curved relationships, then calculate and interpret residuals using the model.	p. 166	5–8
Use residual plots to determine the most appropriate model.	p. 168	9–10

Exercises   *Lesson 2.8*

Mastering Concepts and Skills

1. **Diamonds are forever . . . and costly!** The table
pg 163 gives the price (in dollars) and weight (in carats) for 15 randomly selected round, clear, internally flawless diamonds with excellent cuts.[72]

Carat	Price ($)	Carat	Price ($)
0.50	4070	1.04	24,679
3.61	365,700	1.20	24,855
1.21	30,315	1.10	25,127
1.74	57,282	1.08	24,026
0.31	1404	1.61	47,126
3.56	329,613	1.29	32,315
1.51	45,928	1.13	25,075
3.01	229,509		

(a) Calculate a quadratic model for these data using carat weight as the explanatory variable.

(b) Sketch the scatterplot, along with the quadratic model.

(c) Calculate and interpret the residual for the diamond that weighed 3.61 carats.

2. **Tortoise eggs** A team of researchers measured the carapace (shell) length and clutch size (number of eggs) for female gopher tortoises in Okeeheelee County Park, Florida. Here are the data for 16 tortoises.[73]

Carapace length (mm)	Clutch size	Carapace length (mm)	Clutch size
284	3	310	10
290	2	311	13
290	7	317	9
298	11	320	6
299	12	323	13
302	10	334	2
307	8	334	8
309	9		

(a) Calculate a quadratic model for these data using carapace length as the explanatory variable. *Hint:* If you are using the applet, adjust the preferences to 5 decimal places for this exercise.

(b) Sketch the scatterplot, along with the quadratic model.

(c) Calculate and interpret the residual for the tortoise with carapace length 298 millimeters.

3. **Whip it up** Hall-of-Fame pitcher Bob Gibson of the St. Louis Cardinals was one of the most dominant pitchers during the 1960s. The table shows his age and WHIP (walks + hits per inning pitched) for each year of his career, beginning in 1959.[74] Lower WHIP values are better for pitchers.

Year	Age	WHIP	Year	Age	WHIP
1959	23	1.533	1968	32	0.853
1960	24	1.673	1969	33	1.102
1961	25	1.443	1970	34	1.190
1962	26	1.151	1971	35	1.185
1963	27	1.257	1972	36	1.129
1964	28	1.169	1973	37	1.108
1965	29	1.157	1974	38	1.417
1966	30	1.027	1975	39	1.670
1967	31	1.089			

(a) Calculate a quadratic model for these data using age as the explanatory variable.

(b) Sketch the scatterplot, along with the quadratic model.

(c) Calculate and interpret the residual for his famous 1968 season.

4. **Slow down and save** Fuel efficiency for many cars is related to how fast the car is driven. To explore this relationship, some researchers calculated the mean fuel efficiency for nine different car models at 13 different speeds. Here are the results.

Speed (mph)	Mean fuel efficiency (mpg)	Speed (mph)	Mean fuel efficiency (mpg)
15	24.4	50	32.4
20	27.9	55	32.4
25	30.5	60	31.4
30	31.7	65	29.2
35	31.2	70	26.8
40	31.0	75	24.8
45	31.6		

(a) Calculate a quadratic model for these data using speed as the explanatory variable.

(b) Sketch the scatterplot, along with the quadratic model.

(c) Calculate and interpret the residual for the cars traveling 60 miles per hour.

5. **European population growth** Many populations grow exponentially. Here are the data for the estimated population of Europe (in millions) from 1700 to 2012.[75] The dates are recorded as years since 1700 so that $x = 312$ is the year 2012.

Years since 1700	0	50	100	150	200
Population (in millions)	125	163	203	276	408
Years since 1700	250	299	308	310	312
Population (in millions)	547	729	732	738	740

(a) Calculate an exponential model for these data using years since 1700 as the explanatory variable.

(b) Sketch the scatterplot, along with the exponential model.

(c) Calculate and interpret the residual for the year 2012.

6. **North American population growth** Many populations grow exponentially. Here are the data for the estimated population of North America (in millions) from 1700 to 2012.[76] The dates are recorded as years since 1700 so that $x = 312$ is the year 2012.

Years since 1700	0	50	100	150	200
Population (in millions)	2	2	7	26	82
Years since 1700	250	299	308	310	312
Population (in millions)	172	307	337	345	351

(a) Calculate an exponential model for these data using years since 1700 as the explanatory variable.

(b) Sketch the scatterplot, along with the exponential model.

(c) Calculate and interpret the residual for the year 2012.

7. **Zapping bacteria** Expose marine bacteria to X-rays for time periods from 1 to 15 minutes. Here are the number of surviving bacteria (in hundreds) on a culture plate after each exposure time.[77]

Time t	Count y	Time t	Count y
1	355	9	56
2	211	10	38
3	197	11	36
4	166	12	32
5	142	13	21
6	106	14	19
7	104	15	15
8	60		

(a) Calculate an exponential model for these data using time as the explanatory variable.

(b) Sketch the scatterplot, along with the exponential model.

(c) Calculate and interpret the residual for the time $t = 8$.

8. **Exponential candy** A student opened a bag of M&M'S® Chocolate Candies, dumped them out, and ate all the ones with the "M" on top. When he finished, he put the remaining 30 M&M'S back in the bag and repeated the same process over and over until all the M&M'S were gone. Here are data on the number of M&M'S remaining at the end of each of the first 6 courses.

Course	1	2	3	4	5	6
Number remaining	30	13	10	3	2	1

(a) Calculate an exponential model for these data using course as the explanatory variable.

(b) Sketch the scatterplot, along with the exponential model.

(c) Calculate and interpret the residual for the first course.

9. **Slow down and chill out** In a fascinating study about the pace of life, M. H. Bernstein was able to show a positive relationship between the average walking speed of pedestrians in 15 cities around the world and the population size of the cities.[78] We calculated quadratic and exponential models to predict y = city size from x = walking speed. Here are the residual plots for these models. Based on the residual plots, which model is more appropriate? Explain.

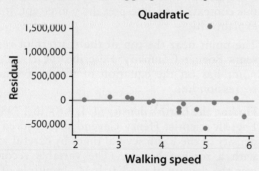

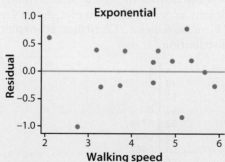

10. **Baseball salaries** The average salary of Major League Baseball players has been growing very rapidly since 1970.[79] A quadratic model and an exponential model were calculated to predict y = average salary from x = years since 1970. Here are the residual plots for each model. Based on the residual plots, which model is more appropriate? Explain.

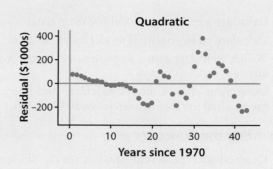

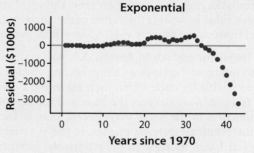

Applying the Concepts

11. **Growth in AP Statistics** The AP® Statistics exam was first given in 1997. The following data show the growth in the number of AP® Statistics exams from 1997 (Year 1) to 2014 (Year 18).[80]

Year	1	2	3	4	5	6
Exams	7667	15,486	25,240	34,118	40,259	49,824
Year	7	8	9	10	11	12
Exams	58,230	65,878	76,786	88,237	98,033	108,284
Year	13	14	15	16	17	18
Exams	116,876	129,899	142,910	153,589	169,508	184,752

(a) Calculate a quadratic model for these data.

(b) Calculate an exponential model for these data.

(c) Which model is more appropriate? Justify your answer.

(d) Using your chosen model, calculate and interpret the residual for 2013 (year = 17).

12. **Light through the water** Some college students collected data on the intensity of light at various depths in a lake. Here are their data.

Depth (m)	Light intensity (lumens)	Depth (m)	Light intensity (lumens)
5	168.00	9	44.34
6	120.42	10	31.78
7	86.31	11	22.78
8	61.87		

pg 168

(a) Calculate a quadratic model for these data.

(b) Calculate an exponential model for these data.

(c) Which model is more appropriate? Justify your answer.

(d) Using your chosen model, calculate and interpret the residual for the measurement at 11 meters.

Extending the Concepts

13. **Quarterbacks' peak year** At what age do NFL quarterbacks peak (have their greatest ability)? Estimate this value by finding the vertex of the quadratic model from page 160, $\hat{y} = -13.65x^2 + 835.5x - 9258$.

14. **Transistor growth** In Lesson 2.5, you learned how to interpret the slope as the predicted change in the y variable for each 1-unit increase in the x variable. You can also interpret the base b of an exponential function in the same way, except that it describes the predicted *percent* change in the y variable for each 1-unit increase in the x variable. Interpret the base ($b = 1.144171$) of the exponential model discussed on page 164, $\hat{y} = 1169.93(1.44171)^x$.

15. **Comparing growth rates** In Exercises 5 and 6, you calculated exponential models for the population growth in Europe and North America. Which continent had a greater growth rate? Justify your answer using the equations of the exponential models.

16. **Power models** In addition to quadratic and exponential models, another common type of model is called a *power model*. Power models are models in the form $\hat{y} = a \cdot x^p$. Here are data on the eight planets of our solar system. Distance from the sun is measured in astronomical units (AU), the average distance Earth is from the sun.[81]

Planet	Distance from sun (astronomical units)	Period of revolution (Earth years)
Mercury	0.387	0.241
Venus	0.723	0.615
Earth	1.000	1.000
Mars	1.524	1.881
Jupiter	5.203	11.862
Saturn	9.539	29.456
Uranus	19.191	84.070
Neptune	30.061	164.810

(a) Use the PwrReg command on your graphing calculator to find a power model relating y = period to x = distance.

(b) Sketch the scatterplot, along with the power model.

(c) Calculate and interpret the residual for Neptune.

Recycle and Review

17. **Past performance (2.5, 2.6)** Can we predict the growth of a company's stock price in one year from its growth during the previous year? The scatterplot gives the annual percent change in 2012 and 2013 for the 30 stocks that make up the Dow Jones Industrial Average.[82] The equation of the least-squares regression line for this relationship is $\hat{y} = 24.90 + 0.396x$, where x = percent change in 2012 and y = percent change in 2013.

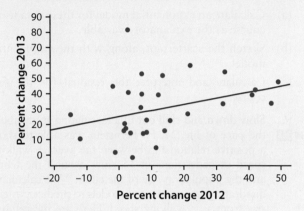

(a) Interpret the slope of the regression line.

(b) Does the value of the y intercept have meaning in this context? If so, interpret the y intercept. If not, explain why.

(c) The point near the top of the scatterplot represents Boeing Company. Describe the effect this point has on the equation of the least-squares regression line.

18. **Finding the Earth's density (1.4, 1.8)** In 1798 the English scientist Henry Cavendish measured the density of the Earth several times by careful work with a torsion balance. The variable recorded was the density of the Earth as a multiple of the density of water. The stemplot shows Cavendish's 29 measurements. Construct a boxplot for this distribution.[83]

```
48 | 8
49 |
50 | 7
51 | 0
52 | 6799
53 | 04469
54 | 2467
55 | 03578
56 | 12358
57 | 59
58 | 5
```
KEY: 48 | 8 = 4.88

Hiro 1775/Getty Images

Chapter 2 RESOLVED

STATS applied!

When will the cherry trees blossom?

In the STATS applied! on page 95, the Teague family was trying to plan a trip to Japan to see the cherry trees blossom. Their research suggested that the trees bloom earlier when the spring is warm and later when the spring is cool. With only a week to visit, when should the Teague family plan to arrive if they want to have the best chance to see the first blossom?

The table shows the average March temperature (in degrees Celsius) and the date of first blossom in April for a 24-year period.

Temperature (°C)	April days	Temperature (°C)	April days
4	14	4.3	13
5.4	8	1.5	28
3.2	11	3.7	17
2.6	19	3.8	19
4.2	14	4.5	10
4.7	14	4.1	17
4.9	14	6.1	3
4	21	6.2	3
4.9	9	5.1	11
3.8	14	5	6
4	13	4.6	9
5.1	11	4	11

1. Make a scatterplot to display the relationship between average March temperature and date of first blossom in April. Describe the association.

2. Calculate the equation of the least-squares regression line and display the line on your scatterplot.

3. Interpret the slope of the least-squares regression line.

4. Calculate and interpret the residual for the year when the average March temperature was 4°C and the first blossom was on April 14.

5. The Teague family learns that the average March temperature in the current year is 5°C. Predict the date of first blossom for the current year. By about how many days should they expect their prediction to be off? Explain.

Main Points   Chapter 2

Analyzing Two-Variable Data

- A **response variable** measures an outcome of a study. An **explanatory variable** may help predict or explain changes in a response variable. Explanatory and response variables can be either categorical or quantitative.

- Two variables have an **association** if knowing the value of one variable helps us predict the value of the other. If knowing the value of one variable does not help us predict the value of the other, there is no association between the variables.

- Even when two variables have an association, you shouldn't automatically conclude that there is a cause-and-effect relationship between them. **Association does not imply causation.**

Displaying Two-Variable Data

- A **segmented bar chart** displays the possible values of a categorical variable as segments of a rectangle, with the area of each segment proportional to the percent of individuals in the corresponding category.

- A **scatterplot** shows the relationship between two quantitative variables measured on the same individuals. The values of one variable appear on the horizontal axis, and the values of the other variable appear on the vertical axis. Each individual in the data appears as a point in the graph.

- To describe a scatterplot, look for the overall pattern and for striking departures from that pattern. You can describe the overall pattern of a scatterplot by the **direction, form,** and **strength** of the relationship. An important kind of departure is an **outlier,** an individual value that falls outside the overall pattern of the relationship.

Correlation

- The **correlation** r is a measure of the strength and direction of a linear relationship between two quantitative variables. The correlation takes values between -1 and 1, where positive values indicate a positive association and negative values indicate a negative association. Values closer to 1 and -1 indicate a stronger linear relationship and values closer to 0 indicate a weaker linear relationship.

- Calculate the correlation using technology or by hand using the formula

$$r = \frac{\sum \left(\dfrac{x_i - \bar{x}}{s_x} \right)\left(\dfrac{y_i - \bar{y}}{s_y} \right)}{n - 1} = \frac{\sum z_x z_y}{n - 1}$$

- The correlation makes no distinction between explanatory and response variables. Correlation has no units. Correlation isn't affected by changes in the units of the explanatory or response variable.

- The correlation isn't resistant—outliers can greatly influence the value of the correlation.

Modeling Linear Associations

- A **regression line** is a line that describes how a response variable y changes as an explanatory variable x changes. Regression lines are expressed in the form $\hat{y} = a + bx$, where $\hat{y}$ is the predicted value of y for a particular value of x.

- **Extrapolation** is the use of a regression line for prediction far outside the interval of x values used to obtain the line. Such predictions often are not accurate.

- A **residual** is the difference between an observed value of y and the value of y predicted by the regression line. That is,

$$\text{residual} = \text{actual } y - \text{predicted } y = y - \hat{y}$$

- The **slope** b of a regression line describes the *predicted* change in the y variable for each 1-unit increase in the x variable. The **y intercept** a is the *predicted* value of y when $x = 0$.

- The **least-squares regression line** is the line that makes the sum of the squared residuals as small as possible.

- Calculate the least-squares regression line using technology or by hand with the following formulas:

$$\text{slope} = b = r\frac{s_y}{s_x}$$

$$y \text{ intercept} = a = \bar{y} - b\bar{x}$$

- The least-squares regression line is not resistant—outliers can greatly influence the equation of the line.

- A **residual plot** is a scatterplot that plots the residuals on the vertical axis and the explanatory variable on the horizontal axis. If a residual plot shows no leftover patterns, the regression model is appropriate. If a residual plot shows a leftover pattern, the regression model is not appropriate.

- The **standard deviation of the residuals** s measures the size of a typical residual. That is, s measures the typical distance between the actual y values and the predicted y values.

- The **coefficient of determination** r^2 measures the percent reduction in the sum of squared residuals when using the least-squares regression line to make predictions rather than the mean value of y. In other words, r^2 measures the percent of the variability in the response variable that is accounted for by the least-squares regression line.

Modeling Nonlinear Associations

- A **quadratic model** is a model in the form $\hat{y} = ax^2 + bx + c$. The graph of a quadratic model is a parabola. Use technology to calculate a quadratic model.

- An **exponential model** is a model in the form $y = a \cdot b^x$. The graph of an exponential model can show either exponential growth or exponential decay. Use technology to calculate an exponential model.

- To decide which model is most appropriate, use the model with the most randomly scattered residual plot.

Chapter 2 Review Exercises begin on page 176

Chapter 2 **Review Exercises**

1. **Middle-school values (2.1)** Researchers carried out a survey of fourth-, fifth-, and sixth-grade students in Michigan. Students were asked if good grades, athletic ability, or being popular was most important to them. The two-way table summarizes the survey data.[84]

		Grade			
		4th grade	5th grade	6th grade	Total
What is most important?	Grades	49	50	69	168
	Athletic	24	36	38	98
	Popular	19	22	28	69
	Total	92	108	135	335

(a) Identify the explanatory and response variables in this context.

(b) Make a segmented bar chart to show the relationship between grade level and which goal was most important to students.

(c) Based on the graph, is there an association between these variables? Explain your reasoning. If there is an association, briefly describe it.

2. **Is crawling seasonal? (2.2, 2.3)** At what age do babies learn to crawl? Does it take longer to learn in the winter, when babies are often bundled in clothes that restrict movement? There might even be an association between babies' crawling age and the average temperature during the month when they first try to crawl (around 6 months after birth). Data were collected from parents who reported the birth month and the age at which their child was first able to creep or crawl a distance of 4 feet within 1 minute. Information was obtained on 414 infants, 208 boys and 206 girls. Average crawling age is given in weeks, and the average temperature (in degrees Fahrenheit) is for the month that is 6 months after the birth month.[85]

(a) Identify the explanatory and response variables. Explain your reasoning.

(b) Make a scatterplot to display the relationship between average 6-month temperature and average crawling age.

(c) Describe the relationship shown in the scatterplot.

(d) Give an estimate of the correlation for this relationship based on the scatterplot. Explain your reasoning.

Birth month	Average temperature (°F)	Average crawling age (weeks)
January	66	29.84
February	73	30.52
March	72	29.70
April	63	31.84
May	52	28.58
June	39	31.44
July	33	33.64
August	30	32.82
September	33	33.83
October	37	33.35
November	48	33.38
December	57	32.32

3. **More calculators (2.3)** The principal of a high school read a study that reported a positive correlation between the number of calculators owned by high school students and their math achievement. Based on this study, he decides to buy each student at his school 2 calculators, hoping to improve their math achievement. Explain the flaw in the principal's reasoning.

4. **The heights of dating (2.2, 2.3, 2.4)** A statistics student wonders if tall women tend to date taller men than do short women. She measures herself, her dormitory roommate, and the women in the adjoining rooms. Then, she measures the next man each woman dates. Here are the data (height in inches).

Women	66	64	66	65	70	65
Men	72	68	70	68	71	65

(a) Make a scatterplot for these data, using women's height as the explanatory variable.

(b) Calculate and interpret the correlation for these data.

(c) What effect does the pair (70, 71) have on the correlation? Explain.

(d) How would the correlation change if the heights of the women were measured in centimeters instead of inches?

5. **It's warmer down here (2.5)** The scatterplot shows the relationship between latitude and mean July temperature (in degrees Fahrenheit) for 12 cities in the United States.[86] The equation of the regression line relating these variables is $\hat{y} = 106.5 - 0.782x$.

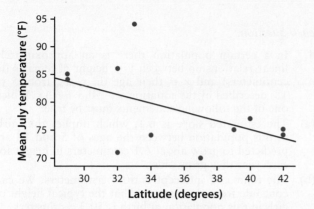

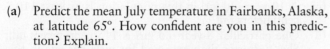

(a) Predict the mean July temperature in Fairbanks, Alaska, at latitude 65°. How confident are you in this prediction? Explain.

(b) Los Angeles, California, is at latitude 34° and has a mean July temperature of 74°. Calculate its residual.

(c) Interpret the slope of the regression line.

(d) Does the value of the y intercept have meaning in this context? If so, interpret the y intercept. If not, explain why.

6. **Crawl before you walk (2.6)** In Exercise 2, we investigated the relationship between the average temperature 6 months after birth (in degrees Fahrenheit) and the average age when babies were able to crawl (in weeks).

(a) Use technology to calculate the equation of the least-squares regression line for these data.

(b) Using your scatterplot from Exercise 2(b), describe what is unusual about the point representing May.

(c) How does the point representing May affect the equation of the least-squares regression line? Explain.

7. **Predicting test scores (2.6)** Each year, students in an elementary school take a standardized math test at the end of the school year. For a class of fourth-graders, the average score was 55.1, with a standard deviation of 12.3. In the third grade, these same students had an average score of 61.7, with a standard deviation of 14.0. The correlation between the two sets of scores is $r = 0.95$. Calculate the equation of the least-squares regression line for predicting a fourth-grade score from a third-grade score.

8. **It's even warmer down here (2.7)** In Exercise 5, we used a least-squares regression line to model the relationship between x = latitude and y = mean July temperature (in degrees Fahrenheit) for a sample of 12 cities in the United States. Here is the residual plot for this model.

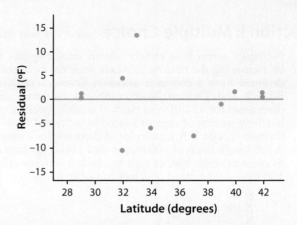

(a) Explain what the residual plot suggests about the appropriateness of the linear model.

(b) The standard deviation of the residuals for this model is $s = 6.4$. Interpret this value.

(c) The value of r^2 for this model is $r^2 = 0.277$. Interpret this value.

9. **Multiplying Microsoft employees (2.8)** For a number of years after it started up, Microsoft Corporation grew quite rapidly. The table shows the number of Microsoft employees and the number of years since Microsoft started up in 1976.[87]

Years since 1976	Employees	Years since 1976	Employees
0	7	6	220
1	9	7	476
2	13	8	608
3	28	9	910
4	40	10	1442
5	128		

(a) Sketch a scatterplot for these data.

(b) Calculate a quadratic model for these data using year as the explanatory variable.

(c) Calculate an exponential model for these data using year as the explanatory variable.

(d) Using each model, calculate and interpret the residual for the year 1981, 5 years after the company started.

(e) Explain how you could use residual plots to determine which model is better.

Section I: Multiple Choice *Select the best answer for each question.*

1. Biologists assess how closely related similar species are by measuring the number of years since the two species diverged from a common ancestor. Researchers Daniel Bolnick and Thomas Near compared the years since divergence for 12 different pairs of sunfish species to the hatching success of eggs produced by a "cross" between the two species.[88] A scatterplot of their results is shown. A hatching success of 100 means the hybrid eggs hatched as often as single-species eggs hatched. A number below 100 means the hybrid eggs hatch less often.

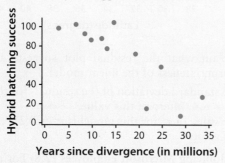

Which of the following is closest to the correlation between these two variables?

(a) −0.7   (b) −0.2   (c) 0.2   (d) 0.7

2. Which of the following statements is *not* true of the correlation r between the length in inches and weight in pounds of a sample of brook trout?

(a) r must take a value between −1 and 1.

(b) r is measured in inches.

(c) If longer trout tend to also be heavier, then $r > 0$.

(d) r would not change if we measured the lengths of the trout in centimeters instead of inches.

3. The scatterplot shows the relationship between life expectancy (in years) and Internet users (per 100 individuals in the population) for 35 countries in North and South America, along with the least-squares regression line. Haiti had a life expectancy of 45 years and 8.37 Internet users per 100 people.[89] What effect would removing this point have on the regression line?

(a) Slope would increase; y intercept would increase.

(b) Slope would increase; y intercept would decrease.

(c) Slope would decrease; y intercept would increase.

(d) Slope would decrease; y intercept would decrease.

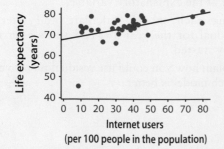

4. In a certain population there is an approximately linear relationship between y = height of females (in centimeters) and x = their age (in years, from 5 to 18) described by the equation $\hat{y} = 50.3 + 6.1x$. Which one of the following statements must be true?

(a) The estimated slope is 6.1, which implies that girls in this population between the ages of 5 and 18 are predicted to grow about 6.1 centimeters in height for each year they grow older.

(b) The estimated intercept is 50.3 centimeters. We can conclude from the intercept that the typical height of girls in this population at birth is 50.3 centimeters.

(c) A 10-year-old girl in this population is at least 110.3 centimeters tall.

(d) One particular 8-year-old girl is 115 centimeters tall. She is shorter than expected for girls her age in this population.

5. Scientists examined the activity level of 7 fish at different temperatures. The least-squares regression equation was calculated for the relationship between y = fish activity (rated as 0 = no activity and 100 = maximal activity) and x = water temperature in degrees Celsius. The regression calculation produced a value of $s = 4.785$. Which of the following gives a correct interpretation of s in this setting?

(a) For every 1°C increase in temperature, fish activity is predicted to increase by 4.785 units.

(b) The typical distance of the temperature readings from their mean is about 4.785°C.

(c) The typical distance of the activity level ratings from the least-squares line is about 4.785 units.

(d) The typical distance of the activity-level ratings from their mean is about 4.785 units.

6. The fraction of the variation in the values of y that is accounted for by the least-squares regression of y on x is

(a) the correlation.

(b) the slope of the least-squares regression line.

(c) the square of the correlation.

(d) the standard deviation of the residuals.

7. There is a positive correlation between the size of a hospital (measured by number of beds) and the median number of days that patients remain in the hospital. Does this mean that you can shorten a hospital stay by choosing to go to a small hospital?

(a) No; a negative correlation would allow that conclusion, but this correlation is positive.

(b) No; the positive correlation probably arises because seriously ill people go to large hospitals.

(c) Yes; the positive correlation establishes that a larger hospital size is the reason for longer stays.

(d) Yes; but only if r is very close to 1.

8. Derek Jeter, the New York Yankees' star shortstop, retired after the 2014 season. A linear regression model and quadratic regression model were calculated to predict y = the number of home runs Jeter hit in a given year from x = his age that year.[90] Here are the residual plots for each model. Which of the following statements is supported by these residual plots?

(a) The linear model is a better fit because there is no pattern in the residuals of the linear model.

(b) The linear model is a better fit because there is an upside-down, U-shaped pattern in the residuals of the linear model.

(c) The quadratic model is a better fit because there is no pattern in the residuals of the quadratic model.

(d) The quadratic model is a better fit because there is a U-shaped pattern in the residuals of the quadratic model.

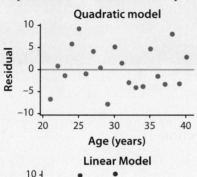

Quadratic model

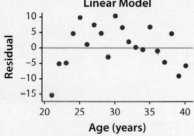

Linear Model

9. China's carbon dioxide emissions from the burning of fossil fuels have grown rapidly for many years. An exponential regression model was calculated to predict $y = CO_2$ emissions from fossil fuels (in metric tons) from x = years since 1902.[91] The regression equation is $\hat{y} = 5,250,100(1.072)^x$. In 1962 ($x = 60$), China's CO_2 emissions totaled 440,319,000 metric tons. Which of the following is the residual for this observation?

(a) $-340,294,317.5$ (c) $100,024,682.5$

(b) $-100,024,682.5$ (d) $340,294,317.5$

10. A school guidance counselor examines the number of extracurricular activities that students do and their grade point average. The guidance counselor says, "The evidence indicates that the correlation between the number of extracurricular activities a student participates in and his or her grade point average is close to zero." A correct interpretation of this statement would be that

(a) active students tend to be students with poor grades, and vice versa.

(b) students with good grades tend to be students who are not involved in many extracurricular activities, and vice versa.

(c) students involved in many extracurricular activities are just as likely to get good grades as bad grades; the same is true for students involved in few extracurricular activities.

(d) there is no linear relationship between number of activities and grade point average for students at this school.

Section II: Free Response

11. A survey was designed to study how business operations vary according to their size. Companies were classified as small, medium, or large. Questionnaires were sent to 200 randomly selected businesses of each size. Because not all questionnaires were returned, researchers decided to investigate the relationship between the response rate and the size of the business. The data are given in the two-way table.

		Business size			
		Small	Medium	Large	Total
Response?	Yes	125	81	40	246
	No	75	119	160	354
	Total	200	200	200	600

(a) Which variable should be used as the explanatory variable? Explain.

(b) Construct a segmented bar chart to summarize the relationship between business size and whether or not the business replied to the survey.

(c) Use your graph from part (b) to determine if there is an association between the size of business and whether or not the business replied to the survey.

12. The scatterplot displays the relationship between height and field-goal percentage for all the basketball players on the 2013 roster of the Brooklyn Nets.[92]

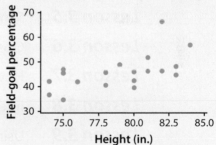

(a) Describe the association shown in the scatterplot.

(b) Use the following summary statistics to calculate the equation of the least-squares regression line: mean height = 78.9 inches, standard deviation of height = 3.29 inches, mean field-goal percentage = 45.3, standard deviation of field-goal percentage = 7.1, and correlation = 0.59.

13. Sarah's parents are concerned that she seems short for her age. Their doctor has the following record of Sarah's height:

Age (months)	36	48	51	54	57	60
Height (cm)	86	90	91	93	94	95

(a) Make a scatterplot of these data.

(b) Use technology to find the equation of the least-squares regression line to predict height from age.

(c) Use your regression line to predict Sarah's height at age 40 years (480 months). Convert your prediction to inches (2.54 centimeters = 1 inch).

(d) The prediction is impossibly large. Explain why this happened.

3

Collecting Data

Lesson 3.1	Introduction to Data Collection	182
Lesson 3.2	Sampling: Good and Bad	187
Lesson 3.3	Simple Random Samples	193
Lesson 3.4	Estimating a Margin of Error	204
Lesson 3.5	Sampling and Surveys	214
Lesson 3.6	Observational Studies and Experiments	220
Lesson 3.7	How to Experiment Well	228
Lesson 3.8	Inference for Experiments	235
Lesson 3.9	Using Studies Wisely	245
Chapter 3	Main Points	253
Chapter 3	Review Exercises	255
Chapter 3	Practice Test	257

180

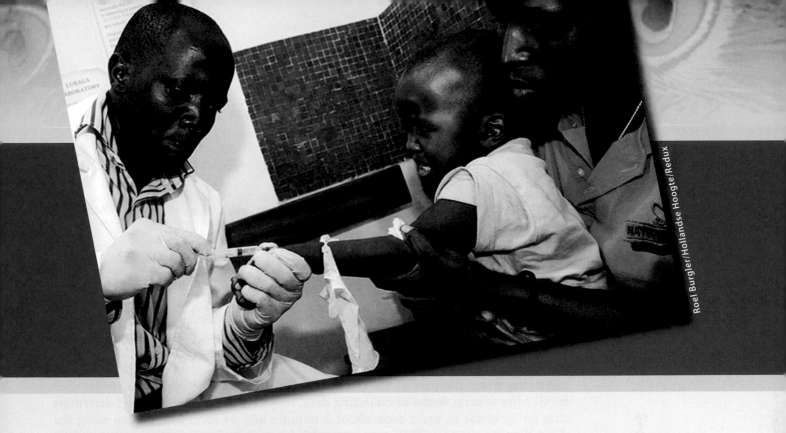

STATS applied!

How can we prevent malaria?

Malaria causes hundreds of thousands of deaths each year, with many of the victims being children. Will regularly screening children for the malaria parasite and treating those who test positive reduce the proportion of children who develop the disease? Researchers worked with children in 101 schools in Kenya, randomly assigning half of the schools to receive regular screenings and follow-up treatments and the remaining schools to receive no regular screening. Children at all 101 schools were tested for malaria at the end of the study.[1]

> If the proportion of children who develop the disease is smaller in the group that received the screening, can we conclude that the screening caused the decrease in malaria? Can we apply the results of this study to children in other countries?

We'll revisit STATS applied! at the end of the chapter, so you can use what you have learned to help answer these questions.

Introduction to Data Collection

L E A R N I N G   T A R G E T S

- Distinguish statistical questions from other types of questions.
- Identify the population and sample in a statistical study.
- Distinguish between an observational study and an experiment.

Tommy and Hannah go off campus to eat lunch every day. Because the lunch period is short, they wonder if it would be faster to go inside their favorite fast-food restaurant to order or use the drive-thru. Each day, they flip a coin to determine which method (inside or drive-thru) to use and record the total length of time it takes from the moment they enter the parking lot to the moment they exit the parking lot with their food. After several weeks of collecting data, they analyze their results and determine that going inside to order took about 2 minutes and 34 seconds less than using the drive-thru, on average. Their conclusion: It's faster to go inside.

Asking Statistical Questions

Tommy and Hannah's study illustrates the four steps of the statistical problem-solving process.

How to Complete the Statistical Problem-Solving Process

- **Ask questions:** Clarify the research problem and ask one or more valid statistical questions.
- **Collect data:** Design and carry out an appropriate plan to collect the data.
- **Analyze data:** Use appropriate graphical and numerical methods to analyze the data.
- **Interpret results:** Draw conclusions based on the data analysis.

A statistics problem starts with a statistical question. Not just any question will do. *A valid statistical question is based on data that vary.* That is, the answer to the question won't be the same each time an observation is recorded. Tommy and Hannah asked a valid statistical question because their answer was based on data that vary—the length of time it takes to order and receive their food, from the moment they enter the parking lot to the moment they leave it.

E X A M P L E

To be or not to be?

Asking valid statistical questions

PROBLEM: Determine if each of the following is a valid statistical question. Justify your answer.

(a) How much money does Mr. Tagawa have in his wallet right now?

(b) How long can students at your school hold their breath?

(c) Do most U.S. adults engage in vigorous exercise at least once a week?

SOLUTION:

(a) Not valid. This question can be answered with a single value that doesn't vary.

(b) Valid. The answer to this question is based on data that vary—the length of time that each student at your school can hold his or her breath.

(c) Valid. The answer to this question is based on data that vary—some adults exercise vigorously and some do not.

Part (a) would be a valid statistical question if it were asked in a slightly different way: "How much money does Mr. Tagawa typically have in his wallet?" The answer to this question is based on data that vary from day to day.

FOR PRACTICE   TRY EXERCISE 1.

We have just discussed the first step of the statistical problem-solving process: ask questions. The rest of this chapter addresses the second step of the statistical problem-solving process: collect data.

Populations and Samples

Suppose we want to find out what percent of young drivers in the United States text while driving. To answer the question, we will survey 16- to 20-year-olds who live in the United States and drive. Ideally, we would ask them all (take a **census**). But contacting every driver in this age group wouldn't be practical. It would take too much time and cost too much money. Instead, we put the question to a **sample** chosen to represent the entire **population** of young drivers.

DEFINITION Population, Census, Sample

The **population** in a statistical study is the entire group of individuals we want information about.

A **census** collects data from every individual in the population.

A **sample** is a subset of individuals in the population from which we collect data.

EXAMPLE

Samples of students and sodas?

Populations and samples

PROBLEM: Identify the population and sample in each of the following settings.

(a) The student government at a high school surveys 100 students at the school to get their opinions about a change to the bell schedule.

(b) The quality-control manager at a bottling company selects 10 cans from the production line every hour to see if the volume of soda is within acceptable limits.

SOLUTION:

(a) Population: All students at the school. Sample: The 100 students surveyed.

(b) Population: All cans filled that hour. Sample: The 10 cans inspected.

FOR PRACTICE   TRY EXERCISE 5.

Observational Studies and Experiments

A sample survey usually aims to gather information about a population without disturbing the population in the process. Sample surveys are one kind of **observational study.** Other observational studies watch the behavior of animals in the wild or track the medical history of patients to look for associations between variables such as diet, exercise, and heart disease.

In contrast to observational studies, **experiments** don't just observe individuals or ask them questions. They actively *impose* some treatment to measure the response. Experiments can answer questions like "Does aspirin reduce the chance of a heart attack?" and "Do plants grow better when classical music is playing?"

DEFINITION Observational study, Experiment

An **observational study** observes individuals and measures variables of interest, but does not attempt to influence the responses.

An **experiment** deliberately imposes some treatment on individuals to measure their responses.

The goal of an observational study can be to describe some group or situation, to compare groups, or to examine relationships between variables. The purpose of an experiment is to determine if the treatment causes a change in the response. An observational study, even one based on a random sample, is a poor way to gauge the effect that changes in one variable have on another variable. To see the response to a change, researchers must impose the change. *When the goal is to understand cause and effect, experiments are the only source of fully convincing data.* For this reason, the distinction between observational study and experiment is one of the most important ideas in statistics.

EXAMPLE

What affects blood pressure and memory?

Observational studies and experiments

PROBLEM: Determine if each of the following settings describes an observational study or an experiment. Explain your reasoning.

(a) Ninety residents of a retirement community were selected at random. Each member of the sample was asked a number of questions, including questions about exercise and blood pressure. Residents who exercised more often were less likely to have high blood pressure.

(b) Students in a statistics class were divided into two groups at random. Students in one group were given a set of words to memorize in a room with music playing. Students in the other group were given the same list of words to memorize, but in a quiet room. Students in the quiet room were able to remember more words, on average.

SOLUTION:

(a) Observational study. There were no treatments imposed on residents of the retirement community. In other words, residents weren't told to exercise a certain number of times per week.

(b) Experiment. Treatments were imposed on the students. Some students were assigned to memorize the words in a room with music playing and others were assigned to memorize the words in a quiet room.

FOR PRACTICE   TRY EXERCISE 11.

LESSON APP 3.1

Do you have dinner plans?

Researchers at Columbia University randomly selected 1000 teenagers in the United States for a survey. According to an ABC News article about the research, "Teenagers who eat with their families at least five times a week are more likely to get better grades in school."[2]

1. What is the statistical question that the researchers were trying to answer?

2. Identify the population and sample.

3. Is this an observational study or an experiment? Explain.

© Ariel Skelley/Blend Images/Corbis

Lesson 3.1

WHAT DID YOU LEARN?

LEARNING TARGET	EXAMPLES	EXERCISES
Distinguish statistical questions from other types of questions.	p. 182	1–4
Identify the population and sample in a statistical study.	p. 183	5–10
Distinguish between an observational study and an experiment.	p. 184	11–16

Exercises *Lesson 3.1*

Mastering Concepts and Skills

1. **Money problems** Determine if each of the following
pg 182 is a valid statistical question. Justify your answer.

(a) How much money do high-school students carry with them?

(b) How many quarters equal $10?

2. **Elm trees** Determine if each of the following is a valid statistical question. Justify your answer.

(a) How tall is the elm tree in Laura's backyard?

(b) How tall are the elm trees on Elm Street?

3. **Political questions** Determine if each of the following is a valid statistical question. Justify your answer.

(a) How many electoral votes did Barack Obama receive in 2012?

(b) What percent of U.S. residents think the president is doing a good job?

4. **Ice cream** Determine if each of the following is a valid statistical question. Justify your answer.

(a) What proportion of New Jersey residents would choose vanilla as their favorite ice cream flavor?

(b) What is the governor's favorite ice cream flavor?

5. **Sampling hardwood** A furniture maker buys hard-
pg 183 wood in large batches. The supplier is supposed to dry the wood before shipping (wood that isn't dry won't hold its size and shape). The furniture maker chooses 5 pieces of wood from each batch and tests their moisture content. If any piece exceeds 12% moisture content, the entire batch is sent back. Identify the population and sample in this setting.

6. **Sampling humans** Each week, the Gallup Poll surveys about 1500 adult U.S. residents to determine current opinions on a wide variety of issues. Identify the population and sample in this setting.

7. **Sampling artifacts** An archaeological dig turns up large numbers of pottery shards, broken stone tools, and other artifacts. Students working on the project classify each artifact and assign it a number. The counts in different categories are important for understanding the site, so the project director chooses 2% of the artifacts at random and checks the students' work. Identify the population and sample in this setting.

8. **Sampling envelopes** A large retailer prepares its customers' monthly credit card bills using a machine that folds the bills, stuffs them into envelopes, and seals the

envelopes for mailing. Are the envelopes completely sealed? Inspectors choose 40 envelopes at random from the 1000 stuffed each hour for visual inspection. Identify the population and sample in this setting.

9. **Sampling customers** A department store mails a customer-satisfaction survey to people who make credit card purchases at the store. This month, 45,000 people made credit card purchases. Surveys are mailed to 1000 of these people, chosen at random, and 137 people return the survey form. Identify the population and sample in this setting.

10. **Sampling orange juice** How much vitamin C does orange juice contain? A nutrition magazine measures the amount of vitamin C in 50 half-gallon containers of a popular brand of orange juice from 10 different grocery stores and concludes that the containers produced by this company do not have as much vitamin C as advertised. Identify the population and sample in this setting.

11. **Teaching biology** An educator wants to compare the effectiveness of computer software for teaching biology with that of a textbook presentation. She gives a biology pretest to each of a group of high school juniors, then randomly divides them into two groups. One group uses the computer, and the other studies the text. At the end of the year, she tests all the students again and compares the increase in biology test scores in the two groups. Is this an observational study or an experiment? Explain your reasoning.
pg 184

12. **Positively chocolate** A University of Helsinki (Finland) study wanted to determine if chocolate consumption during pregnancy had an effect on infant temperament at age 6 months. Researchers began by asking 305 healthy pregnant women to report their chocolate consumption. Six months after birth, the researchers asked mothers to rate their infants' temperament, including smiling, laughter, and fear. The babies born to women who had eaten chocolate daily during pregnancy were found to be more active and "positively reactive"—a measure that the investigators said encompasses traits like smiling and laughter. Is this an observational study or an experiment? Explain your reasoning.[3]

13. **Battery life** Does reducing screen brightness increase battery life in laptop computers? To find out, researchers obtained 30 new laptops of the same brand. They chose 15 of the computers at random and adjusted their screens to the brightest setting. The other 15 laptop screens were left at the default setting—moderate brightness. Researchers then measured how long each machine's battery lasted. Is this an observational study or an experiment? Explain your reasoning.

14. **Effects of child care** A study of child care enrolled 1364 infants and followed them through their sixth year in school. Later, the researchers published an article in which they stated that "the more time children spent in child care from birth to age four-and-a-half, the more adults tended to rate them, both at age four-and-a-half and at kindergarten, as less likely to get along with others, as more assertive, as disobedient, and as aggressive." Is this an observational study or an experiment? Explain your reasoning.[4]

15. **Lack of sleep** Does lack of sleep affect your academic performance? A teacher explores this question by asking everyone in his statistics class to write down on a piece of paper the total number of hours he or she slept on each of the last three nights before taking a test. He divides the students into those who slept for an average of 8 or more hours each night and those who slept for less. Then he compares the average test scores of the two groups. Is this an observational study or an experiment? Explain your reasoning.

16. **Stopping strokes** Aspirin prevents blood from clotting and so helps prevent strokes. The Second European Stroke Prevention Study asked if adding another anticlotting drug named dipyridamole would help. Patients who had already had a stroke were randomly assigned to receive either aspirin only, dipyridamole only, both, or a placebo. Is this an observational study or an experiment? Explain your reasoning.[5]

Applying the Concepts

17. **The buzz about coffee** A recent study of over 400,000 older people in eight states asked participants whether or not they drank coffee, and followed them for 13 years. The researchers found that coffee drinkers tend to live longer than non-coffee drinkers.[6]

(a) What is the statistical question that the researchers were trying to answer?

(b) Identify the population and sample.

(c) Is this an observational study or an experiment? Explain.

18. **Paid to quit** In an effort to reduce health care costs, General Motors sponsored a study to help employees stop smoking. In the study, 439 volunteer subjects were randomly assigned to receive up to $750 for quitting for a year, while the other 439 volunteer subjects were simply encouraged to use traditional methods to stop smoking. After one year, people who had the financial incentive were 3 times more likely to have quit smoking.[7]

(a) What is the statistical question that the researchers were trying to answer?

(b) Identify the population and sample.

(c) Is this an observational study or an experiment? Explain.

Extending the Concepts

19. **Eating habits** Write three different statistical questions that you could ask about the eating habits of college students.

20. Technology use Write three different statistical questions that you could ask about technology use among senior citizens.

Recycle and Review

21. Working hard (1.6) The pie chart shows the distribution of hours in the workweek for a sample of 1271 full-time workers in the United States.

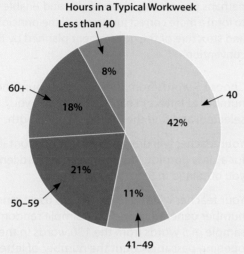

Hours in a Typical Workweek

Less than 40

60+ → 18%

50–59 → 21%

41–49 → 11%

8%

40

42%

Full-time workers put in 47 hours a week, on average.

(a) Estimate the median number of work hours per week in this sample. Explain your reasoning.

(b) These data are a "categorical" version of a quantitative variable. Taking into account that the legend says, "Full-time workers put in an average of 47 hours a week," would you say that the distribution of the underlying quantitative variable is symmetric, skewed right, or skewed left? Justify your answer.

22. Return of the sparrowhawk (2.2, 2.5, 2.6, 2.7) Is there an association between the percent of adult birds in a colony that return from the previous year and the number of new adults that join the colony? Research on 13 colonies of sparrowhawks produced the following data.[8]

Percent return	74	66	81	52	73	62	52
New adults	5	6	8	11	12	15	16
Percent return	45	62	46	60	46	38	
New adults	17	18	18	19	20	20	

(a) Sketch a scatterplot of these data, using percent return as the explanatory variable.

(b) Use technology to calculate a least-squares linear regression equation to predict the number of new adults from the percent of adults returning from the previous year.

(c) Interpret the slope of the least-squares regression line.

(d) Interpret the value $s = 3.67$.

Lesson 3.2
Sampling: Good and Bad

LEARNING TARGETS

- Describe how convenience sampling can lead to bias.
- Describe how voluntary response sampling can lead to bias.
- Explain how random sampling can help to avoid bias.

Many statistical studies use information from a sample to make a conclusion about an entire population. To ensure that these conclusions are accurate, we must be mindful of how the sample is selected.

ACTIVITY

Who wrote the Federalist Papers?

In this activity, you will learn how statistics can be used to help identify the author of an anonymous text.

The Federalist Papers are a series of 85 essays supporting the ratification of the U.S. Constitution. When newspapers in New York published the essays in 1787 and 1788, the identity of the authors was a secret known to just a few people. Later, the authors were identified as Alexander Hamilton, James Madison, and John Jay. The authorship of 73 of the essays is fairly certain, leaving 12 in dispute. Thanks in some part to statistical analysis, most scholars now believe that these 12 essays were written by Madison alone or in collaboration with Hamilton.[9]

There are several ways to use statistics to help determine the authorship of a text. One method is to estimate the average word length in a disputed text and compare it to the average word lengths of works where the authorship is not in dispute.

The following paragraph is the opening of Federalist Paper No. 51, one of the disputed essays. The theme of this essay is the separation of powers among the three branches of government.

> To what expedient, then, shall we finally resort, for maintaining in practice the necessary partition of power among the several departments, as laid down in the Constitution? The only answer that can be given is, that as all these exterior provisions are found to be inadequate, the defect must be supplied, by so contriving the interior structure of the government as that its several constituent parts may, by their mutual relations, be the means of keeping each other in their proper places. Without presuming to undertake a full development of this important idea, I will hazard a few general observations, which may perhaps place it in a clearer light, and enable us to form a more correct judgment of the principles and structure of the government planned by the convention.

1. Choose 5 words from this passage. Count the number of letters in each of the words you selected, and find the average word length.

2. Your teacher will draw and label a horizontal axis for a class dotplot. Plot the average word length you obtained in Step 1 on the graph.

3. Your teacher will show you how to use a random number generator to select a simple random sample of 5 words from the 130 words in the opening passage. Count the number of letters in each of the words you selected, and find the average word length.

4. Your teacher will draw and label another horizontal axis with the same scale for a comparative class dotplot. Plot the average word length you obtained in Step 3 on the graph.

5. How do the dotplots compare? Can you think of any reasons why they might be different? Discuss with your classmates.

How to Sample Poorly: Convenience Samples

Suppose we want to know how much time students at a large high school spent doing homework last week. We might go to the school library and ask the first 30 students we see about their homework time. The sample we get is called a **convenience sample.**

‖ DEFINITION Convenience sample

Choosing individuals from the population who are easy to reach results in a **convenience sample.**

Convenience sampling often produces unrepresentative data. Students who hang out in the library probably spend more time doing homework than a typical student. Our estimate for the average amount of time spent doing homework will be too high. In fact, if we were to repeat this sampling process again and again, we would almost always overestimate the average time spent doing homework in the population of all students. This predictable overestimation is due to **bias** in the sampling method.

> **DEFINITION** Bias
>
> The design of a statistical study shows **bias** if it would consistently underestimate or consistently overestimate the value you want to know when the study is repeated many times.

Bias is not just bad luck in one sample. It's the result of a bad study design that will consistently miss the truth about the population in the same direction.

EXAMPLE

Orange you thirsty?

Convenience samples

PROBLEM: A farmer brings a juice company many crates of oranges each week. A company inspector looks at 10 oranges from the top of each crate before deciding whether to buy all the oranges. Explain why this sampling method is biased. Is the proportion of damaged oranges in the sample likely greater than or less than the proportion of all oranges in the crate that are damaged?

eyespy/Getty Images

SOLUTION:

It is likely that the oranges on top of the crate are in better condition than the oranges in the rest of the crate. Perhaps the farmer put the best oranges on top. Or the oranges on bottom may be damaged due to the weight of the oranges on top of them. This means that the sample proportion of damaged oranges is likely to be less than the proportion of all oranges that are damaged.

FOR PRACTICE TRY EXERCISE 1.

How to Sample Poorly: Voluntary Response Samples

Many websites have polls that ask visitors to express their opinions. These polls, along with other call-in or write-in polls, are not reliable because they use **voluntary response samples**.

> **DEFINITION** Voluntary response sample
>
> A **voluntary response sample** consists of people who choose to be in the sample by responding to a general invitation. Voluntary response samples are sometimes called *self-selected samples*.

Ann Landers once asked the readers of her long-running advice column, "If you had it to do over again, would you have children?" She received nearly 10,000 responses, almost 70% saying "NO!" Can it be true that 70% of parents regret having children? Not at all. People who feel strongly about an issue, particularly people with strong negative feelings, are more likely to take the trouble to respond. Ann Landers's results are misleading—the percent of parents who would not have children again is much higher in her sample than in the population of all parents.

EXAMPLE

Is that bias pushing up the mean?

Voluntary response samples

PROBLEM: The athletic department at El Dorado Community College wants to learn more about the physical fitness of students at the school, including the number of push-ups that students can perform. To obtain their sample, the athletic department sets up a booth in the center of campus with a sign that says "Free Physical Fitness Testing!" Explain why this sampling method is biased. Is the mean number of push-ups for the members of their sample likely to be greater than or less than the mean number of push-ups for the entire student body?

SOLUTION:

It is likely that only physically fit students will volunteer for the testing. Students who aren't in good shape are unlikely to voluntarily take a physical fitness test in the center of campus. As a result, the mean number of push-ups for the members of the sample is likely to be greater than the mean number of push-ups for the entire student body.

FOR PRACTICE   TRY EXERCISE 5.

How to Sample Well: Random Samples

In convenience sampling, the researcher chooses easy-to-reach members of the population. In voluntary response sampling, people decide whether to join the sample. Both sampling methods suffer from bias due to personal choice. The best way to avoid this problem is to let chance choose the sample. That's the idea of **random sampling.**

> **DEFINITION** Random sampling
>
> **Random sampling** involves using a chance process to determine which members of a population are included in the sample.

In the Federalist Papers activity, the second sample you selected was a random sample. Using a random sample helped avoid bias. In random samples of $n = 5$ words, the sample means won't be consistently too high or consistently too low—about half of the sample means should be less than the true mean and about half should be greater than the true mean.

EXAMPLE

Can we get some unbiased push-ups?

Random samples

PROBLEM: Explain how the athletic department at El Dorado Community College can avoid the bias identified in the previous example.

SOLUTION:

Instead of recruiting volunteers in the center of campus, the athletic department could obtain a list of all students and randomly select 100 for physical-fitness testing. Because no personal choice is involved, this sample should be more representative of the entire student body.

FOR PRACTICE   TRY EXERCISE 9.

LESSON APP 3.2

Still on the phone?

In June 2008 *Parade* magazine posed the following question: "Should drivers be banned from using all cell phones?" Readers were encouraged to vote online at www.parade.com. The July 13, 2008, issue of *Parade* reported the results: 2407 (85%) said "Yes" and 410 (15%) said "No."

1. What type of sample did the *Parade* survey obtain?

2. Explain why this sampling method is biased. Is 85% likely to be greater than or less than the percentage of all adults who believe that cell-phone use while driving should be banned? Why?

3. Explain how *Parade* magazine could avoid the bias described in Question 2.

Lesson 3.2

WHAT DID YOU LEARN?

LEARNING TARGET	EXAMPLES	EXERCISES
Describe how convenience sampling can lead to bias.	p. 189	1–4
Describe how voluntary response sampling can lead to bias.	p. 190	5–8
Explain how random sampling can help to avoid bias.	p. 190	9–12

Exercises

Lesson 3.2

Mastering Concepts and Skills

1. **Sampling mall shoppers** You have probably seen the mall interviewer, clipboard in hand, approaching people passing by. Explain why even a large sample of mall shoppers would not provide a trustworthy estimate of the current unemployment rate in the city where the mall is located. Is the proportion of unemployed people in the sample likely greater than or less than the proportion of all people who are unemployed?
   pg 189

2. **Funding for fine arts** The band director at a high school wants to know what percentage of parents supports a decrease in the budget for fine arts. Because many parents attend the school's annual musical, the director surveys the first 30 parents who arrive at the show. Explain why this sampling method is biased. Is the proportion of parents in the sample who support the decrease likely greater than or less than the proportion of all parents in the school who support the decrease?

3. **Students sleeping** How much sleep do high-school students get on a typical school night? An interested student designed a survey to find out. To make data collection easier, the student surveyed the first 100 students to arrive at school on a particular morning. These students reported an average of 7.2 hours of sleep on the previous night. Explain why this sampling method is biased. Is 7.2 hours likely greater than or less than the true average amount of sleep last night for all students at the school?

4. **Quick quality check?** The quality-control department at an automobile factory checks the tightness of motor-mounting bolts installed by assembly-line workers by sampling the first 25 cars produced by the assembly line each day. Explain why this sampling method is biased. Is this method likely to overestimate or underestimate the proportion of bolts that are improperly tightened?

5. **Yelping reviews** Many websites include customer reviews of products, restaurants, hotels, and so on. The manager of a hotel was upset to see that 26% of reviewers on a travel website gave the hotel a "1 star" rating—the lowest possible rating. Explain why this sampling method is biased. Is 26% likely greater
   pg 190

than or less than the percentage of all the hotel's customers who would give the hotel 1 star? Explain.

6. **Explain to the senator** You are on the staff of a member of the U.S. Senate who is considering a bill that would provide government-sponsored insurance for nursing-home care. You report that 1128 letters and emails have been received on the issue, of which 871 (77%) oppose the legislation. "I'm surprised that most of my constituents oppose the bill. I thought it would be quite popular," says the senator. Explain why this sampling method is biased. Is 77% likely greater than or less than the percentage of all the senator's constituents who oppose the legislation?

7. **Letters to the editor** Newspaper readers sometimes write letters to the editor, hoping that their opinions will be published in the paper. The editorial staff at a local newspaper keeps track of the content of each letter that is submitted. After a controversial decision made by the mayor, 90% of the letters to the editor expressed opposition to the mayor's action. Explain why this sampling method is biased. Is 90% likely greater or less than the percentage of all citizens who oppose the mayor's action?

8. **Illegal immigration** TV commentator Lou Dobbs doesn't like illegal immigration. One of his shows was largely devoted to attacking a proposal to offer driver's licenses to illegal immigrants. During the show, Dobbs invited his viewers to loudobbs.com to vote on the question "Would you be more or less likely to vote for a presidential candidate who supports giving driver's licenses to illegal aliens?" The result: 97% of the 7350 people who voted by the end of the show said, "Less likely." Explain why this sampling method is biased. Is 97% likely greater than or less than the percentage of all Americans who would say "Less likely"?

9. **Better sampling at the mall** Explain how to avoid
pg **190** the bias you described in Exercise 1.

10. **Finding funds for fine arts** Explain how to avoid the bias you described in Exercise 2.

11. **Unbiased reviews** Explain how to avoid the bias you described in Exercise 5.

12. **Help the senator** Explain how to avoid the bias you described in Exercise 6.

Applying the Concepts

13. **Who has a driver's license?** Sammy wants to know what percent of students at her high school have a driver's license. She surveys all students in her statistics class and finds that 68% of the students in her sample have a driver's license.

(a) What type of sample did Sammy obtain?

(b) Explain why this sampling method is biased. Is 68% likely to be greater than or less than the percent of all students at her high school who have a driver's license?

(c) Explain how Sammy could avoid the bias described in part (b).

14. **Environmental action** Chris is the president of an environmental club at his school. To advertise, he posts flyers around campus that ask, "Want to save the environment? Make your voice heard Wednesdays at lunch in room 502." At the next meeting, he polls those in attendance and finds that 80% of students want to ban the sale of plastic water bottles on campus.

(a) What type of sample did Chris obtain?

(b) Explain why this sampling method is biased. Is 80% likely to be greater than or less than the percent of all students at Chris's high school who want to ban the sale of plastic water bottles on campus?

(c) Explain how Chris could avoid the bias described in part (b).

15. **Tablets for all** To fund a program to buy electronic tablets for every second-grade student in Springdale public schools, the board of education proposes a 5% increase in property taxes. One local "golden oldies" radio station asks listeners to call in to voice their support for, or opposition to, the proposal. The station finds that 78% of the callers are opposed.

(a) What type of sample did the radio station obtain?

(b) Explain why this sampling method is biased. Is 78% likely to be greater than or less than the percent of all Springdale residents who oppose the proposal?

(c) Explain how the radio station could avoid the bias described in part (b).

16. **Infested hemlocks** A forester studying the woolly adelgid, an invasive species of insect that infests and weakens hemlock trees, wants to estimate the proportion of hemlocks in a large forest that are infested. He starts at a roadside rest area and walks along the road, using the first 50 hemlocks he encounters as his sample. He finds that 72% of the trees are infested by the woolly adelgid.

(a) What type of sample did the forester obtain?

(b) Explain why this sampling method is biased. Is 72% likely to be greater than or less than the percent of all hemlock trees in the forest that are infested?

(c) Explain how the forester could avoid the bias described in part (b).

Extending the Concepts

17. **Voluntary response on the Web** Find an example of an Internet poll that is potentially biased. Discuss potential sources of bias and how the result from the poll might differ from the true value.

Recycle and Review

18. Tennis players' ages (1.8) The parallel boxplots show the distribution of ages for the top 100 male and female professional tennis players in 2014.[10] Use the boxplot to discuss similarities and differences in the two distributions.

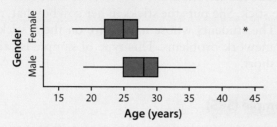

19. Practice and memory (2.7, 2.8) An experiment was conducted to determine the effect of practice time (in seconds) on the percent of unfamiliar words recalled. Here is a scatterplot of the results with the least-squares regression line.

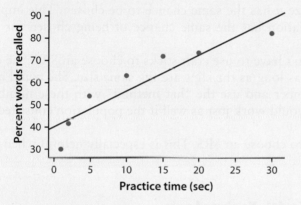

(a) Sketch an approximate residual plot. Be sure to label your axes.

(b) Explain why a linear model is not appropriate for describing the relationship between practice time and percent of words recalled.

(c) Here are residual plots for exponential and quadratic models of this relationship. Is one of these models more appropriate than the other for describing this relationship? Justify your answer.

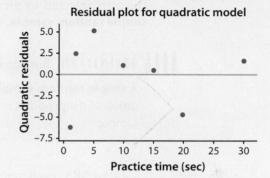

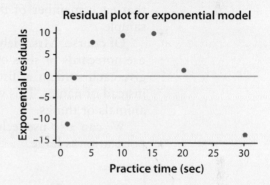

Lesson 3.3

Simple Random Samples

LEARNING TARGETS

- Describe how to obtain a simple random sample using slips of paper or technology.
- Explain the concept of sampling variability and the effect of increasing sample size.
- Use simulation to test a claim about a population proportion.

In Lesson 3.2, you learned that choosing a sample at random is a good way to avoid bias. In this lesson, we start with the details of how to select a random sample.

Choosing a Simple Random Sample

Each day in class, Mrs. Hebert randomly selects 5 students to present homework problems on the white board. To make the selection, she uses a set of craft sticks, with each student's name written on one stick. She puts the sticks in her cowboy hat, mixes them well, and pulls out 5 sticks. The students whose names are on the 5 sticks are the ones selected to present the homework problems. This type of sample is called a **simple random sample,** or **SRS** for short.

> **DEFINITION** Simple random sample (SRS)
>
> A **simple random sample** (**SRS**) of size *n* is a sample chosen in such a way that every group of *n* individuals in the population has an equal chance of being selected as the sample.

In an SRS, each sample of size *n* has the same chance to be chosen. This implies that each member of the population has the same chance of being chosen for the sample.

Of course, Mrs. Hebert doesn't have to use craft sticks to choose an SRS. She can use notecards or slips of paper, as long as the slips are the same size. She could also give each student a distinct number and use the "hat method" with these numbers instead of names. This version would work just as well if the population consisted of animals or things.

We can also use technology to choose an SRS. This is especially helpful when the population is large.

> ## How to Choose an SRS with Technology
>
> 1. **Label.** Give each individual in the population a distinct numerical label from 1 to *N*, where *N* is the number of individuals in the population.
> 2. **Randomize.** Use a random number generator to obtain *n different* integers from 1 to *N*.
> 3. **Select.** Choose the individuals that correspond to the randomly selected integers.

When choosing an SRS, we make the selections *without replacement*. That is, once an individual is selected for a sample, that individual cannot be selected again. Most random number generators sample numbers *with replacement,* so it is important that you explain that repeated numbers should be ignored when using technology to select an SRS.

There are many random number generators available on the Internet, including those at www.random.org. You can also use the random number generator on your calculator.

TECH CORNER

Choosing an SRS on a TI-83/84

In addition to using a website to generate random integers, you can also use a graphing calculator. Let's select an SRS of 5 students from Mrs. Hebert's class using a TI-83/84. She has already given each student a distinct label from 1 to 35.

1. Press MATH , then select PROB(PRB) and randint(.

2. Complete the command randint(1,35) and press ENTER until you have chosen 5 *different* numbers. If you are using OS 2.55 or later, leave the entry for *n* blank.

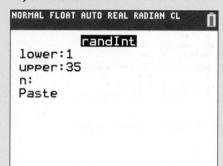

```
NORMAL FLOAT AUTO REAL RADIAN CL
            randInt
 lower:1
 upper:35
 n:
 Paste
```

```
NORMAL FLOAT AUTO REAL RADIAN CL
                              34.
randInt(1,35)
                              32.
randInt(1,35)
                               6.
randInt(1,35)
                              19.
randInt(1,35)
                              15.
```

Note: If you have a TI-84 with OS 2.55 or later, you can use the command RandIntNoRep(1,35) to sort the integers from 1 to 35 in random order. The first 5 numbers listed give the labels of the chosen students. If you have a TI-84 Plus CE or TI-84 Plus C Silver Edition, you can use the command RandIntNoRep(1,35,5) to get a sample of 5 *different* integers from 1 to 35.

EXAMPLE

What do we do if the landfill is full?

Selecting an SRS

PROBLEM: When landfills run out of room, they are sometimes converted into parks. Does the presence of a landfill underneath affect the trees that grow in the park? One such park has 283 trees and researchers want to closely inspect a sample of 20 trees.

(a) Describe how to select an SRS of 20 trees using slips of paper.

(b) Describe how to select an SRS of 20 trees using a random number generator.

SOLUTION:

(a) Give each tree a distinct label from 1 to 283. Number 283 identically sized slips of paper from 1 to 283. Put these slips of paper into a hat and mix well. Select 20 slips of paper and inspect the 20 trees that are labeled with the numbers selected.

(b) Give each tree a distinct label from 1 to 283.

> **1. Label.** Give each individual in the population a distinct numerical label from 1 to *N*, where *N* is the number of individuals in the population.

Randomly generate 20 integers from 1 to 283, ignoring repeated integers.

> **2. Randomize.** Use a random number generator to obtain *n different* integers from 1 to *N*.

Inspect the 20 trees that are labeled with the generated integers.

> **3. Select.** Choose the individuals that correspond to the randomly selected integers.

FOR PRACTICE   TRY EXERCISE 1.

Before there was technology to generate random numbers, statisticians used tables of random digits. These tables have lines of randomly generated integers from 0 to 9. For example, the first line of Table D in the back of the book begins like this:

19223 95034 05756 . . .

To choose an SRS using a random-digit table, give each individual a distinct numerical label with the same number of digits. Then, read consecutive, appropriately sized groups of digits until you have the number of labels you require. In the landfill example, we would give each tree a label from 001 to 283 and go through the random-digit table looking at sets of 3 digits until we found 20 distinct numbers from 001 to 283.

Sampling Variability

When Mr. McCourtney's class did the activity about the Federalist Papers, his students got the results shown below.

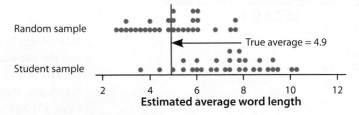

When his students picked the words themselves, the estimates were consistently too large. Using random samples helped them avoid bias by giving each word the same chance to be chosen. Still, none of their estimates was equal to the true mean of $\mu = 4.9$, even though the students used an unbiased sampling method. This isn't a contradiction. Instead, it illustrates the concept of **sampling variability.**

> **DEFINITION** Sampling variability
>
> The fact that different random samples of the same size from the same population produce different estimates is called **sampling variability.**

For example, here is one random sample of 5 words from the activity about the Federalist Papers:

correct other a found by

The mean length of these 5 words is $(7 + 5 + 1 + 5 + 2)/5 = 4$. Other random samples will produce different means. The dotplot in Figure 3.1 displays the mean word lengths for 50 different random samples of size $n = 5$. The sample means vary from as small as 2.2 to as large as 8.0, even though the samples were selected from the same population.

FIGURE 3.1 Distribution of the sample mean word length for 50 SRSs of size $n = 5$.

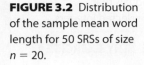

We can reduce sampling variability by using a larger sample size. The mean word lengths for 50 random samples of size $n = 20$ are given in Figure 3.2. Notice that the sample means are now less variable than they were before. In other words, our estimates are now closer to the truth, on average.

FIGURE 3.2 Distribution of the sample mean word length for 50 SRSs of size $n = 20$.

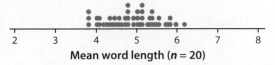

EXAMPLE

What's the median income in Virginia?

Sampling variability

PROBLEM: The median household income in a random sample of 371 Virginia households is $61,784.

(a) Do you think that the median income of *all* Virginia households is equal to $61,784? Explain.

(b) If the sample size was increased to 1000 households, what effect would this have on the estimated median income?

SOLUTION:

(a) No. Because different random samples will produce different medians, it is unlikely this sample provides a median that is exactly correct.

(b) We could expect our estimate to be closer to the true median income of all Virginia households.

FOR PRACTICE   TRY EXERCISE 5.

Inference for Sampling

The purpose of a sample is to give us information about a larger population. The process of drawing conclusions about a population on the basis of sample data is called *inference* because we *infer* information about the population from what we know about the sample.

ACTIVITY

Do students prefer name-brand cookies?

In this activity, you will get a taste of what inference is all about.

Thirty randomly selected high-school students were asked to taste two unlabeled cookies and identify which cookie they preferred. One of these cookies was a name-brand cookie (Chips Ahoy!) and the other was a store-brand cookie (ChipMates). The order in which the two cookies were presented was determined at random using a coin flip. Of these 30 students, 19 preferred the name-brand cookie and 11 preferred the store-brand cookie.

Although more than half of the members of the sample (19/30) preferred the name-brand cookie, it is possible that the two types of cookies are equally preferred at this high school and we got a value this large because of sampling variability. Let's do a simulation to determine if there is *convincing* evidence that the majority of students at this high school prefer the name-brand cookie.

1. The spinner below is divided into two regions: 50% for the name-brand cookie and 50% for the store-brand cookie. Why do you think it was divided into two equal sections?

2. To complete the spinner, get a small paper clip and "unbend" one end so that it has a loop at one end and a pointer at the other end, as shown to the right. Place your pencil tip through the loop of the paper clip and put the tip of the pencil on the dot in the middle of the spinner.

Loop

3. To simulate the preferences of a random sample of 30 students, spin the spinner 30 times and record how often it lands in the name-brand cookie region.

4. Share your results with the class and make a class dotplot.

5. Based on the dotplot, is there convincing evidence that the name-brand cookie is preferred? Or, is it plausible that the cookies are equally preferred at the school? Explain your reasoning.

Mr. Powell's class simulated 100 SRSs of size 30 for the name-brand cookies activity using the *One Categorical Variable* applet at highschool.bfwpub.com/spa3e. Here are their results.

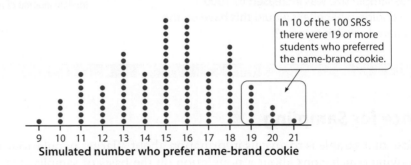

In 10 of the 100 SRSs there were 19 or more students who preferred the name-brand cookie.

Simulated number who prefer name-brand cookie

According to the dotplot, 10 of the 100 simulated SRSs resulted in at least 19 "successes." Because a result of 19 or more successes isn't that unlikely, Mr. Powell's class decided there wasn't convincing evidence that students at the high school preferred the name-brand cookie. In other words, it is plausible that the two brands of cookie are equally preferred and the sample result of 19/30 could have happened because of sampling variability.

EXAMPLE

Is the last chocolate the best chocolate?

Testing a claim

PROBLEM: Do people have a preference for the last thing they taste? Researchers at the University of Michigan designed a study to find out. The researchers gave 22 students five different Hershey's Kisses (milk chocolate, dark chocolate, crème, caramel, and almond) in random order and asked the students to rate each one. Participants were not told how many Kisses they would taste. However, when the fifth and final Kiss was presented, participants were informed that it would be their last one. Of the 22 students, 14 (64%) gave the final Kiss the highest rating. This is greater than 20%—the expected percentage if people don't have a special preference for the last thing they taste.[11]

To determine if these data provide convincing evidence that more than 20% of students prefer the last Kiss they taste, 100 simulated SRSs were selected. Each dot in the graph shows the number of students (out of 22) who preferred the last chocolate, assuming each student had a 20% chance of choosing the last Kiss.

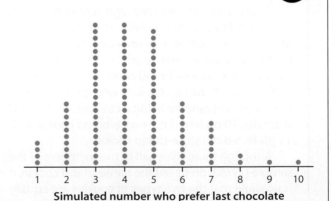

Simulated number who prefer last chocolate

(a) Explain how the graph illustrates the concept of sampling variability.

(b) Based on the results of the simulation, is there convincing evidence that people have a preference for the last Kiss they taste? Explain.

SOLUTION:

(a) If researchers repeatedly take random samples of size 22 from a population of students who show no preference for what they taste last, the number of students in a sample who prefer the last Kiss varies from about 1 to 10.

(b) Yes. In the study, there were 14 people who preferred the last Kiss they tasted. This is much greater than what we would expect to happen by chance alone. In the simulation, the largest number of people choosing the last Kiss they tasted was 10.

FOR PRACTICE   TRY EXERCISE 9.

LESSON APP 3.3

Do you tweet?

What proportion of students at your school use Twitter? To find out, you decide to survey a simple random sample of students from your school.

1. Describe how you could select an SRS of 50 students.

2. Will your sample result be exactly the same as the true population proportion? Explain.

3. Which would be more likely to yield a sample result closer to the true population value: an SRS of 50 students or an SRS of 100 students? Explain.

Steve Debenport/Getty Images

TECH CORNER

Simulating the Number of Successes in a Random Sample

We can use the *One Categorical Variable* applet at **highschool.bfwpub.com/spa3e** to simulate the number of successes in a random sample. We'll illustrate using the data from the cookie example.

1. Enter the variable name "Preference" and the category names ("Name-brand" and "Store-brand") with their corresponding counts, as shown in the screen shot below.

One Categorical Variable

Variable name: Preference
Groups: Single ▾
Input data as: Counts in categories ▾

	Category Name	Frequency	
1	Name-brand	19	−
2	Store-brand	11	−
			+

Begin analysis   Edit inputs   Reset everything

2. Press the "Begin analysis" button. Then, scroll down to Perform Inference and choose Simulate sample count from the drop-down menu. Keep "Name-brand" as the category to indicate as success and leave the third drop-down menu as a hypothesized proportion. Enter 0.5 as a hypothesized value, as shown in the screen shot below.

Perform Inference

Inference procedure: Simulate sample count ▾ Category to indicate as success: Name-brand ▾

Simulate the distribution of the sample count for samples of the original size assuming that the true proportion is equal to a hypothesized value ▾.
Hypothesized proportion: 0.5
Add this many samples: ____   Perform simulation   Clear simulation results

3. Enter 300 for the number of samples and press the "Perform simulation" button. To find out how often a value of 19 or greater would occur by chance alone, change the menu options to match the screen shot.

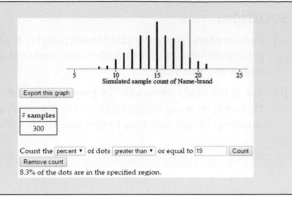

Export this graph

samples
300

Count the [percent ▾] of dots [greater than ▾] or equal to [19]   [Count]
[Remove count]
8.3% of the dots are in the specified region.

Lesson 3.3

WHAT DID YOU LEARN?

LEARNING TARGET	EXAMPLES	EXERCISES
Describe how to obtain a simple random sample using slips of paper or technology.	p. 195	1–4
Explain the concept of sampling variability and the effect of increasing sample size.	p. 197	5–8
Use simulation to test a claim about a population proportion.	p. 198	9–12

Lesson 3.3

Mastering Concepts and Skills

1. **Spring break family-style** The school newspaper is
pg 195 planning an article on family-friendly places to stay over spring break at a nearby beach town. The editors intend to call 4 randomly chosen hotels to ask about their amenities for families with children. They have an alphabetized list of all 28 hotels in the town.

(a) Describe how to select an SRS of 4 hotels using slips of paper.

(b) Describe how to select an SRS of 4 hotels using a random number generator.

2. **Holiday hours** The management company of a local mall with 21 stores plans to survey 3 of the stores to determine the hours they would like to stay open during the holiday season.

(a) Describe how to select an SRS of 3 stores using slips of paper.

(b) Describe how to select an SRS of 3 stores using a random number generator.

3. **Sampling pines** To gather data on a 1200-acre pine forest in Louisiana, the U.S. Forest Service laid a grid of 1410 equally spaced circular plots over a map of the forest. The Forest Service wants to visit a sample of 10% of these plots.[12]

(a) Explain why it would be difficult to select an SRS of 10% of the plots using slips of paper.

(b) Describe how to select an SRS of 10% of the plots using a random number generator.

4. **Sampling for genealogy** The local genealogical society in Coles County, Illinois, has compiled records on all 55,914 gravestones in cemeteries in the county for the years 1825 to 1985. Historians plan to use these records to learn about African Americans in Coles County's history. They first choose an SRS

of 395 records to check their accuracy by visiting the actual gravestones.[13]

(a) Explain why it would be difficult to select an SRS of 395 gravestones using slips of paper.

(b) Describe how to select an SRS of 395 gravestones using a random number generator.

5. **Estimating correlation** In an SRS of $n = 6$ students from Washington High School, the correlation between arm span (in centimeters) and height (in centimeters) is $r = 0.928$.

pg 197

(a) Do you think that the correlation between arm span and height for all students at Washington High School is equal to 0.928? Explain.

(b) If the sample size was increased to 100 students, what effect would this have on the estimated correlation?

6. **Estimating slope** In an SRS of $n = 10$ used Honda CR-Vs, the slope of the least-squares regression line relating $y =$ asking price (in dollars) to $x =$ age (in years) is $b = -892$.

(a) Do you think that the slope of the least-squares regression line relating asking price to age for all used Honda CR-Vs would equal -892? Explain.

(b) If the sample size was increased to 50 used Honda CR-Vs, what effect would this have on the estimated slope?

7. **Estimating a mean** In an SRS of 100 calls to a customer service line, the mean time that callers were placed on hold was 12.5 minutes.

(a) Do you think the mean time that all callers to the customer service line were placed on hold is 12.5 minutes? Explain.

(b) Suppose the customer-service manager only had time to collect an SRS of 25 calls before reporting to a managers' meeting. What effect would this have on the estimated mean?

8. **Estimating a proportion** During the political crisis in Ukraine in spring 2014, a national polling organization asked an SRS of 2066 Americans to locate Ukraine on a map of the world. The sample proportion who successfully located Ukraine was 0.166.[14]

(a) Do you think that the proportion of all Americans who can locate Ukraine on a map is 0.166? Explain.

(b) Suppose the polling organization had only polled an SRS of 500 Americans. What effect would this have on the estimated proportion?

9. **Spinning heads** When a fair coin is flipped, we all know that the probability the coin lands on heads is 0.50. However, what if a coin is spun? Two Polish math professors and their students spun a Belgian euro coin 250 times. It landed heads 140

pg 198

times. One of the professors concluded that the coin was minted asymmetrically and more likely to land on heads. A representative from the Belgian mint indicated the result was just chance.[15]

To determine if these data provide convincing evidence that this Belgian euro coin is more likely to land on heads, 100 trials of a simulation were conducted. Each dot in the graph shows the number of heads in a random sample of 250 spins, assuming the coin has a 50% chance of landing on heads.

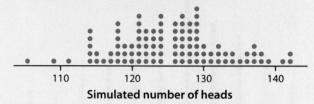

Simulated number of heads

(a) Explain how the graph illustrates the concept of sampling variability.

(b) Based on the results of the simulation, is there convincing evidence that this Belgian euro coin is more likely to land on heads? Explain.

10. **Kissing the right way** According to a newspaper article, "Most people are kissing the 'right way.'" That is, according to a study, the majority of couples prefer to tilt their heads to the right when kissing. In the study, a researcher observed a random sample of 124 kissing couples and found that 83/124 (66.9%) of the couples tilted to the right.[16]

To determine if these data provide convincing evidence that couples are more likely to tilt their heads to the right, 100 simulated SRSs were selected. Each dot in the graph shows the number of couples that tilt to the right in a simulated SRS of 124 couples, assuming that each couple has a 50% chance of tilting to the right.

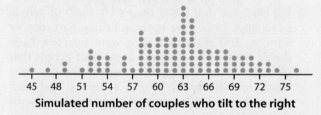

Simulated number of couples who tilt to the right

(a) Explain how the graph illustrates the concept of sampling variability.

(b) Based on the results of the simulation, is there convincing evidence that couples prefer to kiss the "right way"? Explain.

11. **More lefties?** In the population of people in the United States, about 10% are left-handed. After bumping elbows at lunch with several left-handed students, Simon wondered if more than 10% of students at his school are left-handed. To investigate, he selected an SRS of 50 students and found 7 lefties.

To determine if these data provide convincing evidence that more than 10% of the students at Simon's school are left-handed, 100 simulated SRSs were selected. Each dot in the graph shows the number of students that are left-handed in a simulated SRS of 50 students, assuming that each student has a 10% chance of being left-handed.

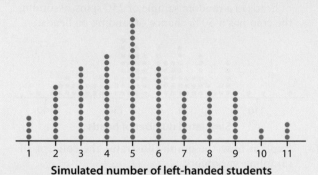

Simulated number of left-handed students

(a) Explain how the graph illustrates the concept of sampling variability.

(b) Based on the results of the simulation, is there convincing evidence that more than 10% of students at Simon's school are left-handed? Explain.

12. **Weekend birthdays** Over the years, the percentage of births that are planned caesarian sections has been rising. Because doctors can schedule these deliveries, there might be more children born during the week and fewer born on the weekend than if births were uniformly distributed throughout the week. To investigate, Mrs. McDonald and her class selected an SRS of 73 people born since 1993. Of these people, 24 were born on Friday, Saturday, or Sunday.

To determine if these data provide convincing evidence that fewer than 43% (3/7) of people born since 1993 were born on Friday, Saturday, or Sunday, 100 simulated SRSs were selected. Each dot in the graph shows the number of people that were born on Friday, Saturday, or Sunday in a simulated SRS of 73 people, assuming that each person had a 43% chance of being born on one of these three days.

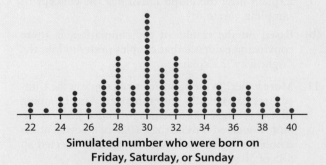

Simulated number who were born on Friday, Saturday, or Sunday

(a) Explain how the graph illustrates the concept of sampling variability.

(b) Based on the results of the simulation, is there convincing evidence that fewer than 43% of people born since 1993 were born on Friday, Saturday, or Sunday? Explain.

Applying the Concepts

13. **How far from home?** A college administrator wants to estimate the mean distance that students at a large community college live from campus. To find out, she obtains a list of all students from the registrar's database.

(a) Describe how you could select an SRS of 100 students.

(b) Will your sample result be exactly the same as the true population mean? Explain.

(c) Which would be more likely to yield a sample result closer to the true population mean: an SRS of 50 students or an SRS of 100 students? Explain.

14. **Who banks online?** A bank wants to estimate the proportion of its checking account customers who prefer to deposit checks in their account using a smartphone rather than making deposits at the local branch bank.

(a) Describe how the bank could select an SRS of 100 account holders.

(b) Will the sample result be exactly the same as the true population proportion? Explain.

(c) Which would be more likely to yield a sample result closer to the true population proportion: an SRS of 50 account holders or an SRS of 100 account holders? Explain.

Extending the Concepts

15. **Mostly on time?** An airline claims that 95% of its flights arrive on time. To check this claim, an SRS of 25 flights was selected and only 22 were on time. Do these data provide convincing evidence that less than 95% of all of this airline's flights arrive on time?

(a) Describe how you could use a spinner to simulate the number of on-time flights in an SRS of 25 flights, assuming that 95% of all flights arrive on time.

(b) Select at least 20 simulated SRSs, using a spinner or the *One Categorical Variable* applet at highschool. bfwpub.com/spa3e. Record the results of your simulation on a dotplot.

(c) Based on the results of the simulation, is there convincing evidence that less than 95% of all of this airline's flights arrive on time? Explain.

16. Can you trust the WWW? You want to ask a sample of high-school students the question "How much do you trust information about health that you find on the Internet—a great deal, somewhat, not much, or not at all?" You try out this and other questions on a pilot group of 5 students chosen from your class. The class members are listed below.

Anderson	Deng	Glaus	Nguyen	Samuels
Arroyo	De Ramos	Helling	Palmiero	Shen
Batista	Drasin	Husain	Percival	Tse
Bell	Eckstein	Johnson	Prince	Velasco
Burke	Fernandez	Kim	Puri	Wallace
Cabrera	Fullmer	Molina	Richards	Washburn
Calloway	Gandhi	Morgan	Rider	Zabidi
Delluci	Garcia	Murphy	Rodriguez	Zhao

(a) Explain how you would use the lines of random digits shown below to choose an SRS of 5 students from the class list.

82739 57890 20807 47511
81676 55300 94383 14833

(b) Use the lines of random digits to select the sample. Show how you use each of the digits.

17. Apartment complexes You are planning a report on apartment living in a college town. You decide to select three apartment complexes at random for in-depth interviews with residents. The list of all apartment complexes in the college town is shown below.

Ashley Oaks	Chauncey Village	Franklin Park	Richfield
Bay Pointe	Country Squire	Georgetown	Sagamore Ridge
Beau Jardin	Country View	Greenacres	Salem Courthouse
Bluffs	Country Villa	Lahr House	Village Manor
Brandon Place	Crestview	Mayfair Village	Waterford Court
Briarwood	Del-Lynn	Nobb Hill	Williamsburg
Brownstone	Fairington	Pemberly Courts	
Burberry	Fairway Knolls	Peppermill	
Cambridge	Fowler	Pheasant Run	

(a) Explain how you would use the lines of random digits shown below to choose an SRS of 3 apartment complexes from the list.

38167 98532 62183 70632
23417 26185 41448 75532

(b) Use the lines of random digits to select the sample. Show how you use each of the digits.

Recycle and Review

18. Get fit! (2.3, 3.1) A college fitness center offers an exercise program for staff members who choose to participate. The program assesses each participant's fitness using a treadmill test and also administers a personality questionnaire. There is a moderately strong positive correlation between fitness score and score for self-confidence.

(a) Identify the population in this study.

(b) Is this an experiment or an observational study? Explain.

(c) Can we conclude from this study that a fitness program will boost self-confidence? Explain.

19. The mystery of the missing grade (2.5, 2.6) In Professor Friedman's economics course, the correlation between the students' total scores before the final examination and their final examination scores is $r = 0.6$. The pre-exam totals for all students in the course have mean 280 and standard deviation 30. The final-exam scores have mean 75 and standard deviation 8. Professor Friedman has lost Julie's final exam, but knows that her total before the exam was 300. He decides to predict her final exam score from her pre-exam total.

(a) Find the equation for the appropriate least-squares regression line for Professor Friedman's prediction.

(b) Use the least-squares regression line to predict Julie's final-exam score.

(c) Explain the meaning of the phrase "least squares" in the context of this question.

Lesson 3.4
Estimating a Margin of Error

In Lesson 3.3, you learned that the purpose of random sampling is to provide information about a larger population. You also learned that different random samples from the same population will produce different estimates for a population proportion or a population mean. In this lesson, you will learn how to approximate and interpret the margin of error for a sample proportion and a sample mean.

Margin of Error: Estimating a Population Proportion

One of the most common uses of statistics is to estimate the proportion of individuals in a population that share some characteristic. For example, what proportion of students at your school text during class? What proportion of U.S. residents approve of the president's job performance?

To make a good estimate of a population proportion, start with a random sample from the population of interest. Then, use the proportion of successes in the sample as the estimate for the proportion of successes in the population. To investigate how far the estimate might be from the truth, we can perform a simulation to estimate the **margin of error.**

> **DEFINITION   Margin of error**
>
> The **margin of error** of an estimate describes how far, at most, we expect the estimate to vary from the true population value.

Carolina selected a simple random sample of 50 students from the 2000 students at her high school. She asked each student to fill out an anonymous survey that included a question about texting during class. Overall, 64% (32/50) of the students in the sample admitted to texting during class. Thus, her estimate of the proportion of *all* students at her school that would admit to texting during class is 0.64.

To approximate the margin of error for Carolina's estimate, we can use simulation to investigate how much sample proportions tend to vary from the true proportion in random samples of size 50. Unfortunately, we don't know the true proportion of students who would admit to texting at Carolina's school! Because Carolina's sample proportion of 0.64 should be fairly close to the true proportion, we will temporarily assume that the true proportion is 0.64 for the purposes of the simulation.

204

Estimating the Margin of Error for a Proportion

We can use the *One Categorical Variable* applet at **highschool.bfwpub.com/spa3e** to help estimate the margin of error for a proportion. We'll illustrate using Carolina's texting data.

1. Enter the variable name "Response to texting question" and the category names ("Yes" and "No") with their corresponding counts, as shown in the screen shot below.

One Categorical Variable

Variable name: Response to texting question
Groups: Single ▼
Input data as: Counts in categories ▼

	Category Name	Frequency	
1	Yes	32	-
2	No	18	-
			+

Begin analysis | Edit inputs | Reset everything

2. Press the "Begin analysis" button. Then, scroll down to Perform Inference and choose Simulate sample proportion from the drop-down menu. Keep "Yes"

as the category to indicate as success and change the third drop-down menu to the observed proportion, as shown in the screen shot below.

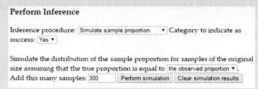

3. Enter 300 for the number of samples and press the "Perform simulation" button. Your output should look similar to the screen shot below. The standard deviation is reported below the dotplot.

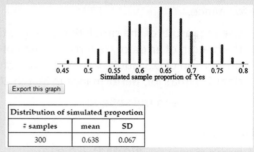

Export this graph

Distribution of simulated proportion		
# samples	mean	SD
300	0.638	0.067

In the tech corner, we used the *One Categorical Variable* applet to simulate taking 300 random samples of size 50, assuming that 64% of all students at the school would admit to texting during class. The dotplot shows the proportion of students who would admit to texting in class for each of the 300 samples. The standard deviation measures the typical distance between a sample proportion and the true population proportion. In this case, the sample proportions typically vary from the assumed population value by about 0.067.

For nearly all samples, it is also true that the distance between the sample proportion and the true proportion will be less than 2 standard deviations. For example, 95.3% (286/300) of the values in the simulation were within 2 standard deviations of 0.64. Thus, to approximate the margin of error for Carolina's estimate of 0.64, we double the standard deviation from the simulation:

$$\text{margin of error} \approx 2(0.067) = 0.134$$

We expect the true proportion of all students at Carolina's school who would admit to texting to be at most 0.134 from the estimate of 0.64.

Estimating the Margin of Error for a Proportion

To estimate the margin of error for a sample proportion resulting from a random sample of size *n*:

1. Simulate the distribution of the sample proportion for many random samples of size *n* assuming that the true population value is equal to the sample proportion from the original sample.

2. Multiply the standard deviation of the simulated distribution by 2 to get the margin of error.

When using this method, the distance between the estimate and the true population value will be less than the margin of error for about 95% of samples. We'll see why this is the case in Lesson 7.2.

EXAMPLE

Yuri Gripas/AFP/Getty Images

Do people approve of the president's job performance?

Margin of error for a proportion

PROBLEM: In March 2014 the Gallup polling organization surveyed a random sample of 1500 U.S. adults. Overall, 42% of the people in the sample approved of the way Barack Obama was handling his job as president. The dotplot shows the proportion who approved of the president's job performance in each of 300 random samples of size 1500 from a population where 42% approved of his job performance.

(a) Use the results of the simulation to approximate the margin of error for Gallup's estimate of the proportion of U.S. adults who approved of President Obama's job performance in March 2014.

(b) Interpret the margin of error.

SOLUTION:

(a) margin of error ≈ 2(0.013) = 0.026

(b) We expect the true proportion of all U.S. adults who approved of President Obama's job performance in March 2014 to be at most 0.026 from the estimate of 0.42.

Simulated proportion who approved of the president's job performance

0.38 0.39 0.4 0.41 0.42 0.43 0.44 0.45 0.46

Distribution of simulated proportion		
# samples	mean	SD
300	0.42	0.013

To calculate the margin of error, multiply the standard deviation from the simulation by 2.

FOR PRACTICE TRY EXERCISE 1.

Notice that the margin of error for Gallup's estimate of President Obama's job approval (0.026) is much smaller than Carolina's estimate of the proportion of students who text during class (0.134). This is mainly because Gallup's sample size (1500) was so much larger than Carolina's (50).

It is common in the media to report the margin of error of an estimate in the "fine print" at the bottom of an article or graphic. For example, an article might state that "the margin of error is ±2.6 percentage points" or might report the estimate and margin of error together as "0.42 ± 0.026."

Margin of Error: Estimating a Population Mean

Carolina also wanted to know the average number of hours of sleep that students at her school typically get on a school night. The dotplot in Figure 3.3 displays the typical sleep time (in hours) for the 50 students she randomly selected from her school.

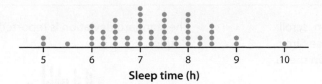

The mean amount of sleep time for the students in the sample is $\bar{x} = 7.24$ hours. Thus, Carolina estimates that the true mean amount of time *all* students at her school typically sleep on a school night is 7.24 hours.

To approximate the margin of error for this estimate, we can use a simulation to investigate how much the sample mean $\bar{x}$ typically varies from the true mean. To do this, we should take many samples of size 50 from the population, calculate the sample mean $\bar{x}$ for each sample, and graph the distribution of $\bar{x}$ on a dotplot. However, we don't know the shape, mean, or standard deviation of the population distribution!

Fortunately, Carolina's random sample of size 50 gives us a fairly good idea of what the distribution would look like for the entire population of 2000 students. The dotplot in Figure 3.4 shows this approximate "population" distribution by replicating the original sample 40 times to get $50 \times 40 = 2000$ students.

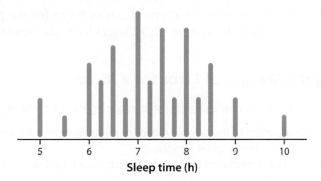

FIGURE 3.4 Distribution of typical school-night sleep time (in hours) for a "population" of 2000 students from Carolina's school.

We can take random samples of size 50 from this approximate "population" to explore the variability of $\bar{x}$. Of course, we don't have to create a new "population" when we want to estimate the margin of error for a mean. We could accomplish the same goal by selecting random samples of size 50 from the original sample, *with replacement*.

TECH CORNER

Estimating the Margin of Error for a Mean

We can use the *One Quantitative Variable* applet at **highschool.bfwpub.com/spa3e** to estimate the margin of error for a mean. We'll illustrate using Carolina's sleep time data, shown in the dotplot below.

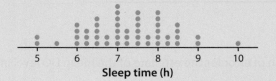

1. Enter the variable name "Sleep time (hours)" and the data as shown in the screen shot below.

One Quantitative Variable

Variable name: Sleep time (hours)
Number of groups: 1 ▼
Input: Raw data ▼

Input data separated by commas or spaces.
Group 1 data: 5.00 5.00 5.50 6.00 6.00 6.25 6.25 6.50 6.25 7.00 6.5

Begin analysis Edit inputs Reset everything

2. Press the "Begin analysis" button. Then, scroll down to Perform Inference and choose Simulate sample mean from the drop-down menu.

The standard deviation is reported below the dotplot.

Distribution of Simulated Mean		
# samples	mean	SD
300	7.222	0.146

3. Enter 300 for the number of samples and press the "Perform simulation" button. Your output should look similar to the final screen shot.

To approximate the margin of error for Carolina's estimate of 7.24 hours, we double the standard deviation from our simulated distribution of $\bar{x}$:

$$\text{margin of error} \approx 2(0.146) = 0.292 \text{ hours}$$

We expect the true mean sleep time on a typical school night for the population of all students at Carolina's school to be at most 0.292 hours from the estimate of 7.24 hours.

Estimating the Margin of Error for a Mean

To estimate the margin of error for a sample mean resulting from a random sample of size n:

1. Simulate the distribution of the sample mean for many random samples of size n taken *with replacement* from the original sample.

2. Multiply the standard deviation of the simulated distribution by 2 to get the margin of error.

EXAMPLE

Is this stream oxygenated?

Margin of error for a mean

PROBLEM: The level of dissolved oxygen (DO) in a stream or river is an important indicator of the water's ability to support aquatic life. A researcher measures the DO level at 15 randomly chosen locations along a stream. Here are the results in milligrams per liter (mg/l):

4.53	5.04	3.29	5.23	4.13	5.50	4.83	4.40
5.42	6.38	4.01	4.66	2.87	5.73	5.55	

The sample mean for these data is $\bar{x} = 4.77$ mg/l.

The dotplot shows the distribution of the sample mean DO level for 300 random samples of size 15 taken with replacement from the original sample.

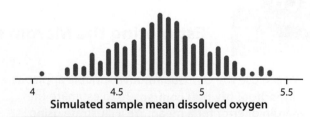

Distribution of simulated mean		
# samples	mean	SD
300	4.778	0.246

(a) Use the results of the simulation to approximate the margin of error for the estimate of the mean DO level in this stream.

(b) Interpret the margin of error.

SOLUTION:

(a) margin of error $\approx 2(0.246) = 0.492$

(b) We expect the true mean DO level in this stream to be at most 0.492 mg/l from the estimate of 4.77 mg/l.

> To calculate the margin of error, multiply the standard deviation from the simulation by 2.

> Don't forget to include units when appropriate.

FOR PRACTICE **TRY EXERCISE 5.**

In a stream, a dissolved oxygen (DO) level below 5 mg/l puts aquatic life at risk. Should the researcher in the previous example be concerned about the stream she is investigating? There is some evidence that the aquatic life is in danger because the sample mean DO level of 4.77 mg/l is less than 5 mg/l. But 4.77 is just an estimate of the mean DO level in the stream—and this estimate has a margin of error of 0.492. Therefore, the true mean could be as high as $4.77 + 0.492 = 5.262$ mg/l. Because there are plausible values for the true mean that are greater than 5 mg/l, the researcher does not have convincing evidence that the true mean DO level in the stream is less than 5 mg/l.

LESSON APP 3.4

Can you roll your tongue?

Many people can roll their tongues, but some can't. Javier is interested in determining the proportion of students at his school that can roll their tongue. In a random sample of 100 students, Javier determines that 70 can roll their tongue. The dotplot below shows the proportion of people who can roll their tongue in each of 500 random samples of size 100 from a population where 70% can roll their tongue.

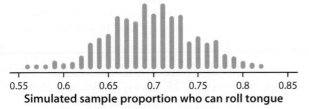

Simulated sample proportion who can roll tongue

Distribution of simulated proportion		
# samples	mean	SD
500	0.701	0.048

1. Use the results of the simulation to approximate the margin of error for Javier's estimate of the proportion of students at his school that can roll their tongue.

2. Interpret the margin of error.

3. Javier's biology teacher claims that 75% of people can roll their tongue. According to Javier's study, is this claim plausible? Explain.

4. Explain how Javier could decrease the margin of error.

Lesson 3.4

WHAT DID YOU LEARN?

LEARNING TARGET	EXAMPLES	EXERCISES
Use simulation to approximate the margin of error for a sample proportion and interpret the margin of error.	p. 206	1–4
Use simulation to approximate the margin of error for a sample mean and interpret the margin of error.	p. 208	5–8

Exercises

Lesson 3.4

Mastering Concepts and Skills

1. **Satisfaction** A recent Gallup Poll asked a random sample of 1025 adults, "In general, are you satisfied or dissatisfied with the way things are going in the United States at this time?" In all, 25% said that they were satisfied. The dotplot shows the proportion who said they were satisfied in each of 500 random samples of size 1025 from a population where 25% are satisfied.

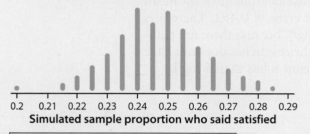

Simulated sample proportion who said satisfied

Distribution of simulated proportion		
# samples	mean	SD
500	0.25	0.014

(a) Use the results of the simulation to approximate the margin of error for Gallup's estimate of the proportion of U.S. adults who were satisfied with the way things were going in the United States at the time of the poll.

(b) Interpret the margin of error.

2. **Reporting on cheats** What proportion of students are willing to report cheating by other students? A student project put this question to an SRS of 172 undergraduates at a large university: "You witness two students cheating on a quiz. Do you go to the professor?" Only 11% answered "Yes." The dotplot shows the proportion who would go to the professor in each of 1000 random samples of size 172 from a population where 11% would go to the professor.[17]

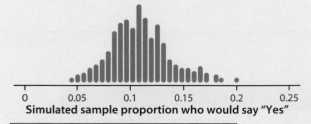

Simulated sample proportion who would say "Yes"

Distribution of simulated proportion		
# samples	mean	SD
1000	0.11	0.024

(a) Use the results of the simulation to approximate the margin of error for the estimate of the proportion of undergrads at the large university who would go to the professor.

(b) Interpret the margin of error.

3. **Twittering away** The Pew Internet and American Life Project asked a random sample of 2253 U.S. adults, "Do you ever . . . use Twitter or another service to share updates about yourself or to see updates about others?" In the sample, 19% said "Yes." The dotplot shows the proportion who would say "Yes" to this question in each of 1000 random samples of size 2253 from a population where 19% would say "Yes."[18]

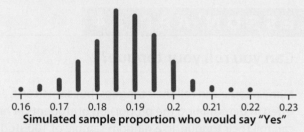

Simulated sample proportion who would say "Yes"

Distribution of simulated proportion		
# samples	mean	SD
1000	0.19	0.009

(a) Use the results of the simulation to approximate the margin of error for the estimate of the proportion of U.S. adults who would say "Yes" to this question.

(b) Interpret the margin of error.

4. **Size of the labor force** In a random sample of 320 U.S. residents over 25 years old, 63% said they are currently employed. The dotplot shows the proportion who are currently employed in each of 500 random samples of size 320 from a population where 63% are employed.

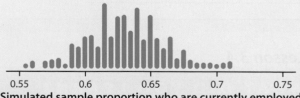

Simulated sample proportion who are currently employed

Distribution of simulated proportion		
# samples	mean	SD
500	0.632	0.026

(a) Use the results of the simulation to approximate the margin of error for the estimate of the proportion of U.S. adult residents over 25 years old who are currently employed.

(b) Interpret the margin of error.

5. **Vitamin C content** Several years ago, the U.S. Agency for International Development provided 238,300 metric tons of corn–soy blend (CSB) for emergency relief in countries throughout the world. CSB is a highly nutritious, low-cost fortified food. The following data are the amounts of vitamin C, measured in milligrams per 100 grams (mg/100 g) of blend, for a random sample of size 8 from one production run:

26 31 23 22 11 22 14 31

The sample mean for these data is $\overline{x} = 22.5$ mg/100 g.

The dotplot shows the distribution of the sample mean vitamin C content for 300 random samples of size 8 taken with replacement from the original sample.[19]

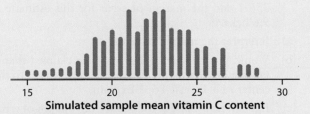

Simulated sample mean vitamin C content

Distribution of simulated mean		
# samples	mean	SD
300	22.356	2.405

(a) Use the results of the simulation to approximate the margin of error for the estimate of the mean vitamin C content in the blend for this production run.

(b) Interpret the margin of error.

6. **Screen tension** A manufacturer of high-resolution video terminals must control the tension on the mesh of fine wires that lies behind the surface of the viewing screen. Too much tension tears the mesh, and too little allows wrinkles. The tension is measured by an electrical device with output readings in millivolts (mV). Here are the tension readings from a random sample of 20 screens from a single day's production:

269.5 297.0 269.6 283.3 304.8 280.4 233.5 257.4 317.5 327.4

264.7 307.7 310.0 343.3 328.1 342.6 338.8 340.1 374.6 336.1

The sample mean for these data is $\overline{x} =$ 306.32 mV.

The dotplot shows the distribution of the sample mean tension readings for 300 random samples of size 20 taken with replacement from the original sample.

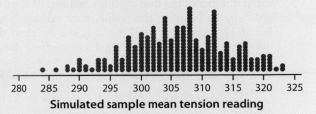

Simulated sample mean tension reading

Distribution of simulated mean		
# samples	mean	SD
300	306.673	7.734

(a) Use the results of the simulation to approximate the margin of error for the estimate of the mean tension reading for screens produced this day.

(b) Interpret the margin of error.

7. **Pepperoni pizza** Melissa and Madeline love pepperoni pizza, but sometimes they are disappointed with the small number of pepperonis on their pizza. To investigate, they went to their favorite pizza restaurant at 10 random times during the week and ordered a large pepperoni pizza. Here are the number of pepperonis on each pizza:

47 36 25 37 46 36 49 32 32 34

The sample mean for these data is $\overline{x} = 37.4$ pepperonis.[20]

The dotplot shows the distribution of the sample mean number of pepperonis for 500 random samples of size 10 taken with replacement from the original sample.

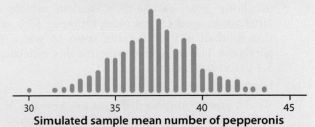

Simulated sample mean number of pepperonis

Distribution of simulated mean		
# samples	mean	SD
500	37.481	2.127

(a) Use the results of the simulation to approximate the margin of error for the estimate of the mean number of pepperonis on a large pizza from this restaurant.

(b) Interpret the margin of error.

8. **Snacking on Goldfish** Carly and Maysem plan to be pre-school teachers after they graduate from college. To prepare for snack time, they want to know the mean number of goldfish crackers in a bag of original goldfish. To estimate this value, they randomly selected 12 bags of original goldfish and counted the number of crackers in each bag. Here are their data:

317 330 325 323 332 337 324 342 330 349 335 333

The sample mean for these data is $\bar{x}$ = 331.4 goldfish.[21]

The dotplot shows the distribution of the sample mean number of goldfish for 500 random samples of size 12 taken with replacement from the original sample.

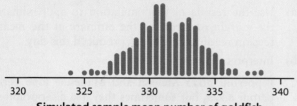

Simulated sample mean number of goldfish

Distribution of simulated mean		
# samples	mean	SD
500	331.545	2.395

(a) Use the results of the simulation to approximate the margin of error for the estimate of the mean number of goldfish in a bag of original goldfish.

(b) Interpret the margin of error.

Applying the Concepts

9. **Football on TV** A recent Gallup poll conducted telephone interviews with a random sample of 1000 adults aged 18 and older. Of these, 37% said that football is their favorite sport to watch on television. The margin of error for this estimate is 3.1 percentage points.

(a) Interpret the margin of error.

(b) Would you be surprised if a census revealed that the population proportion was 0.50? Explain.

(c) Explain how Gallup could decrease the margin of error.

10. **Car colors in Miami** Using a webcam, a traffic analyst selected a random sample of 800 cars traveling on I-195 in Miami on a weekday morning. Among the 800 cars in the sample, 24% were white. The margin of error for this estimate is 3.0 percentage points.

(a) Interpret the margin of error.

(b) Would you be surprised if a census revealed that the population proportion was 0.26? Explain.

(c) Explain how the traffic analyst could decrease the margin of error.

11. **Commuting to work** A study of commuting times reports the travel times to work of a random sample of 20 employed adults in New York State. The mean travel time is $\bar{x}$ = 31.25 minutes and the margin of error for this estimate is 4.47 minutes.

(a) Interpret the margin of error.

(b) Would you be skeptical if the governor claimed that the average travel time to work for all employed adults in New York State is less than 30 minutes? Explain.

(c) Explain how researchers could reduce the margin of error for this estimate.

12. **How many licks?** Many people have asked the question, but few have been patient enough to collect the data. How many licks does it take to get to the center of a Tootsie Pop? Researcher Corey Heid decided to find out. He instructed a random sample of 92 students to lick a Tootsie Pop along the non-banded side until they could taste the chocolate center. The mean number of licks was 356.1 and the margin of error for this estimate is 9.6 licks.[22]

(a) Interpret the margin of error.

(b) Would you be skeptical if someone claimed that it takes more than 400 licks, on average, to get to the center of a Tootsie Pop? Explain.

(c) Explain how Heid could reduce the margin of error for his estimate.

13. **Football on TV** Refer to Exercise 9. Suppose that the researchers wanted to make separate estimates for males and females using the original sample of 1000 adults. How does the margin of error for each of these estimates compare to the margin of error from Exercise 9? Explain.

14. **More colorful cars in Miami** Refer to Exercise 10. Suppose that we wanted to make separate estimates for eastbound and westbound cars using the original sample of 800. How does the margin of error for each of these estimates compare to the margin of error from Exercise 10? Explain.

15. **Running a red light** A random digit dialing telephone survey of 880 drivers asked, "Recalling the last ten traffic lights you drove through, how many of them were red when you entered the intersections?" Of the 880 respondents, 171 admitted that at least one light had been red.[23]

(a) Based on this sample, estimate the proportion of all drivers who would say that at least one light was red.

(b) Use the *One Categorical Variable* applet at highschool.bfwpub.com/spa3e to approximate and interpret the margin of error for this estimate.

16. **Can you spare a square?** Christina and Rachel randomly selected 18 rolls of a generic brand of toilet paper to measure how well this brand could absorb water. To do this, they poured $\frac{1}{4}$ cup of water onto a hard surface and counted how many squares of toilet paper it took to completely absorb the water. Here are the results from their 18 rolls:

29 20 25 29 21 24 27 25 24 29 24 27 28 21 25 26 22 23

   (a) Based on this sample, estimate the mean number of squares of generic brand toilet paper it takes to completely absorb $\frac{1}{4}$ cup of water.

   (b) Use the *One Quantitative Variable* applet at high-school.bfwpub.com/spa3e to approximate and interpret the margin of error for this estimate.[24]

Extending the Concepts

17. **Going to prom** Mackenzie randomly selected 50 seniors at her school to estimate the proportion of all seniors who are planning to go to the prom. Her estimate is 0.58 and the margin of error for her estimate is 0.14. If there are 600 seniors at her school, what is the best estimate for the total number of seniors who are planning to go to the prom? What is the margin of error for this estimate?

18. **A wordy novel** Sanjung wants to estimate the total number of words in her favorite novel. Instead of counting every word on every page, she estimates the mean number of words on each page by randomly selecting 20 pages and counting the number of words on those pages. Her estimate is 382 words per page and the margin of error for her estimate is 28 words. If there are 520 pages in the novel, what is the best estimate for the total number of words in the novel? What is the margin of error for this estimate?

Recycle and Review

19. **Still living at home?** (1.2) What kinds of living arrangements are most prevalent among young adults? Have things changed since the 1960s? The bar graph compares the living arrangements of adults aged 18 to 31 years in 1968 and in 2012.

Write a few sentences comparing the distribution of living arrangements for the two years.[25]

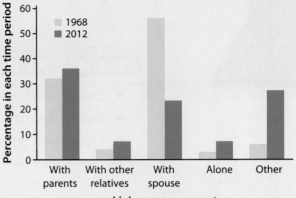

20. **Everybody loves a winner** (2.2, 2.3, 2.4) The scatterplot shows the relationship between wins and average attendance at home games during the 2014–2015 season for the 30 teams in the National Basketball Association.[26]

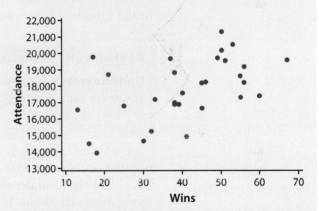

   (a) Describe the relationship between wins and home attendance.

   (b) Which of the following values is closest to the correlation between these two variables: –1.0, –0.5, 0, 0.5, or 1.0? Explain your reasoning.

   (c) If we plotted wins on the y-axis and attendance on the x-axis, how would the correlation change? Explain.

Lesson 3.5

Sampling and Surveys

LEARNING TARGETS

- Explain how undercoverage can lead to bias.
- Explain how nonresponse can lead to bias.
- Explain how other aspects of a sample survey can lead to bias.

Even when samples are selected at random, there are other potential sources of bias in a survey.

Undercoverage

Some samples are selected by randomly calling landline telephones. Unfortunately, this means people who only use a cell phone (or no phone at all) cannot be included in the sample. These types of samples suffer from **undercoverage.**

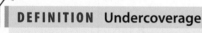

DEFINITION Undercoverage

Undercoverage occurs when some members of the population are less likely to be chosen or cannot be chosen for the sample.

Undercoverage leads to bias when the "undercovered" individuals differ from the population in ways that affect their responses. For example, randomly dialing landlines is likely to under-represent young people because many younger people exclusively use a cell phone. If a landline survey is about a tax increase that will help reduce the cost of college, you can bet that the proportion in the sample who favor the proposal will be less than the proportion in the entire population.

EXAMPLE

How long can you sleep in?

Undercoverage

PROBLEM: How long before school starts do students get out of bed, on average? Administrators survey a random sample of students on each school bus one morning. Describe how undercoverage might lead to bias in this study. Explain the likely direction of the bias.

SOLUTION:

Students who take the bus probably have to wake up earlier than other students, so they don't miss the bus. Because the sample includes only students who take the bus and not other students, the administrator's estimate of the average time is likely to be too large.

FOR PRACTICE   TRY EXERCISE 1.

To avoid the bias caused by undercoverage in the previous example, the administrators should select a sample at random using a list of every student in the school. The list of individuals from which a sample will be drawn is called the *sampling frame*. If the sampling frame doesn't contain every member of the population, there will be undercoverage.

Nonresponse

Even if every member of the population is equally likely to be selected for a sample, not all members of the population are equally likely to provide a response. Some people are never at home and cannot be reached by pollsters on the phone or in person. Other people see an unfamiliar phone number on their caller ID and never pick up the phone or quickly hang up when they don't recognize the voice of the caller. These are examples of **nonresponse,** another major source of bias in surveys.

DEFINITION Nonresponse

Nonresponse occurs when an individual chosen for the sample can't be contacted or refuses to participate.

Nonresponse leads to bias when the individuals who can't be contacted or refuse to participate would respond differently from those who do participate. Consider a telephone survey that asks people how many hours of television they watch per day. People who are selected but are out of the house won't be able to respond. Because these people probably watch less television than the people who are at home when the phone call is made, the mean number of hours obtained in the sample is likely to be greater than the mean number of hours for the entire population.

How bad is nonresponse? According to polling guru Nate Silver, "Response rates to political polls are dismal. Even polls that make every effort to contact a representative sample of voters now get no more than 10 percent to complete their surveys—down from about 35 percent in the 1990s."[26]

EXAMPLE

Can they keep the weight off?

Nonresponse

PROBLEM: Three hundred people participated in a free 12-week weight-loss course at a community health clinic. After one year, administrators emailed each of the 300 participants to see how much weight they had lost since the end of the course. Only 56 participants responded to the survey. The mean weight loss for this sample was $\bar{x} = 13.6$ pounds. Describe how nonresponse might lead to bias in this study. Explain the likely direction of the bias.

SOLUTION:

People who haven't lost weight might be embarrassed and choose not to reply to the survey. On the other hand, people who have lost weight are probably happy to share their success with the administrators. This means that the estimate from the sample is likely to be larger than the mean weight loss for all 300 participants in the course.

FOR PRACTICE   TRY EXERCISE 5.

The best way to reduce the effect of nonresponse is to follow up with people who don't respond the first time. In the weight-loss example, administrators could call people who didn't reply to the email, send them a letter, or visit the participants at their home. In other types of surveys, polling companies encourage people to respond by offering a small payment.

Be sure you don't use the term "voluntary response" instead of "nonresponse" to explain why certain individuals don't respond after being selected for a sample survey. Think about it this way: Nonresponse can occur only *after* a sample has been selected. In a voluntary response sample, every individual has opted to take part, so there won't be any nonresponse.

Other Sources of Bias

Even when a researcher can avoid undercoverage and nonresponse, it is still possible for bias to affect the results. Characteristics of the interviewer, wording of the question, respondents being untruthful, lack of anonymity, and many other factors can lead to **response bias**.

DEFINITION Response bias

Response bias occurs when there is a consistent pattern of inaccurate responses to a survey question.

Is cheerleading a sport? It depends on who's asking. When Hailey surveyed students at her school, 80% said "Yes" when she was wearing her cheerleading outfit. When she was dressed in regular school clothes, only 28% said "Yes."[28]

Do Americans doubt the Holocaust occurred? A badly worded question seemed to indicate that nearly one-third of all Americans had some doubt that the Holocaust occurred. Here was the question that the Roper polling organization asked in 1992: "Does it seem possible or does it seem impossible to you that the Nazi extermination of the Jews never happened?" Only 65% said it was impossible that it never happened. But the wording of the question included a double negative, making it difficult to understand. When the question was revised, the percentage who were certain that the Holocaust occurred dramatically increased.[29]

EXAMPLE

Did you wash your hands?

Response bias

PROBLEM: What percent of Americans wash their hands after using the bathroom? It depends on how you collect the data. In a telephone survey of 1006 U.S adults, 96% said they always wash their hands after using a public restroom. An observational study of 6028 adults in public restrooms told a different story: Only 85% of those observed washed their hands after using the restroom. Explain why the results of the two studies are so different.[30.]

SOLUTION:

When asked in person, many people might lie about always washing their hands because they want to appear like they have good hygiene. When people are only observed and not asked directly, the percentage who wash their hands will be smaller—and much closer to the truth.

FOR PRACTICE   TRY EXERCISE 9.

In the preceding example, the 85% figure was obtained by stationing an "observer" in the bathroom who was brushing teeth or putting on make-up. Imagine how much lower the percentage of hand washers might be if no one else was present in the bathroom to encourage good behavior!

How to Survey Well

Completely avoiding bias when conducting a survey is likely impossible. But there are ways that thoughtful researchers can keep bias to a minimum. Selecting a random sample is a good start. And taking steps to avoid undercoverage, nonresponse, and response bias can make surveys even better. Finally, it is a good idea to test out the survey on a small group of people before administering it to the entire sample. This type of pilot study can help discover any problems with question wording or other unforeseen issues.

LESSON APP 3.5

Who did you say is calling? *Literary Digest?*

One of the most famous flops in survey history occurred in 1936. To predict the outcome of the presidential election between Republican Alf Landon and Democrat Franklin D. Roosevelt, the magazine *Literary Digest* sent over 10,000,000 "ballots" to its subscribers. They also sent "ballots" to registered owners of an automobile or telephone. About 2,400,000 of the ballots were returned, with a large majority (57%) favoring Landon. The election turned out to be a landslide, but for Roosevelt (61%) instead of Landon.[31]

1. Explain how undercoverage might have led to bias in this survey.

2. Explain how nonresponse might have led to bias in this survey.

3. If the magazine followed up with people who didn't return their ballots and was able to obtain responses, would this eliminate the bias described in part (a) or (b)? Explain.

Lesson 3.5

WHAT DID YOU LEARN?

LEARNING TARGET	EXAMPLES	EXERCISES
Explain how undercoverage can lead to bias.	p. 214	1–4
Explain how nonresponse can lead to bias.	p. 215	5–8
Explain how other aspects of a sample survey can lead to bias.	p. 216	9–14

Exercises · Lesson 3.5

Mastering Concepts and Skills

1. You paid what for those tickets? Suppose you want to know the average amount of money spent by the fans attending opening day for the Cleveland Indians baseball season. You get permission from the team's management to conduct a survey at the stadium, but they will not allow you to bother the fans in the club seating or box seats (the most expensive seating). Using a random number generator, you select 500 seats from the rest of the stadium. During the game, you ask the fans in those seats how much they spent that day. Describe how undercoverage might lead to bias in this study. Explain the likely direction of the bias.

pg 214

2. Eating on campus The director of student life at a university wants to know what percent of students eat regularly on campus. To find out, the director selects an SRS of 300 students who live in the dorms. Describe how undercoverage might lead to bias in this study. Explain the likely direction of the bias.

3. Immigration reform A news organization wants to know what percent of U.S. residents support a "pathway to citizenship" for people who came to the United States illegally. The news organization randomly selects registered voters for the survey. Describe how undercoverage might lead to bias in this study. Explain the likely direction of the bias.

4. Mowing the grass The manager of a home improvement store wants to know the percent of residents in his town who mow their own lawns. To find out, he selects a random sample of 100 customers and asks each one if he or she mows the lawn. Describe how undercoverage might lead to bias in this study. Explain the likely direction of the bias.

5. Improve your golf scores A website offers "the secret technique to improving your golf game" for a mere $49.95. To provide evidence of the technique's

pg 215

success, the seller sends out emails to everyone who has paid for the secret to report on how much it has lowered their scores (low golf scores are better than high scores). The mean reduction in score for the golfers who respond is 2.4 strokes. Describe how nonresponse might lead to bias in this study. Is the mean reduction of 2.4 strokes likely greater than or less than the actual mean reduction for all purchasers of the secret?

6. Effective SAT prep? Suppose an SAT preparation course makes the following statement in their advertising brochure: "We asked our graduates to report on their SAT scores before and after taking our course. The mean increase in SAT scores for the 500 graduates who responded was 210 points." Describe how nonresponse might lead to bias in this study. Is the mean of 210 points likely greater than or less than the actual mean increase for all graduates of the course?

7. Driving distance A survey of drivers began by randomly sampling all listed residential telephone numbers in the United States. Of 45,956 calls to these numbers, 5029 were completed. The goal of the survey was to estimate how far people drive, on average, per day. Describe how nonresponse might lead to bias in this study. Explain the likely direction of the bias. [32]

8. Who's at home? A local news agency conducted a survey about unemployment by randomly dialing phone numbers until it had gathered responses from 1000 adults in its state. In the survey, 19% of those who responded said they were not currently employed. Describe how nonresponse might lead to bias in this study. Explain the likely direction of the bias.

9. Did you use your seat belt? A study in El Paso, Texas, looked at seat-belt use by drivers. Drivers were observed at randomly chosen convenience stores. After they left their cars, they were invited

pg 216

to answer questions that included questions about seat-belt use. In all, 75% said they always used seat belts, yet only 61.5% were wearing seat belts when they pulled into the store parking lots. Explain why the two percentages are so different.[33]

10. **A picture of health** Emma asked 80 randomly selected people if free health care should be provided to the homeless. Half of the people were shown a picture of a homeless woman with a small child. When shown the picture, 67.5% agreed that free health care should be provided to the homeless. When the picture was not shown, only 45% agreed with this statement. Explain why the two percentages are so different.[34]

11. **Boys don't cry?** Two female statistics students asked a random sample of 60 high-school boys if they have ever cried during a movie. Thirty of the boys were asked directly and the other 30 were asked anonymously by means of a "secret ballot." When the responses were anonymous, 63% of the boys said "Yes," while only 23% of the other group said "Yes." Explain why the two percentages are so different.

12. **Weight? Wait what?** Marcos asked a random sample of 50 mall shoppers for their weight. Twenty-five of the shoppers were asked directly and the other 25 were asked anonymously by means of a "secret ballot." The mean reported weight was 13 pounds heavier for the anonymous group. Explain why the two means were so different.[35]

13. **Throwing away diapers** A survey paid for by makers of disposable diapers found that 84% of the sample opposed banning disposable diapers. Here is the actual question:

It is estimated that disposable diapers account for less than 2% of the trash in today's landfills. In contrast, beverage containers, third-class mail, and yard wastes are estimated to account for about 21% of the trash in landfills. Given this, in your opinion, would it be fair to ban disposable diapers?

Explain how the wording of the question could result in bias. Be sure to specify the direction of the bias.[36]

14. **National health insurance** A simple random sample of 1200 adult Americans is selected, and each person is asked the following question:

In light of the huge national deficit, should the government at this time spend additional money to establish a national system of health insurance?

Explain how the wording of the question could result in bias. Be sure to specify the direction of the bias.

Applying the Concepts

15. **Blowing through red lights** A sample of 880 drivers was asked the following question: "Recalling the last ten traffic lights you drove through, how many of them were red when you entered the intersections?" Of the 880 respondents, 171 admitted that at least one light had been red. The drivers were selected by randomly sampling all listed residential telephone numbers.[37]

(a) Explain how undercoverage might have led to bias in this survey.

(b) Explain how response bias might affect the results of this survey.

(c) Would increasing the sample size eliminate the bias described in part (a) or (b)? Explain.

16. **Still another Facebook poll** A U.S. senator sends out a poll to people who have "liked" him on Facebook, asking, "Do you approve or disapprove of the U.S. Supreme Court decision in *Citizens United* that let corporations and wealthy individuals spend unlimited amounts on campaigns?" Ninety-nine percent of the responses were "disapprove."

(a) Explain how undercoverage might have led to bias in this survey.

(b) Explain how the wording of the question might affect the results of this survey.

(c) Would increasing the sample size eliminate the bias described in part (a) or (b)? Explain.

17. **Paper or plastic?** Some cities are proposing a ban on plastic grocery bags. Write two versions of a question that ask respondents about this issue. One version should be neutral, while the other should encourage people to either support or oppose the ban.

18. **Homework time?** The principal of a high school wants to estimate the mean time that students spend on homework each night. Write two versions of a question that asks students to estimate the typical amount of time they spend on homework each night. One version should be neutral, while the other should encourage them to either overstate or understate the time.

Extending the Concepts

19. **Ring, ring, ring** A common form of nonresponse in telephone surveys is "ring-no-answer." That is, a call is made to an active number but no one answers. The Italian National Statistical Institute looked at nonresponse to a government survey of households in Italy during the periods January 1 to Easter and July 1 to August 31. All calls were made between 7:00 and 10:00 P.M., but 21.4% gave "ring-no-answer" in one period versus 41.5% "ring-no-answer" in the other period.

(a) Which period do you think had the higher rate of no answers? Why?

(b) Write a question where you would expect the results to differ in the two time periods.

Recycle and Review

20. **Mean versus median** (1.6) For each of the following settings, explain how the mean and median would compare and which measure of center is the more appropriate choice.

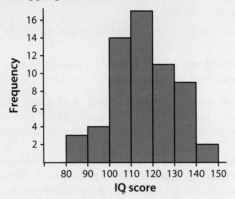

(a) The histogram displays data on the IQ scores of 60 randomly selected fifth-grade students from one school.

(b) In Major League Baseball, most players earn close to the minimum salary (which was $500,000 in 2013), while a few earn more than $10 million.[38]

21. **Olympic times** (2.2, 2.5) The scatterplot shows the winning times in the Men's Olympic 100-meter dash from 1900 through 2012.

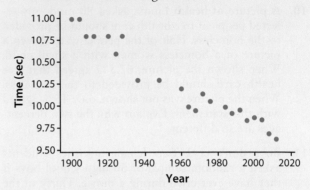

(a) Describe the relationship between the winning time and year.

(b) The equation of the least-squares linear regression line is time = 32.84 − 0.01153 (year). Interpret the slope of this line in the context of the problem.

(c) Predict the winning time in the 2100 Olympics. Is this realistic? Explain.

Lesson 3.6

Observational Studies and Experiments

LEARNING TARGETS

- Explain the concept of confounding and how it limits the ability to make cause-and-effect conclusions.
- Explain the purpose of comparison in an experiment.
- Describe the placebo effect and the purpose of blinding in an experiment.

In Lesson 3.1, you learned how to distinguish observational studies from experiments. The goal of an observational study can be to describe some group or situation, to compare groups, or to examine relationships between variables. Unfortunately, observational studies make it very difficult to identify cause-and-effect relationships.

Confounding

In Lesson App 3.1, you read about an observational study that revealed teenagers who eat with their families at least 5 times a week are more likely to get better grades in school. Unfortunately, it's not appropriate to conclude that increasing the number of meals you eat with your family will *cause* your grades to improve. As shown in the table below, there are many differences between teens who eat often with their families and teens who don't. Any of these differences could be the cause of the difference in grades.

Perhaps the teens who eat more meals with their families are less likely to have part-time jobs or participate in after-school activities, allowing more time for studying—and for dinner with their families. In this case, we wouldn't know which variable was causing the good grades—the number of meals with their family or the amount of available study time.

Families who eat dinner together often	Families who do *not* eat dinner together often
Students are *less* likely to have a part-time job.	Students are *more* likely to have a part-time job.
Students are *less* likely to participate in extracurricular activities.	Students are *more* likely to participate in extracurricular activities.
Students have *more* time to study.	Students have *less* time to study.
Students have *better* grades.	**Students have *worse* grades.**

Recall that a response variable measures an outcome of a study. An explanatory variable may help predict or explain changes in a response variable. When it is impossible to tell if the explanatory variable or some other variable is causing a change in a response variable, we say the two variables are **confounded**.

DEFINITION Confounding

Confounding occurs when two variables are associated in such a way that their effects on a response variable cannot be distinguished from each other.

Observational studies cannot definitively show a cause-and-effect relationship between an explanatory variable and a response variable because of confounding.

CAUTION

EXAMPLE

Can ADHD be linked to mothers who smoke?

Confounding

PROBLEM: In a study of more than 4700 children, researchers from Cincinnati Children's Hospital Medical Center found that those children whose mothers smoked during pregnancy were more than twice as likely to develop ADHD as children whose mothers had not smoked.[39] Based on this study, is it reasonable to conclude that a mother's smoking during pregnancy causes an increase in the risk of ADHD in her children? Explain.

SOLUTION:

No. Although smoking during pregnancy may increase the risk of ADHD in children, it is possible that the mothers who smoked during pregnancy were also more likely to have unhealthy diets. Perhaps it was the unhealthy diets that caused the increase in ADHD risk, not the smoking.

FOR PRACTICE   TRY EXERCISE 1.

The easiest way to identify confounding in an observational study is to think about other variables that are associated with the explanatory variable that might cause a change in the response variable. In the smoking and ADHD study, there were many potential differences between the group of mothers who smoked during pregnancy and the group of mothers who didn't. Likewise, there are many potential differences between the group of teens who eat often with their families and the group of teens who don't. Any of these differences could be the cause of the change in the response variable.

Experiments: Comparison

Conducting a well-designed experiment is one of the best ways to prevent confounding. Recall that an experiment deliberately imposes some **treatment** on individuals to measure their responses. These individuals are called **experimental units,** or **subjects.**

> **DEFINITION** Treatment, Experimental unit, Subject
>
> A **treatment** is a specific condition applied to the individuals in an experiment. The **experimental units** are the smallest collection of individuals to which treatments are randomly assigned. When the units are human beings, they often are called **subjects.**

To learn what it means to be a "well-designed" experiment, let's start with a poorly designed experiment. Mr. Luckow's class wants to investigate if caffeine affects pulse rates. One of his students proposes the following plan:

1. Have each student measure his or her initial pulse rate.
2. Give each student 12 ounces of cola.
3. Wait for 15 minutes.
4. Have each student measure his or her final pulse rate.
5. Compare initial and final pulse rates.

There are a number of problems with this proposed experiment. If the cola has sugar as well as caffeine, we wouldn't be able to tell if the increase in pulse rates was due to the presence of caffeine or the presence of sugar. (Can you say confounding?) Likewise, there could be other events between the initial and final pulse rate measurements that could be the cause of an increase in pulse rates, such as a fire drill or an exciting statistics lesson.

The solution for these problems is to have *two groups* in the experiment: one group that receives the caffeine and a **control group** that doesn't receive the caffeine.

> **DEFINITION** Control group
>
> In an experiment, a **control group** is a group used to provide a baseline for comparing the effects of other treatments. Depending on the purpose of the experiment, a control group may be given an inactive treatment or an active treatment.

In all other ways, these groups should be treated exactly the same so that the only difference is the caffeine. That way, if there is convincing evidence of a difference in the average increase in pulse rates, we can conclude it was *caused* by the caffeine. This means that one group could get regular cola with caffeine, while the control group gets caffeine-free cola. Both groups would get the same amount of sugar, so sugar consumption would no longer be confounded with caffeine intake. Likewise, both groups would experience the same events during the experiment so what happens during the experiment won't be confounded with caffeine intake either. *Using a design that compares two or more treatments is the first step in designing a good experiment.*

What devices get customers to conserve electricity?

Comparison

PROBLEM: Many utility companies have introduced programs to encourage energy conservation among their customers. An electric company considers placing small digital displays in households to show current electricity use along with a projected monthly cost. Will the displays reduce electricity use? One cheaper approach is to give customers a chart and information about monitoring their electricity use from their outside meter. Would this method work almost as well? The company decides to conduct an experiment to compare these two approaches (display, chart) with a control group of customers who receive information about energy consumption but no help in monitoring electricity use. Explain why it was necessary to include the control group.

SOLUTION:

The purpose of the control group is to provide a baseline for comparing the effects of the other treatments. Otherwise, we wouldn't be able to tell if the displays or charts caused a reduction in electricity use.

FOR PRACTICE **TRY EXERCISE 5.**

Comparing the group that received the display with the group that received the chart helps us decide if one of these methods is better than the other. Including a control group helps us determine if these two methods are better than providing customers with no information at all. Perhaps the weather this year is less extreme than in the past, allowing households to use less electricity. If this was the case, use of the displays and charts would be confounded with the weather—we wouldn't know if electricity use was lower because of the weather or because of the displays and charts. But having a control group allows us to account for the effect of the weather.

Experiments: The Placebo Effect

In the caffeine experiment, we used comparison to help prevent confounding. But even when there is comparison, confounding is still possible. If each student in Mr. Luckow's class knows what type of soda he or she is receiving, the expectations of the two groups will be different. The knowledge that a subject is receiving caffeine may increase his or her pulse rate, apart from the caffeine itself. This is an example of the **placebo effect.**

DEFINITION Placebo effect

The **placebo effect** describes the fact that some subjects in an experiment will respond favorably to any treatment, even an inactive treatment.

To make sure the expectations of all subjects are the same in Mr. Luckow's caffeine experiment, we give one group cola with caffeine and the other group a **placebo**—a cola that looks and tastes identical, but without caffeine.

DEFINITION Placebo

A **placebo** is a treatment that has no active ingredient, but is otherwise like other treatments.

Placebos are often "sugar pills," though they can also be any other type of substance as long as it looks, tastes, smells, and feels the same as the active treatment. Placebos can even be a procedure such as placebo surgery!

Of course, a placebo isn't that useful if a subject knows he or she is getting a placebo. It is important that the placebo look, feel, and taste just like the other treatments, so the subjects will not know which treatment they are receiving. Otherwise, the expectations of the subjects receiving the placebo will differ from the expectations of other subjects, creating confounding between the explanatory variable and the expectations of the subjects.

THINK ABOUT IT **Do placebos work?** One study found that 42% of balding men maintained or increased the amount of hair on their heads when they took a placebo. In another study, researchers zapped the wrists of 24 test subjects with a painful jolt of electricity. Then, they rubbed a cream with no active medicine on subjects' wrists and told them the cream should help soothe the pain. When researchers shocked them again, 8 subjects said they experienced significantly less pain.[40] With ailments like depression, some experts think that the placebo effect accounts for about three-quarters of the effect of the most widely used drugs.[41] Others disagree. In any case, "placebos work" is a good place to start when you think about planning medical experiments.

EXAMPLE

Do magnets repel pain?

The placebo effect

PROBLEM: Early research showed that magnetic fields affected living tissue in humans. Some doctors have begun to use magnets to treat patients with chronic pain. Scientists wondered if this type of therapy really worked. They designed a study to find out.

Fifty patients with chronic pain were recruited for the study. A doctor identified a painful site on each patient and asked him or her to rate the pain on a scale from 0 (mild pain) to 10 (severe pain). Then, the doctor selected a sealed envelope containing a magnet at random from a box with a mixture of active and inactive magnets. The chosen magnet was applied to the site of the pain for 45 minutes. After being treated, each patient was again asked to rate the level of pain from 0 to 10.[42]

In this experiment, the patient didn't know which type of magnet was being used. Explain why this is an important consideration.

SOLUTION:

If subjects know they are receiving an active treatment, researchers wouldn't know if any improvement was due to the magnets or to the expectation of getting better (the placebo effect).

FOR PRACTICE **TRY EXERCISE 9.**

Besides the subjects, the doctor in the magnet study was also unaware of which patients were receiving the active treatment. This type of experiment is called **double-blind.** Other experiments are **single-blind.**

DEFINITION Double-blind, Single-blind

In a **double-blind** experiment, neither the subjects nor those who interact with them and measure the response variable know which treatment a subject received.

In a **single-blind** experiment, either the subjects don't know which treatment they are receiving or the people who interact with them and measure the response variable don't know which subjects are receiving which treatment.

If the doctor knew who was receiving the active magnets, she might consciously or subconsciously treat the subjects differently. To avoid this potential source of confounding, it's best if the experiment is double-blind.

LESSON APP 3.6

What happens when physicians study themselves?

Does regularly taking aspirin help protect people against heart attacks? The Physicians' Health Study I was a medical experiment that helped answer this question. The subjects in this experiment were 21,996 male physicians. Half of these subjects took an aspirin tablet every other day and the remaining subjects took a dummy pill that looked and tasted like the aspirin but had no active ingredient. After several years, 239 of the control group but only 139 of the aspirin group had suffered heart attacks. This difference is large enough to provide convincing evidence that taking aspirin does reduce heart attacks.[43]

1. Why was it necessary to perform an experiment rather than simply asking the doctors whether or not they take aspirin regularly?

2. Explain why it was necessary for the experiment to include a control group that didn't receive aspirin.

3. Was blinding used in this experiment? Explain why this is an important consideration.

© Science Photo Library/age fotostock

Lesson 3.6

WHAT DID YOU LEARN?

LEARNING TARGET	EXAMPLES	EXERCISES
Explain the concept of confounding and how it limits the ability to make cause-and-effect conclusions.	p. 221	1–4
Explain the purpose of comparison in an experiment.	p. 223	5–8
Describe the placebo effect and the purpose of blinding in an experiment.	p. 224	9–12

Exercises — Lesson 3.6

Mastering Concepts and Skills

1. **Good for the gut?** Is fish good for the gut? Researchers tracked 22,000 male physicians for 22 years. Those who reported eating seafood of any kind at least 5 times per week had a 40% lower risk of colon cancer than those who said they ate seafood less than once a week. Based on this study, is it reasonable to conclude that eating seafood causes a reduction in the risk of colon cancer? Explain.[44]
   pg 221

2. **Move over carrots?** In a long-term study, researchers collected information about the diets of participants and whether or not they developed the eye condition called age-related macular degeneration (AMD). People who ate lots of green leafy vegetables, which are high in lutein and zeaxanthin, had a lower risk of AMD. Based on this study, is it reasonable to conclude that eating green leafy vegetables causes a reduction in the risk of AMD? Explain.[45]

3. **It pays to be mean?** A recent study based on data from three different surveys spanning 20 years found that men who scored below average on "agreeableness" earned, on average, 18% more than those who scored above average. Though the gap was smaller, a similar result was found for women. Based on this study, is it reasonable to conclude that men are likely to earn a higher salary if they act less agreeable? Explain.[46]

4. **Straight As now, healthy later** A study by Pamela Herd of the University of Wisconsin-Madison found a link between high-school grades and health. Analyzing data from the Wisconsin Longitudinal Study, which has tracked the lives of thousands of Wisconsin high-school graduates from the class of 1957, Herd found that students with higher grade-point averages were more likely to say they were in excellent or very good health in their early 60s. Does this mean people will live healthier lives if they increase their GPA? Explain.[47]

5. **More rain?** The changing climate might affect the amount of rain that falls during different seasons. Eighteen plots of open grassland, each with an area 70 square meters, were available for a study. One response variable is total plant biomass produced in a plot over a year. Kenwyn Suttle of the University of California at Berkeley and his coworkers wanted to compare the effects of three treatments: added water equal to 20% of annual rainfall either during January to March (winter) or during April to June (spring), and no added water (control). Explain why it was necessary to include a control group that didn't get additional water.[48]

pg 223

6. **Getting the teacher to show up** Elementary schools in rural India are usually small, with a single teacher. The teachers often fail to show up for work. Here is an idea for improving attendance: Offer the teacher better pay for good attendance. Give the teacher a digital camera with a tamper-proof time and date stamp, and ask a student to take a photo of the teacher and class at the beginning and end of the day to verify attendance. Will this work? Researchers obtained permission to use 120 rural schools in Rajasthan for an experiment to find out. Explain why it is necessary to include a control group that doesn't get a financial incentive.[49]

7. **Career education** Surveys of recent high-school dropouts revealed that many of these students saw little connection between what they were studying in school and their future plans. To change this perception, researchers developed a program called CareerStart, in which teachers show students how the topics they learn about are used in specific careers. Seven of the 14 schools in Forsyth County, North Carolina, were selected at random to use CareerStart along with the district's standard curriculum. The other seven schools just followed the standard curriculum. Researchers followed both groups of students for several years, collecting data on students' attendance, behavior, standardized test scores, level of engagement in school, and whether or not the students graduated from high school. *Results:* Students at schools that used CareerStart typically did better in each of these areas. Explain why it was necessary to include a control group of schools that didn't use the CareerStart program.[50]

8. **Yum, jellyfish!** Research by Andrew Sweetman of Norway's International Research Institute of Stavanger focused on whether or not deep-sea scavengers consume dead jellyfish. His team lowered platforms piled with dead jellyfish and other platforms piled with dead mackerel more than 4000 feet into Norway's largest fjord, and they found that hagfish, crabs, and other scavengers consumed the jellyfish in a few hours—faster than they consumed the mackerel. Explain why it was necessary to include platforms containing mackerel in the experiment, even though the study was focused on the consumption of jellyfish.[51]

9. **Oils to make us limber?** The extracts of avocado and soybean oils have been shown to slow cell inflammation in test tubes. Will taking avocado and soybean unsaponifiables (ASU, i.e., components of the oily mixture) help relieve pain for subjects with joint stiffness due to arthritis? In an experiment, 345 men and women were randomly assigned to receive either 300 milligrams of ASU daily or a daily placebo for three years. Could blinding be used in this experiment? Explain. Why is blinding an important consideration in this experiment?[52]

pg 224

10. **Supplements for testosterone** As men age, their testosterone levels gradually decrease. This may cause a reduction in energy, an increase in fat, and other undesirable changes. Do testosterone supplements reverse some of these effects? A study in the Netherlands assigned 237 men aged 60 to 80 with low or low-normal testosterone levels to either a testosterone supplement or a placebo. Could blinding be used in this experiment? Explain. Why is blinding an important consideration in this experiment?[53]

11. **Meditate to lessen anxiety** An experimenter interviewed subjects and rated their level of anxiety. Then, the subjects were randomly assigned to two groups. The experimenter taught one group how to meditate, and they meditated daily for a month. The other group was simply told to relax more. At the end of the month, the experimenter interviewed all the subjects again and rated their anxiety level. The meditation group now had less anxiety. Psychologists said that the results were suspect because the ratings were not blind. Explain what this means and why this is an important consideration in this experiment.

12. **Side effects** Even if an experiment is double-blind, the blinding might be compromised if side effects of the treatments differ. For example, suppose researchers at a skin-care company are comparing their new acne treatment against the leading competitor. Fifty subjects are assigned at random to each treatment, and the company's researchers will rate the improvement for each of the 100 subjects. The researchers aren't told which subjects received which treatments, but they know that their new acne treatment causes a slight reddening of the skin. How might this knowledge compromise the blinding? Explain why this is an important consideration in this experiment.

Applying the Concepts

13. **Ultrasounds** Researchers examined the effect of ultrasound on birth weight. Pregnant women participating in the study were randomly assigned to one of two groups. The first group of women received an ultrasound; the second group did not. When the subjects' babies were born, their birth weights were recorded. The women who received the ultrasounds had heavier babies, on average.[54]

(a) Why was it necessary to perform an experiment rather than simply asking the women whether or not they had an ultrasound during pregnancy?

(b) Explain why it was important for the experiment to include a control group that didn't receive an ultrasound.

(c) Could blinding be used in this experiment? Explain. Why is blinding an important consideration in this experiment?

14. **Flu shots and heart attacks?** In a series of experiments, subjects were randomly assigned to receive a flu vaccine or a placebo. In addition to recording whether or not subjects contracted the flu, researchers also recorded whether or not subjects suffered from "major adverse cardiovascular events" such as heart attacks. The subjects that received the flu vaccine were less likely to have a cardiovascular event.[55]

(a) Why was it necessary to perform an experiment rather than simply asking people if they have received a flu vaccine?

(b) Explain why it was important for the experiment to include a control group that didn't receive a flu vaccine.

(c) Could blinding be used in this experiment? Explain. Why is blinding an important consideration in this experiment?

15. **Cocoa in the blood** A study involved 27 healthy people aged 18 to 72 who consumed a cocoa beverage containing 900 mg of flavonols (a class of flavonoids) daily for 5 days. Using a finger cuff, blood flow was measured on the first and fifth days of the study. After 5 days, researchers measured what they called "significant improvement" in blood flow and the function of the cells that line the blood vessels. What flaw in the design of this experiment makes it impossible to say if the cocoa really caused the improved blood flow? Explain.[56]

16. **Back to work** Will cash bonuses speed the return to work of unemployed people? A state department of labor notes that last year 68% of people who filed claims for unemployment insurance found a new job within 15 weeks. This year, the state offers $500 to people filing unemployment claims if they find a job within 15 weeks. The percent increases to 77%. What flaw in the design of this experiment makes it impossible to say if the bonus really caused the increase? Explain.

17. **Do you get what you pay for?** In a recent study, researchers had volunteers rate the pain of an electric shock before and after taking a new medication. However, half of the subjects were told the medication cost $2.50 per dose, while the other half were told the medication cost $0.10 per dose. In reality, both medications were placebos. Of the "cheap" placebo users, 61% experienced pain relief, while 85% of the "expensive" placebo users experienced pain relief. Explain how the results of this study support the idea of a placebo effect.[57]

18. **Poison ivy** Researchers in Japan conducted an experiment on 13 individuals who were extremely allergic to poison ivy. On one arm, each subject was rubbed with a poison ivy leaf and told the leaf was harmless. On the other arm, each subject was rubbed with a harmless leaf and told it was poison ivy. All the subjects developed a rash on the arm where the harmless leaf was rubbed. Of the 13 subjects, 11 did not have any reaction to the real poison ivy leaf. Explain how the results of this study support the idea of a placebo effect.[58]

Extending the Concepts

19. **Piglet chow** A well-designed experiment should not be subject to confounding, but a *poorly* designed experiment might be. Suppose you are conducting an experiment to determine the weight gain of piglets on a newly designed piglet food. You have two litters of piglets and an abundant supply of both the new piglet food and another type of piglet food that is commonly used. Design an experiment that would be subject to confounding, and explain why it is.

Recycle and Review

20. Reading and IQ related? (2.2, 2.4) Here is a scatterplot of reading test scores against IQ test scores for 14 fifth-grade children.

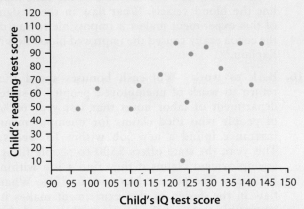

(a) Describe the relationship between reading score and IQ for these children.

(b) There is one low outlier in the plot. What effect does this low outlier have on the correlation?

21. Survival times (1.5, 1.7, 1.8) Here are the survival times in days of 72 guinea pigs after they were injected with infectious bacteria in a medical experiment. Survival times usually have distributions that are skewed to the right.[59]

43	45	53	56	56	57	58	66	67	73	74	79
80	80	81	81	81	82	83	83	84	88	89	91
91	92	92	97	99	99	100	100	101	102	102	102
103	104	107	108	109	113	114	118	121	123	126	128
137	138	139	144	145	147	156	162	174	178	179	184
191	198	211	214	243	249	329	380	403	511	522	598

(a) Make a histogram of the data and describe its main features. Does it show the expected right skew?

(b) Now make a boxplot of the data. Be sure to check for outliers.

(c) Which measure of center and spread would you use to summarize the distribution—the mean and standard deviation or the median and *IQR*? Justify your answer.

Lesson 3.7

How to Experiment Well

LEARNING TARGETS

- Describe how to randomly assign treatments using slips of paper or technology.
- Explain the purpose of random assignment in an experiment.
- Identify other sources of variability in an experiment and explain the benefits of keeping these variables the same for all experimental units.

In Lesson 3.6, you read about an experiment to determine if caffeine affects pulse rates. When planning this experiment, Mr. Luckow's class eliminated many sources of confounding by comparing two groups of subjects—one that got cola with caffeine and the other that got cola with no caffeine. However, comparison alone isn't enough to produce results we can trust.

Many other variables affect pulse rates besides caffeine. Some of these variables, such as caffeine tolerance, weight, and recent caffeine consumption, describe characteristics of the subjects. Other variables, such as sugar content, amount of soda consumed, and temperature, describe characteristics of the experimental process. To experiment well, we want to create groups that are identical to each other and treated in exactly the same way, other than the caffeine.

Experiments: Random Assignment

If treatments are given to groups that differ greatly when the experiment begins, confounding is likely. If Mr. Luckow lets students choose which group to join, those who rarely consume caffeine may be more likely to choose the placebo. Then, the two groups will differ with respect to additional variables that might affect pulse rates, such as caffeine tolerance and recent consumption of caffeine. This prevents us from being able to conclude that caffeine is the cause of a change in pulse rates. The solution to this problem is **random assignment**.

DEFINITION Random assignment

In an experiment, **random assignment** means that experimental units are assigned to treatments using a chance process.

EXAMPLE

Who gets the caffeine?

Random assignment methods

PROBLEM: In Mr. Luckow's class, 20 students volunteer to participate in the caffeine experiment. Describe how you would randomly assign 10 students to each treatment:

(a) Using identical slips of paper

(b) Using a random number generator

SOLUTION:

(a) On 10 slips of paper, write the letter "A"; on the remaining 10 slips, write the letter "B." Shuffle the slips of paper and hand out one slip of paper to each volunteer. Students who get an "A" slip will receive the cola with caffeine and students who get a "B" slip will receive the cola without caffeine.

(b) Label each student with a different integer from 1 to 20. Then, randomly generate 10 *different* integers from 1 to 20. The students with these labels will receive the cola with caffeine. The remaining 10 students will receive the cola without caffeine.

FOR PRACTICE   TRY EXERCISE 1.

If you use a random-number generator to assign the treatments, make sure your description addresses the possibility of repeated integers. Do this by explicitly stating "ignore repeated integers" or by saying you will generate a certain number of "different" integers. Regardless of the method used for random assignment, make sure students aren't told which treatment they are receiving!

Random assignment should distribute the students who have a caffeine tolerance in roughly equal numbers to each group. It should also roughly balance out the number of students who have had caffeine recently and make the average weight of subjects in each group roughly the same. In fact, the random assignment should create two groups that are roughly equivalent with respect to *every* variable that might affect the response.

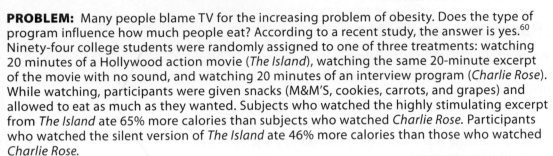

EXAMPLE

Does watching TV make you snack?

The purpose of random assignment

PROBLEM: Many people blame TV for the increasing problem of obesity. Does the type of program influence how much people eat? According to a recent study, the answer is yes.[60] Ninety-four college students were randomly assigned to one of three treatments: watching 20 minutes of a Hollywood action movie (*The Island*), watching the same 20-minute excerpt of the movie with no sound, and watching 20 minutes of an interview program (*Charlie Rose*). While watching, participants were given snacks (M&M'S, cookies, carrots, and grapes) and allowed to eat as much as they wanted. Subjects who watched the highly stimulating excerpt from *The Island* ate 65% more calories than subjects who watched *Charlie Rose*. Participants who watched the silent version of *The Island* ate 46% more calories than those who watched *Charlie Rose*.

Explain the purpose of the random assignment in the context of this experiment.

SOLUTION:

The purpose of random assignment is to create roughly equivalent groups at the beginning of the experiment. For example, if you let people choose which show they will watch, it is possible that more males would choose the action movie. Because males typically eat more than females, we wouldn't know if any difference in calorie consumption was due to gender or the type of programming.

FOR PRACTICE **TRY EXERCISE 5.**

As noted in the example, *the purpose of random assignment is to create groups that are roughly equivalent at the beginning of an experiment.* Random assignment works best when there are many subjects in the experiment. If only 6 subjects participated in the TV-watching experiment, the three groups are likely to be quite different, even if they were randomly assigned. For example, if 4 of the 6 volunteers are female, at least one of the treatment groups will be 100% female! In the actual study, there were 57 females and 37 males. With this many subjects, it is almost impossible that one of the treatment groups will end up as 100% female. The idea that we should use enough subjects to create roughly equivalent groups is sometimes called *replication*.

THINK ABOUT IT Does there have to be random assignment in an experiment? You could try to use only subjects that were almost the same to begin with, but several problems quickly arise. First, if you restrict your subjects to those with certain characteristics, you limit your ability to learn about other types of people. For example, if you only allow non-caffeine-drinking males that are between 160 and 170 lb into the experiment, you won't learn anything about females, caffeine drinkers, or people who weigh less than 160 lb or more than 170 lb. Second, it will be very difficult to find enough subjects to participate if there are very specific requirements for participation, making replication difficult. Third, even if you are using experimental units that seem identical, such as pea seeds or bottles of water, these units will still vary in other ways that might affect the response variable, whether you realize it or not. Because there will always be some differences between experimental units, you must always use random assignment to make the treatment groups roughly equivalent at the beginning of the study.

Experiments: Other Sources of Variability

Although random assignment should create groups of experimental units that are roughly equivalent at the beginning of an experiment, there are other variables that might have an impact on the response variable. In Mr. Luckow's caffeine experiment, the sugar content, amount of soda consumed, and temperature of the room during the experiment will likely affect pulse rates.

Because sugar almost certainly affects pulse rates, it is important that each subject receive the same amount of sugar. If one group received regular cola with caffeine and the other group received caffeine-free *diet* cola with no sugar, then sugar consumption and caffeine intake would be confounded. *One reason to keep other variables the same for each subject is to prevent confounding.*

We should also make sure that each student has the same amount of cola to drink. If each student was able to drink as much or as little as he or she wanted, the changes in pulse rates would be more variable than they would be otherwise. This increase in variation makes it harder to see if there is a difference in the average pulse rate increase for the two groups.

The dotplots in Figure 3.5(a) show the results of an experiment in which the amount of cola was the same for all participating students. From these graphs, it seems clear that the caffeine does increase pulse rates. The dotplots in Figure 3.5(b) show the results of an experiment in which the students were permitted to choose the amount of soda they drank. Notice that the centers of the distributions haven't changed, but the distributions on the right are much more variable. The increased overlap in the graphs makes the evidence supporting the effect of caffeine less convincing.

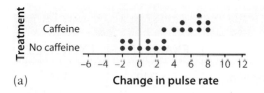

(a)

(b)

FIGURE 3.5 Dotplots showing the results of the caffeine experiment when (a) the amount of soda is kept the same and (b) when the amount of soda is allowed to vary.

The second reason we keep other variables the same is to reduce the variability in the response variable, making it easier to determine if one treatment is more effective than another.

EXAMPLE

More distractions, more calories?

Other sources of variability

PROBLEM: In the TV-and-snacks example from earlier in this lesson, researchers determined that people typically consume more calories when watching TV shows with more distractions. Identify one variable that the researchers kept the same for all subjects during the experiment. Provide two reasons why it was important for the researchers to keep this variable the same.

SOLUTION:

Amount of time watching TV. If one of the treatment groups was allowed to watch TV for a longer period of time, we wouldn't know if the amount of time or the type of program was the cause of a difference in calories. Also, if the researchers let subjects within each group watch TV for as little or as long as they wanted, the number of calories consumed would be much more variable than it would be otherwise, making it harder to tell if the type of program made a difference.

FOR PRACTICE   TRY EXERCISE 9.

LESSON APP 3.7

Multitasking? Or multiple distractions?

Researchers in Canada performed an experiment with university students to examine the effects of multitasking on student learning. The 40 participants in the study were asked to attend a lecture and take notes with their laptops. Half of the participants were randomly assigned to complete other online tasks not related to the lecture during that time. These tasks were meant to imitate typical student Web browsing during classes. The remaining students simply took notes with their laptops.

At the end of the lecture, all participants took a comprehension test to measure how much they learned from it. The results: students who were assigned to multitask did significantly worse (11%) than students who were not assigned to multitask.[61]

dgmata/Shutterstock

1. Describe how the researchers could have carried out the random assignment.

2. Why was it important that the researchers randomly assigned treatments to the students?

3. Identify one variable that the researchers kept the same for all subjects. Provide two reasons why it was important for the researchers to keep this variable the same.

Lesson 3.7

WHAT DID YOU LEARN?

LEARNING TARGET	EXAMPLES	EXERCISES
Describe how to randomly assign treatments using slips of paper or technology.	p. 229	1–4
Explain the purpose of random assignment in an experiment.	p. 230	5–8
Identify other sources of variability in an experiment and explain the benefits of keeping these variables the same for all experimental units.	p. 231	9–12

Exercises   Lesson 3.7

Mastering Concepts and Skills

1. **Pricey pizza?** The cost of a meal might affect how customers evaluate and appreciate food. To investigate, researchers worked with an Italian all-you-can-eat buffet to perform an experiment. A total of 139 subjects were randomly assigned to pay either $4 or $8 for the buffet and then asked to rate the quality of the pizza on a 9-point scale. Subjects who paid $8 rated the pizza 11% higher than those who paid only $4.[62] Describe how the researchers could have randomly assigned the subjects to treatments using (a) identical slips of paper and (b) a random-number generator.
pg 229

2. **Power and speech** Recent research suggests that people's speech patterns are influenced by how much power they think they have in a particular negotiation. In one study, 161 college students were randomly assigned to two groups. The subjects in

one group were told they had higher status or inside information, giving them the sense that they had more power than those they were negotiating with. Subjects in the other group were told something that lowered their perceived power. All subjects were given the same statement to read. The voices of the subjects in the "high power" group tended to go up in pitch and become more variable in loudness than the voices of students in the "low-power" group.[63] Describe how the researchers could have randomly assigned the subjects to treatments using (a) identical slips of paper and (b) a random-number generator.

3. **Effects of layoffs** Workers who survive a layoff of other employees at their location may suffer from "survivor guilt." A study used as subjects 120 students who were offered an opportunity to earn extra course credit by doing proofreading. Each subject worked in the same cubicle as another student, who was an accomplice of the experimenters. At a break midway through the work, one of three things happened:

Treatment 1: The accomplice was told to leave; it was explained that this was because he or she performed poorly.

Treatment 2: It was explained that unforeseen circumstances meant there was only enough work for one person. By "chance," the accomplice was chosen to be laid off.

Treatment 3: Both students continued to work after the break.

The subjects' work performance after the break was compared with performance before the break.[64] Describe how you would randomly assign the subjects to the treatments using (a) identical slips of paper and (b) a random-number generator.

4. **Headaches compared** Doctors identify "chronic tension-type headaches" as headaches that occur almost daily for at least six months. Can antidepressant medications or stress-management training reduce the number and severity of these headaches? Are both together more effective than either alone? Researchers want to compare four treatments: antidepressant alone, placebo alone, antidepressant plus stress management, and placebo plus stress management. Describe how you would randomly assign 36 volunteer subjects to the treatments using (a) identical slips of paper and (b) a random number generator.

5. **Even more pizza** Explain the purpose of the random assignment in the experiment described in Exercise 1.
pg 230

6. **More power, more speech?** Explain the purpose of the random assignment in the experiment described in Exercise 2.

7. **More layoffs** Explain the purpose of the random assignment in the experiment described in Exercise 3.

8. **More headache relief** Explain the purpose of the random assignment in the experiment described in Exercise 4.

9. **Pay to play?** Are politicians more available to people who donate money to their campaigns? Researchers had members of a political organization attempt to schedule an appointment with members of Congress to discuss the banning of a particular chemical. Callers were identified as either a "local campaign donor" or a "local constituent," with the identification determined at random for each member of Congress contacted. Otherwise, the protocol for the calls was identical, including the details of the meeting request, which were delivered by email. *Results*: "Donors" were more successful obtaining meetings and were given better access to higher level of staff, including the member of Congress.[65]
pg 231

(a) Identify one variable that researchers kept the same in this experiment.

(b) Explain how keeping this variable the same helps prevent confounding.

10. **Vitamin D and asthma** Researchers randomly assigned 408 adults with persistent asthma to take one of two treatments: an inhaled medicine with a vitamin D supplement or the inhaled medicine with a placebo. *Results*: After seven months, the vitamin D users were no less likely to experience reduced air flow, hospitalization, or the need for additional medicine.[66]

(a) Identify one variable that researchers kept the same in this experiment.

(b) Explain how keeping this variable the same helps prevent confounding.

11. **An apple a day?** In an experiment to determine how calories in beverages affect appetite, researchers randomly assigned volunteers to one of four treatments: 150 calories of apple slices, 150 calories of applesauce, 150 calories of apple juice, and 150 calories of apple juice with added fiber. Fifteen minutes later, researchers provided each participant with a large bowl of pasta and measured how many calories each person consumed. The apple slice group ate 190 fewer calories, on average, compared to the other three groups.[67]

(a) Identify one variable that researchers kept the same in this experiment.

(b) Other than to avoid confounding, explain another benefit of keeping this variable the same.

12. **Radishes and light** Mr. Mills's biology class conducted an experiment to investigate the impact of different light regimens on the growth of radish

plants. They randomly assigned 60 2-day-old plants in identical pots to three treatments. Twenty plants were exposed to artificial "grow-light" lamps for 8 hours per 24-hour period, 20 plants were exposed to the same light for 16 hours, and 20 plants were exposed to light for the entire 24-hour period. All plants were grown in the same room at 24°C, and all were given the same amount of water every other day. After 20 days, the mean weight of the 16-hour plants was higher than the mean weight of the 8-hour and 24-hour plants.

(a) Identify one variable that the class kept the same in this experiment.

(b) Other than to avoid confounding, explain another benefit of keeping this variable the same.

Applying the Concepts

13. **Which diets work?** Dr. Linda Stern and her colleagues recruited 132 obese adults at the Philadelphia Veterans Affairs Medical Center in Pennsylvania. Half the participants were randomly assigned to a low-carbohydrate diet and the other half to a low-fat diet. Researchers measured each participant's change in weight and cholesterol level after six months and again after one year.

(a) Describe how the researchers could have carried out the random assignment.

(b) Why was it important that the researchers randomly assigned treatments to the subjects?

(c) Identify one variable that the researchers kept the same for all subjects. Why was it important that the researchers kept this variable the same?

14. **The future of preschool** Does preschool help low-income children stay in school and hold good jobs later in life? The Carolina Abecedarian Project (the name suggests the ABCs) has followed a group of 111 children since 1972. Back then, these individuals were all healthy, low-income black infants in Chapel Hill, North Carolina. All the infants received nutritional supplements and help from social workers. Half were also assigned at random to an intensive preschool program.[68]

(a) Describe how the researchers could have carried out the random assignment.

(b) Why was it important that the researchers randomly assigned treatments to the children?

(c) Identify one variable that the researchers kept the same for all subjects. Why was it important that the researchers kept this variable the same?

15. **Building strength** A high school football coach hears that a new exercise program will increase upper-body strength better than lifting weights. In the off-season, the coach lets his players choose which of the two treatments they will undergo for 3 weeks— exercise or weight lifting. He will use the number of push-ups a player can do at the end of the experiment as the response variable. What flaw in this study makes it impossible to determine if the new program is effective? Explain.

16. **Comparing treatments** A large study used records from Canada's national health care system to compare the effectiveness of two ways to treat prostate disease. The two treatments are traditional surgery and a new method that does not require surgery. The records described many cases where doctors had chosen the method of treatment for their patients. The study found that patients treated by the new method were significantly more likely to die within 8 years. What flaw in this study makes it impossible to determine if the new method is more dangerous? Explain.[69]

Extending the Concepts

17. **The power of milk?** An article from the *Arizona Daily Star* tells the story of a gym trainer in London who swears by the power of milk to help him recover from workouts. The article describes how milk includes many of the same ingredients as sports drinks, as well as other proteins that help rebuild muscles. Design an experiment to determine if milk is better than sports drinks for helping football players at your school recover after weight training. Make sure to clearly explain how you use random assignment and what variables should be kept the same for all subjects.[70]

18. **Random headache relief** In Exercise 4, you described how to use slips of paper and a random number generator to randomly assign 36 subjects to one of four treatments: (1) antidepressant alone, (2) placebo alone, (3) antidepressant plus stress management, and (4) placebo plus stress management.

(a) Describe how you could use the table of random digits below to perform the random assignment.

(b) Use the random digits below to assign 9 subjects to the antidepressant alone group.

36009 19365 15412 39636 85453 46816 83485 41979

38448 48789 18338 24697 39364 42006 76688 08708

81486 69487 60513 09297 00412 71238 27649 39950

19. **"Natural" experiment?** A recent study compared the rate of crashes involving 16- to 18-year-old drivers in two adjacent Virginia counties. For two consecutive years, there were more crashes per person for drivers in this age group in Chesterfield County, where school starts at 7:20 A.M., than in Henrico County, where school starts at 8:45 A.M.

The study was described as a "natural experiment," because the two counties are nearly identical in socioeconomic characteristics and in the percentage of roads with traffic congestion. Explain why it is harder to establish a cause-and-effect conclusion with this type of "natural experiment" than an experiment that uses random assignment.[71]

Recycle and Review

20. **Voice vote (3.2)** On the popular reality-TV show *The Voice*, viewers can vote for their favorite contestant by calling a toll-free number, by voting online, by texting, by using a smartphone app, or by purchasing a song on iTunes. The online rules describe the following restriction: "In each voting period, there is a limit of 10 votes per artist per originating phone number via phone voting, 10 votes per artist per email address for online voting, 10 votes per artist per email address for App voting, 10 votes per artist per originating phone number for text voting, and 1 vote per artist per

eligible song for valid iTunes voting." Explain why the winning contestant may *not* be the one that the majority of the show's viewers prefer.[72]

21. **Age of Facebook users (2.1)** Is there a relationship between Facebook use and age among college students? The following two-way table displays data for the 219 students who responded to a survey.

		Age			
		Younger (18–22)	Middle (23–27)	Older (28 and up)	Total
Facebook user?	Yes	78	49	21	148
	No	4	21	46	71
	Total	82	70	67	219

Use the information in the two-way table to make a segmented bar chart to show the relationship between age and Facebook use. Is there an association between age and Facebook use for the members of this sample? Explain.[73]

Lesson 3.8

Inference for Experiments

LEARNING TARGETS

- Outline an experiment that uses a completely randomized design.
- Explain the concept of statistical significance in the context of an experiment.
- Use simulation to determine if the difference between two means or two proportions in an experiment is significant.

In the preceding two lessons, you learned how to properly design an experiment. In this lesson, you will learn how to outline a completely randomized experiment and how to analyze the results of an experiment that compares two treatments.

Completely Randomized Designs

The diagram in Figure 3.6 presents the details of the caffeine experiment from Lesson 3.7: random assignment, the sizes of the groups and which treatment they receive, and the response variable. This type of experimental design is called a **completely randomized design.**

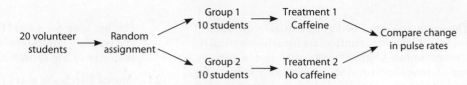

FIGURE 3.6 Outline of a completely randomized design to investigate the effects of caffeine on pulse rates.

||||| **DEFINITION** Completely randomized design

In a **completely randomized design,** the experimental units are assigned to the treatments completely by chance.

Random assignment to treatments is absolutely essential in any experiment. There is no requirement that treatment groups be exactly the same size, although there are some statistical advantages if they are roughly equal in size.

EXAMPLE

What distracts drivers?

Completely randomized designs

PROBLEM: Is talking on a cell phone while driving more distracting than talking to a passenger? David Strayer and his colleagues at the University of Utah used 48 undergraduate students as subjects in an experiment. The researchers randomly assigned half of the subjects to drive in a simulator while talking on a cell phone and the other half to drive in the simulator while talking to a passenger. One response variable was whether or not the driver stopped at a rest area that was specified by researchers before the simulation started. Outline a completely randomized design for this experiment.[74]

SOLUTION:

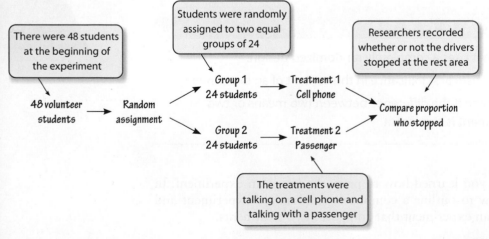

FOR PRACTICE **TRY EXERCISE 1.**

Statistical Significance

In Lesson App 3.6, you read about an experiment that investigated if magnets can help patients with chronic pain. A doctor identified a painful site on each patient and asked him or her to rate the pain on a scale from 0 (mild pain) to 10 (severe pain). Then, the doctor selected a sealed envelope containing a magnet at random from a box with a mixture of active and inactive magnets. The chosen magnet was applied to the site of the pain for 45 minutes. After being treated, each patient was again asked to rate the level of pain from 0 to 10 and the improvement in pain level was recorded.

In this experiment, the treatments were assigned by having the doctor choose an envelope at random. The purpose of this random assignment was to create two groups of patients who were roughly equivalent at the beginning of the experiment. With only 50 patients, however, it is unlikely that the two groups were exactly equivalent. By chance, it is possible that the active magnet group included a higher proportion of patients who were going to feel better regardless of which treatment they received. This means that the average improvement for the active magnet group might be great-er than the average improvement for the inactive magnet group, *even if the active magnet had no beneficial effect.*

Before making any firm conclusions, we must account for the possible differences in response that could occur *only* to the random assignment of treatments. That is, we must make sure that the difference in response for the groups in an experiment is larger than what we would expect to occur simply due to chance variation in the random assign-ment. If this is true, the results of the experiment are **statistically significant.**

DEFINITION   Statistically significant

When an observed difference in responses between the groups in an experiment is too large to be explained by chance variation in the random assignment, we say that the result is **statistically significant.**

The dotplots in Figure 3.7 show the results of the magnet experiment. The mean improvement for patients with the active magnet is $\bar{x}_A = 5.24$ and the mean improve-ment for the patients with inactive magnets is $\bar{x}_I = 1.10$, resulting in a difference of $\bar{x}_A - \bar{x}_I = 5.24 - 1.10 = 4.14$.

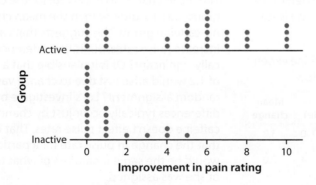

FIGURE 3.7 Improve-ment in pain rating for subjects receiving active and inactive magnets.

The researchers in this study determined that the difference of 4.14 was statisti-cally significant. In other words, a difference in mean improvement of 4.14 was too large to occur simply due to chance variability in the random assignment.

EXAMPLE

How did the doctors do?

Statistical significance

PROBLEM: In Lesson App 3.6, you read about the Physicians' Health Study. Nearly 22,000 doctors were randomly assigned to take aspirin or a placebo every odd-numbered day, hoping to find that aspirin helps prevent heart attacks. These doctors were also randomly assigned to take beta-carotene or a placebo every even-numbered day, hoping to find that beta-carotene helps prevent cancer.

(a) The proportion of doctors taking aspirin who had a heart attack was smaller than the proportion of doctors taking a placebo who had a heart attack, and the difference was statistically significant. Explain what it means that the difference was statistically significant.

(b) The difference in the proportion of doctors taking beta-carotene who developed cancer and the proportion of doctors taking a placebo who developed cancer was not statistically significant. Explain what it means that the difference was not statistically significant.

SOLUTION:

(a) The difference in the proportion of doctors who had a heart attack in the two groups was too large to be due only to chance variation in the random assignment to treatments.

(b) The difference in the proportion of doctors who developed cancer in the two groups was small enough that it could be due to chance variation in the random assignment to treatments.

FOR PRACTICE   TRY EXERCISE 5.

Determining Statistical Significance with Simulation

We can use simulation to determine if the difference in two statistics is significant. The following activity shows you how.

ACTIVITY

Drawing conclusions from the caffeine experiment

In this activity, we will use a simulation to determine if the difference in mean change in pulse rate is statistically significant.

Mr. Luckow's class performed the caffeine experiment and obtained the results shown below.

	Change in pulse rate (Final pulse rate – Initial pulse rate)										Mean change
Caffeine	8	3	5	1	4	0	6	1	4	0	3.2
No caffeine	3	–2	4	–1	5	5	1	2	–1	4	2.0

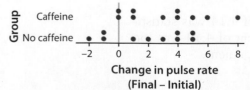

The mean change in pulse rate for the caffeine group was 1.2 greater than the mean change for the no-caffeine group. This suggests that caffeine does increase pulse rates. But is the difference statistically significant? Or is it plausible that a difference of 1.2 would arise just due to chance variation in the random assignment? Let's investigate by seeing what differences typically occur just by chance, assuming caffeine doesn't affect pulse rates. That is, assuming that the change in pulse rate for a particular student would be the same regardless of what treatment he or she was assigned.

1. Gather 20 index cards to represent the 20 students in this experiment. On each card, write one of the 20 outcomes listed in the table. For example, write "8" on the first card, "3" on the second card, and so on.

2. Shuffle the cards and deal two piles of 10 cards each. This represents randomly assigning the 20 students to the two treatments, *assuming that the treatment received doesn't affect the change in pulse rate.* The first pile of 10 cards represents the caffeine group, and the second pile of 10 cards represents the no caffeine group.

3. Find the mean change for each group and subtract the means (Caffeine – No caffeine).

Note: It is possible to get a negative difference.

4. Your teacher will draw and label an axis for a class dotplot. Plot the difference you got in Step 3 on the graph.

5. In Mr. Luckow's class, the observed difference in means was 1.2. Is a difference of 1.2 statistically significant? Discuss with your classmates.

We used the *One Quantitative Variable* applet at highschool.bfwpub.com/spa3e to do 100 trials of the simulation described in the activity. The dotplot in Figure 3.8 shows that getting a difference of 1.2 isn't that unusual. In 19 of the 100 trials, we obtained a difference of 1.2 or more simply due to chance variation in the random assignment.

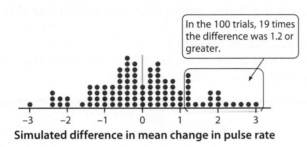

In the 100 trials, 19 times the difference was 1.2 or greater.

Simulated difference in mean change in pulse rate

FIGURE 3.8 Dotplot showing the differences in means that occurred in 100 simulated random assignments, assuming that caffeine has no effect on pulse rates.

Because the difference of 1.2 or greater is somewhat likely to occur by chance alone, the results of Mr. Luckow's class experiment aren't statistically significant.

EXAMPLE

What is distracting these drivers?

Determining statistical significance

PROBLEM: Here are the results of the distracted-driver experiment described in the first example of this lesson.

		Treatment		
		Cell phone	Passenger	Total
Response	Stopped at rest area	12	21	33
	Didn't stop	12	3	15
	Total	24	24	48

(a) Calculate the difference in the proportion of students who stopped at the rest area in the two groups (Passenger − Cell phone).

(b) One hundred trials of a simulation were performed to see what differences in proportions would occur due only to chance variation in the random assignment, assuming that the type of distraction doesn't matter. That is, 33 "stoppers" and 15 "non-stoppers" were randomly assigned to two groups of 24. Use the results of the simulation in the dotplot on the next page to determine if the difference in proportions from part (a) is statistically significant. Explain your reasoning.

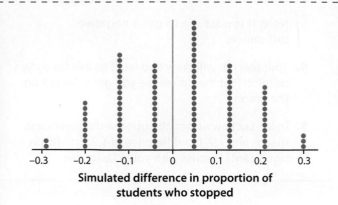

Simulated difference in proportion of
students who stopped

SOLUTION:

(a) Difference in proportions = 21/24 − 12/24 =
0.875 − 0.500 = 0.375.

(b) Because a difference of 0.375 or greater never occurred in
the simulation, the difference is statistically significant. It
is extremely unlikely to get a difference this big simply due
to chance variation in the random assignment.

FOR PRACTICE   TRY EXERCISE 9.

THINK ABOUT IT   **How unusual does a difference have to be before it is considered statistically significant?** In the caffeine example, we said that a difference in means of 1.2 *was not unusual* because a difference this big or bigger occurred 19% of the time by chance alone. In the distracted-drivers example, we said that a difference in proportions of 0.375 *was unusual* because a difference this big or bigger occurred 0% of the time by chance alone. So the boundary between "not unusual" and "unusual" must be somewhere between 0% and 19%. For now, we recommend using a boundary of 5% so that differences that would occur less than 5% of the time by chance alone are considered statistically significant.

LESSON APP 3.8

Does fish oil affect blood pressure?

To see if fish oil can help reduce blood pressure, males with high blood pressure were recruited and randomly assigned to different treatments. Seven of the men were randomly assigned to a 4-week diet that included fish oil. Seven other men were assigned to a 4-week diet that included a mixture of oils that approximated the types of fat in a typical diet. At the end of the 4 weeks, each volunteer's blood pressure was measured again and the reduction in diastolic blood pressure was recorded. These reductions are shown in the table. Note that a negative value means that the subject's blood pressure *increased*.[75]

| Fish oil | 8 | 12 | 10 | 14 | 2 | 0 | 0 |
| Mixture | −6 | 0 | 1 | 2 | −3 | −4 | 2 |

1. Outline a completely randomized design for this experiment.

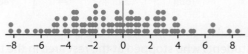

2. Calculate the mean reduction for each group and the difference in mean reduction (Fish oil − Mixture).

3. One hundred trials of a simulation were performed to see what differences in means are likely to occur due only to chance variation in the random assignment, assuming that the type of oil doesn't matter. Use the results of the simulation below to determine if the difference in means from Question 2 is statistically significant. Explain your reasoning.

Simulated difference in mean reduction

Lesson 3.8

WHAT DID YOU LEARN?

LEARNING TARGET	EXAMPLES	EXERCISES
Outline an experiment that uses a completely randomized design.	p. 236	1–4
Explain the concept of statistical significance in the context of an experiment.	p. 238	5–8
Use simulation to determine if the difference between two means or two proportions is significant.	p. 239	9–12

Exercises

Lesson 3.8

Mastering Concepts and Skills

1. **Internet telephone calls** One of the most popular Web-based long-distance services is Skype. How will the appearance of ads during calls affect the use of this service? Researchers design an experiment to find out. They recruit 300 people who have not used Skype before to participate. Some people get a version of Skype with no ads. Others see ads whenever they make calls. The researchers are interested in frequency and length of phone calls. Outline a completely randomized design for this experiment.

   pg 236

2. **Personal stories** Timothy Wilson, a psychology professor at the University of Virginia, studies the effect of "personal narratives" on people's behavior. For example, if the "story" you have about your academic ability is positive, you do better in school than if your "story" is negative. Something as simple as hearing older students describe overcoming challenges similar to yours can help you change your story and improve your performance. Suppose you have 40 college freshmen who have volunteered to be part of an experiment. Some students will hear older students talking about overcoming challenges and others will not. Outline a completely randomized design for this experiment.[76]

3. **Growing in the shade** Adaptation to growing in shade may help pines found in the dry forests of Arizona to resist drought. How well do these pines grow in shade? Investigators planted pine seedlings in a greenhouse in either full light, light reduced to 25% of normal by shade cloth, or light reduced to 5% of normal. At the end of the study, they dried the young trees and weighed them. Outline a completely randomized design for this experiment.

4. **Eating and exercising better** Most American middle-school students don't eat well and don't exercise enough. Investigators designed a "physical activity intervention" to increase activity in physical education classes and during leisure periods throughout the school day. They also designed a "nutrition intervention" that improved school lunches and offered ideas for healthy home-packed lunches. Each participating school was assigned to one of the interventions, both interventions, or no intervention. The investigators observed physical activity and lunchtime consumption of fat. Outline a completely randomized design for this experiment.

5. **A cure for the common cold?** An experiment randomly assigned over 350 subjects to echinacea (a popular herbal supplement) or a placebo to see if taking echinacea reduced the length or severity of a cold. *Results:* The colds lasted longer (6.87 days) for subjects who received the placebo than for subjects who received the echinacea (6.34 days), but the difference wasn't statistically significant. Explain what it means that the difference was not statistically significant.[77]

   pg 238

6. **Acupuncture and fertility** A study sought to determine if the ancient Chinese art of acupuncture could help infertile women become pregnant. A total of 160 healthy women undergoing assisted reproductive therapy were recruited for the study. Half of the subjects were randomly assigned to receive acupuncture treatment 25 minutes before embryo transfer and again 25 minutes after the transfer. The remaining 80 subjects were instructed to lie still for 25 minutes after the embryo transfer. *Results:* The difference in the percent of women who became pregnant in the two groups is statistically significant, with 34 pregnancies in the acupuncture group and only 21 pregnancies in the control group. Explain what it means that the difference was statistically significant.[78]

7. **Tai chi for balance** People with Parkinson's disease often suffer from tremors and stiff, jerky movements. Can tai chi, with its emphasis on slow, graceful movements, help with these symptoms? In an experiment, 195 people with Parkinson's disease

were randomly assigned to twice-a-week classes of tai chi, stretching, or resistance training. *Results:* After six months, the subjects in the tai chi group did better in tests of balance than the subjects in the other groups, and the differences were statistically significant. Explain what it means that these differences were statistically significant.[79]

8. **Stopping sinus infections** In an experiment, 166 adults from the St. Louis area were recruited and randomly assigned to receive one of two treatments for a sinus infection. Half of the subjects received an antibiotic (amoxicillin) and the other half received a placebo. At different stages during the next month, all subjects took the Sino-Nasal Outcome Test. After 10 days, the difference in average test scores was not statistically significant. Explain what it means that this difference was not statistically significant.[80]

9. **Leaking helium** After buying many helium balloons only to see them deflate within a couple of days, Erin and Jenna decided to test if helium-filled balloons deflate faster than air-filled balloons. They bought 60 balloons and randomly divided them into two piles of 30, filling the balloons in the first pile with helium and the balloons in the second pile with air. Then, they measured the circumference of each balloon immediately after it was filled and again three days later. The average decrease in circumference of the helium-filled balloons was 26.5 centimeters and the average decrease of the air-filled balloons was 2.1 centimeters.[81]

pg 239

(a) Calculate the difference in the mean decrease in diameter for the two groups (Helium – Air).

(b) One hundred trials of a simulation were performed to see what differences in means would occur due only to chance variation in the random assignment, assuming that the type of gas doesn't matter. Use the results of the simulation in the dotplot to determine if the difference in means from part (a) is statistically significant. Explain your reasoning.

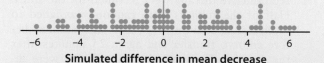

Simulated difference in mean decrease

10. **I work out a lot** Are people influenced by what others say? Michael conducted an experiment in front of a popular gym. As people entered, he asked them how many days they typically work out per week. As he asked the question, he showed the subjects one of two clipboards, determined at random. Clipboard A had the question and several responses written down, where the majority of responses were 6 or 7 days per week. Clipboard B was the same, except most of the responses were 1 or 2 days per week. The mean response for the Clipboard A group was 4.68 and the mean response for the Clipboard B group was 4.21.[82]

(a) Calculate the difference in the mean number of days for the two groups (A – B).

(b) One hundred trials of a simulation were performed to see what differences in means would occur due only to chance variation in the random assignment, assuming that the responses on the clipboard don't matter. Use the results of the simulation in the dotplot to determine if the difference in means from part (a) is statistically significant. Explain your reasoning.

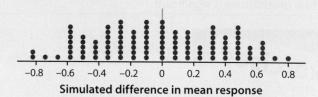

Simulated difference in mean response

11. **Texting and driving** Does providing additional information affect responses to a survey question? Two statistics students decided to investigate by asking different versions of a question about texting and driving. Fifty mall shoppers were divided into two groups of 25 at random. The first group was asked Version A and the other half were asked Version B.

• Version A: A lot of people text and drive. Are you one of them?

• Version B: About 6000 deaths occur per year due to texting and driving. Knowing the potential consequences, do you text and drive?

Of the 25 shoppers assigned to Version A, 16 admitted to texting and driving. Of the shoppers assigned to Version B, only 12 admitted to texting and driving.

(a) Calculate the difference in the proportion of shoppers who admitted to texting and driving in the two groups (A – B).

(b) One hundred trials of a simulation were performed to see what differences in proportions would occur due only to chance variation in the random assignment, assuming that the wording of the question doesn't matter. Use the results of the simulation in the dotplot to determine if the difference in proportions from part (a) is statistically significant. Explain your reasoning.

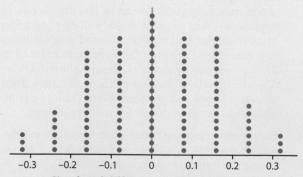

Simulated difference in proportion who admit texting and driving

12. **A louse-y situation** A study published in the *New England Journal of Medicine* compared two medicines to treat head lice: an oral medication called ivermectin and a topical lotion containing malathion. Researchers studied 812 people in 376 households in seven areas around the world. Of the 185 households randomly assigned to ivermectin, 171 were free from head lice after 2 weeks compared with only 151 of the 191 households randomly assigned to malathion.[83]

(a) Calculate the difference in the proportion of households that were free from head lice in the two groups (Ivermectin – Malathion).

(b) One hundred trials of a simulation were performed to see what differences in proportions would occur due only to chance variation in the random assignment, assuming that the type of medication doesn't matter. Use the results of the simulation in the dotplot to determine if the difference in proportions from part (a) is statistically significant. Explain your reasoning.

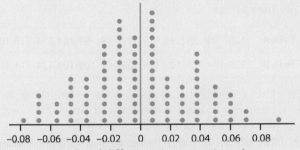

Simulated difference in proportion who were free from head lice

Applying the Concepts

13. **Wet popcorn** Does soaking popcorn kernels before popping increase the percentage of kernels that pop? To find out, Jantzen randomly assigned 10 cups of kernels to be soaked in water before popping and 10 cups of kernels to be popped without soaking. After popping each cup, she calculated the percentage of kernels that popped.[84] Here are her results.

Percent popped (soaked)	83	91	88	86	96	97	94	92	86	93
Percent popped (unsoaked)	89	89	81	77	84	79	82	96	91	93

(a) Outline a completely randomized design for this experiment.

(b) Calculate the mean percentage of popped kernels for each group and the difference in mean percentage (Soaked – Unsoaked).

(c) One hundred trials of a simulation were performed to see what differences in means are likely to occur due only to chance variation in the random

assignment, assuming that soaking doesn't matter. Use the results of the simulation below to determine if the difference in means from part (b) is statistically significant. Explain your reasoning.

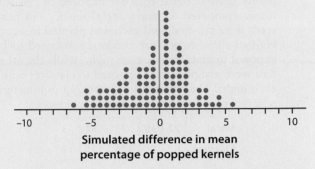

Simulated difference in mean percentage of popped kernels

14. **Crime metaphors** A recent study examined the impact of metaphors on the way people think about complex issues. In one part of the study, subjects were divided randomly into two groups. One group read a passage that described crime in a fictional city as a "beast" ravaging the city. The second group read an identical passage with one change—the word "beast" was replaced with the word "virus." After reading the passage, all the subjects were asked what the town should do in response to crime. Responses were categorized as either suggesting more enforcement or more social programs.[85] Here are the results.

		Treatment		
		"Beast"	"Virus"	Total
Response	Enforcement	80	72	152
	Social	30	61	91
	Total	110	133	243

(a) Outline a completely randomized design for this experiment.

(b) Calculate the proportion of subjects who suggested "enforcement" in each of the two treatment groups and the difference in proportions (Beast – Virus).

(c) One hundred trials of a simulation were performed to see what differences in proportions are likely to occur due only to chance variation in the random assignment, assuming that the type of metaphor didn't matter. Use the results of the simulation to determine if the difference in proportions from part (b) is statistically significant. Explain your reasoning.

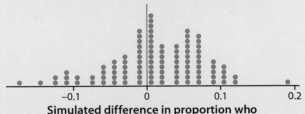

Simulated difference in proportion who suggested enforcement

Extending the Concepts

15. **Bached beans?** For their final statistics project, two students performed an experiment to determine if plants grow better if they are exposed to classical music compared to heavy metal music. Ten bean seeds were selected, and each was planted in a cup. Half of these cups were randomly assigned to be exposed to metal music each night, while the other half were assigned to be exposed to classical music each night. The amount of growth, in millimeters, was recorded for each plant after two weeks.

Metal	22	36	73	57	3
Classical	87	78	124	121	19

(a) Calculate the difference in mean growth (Classical – Metal).

(b) By hand or using the *One Quantitative Variable* applet at highschool.bfwpub.com/spa3e, conduct at least 20 trials of a simulation to determine what differences are likely to occur due to chance variation in the random assignment, assuming that the type of music doesn't matter. Record your results on a dotplot.

(c) Using the results of the simulation, determine if the difference in means from part (a) is statistically significant. Explain your reasoning.

16. **Color my world** Many people believe that color can influence mood, but can mood influence a person's choice of color? Maddi and Natasha randomly assigned 40 high-school students to one of two surveys. The "happy" survey asked questions that were designed to make the subjects happy. The "angry" survey asked questions that were designed to make the subjects angry. After the surveys were completed, each subject was offered a reward. The rewards were identical pieces of candy, but half of the candies were in a yellow box and half were in a red box. Maddi and Natasha suspected that people who had the "angry" survey would be more likely to choose the red box.[86] Here are their results.

		Survey type		
		Angry	Happy	Total
Color of box	Red	13	6	19
	Yellow	7	14	21
	Total	20	20	40

(a) Calculate the difference in the proportion of students who chose a candy from the red box (Angry – Happy).

(b) By hand or using the *One Categorical Variable* applet at highschool.bfwpub.com/spa3e, conduct at least 20 trials of a simulation to determine what differences are likely to occur due to chance variation in the random assignment, assuming that the type of survey doesn't matter. Record your results on a dotplot.

(c) Using the results of the simulation, determine if the difference in proportions from part (a) is statistically significant. Explain your reasoning.

Recycle and Review

17. **Separated at birth (1.3, 2.2)** A researcher studied a group of identical twins who had been separated and adopted at birth. In each case, one twin (Twin A) was adopted by a low-income family and the other (Twin B) by a high-income family. Both twins were given an IQ test as adults.[87] Here are their scores.

Twin A	120	99	99	94	111	97	99	94	104	114	113	100
Twin B	128	104	108	100	116	105	100	100	103	124	114	112

(a) Use a scatterplot to display the relationship between IQ scores for these sets of twins. How well does one twin's IQ predict the other's?

(b) Do identical twins living in low-income homes tend to have lower IQs later in life than their twins who live in high-income homes? Calculate the difference in IQ for each set of twins, make a dotplot to display the differences, and use the dotplot to answer the question.

18. **Initial public offerings (1.6)** The business magazine *Forbes* reports that 4567 companies sold their first stock to the public between 1990 and 2000. The *mean* change in the stock price of these companies since the first stock was issued was +111%. The *median* change was −31%. Explain how this could happen.[88]

Lesson 3.9

Using Studies Wisely

Researchers who conduct statistical studies often want to draw conclusions (make inferences) from the data they produce. Here are two examples.

- The U.S. Census Bureau carries out a monthly Current Population Survey of about 60,000 households. One goal is to use data from these randomly selected households to estimate the percent of unemployed individuals in the population.

- Scientists performed an experiment that randomly assigned 21 volunteer subjects to one of two treatments: sleep deprivation for one night or unrestricted sleep. The experimenters hoped to show that sleep deprivation causes a decrease in performance two days later. [89]

The Scope of Inference

The type of inference that can be made from a study depends on how the study was designed.

In the Census Bureau's sample survey, the individuals who responded were *chosen at random* from the population of interest. Because the sample was randomly selected, it should be representative of the entire population. Thus, random sampling helps to avoid bias and produces reliable estimates of the truth about the population. The Census Bureau should be safe making an *inference about the population* based on the results of the sample.

In the sleep deprivation experiment, subjects were *randomly assigned* to the sleep deprivation and unrestricted sleep treatments. Random assignment helps ensure that the two groups of subjects are as alike as possible before the treatments are imposed. If the unrestricted sleep group performs significantly better than the sleep deprivation group, the scientists could safely conclude that sleep deprivation caused the decrease in performance. That is, they can make an *inference about cause and effect*. However, because the experiment used volunteer subjects, the scientists cannot apply this conclusion to a larger population.

The Scope of Inference

- Random selection of individuals allows inference about the population.
- Random assignment of individuals to groups permits inference about cause and effect.

EXAMPLE

Feeling sleepy?

The scope of inference

PROBLEM: The headline of a recent article declared: "Study: Teens Getting Less Sleep; Social Media Could Be a Reason." Using random samples of teens from 1991 and 2012, the study found that teens are getting less sleep than they did 20 years earlier. For example, just over half of 15-year-olds reported at least 7 hours of sleep in 1991, but only 43% of 15-year-olds reported this much sleep in 2012.[90]

(a) Can the results of this study be generalized to all teens in the years 1991 and 2012? Explain.

(b) Based on the study, is it reasonable to say that use of social media is the cause of the reduction in sleep? Explain.

SOLUTION:

(a) Yes, because the study used random samples of teens from 1991 and 2012.

(b) No, because teens weren't randomly assigned to use social media or assigned to avoid using social media.

FOR PRACTICE   TRY EXERCISE 1.

The following chart summarizes the appropriate scope of inference based on the design of a study.

		Were individuals randomly assigned to groups?	
		Yes	No
Were individuals randomly selected?	Yes	Inference about the population: YES Inference about cause and effect: YES	Inference about the population: YES Inference about cause and effect: NO
	No	Inference about the population: NO Inference about cause and effect: YES	Inference about the population: NO Inference about cause and effect: NO

Here is another example that looks at all four possible study designs.

EXAMPLE

Is silence golden?

The scope of inference

PROBLEM: Many students insist that they study better when listening to music. A teacher doubts this claim and suspects that listening to music hurts academic performance. Here are four possible study designs to address this question at your school. In each case, the response variable will be the students' GPA at the end of the semester.

1. Get all the students in your statistics class to participate in a study. Ask them whether or not they study with music on and divide them into two groups based on their answer to this question.

2. Select a random sample of students from your school to participate in a study. Ask them whether or not they study with music on and divide them into two groups based on their answer to this question.

3. Get all the students in your statistics class to participate in a study. Randomly assign half of the students to listen to music while studying for the entire semester and have the remaining half abstain from listening to music while studying.

4. Select a random sample of students from your school to participate in a study. Randomly assign half of the students to listen to music while studying for the entire semester and have the remaining half abstain from listening to music while studying.

What conclusion can we draw from each study, assuming the mean GPA for students who listen to music while studying is significantly lower than the mean GPA of students who didn't listen to music while studying? Explain.

SOLUTION:

1. We can conclude that the students in the statistics class who listen to music while studying have lower GPAs, on average, than those who do not listen to music while studying, but we can't say listening to music is the cause.

   > With no random selection, the results of the study should be applied only to the statistics students in the study. With no random assignment, we cannot make any inferences about cause and effect.

2. We can conclude that students at the school who listen to music while studying have lower GPAs than those who do not listen to music while studying, on average, but we can't say listening to music is the cause.

   > With random selection, the results of the study can be applied to the entire population—in this case, all the students at the school. With no random assignment, we cannot make any inferences about cause and effect.

3. We can conclude that listening to music while studying causes students to have lower GPAs, on average, but only for the statistics students who took part in the study.

   > With no random selection, the results of the study should be applied only to the statistics students in the study. With random assignment, we can make an inference about cause and effect.

4. We can conclude that listening to music while studying causes students to have lower GPAs, on average, for all the students at the school.

   > With random selection, the results of the study can be applied to the entire population—in this case, all the students at the school. With random assignment, we can make an inference about cause and effect.

FOR PRACTICE   TRY EXERCISE 3.

Data Ethics

Although randomized experiments are the best way to make an inference about cause and effect, in some cases it isn't ethical to do an experiment. Consider these important questions:

- Does texting while driving increase the risk of having an accident?
- Does going to church regularly help people live longer?
- Does smoking cause lung cancer?

Although a well-designed experiment would help answer these questions, it would be unethical to randomly assign an individual to text while driving, attend church, or smoke cigarettes! It is sometimes possible to build a strong case for cause and effect even without an experiment, but the evidence must meet a strict set of criteria. For example, the evidence that smoking causes cancer is about as strong as nonexperimental evidence can be.

The most complicated ethical issues arise when we collect data from people. Trials of new medical treatments, for example, can do harm as well as good to their subjects. Likewise, there are ethical issues to consider when using a sample survey, even though no treatments are imposed.

Here are some basic standards of data ethics that must be obeyed by all studies that gather data from human subjects, both observational studies and experiments.

Basic Principles of Data Ethics

- All planned studies must be reviewed in advance by an *institutional review board* charged with protecting the safety and well-being of the subjects.

- All individuals who are subjects in a study must give their *informed consent* before data are collected.

- All individual data must be kept *confidential*. Only statistical summaries for groups of subjects may be made public.

The law requires that studies carried out or funded by the federal government obey these principles.[91] But neither the law nor the consensus of experts is completely clear about the details of their application.

The purpose of an institutional review board is not to decide if a proposed study will produce valuable information or if it is statistically sound. The board's purpose is, in the words of one university's board, "to protect the rights and welfare of human subjects (including patients) recruited to participate in research activities." An institutional review board would certainly reject an experiment that required subjects to smoke cigarettes or text while driving.

Both words in the phrase "informed consent" are important, and both can be controversial. Subjects must be *informed* in advance about the nature of a study and any risk of harm it may bring. People who are asked to answer survey questions should be told what kinds of questions the survey will ask and about how much of their time it will take. Experimenters must tell subjects the nature and purpose of the study and outline possible risks. Subjects must then *consent*, or agree, in writing.

It is important to protect individuals' privacy by keeping all data about them *confidential*. The report of an opinion poll may say what percent of the 1200 respondents believed that legal immigration should be increased. It may not report what *you* said about this or any other issue. Confidentiality is not the same as *anonymity*. Anonymity means that the names of individuals are not known, even to the director of the study.

EXAMPLE

What would you do?

Data ethics

PROBLEM: Will people try to stop someone from driving drunk? A television news program hired an actor to play a drunk driver and used a hidden camera to record the behavior of individuals who encountered the driver. Was this study carried out ethically? Explain.

SOLUTION:

No. Participants in the study didn't provide informed consent before the study was conducted. Also, because the participants were filmed for a television program, the results are not confidential.

FOR PRACTICE **TRY EXERCISE 7.**

The drunk-driving study in this example doesn't meet the basic standards of data ethics. But asking participants to provide informed consent before the study would almost certainly make the results of the study useless. If people knew they were being filmed, their behavior might be much different! For many interesting questions like this one, it can be a challenge to collect meaningful data in an ethical way.

LESSON APP 3.9

Is foster care better for children than an orphanage?

Do abandoned children placed in foster homes do better than similar children placed in an institution? The Bucharest Early Intervention Project found that the answer is a clear "Yes." The subjects were 136 young children abandoned at birth and living in orphanages in Bucharest, Romania. Half of the children, chosen at random, were placed in foster homes. The other half remained in the orphanages. (Foster care was not easily available in Romania at the time and so was paid for by the study.)[92]

BSIP/Getty Images

1. What conclusion can we draw from this study? Explain.

2. The children in this study were too young to provide informed consent. Does this make this study unethical? Explain.

Lesson 3.9

WHAT DID YOU LEARN?

LEARNING TARGET	EXAMPLES	EXERCISES
Identify when it is appropriate to use information from a sample to make an inference about a population and when it is appropriate to make an inference about cause and effect.	p. 246	1–6
Evaluate if a statistical study has been carried out in an ethical manner.	p. 248	7–10

Exercises   Lesson 3.9

Mastering Concepts and Skills

1. **Batteries on ice** Will storing batteries in a freezer make them last longer? A company that produces batteries takes a random sample of 100 AA batteries from its warehouse. The company statistician randomly assigns 50 batteries to be stored in the freezer and the other 50 to be stored at room temperature for 3 years. At the end of that time period, each battery's charge is tested. *Result:* Batteries stored in the freezer had a higher average charge, and the difference between the groups was statistically significant.

   pg 246

   (a) Can the results of this study be generalized to all AA batteries in the warehouse? Explain.

   (b) Based on the study, is it reasonable to conclude that storing the batteries in the freezer was the cause of the higher average charge? Explain.

2. **Sugar and gum disease** The National Health and Nutrition Examination Survey is a long-term research program that uses random sampling methods to examine the health and nutrition among adults and children in the United States. One study of 2400 young adults found that subjects who consumed sugary foods at least 5 times per week had a 73% higher risk of developing periodontal (gum) disease in at least two teeth than those who never ate such foods.[93]

   (a) Can the results of this study be generalized to all young adults in the United States? Explain.

   (b) Based on the study, is it reasonable to conclude that consuming sugary foods at least 5 times per week causes periodontal (gum) disease? Explain.

3. **Church and long life** One of the better studies of the effect of regular attendance at religious services gathered data from a random sample of 3617 adults. The researchers then measured many variables, not just the explanatory variable (religious activities) and the response variable (length of life). A news article said: "Churchgoers were more likely to be nonsmokers, physically active, and at their right weight. But even after health behaviors were taken into account, those not attending religious services regularly still were about 25% more likely to have died." What conclusion can we draw from this study? Explain.[94]

   pg 246

4. **Exercise and memory** A study of strength training and memory randomly assigned 46 young adults to two groups. After both groups were shown 90 pictures, one group had to bend and extend one leg against heavy resistance 60 times. The other group stayed relaxed, while the researchers used the same exercise machine to bend and extend their legs with no resistance. Two days later, each subject was shown 180 pictures—the original 90 pictures plus 90 new pictures and asked to identify which pictures were shown two days earlier. The resistance group had significantly more success identifying these pictures than the relax group. What conclusions can we draw from this study? Explain.[95]

5. **Berry good** Eating blueberries and strawberries might improve heart health, according to a long-term study of 93,600 women who volunteered to take part. These berries are high in anthocyanins due to their pigment. Women who reported consuming the most anthocyanins had a significantly smaller risk of heart attack compared to the women who reported consuming the least. What conclusion can we draw from this study? Explain.[96]

6. **Are women chattier than men?** According to Louann Brizendine, author of *The Female Brain,* women say nearly 3 times as many words per day as men. Skeptical researchers devised a study to test this claim. They used electronic devices to record the talking patterns of 396 university students who volunteered to participate. The device was programmed to record 30 seconds of sound every 12.5 minutes without the carrier's knowledge. According to a published report in *Scientific American,* "Men showed a slightly wider variability in words uttered. . . . But in the end, the sexes came out just about even in the daily averages: women at 16,215 words and men at 15,669." This difference was not statistically significant. What conclusion can we draw from this study? Explain.[97]

7. **Unethical hepatitis studies** In the 1960s, children entering the Willowbrook State School, an institution for the intellectually disabled on Staten Island in New York, were deliberately infected with hepatitis. The researchers argued that almost all children in the institution quickly became infected anyway. The studies showed for the first time that different strains of hepatitis existed. This finding contributed to the development of effective vaccines. Despite these valuable results, explain why this study was unethical.

pg 248

8. **Facebook emotions** In cooperation with researchers from Cornell University, Facebook randomly selected almost 700,000 users for an experiment in "emotional contagion." Users' news feeds were manipulated (without their knowledge) to selectively show postings from their friends that were either more positive or more negative in tone, and the emotional tone of their own subsequent postings was measured. The researchers found evidence that people who read emotionally negative postings were more likely to post messages with a negative tone and those reading positive messages were more likely to post messages with a positive tone. The research was widely criticized for being unethical. Explain why.[98]

9. **Monitoring tax returns** A government agency takes a random sample of income tax returns to obtain information on the average income of people in different occupations. Only the incomes and occupations are recorded from the returns, not the names. Should this study require informed consent? Explain your reasoning.

10. **Religious behavior** A social psychologist attends public meetings of a religious group to study the behavior patterns of members. Should this study require informed consent? Explain your reasoning. Would your answer change if the social psychologist pretends to be converted and attends private meetings to study the behavior patterns of members?

Applying the Concepts

11. **Ah nuts!** Can eating nuts during pregnancy help children avoid nut allergies? Researchers studied over 8000 children who were born in the early 1990s to mothers who were part of the Nurses' Health Study II. Children whose mothers ate the most nuts during pregnancy (at least 5 times per week) were significantly less likely to develop nut allergies than children whose mothers ate the least amount of nuts during pregnancy (less than once per month).[99]

(a) Explain why we shouldn't conclude that eating nuts during pregnancy *causes* a reduced risk of nut allergies in children, based on this study.

(b) Would it be ethical to conduct an experiment to answer this question? Explain.

12. **Feeling upbeat?** Are people who report being cheerful and happy less likely to get sick than people who report being sad and angry? Yes, according to a study where researchers exposed 193 healthy volunteers aged 21–55 to a cold virus and recorded who got sick. Cheerful people also had milder symptoms when they did get sick.[100]

(a) Based on this study, is it appropriate to conclude that being happy and cheerful reduces the risk of catching a cold? Explain.

(b) Was this study carried out in an ethical manner? Explain.

13. **Who can review?** Government regulations require that institutional review boards consist of at least 5 people, including at least 1 scientist, 1 nonscientist, and 1 person from outside the institution. Most boards are larger, but many contain just 1 outsider.

(a) Why should review boards contain people who are not scientists?

(b) Do you think that 1 outside member is enough? How would you choose that member? For example, would you prefer a medical doctor? A religious leader? An activist for patients' rights?

14. **Anonymous or confidential?** One of the most important nongovernment surveys in the United States is the National Opinion Research Center's General Social Survey. The GSS regularly monitors public opinion on a wide variety of political and social issues. Interviews are conducted in person in the subject's home. Are a subject's responses to GSS questions anonymous, confidential, or both? Explain your answer.

Extending the Concepts

15. **Randomized response** When asked sensitive questions, many people give untruthful responses—especially if the survey is not anonymous. To encourage honest answers, researchers developed the "randomized response" method. For example, ask students if they have cheated on an exam this year. Before they answer, have each student flip a coin. If the coin lands on heads, answer truthfully. If it lands on tails, answer "Yes."

(a) Explain why this method might encourage students to answer the question honestly.

(b) Suppose that 100 students used this method to answer the question about cheating on an exam this year. Estimate the proportion of students that have cheated if 63 students said "Yes" and 37 students said "No."

Recycle and Review

16. **Mercury concentration in tuna (1.5, 1.7, 1.8)** What is the typical mercury concentration in cans of tuna sold in stores? A study conducted by Defenders of Wildlife set out to answer this question. Defenders collected a sample of 164 cans of tuna from stores across the United States. They sent the selected cans to a laboratory that is often used by the Environmental Protection Agency for mercury testing. A histogram and some computer output provide information about the mercury concentration in the sampled cans (in parts per million, ppm).[101]

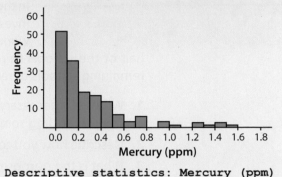

Descriptive statistics: Mercury (ppm)

Variable	N	Mean	StDev	Min
Mercury	164	0.285	0.300	0.012

Variable	Q1	Med	Q3	Max
Mercury	0.071	0.180	0.380	1.500

(a) Interpret the standard deviation in context.

(b) Determine if there are any outliers.

(c) Describe the shape, center, and variability of the distribution.

17. **A drop in crime? (3.6)** An article in the *New York Times Magazine* on Manhattan District Attorney Cyrus Vance, Jr. cites a substantial decrease in crime during Vance's tenure. But the author cautions, "It's hard to know how much one person or policy can affect the crime rate, and consequently how much credit or blame should be assigned when things go well or don't. . . . The sun comes up, the roosters preen." Explain how confounding makes it difficult to give Vance credit for the decrease in crime in Manhattan.[102]

Alexander Joe/AFP/Getty Images

Chapter 3

STATS applied!

How can we prevent malaria?

In the STATS applied! on page 181, we described an experiment that investigated the effect of regular screening for malaria. Researchers worked with children in 101 schools in Kenya, randomly assigning half of the schools to receive regular screenings and follow-up treatments and the remaining schools to receive no regular screening.

1. Why was it necessary to include a group of schools that didn't receive the screening? Does excluding some schools from screening raise any ethical concerns?

2. Describe how you could randomly assign the 101 schools into groups of 51 and 50.

3. What is the purpose of random assignment in this experiment?

4. If the researchers found convincing evidence that the proportion with malaria is lower for children in schools that are regularly screened, would it be reasonable to say that screening caused the reduction in malaria? Would these results apply to all schools in Africa? Explain.

5. Unfortunately, the results of the study were not statistically significant. Explain what this means in the context of this study.

Main Points   *Chapter 3*

The Statistical Problem-Solving Process

- The **statistical problem-solving process** involves four steps: ask questions, collect data, analyze data, and interpret results.

- A valid **statistical question** is based on data that vary.

- There are many **data collection methods,** including sample surveys, observational studies, and experiments.

- The **scope of inference** determines how we interpret results.

  - **Random selection** of individuals allows inference about the population.

  - **Random assignment** of individuals to treatments permits inference about cause and effect.

- All data collection methods should follow basic principles of **data ethics.** This includes the use of an institutional review board, informed consent, and confidentiality.

Sampling and Surveys

- The **population** in a statistical study is the entire group of individuals we want information about. A **census** collects data from every individual in the population.

- A **sample** is a subset of individuals in the population from which we collect data.

- **Random sampling** involves using a chance process to determine which members of a population are included in the sample. We can use the data collected from a random sample to make inferences about the population from which the sample was selected.

- A **simple random sample (SRS)** of size n is chosen in such a way that every group of n individuals in the population has an equal chance to be selected as the sample.

- The fact that different random samples of the same size from the same population produce different estimates is called **sampling variability.** Sampling variability can be reduced by increasing the sample size.

- When making an estimate, we can account for sampling variability by including a **margin of error** that describes how far, at most, we expect the estimate to vary from the true population value.

- The design of a statistical study shows **bias** if it would consistently underestimate or consistently overestimate the value you want to know when the study is repeated many times.

  - Choosing individuals from the population who are easy to reach results in a **convenience sample.** This method of selecting a sample is biased because the individuals chosen are typically not representative of the population.

  - A **voluntary response sample** consists of people who choose to be in the sample by responding to a general invitation. This method of selecting a sample is biased because the individuals in the sample are typically not representative of the population.

  - **Undercoverage** occurs when some members of the population are less likely to be chosen for the sample. Sampling methods that suffer from undercoverage can show bias if the less-likely individuals differ in relevant ways from the other members of the population.

  - **Nonresponse** occurs when an individual chosen for the sample can't be contacted or refuses to participate. Sampling methods that suffer from nonresponse can show bias if the individuals who don't respond differ in relevant ways from the other members of the population.

  - **Response bias** occurs when there is a consistent pattern of inaccurate responses to a survey question. Response bias can be caused by the wording of questions, characteristics of the interviewer, lack of anonymity, and other factors.

Observational Studies

■ An **observational study** observes individuals and measures variables of interest but does not attempt to influence the responses.

■ **Confounding** occurs when two variables are associated in such a way that their effects on a response variable cannot be distinguished from each other.

■ Observational studies cannot definitively show a cause-and-effect relationship between an explanatory variable and a response variable because of confounding.

Experiments

■ An **experiment** deliberately imposes some treatment on individuals to measure their responses. We can use the results of an experiment to make inference about cause and effect.

■ A **treatment** is a specific condition applied to the individuals in an experiment. The **experimental units** are the smallest collection of individuals to which treatments are applied. When the units are human beings, they are often called **subjects.**

■ The **placebo effect** describes the fact that some subjects in an experiment will respond favorably to any treatment, even an inactive treatment. A **placebo** is a treatment that has no active ingredient but is otherwise like other treatments.

■ In a **double-blind** experiment, neither the subjects nor those who interact with them and measure the response variable know which treatment a subject received. In a **single-blind** experiment, either the subjects don't know which treatment they are receiving or the people who interact with them and measure the response variable don't know which subjects are receiving which treatment.

■ In an experiment, a **control group** is used to provide a baseline for comparing the effects of other treatments. Depending on the purpose of the experiment, a control group may be given a placebo or an active treatment.

■ **Random assignment** means that experimental units are assigned to treatments using a chance process. The purpose of random assignment is to avoid confounding by creating treatment groups that are roughly equivalent at the beginning of the experiment.

■ In a **completely randomized design,** the experimental units are assigned to the treatments completely by chance.

■ In an experiment, it is important to keep all variables the same, other than the explanatory variable. Then, these additional variables won't be confounded with the explanatory variable or add variability to the response variable, making it easier to detect a difference in the effects of the treatments.

■ When an observed difference in responses between the groups in an experiment is too large to be explained by chance variation in the random assignment, we say that the result is **statistically significant.** We can use simulation to investigate what differences are likely to occur due to the chance variation in random assignment.

Chapter 3 Review Exercises

1. **Statistical questions (3.1)** Indicate which of the following are valid statistical questions. Explain your reasoning.

(a) How many people visited Acadia National Park last Tuesday?

(b) How many people visit Acadia National Park on a typical weekday in August?

(c) What proportion of soda bottles produced by a particular manufacturer contain less soda than the label on the bottle indicates?

2. **Nurses are the best (3.1, 3.3, 3.4)** A recent random sample of $n = 805$ adult U.S. residents found that the proportion who rated the honesty and ethical standards of nurses as very high or high is 0.85. This is 0.15 higher than the proportion recorded for doctors, the next highest-ranked profession.[103]

(a) Identify the sample and the population in this setting.

(b) Do you think that the proportion of *all* U.S. residents who rate the honesty and ethical standards of nurses as very high or high is exactly 0.85? Explain.

(c) To approximate the margin of error for this estimate, a simulation was conducted. The dotplot shows the proportion who rated the honesty and ethical standards of nurses as very high or high in each of 500 random samples of size 805 from a population where 85% would give this rating. Use the results of the simulation to approximate *and* interpret the margin of error for the estimate of the proportion of all U.S. residents who rate the honesty and ethical standards of nurses as very high or high.

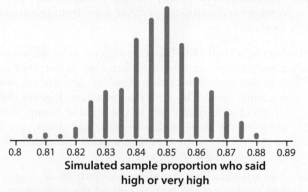

Simulated sample proportion who said high or very high

Distribution of simulated proportion		
# samples	mean	SD
500	0.85	0.012

(d) If the size of the sample in the poll was increased to 1600 residents, what effect would this have on the margin of error? Explain.

3. **Good and bad sampling (3.2, 3.3)** The administration at a high school with 1800 students wants to gather student opinion about parking for students on campus. It isn't practical to contact all students.

(a) Give an example of a way to choose a voluntary response sample of students. Explain how this method could lead to bias.

(b) Give an example of a way to choose a convenience sample of students. Explain how this method could lead to bias.

(c) Describe how to select an SRS of 50 students from the school using technology.

(d) Explain how the method you described in part (c) avoids the biases you described in parts (a) and (b).

4. **Flipping the disk (3.3)** Is flipping a flying disk as fair as flipping a coin? Hailey flips a disk 40 times and it lands right side up only 16 times. She suspects that the disk is more likely to land upside down. To determine if these data provide convincing evidence in support of Hailey's conclusion, 100 trials of a simulation were conducted. Each dot in the graph shows the number of right-side-up flips in a random sample of 40 flips, assuming that each flip has a 50% chance of landing right-side up.

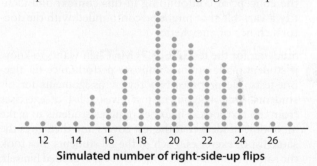

Simulated number of right-side-up flips

(a) Explain how the graph illustrates the concept of sampling variability.

(b) Based on the results of the simulation, is there convincing evidence that flying disks are more likely to land upside down? Explain.

5. **Been to the movies lately? (3.5)** An opinion poll calls 2000 randomly chosen residential telephone numbers, then asks to speak with an adult member of the household. The interviewer asks, "Box-office revenues are at an all-time high. How many movies have you watched in a movie theater in the past 12 months?" In all, 1131 people responded. The researchers used the responses to estimate the mean number of movies adults have watched in a movie theater in the past 12 months.

(a) Describe a potential source of bias related to the wording of the question. Suggest a change that would help fix this problem.

(b) Describe how using only residential phone numbers might lead to bias and how this will affect the estimate.

(c) Describe how nonresponse might lead to bias and how this will affect the estimate.

6. **How safe are anesthetics?** (3.1, 3.6) The National Halothane Study was a major investigation of the safety of anesthetics. Performed in 34 major hospitals, the study showed the following death rates for four common anesthetics:

Anesthetic	A	B	C	D
Death rate	1.7%	1.7%	3.4%	1.9%

There seems to be a clear association between the anesthetic used and the death rate of patients. Anesthetic C appears to be more dangerous.[104]

(a) Explain why we call the National Halothane Study an observational study rather than an experiment, even though it compared the results of using different anesthetics in actual surgery.

(b) When the study looked at other variables that are related to a doctor's choice of anesthetic, it found that Anesthetic C was not causing extra deaths. Explain the concept of confounding in this context and identify a variable that might be confounded with the doctor's choice of anesthetic.

7. **Studying for the test** (3.6, 3.7) Mr. Chen wants to know if student-made outlines improve performance on algebra tests. He prepares two review assignments for his students. One assignment just gives a list of exercises from the textbook. The other requires students to make an outline of ideas that the test covers and then gives the same list of exercises. Each of the 28 students then took the same 50-minute test, which Mr. Chen graded himself.

(a) Explain why Mr. Chen included a group that didn't prepare an outline.

(b) Can this be a double-blind experiment? Explain.

(c) Identify two variables that Mr. Chen kept the same during the experiment. Explain why it was important for him to keep these variables the same.

(d) Describe how you would randomly assign the students to treatments using identical slips of paper. Explain the purpose of the random assignment in this context.

8. **Studying for the test, Part 2** (3.8) Using the results from the experiment described in Exercise 7, Mr. Chen found that the mean score for the group who prepared an outline was 87.9 and the mean score for the other group was 80.4.

(a) Calculate the difference in mean score for the two groups (Outline – No outline).

(b) Two hundred trials of a simulation were performed to see what differences in means would occur due only to chance variation in the random assignment, assuming that preparing an outline did not affect test scores. Use the results of the simulation to determine if the difference in means from part (a) is statistically significant. Explain your reasoning.

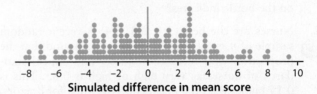

Simulated difference in mean score

9. **Bedtime reading** (3.8, 3.9) Is there a difference between reading an ebook and a printed book before going to bed? Researchers randomly assigned 12 young adults to read either an electronic book or a printed book for the same amount of time. They found that those who read the ebook had decreased sleepiness, reduced REM sleep, and a smaller increase in blood levels of melatonin, the hormone that regulates the sleep–wake cycle. Suppose you want to repeat this research with high-school-aged subjects, and you have 50 volunteers ready to participate.[105]

(a) Outline an experiment using a completely randomized design to examine the effect of ereaders on sleep.

(b) If the difference in sleep quality is statistically significant, what conclusions can you draw from this study? Explain.

10. **Staging a theft** (3.9) Students sign up to be subjects in a psychology experiment. When they arrive, they are told that interviews are running late and are taken to a waiting room. The experimenters then stage the theft of a valuable object left in the waiting room. Some subjects are alone with the thief, and others are in pairs—these are the treatments being compared. Will the subject report the theft?

(a) The students had agreed to take part in an unspecified study, and the true nature of the experiment is explained to them afterward. Does this meet the requirement of informed consent? Explain.

(b) What two other ethical principles should be followed in this study?

Chapter 3 Practice Test

Section I: Multiple Choice *Select the best answer for each question.*

1. Which one of the following is a valid statistical question?
(a) How tall is the Empire State Building?
(b) How many points did LeBron James score in the 2014 NBA playoffs?
(c) When you call your cell-phone provider's customer service number, how long will you be on hold?
(d) How much money did Ivan's family spend on groceries last month?

2. A sportswriter wants to know how strongly Albuquerque residents support the local minor league baseball team, the Isotopes. She stands outside the stadium before a game and interviews the first 20 people who enter the stadium, asking them to rate their enthusiasm for the team on a 1 (lowest) to 5 (highest) scale. Which of the following best describes the results of this survey?
(a) Because this is a random sample, there will be some sampling variability, but we can expect it to produce quite accurate results.
(b) This is a random sample, but the size of the sample is too small to produce reliable results.
(c) This is a voluntary response sample and is likely to underestimate support for the team.
(d) This is a convenience sample and is likely to overestimate the level of support for the team.

3. The Web portal AOL places opinion poll questions next to many of its news stories. Simply click your response to join the sample. One recent question was "Do you plan to diet this year?" More than 30,000 people responded, with 68% saying "Yes." You can conclude that
(a) about 68% of Americans planned to diet.
(b) the poll used a convenience sample, so the results tell us little about the population of all adults.
(c) the poll uses voluntary response, so the results tell us little about the population of all adults.
(d) the sample is too small to draw any conclusion.

4. A local news agency conducted a survey about unemployment by randomly dialing phone numbers until it had gathered responses from 1000 adults in the state. In the survey, 19% of those who responded said they were not currently employed. In reality, only 6% of the adults in the state were not currently employed at the time of the survey. Which of the following best explains the difference in the two percentages?
(a) The difference is due to sampling variability. We shouldn't expect the results of a random sample to match the truth about the population every time.

(b) The difference is due to response bias. Adults who are employed are likely to lie and say that they are unemployed.
(c) The difference is due to undercoverage. The survey included only adults and did not include teenagers who are eligible to work.
(d) The difference is due to nonresponse. Adults who are employed are less likely to be available for the sample than adults who are unemployed.

5. A study of treatments for angina (pain due to low blood supply to the heart) compared bypass surgery, angioplasty, and use of drugs. The study looked at the medical records of thousands of angina patients whose doctors had chosen one of these treatments. It found that the average survival time of patients given drugs was the highest. What do you conclude?
(a) This study provides convincing evidence that drugs prolong life and should be the treatment of choice.
(b) We can't conclude that drugs prolong life because the patients were volunteers.
(c) We can't conclude that drugs prolong life because this was an observational study.
(d) We can't conclude that drugs prolong life because no placebo was used.

6. Some studies of the relationship between car color and frequency of accidents have found that red cars are more likely to be in accidents than black cars, despite how visible they are. Some experts warn that we should not conclude red cars are less safe than black cars because of possible confounding. Which of the following best describes what this means?
(a) There are too many variables involved in traffic accidents to isolate the effect of car color.
(b) The accidents used in the study might not be a random sample of all accidents.
(c) It is not possible to separate the effect of car color from the type of people who choose to drive red cars.
(d) It can be hard to determine if red cars or black cars are more visible on modern highways.

7. A new headache remedy was given to a group of 25 subjects who had headaches. Four hours after taking the new remedy, 20 of the subjects reported that their headaches had disappeared. From this information, you conclude
(a) that the remedy is effective for the treatment of headaches.
(b) nothing because the sample size is too small.

(c) nothing because there is no control group for comparison.

(d) that the new treatment is better than aspirin.

8. One hundred volunteers who suffer from attention deficit hyperactivity disorder (ADHD) are available for a study. Fifty are randomly assigned to receive a new drug that is thought to be particularly effective in treating ADHD. The other 50 are given a commonly used drug. A psychiatrist evaluates the symptoms of all volunteers after four weeks to determine if there has been substantial improvement in symptoms. The study would be double-blind if

(a) neither drug had any identifying marks on it.

(b) none of the volunteers were allowed to see the psychiatrist, and the psychiatrist is also not allowed to see the volunteers during the evaluation session.

(c) neither the volunteers nor the psychiatrist knew which treatment any person had received.

(d) the patients were given a placebo.

9. Consider an experiment to investigate the effectiveness of different insecticides in controlling pests and their impact on the productivity of tomato plants. What is the best reason for randomly assigning treatments (spraying or not spraying) to the farms?

(a) Random assignment allows researchers to generalize conclusions about the effectiveness of the insecticides to all farms.

(b) Random assignment will tend to average out all other variables, such as soil fertility, so that they are not confounded with the treatment effects.

(c) Random assignment eliminates the effects of other variables, like soil fertility.

(d) Random assignment eliminates chance variation in the responses.

10. Which of the following is *not* a benefit of keeping other variables the same in an experiment?

(a) Keeping other variables the same helps prevent confounding.

(b) Keeping other variables the same reduces variability in the response variable.

(c) Keeping other variables the same makes it easier to get statistically significant results if one treatment is more effective than the other.

(d) Keeping other variables the same eliminates the need for random assignment.

11. Every hour, the quality control manager at a factory making tortilla chips selects a random sample of 20 bags of chips and weighs the contents. If the manager is convinced that the mean weight of all bags produced that hour differs from the target of 16 ounces, the production line will be stopped so the problem can be identified. In the current sample of 20 bags, the mean weight is 15.94 ounces, and this estimate has a margin of error of 0.12 ounces. Which of the following conclusions is best?

(a) The manager should stop the production line because the mean of 15.94 is less than 16.

(b) The manager should stop the production line because the margin of error is greater than 0.

(c) The manager should not stop the production line because the difference between 15.94 and 16 is less than the margin of error.

(d) The manager should not stop the production line because the margin of error is very small.

Section II: Free Response

12. To examine the impact of question wording on survey results, Sam and Joshan randomly selected 60 students from their school's student list and divided them at random into two groups of 30. By email, they asked the students the following questions about a nearby convenience store, the Quick Shop:

Group 1 Question ("Markup"): Do you think the prices at the Quick Shop are fair, given that the typical markup is 25%?

Group 2 Question ("Convenience"): Keeping in mind the convenient location of the Quick Shop, do you think the prices are fair, given that the typical markup is 25%? Here are the results of their survey.

		Question Wording		
		Markup	Convenience	Total
Response	Fair	7	18	25
	Not fair	23	12	35
	Total	30	30	60

(a) Is this an experiment or an observational study? Justify your answer.

(b) Calculate the proportion of subjects who responded "fair" in each of the two treatment groups and the difference in proportions (Convenience – Markup).

(c) One hundred trials of a simulation were performed to see what differences in proportions would happen due only to chance variation in the random assignment, assuming that the wording of the question didn't matter. Use the results of the simulation on the next page to determine if the difference in proportions from part (b) is statistically significant. Explain your reasoning.

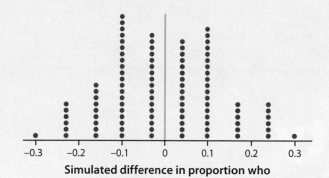

Simulated difference in proportion who responded fair

(d) Based on the design of the study, what conclusions can we draw? Explain.

13. The director of buildings and grounds at a university wants to evaluate the satisfaction level of students living in one of the dormitories on campus. She plans to interview a random sample of 50 of the 816 residents of the building.

(a) Explain how you could use a random number generator to select a simple random sample of 50 residents of the dormitory.

(b) During each interview, the director asks, "If you could make the choice again, would you still choose this dorm?" In the sample of 50, 66% of the residents answered "Yes." If the director took a second random sample of 50 students, would this percent be the same? Explain.

(c) Explain to the director why response bias might make the estimate from part (b) unreliable. Would increasing the sample size solve this problem? Explain.

14. Elephants sometimes damage trees in Africa. It turns out that elephants dislike bees. They recognize beehives and avoid them. Can this information be used to keep elephants away from trees? Will elephant damage be less in trees with hives? Will empty hives even keep elephants away? Researchers want to design an experiment to answer these questions using 72 acacia trees.[106]

(a) Identify the experimental units, treatments, and the response variable.

(b) Describe how the researchers could carry out a completely randomized design for this experiment. Include a description of how the treatments should be assigned.

15. In the early 1960s, Stanley Milgram ran an experiment in which subjects were told they were "teachers" and were instructed to give electric shocks to other subjects ("students") when the latter made mistakes on a word-matching test. What the "teachers" didn't know was that the "students" were Milgram's assistants, who faked their discomfort at receiving nonexistent shocks. Milgram showed that people were willing to administer what they thought were serious shocks rather than disobey someone in authority. In what way did this experiment violate the principles of ethical research?[107]

4

Probability

Lesson 4.1   Randomness, Probability, and Simulation   262

Lesson 4.2   Basic Probability Rules   270

Lesson 4.3   Two-Way Tables and Venn Diagrams   277

Lesson 4.4   Conditional Probability and Independence   286

Lesson 4.5   The General Multiplication Rule and Tree Diagrams   294

Lesson 4.6   The Multiplication Rule for Independent Events   302

Lesson 4.7   The Multiplication Counting Principle and Permutations   309

Lesson 4.8   Combinations and Probability   317

Chapter 4   Main Points   325

Chapter 4   Review Exercises   327

Chapter 4   Practice Test   328

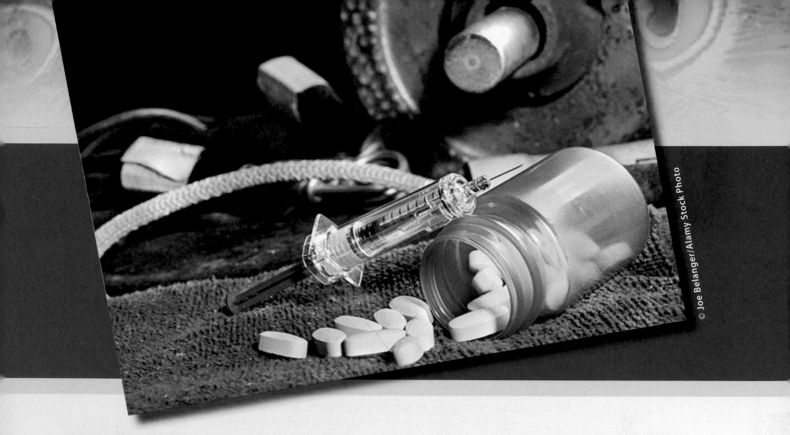

© Joe Belanger/Alamy Stock Photo

STATS applied!

Should an athlete who fails a drug test be suspended?

Many high schools now have drug-testing programs for athletes. The main goal of these programs Is to reduce the use of banned substances by students who play sports. It is not practical to test every athlete for drug use regularly. Instead, school administrators give drug tests to randomly selected student athletes at unannounced times during the school year. Students who test positive face serious consequences, including letters to their parents, mandatory counseling, and suspension from athletic participation.

Drug tests aren't perfect. Sometimes the tests say that athletes took a banned substance when they did not. This is known as a *false positive*. Other times, drug tests say that athletes are "clean" when they did, in fact, take a banned substance. This is called a *false negative*.

Suppose that 16% of the high school athletes in a large school district have taken a banned substance. The drug test used by this district has a false positive rate of 5% and a false negative rate of 10%. If a randomly chosen athlete tests positive, what's the chance that the student took a banned substance? Should an athlete who tests positive be suspended from athletic competition for a year?

We'll revisit STATS applied! at the end of the chapter, so you can use what you have learned to help answer these questions.

Lesson 4.1

Randomness, Probability, and Simulation

Chance is all around us. You and your friend play rock-paper-scissors to determine who gets the last slice of pizza. A coin toss decides which team gets to receive the ball first in a football game. People young and old play games of chance involving cards, dice, or spinners. The traits that children inherit—gender, hair and eye color, blood type, handedness, dimples, whether or not they can roll their tongues—are determined by the chance involved in which genes their parents pass along.

The mathematics of chance behavior is called *probability*. Probability is the topic of this chapter.

The Idea of Probability

A C T I V I T Y

What is probability?

If you toss a fair coin, what's the probability that it shows heads? It's 1/2, or 0.5, right? But what does a probability of 1/2 really mean? In this activity, you will investigate by flipping a coin several times.

1. Toss your coin once. Record whether you get heads or tails.

2. Toss your coin a second time. Record whether you get heads or tails. What proportion of your first two tosses is heads?

3. Toss your coin 8 more times, so that you have 10 tosses in all. Record whether you get heads or tails on each toss in a table like the one that follows.

4. Calculate the proportion of heads after each toss and record these values in the bottom row of the table. For instance, suppose you got tails on the first toss and heads on the second toss. Then your proportions of heads would be 0/1 = 0 after the first toss and 1/2 = 0.50 after the second toss.

Toss	1	2	3	4	5
Result (H or T)					
Proportion of heads					
Toss	6	7	8	9	10
Result (H or T)					
Proportion of heads					

5. Launch our *The Idea of Probability* applet at highschool.bfwpub.com/spa3e. Set the number of tosses at 10 and click "Toss." What proportion of the tosses were heads? Click "Reset" and toss the coin 10 more times. What proportion of heads did you get this time? Repeat this process several more times. What do you notice?

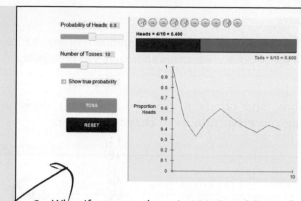

7. Keep on tossing without hitting "Reset." What happens to the proportion of heads?

8. As a class, discuss what the following statement means: "If you toss a fair coin, the probability of heads is 0.5."

9. If you toss a coin, it can land heads or tails. If you "toss" a thumbtack, it can land with the point sticking up or with the point down. Does that mean the probability of a tossed thumbtack landing point up is 0.5? How can you find out? Discuss with your classmates.

6. What if you toss the coin 100 times? Reset the applet and have it do 100 tosses. Is the proportion of heads exactly equal to 0.5? Close to 0.5?

Figure 4.1 shows some results from the activity. The proportion of tosses that land heads varies from 0.30 to 1.00 in the first 10 tosses. As we make more and more tosses, however, the proportion of heads gets closer to 0.5 and stays there.

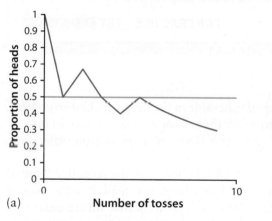

(a)

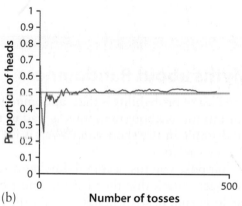

(b)

FIGURE 4.1 (a) The proportion of heads in the first 10 tosses of a coin. (b) The proportion of heads in the first 500 tosses of a coin.

When we watch coin tosses or the results of random sampling and random assignment closely, a remarkable fact emerges: *Chance behavior is unpredictable in the short run but has a regular and predictable pattern in the long run.* This is the basis for the idea of **probability.**

DEFINITION Probability

The probability of any outcome of a chance process is a number between 0 and 1 that describes the proportion of times the outcome would occur in a very large number of repetitions.

Outcomes that never occur have probability 0. An outcome that happens on every repetition has probability 1. An outcome that happens half the time in a very long series of trials has probability 0.5.

The fact that the proportion of heads in many tosses eventually closes in on 0.5 is guaranteed by the **law of large numbers.**

> **DEFINITION** Law of large numbers
>
> The **law of large numbers** says that if we observe more and more repetitions of any chance process, the proportion of times that a specific outcome occurs approaches its probability.

Life-insurance companies, casinos, and others who make important decisions based on probability rely on the long-run predictability of chance behavior.

EXAMPLE

Who eats breakfast?

Interpreting probability

PROBLEM: According to *The Book of Odds*, the probability that a randomly selected U.S. adult usually eats breakfast is 0.61.

(a) Explain what probability 0.61 means in this setting.

(b) Does this probability say that if 100 U.S. adults are chosen at random, exactly 61 of them usually eat breakfast? Explain.

SOLUTION:

(a) If you take a very large random sample of U.S. adults, about 61% of them will be people who usually eat breakfast.

(b) No. Probability describes what happens in many, many repetitions (way more than 100) of a chance process. We would expect to get *about* 61 people who usually eat breakfast in a random sample of 100 U.S. adults.

FOR PRACTICE   TRY EXERCISE 1.

Myths about Randomness

The idea of probability is that randomness is predictable in the long run. Unfortunately, our intuitions about randomness lead us to think that chance behavior should also be predictable in the short run. When it isn't, we look for some explanation other than chance variation.

Suppose you toss a coin 6 times and get TTTTTT. Believers in the so-called "law of averages" think that the next toss must be more likely to give a head. It's true that in the long run, heads will appear half the time. What is a myth is that future outcomes must make up for an imbalance like six straight tails.

Coins and dice have no memories. A coin doesn't know that the first 6 outcomes were tails, and it can't try to get a head on the next toss to even things out. Of course, things do even out *in the long run*. That's the law of large numbers in action. After 10,000 tosses, the results of the first six tosses don't matter. They are overwhelmed by the results of the next 9994 tosses.

EXAMPLE

Aren't we due for a boy?

Beware of the "law of averages"

PROBLEM: A husband and wife decide to have children until they have at least one child of each gender. The couple had seven girls in a row. Their doctor assured them that they were much more likely to have a boy for their next child after all those girls. Explain why the doctor is wrong.

SOLUTION:

The doctor's claim is based on the "law of averages." This couple is just as likely to have a girl as a boy for their next child.

> Having children is like tossing coins. There is about a 1/2 probability that any child born to them will be a boy.

FOR PRACTICE   TRY EXERCISE 5.

When asked to predict the gender—boy (B) or girl (G)—of the next seven babies born in a local hospital, most people will guess something like B-G-B-G-B-G-G. Few people would say G-G-G-B-B-B-G because this sequence of outcomes doesn't "look random." In fact, these two sequences of births are equally likely. "Runs" consisting of several of the same outcomes in a row are surprisingly common in chance behavior.

Simulation

We can model chance behavior and estimate probabilities with a **simulation.**

> **DEFINITION** Simulation
>
> **Simulation** is the imitation of chance behavior, based on a model that accurately reflects the situation.

You already have some experience with simulations. The "1 in 6 wins" game in Lesson 1.1 (page 4) had you roll a die several times to imitate buying 20-ounce sodas and looking under the cap. Lesson 3.3's Name-Brand Cookie activity (page 197) asked you to use a spinner to simulate the choice of cookies. Lesson 3.7's Caffeine Experiment activity (page 238) asked you to shuffle and deal piles of index cards to mimic the random assignment of subjects to treatments.

These simulations involved different "chance devices"—dice, a spinner, and index cards. But the same basic strategy was followed in each simulation.

> ## How to Perform a Simulation
>
> **STATE:** Ask a question about some chance process.
>
> **PLAN:** Describe how to use a chance device to imitate one repetition of the process. Tell what you will record at the end of each repetition.
>
> **DO:** Perform many repetitions.
>
> **CONCLUDE:** Use the results of your simulation to help answer the question.

So far, we have used physical devices for our simulations. Technology provides another option: a random number generator.

EXAMPLE

How many boxes of cereal to collect them all?

Simulations with technology

PROBLEM: In an attempt to increase sales, a breakfast cereal company decides to offer a NASCAR promotion. Each box of cereal will contain a collectible card featuring one of these NASCAR drivers: Jeff Gordon, Dale Earnhardt, Jr., Kasey Kahne, Danica Patrick, or Jimmie Johnson. The company claims that each of the 5 cards is equally likely to appear in any box of cereal. A NASCAR fan decides to keep buying boxes of the cereal until she has all 5 drivers' cards. She is surprised when it takes her 23 boxes to get the full set of cards.

Design and carry out a simulation to estimate the probability that it would take 23 or more boxes to get all 5 drivers' cards if the company's claim is true. What conclusion should the fan draw?

Daniel SHirey/Getty Images

SOLUTION:

STATE: What's the probability that it will take 23 or more boxes to get a full set of 5 NASCAR collectible cards if the company's claim is true?

PLAN:

- Let 1 = Jeff Gordon; 2 = Dale Earnhardt, Jr.; 3 = Kasey Kahne; 4 = Danica Patrick; and 5 = Jimmie Johnson.
- Generate a random integer from 1 to 5 to simulate buying one box of cereal and looking at which card is inside.
- Keep generating integers until all five of the labels from 1 to 5 appear. Record the number of boxes it took.

DO: I will perform 50 repetitions.

Rep 1: <u>3</u> <u>5</u> <u>2</u> <u>1</u> 5 2 3 5 <u>4</u>   9 boxes

Rep 2: <u>5</u> <u>1</u> <u>2</u> 5 1 <u>4</u> 1 4 1 2 2 2 4 4 5 <u>3</u>   16 boxes

Rep 3: 5 5 5 <u>2</u> <u>4</u> <u>1</u> 2 1 <u>5</u> <u>3</u>   10 boxes

...

CONCLUDE: In 50 repetitions of the simulation, it never took more than 22 boxes to get the full set of NASCAR drivers' cards. The estimated probability that it would take 23 or more boxes if the company's claim is true is approximately 0/50 = 0. The NASCAR fan has convincing evidence that the cereal company's claim isn't true.

STATE: Ask a question about some chance process.

PLAN: Describe how to use a chance device to imitate one repetition of the process. Tell what you will record at the end of each repetition.

DO: Perform many repetitions.

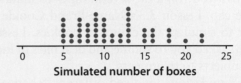

The dotplot shows the number of boxes it took to get all 5 drivers' cards in 50 repetitions.

Simulated number of boxes

CONCLUDE: Use the results of your simulation to help answer the question.

FOR PRACTICE   TRY EXERCISE 9.

LESSON APP 4.1

Will the train arrive on time?

New Jersey Transit claims that its 8:00 A.M. train from Princeton to New York has probability 0.9 of arriving on time. Assume for now that this claim is true.

1. Explain what probability 0.9 means in this setting.

2. The 8:00 A.M. train has arrived on time 5 days in a row. What's the probability that it will arrive on time tomorrow? Explain.

3. A businessman takes the 8:00 A.M. train to work on 20 days in a month. He is surprised when the train arrives late in New York on 3 of the 20 days. Should he be surprised? Describe how you would carry out a simulation to estimate the probability that the train would arrive late on 3 or more of 20 days if New Jersey Transit's claim is true. Do not perform the simulation.

4. The following dotplot shows the number of days on which the train arrived late in 100 repetitions of the simulation. What is the resulting estimate of the probability described in Question 3? Should the businessman be surprised?

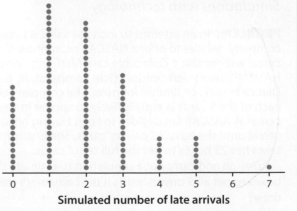

Simulated number of late arrivals

Andrew Burton/Getty Images

Lesson 4.1

WHAT DID YOU LEARN?

LEARNING TARGET	EXAMPLES	EXERCISES
Interpret probability as a long-run relative frequency.	p. 264	1–4
Dispel common myths about randomness.	p. 264	5–8
Use simulation to model chance behavior.	p. 265	9–12

Exercises   Lesson 4.1

Mastering Concepts and Skills

1. **Genetic testing** Suppose a married man and woman
pg 264 both carry a gene for cystic fibrosis but don't have the disease themselves. According to the laws of genetics, the probability that any child they have will develop cystic fibrosis is 0.25.

(a) Explain what this probability means.

(b) If the couple has 4 children, is one of them guaranteed to get cystic fibrosis? Explain.

2. **The coffee majority** According to *The Book of Odds*, the probability that a randomly selected U.S. adult drinks coffee on a given day is 0.56.

(a) Explain what this probability means.

(b) If a researcher surveys 100 U.S. adults at random on the same day, will exactly 56 of them have consumed coffee that day? Explain.

3. **Red light!** Pedro drives the same route to work on Monday through Friday. His route includes one traffic light. According to the local traffic department, there is a 55% chance that the light will be red. Explain what this probability means.

4. **Take the umbrella?** A local weather forecast says that there is a 20% chance of rain tomorrow. Explain what this probability means.

5. **Predicting a hit** A very good professional baseball
pg 264 player gets a hit about 35% of the time over an entire season. After the player failed to hit safely in six straight at-bats, a TV commentator said, "He is due for a hit by the law of averages." Explain why the commentator is wrong.

6. **Hot hand?** A college basketball player makes about 75% of her free throws. In one game, the player makes her first 5 free-throw attempts. When the player steps to the free-throw line for her sixth attempt, a radio announcer says that she is "due for a miss." Explain why the announcer is wrong.

7. **Heads up** Imagine tossing a coin 6 times and recording heads (H) or tails (T) on each toss. Which of the following outcomes is more likely: HTHTTH or TTTHHH? Justify your answer.

8. **No dice** Imagine rolling a die 12 times and recording the result of each roll. Which of the following outcomes is more likely: 1 2 3 4 5 6 6 5 4 3 2 1 or 1 5 4 5 2 4 3 3 6 1 2 6? Justify your answer.

9. **Color-blindness in men** A randomly selected U.S.
pg 265 adult male has probability about 0.07 of having some form of red–green color blindness. Suppose we choose 4 U.S. adult males at random. What's the probability that at least one of them is red–green color-blind? Design and carry out a simulation to help answer this question.

10. **Whose book is this?** Suppose that 4 friends get together to study at Leigh's house for their next statistics test. When they go for a snack in the kitchen, Leigh's 3-year-old brother makes a tower using their textbooks. Unfortunately, none of the students has written his or her name in the books, so when they leave, each student takes one of the books at random. When the students return the books at the end of the year and the clerk scans their barcodes, the students are surprised to learn that none of the 4 had their own book. What's the probability that none of the 4 students ended up with the correct book? Design and carry out a simulation to help answer this question.

11. **Streaky shooting** A basketball announcer suggests that a certain player is streaky. That is, the announcer believes that if the player makes a shot, the player is more likely to make the next shot. As evidence, the announcer points to a recent game where the player took 30 shots and had a streak of 10 made shots in a row. Is this convincing evidence of streaky shooting by the player? Assume that this player makes 50% of the shots and that the results of a shot don't depend on previous shots.

(a) Describe how you would carry out a simulation to estimate the probability that a 50% shooter who takes 30 shots in a game would have a streak of 10 or more made shots. Do not perform the simulation.

(b) The dotplot displays the results of 50 simulated games in which this player took 30 shots. What conclusion would you draw about whether this player was streaky? Explain.

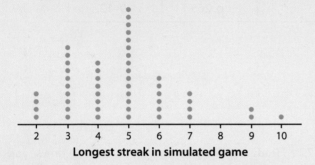

Longest streak in simulated game

12. **Double fault!** A professional tennis player claims to get 90% of her second serves in. In a recent match, the player missed 5 of her first 20 second serves. Is this a surprising result if the player's claim is true? Assume that the player has a 0.90 probability of getting each second serve in.

(a) Describe how you would carry out a simulation to estimate the probability that a player who gets 90% of her second serves in would miss 5 or more of her first 20 second serves. Do not perform the simulation.

(b) The dotplot displays the results of the first 20 second serves by this player in 100 simulated matches. What conclusion would you draw about the player's claim that she makes 90% of her second serves? Explain.

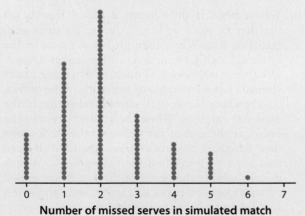

Number of missed serves in simulated match

Applying the Concepts

13. **Virtual three pointers** The figure shows the results of a virtual basketball player shooting many 3-point shots. Explain what this graph tells you about chance behavior in the short run and long run.

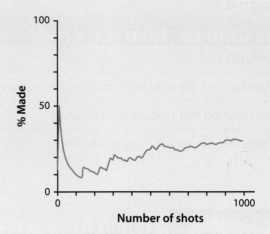

Number of shots

14. **Keep on tossing** The figure shows the results of two different sets of 5000 coin tosses. Explain what this graph tells you about chance behavior in the short run and the long run.

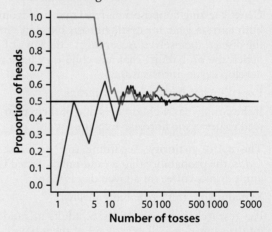

Number of tosses

15. **Do most teens recycle?** A statistics class asked an SRS of 100 students at their school whether they regularly recycle or not. In the sample, 55 students said that they recycle. Is this convincing evidence that more than half of the students at the school would say they regularly recycle? The dotplot shows the results of taking 200 SRSs of 100 students from a population in which the true proportion who recycle is 0.50.

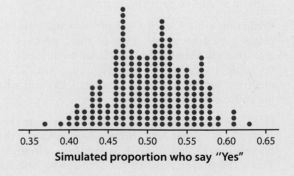

Simulated proportion who say "Yes"

(a) Explain why the sample result (55 out of 100 said "Yes") does not give convincing evidence that more than half of the school's students recycle.

(b) Suppose instead that 63 students in the class's sample had said "Yes." Explain why this result would give convincing evidence that a majority of the school's students recycle.

16. **Brushing teeth, wasting water?** A recent study reported that fewer than half of young adults turn off the water while brushing their teeth. Is the same true for teenagers? To find out, a group of statistics students asked an SRS of 60 students at their school if they usually brush with the water off. In the sample, 27 students said "Yes." The dotplot shows the results of taking 200 SRSs of 60 students from a population in which the true proportion who brush with the water off is 0.50.

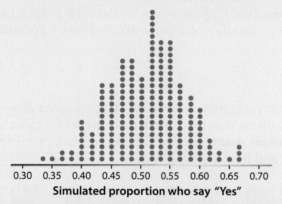

Simulated proportion who say "Yes"

(a) Explain why the sample result (27 out of 60 said "Yes") does not give convincing evidence that fewer than half of the school's students brush their teeth with the water off.

(b) Suppose instead that 18 of the 60 students in the class's sample had said "Yes." Explain why this result would give convincing evidence that fewer than 50% of the school's students brush their teeth with the water off.

Extending the Concepts

Some simulations involve imitating the process of choosing a random sample. It is important to avoid selecting the same individual more than once when performing such simulations.

17. **All vowels?** In the game of Scrabble, the first player draws 7 tiles at random from a bag containing 100 tiles. There are 42 vowels, 56 consonants, and 2 blank tiles in the bag. Cait draws first and is surprised to discover that all 7 tiles are vowels. Should she be surprised? Design and carry

out a simulation to estimate the probability of getting all vowels when selecting a random sample of 7 tiles from the bag.

18. **Extra security?** The Transportation Security Administration (TSA) is responsible for airport safety. On some flights, TSA officers randomly select passengers for an extra security check before boarding. One such flight had 76 passengers—12 in first class and 64 in coach class. Some passengers were surprised when none of the 10 passengers chosen for screening was seated in first class. Should they be surprised? Design and carry out a simulation to estimate the probability that none of 10 randomly selected passengers from such a flight are seated in first class.

Recycle and Review

19. **Wake up and work!** (1.3, 3.9) Mr. Brown has two statistics classes, one that meets at 8:00 A.M. and one that meets at 10:00 A.M. He wonders if student performance differs in the two classes. Here are the scores earned by the students in both classes on a recent quiz.

10 A.M.	24	28	28	24	11	19	9	28
	24	28	26	23	23	27	20	
8 A.M.	22	21	22	24	23	21	22	18
	21	17	22	23	23	22	21	20

(a) Make parallel dotplots of the data for the two classes.

(b) Use your graph from part (a) to compare the quiz score distributions.

(c) Can Mr. Brown conclude that any difference in student performance for the two classes is caused by the times that the classes meet? Explain.

20. **AARP and Medicare** (3.5) To find out what proportion of Americans support proposed Medicare legislation to help pay medical costs, the AARP conducted a survey of their members (people over age 50 who pay membership dues). One of the questions was: "Even if this plan won't affect you personally either way, do you think it should be passed so that people with low incomes or people with high drug costs can be helped?" Of the respondents, 75% answered yes.[1]

(a) Describe how undercoverage might lead to bias in this study. Explain the likely direction of the bias.

(b) Describe how the wording of the question might lead to bias in this study. Explain the likely direction of the bias.

Lesson 4.2

Basic Probability Rules

LEARNING TARGETS

- Give a probability model for a chance process with equally likely outcomes and use it to find the probability of an event.
- Use the complement rule to find probabilities.
- Use the addition rule for mutually exclusive events to find probabilities.

In Lesson 4.1, we used simulation to imitate chance behavior. Do we always have to repeat a chance process—rolling two dice, flipping a coin, or drawing a name from a hat—many times to determine the probability of a particular outcome? Fortunately, the answer is no.

Probability Models

Many board games involve rolling dice. Imagine rolling two fair, six-sided dice—one that's red and one that's blue. How do we develop a **probability model** for this chance process? Figure 4.2 displays the **sample space**. Because the dice are fair, each of these 36 outcomes will be equally likely and have probability 1/36.

why rolling a 7 is not 1/11

FIGURE 4.2 The 36 possible outcomes from rolling two dice, one red and one blue.

▌▌▌ **DEFINITION** Probability model, Sample space

A **probability model** is a description of some chance process that consists of two parts: a list of all possible outcomes and the probability for each outcome.

The list of all possible outcomes is called the **sample space.**

A probability model does more than just assign a probability to each outcome. It allows us to find the probability of an **event.**

270

> **DEFINITION** Event
>
> An **event** is any collection of outcomes from some chance process.

Events are usually designated by capital letters, like A, B, C, and so on. For rolling two 6-sided dice, we can define event A as getting a sum of 5. We write the probability of event A as $P(A)$ or $P(\text{sum is } 5)$.

It is fairly easy to find the probability of an event in the case of equally likely outcomes. There are 4 outcomes in event A:

The probability that event A occurs is therefore

$$P(A) = \frac{\text{number of outcomes in event A}}{\text{total number of outcomes in sample space}} = \frac{4}{36} = 0.111$$

Finding Probabilities: Equally Likely Outcomes

If all outcomes in the sample space are equally likely, the probability that event A occurs can be found using the formula

$$P(A) = \frac{\text{number of outcomes in event A}}{\text{total number of outcomes in sample space}}$$

EXAMPLE

Will you get a head?

Probability models: Equally likely outcomes

PROBLEM: Suppose you flip a fair coin 3 times.

(a) Give a probability model for this chance process.

(b) Define event A as getting exactly two heads. Find $P(A)$.

SOLUTION:

(a) The sample space is

HHH   HHT   HTH   THH   HTT   THT   TTH   TTT

Because the coin is fair, each of these 8 outcomes will be equally likely and have probability 1/8.

(b) There are 3 outcomes—HHT, HTH, THH—with exactly two heads. So

$$P(A) = \frac{3}{8} = 0.375$$

$$P(A) = \frac{\text{number of outcomes in event A}}{\text{total number of outcomes in sample space}}$$

FOR PRACTICE   TRY EXERCISE 1.

Basic Probability Rules

Our work so far suggests two common sense rules that a valid probability model must obey:

- **The probability of any event is a number between 0 and 1.** This rule follows from the definition of probability: the proportion of times the event would occur in many repetitions of the chance process.

- **All possible outcomes together must have probabilities that add up to 1.** Any time we observe a chance process, some outcome must occur.

Here's one more rule that follows from the previous two:

- **The probability that an event does *not* occur is 1 minus the probability that the event does occur.** Earlier, we found that the probability of getting a sum of 5 when rolling two fair, six-sided dice is 4/36. What's the probability that the sum is *not* 5?

$$P(\text{sum is not } 5) = 1 - P(\text{sum is } 5) = 1 - \frac{4}{36} = \frac{32}{36} = 0.889$$

We refer to the event "not A" as the **complement** of A and denote it by A^C. For that reason, this handy result is known as the **complement rule.** Using the complement rule in this setting is much easier than counting all 32 possible ways to get a sum that isn't 5.

> **DEFINITION** Complement rule, Complement
>
> The **complement rule** says that $P(A^C) = 1 - P(A)$ where A^C is the **complement** of event A; that is, the event that A does not happen.

EXAMPLE

Can you avoid the blue M&M'S?

Complement rule

PROBLEM: Suppose you tear open the corner of a bag of M&M'S® Milk Chocolate Candies, pour one candy into your hand, and observe the color. According to Mars, Incorporated, the maker of M&M'S, the probability model is

Color	Blue	Orange	Green	Yellow	Red	Brown
Probability	0.24	0.20	0.16	0.14	0.13	0.13

(a) Explain why this is a valid probability model.

(b) Find the probability that you don't get a blue M&M.

© Niels Poulsen std/Alamy Stock Photo

SOLUTION:

(a) The probability of each outcome is a number between 0 and 1, and $0.24 + 0.20 + 0.16 + 0.14 + 0.13 + 0.13 = 1$

(b) $P(\text{not blue}) = 1 - P(\text{blue}) = 1 - 0.24 = 0.76.$

> There is a 76% chance that the candy isn't blue.

FOR PRACTICE   TRY EXERCISE 5.

What's the probability that you get a green or a red M&M? It's

$$P(\text{green or red}) = P(\text{green}) + P(\text{red}) = 0.16 + 0.13 = 0.29$$

Why does this formula work? Because the events "getting a green" and "getting a red" have no outcomes in common—that is, there are no M&M'S that are both green and red. We say that these two events are **mutually exclusive.** As a result, this intuitive formula is known as the **addition rule for mutually exclusive events.**

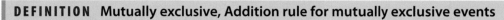

> **DEFINITION** Mutually exclusive, Addition rule for mutually exclusive events
>
> Two events A and B are **mutually exclusive** if they have no outcomes in common and so can never occur together—that is, if $P(A \text{ and } B) = 0$.
>
> The **addition rule for mutually exclusive events** A and B says that
> $$P(A \text{ or } B) = P(A) + P(B)$$

Note that this rule only works for mutually exclusive events. We will develop a more general rule for finding $P(A \text{ or } B)$ that works for *any* two events in Lesson 4.3.

CAUTION

EXAMPLE

Is it easy to pass the AP® exam?

Addition rule for mutually exclusive events

PROBLEM: Randomly select a student who took the 2015 AP® Statistics exam and record the student's score. Here is the probability model according to the College Board:

Score	1	2	3	4	5
Probability	0.236	0.186	0.252	0.191	0.135

Many people consider scores of 3, 4, or 5 as "passing scores" because many colleges award credit or placement to students who earn these scores.

(a) Find the probability that the chosen student scored less than a 3.

(b) Find the probability that the chosen student earned a passing score.

SOLUTION:

(a) By the addition rule for mutually exclusive events,
$$P(\text{scored less than 3}) = P(\text{scored 1 or 2}) = P(\text{scored 1}) + P(\text{scored 2}) = 0.236 + 0.186 = 0.422$$

(b) By the complement rule,
$$P(\text{earned passing score}) = 1 - P(\text{scored less than 3}) = 1 - 0.422 = 0.578$$

FOR PRACTICE   TRY EXERCISE 9.

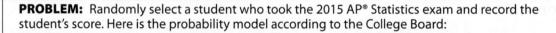

Note that you could also find the probability in part (b) of the example using the addition rule for mutually exclusive events:

$$
\begin{aligned}
P(\text{earned passing score}) &= P(\text{scored 3 or 4 or 5}) \\
&= P(\text{scored 3}) + P(\text{scored 4}) + P(\text{scored 5}) \\
&= 0.252 + 0.191 + 0.135 \\
&= 0.578
\end{aligned}
$$

LESSON APP 4.2

How prevalent is high cholesterol?

Choose an American adult at random. Define two events:

A = the person has a cholesterol level of 240 milligrams per deciliter of blood (mg/dl) or above (high cholesterol)
B = the person has a cholesterol level of 200 to <240 mg/dl (borderline high cholesterol)

According to the American Heart Association, $P(A) = 0.16$ and $P(B) = 0.29$.

1. Explain why events A and B are mutually exclusive.

2. Say in plain language what the event "A or B" is. Find $P(A \text{ or } B)$.

Westend61/Getty Images

3. Let C be the event that the person chosen has a cholesterol level below 200 mg/dl (normal cholesterol). Find $P(C)$.

Lesson 4.2

WHAT DID YOU LEARN?

LEARNING TARGET	EXAMPLES	EXERCISES
Give a probability model for a chance process with equally likely outcomes and use it to find the probability of an event.	p. 271	1–4
Use the complement rule to find probabilities.	p. 272	5–8
Use the addition rule for mutually exclusive events to find probabilities.	p. 273	9–12

Exercises · Lesson 4.2

Mastering Concepts and Skills

1. **Four-sided dice** A four-sided die is a pyramid whose four faces are labeled with the numbers 1, 2, 3, and 4 (see image). Imagine rolling two fair, four-sided dice—one blue and one yellow—and recording the number on the bottom of each pyramid.

   pg **271**

   (a) Give a probability model for this chance process.
   (b) What is the probability that the sum of the numbers is 5? Show your work.

2. **Toss 4 times** Imagine tossing a fair coin 4 times.
   (a) Give a probability model for this chance process.
   (b) Define event B as getting exactly three tails. Find $P(B)$.

3. **Rock smashes scissors** Almost everyone has played the rock-paper-scissors game at some point. Two players face each other and, at the count of 3, make a fist (rock), an extended hand, palm side down (paper), or a "V" with the index and middle fingers (scissors). The winner is determined by these rules: rock smashes scissors; paper covers rock; and scissors cut paper. If both players choose the same object, then the game is a tie. Suppose that Player 1 and Player 2 are both equally likely to choose rock, paper, or scissors.
   (a) Give a probability model for this chance process.
   (b) Find the probability that Player 1 wins the game on the first throw.

4. **Who's paying?** Ari, Betty, Charlie, Daniela, and Ethel go to the bagel shop for lunch every Thursday. Each time, they randomly pick 2 of the group to pay for lunch by drawing names from a hat.

(a) Give a probability model for this chance process.

(b) Find the probability that Charlie or Daniela (or both) ends up paying for lunch.

5. **Blood types vary** All human blood can be typed as one of O, A, B, or AB, but the distribution of the types varies with race. Here is the distribution of blood types of black Americans. Suppose we choose one black American at random.

Blood type	O	A	B	AB
Probability	0.49	0.27	0.20	?

(a) What is the probability that the chosen person has type AB blood? Why?

(b) Find the probability that the chosen person does not have type AB blood.

6. **Speaking Canada's languages** Canada has two official languages, English and French. Choose a Canadian at random and ask, "What is your mother tongue?" Here is the distribution of responses, combining many separate languages from the broad Asia/Pacific region.[2]

Language	English	French	Asian/Pacific	Other
Probability	0.63	0.22	0.06	?

(a) What is the probability that this person's mother tongue would be classified as "Other"? Why?

(b) Find the probability that this Canadian's mother tongue is not English.

7. **Household size** In government data, a household consists of all occupants of a dwelling unit. Choose an American household at random and count the number of people it contains. Here is the assignment of probabilities for the outcome.

Number of persons	1	2	3	4	5	6	7+
Probability	0.25	0.32	?	?	0.07	0.03	0.01

The probability of finding 3 people in a household is the same as the probability of finding 4 people.

(a) What probability should replace "?" in the table? Why?

(b) Find the probability that the chosen household contains more than 1 person.

8. **When did you leave?** The National Household Travel Survey gathers data on the time of day when people begin a trip in their car or other vehicle. Choose a trip at random and record the time at which the trip started.[3] Here is an assignment of probabilities for the outcome.

Time of day	10 P.M.–12:59 A.M.	1 A.M.–5:59 A.M.	6 A.M.–8:59 A.M.	9 A.M.–12:59 P.M.
Probability	0.040	0.033	0.144	0.234
Time of day	1 P.M.–3:59 P.M.	4 P.M.–6:59 P.M.	7 P.M.–9:59 P.M.	
Probability	0.208	?	0.123	

(a) What probability should replace "?" in the table? Why?

(b) Find the probability that the chosen trip did not begin between 9 A.M. and 12:59 P.M.

9. **Blood type compatibility** Refer to Exercise 5. Rolanda has type B blood. She can safely receive blood transfusions from people with blood types O and B.

(a) Find the probability that a randomly chosen black American can donate blood to Rolanda.

(b) Find the probability that a randomly chosen black American cannot donate blood to Rolanda.

10. **Which language of Canada?** Refer to Exercise 6.

(a) Find the probability that the chosen Canadian's mother tongue is English or French.

(b) Find the probability that the chosen Canadian's mother tongue is a language other than English or French.

11. **Specific household size** Refer to Exercise 7. What's the probability that a randomly selected American household has 5 or fewer people?

12. **What time did you leave?** Refer to Exercise 8. What's the probability that a randomly chosen trip began between 6 A.M. and 9:59 P.M.?

Applying the Concepts

13. **Education among young adults** Choose a young adult (aged 25 to 29) at random. The probability is 0.13 that the person chosen did not complete high school, 0.29 that the person has a high school diploma but no further education, and 0.30 that the person has at least a bachelor's degree.

(a) What must be the probability that a randomly chosen young adult has some education beyond high school but does not have a bachelor's degree? Why?

(b) Find the probability that the young adult completed high school. Which probability rule did you use to find the answer?

(c) Find the probability that the young adult has further education beyond high school. Which probability rule did you use to find the answer?

14. **Preparing for the GMAT** A company that offers courses to prepare students for the Graduate Management Admission Test (GMAT) has the following information about its customers: 20% are currently undergraduate students in business, 15% are currently undergraduate students in other fields of study, and 60% are college graduates who are currently employed. Choose a customer at random.

(a) What must be the probability that the customer is a college graduate who is not currently employed? Why?

(b) Find the probability that the customer is currently an undergraduate. Which probability rule did you use to find the answer?

(c) Find the probability that the customer is not an undergraduate business student. Which probability rule did you use to find the answer?

Extending the Concepts

15. **Rock, paper, scissors, lizard, Spock!** The characters on the television show *Big Bang Theory* invented a new version of the traditional rock, paper, scissors game described in Exercise 3. Do some research to find out how the game works. Suppose that Player 1 and Player 2 are both equally likely to choose rock, paper, scissors, lizard, or Spock. Find the probability that Player 2 wins.

Recycle and Review

16. **Is it random?** (3.3) Charles claims he can act as a random number generator. Samantha has doubts. She tells Charles to write down 50 random digits, suspecting that he won't write down zero as often

as he should. Charles includes only 2 zeros. Do these data provide convincing evidence that Charles produces fewer zeros than the expected 10%? Samantha simulated 100 trials of choosing 50 random digits, assuming that zero has a 1/10 chance of occurring each time. The dotplot shows the number of zeros in each trial.

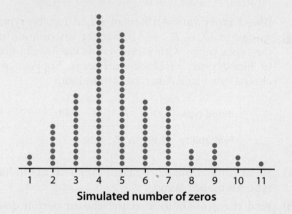

Simulated number of zeros

(a) Explain how the graph illustrates chance variation.

(b) Based on the results of the simulation, is there convincing evidence that Charles doesn't choose zero often enough when he's trying to generate random numbers? Explain.

17. **Owed some sleep?** (2.5) A researcher reported that the typical teenager needs 9.3 hours of sleep per night but gets only 6.3 hours.[4] By the end of a 5-day school week, a teenager would accumulate about 15 hours of "sleep debt." Students in a high school statistics class were skeptical, so they gathered data on the amount of sleep debt (in hours) accumulated over time (in days) by a random sample of 25 high school students. The resulting least-squares regression equation for their data is

Predicted sleep debt = 2.23 + 3.17(days)

(a) Interpret the slope of the regression line in context.

(b) Are the students' results consistent with the researcher's report? Explain.

Lesson 4.3

Two-Way Tables and Venn Diagrams

So far, you have learned how to model chance behavior and some basic rules for finding the probability of an event. What if you're interested in finding probabilities involving two events? For instance, a 2014 survey suggests that 71% of U.S. teenagers use Facebook, 33% use Twitter, and 15% do both.[5] Suppose we select a U.S. teenager at random. What's the probability that the student uses Facebook *or* uses Twitter? How about *P*(does not use Facebook *and* does not use Twitter)? In this lesson, we will introduce several methods to help you answer questions like these.

Two-Way Tables and the General Addition Rule

There are two different uses of the word "or" in everyday life. In a restaurant, when you are asked if you want "soup or salad," the waiter wants you to choose one or the other, but not both. However, when you order coffee and are asked if you want "cream or sugar," it's OK to ask for one or the other or both. This same issue arises in statistics.

Mutually exclusive events A and B cannot both happen at the same time. For such events, "A or B" means that only event A happens or only event B happens. You can find *P*(A or B) with the addition rule for mutually exclusive events:

$$P(\text{A or B}) = P(\text{A}) + P(\text{B})$$

How can we find *P*(A or B) when the two events are *not* mutually exclusive? Now we have to deal with the fact that "A or B" means one or the other or both. For instance, "uses Facebook or uses Twitter" in the scenario above includes U.S. teenagers who do both.

When you're trying to find probabilities involving two events, like *P*(A or B), a two-way table can display the sample space in a way that makes probability calculations easier.

EXAMPLE

How many men or women have a pierced ear?

Two-way tables and probability

Students in a college statistics class wanted to find out how common it is for young adults to have their ears pierced. They recorded data on two variables—gender and whether or not the student had a pierced ear—for all 178 people in the class. The two-way table summarizes the data.

		Gender		
		Male	Female	Total
Pierced ear	Yes	19	84	103
	No	71	4	75
	Total	90	88	178

PROBLEM: Suppose we choose a student from the class at random. Define event A as getting a male student and event B as getting a student with a pierced ear.

(a) Find $P(B)$. Interpret this value in context.

(b) Find P(male and pierced ear).

(c) Find $P(A \text{ or } B)$.

SOLUTION:

(a) $P(B) = P(\text{pierced ear}) = \dfrac{103}{178} = 0.579$. There is about a 58% chance that a randomly selected student from this class will have a pierced ear.

(b) $P(\text{male and pierced ear}) = \dfrac{19}{178} = 0.107$

(c) $P(A \text{ or } B) = P(\text{male or pierced ear})$

$$= \frac{71 + 84 + 19}{178} = \frac{174}{178} = 0.978$$

> In statistics, "or" means one or the other or both. So "male or pierced ear" includes (i) male but no pierced ear; (ii) pierced ear but not male; and (iii) male and pierced ear.

FOR PRACTICE   TRY EXERCISE 1.

When we found P(male and pierced ear) in part (b) of the example, we could have described this as either $P(A \text{ and } B)$ or $P(B \text{ and } A)$. Why? Because "A and B" describes the same event as "B and A." Likewise, $P(A \text{ or } B)$ is the same as $P(B \text{ or } A)$. Don't get so caught up in the notation that you lose sight of what's really happening!

Part (c) of the example reveals an important fact about finding the probability $P(A \text{ or } B)$: We can't use the addition rule for mutually exclusive events unless events A and B have no outcomes in common. In this case, there are 19 outcomes that are shared by events A and B—the students who are male and have a pierced ear. If we did add the probabilities of A and B, we'd get $90/178 + 103/178 = 193/178$. This is clearly wrong because the probability is bigger than 1! As Figure 4.3 illustrates, outcomes common to both events are counted twice when we add the probabilities of these two events.

FIGURE 4.3 Two-way table showing events A and B from the pierced-ear example. These events are *not* mutually exclusive, so we can't find $P(A \text{ or } B)$ by just adding the probabilities of the two events.

$P(A \text{ and } B) = \dfrac{19}{178}$

Outcomes here are double-counted by $P(A) + P(B)$

Gender

Pierced ear		Male	Female	Total	
	Yes	19	84	103	$P(B) = \dfrac{103}{178}$
	No	71	4	75	
	Total	90	88	178	

$P(A) = \dfrac{90}{178}$

We can fix the double-counting problem illustrated in the two-way table by subtracting the probability P(male and pierced ear) from the sum. That is,

$P(\text{male or pierced ear}) = P(\text{male}) + P(\text{pierced ear}) - P(\text{male and pierced ear})$

$\qquad = 90/178 + 103/178 - 19/178$

$\qquad = 174/178$

This result is known as the **general addition rule.**

> **DEFINITION** The general addition rule for two events
>
> If A and B are any two events resulting from some chance process, the **general addition rule** says that
>
> $$P(A \text{ or } B) = P(A) + P(B) - P(A \text{ and } B)$$

EXAMPLE

Which do teens prefer, Facebook or Twitter?

General addition rule

PROBLEM: A 2014 survey suggests that 71% of U.S. teenagers use Facebook, 33% use Twitter, and 15% do both. Suppose we select a U.S. teenager at random. What's the probability that the student uses Facebook or uses Twitter?

SOLUTION:

Let F = uses Facebook and T = uses Twitter.

$$P(F \text{ or } T) = P(F) + P(T) - P(F \text{ and } T)$$
$$= 0.71 + 0.33 - 0.15$$
$$= 0.89$$

> Use the general addition rule.
> $$P(A \text{ or } B) = P(A) + P(B) - P(A \text{ and } B)$$

FOR PRACTICE   TRY EXERCISE 5.

As the example suggests, it is sometimes easier to designate events with letters that relate to the context, like F for "uses Facebook" and T for "uses Twitter."

THINK ABOUT IT   **What happens if we use the general addition rule for two mutually exclusive events A and B?** In that case, $P(A \text{ and } B) = 0$, and the formula reduces to $P(A \text{ or } B) = P(A) + P(B)$. In other words, the addition rule for mutually exclusive events is just a special case of the general addition rule.

Venn Diagrams and Probability

We have seen that two-way tables can be used to illustrate the sample space of a chance process involving two events. So can **Venn diagrams.**

> **DEFINITION** Venn diagram
>
> A **Venn diagram** consists of one or more circles surrounded by a rectangle. Each circle represents an event. The region inside the rectangle represents the sample space of the chance process. See Figure 4.4.

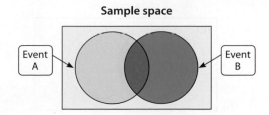

FIGURE 4.4 A typical Venn diagram that shows the sample space and the relationship between two events A and B.

In an earlier example, we looked at data from a survey on gender and ear piercings for a large group of college students. The chance process was selecting a student in the class at random. Our events of interest were A: is male and B: has a pierced ear. Here is the two-way table that summarizes the data.

		Gender		
		Male	Female	Total
Pierced ear	Yes	19	84	103
	No	71	4	75
	Total	90	88	178

The Venn diagram in Figure 4.5 displays the sample space in a slightly different way. There are four distinct regions in the Venn diagram. These regions correspond to the four cells in the two-way table as follows.

Region in Venn diagram	In words	Count
In the intersection of two circles	Male and pierced ear	19
Inside circle A, outside circle B	Male and no pierced ear	71
Inside circle B, outside circle A	Female and pierced ear	84
Outside both circles	Female and no pierced ear	4

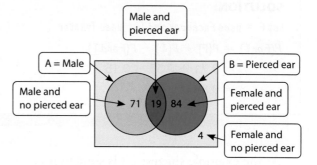

FIGURE 4.5 The completed Venn diagram for the large group of college students. The circles represent the two events A = male and B = has a pierced ear.

Some standard vocabulary and notation have been developed to make our work with Venn diagrams a bit easier.

- We introduced the *complement* of an event earlier. In Figure 4.6(a), the complement A^C contains the outcomes that are not in A.

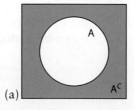

(a)

- Figure 4.6(b) shows the event "A and B." You can see why this event is also called the **intersection** of A and B. The corresponding notation is $A \cap B$.

$$A \cap B$$

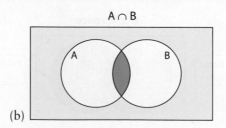

(b)

- The event "A or B" is shown in Figure 4.6(c). This event is also known as the **union** of A and B. The corresponding notation is A ∪ B.

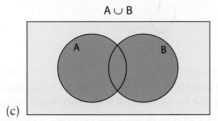

$A \cup B$

(c)

FIGURE 4.6 The green shaded region in each Venn diagram shows (a) the *complement* A^C of event A, (b) the *intersection* of events A and B, and (c) the *union* of events A and B.

> **DEFINITION** Intersection, Union
>
> The event "A and B" is called the **intersection** of events A and B. It consists of all outcomes that are common to both events, and is denoted A ∩ B.
>
> The event "A or B" is called the **union** of events A and B. It consists of all outcomes that are in event A or event B, or both, and is denoted A ∪ B.

With this new notation, we can rewrite the general addition rule in symbols as

$$P(A \cup B) = P(A) + P(B) - P(A \cap B)$$

This Venn diagram shows why the formula works in the pierced-ear example.

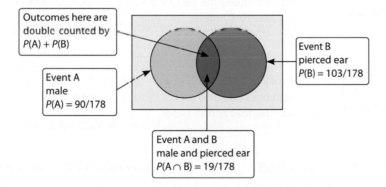

Outcomes here are double counted by $P(A) + P(B)$

Event A
male
$P(A) = 90/178$

Event B
pierced ear
$P(B) = 103/178$

Event A and B
male and pierced ear
$P(A \cap B) = 19/178$

For mutually exclusive events A and B, the two events have no outcomes in common. So the corresponding Venn diagram consists of two non-overlapping circles. You can see from the figure below why the general addition rule reduces to

$$P(A \cup B) = P(A) + P(B)$$

in this special case.

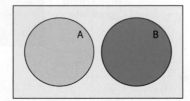

EXAMPLE

Which paper do the neighbors read?

Venn diagrams and probability

PROBLEM: In a large apartment complex, 40% of residents read *USA Today*. Only 25% read the *New York Times*. Five percent of residents read both papers. Suppose we select a resident of the apartment complex at random and record which of the two papers the person reads.

(a) Make a Venn diagram to display the sample space of this chance process using the events A: reads *USA Today* and B: reads *New York Times*.

(b) Find the probability that the person reads exactly one of the two papers.

SOLUTION:

(a)

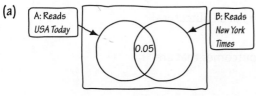

Start with the intersection. Because 5% of residents read both papers, $P(A \cap B) = 0.05$.

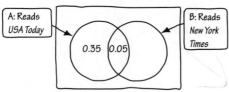

We know that 40% of residents read *USA Today*. That figure includes the 5% who read both papers. So 35% read only *USA Today*.

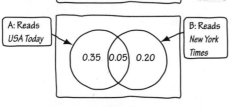

We know that 25% of residents read the *New York Times*. That figure includes the 5% who read both papers. So 20% read only the *New York Times*.

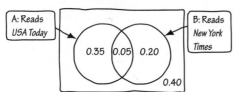

A total of $0.35 + 0.05 + 0.20 = 0.60$ (or 60%) of residents read at least one of the two papers. By the complement rule, $1 - 0.60 = 0.40$ (or 40%) read neither paper.

(b) $P(\text{reads exactly one of the two papers}) = 0.35 + 0.20 = 0.55$

FOR PRACTICE   **TRY EXERCISE 9.**

LESSON APP 4.3

Who owns a home?

What is the relationship between educational achievement and home ownership? A random sample of 500 U.S. adults was selected. Each member of the sample was identified as a high school graduate (or not) and as a homeowner (or not). The two-way table displays the data.

		High school graduate	
		Yes	No
Homeowner	Yes	221	119
	No	89	71

Andresr/Shutterstock

Suppose we choose a member of the sample at random. Define events G: is a high school graduate and H: is a homeowner.

1. Explain why $P(G \text{ or } H) \neq P(G) + P(H)$. Then find $P(G \text{ or } H)$.

2. Make a Venn diagram to display the sample space of this chance process.

3. Write the event "is not a high school graduate but is a homeowner" in symbolic form.

Lesson 4.3

WHAT DID YOU LEARN?

LEARNING TARGET	EXAMPLES	EXERCISES
Use a two-way table to find probabilities.	p. 277	1–4
Calculate probabilities with the general addition rule.	p. 279	5–8
Use a Venn diagram to find probabilities.	p. 282	9–12

Exercises   *Lesson 4.3*

Mastering Concepts and Skills

1. **Breakfast every day?** Students in an urban school
pg 277 were curious about how many children regularly
eat breakfast. They conducted a survey, asking,
"Do you eat breakfast regularly?" All 595 students
in the school responded to the survey. The resulting
data are summarized in the two-way table.[6]

Gender

		Male	Female	Total
Eats breakfast regularly	Yes	190	110	300
	No	130	165	295
	Total	320	275	595

Suppose we select a student from the school at
random. Define event F as getting a female student
and event B as getting a student who eats breakfast
regularly.

(a) Find $P(B^C)$. Interpret this value in context.

(b) Find $P(\text{female and doesn't eat breakfast regularly})$.

(c) Find $P(F \text{ or } B^C)$.

2. **Is this your card?** A standard deck of playing cards
(with jokers removed) consists of 52 cards in four
suits—clubs, diamonds, hearts, and spades. Each
suit has 13 cards, with denominations ace, 2, 3, 4, 5,
6, 7, 8, 9, 10, jack, queen, and king. The jack, queen,
and king are referred to as "face cards." Imagine
that we shuffle the deck thoroughly and deal one
card. Let's define events F: getting a face card and

H: getting a heart. The two-way table summarizes
the sample space for this chance process.

Card

		Face card	Nonface card	Total
Suit	Heart	3	10	13
	Nonheart	9	30	39
	Total	12	40	52

(a) Find $P(H^C)$. Interpret this value in context.

(b) Find $P(\text{face card and not a heart})$.

(c) Find $P(H^C \text{ or } F)$.

3. **Casualties of the *Titanic*** In 1912 the *Titanic* struck
an iceberg and sank on its first voyage. Some passengers got off the ship in lifeboats, but many died.
The following two-way table gives information
about adult passengers who survived and who
died, by class of travel.

Class

		First	Second	Third
Survived	Yes	197	94	151
	No	122	167	476

Suppose we randomly select one of the adult passengers who rode on the *Titanic*. Define event D as
getting a person who died and event F as getting a
passenger in first class.

(a) Find $P(F^C)$.

(b) Find P(not a passenger in first class and survived).

(c) Find P(not a passenger in first class or survived).

4. **Python nests** How is the hatching of water python eggs influenced by the temperature of the snake's nest? Researchers randomly assigned newly laid eggs to one of three water temperatures: hot, neutral, or cold. Hot duplicates the extra warmth provided by the mother python, and cold duplicates the absence of the mother.

		Nest temperature		
		Cold	Neutral	Hot
Hatching status	Hatched	16	38	75
	Didn't hatch	11	18	29

Suppose we select one of the eggs at random. Define events C as getting an egg that was assigned to cold water temperature and H as getting an egg that hatched.

(a) Find $P(H)$.

(b) Find P(not cold water temperature and hatched).

(c) Find $P(H^C$ or $C)$.

5. **Phone choices** According to the National Center for Health Statistics, in December 2012, 60% of U.S. households had a traditional landline telephone, 89% of households had cell phones, and 51% had both.[7] Suppose we randomly selected a household in December 2012. What's the probability that the household has a traditional landline telephone or a cell phone?

6. **Facebook or Instagram?** A December 2013 Pew Research Center poll of adults who use the Internet found that 71% of online adults use Facebook, 17% use Instagram, and 16% use both.[8] Suppose we randomly select a person who responded to this poll. What's the probability that he or she uses Facebook or Instagram?

7. **Mac or PC?** A recent census at a major university revealed that 60% of its students mainly used Macs. The rest mainly used PCs. At the time of the census, 67% of the school's students were undergraduates. The rest were graduate students. In the census, 23% of respondents were graduate students who said that they used Macs as their main computers. Suppose we select a student at random from among those who were part of the census. What's the probability that this person is a graduate student or mainly uses a Mac?

8. **Gender and political party** In March 2015, 57% of the voting members of the U.S. House of Representatives were Republicans and the rest were Democrats. Nineteen percent of the House members were women, and 36% of the House members were men who identified themselves as Democrats.[9] Suppose we select a representative at random. What is the probability that this person is male or a Democrat?

9. **Dropping the landline?** Refer to Exercise 5.

(a) Construct a Venn diagram to represent the outcomes of this chance process using the events T: has a traditional landline phone and C: has a cell phone.

(b) Find the probability that the household has a cell phone only.

10. **Friend or follower?** Refer to Exercise 6.

(a) Construct a Venn diagram to represent the outcomes of this chance process using the events F: uses Facebook and I: uses Instagram.

(b) Find the probability that the person uses Facebook but not Instagram.

11. **Mac world** Refer to Exercise 7.

(a) Construct a Venn diagram to represent the outcomes of this chance process using the events G: is a graduate student and M: mainly uses a Mac.

(b) Find $P(G^C \cap M^C)$. Interpret this value in context.

12. **Gender and the Republican Party** Refer to Exercise 8.

(a) Construct a Venn diagram to represent the outcomes of this chance process using the events M: is male and D: is a Democrat.

(b) Find $P(M^C \cap D^C)$. Interpret this value in context.

Applying the Concepts

13. **Middle school values** Researchers carried out a survey of fourth-, fifth-, and sixth-grade students in Michigan. Students were asked whether good grades, athletic ability, or being popular was most important to them. This two-way table summarizes the survey data.[10]

		Grade			
		4th grade	5th grade	6th grade	Total
Most important	Grades	49	50	69	168
	Athletic	24	36	38	98
	Popular	19	22	28	69
	Total	92	108	135	335

Suppose we select one of these students at random. What's the probability that:

(a) The student is a sixth grader or a student who rated good grades as important?

(b) The student is not a sixth grader and did not rate good grades as important?

14. Mobile systems The Pew Research Center asked a random sample of 2024 adult cell-phone owners from the United States their age and which type of cell phone they own: iPhone, Android, or other (including non-smartphones). The two-way table summarizes the data.

		Age			
		18–34	35–54	55+	Total
Type of cell phone	iPhone	169	171	127	467
	Android	214	189	100	503
	Other	134	277	643	1054
	Total	517	637	870	2024

Suppose we select one of the survey respondents at random. What's the probability that:

(a) The person is not age 18 to 34 and does not own an iPhone?

(b) The person is age 18 to 34 or owns an iPhone?

15. Disks of four colors A jar contains 36 disks: 9 each of four colors—red, green, blue, and yellow. Each set of disks of the same color is numbered from 1 to 9. Suppose you draw one disk at random from the jar. Define events B: get a blue disk, and E: get a disk with the number 8.

(a) Make a two-way table that describes the sample space in terms of events B and E.

(b) Find $P(B)$ and $P(E)$.

(c) Write the event "blue eight" in symbolic form. Then find the probability of this event.

(d) Explain why $P(B \cup E) \neq P(B) + P(E)$. Then use the general addition rule to compute $P(B \cup E)$.

16. Jack of hearts? Shuffle a standard deck of playing cards and deal one card. (Refer to Exercise 2 for the make-up of a deck of cards.) Define events J: get a jack, and R: get a red card.

(a) Make a two-way table that describes the sample space in terms of events J and R.

(b) Find $P(J)$ and $P(R)$.

(c) Write the event "red jack" in symbolic form. Then find the probability of this event.

(d) Explain why $P(J \cup R) \neq P(J) + P(R)$. Then use the general addition rule to compute $P(J \cup R)$.

Extending the Concepts

17. Mutually exclusive versus complementary Classify each of the following statements as true or false. Justify your answer.

(a) If one event is the complement of another event, the two events are mutually exclusive.

(b) If two events are mutually exclusive, one event is the complement of the other.

18. Who goes here? At a large university, 49% of the students are female, 69% live in the dorms, and 42% are in-state residents. Further, 34% of the students are females who live in the dorms, 19% are females who are in-state residents, and 16% are in-state residents who live in the dorms. Finally, 7% of the students are females who live in the dorms and are in-state residents. Pick a student at random from this university. What's the probability that the chosen student is a male who does not live in the dorm and is not an in-state resident? (*Hint*: It might be helpful to make a Venn diagram with three circles.)

Recycle and Review

19. Wedding bells? (2.1) A national survey interviewed several thousand teens in grades 7 through 12. One question was "What do you think are the chances you will be married in the next 10 years?" Here is a segmented bar chart of the responses by gender.[11] Use the graph to discuss if there is an association between gender and responses to this question.

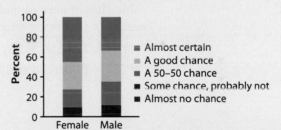

20. Mean Facebook friends (3.2, 3.3, 3.5) Karla, a senior at Springfield High School, wants to estimate the mean number of Facebook friends that students at the school have. To gather data, she makes an announcement one morning on the school's P.A. system asking students to find her during the 12th-grade lunch period and report the number of Facebook friends they have on a slip of paper.

(a) Identify a possible source of bias in Karla's data collection method. Is Karla's method likely to overestimate or underestimate the mean number of Facebook friends students at this school have?

(b) Describe an unbiased sampling method that Karla could use to obtain a better estimate.

Lesson 4.4

Conditional Probability and Independence

LEARNING TARGETS

- Find and interpret conditional probabilities using two-way tables.
- Use the conditional probability formula to calculate probabilities.
- Determine whether two events are independent.

The probability of an event can change if we know that some other event has occurred. For instance, suppose you toss a fair coin twice. The probability of getting two heads is 1/4 because the sample space consists of the 4 equally likely outcomes

HH   HT   TH   TT

Suppose that the first toss lands tails. Now what's the probability of getting two heads? It's 0. Knowing that the first toss is a tail changes the probability that you get two heads.

This idea is the key to many applications of probability.

What Is Conditional Probability?

Let's return to the college statistics class from Lesson 4.3. Earlier, we used the two-way table shown here to find probabilities involving events A: is male and B: has a pierced ear for a randomly selected student. Here is a summary of our previous results.

$P(A) = P(\text{male}) = 90/178$ $P(A \cap B) = P(\text{male and pierced ear}) = 19/178$

$P(B) = P(\text{pierced ear}) = 103/178$ $P(A \cup B) = P(\text{male or pierced ear}) = 174/178$

		Gender		
		Male	Female	Total
Pierced ear	Yes	19	84	103
	No	71	4	75
	Total	90	88	178

Now let's turn our attention to some other interesting probability questions.

1. **If we know that a randomly selected student has a pierced ear, what is the probability that the student is male?** There are 103 students in the class with a pierced ear. We can restrict our attention to this group, since we are told that the chosen student has a pierced ear. Because there are 19 males among the 103 students with a pierced ear, the desired probability is

   P(male given pierced ear) = 19/103, or about 18.4%

2. **If we know that a randomly selected student is male, what's the probability that the student has a pierced ear?** This time, our attention is focused on the males in the class. Because 19 of the 90 males in the class have a pierced ear,

   P(pierced ear given male) = 19/90, or about 21.1%

These two questions sound alike, but they're asking about two very different things. Each of these probabilities is an example of a **conditional probability.**

DEFINITION Conditional probability

The probability that one event happens given that another event is known to have happened is called a **conditional probability.** The conditional probability that event B happens given that event A has happened is denoted by $P(B \mid A)$.

With this new notation, we can restate the answers to the two questions on page 286 as

$$P(\text{male} \mid \text{pierced ear}) = P(A \mid B) = 19/103$$

and

$$P(\text{pierced ear} \mid \text{male}) = P(B \mid A) = 19/90$$

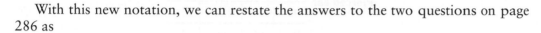

EXAMPLE

Who rides snowmobiles in Yellowstone?

Conditional probabilities and two-way tables

PROBLEM: Yellowstone National Park surveyed a random sample of 1526 winter visitors to the park. They asked each person whether he or she owned, rented, or had never used a snowmobile. Respondents were also asked whether or not they belonged to an environmental organization (like the Sierra Club). The two-way table summarizes the survey responses.

Suppose we randomly select one of the survey respondents. Define events E: environmental club member, S: snowmobile owner, and N: never used.

(a) Find $P(N \mid E)$. Interpret this value in context.

(b) Given that the chosen person is not a snowmobile owner, what's the probability that she or he is an environmental club member? Write your answer as a probability statement using correct symbols for the events.

		Environmental club		
		No	Yes	Total
Snowmobile experience	Never used	445	212	657
	Renter	497	77	574
	Owner	279	16	295
	Total	1221	305	1526

SOLUTION:

(a) $P(N \mid E) = P(\text{never used} \mid \text{environmental club member})$

$\qquad = 212/305 = 0.695$

Given that the randomly chosen person is an environmental club member, there is about a 69.5% chance that she or he has never used a snowmobile.

> To answer part (a), only consider values in the "Yes" column.

(b) $P(\text{environmental club member} \mid \text{not snowmobile owner})$

$= P(E \mid S^c) = \dfrac{212 + 77}{657 + 574} = \dfrac{289}{1231} = 0.235$

> To answer part (b), only consider values in the "Never used" and "Renter" rows.

FOR PRACTICE   TRY EXERCISE 1.

Daniel Milchev/Getty Images

Let's look more closely at how conditional probabilities are calculated. From the following two-way table, we see that

$$P(\text{male} \mid \text{pierced ear}) = \frac{19}{103} = \frac{\text{number of students who are male and have pierced ear}}{\text{number of students with pierced ear}}$$

		Gender		
		Male	Female	Total
Pierced ear	Yes	19	84	103
	No	71	4	75
	Total	90	88	178

What if we focus on probabilities instead of numbers of students? Notice that

$$\frac{P(\text{male and pierced ear})}{P(\text{pierced ear})} = \frac{\dfrac{19}{178}}{\dfrac{103}{178}} = \frac{19}{103} = P(\text{male} \mid \text{pierced ear})$$

This observation leads to a general formula for computing a conditional probability.

> ## Calculating Conditional Probabilities
>
> To find the conditional probability $P(A \mid B)$, use the formula
>
> $$P(A \mid B) = \frac{P(A \text{ and } B)}{P(B)} = \frac{P(A \cap B)}{P(B)} = \frac{P(\text{both events occur})}{P(\text{given event occurs})}$$

By the same reasoning,

$$P(B \mid A) = \frac{P(B \text{ and } A)}{P(A)} = \frac{P(B \cap A)}{P(A)}$$

EXAMPLE

Do our neighbors read both newspapers?

Calculating conditional probability

PROBLEM: In Lesson 4.3, we classified the residents of a large apartment complex based on the events A: reads *USA Today*, and B: reads the *New York Times*. The completed Venn diagram is reproduced here.

What's the probability that a randomly selected resident who reads *USA Today* also reads the *New York Times*?

A: Reads USA Today   B: Reads New York Times
0.35 0.05 0.20
0.40

SOLUTION:

$$P(\text{reads New York Times} \mid \text{reads USA Today}) = P(B \mid A) = \frac{P(B \cap A)}{P(A)} = \frac{0.05}{0.35 + 0.05} = 0.125$$

FOR PRACTICE   TRY EXERCISE 5.

Conditional Probability and Independence

Suppose you toss a fair coin twice. Define events A: first toss is a head, and B: second toss is a head. We know that $P(A) = 1/2$ and $P(B) = 1/2$.

- What's $P(B \mid A)$? It's the conditional probability that the second toss is a head given that the first toss was a head. The coin has no memory, so $P(B \mid A) = 1/2$.

- What's $P(B \mid A^C)$? It's the conditional probability that the second toss is a head given that the first toss was not a head. Getting a tail on the first toss does not change the probability of getting a head on the second toss, so $P(B \mid A^C) = 1/2$.

In this case, $P(B \mid A) = P(B \mid A^C) = P(B)$. Knowing whether or not the first toss was a head does not change the probability that the second toss is a head. We say that A and B are **independent events.**

> ### DEFINITION   Independent events
>
> A and B are **independent events** if knowing whether or not one event has occurred does not change the probability that the other event will happen. In other words, events A and B are independent if
>
> $$P(A \mid B) = P(A \mid B^C) = P(A) \quad \text{and} \quad P(B \mid A) = P(B \mid A^C) = P(B)$$

Let's contrast the coin-toss scenario with our earlier pierced-ear example. In that case, the chance process involved randomly selecting a student from a college statistics class. The events of interest were A: is male, and B: has a pierced ear. Are these two events independent?

Gender

Pierced ear		Male	Female	Total
	Yes	19	84	103
	No	71	4	75
	Total	90	88	178

- Suppose that the chosen student is male. We can see from the two-way table that $P(\text{pierced ear} \mid \text{male}) = P(B \mid A) = 19/90 = 0.211$.

- Suppose that the chosen student is female. From the two-way table, we see that $P(\text{pierced ear} \mid \text{female}) = P(\text{pierced ear} \mid \text{not male}) = P(B \mid A^C) = 84/88 = 0.955$.

Knowing that the chosen student is a male changes (greatly reduces) the probability that the student has a pierced ear. So these two events are not independent.

Another way to determine whether two events A and B are independent is to compare $P(A \mid B)$ to $P(A)$ or $P(B \mid A)$ to $P(B)$. For the pierced-ear setting,

$$P(\text{pierced ear} \mid \text{male}) = P(B \mid A) = 19/90 = 0.211$$

The unconditional probability that the chosen student has a pierced ear is

$$P(\text{pierced ear}) = P(B) = 103/178 = 0.579$$

Again, knowing that the chosen student is male changes (reduces) the probability that this person has a pierced ear. So these two events are not independent.

EXAMPLE

Are more males or females left-handed?

Checking for independence

PROBLEM: Is there a relationship between gender and handedness? To find out, we chose an SRS of 100 Australian high school students who completed a survey. The two-way table summarizes the relationship between the gender and dominant hand of each student.

Suppose we choose one of the students in the sample at random. Are the events "male" and "left-handed" independent? Justify your answer.

		Gender		
		Male	Female	Total
Dominant hand	Right	39	51	90
	Left	7	3	10
	Total	46	54	100

SOLUTION:

$P(\text{left-handed} \mid \text{male}) = 7/46 = 0.152$
$P(\text{left-handed} \mid \text{female}) = 3/54 = 0.056$

> Does knowing whether or not the student is male change the probability of left-handedness?

Because these probabilities are not equal, the events "male" and "left-handed" are not independent. Knowing that the student is male increases the probability that the student is left-handed.

FOR PRACTICE   TRY EXERCISE 9.

In the example, we could have also determined that the two events are not independent by showing that

$$P(\text{left-handed} \mid \text{male}) = 7/46 = 0.152 \neq P(\text{left-handed}) = 10/100 = 0.10$$

Or we could have focused on whether knowing that the chosen student is left-handed changes the probability that the person is male. Because

$$P(\text{male} \mid \text{left-handed}) = 7/10 = 0.70 \neq P(\text{male} \mid \text{right-handed}) = 39/90 = 0.433$$

The events "male" and "left-handed" are not independent.

THINK ABOUT IT   **Is there a connection between independence of events and association between two variables?** Yes! In the previous example, we found that the events "male" and "left-handed" were not independent for the sample of 100 Australian high school students. Knowing a student's gender helped us predict his or her dominant hand. By what you learned in Lesson 2.1, there is an association between gender and handedness for the students in the sample. The segmented bar chart shows the association in picture form.

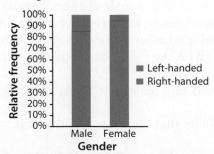

Does that mean there is an association between gender and handedness in the larger population? Maybe or maybe not. We'll discuss this issue further in Chapter 10.

LESSON APP 4.4

Who earns A's in college?

Students at the University of New Hampshire received 10,000 course grades in a recent semester. The following two-way table breaks down these grades by which school of the university taught the course. The schools are Liberal Arts, Engineering and Physical Sciences (EPS), and Health and Human Services.[12]

		School		
		Liberal Arts	Engineering and Physical Sciences	Health and Human Services
	A	2142	368	882
Grade	B	1890	432	630
	Lower than B	2268	800	588

Choose a University of New Hampshire course grade at random. Consider the two events E: the grade comes from an EPS course, and L: the grade is lower than a B.

1. Find $P(L \mid E)$. Interpret this probability in context.

2. Are events L and E independent? Justify your answer.

Lesson 4.4

WHAT DID YOU LEARN?

LEARNING TARGET	EXAMPLES	EXERCISES
Find and interpret conditional probabilities using two-way tables.	p. 287	1–4
Use the conditional probability formula to calculate probabilities.	p. 288	5–8
Determine whether two events are independent.	p. 290	9–12

Exercises   *Lesson 4.4*

Mastering Concepts and Skills

1. **Finger length and gender** Is there a relationship between gender and relative finger length? To find out, we randomly selected 452 U.S. high school students who completed a survey. The two-way table summarizes the relationship between gender and which finger was longer on the left hand (index finger or ring finger).

pg 287

		Gender		
		Female	Male	Total
Longer finger	Index finger	78	45	123
	Ring finger	82	152	234
	Same length	52	43	95
	Total	212	240	452

Suppose we randomly select one of the survey respondents. Define events R: ring finger longer and F: female.

(a) Find $P(R \mid F)$. Interpret this value in context.

(b) Given that the chosen student does not have a longer ring finger, what's the probability that this person is male? Write your answer as a probability statement using correct symbols for the events.

2. **The young and the rich** A survey of 4826 randomly selected young adults (aged 19 to 25) asked, "What do you think are the chances you will have much more than a middle-class income at age 30?" The two-way table summarizes the responses.[13]

Opinion		Gender		
		Female	Male	Total
	Almost no chance	96	98	194
	Some chance but probably not	426	286	712
	A 50-50 chance	696	720	1416
	A good chance	663	758	1421
	Almost certain	486	597	1083
	Total	2367	2459	4826

Choose a survey respondent at random. Define events G: a good chance, M: male, and N: almost no chance.

(a) Find $P(G \mid M)$. Interpret this value in context.

(b) Given that the chosen student didn't say "almost no chance," what's the probability that this person is female? Write your answer as a probability statement using correct symbols for the events.

3. **Third-class passengers** The *Titanic* struck an iceberg and sank on its first voyage across the Atlantic in 1912. Some passengers got off the ship in lifeboats, but many died. The two-way table gives information about adult passengers who survived and who died, by class of travel.

Survived		Class		
		First	Second	Third
	Yes	197	94	151
	No	122	167	476

Suppose we randomly select one of the adult passengers who rode on the *Titanic*.

(a) Given that the person selected was in first class, what's the probability that he or she survived?

(b) If the person selected survived, what's the probability that he or she was not a third-class passenger?

4. **Temperature and hatching** How is the hatching of water python eggs influenced by the temperature of the snake's nest? Researchers randomly assigned newly laid eggs to one of three water temperatures: hot, neutral, or cold. Hot duplicates the extra warmth provided by the mother python, and cold duplicates the absence of the mother.

Hatching status		Nest temperature		
		Cold	Neutral	Hot
	Hatched	16	38	75
	Didn't hatch	11	18	29

Suppose we select one of the eggs at random.

(a) Given that the chosen egg was assigned to hot water, what is the probability that it hatched?

(b) If the chosen egg hatched, what is the probability that it was not assigned to hot water?

5. **Who has landlines?** The National Center for Health Statistics conducted a survey of phone usage in a random sample of U.S. households in December 2012. Each household was identified as having a landline telephone (or not) and having a cell phone (or not).[14] Define event L as having a landline phone and event C as having a cell phone. The Venn diagram summarizes the data based on these two events. What is the probability that a randomly selected household with a landline also has a cell phone?

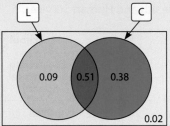

6. **Homeowner education** What is the relationship between educational achievement and home ownership? A random sample of 500 U.S. adults was selected. Each member of the sample was identified as a high school graduate (or not) and as a homeowner (or not). Define event G as being a high school graduate and event H as being a homeowner. The Venn diagram summarizes the data based on these two events. What's the probability that a randomly selected person who owns a home is also a high school graduate?

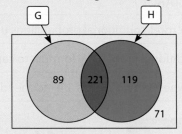

7. **140 characters** A 2014 survey suggests that 71% of U.S. teenagers use Facebook, 33% use Twitter, and 15% do both. Suppose we select a U.S. teenager at random and learn that the student uses Facebook. Find the probability that the student uses Twitter.

8. **Mac grads?** A recent census at a major university revealed that 60% of its students mainly used Macs. The rest primarily used PCs. At the time of the census, 67% of the school's students were undergraduates. The rest were graduate students. In the census, 23% of respondents were graduate students who said that they used Macs as their primary computers. Suppose we select a student at random from among those who were part of the census and learn that the student mainly uses a Mac. Find the probability that this person is a graduate student.

9. **Give me a hand** The 28 students in Mr. Tabor's statistics class completed a brief survey. One of the questions asked whether each student was

right- or left-handed. The two-way table summarizes the class data. Choose a student from the class at random. Are the events "female" and "right-handed" independent? Justify your answer.

		Gender	
		Female	Male
Handedness	Left	3	1
	Right	18	6

10. **Big Papi** Baseball star David Ortiz—nicknamed "Big Papi"—is known for his ability to deliver hits in high-pressure situations. Here is a two-way table of his hits, walks, and outs in all of his regular-season and post-season plate appearances from 1997 through 2014.[15] Choose a plate appearance at random. Are the events "Hit" and "Post-season" independent? Justify your answer.

		At-bat		
		Hit	Walk	Out
Season	Regular	2023	1474	5034
	Post	87	57	208

11. **Preferred passengers?** Refer to Exercise 3. Are the events "Survived" and "First class" independent? Justify your answer.

12. **Hatching temperature** Refer to Exercise 4. Are the events "Hot water" and "Hatched" independent? Justify your answer.

Applying the Concepts

13. **Language learners** Choose a student in grades 9 to 12 at random and ask if he or she is studying a language other than English. Here is the distribution of results. What is the probability that a student is studying Spanish given that he or she is studying some language other than English?

Language	Spanish	French	German	All others	None
Probability	0.26	0.09	0.03	0.03	0.59

14. **Higher income** Here is the distribution of the adjusted gross income (in thousands of dollars) reported on individual federal income tax returns in a recent year. Given that a return shows an income of at least $50,000, what is the probability that the income is at least $100,000?

Income	<25	25–49	50–99	100–499	≥500
Probability	0.431	0.248	0.215	0.100	0.006

15. **Do you play basketball?** Select an adult at random. Define events T: person is over 6 feet tall, and B: person is a professional basketball player. Rank the following probabilities from smallest to largest. Justify your answer.

$$P(T) \quad P(B) \quad P(T \mid B) \quad P(B \mid T)$$

16. **Learn before you teach** Select an adult at random. Define events A: person has earned a college degree, and T: person has chosen teaching as his or her career. Rank the following probabilities from smallest to largest. Justify your answer.

$$P(A) \quad P(T) \quad P(A \mid T) \quad P(T \mid A)$$

17. **Independent dice?** Suppose you roll two fair, six-sided dice—one red and one blue. Are the events "Sum is 7" and "Blue die shows a 4" independent? Justify your answer.

18. **More independent dice?** Suppose you roll two fair, six-sided dice—one red and one blue. Are the events "Sum is 8" and "Blue die shows a 4" independent? Justify your answer.

Extending the Concepts

19. **Independence and association** In a random sample of 200 children ages 9–17 from the United Kingdom, the gender of each student was recorded along with which superpower they would most like to have: invisibility, super strength, telepathy (ability to read minds), ability to fly, or ability to freeze time. The segmented bar chart shows the percents of males and females who chose each superpower.

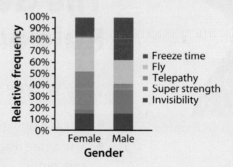

Suppose we choose one of the students from the sample at random.

(a) Explain why the events "Male" and "Fly" are not independent.

(b) Name two events that do appear to be independent.

(c) Use your work from part (a) or (b) to explain whether there is an association between gender and superpower preference for the students in the sample.

Recycle and Review

20. **What kind of study?** (3.1) For each study, determine whether it is an observational study or an experiment. Justify your answer.

(a) A researcher wants to study the effect of price promotions on consumers' expectations. The researcher makes up two different histories of the store price of a video game for the past year. Students in an economics course view one or the other price history on a computer. Some students see a steady price, while others see regular promotions that temporarily cut the price. Then the students are asked what price they would expect to pay for the video game.

(b) In a study of the effect of living in public housing on family stability in low-income households, a list of applicants accepted for public housing was obtained, together with a list of families who applied but were rejected by the housing authorities. A random sample was drawn from each list. Researchers followed up with the families in both groups several times over a period of 20 years.

(c) A doctor at a veterans hospital examines all of its patient records over a 9-year period and finds that twice as many men as women fell out of their hospital beds during their stay. The doctor asserts that this is evidence men are clumsier than women.

21. **Standard deviation by eye (1.7)** For the four small data sets given here, indicate which has the smallest standard deviation and which has the largest without performing any calculations. Justify your answer.

Set 1: 1, 1, 1, 9, 9, 9

Set 2: 1, 1, 1, 6, 6, 6, 9, 9, 9

Set 3: 3, 4, 5, 6, 7, 8, 9

Set 4: 3, 4, 5, 6, 6, 6, 7, 8, 9

Lesson 4.5

The General Multiplication Rule and Tree Diagrams

LEARNING TARGETS

- Use the general multiplication rule to calculate probabilities.
- Use a tree diagram to model a chance process involving a sequence of outcomes.
- Calculate conditional probabilities using tree diagrams.

Suppose that A and B are two events resulting from the same chance process. We can find the probability $P(A \text{ or } B)$ with the general addition rule:

$$P(A \text{ or } B) = P(A) + P(B) - P(A \text{ and } B)$$

How do we find the probability that both events happen, $P(A \text{ and } B)$?

The General Multiplication Rule

About 55% of high school students participate in a school athletic team at some level. Roughly 5% of these athletes go on to play on a college team in the NCAA.[16] What percent of high school students play a sport in high school *and* go on to play on an NCAA team? About 5% of 55%, or roughly 2.75%.

Let's restate the situation in probability language. Suppose we select a high school student at random. What's the probability that the student plays a sport in high school and goes on to play on an NCAA team? The given information suggests that

P(high school sport) = 0.55 and P(NCAA team | high school sport) = 0.05

By the logic stated above,

P(high school sport and NCAA team)

$= P$(high school sport) $\cdot P$(NCAA team | high school sport)

$= (0.55)(0.05) = 0.0275$

This is an example of the **general multiplication rule.**

DEFINITION **General multiplication rule**

For any chance process, the probability that events A and B both occur can be found using the **general multiplication rule:**

$$P(A \text{ and } B) = P(A) \cdot P(B \mid A)$$

The general multiplication rule says that for both of two events to occur, first one must occur. Then, given that the first event has occurred, the second must occur. To confirm that this result is correct, start with the conditional probability formula

$$P(B \mid A) = \frac{P(B \text{ and } A)}{P(A)}$$

The numerator gives the probability we want because P(B and A) is the same as P(A and B). Multiply both sides of the above equation by P(A) to get

$$P(A) \cdot P(B \mid A) = P(A \text{ and } B)$$

EXAMPLE

Do teenagers post on social-networking sites?

The general multiplication rule

PROBLEM: The Pew Internet and American Life Project found that 93% of teenagers (ages 12 to 17) use the Internet and that 55% of online teens have posted a profile on a social-networking site.[17] Find the probability that a randomly selected teen uses the Internet and has posted a profile. Show your work.

SOLUTION:

P(Internet and has profile) $= P$(Internet) $\cdot P$(profile | Internet)

$= (0.93)(0.55) = 0.5115$

FOR PRACTICE **TRY EXERCISE 1.**

Tree Diagrams and Conditional Probability

Shannon hits the snooze button on her alarm on 60% of school days. If she hits snooze, there is a 0.70 probability that she makes it to her first class on time. If she doesn't hit snooze and gets up right away, there is a 0.90 probability that she makes it to class on time. Suppose we select a school day at random and record whether Shannon hits the snooze button and whether she arrives in class on time. Figure 4.7 shows a **tree diagram** for this chance process.

FIGURE 4.7 A tree diagram displaying the sample space of randomly choosing a school day and noting if Shannon hits the snooze button and whether or not she gets to her first class on time.

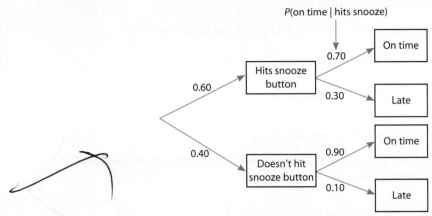

There are only two possible outcomes at the first "stage" of this chance process: Shannon hits the snooze button or she doesn't. The first set of branches in the tree diagram displays these outcomes with their probabilities. The second set of branches shows the two possible results at the next "stage" of the process—Shannon gets to her first class on time or late—and the probability of each result based on whether or not she hit the snooze button. Note that the probabilities on the second set of branches are conditional probabilities, like $P(\text{on time} \mid \text{hits snooze}) = 0.70$.

> ### DEFINITION Tree diagram
>
> A **tree diagram** shows the sample space of a chance process involving multiple stages. The probability of each outcome is shown on the corresponding branch of the tree. All probabilities after the first stage are conditional probabilities.

What is the probability that Shannon hits the snooze button and is late for class on a randomly selected school day? The general multiplication rule provides the answer:

$$P(\text{hits snooze and late}) = P(\text{hits snooze}) \cdot P(\text{late} \mid \text{hits snooze})$$
$$= (0.60)(0.30) = 0.18$$

The previous calculation amounts to multiplying probabilities along the branches of the tree diagram.

What's the probability that Shannon is late to class on a randomly selected school day? Figure 4.8 (on the next page) illustrates two ways this can happen: Shannon hits the snooze button and is late or she doesn't hit snooze and is late. Because these outcomes are mutually exclusive,

$$P(\text{late}) = P(\text{hits snooze and late}) + P(\text{doesn't hit snooze and late})$$

The general multiplication rule tells us that

$P(\text{doesn't hit snooze and late}) = P(\text{doesn't hit snooze}) \cdot P(\text{late} \mid \text{doesn't hit snooze})$
$= (0.40)(0.10) = 0.04$

So $P(\text{late}) = 0.18 + 0.04 = 0.22$. There is about a 22% chance that Shannon will be late to class.

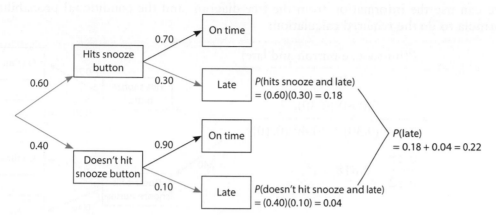

FIGURE 4.8 Tree diagram showing the two possible ways that Shannon can be late to class on a randomly selected day.

EXAMPLE

Does media use relate to good grades?

Tree diagrams

PROBLEM: The Kaiser Family Foundation recently released a study about the influence of media in the lives of young people aged 8–18.[18] In the study, 17% of the youth were classified as light media users, 62% were classified as moderate media users, and 21% were classified as heavy media users. Of the light users who responded, 74% described their grades as good (A's and B's), while only 68% of the moderate users and 52% of the heavy users described their grades as good. Suppose that we select one young person at random.

(a) Draw a tree diagram to model this chance process.

(b) Find the probability that this person describes his or her grades as good.

SOLUTION:

(a)

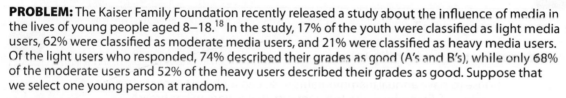

(b) $P(\text{good grades}) = (0.17)(0.74) + (0.62)(0.68) + (0.21)(0.52)$
$= 0.1258 + 0.4216 + 0.1092 = 0.6566$

$P(\text{good grades}) = P(\text{light and good OR moderate and good OR heavy and good}).$

FOR PRACTICE   TRY EXERCISE 5.

Some interesting conditional probability questions involve "going in reverse" on a tree diagram. For instance, suppose that Shannon is late for class on a randomly chosen school day. What is the probability that she hit the snooze button that morning? To find this probability, we start with the given information that Shannon is late, which is displayed on the second set of branches in the following tree diagram, and ask whether she hit the snooze button, which is shown on the first set of branches. We can use the information from the tree diagram and the conditional probability formula to do the required calculation:

$$P(\text{hit snooze button} \mid \text{late}) = \frac{P(\text{hit snooze button and late})}{P(\text{late})}$$

$$= \frac{(0.60)(0.30)}{(0.60)(0.30) + (0.40)(0.10)}$$

$$= \frac{0.18}{0.22} = 0.818$$

When Shannon is late, there is a 0.818 probability that she hit the snooze button that morning.

EXAMPLE

How reliable are mammograms?

Tree diagrams and conditional probability

PROBLEM: Many women choose to have annual mammograms to screen for breast cancer after age 40. A mammogram isn't foolproof. Sometimes, the test suggests that a woman has breast cancer when she really doesn't (a "false positive"). Other times, the test says that a woman doesn't have breast cancer when she actually does (a "false negative").

Suppose that we know the following information about breast cancer and mammograms in a particular population:

- One percent of the women aged 40 or over in this population have breast cancer.
- For women who have breast cancer, the probability of a negative mammogram is 0.03.
- For women who don't have breast cancer, the probability of a positive mammogram is 0.06.

A randomly selected woman aged 40 or over from this population tests positive for breast cancer in a mammogram. Find the probability that she has breast cancer.

SOLUTION:

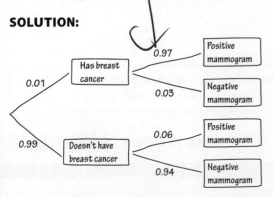

Start by making a tree diagram to summarize the possible outcomes.

- Because 1% of women in this population have breast cancer, 99% don't.
- Of those women who do have breast cancer, 3% would test negative on a mammogram. The remaining 97% would (correctly) test positive.
- Among the women who don't have breast cancer, 6% would test positive on a mammogram. The remaining 94% would (correctly) test negative.

$$P(\text{breast cancer} \mid \text{positive mammogram}) = \frac{P(\text{breast cancer and positive mammogram})}{P(\text{positive mammogram})}$$

> Use the conditional probability formula.

$$= \frac{(0.01)(0.97)}{(0.01)(0.97) + (0.99)(0.06)}$$

$$= \frac{0.0097}{0.0691} = 0.14$$

FOR PRACTICE   TRY EXERCISE 9.

Are you surprised by the final result of the example: Given that a randomly selected woman from the population has a positive mammogram, there is only about a 14% chance that she has breast cancer? Most people are. Sometimes, a two-way table that includes counts is more convincing. To make calculations simple, we'll suppose that there are exactly 10,000 women aged 40 or over in this population, and that exactly 100 have breast cancer (that's 1% of the women).

How many of those 100 would have a positive mammogram? It would be 97% of 100, or 97 of them. That leaves 3 who would test negative. How many of the 9900 women who don't have breast cancer would get a positive mammogram? Six percent of them, or (9900)(0.06) = 594 women. The remaining 9900 − 594 = 9306 would test negative. In total, 97 + 594 = 691 women would have positive mammograms and 3 + 9306 = 9309 women would have negative mammograms. This information is summarized in the two-way table.

		Has breast cancer?		
		Yes	No	Total
Mammogram result	Positive	97	594	691
	Negative	3	9,306	9,309
	Total	100	9,900	10,000

Given that a randomly selected woman has a positive mammogram, the two-way table shows that the conditional probability $P(\text{breast cancer} \mid \text{positive mammogram})$ = 97/691 = 0.14.

This example illustrates an important fact when considering proposals for widespread testing for serious diseases or illegal drug use: If the condition being tested is uncommon in the population, many positives will be false positives. The best remedy is to retest any individual who tests positive.

LESSON APP 4.5

Not milk?

Lactose intolerance causes difficulty in digesting dairy products that contain lactose (milk sugar). It is particularly common among people of African and Asian ancestry. In the United States (not including other groups and people who consider themselves to belong to more than one race), 82% of the population is white, 14% is black, and 4% is Asian. Moreover, 15% of whites, 70% of blacks, and 90% of Asians are lactose intolerant.[19] Suppose we select a U.S. person at random.

1. Construct a tree diagram to represent this situation.

2. Find the probability that the person is lactose intolerant.

3. Given that the chosen person is lactose intolerant, what is the probability that he or she is Asian?

Snorg Tees

Lesson 4.5

WHAT DID YOU LEARN?

LEARNING TARGET	EXAMPLES	EXERCISES
Use the general multiplication rule to calculate probabilities.	p. 295	1–4
Use a tree diagram to model a chance process involving a sequence of outcomes.	p. 297	5–8
Calculate conditional probabilities using tree diagrams.	p. 298	9–12

Exercises Lesson 4.5

Mastering Concepts and Skills

1. **Whose copyright?** Illegal music downloading has
pg 295 become a big problem: 29% of Internet users download music files, and 67% of downloaders say they don't care if the music is copyrighted.[20] Find the probability that a randomly selected Internet user downloads music and doesn't care if it's copyrighted.

2. **Hanging out at the gym** Suppose that 10% of adults belong to health clubs, and 40% of these health club members go to the club at least twice a week. Find the probability that a randomly selected adult belongs to a health club and goes there at least twice a week.

3. **Gump probabilities** According to Forrest Gump, "Life is like a box of chocolates. You never know what you're gonna get." Suppose a candy maker offers a special "Gump box" with 20 chocolate candies that look alike. In fact, 14 of the candies have soft centers and 6 have hard centers. Choose 2 of the candies from a Gump box at random. Find the probability that both candies have soft centers.

4. **Sampling students** A statistics class with 30 students has 10 males and 20 females. Choose 2 of the students in the class at random. Find the probability that both are female.

5. **Putting it on the card** In a recent month, 88% of
pg 297 automobile drivers filled their vehicles with regular gasoline, 2% purchased midgrade gas, and 10% bought premium gas.[21] Of those who bought regular gas, 28% paid with a credit card. Of customers who bought midgrade and premium gas, 34% and 42%, respectively, paid with a credit card. Suppose we select a customer at random.

   (a) Draw a tree diagram to model this chance process.

   (b) Find the probability that the customer paid with a credit card.

6. **Shipping you a laptop?** A computer company makes desktop and laptop computers at factories in three states—California, Texas, and New York. The California factory produces 40% of the company's computers, the Texas factory makes 25%, and the remaining 35% are manufactured in New York. Of the computers made in California, 75% are laptops. Of those made in Texas and New York, 70% and 50%, respectively, are laptops. All computers are first shipped to a distribution center in Missouri before being sent out to stores. Suppose we select a computer at random from the distribution center.[22]

   (a) Draw a tree diagram to model this chance process.

   (b) Find the probability that the computer is a laptop.

7. **Gump-ology** Refer to Exercise 3.

   (a) Draw a tree diagram to model this chance process.

   (b) Find the probability that one of the chocolates has a soft center and the other one doesn't.

8. **Sampling students** Refer to Exercise 4.

   (a) Draw a tree diagram to model this chance process.

   (b) Find the probability that one of the students is female and the other is male.

9. **The first serve** Tennis great Roger Federer made
pg 298 63% of his first serves in a recent season. When Federer made his first serve, he won 78% of the points. When Federer missed his first serve and had to serve again, he won only 57% of the points.[23] Suppose you randomly choose a point on which Federer served. You get distracted before seeing his first serve but look up in time to see Federer win the point. What's the probability that he missed his first serve?

10. **Eureka?** A boy uses a homemade metal detector to look for valuable metal objects on a beach. The machine isn't perfect—it identifies only 98% of the metal objects over which it passes, and it identifies 4% of the nonmetallic objects over which it passes.

Suppose that 25% of the objects that the machine passes over are metal. Choose an object from this beach at random. If the machine gives a signal when it passes over this object, find the probability that the boy has found a metal object.

11. **Paying for premium** Refer to Exercise 5. Given that the customer paid with a credit card, find the probability that she or he bought premium gas.

12. **Laptop from California?** Refer to Exercise 6. Given that a laptop is selected, what is the probability that it was made in California?

Applying the Concepts

13. **False positives, false negatives** Enzyme immunoassay (EIA) tests are used to screen blood specimens for the presence of antibodies to HIV, the virus that causes AIDS. Antibodies indicate the presence of the virus. The test is quite accurate but is not always correct. A *false positive* occurs when the test gives a positive result but no HIV antibodies are actually present in the blood. A *false negative* occurs when the test gives a negative result but HIV antibodies are present in the blood. Here are approximate probabilities of positive and negative EIA outcomes when the blood tested does and does not actually contain antibodies to HIV.[24]

		Test result	
		+	−
Truth	Antibodies present	0.9985	0.0015
	Antibodies absent	0.006	0.994

Suppose that 1% of a large population carries antibodies to HIV in their blood. Imagine choosing a person from this population at random.

(a) Draw a tree diagram to model this chance process.

(b) Find the probability that the EIA test result is positive.

(c) Given that the EIA test is positive, find the probability that the person has the HIV antibody.

14. **Does the new hire use drugs?** Many employers require prospective employees to take a drug test. A positive result on this test suggests that the prospective employee uses illegal drugs. However, not all people who test positive use illegal drugs. The test result could be a *false positive*. A negative test result could be a *false negative* if the person really does use illegal drugs. Suppose that 4% of prospective employees use drugs, and that the drug test has a false positive rate of 5%, and a false negative rate of 10%.[25] Imagine choosing a prospective employee at random.

(a) Draw a tree diagram to model this chance process.

(b) Find the probability that the drug test result is positive.

(c) Given that the drug test result is positive, find the probability that the prospective employee uses illegal drugs.

Extending the Concepts

15. **Fundraising by telephone** Tree diagrams can organize problems having more than two stages. The following figure shows probabilities for a charity calling potential donors by telephone.[26] Each person called is either a recent donor, a past donor, or a new prospect. At the next stage, the person called either does or does not pledge to contribute, with conditional probabilities that depend on the donor class to which the person belongs. Finally, those who make a pledge either do or don't make a contribution. Suppose we randomly select a person who is called by the charity.

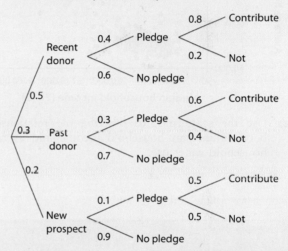

(a) What is the probability that the person contributed to the charity?

(b) Given that the person contributed, find the probability that he or she is a recent donor.

16. **Matching suits** A standard deck of playing cards consists of 52 cards with 13 cards in each of four suits: spades, diamonds, clubs, and hearts. Suppose you shuffle the deck thoroughly and deal 5 cards face-up onto a table.

(a) What is the probability of dealing five spades in a row?

(b) Find the probability that all five cards on the table have the same suit.

17. **HIV and confirmation testing** Refer to Exercise 13. Many of the positive results from EIA tests are false positives. It is therefore common practice to perform a second EIA test on another blood sample from a person whose initial specimen tests positive. Assume that the false positive and false negative rates remain the same for a person's second test. Find the probability that a person who gets a positive result on both EIA tests has HIV antibodies.

Recycle and Review

18. **Water and wealth (2.2, 2.6)** Water is an expensive commodity in the U.S. Southwest. A recent article in the *Arizona Daily Star* reported on the relationship between average household water consumption per year and median household income for 186 different census tracts served by Tucson Water. The scatterplot summarizes this relationship.

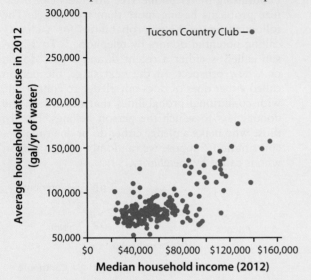

(a) Use the scatterplot to describe the relationship between median household income and average household water use.

(b) The single dot in the upper right region of the scatterplot is from the Tucson Country Club Estates development. Describe the characteristics of this neighborhood.

(c) How would the slope of the least-squares regression line change if the Tucson Country Club Estates point were removed from the data set?

19. **Gas up the roadster! (1.8)** Interested in a sporty car? Worried that it might use too much gas? The Environmental Protection Agency lists most such vehicles in its "two-seater" or "minicompact" categories. The figure shows boxplots for both city and highway gas mileages for these two groups of cars.[28] Write a few sentences comparing their distributions.

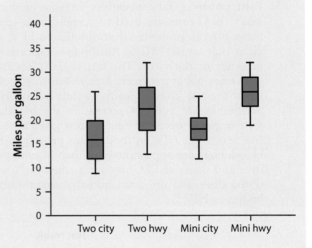

Lesson 4.6

The Multiplication Rule for Independent Events

LEARNING TARGETS

- Use the multiplication rule for independent events to calculate probabilities.
- Calculate *P*(at least one) using the complement rule and the multiplication rule for independent events.
- Determine if it is appropriate to use the multiplication rule for independent events in a given setting.

What happens to the general multiplication rule in the special case when events A and B are independent? In that case, $P(B \mid A) = P(B)$ because knowing that event A occurred doesn't change the probability that event B occurs. We can simplify the general multiplication rule as follows:

$$P(A \text{ and } B) = P(A) \cdot P(B \mid A)$$
$$= P(A) \cdot P(B)$$

This result is known as the **multiplication rule for independent events.**

> **DEFINITION** Multiplication rule for independent events
>
> If A and B are independent events, the probability that A and B both occur is
>
> $$P(A \text{ and } B) = P(A) \cdot P(B)$$

Note that this rule applies *only* to independent events.

CAUTION

Calculating Probabilities with the Multiplication Rule for Independent Events

Suppose that Pedro drives the same route to work on Monday through Friday. His route includes one traffic light. The probability that the light will be green is 0.42, yellow is 0.03, and red is 0.55.

1. **What's the probability that the light is green on Monday and red on Tuesday?** Let event A be green light on Monday and event B be red light on Tuesday. These two events are independent because knowing whether or not the light was green on Monday doesn't help us predict the color of the light on Tuesday. By the multiplication rule for independent events,

$$P(\text{green on Monday and red on Tuesday})$$
$$= P(A \text{ and } B) = P(A) \cdot P(B)$$
$$= (0.42)(0.55) = 0.231$$

There is about a 23% chance that the light will be green on Monday and red on Tuesday.

2. **What's the probability that Pedro finds the light red on Monday through Friday?** We can extend the multiplication rule for independent events to more than two events.

$P(\text{red Monday } and \text{ red Tuesday } and \text{ red Wednesday } and \text{ red Thursday } and \text{ red Friday})$
$= P(\text{red Monday}) \cdot P(\text{red Tuesday}) \cdot P(\text{red Wednesday}) \cdot P(\text{red Thursday}) \cdot P(\text{red Friday})$
$= (0.55)(0.55)(0.55)(0.55)(0.55)$
$= (0.55)^5$
$= 0.0503$

There is about a 5% chance that Pedro will encounter a red light on all five days in a workweek.

EXAMPLE

What factors led to the Challenger disaster?

Multiplication rule for independent events

PROBLEM: On January 28, 1986, Space Shuttle *Challenger* exploded on takeoff, killing all seven crew members aboard. Afterward, scientists and statisticians helped analyze what went wrong. They determined that the failure of O-ring joints in the shuttle's booster rockets was to blame. Experts estimated that the probability that an individual O-ring joint would function properly under the cold conditions that day was 0.977. But there were six of these O-ring joints, and all six had to function properly for the shuttle to launch safely. Assuming that O-ring joints succeed or fail independently, find the probability that the shuttle would launch safely under similar conditions.

Thom Baur/AP Photo

SOLUTION:

P(O-ring 1 OK and O-ring 2 OK and . . . and O-ring 6 OK)

$= P$(O-ring 1 OK) $\cdot$ P(O-ring 2 OK) $\cdot$ . . . $\cdot$ P(O-ring 6 OK)

$= (0.977)(0.977)(0.977)(0.977)(0.977)(0.977)$

$= (0.977)^6$

$= 0.87$

> For the shuttle to launch safely, all six O-ring joints must function properly.

FOR PRACTICE   TRY EXERCISE 1.

The multiplication rule for independent events can also be used to help find P(at least one). In the previous example, the shuttle would *not* launch safely under similar conditions if 1, 2, 3, 4, 5, or all 6 O-ring joints fail—that is, if *at least one* O-ring fails. The only possible number of O-ring failures excluded is 0. So the events "at least one O-ring joint fails" and "no O-ring joints fail" are complementary events. By the complement rule,

$$P(\text{at least one O-ring fails}) = 1 - P(\text{no O-ring fails})$$
$$= 1 - 0.87 = 0.13$$

That's a very high chance of failure! As a result of this analysis following the *Challenger* disaster, NASA made important safety changes to the design of the shuttle's booster rockets.

EXAMPLE

Trading speed for accuracy in HIV testing?

Finding the probability of "at least 1"

PROBLEM: Many people who come to clinics to be tested for HIV, the virus that causes AIDS, don't come back to learn the test results. Clinics now use "rapid HIV tests" that give a result while the client waits. In a clinic in Malawi, use of rapid tests increased the percent of clients who learned their test results from 69% to 99.7%.

The trade-off for fast results is that rapid tests are often less accurate than slower laboratory tests. Applied to people who have no HIV antibodies, one rapid test has a probability of about 0.004 of producing a false positive (i.e., of falsely indicating that HIV antibodies are present).[29] If a clinic tests 200 randomly selected people who are free of HIV antibodies, what is the chance that at least one false positive will occur? Assume that test results for different individuals are independent.

SOLUTION:

$$P(\text{no false positives}) = P(\text{all 200 tests negative})$$

> The probability that any individual test result is negative is $1 - 0.004 = 0.996$

$$= (0.996)(0.996) \cdots (0.996)$$

> Start by finding P(no false positives).

$$= 0.996^{200}$$

$$= 0.4486$$

> By the complement rule, P(at least one false positive) $= 1 - P$(no false positives).

$$P(\text{at least one false positive}) = 1 - 0.4486 = 0.5514$$

FOR PRACTICE   TRY EXERCISE 5.

Use the Multiplication Rule for Independent Events Wisely

The multiplication rule $P(A \text{ and } B) = P(A) \cdot P(B)$ holds if A and B are *independent* but not otherwise. The addition rule $P(A \text{ or } B) = P(A) + P(B)$ holds if A and B are *mutually exclusive* but not otherwise. Resist the temptation to use these simple rules when the conditions that justify them are not met.

CAUTION

Hagar the Horrible

World rights reserved.

VERY INTERESTING!

WHAT ARE THE ODDS OF **TWO ROPES** BREAKING AT THE EXACT SAME TIME?!

©2008 by King Features Syndicate, Inc.

TWANG!

TWANG!

CHRIS BROWNE

12-22

EXAMPLE

Are there many old college students?

Beware lack of independence!

PROBLEM: Government data show that 8% of adults are full-time college students and that 30% of adults are age 55 or older. If we randomly select an adult, is P(full-time college student and over 55) $= (0.08)(0.30) = 0.024$? Why or why not?

SOLUTION:

No, because being a full-time college student and being 55 or older are not independent events. Knowing that the chosen adult is a full-time college student makes it much less likely that she or he is 55 or older.

FOR PRACTICE   TRY EXERCISE 9.

The multiplication rule $P(A \text{ and } B) = P(A) \cdot P(B)$ gives us another way to determine if two events are independent. Let's return to the pierced-ear example from earlier in the chapter. The following two-way table summarizes data from a college statistics class.

		Gender		
		Male	Female	Total
Pierced ear	Yes	19	84	103
	No	71	4	75
	Total	90	88	178

Our events of interest were A: is male and B: has a pierced ear. Are these two events independent? No, because

$$P(\text{A and B}) = P(\text{male and pierced ear}) = \frac{19}{178} = 0.107$$

is not equal to

$$P(\text{A}) \cdot P(\text{B}) = P(\text{male}) \cdot P(\text{pierced ear}) = \frac{90}{178} \cdot \frac{103}{178} = 0.293$$

THINK ABOUT IT   **Is there a connection between mutually exclusive and independent?** Let's start with a new chance process. Choose a U.S. adult at random. Define event A: the person is male, and event B: the person is pregnant. It's fairly clear that these two events are mutually exclusive (can't happen together)! Are they also independent?

If you know that event A has occurred, does this change the probability that event B happens? Of course! If we know the person is male, the chance that the person is pregnant is 0. Because $P(\text{B} \mid \text{A}) \neq P(\text{B})$, the two events are not independent.

Two mutually exclusive events (with nonzero probabilities) can *never* be independent, because if one event happens, the other event is guaranteed not to happen.

LESSON APP 4.6

How should we interpret genetic screening?

The First Trimester Screen is a test given during the first trimester of pregnancy to determine if there are specific chromosomal abnormalities in the fetus. According to a study published in the *New England Journal of Medicine*,[30] approximately 5% of normal pregnancies will receive a positive result. Assume that test results for individual women are independent.

1. Suppose that two unrelated women who are having normal pregnancies, Devondra and Miranda, are given the First Trimester Screen. What is the probability that Devondra gets a positive result and Miranda gets a negative result?

2. If 100 unrelated women with normal pregnancies are tested with the First Trimester Screen, what is the probability that at least 1 woman will receive a positive result?

Lesson 4.6

WHAT DID YOU LEARN?

LEARNING TARGET	EXAMPLES	EXERCISES
Use the multiplication rule for independent events to calculate probabilities.	p. 304	1–4
Calculate P(at least one) using the complement rule and the multiplication rule for independent events.	p. 304	5–8
Determine if it is appropriate to use the multiplication rule for independent events in a given setting.	p. 305	9–12

Exercises   Lesson 4.6

Mastering Concepts and Skills

1. **Merry and bright?** A string of Christmas lights contains 20 lights. The lights are wired in series, so that if any light fails, the whole string will go dark. Each light has probability 0.98 of working for a 3-year period. The lights fail independently of each other. Find the probability that the string of lights will remain bright for 3 years.
   pg 304

2. **Pitching a perfect game** In baseball, a perfect game is when a pitcher doesn't allow any hitters to reach base in all 9 innings. Historically, pitchers throw a perfect inning—an inning where no hitters reach base—about 40% of the time.[31] So to throw a perfect game, a pitcher must have 9 perfect innings in a row. Find the probability that a pitcher throws 9 perfect innings in a row, assuming the pitcher's performance in any one inning is independent of his performance in other innings.

3. **Liar, liar!** From experience, we know that a certain lie detector will show a positive reading (indicating a lie) 10% of the time when a person is telling the truth. Suppose that a random sample of 5 suspects is subjected to a lie detector test. Find the probability of observing no positive readings if all suspects are telling the truth.

4. **Making a reservation** An airline estimates that the probability a randomly selected call to its reservation phone line results in a reservation being made is 0.31. Find the probability that none of 4 randomly selected calls to the phone line results in a reservation.

5. **O-Negative** People with type O-negative blood are universal donors. Any patient can receive a transfusion of O-negative blood. Only 7.2% of the American population has O-negative blood. If we choose
   pg 304

10 Americans at random, what is the probability that at least 1 of them is a universal donor?

6. **Playing with Canadians** If you buy 1 ticket in Canada's Lotto 6/49 lottery game, the probability that you will win a prize is 0.15. Given the nature of lotteries, the probability of winning is independent from month to month. If you buy 1 ticket each month for five months, what is the probability that you will win at least one prize?

7. **On a roll** Suppose that you roll a fair, six-sided die 10 times. What's the probability that you get at least one 6?

8. **Is the package late?** A shipping company claims that 90% of its shipments arrive on time. Suppose this claim is true. If we take a random sample of 20 shipments made by the company, what's the probability that at least 1 of them arrives late?

9. **Late to the gate?** An airline reports that 85% of its flights arrive on time. To find the probability that its next 4 flights into LaGuardia Airport all arrive on time, can we multiply (0.85)(0.85)(0.85)(0.85)? Why or why not?
   pg 305

10. **Late shows** Some TV shows begin after their scheduled times when earlier programs run late. According to a network's records, about 3% of its shows start late. To find the probability that 3 consecutive shows on this network start on time, can we multiply (0.97)(0.97)(0.97)? Why or why not?

11. **Independent news** In a large apartment complex, 40% of residents read *USA Today*, 25% read the *New York Times*, and 5% read both papers. Suppose we select a resident of the apartment complex at random. Are the events A: reads *USA Today* and B: reads *New York Times* independent? Justify your answer.

12. **By land or by cell?** According to the National Center for Health Statistics, in December 2012, 60% of U.S. households had a traditional landline telephone, 89% of households had cell phones, and 51% had both.[32] Suppose we randomly selected a household in December 2012. Are the events A: had a landline phone and B: had a cell phone independent? Justify your answer.

Applying the Concepts

13. **Fire or medical?** Many fire stations handle more emergency calls for medical help than for fires. At one fire station, 81% of incoming calls are for medical help. Suppose we choose 4 incoming calls to the station at random.

    (a) Find the probability that all 4 calls are for medical help.

    (b) What's the probability that at least 1 of the calls is not for medical help?

    (c) Explain why the calculation in part (a) may not be valid if we choose 4 consecutive calls to the station.

14. **Broken links** Internet sites often vanish or move, so that references to them can't be followed. In fact, 13% of Internet sites referred to in major scientific journals are lost within two years of publication.[33] Suppose we randomly select 7 Internet references from scientific journals.

    (a) Find the probability that all 7 references still work two years later.

    (b) What's the probability that at least 1 of them doesn't work two years later?

    (c) Explain why the calculation in part (a) may not be valid if we choose 7 Internet references from one issue of the same journal.

Extending the Concepts

15. **Mutually exclusive versus independent** The two-way table summarizes data on the gender and eye color of students in a college statistics class. Imagine choosing a student from the class at random. Define event A: student is male and event B: student has blue eyes.[34]

		Gender		
		Male	Female	Total
Eye color	Blue			10
	Brown			40
	Total	20	30	50

    (a) Copy and complete the two-way table so that events A and B are mutually exclusive.

    (b) Copy and complete the two-way table so that events A and B are independent.

Recycle and Review

16. **Snappy dressers (3.8, 3.9)** Matt and Diego suspect that people are more likely to agree to participate in a survey if the interviewers are dressed up. To test this, they went to the local grocery store on two consecutive Saturday mornings at 10 A.M. On the first Saturday, they wore casual clothing (tank tops and jeans). On the second Saturday, they dressed in button-down shirts and nicer slacks. Each day, they asked every fifth person who walked into the store to participate in a survey. Their response variable was whether or not the person agreed to participate. Here are their results.

		Clothing	
		Casual	Nice
Participation	Agreed	14	27
	Declined	36	23

    (a) Calculate the difference in the proportion of subjects that agreed to participate in the survey in the two groups (Casual − Nice).

    (b) Assume the study design is equivalent to randomly assigning shoppers to the "casual" or "nice" groups. A total of 100 trials of a simulation were performed to see what differences in proportions would occur due only to chance variation in this random assignment. Use the results of the simulation in the following dotplot to determine if the difference in proportions from part (a) is statistically significant. Explain your reasoning.

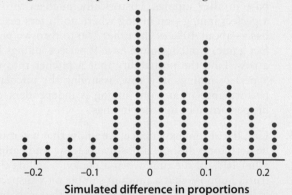

    (c) What flaw in the design of this experiment would prevent Matt and Diego from drawing a cause-and-effect conclusion about the impact of an interviewers' attire on nonresponse in a survey?

17. **Butter side down (4.1)** Researchers at Manchester Metropolitan University in England determined experimentally that if a piece of toast is dropped from a 2.5-foot-high table, the probability that it lands butter side down is 0.81.

    (a) Explain what this probability means.

    (b) If you dropped 100 pieces of toast, will exactly 81 of them land butter side down? Explain.

(c) Maria decides to test this probability and drops 10 pieces of toast from a 2.5-foot table. Only 4 of them land butter side down. Should she be surprised? Describe how you would carry out a simulation to answer this question, assuming the probability that the toast lands butter side down is 0.81. Do not perform the simulation.

(d) The dotplot displays the results of 50 simulated trials of dropping 10 pieces of toast. Does the simulation suggest that Maria should be surprised? Explain.

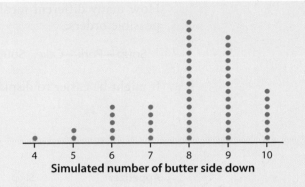

Simulated number of butter side down

Lesson 4.7

The Multiplication Counting Principle and Permutations

LEARNING TARGETS

- Use the multiplication counting principle to determine the number of ways to complete a process involving several steps.
- Use factorials to count the number of permutations of a group of individuals.
- Compute the number of permutations of n individuals taken k at a time.

Finding the probability of an event often involves counting the number of possible outcomes of some chance process. In this lesson, we will show you two techniques for determining the number of ways that a multistep process can happen when the order of the steps matters.

The Multiplication Counting Principle

The Agricola Restaurant offers a three-course dinner menu. Customers who order from this menu must choose one appetizer, one main dish, and one dessert. Here are the options for each course.

Appetizer	Main dish	Dessert
Butternut squash soup	Grilled pork chop	Chocolate cake
Green salad	Ribeye steak	Apple and cranberry tart
Caesar salad	Roasted chicken breast	
	Poached salmon	

How many different meals can be ordered from this menu? We could try to list all possible orders:

Soup—Pork—Cake, Soup—Pork—Tart, Soup—Steak—Cake, Soup—Steak—Tart,...

It might be easier to display all of the options in a diagram, like the one shown here.

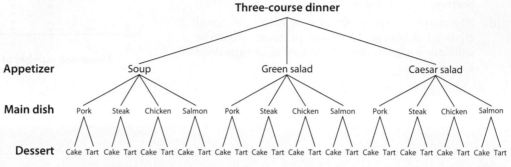

Three-course dinner

Appetizer Soup	Green salad	Caesar salad
Main dish Pork Steak Chicken Salmon	Pork Steak Chicken Salmon	Pork Steak Chicken Salmon
Dessert Cake Tart Cake Tart Cake Tart Cake Tart	Cake Tart Cake Tart Cake Tart Cake Tart	Cake Tart Cake Tart Cake Tart Cake Tart

From the diagram, we can see that for each of the three choices of appetizer, there are four choices of main dish, and for each of those main dish choices, there are two dessert choices. So there are

$$\underbrace{3}_{\text{Appetizer}} \cdot \underbrace{4}_{\text{Main dish}} \cdot \underbrace{2}_{\text{Dessert}} = 24$$

different meals that can be ordered from the three-course dinner menu. This is an example of the **multiplication counting principle.**

DEFINITION Multiplication counting principle

For a process involving multiple (k) steps, suppose that there are n_1 ways to do Step 1, n_2 ways to do Step 2, ..., and n_k ways to do Step k. The total number of different ways to complete the process is

$$n_1 \cdot n_2 \cdot \ldots \cdot n_k$$

This result is called the **multiplication counting principle.**

EXAMPLE

How many license plates in the Golden State?

The multiplication counting principle

PROBLEM: The standard license plate for California passenger cars has one digit, followed by three letters, and then three more digits. The first digit cannot be a 0. The first and third letters cannot be I, O, or Q. How many possible license plates are there?

SOLUTION:

By the multiplication counting principle, there are

$$\underset{\substack{\text{Digit}\\\text{Not 0}}}{9} \cdot \underset{\substack{\text{Letter}\\\text{Not I, O, Q}}}{23} \cdot \underset{\text{Letter}}{26} \cdot \underset{\substack{\text{Letter}\\\text{Not I, O, Q}}}{23} \cdot \underset{\text{Digit}}{10} \cdot \underset{\text{Digit}}{10} \cdot \underset{\text{Digit}}{10}$$

$$= 123,786,000 \text{ possible license plates}$$

There are 26 letters and 10 digits from 0 to 9.

Notice the restrictions: The first digit can't be 0 and the first and third letters can't be I, O, or Q.

FOR PRACTICE   TRY EXERCISE 1.

Permutations

The multiplication counting principle can also help us determine how many ways there are to arrange a group of people, animals, or things. For example, suppose you have 5 framed photographs of different family members that you want to arrange in a line on top of your dresser. In how many ways can you do this? Let's count the options moving from left to right across the dresser. There are 5 options for the first photo, 4 options for the next photo, and so on. By the multiplication counting principle, there are

$$\underbrace{5}_{\text{Photo \#1}} \cdot \underbrace{4}_{\text{Photo \#2}} \cdot \underbrace{3}_{\text{Photo \#3}} \cdot \underbrace{2}_{\text{Photo \#4}} \cdot \underbrace{1}_{\text{Photo \#5}} = 120$$

different photo arrangements. We call arrangements like this, where the order matters, **permutations.**

DEFINITION Permutation

A **permutation** is a distinct arrangement of some group of individuals.

Expressions like $5 \cdot 4 \cdot 3 \cdot 2 \cdot 1$ occur often enough in counting problems that mathematicians invented a special name and notation for them. We write $5 \cdot 4 \cdot 3 \cdot 2 \cdot 1 = 5!$, read as "5 **factorial.**"

DEFINITION Factorial

For any positive integer n, we define $n!$ (read "n **factorial**") as

$$n! = n(n-1)(n-2)\ldots \cdot 3 \cdot 2 \cdot 1$$

That is, n factorial is the product of the numbers starting with n and going down to 1.

EXAMPLE

How many different batting orders?

Permutations and factorials

PROBLEM: The manager of a youth baseball team has picked 9 players to start an upcoming playoff game. How many different ways are there for the manager to arrange these 9 players to make up the team's batting order?

SOLUTION:

By the multiplication counting principle, there are

$$\underbrace{9}_{\text{Batter \#1}} \cdot \underbrace{8}_{\text{Batter \#2}} \cdot \underbrace{7}_{\text{Batter \#3}} \cdot \underbrace{6}_{\text{Batter \#4}} \cdot \underbrace{5}_{\text{Batter \#5}} \cdot \underbrace{4}_{\text{Batter \#6}} \cdot \underbrace{3}_{\text{Batter \#7}} \cdot \underbrace{2}_{\text{Batter \#8}} \cdot \underbrace{1}_{\text{Batter \#9}} = 9!$$

or 362,880 possible batting orders.

FOR PRACTICE   TRY EXERCISE 5.

So far, we have shown how to count the number of distinct arrangements of *all* the individuals in a group of people, animals, or things. Sometimes, we want to determine how many ways there are to select and arrange only *some* of the individuals in a group.

Mr. Wilcox likes to get the students in his statistics class involved in the action. But he doesn't want to play favorites. Each day, Mr. Wilcox puts the names of all 28 of his students in a hat and mixes them up. He then draws out 3 names, one at a time. The student whose name is picked first gets to operate the display calculator for the day. The second student selected is in charge of reading the answers to the even-numbered homework problems. The third student picked writes class notes on the interactive whiteboard. In how many different ways can Mr. Wilcox fill these three jobs?

By the multiplication counting principle, there are

$$\underset{\text{Job 1}}{28} \cdot \underset{\text{Job 2}}{27} \cdot \underset{\text{Job 3}}{26} = 19{,}656$$

ways for Mr. Wilcox to fill the three different jobs. We can describe this result as the number of permutations of 28 people taken 3 at a time. In symbols, we'll write this as $_{28}P_3$.

Denoting Permutations: $_nP_k$

If there are n individuals, the notation $_nP_k$ represents the number of different permutations of k individuals selected from the entire group of n.

EXAMPLE

How many ways can you arrange the batting order?

Finding the number of permutations

PROBLEM: A youth baseball team has 15 players. How many different ways are there for the team's manager to select and arrange 9 of these players to make up the team's batting order?

SOLUTION:

There are

$$\underset{\text{Batter \#1}}{15} \cdot \underset{\text{Batter \#2}}{14} \cdot \underset{\text{Batter \#3}}{13} \cdot \underset{\text{Batter \#4}}{12} \cdot \underset{\text{Batter \#5}}{11} \cdot \underset{\text{Batter \#6}}{10} \cdot \underset{\text{Batter \#7}}{9} \cdot \underset{\text{Batter \#8}}{8} \cdot \underset{\text{Batter \#9}}{7}$$

$= {}_{15}P_9 = 1{,}816{,}214{,}400$ possible batting orders

FOR PRACTICE   TRY EXERCISE 9.

With a little clever math, we can rewrite $_{15}P_9$ as follows:

$$\begin{aligned}
{}_{15}P_9 &= 15 \cdot 14 \cdot 13 \cdot 12 \cdot 11 \cdot 10 \cdot 9 \cdot 8 \cdot 7 \\
&= \frac{15 \cdot 14 \cdot 13 \cdot 12 \cdot 11 \cdot 10 \cdot 9 \cdot 8 \cdot 7 \cdot 6 \cdot 5 \cdot 4 \cdot 3 \cdot 2 \cdot 1}{6 \cdot 5 \cdot 4 \cdot 3 \cdot 2 \cdot 1} \\
&= \frac{15!}{6!} \\
&= \frac{15!}{(15 - 9)!}
\end{aligned}$$

This method leads to a general formula for $_nP_k$.

Two Ways to Compute Permutations

You can calculate the number of permutations of n individuals taken k at a time (where $k \leq n$) using the multiplication counting principle or with the formula

$$_nP_k = \frac{n!}{(n-k)!}$$

By definition, $0! = 1$.

THINK ABOUT IT **Why do we define $0! = 1$?** To make the formula for $_nP_k$ work for all possible values of k, consider $_{28}P_{28}$. This is the number of different arrangements of 28 individuals taken 28 at a time. By the multiplication counting principle, there are $28 \cdot 27 \cdot 26 \cdot \ldots \cdot 3 \cdot 2 \cdot 1 = 28!$ such arrangements. Using the formula for $_nP_k$, we get

$$_{28}P_{28} = \frac{28!}{(28-28)!} = \frac{28!}{0!}$$

If we define $0! = 1$, the values obtained by the two methods will agree.

LESSON APP 4.7

Do you scream for ice cream?

The local ice cream shop in Dontrelle's town is called 21 Choices. Why? Because they offer 21 different flavors of ice cream. Dontrelle likes all but three of the flavors that 21 Choices offers: bubble gum, butter pecan, and pistachio.

1. A 21 Choices "basic sundae" comes in three sizes—small, medium, or large—and includes one flavor of ice cream and one of 12 toppings. Dontrelle has enough money for a small or medium basic sundae. How many different sundaes could Dontrelle order that include only flavors that he likes?

2. Dontrelle could order a cone with three scoops of ice cream instead of a sundae. He prefers to

Eric Futran/Chefshots/Getty Images

have three different flavors (for variety) and he considers the order of the flavors on his cone to be important. How many three-scoop cones with three different flavors that Dontrelle likes are possible at 21 Choices? Give your answer as a number and using $_nP_k$ notation.

Calculating Factorials and Permutations

You can use an applet or a TI-83/84 to compute the number of distinct arrangements of a group of individuals.

Applet

1. Go to highschool.bfwpub.com/spa3e and launch the *Probability* applet.

2. Select type "Counting methods" and click on "Begin analysis."

3. To evaluate 9! from the first "Batter up!" example, enter $n = 9$. Then click on "Calculate n!."

Probability

Select type: Counting methods ▾ | Begin analysis

$n = $ 9 $r = $

Calculate n! | Calculate nPr | Calculate nCr
9! = 362880

4. To compute the number of possible batting orders for the youth baseball team with 15 players from the most recent example, enter $n = 15$ and $r = 9$. Then click on "Calculate $_nP_r$."

Probability

Select type: Counting methods ▾ | Begin analysis

$n = $ 15 $r = $ 9

Calculate n! | Calculate nPr | Calculate nCr
$_{15}P_9 = 1816214400$

TI-83/84

1. We can use the factorial command to evaluate 9! from the first "Batter up!" example.

● Type 9. Then press MATH, arrow to PROB, choose !, and press ENTER.

2. We can use the $_nP_r$ command to compute the number of possible batting orders for the youth baseball team with 15 players from the most recent example.

● Type 15. Then press MATH, choose PROB, and $_nP_r$. Complete the command $_{15}P_9$ (Older OS: 15 $_nP_r$ 9) and press ENTER.

```
NORMAL FLOAT AUTO REAL RADIAN MP
9!
                              362880
15P9
                          1816214400
```

Lesson 4.7

WHAT DID YOU LEARN?

LEARNING TARGET	EXAMPLES	EXERCISES
Use the multiplication counting principle to determine the number of ways to complete a process involving several steps.	p. 310	1–4
Use factorials to count the number of permutations of a group of individuals.	p. 311	5–8
Compute the number of permutations of n individuals taken k at a time.	p. 312	9–12

Exercises · Lesson 4.7

Mastering Concepts and Skills

1. **Alphanumeric NJ plates** A standard-issue 2014 New Jersey license plate had three digits followed by three letters.

pg 310

New Jersey
100 SPA
Garden State

These plates were not allowed to have letters D, T, or X in the fourth position, or the letters I, O, or Q in *any* position. With these restrictions, how many different license plates were possible in 2014?

2. **National insurance code** In the United Kingdom, every resident is issued a national insurance "number" for the national health system and social security. The format of the "number" is two letters, followed by six digits, and then a third letter. The first two letters cannot be D, F, I, Q, U, or V, and the second letter also cannot be O. The last letter is either A, B, C, or D. All six digits can be 0 through 9, with no restrictions. How many national insurance numbers can be issued with these restrictions?

3. **Three-letter call signs** In 1912 the U.S. government began issuing licenses to radio stations. Each station was given a unique three-letter "call sign." By international agreement, the United States received rights to all call signs beginning with the letters W, N, and K. Radio stations in the western United States were given call signs starting with K. Stations in the East were given call signs starting with W. (N was reserved for use by the U.S. Navy.)

(a) How many three-letter call signs (like WGO) start with the letter W?

(b) How many three-letter call signs starting with W or K were available for U.S. radio stations?

4. **Four-letter call signs** Refer to Exercise 3. By 1922 there were more applications for radio station licenses than the number of three-letter call signs available. A radio station in New Orleans applied for and was granted the call sign WAAB.

(a) How many four-letter call signs starting with W are possible?

(b) How many four-letter call signs starting with W or K are possible?

5. **Do your homework!** Suppose you have 6 homework assignments to complete one night. In how many different orders can you complete all of the assignments?

pg 311

6. **Bookshelf** Mario has 10 science fiction novels on a shelf in his bedroom. How many different ways are there for him to arrange the books on the shelf?

7. **Line up** At the beginning of class, Charise's teacher makes a special offer. She will dismiss the class early if they can make all possible arrangements with the 15 students in the class in a single-file line. How many such arrangements are there? Should the class take the offer? Explain.

8. **Assigned seating forever!** Ms. Kreppel has 28 desks in her classroom. She numbers the desks from 1 to 28. On the first day of class, Ms. Kreppel places identical slips of paper numbered 1 to 28 in a hat. Each of the 28 students in her statistics class draws a slip from the hat upon entering the classroom to determine his or her assigned seat. How many possible seating assignments are there?

9. **iPod shuffle** Declan has 100 songs on his MP3 player. He is going for a run and has time to listen to 8 songs while he runs. Declan decides to use the MP3 player's random shuffle feature to determine which songs will play, and in what order. (The random shuffle feature does not allow any song to be played twice.) How many different lists of 8 songs are possible?

pg 312

10. **Random test generation** Mr. Vellman has a test bank with 75 multiple-choice questions on Lesson 4.7. The test bank comes with a random generator that will select and arrange questions to make different versions of a test. How many different versions of a 10-question multiple-choice test on this lesson could Mr. Vellman make?

11. **Penalty kicks** A soccer team has 11 players on the field at the end of a scoreless game. According to league rules, the coach must select 5 of the players and designate an order in which they will take penalty kicks. How many different ways are there for the coach to do this?

12. **Open up!** "Letter lock" padlocks open when a correct sequence of 3 letters is selected on the lock's dial. If the dial has 20 letters on it and letters cannot be repeated, how many different sequences of letters are possible?

Applying the Concepts

13. **ATM passwords** Many banks require customers who use the automated teller machine (ATM) to enter a four-digit password before they begin a transaction.

    (a) How many possible four-digit passwords are there?

    (b) How many four-digit passwords contain no 3s?

14. **Random music** The pentatonic musical scale contains 5 notes in an octave: C, D, E, G, and A. Miles decides to look for new musical themes by playing random sequences of 4 notes from the pentatonic scale.

    (a) How many possible four-note sequences can be played?

    (b) How many possible four-note sequences contain no Gs?

15. **Call on me** Mrs. Mortlock uses a calculator program to select students at random to answer questions during class. Each of the 28 students in her class is equally likely to be chosen every time she runs the program. Suppose Mrs. Mortlock runs the program for five different questions.

    (a) In how many different ways can the program select which students will answer the five questions?

    (b) How many of the selections in part (a) include 5 different students?

16. **Five on a roll** Suppose you roll 5 fair, six-sided dice one after the other.

    (a) How many different possible outcomes are there?

    (b) In how many of the outcomes from part (a) do all 5 dice show the same number?

Extending the Concepts

17. **Call on us five** Refer to Exercise 15. Find the probability that the program will select 5 different students to answer the five questions.

18. **Doubling passwords?** Refer to Exercise 13. How many four-digit passwords contain at least two of the same digit?

Recycle and Review

19. **Bagel nutrition (2.3, 2.4)** The scatterplot shows the relationship between calories and total carbohydrates (in grams) for 19 varieties of bagels sold by a national chain of bagel stores.[36]

    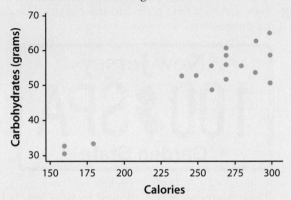

    (a) The correlation for these data is $r = 0.915$. Interpret this value in context.

    (b) How would the correlation be affected if carbohydrates were measured in ounces instead of grams? (1 gram = 0.035 ounces.)

    (c) The three data points in the lower left represent three special low-calorie bagels. How would the correlation be affected if these three bagels were removed from the data set?

20. **Workplace injuries (4.2, 4.4)** Pick a random nonfatal workplace injury in private industry in 2013. The table gives the probability model for the day of the week that the injury took place.[37]

Day	Sunday	Monday	Tuesday	Wednesday
Probability	0.06	0.18	0.17	0.17
Day	Thursday	Friday	Saturday	
Probability	x	0.15	0.09	

    (a) Find the probability that the randomly chosen injury took place on a Thursday.

    (b) Find the probability that the injury took place on a weekday (Monday through Friday).

    (c) Given that an injury took place on a weekday, what is the probability that it took place on a Monday?

Lesson 4.8

Combinations and Probability

LEARNING TARGETS

- Compute the number of combinations of *n* individuals taken *k* at a time.
- Use combinations to calculate probabilities.
- Use the multiplication counting principle and combinations to calculate probabilities.

Recall from Lesson 4.7 that a permutation is a distinct arrangement of some group of individuals. With permutations, order matters. Sometimes, we're just interested in finding how many ways there are to choose some number of individuals from a group, but we don't care about the order in which the individuals are selected.

Combinations

Mr. Wilcox decides to randomly select 3 students' homework papers from his statistics class to grade each day. He puts all 28 names in a hat, mixes them up, and draws out 3 names, one at a time. In how many different ways can he choose 3 students' papers to grade? It's tempting to say that there are

$$\underline{28} \cdot \underline{27} \cdot \underline{26} = 19{,}656$$

ways for Mr. Wilcox to do this. That's not correct, however.

Suppose Mr. Wilcox picks Lucretia, Tim, and Kiran—in that order. That's really no different from getting Tim, then Lucretia, then Kiran. Or Kiran, then Lucretia, then Tim. How many selections consist of these same 3 students? We can list all the possibilities:

Lucretia—Tim—Kiran	Lucretia—Kiran—Tim	Tim—Lucretia—Kiran
Tim—Kiran—Lucretia	Kiran—Lucretia—Tim	Kiran—Tim—Lucretia

Or we could use the multiplication counting principle: there are 3 possibilities for the first pick, 2 options for the second pick, and only 1 option for the last pick. So there are

$$\underline{3} \cdot \underline{2} \cdot \underline{1} = 3! = 6$$

arrangements that consist of these 3 students. This same argument applies for any 3 students whom Mr. Wilcox selects.

The order in which Mr. Wilcox chooses the 3 papers to grade doesn't matter. To avoid counting the same group of 3 students 6 times, we have to divide our original (wrong) answer by $3! = 6$:

$$\text{number of ways to choose 3 homework papers out of 28} = \frac{28 \cdot 27 \cdot 26}{3 \cdot 2 \cdot 1} = 3276$$

This result gives the number of **combinations** of 28 students taken 3 at a time, which we'll write as $_{28}C_3$. Some people prefer to write this as $\binom{28}{3}$ instead.

> **│││ DEFINITION** Combination, $_nC_k$
>
> A **combination** is a selection of individuals from some group in which the order of selection doesn't matter. If there are n individuals, then the notation $_nC_k$ represents the number of different combinations of k individuals chosen from the entire group of n.

We can rewrite the above answer as

$$_{28}C_3 = \frac{28 \cdot 27 \cdot 26}{3 \cdot 2 \cdot 1} = \frac{_{28}P_3}{3!}$$

which shows an important connection between the number of permutations and the number of combinations in this setting. With a little fancy math, we can also think of this result as

$$_{28}C_3 = \frac{_{28}P_3}{3!} = \frac{\left(\dfrac{28!}{(28-3)!}\right)}{3!} = \frac{\left(\dfrac{28!}{25!}\right)}{3!} = \frac{28!}{3!\,25!}$$

How to Compute Combinations

You can calculate the number of combinations of n individuals taken k at a time (where $k \leq n$) using the multiplication counting principle, with the formula

$$_nC_k = \frac{_nP_k}{k!}$$

or with the formula

$$_nC_k = \frac{n!}{k!(n-k)!}$$

EXAMPLE

Which six to pick?

Combinations

▶

PROBLEM: In the New Jersey "Pick Six" lotto game, a player chooses 6 different numbers from 1 to 49. The 6 winning numbers for the lottery are chosen at random. If the player matches all 6 numbers, she wins the jackpot, which starts at $2 million. How many different possible sets of winning numbers are there for the New Jersey Pick Six lotto game?

SOLUTION:

There are

$$_{49}C_6 = \frac{_{49}P_6}{6!} = \frac{49 \cdot 48 \cdot 47 \cdot 46 \cdot 45 \cdot 44}{6 \cdot 5 \cdot 4 \cdot 3 \cdot 2 \cdot 1} = 13{,}983{,}816$$

different sets of winning numbers for this game.

> All that matters is which 6 numbers are picked—the order of selection doesn't matter. That calls for combinations!
>
> $$_nC_k = \frac{_nP_k}{k!}$$

FOR PRACTICE   TRY EXERCISE 1.

Counting and Probability

The focus of this chapter is probability. Recall that when a chance process results in equally likely outcomes, the probability that event A occurs is

$$P(A) = \frac{\text{number of outcomes in event A}}{\text{total number of outcomes in sample space}}$$

You can use the multiplication counting principle and what you have learned about permutations and combinations to help count the number of outcomes.

Consider New Jersey's "Pick Six" lotto game from the previous example. What's the probability that a player wins the jackpot by matching all 6 winning numbers? Because the winning numbers are randomly selected, any set of 6 numbers from 1 to 49 is equally likely to be chosen. So we can use the above formula to calculate

$$P(\text{win the jackpot}) = \frac{\text{number of ways to choose all 6 winning numbers}}{\text{total number of ways to choose 6 numbers from 1 to 49}}$$

$$= \frac{{}_6C_6}{{}_{49}C_6}$$

$$= \frac{1}{13{,}983{,}816} = 0.0000000715$$

Your calculator may give the probability in scientific notation as 7.15E-08, which is $7.15 \cdot 10^{-8}$. However you write it, the player's chance of winning the jackpot is very small!

EXAMPLE

Is this the girls' lucky day?

From counting to probability

PROBLEM: There are 28 students in Mr. Wilcox's statistics class—16 boys and 12 girls. Suppose Mr. Wilcox randomly selects 3 students' homework papers to grade. Use combinations to find the probability that all 3 students chosen are girls.

SOLUTION:

$P(3 \text{ girls selected}) = \dfrac{\# \text{ of ways to choose 3 girls out of 12}}{\text{Total } \# \text{ of ways to choose 3 students from the class}}$

$$= \frac{{}_{12}C_3}{{}_{28}C_3}$$

$$= \frac{\dfrac{12 \cdot 11 \cdot 10}{3 \cdot 2 \cdot 1}}{\dfrac{28 \cdot 27 \cdot 26}{3 \cdot 2 \cdot 1}}$$

$$= \frac{220}{3276} = 0.067$$

Because each subgroup of 3 students in the class is equally likely to be chosen, you can use the formula

$P(A) = \dfrac{\text{number of outcomes in event A}}{\text{total number of outcomes in sample space}}$,

where event A = 3 girls are chosen.

There's about a 7% chance that Mr. Wilcox will randomly select 3 girls' homework papers to grade.

FOR PRACTICE   TRY EXERCISE 5.

We could also have used the general multiplication rule from Lesson 4.5 to find the desired probability in the example:

P(three girls selected)

$= P$(1st student female *and* 2nd student female *and* 3rd student female)

$= P$(1st student female) $\cdot$ P(2nd student female | 1st student female) $\cdot$

P(3rd student female | 1st two students female)

$$= \frac{12}{28} \cdot \frac{11}{27} \cdot \frac{10}{26} = 0.067$$

Note that these two different methods give the same result.

EXAMPLE

Is this lottery fair to all pilots?

Finding probabilities with combinations

PROBLEM: An airline has just finished training 25 junior pilots—15 male and 10 female—to become captains. Unfortunately, only 8 captain positions are available right now. Airline managers announce that they will use a lottery process to determine which pilots will fill the available positions. The names of all 25 pilots will be written on identical slips of paper, placed in a hat, mixed thoroughly, and drawn out one at a time until all 8 captains have been identified.

A day later, managers announce the results of the lottery. Of the 8 captains chosen, 5 are female and only 3 are male. Some of the male pilots who were not selected suspect that the lottery was not carried out fairly.

(a) Find the number of ways in which a fair lottery can result in 5 female and 3 male pilots being selected.

(b) Find the probability that a fair lottery would result in the selection of 5 female and 3 male pilots.

(c) Based on your answer to part (b), is there convincing evidence that the lottery wasn't carried out fairly? Explain.

SOLUTION:

(a) $\dfrac{{}_{10}C_5}{\text{Females}} \cdot \dfrac{{}_{15}C_3}{\text{Males}}$

$$= \left(\frac{10 \cdot 9 \cdot 8 \cdot 7 \cdot 6}{5 \cdot 4 \cdot 3 \cdot 2 \cdot 1}\right) \cdot \left(\frac{15 \cdot 14 \cdot 13}{3 \cdot 2 \cdot 1}\right)$$

$$= \left(\frac{30,240}{120}\right) \cdot \left(\frac{2730}{6}\right)$$

$$= 252 \cdot 455$$

$$= 114,660$$

> The number of ways to select 5 of the 10 female pilots is ${}_{10}C_5$. The number of ways to select 3 of the 15 male pilots is ${}_{15}C_3$. Now use the multiplication counting principle to find the number of ways in which the lottery yields 5 female pilots and 3 male pilots.

(b) P(5 female and 3 male pilots selected in fair lottery)

$$= \frac{\text{number of ways to get 5 female and 3 male pilots}}{\text{total number of ways to select 8 pilots}}$$

$$= \frac{{}_{10}C_5 \cdot {}_{15}C_3}{{}_{25}C_8}$$

$$= \frac{114,660}{1,081,575} = 0.106$$

> $${}_{25}C_8 = \frac{25 \cdot 24 \cdot 23 \cdot 22 \cdot 21 \cdot 20 \cdot 19 \cdot 18}{8 \cdot 7 \cdot 6 \cdot 5 \cdot 4 \cdot 3 \cdot 2 \cdot 1}$$
>
> $$= \frac{43,609,104,000}{40,320} = 1,081,575$$

> Each of these 1,081,575 possible sets of 8 pilots is equally likely to be selected.

(c) No. There is more than a 10% chance that a fair lottery would result in 5 female and 3 male pilots being selected.

> Recall from Lesson 3.7 that researchers often identify results that occur at least 5% of the time by chance alone as "not statistically significant."

FOR PRACTICE   TRY EXERCISE 9.

LESSON APP 4.8

How many ways can you set up an iPod play list?

Janine wants to set up a play list with 8 songs on her iPod. She has 50 songs to choose from, including 15 songs by One Direction. Janine's iPod won't allow any song to appear more than once in a play list.

1. How many different sets of 8 songs are possible for Janine's play list? Assume that the order of the songs doesn't matter.

2. How many 8-song play lists contain no songs by One Direction?

Suppose Janine decides to let her iPod select an 8-song play list at random.

3. What's the probability that none of the 8 songs is by One Direction?

4. Find the probability that exactly 2 of the songs on the play list are by One Direction.

Ann Heath

Calculating Combinations

You can use an applet or a TI-83/84 to compute the number of combinations of *n* individuals taken *r* at a time.

Applet

1. Go to highschool.bfwpub.com/spa3e and launch the *Probability* applet.

2. Select type "Counting methods" and click on "Begin analysis."

3. To calculate the number of ways in which Mr. Wilcox can randomly select 3 of his 28 students' homework papers to grade, enter $n = 28$ and $r = 3$. Then click on "Calculate $_nC_r$."

Probability

Select type: Counting methods ▼ | Begin analysis

$n =$ 28 $r =$ 3

Calculate n! | Calculate nPr | Calculate nCr

$_{28}C_3 = 3276$

TI-83/84

We can use the $_nC_r$ command to compute the number of ways in which Mr. Wilcox can randomly select 3 of his 28 students' homework papers to grade.

● Type 28. Then press $\boxed{\text{MATH}}$, choose PROB, and $_nC_r$. Complete the command $_{28}C_3$ (Older OS: 28 $_nC_r$ 3) and press $\boxed{\text{ENTER}}$.

```
NORMAL FLOAT AUTO REAL RADIAN MP

28C3
                              3276
```

Lesson 4.8

WHAT DID YOU LEARN?

LEARNING TARGET	EXAMPLES	EXERCISES
Compute the number of combinations of *n* individuals taken *k* at a time.	p. 318	1–4
Use combinations to calculate probabilities.	p. 319	5–8
Use the multiplication counting principle and combinations to calculate probabilities.	p. 320	9–12

Exercises · Lesson 4.8

Mastering Concepts and Skills

1. Papa's Pizza On Monday nights, Papa's Pizza of-
pg 318 fers a special two-for-one deal on its medium two-topping pizzas. All pizzas come with sauce and cheese. The customer can choose any 2 of Papa's 25 toppings to complete the order. How many different possible pizzas are there, assuming that the customer must choose exactly 2 different toppings?

2. Five-card hands A standard deck has 52 playing cards. Suppose you shuffle the deck well and deal out 5 cards. How many different possible 5-card hands are there?

3. Random raffle At the end of a weeklong seminar, the presenter decides to give away signed copies of his book to 4 randomly selected people in the audience. How many different ways can this be done if 30 people are present at the seminar?

4. Two scoops The local ice cream shop in Dontrelle's town is called 21 Choices. Why? Because they offer 21 different flavors of ice cream. Dontrelle's sister Emogene wants to order a small cup with 2 scoops of different flavors. In how many ways can she do this? (Note that the order of the two flavors doesn't matter.)

5. Sleepy batteries? Almost everyone has one—a
pg 319 drawer that holds miscellaneous batteries of all sizes. Suppose that your drawer contains 8 AAA batteries but only 6 of them are good. You have to choose 4 for your graphing calculator. If you randomly select 4 batteries, what is the probability that all 4 of them will work? Use combinations to help answer this question.

6. Seven vowels? In the game of Scrabble, each player begins by drawing 7 tiles from a bag initially containing 100 tiles. There are 42 vowels, 56 consonants, and 2 blank tiles in the bag. Suppose you are playing Scrabble and get to go first. If you randomly select 7 tiles from the bag, what's the probability that all of them are vowels? Use combinations to help answer this question.

7. Parking tickets At a local high school, 95 students have permission to park on campus. Each month, the student council holds a "golden ticket parking lottery." The three lucky winners are given reserved parking spots next to the main entrance. Last month, the winning tickets were drawn by a student council member who is in Mr. Wilder's statistics class. When all three golden tickets went to members of that class, some people thought the lottery had been rigged. There are 28 students in the statistics class, all of whom are eligible to park on campus.

(a) Use combinations to find the probability that a fair lottery would result in all three golden tickets going to members of Mr. Wilder's statistics class.

(b) Based on your answer to part (a), do you think the lottery was carried out fairly? Explain.

8. Airport security The Transportation Security Administration (TSA) is responsible for airport safety. On some flights, TSA officers randomly select passengers for an extra security check prior to boarding. One such flight had 76 passengers—12 in first class and 64 in economy class. Some passengers were suspicious when all 10 passengers chosen for screening were seated in economy class.

(a) Use combinations to find the probability that all 10 passengers chosen at random for security screening are in economy class.

(b) Based on your answer to part (a), should passengers be suspicious about this result? Explain.

9. Statistical ringers? At a university's annual picnic,
pg 320 18 students in the mathematics/statistics department decide to play a softball game. Twelve of the 18 students are math majors and 6 are statistics majors. To divide into two teams of 9, one of the professors put all the players' names into a hat and drew out 9 players to form team Fisher, with the remaining 9 players forming team Newton. The players were surprised when all of the statistics majors ended up on team Fisher.

(a) Find the number of ways in which a random selection of 9 students for team Fisher would result in all 6 statistics majors (and 3 math majors) being selected.

(b) Find the probability that a random selection of 9 students for team Fisher would result in all 6 statistics majors being selected.

(c) Is there convincing evidence that the professor didn't mix the names well before drawing them out of the hat? Explain.

10. **You're fired!** Recently, a company had to lay off 10 employees. The company claimed that it selected the employees who were fired at random. Of the 10 employees fired, 5 were women (50%). However, only 10 of the company's 30 employees were women (33%).

(a) Find the number of ways in which a random selection of employees from the company would result in 5 women and 5 men being selected.

(b) Find the probability that a random selection of employees from the company would result in the selection of 5 women and 5 men.

(c) Does the result in part (b) give convincing evidence that the company did not select the employees who were fired at random? Explain.

11. **Random assignment** Researchers recruited 20 volunteers—8 men and 12 women—to take part in an experiment involving a new drug and a placebo. They randomly assigned the subjects into two groups of 10 people each. To their surprise, 6 of the 8 men were randomly assigned to the new drug treatment. Should they be surprised? Find the probability that the random assignment would put 6 men in the new drug treatment group purely by chance.

12. **Three of a kind** A standard deck has 52 playing cards—4 cards each of 13 different denominations (ace, two, three, . . . , ten, jack, queen, and king). You and a friend decide to play a game in which you each start with 5 cards. Suppose you shuffle the deck well and deal out 5 cards to you and your friend. You are surprised when you look at your hand and see three aces (along with two non-aces). Should you be surprised? Find the probability of dealing yourself three aces in a 5-card hand from a well-shuffled deck.

Applying the Concepts

13. **Not even one match?** Refer to the example on page 318 about New Jersey's Pick Six lotto game. Use combinations to find the probability that a player matches none of the 6 winning numbers.

14. **Selected students** Refer to the example on page 319 about Mr. Wilcox randomly selecting students' homework papers from his class. Use combinations to find the probability that none of the 3 students chosen are girls.

15. **Balancing the starting lineup** A certain basketball team has 12 players. Five of these players are classified as guards and 7 of the players are classified as forwards/centers.

(a) If the coach wanted to choose her 5 starters at random by drawing names from a hat, how many possible groups of 5 starters could she choose?

(b) To avoid an unbalanced lineup, the coach wants to choose 2 of her guards and 3 of her forwards/centers. To do this, she places the names of her 5 guards in one hat and the names of her 7 forwards/centers in a second hat. Then, she will randomly select 2 guards from the first hat and 3 forwards/centers from the second hat. How many different lineups are possible?

(c) If the coach chooses her 5 starters at random by drawing names from a hat, find the probability that she selects an ideal lineup with 2 guards and 3 forwards/centers.

16. **Random gifts of candy** At the end of each week, a generous teacher gives some candy to a few randomly selected students. There are 30 students in this teacher's class—16 girls and 14 boys.

(a) If the teacher wanted to give candy to 6 of the students, how many possible groups of 6 students can she choose?

(b) To avoid appearing to favor one gender, the teacher will randomly choose 3 of the 16 girls and 3 of the 14 boys. How many possible groups of 6 can she choose?

(c) Find the probability that the teacher gives candy to 3 boys and 3 girls if she chooses 6 students from the class at random.

Extending the Concepts

17. **Tuesdays at Papa's** Refer to Exercise 1. On Tuesday nights, Papa's Pizza allows a customer to choose from 0 to 8 toppings for the same price on its medium pizzas. How many different pizzas are possible?

18. **Sampling solutions** Mr. Tabor has 32 students in his statistics class—20 seniors, 7 juniors, and 5 sophomores. Suppose Mr. Tabor chooses 6 students at random to present solutions to the six homework problems he assigned last night. What's the probability that 3 seniors, 2 juniors, and 1 sophomore are selected?

Recycle and Review

19. **Home runs (1.4, 1.8)** The following stemplot shows the number of home runs hit by each of the 30 Major League Baseball teams in 2014.[38] Construct a boxplot for this distribution. Be sure to identify any outliers.

```
 9 | 5
10 | 5 9
11 | 1 7 8
12 | 2 3 3 5 5 8
13 | 1 2 4 6
14 | 2 6 7
15 | 0 2 5 5 5 6 7
16 | 3
17 | 7
18 | 6        KEY: 12 | 3 = 123 Home runs
19 |
20 |
21 | 1
```

20. **Life at work (4.4)** The University of Chicago's General Social Survey asked a representative sample of adults this question: "Which of the following statements best describes how your daily work is organized? (1) I am free to decide how my daily work is organized. (2) I can decide how my daily work is organized, within certain limits. (3) I am not free to decide how my daily work is organized." Here is a two-way table of the responses for three levels of education.[39]

		Highest degree completed		
		Less than high school	High school	Bachelor's
Response	1	31	161	81
	2	49	269	85
	3	47	112	14

(a) Choose an individual from the sample at random. If the individual's highest degree was a high school diploma, what is the probability that his or her response was "I am free to decide how my daily work is organized"?

(b) Show that the events "Response 3" and "Less than High School" are not independent.

John Lund/Sam Diephuis/Getty Images

Chapter 4

RESOLVED

STATS applied!

Should an athlete who fails a drug test be suspended?

In the STATS applied! on page 261, we described drug-testing programs for high school athletes. Suppose that 16% of the high school athletes in a large school district have taken a banned substance. The drug test used by this district has a false positive rate of 5% and a false negative rate of 10%. Use what you have learned in this chapter to help answer the following questions about the district's drug-testing program.

1. What's the probability that a randomly chosen athlete tests positive for banned substances?

2. If two athletes are randomly selected, what's the probability that at least one of them tests positive?

3. If a randomly chosen athlete tests positive, what's the probability that the student did not take a banned substance? Based on your answer, do you think that an athlete who tests positive should be suspended from athletic competition for a year? Why or why not?

4. If a randomly chosen athlete tests negative, what's the probability that the student took a banned substance? Explain why it makes sense for the drug-testing process to be designed so that this probability is less than the one you found in Question 3.

Challenge: The district decides to immediately "re-test" any athlete who tests positive. Assume that the results of an athlete's two tests are independent. Find the probability that a student who gets a positive result on both tests took a banned substance. Based on your answer, do you think that an athlete who tests positive twice should be suspended from athletic competition for a year? Why or why not?

Main Points — Chapter 4

Randomness, Probability, and Simulation

■ Chance behavior is unpredictable in the short run but has a regular and predictable pattern in the long run.

■ The long-run relative frequency of a chance outcome is its **probability**. A probability is a number between 0 (never occurs) and 1 (always occurs).

■ The **law of large numbers** says that in many repetitions of the same chance process, the proportion of times that a particular outcome occurs will approach its probability.

■ **Simulation** can be used to imitate chance behavior and to estimate probabilities. To perform a simulation:

■ **State:** Ask a question about some chance process.

■ **Plan:** Describe how to use a chance device to imitate one repetition of the process. Tell what you will record at the end of each repetition.

■ **Do:** Perform many repetitions of the simulation.

■ **Conclude:** Use the results of your simulation to answer the question.

Probability Models and Probability Rules

■ A **probability model** describes a chance process by listing all possible outcomes in the **sample space** and giving the probability of each outcome. A valid probability model requires that all possible outcomes have probabilities that add up to 1.

■ An **event** is a collection of possible outcomes from the sample space. The probability of any event is a number between 0 and 1.

■ To find the probability that an event occurs, we can use some basic rules:

■ If all outcomes in the sample space are equally likely,

$$P(A) = \frac{\text{number of outcomes in event A}}{\text{total number of outcomes in sample space}}$$

■ **Complement rule:** $P(A^C) = 1 - P(A)$, where A^C is the **complement** of event A; that is, the event that A does not happen.

■ **General addition rule:** For any two events A and B,

$$P(A \text{ or } B) = P(A) + P(B) - P(A \text{ and } B)$$

■ **Addition rule for mutually exclusive events:** Events A and B are **mutually exclusive** if they have no outcomes in common. If A and B are mutually exclusive, $P(A \text{ or } B) = P(A) + P(B)$.

■ A **two-way table** or **Venn diagram** can be used to display the sample space and to help find probabilities for a chance process involving two events.

■ The event "A or B" is known as the **union** of A and B, denoted by $A \cup B$. It consists of all outcomes in event A, event B, or both.

■ The event "A and B" is known as the **intersection** of A and B, denoted by $A \cap B$. It consists of all outcomes that are common to both events.

Conditional Probability, Multiplication Rules, and Independence

■ A **conditional probability** describes the probability that one event happens given that another event is already known to have happened. To calculate the conditional probability $P(A|B)$ that event A occurs given that event B has occurred, use the formula

$$P(A|B) = \frac{P(A \text{ and } B)}{P(B)} = \frac{P(\text{both events occur})}{P(\text{given event occurs})}$$

■ Use the **general multiplication rule** to calculate the probability that events A and B both occur:

$$P(A \text{ and } B) = P(A) \cdot P(B|A)$$

■ When a chance process involves multiple stages, a **tree diagram** can be used to display the sample space and to help answer questions involving conditional probability.

■ When knowing whether or not one event has occurred does not change the probability that another event happens, we say that the two events are **independent**. For independent events A and B,

$$P(A|B) = P(A|B^C) = P(A) \text{ and}$$
$$P(B|A) = P(B|A^C) = P(B)$$

■ In the special case of independent events, the multiplication rule becomes

$$P(A \text{ and } B) = P(A) \cdot P(B)$$

Counting and Probability

■ When a chance process involves several steps, the **multiplication counting principle** can be used to determine the total number of possible outcomes. For a process involving k steps, if there are n_1 ways to do Step 1, n_2 ways to do Step 2, . . . , and n_k ways to do Step k, the total number of different ways to complete the process is $n_1 \cdot n_2 \cdot \ldots \cdot n_k$.

■ The multiplication counting principle also helps us count the number of distinct arrangements of some group of individuals. We call such arrangements, where the order of selection matters, **permutations.**

■ The number of permutations of an entire group of n individuals is n **factorial:**

$$n! = n(n-1)(n-2)\ldots 3 \cdot 2 \cdot 1$$

■ The number of permutations of k individuals selected from a group of n individuals (where $k \le n$) is denoted by $_nP_k$ and can be computed using the multiplication counting principle, with technology, or with the formula

$$_nP_k = \frac{n!}{(n-k)!}$$

■ A **combination** is a selection of individuals from some group in which the order of selection doesn't matter. If there are n individuals, $_nC_k$ represents the number of different combinations of k individuals chosen from the entire group of n. You can calculate $_nC_k$ using the multiplication counting principle, with technology, or with the formula

$$_nC_k = \frac{_nP_k}{k!} = \frac{n!}{k!(n-k)!}$$

■ The multiplication counting principle, permutations, and combinations are useful tools in helping to calculate probabilities.

Chapter 4 Review Exercises

1. **Live long and prosper (4.1, 4.6)** According to an article on *The Guardian* website, a British woman born in 2003 has probability 0.27 of living at least 100 years.[40]

   (a) Suppose that you randomly select 200 women who were born in 2003. Will exactly 54 of them live at least 100 years? Explain.

   (b) Five high school friends all happen to be females born in 2003. Assuming that the survival of each one is independent of the others, what is the probability that at least one of them lives to be 100 years old?

2. **What kind of vehicle? (4.1, 4.2, 4.4)** Choose a new vehicle sold in the United States in November 2014 at random.[41] The probability distribution for the type of vehicle chosen is given here.

Vehicle type	Passenger car	Pickup truck	SUV	Crossover	Minivan
Probability	0.46	0.15	0.10	?	0.05

   (a) What is the probability that the vehicle is a crossover? How do you know?

   (b) Given that the vehicle is not a passenger car, what is the probability that it is a pickup truck?

   (c) What is the probability that the vehicle is a pickup truck, SUV, or minivan?

   (d) Suppose you select 3 vehicles at random. You want to know the probability that the vehicles are of three different types. Describe how you would carry out a simulation to help answer this question. Do not actually perform the simulation.

3. **Taste test (4.3, 4.4)** Ivy conducted a taste test for 4 different brands of chocolate chip cookies. Here is a two-way table that describes which cookie each subject preferred and that person's gender.

		Cookie brand			
		A	B	C	D
Gender	Female	4	6	13	13
	Male	22	11	11	14

   Suppose one subject from this experiment is selected at random.

   (a) Find the probability that the selected subject preferred Brand C.

   (b) Find the probability that the selected subject preferred Brand C or is female.

   (c) Find the probability that the selected subject preferred Brand C, given that she is female.

4. **Mike's pizza (4.3, 4.4)** You work at Mike's pizza shop. You have the following information about the 7 pizzas in the oven: 3 of the 7 have thick crust and 2 of the 3 thick-crust pizzas have mushrooms. Of the remaining 4 pizzas, 2 have mushrooms. Choose a pizza at random from the oven.

   (a) Make a Venn diagram to model this chance process.

   (b) Are the events "getting a thick-crust pizza" and "getting a pizza with mushrooms" independent? Explain.

   (c) You add an eighth pizza to the oven. This pizza has thick crust with only cheese. Now are the events "getting a thick-crust pizza" and "getting a pizza with mushrooms" independent? Explain.

5. **Matching socks (4.5 or 4.8)** Gabriel is not one for pairing up his socks before he throws them into his sock drawer. One dark morning, his drawer contains 8 blue socks, 6 brown socks, and 4 gray socks. He pulls 2 socks at random from his drawer without looking at them.

   (a) What is the probability that both socks are blue?

   (b) What is the probability that the 2 socks are the same color?

6. **Detecting steroids (4.5)** A company has developed a drug test to detect steroid use by athletes. The test is accurate 95% of the time when an athlete has taken steroids. It is 97% accurate when an athlete hasn't taken steroids. Suppose that the drug test will be used in a population of athletes in which 10% have taken steroids. Let's choose an athlete at random and administer the drug test.

   (a) Make a tree diagram showing the sample space of this chance process.

   (b) What's the probability that the randomly selected athlete tests positive?

   (c) Suppose that the chosen athlete tests positive. What's the probability that he or she used steroids?

7. **Many-sided die (4.6, 4.7)** The outcome of chance events in a fantasy role-playing game is determined by rolling polyhedral dice with anywhere from 4 to 20 sides. Suppose you roll a 16-sided die 5 times and observe the number on the top of the die.

   (a) How many possible outcomes are there for these 5 rolls?

   (b) In how many of the outcomes in part (a) do the 5 rolls produce five different numbers?

   (c) What is the probability that at least 2 of the rolls are the same?

8. **Dream team (4.8)** Mr. Bakare often uses a computer program to randomly select 4 students from his class of 24 students to solve a problem on the board.

   (a) How many possible teams of 4 students can be created from this class of 24?

   (b) One day Mr. Bakare poses a particularly challenging problem, and his random name generator produces a team of 4 students that includes 2 of the 3 top students in the class. What is the probability that a randomly selected group of 4 students will include 2 of the 3 strongest students?

   (c) Based on your answer to part (b), do you suspect that Mr. Bakare rigged his random name generator? Explain.

Chapter 4 Practice Test

Section I: Multiple Choice *Select the best answer for each question.*

1. Dr. Stats plans to toss a fair coin 10,000 times in the hope that it will lead him to a deeper understanding of the laws of probability. Which of the following statements is true?

(a) It is unlikely that Dr. Stats will get more than 5000 heads.

(b) Whenever Dr. Stats gets a string of 15 tails in a row, it becomes more likely that the next toss will be a head.

(c) The chance that the 100th toss will be a head depends somewhat on the results of the first 99 tosses.

(d) It is likely that Dr. Stats will get about 50% heads.

2. Choose a U.S. household at random and record the number of vehicles. Here is the probability model if we ignore the few households that own more than 5 cars:

Number of cars	0	1	2	3	4	5
Probability	0.09	0.36	0.35	0.13	0.05	0.02

A housing company builds houses with two-car garages. What percent of households have more cars than the garage can hold?

(a) 7% (b) 13% (c) 20% (d) 45%

3. When three-digit telephone area codes were first put in place in the United States and Canada, the first number could be any digit from 2 through 9, the second number could only be 0 or 1, and the third number could be any digit from 0 through 9. How many possible area codes can be created with this format?

(a) 24 (b) 126 (c) 140 (d) 160

4. Wilt is a fine basketball player, but his free-throw shooting could use some work. For the past three seasons, he has made only 56% of his free throws. His coach sends him to a summer clinic to work on his shot, and when he returns, his coach has him step to the free-throw line and take 50 shots. He makes 34 shots. Is this result convincing evidence that Wilt's shooting has improved? The dotplot displays the number of made shots in 100 simulated sets of 50 free throws by someone who makes 56% of his shots.

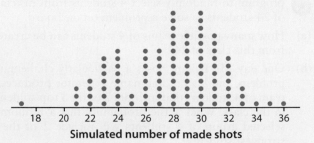

Simulated number of made shots

Which of the following is an appropriate statement about Wilt's free-throw shooting, based on this dotplot?

(a) If Wilt were still only a 56% shooter, the probability that he would make at least 34 of his shots is about 0.03.

(b) If Wilt were still only a 56% shooter, the probability that he would make at least 34 of his shots is about 0.97.

(c) If Wilt is now shooting better than 56%, the probability that he would make at least 34 of his shots is about 0.03.

(d) If Wilt is now shooting better than 56%, the probability that he would make at least 34 of his shots is about 0.97.

Questions 5–8 refer to the following setting. A random sample of 88 U.S. 11th- and 12th-graders was selected. The two-way table summarizes the gender of the students and their response to the question "Do you have allergies?" Suppose we choose a student from this group at random.

		Gender		
		Female	Male	Total
Allergies	Yes	19	15	34
	No	24	30	54
	Total	43	45	88

5. What is the probability that the student has allergies?

(a) $\dfrac{19}{88}$ (b) $\dfrac{34}{88}$ (c) $\dfrac{54}{88}$ (d) $\dfrac{34}{54}$

6. What is the probability that the student is female or has allergies?

(a) $\dfrac{19}{88}$ (b) $\dfrac{39}{88}$ (c) $\dfrac{58}{88}$ (d) $\dfrac{77}{88}$

7. If the student has allergies, what is the probability that the student is female?

(a) $\dfrac{34}{43}$ (b) $\dfrac{19}{88}$ (c) $\dfrac{19}{43}$ (d) $\dfrac{19}{34}$

8. Which of the following is true about the events "Student is female" and "Student has allergies"?

(a) The events are not mutually exclusive, but they are independent.

(b) The events are mutually exclusive, but they are not independent.

(c) The events are not mutually exclusive, nor are they independent.

(d) The events are both mutually exclusive and independent.

9. The Venn diagram displays the sample space for the chance process of randomly selecting a teacher from a local high school and recording whether or not the teacher regularly uses the Internet as a source of news (E) and whether or not he or she regularly uses print media as a source of news (P).

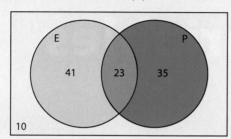

Which of the following two-way tables conveys the same information?

(a)

		Print news		
		P	P^C	Total
Internet news	E	23	17	40
	E^C	35	10	45
	Total	58	27	

(b)

		Print news		
		P	P^C	Total
Internet news	E	23	41	64
	E^C	35	10	45
	Total	58	51	

(c)

		Print news		
		P	P^C	Total
Internet news	E	10	41	51
	E^C	35	23	58
	Total	45	64	

(d)

		Print news		
		P	P^C	Total
Internet news	E	23	18	41
	E^C	41	10	51
	Total	64	28	

10. There are 12 students running for three student council positions: president, vice president, and secretary. Which of the following expressions represents the number of ways in which these three positions can be filled by the 12 candidates?

(a) $12 \cdot 12 \cdot 12$

(b) $12 \cdot 11 \cdot 10$

(c) $\dfrac{12 \cdot 11 \cdot 10}{3 \cdot 2 \cdot 1}$

(d) $12!$

Section II: Free Response

11. Three machines—A, B, and C—are used to produce a large quantity of identical parts at a factory. Machine A produces 60% of the parts, while Machines B and C produce 30% and 10% of the parts, respectively. Historical records indicate that 10% of the parts produced by Machine A are defective, compared with 30% for Machine B and 40% for Machine C. Suppose we randomly select a part produced by one of these three machines at random.

(a) Draw a tree diagram to represent this chance process.

(b) What's the probability that the part is defective?

(c) If the part is inspected and found to be defective, what's the probability that it was produced by Machine C?

12. Researchers are interested in the relationship between cigarette smoking and lung cancer. Suppose an adult male is randomly selected from a particular population. The following table shows the probabilities of some events related to this chance process.

Event	Probability
Smokes	0.25
Smokes and gets cancer	0.08
Does not smoke and does not get cancer	0.71

(a) Make a two way table or a Venn diagram to display the sample space.

(b) Find the probability that the individual gets cancer given that he is a smoker.

(c) Find the probability that the individual smokes or gets cancer.

(d) Two adult males are selected at random. Find the probability that at least one of the two gets cancer.

13. Fourteen musical acts have asked to perform at Springfield High School's winter talent show. Eight of them play acoustic music and 6 play electric music. Unfortunately, there is only room on the program for 10 acts—the others will have to wait until the spring show.

(a) How many different groups of 10 acts can be selected to play in the winter show?

(b) If the 10 musical acts for the winter show are randomly selected, what is the probability that all of the electric acts get to play in the show?

(c) If the 10 musical acts for the winter show are randomly selected, what is the probability that 5 acoustic acts and 5 electric acts get to play in the show?

5

Random Variables

Lesson 5.1	Two Types of Random Variables	332
Lesson 5.2	Analyzing Discrete Random Variables	338
Lesson 5.3	Binomial Random Variables	348
Lesson 5.4	Analyzing Binomial Random Variables	356
Lesson 5.5	Continuous Random Variables	365
Lesson 5.6	The Standard Normal Distribution	375
Lesson 5.7	Normal Distribution Calculations	384
Chapter 5	Main Points	393
Chapter 5	Review Exercises	395
Chapter 5	Practice Test	396

STATS applied!

How did that vending machine go wrong?

Have you ever purchased coffee from a vending machine? The intended sequence of events runs something like this. You insert your money into the machine and select your preferred beverage. A cup falls out of the machine, landing upright. Coffee pours out until the cup is nearly full. You reach in, grab the piping hot beverage, and drink happily.

Sometimes, things go wrong. The machine might swipe your money. Or the cup might fall over. More often, everything goes smoothly until the coffee begins to flow. It might stop flowing when the cup is only half full. Or the coffee might keep coming until your cup overflows. Neither of these results leaves you satisfied.

> How should the company set the machine's mean amount of coffee dispensed to satisfy both concerns? If a quality-control inspector measures the amount of liquid in 10 randomly selected cups, how can she determine if the machine is operating correctly?

We'll revisit STATS applied! at the end of the chapter, so you can use what you have learned to help answer these questions.

Lesson 5.1

Two Types of Random Variables

- Verify that the probability distribution of a discrete random variable is valid.
- Calculate probabilities involving a discrete random variable.
- Classify a random variable as discrete or continuous.

Suppose you toss a fair coin 3 times. The sample space for this chance process is

HHH   HHT   HTH   THH   HTT   THT   TTH   TTT

Because there are 8 equally likely outcomes, the probability is 1/8 for each possible outcome.

Define the **random variable** X = the number of heads obtained in 3 tosses. The value of X will vary from one set of tosses to another, but it will always be one of the numbers 0, 1, 2, or 3. How likely is X to take each of those values? It will be easier to answer this question if we group the possible outcomes by the number of heads obtained:

$X = 0$: TTT $X = 1$: HTT THT TTH $X = 2$: HHT HTH THH $X = 3$: HHH

We can summarize the **probability distribution** of X in a table:

Value	0	1	2	3
Probability	1/8	3/8	3/8	1/8

DEFINITION   Random variable, Probability distribution

A **random variable** takes numerical values that describe the outcomes of a chance process. The **probability distribution** of a random variable gives its possible values and their probabilities.

There are two main types of probability distributions, corresponding to two types of random variables: *discrete* and *continuous*.

Discrete Random Variables

The variable X in the coin-tossing example is a **discrete random variable**.

DEFINITION   Discrete random variable

A **discrete random variable** X takes a fixed set of possible values with gaps between. The probability distribution of a discrete random variable X lists the values x_i and their probabilities p_i:

Value	x_1	x_2	x_3	…
Probability	p_1	p_2	p_3	…

In a valid probability distribution, the probabilities p_i must satisfy two requirements:

1. Every probability p_i is a number between 0 and 1, inclusive.

2. The sum of the probabilities is 1: $p_1 + p_2 + p_3 + \ldots = 1$.

We can list the possible values of X = the number of heads obtained in 3 tosses as 0, 1, 2, 3, as shown in this probability distribution.

Value	0	1	2	3
Probability	1/8	3/8	3/8	1/8

Note that there are gaps between these values on a number line. For example, a gap exists between $X = 1$ and $X = 2$ because X cannot take values such as 1.2 or 1.84. From the probability distribution, we can see that the probabilities are all between 0 and 1, and their sum is

$$1/8 + 3/8 + 3/8 + 1/8 = 1$$

EXAMPLE

How can we quickly assess babies' health at birth?

Valid probability distributions

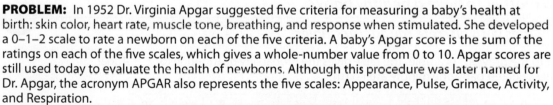

PROBLEM: In 1952 Dr. Virginia Apgar suggested five criteria for measuring a baby's health at birth: skin color, heart rate, muscle tone, breathing, and response when stimulated. She developed a 0–1–2 scale to rate a newborn on each of the five criteria. A baby's Apgar score is the sum of the ratings on each of the five scales, which gives a whole-number value from 0 to 10. Apgar scores are still used today to evaluate the health of newborns. Although this procedure was later named for Dr. Apgar, the acronym APGAR also represents the five scales: Appearance, Pulse, Grimace, Activity, and Respiration.

What Apgar scores are typical? To find out, researchers recorded the Apgar scores of more than 2 million newborn babies in a single year.[1] Imagine selecting one of these newborns at random. (That's our chance process.) Define the random variable X = Apgar score of a randomly selected baby one minute after birth. The table gives the probability distribution for X. Show that the probability distribution for X is valid.

Value	0	1	2	3	4	5	6	7	8	9	10
Probability	0.001	0.006	0.007	0.008	0.012	0.020	0.038	0.099	0.319	0.437	0.053

SOLUTION:

* The probabilities are all between 0 and 1.

* The sum of the probabilities is $0.001 + 0.006 + 0.007 + 0.008 + 0.012 + 0.020 + 0.038 + 0.099 + 0.319 + 0.437 + 0.053 = 1$.

FOR PRACTICE **TRY EXERCISE 1.**

Finding Probabilities for Discrete Random Variables

We can use the probability distribution of a discrete random variable X to find the probability of an event. For instance, the probability distribution for $X =$ the number of heads obtained when tossing a fair coin 3 times is shown here.

Value	0	1	2	3
Probability	1/8	3/8	3/8	1/8

What's the probability that we get at least one head in 3 tosses of the coin? In symbols, we want to find $P(X \geq 1)$. We know that

$$P(X \geq 1) = P(X = 1 \text{ or } X = 2 \text{ or } X = 3)$$

Because the events $X = 1$, $X = 2$, and $X = 3$ are mutually exclusive, we can add their probabilities to get the answer:

$$P(X \geq 1) = P(X = 1) + P(X = 2) + P(X = 3)$$
$$= 1/8 + 3/8 + 3/8 = 7/8$$

Or we could use the complement rule from Lesson 4.2:

$$P(X \geq 1) = 1 - P(X < 1) = 1 - P(X = 0)$$
$$= 1 - 1/8 = 7/8$$

EXAMPLE

How likely is a mother to have a healthy baby?

Finding probabilities with discrete random variables

PROBLEM: Refer to the previous example. The table gives the probability distribution for $X =$ Apgar score of a randomly selected baby one minute after birth.

Value	0	1	2	3	4	5	6	7	8	9	10
Probability	0.001	0.006	0.007	0.008	0.012	0.020	0.038	0.099	0.319	0.437	0.053

Doctors decided that Apgar scores of 7 or higher indicate a healthy baby. What's the probability that a randomly selected baby is healthy?

SOLUTION:

$P(X \geq 7) = 0.099 + 0.319 + 0.437 + 0.053$
$\qquad = 0.908$

The probability of choosing a healthy baby is

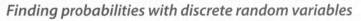

$P(X \geq 7) = P(X = 7) + P(X = 8) + P(X = 9) + P(X = 10)$

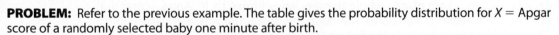

Value	0	...	7	8	9	10
Probability	0.001	...	0.099	0.319	0.437	0.053

FOR PRACTICE   TRY EXERCISE 5.

Note that the probability of randomly selecting a newborn whose Apgar score is *at least* 7 is not the same as the probability that the baby's Apgar score is *greater than* 7. The latter probability is

$$P(X > 7) = P(X = 8) + P(X = 9) + P(X = 10)$$
$$= 0.319 + 0.437 + 0.053 = 0.809$$

The outcome $X = 7$ is included in "greater than or equal to" but is not included in "greater than." Be sure to consider whether to include the boundary value in your calculations when dealing with discrete random variables.

CAUTION

Continuous Random Variables

Suppose we want to randomly select a number between 0 and 1, allowing *any* number between 0 and 1 as the outcome (like 0.84522 or 0.1111119). Calculator and computer random number generators will do this. The sample space of this chance process is the entire interval of values between 0 and 1 on a number line. If we define $Y =$ the outcome of the random number generator, then Y is a **continuous random variable.**

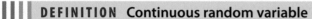

> **DEFINITION Continuous random variable**
>
> A **continuous random variable** X can take any value in an interval on the number line.

We'll show you how to find probabilities involving continuous random variables in Lesson 5.5. For now, you should focus on distinguishing between the two types of random variables.

Most discrete random variables result from counting something, like the number of siblings that a randomly selected student has. Continuous random variables typically result from measuring something, like the height or time to run a mile for a randomly selected student.

EXAMPLE

Feet, gas pumps, and tennis rackets: How do they vary?

Discrete or continuous random variables

PROBLEM: Classify each of the following random variables as discrete or continuous.

(a) $X =$ foot length of a randomly selected student at your school

(b) $G =$ the number of pumps (out of 12) in use at a gas station at a randomly selected time of day

(c) $T =$ string tension of a randomly selected tennis racket

Matthew Ward/Getty Images

SOLUTION:

(a) Continuous

(b) Discrete

(c) Continuous

> X could take any value in an interval from about 9 inches to 15 inches.

> G could take any of the values 0, 1, 2, . . . , 12 but not values like 2.387465819.

> T could take any value in the interval from about 50 pounds to 70 pounds.

FOR PRACTICE TRY EXERCISE 9.

LESSON APP 5.1

Making the grade?

Indiana University Bloomington posts the grade distributions for its courses online.[2] In a recent semester of a Business Statistics course, 45.7% of students received A's, 36.2% B's, 13.8% C's, 3.2% D's, and 1.1% F's. Choose a Business Statistics student at random. The student's grade on a 4-point scale (with A = 4) is a random variable X with this probability distribution:

Value	0	1	2	3	4
Probability	0.011	0.032	0.138	0.362	0.457

1. Is X a discrete or a continuous random variable? Explain.

2. Show that the probability distribution of X is valid.

3. Explain in words what $P(X \geq 3)$ means. What is this probability?

4. Write the event "the student got a grade worse than C" using probability notation. What is the probability of this event?

Lesson 5.1

WHAT DID YOU LEARN?

LEARNING TARGET	EXAMPLES	EXERCISES
Verify that the probability distribution of a discrete random variable is valid.	p. 333	1–4
Calculate probabilities involving a discrete random variable.	p. 334	5–8
Classify a random variable as discrete or continuous.	p. 335	9–12

Lesson 5.1

Mastering Concepts and Skills

1. **Benford's law** Faked numbers on tax returns, invoices, or expense account claims often display patterns that aren't present in legitimate records. Some patterns, like too many round numbers, are obvious and easily avoided by a clever crook. Others are more subtle. It is a striking fact that the first digits of numbers in legitimate records often follow a model known as Benford's law.[3] Call the first digit of a randomly chosen record X for short. Benford's law gives this probability model for X (note that a first digit can't be 0). Show that this is a valid probability distribution.

pg 333

First digit	1	2	3	4	5
Probability	0.301	0.176	0.125	0.097	0.079
First digit	6	7	8	9	
Probability	0.067	0.058	0.051	0.046	

2. **Working out** Choose a person aged 19 to 25 years at random and ask, "In the past seven days, how many times did you go to an exercise or fitness center or work out?" Call the response Y for short. Based on a large sample survey, here is a probability model for the answer you will get.[4] Show that this is a valid probability distribution.

Days	0	1	2	3	4	5	6	7
Probability	0.68	0.05	0.07	0.08	0.05	0.04	0.01	0.02

3. **Kids and toys** In an experiment on the behavior of young children, each subject is placed in an area with 5 toys. Past experiments have shown that the probability distribution of the number X of toys played with by a randomly selected subject is as follows. Find the missing value for $P(X = 5)$ in the probability distribution.

Number of toys	0	1	2	3	4	5
Probability	0.03	0.16	0.30	0.23	0.17	

4. **Spell check?** Spell-checking software catches "non-word errors," which result in a string of letters that is not a word, as when "the" is typed as "teh." When undergraduates are asked to write a 250-word essay (without spell-checking), the number X of nonword errors in a randomly selected essay has the following distribution. Find the missing value for $P(X = 2)$ in the probability distribution.

Value	0	1	2	3	4
Probability	0.1	0.2		0.3	0.1

5. **Large first digits** Refer to Exercise 1. Describe the event $X \geq 6$ in words. What is $P(X \geq 6)$?
pg 334

6. **Not working out** Refer to Exercise 2. Describe the event $Y < 3$ in words. What is $P(Y < 3)$?

7. **Playing with few toys** Refer to Exercise 3.

(a) Write the event "plays with at most 2 toys" in terms of X. What is the probability of this event?

(b) Find $P(X < 2)$. Explain why this answer is different from the one you found in part (a).

8. **Nonword errors** Refer to Exercise 4.

(a) Write the event "at least 1 nonword error" in terms of X. What is the probability of this event?

(b) Find $P(X > 1)$. Explain why this answer is different from the one you found in part (a).

9. **Discrete or continuous?** Classify each of the following random variables as discrete or continuous.
pg 335

(a) X = the pH of a water sample that has been randomly selected from a stream

(b) Y = the number correct on a recent multiple-choice test for a randomly selected student in your class

(c) W = the exact amount of sleep that a randomly selected student from your school got last night

10. **Which type?** Classify each of the following random variables as discrete or continuous.

(a) H = the number of homework problems assigned by your statistics teacher on a randomly selected class day

(b) R = the temperature of the turkey meat in a randomly chosen location in a cooked turkey

(c) X = the reported score of a randomly selected senior at your school on the SAT Math test

11. **On a roll** Classify each of the following random variables as discrete or continuous.

(a) R = the number of times you have to roll a fair, six-sided die to get a 1

(b) T = winning time in the men's 100-meter dash at a randomly selected international track meet

12. **Sizing them up** Classify each of the following random variables as discrete or continuous.

(a) X = age (in years) of the oldest person that a randomly selected adult has ever met

(b) Y = head circumference of a randomly selected female student at a large high school

Applying the Concepts

13. **How many languages?** Imagine selecting a U.S. high school student at random. Define the random variable X = number of languages spoken by the student. The table gives the probability distribution of X.[5]

Languages	1	2	3	4	5
Probability	0.630	0.295	0.065	0.008	0.002

(a) Is X a discrete or a continuous random variable? Explain.

(b) Show that the probability distribution of X is valid.

(c) Explain in words what $P(X \geq 3)$ means. What is this probability?

14. **Mice at night** The following probability distribution is for the random variable X = number of mice caught in traps during a single night in a small apartment building.

Mice	0	1	2	3	4	5
Probability	0.12	0.20	0.31	0.14	0.16	0.07

(a) Is X a discrete or a continuous random variable? Explain.

(b) Show that the probability distribution of X is valid.

(c) Express the event "trapping at least 1 mouse" in terms of X and find its probability.

15. **Toss 4 times** Suppose you toss a fair coin 4 times. Let X = the number of heads you get.

(a) Find the probability distribution of X.

(b) Find $P(X \leq 3)$ and interpret the result.

16. **Pair-a-dice** Suppose you roll a pair of fair, six-sided dice. Let T = the sum of the spots showing on the up-faces.

(a) Find the probability distribution of T.

(b) Find $P(T \geq 5)$ and interpret the result.

Recycle and Review

17. **Did my phone just vibrate?** (4.6) According to a recent survey, 67% of cell-phone owners find themselves checking their phone for messages, alerts, or calls—even when they don't notice their phone ringing or vibrating.[6] Suppose we randomly select 5 cell-phone owners at random.

(a) Find the probability that all 5 cell-phone owners check their phones even when the phone doesn't ring or vibrate.

(b) What's the probability that at least 1 of them doesn't check the phone?

18. **Fluoride varnish (3.6, 3.7)** In an experiment to measure the effect of fluoride "varnish" on the incidence of tooth cavities, thirty-four 10-year-old girls whose parents volunteered them for the study were randomly assigned to two groups. One group was given fluoride varnish annually for 4 years, along with standard dental hygiene; the other group followed only the standard dental hygiene regimen. The mean number of cavities in the two groups was compared at the end of the treatments.

(a) Is this experiment subject to the placebo effect? Explain.

(b) Describe how you could alter this experiment to make it double-blind.

(c) Explain the purpose of the random assignment in this experiment.

Lesson 5.2

Analyzing Discrete Random Variables

LEARNING TARGETS

- Make a histogram to display the probability distribution of a discrete random variable and describe its shape.
- Calculate and interpret the mean (expected value) of a discrete random variable.
- Calculate and interpret the standard deviation of a discrete random variable.

When we analyzed distributions of quantitative data in Chapter 1, we made it a point to discuss their shape, center, and variability. We'll do the same with probability distributions of random variables. This lesson focuses on discrete random variables. Lesson 5.5 examines continuous random variables.

Displaying Discrete Probability Distributions: Histograms and Shape

In Lesson 5.1, we considered the discrete random variable X = Apgar score of a randomly selected baby one minute after birth. The table gives the probability distribution of X.

Value	0	1	2	3	4	5	6	7	8	9	10
Probability	0.001	0.006	0.007	0.008	0.012	0.020	0.038	0.099	0.319	0.437	0.053

We can display the probability distribution in a histogram. Values of the variable go on the horizontal axis and probabilities go on the vertical axis. There is one bar in the histogram for each value of X. The height of each bar gives the probability for the corresponding value of the variable.

Figure 5.1 shows a histogram of the probability distribution for X. This distribution is skewed to the left and single-peaked.

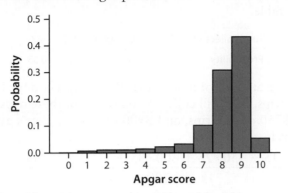

FIGURE 5.1 Histogram of the probability distribution for the random variable X = Apgar score of a randomly selected baby one minute after birth.

A probability distribution histogram is really just a relative frequency histogram because probabilities are long-run relative frequencies.

EXAMPLE

How many passengers for Pete's Jeeps?

Displaying a probability distribution

PROBLEM: Pete's Jeep Tours offers a popular half-day trip in a tourist area. There must be at least 2 passengers for the trip to run, and the vehicle will hold up to 6 passengers. Pete charges $150 per passenger. Let C = the total amount of money that Pete collects on a randomly selected trip. The probability distribution of C is given in the table. Make a histogram of the probability distribution. Describe its shape.

Jucy M. Starnes

Total collected	300	450	600	750	900
Probability	0.15	0.25	0.35	0.20	0.05

SOLUTION:

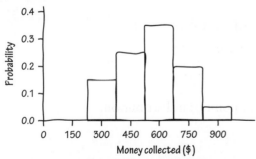

Remember: Values of the variable go on the horizontal axis and probabilities go on the vertical axis. Don't forget to properly label and scale each axis!

The graph is roughly symmetric and has a single peak at $600.

FOR PRACTICE **TRY EXERCISE 1.**

Measuring Center: The Mean (Expected Value) of a Discrete Random Variable

In Chapter 1, you learned about summarizing the center of a distribution of quantitative data. For discrete random variables, the mean is typically used to summarize the center of a probability distribution. But how do we find the **mean of a discrete random variable?**

Consider the random variable C = the total amount of money that Pete collects on a randomly selected jeep tour from the previous example. The probability distribution of C is given in the table.

Total collected	300	450	600	750	900
Probability	0.15	0.25	0.35	0.20	0.05

What's the average amount of money that Pete collects on his jeep tours? Imagine a hypothetical 100 trips. Pete will collect $300 on 15 of these trips, $450 on 25 trips, $600 on 35 trips, $750 on 20 trips, and $900 on 5 trips. Pete's average amount collected for these trips is

$$\mu_C = \frac{300 \cdot 15 + 450 \cdot 25 + 600 \cdot 35 + 750 \cdot 20 + 900 \cdot 5}{100}$$

$$= \frac{300 \cdot 15}{100} + \frac{450 \cdot 25}{100} + \frac{600 \cdot 35}{100} + \frac{750 \cdot 25}{100} + \frac{900 \cdot 5}{100}$$

$$= 300(0.15) + 450(0.25) + 600(0.35) + 750(0.20) + 900(0.05)$$

$$= \$562.50$$

This is also known as the **expected value** of the random variable C, denoted by $E(C)$.

The mean (expected value) of any discrete random variable is found in a similar way. It is an average of the possible outcomes, but a weighted average in which each outcome is weighted by its probability.

DEFINITION   Mean (expected value) of a discrete random variable

The **mean (expected value) of a discrete random variable** is its long-run average value over many, many repetitions of the same chance process.

Suppose that X is a discrete random variable with probability distribution

Value	x_1	x_2	x_3	...
Probability	p_1	p_2	p_3	...

To find the mean (expected value) of X, multiply each possible value of X by its probability, then add all the products:

$$\mu_X = E(X) = x_1 p_1 + x_2 p_2 + x_3 p_3 + \dots$$

$$= \sum x_i p_i$$

Recall that the mean is the balance point of a distribution.

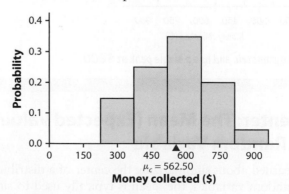

For Pete's distribution of money collected on a randomly selected jeep tour, the histogram balances at $\mu_C = 562.50$. How do we interpret this value? If we randomly select many, many jeep tours, Pete will make about $562.50 per trip, on average.

EXAMPLE

Apgar scores: What's expected?

Finding and interpreting the mean

PROBLEM: Earlier, we defined the random variable X to be the Apgar score of a randomly selected baby one minute after birth. The table gives the probability distribution for X once again. Calculate and interpret the expected value of X.

Value	0	1	2	3	4	5	6	7	8	9	10
Probability	0.001	0.006	0.007	0.008	0.012	0.020	0.038	0.099	0.319	0.437	0.053

SOLUTION:

$$\mu_X = E(X) = 0(0.001) + 1(0.006) + 2(0.007) + \cdots + 10(0.053)$$
$$= 8.128$$

$$\mu_X = E(X) = x_1 p_1 + x_2 p_2 + x_3 p_3 + \cdots$$

If many, many newborns are randomly selected, their average Apgar score one minute after birth will be about 8.128.

FOR PRACTICE   TRY EXERCISE 5.

Notice that the mean Apgar score, 8.128, is not a possible value of the random variable X. It's also not an integer. These facts shouldn't bother you if you think of the mean (expected value) as a long-run average over many repetitions.

THINK ABOUT IT **How can we find the median of a discrete random variable?** In Lesson 1.6, we defined the median as "the midpoint of a distribution, the number such that about half the observations are smaller and about half are larger." The median of a discrete random variable is the 50th percentile of its probability distribution. We can find the median by adding a cumulative probability row to the probability distribution table, and then locating the smallest value for which the cumulative probability equals or exceeds 0.5. For the distribution of amount of money collected on Pete's jeep tour, we see that the median is $600.

Total collected	300	450	600	750	900
Probability	0.15	0.25	0.35	0.20	0.05
Cumulative probability	0.15	0.40	0.75	0.95	1.00

Measuring Variability: The Standard Deviation of a Discrete Random Variable

With the mean as our measure of center for a discrete random variable, it shouldn't surprise you that we'll use the standard deviation as our measure of variability. In Lesson 1.7, we defined the standard deviation s_x of a distribution of quantitative data as the typical distance of the values in the data set from the mean. To get the standard deviation, we started by "averaging" the squared deviations from the mean and then took the square root:

$$s_x = \sqrt{\frac{(x_1 - \overline{x})^2 + (x_2 - \overline{x})^2 + \cdots + (x_n - \overline{x})^2}{n - 1}}$$

We can modify this approach to calculate the **standard deviation of a discrete random variable** X. Start by finding a weighted average of the squared deviations $(x_i - \mu_X)^2$ of the values of the variable X from its mean μ_X. The probability distribution gives the appropriate weight for each squared deviation. (We call this weighted average the *variance* of X, denoted by σ_x^2.) Then take the square root to get the standard deviation σ_X.

DEFINITION   Standard deviation of a discrete random variable

The **standard deviation of a discrete random variable** measures how much the values of the variable typically differ from the mean.

Suppose that X is a discrete random variable with probability distribution

Value	x_1	x_2	x_3	...
Probability	p_1	p_2	p_3	...

and that μ_X is the mean of X. The standard deviation of X is

$$\sigma_X = \sqrt{(x_1 - \mu_X)^2\, p_1 + (x_2 - \mu_X)^2\, p_2 + (x_3 - \mu_X)^2\, p_3 + \cdots}$$
$$= \sqrt{\sum (x_i - \mu_X)^2\, p_i}$$

Let's return to the random variable $C =$ the total amount of money that Pete collects on a randomly selected jeep tour. The left two columns of the following table give the probability distribution. Recall that the mean of C is $\mu_C = 562.50$. The third column of the table shows the squared deviation of each value from the mean. The fourth column gives the weighted squared deviations.

Total collected	Probability	Squared deviation from the mean	Weighted squared deviation
300	0.15	$(300 - 562.50)^2$	$(300 - 562.50)^2\,(0.15)$
450	0.25	$(450 - 562.50)^2$	$(450 - 562.50)^2\,(0.25)$
600	0.35	$(600 - 562.50)^2$	$(600 - 562.50)^2\,(0.35)$
750	0.20	$(750 - 562.50)^2$	$(750 - 562.50)^2\,(0.20)$
900	0.05	$(900 - 562.50)^2$	$(900 - 562.50)^2\,(0.05)$
			Total = 26,718.75

You can see from the fourth column that the weighted average of the squared deviations is 26,718.75. The standard deviation of C is the square root of this weighted average:

$$\sigma_C = \sqrt{(300 - 562.50)^2 (0.15) + (450 - 562.50)^2 (0.25) + \cdots + (900 - 562.50)^2 (0.05)}$$

$$= \sqrt{26,718.75} = \$163.46$$

The amount of money that Pete collects on a randomly selected day typically differs from the mean of \$562.50 by about \$163.46.

EXAMPLE

How much do Apgar scores vary?

Finding and interpreting the standard deviation

PROBLEM: Earlier, we defined the random variable X to be the Apgar score of a randomly selected baby one minute after birth. The table gives the probability distribution for X once again. In the last example, we calculated the mean Apgar score of a randomly chosen newborn to be $\mu_x = 8.128$. Calculate and interpret the standard deviation of X.

Value	0	1	2	3	4	5	6	7	8	9	10
Probability	0.001	0.006	0.007	0.008	0.012	0.020	0.038	0.099	0.319	0.437	0.053

SOLUTION:

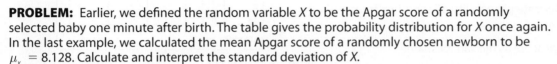

$$\sigma_X = \sqrt{(x_1 - \mu_X)^2 p_1 + (x_2 - \mu_X)^2 p_2 + \cdots}$$

$$\sigma_X = \sqrt{(0 - 8.128)^2 (0.001) + (1 - 8.128)^2 (0.006) + \cdots + (10 - 8.128)^2 (0.053)}$$

$$= \sqrt{2.066} = 1.437$$

A randomly selected baby's Apgar score one minute after birth will typically differ from the mean (8.128) by about 1.4 units.

FOR PRACTICE   TRY EXERCISE 9.

LESSON APP 5.2

How much do college grades vary?

Indiana University Bloomington posts the grade distributions for its courses online.[7] In a recent semester of a Business Statistics course, 45.7% of students received A's, 36.2% B's, 13.8% C's, 3.2% D's, and 1.1% F's. Choose a statistics student at random. The student's grade on a 4-point scale (with A = 4) is a random variable X with this probability distribution:

Grade	0	1	2	3	4
Probability	0.011	0.032	0.138	0.362	0.457

1. Make a histogram of the probability distribution. Describe its shape.

2. Calculate and interpret the mean of X.

3. Calculate and interpret the standard deviation of X.

TECH CORNER

Analyzing Discrete Random Variables with Technology

You can use an applet or a graphing calculator to display the probability distribution of a discrete random variable and to calculate its mean and standard deviation. For the random variable $X =$ Apgar score of a randomly selected newborn one minute after birth (see page 333), use either of the following.

Applet

1. Go to highschool.bfwpub.com/spa3e and launch the *Probability* applet.

2. Choose "Discrete probability distribution" from the drop-down menu.

3. Enter Apgar score as the variable name.

4. Type in the values of the variable and the corresponding probabilities. Click on the "+" sign in the bottom right corner of the table to add rows.

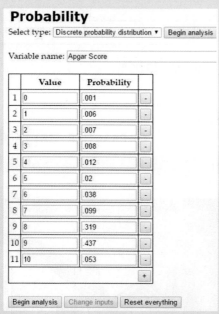

5. Click "Begin analysis." A histogram of the probability distribution should appear, along with the mean and standard deviation.

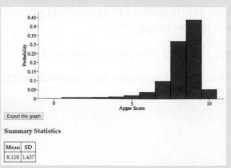

TI-83/84

1. Enter the values of the random variable in list L_1 and the corresponding probabilities in list L_2.

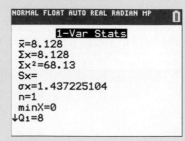

2. To graph a histogram of the probability distribution:

● Set up a statistics plot to be a histogram with Xlist: L1 and Freq: L2.

● Adjust your window settings as follows: Xmin $= -1$, Xmax $= 11$, Xscl $= 1$, Ymin $= -0.1$, Ymax $= 0.5$, Yscl $= 0.1$.

● Press GRAPH.

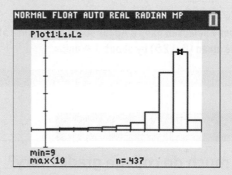

3. To calculate the mean and standard deviation of the random variable, use one-variable statistics with the values in L1 and the probabilities (relative frequencies) in L2. Press STAT ▶ (CALC) and choose 1-Var Stats.

OS 2.55 or later: In the dialog box, specify List: L1 and FreqList: L2. Then choose Calculate. **Older OS:** Execute the command 1-Var Stats L1,L2.

Lesson 5.2

WHAT DID YOU LEARN?

LEARNING TARGET	EXAMPLES	EXERCISES
Make a histogram to display the probability distribution of a discrete random variable and describe its shape.	p. 339	1–4
Calculate and interpret the mean (expected value) of a discrete random variable.	p. 341	5–8
Calculate and interpret the standard deviation of a discrete random variable.	p. 343	9–12

Exercises   Lesson 5.2

Mastering Concepts and Skills

1. **College costs** El Dorado Community College considers a student to be full-time if he or she is taking between 12 and 18 units. The tuition charge for a student is $50 per unit. Suppose we choose a full-time El Dorado Community College student at random. The probability distribution for the random variable T = tuition charge for the chosen student is shown here. Make a histogram of the probability distribution. Describe its shape.

pg 339

Tuition charge	$600	$650	$700	$750	$800	$850	$900
Probability	0.25	0.10	0.05	0.30	0.10	0.05	0.15

2. **Dedicated to Skee Ball** Ana is a dedicated Skee Ball player who always rolls for the 50-point slot. The probability distribution of Ana's score X on a randomly selected roll of the ball is shown here. Make a histogram of the probability distribution. Describe its shape.

Score	10	20	30	40	50
Probability	0.32	0.27	0.19	0.15	0.07

© Peter Carroll/Alamy Stock Photo

3. **Benford's law distribution** Faked numbers in tax returns, invoices, or expense account claims often display patterns that aren't present in legitimate records. Some patterns, like too many round numbers, are obvious and easily avoided by a clever crook. Others are more subtle. It is a striking fact that the first digits of numbers in legitimate records often follow a model known as Benford's law.[8] Call the first digit of a randomly chosen record X for short. The probability distribution for X is shown here (note that a first digit can't be 0). Make a histogram of the probability distribution. Describe its shape.

First digit	1	2	3	4	5
Probability	0.301	0.176	0.125	0.097	0.079
First digit	6	7	8	9	
Probability	0.067	0.058	0.051	0.046	

4. **Get on the boat!** A small ferry runs every half hour from one side of a large river to the other. The probability distribution for the random variable Y = money collected on a randomly selected ferry trip is shown here. Make a histogram of the probability distribution. Describe its shape.

Money collected	0	5	10	15	20	25
Probability	0.02	0.05	0.08	0.16	0.27	0.42

5. **Mean tuition** Refer to Exercise 1. Calculate and interpret the mean of T.

pg 341

6. **Average at Skee Ball** Refer to Exercise 2. Calculate and interpret the mean of X.

7. **Benford's law and expected value** Refer to Exercise 3. Compute and interpret the expected value of X.

8. **Expecting to get on the boat!** Refer to Exercise 4. Compute and interpret the expected value of Y.

9. **College costs vary** Refer to Exercises 1 and 5.
pg **343** Calculate and interpret the standard deviation of T.

10. **Skee Ball variation** Refer to Exercises 2 and 6. Calculate and interpret the standard deviation of X.

11. **First digits vary: Benford's law** Refer to Exercises 3 and 7. Calculate and interpret the standard deviation of X.

12. **Ferry income varies** Refer to Exercises 4 and 8. Calculate and interpret the standard deviation of Y.

Applying the Concepts

13. **Size of U.S. households** In government data, a household consists of all occupants of a dwelling unit, while a family consists of 2 or more persons who live together and are related by blood or marriage. So all families form households, but some households are not families. Here are the distributions of household size and family size in the United States.

	Number of people						
	1	2	3	4	5	6	7
Household probability	0.25	0.32	0.17	0.15	0.07	0.03	0.01
Family probability	0	0.42	0.23	0.21	0.09	0.03	0.02

Let H = the number of people in a randomly selected U.S. household and F = the number of people in a randomly chosen U.S. family.

(a) Make histograms suitable for comparing the probability distributions of H and F. Describe any differences that you observe.

(b) Find the expected value of each random variable. Explain why this difference makes sense.

(c) The standard deviations of the 2 random variables are $\sigma_H = 1.421$ and $\sigma_F = 1.249$. Explain why this difference makes sense.

14. **Owning or renting** How do rented housing units differ from units occupied by their owners? Here are the distributions of the number of rooms for owner-occupied units and renter-occupied units in San Jose, California.[9]

	Number of rooms				
	1	2	3	4	5
Owned probability	0.003	0.002	0.023	0.104	0.210
Rented probability	0.008	0.027	0.287	0.363	0.164
	6	7	8	9	10
Owned probability	0.224	0.197	0.149	0.053	0.035
Rented probability	0.093	0.039	0.013	0.003	0.003

Let X = the number of rooms in a randomly selected owner-occupied unit and Y = the number of rooms in a randomly chosen renter-occupied unit.

(a) Make histograms suitable for comparing the probability distributions of X and Y. Describe any differences that you observe.

(b) Find the expected number of rooms for both types of housing unit. Explain why this difference makes sense.

(c) The standard deviations of the 2 random variables are $\sigma_X = 1.640$ and $\sigma_Y = 1.308$. Explain why this difference makes sense.

15. **How life insurance works** A life insurance company sells a term insurance policy to 21-year-old males that pays $100,000 if the insured dies within the next 5 years. The probability that a randomly chosen male will die each year can be found in mortality tables. The company collects a premium of $250 each year as payment for the insurance. The amount Y that the company earns on a randomly selected policy of this type is $250 per year, less the $100,000 that it must pay if the insured dies. Here is the probability distribution of Y.

Age at death	21	22	23
Profit	−$99,750	−$99,500	−$99,250
Probability	0.00183	0.00186	0.00189
Age at death	24	25	26 or more
Profit	−$99,000	−$98,750	$1250
Probability	0.00191	0.00193	0.99058

(a) Explain why the company suffers a loss of $98,750 on such a policy if a customer dies at age 25.

(b) Compute the expected value of Y. Explain what this result means for the insurance company.

(c) Compute the standard deviation of Y. Explain what this result means for the insurance company.

16. **Insuring against fire** Suppose a homeowner spends $300 for a home insurance policy that will pay out $200,000 if the home is destroyed by fire. Let P = the profit made by the company on a single policy. From previous data, the probability that a home in this area will be destroyed by fire is 0.0002.

(a) Make a table that shows the probability distribution of P.

(b) Compute the expected value of P. Explain what this result means for the insurance company.

(c) Compute the standard deviation of P. Explain what this result means for the insurance company.

Extending the Concepts

17. **Using Benford's law against fraud** A not-so-clever employee decided to fake his monthly expense report. He believed that the first digits of his expense amounts should be equally likely to be any of the numbers from 1 to 9. In that case, the first digit Y of a randomly selected expense amount would have the probability distribution shown in the histogram.

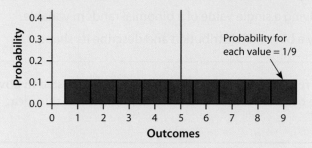

Probability for each value = 1/9

(a) Explain why the mean of the random variable Y is located at the solid red line in the figure.

(b) The first digits of randomly selected expense amounts follow Benford's law (Exercise 3). According to Benford's law, what's the expected value of the first digit? Explain how this information could be used to detect a fake expense report.

(c) Calculate the standard deviation σ_Y. This gives us an idea of how much variation we'd expect in the employee's expense records if he assumed that first digits from 1 to 9 were equally likely.

(d) In Exercise 11, you calculated the standard deviation of first digits that follow Benford's law to be $\sigma_X = 2.46$. Would using standard deviations be a good way to detect fraud? Explain.

Recycle and Review

18. **Win with Warren (2.8)** If you had bought one share of Warren Buffett's company Berkshire Hathaway at the market price of $21 in 1967, it would have been worth $177,900 in 2012. Here is the December stock price for Berkshire Hathaway in five-year intervals from 1967.

Year since 1967	Stock price	Year since 1967	Stock price
0	21	25	11,750
5	80	30	46,000
10	138	35	72,750
15	775	40	141,600
20	2950	45	177,900

(a) Calculate an exponential model for these data using years since 1967 as the explanatory variable.

(b) Sketch the scatterplot, along with the exponential model.

(c) Calculate and interpret the residual for the year 2012.

19. **Late for work (4.5)** Some days, Ramon drives to work. The rest of the time he rides his bike. Suppose we choose a workday at random. The table gives the probabilities of several events involving Ramon.

Event	Probability
Drives to work	0.20
Drives to work and is late	0.05
Late for work, given that he bikes	0.30

(a) Find the probability that Ramon is late for work, given that he drives.

(b) Draw a tree diagram to summarize the probabilities.

(c) On a randomly selected workday, Ramon was late for work. What is the probability that he biked?

Lesson 5.3

Binomial Random Variables

LEARNING TARGETS

- Determine whether or not a given scenario is a binomial setting.
- Calculate probabilities involving a single value of a binomial random variable.
- Make a histogram to display a binomial distribution and describe its shape.

When the same chance process is repeated several times, we are often interested in how many times a particular outcome occurs. Here's an activity that illustrates this idea.

ACTIVITY

Pop quiz!

It's time for a pop quiz! We hope you are ready. The quiz consists of 10 multiple-choice questions. Each question has five answer choices, labeled A through E. Now for the bad news: You will not get to see the questions. You just have to guess the answer for each one!

1. Get out a blank sheet of paper. Write your name at the top. Number your paper from 1 to 10. Then guess the answer to each question: A, B, C, D, or E. Do not look at anyone else's paper! You have 2 minutes.

2. Now it's time to grade the quizzes. Exchange papers with a classmate. Your teacher will display the answer key. The correct answer for each of the 10 questions was determined randomly so that A, B, C, D, or E was equally likely to be chosen.

3. How did you do on your quiz? Make a class dot-plot that shows the number of correct answers for each student in your class. As a class, describe what you see.

Binomial Settings

In the Pop Quiz activity, students are performing repeated *trials* of the same chance process: guessing the answer to a multiple-choice question. We're interested in the number of times that a specific event occurs: getting a correct answer (which we'll call a "success"). Knowing the outcome of one question (right or wrong guess) tells us nothing about the outcome of any other question. That is, the trials are independent. The number of trials is fixed in advance: $n = 10$. And a student's probability of getting a "success" is the same on each trial: $p = 0.2$. When these conditions are met, we have a **binomial setting.**

||| **DEFINITION** Binomial setting

A **binomial setting** arises when we perform n independent trials of the same chance process and count the number of times that a particular outcome (called a "success") occurs.

The four conditions for a binomial setting are

- **B**inary? The possible outcomes of each trial can be classified as "success" or "failure."
- **I**ndependent? Trials must be independent. That is, knowing the result of one trial must not tell us anything about the result of any other trial.
- **N**umber? The number of trials n of the chance process must be fixed in advance.
- **S**ame probability? There is the same probability of success p on each trial.

The boldface letters in the box give you a helpful way to remember the conditions for a binomial setting: just check the BINS!

When checking the Binary condition, note that there can be more than two possible outcomes per trial—in the Pop Quiz activity, each question (trial) had five possible answer choices: A, B, C, D, or E. If we define "success" as guessing the correct answer to a question, then "failure" occurs when the student guesses any of the four incorrect answer choices.

EXAMPLE

Is blood type like a card game?

Identifying binomial settings

PROBLEM: Determine whether or not the given scenario describes a binomial setting. Justify your answer.

(a) Genetics says that the genes children receive from their parents are independent from one child to another. Each child of a particular set of parents has probability 0.25 of having type O blood. Suppose these parents have 5 children. Count the number of children with type O blood.

(b) Shuffle a standard deck of 52 playing cards. Turn over the first 10 cards, one at a time. Record the number of aces you observe.

SOLUTION:

(a) • Binary? "Success" = has type O blood.

"Failure" = doesn't have type O blood.

> Check the BINS! Note that each trial consists of observing the blood type for one of these parents' children.

• Independent? Knowing one child's blood type tells you nothing about another child's because each of them inherits genes independently from their parents.

• Number? $n = 5$

• Same probability? $p = 0.25$

This is a binomial setting.

> All the conditions are met, and we are counting the number of successes (children with type O blood).

(b) • Binary? "Success" = get an ace. "Failure" = don't get an ace.

> Check the BINS! A trial consists of turning over a card from the deck.

• Independent? No. If the first card you turn over is an ace, the next card is less likely to be an ace because you're not replacing the top card in the deck. If the first card isn't an ace, the second card is more likely to be an ace. This is not a binomial setting because the independent condition is not met.

> To check for independence, you could also write P(2nd card ace | 1st card ace) = 3/51 and P(2nd card ace | 1st card not ace) = 4/51. Because the two probabilities are not equal, the trials are not independent.

FOR PRACTICE   TRY EXERCISE 1.

The blood type scenario in part (a) of the example is a binomial setting. If we let X = the number of children with type O blood, then X is a **binomial random variable.**

DEFINITION Binomial random variable

The count of successes X in a binomial setting is a **binomial random variable.** The possible values of X are 0, 1, 2, . . . , n.

In the Pop Quiz activity at the beginning of the lesson, the binomial random variable is X = the number of correct answers.

Calculating Binomial Probabilities

How can we calculate probabilities involving binomial random variables? Let's return to the scenario from part (a) of the previous example:

> Genetics says that the genes children receive from their parents are independent from one child to another. Each child of a particular set of parents has probability 0.25 of having type O blood. Suppose these parents have 5 children. Count the number of children with type O blood.

In this binomial setting, a child with type O blood is a "success" (S) and a child with another blood type is a "failure" (F). The count X of children with type O blood is a binomial random variable with $n = 5$ trials and probability $p = 0.25$ of success on each trial.

• What's $P(X = 0)$? That is, what's the probability that *none* of the 5 children has type O blood? It's the chance that all 5 children *don't* have type O blood. The probability that any one of this couple's children doesn't have type O blood is $1 - 0.25 = 0.75$ (complement rule). By the multiplication rule for independent events (Lesson 4.6),

$$P(X = 0) = P(\text{FFFFF}) = (0.75)(0.75)(0.75)(0.75)(0.75) = (0.75)^5 = 0.2373$$

• How about $P(X = 1)$? There are several different ways in which exactly 1 of the 5 children could have type O blood. For instance, the first child born might have type O blood, while the remaining 4 children don't have type O blood. The probability that this happens is

$$P(\text{SFFFF}) = (0.25)(0.75)(0.75)(0.75)(0.75) = (0.25)^1(0.75)^4$$

Or Child 2 could be the one that has type O blood. The corresponding probability is

$$P(\text{FSFFF}) = (0.75)(0.25)(0.75)(0.75)(0.75) = (0.25)^1(0.75)^4$$

There are three more possibilities to consider—those in which Child 3, Child 4, and Child 5 are the only ones to inherit type O blood. Of course, the probability will be the same for each of those cases. In all, there are five different ways in which exactly 1 child would have type O blood, each with the same probability of occurring. As a result,

$$P(X = 1) = P(\text{exactly 1 child with type O blood})$$
$$= 5(0.25)^1(0.75)^4 = 0.3955$$

Where did the 5 come from in the above formula? It's the number of ways to choose which one of the 5 children inherits type O blood. In Lesson 4.8, we wrote this as $_5C_1$, the number of combinations of 5 individuals taken 1 at a time. So we can rewrite the formula as

$$P(X = 1) = {_5C_1}(0.25)^1(0.75)^4 = 0.3955$$

The pattern of this calculation works for any binomial probability.

> ## Binomial Probability Formula
>
> Suppose that X is a binomial random variable with n trials and probability p of success on each trial. The probability of getting exactly k successes in n trials ($k = 0, 1, 2, ..., n$) is
>
> $$P(X = k) = {_nC_k} \cdot p^k(1 - p)^{n-k}$$
>
> where
>
> $$_nC_k = \frac{n!}{k!(n - k)!}$$

The binomial probability formula looks complicated, but it is fairly straightforward if you know what each part of the formula represents:

$$P(X = k) = (\text{\# of ways to get } k \text{ successes in } n \text{ trials}) \cdot (\text{success probability})^k (\text{failure probability})^{n-k}$$

EXAMPLE

How do we inherit blood type?

Using the binomial probability formula

PROBLEM: Genetics says that the genes children receive from their parents are independent from one child to another. Each child of a particular set of parents has probability 0.25 of having type O blood. Suppose these parents have 5 children. Find the probability that exactly 3 of the children have type O blood.

SOLUTION:

Let $X =$ the number of children with type O blood. X is a binomial random variable with $n = 5$ and $p = 0.25$.

$$P(X = 3) = {}_5C_3 \cdot (0.25)^3 \cdot (1 - 0.25)^{5-3}$$
$$= 10(0.25)^3(0.75)^2$$
$$= 0.0879$$

Use the binomial probability formula to find $P(X = 3)$:
$$P(X = k) = {}_nC_K \cdot p^k(1 - p)^{n-k}$$

Use one of the methods presented in Lesson 4.8 to find ${}_5C_3$:
$${}_5C_3 = \frac{{}_5P_3}{3!} = \frac{5 \cdot 4 \cdot 3}{3 \cdot 2 \cdot 1} = 10$$

FOR PRACTICE   TRY EXERCISE 5.

Binomial Distributions and Shape

What does the probability distribution of a binomial random variable look like? The table shows the possible values and corresponding probabilities for $X =$ the number of children with type O blood from the previous example. The probability distribution of X is called a **binomial distribution.**

Value	0	1	2	3	4	5
Probability	0.23730	0.39551	0.26367	0.08789	0.01465	0.00098

DEFINITION   Binomial distribution

The probability distribution of a binomial random variable is a **binomial distribution.** Any binomial distribution is completely specified by two numbers: the number of trials n of the chance process and the probability p of success on any trial.

A graph of the probability distribution of $X =$ the number of children with type O blood is shown in Figure 5.2 (on the next page). This binomial distribution with $n = 5$ and $p = 0.25$ has a clear right-skewed shape.

FIGURE 5.2 Histogram showing the probability distribution of the binomial random variable *X* = number of children with type O blood in a family with 5 children.

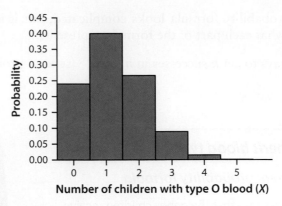

You can use the *Probability* applet at highschool.bfwpub.com/spa3e or a TI-83/84 to graph the probability distribution of a binomial random variable. See the Tech Corner at the end of the lesson for details.

EXAMPLE

Getting a head with a fair coin?

Graphing a binomial distribution

PROBLEM: Imagine that you toss a fair coin 10 times. Let *Y* = the number of heads you get. Make a histogram of the probability distribution of *Y*. Describe its shape.

SOLUTION:

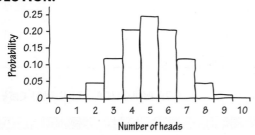

Use the *Probability* applet to make the graph. Enter *n* = 10 and *p* = 0.5. Then click "Plot distribution."

The graph is symmetric with a single peak at *Y* = 5.

FOR PRACTICE **TRY EXERCISE 9.**

LESSON APP 5.3

Is the train binomial?

According to New Jersey Transit, the 8:00 A.M. weekday train from Princeton to New York City has a 90% chance of arriving on time on a randomly selected day. Suppose this claim is true. Choose 6 days at random. Let *X* = the number of days on which the train arrives on time.

1. Explain why this is a binomial setting.

2. Calculate and interpret *P*(*X* = 4).

3. Make a histogram of the probability distribution of *X*. Describe its shape.

Binomial Distributions with Technology

You can use an applet or a TI-83/84 to graph a binomial distribution and to calculate binomial probabilities. Let's use technology to do both of these things for the binomial random variable $X =$ the number of children with type O blood from the example on page 351. Recall that we were trying to find the probability that exactly 3 of the family's 5 children have type O blood.

Applet

1. Launch the *Probability* applet at highschool.bfwpub.com/spa3e.

2. Choose "Binomial distribution" from the drop-down menu and click on "Begin analysis."

3. Enter $n = 5$ and $p = 0.25$. Then click on "Plot distribution" to see a graph of the appropriate binomial distribution.

4. Choose "exactly" from the drop-down menu, type "3," and click "Go!" to find the probability of *exactly* 3 successes. The desired probability is given and illustrated in the graph.

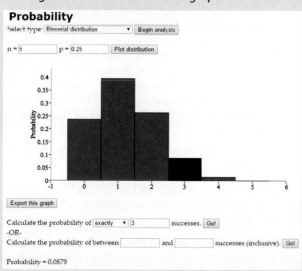

TI-83/84

The TI-84 command binompdf(n,p,k) computes $P(X = k)$ in a setting with n trials and success probability p on each trial.

1. To calculate $P(X = 3)$, press 2nd VARS (DISTR) and choose binompdf(.

OS 2.55 or later: In the dialog box, enter these values: trials:5, p:0.25, x value:3, choose Paste, and then press ENTER.

Older OS: Complete the command binompdf (5,0.25,3) and press ENTER.

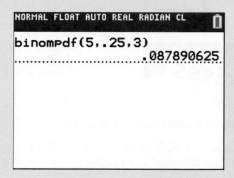

2. To graph the binomial probability distribution for $n = 5$ and $p = 0.25$:

● Type the possible values of the random variable X into list L_1: 0, 1, 2, 3, 4, and 5.

● Highlight L_2 with your cursor. Enter the command binompdf(5,0.25) and press ENTER.

● Make a histogram of the probability distribution using the method shown in Lesson 5.2's Tech Corner (page 344).

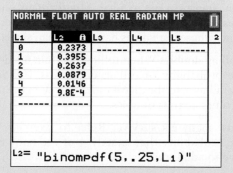

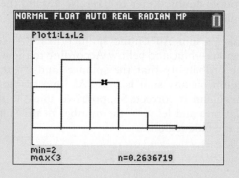

Lesson 5.3

WHAT DID YOU LEARN?

LEARNING TARGET	EXAMPLES	EXERCISES
Determine whether or not a given scenario is a binomial setting.	p. 349	1–4
Calculate probabilities involving a single value of a binomial random variable.	p. 351	5–8
Make a histogram to display a binomial distribution and describe its shape.	p. 352	9–12

Exercises | Lesson 5.3

Mastering Concepts and Skills

In Exercises 1–4, determine whether the given scenario describes a binomial setting. Justify your answer.

1. **Baby elk** Biologists estimate that a randomly
pg 349 selected baby elk has a 44% chance of surviving to adulthood. Assume this estimate is correct. Suppose researchers choose 7 baby elk at random to monitor. Let X = the number that survive to adulthood.

2. **Bullseye!** Lawrence likes to shoot a bow and arrow in his free time. On any shot, he has about a 10% chance of hitting the bullseye. As a challenge one day, Lawrence decides to keep shooting until he gets a bullseye. Let X = the number of shots he takes.

3. **Long or short?** Put the names of all the students in your statistics class in a hat. Mix them up, and draw 4 names without looking. Let Y = the number of students whose last names have more than six letters.

4. **Rolling doubles** When rolling two fair, 6-sided dice, the probability of rolling doubles is 1/6. Suppose that Elias rolls the dice 4 times. Let W = the number of times he rolls doubles.

5. **How many elk survive?** Refer to Exercise 1. Find
pg 351 the probability that exactly 4 of the baby elk survive to adulthood.

6. **Double double** Refer to Exercise 4. Find the probability that Elias rolls doubles twice.

7. **Take a spin** An online spinner has two colored regions—blue and yellow. According to the website, the probability that the spinner lands in the blue region on any spin is 0.80. Assume for now that this claim is correct. Suppose we spin the spinner 12 times and let X = the number of times it lands in the blue region.

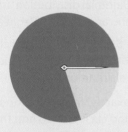

(a) Explain why X is a binomial random variable.

(b) Find $P(X = 8)$. Interpret this value in context.

8. **Red light!** Pedro drives the same route to work on Monday through Friday. His route includes one traffic light. According to the local traffic department, there is a 55% chance that the light will be red when he arrives at the intersection on a randomly selected workday. Suppose we choose 10 of Pedro's workdays at random and let Y = the number of times that the light is red.

(a) Explain why Y is a binomial random variable.

(b) Find $P(Y = 8)$. Interpret this value in context.

9. **Elk distribution** Refer to Exercise 1. Make a histo-
pg 352 gram of the probability distribution of X. Describe its shape.

10. **Double distribution** Refer to Exercise 4. Make a histogram of the probability distribution of W. Describe its shape.

11. **Blue distribution** Refer to Exercise 7. Make a histogram of the probability distribution of X. Describe its shape.

12. **Red light distribution** Refer to Exercise 8. Make a histogram of the probability distribution of Y. Describe its shape.

LESSON 5.3 • **Binomial Random Variables** **355**

Applying the Concepts

13. **Bag check** Thousands of travelers pass through the airport in Guadalajara, Mexico, each day. Before leaving the airport, each passenger must go through the customs inspection area. Customs agents want to be sure that passengers do not bring illegal items into the country, but they do not have time to search every traveler's luggage. Instead, they require each person to press a button. Either a red or a green bulb lights up. If a red light flashes, the passenger will be searched by customs agents. A green light means it is OK for the passenger to "go ahead." Customs agents claim that the light has probability 0.30 of showing red on any push of the button. Assume for now that this claim is true. Suppose we watch 20 passengers press the button. Let R = the number who get a red light.

(a) Explain why R is a binomial random variable.

(b) Make a histogram of the probability distribution of R. Describe its shape.

(c) Find the probability that exactly 6 of the 20 passengers get a red light.

14. **Cranky mower** A company has developed an "easy-start" mower that cranks the engine with the push of a button. The company claims that the probability the mower will start on any push of the button is 0.9. Assume for now that this claim is true. On the next 30 uses of the mower, let Y = the number of times it starts on the first push of the button.

(a) Explain why Y is a binomial random variable.

(b) Make a histogram of the probability distribution of Y. Describe its shape.

(c) Find the probability that the mower starts exactly 27 times in 30 uses.

Extending the Concepts

15. **Sampling without replacement** An airline has just finished training 25 junior pilots—15 male and 10 female—to become captains. Unfortunately, only 8 captain positions are available right now. Airline managers announce that they will use a lottery process to determine which pilots will fill the available positions. The names of all 25 pilots will be written on identical slips of paper, placed in a hat, mixed thoroughly, and drawn out one at a time until all 8 captains have been identified.

A day later, managers announce the results of the lottery. Of the 8 captains chosen, 5 are female and only 3 are male. Explain why the probability that 5 female pilots are chosen in a fair lottery is *not* $P(X = 5) = {}_8C_5(0.40)^5(0.60)^3 = 0.124$.

16. **Exploring shape** In this exercise, you will use the *Probability* applet to investigate what happens to the shape of a binomial distribution with $p = 0.2$ as the sample size increases. For each of the following sample sizes, use the applet to make a graph of the probability distribution and describe its shape: $n = 4$, $n = 20$, $n = 100$.

Recycle and Review

17. **Gluten-free?** (3.6, 3.8) According to a recent article on the website www.fivethirtyeight.com, 30% of Americans say they're trying to reduce or eliminate gluten in their diets.[10] Yet only about 1 percent of the population has an autoimmune response to gluten.

(a) Many people wonder if a gluten-free diet offers benefits to more than just those with a diagnosed medical condition. The article points out that it can be difficult to establish this because of the placebo effect. Explain why.

(b) Suppose you have 50 volunteers with gastrointestinal problems that are often associated with "gluten sensitivity." There are dietary supplements available that contain gluten and similar ones that do not. Outline a completely randomized design for an experiment to test the impact of a gluten-free diet on people with these gastrointestinal problems.

18. **Super Bowl Nielsen ratings** (1.3, 1.4, 1.5) The Nielsen company collects data on the television-viewing habits of American households. One measurement of a show's popularity is the "share," which looks only at the households in which at least one television is turned on, and calculates the percentage of those televisions that are tuned to a particular show. The Nielsen share scores for every NFL Super Bowl from 1967 through 2015 follow.[11] Make an appropriate graph to display these data. Describe what you see.

71	68	62	63	62	69	64	63	61	66	69	65	61
67	66	71	64	61	65	69	62	61	68	61	63	63
68	62	66	70	63	71	69	73	63	67	74	67	73
78	72	73	72	74	75	69	70	68	79		

Lesson 5.4

Analyzing Binomial Random Variables

LEARNING TARGETS

- Calculate and interpret the mean and standard deviation of a binomial distribution.
- Find probabilities involving several values of a binomial random variable.
- Use technology to calculate cumulative binomial probabilities.

In Lesson 5.3, you learned how to check the conditions for a binomial setting and how to calculate binomial probabilities. You also used technology to graph the probability distribution of a binomial random variable. It is easy to describe the shape of a binomial distribution from its graph. In this lesson, you will learn a simple way to find the mean and standard deviation of a binomial random variable.

The Mean and Standard Deviation of a Binomial Distribution

In Lesson 5.3, we considered this scenario: According to the science of genetics, the genes children receive from their parents are independent from one child to another. Each child of a particular set of parents has probability 0.25 of having type O blood. Suppose these parents have 5 children. Let $X =$ the number of children with type O blood.

We determined that X is a binomial random variable with $n = 5$ and $p = 0.25$. Its probability distribution is shown in the table and the histogram.

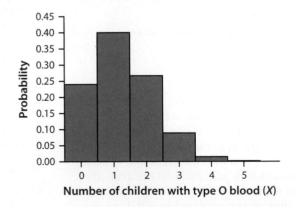

Value	0	1	2	3	4	5
Probability	0.23730	0.39551	0.26367	0.08789	0.01465	0.00098

Because X is a discrete random variable, we can calculate its mean using the formula

$$\mu_X = E(X) = x_1p_1 + x_2p_2 + x_3p_3 + \cdots$$

from Lesson 5.2. We get

$$\mu_X = (0)(0.23730) + (1)(0.39551) + (2)(0.26367) + (3)(0.08789) + (4)(0.01465) + (5)(0.00098) = 1.25$$

So the expected number of children with type O blood in families like this one with 5 children is 1.25.

Did you think about why the mean is $\mu_X = 1.25$? Because each child has a 0.25 chance of inheriting type O blood, we'd expect one-fourth of the 5 children to have this blood type. In other words,

$$\mu_X = 5(0.25) = 1.25$$

This method can be used to find the mean of any binomial random variable.

Calculating the Mean of a Binomial Random Variable

If a count X of successes has a binomial distribution with number of trials n and probability of success p, the mean of X is

$$\mu_X = np$$

To calculate the standard deviation of X, we use the formula

$$\sigma_X = \sqrt{(x_1 - \mu_X)^2 p_1 + (x_2 - \mu_X)^2 p_2 + (x_3 - \mu_X)^2 p_3 + \cdots}$$

from Lesson 5.2 with $\mu_X = 1.25$. We get

$$\sigma_X = \sqrt{(0 - 1.25)^2(0.23720) + (1 - 1.25)^2(0.39551) + \cdots + (5 - 1.25)^2(0.00098)}$$

$$= \sqrt{0.9375} \approx 0.968$$

The number of children with type O blood will typically differ from the mean of 1.25 by about 0.968 children in families like this one with 5 children.

There is a simple formula for the standard deviation of a binomial random variable, but it isn't easy to explain. For our family with $n = 5$ children and $p = 0.25$ of type O blood, the *variance* of X is

$$5(0.25)(0.75) = 0.9375$$

To get the standard deviation, we just take the square root:

$$\sigma_X = \sqrt{5(0.25)(0.75)} \approx 0.968$$

This method works for any binomial random variable.

Calculating the Standard Deviation of a Binomial Random Variable

If a count X of successes has a binomial distribution with number of trials n and probability of success p, the standard deviation of X is

$$\sigma_X = \sqrt{np(1 - p)}$$

Remember that these formulas for the mean and standard deviation work *only* for binomial distributions.

CAUTION

EXAMPLE

How many correct on the pop quiz?

Mean and SD of a binomial distribution

PROBLEM: To introduce her class to binomial distributions, Mrs. Desai does the Pop Quiz activity from Lesson 5.3 (page 348). Each student in the class guesses an answer from A through E on each of the 10 multiple-choice questions. Mrs. Desai determines the "correct" answer for each of the 10 questions randomly so that A, B, C, D, or E was equally likely to be chosen. Let X = the number of correct answers that a student gets on the quiz.

(a) Calculate and interpret the mean of X.

(b) Calculate and interpret the standard deviation of X.

SOLUTION:

The random variable X has a binomial distribution with $n = 10$ and $p = 1/5 = 0.2$.

(a) $\mu_X = 10(0.2) = 2$. If many students took the quiz, we'd expect students to get about 2 answers correct, on average.

> The mean of a binomial random variable is $\mu_X = np$.

(b) $\sigma_X = \sqrt{10(0.2)(0.8)} = 1.265$. If many students took the quiz, their scores would typically vary from the mean (2) by about 1.265 correct answers.

> The standard deviation of a binomial random variable is $\sigma_X = \sqrt{np(1 - p)}$.

FOR PRACTICE TRY EXERCISE 1.

Cumulative Binomial Probabilities

Let's return to the parents with 5 children one more time. Recall that X = the number of children with type O blood is a binomial random variable with $n = 5$ and $p = 0.25$. Its probability distribution is shown in the table.

Value	0	1	2	3	4	5
Probability	0.23730	0.39551	0.26367	0.08789	0.01465	0.00098

What's the probability that *at most 1* of the children has type O blood? In symbols, it's $P(X \le 1)$. We can compute this cumulative binomial probability using the fact that

$$P(X \le 1) = P(X = 0) + P(X = 1)$$
$$= 0.23730 + 0.39551$$
$$= 0.63281$$

What if we want to find the probability that *at least 2* of the couple's 5 children have type O blood? In symbols, that's $P(X \ge 2)$. We could compute this probability using the fact that

$$P(X \ge 2) = P(X = 2) + P(X = 3) + P(X = 4) + P(X = 5)$$
$$= 0.26367 + 0.08789 + 0.01465 + 0.00098$$
$$= 0.36719$$

This approach requires us to add four separate binomial probabilities.

A clever alternative is to use the complement rule:

$$P(X \geq 2) = 1 - P(X \leq 1)$$
$$= 1 - [P(X = 0) + P(X = 1)]$$
$$= 1 - [0.23730 + 0.39551]$$
$$= 1 - 0.63281$$
$$= 0.36719$$

Notice that this strategy only requires us to add two binomial probabilities!

Here's a helpful tip to avoid making mistakes on probability questions that involve an interval of values for a binomial random variable: Write out the possible values of the variable, circle the ones you want to find the probability of, and cross out the rest. In this scenario, X can take values from 0 to 5 and we want to find $P(X \geq 2)$:

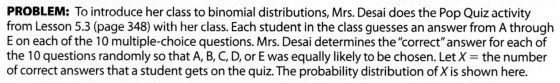

Crossing out the values for 0 and 1 shows why the correct calculation is $1 - P(X \leq 1)$.

EXAMPLE

Will Patti pass the pop quiz?

Finding cumulative binomial probabilities

PROBLEM: To introduce her class to binomial distributions, Mrs. Desai does the Pop Quiz activity from Lesson 5.3 (page 348) with her class. Each student in the class guesses an answer from A through E on each of the 10 multiple-choice questions. Mrs. Desai determines the "correct" answer for each of the 10 questions randomly so that A, B, C, D, or E was equally likely to be chosen. Let $X =$ the number of correct answers that a student gets on the quiz. The probability distribution of X is shown here.

Number correct	0	1	2	3	4	5
Probability	0.1074	0.2684	0.3020	0.2013	0.0881	0.0264
Number correct	6	7	8	9	10	
Probability	0.0055	0.00079	0.00007	0.000004	0.0000001	

Patti is one of the students in this class. Find the probability that Patti gets more than 2 correct answers on the pop quiz.

SOLUTION:

$$P(X > 2) = 1 - P(X \leq 2)$$
$$= 1 - [P(X = 0) + P(X = 1) + P(X = 2)]$$
$$= 1 - [0.1074 + 0.2684 + 0.3020]$$
$$= 1 - 0.6778$$
$$= 0.3222$$

You want to find $P(X > 2)$.

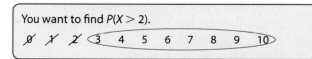

FOR PRACTICE   TRY EXERCISE 5.

What if the probability distribution of X had not been provided in the example? You could use the binomial probability formula from Lesson 5.3:

$$P(X = k) = {}_nC_k \cdot p^k(1 - p)^{n-k}$$

to find the probabilities for $X = 0$, $X = 1$, and $X = 2$. Then you would add together those probabilities and subtract the result from 1. In symbols,

$$P(X > 2) = 1 - P(X \le 2)$$
$$= 1 - [{}_{10}C_0 \cdot (0.2)^0(0.8)^{10} + {}_{10}C_1 \cdot (0.2)^1(0.8)^9 + {}_{10}C_2 \cdot (0.2)^2(0.8)^8]$$
$$= 1 - 0.6778$$
$$= 0.3222$$

USING TECHNOLOGY

It is tedious to use the binomial probability formula for calculations involving several values of a binomial random variable. Technology is a much more practical alternative.

Cumulative Binomial Probabilities with Technology

You can use the *Probability* applet or a TI-83/84 to calculate binomial probabilities. Let's use technology to confirm our answer to the previous example. Recall that we were trying to find the probability that Patti gets more than 2 correct answers on Mrs. Desai's pop quiz.

Applet

1. Launch the *Probability* applet at highschool. bfwpub.com/spa3e.

2. Choose "Binomial distribution" from the drop-down menu and click on "Begin analysis."

3. Enter $n = 10$ and $p = 0.2$. Then click on "Plot distribution." A graph of the appropriate binomial distribution is shown.

4. To find $P(X > 2)$, choose "more than" from the drop-down menu, type 2 in the adjacent box, and click "Go!" The desired probability is illustrated in the graph and displayed below the graph.

TI-83/84

The TI-83/84 command binomcdf(n,p,k) computes $P(X \le k)$ in a setting with n trials and success probability p on each trial.

● Press [2nd] [VARS] (DISTR) and choose binomcdf(.

OS 2.55 or later: In the dialog box, enter these values: trials:10, p:0.2, x value:2, choose Paste, and then press [ENTER]. Subtract this result from 1 to get the answer.

Older OS: Complete the command binomcdf(10,0.2,2) and press [ENTER]. Subtract this result from 1 to get the answer.

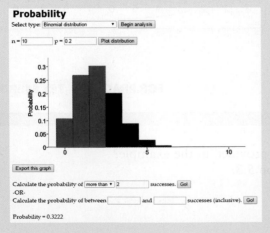

EXAMPLE

Can you tell bottled water from tap water?

Cumulative binomial probabilities with technology

PROBLEM: Mr. Bullard's statistics class did an activity to see if students could tell bottled water from tap water. There were 21 students in the class. Each student was given 3 cups—2 that contained tap water and 1 that contained bottled water—in a random order. The student was asked to drink the cups of water in the specified order and then to identify which cup contained bottled water. If we assume that the students in Mr. Bullard's class could not tell tap water from bottled water, then each one was guessing, with a 1/3 chance of being correct. Let X = the number of students who correctly identify the cup containing bottled water when guessing.

Of the 21 students in the class, 13 made correct identifications. Are you convinced that Mr. Bullard's students could tell bottled water from tap water? Compute $P(X \geq 13)$ with technology and use this result to support your answer.

SOLUTION:

X is a binomial random variable with $n = 21$ and $p = 1/3$.

$$P(X \geq 13) = 1 - P(X \leq 12)$$
$$= 1 - 0.9932$$
$$= 0.0068$$

> Assuming that students were just guessing, the activity consisted of 21 trials of a chance process with success probability 1/3 on each trial. Use the TI-83/84 binomcdf command $1 -$ binomcdf(21, 1/3, 12) or the *Probability* applet to do the calculation.

The students had less than a 1% chance of getting so many right if all of them were just guessing. Because this is unlikely to happen by chance alone, there is convincing evidence that some of the students in the class could tell bottled water from tap water.

FOR PRACTICE   TRY EXERCISE 9.

LESSON APP 5.4

Free lunch?

A local fast-food restaurant is running a "Draw a three, get it free" lunch promotion. After each customer orders, a touch-screen display shows the message, "Press here to win a free lunch." A computer program then simulates one card being drawn at random from a standard deck of playing cards. If the chosen card is a 3, the customer's order is free. (Note that the probability of drawing a 3 from a standard deck of playing cards is 4/52.) Otherwise, the customer must pay the bill. Suppose that 250 customers place lunch orders on the first day of the promotion. Let X = the number of people who win a free lunch.

1. Explain why X is a binomial random variable.

2. Find the mean of X. Interpret this value in context.

3. Find the standard deviation of X. Interpret this value in context.

4. One of the customers is surprised when only 10 people win a free lunch. Should the customer be surprised? Find the probability that 10 or fewer people win a free lunch by chance alone and use this result to support your answer.

Lesson 5.4

WHAT DID YOU LEARN?

LEARNING TARGET	EXAMPLES	EXERCISES
Calculate and interpret the mean and standard deviation of a binomial distribution.	p. 358	1–4
Find probabilities involving several values of a binomial random variable.	p. 359	5–8
Use technology to calculate cumulative binomial probabilities.	p. 361	9–12

Exercises   Lesson 5.4

Mastering Concepts and Skills

1. **Random digit dialing** When a polling company
pg 358 calls a telephone number at random, there is only a 9% chance that the call reaches a live person and the survey is successfully completed.[12] Suppose the random digit dialing machine makes 15 calls. Let X = the number of calls that result in a completed survey.

   (a) Calculate and interpret the mean of X.

   (b) Calculate and interpret the standard deviation of X.

2. **Detecting lies** A federal report finds that lie detector tests given to truthful persons have probability 0.2 of suggesting that the person is deceptive.[13] A company asks 12 job applicants about thefts from previous employers, using a lie detector to assess their truthfulness. Suppose that all 12 answer truthfully. Let X = the number of people whom the lie detector identifies as being deceptive.

   (a) Calculate and interpret the mean of X.

   (b) Calculate and interpret the standard deviation of X.

3. **Take a virtual spin** An online spinner has 2 colored regions—blue and yellow. According to the website, the probability that the spinner lands in the blue region on any spin is 0.80. Assume for now that this claim is correct. Suppose we spin the spinner 12 times and let X = the number of times it lands in the blue region.

   (a) Calculate and interpret μ_X.

   (b) Calculate and interpret σ_X.

4. **How many red lights?** Pedro drives the same route to work on Monday through Friday. His route includes one traffic light. According to the local traffic department, there is a 55% chance that the light will be red when he arrives at the intersection on a randomly selected workday. Suppose we choose 10 of Pedro's workdays at random and let X = the number of times that the light is red.

   (a) Calculate and interpret μ_X.

   (b) Calculate and interpret σ_X.

5. **Baby elk and their chances** Biologists estimate that
pg 359 a randomly selected baby elk has a 44% chance of surviving to adulthood. Assume this estimate is correct. Suppose researchers choose 7 baby elk at random to monitor. Let X = the number who survive to adulthood. The probability distribution of X is shown here. Find the probability that fewer than 3 of the elk survive to adulthood.

Value	0	1	2	3
Probability	0.0173	0.0950	0.2239	0.2932
Value	4	5	6	7
Probability	0.2304	0.1086	0.0284	0.0032

6. **Rolling doubles** When rolling two fair, 6-sided dice, the probability of rolling doubles is 1/6. Suppose Elias rolls the dice 4 times. Let W = the number of times he rolls doubles. The probability distribution of W is shown here. Find the probability that Elias rolls doubles more than twice.

Value	0	1	2	3	4
Probability	0.482	0.386	0.116	0.015	0.001

7. **Please try again** As a special promotion for its 20-ounce bottles of soda, a soft drink company printed a message on the inside of each cap. Some of the caps said, "Please try again," while others said, "You're a winner!" The company advertised the promotion with the slogan "1 in 6 wins a prize." Suppose the company is telling the truth and that every 20-ounce bottle of soda it fills has a 1-in-6 chance of being a winner. Seven friends each buy one 20-ounce bottle of the soda at a local convenience store. Let X = the number who win a prize. The probability distribution of X is shown here.

Winners	0	1	2	3
Probability	0.2791	0.3907	0.2344	0.0781
Winners	4	5	6	7
Probability	0.0156	0.0019	0.0001	0.000004

The store clerk is surprised when 3 of the friends win a prize. Is this group of friends just lucky, or is the company's 1-in-6 claim inaccurate? Find the probability that at least 3 of a group of 7 friends would win a prize and use the result to justify your answer.

8. **How many white cars?** About 20% of cars sold in North America are white. Let's assume that the color of any car on the road is independent of cars that come before or after it and that the proportion of white cars is the same throughout North America. The probability distribution of X = the number of white cars among 6 randomly selected cars is given here.

White cars	0	1	2	3
Probability	0.2621	0.3932	0.2458	0.0819
White cars	4	5	6	
Probability	0.0154	0.0015	0.0001	

You are standing on a bridge over an interstate highway, and 4 of the next 6 cars that pass by are white. Does this suggest that the proportion of white cars that use this particular highway is greater than 0.20? Find the probability that at least 4 cars in randomly selected groups of 6 cars are white and use this to justify your answer.

9. **The last Kiss** Do people have a preference for
pg 361 the last thing they taste? Researchers at the University of Michigan designed a study to find out. The researchers gave 22 students five different Hershey's Kisses (milk chocolate, dark chocolate, crème, caramel, and almond) in random order and asked the student to rate each candy.

Participants were not told how many Kisses they would taste. However, when the 5th and final Kiss was presented, participants were told that it would be their last one.[14] Assume that the participants in the study don't have a special preference for the last thing they taste, that is, the probability of preferring the last Kiss tasted is $p = 0.20$. Let X = the number of participants who choose the last Kiss.

Of the 22 students, 14 gave the final Kiss the highest rating. Does this give convincing evidence that the participants have a preference for the last thing they taste? Compute $P(X \geq 14)$ with technology and use this result to support your answer.

10. **Tastes as good as the real thing?** The makers of a diet cola claim that its taste is indistinguishable from the taste of the full-calorie version of the same cola. To investigate, a statistics student named Emily prepared small samples of each type of soda in identical cups. Then she had volunteers taste each cola in random order and try to identify which was the diet cola and which was the regular cola. If we assume that the volunteers couldn't tell the difference, each one was guessing, with a 1/2 chance of being correct. Let X = the number of volunteers who correctly identify the colas.

Of the 30 volunteers, 23 made correct identifications. Does this give convincing evidence that the volunteers can taste the difference between the diet and regular colas? Compute $P(X \geq 23)$ with technology and use this result to support your answer.

11. **Spin again** Refer to Exercise 3. If only 7 of the 12 spins land in the blue region, do we have convincing evidence that the website's claim is false? Compute $P(X \leq 7)$ with technology and use this result to support your answer.

12. **Red light!** Refer to Exercise 4. If the light is red on 7 of the 10 days, do we have convincing evidence that the traffic department's claim is false? Compute $P(X \geq 7)$ with technology and use this result to support your answer.

Applying the Concepts

13. **Bag check** Thousands of travelers pass through the airport in Guadalajara, Mexico, each day. Mexican customs agents want to be sure that passengers do not bring in illegal items, but do not have time to search every traveler's luggage. Instead, customs requires each person to press a button. Either a red or a green bulb lights up. If red, the passenger will be searched by customs agents. Green means "go ahead." Customs agents

claim that the light has probability 0.30 of showing red on any push of the button. Assume for now that this claim is true. Suppose we watch 20 passengers press the button. Let R = the number who get a red light. Exercise 13 in Lesson 5.3 (page 355) asked you to show that R is a binomial random variable.

(a) Find the mean and standard deviation of R.

(b) Suppose that only 3 of the 20 passengers get a red light after pressing the button. Does this give convincing evidence that the customs agents' claimed value of $p = 0.3$ is too high? Find the probability that at most 3 people out of 20 would get a red light if the agents' claim is true, and use this result to support your answer.

14. **Still cranky?** A company has developed an "easy-start" mower that cranks the engine with the push of a button. The company claims that the probability the mower will start on any push of the button is 0.9. Assume for now that this claim is true. On the next 30 uses of the mower, let Y = the number of times it starts on the first push of the button. Exercise 14 in Lesson 5.3 (page 355) asked you to show that Y is a binomial random variable.

(a) Find the mean and standard deviation of Y.

(b) Suppose that the mower only starts on 23 of the 30 attempts. Does this give convincing evidence that the company's claim is exaggerated? Find the probability of at most 23 starts in 30 attempts if the company's claim is true, and use this result to support your answer.

Extending the Concepts

15. **No answer** Refer to Exercise 1. Let Y = the number of calls that *don't* result in a completed survey.

(a) Find the mean of Y. How is it related to the mean of X? Explain why this makes sense.

(b) Find the standard deviation of Y. How is it related to the standard deviation of X? Explain why this makes sense.

16. **Truth or lie?** Refer to Exercise 2. Let Y = the number of people whom the lie detector identifies as *telling the truth*. Use technology to find $P(Y \geq 10)$ and $P(X \leq 2)$. Explain the relationship between these two values.

Recycle and Review

17. **Which 'wich?** (4.7) Sam's Sandwich Shop lets you design your own sandwich. There are 5 choices for bread, 6 choices for meat, and 4 choices for cheese. You can also choose to include (or not include) each of the following items by request: lettuce, tomato, hot peppers, mayonnaise, or mustard.

(a) A standard sandwich has 1 type of meat and 1 type of cheese (along with your requested items). How many different standard sandwiches can be created?

(b) A "doubles" sandwich contains 2 different kinds of meat, 2 different kinds of cheese (plus your requested items). How many different "doubles" sandwiches can be created?

18. **Statistics for investing** (2.3) Joe's retirement plan invests in stocks through an "index fund" that follows the behavior of the stock market as a whole, as measured by the Standard & Poor's (S&P) 500 stock index. Joe wants to buy a mutual fund that does not track the index closely. He reads that monthly returns from Fidelity Technology Fund have correlation $r = 0.77$ with the S&P 500 index and that Fidelity Real Estate Fund has correlation $r = 0.37$ with the index.

(a) Which of these funds has the closer relationship to returns from the stock market as a whole? How do you know?

(b) Does the information given tell Joe anything about which fund has had higher returns?

Lesson 5.5

Continuous Random Variables

Figure 5.3 shows a relative frequency histogram of the scores of all seventh-grade students in Gary, Indiana, on the vocabulary part of the Iowa Test of Basic Skills (ITBS).[15] The scores are grade-level equivalents. So a score of 6.3 indicates that the student's performance is typical for a student in the third month of grade 6.

The histogram is roughly symmetric, and both tails fall off smoothly from a single center peak. There are no large gaps or obvious outliers.

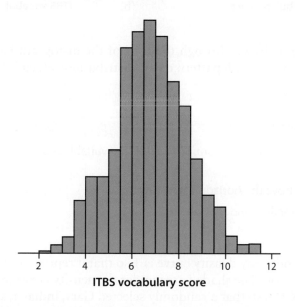

ITBS vocabulary score

FIGURE 5.3 Histogram of the Iowa Test of Basic Skills (ITBS) vocabulary scores of all seventh-grade students in Gary, Indiana.

Suppose that we choose a Gary seventh-grader at random. Let X = the student's ITBS vocabulary score. Although ITBS scores are usually reported to the nearest tenth, X can take any value in the interval from 1 (beginning of first grade) to 13 (high school graduate). From what you learned in Lesson 5.1, X would be classified as a continuous random variable.

This lesson examines probability distributions for continuous random variables.

Finding Probabilities for Continuous Random Variables

What's $P(X < 6.0)$? It's the probability that a randomly selected student earned a vocabulary score less than the sixth-grade equivalent on the ITBS test. The area of the shaded bars in the relative frequency histogram in part (a) (see the figure that follows) represents the proportion of students with vocabulary scores less than 6.0. From the raw data, we can determine that this proportion is $287/947 = 0.303$. So $P(X < 6.0) = 0.303$.

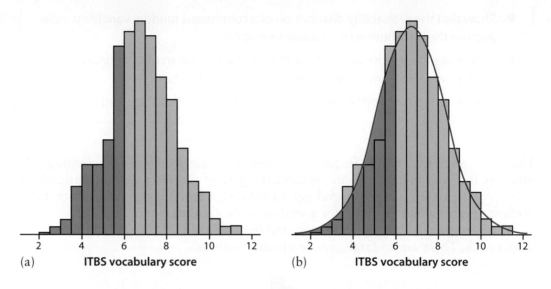

(a) **ITBS vocabulary score** (b) **ITBS vocabulary score**

The smooth curve drawn through the tops of the histogram bars in part (b) is a good description of the overall pattern of the distribution. We call this curve a **density curve.**

DEFINITION Density curve

The probability distribution of a continuous random variable is described by a **density curve,** a curve that

- is always on or above the horizontal axis, and
- has an area of exactly 1 underneath it.

The total area under any density curve is 1 so that it represents 100% of the observations in a distribution. The shaded area under the density curve in part (b) estimates the probability $P(X < 6.0)$ that a randomly selected Gary, Indiana, seventh-grader has an ITBS vocabulary score lower than 6.0. We used technology to find that this area is 0.293, only 0.010 away from the actual probability 0.303.

How to Find Probabilities for a Continuous Random Variable

The probability of any event involving a continuous random variable X is the area under the density curve and directly above the values of X on the horizontal axis that make up the event.

EXAMPLE

Uniform and random decimals?

Continuous random variables and density curves

PROBLEM: Suppose you use a calculator or computer random-number generator to produce a number *Y* between 0 and 1 (like 0.84522 or 0.1111119). The random-number generator will spread its output uniformly across the entire interval from 0 to 1 as we allow it to generate a long sequence of random numbers. The probability distribution of *Y* is the density curve of a *uniform distribution*, shown in purple here.

(a) Explain why the probability distribution of *Y* is valid.

(b) Find and interpret $P(0.3 \leq Y \leq 0.7)$.

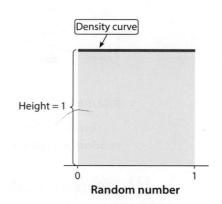

SOLUTION:

(a) The density curve is entirely above the horizontal axis and the area under the density curve is length × width = 1 × 1 = 1.

(b) The probability that the random number generator produces a number *Y* between 0.3 and 0.7 is $P(0.3 \leq Y \leq 0.7) = 0.4$.

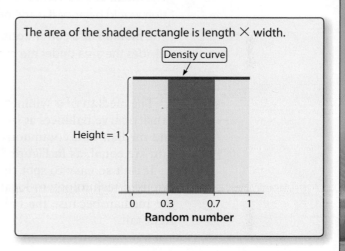

The area of the shaded rectangle is length × width.

FOR PRACTICE **TRY EXERCISE 1.**

THINK ABOUT IT What's the probability that a continuous random variable takes a specific value? Consider a specific outcome from the random number generator of the previous example, such as $P(Y = 0.7)$. The probability of this event is the area under the density curve that's directly above the point 0.70000 . . . on the horizontal axis. But this vertical line segment has no width, so the area is 0. In fact, all continuous probability models assign probability 0 to every individual outcome. For that reason,

$$P(0.3 \leq Y \leq 0.7) = P(0.3 \leq Y < 0.7) = P(0.3 < Y \leq 0.7) = P(0.3 < Y < 0.7) = 0.4$$

Remember: The probability distribution for a continuous random variable assigns probabilities to *intervals* of outcomes rather than to individual outcomes.

The Mean and Median of a Continuous Random Variable

Recall that the mean is the "balance point" of a distribution. Figure 5.4 illustrates this idea for the **mean of a continuous random variable.**

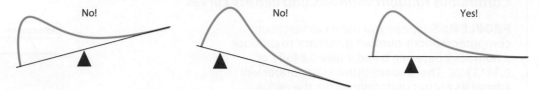

FIGURE 5.4 The mean of a continuous random variable is the balance point of a density curve.

Areas under a density curve represent probabilities. The **median of a continuous random variable** is the 50th percentile of its probability distribution.

> **DEFINITION** Mean and median of a continuous random variable
>
> The **mean of a continuous random variable** is the point at which its probability distribution would balance if made of solid material.
>
> The **median of a continuous random variable** is the equal-areas point, the point that divides the area under the probability distribution in half.

The median of a symmetric probability distribution is at its midpoint. A symmetric density curve balances at its midpoint because the two sides are identical. So the mean and median of a continuous random variable with a symmetric probability distribution are equal, as in Figure 5.5(a).

It isn't so easy to spot the equal-areas point on a skewed probability distribution. We used technology to locate the median in Figure 5.5(b). The mean is greater than the median because the balance point of the distribution is pulled toward the long right tail.

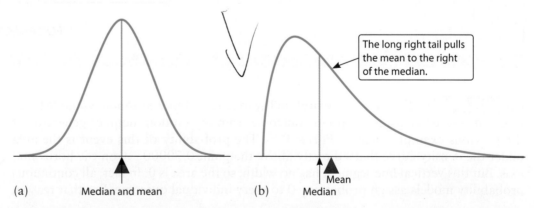

The long right tail pulls the mean to the right of the median.

(a) Median and mean   (b) Mean Median

FIGURE 5.5 (a) The median and mean of a symmetric probability distribution both lie at the point of symmetry. (b) The median and mean of a right-skewed probability distribution. The mean is pulled away from the median toward the long tail.

What does the left skew do?

Mean versus median

PROBLEM: The probability distribution of a continuous random variable is shown. Identify the location of the mean and median by letter. Justify your answers.

SOLUTION:

Median = B, Mean = A. B is the equal-areas point of the distribution. The mean will be less than the median due to the left-skewed shape.

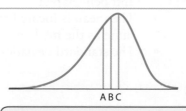

A B C

Even though C is directly under the peak of the curve, more than half of the area is to its left, so it cannot be the median.

FOR PRACTICE   **TRY EXERCISE 5.**

The mean μ of a continuous random variable has the same interpretation as for a discrete random variable. It is the long-run average value (the expected value) of the variable if the chance process is repeated many, many times. The standard deviation σ of a continuous random variable measures how much the values of the variable typically differ from the mean.

We can roughly locate the mean of a continuous random variable by eye, as the balance point of its density curve. No easy way exists to estimate the standard deviation for density curves in general. But there is one family of probability distributions for which we *can* estimate the standard deviation by eye.

Normal Distributions

Earlier, we defined the random variable $X =$ the ITBS grade-equivalent vocabulary score of a randomly selected Gary, Indiana, seventh-grader. Its density curve (shown in red in the figure) has a distinctive "bell" shape. The probability distribution of X is called a **normal distribution.**

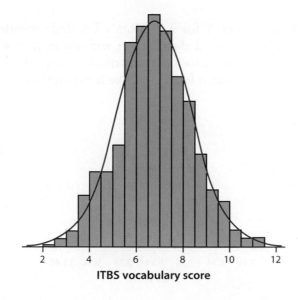

ITBS vocabulary score

Look at the two normal distributions in Figure 5.6. They illustrate several important facts:

● All normal distributions have the same overall shape: symmetric, single-peaked, and bell-shaped.

● The mean is located at the midpoint of the symmetric density curve and is the same as the median.

● The standard deviation σ measures the variability (width) of a normal distribution.

FIGURE 5.6 Two normal distributions, showing the mean μ and standard deviation σ.

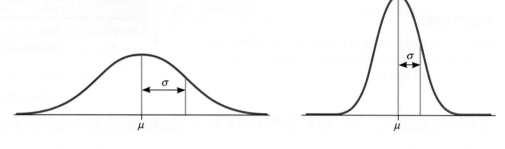

You can estimate σ by eye on a normal density curve. Here's how. Imagine that you are skiing down a mountain that has the shape of a normal distribution. At first, you descend at an increasingly steep angle as you go out from the peak. Fortunately, before you find yourself going straight down, the slope begins to get flatter rather than steeper as you go out and down.

The points at which this change of curvature takes place are located at a distance σ on either side of the mean μ. (Advanced math students know these as "inflection points.") You can feel the change in curvature as you run a pencil along a normal distribution, which will allow you to estimate the standard deviation.

> **DEFINITION** Normal distribution
>
> A **normal distribution** is described by a symmetric, single-peaked, bell-shaped density curve. Any normal distribution is completely specified by two numbers: its mean μ and standard deviation σ.

The ITBS vocabulary score X for a randomly selected seventh-grade student in Gary is modeled well by a normal distribution with mean $\mu = 6.84$ and standard deviation $\sigma = 1.55$. The figure shows this distribution with the points 1, 2, and 3 standard deviations from the mean labeled on the horizontal axis.

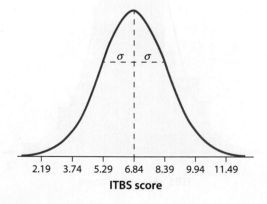

EXAMPLE

Is it a hit?

Graphing a normal distribution

PROBLEM: In baseball, a player's batting average is the proportion of times the player gets a hit out of his total number of times at bat. Suppose we select a Major League Baseball player at random and let X = the player's batting average. The probability distribution of X is approximately normal, with mean $\mu = 0.261$ and standard deviation $\sigma = 0.034$. Sketch this probability distribution. Label the mean and the points that are 1, 2, and 3 standard deviations from the mean.

SOLUTION:

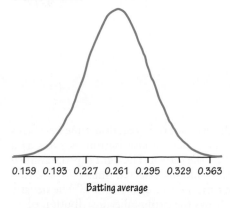

0.159 0.193 0.227 0.261 0.295 0.329 0.363
Batting average

The mean is at the midpoint of the bell-shaped density curve. The standard deviation (0.034) is the distance from the center to the change-of-curvature points on either side. Label the mean (0.261) and the points 1, 2, and 3 SDs from the mean.

1 SD: $0.261 - 1(0.034) = 0.227$ $0.261 + 1(0.034) = 0.295$
2 SD: $0.261 - 2(0.034) = 0.193$ $0.261 + 2(0.034) = 0.329$
3 SD: $0.261 - 3(0.034) = 0.159$ $0.261 + 3(0.034) = 0.363$

FOR PRACTICE   TRY EXERCISE 9.

Remember that μ and σ alone do not specify the appearance of most distributions. The shape of density curves, in general, does not reveal σ. These are special properties of normal distributions.

CAUTION

LESSON APP 5.5

Still waiting for the server?

How does your Web browser get a file from the Internet? Your computer sends a request for the file to a Web server, and the Web server sends back a response. For one particular Web server, the time X (in seconds) after the start of an hour at which a randomly selected request is received has the uniform distribution shown in the figure.

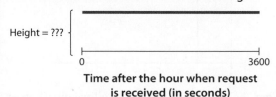

Height = ???

0   3600
**Time after the hour when request
is received (in seconds)**

1. What height must the probability distribution have? Justify your answer.

2. Find the probability that the request is received within the first 300 seconds (5 minutes) after the hour.

3. What is the mean of X? Explain.

4. What is the median of X? Explain.

Masson/Shutterstock

Lesson 5.5

WHAT DID YOU LEARN?

LEARNING TARGET	EXAMPLES	EXERCISES
Show that the probability distribution of a continuous random variable is valid and use the distribution to calculate probabilities.	p. 367	1–4
Determine the relative locations of the mean and median of a continuous random variable from the shape of its probability distribution.	p. 369	5–8
Draw a normal probability distribution with a given mean and standard deviation.	p. 371	9–12

Exercises Lesson 5.5

Mastering Concepts and Skills

1. Biking accidents Accidents on a level, 3-mile bike
pg 367 path occur uniformly along the length of the path.
Let Y represent the distance from the start of the
bike path at which a randomly selected accident
occurs. The figure displays the density curve that
describes the probability distribution of Y.

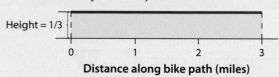

Distance along bike path (miles)

(a) Explain why the probability distribution of Y is
valid.

(b) Aaliyah's property adjoins the bike path between
the 0.8-mile mark and the 1.1-mile mark. Find and
interpret $P(0.8 \le Y \le 1.1)$.

2. Where's the bus? Jayden takes the same bus
to work every morning. The amount of time Y
(in minutes) that he has to wait for the bus to
arrive on a randomly selected workday morning
is described by the uniform density curve that
follows.

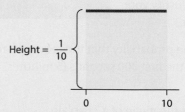

(a) Explain why the probability distribution of Y is
valid.

(b) Find and interpret $P(2.5 \le Y \le 5.3)$.

3. Quick, click! An Internet reaction time test asks
subjects to click their mouse button as soon as a
light flashes on the screen. The light is programmed
to go on at a randomly selected time T from 2 to 5
seconds after the subject clicks "Start." The density
curve here shows the probability distribution of T.

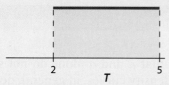

(a) What height must the probability distribution
have? Justify your answer.

(b) Find the probability that the light turns on between
2.5 and 4 seconds after the subject clicks "Start."

4. Class is over! Mr. Wilder does not always let his statistics class out when the bell rings. In fact, he seems
to end class according to his own "internal clock."
The random variable D describes the amount of
time after the bell rings (in minutes) when Mr.
Wilder dismisses the class on a randomly selected
day. (A negative value of D indicates he dismissed
his class before the bell rang.) The density curve
that follows gives the probability distribution of D.

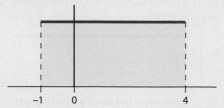

(a) What height must the probability distribution
have? Justify your answer.

(b) Find the probability that Mr. Wilder ends class within 1 minute (before or after) of when the bell rings.

5. Which is the mean? The probability distribution of a continuous random variable is shown here. Identify the location of the mean and median by letter. Justify your answers.

pg 369

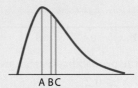

6. Which is the median? The probability distribution of a continuous random variable is shown here. Identify the location of the mean and median by letter. Justify your answers.

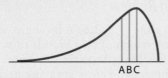

7. Bimodal mean and median The probability distribution of a continuous random variable is shown here. Identify the location of the mean and median by letter. Justify your answers.

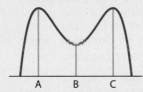

8. Bell-shaped mean and median The probability distribution of a continuous random variable is shown here. Identify the location of the mean and median by letter. Justify your answers.

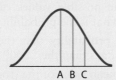

9. Nine ounces of chips? The distribution of weights of 9-ounce bags of a particular brand of potato chips is approximately normal, with mean $\mu = 9.12$ ounces and standard deviation $\sigma = 0.05$ ounce. Sketch this probability distribution. Label the mean and the points that are 1, 2, and 3 standard deviations from the mean.

pg 371

10. Men's heights The distribution of heights of adult American men is approximately normal, with mean $\mu = 69$ inches and standard deviation $\sigma = 2.5$ inches.

Sketch this probability distribution. Label the mean and the points that are 1, 2, and 3 standard deviations from the mean.

11. Rafa serves! Tennis superstar Rafael Nadal's first-serve speeds (in miles per hour) in a recent season can be modeled by the normal distribution shown here. The mean and the points that are 1, 2, and 3 standard deviations from the mean are labeled on the horizontal axis. Identify the mean and the standard deviation.

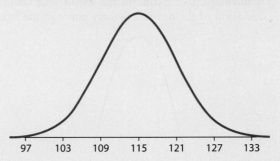

12. Cholesterol levels High levels of cholesterol in the blood increase the risk of heart disease. Cholesterol levels for 14-year-old boys (in milligrams per deciliter) can be modeled by the normal distribution shown here. The mean and the points that are 1, 2, and 3 standard deviations from the mean are labeled on the horizontal axis. Identify the mean and the standard deviation.

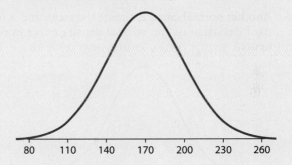

Applying the Concepts

13. Set the alarm Old-fashioned mechanical alarm clocks were not all that accurate. Suppose that the alarm on one such clock is equally likely to go off at any time from 2 minutes before to 2 minutes after the alarm's setting. Let the random variable X = the amount of time (in minutes) from when the alarm is set to when it goes off. Note that X will be negative if the alarm goes off early.

(a) Sketch a graph of the uniform probability distribution of X.

(b) Find the probability that the alarm goes off within 10 seconds of the time for which it is set on a randomly selected day.

14. **Preheat before baking** The time X it takes a certain oven to preheat to 350°F is equally likely to be any value from 7 to 12 minutes.

(a) Sketch a graph of the uniform probability distribution of X.

(b) Find the probability that the oven preheats in less than 8 minutes and 45 seconds on a randomly selected day.

15. **A normal curve** Estimate the mean and standard deviation of the normal density curve in the figure.

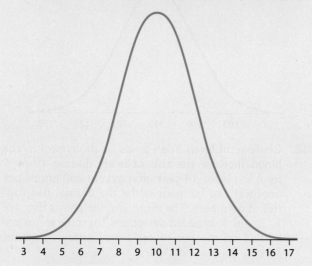

16. **Another normal curve** Estimate the mean and standard deviation of the normal density curve in the figure.

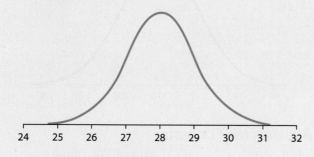

Extending the Concepts

17. **A weird density curve** The figure shows the density curve that describes the probability distribution of a random variable X.

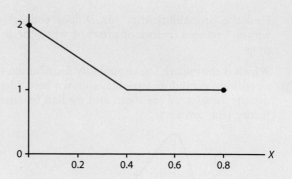

(a) Show that the probability distribution of X is valid.

(b) Find $P(0 \leq X \leq 0.2)$.

(c) The median of X is between 0.2 and 0.4. Explain why.

(d) Is the mean of X less than, equal to, or greater than the median of X? Justify your answer.

Recycle and Review

18. **Weed be gone** (3.5, 3.6) A biologist would like to determine which of two brands of weed killer, X or Y, is less likely to harm the plants in a garden at the university. Before spraying near the plants, the biologist decides to conduct an experiment using 24 individual plants. The biologist chooses the 12 healthiest-looking plants and then flips a coin. If it lands heads up, apply Brand X weed killer to these plants and Brand Y weed killer to the remaining 12 plants. If it lands tails up, do the opposite. What flaw in this study makes it impossible to determine which brand of weed killer is better? Explain.

19. **Buying stock** (4.6, 5.2) You purchase a hot stock for $1000. The stock either gains 30% or loses 25% each day, each with probability 0.5. Its returns on consecutive days are independent of each other. You plan to sell the stock after 2 days.

(a) What are the possible values of the stock after 2 days, and what is the probability for each value?

(b) What is the probability that the stock is worth more after 2 days than the $1000 you paid for it?

(c) What is the mean value of the stock after 2 days?

(*Comment:* The criteria in parts (b) and (c) give different answers to the question "Should I invest?")

Lesson 5.6

The Standard Normal Distribution

Why are normal distributions important in statistics? Here are three reasons. First, normal distributions are good descriptions for some distributions of real data. Distributions that are often close to normal include:

- Scores on tests taken by many people (such as SAT exams and IQ tests)
- Repeated careful measurements of the same quantity (like the diameter of a tennis ball)
- Characteristics of biological populations (such as lengths of crickets and yields of corn)

Second, normal distributions are good approximations to the results of many kinds of chance outcomes, like the number of heads in many tosses of a fair coin. Third, and most important, we will see that many of the inference methods in Chapters 7–10 are based on normal distributions.

The 68–95–99.7 Rule

In Lesson 5.5, we defined the continuous random variable X = the ITBS grade-equivalent vocabulary score for a randomly selected seventh-grade student from Gary, Indiana. Suppose the probability distribution of this random variable is approximately normal, with mean $\mu = 6.84$ and standard deviation $\sigma = 1.55$. Figure 5.7 shows the normal density curve for this distribution with the points 1, 2, and 3 standard deviations from the mean labeled on the horizontal axis.

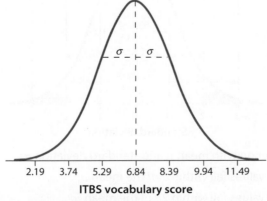

ITBS vocabulary score

FIGURE 5.7 The random variable X = ITBS vocabulary score of a randomly selected Gary, Indiana, seventh-grader follows a normal distribution, with $\mu = 6.84$ and $\sigma = 1.55$.

How unusual is it for a Gary seventh-grader to get an ITBS score below 3.74? That is, what's $P(X < 3.74)$? The following activity should help shed some light on this question.

ACTIVITY

What's so special about normal distributions?

In this activity, you will use the *Probability* applet at the book's website (**highschool.bfwpub.com/spa3e**) to explore an interesting property of normal distributions.

- Choose "Normal distribution" in the drop-down menu and click on "Begin analysis."
- To generate a *standard normal distribution*, enter 0 for the mean and 1 for the standard deviation.
- Click on "Plot distribution" to display the graph.

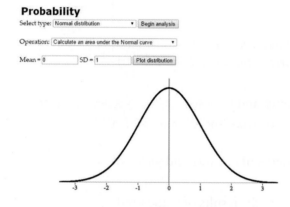

Use the applet to help you answer the questions.

1. What percent of the area in a standard normal distribution lies between −1 and 1?

2. What percent of the area in a standard normal distribution lies within 2 standard deviations of the mean?

3. Use the applet to confirm that about 99.7% of the area in a standard normal distribution lies within 3 standard deviations of the mean.

4. Change the mean to 6.84 and the standard deviation to 1.55. (These values are from the ITBS vocabulary scores in Gary, Indiana.) What percent of the area in this normal distribution lies within 1, 2, and 3 standard deviations of the mean? See Figure 5.7 on page 375 for the boundary values.

5. Change the mean and standard deviation to two new values. What percent of the area in this normal distribution lies within 1, 2, and 3 standard deviations of the mean?

6. *Summarize:* Complete this sentence: "For any normal distribution, the area under the curve within 1, 2, and 3 standard deviations of the mean is about ___%, ___%, and ___%."

Although there are many normal distributions, they all have properties in common. In particular, all normal distributions obey the **68–95–99.7 rule.**

DEFINITION The 68–95–99.7 rule

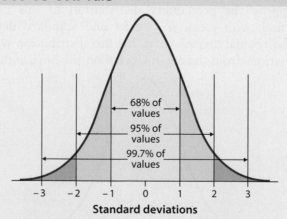

In any normal distribution with mean μ and standard deviation σ:

- About **68%** of the values fall within σ of the mean μ.
- About **95%** of the values fall within 2σ of the mean μ.
- About **99.7%** of the values fall within 3σ of the mean μ.

This important result is known as the **68–95–99.7 rule.**

Some people refer to the 68–95–99.7 rule result as the "empirical rule" (*empirical* means "learned from experience or by observation"). By remembering these three numbers, you can quickly estimate some probabilities involving normal distributions and recognize when an observation is unusual.

EXAMPLE

How can you interpret ITBS vocabulary scores?

Using the 68–95–99.7 rule

PROBLEM: The random variable $X =$ ITBS vocabulary score of a randomly selected Gary, Indiana, seventh-grader can be modeled by a normal distribution with $\mu = 6.84$ and $\sigma = 1.55$. Use the 68–95–99.7 rule to approximate:

(a) $P(X < 3.74)$

(b) The probability that a randomly selected Gary seventh-grader gets an ITBS vocabulary score between 5.29 and 9.94

SOLUTION:

(a)

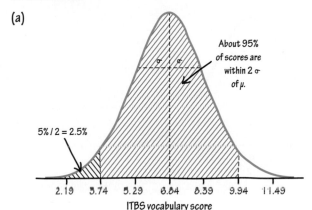

$P(X < 3.74) \approx 0.025$

Draw a picture of the normal probability distribution with the mean and the points 1, 2, and 3 standard deviations from the mean labeled.

Shade the area under the normal density curve that corresponds to the desired probability.

By the 68–95–99.7 rule, about 95% of all scores fall between 3.74 and 9.94. The other 5% of scores are outside this range. Because normal distributions are symmetric, half of these scores are lower than 3.74 and half are higher than 9.94.

(b)

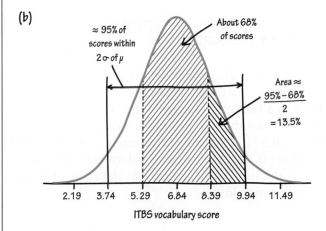

$P(5.29 \le X \le 9.94) \approx 0.68 + 0.135 = 0.815$

About 68% of ITBS scores fall between 5.29 and 8.39. Now we add the percent of scores that are between 8.39 and 9.94.

About 95% of ITBS scores fall between 3.74 and 9.94. So about 95% − 68% = 27% of scores are between 1 and 2 standard deviations from the mean. By the symmetry of normal distributions, half of these scores (27% / 2 = 13.5%) are between 8.39 and 9.94.

So about 68% + 13.5% = 81.5% of ITBS scores fall between 5.29 and 9.94.

FOR PRACTICE   TRY EXERCISE 1.

CAUTION

Note that the 68–95–99.7 rule applies *only* to normal distributions.

The Standard Normal Distribution

As the 68–95–99.7 rule suggests, all normal distributions share many properties. In fact, all normal distributions are the same if we measure in units of size σ from the mean μ as center. Changing to these units requires us to standardize, just as we did in Lesson 1.9:

$$z = \frac{\text{value} - \text{mean}}{\text{standard deviation}} = \frac{x - \mu}{\sigma}$$

Suppose that we standardize a continuous random variable X that has a normal distribution. The resulting random variable Z has a normal distribution with mean $\mu = 0$ and standard deviation $\sigma = 1$. This distribution is known as the **standard normal distribution.**

▌▌▌ **DEFINITION** Standard normal distribution

The **standard normal distribution** is the normal distribution with mean 0 and standard deviation 1.

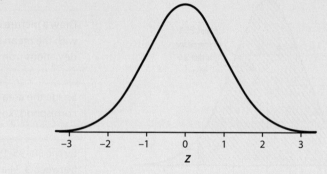

In a standard normal distribution, the 68–95–99.7 rule tells us that about 68% of the values fall between $z = -1$ and $z = 1$ (i.e., within 1 standard deviation of the mean). What if we want to find $P(-1.25 \leq Z \leq 1.25)$? The 68–95–99.7 rule can't help us.

FINDING PROBABILITIES IN THE STANDARD NORMAL DISTRIBUTION

Because all normal distributions are the same when we standardize, we can find probabilities for any normal distribution using areas from the standard normal distribution. Table A in the back of the book is a table of standard normal probabilities. The table entry for each z-score is the area under the curve to the left of z.

Suppose we wanted to find $P(Z < 0.81)$. To find the area to the left of $z = 0.81$ in a standard normal distribution, locate 0.8 in the left-hand column of Table A, then locate the remaining digit 1 as .01 in the top row. The entry to the right of 0.8 and under .01 is .7910. This is the desired probability.

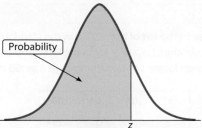

Table entry for *z* is the area under the standard normal curve to the left of *z*.

Table A Standard normal probabilities (continued)										
z	.00	.01	.02	.03	.04	.05	.06	.07	.08	.09
0.0	.5000	.5040	.5080	.5120	.5160	.5199	.5239	.5279	.5319	.5359
0.1	.5398	.5438	.5478	.5517	.5557	.5596	.5636	.5675	.5714	.5753
0.2	.5793	.5832	.5871	.5910	.5948	.5987	.6026	.6064	.6103	.6141
0.3	.6179	.6217	.6255	.6293	.6331	.6368	.6406	.6443	.6480	.6517
0.4	.6554	.6591	.6628	.6664	6700	.6736	.6772	.6808	.6844	.6879
0.5	.6915	.6950	.6985	.7019	.7054	.7088	.7123	.7157	.7190	.7224
0.6	.7257	.7291	.7324	.7357	.7389	.7422	.7454	.7486	.7517	.7549
0.7	.7580	.7611	.7642	.7673	.7704	.7734	.7764	.7794	.7823	.7852
0.8	.7881	.7910	.7939	.7967	.7995	.8023	.8051	.8078	.8106	.8133
0.9	.8159	.8186	.8212	.8238	.8264	.8289	.8315	.8340	.8365	.8389

Figure 5.8 illustrates the relationship between the value $z = 0.81$ and the area 0.7910. About 79% of the values in a normal distribution will have a *z*-score less than $z = 0.81$.

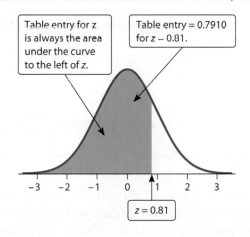

Table entry for z is always the area under the curve to the left of z.

Table entry = 0.7910 for z – 0.81.

z = 0.81

FIGURE 5.8 The area in a standard normal distribution to the left of the point $z = 0.81$ is 0.7910.

EXAMPLE

How do you find probabilities in a standard normal distribution?

Using Table A

PROBLEM: Draw a standard normal distribution with the area of interest shaded. Use Table A to find the indicated probability.

(a) The probability that *Z* is greater than -1.78 (b) $P(-1.25 \le Z \le 0.81)$

SOLUTION:

(a)

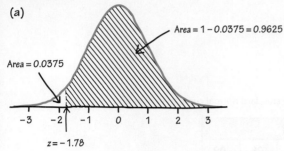

Area = 0.0375

Area = 1 − 0.0375 = 0.9625

z = −1.78

$$P(Z > -1.78) = 1 - 0.0375 = 0.9625$$

(b)

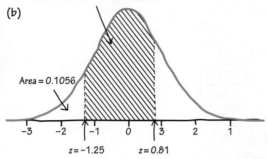

Area = 0.7910 − 0.1056 = 0.6854

Area = 0.1056

z = −1.25   z = 0.81

$$P(-1.25 \le Z \le 0.81) = 0.7910 - 0.1056 = 0.6854$$

Find the area to the *left* of $z = -1.78$ in the standard normal distribution. Locate -1.7 in the left-hand column of Table A, then locate the remaining digit 8 as .08 in the top row.

z	.07	.08	.09
−1.8	.0307	.0301	.0294
−1.7	.0384	.0375	.0367
−1.6	.0475	.0465	.0455

Use the fact that the total area in the standard normal distribution is 1 to find the area to the *right* of $z = -1.78$.

Use Table A to find the area to the left of $z = 0.81$ in the standard normal distribution.

Subtract the area to the left of $z = -1.25$ in the standard normal distribution.

FOR PRACTICE   TRY EXERCISE 5.

The figure that follows shows why the method used in part (b) of the example works.

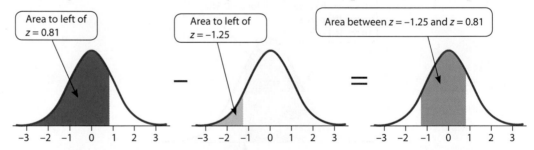

Area to left of $z = 0.81$

Area to left of $z = -1.25$

Area between $z = -1.25$ and $z = 0.81$

Table A does not go beyond $z = -3.50$ and $z = 3.50$ because it is highly unusual for a value to be more than 3.5 standard deviations from the mean in a normal distribution. For practical purposes, we can act as if there is approximately zero probability outside the range of Table A.

WORKING BACKWARD: FROM PROBABILITIES (AREAS) TO Z-SCORES

So far, we have used Table A to find probabilities in a standard normal distribution from z-scores. What if we want to find the z-score that corresponds to a particular area or probability?

For example, let's find the 90th percentile of the standard normal distribution. We're looking for the z-score that has 90% of the area to its left, as shown in Figure 5.9.

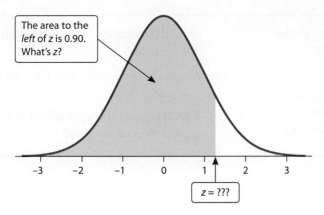

The area to the *left* of *z* is 0.90. What's *z*?

FIGURE 5.9 The *z*-score with area 0.90 to its left under the standard normal curve.

Because Table A gives areas to the left of a specified *z*-score, all we have to do is find the value closest to 0.90 in the middle of the table. From the reproduced portion of Table A, you can see that the desired value is $z = 1.28$. That is, $P(Z < 1.28) \approx 0.90$.

z	.07	.08	.09
1.1	.8790	.8810	.8830
1.2	.8980	.8997	.9015
1.3	.9147	.9162	.9177

EXAMPLE

How do you convert probabilities to z-scores in a standard normal distribution?

Using Table A in reverse

PROBLEM: Use Table A to find the value of *z* for which $P(Z > \_\_\_) = 0.20$. Draw a standard normal distribution with the area of interest shaded.

SOLUTION:

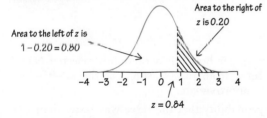

Area to the left of z is
$1 - 0.20 = 0.80$

Area to the right of z is 0.20

z = 0.84

Use Table A to look up the area to the *left* of the desired *z*-score in the standard normal distribution. The closest area to 0.8000 is 0.7995, which corresponds to a *z*-score of $z = 0.84$.

From Table A, $z = 0.84$.

FOR PRACTICE   TRY EXERCISE 9.

LESSON APP 5.6

What's a good batting average?

1. In baseball, a player's batting average is the proportion of times that the player gets a hit out of his total number of times at bat. Suppose we select a Major League Baseball player at random. The random variable $X =$ the player's batting average can be modeled by a normal distribution with mean $\mu = 0.261$ and standard deviation $\sigma = 0.034$. Use the 68–95–99.7 rule to approximate:

(a) The probability that a randomly selected player has a batting average greater than 0.329

(b) $P(0.193 \leq X \leq 0.295)$

2. Suppose we convert the randomly selected player's batting average to a z-score. Use Table A to find each of the following. Draw a standard normal distribution with the desired area shaded in each case.

(a) What's the probability that the z-score is between −0.58 and 1.79?

(b) 45% of batting averages will have a z-score greater than what value?

Lesson 5.6

WHAT DID YOU LEARN?

LEARNING TARGET	EXAMPLES	EXERCISES
Use the 68–95–99.7 rule to find approximate probabilities in a normal distribution.	p. 377	1–4
Use Table A to find a probability (area) from a z-score in the standard normal distribution.	p. 379	5–8
Use Table A to find a z-score from a probability (area) in the standard normal distribution.	p. 381	9–12

Exercises   Lesson 5.6

Mastering Concepts and Skills

1. **How much is 9 ounces?** The weights of 9-ounce bags of a particular brand of potato chips can be modeled by a normal distribution with mean $\mu = 9.12$ ounces and standard deviation $\sigma = 0.05$ ounce. Let $X =$ the weight (in ounces) of a randomly selected bag. Use the 68–95–99.7 rule to approximate:
   pg 377

(a) $P(X < 9.02)$

(b) The probability that the bag weighs between 8.97 and 9.17 ounces

2. **Modeling men's heights** The distribution of heights of adult American men can be modeled by a normal distribution with mean $\mu = 69$ inches and standard deviation $\sigma = 2.5$ inches. Let $X =$ the height (in inches) of a randomly selected adult American man. Use the 68–95–99.7 rule to approximate:

(a) $P(X > 74)$

(b) The probability that the man has a height between 64 and 71.5 inches

3. **Rafa serves again!** Tennis superstar Rafael Nadal's first-serve speeds in a recent season can be modeled by a normal distribution with mean 115 mph and standard deviation 6 mph. Let X be the speed (in miles per hour) of a randomly selected Nadal first serve from that season. Use the 68–95–99.7 rule to approximate:

(a) The probability that the serve was faster than 121 mph

(b) $P(109 \leq X \leq 133)$

4. **Cholesterol modeled** Cholesterol levels for 14-year-old boys can be modeled by a normal distribution with mean 170 mg/dl and standard deviation 30 mg/dl. Let $X =$ the cholesterol level (in milligrams per deciliter) of a randomly selected 14-year-old boy. Use the 68–95–99.7 rule to approximate:

(a) The probability that his cholesterol level is less than 140 mg/dl

(b) $P(200 \leq X \leq 230)$

For Exercises 5–8, draw a standard normal distribution with the area of interest shaded. Use Table A to find the indicated probability.

5. **Table A practice**
   pg 379

(a) The probability that Z is greater than −2.15

(b) $P(-0.56 \leq Z \leq 1.87)$

6. **Table A practice**

(a) The probability that Z is greater than 1.53

(b) $P(-1.66 \leq Z \leq 2.85)$

7. **More Table A practice**

(a) $P(Z < 2.46)$

(b) The probability that Z is between 0.50 and 1.79

8. **More Table A practice**

(a) $P(Z < -1.39)$

(b) The probability that Z is between -1.11 and -0.32

9. **Working backward** Use Table A to find the value
pg **381** of z for which $P(Z > \underline{\quad}) = 0.34$. Draw a standard normal distribution with the area of interest shaded.

10. **Working backward** Use Table A to find the value of z for which $P(Z > \underline{\quad}) = 0.75$. Draw a standard normal distribution with the area of interest shaded.

11. **Quartiles** Find the 25th percentile (Q_1) and the 75th percentile (Q_3) of the standard normal distribution.

12. **Deciles** The deciles of any distribution are the values at the 10th, 20th, ..., 90th percentiles. The first and last deciles are the 10th and the 90th percentiles, respectively. What are the first and last deciles of the standard normal distribution?

Applying the Concepts

13. **Women's heights: Mean and SD** The distribution of heights in a population of women is approximately normal. Sixteen percent of the women have heights less than 62 inches. About 97.5% of the women have heights less than 70 inches. Use the 68–95–99.7 rule to estimate the mean and standard deviation of the heights in this population.

14. **Quiz scores: Mean and SD** The distribution of scores on a recent quiz in a large college statistics class is approximately normal. About 2.5% of the students scored below 25 on the quiz. Eighty-four percent of the students scored below 40 on the quiz. Use the 68–95–99.7 rule to estimate the mean and standard deviation of the quiz scores in this class.

Extending the Concepts

15. **Always outliers** Refer to Exercise 11. The percent of the values that are classified as outliers by the $1.5 \times IQR$ rule is the same in any normal distribution. What is this percent? Show your method clearly.

16. **Chebyshev's inequality** An interesting result known as *Chebyshev's inequality* says that in *any* distribution, at least $100\left(1 - \dfrac{1}{k^2}\right)\%$ of the values are within k standard deviations of the mean. If $k = 2$, for example, Chebyshev's inequality tells us that at least $100\left(1 - \dfrac{1}{2^2}\right) = 100\left(\dfrac{3}{4}\right) = 75\%$ of the values in any distribution are within 2 standard deviations of the mean. For normal distributions, we know that about 95% of the values are within 2 standard deviations of the mean by the 68–95–99.7 rule.

(a) Make a table that shows what percent of observations must fall within 1, 2, 3, 4, and 5 standard deviations of the mean in any distribution.

(b) Explain why values 5 or more standard deviations from the mean in any distribution could be considered unusual.

Recycle and Review

17. **Squirrels and food supply (2.7)** Animal species produce more offspring when their supply of food goes up. Some animals appear able to anticipate unusual food abundance. Red squirrels eat seeds from pinecones, a food source that sometimes has very large crops. Researchers collected data on an index of the abundance of pinecones and the average number of offspring per female over 16 years. The least-squares regression line calculated from these data was

$$\overline{\text{offspring per female}} = 1.4146 + 0.4399 \text{ (cone index)}$$

A residual plot follows.

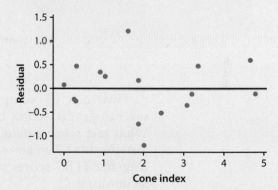

(a) Is the linear model appropriate for these data? Explain.

(b) Estimate the actual number of offspring per female for the year with a cone index of 2.01.

18. **Standard deviations (1.7, 5.5)** Continuous random variables A, B, and C all take values between 0 and 10. Their density curves, drawn on the same horizontal scales, are shown here. Rank the standard deviations of the three random variables from lowest to highest. Justify your answer.

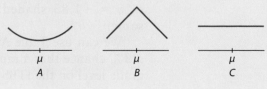

Lesson 5.7

Normal Distribution Calculations

LEARNING TARGETS

- Calculate the probability that a value falls within a given interval in a normal distribution.
- Find a value corresponding to a given probability (area) in a normal distribution.

In Lessons 5.5 and 5.6, we examined the continuous random variable X = the ITBS grade-equivalent vocabulary score for a randomly selected seventh-grade student in Gary, Indiana. Figure 5.10 shows the probability distribution of this random variable. The distribution is approximately normal with mean $\mu = 6.84$ and standard deviation $\sigma = 1.55$.

FIGURE 5.10 The random variable X = ITBS vocabulary score of a randomly selected Gary seventh-grader follows a normal distribution with $\mu = 6.84$ and $\sigma = 1.55$.

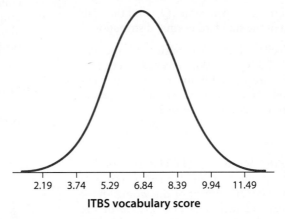

ITBS vocabulary score

How can we compute the probability $P(X < 4)$ that a randomly selected seventh-grader scores below the fourth-grade level on the ITBS vocabulary test? What test score would place a Gary seventh-grader at the 90th percentile of the distribution? The answers to both of these questions can be found by standardizing the ITBS scores and using the standard normal distribution or by using technology.

Calculating Probabilities

Figure 5.11(a) shows $P(X < 4)$ for a randomly selected Gary seventh-grader. Recall from Lesson 5.6 that if we standardize the ITBS vocabulary scores, we get a standard normal distribution. The boundary value $X = 4$ corresponds to

$$z = \frac{4 - 6.84}{1.55} = -1.83$$

Figure 5.11(b) shows the standard normal distribution with the area to the left of $z = -1.83$ shaded. Notice that the shaded areas in the two figures are the same.

We can use Table A to find that $P(Z < -1.83) = 0.0336$. That is, there's about a 3% chance that a randomly selected Gary seventh-grader scores below the fourth-grade level on the ITBS vocabulary test.

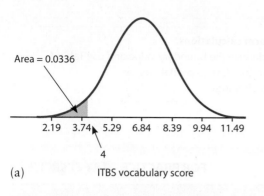

Area = 0.0336

2.19 3.74 5.29 6.84 8.39 9.94 11.49

4

(a)   ITBS vocabulary score

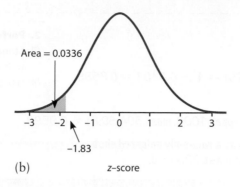

Area = 0.0336

−3   −2   −1   0   1   2   3

−1.83

(b)   z–score

FIGURE 5.11 (a) Normal distribution showing the probability that a randomly selected Gary seventh-grader gets an ITBS vocabulary score below the fourth-grade level. (b) The corresponding area in the standard normal distribution.

It's possible to find $P(X < 4)$ directly from the original (unstandardized) normal distribution using technology. See the Tech Corner at the end of the lesson for details. Check with your teacher to see which method will be used in your class.

How to Find Probabilities (Areas) in Any Normal Distribution

1. **Draw a normal distribution** with the horizontal axis labeled and scaled using the mean and standard deviation, the boundary value(s) clearly identified, and the area of interest shaded.

2. **Perform calculations.** Do one of the following:

   (i)   Standardize each boundary value and use Table A to find the desired probability (area) under the standard normal curve; or

   (ii) use technology to find the desired probability (area) without standardizing.

   Be sure to answer the question that was asked.

EXAMPLE

Is Rory McIlroy a long driver?

Calculating probabilities in a normal distribution

PROBLEM: On the driving range, Rory McIlroy practices his swing with a particular club by hitting many, many golf balls. Let X = the distance the ball travels on a randomly selected shot with his driver. Suppose that X follows a normal distribution with mean 304 yards and standard deviation 8 yards. Find and interpret $P(X \geq 290)$.

SOLUTION:

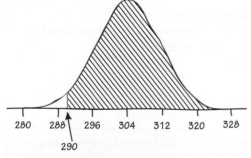

280   288   296   304   312   320   328

290

Distance traveled (yards)

1. **Draw a normal distribution.**

Be sure to:
● Scale the horizontal axis with the mean and the points 1, 2, and 3 standard deviations from the mean.
● Label the horizontal axis with the variable name.
● Clearly identify the boundary value(s).
● Shade the area of interest.

Madlie Meyer/Getty Images

$$z = \frac{290 - 304}{8} = -1.75$$

$$P(X \geq 290) = P(Z \geq -1.75) = 1 - 0.0401 = 0.9599$$

Using technology:

Applet/normalcdf (lower:290, upper:1000, mean:304, SD:8) = 0.9599

There's about a 96% chance that a randomly selected shot with Rory's driver will travel at least 290 yards.

2. Perform calculations.
(i) Standardize the boundary value and use Table A to find the desired probability; or
(ii) Use technology.

FOR PRACTICE   TRY EXERCISE 1.

THINK ABOUT IT   What proportion of Rory McIlroy's drives go *exactly* 290 yards? There is no area under the normal density curve in the previous example exactly over the point 290.000000000. . . . So the answer to our question based on the normal model is 0. One more thing: the areas under the curve with $X \geq 290$ and $X > 290$ are the same. According to the normal model, the proportion of Rory's drives that travel at least 290 yards is the same as the proportion that travel more than 290 yards.

EXAMPLE

Can Rory hit it that far?

Calculating more complicated probabilities

PROBLEM: Refer to the previous example. What percent of Rory's drives travel between 305 and 325 yards?

SOLUTION:

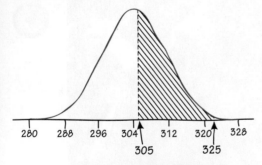

Distance traveled (yards)

1. Draw a normal distribution.

Be sure to:
● Scale the horizontal axis with the mean and the points 1, 2, and 3 standard deviations from the mean.
● Label the horizontal axis with the variable name.
● Clearly identify the boundary value(s).
● Shade the area of interest.

$$z = \frac{305 - 304}{8} = 0.125 \approx 0.13$$

$$z = \frac{325 - 304}{8} = 2.625 \approx 2.63$$

2. Perform calculations.
(i) Standardize the boundary values and use Table A to find the desired probability; or
(ii) Use technology.

$$P(305 \leq X \leq 325) = P(0.13 \leq Z \leq 2.63) = 0.9957 - 0.5517 = 0.4440$$

Using technology:

Applet/normalcdf (lower:305, upper:325, mean:304, SD:8)

$= 0.4459$

The slight difference between the answers using the two methods comes from using z-scores rounded to two decimal places for Table A.

FOR PRACTICE   TRY EXERCISE 5.

Finding Values from Probabilities

Sometimes, we want to find the value in a normal distribution that corresponds to a given area or probability. For instance, the probability distribution of X = ITBS vocabulary score for a randomly selected Gary, Indiana, seventh-grader is approximately normal with mean $\mu = 6.84$ and standard deviation $\sigma = 1.55$. What score x would a student have to earn to be at the 90th percentile of the distribution? Figure 5.12(a) shows what we are trying to find.

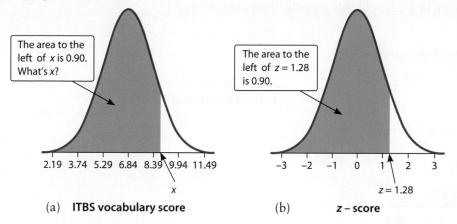

The area to the left of x is 0.90. What's x?

The area to the left of $z = 1.28$ is 0.90.

2.19 3.74 5.29 6.84 8.39 9.94 11.49

x

(a) **ITBS vocabulary score**

−3 −2 −1 0 1 2 3

$z = 1.28$

(b) **z – score**

FIGURE 5.12 (a) Normal distribution showing the 90th percentile of ITBS vocabulary scores for Gary, Indiana, seventh-graders. (b) The 90th percentile in the standard normal distribution is $z = 1.28$.

You can use Table A to find the z-score that corresponds to the 90th percentile in a standard normal distribution. It's $z = 1.28$. See Figure 5.12(b). Earlier, you used the formula

$$z = \frac{\text{value} - \text{mean}}{\text{standard deviation}} = \frac{x - \mu}{\sigma}$$

to compute the standardized score (z-score) for a given value in a distribution. This formula can also be used to find the value in a distribution that corresponds to a given z-score. To calculate the ITBS vocabulary score x that corresponds to $z = 1.28$:

$$1.28 = \frac{x - 6.84}{1.55} \Rightarrow 1.28(1.55) = x - 6.84 \Rightarrow 1.28(1.55) + 6.84 = x \Rightarrow 8.824 = x$$

It's possible to find the 90th percentile of the seventh-grade ITBS vocabulary scores in Gary directly from the original (unstandardized) normal distribution using technology. The Tech Corner at the end of the lesson provides the details. Check with your teacher to see which method will be used in your class.

How to Find Values from Probabilities (Areas) in Any Normal Distribution

1. **Draw a normal distribution** with the horizontal axis labeled and scaled using the mean and standard deviation, the area of interest shaded, and unknown boundary value clearly marked.

2. **Perform calculations.** Do one of the following:

   (i) Use Table A to find the value of z with the indicated area under the standard normal curve to the left of the boundary, then "unstandardize" to transform back to the original distribution; or

   (ii) use technology to find the boundary value from the probability (area) to the left of the boundary without standardizing.

   Be sure to answer the question that was asked.

EXAMPLE

How tall are 3-year-old girls?

Finding a value from a probability (area)

PROBLEM: According to www.cdc.gov/growthcharts/, the heights of 3-year-old females are approximately normally distributed with a mean of 94.5 centimeters and a standard deviation of 4 centimeters. Seventy-five percent of 3-year-old girls are taller than what height?

SOLUTION:

Let $X =$ the height of a randomly selected 3-year-old girl.
I want to find the value of X for which $P(X \geq \underline{\hspace{1cm}}) = 0.75$.

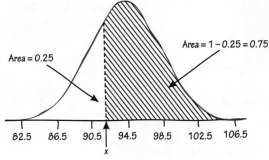

Area = 0.25
Area = 1 − 0.25 = 0.75

82.5   86.5   90.5   94.5   98.5   102.5   106.5
x

Heights of 3-year-old girls (cm)

1. **Draw a normal distribution.**

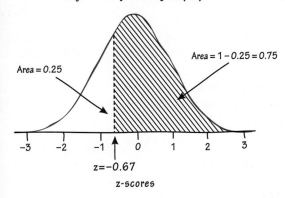

Area = 0.25
Area = 1 − 0.25 = 0.75

−3   −2   −1   0   1   2   3
$z = -0.67$

z-scores

2. **Perform calculations.**
   (i) Use Table A to find the value of z with the indicated area under the standard normal curve to the left of the boundary, then "unstandardize"; or
   (ii) Use technology.

The slight difference in the answers comes from using a z-score rounded to two decimal places for Table A.

$$-0.67 = \frac{x - 94.5}{4}$$

$$-0.67(4) = x - 94.5$$

$$-0.67(4) + 94.5 = x$$

$$91.82 = x$$

Using technology:

Applet/invNorm(area:0.25, mean:94.5, SD:4) = 91.80

About 75% of 3-year-old girls are taller than 91.80 centimeters.

FOR PRACTICE   TRY EXERCISE 9.

LESSON APP 5.7

What cholesterol levels are unhealthy for teen boys?

High levels of cholesterol in the blood increase the risk of heart disease. For 14-year-old boys, the distribution of blood cholesterol is approximately normal with mean $\mu = 170$ milligrams of cholesterol per deciliter of blood (mg/dl) and standard deviation $\sigma = 30$ mg/dl.[16] Let $X =$ the cholesterol level of a randomly selected 14-year-old boy.

1. Cholesterol levels above 240 mg/dl may require medical attention. Find and interpret $P(X > 240)$.

2. People with cholesterol levels between 200 and 240 mg/dl are at considerable risk for heart disease. What percent of 14-year-old boys have blood cholesterol between 200 and 240 mg/dl?

3. What cholesterol level would place a 14-year-old boy at the 10th percentile of the distribution?

Finding Values from Probabilities in Any Normal Distribution, and Vice Versa

You can use technology to find probabilities from values and to find values from probabilities in *any* normal distribution. Let's use an applet and the TI-83/84 to confirm our earlier results for the random variable $X =$ the ITBS vocabulary score of a randomly selected Gary, Indiana, seventh-grader. The distribution of X is approximately normal with mean $\mu = 6.84$ and standard deviation $\sigma = 1.55$.

(a) Find $P(X < 4)$.

Applet

1. Launch the *Probability* applet at highschool. bfwpub.com/spa3e.

2. Choose "Normal distribution" from the drop-down menu and click "Begin analysis."

3. Select the option to "Calculate an area under the normal curve."

4. Enter 6.84 for the mean and 1.55 for the standard deviation. Then click "Plot distribution" to see the graph of the appropriate normal distribution.

5. To find $P(X < 4)$, select "Calculate the area to the left of a value" from the pull-down menu, type 4 for the Value, and click "Calculate area."

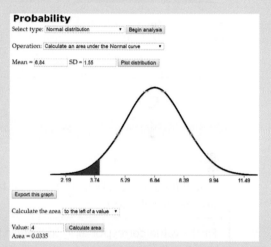

TI-83/84

1. Press [2nd] [VARS] (DISTR) and choose normalcdf(.

OS 2.55 or later: In the dialog box, enter these values: lower: –1000, upper: 4, μ:6.84, σ:1.55, choose Paste, and then press [ENTER].

Older OS: Complete the command normalcdf (–1000,4,6.84,1.55) and press [ENTER].

```
NORMAL FLOAT AUTO REAL RADIAN MP      []
normalcdf(-1000,4,6.84,1.▶
                   .0334564318
```

(b) What score *x* is the 90th percentile of the distribution?

Applet

1. Launch the *Probability* applet. Choose "Normal distribution" from the drop-down menu and click "Begin analysis."

2. Select the option to calculate a value corresponding to a left-tail area.

3. Enter 6.84 for the mean and 1.55 for the standard deviation. Then click "Plot distribution" to see the graph of the appropriate normal distribution.

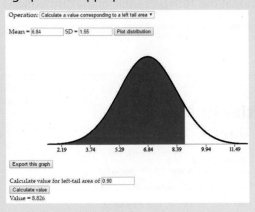

4. To find the 90th percentile of the distribution, after "Calculate value for left-tail area of," type 0.90 in the box. Then click "Calculate value."

TI-83/84

1. Press [2nd] [VARS] (DISTR) and choose invNorm(.

OS 2.55 or later: In the dialog box, enter these values: area:0.90,μ:6.84,σ:1.55, choose Paste, and then press [ENTER].

Older OS: Complete the command invNorm(0.90, 6.84,1.55) and press [ENTER].

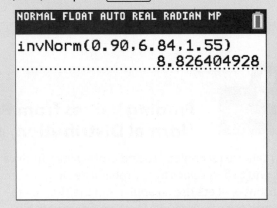

Lesson 5.7

WHAT DID YOU LEARN?

LEARNING TARGET	EXAMPLES	EXERCISES
Calculate the probability that a value falls within a given interval in a normal distribution.	pp. 385, 386	1–8
Find a value corresponding to a given probability (area) in a normal distribution.	p. 388	9–12

Exercises

Lesson 5.7

Mastering Concepts and Skills

1. **Weighing potato chips** The weights of 9-ounce bags
pg 385 of a particular brand of potato chips can be modeled by a normal distribution with mean $\mu = 9.12$ ounces and standard deviation $\sigma = 0.05$ ounce. Let X = the weight (in ounces) of a randomly selected bag. Find and interpret $P(X < 9)$.

2. **Blood pressure** Suppose we randomly select an adult. The random variable X = the person's diastolic blood pressure (in millimeters of mercury) can be modeled by a normal distribution with mean 70 and standard deviation 20. A healthy diastolic pressure for an adult is less than 80. Find and interpret $P(X < 80)$.

3. **Watch the salt!** A study investigated about 3000 meals ordered from Chipotle restaurants using the online site GrubHub. Researchers calculated the sodium content, in milligrams (mg), for each order based on Chipotle's published nutrition information. Let X = the sodium content for a randomly selected meal. The distribution of X is approximately normal with mean 2000 mg and standard deviation 500 mg.[17] What percent of the meals ordered exceeded the recommended daily allowance of 2400 mg of sodium?

4. **Batter up!** In baseball, a player's batting average is the proportion of times the player gets a hit out of his total number of times at bat. Select a Major League Baseball player at random. The random variable X = the player's batting average can be modeled by a normal distribution with mean $\mu = 0.261$ and standard deviation $\sigma = 0.034$. What percent of players have a batting average of 0.300 or higher?

5. **Medium salt!** Refer to Exercise 3. What percent of the meals ordered had between 1200 mg and 1800 mg of sodium?

6. **Borderline blood pressure** Refer to Exercise 2. A diastolic blood pressure between 80 and 90 indicates borderline high blood pressure. What percent of adults have borderline high blood pressure?

7. **Put a lid on it!** At fast-food restaurants, the lids for drink cups are made with a small amount of flexibility, so they can be stretched across the mouth of the cup and then snugly secured. When lids are too small or too large, customers can get frustrated, especially if they end up spilling their drinks. At one restaurant, large drink cups require lids with a diameter of between 3.95 and 4.05 inches. The restaurant's lid supplier claims that the diameter of the large lids follows a normal distribution with mean 3.98 inches and standard deviation 0.02 inches. Assume that the supplier's claim is true. Let X = the diameter of a randomly selected large lid from this supplier. Find and interpret $P(3.95 \leq X \leq 4.05)$.

8. **Egg weights** In the United States, egg sizes are set by the Department of Agriculture. A "large" egg, for example, weighs between 57 and 64 grams. Suppose the weights of eggs produced by hens owned by a particular farmer are normally distributed with a mean of 55.8 grams and a standard deviation of 7.5 grams. Let X = the weight of a randomly selected egg from this farmer's hens. Find and interpret $P(57 \leq X \leq 64)$.

9. **Fire!** A fire department in a rural county reports that its response time to fires is approximately normally distributed with a mean of 22 minutes and a standard deviation of 11.9 minutes. Assume that this claim is true. Ninety-nine percent of the time, the fire department responds to a fire within how many minutes?

10. **Mail the letter** A local post office weighs outgoing mail and finds that the weights of first-class letters are approximately normally distributed with a mean of 0.69 ounce and a standard deviation of 0.16 ounce. Find the 60th percentile of first-class letter weights.

11. **Get a Harley!** The average sale price (online) for 4-year-old Harley Davidson touring motorcycles is approximately normally distributed with a mean of $14,000 and a standard deviation of $4000. Find the 30th percentile for the prices of this type of motorcycle.

12. **Helmet sizes** The army reports that the distribution of head circumference among soldiers is approximately normal with mean 22.8 inches and standard deviation 1.1 inches. Helmets are mass-produced for all except the smallest 5% and the largest 5% of head sizes. Soldiers in the smallest or largest 5% get custom-made helmets. What head sizes get custom-made helmets?

Applying the Concepts

13. **IQ test scores** Scores on the Wechsler Adult Intelligence Scale (a standard IQ test) for the 20-to-34 age group are approximately normally distributed with $\mu = 110$ and $\sigma = 25$.

(a) What percent of people aged 20 to 34 have IQs between 125 and 150?

(b) MENSA is an elite organization that only admits as members people who score in the top 2% on IQ tests. What score on the Wechsler Adult Intelligence Scale would an individual aged 20 to 34 have to earn to qualify for MENSA membership?

14. **Low birth weight** Researchers in Norway analyzed data on the birth weights of 400,000 newborns over a 6-year period. The distribution of birth weights is approximately normal with a mean of 3668 grams and a standard deviation of 511 grams.[18] Babies who weigh less than 2500 grams at birth are classified as "low birth weight."

(a) What percent of babies will be identified as low birth weight?

(b) Find the first and third quartiles of the birth weight distribution.

Extending the Concepts

15. **Put a bigger lid on it!** Refer to Exercise 7. The supplier is considering two changes to reduce the percent of its large-cup lids that are too small to 1%. One way is to adjust the mean diameter of its lids. Another option is to alter the production process to decrease the standard deviation of the lid diameters.

(a) If the standard deviation remains at $\sigma = 0.02$ inch, at what value should the supplier set the mean diameter of its large-cup lids so that only 1% are too small to fit? Show your method.

(b) If the mean diameter stays at $\mu = 3.98$ inches, what value of the standard deviation will result in only 1% of lids that are too small to fit? Show your method.

16. Brush your teeth The amount of time Riccardo spends brushing his teeth follows a normal distribution with unknown mean and standard deviation. Riccardo spends less than 1 minute brushing his teeth about 40% of the time. He spends more than 2 minutes brushing his teeth 2% of the time. Use this information to determine the mean and standard deviation of this distribution.

Recycle and Review

17. Binomial settings (5.3) Determine whether or not each random variable described satisfies the conditions for a binomial setting. Justify your answers.

(a) Deal 7 cards from a standard deck of 52 cards. Let H = the number of hearts dealt.

(b) A high school principal goes to 10 different classrooms and randomly selects 1 student from each class. Let X = the number of female students in his group of 10 students.

(c) You are on Interstate 80 in Pennsylvania, counting the occupants in every 5th car you pass. Let Z = the number of cars you pass before you see one with more than 2 occupants.

18. Faulty forecasts? Nate Silver is famous for making predictions. In his book *The Signal and the Noise: Why So Many Predictions Fail—But Some Don't*, Silver claimed that The Weather Channel deliberately makes poor predictions. When The Weather Channel states that there is a 20% chance of rain, it should rain 20% of the time for their predictions to be considered "calibrated." However, a study found that of 113 days that The Weather Channel predicted a 20% chance of rain, it rained only 6 times. To determine if these data provide convincing evidence that The Weather Channel deliberately makes mistakes, Chiara performed 100 repetitions of a simulation using an online spinner with 20% blue area. On each repetition, she spun the spinner 113 times and recorded the number of times the spinner landed on blue. The dotplot shows the results.

Based on the results of the simulation, is there convincing evidence that The Weather Channel deliberately makes mistakes? Explain.

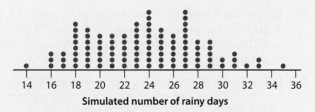

Chapter 5   RESOLVED

STATS applied!

How did that vending machine go wrong?

The STATS applied! on page 331 described a vending machine that dispenses coffee. Here is a summary of important facts about the machine:

- Cups will hold 8 fluid ounces.
- The amount of coffee dispensed varies according to a normal distribution centered at the mean μ that is set in the machine and with standard deviation $\sigma = 0.22$ ounce.

If a cup contains more than 7.4 ounces of coffee, a customer may get burned from a spill. This could result in an expensive lawsuit for the company. On the other hand, customers may become irritated if they get a cup with less than 6.75 ounces of coffee from the machine. Suppose the company sets the machine's mean at $\mu = 7$ ounces.

1. What percent of cups will get an acceptable amount of coffee—between 6.75 and 7.4 ounces?

2. A quality-control inspector measures the volume of coffee in 10 randomly selected cups. Let X = the number of cups that have an acceptable amount of coffee.

(a) Justify that X is a binomial random variable.

(b) Find $P(X \leq 5)$. Should the inspector conclude that the machine is working properly if only 5 cups have an acceptable amount of coffee? Explain.

3. The company's lawyer insists that, at most, 0.1% of cups get more than 7.4 ounces of coffee. What is the largest value of μ that meets the lawyer's requirement?

Main Points

Chapter 5

Analyzing Random Variables

- A **random variable** takes numerical values determined by the outcome of a chance process. The **probability distribution** of a random variable gives its possible values and their probabilities. There are two types of random variables: *discrete* and *continuous*.

- A **discrete random variable** has a fixed set of possible values with gaps between them.

  - A valid probability distribution assigns each of these values a probability between 0 and 1 such that the sum of all the probabilities is exactly 1.

  - We can display the probability distribution as a histogram, with the values of the variable on the horizontal axis and the probabilities on the vertical axis.

  - The probability of any event is the sum of the probabilities of all the values that make up the event.

- A **continuous random variable** takes all possible values in an interval on the number line.

  - A valid probability distribution for a continuous random variable is described by a **density curve,** a curve that is always on or above the horizontal axis, and has area exactly 1 underneath it.

- The probability of any event is the area under the density curve and directly above the values on the horizontal axis that make up the event.

- We can describe the *shape* of a probability distribution histogram or density curve in the same way as we did a distribution of quantitative data—by identifying symmetry or skewness and any major peaks.

- Use the mean to summarize the *center* of a probability distribution. The **mean of a random variable** μ_X is the balance point of the probability distribution histogram or density curve.

  - The mean is the long-run average value of the variable after many repetitions of the chance process. It is also known as the **expected value** of the random variable, $E(X)$.

  - If X is a discrete random variable, the mean is the average of the values of X, each weighted by its probability:

$$\mu_X = E(X) = x_1 p_1 + x_2 p_2 + x_3 p_3 + \cdots = \sum x_i p_i$$

- Use the standard deviation to summarize the *variability* of a probability distribution. The **standard deviation of a random variable** σ_X measures how much the values of the variable typically differ from the mean.

- If X is a discrete random variable, the standard deviation of X is

$$\sigma_X = \sqrt{(x_1 - \mu_X)^2 p_1 + (x_2 - \mu_X)^2 p_2 + \ldots}$$

$$= \sqrt{\sum (x_i - \mu_X)^2 p_i}$$

Binomial Random Variables

- A **binomial setting** arises when we perform n independent trials of the same chance process and count the number of times that a particular outcome (a "success") occurs. The conditions for a binomial setting are:

  - **Binary?** The possible outcomes of each trial can be classified as "success" or "failure."

  - **Independent?** Trials must be independent. That is, knowing the result of one trial must not tell us anything about the result of any other trial.

  - **Number?** The number of trials n of the chance process must be fixed in advance.

  - **Same probability?** There is the same probability of success p on each trial.

- The count of successes X in a binomial setting is a special type of discrete random variable known as a **binomial random variable.** Its probability distribution is a **binomial distribution.** Any binomial distribution is completely specified by two numbers: the number of trials n of the chance process and the probability of success p on any trial.

- To calculate probabilities involving a binomial random variable, use the binomial probability formula or technology.

  - *Binomial probability formula:* The probability of getting exactly k successes in n trials ($k = 0, 1, 2, \ldots, n$) is

$$P(X = k) = {}_nC_k \cdot p^k (1 - p)^{n-k}$$

where

$${}_nC_k = \frac{n!}{k!(n - k)!}$$

  - The *Probability* applet and the TI-84 will calculate binomial probabilities. The TI-84 command binompdf(n,p,k) computes $P(X = k)$ in a setting with n trials and success probability p on each trial. The TI-84 command binomcdf(n,p,k) computes $P(X \le k)$ in a setting with n trials and success probability p on each trial.

- The mean and standard deviation of a binomial random variable are

$$\mu_X = np \quad \text{and} \quad \sigma_X = \sqrt{np(1 - p)}$$

Normal Distributions

- Some continuous random variables have probability distributions that are described by symmetric, single-peaked, bell-shaped density curves called **normal distributions.**

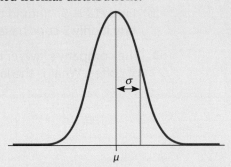

- Any normal distribution is completely specified by two numbers: its mean μ and standard deviation σ. The mean is the center of the curve, and σ is the distance from μ to the change-of-curvature points on either side.

- The **68–95–99.7 rule** describes the approximate percent of values in any normal distribution that fall within 1, 2, and 3 standard deviations of the mean.

- All normal distributions are the same when values are standardized. If X follows a normal distribution with mean μ and standard deviation σ, we can standardize using

$$z = \frac{\text{value} - \text{mean}}{\text{standard deviation}} = \frac{x - \mu}{\sigma}$$

The random variable Z has the **standard normal distribution** with mean 0 and standard deviation 1.

- Table A in the back of the book is a table of standard normal probabilities. The table entry for each z-score is the area under the curve to the left of z. We can use Table A to find a probability (area) from a z-score or to find a z-score from a probability (area) in the standard normal distribution.

- For *any* normal distribution, we can standardize and use Table A to:

  - Calculate the probability that a value falls in a given interval.

  - Find the value corresponding to a given probability.

- The *Probability* applet and the TI-83/84 will perform normal calculations. The TI-83/84 command normalcdf(lower,upper,mean,SD) computes the probability for an interval of values in a normal distribution. The command invNorm(area,mean,SD) finds the value in a normal distribution that has a given area to the left.

Chapter 5 Review Exercises

1. **Televisions (5.1)** If X = the number of televisions in a randomly selected U.S. household, the probability distribution of X is given here.[19]

Number of TVs	0	1	2	3	4	5
Probability	0.01	0.21	0.33	0.23	0.13	0.09

(a) Explain why this is a valid probability distribution.

(b) Is X a discrete or a continuous random variable? Explain.

(c) What is the probability that a randomly selected household has at least 2 televisions?

2. **Televisions (5.2)** Refer to the random variable X described in Exercise 1.

(a) Make a histogram to display the probability distribution of X.

(b) Calculate and interpret the expected value of X.

(c) Calculate and interpret the standard deviation of X.

3. **ESP (5.3, 5.4)** To test whether someone has extra-sensory perception (ESP), choose 1 of 4 cards at random—a star, wave, cross, or circle. Ask the person to identify the card without seeing it. Do this a total of 20 times and see how many cards the person identifies correctly. Let X = the number of correct identifications, assuming that the person does not have ESP and is just guessing on each card.

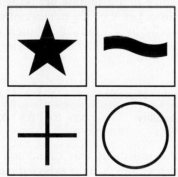

(a) Explain why X is a binomial random variable.

(b) Find and interpret the mean of X.

(c) Find and interpret the standard deviation of X.

4. **ESP (5.3, 5.4)** Refer to Exercise 3.

(a) Use the binomial probability formula to calculate $P(X = 5)$. Interpret this result in context.

(b) Alec makes 8 correct identifications out of 20 cards. Does this provide convincing evidence that Alec has ESP? Find the probability that a person without ESP identifies 8 or more cards correctly and use the result to answer the question.

5. **Fetch, Bucket! (5.5)** Kristen likes to throw tennis balls for her dog Bucket. Let T = the time it takes for Bucket to chase down a ball, return to Kristen, and drop the ball at her feet on a randomly selected throw. T is equally likely to take any value in the interval from 8 seconds to 15 seconds.

(a) Sketch a graph of the probability distribution of T. Be sure to include a vertical scale.

(b) Find the probability that, on a randomly selected throw, Bucket returns the ball within 13 seconds.

(c) Find the mean of T.

6. **Horse pregnancies (5.5, 5.6)** Bigger animals tend to carry their young longer before birth. The length of horse pregnancies from conception to birth varies according to a roughly normal distribution with mean 336 days and standard deviation 6 days.

(a) Sketch a graph of the distribution of lengths of horse pregnancies. Label the mean and the points that are 1, 2, and 3 standard deviations from the mean.

(b) Use the 68–95–99.7 rule to estimate the percentage of horse pregnancies that are longer than 342 days.

7. **Standard normal distribution (5.6)** Use Table A to find each of the following. In each case, draw a standard normal distribution with the area of interest shaded.

(a) $P(Z > 1.77)$

(b) The value z such that 35% of all observations are greater than z

8. **Ketchup (5.7)** A fast-food restaurant has just installed a new automatic ketchup dispenser for use in preparing its burgers. The amount of ketchup dispensed by the machine follows a normal distribution with mean 1.05 fluid ounces and standard deviation 0.08 fluid ounce.

(a) If the restaurant's goal is to put between 1 and 1.2 fluid ounces of ketchup on each burger, what percent of the time will this happen? Show your work.

(b) Find the 99th percentile of the distribution of amount of ketchup dispensed by the machine.

Chapter 5 Practice Test

Section I: Multiple Choice *Select the best answer for each question.*

1. An ecologist studying starfish populations records values for each of the following variables from randomly selected plots on a rocky coastline.

X = The number of starfish in the plot
Y = The total weight of starfish in the plot
W = The percent of area in the plot that is covered by barnacles (a popular food for starfish)
T = The primary type of rock in the plot

How many of these are continuous random variables and how many are discrete random variables?

(a) Two continuous, two discrete
(b) One continuous, three discrete
(c) Two continuous, one discrete, and one that is not a random variable
(d) One continuous, two discrete, and one that is not a random variable

2. Which of the following is closest to the 28th percentile in the standard normal distribution?

(a) –0.58 (b) –0.50 (c) 0.39 (d) 0.61

3. A hat holds a large number of slips of paper, each with a single digit from 1 to 4 on it. Which one of the following is a possible probability distribution for X = the number on a randomly selected slip of paper?

(a)

x_i	p_i
0	0.2
1	0.2
2	0.1
3	0.1
4	0.1

(b)

x_i	p_i
0	0.25
1	0.05
2	0.3
3	0.2
4	0.2

(c)

x_i	p_i
1	0.25
2	0.15
3	0.15
4	0.15

(d)

x_i	p_i
1	0.05
2	0.15
3	0.5
4	0.3

4. A private school has 25 students in each of grades 1 through 12. The principal randomly selects 1 student from each grade and records whether or not the student is over 5 feet tall. Let X = the number of students in the sample who are over 5 feet tall. Which of the following requirements for a binomial setting is *violated* in this case?

(a) There are two possible outcomes for each trial.
(b) The trials are independent.
(c) The probability of "success" is the same for each trial.
(d) The number of trials is fixed.

5. The figure shows the density curve that is the probability distribution of a continuous random variable. Seven values are marked on the density curve. Which of the following statements is true?

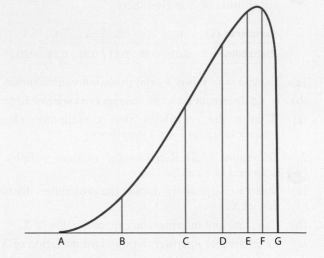

(a) The mean of the distribution is E.
(b) The median of the distribution is C.
(c) The third quartile of the distribution is D.
(d) The area under the curve between A and G is 1.

Questions 6 and 7 refer to the following setting. A psychologist studied the number of puzzles that subjects were able to solve in a 5-minute period while listening to soothing music. Let X be the number of puzzles completed successfully by a randomly chosen subject. The psychologist found that X had the following probability distribution:

Number of puzzles	1	2	3	4
Probability	0.2	0.4	0.3	0.1

6. What is the probability that a randomly chosen subject completes more than the expected number of puzzles in the 5-minute period while listening to soothing music?

(a) 0.1 (b) 0.4 (c) 0.5 (d) 0.8

7. The standard deviation of X is 0.9. Which of the following is the best interpretation of this value?

(a) About 95% of subjects solved between 1.8 puzzles less and 1.8 puzzles more than the mean.
(b) The typical subject solved an average of 0.9 puzzles.
(c) The number of puzzles solved by subjects typically differed from the mean by about 0.9 puzzles.
(d) The number of puzzles solved by subjects typically differed from one another by about 0.9 puzzles.

8. For the normal distribution shown, the standard deviation is closest to

(a) 1   (b) 2   (c) 3   (d) 6

9. In the game Pass the Pigs, a small plastic pig is rolled, and its orientation when it comes to a stop determines the number of points a player scores. Approximately 9% of rolls produce a "trotter"—that is, the pig lands right side up, standing on its four feet.[20] Suppose you roll a pig 10 times. Which of the following expressions represents the probability that you roll exactly 2 trotters?

(a) $(0.09)^2(0.91)^8$

(b) $_{10}C_2 \cdot (0.09)^2(0.91)^8$

(c) $(0.09)^2(0.91)^{10}$

(d) $_8C_2 \cdot (0.09)^2(0.91)^8$

10. If the heights of a population of men follow a normal distribution, and 99.7% have heights between 5 feet 0 inches and 7 feet 0 inches, what is your estimate of the standard deviation of the heights in this population?

(a) 1 inch   (c) 4 inches

(b) 3 inches   (d) 6 inches

Section II: Free Response

11. Let Y denote the number of broken eggs in a randomly selected carton of one dozen "store brand" eggs at a local supermarket. Suppose that the probability distribution of Y is as follows.

Value	0	1	2	3	4
Probability	0.77	0.11	0.08	0.03	0.01

(a) What is the probability that at least 2 eggs in a randomly selected carton are broken?

(b) Calculate and interpret the mean of Y.

(c) Calculate and interpret the standard deviation of Y.

12. An airline claims that its 7:00 A.M. New York to Los Angeles flight has an 85% chance of arriving on time on a randomly selected day. Assume for now that this claim is true. Suppose we take a random sample of 20 of these flights. Let X = the number of flights that arrive on time.

(a) Explain why X is a binomial random variable.

(b) Use the binomial probability formula to find $P(X = 19)$. Interpret this value in context.

(c) Calculate the mean and standard deviation of X.

(d) If only 14 of the 20 flights arrive on time, do we have convincing evidence that the airline's claim is false? Calculate $P(X \le 14)$ if the company's claim is true and use this result to support your answer.

13. Professional tennis player Novak Djokovic hits the ball extremely hard. His first-serve speeds follow a normal distribution with mean 112 miles per hour (mph) and standard deviation 6 mph. Choose one of Djokovic's first serves at random. Let Y = its speed, measured in miles per hour.

(a) A first serve with a speed less than 100 miles per hour is considered "slow." What percent of Djokovic's first serves are slow?

(b) Calculate $P(120 \le Y \le 125)$ and interpret the result.

(c) Find the 15th percentile of Novak Djokovic's first-serve speed distribution.

6

Sampling Distributions

Lesson 6.1 What Is a Sampling Distribution?   400

Lesson 6.2 Sampling Distributions: Center and Variability   409

Lesson 6.3 The Sampling Distribution of a Sample Count
(The Normal Approximation to the Binomial)   417

Lesson 6.4 The Sampling Distribution of a Sample Proportion   424

Lesson 6.5 The Sampling Distribution of a Sample Mean   432

Lesson 6.6 The Central Limit Theorem   439

Chapter 6   Main Points   445

Chapter 6   Review Exercises   447

Chapter 6   Practice Test   448

398

Steve Gorton and Gary Ombler/Getty Images

STATS applied!

How can we build "greener" batteries?

Kids love getting toys for their birthdays, especially electronic ones that have flashing lights and make loud noises. But these devices require lots of power and can drain batteries quickly. Battery manufacturers are constantly searching for ways to build longer-lasting batteries.

When the manufacturing process is working correctly, AA batteries from a particular company should last an average of 17 hours, with a standard deviation of 0.8 hours. Also, at least 73% of the batteries should last 16.5 hours or more.

Quality-control inspectors select a random sample of 50 batteries during each hour of production and then drain them under conditions that mimic normal use. The graph and summary statistics describe the distribution of the lifetimes (in hours) of the batteries from one sample of 50 AA batteries.

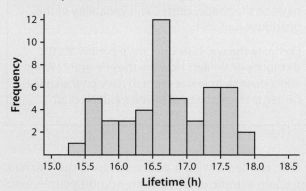

n	Mean	SD	min	Q_1	med	Q_3	max
50	16.718	0.66	15.46	16.31	16.7	17.28	17.98

> Do these data suggest that the production process isn't working properly? Or is it safe for plant managers to send out all the batteries produced in this hour for sale?

We'll revisit STATS applied! at the end of the chapter, so you can use what you have learned to help answer these questions.

Lesson 6.1

What Is a Sampling Distribution?

LEARNING TARGETS

- Distinguish between a parameter and a statistic.
- Create a sampling distribution using all possible samples from a small population.
- Use the sampling distribution of a statistic to evaluate a claim about a parameter.

ACTIVITY

A penny for your thoughts?

In this activity, your class will investigate how the mean year $\bar{x}$ and the proportion of pennies from the 2000s $\hat{p}$ vary from sample to sample, using a large population of pennies of various ages.[1]

1. Have each member of the class randomly select 1 penny from the population and record the year of the penny with an "X" on the dotplot provided by your teacher. Return the penny to the population. Repeat this process until at least 100 pennies have been selected and recorded. This graph gives you an idea of what the population distribution of penny years looks like.

2. Have each member of the class take an SRS of 5 pennies from the population and note the year on each penny.

    - Record the average year of these 5 pennies with an "$\bar{x}$" on a new class dotplot. Make sure this dotplot is on the same scale as the dotplot in Step 1 above.

    - Record the proportion of pennies from the 2000s with a "$\hat{p}$" on a different dotplot provided by your teacher.

    Return the pennies to the population. Repeat this process until there are at least 100 $\bar{x}$'s and 100 $\hat{p}$'s.

3. Repeat Step 2 with SRSs of size $n = 20$. Make sure these dotplots are on the same scale as the corresponding dotplots from Step 2 above.

4. Compare the distribution of X (year of penny) with the two distributions of $\bar{x}$ (mean year). How are the distributions similar? How are they different? What effect does sample size seem to have on the shape, center, and variability of the distribution of $\bar{x}$?

5. Compare the two distributions of $\hat{p}$. How are the distributions similar? How are they different? What effect does sample size seem to have on the shape, center, and variability of the distribution of $\hat{p}$?

To estimate the mean income of U.S. residents with a college degree, the Current Population Survey (CPS) selected a random sample of more than 60,000 people with at least a bachelor's degree. The mean income *in the sample* was $69,609.[2] How close is this estimate to the mean income for *all* members of the population? To find out how an estimate varies from sample to sample, we want to gain some understanding of *sampling distributions*.

Parameters and Statistics

For the sample of college graduates contacted by the CPS, the mean income was $\bar{x}$ = $69,609. The number $69,609 is a **statistic** because it describes this one *sample*. The population that the researchers want to draw conclusions about is all U.S. college graduates. In this case, the **parameter** of interest is the mean income μ of the *population* of all college graduates.

DEFINITION Statistic, Parameter

A **statistic** is a number that describes some characteristic of a sample.

A **parameter** is a number that describes some characteristic of the population.

Because we can't examine the entire population, the value of a parameter is usually unknown. To estimate the value of the parameter, we use a statistic calculated using data from a random sample of the population.

Remember **s** and **p**: statistics come from samples, and parameters come from populations. The notation we use should reflect this distinction. For example, we write μ (the Greek letter mu) for the population mean and $\bar{x}$ for the sample mean. The table lists some additional examples of statistics and their corresponding parameters.

Sample statistic		Population parameter
$\bar{x}$ (the sample mean)	estimates	μ (the population mean)
$\hat{p}$ (the sample proportion)	estimates	p (the population proportion)
s (the sample SD)	estimates	σ (the population SD)

EXAMPLE

How are teens different from turkeys?

Parameters and statistics

PROBLEM: Identify the population, the parameter, the sample, and the statistic in each of the following settings:

(a) A Pew Research Center poll asked 1102 12- to 17-year-olds in the United States if they have a cell phone. Of the respondents, 71% said "Yes."[3]

(b) Tom is roasting a large turkey breast for a holiday meal. He wants to be sure that the turkey is safe to eat, which requires a minimum internal temperature of 165°F. Tom uses a thermometer to measure the temperature of the turkey breast at four randomly chosen points. The minimum reading he gets is 170°F.

SOLUTION:

(a) Population: all 12- to 17-year-olds in the United States. Parameter: p = the proportion of all 12- to 17-year-olds with cell phones. Sample: the 1102 12- to 17-year-olds contacted. Statistic: the sample proportion with a cell phone, $\hat{p} = 0.71$.

(b) Population: all possible locations in the turkey breast. Parameter: the true minimum temperature in all possible locations. Sample: the four randomly chosen locations. Statistic: the sample minimum, 170°F.

FOR PRACTICE TRY EXERCISE 1.

While some parameters and statistics have special symbols (such as p for the population proportion and $\hat{p}$ for the sample proportion), many parameters and statistics do not have their own symbol. To distinguish between a parameter and statistic, use descriptors such as "true" minimum and "sample" minimum as we did in the turkey example.

Sampling Distributions

In the Penny for Your Thoughts Activity, you encountered sampling variability—meaning that different random samples of the same size from the same population produce different values of a statistic. The statistics that come from these samples form a **sampling distribution.**

> **DEFINITION** Sampling distribution
>
> The **sampling distribution** of a statistic is the distribution of values taken by the statistic in all possible samples of the same size from the same population.

Remember that a distribution describes the possible values of a variable and how often these values occur. The easiest way to picture a distribution is with a graph, such as a dotplot or histogram.

EXAMPLE

Just how tall are their sons?

Sampling distributions

PROBLEM: John and Carol have four grown sons. Their heights (in inches) are 71, 75, 72, and 68. List all 6 possible SRSs of size $n = 2$, calculate the mean height for each sample, and display the sampling distribution of the sample mean on a dotplot.

SOLUTION:

Sample 1: 71, 75 $\bar{x} = 73$ Sample 4: 75, 72 $\bar{x} = 73.5$
Sample 2: 71, 72 $\bar{x} = 71.5$ Sample 5: 75, 68 $\bar{x} = 71.5$
Sample 3: 71, 68 $\bar{x} = 69.5$ Sample 6: 72, 68 $\bar{x} = 70$

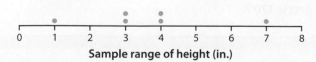

FOR PRACTICE   TRY EXERCISE 5.

Every statistic has its own sampling distribution. For example, Figure 6.1 shows the sampling distribution of the sample range of height for SRSs of size $n = 2$ from John and Carol's four sons.

Sample 1: 71, 75   sample range = 4   Sample 4: 75, 72   sample range = 3

Sample 2: 71, 72   sample range = 1   Sample 5: 75, 68   sample range = 7

Sample 3: 71, 68   sample range = 3   Sample 6: 72, 68   sample range = 4

FIGURE 6.1 Dotplot showing the sampling distribution of the sample range of height for SRSs of size $n = 2$.

Be specific when you use the word "distribution." There are three different types of distributions in this setting:

1. The distribution of height in the population (the four heights):

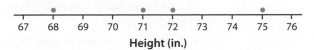

Height (in.)

2. The distribution of height in a particular sample (two of the heights):

Height (in.)

3. The sampling distribution of the sample range for all possible samples (the six sample ranges):

Sample range of height (in.)

Notice that the first two distributions consist of heights (data values), while the third distribution consists of ranges (statistics). Lesson: Always use "the distribution of __" and never just "the distribution."

Using Sampling Distributions to Evaluate Claims

Sampling distributions are the foundation for the methods of statistical inference you will learn about in Chapters 7–10. Knowing the sampling distribution of a statistic will help us know how much the statistic tends to vary from its corresponding parameter and what values of the statistic should be considered unusual.

EXAMPLE

How long will we bead doing the homework?

Evaluating a claim

PROBLEM: At the beginning of class, Mrs. Chauvet shows her class a box filled with black and white beads. She claims that the proportion of black beads in the box is $p = 0.50$. To determine the number of homework exercises she will assign that evening, she invites a student to select an SRS of $n = 30$ beads from the box. The number of black beads selected will be the number of homework exercises assigned. When the student selects 19 black beads ($\hat{p} = 19/30 = 0.63$), the students groan and suggest that Mrs. Chauvet included more than 50% black beads in the box.

 To determine if a sample proportion of $\hat{p} = 0.63$ provides convincing evidence that Mrs. Chauvet cheated, the class simulated 100 SRSs of size $n = 30$, assuming that she was telling the truth. That is, they sampled from a population with 50% black beads. For each sample, they recorded the sample proportion of black beads. The results of the simulation are shown on the next page.

© Monalyn Gracia/Corbis

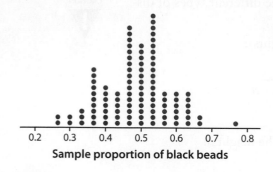

Sample proportion of black beads

(a) There is one dot on the graph at $\hat{p} = 0.77$. Explain what this dot represents.

(b) Would it be unusual to get a sample proportion of 0.63 or higher in a sample of size 30 when $p = 0.50$? Explain.

(c) Based on your answer to part (b), is there convincing evidence that Mrs. Chauvet lied about the contents of the box?

SOLUTION:

(a) In one simulated SRS of size $n = 30$, 77% of the beads were black.

(b) No. In the 100 simulated samples, 9 of the SRSs included at least 63% black beads.

(c) No. Because the probability from part (b) isn't that small, it is plausible that the proportion of black beads in the box is $p = 0.50$ and the student got a sample proportion of $\hat{p} = 0.63$ by chance alone.

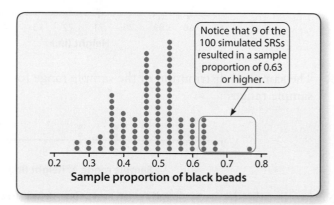

Notice that 9 of the 100 simulated SRSs resulted in a sample proportion of 0.63 or higher.

Sample proportion of black beads

FOR PRACTICE   TRY EXERCISE 9.

We used 100 simulated samples to produce the dotplot of sample proportions in this example. Because it doesn't include all possible samples of size 30, it is only an approximation of the actual sampling distribution of $\hat{p}$. Thankfully, the simulated sampling distribution should be a good approximation as long as we use a large number of samples in the simulation.

LESSON APP 6.1

How cold is it inside the cabin?

During the winter months, outside temperatures at the Starneses' cabin in Colorado can stay well below freezing (32°F, or 0°C) for weeks at a time. To prevent the pipes from freezing, Mrs. Starnes sets the thermostat at 50°F. The manufacturer claims that the thermostat allows variation in home temperature that follows a normal distribution with $\sigma = 3$°F. To test this claim, Mrs. Starnes programs her digital thermostat to take an SRS of $n = 10$ readings during a 24-hour period. The standard deviation of the results is $s_x = 5$°F.

1. Identify the population, the parameter, the sample, and the statistic in this context.

Suppose the thermostat is working properly and that the temperatures in the cabin vary according to a normal distribution with mean $\mu = 50$°F and standard deviation $\sigma = 3$°F. The dotplot shows the distribution of the sample standard deviation in 100 simulated SRSs of size $n = 10$ from this distribution.

Quasarphoto/Getty Images

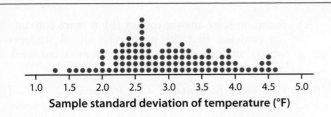
Sample standard deviation of temperature (°F)

2. Would it be unusual to get a sample standard deviation of $s_x = 5$°F or higher in a sample of size $n = 10$ when $\sigma = 3$°F? Explain.

3. Based on your answer to Question 2, is there convincing evidence that the thermometer is more variable than the manufacturer claims? Explain.

Lesson 6.1

WHAT DID YOU LEARN?

LEARNING TARGET	EXAMPLES	EXERCISES
Distinguish between a parameter and a statistic.	p. 401	1–4
Create a sampling distribution using all possible samples from a small population.	p. 402	5–8
Use the sampling distribution of a statistic to evaluate a claim about a parameter.	p. 403	9–12

Lesson 6.1

Mastering Concepts and Skills

For Exercises 1–4, identify the population, the parameter, the sample, and the statistic in each setting.

1. **Smoking and height**

   (a) From a large group of people who signed a card saying they intended to quit smoking, a random sample of 1000 people was selected. It turned out that 210 (21%) of the sampled individuals had not smoked over the past 6 months.

   (b) A pediatrician wants to know the 75th percentile for the distribution of heights of 10-year-old boys, so she selects a sample of 50 10-year-old male patients and calculates that the 75th percentile in the sample is 56 inches.

2. **Unemployment and gas prices**

   (a) Each month, the Current Population Survey interviews a random sample of individuals in about 60,000 U.S. households. One of its goals is to estimate the national unemployment rate. In January 2015, 5.7% of those interviewed were unemployed.

   (b) How much do gasoline prices vary in a large city? To find out, a reporter records the price per gallon of regular unleaded gasoline at a random sample of 10 gas stations in the city on the same day.

The range (Maximum − Minimum) of the prices in the sample is 25 cents.

3. **Tea and screening**

   (a) On Tuesday, the bottles of iced tea filled in a plant were supposed to contain an average of 20 ounces of iced tea. Quality-control inspectors sampled 50 bottles at random from the day's production. These bottles contained an average of 19.6 ounces of iced tea.

   (b) On a New York–Denver flight, 8% of the 125 passengers were selected for random security screening before boarding. According to the Transportation Security Administration, 10% of passengers at this airport are supposed to be chosen for random screening.

4. **Bearings and thermostats**

   (a) A production run of ball bearings is supposed to have a mean diameter of 2.5000 centimeters. An inspector chooses a random sample of 100 bearings from the container and calculates a mean diameter of 2.5009 centimeters.

   (b) During the winter months, Mrs. Starnes sets the thermostat at 50°F to prevent the pipes from freezing in her cabin. She wants to know how low the interior temperature gets. A digital thermometer records the indoor temperature at 20 randomly chosen times during a given day. The minimum reading is 38°F.

Exercises 5–8 refer to the following population of 2 male students and 3 female students, along with their quiz scores:

> Abigail 10   Bobby 5   Carlos 10   DeAnna 7   Emily 9

5. **Sample means** List all 10 possible SRSs of size $n = 2$, calculate the mean quiz score for each sample, and display the sampling distribution of the sample mean on a dotplot.
   [pg 402]

6. **Sample ranges** List all 10 possible SRSs of size $n = 3$, calculate the range of quiz scores for each sample, and display the sampling distribution of the sample range on a dotplot.

7. **Sample proportions** List all 10 possible SRSs of size $n = 2$, calculate the proportion of females for each sample, and display the sampling distribution of the sample proportion on a dotplot.

8. **Sample medians** List all 10 possible SRSs of size $n = 3$, calculate the median quiz score for each sample, and display the sampling distribution of the sample median on a dotplot.

9. **Who does their homework?** A school newspaper article claims that 60% of the students at a large high school completed their assigned homework last week. Some statistics students want to investigate if this claim is true, so they choose an SRS of 100 students from the school to interview. When they found that only 45 of the 100 students completed their assigned homework last week, they suspected that the proportion of all students who completed their assigned homework last week is less than the 60% claimed by the newspaper.
   [pg 403]

   To determine if a sample proportion of $\hat{p} = 0.45$ provides convincing evidence that the true proportion is less than $p = 0.60$, the class simulated 250 SRSs of size $n = 100$ from a population in which $p = 0.60$. Here are the results of the simulation.

   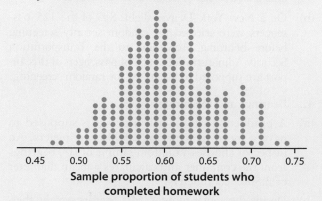

   Sample proportion of students who completed homework

   (a) There is one dot on the graph at 0.73. Explain what this dot represents.

   (b) Would it be surprising to get a sample proportion of 0.45 or less in an SRS of size 100 when $p = 0.60$? Explain.

(c) Based on your answer to part (b), is there convincing evidence that the proportion of all students who completed their assigned homework last week is less than $p = 0.60$? Explain.

10. **First-serve percentage** One important aspect of a tennis player's effectiveness is her first-serve percentage—the proportion of the time the first of her two attempts to serve the ball to her opponent is successful. For her first three years on the tennis team, Shruti's first-serve percentage is 53%. Hoping to improve, Shruti works over the summer with a coach who specializes in serves. In her first match of the next season, Shruti's first serve is successful 42 times in 60 attempts, a first-serve percentage of 70%.

    Suppose we treat Shruti's first 60 attempts as an SRS of her serves after working with the new coach. To determine if a sample proportion of $\hat{p} = 0.70$ provides convincing evidence that the true proportion is greater than $p = 0.53$, we simulate 200 SRSs of size $n = 60$ from a population in which $p = 0.53$. Here are the results of the simulation.

    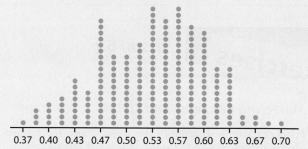

    Sample proportion of successful first serves

    (a) There is one dot on the graph at 0.37. Explain what this dot represents.

    (b) Would it be surprising to get a sample proportion of 0.70 or more in an SRS of size 60 when $p = 0.53$? Explain.

    (c) Based on your answer to part (b), is there convincing evidence that Shruti's first-serve percentage has improved since working with the new coach? Explain.

11. **Are we taller?** According to the National Center for Health Statistics, the distribution of heights for 16-year-old females is modeled well by a normal distribution with mean $\mu = 64$ inches and standard deviation $\sigma = 2.5$ inches. To see if this distribution applies at their high school, a statistics class takes an SRS of 20 of the 300 16-year-old females at the school and measures their heights. When they calculate a sample mean of 64.7 inches, they wonder if the population of 16-year-old girls at their school has a mean height greater than 64 inches.

    To determine if a sample mean of $\bar{x} = 64.7$ inches provides convincing evidence that the average height of 16-year-old girls at the school is taller than 64 inches, the class simulated 200 SRSs of size $n = 20$ from a normal population with mean $\mu = 64$ inches and standard deviation $\sigma = 2.5$ inches. Here are the results of the simulation.

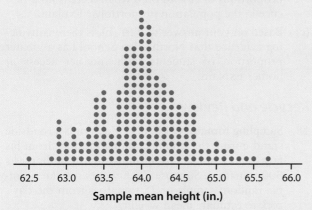

Sample mean height (in.)

(a) There is one dot on the graph at 62.5. Explain what this dot represents.

(b) Would it be unusual to get a sample mean of 64.7 or more in a sample of size 20 when $\mu = 64$? Explain.

(c) Based on your answer to part (b), is there convincing evidence that the mean height of the population of 16-year-old girls at this school is greater than 64 inches? Explain.

12. **Relying on bathroom scales** A manufacturer of bathroom scales says that when a 150-pound weight is placed on a scale produced in the factory, the weight indicated by the scale is normally distributed with a mean of 150 pounds and a standard deviation of 2 pounds. A consumer-advocacy group acquires an SRS of 12 scales from the manufacturer and places a 150-pound weight on each one. The group gets a mean weight of 149.1 pounds, which makes them suspect that the scales underestimate the true weight. To test this, they use a computer to simulate 200 samples of 12 scales from a population with a mean of 150 pounds and standard deviation of 2 pounds. Here is a dotplot of the means from these 200 samples.

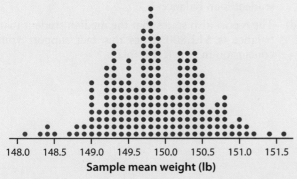

Sample mean weight (lb)

(a) There is one dot on the graph at 151.2. Explain what this dot represents.

(b) Would it be unusual to get a sample mean of 149.1 or less in a sample of size 12 when $\mu = 150$? Explain.

(c) Based on your answer to part (b), is there convincing evidence that the scales produced by this manufacturer underestimate true weight? Explain.

Applying the Concepts

13. **Instant winners** A fast-food restaurant promotes certain food items by giving a game piece with each item. Advertisements proclaim that "25% of the game pieces are Instant Winners!" To test this claim, a frequent diner collects 20 game pieces and gets only 3 instant winners.

(a) Identify the population, the parameter, the sample, and the statistic in this context.

Suppose the advertisements are correct and $p = 0.25$. The dotplot shows the distribution of the sample proportion of instant winners in 100 simulated SRSs of size $n = 20$.

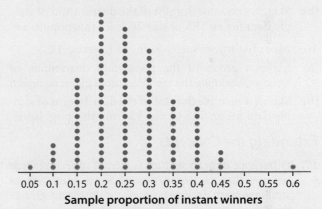

Sample proportion of instant winners

(b) Would it be unusual to get a sample proportion of $\hat{p} = 3/20 = 0.15$ or less in a sample of size $n = 20$ when $p = 0.25$? Explain.

(c) Based on your answer to part (b), is there convincing evidence that fewer than 25% of all game pieces are instant winners? Explain.

14. **Puny guppies?** A large pet store that specializes in tropical fish has several thousand guppies. The store claims that the lengths of its guppies are approximately normally distributed with a mean of 5 centimeters and a standard deviation of 0.5 centimeter. You come to the store and buy 10 randomly selected guppies and find that the mean length of your 10 guppies is only 4.8 centimeters.

(a) Identify the population, the parameter, the sample, and the statistic in this context.

Suppose the store's description of the lengths of its guppies is true. The dotplot on the next page shows the distribution of sample means from 200 simulated SRSs of size $n = 10$ from a normally distributed population with $\mu = 5$ centimeters and $\sigma = 0.5$ centimeter.

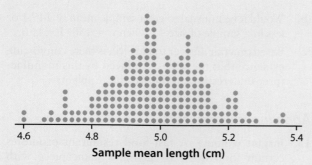

Sample mean length (cm)

(b) Would it be unusual to get a sample mean of $\bar{x} = 4.8$ centimeters or less in a sample of size $n = 10$ from this population? Explain.

(c) Based on your answer to part (b), is there convincing evidence that the mean length of guppies at this store is less than 5 centimeters? Explain.

15. More tall girls Refer to Exercise 11.

(a) Make a graph of the population distribution of heights for 16-year-old females.

(b) Sketch a possible dotplot of the distribution of sample data for an SRS of size 20 from this population.

16. More bathroom scales Refer to Exercise 12.

(a) Make a graph of the population distribution of weights, assuming the manufacturer's claim is correct.

(b) Sketch a possible dotplot of the distribution of sample data for an SRS of size 12 from this population.

Extending the Concepts

17. Difference of proportions A school superintendent believes that the proportion of North High School students with Internet access at home is greater than the proportion of South High School students with Internet access at home. To investigate, she selects SRSs of size $n = 50$ from each school and finds $\hat{p}_N = 46/50 = 0.92$ and $\hat{p}_S = 36/50 = 0.72$.

To determine if a difference in proportions of 0.20 provides convincing evidence that North High School has a greater proportion of students with Internet access at home, we simulated two random samples of size $n = 50$ from populations with the same proportion of students with Internet access. Then, we subtracted the sample proportions. Here are the results from repeating this process 100 times.

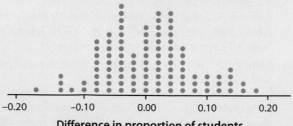

Difference in proportion of students with Internet access at home

(a) There are ten dots at 0. Explain what these dots represent.

(b) Would it be unusual to get a difference in sample proportions of at least 0.20 when there is no difference in the population proportions? Explain.

(c) Based on your answer to part (b), is there convincing evidence that North High School has a greater proportion of students with Internet access at home? Explain.

Recycle and Review

18. Sampling tomatoes (4.8, 6.1) Zach runs a roadside stand during the summer, selling produce from his farm. On a single day in mid-August, he harvests 300 tomatoes. Suppose Zach wants to take a simple random sample of 25 tomatoes from the day's pick to estimate mean weight.

(a) How many possible sets of 25 tomatoes could be sampled from the 300 tomatoes in the day's crop?

(b) What does this say about the practicality of examining the complete sampling distribution of the sample mean for samples of size 25 from this population?

19. College debt (5.7) A report published by the Federal Reserve Bank of New York in 2012 reported the results of a nationwide study of college student debt. Researchers found that the average student loan balance per borrower is $23,300. They also reported that about one-quarter of borrowers owe more than $28,000.[4]

(a) Assuming that the distribution of student loan balances is approximately normal, estimate the standard deviation of the distribution of student loan balances.

(b) Assuming that the distribution of student loan balances is approximately normal, use your answer to part (a) to estimate the proportion of borrowers who owe more than $54,000.

(c) In fact, the report states that about 10% of borrowers owe more than $54,000. What does this fact indicate about the shape of the distribution of student loan balances?

(d) The report also states that the median student loan balance is $12,800. Does this fact support your conclusion in part (c)? Explain.

Lesson 6.2

Sampling Distributions: Center and Variability

ACTIVITY

How many craft sticks are in the bag?

In this activity, you will create a statistic for estimating the total number of craft sticks in a bag (N). The sticks are numbered 1, 2, 3, . . . , N. Near the end of the activity, your teacher will select a random sample of $n = 7$ sticks and read the number on each stick to the class. The team that has the best estimate for the total number of sticks will win a prize.

1. Form teams of three or four students. As a team, spend about 10 minutes brainstorming different ways to estimate the total number of sticks. Try to come up with at least three different statistics.

2. Before your teacher provides the sample of sticks, use simulation to investigate the sampling distribution of each statistic. For the simulation, assume that there are $N = 100$ sticks in the bag and that you will be selecting samples of size $n = 7$.

- Using your TI-83/84 calculator, select an SRS of size 7 using the command RandIntNoRep(lower: 1,upper:100,n:7). [With older OS, use the command RandInt(lower:1,upper:100,n:7) and verify that there are no repeated numbers. If there are repeats, press [ENTER] to get a new sample.]

- For each sample, calculate the value of each of your three statistics.

- Graph these values on a set of dotplots like those shown here.

- Perform as many trials of the simulation as possible.

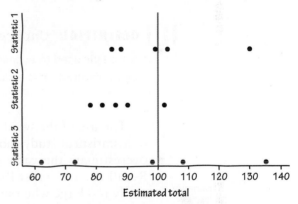

3. Based on the simulated sampling distributions, which of your statistics is likely to produce the best estimate? Discuss as a team.

4. Your teacher will now draw a sample of $n = 7$ sticks from the bag. On a piece of paper, write the names of your group members, your group's estimate for the number of sticks in the bag (a number), and the statistic you used to calculate your estimate (a formula).

Unbiased Estimators

In the craft sticks activity, the goal was to estimate the maximum value in a population, with the assumption that the members of the population are numbered 1, 2, . . . , N. Two possible statistics that might be used to estimate N are the sample maximum (max) and twice the sample median (2 × median).

Assuming that the population has $N = 100$ members and we use SRSs of size $n = 7$, Figure 6.2 shows the simulated sampling distributions of the sample maximum and twice the sample median.

409

FIGURE 6.2 Simulated sampling distributions of the sample maximum and twice the sample median for samples of size $n = 7$ from a population with $N = 100$.

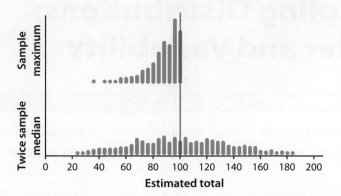

These simulated sampling distributions look quite different. The sampling distribution of the sample maximum is skewed left, while the sampling distribution of twice the sample median is roughly symmetric.

The values of the sample maximum are consistently less than the population maximum N. However, the values of twice the sample median aren't consistently less than or consistently greater than the population maximum N. It appears that twice the sample median might be an **unbiased estimator** of the population maximum, while the sample maximum is clearly biased.

> ||| **DEFINITION** Unbiased estimator
>
> A statistic used to estimate a parameter is an **unbiased estimator** if the mean of its sampling distribution is equal to the value of the parameter being estimated.

The use of the word "bias" here is consistent with its use in Chapter 3. The design of a statistical study shows bias if it would consistently underestimate or consistently overestimate the value you want to know when you repeat the study many times. Recall the Federalist Papers activity (page 188) in which the estimates were consistently too large when students were allowed to choose the words in the sample. Don't trust an estimate that comes from a biased sampling method.

CAUTION

EXAMPLE

Why do we divide by n − 1?

Unbiased estimators

PROBLEM: In Chapter 1, you learned to calculate the standard deviation of sample data using the formula

$$s_x = \sqrt{\frac{\sum (x_i - \bar{x})^2}{n - 1}}$$

What if you divided by n instead of $n - 1$? Let's simulate the sampling distributions of two statistics that can be used to estimate the variance of a distribution, where the variance is the square of the standard deviation (variance = standard deviation²).

Statistic 1: $\dfrac{\sum (x_i - \bar{x})^2}{n - 1}$ Statistic 2: $\dfrac{\sum (x_i - \bar{x})^2}{n}$

These simulated sampling distributions are based on 1000 SRSs of size $n = 3$ from a population with variance = 25. The mean of each distribution is indicated by a blue line segment.

Is either of these statistics an unbiased estimator of the population variance? Explain your reasoning.

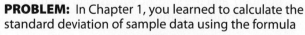

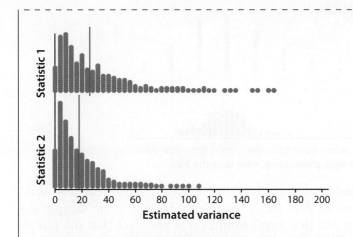

SOLUTION:

Statistic 1 appears to be unbiased because the mean of its sampling distribution is very close to 25, the value of the population variance. Statistic 2 appears to be biased because the mean of its sampling distribution is clearly less than 25, the value of the population variance.

FOR PRACTICE   TRY EXERCISE 1.

We divide by $n - 1$ when calculating the sample variance so it will be an unbiased estimator of the population variance. If we divided by n instead, our estimates would be consistently too small. Likewise, it is better to divide by $n - 1$ instead of n when calculating the standard deviation for a distribution of sample data.

Sampling Variability

Another possible statistic that could be used in the craft sticks activity is twice the sample mean. Figure 6.3 shows the simulated sampling distributions of twice the sample mean and twice the sample median.

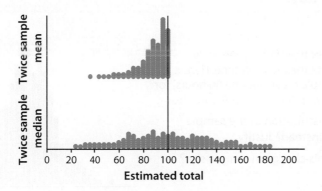

FIGURE 6.3 Simulated sampling distributions of twice the sample mean and twice the sample median for samples of size $n = 7$ from a population with $N = 100$.

Both statistics appear to be unbiased estimators because the mean of each sampling distribution is around 100. However, the sampling distribution of twice the sample mean (standard deviation ≈ 22) is less variable than the sampling distribution of twice the sample median (standard deviation ≈ 34). In general, we prefer statistics that are less variable because they produce estimates that tend to be closer to the value of the parameter.

For some parameters, there is an obvious choice for a statistic. For example, to estimate the proportion of successes in a population p, we use the proportion of successes in the sample, $\hat{p}$. Fortunately, $\hat{p}$ is an unbiased estimator of p. And as we learned in Lesson 3.3, we can reduce the variability of an estimate by increasing the sample size.

Figure 6.4 on the next page shows the simulated sampling distributions for $\hat{p} =$ the proportion of students in the sample who take the bus to school when taking SRSs of size $n = 10$ and SRSs of size $n = 50$ from a population in which the proportion of all students who take the bus to school is $p = 0.70$.

FIGURE 6.4 Simulated sampling distributions of the sample proportion $\hat{p}$ for samples of size $n = 10$ and samples of size $n = 50$ from a population with $p = 0.7$.

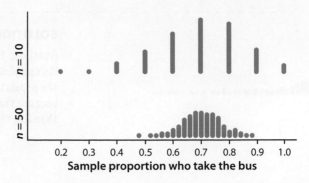

As expected, both simulated sampling distributions have means near $p = 0.70$. Also, the sampling distribution of $\hat{p}$ is much more variable when the sample size is $n = 10$, compared with $n = 50$. In a small sample, it is plausible that the sample proportion could be much smaller or much larger than the parameter, just by chance. However, when the sample size gets bigger, we expect the sample proportion to be fairly close to the value of the parameter.

Decreasing sampling variability

The sampling distribution of any statistic will have less variability when the sample size is larger.

EXAMPLE

The lifetime of batteries: Hours or days?

Sampling variability

PROBLEM: For quality control, workers at a battery factory regularly select random samples of batteries to estimate the mean lifetime. Here is a simulated sampling distribution of $\bar{x}$, the sample mean lifetime (in hours) for 1000 random samples of size $n = 100$ from a population of AAA batteries.

(a) What would happen to the sampling distribution of the sample mean $\bar{x}$ if the sample size were $n = 50$ instead? Justify.

(b) What is the practical consequence of this change in sample size?

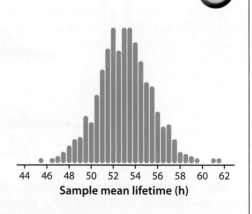

SOLUTION:

(a) The sampling distribution of the sample mean $\bar{x}$ will be more variable because the sample size is smaller.

(b) The estimated mean lifetime will typically be farther away from the true mean lifetime. In other words, the estimate will be less precise.

FOR PRACTICE   TRY EXERCISE 5.

Putting It All Together: Center and Variability

We can think of the true value of the population parameter as the bullseye on a target and of the sample statistic as an arrow fired at the target. Both bias and variability describe what happens when we take many shots at the target.

• *Bias* means that our aim is off and we consistently miss the bullseye in the same direction. That is, our sample values do not center on the population value.

- *High variability* means that repeated shots are widely scattered on the target. In other words, repeated samples do not give very similar results.

Figure 6.5 shows this target illustration of bias and variability. Notice that low variability (shots are close together) can accompany high bias (shots are consistently away from the bullseye in one direction). And low or no bias (shots center on the bullseye) can accompany high variability (shots are widely scattered). Ideally, we'd like our estimates to be *accurate* (unbiased) and *precise* (have low variability).

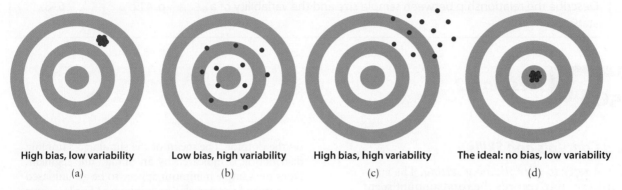

| High bias, low variability | Low bias, high variability | High bias, high variability | The ideal: no bias, low variability |
| (a) | (b) | (c) | (d) |

FIGURE 6.5 Bias and variability. (a) High bias, low variability. (b) Low bias, high variability. (c) High bias, high variability. (d) The ideal: no bias, low variability.

LESSON APP 6.2

How many tanks does the enemy have?

During World War II, the Allies captured many German tanks. Each tank had a serial number on it. Allied commanders wanted to know how many tanks the Germans had so that they could allocate their forces appropriately. They sent the serial numbers of the captured tanks to a group of mathematicians in Washington, D.C., and asked for an estimate of the total number of German tanks *N*.

Here are simulated sampling distributions for three statistics that the mathematicians considered, using samples of size *n* = 7. The blue line marks *N*, the total number of German tanks. The shorter red line segments mark the mean of each simulated sampling distribution.

© Bettmann/Corbis

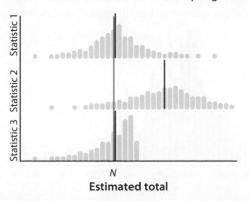

1. Do any of these statistics appear to be unbiased? Justify.

2. Which of these statistics do you think is best? Explain your reasoning.

3. Explain how the Allies could get a more precise estimate of the number of German tanks using the statistic you chose in Question 2.

Lesson 6.2

WHAT DID YOU LEARN?

LEARNING TARGET	EXAMPLES	EXERCISES
Determine if a statistic is an unbiased estimator of a population parameter.	p. 410	1–4
Describe the relationship between sample size and the variability of a statistic.	p. 412	5–8

Exercises   Lesson 6.2

Mastering Concepts and Skills

Exercises 1–4 refer to the following setting. The manager of a grocery store records the total amount spent (in dollars) for each customer who makes a purchase at his store during a week. The values in the table summarize the distribution of amount spent for this population:

N	mean	SD	Min	Q_1	med	Q_3	Max
749	29.85	24.63	0.20	12.29	22.96	39.93	153.73

1. Is the median unbiased? To investigate if the sample median is an unbiased estimator of the population median, 1000 SRSs of size $n = 10$ were selected from the population described. The sample median for each of these samples was recorded on the dotplot. The mean of the simulated sampling distribution is indicated by an orange line segment. Does the sample median appear to be an unbiased estimator of the population median? Explain your reasoning.

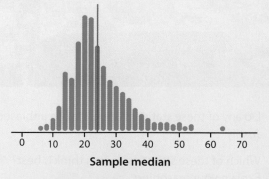

Sample median

2. Is the minimum unbiased? To investigate if the sample minimum is an unbiased estimator of the population minimum, 1000 SRSs of size $n = 10$ were selected from the population described. The sample minimum for each of these samples was recorded

on the dotplot. The mean of the simulated sampling distribution is indicated by an orange line segment. Does the sample minimum appear to be an unbiased estimator of the population minimum? Explain your reasoning.

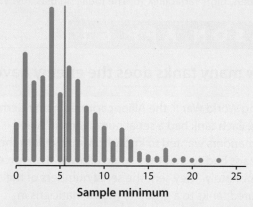

Sample minimum

3. Is the range unbiased? To investigate if the sample range is an unbiased estimator of the population range, 1000 SRSs of size $n = 10$ were selected from the population described. The sample range for each of these samples was recorded on the dotplot. The mean of the simulated sampling distribution is indicated by an orange line segment. Does the sample range appear to be an unbiased estimator of the population range? Explain your reasoning.

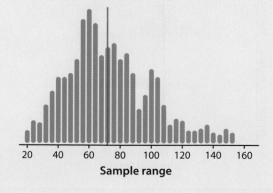

Sample range

4. Is the *IQR* unbiased? To investigate if the sample *IQR* is an unbiased estimator of the population *IQR*, 1000 SRSs of size $n = 10$ were selected from the population described. The sample *IQR* for each of these samples was recorded on the dotplot. The mean of the simulated sampling distribution is indicated by an orange line segment. Does the sample *IQR* appear to be an unbiased estimator of the population *IQR*? Explain your reasoning.

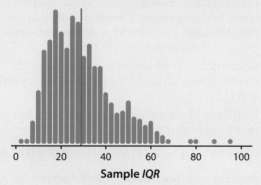

Sample *IQR*

5. More about medians Refer to Exercise 1.

pg 412
(a) What would happen to the sampling distribution of the sample median if the sample size were $n = 50$ instead? Justify.

(b) What is the practical consequence of this change in sample size?

6. More about minimums Refer to Exercise 2.

(a) What would happen to the sampling distribution of the sample minimum if the sample size were $n = 50$ instead? Justify.

(b) What is the practical consequence of this change in sample size?

7. More about ranges Refer to Exercise 3.

(a) What would happen to the sampling distribution of the sample range if the sample size were $n = 5$ instead? Justify.

(b) What is the practical consequence of this change in sample size?

8. More about *IQRs* Refer to Exercise 4.

(a) What would happen to the sampling distribution of the sample *IQR* if the sample size were $n = 5$ instead? Justify.

(b) What is the practical consequence of this change in sample size?

Applying the Concepts

9. Cholesterol in teens A study of the health of teenagers plans to measure the blood cholesterol levels of an SRS of 13- to 16-year-olds. The researchers will report the mean $\overline{x}$ from their sample as an estimate of the mean cholesterol level μ in this population. Explain to someone who knows little about statistics what it means to say that $\overline{x}$ is an unbiased estimator of μ.

10. Predict the election A polling organization plans to ask a random sample of likely voters who they will vote for in an upcoming election. The researchers will report the sample proportion $\hat{p}$ that favors the incumbent as an estimate of the population proportion p that favors the incumbent. Explain to someone who knows little about statistics what it means to say that $\hat{p}$ is an unbiased estimator of p.

11. Sampling more teens Refer to Exercise 9. The sample mean $\overline{x}$ is an unbiased estimator of the population mean μ no matter what size SRS the study chooses. Explain to someone who knows nothing about statistics why a large random sample will give more reliable results than a small random sample.

12. Sampling more voters Refer to Exercise 10. The sample proportion $\hat{p}$ is an unbiased estimator of the population proportion p no matter what size random sample the polling organization chooses. Explain to someone who knows nothing about statistics why a large random sample will give more trustworthy results than a small random sample.

13. Housing prices In a residential neighborhood, the median value of a house is $200,000. For which of the following sample sizes, $n = 10$ or $n = 100$, is the sample median most likely to be greater than $250,000? Explain.

14. Houses with basements In a particular city, 74% of houses have basements. For which of the following sample sizes, $n = 10$ or $n = 100$, is the sample proportion of houses with a basement *more* likely to be greater than 0.70? Explain.

15. Bias and variability The histograms show sampling distributions for four different statistics intended to estimate the same parameter.

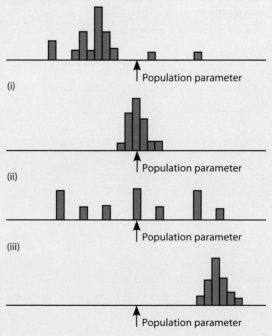

(i)

(ii)

(iii)

(iv)

(a) Which statistics are unbiased estimators? Justify your answer.

(b) Which statistic does the best job of estimating the parameter? Explain.

Extending the Concepts

16. **More about means** In the Exercises for Lesson 6.1, you were introduced to the following population of 2 male students and 3 female students, along with their quiz scores:

    Abigail 10   Bobby 5   Carlos 10   DeAnna 7   Emily 9

(a) Calculate the mean quiz score for the entire population.

(b) List all 10 possible SRSs of size $n = 2$, calculate the mean quiz score for each sample, and display the sampling distribution of the sample mean in a dotplot.

(c) Calculate the mean of the sampling distribution from part (b). Is the sample mean an unbiased estimator of the population mean? Explain.

17. **More about proportions** In the Exercises for Lesson 6.1, you were introduced to the following population of 2 male students and 3 female students, along with their quiz scores:

    Abigail 10   Bobby 5   Carlos 10   DeAnna 7   Emily 9

(a) Calculate the proportion of females in the entire population.

(b) List all 10 possible SRSs of size $n = 2$, calculate the proportion of females for each sample, and display the sampling distribution of the sample proportion in a dotplot.

(c) Calculate the mean of the sampling distribution from part (b). Is the sample proportion an unbiased estimator of the population proportion? Explain.

Recycle and Review

18. **Students and housing (4.3, 4.4)** There are 104 students in Professor Negroponte's statistics class, 49 males and 55 females. Sixty of the students live in the dorms and the rest live off campus. Twenty of the males live off-campus. Choose a student at random from this class. Let Event M = the student is male and Event D = the student lives in the dorms.

(a) Construct a Venn diagram to represent the outcomes of this chance process using the events M and D.

(b) Find each of the following probabilities and interpret them in context.

   (i) $P(M \cup D)$ **(ii)** $P(M^C \cap D)$ **(iii)** $P(D \mid M)$

19. **Students and homework (5.3, 5.4)** Refer to Exercise 18. At the beginning of each day that Professor Negroponte's class meets, he randomly selects a member of the class to present the solution to a homework problem. Suppose the class meets 40 times during the semester and the selections are made with replacement. Let X = the number of times a female student is selected to present a solution.

(a) Is X a binomial random variable? Justify your answer.

(b) Calculate the mean and standard deviation of X.

(c) For the first 10 meetings of the class, Professor Negroponte selects only 1 female student to solve a problem. Is there convincing evidence that his selection process is not really random? Support your answer with an appropriate probability calculation.

Lesson 6.3

The Sampling Distribution of a Sample Count (The Normal Approximation to the Binomial)

In many cases, we are interested in the number of successes X in a random sample from some population. For example, $X =$ the number of defective flash drives in a random sample of 10 flash drives or $X =$ the number of Democrats in a random sample of 1000 registered voters. To do probability calculations involving X, we want an understanding of the **sampling distribution of the sample count X**.

> **DEFINITION** Sampling distribution of the sample count X
>
> The **sampling distribution of the sample count X** describes the distribution of values taken by the sample count X in all possible samples of the same size from the same population.

The sampling distribution of X is closely related to the binomial distributions that you learned about in Lessons 5.3 and 5.4.

Suppose that a supplier inspects an SRS of 10 flash drives from a shipment of 10,000 flash drives in which 200 are defective. Let $X =$ the number of bad flash drives in the sample. This is not quite a binomial setting. Because we are sampling without replacement, the independence condition is violated. The conditional probability that the second flash drive chosen is bad changes when we know whether the first is good or bad: $P(\text{second is bad} \mid \text{first is good}) = 200/9999 = 0.0200$ but $P(\text{second is bad} \mid \text{first is bad}) = 199/9999 = 0.0199$. These probabilities are very close because removing 1 flash drive from a shipment of 10,000 changes the makeup of the remaining 9999 flash drives very little. The distribution of X is very close to the binomial distribution with $n = 10$ and $p = 0.02$.

In practice, we can ignore the violation of the independence condition caused by sampling without replacement whenever the sample size is relatively small compared to the population size. Specifically, we can assume that the sampling distribution of a sample count X is approximately binomial when the sample size is less than 10% of the population size.

Center and Variability

Because the sampling distribution of a sample count X is approximately binomial when the sample is a small fraction of the population, we can use the formulas from Lesson 5.4 to calculate the mean and standard deviation of X.

> ### How to Calculate μ_x and σ_x for a Binomial Distribution
> Suppose X is the number of successes in a random sample of size n from a large population with proportion of successes p. Then:
>
> - The **mean** of the sampling distribution of X is $\mu_X = np$.
>
> - The **standard deviation** of the sampling distribution of X is $\sigma_X = \sqrt{np(1 - p)}$.

The formula for the mean is always correct, even if we are sampling without replacement. However, the formula for the standard deviation is not appropriate to use when the sample size is more than 10% of the population size.

EXAMPLE

How many flash drives are defective?

Mean and SD of the sampling distribution of X

PROBLEM: Two percent of the flash drives in a shipment of 10,000 flash drives are defective. An inspector randomly selects 10 flash drives from the shipment and records $X =$ the number of defective flash drives in the sample.

(a) Calculate the mean and standard deviation of the sampling distribution of X.

(b) Interpret the standard deviation from part (a).

SOLUTION:

(a) $\mu_x = 10(0.02) = 0.2$ flash drives

$\sigma_x = \sqrt{10(0.02)(1 - 0.02)} = 0.44$ flash drives

(b) If the inspector took many samples of size 10, the number of defective flash drives would typically vary by about 0.44 from the mean of 0.2.

> Because the sample size is less than 10% of the population size, the distribution of X is approximately binomial with $n = 10$ and $p = 0.2$. The mean is $\mu_X = np$ and the standard deviation is $\sigma_X = \sqrt{np(1 - p)}$.

FOR PRACTICE   TRY EXERCISE 1.

Shape

As you learned in Lesson 5.3, the shape of a binomial distribution can be skewed to the right, skewed to the left, or roughly symmetric. The histogram in Figure 6.6 shows the sampling distribution of X = the number of defective flash drives from the previous example. It is clearly skewed to the right.

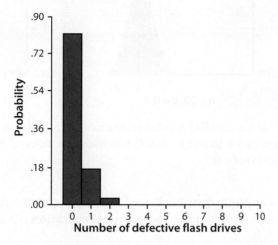

FIGURE 6.6 Probability histogram of X = the number of defective flash drives in a sample of size $n = 10$ from a population in which $p = 0.02$.

The following activity explores the shape of the sampling distribution of a sample count for various combinations of n and p.

ACTIVITY

Simulating with the *Normal Approximation to Binomial Distributions* applet

In this activity, you will explore the shape of the sampling distribution of a sample count X using an applet from the book's website.

1. Launch the *Normal Approximation to the Binomial Distributions* applet at **highschool.bfwpub.com/spa3e**. You will see a histogram with a normal curve overlaid.

2. Using the sliders, set the number of trials to $n = 10$ and the probability of success to $p = 0.02$. *Hint:* You can also use the arrow keys on your computer's keyboard to move the sliders. The normal curve has the same mean and standard deviation as the histogram, but it doesn't model the histogram very well.

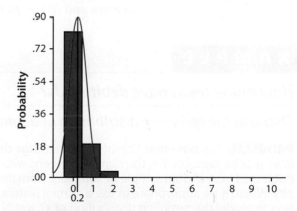

3. Use the slider (or the arrow keys) to gradually change the probability from $p = 0.02$ to $p = 1.00$ while keeping the number of trials the same. Does the normal curve fit well when p is close to 0? Close to 0.5? Close to 1?

4. Keep the number of trials set to $n = 10$ and change the probability to $p = 0.1$. Use the slider (or the arrow keys) to gradually increase the sample size from $n = 10$ to $n = 100$. Does the normal curve fit the histogram better when n is smaller or larger?

5. Under what conditions will the distribution of X be approximately normal? Under what conditions will the distribution of X not be approximately normal?

As you learned in the activity, the shape of the sampling distribution of X will be approximately normal when the sample size is large enough. You also learned that "large enough" depends on the value of *p*. The farther *p* is from 0.5, the larger the sample size needs to be, as shown in Figure 6.7.

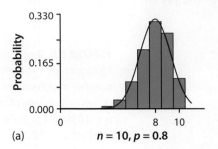
(a) $n = 10, p = 0.8$

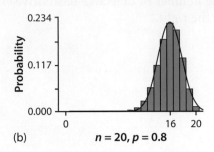
(b) $n = 20, p = 0.8$

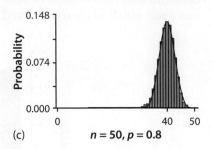
(c) $n = 50, p = 0.8$

FIGURE 6.7 Histograms of the sampling distribution of a sample count X with (a) $n = 10$ and $p = 0.8$, (b) $n = 20$ and $p = 0.8$, and (c) $n = 50$ and $p = 0.8$. As n increases, the shape of the sampling distribution gets closer and closer to normal.

In practice, the sampling distribution of a sample count will have an approximately normal distribution when the **Large Counts condition** is met.

> **DEFINITION**   **The Large Counts condition**
>
> Suppose X is the number of successes in a random sample of size n from a population with proportion of successes p. The **Large Counts condition** says that the distribution of X will be approximately normal when
>
> $$np \geq 10 \quad \text{and} \quad n(1 - p) \geq 10$$

This condition is called "large counts" because *np* is the expected (mean) count of successes and $n(1 - p)$ is the expected (mean) count of failures.

EXAMPLE

How many teens have debit cards?

Shape of the sampling distribution of a sample count

PROBLEM: Suppose that 12% of teens in a large city have a debit card. Let X = the number of teens with a debit card in a random sample of 500 teens from this city. Would it be appropriate to use a normal distribution to model the sampling distribution of X? Justify your answer.

Check the Large Counts condition to determine if X will have an approximately normal distribution.

SOLUTION:

Because $np = 500(0.12) = 60 \geq 10$ and $n(1 - p) = 500(1 - 0.12) = 440 \geq 10$, the sampling distribution of X is approximately normal.

In this context, 60 is the expected (mean) count of teens with a debit card and 440 is the expected (mean) count of teens without a debit card.

FOR PRACTICE   TRY EXERCISE 5.

Finding Probabilities Involving *X*

When the Large Counts condition is met, we can use a normal distribution to calculate probabilities involving X = the number of successes in a random sample of size n.

EXAMPLE

Is it fun to shop anymore?

Probabilities involving X

PROBLEM: Sample surveys show that fewer people enjoy shopping than in the past. A survey asked a nationwide random sample of 2500 adults if they agreed or disagreed with the statement "I like buying new clothes, but shopping is often frustrating and time-consuming."[5] Suppose that exactly 60% of all adult U.S. residents would say "Agree" if asked the same question. Calculate the probability that at least 1520 members of the sample would say "Agree."

Chris Hondros/Getty Images

Let X = the number who would say "Agree." The sampling distribution of X is approximately binomial with $n = 2500$ and $p = 0.60$.

To use a normal approximation to calculate probabilities involving X, we need to know the mean, standard deviation, and shape of the sampling distribution of X. Recall that the mean is $\mu_X = np$ and the standard deviation is $\sigma_X = \sqrt{np(1 - p)}$.

SOLUTION:

- Mean: $\mu_X = 2500(0.60) = 1500$

- SD: $\sigma_X = \sqrt{2500(0.6)(1 - 0.6)} = 24.49$

- Shape: Approximately normal because
  $np = 2500(0.60) = 1500 \geq 10$ and
  $n(1 - p) = 2500(1 - 0.6) = 1000 \geq 10$

1. Draw a normal distribution.

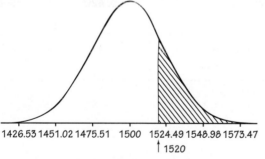

| 1426.53 | 1451.02 | 1475.51 | 1500 | 1524.49 | 1548.98 | 1573.47 |

↑ 1520

Sample count who would say "Agree"

Using Table A: $Z = \dfrac{1520 - 1500}{24.49} = 0.82$

$P(Z \geq 0.82) = 1 - 0.7939 = 0.2061$

Using technology: Applet/normalcdf (lower:1520, upper:100000, mean:1500, SD: 24.49) = 0.2071

2. Perform calculations.
(i) Standardize the boundary value and use Table A to find the desired probability; or
(ii) Use technology.

FOR PRACTICE   TRY EXERCISE 9.

LESSON APP 6.3

How can we check for bias in a survey?

One way of checking the effect of undercoverage, nonresponse, and other sources of bias in a sample survey is to compare the sample with known facts about the population. About 12% of American adults identify themselves as black. Suppose we take an SRS of 1500 American adults and let X be the number of blacks in the sample.

Rawpixel Ltd/Getty Images

1. Calculate the mean and standard deviation of the sampling distribution of X. Interpret the standard deviation.

2. Justify that the sampling distribution of X is approximately normal.

3. Calculate the probability that an SRS of 1500 American adults will contain between 155 and 205 blacks.

4. Explain how a polling organization could use the results from the previous question to check for undercoverage and other sources of bias.

Lesson 6.3

WHAT DID YOU LEARN?

LEARNING TARGET	EXAMPLES	EXERCISES
Calculate the mean and the standard deviation of the sampling distribution of a sample count and interpret the standard deviation.	p. 418	1–4
Determine if the sampling distribution of a sample count is approximately normal.	p. 420	5–8
If appropriate, use the normal approximation to the binomial distribution to calculate probabilities involving a sample count.	p. 421	9–12

Exercises   *Lesson 6.3*

Mastering Concepts and Skills

1. **Lefties** Eleven percent of students at a large high school are left-handed. A statistics teacher selects a random sample of 100 students and records $X = $ the number of left-handed students in the sample.

   pg 418

   (a) Calculate the mean and standard deviation of the sampling distribution of X.

   (b) Interpret the standard deviation from part (a).

2. **Hip dysplasia** Dysplasia is a malformation of the hip socket that is very common in certain dog breeds and causes arthritis as a dog gets older. According to the Orthopedic Foundation for Animals, 11.6% of all Labrador retrievers have hip dysplasia.[6] A veterinarian tests a random sample of 50 Labrador retrievers and records $Y = $ the number of Labs with dysplasia in the sample.

   (a) Calculate the mean and standard deviation of the sampling distribution of Y.

   (b) Interpret the standard deviation from part (a).

3. **NASCAR cards and cereal boxes** In an attempt to increase sales, a breakfast cereal company decides to offer a promotion. Each box of cereal will contain a collectible card featuring one NASCAR driver: Kyle Busch; Dale Earnhardt, Jr.; Kasey Kahne; Danica Patrick; or Jimmie Johnson. The company says that each of the 5 cards is equally likely to

appear in the 100,000 boxes of cereal that are part of this promotion. You buy 12 boxes and let X = the number of Kyle Busch cards in the sample.

(a) Calculate the mean and standard deviation of the sampling distribution of X.

(b) Interpret the standard deviation from part (a).

4. **What, me marry?** In the United States, 20% of adults ages 25 and older have never been married, more than double the figure recorded for 1960.[7] Select a random sample of 50 U.S. adults ages 25 and older and let Y = the number of individuals in the sample who have never married.

(a) Calculate the mean and standard deviation of the sampling distribution of Y.

(b) Interpret the standard deviation from part (a).

5. **Are lefties normal?** Refer to Exercise 1. Would it be
pg 420   appropriate to use a normal distribution to model the sampling distribution of X = the number of left-handed students in the sample? Justify your answer.

6. **Is hip dysplasia normal?** Refer to Exercise 2. Would it be appropriate to use a normal distribution to model the sampling distribution of Y = the number of Labs with dysplasia in the sample? Justify your answer.

7. **Is NASCAR normal?** Refer to Exercise 3. Would it be appropriate to use a normal distribution to model the sampling distribution of X = the number of Kyle Busch cards in the sample? Justify your answer.

8. **A normal marriage?** Refer to Exercise 4. Would it be appropriate to use a normal distribution to model the sampling distribution of Y = the number of individuals in the sample who have never married? Justify your answer.

9. **Lefties are all right** Refer to Exercises 1 and 5.
pg 421   Calculate the probability that at least 15 of the members of the sample are left-handed.

10. **Never been married** Refer to Exercises 4 and 8. Calculate the probability that at most 5 of the individuals in the sample have never been married.

11. **Public transportation** In a large city, 34% of residents use public transportation at least once per week. If the mayor selects a random sample of 200 residents, calculate the probability that at most 60 residents in the sample use public transportation at least once per week.

12. **U.S. quarters** According to www.usmint.gov, 54% of the quarters minted in 2014 were produced by the U.S. Mint in Denver, Colorado (the rest were produced in Philadelphia). In a random sample of 200 quarters, what is the probability that at least 115 of them were minted in Denver?

Applying the Concepts

13. **Tasty chips** For a statistics project, Zenon decided to investigate if students at his school prefer name-brand potato chips to store-brand potato chips. He prepared two identical bowls of chips, filling one with name-brand chips and the other with store-brand chips. Then, he selected a random sample of 30 students, had each student try both types of chips in random order, and recorded which type of chip each student preferred. Assume that 50% of students at Zenon's school prefer the name-brand chips. Let X = the number of students in the sample that prefer the name-brand chips.[8]

(a) Calculate the mean and standard deviation of the sampling distribution of X. Interpret the standard deviation.

(b) Justify that the distribution of X is approximately normal.

(c) Calculate the probability that 19 or more of the students will prefer the name-brand chips.

14. **Blood types** About 10% of people in the United States have type B blood. Suppose we take a random sample of 120 U.S. residents, and let X = the number of residents in the sample who have type B blood.

(a) Calculate the mean and standard deviation of the sampling distribution of X. Interpret the standard deviation.

(b) Justify that the distribution of X is approximately normal.

(c) Calculate the probability that 16 or more individuals in the sample have type B blood.

15. **More chips!** Refer to Exercise 13. In Zenon's study, 19 of the 30 students chose the name-brand chips. Based on your answer to Exercise 13(c), does this provide convincing evidence that more than half of the students at Zenon's school prefer name-brand potato chips? Explain.

16. **More on blood type** Refer to Exercise 14. Some people believe that one's blood type has an impact on personality. For example, people with type B blood are supposed to be more creative, active, and passionate. To test this hypothesis, Jason selects a random sample of 120 art, music, and drama majors at his college and finds that 16 of them have type B blood. Based on your answer to Exercise 14(c), does this provide convincing evidence that art, music, and drama majors at Jason's college are more likely than the general population to have type B blood? Explain.

Extending the Concepts

17. **Binomial transportation** Refer to Exercise 11. Use a binomial distribution to calculate the probability that at most 60 residents in the sample use public transportation at least once per week. *Hint:* See Lesson 5.4.

Recycle and Review

18. Summer reading (3.6, 3.8) A group of educational researchers studied the impact of summer reading with a randomized experiment involving second- and third-graders in North Carolina. Students were randomly assigned to either a group that was mailed one book a week for 10 weeks or a control group that was not mailed any books. Both groups were given a reading comprehension test at the start and end of the summer. Third-grade girls who were mailed books showed a statistically significant increase in reading ability, but third-grade boys and second-graders of both genders did not.[9]

(a) Explain the purpose of including a control group in this experiment.

(b) Explain what is meant by "statistically significant increase" in the last sentence.

19. Sisters and brothers (2.2, 2.5, 2.7) How strongly do physical characteristics of sisters and brothers correlate? Here are data on the heights (in inches) of 11 adult pairs:[10]

Brother	71	68	66	67	70	71	70	73	72	65	66
Sister	69	64	65	63	65	62	65	64	66	59	62

(a) Construct a scatterplot using brother's height as the explanatory variable. Describe what you see.

(b) Use technology to compute the least-squares regression line for predicting sister's height from brother's height. Interpret the slope in context.

(c) Damien is 70 inches tall. Predict the height of his sister Tonya.

(d) The standard deviation of residuals for this model is $s = 2.247$. Interpret this value in context.

Lesson 6.4

The Sampling Distribution of a Sample Proportion

LEARNING TARGETS

- Calculate the mean and standard deviation of the sampling distribution of a sample proportion $\hat{p}$ and interpret the standard deviation.
- Determine if the sampling distribution of $\hat{p}$ is approximately normal.
- If appropriate, use a normal distribution to calculate probabilities involving $\hat{p}$.

What proportion of U.S. teens know that 1492 was the year in which Columbus "discovered" America? A Gallup poll found that 210 out of a random sample of 501 American teens aged 13 to 17 knew this historically important date.[11] The sample proportion $\hat{p} = 210/501 = 0.42$ is the statistic that we use to estimate the unknown population proportion p. Because a random sample of 501 teens is unlikely to perfectly represent all teens, we can only say that "about" 42% of U.S. teenagers know that Columbus voyaged to America in 1492.

To understand how much $\hat{p}$ varies from p and what values of $\hat{p}$ are likely to happen by chance, we want an understanding of the **sampling distribution of the sample proportion $\hat{p}$**.

DEFINITION Sampling distribution of the sample proportion $\hat{p}$

The **sampling distribution of the sample proportion $\hat{p}$** describes the distribution of values taken by the sample proportion $\hat{p}$ in all possible samples of the same size from the same population.

When Mr. Ramirez's class did the Penny for Your Thoughts activity at the beginning of the chapter, his students produced the "dotplot" in Figure 6.8 showing the simulated sampling distribution of $\hat{p}$ = the sample proportion of pennies from the 2000s in 50 samples of size $n = 20$.

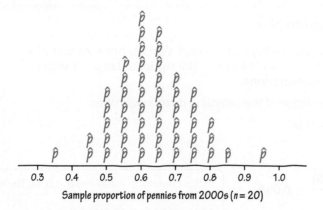

FIGURE 6.8 Simulated sampling distribution of the sample proportion of pennies in 50 samples of size $n = 20$ from a population of pennies.

This distribution is roughly symmetric, with a mean of about 0.65 and a standard deviation of about 0.10. By the end of this lesson, you should be able to anticipate the shape, center, and variability of distributions like this one without getting your hands dirty in a jar of pennies.

Center and Variability

When we select random samples of size n from a population with proportion of successes p, the value of $\hat{p}$ will vary from sample to sample. As with the sampling distribution of the sample count X, there are formulas that describe the center and variability of the sampling distribution of $\hat{p}$.

How to Calculate $\mu_{\hat{p}}$ and $\sigma_{\hat{p}}$

Suppose that $\hat{p}$ is the proportion of successes in an SRS of size n drawn from a large population with proportion of successes p. Then:

- The **mean** of the sampling distribution of $\hat{p}$ is $\mu_{\hat{p}} = p$.

- The **standard deviation** of the sampling distribution of $\hat{p}$ is $\sigma_{\hat{p}} = \sqrt{\dfrac{p(1-p)}{n}}$.

Here are some important facts about the mean and standard deviation of the sampling distribution of the sample proportion $\hat{p}$:

- The sample proportion $\hat{p}$ is an *unbiased estimator* of the population proportion p. This is because the mean of the sampling distribution $\mu_{\hat{p}}$ is equal to the population proportion p.

- The standard deviation of the sampling distribution of $\hat{p}$ describes the typical distance between $\hat{p}$ and the population proportion p.

- The sampling distribution of $\hat{p}$ is less variable for larger samples. This is indicated by the $\sqrt{n}$ in the denominator of the standard deviation formula.
- The formula for the standard deviation of the distribution of $\hat{p}$ requires that the observations are independent. In practice, we are safe assuming independence when sampling without replacement as long as the sample size is less than 10% of the population size.

EXAMPLE

What proportion of students have a smartphone?

Mean and SD of the sampling distribution of $\hat{p}$

PROBLEM: Suppose that 43% of students at a large high school own a smartphone. As part of a schoolwide technology study, the principal surveys an SRS of $n = 100$ students. Let $\hat{p}$ = the proportion of students in the sample who own a smartphone.

(a) Calculate the mean and the standard deviation of the sampling distribution of $\hat{p}$.

(b) Interpret the standard deviation from part (a).

SOLUTION:

(a) $\mu_{\hat{p}} = 0.43$ and $\sigma_{\hat{p}} = \sqrt{\dfrac{0.43(1 - 0.43)}{100}} = 0.050$

> In this context, $n = 100$ and $p = 0.43$. The mean is $\mu_{\hat{p}} = p$ and the standard deviation is $\sigma_{\hat{p}} = \sqrt{\dfrac{p(1 - p)}{n}}$.

(b) In SRSs of size $n = 100$, the sample proportion of students who own a smartphone will typically vary by about 0.050 from the true proportion of $p = 0.43$.

FOR PRACTICE   TRY EXERCISE 1.

THINK ABOUT IT   Is the sampling distribution of $\hat{p}$ (the sample proportion of successes) related to the sampling distribution of X (the sample count of successes)?

Yes!

$$\hat{p} = \frac{\text{number of successes in sample}}{\text{sample size}} = \frac{X}{n}$$

For example, here are dotplots showing the simulated sampling distribution of X = the number of pennies from the 2000s and $\hat{p}$ = the proportion of pennies from the 2000s for samples of size 20 in Mr. Ramirez's class. The distributions are exactly the same, other than the scale on the axis.

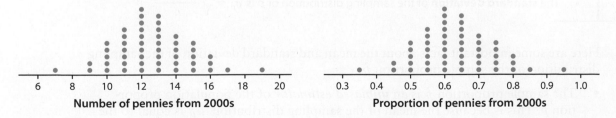

Also, the formulas for the mean and standard deviation of the sampling distribution of $\hat{p}$ are derived from the formulas we learned in the previous lesson:

$$\mu_{\hat{p}} = \frac{\mu_X}{n} = \frac{np}{n} = p$$

$$\sigma_{\hat{p}} = \frac{\sigma_X}{n} = \frac{\sqrt{np(1-p)}}{n} = \sqrt{\frac{np(1-p)}{n^2}} = \sqrt{\frac{p(1-p)}{n}}$$

Shape

Both the sample size and the proportion of successes in the population affect the shape of the sampling distribution of the sample proportion $\hat{p}$. The following activity helps you explore the effect of these two factors.

ACTIVITY

Sampling from the candy machine

Imagine a very large candy machine filled with orange, brown, and yellow candies. When you insert money, the machine dispenses a sample of candies. In this activity, you will use an applet to investigate the sample-to-sample variability in the proportion of orange candies dispensed by the machine.

1. Launch the *Reese's Pieces*® applet at **www.rossmanchance.com/applets**. Click the button for "Proportion of orange" to have the applet calculate and record the value of $\hat{p} =$ the sample proportion of orange candies.

2. Click on the "Draw Samples" button. An animated simple random sample of $n = 25$ candies should be dispensed. The screen shot shows the results of one such sample. Was your sample proportion of orange candies close to the actual population proportion, $p = 0.50$?

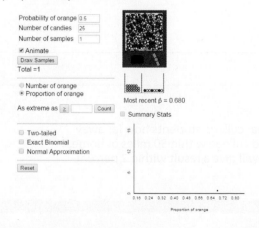

3. Click "Draw Samples" 9 more times, so that you have a total of 10 sample proportions. Look at the dotplot of your $\hat{p}$ values. Does the distribution have a recognizable shape?

4. To take many more samples quickly, enter 990 in the "number of samples" box. Click on the "Animate" box to turn off the animation. Then click "Draw Samples." You have now taken a total of 1000 samples of $n = 25$ candies from the machine. Describe the shape of the simulated sampling distribution of $\hat{p}$ shown in the dotplot.

5. How does the shape of the sampling distribution of $\hat{p}$ change if the proportion of orange candies in the machine is $p = 0.10$ instead of $p = 0.50$? Set the probability of orange candies to $p = 0.10$ and draw 1000 samples of size $n = 25$. What if $p = 0.90$? Describe how the value of p affects the shape of the sampling distribution of $\hat{p}$.

6. How does the shape of the sampling distribution of $\hat{p}$ change if the sample size increases? Set the probability of orange to $p = 0.90$ and the number of candies to $n = 25$ and draw 1000 samples. Then, repeat with sample sizes of $n = 100$ and $n = 500$. Describe how the value of n affects the shape of the sampling distribution of $\hat{p}$.

As you learned in the activity, the shape of the sampling distribution of $\hat{p}$ will be closer to normal when the value of p is closer to 0.5 and the sample size is larger. These relationships are the same as those you discovered in the previous lesson. And the **Large Counts condition** is the same as well.

> **DEFINITION**   The Large Counts condition
>
> Suppose $\hat{p}$ is the proportion of successes in a random sample of size n from a population with proportion of successes p. The **Large Counts condition** says that the distribution of $\hat{p}$ will be approximately normal when
>
> $$np \geq 10 \quad \text{and} \quad n(1-p) \geq 10$$

EXAMPLE

A penny for your thoughts?

Shape of the sampling distribution of $\hat{p}$

PROBLEM: Mr. Ramirez's class did the Penny for Your Thoughts Activity from the beginning of this chapter. In his population of pennies, the proportion of pennies from the 2000s is $p = 0.627$. Would it be appropriate to use a normal distribution to model the sampling distribution of $\hat{p}$ for samples of size $n = 16$? Justify your answer.

Andrew Unangst/Getty Images

SOLUTION:

No. Because $n(1-p) = 16(1-0.627) = 5.968 < 10$, the sampling distribution of $\hat{p}$ is not approximately normal.

> Check the Large Counts condition to determine if $\hat{p}$ will have an approximately normal distribution.

FOR PRACTICE   TRY EXERCISE 5.

Finding Probabilities Involving $\hat{p}$

When the Large Counts condition is met, we can use a normal distribution to calculate probabilities involving $\hat{p} =$ the proportion of successes in a random sample of size n.

EXAMPLE

How far from home do you attend college?

Normal calculations involving $\hat{p}$

PROBLEM: A polling organization asks an SRS of 1500 first-year college students how far away their home is. Suppose that 35% of all first-year students attend college within 50 miles of home. Find the probability that the random sample of 1500 students will give a result within 2 percentage points of this true value.

SOLUTION:

- Mean: $\mu_{\hat{p}} = 0.35$

- SD: $\sigma_{\hat{p}} = \sqrt{\dfrac{0.35(1-0.35)}{1500}} = 0.0123$

- Shape: Approximately normal because $np = 1500(0.35)$
$= 525 \geq 10$ and $n(1-p) = 1500(1-0.35) = 975 \geq 10$

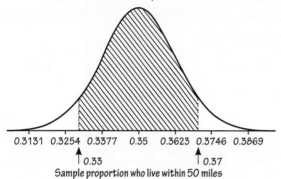

0.3131   0.3254   0.3377   0.35   0.3623   0.3746   0.3869
         ↑0.33   ↑0.37
Sample proportion who live within 50 miles

Using Table A:

$$z = \dfrac{0.33 - 0.35}{0.0123} = -1.63 \text{ and } z = \dfrac{0.37 - 0.35}{0.0123} = 1.63$$

$$P(-1.63 \leq Z \leq 1.63) = 0.9484 - 0.0516 = 0.8968$$

Using technology: Applet/normalcdf (lower:0.33, upper:0.37, mean:0.35, SD: 0.0123) = 0.8961

> Let $\hat{p}$ = the proportion in the sample who attend college within 50 miles of home, where $p = 0.35$ and $n = 1500$.
> To use a normal distribution to calculate probabilities involving $\hat{p}$, we have to know the mean, standard deviation, and shape of the sampling distribution of $\hat{p}$. Recall that the mean is $\mu_{\hat{p}} = p$ and the standard
> deviation is $\sigma_{\hat{p}} = \sqrt{\dfrac{p(1-p)}{n}}$.

> **1. Draw a normal distribution.**

> **2. Perform calculations.**
> (i) Standardize the boundary value and use Table A to find the desired probability; or
> (ii) Use technology.

FOR PRACTICE   TRY EXERCISE 9.

LESSON APP 6.4

What's that spot on my potato chip?

A potato-chip producer and its main supplier agree that each shipment of potatoes must meet certain quality standards. If the producer finds convincing evidence that more than 8% of the potatoes in the entire shipment have "blemishes," the truck will be sent away to get another load of potatoes from the supplier. Otherwise, the entire truckload will be used to make potato chips. To make the decision, a supervisor will inspect a random sample of potatoes from the shipment. Suppose that the proportion of blemished potatoes in the entire shipment is $p = 0.08$ and that the supervisor randomly selects $n = 500$ potatoes for inspection.

1. Calculate the mean and standard deviation of the sampling distribution of $\hat{p}$. Interpret the standard deviation.

2. Justify that the sampling distribution of $\hat{p}$ is approximately normal.

3. Calculate the probability that at least 11% of the potatoes in the sample are blemished.

4. Based on your answer to Question 3, what should the supervisor conclude if he selects an SRS of size $n = 500$ and finds $\hat{p} = 0.11$? Explain.

Lesson 6.4

WHAT DID YOU LEARN?

LEARNING TARGET	EXAMPLES	EXERCISES
Calculate the mean and standard deviation of the sampling distribution of a sample proportion $\hat{p}$ and interpret the standard deviation.	p. 426	1–4
Determine if the sampling distribution of $\hat{p}$ is approximately normal.	p. 428	5–8
If appropriate, use a normal distribution to calculate probabilities involving $\hat{p}$.	p. 428	9–12

Exercises   Lesson 6.4

Mastering Concepts and Skills

1. Orange Skittles® The makers of Skittles claim that
pg 426 20% of Skittles candies are orange. You select a random sample of 20 Skittles from a large bag. Let $\hat{p}$ = the proportion of orange Skittles in the sample.

(a) Calculate the mean and the standard deviation of the sampling distribution of $\hat{p}$.

(b) Interpret the standard deviation from part (a).

2. Registered voters In a congressional district, 55% of registered voters are Democrats. A polling organization selects a random sample of 500 registered voters from this district. Let $\hat{p}$ = the proportion of Democrats in the sample.

(a) Calculate the mean and the standard deviation of the sampling distribution of $\hat{p}$.

(b) Interpret the standard deviation from part (a).

3. On-time shipping A large mail-order company advertises that it ships 90% of its orders within 3 working days. You select an SRS of 100 of the orders received in the past week for an audit. Let $\hat{p}$ = the proportion of orders in the last week that were shipped within 3 working days.

(a) Calculate the mean and the standard deviation of the sampling distribution of $\hat{p}$.

(b) Interpret the standard deviation from part (a).

4. Married with children According to a recent U.S. Bureau of Labor Statistics report, the proportion of married couples with children in which both parents work outside the home is 59%.[12] You select an SRS of 50 married couples with children and let $\hat{p}$ = the sample proportion of couples in which both parents work outside the home.

(a) Calculate the mean and the standard deviation of the sampling distribution of $\hat{p}$.

(b) Interpret the standard deviation from part (a).

5. Airport safety The Transportation Security Admin-
pg 428 istration (TSA) is responsible for airport safety. On some flights, TSA officers randomly select passengers for an extra security check before boarding. One such flight had 120 passengers—16 in first class and 104 in coach class. TSA officers selected an SRS of 10 passengers for screening. Would it be appropriate to use a normal distribution to model the sampling distribution of $\hat{p}$ = the proportion of first-class passengers in the sample? Justify your answer.

6. Only vowels? In the game of Scrabble, each player begins by drawing 7 tiles from a bag containing 100 tiles. There are 42 vowels, 56 consonants, and 2 blank tiles in the bag. Cait chooses an SRS of 7 tiles. Would it be appropriate to use a normal distribution to model the sampling distribution of $\hat{p}$ = the proportion of vowels in her sample? Justify your answer.

7. Model shipping? Refer to Exercise 3. Would it be appropriate to use a normal distribution to model the sampling distribution of $\hat{p}$ = the proportion of orders in the last week that were shipped within 3 working days? Justify your answer.

8. A model marriage? Refer to Exercise 4. Would it be appropriate to use a normal distribution to model the sampling distribution of $\hat{p}$ = the sample proportion of couples in which both parents work outside the home? Justify your answer.

9. Women on diets Suppose that 70% of all col-
pg 428 lege women have been on a diet within the past 12 months. A sample survey interviews an SRS of 267 college women. Find the probability that 75% or more of the women in the sample have been on a diet.

10. Percentage of Harleys Harley-Davidson motorcycles make up 14% of all the motorcycles registered in the United States. You plan to interview an SRS of 500 motorcycle owners. Find the probability that 20% or more of the motorcycle owners in the sample own Harleys.

11. **Success on Kickstarter** The fundraising site Kickstarter regularly tracks the success rate of projects that seek funding on its site. Recently, the percentage of projects that were successfully funded was 38.7%.[13] You select a random sample of 50 Kickstarter projects. What is the probability that less than 30% of them were successfully funded?

12. **Parlez-vous français?** Quebec is the only province in Canada where the one official language is French. According to a recent census, 79.7% of Quebec residents identify French as their mother tongue. You select an SRS of 165 Quebec residents. What is the probability that less than 80% of them identify French as their mother tongue?

Applying the Concepts

13. **Drinking the cereal milk?** A *USA Today* poll asked a random sample of 1012 U.S. adults what they do with the milk in their cereal bowl after they have eaten. Let $\hat{p}$ be the proportion of people in the sample who drink the cereal milk. A spokesman for the dairy industry claims that 70% of all U.S. adults drink the cereal milk. Suppose this claim is true.

(a) Calculate the mean and standard deviation of the sampling distribution of $\hat{p}$. Interpret the standard deviation.

(b) Justify that the sampling distribution of $\hat{p}$ is approximately normal.

(c) Calculate the probability that at most 67% of the people in the sample drink the cereal milk.

14. **Who goes to church?** A Gallup poll asked a random sample of 1785 adults if they attended church during the past week. Let $\hat{p}$ be the proportion of people in the sample who attended church. A newspaper report claims that 40% of all U.S. adults went to church last week. Suppose this claim is true.

(a) Calculate the mean and standard deviation of the sampling distribution of $\hat{p}$. Interpret the standard deviation.

(b) Justify that the sampling distribution of $\hat{p}$ is approximately normal.

(c) Calculate the probability that at least 44% of the people in the sample attended church.

15. **Who drinks the cereal milk?** Refer to Exercise 13. Of the poll respondents, 67% said that they drink the cereal milk. Based on your answer to part (c), does this poll give convincing evidence that less than 70% of all U.S. adults drink the cereal milk? Explain.

16. **Do you go to church?** Refer to Exercise 14. Of the poll respondents, 44% said they attended church last week. Based on your answer to part (c), does this poll give convincing evidence that more than 40% of all U.S. adults attended church last week? Explain.

Extending the Concepts

17. **More milk drinkers** Refer to Exercise 13. What sample size would be required to reduce the standard deviation of the sampling distribution to one-half the value you found in part (a)? Justify your answer.

18. **Off to college** The example on page 428 used the sampling distribution of the sample proportion $\hat{p}$ to find the probability that the random sample of 1500 students from a population where $p = 0.35$ will give a $\hat{p}$ between 0.33 and 0.37. You can also find this probability using the sampling distribution of the sample count X, where X = the number of students in the sample who attend college within 50 miles of their home.

(a) Find the mean and standard deviation of the sampling distribution of the sample count X.

(b) Justify that X has an approximately normal distribution.

(c) Find the values of X that would result in sample proportions of $\hat{p} = 0.33$ and $\hat{p} = 0.37$.

(d) Calculate the probability that X is between the two values from part (c).

Recycle and Review

19. **Waiting with intent (1.8, 3.9)** Do drivers take longer to leave their parking spaces when another car is waiting? Researchers hung out in a parking lot and collected some data. The graphs and numerical summaries display information about how long it took drivers to exit their spaces.

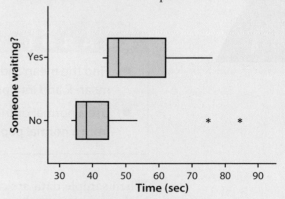

Descriptive Statistics: Time

Waiting	N	Mean	StDev	Min	Q_1	Median	Q_3	Max
No	20	44.42	14.10	33.76	35.61	39.56	48.48	84.92
Yes	20	54.11	14.39	41.61	43.41	47.14	66.44	85.97

(a) Write a few sentences comparing these distributions.

(b) Can we conclude that a waiting car causes drivers to leave their spaces more slowly? Why or why not?

20. **Those baby blues (2.1, 4.4)** The two-way table summarizes information about eye color and gender in a random sample of 200 high school students.

		Gender		
		Male	Female	Total
Eye color	Blue	21	29	50
	Brown	35	40	75
	Green	14	21	35
	Hazel	12	23	35
	Other	2	3	5
	Total	84	116	200

(a) Is there an association between eye color and gender in this group of students? Support your answer with an appropriate graphical summary of the data.

(b) Select a student at random. Are the events "Student is male" and "Student has blue eyes" independent? Justify your answer.

Lesson 6.5

The Sampling Distribution of a Sample Mean

LEARNING TARGETS

- Find the mean and standard deviation of the sampling distribution of a sample mean $\bar{x}$ and interpret the standard deviation.
- Use a normal distribution to calculate probabilities involving $\bar{x}$ when sampling from a normal population.

When sample data are categorical, we often use the count or proportion of successes in the sample to make an inference about a population. When sample data are quantitative, we often use the sample mean $\bar{x}$ to estimate the mean μ of a population. When we select random samples from a population, the value of $\bar{x}$ will vary from sample to sample. To understand how much $\bar{x}$ varies from μ and what values of $\bar{x}$ are likely to happen by chance, we want to understand the **sampling distribution of the sample mean $\bar{x}$.**

> ▐▐▐▐ **DEFINITION** Sampling distribution of the sample mean $\bar{x}$
>
> The **sampling distribution of the sample mean** $\bar{x}$ describes the distribution of values taken by the sample mean $\bar{x}$ in all possible samples of the same size from the same population.

When Mr. Ramirez's class did the Penny for Your Thoughts activity at the beginning of the chapter, his students produced the "dotplot" in Figure 6.9 showing the simulated sampling distribution of $\bar{x}$ = the sample mean year of pennies in 50 samples of size $n = 5$.

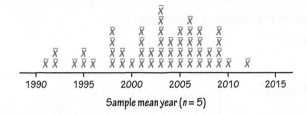

Sample mean year ($n = 5$)

FIGURE 6.9 Simulated sampling distribution of the sample mean year $\bar{x}$ in 50 samples of size $n = 5$ from a population of pennies.

This distribution is slightly skewed to the left, with a mean of about 2002 and a standard deviation of about 5 years. By the end of Lesson 6.6, you should be able to anticipate the shape, center, and variability of distributions like this one without having to do a simulation.

Center and Variability

When we select random samples of size n from a population with mean μ and standard deviation σ, the value of $\bar{x}$ will vary from sample to sample. As with the sampling distribution of $\hat{p}$, there are formulas that describe the center and variability of the sampling distribution of $\bar{x}$.

> **How to Calculate $\mu_{\bar{x}}$ and $\sigma_{\bar{x}}$**
>
> Suppose that $\bar{x}$ is the mean of an SRS of size n drawn from a large population with mean μ and standard deviation σ. Then:
>
> • The **mean** of the sampling distribution of $\bar{x}$ is $\mu_{\bar{x}} = \mu$.
>
> • The **standard deviation** of the sampling distribution of $\bar{x}$ is $\sigma_{\bar{x}} = \dfrac{\sigma}{\sqrt{n}}$.

The behavior of $\bar{x}$ in repeated samples is much like that of the sample proportion $\hat{p}$:

• The sample mean $\bar{x}$ is an *unbiased estimator* of the population mean μ. This is because the mean of the sampling distribution $\mu_{\bar{x}}$ is equal to the mean of the population μ.

• The standard deviation of the sampling distribution of $\bar{x}$ describes the typical distance between the sample mean $\bar{x}$ and the population mean μ.

• The distribution of $\bar{x}$ is less variable for larger samples. This is indicated by the $\sqrt{n}$ in the denominator of the standard deviation formula.

• The formula for the standard deviation of the distribution of $\bar{x}$ requires that the observations be independent. In practice, we are safe assuming independence when we are sampling without replacement as long as the sample size is less than 10% of the population size.

These facts about the mean and standard deviation of $\bar{x}$ are true *no matter what shape the population distribution has.*

EXAMPLE

Seen any good movies lately?

Mean and standard deviation of the sampling distribution of $\bar{x}$

PROBLEM: The number of movies viewed in the last year by students at a large high school has a mean of 19.3 movies with a standard deviation of 15.8 movies. Suppose we take an SRS of 100 students from this school and calculate the mean number of movies viewed by the members of the sample.

(a) Calculate the mean and standard deviation of the sampling distribution of $\bar{x}$.

(b) Interpret the standard deviation from part (a).

SOLUTION:

(a) $\mu_{\bar{x}} = 19.3$ movies and $\sigma_{\bar{x}} = \dfrac{15.8}{\sqrt{100}} = 1.58$ movies

(b) In SRSs of size $n = 100$, the sample mean number of movies will typically vary by about 1.58 movies from the population mean of 19.3 movies.

> Recall that $\mu_{\bar{x}} = \mu$ and $\sigma_{\bar{x}} = \dfrac{\sigma}{\sqrt{n}}$.

FOR PRACTICE **TRY EXERCISE 1.**

Shape

The shape of the sampling distribution of the sample mean $\bar{x}$ depends on the shape of the population distribution. In the following activity, you will explore what happens when sampling from a normal population.

ACTIVITY

Sampling from a normal population

Professor David Lane of Rice University has developed a wonderful applet for investigating the sampling distribution of $\bar{x}$. In this activity, you'll use Professor Lane's applet to explore the shape of the sampling distribution when the population is normally distributed.

1. Go to **http://onlinestatbook.com/stat_sim/ sampling_dist/** or search for "online statbook sampling distributions applet" and go to the website. When the BEGIN button appears on the left side of the screen, click on it. You will then see a yellow page entitled "Sampling Distributions" like the one in the screen shot.

2. There are choices for the population distribution: normal, uniform, skewed, and custom. The

default is normal. Click the "Animated" button. What happens? Click the button several more times. What do the black boxes represent? What is the blue square that drops down onto the plot below?

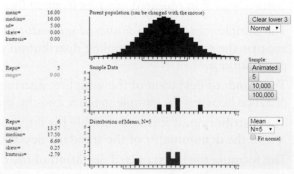

3. Click on "Clear lower 3" to start clean. Then click on the "100,000" button under "Sample:" to simulate taking 100,000 SRSs of size $n = 5$ from the population. Answer these questions:

  • Does the simulated sampling distribution (blue bars) have a recognizable shape? Click the box next to "Fit normal."

  • To the left of each distribution is a set of summary statistics. Compare the mean of the simulated sampling distribution with the mean of the population.

  • How is the standard deviation of the simulated sampling distribution related to the standard deviation of the population?

4. Click "Clear lower 3." Use the drop-down menus to set up the bottom graph to display the mean for samples of size $n = 20$. Then sample 100,000 times. How do the two distributions of $\bar{x}$ compare: shape, center, and variability?

5. What have you learned about the shape of the sampling distribution of $\bar{x}$ when the population has a normal shape?

As the activity demonstrates, if the population distribution is normal, so is the sampling distribution of $\bar{x}$. *This is true no matter what the sample size is.*

Describing a Sampling Distribution of a Sample Mean When Sampling from a Normal Population

Suppose that a population is normally distributed with mean μ and standard deviation σ. Then the sampling distribution of $\bar{x}$ for SRSs of size n has a normal distribution with mean μ and standard deviation $\sigma/\sqrt{n}$.

In the next lesson, you will learn what happens when sampling from a non-normal population.

Finding Probabilities Involving $\bar{x}$

Now we have enough information to calculate probabilities involving $\bar{x}$ when the population distribution is normal.

EXAMPLE

Are those peanuts underweight?

Probabilities involving $\bar{x}$

PROBLEM: At the P. Nutty Peanut Company, dry-roasted, shelled peanuts are placed in jars labeled "16 ounces" by a machine. The distribution of weights in the jars is approximately normal with a mean of 16.1 ounces and a standard deviation of 0.15 ounces. Find the probability that the mean weight of 10 randomly selected jars is less than the advertised weight of 16 ounces.

Charles Nesbit/Getty Images

SOLUTION:

• Mean: $\mu_{\bar{x}} = 16.1$ ounces

• SD: $\sigma_{\bar{x}} = \dfrac{0.15}{\sqrt{10}} = 0.047$ ounces

Let $\bar{x}$ = the sample mean weight of 10 randomly selected jars. To find $P(\bar{x} \leq 16)$, we have to know the mean, standard deviation, and shape of the sampling distribution of $\bar{x}$. Recall that $\mu_{\bar{x}} = \mu$ and $\sigma_{\bar{x}} = \dfrac{\sigma}{\sqrt{n}}$.

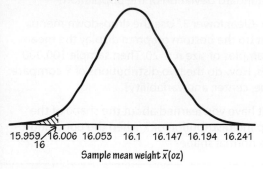

• Shape: Approximately normal because the population distribution is approximately normal

Sample mean weight x̄ (oz)

15.959 16.006 16.053 16.1 16.147 16.194 16.241
16

Using Table A:

$$z = \frac{16 - 16.1}{0.047} = -2.13$$

$$P(Z \le -2.13) = 0.0166$$

Using technology: Applet/normalcdf(lower: −1000, upper: 16, mean: 16.1, SD: 0.047) = 0.0167

1. Draw a normal curve.

2. Perform calculations.
(i) Standardize the boundary value and use Table A to find the desired probability; or
(ii) Use technology.

FOR PRACTICE TRY EXERCISE 5.

Individual values vary more than averages, so randomly selecting a single jar that is under the advertised weight is more likely than getting a sample mean for 10 jars that is less than the advertised weight. This is illustrated in Figure 6.10.

FIGURE 6.10 Normal curves showing the distribution of weights for individual jars of peanuts (purple curve) and distribution of sample mean weights for SRSs of 10 jars (blue curve).

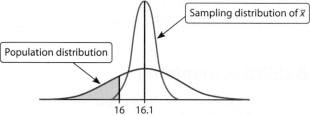

Sampling distribution of x̄

Population distribution

16   16.1

The fact that averages of several observations are less variable than individual observations is important in many settings. For example, it is common practice in science and medicine to repeat a measurement several times and report the average of the results.

LESSON APP 6.5

Are college women taller?

The heights of young women follow a normal distribution with mean $\mu = 64.5$ inches and standard deviation $\sigma = 2.5$ inches.

1. Calculate the mean and standard deviation of the sampling distribution of $\bar{x}$ for SRSs of size 15.

2. Interpret the standard deviation from Question 1.

3. Find the probability that the mean height of an SRS of 15 young women exceeds 66.5 inches.

4. Suppose that the mean height in a sample of $n = 15$ young women from a local college is $\bar{x} = 66.5$. Based on your answer to Question 3, what would you conclude about the mean height for all young women at this college?

Lesson 6.5

WHAT DID YOU LEARN?

LEARNING TARGET	EXAMPLES	EXERCISES
Find the mean and standard deviation of the sampling distribution of a sample mean $\bar{x}$ and interpret the standard deviation.	p. 434	1–4
Use a normal distribution to calculate probabilities involving $\bar{x}$ when sampling from a normal population.	p. 435	5–8

Exercises   Lesson 6.5

Mastering Concepts and Skills

1. **Short songs?** David's iPod has about 10,000 songs. pg 434 The distribution of the play times for these songs is heavily skewed to the right with a mean of 225 seconds and a standard deviation of 60 seconds. Suppose we choose an SRS of 10 songs from this population and calculate the mean play time $\bar{x}$ of these songs.

   (a) Calculate the mean and standard deviation of the sampling distribution of $\bar{x}$.

   (b) Interpret the standard deviation from part (a).

2. **Grinding auto parts** A grinding machine in an auto-parts factory prepares axles with a target diameter of $\mu = 40.125$ millimeters (mm). The machine has some variability, so the standard deviation of the diameters is $\sigma = 0.002$ millimeter. The machine operator inspects a random sample of 4 axles each hour for quality-control purposes and records the sample mean diameter $\bar{x}$.

   (a) Calculate the mean and standard deviation of the sampling distribution of $\bar{x}$.

   (b) Interpret the standard deviation from part (a).

3. **Fresh tomatoes** A local garden center says that a certain variety of tomato plant produces tomatoes with a mean weight of 250 grams and a standard deviation of 42 grams. You take a random sample of 20 tomatoes produced by these plants and calculate their mean weight $\bar{x}$.

   (a) Calculate the mean and standard deviation of the sampling distribution of $\bar{x}$.

   (b) Interpret the standard deviation from part (a).

4. **Fuel efficiency** Driving styles differ, so there is variability in the fuel efficiency of the same model automobile driven by different people. For a certain car model, the mean fuel efficiency is 23.6 miles per gallon with a standard deviation of 2.5 miles per gallon.[14] Take a simple random sample of 25 owners of this model and calculate the sample mean fuel efficiency $\bar{x}$.

   (a) Calculate the mean and standard deviation of the sampling distribution of $\bar{x}$.

   (b) Interpret the standard deviation from part (a).

5. **How much cereal?** A company's cereal boxes advertise pg 435 that 9.65 ounces of cereal are contained in each box. In fact, the amount of cereal in a randomly selected box follows a normal distribution with mean $\mu = 9.70$ ounces and standard deviation $\sigma = 0.03$ ounce. What is the probability that the mean amount of cereal $\bar{x}$ in 5 randomly selected boxes is at most 9.65?

6. **Finch beaks** One dimension of bird beaks is "depth"—the height of the beak where it arises from the bird's head. During a research study on one island in the Galapagos archipelago, the beak depth of all Medium Ground Finches on the island was found to be normally distributed with mean $\mu = 9.5$ millimeters and standard deviation $\sigma = 1.0$ millimeter.[15] What is the probability that the mean depth $\bar{x}$ in 10 randomly selected Medium Ground Finches is at least 10 millimeters?

7. **Estimating cholesterol** Suppose that the blood cholesterol level of all men aged 20 to 34 follows a normal distribution with mean $\mu = 188$ milligrams per deciliter (mg/dl) and standard deviation $\sigma = 41$ mg/dl. In an SRS of size 100, find the probability that $\bar{x}$ estimates μ to within ± 3 mg/dl.

8. **Bottlers at work** A company uses a machine to fill plastic bottles with cola. The contents of the bottles vary according to a normal distribution with mean $\mu = 298$ milliliters and standard deviation $\sigma = 3$ milliliters. In an SRS of size 16, find the probability that $\bar{x}$ estimates μ to within ± 1 milliliter.

Applying the Concepts

9. **Why won't the car start?** An automaker has found that the lifetime of its batteries varies from car to car according to a normal distribution with mean $\mu = 48$ months and standard deviation $\sigma = 8.2$ months. The company installs a new battery on an SRS of 8 cars.

(a) Calculate the mean and standard deviation of the sampling distribution of $\bar{x}$ for SRSs of size 8.

(b) Interpret the standard deviation from part (a).

(c) Find the probability that the sample mean life is less than 42.2 months.

10. **Birth weights** The birth weights of males born full term are normally distributed with mean $\mu = 3.4$ kilograms and standard deviation $\sigma = 0.5$ kilogram.[16] A large city hospital selects a random sample of 15 full-term males born in the last six months.

(a) Calculate the mean and standard deviation of the sampling distribution of $\bar{x}$ for SRSs of size 15.

(b) Interpret the standard deviation from part (a).

(c) Find the probability that the sample mean weight is greater than 3.55 kilograms.

11. **Could it be the battery?** Refer to Exercise 9. Suppose that the average life of the batteries on these 8 cars turns out to be $\bar{x} = 42.2$ months. Based on your answer to Exercise 9, is there convincing evidence that the population mean μ is really less than 48 months? Explain.

12. **More birth weights** Refer to Exercise 10. Suppose that the average birth weight of the 15 babies turns out to be 3.55 kilograms. Based on your answer to Exercise 10, is there convincing evidence that the population mean μ is really more than 3.4 kilograms? Explain.

13. **One man's cholesterol** In Exercise 7, you calculated the probability that $\bar{x}$ would estimate the true mean cholesterol level within ±3 mg/dl of μ in samples of size 100.

(a) If you randomly selected one 20- to 34-year-old male instead of 100, would he be more likely, less likely, or equally likely to have a cholesterol level within ±3 mg/dl of μ? Explain this without doing any calculations.

(b) Calculate the probability of the event described in part (a) to confirm your answer.

14. **One bottle of cola** In Exercise 8, you calculated the probability that $\bar{x}$ would estimate the true mean amount of cola within ±1 milliliter of μ in samples of size 16.

(a) If you randomly selected one bottle instead of 16, would it be more likely, less likely, or equally likely to contain an amount of cola within ±1 milliliter of μ? Explain this without doing any calculations.

(b) Calculate the probability of the event described in part (a) to confirm your answer.

15. **Sampling music** Refer to Exercise 1. How many songs would you have to sample if you wanted the standard deviation of the sampling distribution of $\bar{x}$ to be 30 seconds?

16. **Sampling auto parts** Refer to Exercise 2. How many axles would you have to sample if you wanted the standard deviation of the sampling distribution of $\bar{x}$ to be 0.0005 millimeter?

Extending the Concepts

17. **Orange overage** Mandarin oranges from a certain grove have weights that follow a normal distribution with mean $\mu = 3$ ounces and standard deviation $\sigma = 0.5$ ounce. Bags are filled with an SRS of 20 mandarin oranges. What is the probability that the total weight of oranges in a bag is greater than 65 ounces? *Hint:* Re-express the total weight of 20 oranges in terms of the average weight $\bar{x}$.

Recycle and Review

18. **Let's text** (1.2) We used Census at School's "Random Data Selector" to choose a sample of 50 Canadian students who completed the survey in a recent year. The bar graph displays data on students' responses to the question "Which of these methods do you most often use to communicate with your friends?"

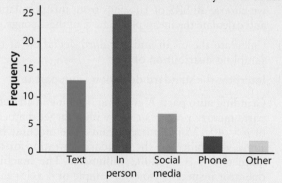

Method of communication

(a) Would it be appropriate to make a pie chart for these data? Why or why not?

(b) Jerry says that he would describe this bar graph as skewed to the right. Explain why Jerry is wrong.

19. **Shut it down and go to sleep!** (5.1, 5.2) A National Sleep Foundation survey of 1103 parents asked, among other questions, how many electronic devices (TVs, video games, smartphones, computers, MP3 players, and so on) children had in their bedrooms.[17] Let X = the number of devices in a randomly chosen child's bedroom. Here is the probability distribution of X.

Number of devices	0	1	2	3	4	5
Probability	0.28	0.27	0.18	0.16	0.07	0.04

(a) Show that this is a legitimate probability distribution.

(b) What is the probability that a randomly chosen child has at least 1 electronic device in her bedroom?

(c) Calculate the expected value and standard deviation of X.

Lesson 6.6
The Central Limit Theorem

- Determine if the sampling distribution of $\bar{x}$ is approximately normal when sampling from a non-normal population.

- If appropriate, use a normal distribution to calculate probabilities involving $\bar{x}$.

In Lesson 6.5, you learned about the sampling distribution of the sample mean $\bar{x}$ when sampling from a normally distributed population. The following activity will help you explore what happens when you sample from non-normal populations.

ACTIVITY

Sampling from a non-normal population

In this activity, we will use an applet to investigate the sampling distribution of the sample mean $\bar{x}$ when sampling from a non-normal population.

1. Go to **http://onlinestatbook.com/stat_sim/sampling_dist/** or search for "online statbook sampling distributions applet" and go to the website. Launch the applet and select the "Skewed" population. Set the bottom two graphs to display the mean—one

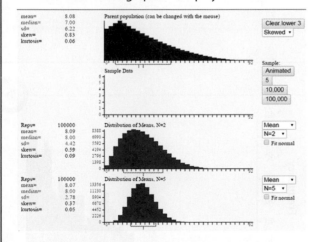

for samples of size 2 and the other for samples of size 5. Click the "Animated" button a few times to be sure you see what's happening. Then "Clear lower 3" and take 100,000 SRSs. Describe what you see.

2. Change the sample sizes to $n = 10$ and $n = 16$ and repeat Step 1. What do you notice?

3. Now change the sample sizes to $n = 20$ and $n = 25$ and take 100,000 samples. Did this confirm what you saw in Step 2?

4. Clear the page, and select "Custom" distribution from the drop-down menu at the top of the page. Click on a point on the horizontal axis, and drag up to create a bar. Make a distribution that looks as strange as you can. (*Note:* You can shorten a bar or get rid of it completely by clicking on the top of the bar and dragging down to the axis.) Then repeat Steps 1 to 3 for your custom distribution. Cool, huh?

5. Summarize what you learned about the shape of the sampling distribution of $\bar{x}$.

The Central Limit Theorem

It is a remarkable fact that as the sample size increases, the sampling distribution of $\bar{x}$ changes shape: It looks less like that of the population and more like a normal distribution. This is true no matter what shape the population distribution has. This famous fact of probability theory is called the **central limit theorem** (sometimes abbreviated as **CLT**).

> **DEFINITION** Central limit theorem (CLT)
>
> Draw an SRS of size n from any population with mean μ and finite standard deviation σ. The **central limit theorem (CLT)** says that when n is large, the sampling distribution of the sample mean $\bar{x}$ is approximately normal.

How large a sample size n is needed for the sampling distribution of $\bar{x}$ to be close to normal depends on the population distribution. A larger sample size is required if the shape of the population distribution is far from normal. In that case, the sampling distribution of $\bar{x}$ will also be far from normal if the sample size is small. To use a normal distribution to calculate probabilities involving $\bar{x}$, check the **Normal/Large Sample condition.**

> **DEFINITION** Normal/Large Sample condition
>
> The Normal/Large Sample condition says that the distribution of $\bar{x}$ will be approximately normal when *either* of the following is true:
>
> • The population distribution is approximately normal. This is true no matter what the sample size n is.
>
> • The sample size is large. If the population distribution is not normal, the sampling distribution of $\bar{x}$ will be approximately normal in most cases if $n \geq 30$.

EXAMPLE

A few more pennies for your thoughts?

Sampling from a non-normal population

PROBLEM: Mr. Ramirez's class did the Penny for Your Thoughts Activity from the beginning of this chapter. The histogram shows the distribution of ages for the 2341 pennies in their collection.

(a) Describe the shape of the sampling distribution of $\bar{x}$ for SRSs of size $n = 2$ from the population of pennies. Justify your answer.

(b) Describe the shape of the sampling distribution of $\bar{x}$ for SRSs of size $n = 50$ from the population of pennies. Justify your answer.

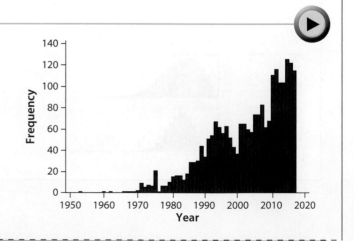

SOLUTION:

(a) Because $n = 2 < 30$, the sampling distribution of $\bar{x}$ will be skewed to the left, but not quite as strongly as the population.

(b) Because $n = 50 \geq 30$, the sampling distribution of $\bar{x}$ will be approximately normal by the central limit theorem.

FOR PRACTICE   TRY EXERCISE 1.

The dotplots in Figure 6.11 show the simulated sampling distributions of the sample mean for (a) 500 SRSs of size $n = 2$ and (b) 500 SRSs of size $n = 50$.

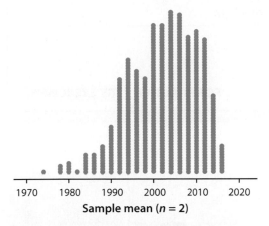

Sample mean ($n = 2$)

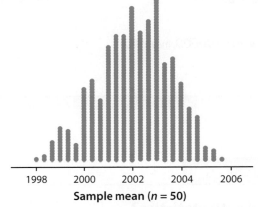

Sample mean ($n = 50$)

FIGURE 6.11 Simulated sampling distributions of the sample mean age for (a) 500 SRSs of size $n = 2$ and (b) 500 SRSs of size $n = 50$ from a population of pennies.

As expected, the simulated sampling distribution of $\bar{x}$ for SRSs of size $n = 2$ is skewed to the left and the simulated sampling distribution of $\bar{x}$ for SRSs of size $n = 50$ is approximately normal—thanks to the central limit theorem.

Probabilities Involving $\bar{x}$

Using the central limit theorem, we can do probability calculations involving $\bar{x}$ even when the population is non-normal.

EXAMPLE

Mean texts?

Probabilities involving $\bar{x}$

PROBLEM: Suppose that the number of texts sent during a typical day by the population of students at a large high school follows a right-skewed distribution with a mean of 45 and a standard deviation of 35. How likely is it that a random sample of 100 students will average at least 50 texts per day?

SOLUTION:

- Mean: $\mu_{\bar{x}} = 45$ texts

- SD: $\sigma_{\bar{x}} = \dfrac{35}{\sqrt{100}} = 3.5$ texts

- Shape: Approximately normal by the CLT because $n = 100 \geq 30$

> Let $\bar{x}$ = the sample mean number of texts. To find $P(\bar{x} \geq 50)$, we have to know the mean, standard deviation, and shape of the sampling distribution of $\bar{x}$.
>
> Recall that $\mu_{\bar{x}} = \mu$ and $\sigma_{\bar{x}} = \dfrac{\sigma}{\sqrt{n}}$.

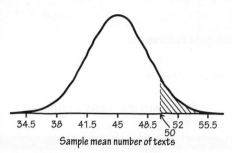

34.5 38 41.5 45 48.5 52 55.5
50
Sample mean number of texts

Using Table A:

$$z = \frac{50 - 45}{3.5} = 1.43$$

$$P(Z \geq 1.43) = 1 - 0.9236 = 0.0764$$

Using technology:
Applet/normalcdf (lower: 50, upper: 100000, mean: 45, SD: 3.5) = 0.0766

1. **Draw a normal curve.**

2. **Perform calculations.**
   (i)   Standardize the boundary value and use Table A to find the desired probability; or
   (ii) Use technology.

FOR PRACTICE   TRY EXERCISE 5.

LESSON APP 6.6

Keeping things cool with statistics?

Your company has a contract to perform preventive maintenance on thousands of air-conditioning units in a large city. Based on service records from the past year, the time (in hours) that a technician requires to complete the work follows a strongly right-skewed distribution with $\mu = 1$ hour and $\sigma = 1.5$ hours. As a promotion, your company will provide service to a random sample of 70 air-conditioning units free of charge. You plan to budget an average of 1.1 hours per unit for a technician to complete the work. Will this be enough time?

1. What is the shape of the sampling distribution of $\bar{x}$ for samples of size $n = 70$ from this population? Justify.

2. Calculate the probability that the average maintenance time $\bar{x}$ for 70 units exceeds 1.1 hours.

3. Based on your answer to the previous problem, did the company budget enough time? Explain.

simazoran/iStock/Getty Images

Lesson 6.6

WHAT DID YOU LEARN?

LEARNING TARGET	EXAMPLES	EXERCISES
Determine if the sampling distribution of $\bar{x}$ is approximately normal when sampling from a non-normal population.	p. 440	1–4
If appropriate, use a normal distribution to calculate probabilities involving $\bar{x}$.	p. 441	5–8

Exercises   *Lesson 6.6*

Mastering Concepts and Skills

1. **Songs on an iPod** David's iPod has about 10,000 songs. The distribution of the play times for these songs is heavily skewed to the right with a mean of 225 seconds and a standard deviation of 60 seconds.

pg 440

(a) Describe the shape of the sampling distribution of $\bar{x}$ for SRSs of size $n = 5$ from the population of songs on David's iPod. Justify your answer.

(b) Describe the shape of the sampling distribution of $\bar{x}$ for SRSs of size $n = 100$ from the population of songs on David's iPod. Justify your answer.

2. **Insurance claims** An insurance company claims that in the entire population of homeowners, the mean annual loss from fire is $\mu = \$250$ with a standard deviation of $\sigma = \$5000$. The distribution of losses is strongly right-skewed: Many policies have $0 loss, but a few have large losses.

(a) Describe the shape of the sampling distribution of $\bar{x}$ for SRSs of size $n = 15$ from the population of homeowners. Justify your answer.

(b) Describe the shape of the sampling distribution of $\bar{x}$ for SRSs of size $n = 1000$ from the population of homeowners. Justify your answer.

3. **How many in a car?** A study of rush-hour traffic in San Francisco counts the number of people in each car entering a freeway at a suburban interchange. Suppose that the number of people per car in the population of all cars that enter at this interchange during rush hours has a mean of $\mu = 1.5$ and a standard deviation of $\sigma = 0.75$.

(a) Could the distribution of the number of people per car be normal for the population of all cars entering the interchange during rush hours? Explain.

(b) Describe the shape of the sampling distribution of $\bar{x}$ for SRSs of size $n = 100$ from the population of all cars that enter this interchange during rush hours. Justify your answer.

4. **Flawed carpets** A supervisor at a carpet factory randomly selects 1-square-yard pieces of carpet and counts the number of flaws in each piece. The number of flaws per square yard in the population of 1-square-yard pieces varies with mean $\mu = 1.6$ and standard deviation $\sigma = 1.2$.

(a) Could the distribution of the number of flaws be normal for the population of all 1-square-yard pieces of carpet? Explain.

(b) Describe the shape of the sampling distribution of $\bar{x}$ for SRSs of size $n = 60$ from the population of all 1-square-yard pieces of carpet. Justify your answer.

5. **More songs on an iPod** Refer to Exercise 1. What is the probability that the mean length in a random sample of 100 songs is less than 4 minutes (240 seconds)?

pg 441

6. **More insurance claims** Refer to Exercise 2. Suppose that the insurance company charges $300 for each policy. What is the probability that the insurance company will make money on a random sample of 1000 homeowners? That is, what is the probability that the mean loss for a random sample of homeowners is less than $300?

7. **More people in a car** Refer to Exercise 3. What is the probability that the mean number of people in a random sample of 100 cars that enter at this interchange during rush hours is at least 1.7?

8. **More flawed carpets** Refer to Exercise 4. What is the probability that the mean number of flaws in a random sample of sixty 1-square-yard pieces of carpet is at least 1.7?

Applying the Concepts

9. **Where does lightning strike?** The number of lightning strikes on a square kilometer of open ground in a year has a mean of 6 and standard deviation of 2.4. The National Lightning Detection Network (NLDN) uses automatic sensors to watch for lightning in a random sample of fifty 1-square-kilometer plots of land. Let $\bar{x}$ be the average number of lightning strikes in the sample.

(a) What is the shape of the sampling distribution of $\bar{x}$ for samples of size $n = 50$ from this population? Justify.

(b) Calculate the probability that the average number of lightning strikes per square kilometer $\bar{x}$ is less than 5.

10. **Please hold** The customer care manager at a cell phone company keeps track of how long each help-line caller spends on hold before speaking to a customer service representative. He finds that the distribution of wait times for all callers has a mean of 12.8 minutes with a standard deviation of 7.2 minutes. The distribution is moderately skewed to the right. Suppose the manager takes a random sample of 45 callers and calculates their mean wait time $\bar{x}$.

(a) What is the shape of the sampling distribution of $\bar{x}$ for samples of size $n = 45$ from this population? Justify.

(b) Calculate the probability that the mean wait time $\bar{x}$ is more than 15 minutes.

11. **Lightning strikes twice?** Refer to Exercise 9.

(a) Explain why you cannot calculate the probability that the average number of lightning strikes per square kilometer $\bar{x}$ is less than 5 for samples of size $n = 10$.

(b) Will the probability referred to in part (a) be less than, greater than, or about the same as the probability in Exercise 9(b)? Explain.

12. **Please continue to hold** Refer to Exercise 10.

(a) Explain why you cannot calculate the probability that the average wait time for customer service $\bar{x}$ is more than 15 minutes for samples of size $n = 5$.

(b) Will the probability referred to in part (a) be less than, greater than, or about the same as the probability in Exercise 10(b)? Explain.

13. **Even more people in a car** Refer to Exercise 7. A researcher selects a random sample of 100 cars that enter this interchange on a Sunday and finds $\bar{x} = 1.7$ people per car. Because the sample mean is greater than 1.5, the researcher concludes that people are more likely to drive with other people in the car on Sundays. Based on your answer to Exercise 7, what would you say to the researcher?

14. **Even more flaws in carpets** Refer to Exercise 8. A supervisor selects a random sample of sixty 1-square-yard pieces of carpet and finds that $\bar{x} = 1.7$ flaws. Because the sample mean is more than the expected mean of 1.6 flaws, the supervisor is thinking about shutting down the machine for inspection. Based on your answer to Exercise 8, what would you say to the supervisor?

15. **What does the CLT say?** Asked what the central limit theorem says, a student replies, "As you take larger and larger samples from a population, the graph of the sample values looks more and more normal." Is the student right? Explain your answer.

16. **Is this what the CLT says?** Asked what the central limit theorem says, a student replies, "As you take larger and larger samples from a population, the variability of the sampling distribution of the sample mean decreases." Is the student right? Explain your answer.

Extending the Concepts

17. **Cost of textbooks** The cost of textbooks for students at a particular college has a mean of $\mu = \$836$ per year with a standard deviation of $\sigma = \$388$. What is the probability that a random sample of 50 students spends a total of more than $45,000 on books this year? *Hint:* Re-express the total cost in terms of the average cost per student $\bar{x}$.

Recycle and Review

18. **Are rich people mean?** (3.1, 3.6) Psychologist Paul Piff from the University of California, Berkeley, studies the relationship between wealth and lawful behavior. In one such study, he had assistants cross a road at a crosswalk and recorded if drivers obeyed the law and stopped to let the person cross or kept driving and cut off the pedestrian. He compared the response of people driving expensive cars and inexpensive cars. Here are his results.[18]

		Type of Car	
		Expensive car	Inexpensive car
Driver behavior	Yielded to pedestrian	32	67
	Cut off pedestrian	26	27

(a) The report on this study stated that the researcher who determined if the cars could be classified as expensive or inexpensive was "blind to the hypothesis of the study." Explain what this means.

(b) Is this an observational study or an experiment? Justify your answer.

19. **How do rich people drive?** (4.3, 4.4) Suppose we choose a driver at random from the results of the study in Exercise 18. Show that the events "Yielded to pedestrians" and "Expensive car" are not independent.

Chapter 6

STATS applied!
How can we build "greener" batteries?

Refer to the STATS applied! on page 399. When the manufacturing process is working properly, the distribution of battery lifetimes has mean $\mu = 17$ hours with standard deviation $\sigma = 0.8$ hour, and 73% last at least 16.5 hours.

1. Assume that the manufacturing process is working properly, and let $\hat{p} =$ the sample proportion of batteries that last at least 16.5 hours. Calculate the mean and standard deviation of the sampling distribution of $\hat{p}$ for random samples of 50 batteries.

2. Describe the shape of the sampling distribution of $\hat{p}$ for random samples of 50 batteries. Justify your answer.

3. In the sample of 50 batteries, only 68% lasted at least 16.5 hours. Find the probability of obtaining a random sample of 50 batteries where $\hat{p}$ is 0.68 or less if the manufacturing process is working properly.

4. Assume that the process is working properly, and let $\bar{x} =$ the sample mean lifetime (in hours). Calculate the mean and standard deviation of the sampling distribution of $\bar{x}$ for random samples of 50 batteries.

5. Describe the shape of the sampling distribution of $\bar{x}$ for random samples of 50 batteries. Justify your answer.

6. In the sample of 50 batteries, the mean lifetime was only 16.718 hours. Find the probability of obtaining a random sample of 50 batteries with a mean lifetime of 16.718 hours or less if the manufacturing process is working properly.

7. Based on your answers to Questions 3 and 6, should the company be worried that the manufacturing process isn't working properly? Explain.

Main Points   *Chapter 6*

The Idea of a Sampling Distribution

- A **parameter** is a number that describes some characteristic of the population. A **statistic** is a number that describes some characteristic of a sample. We use statistics to estimate parameters.

- The **sampling distribution** of a statistic is the distribution of values taken by the statistic in all possible samples of the same size from the same population.

- To determine a sampling distribution, list all possible samples of a particular size, calculate the value of the statistic for each sample, and graph the distribution of the statistic. If there are many possible samples, use simulation to approximate the sampling distribution: Repeatedly select random samples of a particular size, calculate the value of the statistic for each sample, and graph the distribution of the statistic.

- We can use sampling distributions to determine what values of a statistic are likely to happen by chance alone and how much a statistic typically varies from the parameter it is trying to estimate.

- A statistic used to estimate a parameter is an **unbiased estimator** if the mean of its sampling distribution is equal to the value of the parameter being estimated. That is, the statistic doesn't consistently overestimate or consistently underestimate the value of the parameter when many random samples are selected.

- The sampling distribution of any statistic will have less variability when the sample size is larger. That is, the statistic will be a more precise estimator of the parameter with larger sample sizes.

Sample Counts and Sample Proportions

- Let X = the number of successes in a random sample of size n from a large population with proportion of successes p. The **sampling distribution of a sample count X** describes the distribution of values taken by the sample count X in all possible samples of the same size from the same population.

  - The **mean** of the sampling distribution of X is $\mu_X = np$. The mean describes the average value of X in repeated random samples.

  - The **standard deviation** of the sampling distribution of X is $\sigma_X = \sqrt{np(1-p)}$. The standard deviation describes how far the values of X typically vary from μ_X in repeated random samples.

  - The shape of the sampling distribution of X will be approximately normal when the **Large Counts condition** is met: $np \geq 10$ and $n(1-p) \geq 10$.

- Let $\hat{p}$ = the proportion of successes in a random sample of size n from a large population with proportion of successes p. The **sampling distribution of a sample proportion $\hat{p}$** describes the distribution of values taken by the sample proportion $\hat{p}$ in all possible samples of the same size from the same population.

  - The **mean** of the sampling distribution of $\hat{p}$ is $\mu_{\hat{p}} = p$. The mean describes the average value of $\hat{p}$ in repeated random samples.

  - The **standard deviation** of the sampling distribution of $\hat{p}$ is $\sigma_{\hat{p}} = \sqrt{\dfrac{p(1-p)}{n}}$. The standard deviation describes how far the values of $\hat{p}$ typically vary from p in repeated random samples.

  - The shape of the sampling distribution of $\hat{p}$ will be approximately normal when the **Large Counts condition** is met: $np \geq 10$ and $n(1-p) \geq 10$.

Sample Means

- Let $\overline{x}$ = the mean of a random sample of size n from a large population with mean μ and standard deviation σ. The **sampling distribution of a sample mean $\overline{x}$** describes the distribution of values taken by the sample mean $\overline{x}$ in all possible samples of the same size from the same population.

  - The **mean** of the sampling distribution of $\overline{x}$ is $\mu_{\overline{x}} = \mu$. The mean describes the average value of $\overline{x}$ in repeated random samples.

  - The **standard deviation** of the sampling distribution of $\overline{x}$ is $\sigma_{\overline{x}} = \dfrac{\sigma}{\sqrt{n}}$. The standard deviation describes how far the values of $\overline{x}$ typically vary from μ in repeated random samples.

  - The shape of the sampling distribution of $\overline{x}$ will be approximately normal when the **Normal/Large Sample** condition is met: The population is normal or the sample size is large ($n \geq 30$). The fact that the sampling distribution of $\overline{x}$ becomes approximately normal—even when the population is non-normal—as the sample size increases is called the **central limit theorem.**

Probability Calculations

- When the sampling distribution of a statistic is approximately normal, you can use z-scores and Table A or technology to do probability calculations involving the statistic.

- To determine which sampling distribution to use, consider whether the variable of interest is categorical or quantitative. If it is categorical, use the sampling distribution of a sample count X or the sampling distribution of a sample proportion $\hat{p}$. If it is quantitative, use the sampling distribution of a sample mean $\overline{x}$.

Chapter 6 Review Exercises

1. **Bad eggs (6.1)** People who eat eggs that are contaminated with salmonella can get food poisoning. A large egg producer takes an SRS of 200 eggs from all the eggs shipped in one day. The laboratory reports that 9 of these eggs had salmonella contamination. Identify the population, the parameter, the sample, and the statistic.

2. **Five books (6.1, 6.2)** An author has written 5 children's books. The number of pages in these books are 64, 66, 71, 73, and 76.

(a) List all 10 possible SRSs of size $n = 3$, calculate the median number of pages for each sample, and display the sampling distribution of the sample median on a dotplot.

(b) Show that the sample median is a biased estimator of the population median for this population.

(c) Describe how the variability of the sampling distribution of the sample median would change if the sample size was increased to $n = 4$.

3. **Kids these days (6.3)** According to the 2010 U.S. Census, 24% of U.S. residents are under 18 years old. Suppose we take a random sample of 500 U.S. residents. Let $X =$ the number of people in the sample who are under 18 years old.

(a) Calculate the mean and standard deviation of the sampling distribution of X.

(b) Interpret the standard deviation from part (a).

(c) Justify that it is appropriate to use a normal distribution to model the sampling distribution of X.

(d) Using a normal distribution, calculate the probability that the number of people under 18 years old in a random sample of size 500 is between 100 and 110.

4. **Five-second rule (6.1, 6.4)** A report claimed that 20% of respondents subscribe to the "5-second rule." That is, they would eat a piece of food that fell onto the kitchen floor if it was picked up within 5 seconds. Assume this figure is accurate for the population of U.S. adults. Let $\hat{p} =$ the proportion of people who subscribe to the 5-second rule in an SRS of size 80 from this population.

(a) Calculate the mean and the standard deviation of the sampling distribution of $\hat{p}$.

(b) Interpret the standard deviation from part (a).

(c) Justify that it is appropriate to use a normal distribution to model the sampling distribution of $\hat{p}$.

(d) In an SRS of size 80, only 10% subscribed to the 5-second rule. Does this result provide convincing evidence that the proportion of all U.S. adults who subscribe to the 5-second rule is less than 0.20? Calculate $P(\hat{p} \le 0.10)$ and use this result to support your answer.

5. **Normal IQ scores? (6.5, 6.6)** The Wechsler Adult Intelligence Scale (WAIS) is a common "IQ test" for adults. The distribution of WAIS scores for persons over 16 years of age is approximately normal with mean 100 and standard deviation 15. Let $\bar{x} =$ the mean WAIS score in a random sample of 10 people over 16 years of age.

(a) Calculate the mean and standard deviation of the sampling distribution of $\bar{x}$.

(b) Interpret the standard deviation from part (a).

(c) What is the probability that the average WAIS score is 105 or greater for a random sample of 10 people over 16 years of age? Show your work.

(d) Would your answers to any of parts (a), (b), or (c) be affected if the distribution of WAIS scores in the adult population were distinctly non-normal? Explain.

6. **Watching for gypsy moths (6.1, 6.6)** The gypsy moth is a serious threat to oak and aspen trees. A state agriculture department places traps throughout the state to detect the moths. Each month, an SRS of 50 traps is inspected, the number of moths in each trap is recorded, and the mean number of moths is calculated. Based on years of data, the distribution of moth counts is strongly skewed with mean 0.5 and standard deviation 0.7.

(a) Explain why it is reasonable to use a normal distribution to approximate the sampling distribution of $\bar{x}$ for SRSs of size 50.

(b) Calculate the probability that the mean number of moths in a sample of size 50 is at least 0.6 moths.

(c) In a recent month, the mean number of moths in an SRS of size 50 was 0.6. Based on this result, should the state agricultural department be worried that the moth population is getting larger in their state? Explain.

Chapter 6 Practice Test

Section I: Multiple Choice *Select the best answer for each question.*

1. A study of voting chose 663 registered voters at random shortly after an election. Of these, 72% said they had voted in the election. Election records show that only 56% of registered voters voted in the election. Which of the following statements is true about these percentages?

   (a) 72% and 56% are both statistics.

   (b) 72% is a statistic and 56% is a parameter.

   (c) 72% is a parameter and 56% is a statistic.

   (d) 72% and 56% are both parameters.

2. Vermont is particularly beautiful in early October when the leaves begin to change color. At that time of year, a large proportion of cars on Interstate 91 near Brattleboro have out-of-state license plates. Suppose a Vermont state trooper randomly selects 50 cars driving past Exit 2 on I-91, records the state identified on the license plate, and calculates the proportion of cars with out-of-state plates. Which of the following describes the sampling distribution of the sample proportion in this context?

   (a) The distribution of state for all cars in the trooper's sample of cars passing this exit

   (b) The distribution of state for all cars passing this exit

   (c) The distribution of the proportion of cars with out-of-state plates in all possible samples of 50 cars passing this exit

   (d) The distribution of the proportion of cars with out-of-state plates in the trooper's sample of 50 cars passing this exit

3. A polling organization wants to estimate the proportion of voters who favor a new law banning smoking in public buildings. The organization decides to increase the size of its random sample of voters from about 1500 people to about 4000 people right before an election. The effect of this increase is to

   (a) reduce the bias of the estimate.

   (b) increase the bias of the estimate.

   (c) reduce the variability of the estimate.

   (d) increase the variability of the estimate.

4. A machine is designed to fill 16-ounce bottles of shampoo. When the machine is working properly, the amount poured into the bottles follows a normal distribution with mean 16.05 ounces and standard deviation 0.1 ounce. Assume that the machine is working properly. If 4 bottles are randomly selected and the number of ounces in each bottle is measured, then there is about a 95% chance that the sample mean will fall in which of the following intervals?

   (a) 16.00 to 16.10 ounces   (c) 15.90 to 16.20 ounces

   (b) 15.95 to 16.15 ounces   (d) 15.85 to 16.25 ounces

5. The central limit theorem is important in statistics because it allows us to use the normal distribution to find probabilities involving the sample mean if the

   (a) sample size is reasonably large for any population shape.

   (b) sample size is reasonably large and the population is normally distributed.

   (c) population size is reasonably large for any population shape.

   (d) population size is reasonably large and the population is normally distributed.

6. At a high school, 85% of students are right-handed. Let X = the number of students who are right-handed in a random sample of 10 students from the school. Which one of the following statements about the mean and standard deviation of the sampling distribution of X is true?

   (a) $\mu_x = 8.5$; $\sigma_x \approx 1.129$

   (b) $\mu_x = 8.5$; $\sigma_x \approx 0.113$

   (c) $\mu_x = 8.5$; $\sigma_x \approx$ cannot be determined from the information given.

   (d) Neither the mean nor the standard deviation can be determined from the information given.

7. The student newspaper at a large university asks an SRS of 250 undergraduates, "Do you favor eliminating the carnival from the end-of-term celebration?" In the sample, 150 of the 250 undergraduates are in favor. Suppose that 55% of all undergraduates favor eliminating the carnival. If you took a very large number of SRSs of size $n = 250$ from this population, the sampling distribution of the sample proportion $\hat{p}$ would have which of the following characteristics?

   (a) Mean 0.55, standard deviation 0.03, shape unknown

   (b) Mean 0.55, standard deviation 0.03, approximately normal

   (c) Mean 0.60, standard deviation 0.03, shape unknown

   (d) Mean 0.60, standard deviation 0.03, approximately normal

8. Scores on the mathematics part of the SAT exam in a recent year followed a normal distribution with mean 515 and standard deviation 114. You choose an SRS of 100 students and calculate $\bar{x}$ = mean SAT Math score. Which of the following are the mean and standard deviation of the sampling distribution of $\bar{x}$?

(a) Mean = 515, SD = 114

(b) Mean = 515, SD = $\dfrac{114}{\sqrt{100}}$

(c) Mean = $\dfrac{515}{100}$, SD = $\dfrac{114}{100}$

(d) Mean = $\dfrac{515}{100}$, SD = $\dfrac{114}{\sqrt{100}}$

9. In a congressional district, 55% of the registered voters are Democrats. Which of the following is closest to the probability of getting less than 50% Democrats in a random sample of size 100?

(a) 0.157   (b) 0.496   (c) 0.504   (d) 0.843

10. A statistic is an unbiased estimator of a parameter when

(a) the statistic is calculated from a random sample.

(b) in all possible samples of a specific size, the distribution of the statistic has a shape that is approximately normal.

(c) in all possible samples of a specific size, the values of the statistic are very close to the value of the parameter.

(d) in all possible samples of a specific size, the values of the statistic are centered at the value of the parameter.

Section II: Free Response

11. Here are histograms of the values taken by three sample statistics in several hundred samples from the same population. The true value of the population parameter is marked with an arrow on each histogram. Which statistic would provide the best estimate of the parameter? Explain.

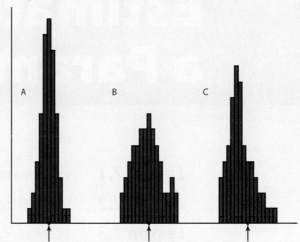

12. The amount that households pay service providers for access to the Internet varies quite a bit, but the mean monthly fee is $48 and the standard deviation is $20. The distribution is not normal: Many households pay a base rate for low-speed access, but some pay much more for faster connections. A sample survey asks an SRS of 500 households with Internet access how much they pay per month. Let $\bar{x}$ be the mean amount paid by the members of the sample.

(a) Calculate the mean and standard deviation of the sampling distribution of $\bar{x}$. Interpret the standard deviation.

(b) What is the shape of the sampling distribution of $\bar{x}$? Justify.

(c) Find the probability that the average amount paid by the sample of households exceeds $50.

13. According to government data, 22% of American children under the age of 6 live in households with incomes less than the official poverty level. A study of learning in early childhood chooses an SRS of 300 children.

(a) Let X = the count of children in this sample who live in households with incomes less than the official poverty level. What is the shape of the sampling distribution of X? Justify your answer.

(b) Find the probability that more than 20% of the sample are from poverty-level households.

7

Estimating a Parameter

Lesson 7.1	The Idea of a Confidence Interval	452
Lesson 7.2	What Affects the Margin of Error?	458
Lesson 7.3	Estimating a Proportion	464
Lesson 7.4	Confidence Intervals for a Proportion	470
Lesson 7.5	Estimating a Mean	476
Lesson 7.6	Confidence Intervals for a Mean	484

Chapter 7	Main Points	492
Chapter 7	Review Exercises	494
Chapter 7	Practice Test	495

STATS applied!

How long will I stay on hold?

If your credit card is denied, you call the number on the back of the card to get it fixed. Does a real person answer your call? These days, probably not. It is far more likely that you will get an automated response. Customers become frustrated if they have to wait too long before speaking to a live person. So companies try hard to minimize the time required to connect to a customer service representative.

A large bank decided to study the call-response times in its credit-card customer service department. The bank's goal was to have a representative answer an incoming call in less than 30 seconds. The histogram in Figure 7.1 shows the distribution of response time in a random sample of 241 calls to the bank's customer service center. What does the graph suggest about how well the bank is meeting its goal?

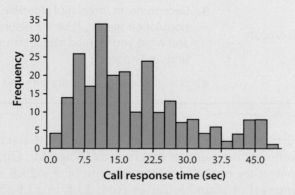

FIGURE 7.1 Histogram showing the distribution of response time (in seconds) at a bank's customer service center for a random sample of 241 calls.

We'll revisit STATS applied! at the end of the chapter, so you can use what you have learned to help answer this question.

Lesson 7.1

The Idea of a Confidence Interval

LEARNING TARGETS

- Interpret a confidence interval in context.
- Determine the point estimate and margin of error from a confidence interval.
- Use confidence intervals to make decisions.

How long does a battery last on the newest iPhone, on average? What proportion of college undergraduates attended all of their classes last week? How much does the weight of a quarter-pound hamburger at a fast-food restaurant vary after cooking? These are the types of questions we would like to answer.

It wouldn't be practical to determine the lifetime of *every* iPhone battery, to ask *all* undergraduates about their attendance, or to weigh *every* burger after cooking. Instead, we choose a random sample of individuals (batteries, undergraduates, burgers) to represent the population and collect data from those individuals.

Chapter 7 begins the formal study of statistical inference—using information from a sample to draw conclusions about a population parameter such as p or μ. This is an important transition from Chapter 6, where you were given information about a population and were asked questions about the distribution of a sample statistic such as $\hat{p}$ or $\overline{x}$. The following activity gives you an idea of what lies ahead.

ACTIVITY

What's the mystery mean?

In this activity, each team of three or four students will try to estimate the mystery value of the population mean μ that your teacher chose before class.[1]

1. Before class, your teacher stored a value of μ (represented by M) in the display calculator. Your teacher then cleared the home screen, so you can't see the value of M.

2. With the class watching, the teacher will execute the following command:

    mean(randNorm(M,20,16))

This tells the calculator to choose an SRS of 16 values from a normal population with mean M and standard deviation 20, and then compute the mean $\overline{x}$ of those 16 sample values. Is the sample mean shown likely to be equal to the mystery mean M? Why or why not?

3. Determine an interval of *plausible* values for the population mean μ. Use the result from Step 2 and what you learned about sampling distributions in Chapter 6.

4. Share your team's results with the class.

This activity illustrates one of the most common tasks of a statistician—using a sample statistic to estimate the value of a parameter. When Mr. Girard's class did the Mystery Mean activity, they obtained a sample mean of $\overline{x} = 23.8$ and used this estimate to create an interval of plausible values from 13.8 to 33.8. When the estimate of a parameter is reported as an interval of values, it is called an interval estimate, or **confidence interval.**

DEFINITION Confidence interval

A **confidence interval** gives an interval of plausible values for a parameter.

"Plausible" means that we shouldn't be surprised if any one of the values in the interval is equal to the value of the parameter. We use an interval of plausible values rather than a single-value estimate to increase our confidence that we have a correct value for the parameter. Of course, as the cartoon illustrates, there is a trade-off between the amount of confidence we have that our estimate is correct and how much information the interval provides.

Interpreting Confidence Intervals

Confidence intervals are constructed so that we know *how much* confidence we should have in the interval. The most common **confidence level** is 95%. You will learn more about how to interpret confidence levels in Lesson 7.2.

DEFINITION Confidence level

The **confidence level** *C* gives the long-run success rate of confidence intervals calculated with *C*% confidence. That is, in *C*% of all possible samples, the interval computed from the sample data will capture the true parameter value.

The Gallup polling organization recently asked a random sample of U.S. adults about the most important financial problem facing their family today. There were many different options, with "health care costs" leading the way with 14% of responses.[2] Based on this sample, the 95% confidence interval for the population proportion is 0.10 to 0.18. That is, we are 95% confident that the interval from 0.10 to 0.18 captures the proportion of *all* U.S. adults who would say that health care costs are the biggest financial problem facing their family today.

How to Interpret a Confidence Interval

To interpret a *C*% confidence interval for an unknown parameter, say, "We are *C*% confident that the interval from____to____captures the [parameter in context]."

When interpreting a confidence interval, make sure that you are describing the parameter and not the statistic. It's wrong to say that we are 95% confident that the interval from 0.10 to 0.18 captures the proportion of U.S. adults who *said* health care costs were the biggest financial problem facing their family today. The "proportion who *said* health care" is the sample proportion, which is known to be 0.14. The interval gives plausible values for the proportion who *would say* health care if asked.

CAUTION

EXAMPLE

How many steps?

Interpreting a confidence interval

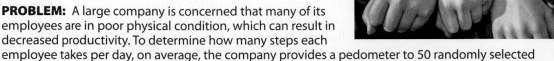

PROBLEM: A large company is concerned that many of its employees are in poor physical condition, which can result in decreased productivity. To determine how many steps each employee takes per day, on average, the company provides a pedometer to 50 randomly selected employees to use for one 24-hour period. After collecting the data, the company statistician reports a 95% confidence interval of 4547 steps to 8473 steps. Interpret this confidence interval.

SOLUTION:

We are 95% confident that the interval from 4547 to 8473 captures the true mean number of steps taken per day for all employees at this company.

FOR PRACTICE   TRY EXERCISE 1.

Sometimes, confidence intervals are reported in the form

(lower boundary, upper boundary)

For example, the confidence interval in the preceding example could be reported as (4547, 8473) instead of "4547 to 8473."

Building a Confidence Interval

To create an interval of plausible values for a parameter, we need two components: a **point estimate** to use as the midpoint of the interval and a **margin of error** to account for sampling variability. The structure of a confidence interval is

point estimate ± margin of error

We can visualize a C% confidence interval like this:

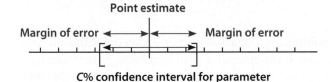

Earlier, we learned that the 95% confidence interval for the proportion of all U.S. adults who would say that health care costs are the biggest financial problem facing their family today is 0.10 to 0.18. This interval could also be expressed as

$$0.14 \pm 0.04$$

Confidence intervals reported in the media are often presented in this format.

> **▌▌ DEFINITION Point estimate, Margin of error**
>
> A **point estimate** is a single-value estimate of a population parameter.
>
> The **margin of error** of an estimate describes how far, at most, we expect the estimate to vary from the true population value. That is, in a C% confidence interval, the distance between the point estimate and the true parameter value will be less than the margin of error in C% of all samples.

The point estimate is our best guess for the value of the parameter. Unfortunately, the point estimate is unlikely to be correct because of the variability introduced by random sampling. We include the margin of error to account for this sampling variability.

EXAMPLE

How many more steps?

Determining the point estimate and margin of error

PROBLEM: In the previous example, we learned that the 95% confidence interval for the true mean number of steps taken per day for all employees at the large company is 4547 to 8473. Calculate the point estimate and margin of error used to create this confidence interval.

SOLUTION:

$$\text{point estimate} = \frac{4547 + 8473}{2} = 6510 \text{ steps}$$

$$\text{margin of error} = 8473 - 6510 = 1963 \text{ steps}$$

> The point estimate is the midpoint of the interval.

> The margin of error is the distance from the point estimate to the endpoints of the interval. You can calculate the margin of error using the lower endpoint as well: $6510 - 4547 = 1963$ steps.

FOR PRACTICE   TRY EXERCISE 5.

Using Confidence Intervals to Make Decisions

Besides estimating the value of a population parameter, we can also use confidence intervals to make decisions about a parameter.

EXAMPLE

Do they need to step it up?

Using confidence intervals to make decisions

PROBLEM: In the preceding examples, we learned that the 95% confidence interval for the true mean number of steps taken per day for all employees at the large company is 4547 to 8473. Recent health guidelines suggest that people aim for 10,000 steps per day. Is there convincing evidence that the employees of this company are not meeting the guideline, on average? Explain.

SOLUTION:

Yes. All the values in the confidence interval are less than 10,000, so there is convincing evidence that the employees of this company are taking fewer than 10,000 steps per day, on average.

FOR PRACTICE   TRY EXERCISE 9.

LESSON APP 7.1

Do you approve of the president's job performance?

Polling organizations regularly survey U.S. adults to ask if they approve of the job the president is doing. On the day of the 2014 midterm elections, the 95% confidence interval for the proportion of all U.S. adults that approved of President Obama's job performance was 0.39 to 0.45.[3]

Steve Debenport/Getty Images

1. Interpret the confidence interval.

2. Calculate the point estimate and margin of error used to create this confidence interval.

3. Is it plausible that on the day of the 2014 midterm elections, a majority of U.S. adults approved of President Obama's job performance? Explain.

Lesson 7.1

WHAT DID YOU LEARN?

LEARNING TARGET	EXAMPLES	EXERCISES
Interpret a confidence interval in context.	p. 454	1–4
Determine the point estimate and margin of error from a confidence interval.	p. 455	5–8
Use confidence intervals to make decisions.	p. 455	9–12

Exercises   Lesson 7.1

Mastering Concepts and Skills

1. **Median U.S. income** According to the U.S. Census Bureau, the 90% confidence interval for the median household income in 2013 is $51,484 to $52,394.[4] Interpret this confidence interval.
   pg 454

2. **Clementines** Tim purchased a random sample of clementines at a local grocery store. The 95% confidence interval for the mean weight of all clementines at this store is 76.6 grams to 90.1 grams. Interpret this confidence interval.

3. **Atmospheric gases** The Pew Research Center and *Smithsonian* magazine recently quizzed a random sample of 1006 U.S. adults on their knowledge of science.[5] One of the questions asked, "Which gas makes up most of the Earth's atmosphere: hydrogen, nitrogen, carbon dioxide, or oxygen?" A 95% confidence interval for the proportion who would correctly answer nitrogen is 0.175 to 0.225. Interpret this confidence interval.

4. **Purpose of a control group** The Pew Research Center and *Smithsonian* magazine recently quizzed a random sample of 1006 U.S. adults on their knowledge of science.[6] One of the questions asked, "Which is the better way to determine whether a new drug is effective in treating a disease? If a scientist has a group of 1000 volunteers with the disease to study, should she (a) Give the drug to all of them and see how many get better or (b) Give the drug to half of them and not to the other half, and compare how many in each group get better?" A 95% confidence interval for the proportion who would correctly answer (b) is 0.723 to 0.777. Interpret this confidence interval.

5. **More about median income** Calculate the point estimate and margin of error used to create the confidence interval in Exercise 1.
   pg 455

6. **More clementines** Calculate the point estimate and margin of error used to create the confidence interval in Exercise 2.

7. **More nitrogen** Calculate the point estimate and margin of error used to create the confidence interval in Exercise 3.

8. **More about control groups** Calculate the point estimate and margin of error used to create the confidence interval in Exercise 4.

9. **Greater median income?** Suppose that $51,759 was the median income for all households in 2012. Does the interval in Exercise 1 provide convincing evidence that the median household income increased in 2013? Explain.
   pg 455

10. **Bigger clementines?** The nutritional label on a bag of clementines says a typical clementine weighs 74 grams. Does the interval in Exercise 2 provide convincing evidence that the mean weight of clementines from this store is larger than the nutritional label suggests? Explain.

11. **Guessing about gases?** Refer to Exercise 3. If people guess one of the four choices at random, about 25% should get the answer correct. Does the interval in Exercise 3 provide convincing evidence that less than 25% of all U.S. adults would answer this question correctly?

12. **Guessing about control groups?** Refer to Exercise 4. If people guess one of the two choices at random, about 50% should get the answer correct. Does the interval in Exercise 4 provide convincing evidence that more than 50% of all U.S. adults would answer this question correctly?

Applying the Concepts

13. **Waiting for pizza** A local pizza shop claims that they deliver pizzas in under 30 minutes, on average. However, it always seems to take longer when you order a pizza from them. To investigate their claim, you randomly select 12 different times during the next week and order a pizza to be delivered to your

home. The 95% confidence interval for the mean delivery time is 26.8 minutes to 41.0 minutes.

(a) Interpret the confidence interval.

(b) Calculate the point estimate and margin of error used to create this confidence interval.

(c) Based on the confidence interval, is it plausible that the true average delivery time is less than 30 minutes? Explain.

14. **Going to the dogs** A magazine reported that 66% of all dog owners usually greet their dog before greeting their spouse or children when they return home at the end of the workday. Researchers select 40 dog owners at random and ask them whom they greet first when returning home. The 95% confidence interval for the proportion of owners who greet their dog first is 0.475 to 0.775.

(a) Interpret the confidence interval.

(b) Calculate the point estimate and margin of error used to create this confidence interval.

(c) Based on the confidence interval, is it plausible that the true proportion of all owners who greet their dog first is 66%? Explain.

Extending the Concepts

15. **Still waiting for pizza?** Refer to Exercise 13. For the 12 pizza orders in the sample, the mean delivery time is 33.9 minutes. How confident are you that the true mean number of minutes for all deliveries is exactly 33.9? Explain.

16. **Counting shoes** The statistics class at Washington High School selected an SRS of 20 female students and an SRS of 20 male students. The selected students were asked how many pairs of shoes they own and the difference in means (Girls – Boys) was calculated. A 95% confidence interval for the difference in the mean number of pairs for girls at the school and the mean number of pairs for boys at the school is 10.9 to 26.5.

(a) Interpret the confidence interval.

(b) Does the confidence interval give convincing evidence that girls at the school own more shoes than boys at the school, on average? Explain.

17. **False profiles** Many teens have posted profiles on sites such as Facebook. A sample survey asked random samples of teens with online profiles if they

included false information in their profiles. Of 170 younger teens (ages 12 to 14) polled, 66.8% said "Yes." Of 317 older teens (ages 15 to 17) polled, 47.9% said "Yes." A 95% confidence interval for the difference in the population proportions (Younger teens – Older teens) is 0.120 to 0.297.[7]

(a) Interpret the confidence interval.

(b) Does the confidence interval give convincing evidence that younger teens are more likely than older teens to include false information on their profiles?

Recycle and Review

18. **Just tall, that's all (1.8)** The 2015 NBA champions, the Golden State Warriors, last won the title in 1975. How has the height of a typical professional basketball player changed over the years? Here are the heights (in inches) of all the players on the two teams that were in the NBA finals in 1975 and 2015.[8]

1975	72	72	73	74	75	75	75	75
	76	76	76	76	76	77	78	78
	78	79	79	79	79	79	80	81
	81	83						

2015	75	75	75	76	77	78	78	78
	78	78	79	79	79	79	80	80
	80	80	80	81	81	82	82	82
	82	83	84	84	85	85		

Draw parallel boxplots using the data from each year. Use the plots to compare the distribution of heights for the two years.

19. **Beagle pups (6.1, 6.2)** Consider a population consisting of the 5 female beagles owned by a dog breeder. The size of the last litter of pups produced by each dog is listed here.

Violet: 9   Patty: 5   Lucy: 4   Sally: 4   Marcie: 8

(a) List all 10 possible SRSs of size $n = 3$ from this population, calculate the mean litter size for each sample, and display the sampling distribution of the sample mean on a dotplot.

(b) Explain how you know that the sample mean is an unbiased estimator of the population mean.

Lesson 7.2

What Affects the Margin of Error?

In Lesson 7.1, you learned that the confidence level gives the overall success rate of the method used to calculate a confidence interval. The following activity will help you explore the idea of confidence level in more depth.

ACTIVITY

Investigating confidence level with the *Confidence Intervals* applet

In this activity, you will use the *Confidence Intervals* applet to learn what it means to say that we are "95% confident" that our confidence interval captures the parameter value.

1. Go to **highschool.bfwpub.com/spa3e** and launch the applet. Use the default settings: confidence level 95% and sample size $n = 20$.

2. Click "Sample" to choose an SRS and display the resulting confidence interval. The 20 values in the sample are marked with yellow dots. The confidence interval is displayed as a horizontal line segment with a dot representing the sample mean $\bar{x}$ in the middle of the interval.

3. Did the interval capture the population mean μ (what the applet calls a "hit")? Do this a total of 10 times. How many of the intervals captured the population mean μ? *Note:* So far, you have used the applet to take 10 SRSs, each of size $n = 20$. Be sure you understand the difference between sample size and the number of samples taken.

4. Reset the applet. Click "Sample 25" twice to choose 50 SRSs and display the confidence intervals based on those samples. How many

intervals captured the parameter μ? Keep clicking "Sample 25" and observe the value of "Percent hit." What do you notice?

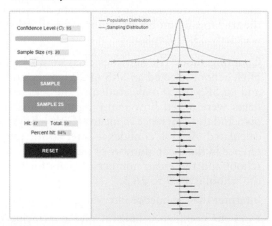

5. Repeat Step 4 using a 90% confidence level.

6. Repeat Step 4 using an 80% confidence level.

7. Summarize what you have learned about the relationship between confidence level and "Percent hit" after taking many samples.

We will investigate the effect of changing the sample size later.

Interpreting Confidence Level

In the preceding lesson, we claimed to be "95% confident that the interval from 0.10 to 0.18 captures the proportion of all U.S. adults who would say that health care costs are the biggest financial problem facing their family today." The claim of "95% confidence" means that we are using a method that successfully captures the parameter 95% of the time. That is, if we were to take many random samples and construct a 95% confidence interval using each sample, about 95% of those intervals would capture p, the true proportion of all U.S. adults who would answer health care costs.

How to Interpret a Confidence Level

To interpret the confidence level C, say, "If we were to select many random samples from a population and construct a C% confidence interval using each sample, about C% of the intervals would capture the [parameter in context]."

Make sure not to interpret the confidence interval when you are asked to interpret a confidence level—and vice versa. Confidence levels describe the process used to make an interval, not a particular interval.

EXAMPLE

Do you use Twitter?

Interpreting confidence level

PROBLEM: The Pew Internet and American Life Project asked a random sample of 2253 U.S. adults, "Do you ever . . . use Twitter or another service to share updates about yourself or to see updates about others?" In the sample, 19% said "Yes." According to Pew, the resulting 95% confidence interval is 0.167 to 0.213. Interpret the confidence level.[9]

SOLUTION:

If the Pew Project took many random samples of U.S. adults and constructed a 95% confidence interval using each sample, about 95% of these intervals would capture the true proportion of all U.S. adults who use Twitter or another service to share updates about themselves or see updates about others.

FOR PRACTICE   **TRY EXERCISE 1.**

The confidence level does not tell us the probability that a particular confidence interval captures the population parameter. The confidence level reveals how likely it is that the method we are using will produce an interval that captures the population parameter *if we use it many times*. However, in practice, researchers usually calculate just a single confidence interval for a given situation.

CAUTION

Factors That Affect the Margin of Error

In general, we prefer short confidence intervals—that is, confidence intervals with a small margin of error. After all, knowing that tomorrow's high temperature will be between −40°F and 200°F won't help you pick out your clothes in the morning! But knowing that tomorrow's high temperature will be between 80°F and 82°F would be quite useful. To reduce the margin of error, we can change two factors: sample size and confidence level. The following activity explores how each of these factors affects the margin of error.

ACTIVITY

Exploring margin of error with the *Confidence Intervals* applet

In this activity, you will use the applet to explore the relationship between the confidence level, the sample size, and the margin of error.

Part 1: Adjusting the Confidence Level

1. Go to highschool.bfwpub.com/spa3e and launch the *Confidence Intervals* applet. Use the default settings: confidence level 95% and sample size $n = 20$. Click "Sample 25" to select 25 SRSs and make 25 confidence intervals.

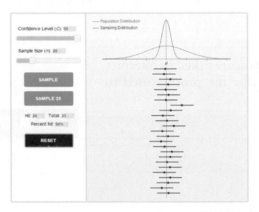

2. Change the confidence level to 99%. What happens to the length of the confidence intervals?

3. Now change the confidence level to 90%. What happens to the length of the confidence intervals?

4. Finally, change the confidence level to 80%. What happens to the length of the confidence intervals?

5. Summarize what you learned about the relationship between the confidence level and the margin of error for a fixed sample size.

Part 2: Adjusting the Sample Size

6. Reset the applet to the default settings: confidence level 95% and sample size $n = 20$. Click "Sample 25" to select 25 SRSs and make 25 confidence intervals.

7. Using the slider, increase the sample size to $n = 100$. What do you notice about the length of the confidence intervals?

8. Using the slider, increase the sample size to $n = 200$. What do you notice about the length of the confidence intervals?

9. Summarize what you learned about the relationship between the sample size and the margin of error for a fixed confidence level.

As the activity illustrates, the price we pay for greater confidence is a wider interval. If we're satisfied with 80% confidence, then our interval of plausible values for the parameter will be much narrower than if we insist on 90%, 95%, or 99% confidence. But intervals constructed at an 80% confidence level will capture the true value of the parameter much less often than intervals that use a 99% confidence level.

The activity also shows that we can get a more precise estimate of a parameter by increasing the sample size. Larger samples yield narrower confidence intervals at any confidence level. However, larger samples cost more time and money to obtain.

EXAMPLE

What's a hashtag?

Changing sample size and confidence level

PROBLEM: In the preceding example, you read about a random sample of 2253 U.S. adults who were surveyed for the Pew Internet and American Life Project. Based on the survey, the 95% confidence interval for the proportion of all U.S. adults who use Twitter or another service to share updates about themselves or to see updates about others is 0.167 to 0.213.

(a) Explain what would happen to the length of the interval if the confidence level were increased to 99%.

(b) Explain what would happen to the length of the original interval if the sample size were increased to 5000.

SOLUTION:

(a) The confidence interval will be wider because increasing the confidence level increases the margin of error.

(b) The confidence interval would be narrower because increasing the sample size decreases the margin of error.

FOR PRACTICE   TRY EXERCISE 5.

What the Margin of Error Doesn't Account For

When we calculate a confidence interval, we include the margin of error because we expect the value of the point estimate to vary somewhat from the parameter. However, the margin of error accounts for *only* the variability we expect from random sampling. It does not account for practical difficulties, such as undercoverage and nonresponse in a sample survey. These problems can produce estimates that are much farther from the parameter than the margin of error would suggest. Remember this unpleasant fact when reading the results of an opinion poll or other sample survey. The margin of error does *not* account for any sources of bias in the data collection process.

CAUTION

!

EXAMPLE

What's your GPA?

Bias and the margin of error

PROBLEM: As part of a project about response bias, Ellery surveyed a random sample of 25 students from her school. One of the questions in the survey required students to state their GPA aloud. Based on the responses, Ellery said she was 90% confident that the interval from 3.14 to 3.52 captures the mean GPA for all students at her school.[10] Describe one potential source of bias in Ellery's study that is not accounted for by the margin of error.

SOLUTION:

The margin of error doesn't account for the fact that many students might lie about their GPAs when having to respond without anonymity. The mean GPA for all students might be even less than 3.14!

FOR PRACTICE   TRY EXERCISE 9.

LESSON APP 7.2

Do you like my photos?

According to a recent survey by the Pew Research Center, a 95% confidence interval for the proportion of Facebook users who comment on other people's photos at least once per day is 0.31 ± 0.035.[11]

1. Interpret the confidence level.

2. Name two things Pew could do to reduce the margin of error. What drawbacks do these actions have?

3. The response rate for this survey was only 9%. That is, Pew was able to obtain responses for 9% of the landline and cell-phone numbers that were randomly selected for the sample. Describe how nonresponse might lead to bias in this survey. Does the stated margin of error account for this possible bias?

Ann Heath

Lesson 7.2

WHAT DID YOU LEARN?		
LEARNING TARGET	EXAMPLES	EXERCISES
Interpret a confidence level in context.	p. 459	1–4
Describe how the confidence level and sample size affect the margin of error.	p. 460	5–8
Explain how practical issues like nonresponse, undercoverage, and response bias can affect the interpretation of a confidence interval.	p. 461	9–12

Exercises   Lesson 7.2

Mastering Concepts and Skills

1. **Time for homework** The principal of a large high
pg 459 school wanted to know how many hours students spend doing homework during a typical week. Based on a random sample of 100 students, a 90% confidence interval for the mean number of hours doing homework for all students is 2.3 hours to 4.1 hours. Interpret the confidence level.

2. **Puzzled** The puzzle editor of a game magazine asked 43 randomly selected subscribers how long it took them to complete a certain crossword puzzle. The 99% confidence interval for the median completion time for all subscribers is 15.2 to 18.6 minutes. Interpret the confidence level.

3. **Prayer in public schools** A *New York Times*/CBS News poll asked a random sample of U.S. adults, "Do you favor an amendment to the Constitution that would permit organized prayer in public schools?" Based on this poll, the 95% confidence interval for the proportion of all U.S. adults who favor such an amendment is 0.63 to 0.69. Interpret the confidence level.

4. **Weighing less** A Gallup poll asked a random sample of U.S. adults, "Would you like to lose weight?" Based on this poll, the 95% confidence interval for the proportion of all U.S. adults who want to lose weight is 0.56 to 0.62. Interpret the confidence level.[12]

5. **More time for homework?** Refer to Exercise 1.
(a) Explain what would happen to the length of the in-
pg 460 terval if the confidence level were increased to 99%.
(b) Explain what would happen to the length of the interval if the sample size were increased to 200 students.

6. **Even more puzzled** Refer to Exercise 2.
(a) Explain what would happen to the length of the interval if the confidence level were decreased to 90%.
(b) Explain what would happen to the length of the interval if the sample size were increased to 100.

7. **More prayer in schools** Refer to Exercise 3.
(a) Explain what would happen to the length of the interval if the confidence level were decreased to 90%.

(b) Explain what would happen to the length of the interval if the sample size was only half as big as in the actual study.

8. **Losing more weight** Refer to Exercise 4.
(a) Explain what would happen to the length of the interval if the confidence level were increased to 99%.
(b) Explain what would happen to the length of the interval if the sample size was only half as big as in the actual study.

9. **Bias and homework** Refer to Exercise 1. After
pg 461 randomly selecting the 100 students, the principal personally interviewed the students about their homework habits, including how many hours they spend doing homework in a typical week. Describe one potential source of bias in the study that is not accounted for by the margin of error.

10. **Puzzled by bias** Refer to Exercise 2. The puzzle editor originally sent requests for completion times to 150 subscribers, only 43 of whom reported their times. Describe one potential source of bias in the study that is not accounted for by the margin of error.

11. **Practical difficulties** Refer to Exercise 3. The news article goes on to say: "The theoretical errors do not take into account additional error resulting from the various practical difficulties in taking any survey of public opinion." List some of the "practical difficulties" that may cause errors not included in the ±3 percentage point margin of error.

12. **Sources of error** Refer to Exercise 4. In the description of its methodology, Gallup states that the 3 percentage point margin of error for this poll includes only sampling variability (what it calls "sampling error"). What other potential sources of error (Gallup calls these "nonsampling errors") could affect the accuracy of the 95% confidence interval?

Applying the Concepts

13. **California's traffic** People love living in California for many reasons, but traffic isn't one of them. Based on a random sample of 572 employed California adults, a 90% confidence interval for average travel

time to work for all employed California adults is 23 minutes to 26 minutes.[13]

(a) Interpret the confidence level.

(b) Name two things you could do to reduce the margin of error. What drawbacks do these actions have?

(c) Describe how nonresponse might lead to bias in this survey. Does the stated margin of error account for this possible bias?

14. **Employment in California** Each month the government releases unemployment statistics. The stated unemployment rate doesn't include people who choose not to be employed, such as retirees. Based on a random sample of 1000 California adults, a 99% confidence interval for the proportion of all California adults employed in the workforce is 0.532 to 0.612.[14]

(a) Interpret the confidence level.

(b) Name two things you could do to reduce the margin of error. What drawbacks do these actions have?

(c) Describe how untruthful answers might lead to bias in this survey. Does the stated margin of error account for this possible bias?

15. **Which confidence level?** The figure shows the result of taking 25 SRSs from a normal population and constructing a confidence interval using each sample. Which confidence level—80%, 90%, 95%, or 99%—do you think was used? Explain.

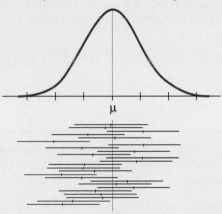

16. **Which confidence level?** The figure shows the result of taking 25 SRSs from a normal population and constructing a confidence interval using each sample. Which confidence level—80%, 90%, 95%, or 99%—do you think was used? Explain.

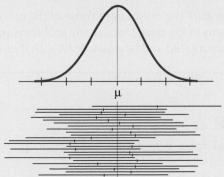

Extending the Concepts

17. **Sample size and confidence** Does increasing the sample size make you more confident? You can investigate this question using the *Confidence Intervals* applet at highschool.bfwpub.com/spa3e.

(a) Using a 95% confidence level and sample size $n = 20$, click "Sample 25" 40 times to select 1000 SRSs and make 1000 confidence intervals. What percent of the intervals captured the true mean μ?

(b) Keeping the 95% confidence level, increase the sample size to $n = 100$ and repeat the process from part (a). What percent of the intervals captured the true mean μ?

(c) Keeping the 95% confidence level, increase the sample size to $n = 200$ and repeat the process from part (a). What percent of the intervals captured the true mean μ?

(d) Based on your answers, does increasing the sample size increase your confidence that a confidence interval will work? Explain.

Recycle and Review

18. **Sources of bias** (3.5) For each of the following studies, describe a possible source of bias. Explain the likely direction of the bias.

(a) In the late 1980s, author Shere Hite distributed 100,000 questionnaires to married women in the United States. Of the 4400 who responded, 84% said that they were not emotionally satisfied by their relationships with men.[16]

(b) A flour company wants to know what percent of local households bake at least twice a week. A company representative calls 500 randomly selected households during the daytime and finds that 156 of the 312 people who responded bake at least twice a week.

19. **Educational attainment** (5.4, 6.4) According to the U.S. Census Bureau's 2014 Current Population Survey, 88% of adult U.S. residents have earned a high school diploma.[15] Suppose we take a random sample of 120 American adults and record the proportion, $\hat{p}$, of individuals in our sample who have a high school diploma.

(a) Describe the shape, center, and spread of the sampling distribution of $\hat{p}$.

(b) Find the probability that the sample proportion of residents who have earned a high school diploma in a random sample of 120 residents is at least 0.90.

(c) If the sample size was only 10, would it still be possible to calculate the probability in part (b)? If so, calculate the probability. If not, explain why not.

Lesson 7.3

Estimating a Proportion

- Check the Random and Large Counts conditions for constructing a confidence interval for a population proportion.

- Determine the critical value for calculating a C% confidence interval for a population proportion using Table A or technology.

- Calculate a C% confidence interval for a population proportion.

In Lessons 7.1 and 7.2, we saw that a confidence interval can be used to estimate an unknown population parameter. We are often interested in estimating the proportion p of some outcome in the population. Here are some examples:

- What proportion of U.S. adults are unemployed right now?
- What proportion of high school students have cheated on a test?
- What proportion of pine trees in a national park are infested with beetles?

In Lesson 3.4, you learned how to approximate the margin of error for a proportion using simulation. In this lesson, you will learn how to calculate a confidence interval for a proportion using a formula. The following activity gives you a feel for what lies ahead.

ACTIVITY

What proportion of the beads are red?

Before class, your teacher will prepare a large population of different-colored beads and put them into a container. In this activity, you and your team will create an interval estimate for the actual proportion of beads in the population that have a particular color (red, for example).

1. As a class, discuss how to use the cup provided to get a simple random sample of beads from the container.

2. Have 1 student take an SRS of beads. Separate the beads into two groups: those that are red and those that aren't. Count the number of beads in each group.

3. Determine the point estimate $\hat{p}$ for the unknown population proportion p of red beads in the container.

4. Now for the challenge: In teams of 3 or 4 students, find a 95% confidence interval for the true proportion of red beads p. *Hint:* Use your knowledge of sampling distributions from Chapter 6 and the 68–95–99.7 rule from Chapter 5.

5. Compare the results with those of the other teams in the class. Discuss any problems you encountered and how you dealt with them.

Conditions for Estimating *p*

Before calculating a confidence interval for a population proportion p, you should check two important conditions:

How to Check the Conditions for Constructing a Confidence Interval for *p*

To construct a confidence interval for *p*, check the following conditions:

* **Random:** The data come from a random sample from the population of interest.

* **Large Counts:** Both $n\hat{p}$ and $n(1 - \hat{p})$ are at least 10.

In Chapter 3, you learned that the scope of inference is limited to the members of the sample if the sample wasn't selected at random from the population of interest. This means that you can't generalize to a larger population if the Random condition isn't met. Remember that the margin of error in a confidence interval doesn't account for bias in the way that the data were collected.

The method we use to calculate a confidence interval for *p* requires that the sampling distribution of $\hat{p}$ be approximately normal. In Lesson 6.4, you learned that this will be true whenever np and $n(1 - p)$ are both at least 10. Because we don't know the value of *p*, we use $\hat{p}$ when checking the Large Counts condition.

EXAMPLE

How many red beads?

Checking conditions

PROBLEM: When Dr. Godlewski's class did the beads activity, they got 107 red beads and 144 other beads. Their point estimate for the proportion of red beads in the container is $\hat{p} = 107/251 = 0.426$. Check if the conditions for calculating a confidence interval for *p* are met.

G. Curt Fiedler/Getty Images

SOLUTION:

Random? The class selected a random sample of beads. ✓

Large Counts? $251\left(\dfrac{107}{251}\right) = 107 \geq 10$ and

$$251\left(1 - \dfrac{107}{251}\right) = 144 \geq 10 \checkmark$$

> To check the Large Counts condition, make sure both $n\hat{p}$ and $n(1 - \hat{p})$ are at least 10.

FOR PRACTICE   TRY EXERCISE 1.

Did you notice something about the values of $n\hat{p}$ and $n(1 - \hat{p})$ in the example? The value $n\hat{p}$ will always equal the observed count of successes in the sample and the value $n(1 - \hat{p})$ will always equal the observed count of failures in the sample.

THINK ABOUT IT **What happens if one of the conditions is violated?** If the data come from a convenience sample or there are other sources of bias in the data collection process, there's no reason to calculate a confidence interval for *p*. You should have *no* confidence in an interval when the Random condition is violated.

Let's use the *Confidence Intervals for Proportions* applet at highschool.bfwpub. com/spa3e to explore violations of the Large Counts condition. We set $p = 0.3$ and $n = 10$ so that $n\hat{p} = 10(0.3) = 3$, which is clearly less than 10. When we generated 1000 "95%" confidence intervals, only 83% of the intervals captured the true proportion—much less than the advertised capture rate of 95%.

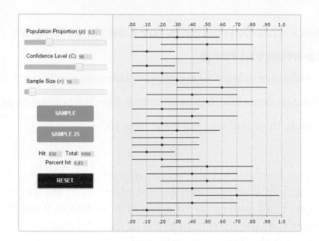

When the Large Counts condition is violated, the actual capture rate will be smaller than the stated capture rate. When we increased the sample size to $n = 50$ so that the Large Counts condition was met, 94.5% of the "95%" intervals captured the true proportion $p = 0.3$. Much better!

Critical Values

You already know that confidence intervals are made up of two parts, the point estimate and the margin of error:

$$\text{point estimate} \pm \text{margin of error}$$

When we are calculating a confidence interval for a population proportion p, the point estimate we use is the sample proportion $\hat{p}$.

The margin of error is more complicated because it is based on two factors: the confidence level and how much the sample proportion typically varies from the population proportion. Fortunately, we can use the facts about the sampling distribution of $\hat{p}$ from Lesson 6.4 as a foundation:

- The standard deviation of the sampling distribution of $\hat{p}$ is $\sigma_{\hat{p}} = \sqrt{\dfrac{p(1 - p)}{n}}$.

- When the Large Counts condition is met, the sampling distribution of $\hat{p}$ is approximately normal.
- According to the 68–95–99.7 rule, $\hat{p}$ will be within 2 standard deviations of p in about 95% of all samples.
- Therefore, in about 95% of all samples, p will be within 2 standard deviations of $\hat{p}$.

Of course, we don't always go out 2 standard deviations in each direction. The number of standard deviations, called a **critical value,** depends on the confidence level.

DEFINITION   Critical value

The **critical value** is a multiplier that makes the interval wide enough to have the stated capture rate.

When the Large Counts condition is met, we can use the standard normal distribution to find the critical value for a specific confidence level. In this example, we will calculate a more precise critical value for a 95% confidence interval.

EXAMPLE

What's the z?

Finding a critical value

PROBLEM: Use Table A or technology to find the critical value z^* for a 95% confidence interval. Assume that the Large Counts condition is met.

SOLUTION:

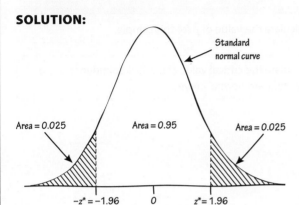

Standard normal curve

Area = 0.025   Area = 0.95   Area = 0.025

$-z^* = -1.96$ 0 $z^* = 1.96$

> To find the critical value for 95% confidence, we have to find the boundaries that capture the middle 95% under the standard normal curve.

> With 95% of the area between the two boundaries, there will be 5% / 2 = 2.5% of the area in each tail.

> Using Table A, search the body of the table for an area closest to 0.025. A z-score of -1.96 corresponds to an area of 0.025.

$z^* = 1.96$

Using technology: Applet/invNorm(left-tail area: 0.025, mean: 0, SD: 1) gives $z = -1.960$, so $z^* = 1.960$.

FOR PRACTICE   TRY EXERCISE 5.

Calculating a Confidence Interval for *p*

In the previous section, we reminded you that the standard deviation of the sampling distribution of $\hat{p}$ is $\sigma_{\hat{p}} = \sqrt{\dfrac{p(1-p)}{n}}$. Unfortunately, we don't know the value of p if we are trying to estimate it using a confidence interval. So we replace p with $\hat{p}$ to calculate the **standard error of $\hat{p}$** or $SE_{\hat{p}}$.

DEFINITION Standard error of $\hat{p}$

The **standard error of $\hat{p}$** is an estimate of the standard deviation of the sampling distribution of $\hat{p}$.

$$SE_{\hat{p}} = \sqrt{\frac{\hat{p}(1-\hat{p})}{n}}$$

The standard error of $\hat{p}$ estimates how much $\hat{p}$ typically varies from p.

We are now ready to calculate a confidence interval for a proportion.

How to Calculate a Confidence Interval for *p*

When the Random and Large Counts conditions are met, a C% confidence interval for the population proportion p is

point estimate $\pm$ margin of error

$$\hat{p} \pm z^* \sqrt{\frac{\hat{p}(1-\hat{p})}{n}}$$

where z^* is the critical value for the standard normal curve with C% of its area between $-z^*$ and z^*.

EXAMPLE

What proportion of the beads are red?

Calculating a confidence interval for p

PROBLEM: Dr. Godlewski's class selected an SRS from her container of beads and got 107 red beads and 144 other beads. Calculate a 90% confidence interval for p.

SOLUTION:

$\hat{p} = 107/251 = 0.426$

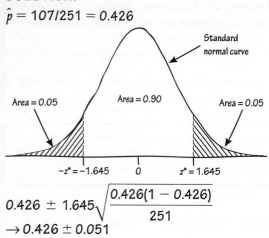

Standard normal curve

Area = 0.05   Area = 0.90   Area = 0.05

$-z^* = -1.645 \qquad 0 \qquad z^* = 1.645$

$0.426 \pm 1.645\sqrt{\dfrac{0.426(1 - 0.426)}{251}}$

$\rightarrow 0.426 \pm 0.051$

$\rightarrow 0.375 \text{ to } 0.477$

> Calculate the value of $\hat{p}$ for this sample.

> Calculate the critical value z^* for 90% confidence using the standard normal curve.

> Compute the confidence interval using the formula
> $$\hat{p} \pm z^*\sqrt{\dfrac{\hat{p}(1 - \hat{p})}{n}}.$$

FOR PRACTICE   TRY EXERCISE 9.

LESSON APP 7.3

Do you know your government?

According to a recent study by the Annenberg Foundation, only 36% of adults in the United States could name all three branches of government. This was based on a survey given to a random sample of 1416 U.S. adults.[17]

1. Show that the conditions for calculating a confidence interval for a proportion are satisfied.

2. Calculate a 99% confidence interval for the proportion of all U.S. adults who could name all three branches of government.

3. Interpret the interval from Question 2.

vichie81/Getty Images

Lesson 7.3

WHAT DID YOU LEARN?

LEARNING TARGET	EXAMPLES	EXERCISES
Check the Random and Large Counts conditions for constructing a confidence interval for a population proportion.	p. 465	1–4
Determine the critical value for calculating a $C\%$ confidence interval for a population proportion using Table A or technology.	p. 467	5–8
Calculate a $C\%$ confidence interval for a population proportion.	p. 468	9–12

Lesson 7.3

Mastering Concepts and Skills

1. **Rating cafeteria food** Latoya wants to estimate $p =$ the proportion of all students at her large boarding high school that like the cafeteria's food. She interviews an SRS of 50 of the students living in the dormitory and finds that 14 think the cafeteria's food is good. Check if the conditions for calculating a confidence interval for p are met.
   pg 465

2. **Finding new customers** An online retailer wants to estimate $p =$ the proportion of all orders in a given week that are from new customers. He takes an SRS of 40 orders and finds that 12 of them are from new customers. Check if the conditions for calculating a confidence interval for p are met.

3. **Contact your representative** Upset with a proposed new law in his state, Ryan emailed his representative in the legislature to express his disapproval. The representative emailed back saying that only 2 of the 200 messages he had received opposed the proposed law. Check if the conditions for calculating a confidence interval for p are met, where $p =$ the proportion of all registered voters in the state who oppose the proposed law.

4. **Whelks and mussels** The small round holes you often see in sea shells were drilled by other sea creatures, who ate the former dwellers of the shells. Whelks often drill into mussels, but this behavior appears to be more or less common in different locations. Researchers collected whelk eggs from the coast of Oregon, raised the whelks in the laboratory, and then put each whelk in a container with some delicious mussels. Only 9 of 98 whelks drilled into a mussel. Check if the conditions for calculating a confidence interval for p are met, where $p =$ the proportion of all Oregon whelks that will spontaneously drill into mussels.[18]

5. **Finding z^*** Use Table A or technology to find the critical value z^* for a 98% confidence interval. Assume that the Large Counts condition is met.
   pg 467

6. **Finding z^*** Use Table A or technology to find the critical value z^* for a 93% confidence interval. Assume that the Large Counts condition is met.

7. **Finding z^*** Use Table A or technology to find the critical value z^* for a 75% confidence interval. Assume that the Large Counts condition is met.

8. **Finding z^*** Use Table A or technology to find the critical value z^* for a 99.9% confidence interval. Assume that the Large Counts condition is met.

9. **Rating the cafeteria** Refer to Exercise 1. Calculate a 90% confidence interval for the proportion of students at Latoya's school who like the cafeteria's food.
   pg 468

10. **Finding more customers** Refer to Exercise 2. Calculate a 99% confidence interval for the proportion of the week's orders that are from new customers.

11. **Spinning the globe** In her first-grade social studies class, Jordan learned that 70% of Earth's surface is covered in water. She wondered if this is true and asked her dad for help. To investigate, he tossed an inflatable globe to her 50 times, being careful to spin the globe each time. When she caught it, he recorded where her right index finger was pointing. In 50 tosses, her finger was pointing to water 33 times.[19] Construct a 95% confidence interval for the proportion of Earth's surface that is covered in water. Assume that the conditions are met.

12. **Hugging the right way** After hearing that the majority of couples tilt to the right when kissing, Stephanie wanted to know the proportion of students who lean to the right when hugging a friend. She observed 51 different hugs and recorded 23 where the huggers leaned to the right. Construct a 95% confidence interval for the proportion of students at Stephanie's school that lean to the right when hugging a friend. Assume the conditions are met.

Applying the Concepts

13. **Reporting a cheat** What proportion of students are willing to report cheating by other students? A student project put this question to an SRS of 172 undergraduates at a large university: "You witness two students cheating on a quiz. Do you go to the professor?" Only 19 answered "Yes."[20]

   (a) Show that the conditions for calculating a confidence interval for a proportion are satisfied.

   (b) Calculate a 99% confidence interval for the proportion of all undergraduate students at this university who would answer "Yes."

   (c) Interpret the interval from part (b).

14. **Sorry!** A recent poll of 738 randomly selected customers of a major U.S. cell-phone carrier found that 170 of them had walked into something or someone while talking on a cell phone.[21]

   (a) Show that the conditions for calculating a confidence interval for a proportion are satisfied.

   (b) Calculate a 90% confidence interval for the proportion of all customers who have walked into something or someone while talking on a cell phone.

   (c) Interpret the interval from part (b).

15. **Reporting cheating** Refer to Exercise 13. Calculate and interpret the standard error of $\hat{p}$, where $\hat{p} =$ the proportion of students in the sample who would answer "Yes."

16. **Still sorry!** Refer to Exercise 14. Calculate and interpret the standard error of $\hat{p}$, where $\hat{p}$ = the proportion of customers who have walked into something or someone while talking on a cell phone.

Extending the Concepts

17. **Equality for women** Have efforts to promote equality for women gone far enough in the United States? A poll on this issue by the cable network MSNBC contacted 1019 adults. A newspaper article about the poll said, "Results have a margin of sampling error of plus or minus 3 percentage points."[22]

(a) The news article said that 65% of men, but only 43% of women, believe that efforts to promote equality have gone far enough. Explain why we do not have enough information to give confidence intervals for men and women separately.

(b) Would the margin of error for women alone be less than 0.03, about equal to 0.03, or greater than 0.03? Why? (You see that the news article's statement about the margin of error for poll results is a bit misleading.)

18. **No critical value** In some journals, researchers report estimates along with the standard error of the estimates, rather than the margin of error of the estimates. For example, the proportion of patients who suffered side effects on a particular medication might be reported as "0.19 (SE = 0.05)." Assuming the conditions are met, what is the confidence level for an interval in the following form: estimate ± SE? Explain.

Recycle and Review

19. **Rolling doubles (4.1, 4.6)** You are playing a board game with some friends. Each turn begins with rolling two dice. In this game, rolling "doubles"—the same number on both dice—is especially beneficial. The probability of rolling doubles is 1/6.

(a) What is the probability of rolling doubles 3 times in a row?

(b) After you roll doubles on your last three turns, one of your friends says, "No way you'll roll doubles this time. It would be nearly impossible." Explain what is wrong with your friend's reasoning.

(c) Given that you have rolled doubles 3 times in a row, what is the probability that you roll doubles on your fourth roll?

20. **You're only young once (1.8)** Here are the percentage of people under 18 in each of the 50 states, according to the 2010 U.S. Census.[23]

20.7	22.3	23.6	24.2	25.1	20.7	22.6	23.6
24.4	25.2	20.9	22.6	23.6	24.4	25.5	21.3
22.9	23.7	24.4	25.5	21.3	22.9	23.7	24.6
25.5	21.7	23.2	23.7	24.7	25.7	21.8	23.4
23.8	24.8	26.4	22.0	23.4	23.9	24.8	27.3
22.3	23.5	23.9	24.9	27.4	22.3	23.5	24.0
25.0	31.5						

(a) Make a boxplot for these data. Be sure to check for outliers.

(b) Describe what you see.

Lesson 7.4

Confidence Intervals for a Proportion

LEARNING TARGETS

● Use the four-step process to construct and interpret a confidence interval for a population proportion.

● Determine the sample size required to obtain a C% confidence interval for a population proportion with a specified margin of error.

In Lesson 7.3, you learned how to check the conditions and calculate a confidence interval for a population proportion. In this lesson, you will use the four-step process introduced in Lesson 4.1 to construct and interpret a confidence interval for a population proportion.

Putting It All Together: The Four-Step Process

When you are constructing and interpreting a confidence interval, follow the four-step process.

How to Use the Four-Step Process: Confidence Intervals

STATE: State the parameter you want to estimate and the confidence level.

PLAN: Identify the appropriate inference method and check conditions.

DO: If the conditions are met, perform calculations.

CONCLUDE: Interpret your interval in the context of the problem.

A confidence interval for a population proportion is often referred to as a *one-sample z interval for p*.

EXAMPLE

Got sleep?

Confidence Interval for p

PROBLEM: Sleep Awareness Week begins in the spring with the release of the National Sleep Foundation's annual poll of U.S. sleep habits and ends with the beginning of daylight saving time, when most people lose an hour of sleep.[24] In the foundation's random sample of 1029 U.S. adults, 48% reported that they "often or always" got enough sleep during the last 7 nights. Calculate and interpret a 95% confidence interval for the proportion of all Americans who would report they "often or always" got enough sleep during the last 7 nights.

SOLUTION:

STATE: We want to estimate p = the proportion of all Americans who would report they "often or always" got enough sleep during the last 7 nights at a 95% confidence level.

> **STATE:** State the parameter you want to estimate and the confidence level.

PLAN: One-sample z interval for p.

Random? Random sample of 1029 U.S. adults ✓

Large Counts? $1029(0.48) \approx 494 \geq 10$ and
$1029(1 - 0.48) \approx 535 \geq 10$ ✓

> **PLAN:** Identify the appropriate inference method and check conditions.

> Because these are the observed counts of successes and failures, they should be rounded to the nearest integer.

DO: $0.48 \pm 1.96 \sqrt{\dfrac{0.48(1 - 0.48)}{1029}} \rightarrow 0.48 \pm 0.031$

$\rightarrow 0.449$ to 0.511

> **DO:** If the conditions are met, perform calculations.

CONCLUDE: We are 95% confident that the interval from 0.449 to 0.511 captures the proportion of all Americans who would report they "often or always" got enough sleep during the last 7 nights.

> **CONCLUDE:** Interpret your interval in the context of the problem.

FOR PRACTICE   TRY EXERCISE 1.

Determining the Sample Size

When planning a study, we may want to choose a sample size that allows us to estimate a population proportion within a given margin of error. The formula for the margin of error (*ME*) in a confidence interval for *p* is

$$ME = z^* \sqrt{\frac{\hat{p}(1 - \hat{p})}{n}}$$

The margin of error calculation includes the sample proportion $\hat{p}$. Unfortunately, we won't know the value of $\hat{p}$ until *after* the study has been conducted. This means we have to guess the value of $\hat{p}$ when choosing *n*. Here are two ways to do this:

1. Use a guess for $\hat{p}$ based on a preliminary study or past experience with similar studies.
2. Use $\hat{p} = 0.5$ as the guess. The margin of error *ME* is largest when $\hat{p} = 0.5$, so this guess is conservative. If we get any other $\hat{p}$ when we do our study, the margin of error will be smaller than planned.

Once you have a guess for $\hat{p}$, the formula for the margin of error can be solved to give the required sample size *n*.

> ### How to Calculate Sample Size for a Desired Margin of Error
>
> To determine the sample size *n* that will yield a *C*% confidence interval for a population proportion *p* with a maximum margin of error *ME*, solve the following inequality for *n*:
>
> $$z^* \sqrt{\frac{\hat{p}(1 - \hat{p})}{n}} \leq ME$$
>
> where $\hat{p}$ is a guessed value for the sample proportion. The margin of error will always be less than or equal to *ME* if you use $\hat{p} = 0.5$.

EXAMPLE

Who has a tattoo?

Determining sample size

PROBLEM: Suppose that you want to estimate *p* = the proportion of all students at your school who have a tattoo with 90% confidence and a margin of error of no more than 0.07. Determine how many students should be surveyed.

SOLUTION:

$$1.645 \sqrt{\frac{0.5(1 - 0.5)}{n}} \leq 0.07$$

> Because we don't have a value of $\hat{p}$ to use from a preliminary or previous study, use $\hat{p} = 0.5$ as the guessed value of $\hat{p}$. Then solve the inequality for *n*.

$$\sqrt{\frac{0.5(1 - 0.5)}{n}} \leq \frac{0.07}{1.645}$$

> Divide each side by 1.645.

$$\frac{0.5(1 - 0.5)}{n} \leq \left(\frac{0.07}{1.645}\right)^2$$

> Square both sides.

$$0.5(1 - 0.5) \leq n \left(\frac{0.07}{1.645}\right)^2$$

> Multiply both sides by *n*.

$$\frac{0.5(1 - 0.5)}{\left(\dfrac{0.07}{1.645}\right)^2} \leq n$$

> Divide both sides by $\left(\dfrac{0.07}{1.645}\right)^2$.

$138.0625 \leq n$

We need to survey at least 139 students.

> Answer the question, making sure to follow the inequality sign when rounding.

FOR PRACTICE   TRY EXERCISE 5.

Even though 138.0625 is closer to 138 than 139, we have to use at least 139 students to *guarantee* that the margin of error will be less than or equal to 0.07. If $\hat{p} = 0.5$ and $n = 138$, the margin of error is 0.07002, which is greater than the desired margin of error of 0.07. In general, we round to the next highest integer when solving for sample size to make sure the margin of error is less than or equal to the desired value.

LESSON APP 7.4

TV in bed?

Many kids aren't getting enough sleep, and TV might be part of the problem. Among 2002 randomly selected 8- to 18-year-olds in the United States, 71% had a TV in their bedroom. [25]

© deux/Corbis

1. Construct and interpret a 99% confidence interval for the proportion of all 8- to 18-year-olds in the United States who have a TV in their bedroom.

2. Based on the interval, is there convincing evidence that a majority of 8- to 18-year-olds have a TV in their bedroom? Explain.

3. If the researchers wanted to reduce the margin of error to at most 1%, about how many *additional* 8- to 18-year-olds do they need to randomly select? Use the value of $\hat{p}$ from the initial study.

TECH CORNER

Confidence Intervals for a Population Proportion

You can use technology to calculate a confidence interval for a population proportion. We'll illustrate using the "Got sleep?" example from earlier. Recall that $n = 1029$ and $\hat{p} = 0.48$, which means that $1029(0.48) \approx 494$ people said that they "often or always" got enough sleep during the last 7 days.

Applet

1. Launch the *One Categorical Variable* applet at highschool.bfwpub.com/spa3e.

2. Enter the variable name "Response" and Category names "Often or always" and "Not." Then, enter 494 for the frequency of "Often or always" and $1029 - 494 = 535$ for "Not."

One Categorical Variable

Variable name: Response
Groups: Single ▾
Input data as: Counts in categories ▾

	Category Name	Frequency	
1	Often or always	494	-
2	Not	535	-
			+

Begin analysis   Edit inputs   Reset everything

3. Press the "Begin analysis" button. Then, scroll down to the "Perform inference" section.

4. Choose the 1-sample *z*-interval for the procedure and 95% for the confidence level. Then, press the "Perform inference" button.

Perform Inference

Inference procedure: 1-sample z interval ▾
Category to indicate as success: Often or always ▾
Confidence level: 95% ▾

Perform inference

Lower Bound	Upper Bound
0.45	0.511

TI-83/84

1. Press STAT then choose TESTS and 1-PropZInt.

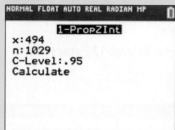
```
NORMAL FLOAT AUTO REAL RADIAN MP
              1-PropZInt
x:494
n:1029
C-Level:.95
Calculate
```

2. When the 1-PropZInt screen appears, enter $x = 494$, $n = 1029$, and confidence level = 0.95.

3. Highlight "Calculate" and press ENTER. The 95% confidence interval for *p* is reported, along with the sample proportion $\hat{p}$ and the sample size, as shown here.

```
NORMAL FLOAT AUTO REAL RADIAN MP
              1-PropZInt
(.44955..5106)
p̂=.4800777454
n=1029
```

Lesson 7.4

WHAT DID YOU LEARN?

LEARNING TARGET	EXAMPLES	EXERCISES
Use the four-step process to construct and interpret a confidence interval for a population proportion.	p. 471	1–4
Determine the sample size required to obtain a C% confidence interval for a population proportion with a specified margin of error.	p. 472	5–8

Exercises   Lesson 7.4

Mastering Concepts and Skills

1. Going to the prom Leticia wants to estimate what proportion of her school's seniors plan to attend the prom. She interviews an SRS of 50 of the 750 seniors in her school and finds that 36 plan to go to the prom. Calculate and interpret a 90% confidence interval for the proportion of all seniors at Leticia's school who plan to go to the prom.

pg 471

2. Car care A large automobile manufacturing plant produces 1200 new cars every day. A quality-control inspector checks a random sample of 90 cars from one day's production and finds that 12 of them have minor paint flaws. Calculate and interpret a 99% confidence interval for the proportion of all cars produced that day with minor paint flaws.

3. **Teens text** The Pew Research Center's Teen Relationship Survey found that 33% of 929 randomly selected teens with cell phones use messaging apps.[26] Calculate and interpret a 95% confidence interval for the proportion of all teens with cell phones who use messaging apps.

4. **Super-size me** A Gallup poll of 1015 randomly selected adults found that 69% would oppose a law that limited the size of soft drinks and other sugary beverages served in restaurants.[27] Calculate and interpret a 95% confidence interval for the proportion of all adults that would oppose a law limiting the size of soft drinks.

5. **The taste of PTC?** PTC is a substance that has a strong bitter taste for some people and is tasteless for others. The ability to taste PTC is inherited. About 75% of Italians can taste PTC, for example. You want to estimate the proportion of Americans of Italian descent who can taste PTC. How large a sample must you test to estimate the proportion of PTC tasters in this population within 0.04 with 90% confidence? Answer this question using the 75% estimate as the guessed value for $\hat{p}$.

pg 472

6. **Getting vouchers** A national opinion poll found that 44% of all American adults agree that parents should be given vouchers that can be applied toward their children's education at any public or private school of their choice. The result was based on a small sample. How large an SRS is required to obtain a margin of error of at most 0.03 in a 99% confidence interval? Answer this question using the previous poll's result as the guessed value for $\hat{p}$.

7. **Mayoral polling** Edgar Martinez and Ingrid Gustafson are the candidates for mayor in a large city. We want to estimate the proportion p of all registered voters in the city who plan to vote for Gustafson with 95% confidence and a margin of error no greater than 0.03. How large a random sample do we need?

8. **Under-age nightclub** A college student organization wants to start a nightclub for students under the age of 21. To assess support for the idea, the organization will select an SRS of students and ask each if he or she would patronize this type of establishment. What sample size is required to obtain a 90% confidence interval with a margin of error of at most 0.04?

Applying the Concepts

9. **Yankees in Connecticut** Connecticut is located between New York and Boston, so baseball fans in the state typically root for either the Yankees or Red Sox. In a recent poll of 803 randomly selected baseball fans in Connecticut, 44% said their favorite team was the Yankees.[28]

(a) Construct and interpret a 95% confidence interval for the proportion of all baseball fans in Connecticut who would say their favorite team is the Yankees.

(b) Based on the interval, is it plausible that the Yankees are the favorite team for a majority of Connecticut baseball fans? Explain.

(c) If the researchers wanted to reduce the margin of error to at most 2%, about how many *additional* baseball fans do they have to randomly select? Use the value of $\hat{p}$ from the initial study.

10. **Life on Mars** A recent CNN poll found that in a random sample of 508 U.S. residents, 31% answered "Yes" to the question "Do you think life has ever existed on Mars?"

(a) Construct and interpret a 99% confidence interval for the proportion of all U.S. residents who believe life ever existed on Mars.

(b) Based on the interval, is it plausible that more than 35% of U.S. residents think life ever existed on Mars? Explain.

(c) If the polltakers wanted to reduce the margin of error to at most 4%, about how many *additional* U.S. residents do they have to randomly select? Use the value of $\hat{p}$ from the initial study.

11. **Text ME** Refer to Exercise 3. How big of a sample would be required to cut the margin of error in half?

12. **Half-size ME** Refer to Exercise 4. How big of a sample would be required to cut the margin of error in half?

Extending the Concepts

13. **At the prom** Use your confidence interval from Exercise 1 to calculate a 90% confidence interval for the *total* number of seniors at Leticia's school who plan to attend the prom.

14. **Paint flaws** Use your confidence interval from Exercise 2 to calculate a 99% confidence interval for the *total* number of cars produced that day with minor paint flaws.

15. **Gambling and college sports** Gambling is an issue of great concern in college athletics. Because of this, the National Collegiate Athletic Association (NCAA) surveyed randomly selected student athletes about gambling-related behaviors. Of the 5594 Division I male athletes in the survey, 3547 reported participation in some form of gambling. This includes playing cards, betting on games of skill, buying lottery tickets, and betting on sports. A report of this study cited a 1% margin of error.[29]

(a) The confidence level was not stated in the report. Use what you have learned to find the confidence level, assuming that the NCAA selected an SRS.

(b) Describe one potential source of bias in the NCAA study that is not accounted for by the margin of error.

Recycle and Review

16. **Bone density (5.6, 5.7)** Osteoporosis is a condition in which the bones become brittle due to loss of minerals. To diagnose osteoporosis, an elaborate apparatus measures bone mineral density (BMD) in standardized form. The standardization is based on a population of healthy young adults. The World Health Organization (WHO) criterion for osteoporosis is a BMD score that is 2.5 standard deviations below the mean for young adults. BMD measurements in a population of people similar in age and gender roughly follow a normal distribution.

(a) What percent of healthy young adults have osteoporosis by the WHO criterion?

(b) Women aged 70 to 79 are, of course, not young adults. The mean BMD in this age group is about -2 on the standard scale for young adults. Suppose that the standard deviation is the same as for young adults. What percent of this older population has osteoporosis?

17. **Buying a house? (6.6)** The distribution of prices for home sales in a certain New Jersey county is skewed right with a mean of \$290,000 and a standard deviation of \$145,000.

(a) Suppose you take a simple random sample of 100 home sales from this large population. What is the probability that the mean of the sample is greater than \$325,000?

(b) Suppose you take a simple random sample of 5 home sales from this population. Explain why you cannot calculate the probability that the mean of this sample is greater than \$325,000.

Lesson 7.5

Estimating a Mean

LEARNING TARGETS

- State and check the Random and Normal/Large Sample conditions for constructing a confidence interval for a population mean.
- Determine critical values for calculating a C% confidence interval for a population mean.
- Calculate a C% confidence interval for a population mean.

In Lessons 7.3 and 7.4, you learned how to construct and interpret a confidence interval for a population proportion. In this lesson, you will learn how to check conditions and calculate a confidence interval for a population mean μ.

Conditions for Estimating μ

Like confidence intervals for a population proportion, there are conditions that need to be met to construct a confidence interval for a population mean.

> ### How to Check the Conditions for Constructing a Confidence Interval for μ
>
> - **Random:** The data come from a random sample from the population of interest.
> - **Normal/Large Sample:** The data come from a normally distributed population or the sample size is large ($n \geq 30$).

When the Random condition is satisfied, we can generalize about the population from which the sample was selected. We check the Normal/Large Sample condition to make sure the distribution of $\bar{x}$ is approximately normal.

EXAMPLE

How good are your grades?

Checking conditions

PROBLEM: To estimate the average GPA of all students at her high school, Quiana randomly selects 30 students from the school. Are the conditions met for calculating a confidence interval for μ? Justify your response.

SOLUTION:

Random? Random sample of students from the school. ✓

Normal/Large Sample? We don't know the shape of the population distribution, but the sample size is large: $n = 30 \geq 30$. ✓

FOR PRACTICE   TRY EXERCISE 1.

The Problem of Unknown σ

In Lesson 7.3, we presented the formula for a confidence interval for a population proportion:

$$\hat{p} \pm z^* \sqrt{\frac{\hat{p}(1 - \hat{p})}{n}}$$

In more general terms, this is

$$\text{statistic} \pm (\text{critical value})(\text{standard error of the statistic})$$

A confidence interval for a population mean has a formula with the same structure. Using $\bar{x}$ as the point estimate for μ and $\sigma/\sqrt{n}$ as the standard deviation of the sampling distribution of $\bar{x}$ gives

$$\bar{x} \pm z^* \frac{\sigma}{\sqrt{n}}$$

Unfortunately, if we don't know the true value of μ, we rarely know the true value of σ either. We can use s_x as an estimate for σ, but things don't work out as nicely as we might like.

ACTIVITY

Confidence interval BINGO!

In this activity, you will investigate the problem caused by replacing σ with s_x when calculating a confidence interval for μ, and how to fix it.

A farmer wants to estimate the mean weight (in grams) of all tomatoes grown on his farm. To do so, he will select a random sample of 4 tomatoes, calculate the mean weight (in grams), and use the sample mean $\bar{x}$ to create a 99% confidence interval for the population mean μ. Suppose that the weights of all tomatoes on his farm are approximately normally distributed, with a mean of 100 grams and a standard deviation of 40 grams.

Let's use an applet to simulate taking an SRS of $n = 4$ tomatoes and calculating a 99% confidence interval for μ using three different methods.

Method 1 (assuming σ is known)

$$\bar{x} \pm z^* \frac{\sigma}{\sqrt{n}} \rightarrow \bar{x} \pm 2.576 \frac{40}{\sqrt{4}}$$

1. Launch the *Simulating Confidence Intervals for a Population Parameter* applet at **www.rossman-chance.com/applets**. Choose "Means" from the drop-down menu and leave the other menus as "Normal" and "z with sigma." Then, enter $\mu = 100$, $\sigma = 40$, $n = 4$, and confidence level = 99%, as shown in the screen shot.

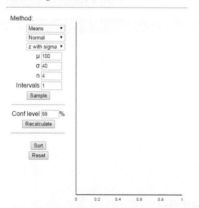

2. Press "Sample." The applet will select an SRS of $n = 4$ tomatoes and calculate a confidence interval for μ. This interval will be displayed as a horizontal line, along with a vertical line at $\mu = 100$. If the interval captures $\mu = 100$, the interval will be green. If the interval misses $\mu = 100$, the interval will be red.

3. Press "Sample" many times, shouting out "BINGO!" whenever you get an interval that misses $\mu = 100$ (i.e., a red interval). Stop when your teacher calls time.

4. How well did Method 1 work? Compare the running total in the lower-left corner with the stated confidence level of 99%.

Method 2 (using s_x as an estimate for σ)

$$\bar{x} \pm z^* \frac{s_x}{\sqrt{n}} \rightarrow \bar{x} \pm 2.576 \frac{s_x}{\sqrt{4}}$$

1. Press the "Reset" button in the lower left. Then, in the third drop-down menu, choose "z with s." Keep everything else the same.

2. Press "Sample" many times, shouting out "BINGO!" whenever you get an interval that misses $\mu = 100$ (i.e., a red interval). Stop when your teacher calls time.

3. Did Method 2 work as well as Method 1? Discuss with your classmates.

4. Now, change the number of intervals to 100 and press "Sample" 10 times, for a total of more than 1000 intervals. Compare the running total in the lower-left corner with the stated confidence level of 99%. What do you notice about the length of the intervals that missed?

To increase the capture rate of the intervals to 99%, we need to make the intervals longer. We can do this by using a different critical value, called a t^* critical value. You'll learn how to calculate this number soon.

Method 3 (using s_x as an estimate for σ and a t^* critical value instead of a z^* critical value)

$$\bar{x} \pm t^* \frac{s_x}{\sqrt{n}} \rightarrow \bar{x} \pm ??? \frac{s_x}{\sqrt{4}}$$

1. Press the "Reset" button in the lower left. Then, in the third drop-down menu, choose "t." Keep everything else the same.

2. Press "Sample" many times, shouting out "BINGO!" whenever you get an interval that misses $\mu = 100$ (i.e., a red interval). Stop when your teacher calls time.

3. Did Method 3 work better than Method 2? How does it compare to Method 1? Discuss with your classmates.

4. Now, change the number of intervals to 100 and press "Sample" 10 times, for a total of at least 1000 intervals. Compare the running total in the lower-left corner with the stated confidence level of 99%. What do you notice about the length of the intervals compared to Method 2?

t* Critical Values

When calculating a confidence interval for a population mean, we use a t^* critical value rather than a z^* critical value whenever we use s_x to estimate σ. This means that we almost always use a t^* critical value when constructing a confidence interval for μ because we seldom know the value of σ.

How to Find t*

1. Using Table B, find the correct confidence level at the bottom of the table.

2. On the left side of the table, find the correct number of *degrees of freedom* (df). For this type of confidence interval, df = $n - 1$.

3. In the body of the table, find the value of t^* that corresponds to the confidence level and df.

4. If the correct df isn't listed, use the greatest df available that is less than the correct df.

In the activity, we calculated 99% confidence intervals with $n = 4$, so df = $4 - 1 = 3$. Here is an excerpt from Table B that shows how to find t^* in this case.

df	Tail probability p			
	0.02	**0.01**	**0.005**	**0.0025**
1	15.89	31.82	63.66	127.3
2	4.849	6.965	9.925	14.09
3	3.482	4.541	5.841	7.453
⋮	⋮	⋮	⋮	⋮
z^*	2.054	2.326	2.576	2.807
	96%	98%	99%	99.5%
	Confidence level C			

For 99% confidence and 3 degrees of freedom, $t^* = 5.841$. That is, the interval should extend 5.841 standard deviations on both sides of the point estimate to have a capture rate of 99%.

The bottom row of Table B gives z^* critical values. That's because the t distributions approach the standard normal distribution as the degrees of freedom approach infinity. If the correct df isn't listed, use the greatest df available that is less than the correct df. "Rounding up" to a larger df will result in confidence intervals that are too narrow. The intervals won't be wide enough to include the true population value as often as suggested by the confidence level.

EXAMPLE

Would you like some t?*

Using Table B

PROBLEM: What critical value t^* from Table B should be used in constructing a confidence interval for the population mean in each of the following settings?

(a) A 95% confidence interval based on an SRS of size $n = 12$

(b) A 90% confidence interval from a random sample of 48 observations

SOLUTION:

(a) $t^* = 2.201$

df	.05	.025	.02	.01
10	1.812	2.228	2.359	2.764
11	1.796	2.201	2.328	2.718
12	1.782	2.179	2.303	2.681
z^*	1.645	1.960	2.054	2.326
	90%	95%	96%	98%

Tail probability p

Confidence level C

> In Table B, we use the column for 95% confidence and move up the column to find the row corresponding to df $= 12 - 1 = 11$.

(b) $t^* = 1.684$

df	.10	.05	.025	.02
30	1.310	1.697	2.042	2.147
40	1.303	1.684	2.021	2.123
50	1.299	1.676	2.009	2.109
z^*	1.282	1.645	1.960	2.054
	80%	90%	95%	96%

Tail probability p

Confidence level C

> With 48 observations, we want to find the t^* critical value for 90% confidence and df $= 48 - 1 = 47$. There is no df $= 47$ row in Table B, so we use the more conservative df $= 40$.

FOR PRACTICE   TRY EXERCISE 5.

Calculating a Confidence Interval for μ

Because we almost never know the population standard deviation σ, we must use the sample standard deviation s_x as an estimate for σ. This means that we must estimate the standard deviation of the sampling distribution of $\bar{x}$ with the **standard error of $\bar{x}$**.

> **||| DEFINITION   Standard error of $\bar{x}$**
>
> The **standard error of $\bar{x}$** is an estimate of the standard deviation of the sampling distribution of $\bar{x}$.
> $$SE_{\bar{x}} = \frac{s_x}{\sqrt{n}}$$
> The standard error of $\bar{x}$ estimates how much $\bar{x}$ typically varies from μ.

Now that we know how to calculate the t^* critical value and the standard error of $\bar{x}$, we can calculate a confidence interval for μ.

> **How to Calculate a Confidence Interval for μ**
>
> When the Random and Normal/Large Sample conditions are met, a $C\%$ confidence interval for the unknown mean μ is
> $$\bar{x} \pm t^* \frac{s_x}{\sqrt{n}}$$
> where t^* is the critical value for a t distribution with df $= n - 1$ and $C\%$ of its area between $-t^*$ and t^*.

Are you confident about your grades?

Calculating a confidence interval for μ

PROBLEM: In the first example of this lesson, we verified that the conditions were met for constructing a confidence interval for μ = the true mean GPA for all students in Quiana's school. In her sample of 30 students, the average GPA was $\bar{x}$ = 3.19 with a standard deviation of s_x = 0.72. Calculate a 99% confidence interval for μ.

SOLUTION:

$df = 30 - 1 = 29$ and $t^* = 2.756$

> Start by finding the t^* critical value using df $= n - 1$.

$3.19 \pm 2.756 \dfrac{0.72}{\sqrt{30}} \rightarrow 3.19 \pm 0.362$

> Use the formula $\bar{x} \pm t^* \dfrac{s_x}{\sqrt{n}}$

$\rightarrow 2.828$ to 3.552

FOR PRACTICE   TRY EXERCISE 9.

LESSON APP 7.5

What does an Oreo weigh?

For their second-semester project, Ann and Tori wanted to estimate the average weight of an Oreo cookie to determine if the average weight was less than advertised. They selected a random sample of 36 cookies and found the weight of each cookie (in grams). The mean weight was $\bar{x}$ = 11.3921 grams, with a standard deviation of s_x = 0.0817 grams.[30]

1. Verify that the conditions are met for constructing a confidence interval for μ.

2. Construct a 95% confidence interval for μ.

3. On the packaging, the stated serving size is 3 cookies (34 grams). Does the interval in Question 2 provide convincing evidence that the average weight of an Oreo cookie is less than advertised? Explain.

Dan Anderson/Recursive Process

Lesson 7.5

WHAT DID YOU LEARN?

LEARNING TARGET	EXAMPLES	EXERCISES
State and check the Random and Normal/Large Sample conditions for constructing a confidence interval for a population mean.	p. 477	1–4
Determine critical values for calculating a C% confidence interval for a population mean.	p. 479	5–8
Calculate a C% confidence interval for a population mean.	p. 481	9–12

Lesson 7.5

Mastering Concepts and Skills

For Exercises 1–4, determine if the conditions for calculating a confidence interval for μ are met in each of the following settings.

1. **Internet use** How much time do students at your school spend on the Internet? You collect data from a random sample of 32 students at your school and calculate the mean amount of time that each student spent on the Internet yesterday.

   pg 477

2. **Athletes' heart rates** The athletic director at a large university records the resting heart rate for 65 randomly selected athletes. We use these data to construct a confidence interval for the mean resting heart rate for all athletes at this university.

3. **Presidents live longer?** We want to estimate the average age at which U.S. presidents have died. So we obtain a list of all U.S. presidents who have died and their ages at death.

4. **Flight capacity** An airline wants to calculate the mean number of passengers on a daily 10:00 A.M. flight from Philadelphia to San Diego in 2015. They use flight manifests to determine the mean number of passengers for all these flights in 2015.

5. **Finding t^*** What critical value t^* from Table B should be used in constructing a confidence interval for the population mean in each of the following settings? Assume the conditions are met.

   pg 479

   (a) A 95% confidence interval based on $n = 10$ randomly selected observations

   (b) A 90% confidence interval based on a random sample of 77 individuals

6. **Finding t^*** What critical value t^* from Table B should be used in constructing a confidence interval for the population mean in each of the following settings? Assume the conditions are met.

   (a) A 90% confidence interval based on $n = 12$ randomly selected observations

   (b) A 99% confidence interval based on a random sample of size 58

7. **Finding t^*** What critical value t^* from Table B should be used in constructing a confidence interval for the population mean in each of the following settings? Assume the conditions are met.

   (a) A 99% confidence interval from an SRS of 20 observations

   (b) A 95% confidence interval based on $n = 85$ randomly selected individuals

8. **Finding t^*** What critical value t^* from Table B should be used in constructing a confidence interval for the population mean in each of the following settings? Assume the conditions are met.

   (a) A 95% confidence interval from an SRS of 30 observations

   (b) A 90% confidence interval based on $n = 162$ randomly selected individuals

9. **Sacrifices for science?** Many people have asked the question, but few have been patient enough to collect the data. How many licks does it take to get to the center of a Tootsie Pop? Researcher Corey Heid decided to find out.[31] He instructed a sample of 92 students to lick a Tootsie Pop along the non-banded side until they could taste the chocolate center. The mean number of licks was 356.1 with a standard deviation of 185.7 licks. Construct a 95% confidence interval for the true mean number of licks to get to the center of a Tootsie Pop. Assume the conditions for inference are met.

   pg 481

10. **Flexible exam times** In Mr. Wright's school district, students are allowed to take as much time as they need to finish the final exam in Algebra II. He asked a random sample of 45 of the 600 students who took the exam one year to record the length of time they spent on the exam. The mean and standard deviation for the 45 students were $\bar{x} = 92.8$ minutes and $s_x = 22.1$ minutes. Construct a 95% confidence interval for the mean completion time of all students who took the exam. Assume the conditions for inference are met.

11. **Valuing houses** The mayor of a small town wants to estimate the average property value for the houses built in the last year. He randomly selects 15 houses and pays an appraiser to determine the value of each house. The mean value of these houses is $183,100 with a standard deviation of $29,200. Construct a 90% confidence interval for the mean value of new houses in this town. Assume the conditions for inference are met.

12. **Oh no! My big toe** A bunion on the big toe is fairly uncommon in youth and often requires surgery. Doctors used X-rays to measure the angle (in degrees) of deformity on the big toe in a random sample of 37 patients under the age of 21 who came to a medical center for surgery to correct a bunion. The angle is a measure of the seriousness of the deformity. For these 37 patients, the mean angle of deformity was 24.76 degrees and the standard deviation was 6.34 degrees. Construct a 90% confidence interval for the mean angle of deformity in the population of patients like these. Assume the conditions for inference are met.[32]

Applying the Concepts

13. **Reading in Atlanta** A particular reading test is scored from 0 to 500. A score of 243 is a "basic" reading level and a score of 281 is "proficient." Scores for a random sample of 1470 eighth-graders in Atlanta had a mean of 240 with a standard deviation of 42.17.[33]

    (a) Verify that the conditions are met for constructing a confidence interval for μ = true mean reading score for Atlanta eighth-graders.

    (b) Construct a 99% confidence interval for μ.

    (c) Based on your interval from part (b), is there convincing evidence that the mean for all Atlanta eighth-graders is less than the basic level? Explain.

14. **Mineral loss during nursing** Breast-feeding mothers secrete calcium into their milk. Some of the calcium may come from their bones, so mothers may lose bone mineral. Researchers measured the percent change in bone mineral content (BMC) of the spines of 47 randomly selected mothers during three months of breast-feeding. The mean change in BMC was −3.587% and the standard deviation was 2.506%.[34]

    (a) Verify that the conditions are met for constructing a confidence interval for μ = true mean change in BMC for breast-feeding mothers.

    (b) Construct a 99% confidence interval for μ.

    (c) Based on your interval from part (b), do these data give convincing evidence that nursing mothers lose bone mineral, on average? Explain.

15. **Measuring blood pressure** A medical study finds that $\bar{x} = 114.9$ and $s_x = 9.3$ for the seated systolic blood pressure of the 27 members of one treatment group. What is the standard error of the mean? Interpret this value in context.

16. **Estimating travel time** A study of commuting times reports the travel times to work of a random sample of 20 employed adults in New York State. The mean is $\bar{x} = 31.25$ minutes, and the standard deviation is $s_x = 21.88$ minutes. What is the standard error of the mean? Interpret this value in context.

Extending the Concepts

17. **Valuing new houses** Refer to Exercise 11. There are a total of 300 new houses in the mayor's town and the town levies a 1% property tax on each house. Use your confidence interval from Exercise 11 to calculate a 90% confidence interval for the total amount of tax revenue generated by the new houses.

18. **A really big toe problem** Refer to Exercise 12. Researchers omitted 1 patient with a deformity angle of 50 degrees from the analysis due to a measurement issue. What effect would including this outlier have on the confidence interval? Justify your answer without doing any calculations.

Recycle and Review

19. **Mail call!** (3.3) Even though electronic communication through email and social media has become increasingly important, a recent study found that 41% of U.S. residents still look forward to opening their U.S. Post Office mailbox each day.[35] A social scientist wonders if the proportion of people in their 20s who look forward to opening their "real" mailbox is less than 41%. He takes a random sample of 50 people in their 20s and finds that only 14 of them say that they look forward to opening their mailbox each day. To determine if these data provide convincing evidence that less than 41% of people in their 20s look forward to opening their mailbox each day, we simulated 100 random samples. Each dot in the graph shows the number of people in their 20s who look forward to opening their mailbox each day in a random sample of 50 people, assuming that the true proportion is 41%.

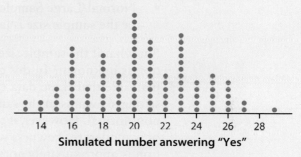

Simulated number answering "Yes"

    (a) Explain how the graph illustrates the concept of sampling variability.

    (b) Based on the results of the simulation, is there convincing evidence that less than 41% of people in their 20s look forward to opening their mailbox each day? Explain.

20. **Mail piles up!** (6.3) Refer to Exercise 19. Let X = the number of people in a random sample of size 50 who say they look forward to opening their mailbox each day. Assume the true proportion of people who look forward to opening their mailbox is 0.41.

    (a) Calculate the mean and standard deviation of the sampling distribution of X. Interpret the standard deviation.

    (b) The mean and standard deviation of the simulated number answering "Yes" in Exercise 19 are 20.46 and 3.54, respectively. How do these compare to your results in part (a)?

Lesson 7.6

Confidence Intervals for a Mean

- Use sample data to check the Normal/Large Sample condition.
- Use the four-step process to construct and interpret a confidence interval for a population mean.

In Lesson 7.5, you learned how to check the conditions and calculate a confidence interval for a mean. In this lesson, we will revisit the Normal/Large Sample condition and practice using the four-step process for constructing and interpreting a confidence interval.

The Normal/Large Sample Condition

In Lesson 7.5, we introduced the Normal/Large Sample condition for constructing a confidence interval for a population mean:

- **Normal/Large Sample:** The data come from a normally distributed population or the sample size is large ($n \geq 30$).

But what if the sample size is small ($n < 30$) and the shape of the population distribution is unknown? In this case, we graph the sample data and ask this question: "Is it plausible that these data came from a normally distributed population?" Outliers or strong skewness in the sample data might indicate that the population distribution is non-normal. However, if the distribution of sample data is roughly symmetric or has only moderate skewness with no outliers, it is believable that the population distribution is approximately normal.

> ### How to Check the Normal/Large Sample Condition
>
> There are three ways that the Normal/Large Sample condition can be met:
>
> 1. The data come from a normally distributed population.
> 2. The sample size is large ($n \geq 30$).
> 3. When the sample size is small and the shape of the population distribution is unknown, a graph of the sample data shows no strong skewness or outliers.

It can be difficult to determine if a graph shows "strong" skewness or only "moderate" skewness. The following example should help you make this distinction.

EXAMPLE

Are the conditions met?

The Normal/Large Sample condition

PROBLEM: Determine if the Normal/Large Sample condition is met in each of the following settings:

(a) How much force does it take to pull wood apart? The stemplot shows the force (in pounds) required to pull apart a random sample of 20 pieces of Douglas fir.

```
23 | 0
24 | 0
25 |
26 | 5
27 |            Key: 31|3 = 313 pounds of force
28 | 7
29 |
30 | 259
31 | 399
32 | 033677
33 | 0236
```

(b) Suppose you want to estimate the mean SAT Math score at a large high school. The boxplot summarizes the distribution of SAT Math scores for a random sample of 12 students at the school.

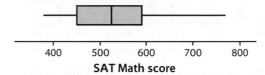

SAT Math score

SOLUTION:

(a) No. The sample size is small ($20 < 30$) and the stemplot shows strong left skewness and possible outliers.

(b) Yes. Although the sample size is small, the boxplot shows no strong skewness or outliers.

FOR PRACTICE   TRY EXERCISE 1.

Putting It All Together: The Four-Step Process

When you construct and interpret a one-sample t interval for a population mean, follow the four-step process from Lesson 7.4: State, Plan, Do, and Conclude.

EXAMPLE

Can you spare a square?

Constructing a confidence interval for μ

PROBLEM: Christina and Rachel randomly selected 18 rolls of a generic brand of toilet paper to measure how well this brand could absorb water. To do this, they poured 1/4 cup of water onto a hard surface and counted how many squares of toilet paper it took to completely absorb the water.[36] Here are the results from their 18 rolls.

29 20 25 29 21 24 27 25 24 29 24 27 28 21 25 26 22 23

Construct and interpret a 99% confidence interval for μ = the true mean number of squares of generic toilet paper needed to absorb 1/4 cup of water.

SOLUTION:

STATE: We want to estimate μ = the true mean number of squares of generic toilet paper needed to absorb 1/4 cup of water with 99% confidence.

PLAN: One-sample t interval for μ

Random? Random sample of 18 generic toilet paper rolls ✓

Normal/Large Sample? The sample size is small, but the dotplot doesn't show any outliers or strong skewness. ✓

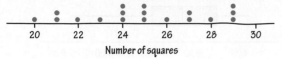

Number of squares

DO: For these data, $\bar{x} = 24.94$, $s_x = 2.86$, and $n = 18$. With 99% confidence and df $= 18 - 1 = 17$, $t^* = 2.898$.

$$24.94 \pm 2.898 \frac{2.86}{\sqrt{18}}$$

$\rightarrow 24.94 \pm 1.95$

$\rightarrow 22.99$ to 26.89

CONCLUDE: We are 99% confident that the interval from 22.99 squares to 26.89 squares captures the true mean number of squares of generic toilet paper needed to absorb 1/4 cup of water.

STATE: State the parameter you want to estimate and the confidence level.

PLAN: Identify the appropriate inference method and check conditions.

DO: If the conditions are met, perform calculations.

Use the sample data to calculate the mean and standard deviation.

CONCLUDE: Interpret your interval in the context of the problem.

Make sure that your conclusion is about a mean and includes units.

FOR PRACTICE TRY EXERCISE 5.

CAUTION

Make sure to include the graph of sample data when the sample size is small and you are checking the Normal/Large Sample condition. It doesn't matter whether you use a dotplot, stemplot, histogram, or boxplot to check for strong skewness or outliers.

THINK ABOUT IT **Is it possible to determine the sample size needed to achieve a specified margin of error when estimating a population mean?** Yes, but it's complicated. The margin of error (*ME*) in the confidence interval for μ is

$$ME = t^* \frac{s_x}{\sqrt{n}}$$

Unfortunately, there are two problems with using this formula to determine the sample size.

1. We don't know the sample standard deviation s_x because we haven't produced the data yet.

2. The critical value t^* depends on the sample size n that we choose.

The second problem is more serious. To get the correct value of t^*, we have to know the sample size. But that's precisely what we're trying to find!

One alternative is to come up with a reasonable estimate for the *population* standard deviation σ from a similar study that was done in the past or from a small-scale preliminary study. By pretending that σ is known, we can use a z^* critical value rather than a t^* critical value.

LESSON APP 7.6

How tense are the video screens?

A manufacturer of high-resolution video terminals must control the tension on the mesh of fine wires that lies behind the surface of the viewing screen. Too much tension will tear the mesh, and too little will allow wrinkles. The tension is measured by an electrical device with output readings in millivolts (mV). Some variation is inherent in the production process. Here are the tension readings from a random sample of 20 screens from a single day's production.

269.5 297.0 269.6 283.3 304.8 280.4 233.5 257.4 317.5 327.4
264.7 307.7 310.0 343.3 328.1 342.6 338.8 340.1 374.6 336.1

1. Construct and interpret a 90% confidence interval for μ = the mean tension of all the screens produced on this day.

Daniel Melekhin/Getty Images

2. The manufacturer's goal is to produce screens with an average tension of 300 millivolts. Based on the interval, is there convincing evidence that the screens produced this day don't meet the manufacturer's goal? Explain.

TECH CORNER

Confidence Intervals for a Population Mean

You can use technology to calculate a confidence interval for a population mean. We'll illustrate using the toilet paper example from earlier. Here are the data.

29 20 25 29 21 24 27 25 24 29 24 27 28 21 25 26 22 23

Applet

1. Launch the *One Quantitative Variable* applet at highschool.bfwpub.com/spa3e.

2. Enter the variable name "Squares" and leave the number of groups as 1. Then, input the data.

One Quantitative Variable

Variable name: Squares
Number of groups: 1 ▾
Input: Raw data   ▾

Input data separated by commas or spaces.
Group 1 data: 29 20 25 29 21 24 27 25 24 29 24 27 28 2

[Begin analysis] [Edit inputs] [Reset everything]

3. Press the "Begin analysis" button. Use the graph to assess if the Normal/Large Sample condition is met.

4. Scroll down to the "Perform inference" section. Choose the 1-sample *t*-interval for the procedure and 99% for the confidence level. Then, press the "Perform inference" button.

Perform Inference

Inference procedure: [1-sample t interval   ▾]
Confidence level: [99% ▾]

[Perform inference]

Lower Bound	Upper Bound	df
22.991	26.897	17

Note: If you don't have the raw data and only have the mean and standard deviation for the sample, choose "Mean and standard deviation" from the Input menu.

TI-83/84

1. Enter the toilet paper data into L1.

2. Press $\boxed{\text{STAT}}$, then choose TESTS and TInterval.

3. When the TInterval screen appears, choose Data as the input method, L1 as the List, and 0.99 as the confidence level.

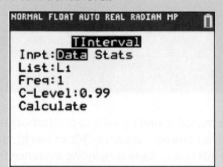

```
NORMAL FLOAT AUTO REAL RADIAN MP
              TInterval
Inpt:Data Stats
List:L1
Freq:1
C-Level:0.99
Calculate
```

4. Highlight "Calculate" and press $\boxed{\text{ENTER}}$. The 99% confidence interval for μ is reported, along with the sample mean $\bar{x}$, sample standard deviation s_x, and sample size n.

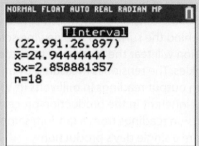

```
NORMAL FLOAT AUTO REAL RADIAN MP
              TInterval
(22.991,26.897)
x̄=24.94444444
Sx=2.858881357
n=18
```

Note: If you don't have the raw data and only have the summary statistics, change the input method to Stats and enter values for the sample mean $\bar{x}$, sample standard deviation s_x, and sample size n.

Lesson 7.6

WHAT DID YOU LEARN?

LEARNING TARGET	EXAMPLES	EXERCISES
Use sample data to check the Normal/Large Sample condition.	p. 485	1–4
Use the four-step process to construct and interpret a confidence interval for a population mean.	p. 485	5–8

Exercises
Lesson 7.6

Mastering Concepts and Skills

Determine if the Normal/Large Sample condition is met in each of the following settings.

1. **Medical jargon** Judy is interested in estimating the reading level of a medical journal using average word length. She records the length of a random sample of 100 words. The histogram displays the data.
pg 485

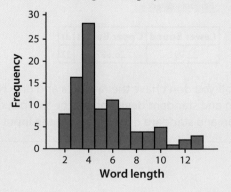

2. **A bad weed** Velvetleaf is an invasive weed in U.S. cornfields, where it produces lots of seeds. How many seeds do velvetleaf plants produce? The histogram shows the counts from a random sample of 28 plants that came up in a cornfield when no herbicide was used.[37]

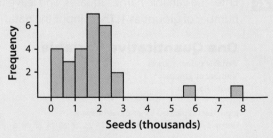

3. **Grand Caravans** How expensive is a used Dodge minivan? The histogram shows the asking price (in thousands of dollars) for a random sample of sixteen 2-year-old Grand Caravans from a popular website listing used cars.

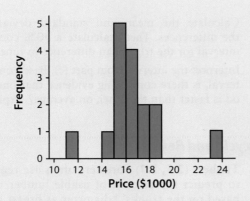

4. **Dissolved oxygen** The level of dissolved oxygen (DO) in a river shows the water's ability to support aquatic life. A researcher collects water samples from 15 randomly chosen locations along a stream and measures the dissolved oxygen. His results are graphed in the boxplot.

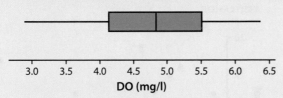

5. **Vitamin C content** Several years ago, the U.S. Agency for International Development provided 238,300 metric tons of corn-soy blend (CSB) for emergency relief in countries throughout the world. CSB is a highly nutritious, low-cost fortified food. As part of a study to evaluate appropriate vitamin C levels in this food, measurements were taken on samples of CSB produced in a factory. The following data are the amounts of vitamin C, measured in milligrams per 100 grams (mg/100 g) of blend, for a random sample of size 8 from one production run.

26 31 23 22 11 22 14 31

Construct and interpret a 95% confidence interval for μ = the mean amount of vitamin C in the CSB from this production run.[38]

6. **Healing newts** Biologists studying the healing of skin wounds measured the rate at which new cells closed a cut made in the skin of an anesthetized newt. Here are data from a random sample of 18 newts, measured in micrometers (millionths of a meter) per hour.[39]

29 27 34 40 22 28 14 35 26 35 12 30 23 18 11 22 23 33

Construct and interpret a 99% confidence interval for μ = the true mean healing rate.

7. **Canadians' wrists** Here are the wrist circumferences (in millimeters) for 19 randomly selected Canadian high-school students.

120 130 135 140 140 140 145 150 150 150
160 160 160 165 170 180 180 190 190

Construct and interpret a 90% confidence interval for μ = the true mean wrist circumference for Canadian high school students.

8. **Crossword tournament** The American Crossword Puzzle Tournament takes place every year. A player's score on a puzzle is based on both completion time and accuracy. In 2015 there were 566 competitors in the tournament. Here are the scores for a random sample of 20 players on the first puzzle that year.[40]

5630 5730 6180 7275 7390 7400 7715 7935 7940 8225
8245 8455 8800 9475 9585 9795 9970 10,460 10,605 10,680

Construct and interpret a 95% confidence interval for μ = the mean score on the first puzzle in the 2015 American Crossword Puzzle Tournament.

Applying the Concepts

9. **Pepperoni pizza** Melissa and Madeline love pepperoni pizza, but sometimes they are disappointed with the small number of pepperonis on their pizza. To investigate, they went to their favorite pizza restaurant at 10 random times during the week and ordered a large pepperoni pizza.[41] Here are the number of pepperonis on each pizza.

47   36   25   37   46   36   49   32   32   34

Construct and interpret a 95% confidence interval for the true mean number of pepperonis on a large pizza at this restaurant.

10. **Catching goldfish for school** Carly and Maysem plan to be preschool teachers after they graduate from college. To prepare for snack time, they want to know the mean number of goldfish crackers in a bag of original-flavored goldfish. To estimate this value, they randomly selected 12 bags of original-flavored goldfish and counted the number of crackers in each bag.[42] Here are their data.

317 330 325 323 332 337 324 342 330 349 335 333

Construct and interpret a 95% confidence interval for the true mean number of crackers in a bag of original-flavored goldfish.

11. **A plethora of pepperoni?** Refer to Exercise 9. According to the manager of the restaurant, there should be an average of 40 pepperonis on a large pizza. Based on the interval, is there convincing evidence that the average number of pepperonis is less than 40? Explain.

12. **A school of fish** Refer to Exercise 10. According to the packaging, there are supposed to be 330 goldfish in each bag of crackers. Based on the interval, is there convincing evidence that the average number of goldfish is less than 330? Explain.

Extending the Concepts

13. **Estimating women's BMI** The body mass index (BMI) of all American young women is believed to follow a normal distribution, with a standard deviation of about 7.5. How large a sample would be needed to estimate the mean BMI μ in this population to within ±1 with 99% confidence?

14. **Homework time** Administrators at your school want to estimate how much time students spend on homework, on average, during a typical week. They want to estimate μ at the 90% confidence level with a margin of error of at most 30 minutes. A pilot study indicated that the standard deviation of time spent on homework per week is about 154 minutes. How many students have to be surveyed to meet the administrators' goal?

15. **Vitamin C variation** Refer to Exercise 5. What percent of the data values are within the 95% confidence interval that you calculated? Why does this percent differ from 95%?

16. **Go in or drive through?** For high school students who can leave campus for lunch, this is a very important question. Ben and Maya decided to investigate it at a local fast-food restaurant. Each time they went, they flipped a coin to determine which of them would go inside and which would use the drive-thru. Both of them ordered the same item, paid with the same amount of cash, and recorded how long it took (in seconds) to wait in line, pay, and receive their item.[43] Here are their data.

Visit	Drive-Thru (sec)	Inside (sec)
1	325	170
2	608	110
3	90	52
4	519	158
5	216	66
6	263	128
7	559	81
8	154	163
9	449	64
10	512	120

(a) Calculate the difference in time (Drive-thru − Inside) for each visit to the restaurant.

(b) Graph the differences using a dotplot. Based on the graph, is the Normal/Large Sample condition satisfied? Explain.

(c) Calculate the mean and standard deviation of the differences. Then, calculate a 90% confidence interval for the true mean difference in time.

(d) Interpret the interval from part (c). Based on the interval, is there convincing evidence that one method is faster than the other, on average? Explain.

Recycle and Review

17. **Timber!** (2.5, 2.7) Foresters often use regression to predict the amount of usable lumber in trees based on the trunks' "diameter at breast height" or DBH. Some data from a U.S. Forest Service research report on 24 Ponderosa Pine Trees were used to calculate a least-squares regression equation for y = tree height (in feet) and x = diameter at breast height (in inches).[44] Here are a residual plot and other information about this regression.

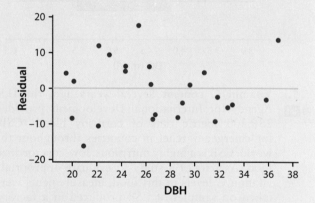

Equation: $\widehat{\text{Height}} = 65.34 + 1.950\,\text{DBH}$, $s = 8.84$, $r^2 = 56.2\%$

(a) Is a linear model appropriate for these data? Explain.

(b) Interpret the slope of the regression line.

(c) One of the trees in the data set had a DBH of 23 inches and a height of 120. Find the residual for this observation.

(d) Would you be willing to use the regression line to predict the height of a tree whose DBH is 15 inches? Explain.

18. **Smoking in the U.K.** (4.5) British government statistics classify adult men by occupation as "managerial and professional" (43% of the population), "intermediate" (34%), or "routine and manual" (23%). A survey finds that 20% of men in managerial and professional occupations smoke, 29% of the intermediate group smoke, and 38% in routine and manual occupations smoke.[45]

(a) Use a tree diagram to find the percent of all adult British men who smoke.

(b) Find the percent of male smokers who have routine and manual occupations.

Chapter 7

STATS applied!

How long will I stay on hold?

The chapter-opening STATS applied! on page 451 described a bank that was studying the call-response times in its credit-card customer service department. The bank manager wants to know whether or not the bank's customer service representatives generally met the goal of answering incoming calls in less than 30 seconds. We can approach this question in two ways: by estimating the proportion p of all calls that were answered within 30 seconds or by estimating the mean response time μ.

Here are a histogram and summary statistics for the random sample of calls.

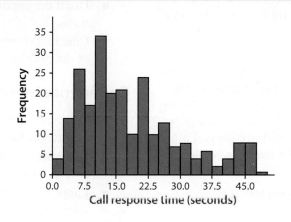

Variable	N	Mean	SE Mean	StDev	Min	Q_1	Med	Q_3	Max
Call response time (sec)	241	18.353	0.758	11.761	1.000	9.000	16.000	25.000	49.000

1. Based on the sample data, a 95% confidence interval for the proportion of calls to the bank's customer service center that are answered in less than 30 seconds is 0.783 to 0.877. Interpret this interval.

2. Calculate the point estimate and margin of error for the interval in the previous question.

3. Explain two ways that the bank could reduce the margin of error for the confidence interval in Question 1, along with any drawbacks to these actions.

4. Construct and interpret a 95% confidence interval for the true mean response time for calls to the bank's customer service center.

5. In this context, what does it mean to be 95% confident?

6. Is the customer service center meeting its goal of answering calls in less than 30 seconds? Explain your reasoning.

Main Points — *Chapter 7*

The Idea of a Confidence Interval

- A **confidence interval** gives an interval of plausible values for a parameter. Sometimes, confidence intervals are used to make decisions about the value of a parameter.

- To interpret a $C\%$ confidence interval for an unknown parameter, say, "We are $C\%$ confident that the interval from _____ to _____ captures the [parameter in context]."

- The **confidence level** C gives the overall success rate of the method for calculating the confidence interval. To interpret the confidence level C, say, "If we were to select many random samples from a population and construct a $C\%$ confidence interval using each sample, about $C\%$ of the intervals would capture the [parameter in context]."

- The structure of a confidence interval is

  point estimate $\pm$ margin of error

  - A **point estimate** is a single-value estimate of a population parameter calculated from sample data.

  - In a $C\%$ confidence interval, the distance between the point estimate and the true parameter value will be less than the **margin of error** in $C\%$ of all samples.

- The margin of error is affected by the confidence level and the sample size. Increasing the confidence level C increases the margin of error. Increasing the sample size decreases the margin of error.

- The margin of error does not account for problems in the data collection process, such as undercoverage, nonresponse, or poorly worded questions. The margin of error only accounts for variability due to random sampling.

- When constructing and interpreting a confidence interval, follow the **four-step process:**
  - **State:** State the parameter you want to estimate and the confidence level.
  - **Plan:** Identify the appropriate inference method and check conditions.
  - **Do:** If the conditions are met, perform calculations.
  - **Conclude:** Interpret your interval in the context of the problem.

Confidence Intervals for a Proportion

- There are two **conditions** for constructing a confidence interval for a population proportion p:
  - **Random:** The data come from a random sample from the population of interest.
  - **Large Counts:** Both $n\hat{p}$ and $n(1 - \hat{p})$ are at least 10.

- When the conditions are met, a **one-sample z interval for p** is

$$\hat{p} \pm z^* \sqrt{\frac{\hat{p}(1 - \hat{p})}{n}}$$

  where $\hat{p}$ is the proportion of successes in the sample, z^* is the critical value, and n is the sample size.

- The **critical value** z^* is a multiplier that makes the interval wide enough to have the stated capture rate. On a standard normal curve, $C\%$ of the area will be between $-z^*$ and z^*. Use Table A or technology to find the value of z^*.

- The **standard error of $\hat{p}$** is an estimate of the standard deviation of the sampling distribution of $\hat{p}$ and estimates how much $\hat{p}$ typically varies from p.

$$SE_{\hat{p}} = \sqrt{\frac{\hat{p}(1 - \hat{p})}{n}}$$

■ To determine the sample size n that will yield a C% confidence interval for a population proportion p with a maximum margin of error ME, solve the following inequality for n:

$$z^*\sqrt{\frac{\hat{p}(1-\hat{p})}{n}} \leq ME$$

where $\hat{p}$ is a guessed value for the sample proportion. The margin of error will always be less than or equal to ME if you use $\hat{p} = 0.5$.

■ You can calculate a one-sample z interval for a proportion using the *One Categorical Variable* applet or the 1-PropZInt command on the TI-83/84.

Confidence Intervals for a Mean

■ There are two **conditions** for constructing a confidence interval for μ:

- **Random:** The data come from a random sample from the population of interest.

- **Normal/Large Sample:** The data come from a normally distributed population or the sample size is large ($n \geq 30$). When the sample size is small and the shape of the population distribution is unknown, a graph of the sample data shows no strong skewness or outliers.

■ When the conditions are met, a **one-sample t interval for** μ is

$$\overline{x} \pm t^*\frac{s_x}{\sqrt{n}}$$

where $\overline{x}$ is the sample mean, t^* is the critical value, s_x is the sample standard deviation, and n is the sample size.

■ The **critical value t^*** is a multiplier that makes the interval wide enough to have the stated capture rate. On the graph of a t distribution with $n - 1$ degrees of freedom, C% of the area will be between $-t^*$ and t^*. Use Table B to find the value of t^*.

■ When calculating a confidence interval for a population mean, we use a t^* critical value rather than a z^* critical value whenever we use s_x to estimate σ.

■ The **standard error of $\overline{x}$** is an estimate of the standard deviation of the sampling distribution of $\overline{x}$ and estimates how much $\overline{x}$ typically varies from μ.

$$SE_{\overline{x}} = \frac{s_x}{\sqrt{n}}$$

■ You can calculate a one-sample t interval for a mean using the *One Quantitative Variable* applet or the TInterval command on the TI-83/84.

Chapter 7 Review Exercises

1. Paying college athletes (7.1, 7.2) A recent ABC News/ *Washington Post* poll asked a random sample of U.S. residents, "Beyond any scholarships they receive, do you support or oppose paying salaries to college athletes?" The 95% confidence interval for the proportion of people who support salaries is 0.295 to 0.365.[46]

(a) Interpret the confidence interval.

(b) Interpret the confidence level.

(c) Calculate the point estimate and margin of error used to create this confidence interval.

(d) Name two ways that the pollsters could reduce the margin of error. What are the drawbacks of these actions?

2. Critical values (7.3, 7.5) Calculate the relevant critical value for each of the following settings, assuming the conditions are met:

(a) You want to calculate a 94% confidence interval for a population proportion based on a sample of size 1200.

(b) You want to calculate a 99% confidence interval for a population mean based on a sample of size 8.

(c) You want to calculate a 90% confidence interval for a population mean using a sample of size 95.

3. Teenage smoking (7.1, 7.3, 7.4) The most recent Youth Risk Behavior Survey found that the percentage of high school students who had smoked a cigarette at least once in the last 30 days was 15.7%. The principal at a large high school wanted to estimate this proportion at her school, so she asked a random sample of 180 students if they had smoked a cigarette in the last 30 days. A total of 21 students in this sample said "Yes."[47]

(a) Verify that the conditions have been met for calculating a confidence interval for p, the proportion of students at this high school who have smoked a cigarette in the last 30 days.

(b) Calculate a 95% confidence interval for p.

(c) Does the interval in part (b) provide convincing evidence that the percentage of students at this school who have smoked in the last 30 days is less than 15.7%? Explain.

4. Still smoking (7.2) The principal in Exercise 3 conducted the survey by randomly selecting students entering the cafeteria for lunch and asking them to fill out a questionnaire and return it to her. Describe one potential source of bias in the study that is not accounted for by the margin of error.

5. Playing games (7.4) You want to estimate, with 99% confidence, the proportion of students at a large high school who play video games at least once a day. How large a sample should you take to ensure that the margin of error is at most 0.05?

6. Smart kids (7.5, 7.6) A school counselor wants to estimate μ = the true mean IQ score for all students at her school. She gets funding from the principal to give an IQ test to an SRS of 60 from among the more than 1000 students. The mean IQ score is 114.98 and the standard deviation is 14.80.[48]

(a) Verify that the conditions for constructing a confidence interval for μ are met.

(b) Construct a 90% confidence interval for the true mean IQ score of all students at the school.

(c) Interpret the interval from part (b).

7. Give it some gas! (7.5, 7.6) A consumer advocacy group randomly selected 20 owners of a particular model of pickup truck. Each owner was asked to report the number of miles per gallon (mpg) for his or her most recent tank of gas. Here are the mpg values for these 20 owners:

| 15.8 | 13.6 | 15.6 | 19.1 | 22.4 | 15.6 | 22.5 | 17.2 | 19.4 | 22.6 |
| 19.4 | 18.0 | 14.6 | 18.7 | 21.0 | 14.8 | 22.6 | 21.5 | 14.3 | 20.9 |

Construct and interpret a 95% confidence interval for μ = the true mean fuel efficiency for this model of pickup truck.

Chapter 7 Practice Test

Section I: Multiple Choice *Select the best answer for each question.*

1. In a recent Gallup poll of randomly selected U.S. adults, 75% said they would vote for a law that imposed term limits on members of the U.S. Congress.[49] The poll's margin of error was 4 percentage points at the 95% confidence level. This means that

(a) the poll used a method that gets an answer within 4 percentage points of the truth about the population 95% of the time.

(b) if Gallup takes another poll on this issue, the results of the second poll will lie between 71% and 79%.

(c) there is convincing evidence that more than 79% of all U.S. adults would vote for term limits.

(d) Gallup can be 95% confident that between 71% and 79% of the adults in the sample would vote for term limits.

2. A political poll was conducted by calling land-line telephones. The researchers conducting the poll are concerned about the possibility of undercoverage because some people do not own a phone or own only a cell phone. Which of the following is the best way for them to correct for this source of bias?

(a) Use a lower confidence level, such as 90%.

(b) Use a higher confidence level, such as 99%.

(c) Select a larger sample.

(d) Throw this sample out and start over again with a better sampling method.

3. A 95% confidence interval for p, the proportion of all shoppers at a large grocery store who purchase cookies, is 0.236 to 0.282. The point estimate and margin of error for this interval are:

(a) Point estimate = 0.236; Margin of error = 0.282.

(b) Point estimate = 0.236; Margin of error = 0.046.

(c) Point estimate = 0.259; Margin of error = 0.046.

(d) Point estimate = 0.259; Margin of error = 0.023.

Questions 4 and 5 refer to the following setting. A quality control manager at a manufacturing plant wants to estimate the mean length of metal rods produced by a certain machine.

4. The manager is deciding between a 95% confidence level and a 99% confidence level. Compared to a 95% confidence interval, a 99% confidence interval will be

(a) narrower and would involve a larger risk of being incorrect.

(b) wider and would involve a smaller risk of being incorrect.

(c) narrower and would involve a smaller risk of being incorrect.

(d) wider and would involve a larger risk of being incorrect.

5. The researcher is deciding between a sample of size $n = 500$ and a sample of size $n = 1000$. Compared to using a sample size of $n = 500$, a 95% confidence interval based on a sample size of $n = 1000$ will be

(a) narrower and would involve a larger risk of being incorrect.

(b) narrower and would involve a smaller risk of being incorrect.

(c) narrower and would involve the same risk of being incorrect.

(d) wider and would involve a smaller risk of being incorrect.

6. Isabel selects a simple random sample of 28 seniors at her school and finds that 20 of them are planning to participate in the school's annual capture-the-flag game. She wants to construct a confidence interval for p = the proportion of all seniors who plan to participate in the game, but she realizes she hasn't met all the conditions for constructing the interval. Which condition for this procedure has she failed to meet?

(a) $n \geq 30$

(b) $n\hat{p} \geq 10$

(c) $n(1 - \hat{p}) \geq 10$

(d) The population must be approximately normal.

7. Most people can roll their tongues, but some can't. Suppose we are interested in determining what proportion of people in a certain population can roll their tongues. We test a random sample of 80 people from this population and find that 64 can roll their tongues. The margin of error for a 95% confidence interval for the true proportion of tongue rollers in this population is closest to

(a) 0.004.

(b) 0.045.

(c) 0.088.

(d) 0.176.

8. Many television viewers express doubts about the validity of certain commercials. In an attempt to answer their critics, Timex Group USA wishes to estimate the proportion of consumers who believe what is shown in Timex television commercials. Let p represent the true proportion of consumers who believe what is shown in Timex television commercials. Which of the following is the smallest number of consumers that Timex can survey to guarantee a margin of error of 0.05 or less at the 99% confidence level?

(a) 600

(b) 650

(c) 700

(d) 750

9. In checking conditions for constructing confidence intervals for a population mean, it's important to plot the distribution of sample data. Each dotplot shows the distribution of a sample from a different population. For which of the three samples would it be safe to construct a one-sample t interval for the population mean?

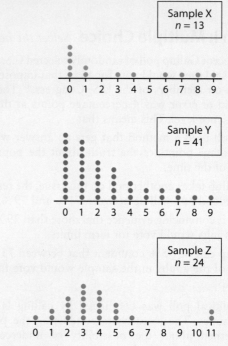

(a) Sample X only

(b) Sample Y only

(c) Samples Y and Z

(d) None of the samples

10. You want to calculate a 98% confidence interval for a population mean from a sample of size $n = 18$. What is the appropriate critical value t^*?

(a) 2.110

(b) 2.224

(c) 2.552

(d) 2.567

Section II: Free Response

11. Pauly's Pizza claims that it takes 30 minutes, on average, to deliver a pizza to dorms at Nat's college. After a long wait one night, Nat decides to test this claim. He randomly selects 15 dormitory residents and asks them to record the time it takes for Pauly's to deliver the next pizza they order. Here are the times (in minutes).

31	38	39	25	26
45	42	32	23	38
42	21	40	37	28

The sample mean is $\bar{x} = 33.8$ minutes and the sample standard deviation is $s_x = 7.72$ minutes.

(a) Construct and interpret a 90% confidence interval for the true mean delivery time.

(b) Does the confidence interval from part (a) provide convincing evidence that the true mean time to deliver a pizza to the college dorms is greater than 30 minutes? Explain.

12. In a recent poll conducted by Marist College, 1104 randomly selected Americans were asked, "This year, are you very likely, somewhat likely, or not likely at all to make a New Year's resolution?" "Not likely at all" was the response of 618 people in the sample.

(a) Construct and interpret a 95% confidence interval for the proportion of all Americans who are not likely at all to make a New Year's resolution.

(b) Explain what is meant by "95% confident" in this context.

8

Testing a Claim

Lesson 8.1	The Idea of a Significance Test	500
Lesson 8.2	Significance Tests and Decision Making	508
Lesson 8.3	Testing a Claim about a Proportion	515
Lesson 8.4	Significance Tests for a Proportion	521
Lesson 8.5	Testing a Claim about a Mean	529
Lesson 8.6	Significance Tests for a Mean	536
	Chapter 8 Main Points	545
	Chapter 8 Review Exercises	547
	Chapter 8 Practice Test	548

STATS applied!

What is normal body temperature?

Several years ago, researchers conducted a study to determine whether the "accepted" value for normal body temperature, 98.6°F, is accurate. They used an oral thermometer to measure the temperatures of a random sample of 130 healthy men and women aged 18 to 40. The dotplot shows the temperature readings.[1] We have added a vertical line at 98.6°F for reference.

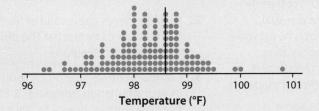

What does the graph reveal about the body temperatures of the 130 people in the sample? Do these data provide convincing evidence that "normal" body temperature in the population of healthy 18- to 40-year-olds is not 98.6°F?

We'll revisit STATS applied! at the end of the chapter, so you can use what you have learned to help answer these questions.

Lesson 8.1

The Idea of a Significance Test

Confidence intervals are one of the two most common types of statistical inference. Use a confidence interval when your goal is to estimate a population parameter. The second common type of inference, called a **significance test,** has a different goal: to test a claim about a parameter.

> **DEFINITION** Significance test
>
> A **significance test** is a formal procedure for using observed data to decide between two competing claims (called *hypotheses*). The claims are statements about a parameter, like the population proportion p or the population mean μ.

Here is an activity that illustrates the reasoning of a significance test.

ACTIVITY

I'm a great free-throw shooter!

In this activity, you and your classmates will perform a simulation to test a claim about a population proportion.

A basketball player claims to make 80% of the free throws that he attempts. That is, he claims $p = 0.80$, where p is the true proportion of free throws he will make in the long run. We suspect that he is exaggerating.

Suppose the player shoots 50 free throws and makes 32 of them. His sample proportion of made shots is $\hat{p} = \dfrac{32}{50} = 0.64$. This result gives *some* evidence that the player makes less than 80% of his free throws in the long run. But do we have *convincing* evidence that $p < 0.80$? Or is it plausible that an 80% shooter can have a performance this poor by chance alone? You can use a simulation to find out.

1. Using the pie chart provided by your teacher, label the 80% region "made shot" and the 20% region "missed shot." Straighten out one of the ends of a paper clip so that there is a loop on one side and a pointer on the other. On a flat surface, place a pencil through the loop and put the tip

of the pencil on the center of the pie chart. Then flick the paper clip and see where the pointed end lands: made shot or missed shot.

2. Flick the paper clip a total of 50 times, and count the number of times that the pointed end lands in the "made shot" region.

3. Compute the sample proportion $\hat{p}$ of made shots in your simulation from Step 2. Plot this value on the class dotplot drawn by your teacher.

4. Repeat Steps 2 and 3 as needed to get at least 40 repetitions of the simulation for your class.

5. Based on the simulation results for the whole class, how likely is it for an 80% shooter to make 64% or less when he shoots 50 free throws?

6. Based on your answer to Question 5, does the observed $\hat{p} = 0.64$ result give convincing evidence that the player is exaggerating? Or is it plausible that an 80% shooter can have a performance this poor by chance alone?

Our reasoning in the activity is based on asking what would happen if the player's claim ($p = 0.80$) were true and we observed many sets of 50 free throws. We used software to simulate 400 sets of 50 shots assuming that the player is really an 80% shooter. Figure 8.1 shows a dotplot of the results. Each dot on the graph represents the sample proportion $\hat{p}$ of made shots in one set of 50 attempts.

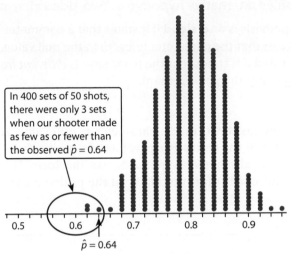

In 400 sets of 50 shots, there were only 3 sets when our shooter made as few as or fewer than the observed $\hat{p} = 0.64$

$\hat{p} = 0.64$

Simulated proportion $\hat{p}$ of made shots

FIGURE 8.1 Dotplot of the simulated sampling distribution of $\hat{p}$, the proportion of free throws made by an 80% shooter in a sample of 50 shots.

The simulation shows that it would be very unlikely for an 80% free-throw shooter to make 32 or fewer out of 50 free throws ($\hat{p} = 0.64$) just by chance. This gives us convincing evidence that the player is less than an 80% shooter.

Be sure that you understand why this evidence is convincing. There are two possible explanations of the fact that the player made only $\hat{p} = 32/50 = 0.64$ of his free throws:

1. The player's claim is true ($p = 0.8$), and purely by chance, a very unlikely outcome occurred.

2. The player's claim is false ($p < 0.8$). That is, the population proportion is less than 0.8, so the sample result is not an unlikely outcome.

Explanation 1 might be correct—the result of our random sample of 50 shots could be due to chance alone. But the probability that such a result would occur by chance is so small (less than 1 in 100) that we are quite confident that Explanation 2 is right.

Stating Hypotheses

A significance test starts with a careful statement of the claims we want to compare. In the activity about the free-throw shooter, the player claims that his long-run proportion of made free throws is $p = 0.80$. This is the claim we seek evidence *against*. We call it the **null hypothesis,** abbreviated H_0.

The claim we hope or suspect to be true instead of the null hypothesis is called the **alternative hypothesis.** We abbreviate the alternative hypothesis as H_a. In this case, we believe the player might be exaggerating, so our alternative hypothesis is H_a: $p < 0.80$.

DEFINITION Null hypothesis H_0, Alternative hypothesis H_a

The claim about the population that we weigh evidence against in a statistical test is called the **null hypothesis (H_0).** The claim about the population that we are trying to find evidence *for* is the **alternative hypothesis (H_a).**

In the free-throw shooter activity, the alternative hypothesis is **one-sided** ($p < 0.80$) because we suspect the player makes less than 80% of his free throws. If you suspect that the true value of a parameter could be either greater than or less than the null value, use a **two-sided** alternative hypothesis.

> |||| **DEFINITION** One-sided alternative hypothesis, Two-sided alternative hypothesis
>
> The alternative hypothesis is **one-sided** if it states that a parameter is *greater than* the null value or if it states that the parameter is *less than* the null value. The alternative hypothesis is **two-sided** if it states that the parameter is *different from* the null value (it could be either greater than or less than).

The null hypothesis has the form H_0: parameter = null value. A one-sided alternative hypothesis has one of the forms H_a: parameter < null value or H_a: parameter > null value. A two-sided alternative hypothesis has the form H_a: parameter ≠ null value. To determine the correct form of H_a, read the problem carefully.

EXAMPLE

Where does the time go?

Stating hypotheses

PROBLEM: For each of the following settings, state appropriate hypotheses for performing a significance test. Be sure to define the parameter of interest.

(a) As part of its 2010 census marketing campaign, the U.S. Census Bureau advertised "10 questions, 10 minutes—that's all it takes." On the census form itself, we read, "The U.S. Census Bureau estimates that, for the average household, this form will take about 10 minutes to complete, including the time for reviewing the instructions and answers." We suspect that the time it takes to complete the form may be longer than advertised.

(b) According to the website sleep-deprivation.com, 85% of teens are getting too little sleep on school nights. Jamila wonders whether this result holds in her large high school. She asks an SRS of 100 students at the school how much sleep they get on a typical school night. In all, 75 of the students are getting less than the recommended amount of sleep.

SOLUTION:

(a) $H_0: \mu = 10$
    $H_a: \mu > 10$

where μ = the true mean amount of time (in minutes) that it takes for U.S. households to complete the census form.

> H_a is one-sided because we suspect that the form takes *longer* to complete than advertised.

(b) $H_0: p = 0.85$
    $H_a: p \neq 0.85$

where p = the proportion of all students at Jamila's high school who get too little sleep on school nights.

> Jamila wonders if the true proportion of students in her high school *differs from* the claimed proportion of $p = 0.85$ on the sleep-deprivation website, so H_a is two-sided.

FOR PRACTICE TRY EXERCISE 1.

CAUTION

The hypotheses should express the hope or suspicion we have *before* we see the data. It is cheating to look at the data first and then frame the alternative hypothesis to fit what the data show. For example, the data for Jamila's sleep study showed that $\hat{p} = 75/100 = 0.75$ for a random sample of 100 students from her high school. She should not change the alternative hypothesis to $H_a: p < 0.85$ after looking at the data.

Hypotheses always refer to a population, not to a sample. Be sure to state H_0 and H_a in terms of population parameters. It is *never* correct to write a hypothesis about a sample statistic, such as $H_0: \hat{p} = 0.85$ or $H_a: \bar{x} > 10$.

Interpreting *P*-Values

In the activity about the free-throw shooter at the beginning of the lesson, a player who claimed to be an 80% free-throw shooter made only $\hat{p} = 32/50 = 0.64$ in a random sample of 50 free throws. This is evidence *against* the null hypothesis that $p = 0.8$ and *in favor of* the alternative hypothesis $p < 0.8$.

But is the evidence convincing? To answer this question, we have to know how likely it is for an 80% shooter to make 64% or less of his free throws by chance alone in a random sample of 50 attempts. This probability is called a **P-value.**

> ### DEFINITION   *P*-value
>
> The **P-value** of a test is the probability of getting evidence for the alternative hypothesis H_a as strong as or stronger than the observed evidence when the null hypothesis H_0 is true.

We used simulation to estimate the *P*-value for our free-throw shooter: $3/400 = 0.0075$. How do we interpret this *P*-value? Assuming that the player makes 80% of his free throws in the long run, there is about a 0.0075 probability of getting a sample proportion of 0.64 or less just by chance in a set of 50 shots.

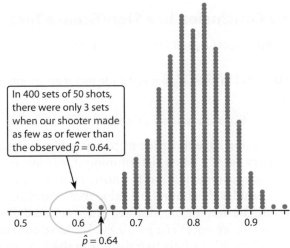

In 400 sets of 50 shots, there were only 3 sets when our shooter made as few as or fewer than the observed $\hat{p} = 0.64$.

$\hat{p} = 0.64$

Simulated proportion $\hat{p}$ of made shots

EXAMPLE

Do teens get enough calcium?

Interpreting a P-value

PROBLEM: Calcium is a vital nutrient for healthy bones and teeth. The National Institutes of Health (NIH) recommend a calcium intake of 1300 milligrams (mg) per day for teenagers. The NIH is concerned that teenagers aren't getting enough calcium, on average. Is this true? Researchers decide to perform a test of

$$H_0: \mu = 1300$$
$$H_a: \mu < 1300$$

where μ is the true mean daily calcium intake in the population of teenagers. They ask a random sample of 20 teens to record their food and drink consumption for 1 day. The researchers then compute the calcium intake for each student. Data analysis reveals that $\bar{x} = 1198$ mg and $s_x = 411$ mg. Researchers performed a significance test and obtained a P-value of 0.1405. Interpret the P-value in context.

SOLUTION:

Assuming that the mean daily calcium intake in the teen population is 1300 mg, there is a 0.1405 probability of getting a sample mean of 1198 mg or less by chance alone in a random sample of 20 teens.

FOR PRACTICE   TRY EXERCISE 5.

Making Conclusions

The final step in performing a significance test is to draw a conclusion about the competing claims being tested. We make a decision based on the strength of the evidence against the null hypothesis (and in favor of the alternative hypothesis) as measured by the P-value. Small P-values give convincing evidence for H_a because they say that the observed result is unlikely to occur when H_0 is true. Large P-values fail to give convincing evidence for H_a because they say that the observed result is likely to occur by chance alone when H_0 is true.

> ### How to Make a Conclusion in a Significance Test
>
> • If the P-value is small, reject H_0 and conclude that there is convincing evidence for H_a (in context).
>
> • If the P-value is large, fail to reject H_0 and conclude that there is not convincing evidence for H_a (in context).

How small does a P-value have to be for us to reject H_0? In Chapter 3, we recommended using a boundary of 5% when determining if a result is *statistically significant*. Keep following this recommendation for now. That is, view a P-value less than 0.05 as "small." We'll consider this boundary value more carefully in Lesson 8.2.

In the activity about the free-throw shooter, the estimated P-value was 0.0075. Because the P-value is small, we reject H_0: $p = 0.80$. We have convincing evidence that the player makes fewer than 80% of his free throws in the long run.

EXAMPLE

Can we tell if teens get enough calcium?

Making conclusions

PROBLEM: Refer to the previous example. What conclusion would you make?

SOLUTION:

The P-value of 0.1405 is large, so we fail to reject H_0. We don't have convincing evidence that teens are getting fewer than 1300 mg of calcium per day, on average.

FOR PRACTICE   TRY EXERCISE 9.

Be careful how you write conclusions when the *P*-value is large. Don't conclude that the null hypothesis is true just because we didn't find convincing evidence for the alternative hypothesis. For example, it would be incorrect to conclude that teens *are* getting 1300 mg of calcium per day, on average. We found *some* evidence that the teens weren't getting enough calcium, but the evidence wasn't convincing enough to reject H_0. Never "accept H_0" or conclude that H_0 is true!

CAUTION

LESSON APP 8.1

Do people kiss the "right" way?

According to an article in the *San Gabriel Valley Tribune,* "Most people are kissing the 'right way.'" That is, according to a study, the majority of couples prefer to tilt their heads to the right when kissing. In the study, a researcher observed a random sample of 124 kissing couples and found that 83/124 ($\hat{p} = 0.669$) of the couples tilted to the right.[2] Do these data provide convincing evidence that more than 50% of kissing couples prefer to tilt their heads to the right?

1. State appropriate hypotheses for performing a significance test. Be sure to define the parameter of interest.

2. The *P*-value for the test in Question 1 is 0.0001. Interpret the *P*-value in context.

3. What conclusion would you make?

Lesson 8.1

WHAT DID YOU LEARN?

LEARNING TARGET	EXAMPLES	EXERCISES
State appropriate hypotheses for a significance test about a population parameter.	p. 502	1–4
Interpret a *P*-value in context.	p. 503	5–8
Make an appropriate conclusion for a significance test based on a *P*-value.	p. 504	9–12

Exercises   Lesson 8.1

Mastering Concepts and Skills

In Exercises 1–4, state appropriate hypotheses for performing a significance test. Be sure to define the parameter of interest.

1. **Yummy pineapples** At the Hawaii Pineapple Company, managers are interested in the size of the pineapples
pg 502
grown in the company's fields. Last year, the mean weight of the pineapples harvested from one large field was 31 ounces. A new irrigation system was installed in this field after the growing season. Managers wonder if the mean weight of pineapples grown in the field this year will be greater than last year. So they take an SRS of 50 pineapples from this year's crop.

2. **Attitude toward school** The Survey of Study Habits and Attitudes (SSHA) is a psychological test that measures students' attitudes toward school and study habits. Scores range from 0 to 200. The mean score for U.S. college students is about 115. A teacher suspects that older students have better attitudes toward school. She gives the SSHA to an SRS of 45 of the more than 1000 students at her college who are at least 30 years of age.

3. **We never argue!** A Gallup Poll report revealed that 72% of teens said they seldom or never argue with their friends.[3] Yvonne wonders whether this result holds true in her large high school. She surveys a random sample of 150 students at her school.

4. **Southpaw reports** Simon reads a newspaper report claiming that 12% of all adults in the United States are left-handed. He wonders if this figure holds true at the large community college he attends. Simon chooses an SRS of 100 students and records whether each student is right- or left-handed.

5. **Juicier pineapples?** In the study of pineapple weights from Exercise 1, the sample mean weight from this year's crop was 31.4 ounces and the sample standard deviation was 2.5 ounces. A significance test yields a *P*-value of 0.1317. Interpret the *P*-value in context.

pg 503

6. **Better attitude** In the study of older students' attitudes from Exercise 2, the sample mean SSHA score was 125.7 and the sample standard deviation was 29.8. A significance test yields a *P*-value of 0.0101. Interpret the *P*-value in context.

7. **Sturdy boards** Lumber companies dry freshly cut wood in kilns before selling it. A certain percentage of the boards become "checked," which means that cracks develop at the ends of the boards during drying. The current drying procedure is known to produce cracks in 16% of the boards. The drying supervisor at a lumber company wants to test a new method to determine if fewer boards crack. She uses the new method on a random sample of 200 boards and finds that the sample proportion of checked boards is 0.11. The supervisor performs a test of $H_0: p = 0.16$ versus $H_a: p < 0.16$, where p is the true proportion of all boards that would crack when dried with the new method. A significance test yields a *P*-value of 0.027. Interpret this *P*-value in context.

8. **What, no homework?!** Mr. Starnes believes that less than 75% of the students at his school did their math homework last night. The math teachers inspect the homework assignments from a random sample of 50 students at the school. Only 68% of the students did their math homework. Mr. Starnes performs a test of $H_0: p = 0.75$ versus $H_a: p < 0.75$,

where p is the true proportion of all students at his school who did their math homework last night. A significance test yields a *P*-value of 0.1265. Interpret this *P*-value in context.

9. **The rest of the pineapples** What conclusion would you make in Exercise 5?

pg 504

10. **Conclusive attitudes** What conclusion would you make in Exercise 6?

11. **Sturdy conclusions** What conclusion would you make in Exercise 7?

12. **No more homework?!** What conclusion would you make in Exercise 8?

Applying the Concepts

13. **An honest loaf** The mean weight of loaves of bread produced at the bakery where you work is supposed to be 1 pound. You are the supervisor of quality control at the bakery, and you are concerned that employees are making loaves that are too light. Suppose you weigh an SRS of 50 bread loaves and find that the mean weight is 0.975 pound.

(a) State appropriate hypotheses for performing a significance test. Be sure to define the parameter of interest.

(b) The *P*-value for the test in part (a) is 0.0806. Interpret the *P*-value in context.

(c) What conclusion would you make?

14. **Philly fanatics?** Nationally, the proportion of red cars on the road is 0.12. A statistically minded fan of the Philadelphia Phillies (whose team color is red) wonders if Phillies fans are more likely to drive red cars. One day during a home game, he takes a random sample of 210 cars parked at Citizens Bank Park (the Phillies home field), and counts 35 red cars.

(a) State appropriate hypotheses for performing a significance test. Be sure to define the parameter of interest.

(b) The *P*-value for the test in (a) is 0.0187. Interpret the *P*-value in context.

(c) What conclusion would you make?

15. **A flawed conclusion** A student performs a test of $H_0: p = 0.5$ versus $H_a: p \neq 0.5$ and gets a *P*-value of 0.63. The student writes: "Because the *P*-value is large, we accept H_0. The data provide convincing evidence that the null hypothesis is true." Explain what is wrong with this conclusion.

16. **An incorrect conclusion** A student performs a test of $H_0: p = 0.75$ versus $H_a: p > 0.75$ and gets a *P*-value of 0.01. The student writes: "Because the *P*-value is small, we reject H_0. The data prove that H_a is true." Explain what is wrong with this conclusion.

Extending the Concepts

17. Who wrote this poem? Statistics can help decide the authorship of literary works. Sonnets by William Shakespeare contain an average of $\mu = 6.9$ new words (words not used in the poet's other works) and a standard deviation of $\sigma = 2.7$ new words. The number of new words is approximately normally distributed. Scholars expect sonnets by other authors to contain more new words. A scholarly sleuth discovers a manuscript with many new sonnets, and a debate erupts over whether the manuscript is Shakespeare's work. Some scholars take a random sample of 5 sonnets from the manuscript and count the number of new words in each. The mean number of new words in these 5 sonnets is $\bar{x} = 9.2$.

Here is a dotplot that shows the results of simulating 200 random samples of size 5 from a normal distribution with a mean of 6.9 and a standard deviation of 2.7, and calculating the mean for each sample.

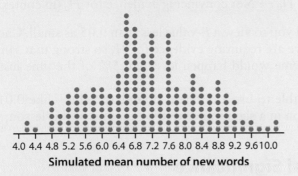

Simulated mean number of new words

(a) State appropriate hypotheses for performing a significance test. Be sure to define the parameter of interest.

(b) Use the simulation results to estimate the P-value of the test in part (a). Interpret the P-value in context.

(c) What conclusion would you make?

18. A better golf club? Mike is an avid golfer who would like to improve his game. A friend suggests getting new clubs and lets Mike try out his 7-iron. Based on years of experience, Mike has established that the mean distance balls travel when hit with his old 7-iron is $\mu = 175$ yards with a standard deviation of $\sigma = 15$ yards. He is hoping that this new club will make his shots with a 7-iron more consistent (less variable), so he goes to the driving range and hits 50 shots with the new 7-iron. The standard deviation of these 50 shots is $s_x = 13.9$ yards.

(a) State appropriate hypotheses for performing a significance test. Be sure to define the parameter of interest.

(b) The P-value for the test in (a) is 0.25. Interpret the P-value in context.

(c) What conclusion would you make?

Recycle and Review

19. Tornadoes in every state (1.5, 1.6) In the years between 1950 and 2011, every U.S. state had at least a few tornadoes, from 3 in Alaska to 7990 in Texas. The histogram shows the distribution for the number of tornadoes reported in each state over that time period.[4]

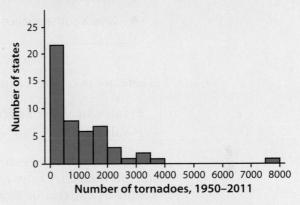

(a) Describe the shape of this distribution. Identify any obvious outliers.

(b) Which measures of center and variability should be used to summarize this distribution? Explain.

20. Animal testing (2.1) The General Social Survey asked a random sample of 1152 adults to react to the following statement. "It is right to use animals for medical testing if it might save human lives." Here is the two-way table of their responses:

		Gender		
		Male	Female	**Total**
Response	Strongly agree	76	59	135
	Agree	270	247	517
	Neither agree nor disagree	87	139	226
	Disagree	61	123	184
	Strongly disagree	22	68	90
	Total	516	636	1152

(a) Make a segmented bar chart to show the relationship between gender and attitude toward using animals in medical testing.

(b) Based on the graph in part (a), is there an association between these variables? Justify your answer.

Lesson 8.2

Significance Tests and Decision Making

There are two types of conclusions you can make in a significance test:

P-value small $\rightarrow$ Reject $H_0 \rightarrow$ Convincing evidence for H_a (in context)
P-value large $\rightarrow$ Fail to reject $H_0 \rightarrow$ Not convincing evidence for H_a (in context)

In Lesson 8.1, we encouraged you to view a P-value less than 0.05 as small. Choosing this boundary value means we are requiring evidence for H_a so strong that sample results as extreme or more extreme would happen less than 5% of the time just by chance when H_0 is true.

Sometimes, it may be preferable to use a different boundary value—like 0.01 or 0.10—when drawing a conclusion in a significance test. By the end of this lesson, you should understand why.

Determining Statistical Significance

To determine if a P-value should be considered small, we compare it to a boundary value called the **significance level.** We denote it by α, the Greek letter alpha.

> **DEFINITION** Significance level
>
> The **significance level** α is the value that we use as a cutoff to decide if an observed result is unlikely to happen by chance alone when the null hypothesis is true.

When our P-value is less than the chosen significance level α in a significance test, we say that the result is "statistically significant at the $\alpha =$ _____ level." In that case, we reject the null hypothesis H_0 and conclude that there is convincing evidence in favor of the alternative hypothesis H_a.

How to Make a Conclusion in a Significance Test

If P-value $< \alpha$:

Reject H_0 and conclude that there is convincing evidence for H_a (in context).

If P-value $\geq \alpha$:

Fail to reject H_0 and conclude that there is not convincing evidence for H_a (in context).

EXAMPLE

Do the company's new batteries last longer?

Determining statistical significance

PROBLEM: A company has developed a new deluxe AAA battery that is supposed to last longer than its regular AAA battery.[5] However, these new batteries are more expensive to produce, so the company would like to be convinced that they really do last longer. Based on years of experience, the company knows that its regular AAA batteries last for 30 hours of continuous use, on average. The company selects an SRS of 15 new batteries and uses them continuously until they are completely drained. The sample mean lifetime is $\bar{x} = 33.9$ hours. A significance test is performed using the hypotheses

$$H_0: \mu = 30$$
$$H_a: \mu > 30$$

where μ is the true mean lifetime (in hours) of the new deluxe AAA batteries. The resulting P-value is 0.0729.

What conclusion would you make for each of the following significance levels?

(a) $\alpha = 0.10$ (b) $\alpha = 0.05$

SOLUTION:

(a) Because the P-value of $0.0729 < \alpha = 0.10$, we reject H_0. We have convincing evidence that the company's deluxe AAA batteries last longer than 30 hours, on average.

> P-value $< \alpha \rightarrow$ Reject H_0
> $\rightarrow$ Convincing evidence for H_a (in context)

(b) Because the P-value of $0.0729 \geq \alpha = 0.05$, we fail to reject H_0. We do not have convincing evidence that the company's deluxe AAA batteries last longer than 30 hours, on average.

> P-value $\geq \alpha \rightarrow$ Fail to reject H_0
> $\rightarrow$ Not convincing evidence for H_a (in context)

FOR PRACTICE   TRY EXERCISE 1.

When a researcher plans to draw a conclusion based on a significance level, α should be stated *before* the data are produced. Otherwise, a deceptive user of statistics might choose α *after* the data have been analyzed in an attempt to manipulate the conclusion. This is just as inappropriate as choosing an alternative hypothesis after looking at the data.

CAUTION

Type I and Type II Errors

When we draw a conclusion from a significance test, we hope our conclusion will be correct. But sometimes it will be wrong. There are two types of errors we can make: a **Type I error** or a **Type II error.**

DEFINITION Type I error, Type II error

A **Type I error** occurs if a test rejects H_0 when H_0 is true. That is, the test finds convincing evidence that H_a is true when it really isn't.

A **Type II error** occurs if a test fails to reject H_0 when H_a is true. That is, the test does not find convincing evidence that H_a is true when it really is.

The possible outcomes of a significance test are summarized in Figure 8.2.

FIGURE 8.2 The two types of errors in significance tests.

Truth about the population

		H_0 true	H_a true
Conclusion based on sample	Reject H_0	Type I error	Correct conclusion
	Fail to reject H_0	Correct conclusion	Type II error

- If H_0 is true:
  - Our conclusion is correct if we don't find convincing evidence that H_a is true.
  - We make a Type I error if we find convincing evidence that H_a is true.

- If H_a is true:
  - Our conclusion is correct if we find convincing evidence that H_a is true.
  - We make a Type II error if we do not find convincing evidence that H_a is true.

Only one error is possible at a time, depending on the conclusion we make.

It is important to be able to describe Type I and Type II errors in the context of a problem. In the better batteries example, the company performed a test of

$$H_0: \mu = 30$$
$$H_a: \mu > 30$$

where μ is the true mean lifetime (in hours) of its new deluxe AAA batteries.

- A Type I error occurs if the company finds convincing evidence that its new AAA batteries last longer than 30 hours, on average, when their true mean lifetime really is 30 hours.

- A Type II error occurs if the company doesn't find convincing evidence that the new batteries last longer than 30 hours, on average, when their true mean lifetime really is greater than 30 hours.

EXAMPLE

Can we make fast food faster?

Type I and Type II errors

PROBLEM: The manager of a fast-food restaurant wants to reduce the proportion of drive-thru customers who have to wait longer than 2 minutes to receive their food after placing an order. Based on store records, the proportion of customers who had to wait longer than 2 minutes was $p = 0.63$. To reduce this proportion, the manager assigns an additional employee to assist with drive-thru orders. During the next month, the manager will collect a random sample of 250 drive-thru times and test the following hypotheses at the $\alpha = 0.10$ significance level:

$$H_0: p = 0.63$$
$$H_a: p < 0.63$$

where p = the true proportion of drive-thru customers who have to wait longer than 2 minutes to receive their food. Describe a Type I error and a Type II error in this setting.

SOLUTION:

Type I error: The manager finds convincing evidence that the true proportion of drive-thru customers who have to wait longer than 2 minutes has decreased from 0.63, when it is really still 0.63.

> A Type I error occurs if a test finds convincing evidence that H_a is true when it really isn't.

Type II error: The manager doesn't find convincing evidence that the true proportion of drive-thru customers who have to wait longer than 2 minutes has decreased from 0.63, when it really is less than 0.63.

> A Type II error occurs if a test does not find convincing evidence that H_a is true when it really is.

FOR PRACTICE   TRY EXERCISE 5.

Which is more serious: a Type I error or a Type II error? That depends on the situation. For the restaurant manager, a Type I error would result in money spent unnecessarily on another employee. A Type II error could lead to upset customers, which might decrease sales. It is important to consider the possible consequences of the two types of error when performing a significance test.

EXAMPLE

Were the new batteries really better?

Consequences of Type I and Type II errors

PROBLEM: Refer to the discussion about Type I and Type II errors for the battery company on page 509. Give a consequence of each type of error in this setting.

SOLUTION:

Type I: The company spends more money to produce the new batteries when they aren't any better than the older, cheaper type.

Type II: The company would not produce the new batteries, even though they last longer.

FOR PRACTICE   TRY EXERCISE 9.

The most common significance levels are $\alpha = 0.05$, $\alpha = 0.01$, and $\alpha = 0.10$. Which is the best choice for a given significance test? That depends on whether a Type I error or a Type II error is more serious.

In the fast-food example, a Type I error occurs if the true proportion of customers who have to wait at least 2 minutes remains $p = 0.63$, but we get a value of $\hat{p}$ small enough to yield a P-value less than $\alpha = 0.10$. When H_0 is true, this will happen 10% of the time just by chance. In other words, $P(\text{Type I error}) = \alpha$.

Determining Type I Error Probability

The probability of making a Type I error in a statistical test is equal to the significance level α.

We can decrease the probability of making a Type I error in a statistical test by using a smaller significance level. For instance, the restaurant manager could use $\alpha = 0.05$ instead of $\alpha = 0.10$. But there is a tradeoff between $P(\text{Type I error})$ and $P(\text{Type II error})$: As one increases, the other decreases. If we make it more difficult to reject H_0 by decreasing α, we increase the probability that we don't find convincing evidence for H_a when it is true. That's why it is important to consider the possible consequences of each type of error before choosing a significance level.

LESSON APP 8.2

Are these potatoes keepers?

A company that makes potato chips requires each shipment of potatoes to meet certain quality standards. If the company finds convincing evidence that more than 8% of the potatoes in the shipment have "blemishes," the truck will be sent back to the supplier to get another load of potatoes. Otherwise, the entire truckload will be used to make potato chips. The producer will perform a significance test using the hypotheses

$$H_0: p = 0.08$$
$$H_a: p > 0.08$$

where p is the true proportion of potatoes with blemishes in a given truckload.

A supervisor selects a random sample of 500 potatoes from the truck and finds that 52 of the potatoes ($\hat{p} = 0.104$) have blemishes. The resulting P-value of the test is 0.0240.

Anatolii-Boida/iStock/Getty Images

1. What conclusion should the supervisor make at $\alpha = 0.05$?

2. Describe a Type I and a Type II error in context.

3. Give a consequence of each type of error in this setting. Which error is more serious for the potato chip producer? Explain.

4. Based on your answer to Question 3, do you agree with the producer's choice of $\alpha = 0.05$? Why or why not?

Lesson 8.2

WHAT DID YOU LEARN?

LEARNING TARGET	EXAMPLES	EXERCISES
Determine if the results of a study are statistically significant and make an appropriate conclusion using a significance level.	p. 509	1–4
Interpret a Type I error and a Type II error in context.	p. 510	5–8
Give a consequence of a Type I error and a Type II error in a given setting.	p. 511	9–12

Exercises   Lesson 8.2

Mastering Concepts and Skills

1. **How much juice?** The label on bottles of one company's grapefruit juice say that they contain 180 milliliters (ml) of liquid. Your friend Jerry suspects that the true mean is less than that, so he takes a random sample of 40 bottles and measures the volume of liquid in each bottle. The mean volume of liquid in the bottles is 179.5 ml and the standard deviation is 1.3 ml. Jerry performs a test of $H_0: \mu = 180$ versus $H_a: \mu < 180$, where μ is the true mean amount of liquid in this company's bottles of grapefruit juice. The test yields a P-value of 0.0098.
pg 509

What conclusion would you make for each of the following significance levels?

(a) $\alpha = 0.05$ (b) $\alpha = 0.01$

2. **Clean water** The Environmental Protection Agency has determined that safe drinking water should contain no more than 1.3 milligrams per liter (mg/l) of copper, on average. To test water from a new source, you collect water in small bottles at each of 30 randomly selected locations. The mean copper content of your bottles is 1.36 mg/l and the standard deviation is 0.18 mg/l. You perform a test of $H_0: \mu = 1.3$ versus $H_a: \mu > 1.3$, where μ is the true mean copper content

of the water from the new source. The test yields a P-value of 0.0391. What conclusion would you make for each of the following significance levels?

(a) $\alpha = 0.05$ (b) $\alpha = 0.01$

3. **Let's not argue!** A Gallup poll report revealed that 72% of teens said they seldom or never argue with their friends.[6] Yvonne wonders if this result holds true in her large high school. So she surveys a random sample of 150 students at her school and finds that 96 of them say they rarely or never argue with friends. She uses the data to perform a test of H_0: $p = 0.72$ versus H_a: $p \neq 0.72$, where p is the true proportion of teens in Yvonne's school who rarely or never argue with their friends. The test yields a P-value of 0.0291. What conclusion would you make for each of the following significance levels?

(a) $\alpha = 0.05$ (b) $\alpha = 0.01$

4. **How many lefties?** Simon reads a newspaper report claiming that 12% of all adults in the United States are left-handed. He wonders if this figure holds true at the large community college he attends. Simon chooses an SRS of 100 students and finds that 16 of them are left-handed. He uses the data to perform a test of H_0: $p = 0.12$ versus H_a: $p \neq 0.12$, where p is the true proportion of lefties at his community college. The test yields a P-value of 0.2184. What conclusion would you make for each of the following significance levels?

(a) $\alpha = 0.10$ (b) $\alpha = 0.05$

5. **Opening a restaurant** You are thinking about opening a restaurant and are searching for a good location. From your research, you know that the mean income of those living near the restaurant must be over $85,000 to support the type of upscale restaurant you wish to open. You decide to take a simple random sample of 50 people living near one potential site. Based on the mean income of this sample, you will perform a test of H_0: $\mu = \$85,000$ versus H_a: $\mu > \$85,000$, where μ is the true mean income in the population of people who live near the restaurant.[7] Describe a Type I error and a Type II error in this setting.

6. **TV reality** Television networks rely heavily on ratings of television shows when deciding whether to renew a show for another season. Suppose the Hyena network has decided that *Miniature Golf with the Stars* will be renewed only if it can be established that more than 12% of U.S. adults watch the show. A polling company asks a random sample of 2000 U.S. adults if they watch *Miniature Golf with the Stars*. The network uses the data to perform a test of H_0: $p = 0.12$ versus H_a: $p > 0.12$, where p is the true proportion of all U.S. adults who watch the show. Describe a Type I error and a Type II error in this setting.

7. **Speed of first responders** Several cities have begun to monitor emergency response times because accident victims with life-threatening injuries generally need medical attention within 8 minutes. In one city, emergency personnel took more than 8 minutes to arrive at 22% of all calls involving life-threatening injuries last year. The city manager then issued guidelines for improving response time to local first responders. After 6 months, the city manager selects an SRS of 400 calls involving life-threatening injuries and examines the response times. She then performs a test of H_0: $p = 0.22$ versus H_a: $p < 0.22$, where p is the true proportion of calls involving life-threatening injuries during this 6-month period for which emergency personnel took more than 8 minutes to arrive. Describe a Type I error and a Type II error in this setting.

8. **High blood pressure?** A company markets a computerized device for detecting high blood pressure. The device measures an individual's blood pressure once per hour at a randomly selected time throughout a 12-hour period. Then it calculates the mean systolic (top number) pressure for the sample of measurements. Based on the sample results, the device performs a test of H_0: $\mu = 130$ versus H_a: $\mu > 130$, where μ is the person's true mean systolic pressure. Describe a Type I error and a Type II error in this setting.

9. **Opening an upscale restaurant** Refer to Exercise 5. pg 511 Give a consequence of each type of error in this setting.

10. **The reality of TV** Refer to Exercise 6. Give a consequence of each type of error in this setting.

11. **Response to accidents** Refer to Exercise 7.

(a) Give a consequence of each type of error in this setting.

(b) Which type of error is more serious? Justify your answer.

12. **Screening blood pressure** Refer to Exercise 8.

(a) Give a consequence of each type of error in this setting.

(b) Which type of error is more serious? Justify your answer.

Applying the Concepts

13. **Light bulbs** A contract between a manufacturer and a consumer of light bulbs specifies that the mean lifetime of the bulbs must be at least 1000 hours. An ordinary testing procedure is difficult because 1000 hours is over 41 days! Because the lifetime of a bulb decreases as the voltage applied increases, a common procedure is to perform an accelerated lifetime test in which the bulbs are

lit using 400 volts (compared to the usual 110 volts). At 400 volts, a 1000-hour bulb is expected to last only 3 hours. This is a well-known procedure, and both sides have agreed that the results from the accelerated test will be a valid indicator of the bulb's lifetime. Let μ = the true mean lifetime of the bulbs. The manufacturer will test the hypotheses $H_0: \mu = 3$ versus $H_a: \mu < 3$ at the $\alpha = 0.05$ level with an SRS of 100 bulbs. In the sample, $\bar{x} = 2.90$. The resulting P-value is 0.04.

(a) What conclusion should we make at $\alpha = 0.05$?

(b) Which type of error—a Type I error or a Type II error—could we have made in part (a)? Explain.

(c) Give a consequence of the type of error you chose in part (b).

14. **Raising the sales tax** Members of the city council want to know if a majority of city residents supports a 1% increase in the sales tax to fund road repairs. To investigate, they survey a random sample of 300 city residents and use the results to test the following hypotheses:

$$H_0: p = 0.50$$
$$H_a: p > 0.50$$

where p is the proportion of all city residents who support a 1% increase in the sales tax to fund road repairs. In the sample, $\hat{p} = 158/300 = 0.527$. The resulting P-value is 0.18.

(a) What conclusion should we make at $\alpha = 0.05$?

(b) Which type of error—a Type I error or a Type II error—could we have made in part (a)? Explain.

(c) Give a consequence of the type of error you chose in part (b).

15. **Responding to accidents** Refer to Exercises 7 and 11. Which significance level—0.10, 0.05, or 0.01—would you choose for the test? Justify your answer.

16. **Screening pressure** Refer to Exercises 8 and 12. Which significance level—0.10, 0.05, or 0.01—would you choose for the test? Justify your answer.

17. **The meaning of "significance"**

(a) Explain why an observed result that is statistically significant at the 1% level must always be significant at the 5% level.

(b) If a result is significant at the 5% level, what can you say about its significance at the 1% level?

18. **Statistical significance explained** Asked to explain the meaning of "statistically significant at the $\alpha = 0.05$ level," a student says, "This means that the probability that the null hypothesis is true is less than 0.05." Is this explanation correct? Why or why not?

Extending the Concepts

The power of a test is the probability that the test will find convincing evidence for H_a at significance level α when a specified alternative value of the parameter is true.

19. **Power** A drug manufacturer claims that less than 10% of patients who take its new drug for treating Alzheimer's disease will experience nausea. To test this claim, a significance test is carried out of

$$H_0: p = 0.10$$
$$H_a: p < 0.10$$

Suppose that the manufacturer's claim about its new drug is correct, and only 8% of patients will experience nausea. The power of this test at the $\alpha = 0.05$ significance level is 0.29.

(a) Explain what "power = 0.29" means in this setting.

(b) Find the probability of a Type II error.

(c) Name one way to increase the power of the test.

Recycle and Review

20. **Paul Bunyan** (3.2, 3.5) Bangor, Maine, is one of several towns that claim to be the birthplace of the legendary lumberjack Paul Bunyan. Bangor is proud of its 31-foot fiberglass statue of Paul, and some townspeople think a second statue should be made of Paul's equally legendary sidekick, Babe the Blue Ox. The *Bangor Daily News* recently conducted an online poll, asking visitors to its website, "Should Bangor add Babe the Blue Ox to the Paul Bunyan statue site?" Of the 1123 people who responded, 864 said "Yes."

(a) Describe how undercoverage might lead to bias in this poll. Explain the likely direction of the bias.

(b) Describe how voluntary response might lead to bias in this poll. Explain the likely direction of the bias.

21. **Buying stocks randomly?** (4.2, 4.6) The "random walk" theory of stock prices holds that price movements in separate time periods are independent of each other. Assume that this theory is correct. Suppose we record only whether the price is up or down each year and that the probability that our portfolio of stocks rises in price in any 1 year is 0.65.

(a) What is the probability that our portfolio goes up for 3 consecutive years?

(b) What is the probability that the portfolio's value moves in the same direction (either up or down) for 3 consecutive years?

(c) What is the probability that the portfolio's value goes down for at least 1 year in 4 consecutive years?

Lesson 8.3

Testing a Claim about a Proportion

LEARNING TARGETS

- Check the Random and Large Counts conditions for performing a significance test about a population proportion.
- Calculate the standardized test statistic for a significance test about a population proportion.
- Find the *P*-value for a one-sided significance test about a population proportion using Table A or technology.

In Lessons 8.1 and 8.2, we saw that a significance test can be used to test a claim about an unknown population parameter. We are often interested in testing a claim about the proportion p of some outcome in the population. For example, a large car dealership claims that the proportion of all car buyers who purchase an extended warranty is $p = 0.60$. An investigative reporter suspects that the true proportion is somewhat lower. This lesson shows you how to check conditions and perform calculations for a significance test about a population proportion.

Conditions for Testing a Claim about p

In Lesson 7.3, we introduced conditions that should be met before we construct a confidence interval for a population proportion. We called them Random and Large Counts. These same conditions must be verified before carrying out a significance test.

The Large Counts condition for proportions requires that both np and $n(1 - p)$ be at least 10. Because we assume H_0 is true when performing a significance test, we use the parameter value specified by the null hypothesis (denoted p_0) when checking the Large Counts condition. In this case, the Large Counts condition says that the *expected* count of successes np_0 and of failures $n(1 - p_0)$ are both at least 10.

> ### How to Check Conditions for Performing a Significance Test about p
>
> - **Random:** The data come from a random sample from the population of interest.
> - **Large Counts:** Both np_0 and $n(1 - p_0)$ are at least 10.

If the data come from a convenience sample or a voluntary response sample, there's no point in carrying out a significance test for p. The same is true if there are other sources of bias in the data-collection process. If the Large Counts condition is violated, a *P*-value calculated from a normal distribution will not be accurate.

EXAMPLE

Is the drive-thru faster?

Checking conditions

PROBLEM: The manager of a fast-food restaurant wants to reduce the proportion of drive-thru customers who have to wait longer than 2 minutes to receive their food after placing an order. Based on store records, the proportion of customers who had to wait longer than 2 minutes was $p = 0.63$. To reduce this proportion, the manager assigns an additional employee to assist with drive-thru orders. The manager would like to carry out a test at the $\alpha = 0.10$ significance level of

$$H_0: p = 0.63$$
$$H_a: p < 0.63$$

where $p =$ the true proportion of drive-thru customers who have to wait longer than 2 minutes to receive their food.

During the next month, the manager collects data on wait times from a random sample of 250 drive-thru orders, and finds that only 141 of the customers have to wait longer than 2 minutes. Check if the conditions for performing the significance test are met.

SOLUTION:

Random? Random sample of 250 orders. ✓

Large Counts? $np_0 = 250(0.63) = 157.5 \geq 10$ and
$\qquad n(1 - p_0) = 250(1 - 0.63) = 92.5 \geq 10$ ✓

> Be sure to use the null value p_0, not the sample proportion $\hat{p}$, when checking the Large Counts condition!

FOR PRACTICE   TRY EXERCISE 1.

Calculations: Standardized Test Statistic and *P*-Value

In the fast-food example, the sample proportion of drive-thru customers who had to wait more than 2 minutes for their order was $\hat{p} = \dfrac{141}{250} = 0.564$. Because this result is less than 0.63, there is *some* evidence against $H_0: p = 0.63$ and in favor of $H_a: p < 0.63$.

But do we have *convincing* evidence that the proportion of all customers who have to wait longer than 2 minutes has decreased? To answer this question, we want to know if it's likely to get a sample proportion of 0.564 or less by chance alone when the null hypothesis is true. In other words, we are looking for a *P*-value.

Suppose for now that the null hypothesis $H_0: p = 0.63$ is true. Consider the sample proportion $\hat{p}$ of customers who have to wait more than 2 minutes for their drive-thru orders in a random sample of size $n = 250$. You learned in Lesson 6.4 that the sampling distribution of $\hat{p}$ will have mean

$$\mu_{\hat{p}} = p = 0.63$$

and standard deviation

$$\sigma_{\hat{p}} = \sqrt{\frac{p(1 - p)}{n}} = \sqrt{\frac{0.63(0.37)}{250}} = 0.031$$

Because the Large Counts condition is met, the sampling distribution of $\hat{p}$ will be approximately normal. Figure 8.3 displays this distribution. We have added the manager's sample result, $\hat{p} = \dfrac{141}{250} = 0.564$.

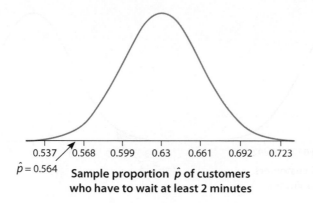

FIGURE 8.3 Normal approximation to the sampling distribution of the sample proportion $\hat{p}$ of drive-thru customers who have to wait at least 2 minutes in random samples of 250 orders when $p = 0.63$.

$\hat{p} = 0.564$

0.537   0.568   0.599   0.63   0.661   0.692   0.723

Sample proportion $\hat{p}$ of customers who have to wait at least 2 minutes

To assess how far the statistic is from the null value, standardize the statistic:

$$z = \frac{0.564 - 0.63}{0.031} = -2.13$$

This value is called the **standardized test statistic.**

> ### DEFINITION Standardized test statistic
>
> A **standardized test statistic** measures how far a sample statistic is from what we would expect if the null hypothesis H_0 were true, in standard deviation units. That is,
>
> $$\text{standardized test statistic} = \frac{\text{statistic} - \text{null value}}{\text{standard deviation of statistic}}$$

EXAMPLE

How do you get that z?

Calculating the test statistic

PROBLEM: Suppose that you want to perform a test of

$$H_0 : p = 0.30$$
$$H_a : p > 0.30$$

An SRS of size 50 from the population of interest yields 19 successes. Calculate the standardized test statistic.

SOLUTION:

Statistic: $\hat{p} = \dfrac{19}{50} = 0.38$

$$z = \frac{0.38 - 0.30}{\sqrt{\dfrac{0.30(1 - 0.30)}{50}}} = 1.23$$

> First calculate the sample statistic.

> Then standardize:
>
> $\text{standardized test statistic} = \dfrac{\text{statistic} - \text{null value}}{\text{standard deviation of statistic}}$

FOR PRACTICE   TRY EXERCISE 5.

You can use the standardized test statistic to find the *P*-value for a significance test. Let's return to the fast-food restaurant. Our test of $H_0 : p = 0.63$ versus $H_a : p < 0.63$ based on the observed result of $\hat{p} = 0.564$ gave a standardized test statistic of $z = -2.13$. The *P*-value is the probability of getting a sample proportion less than or equal to $\hat{p} = 0.564$ by chance alone when $H_0 : p = 0.63$ is true. The shaded area in Figure 8.4(a) on the next page shows this probability. Figure 8.4(b) shows the corresponding area to the left of $z = -2.13$ in the standard normal distribution.

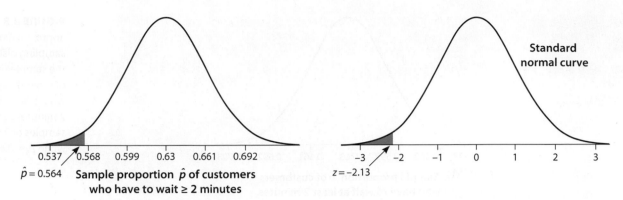

0.537 0.568 0.599 0.63 0.661 0.692

$\hat{p} = 0.564$ **Sample proportion $\hat{p}$ of customers who have to wait ≥ 2 minutes**

Standard normal curve

–3 –2 –1 0 1 2 3

$z = -2.13$

FIGURE 8.4 The shaded area shows the *P*-value for the fast-food example about the proportion of drive-thru customers who had to wait more than 2 minutes to receive their orders (a) on the normal distribution that models the sampling distribution of $\hat{p}$ and (b) on the standard normal curve.

We can find the *P*-value using Table A or technology. Table A gives $P(z \leq -2.13)$ = 0.0166. The *Normal Distribution* applet and the TI-84 command normalcdf(lower: -1000, upper: -2.13, μ:0, σ :1) also give a *P*-value of 0.0166.

EXAMPLE

What are the chances?

Finding the P-value

PROBLEM: Refer to the previous example, where the standardized test statistic was $z = 1.23$. Assume that the conditions for carrying out the test are met. Find the *P*-value using Table A or technology.

SOLUTION:

Using Table A: $P(z \geq 1.23) = 1 - 0.8907 = 0.1093$

Using technology: Applet/normalcdf(lower:1.23, upper:1000, mean:0, SD:1) = 0.1093

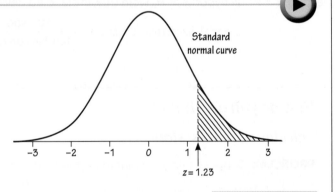

Standard normal curve

–3 –2 –1 0 1 2 3

$z = 1.23$

FOR PRACTICE TRY EXERCISE 9.

You learned how to make a conclusion for a statistical test based on a *P*-value and a significance level in Lesson 8.2. No significance level was stated in the example, so we default to $\alpha = 0.05$. Because the *P*-value of 0.1093 is greater than $\alpha = 0.05$, we fail to reject H_0. We don't have convincing evidence that the population proportion is greater than 0.30.

For the fast-food setting, the manager chose a significance level of $\alpha = 0.10$. Because the *P*-value of 0.0166 is less than $\alpha = 0.10$, we reject H_0. We have convincing evidence that the true proportion of drive-thru customers who have to wait more than 2 minutes is now less than 0.63. We cannot conclude that assigning an additional employee to the drive-thru *caused* the decrease in wait times because this study was not a randomized experiment.

CAUTION

THINK ABOUT IT **What happens when the data don't support H_a?** Suppose the fast-food restaurant manager had obtained a sample proportion of $\hat{p} = 0.66$. This sample result doesn't give *any* evidence to support the alternative hypothesis $H_a: p < 0.63$ because $\hat{p} = 0.66$ is *greater than* 0.63! There's no need to continue with a significance test.

LESSON APP 8.3

Is it better to be last?

On TV shows that feature singing competitions, contestants often wonder if there is an advantage in performing last. To investigate, researchers selected a random sample of 600 college students and showed each student the audition video of 12 different singers. Each student viewed the videos in random order. We would expect approximately 1/12 of the students to prefer the last singer seen, assuming order doesn't matter. In this study, 59 of the 600 students preferred the last singer they viewed. We want to perform a test at the $\alpha = 0.05$ significance level of

$$H_0 : p = \frac{1}{12}$$

$$H_a : p > \frac{1}{12}$$

where $p =$ the true proportion of college students who prefer the last singer they see.

1. Check that the conditions for performing the test are met.

2. Calculate the standardized test statistic.

3. Find the P-value using Table A or technology.

4. What conclusion would you make?

Lesson 8.3

WHAT DID YOU LEARN?

LEARNING TARGET	EXAMPLES	EXERCISES
Check the Random and Large Counts conditions for performing a significance test about a population proportion.	p. 516	1–4
Calculate the standardized test statistic for a significance test about a population proportion.	p. 517	5–8
Find the P-value for a one-sided significance test about a population proportion using Table A or technology.	p. 518	9–12

Exercises

Lesson 8.3

Mastering Concepts and Skills

In Exercises 1–4, check if the conditions for performing the significance test are met.

1. **Computers at home** Jason reads a report that says
 80% of U.S. high school students have a computer at home. He believes the proportion is smaller than 0.80 at his large rural high school. Jason chooses an SRS of 60 students and finds that 41 have a computer at home. He would like to carry out a test at the $\alpha = 0.05$

significance level of $H_0: p = 0.80$ versus $H_a: p < 0.80$, where $p =$ the proportion of all students at Jason's high school who have a computer at home.

2. **No one walks to school** A recent report claimed that 13% of students typically walk to school.[8] DeAnna thinks that the proportion is higher than 0.13 at her large elementary school. She surveys a random sample of 100 students and finds that 17 typically walk to school. DeAnna would like to

carry out a test at the $\alpha = 0.05$ significance level of H_0: $p = 0.13$ versus H_a: $p > 0.13$, where $p =$ the proportion of all students at her elementary school who typically walk to school.

3. **Get rid of Boggs** A college president says, "More than two-thirds of the alumni support my firing of Coach Boggs." The president's statement is based on 200 emails he has received from alumni in the past three months. The college's athletic director wants to perform a test of H_0: $p = 2/3$ versus H_a: $p > 2/3$, where $p =$ the true proportion of the college's alumni who favor firing the coach.

4. **Heads, you win** You want to determine if a coin is fair. You toss it 10 times and record the proportion of tosses that land heads. You would like to perform a test of H_0: $p = 0.5$ versus H_a: $p \neq 0.5$, where $p =$ the proportion of all tosses of the coin that would land heads.

5. **Home computers** Refer to Exercise 1. Calculate the standardized test statistic.
pg 517

6. **Walking to school** Refer to Exercise 2. Calculate the standardized test statistic.

7. **One-sided test** Suppose that you want to perform a test of H_0: $p = 0.9$ versus H_a: $p > 0.9$. An SRS of size 100 from the population of interest yields 96 successes. Assume that the conditions for carrying out the test are met. Calculate the standardized test statistic.

8. **Another one-sided test** Suppose that you want to perform a test of H_0: $p = 0.5$ versus H_a: $p > 0.5$. An SRS of size 75 from the population of interest yields 43 successes. Assume that the conditions for carrying out the test are met. Calculate the standardized test statistic.

9. **Computer P-value** Refer to Exercises 1 and 5. Find the P-value using Table A or technology.
pg 518

10. **Walking P-value** Refer to Exercises 2 and 6. Find the P-value using Table A or technology.

11. **Right-tail test** Refer to Exercise 7. Find the P-value using Table A or technology.

12. **Another right-tail test** Refer to Exercise 8. Find the P-value using Table A or technology.

Applying the Concepts

13. **The chips project** Zenon decided to investigate whether students at his school prefer name-brand potato chips to generic potato chips. He randomly selected 50 students and had each student try both types of chips, in random order. Overall, 32 of the 50 students preferred the name-brand chips. Zenon wants to perform a test at the $\alpha = 0.05$ significance level of H_0: $p = 0.5$ versus H_a: $p > 0.5$, where $p =$ the true proportion of all students at his school who prefer name-brand chips.

(a) Check that the conditions for performing the test are met.

(b) Calculate the standardized test statistic.

(c) Find the P-value using Table A or technology.

(d) What conclusion would you make?

14. **Watching the corn grow** The germination rate of seeds is defined as the proportion of seeds that sprout and grow when properly planted and watered. A certain variety of corn usually has a germination rate of 0.80. As their science project, a group of students wants to see if spraying seeds with a chemical known to increase germination rates will increase the germination rate of this variety of corn. They spray a random sample of 400 seeds with the chemical, and 339 of the seeds germinate. You would like to perform a test at the $\alpha = 0.05$ significance level of H_0: $p = 0.80$ versus H_a: $p > 0.80$, where $p =$ the true proportion of this variety of corn that will germinate when sprayed with the chemical.

(a) Check that the conditions for performing the test are met.

(b) Calculate the standardized test statistic.

(c) Find the P-value using Table A or technology.

(d) What conclusion would you make?

15. **Computer conclusion** Refer to Exercises 1, 5, and 9.

(a) What conclusion would you make?

(b) Which type of error—a Type I error or a Type II error—could you have made in part (a)? Explain.

16. **Walking conclusion** Refer to Exercises 2, 6, and 10.

(a) What conclusion would you make?

(b) Which type of error—a Type I error or a Type II error—could you have made in part (a)? Explain.

Extending the Concepts

17. **Assessing a one-sided test** Suppose that you want to perform a test of

$$H_0: p = 0.9$$
$$H_a: p > 0.9$$

An SRS of size 100 from the population of interest yields 80 successes. Explain why there is really no need to carry out the test in this case.

Recycle and Review

18. **Old enough to vote?** (5.3, 5.4, 6.4) According to the 2010 U.S. Census, 24% of U.S. residents are under 18 years old. Suppose we take a random sample of 10 U.S. residents. Let $X =$ the number of people in the sample who are under 18 years old.

(a) Explain why it is reasonable to use the binomial distribution for probability calculations involving X.

(b) Find the probability that exactly 3 people in the sample are under 18 years old.

(c) Researchers want to perform a test of H_0: $p = 0.24$ versus H_a: $p < 0.24$, where p is the true proportion of North Carolina residents who are under 18 years old. They take a random sample of North Carolina residents; only 1 person is under 18. Assuming H_0 is true, calculate $P(X \leq 1)$. What conclusion should the researchers make?

19. How many miles? (6.5, 6.6) The service department of a large automobile dealership records the odometer readings of cars that it repairs and determines that the distribution of miles driven per year by all of its customers has a mean of 14,000 miles and a standard deviation of 4000 miles. The distribution is skewed to the right. Suppose a random sample of 12 cars is taken from the dealership's service records, and the mean number of miles per year driven by these cars, $\bar{x}$, is calculated.

(a) What is the mean of the sampling distribution of $\bar{x}$?

(b) Calculate and interpret the standard deviation of the sampling distribution of $\bar{x}$.

(c) Describe the shape of the sampling distribution of $\bar{x}$.

Lesson 8.4

Significance Tests for a Proportion

LEARNING TARGETS

- Use the four-step process to perform a one-sided significance test about a population proportion.
- Calculate the P-value for a two-sided significance test about a population proportion using Table A or technology.
- Use the four-step process to perform a two-sided significance test about a population proportion.

In Lesson 8.3, you learned how to check conditions and perform calculations for a one-sided test about a population proportion. We begin this lesson by showing you how to use the four-step process to carry out a one-sided significance test for p. Then we discuss two-sided tests.

Putting It All Together: The Four-Step Process

To perform a significance test, we state hypotheses, check conditions, calculate a test statistic and P-value, and draw a conclusion in the context of the problem. The four-step process is ideal for organizing our work.

How to Use the Four-Step Process: Significance Tests

STATE: State the hypotheses you want to test and the significance level, and define any parameters you use.

PLAN: Identify the appropriate inference method and check conditions.

DO: If the conditions are met, perform calculations.
- Give the sample statistic.
- Calculate the standardized test statistic.
- Find the P-value.

CONCLUDE: Make a conclusion about the hypotheses in the context of the problem.

A significance test for a population proportion is often referred to as a *one-sample z test for p.*

EXAMPLE

Who watches Survivor?

Significance test for a proportion

CBS via Getty Images

PROBLEM: Television executives and companies who advertise on TV are interested in how many viewers watch particular shows. According to Nielsen ratings, *Survivor* was one of the most-watched television shows in the United States during every week that it aired. An avid *Survivor* fan (and textbook author) claims that 35% of all U.S. adults have watched *Survivor.* A skeptical editor believes this figure is too high. He asks a random sample of 200 U.S. adults if they have watched *Survivor;* 60 say, "Yes." Is there convincing evidence at the $\alpha = 0.05$ significance level to confirm the editor's belief?

SOLUTION:

STATE: We want to perform a test at the $\alpha = 0.05$ significance level of

$H_0: p = 0.35$

$H_a: p < 0.35$

where p = the proportion of all U.S. adults who have watched *Survivor.*

PLAN: One-sample z test for p

Random? Random sample of 200 U.S. adults. ✓

Large Counts? $200(0.35) = 70 \geq 10$ and
$\qquad\qquad 200(1 - 0.35) = 130 \geq 10$ ✓

DO:

- $\hat{p} = \dfrac{60}{200} = 0.30$

- $z = \dfrac{0.30 - 0.35}{\sqrt{\dfrac{0.35(1 - 0.35)}{200}}} = -1.48$

- P-value:

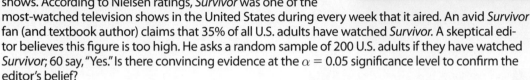

Standard normal curve

$z = -1.48$

Using Table A: $P(z < -1.48) = 0.0694$
Using technology: Applet/normalcdf(lower bound: -1000,
upper bound: -1.48, mean: 0, SD: 1) = 0.0694

CONCLUDE: Because the P-value of 0.0694 is greater than $\alpha = 0.05$, we fail to reject H_0. There is not convincing evidence that the proportion of all U.S. adults who have watched *Survivor* is less than 0.35.

> Follow the four-step process!

> **STATE:** State the hypotheses you want to test and the significance level, and define any parameters you use.

> **PLAN:** Identify the appropriate inference method and check conditions.

> Be sure to use the null value p_0 when checking the Large Counts condition.

> **DO:** If the conditions are met, perform calculations:
> - Identify the sample statistic.
> - Calculate the standardized test statistic.
> - Find the P-value.

> The sample result gives *some* evidence in favor of H_a because $0.30 < 0.35$.

> **CONCLUDE:** Make a conclusion about the hypotheses in the context of the problem.

FOR PRACTICE   TRY EXERCISE 1.

An applet or a calculator will handle the calculations in the "Do" step. See the Tech Corner at the end of the lesson for details.

Two-Sided Tests

The *P*-value in a one-sided test about a population proportion is the area in one tail of a standard normal distribution—the tail specified by H_a. In a two-sided test, the alternative hypothesis has the form $H_a: p \neq p_0$. The *P*-value in such a test is the probability of getting a sample proportion as far as or farther from p_0 *in either direction* than the observed value of $\hat{p}$, assuming the null hypothesis is true. As a result, you have to find the area in both tails of a standard normal distribution to get the *P*-value.

EXAMPLE

What about a two-sided test?

Finding the P-value

PROBLEM: Suppose that you want to perform a test of

$$H_0: p = 0.30$$
$$H_a: p \neq 0.30$$

An SRS of size 50 from the population of interest yields 19 successes. Assume that the conditions for carrying out the test are met. Calculate the standardized test statistic and find the *P*-value using Table A or technology.

SOLUTION:

Statistic: $\hat{p} = \dfrac{19}{50} = 0.38$

> First calculate the sample statistic.

$z = \dfrac{0.38 - 0.30}{\sqrt{\dfrac{0.30(1 - 0.30)}{50}}} = 1.23$

> Then standardize:
>
> standardized test statistic $= \dfrac{\text{statistic} - \text{null value}}{\text{standard deviation of statistic}}$

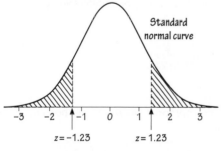

Standard normal curve

$z = -1.23$ $z = 1.23$

> As the graph suggests, the *P*-value for a two-sided test is twice the area in one tail of the standardized distribution.

> Find the *P*-value using Table A or technology.

Using Table A: $P(z \geq 1.23 \text{ or } z \leq -1.23) = 2(0.1093) = 0.2186$

Using technology: Applet/normalcdf(lower:1.23, upper:1000, mean:0, SD:1) $\times 2 = 0.2187$

FOR PRACTICE   TRY EXERCISE 5.

The *P*-value in the example is the probability of getting a sample proportion as far or farther from $p_0 = 0.3$ as the observed $\hat{p} = 0.38$ in either direction when H_0 is true. That is,

$$\text{P-value} = P(\hat{p} \geq 0.38 \text{ or } \hat{p} \leq 0.22 \mid p = 0.30) = 0.2187$$

Figure 8.5 (on the next page) shows this probability as an area under the normal curve that approximates the sampling distribution of $\hat{p}$ when $H_0: p = 0.30$ is true. Notice that the shaded area here is the same as in the standard normal distribution from the example. Because the *P*-value of 0.2187 is greater than our default $\alpha = 0.05$ significance level, we would fail to reject H_0. We do not have convincing evidence that the population proportion differs from 0.30.

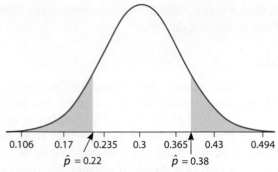

0.106   0.17   0.235   0.3   0.365   0.43   0.494

$\hat{p} = 0.22$ $\hat{p} = 0.38$

Sample proportion $\hat{p}$ of successes

FIGURE 8.5 The shaded area shows the *P*-value for the previous example in the normal distribution that approximates the sampling distribution of $\hat{p}$.

Now you are ready to perform a two-sided test about a population proportion.

EXAMPLE

Are most students nonsmokers?

Two-sided test about a proportion

PROBLEM: According to the Centers for Disease Control and Prevention (CDC) website, 59% of high school students have never smoked a cigarette.[9] Antawn wonders if this national result holds true in his large, urban high school. For his statistics class project, Antawn surveys an SRS of 150 students from his school, and 102 say that they have never smoked a cigarette. Is there convincing evidence that the proportion of all students at Antawn's school differs from the national result?

SOLUTION:

STATE: We want to perform a test of

$H_0: p = 0.59$

$H_a: p \neq 0.59$

> **STATE:** State the hypotheses you want to test and the significance level, and define any parameters you use.

where $p =$ the proportion of all students at Antawn's school who have never smoked a cigarette. Because no significance level was given, we'll use $\alpha = 0.05$.

PLAN: One-sample z test for p

Random? Random sample of 150 students. ✓

Large Counts? $150(0.59) = 88.5 \geq 10$ and
$\qquad 150(1 - 0.59) = 61.5 \geq 10$ ✓

> **PLAN:** Identify the appropriate inference method and check conditions.

DO:

- $\hat{p} = \dfrac{102}{150} = 0.68$

- $z = \dfrac{0.68 - 0.59}{\sqrt{\dfrac{0.59(1 - 0.59)}{150}}} = 2.24$

> **DO:** If the conditions are met, perform calculations:
> - Identify the sample statistic.
> - Calculate the standardized test statistic.
> - Find the *P*-value.

> The sample result gives *some* evidence in favor of H_a because $0.68 \neq 0.59$.

• *P-value:*

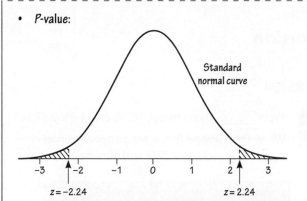

Standard normal curve

$z = -2.24$ $z = 2.24$

Using Table A: $P(z \geq 2.24 \text{ or } z \leq -2.24) = 2(0.0125) = 0.0250$

Using technology: Applet/normalcdf(lower:2.24, upper:1000, mean:0, SD:1) $\times 2 = 0.0251$

CONCLUDE: *Because the P-value of 0.0250 is less than $\alpha = 0.05$, we reject H_0. We have convincing evidence that the proportion of all students at Antawn's school who would say they have never smoked a cigarette differs from the national proportion of 0.59.*

CONCLUDE: Make a conclusion about the hypotheses in the context of the problem.

FOR PRACTICE **TRY EXERCISE 9.**

The result of a significance test begins with a decision to reject H_0 or fail to reject H_0. In Antawn's smoking study, for instance, the data led us to reject H_0: $p = 0.59$ because we found convincing evidence that the proportion of students at his school who would say they have never smoked a cigarette differs from the national value. We're left wondering what the actual proportion p might be. A confidence interval can shed some light on this issue. You learned how to calculate a confidence interval for a population proportion in Lesson 7.3.

A 95% confidence interval for p is

$$0.68 \pm 1.96\sqrt{\frac{0.68(1 - 0.68)}{150}} \rightarrow 0.68 \pm 0.075 \rightarrow 0.605 \text{ to } 0.755$$

This interval gives the values for p that are plausible based on the sample data. We would not be surprised if the proportion of all students at Antawn's school who would say they have never smoked cigarettes was any value between 0.605 and 0.755. Note that the null value of $p = 0.59$ is *not* a plausible value of the parameter. So the 95% confidence interval and the two-sided test at the $\alpha = 0.05$ significance level give consistent results.

LESSON APP 8.4

Who feels job stress?

A news report claims that 75% of restaurant employees feel that work stress has a negative impact on their personal lives.[10] Managers of a large restaurant chain wonder if this claim is valid for their employees. A random sample of 100 employees finds that 68 answer "Yes" when asked, "Does work stress have a negative impact on your personal life?" Do these data provide convincing evidence at the $\alpha = 0.10$ level that the proportion of all employees in this chain who would say "Yes" differs from 0.75?

TECH CORNER

One-sample *z* test for a proportion

You can use technology to perform the calculations for a significance test about a population proportion. We'll illustrate using the *Survivor* example, where we tested $H_0: p = 0.35$ versus $H_a: p < 0.35$. Recall that $n = 200$ and $X = 60$ people said that they have watched *Survivor*.

Applet

1. Launch the *One Categorical Variable* applet at highschool.bfwpub.com/spa3e.

2. Enter the variable name "Watch Survivor?" and Category names "Yes" and "No." Then, enter 60 for the frequency of "Yes" and 200 − 60 = 140 for "No."

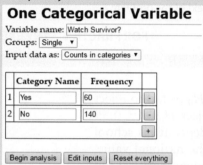

3. Press the "Begin analysis" button. Then, scroll down to the "Perform inference" section. Choose the 1-sample *z* test for the procedure, Yes as the category to indicate success, $p <$ for the alternative hypothesis, and 0.35 for the hypothesized proportion. Then, press the "Perform inference" button.

Perform Inference

Inference procedure: 1-sample z test ▾ Category to indicate as success: Yes ▾
Alternative hypothesis: p < ▾ Hypothesized proportion: 0.35

Perform inference

z	P-value
-1.482	0.069

TI-83/84

1. Press STAT, then choose TESTS and 1-PropZTest.

2. When the 1-PropZTest screen appears, enter $p_0 = .35$, $X = 60$, $n = 200$, and choose prop $< p_0$.

3. Highlight Calculate and press ENTER.

Note: If you select the Draw option, you will get a picture of the standard normal distribution with the area of interest shaded and the standardized test statistic and *P*-value labeled.

Lesson 8.4

WHAT DID YOU LEARN?

LEARNING TARGET	EXAMPLES	EXERCISES
Use the four-step process to perform a one-sided significance test about a population proportion.	p. 522	1–4
Calculate the *P*-value for a two-sided significance test about a population proportion using Table A or technology.	p. 523	5–8
Use the four-step process to perform a two-sided significance test about a population proportion.	p. 524	9–12

Lesson 8.4

Mastering Concepts and Skills

1. Improved parking A local high school makes a
pg 522 change to improve student satisfaction with parking. Before the change, 37% of the school's students approved of parking that was provided. After
the change, the principal surveys an SRS of 200 students at the school. In all, 83 students say that they
approve of the new parking arrangement. Do the
data provide convincing evidence at the $\alpha = 0.05$
significance level that the change was effective?

2. Research into side effects A drug manufacturer
claims that less than 10% of patients who take
its new drug for treating Alzheimer's disease will
experience nausea. To test this claim, researchers conduct an experiment. They give the new
drug to a random sample of 300 Alzheimer's
patients whose families have given informed consent for the patients to participate in the study.
In all, 25 of the subjects experience nausea. Do
these data provide convincing evidence at the
$\alpha = 0.05$ significance level to support the drug
manufacturer's claim?

3. Think before you post! According to the National
Campaign to Prevent Teen and Unplanned Pregnancy, 20% of teens aged 13 to 19 say that they
have electronically sent or posted suggestive images
of themselves.[11] The counselor at a large high school
worries that the actual figure might be higher at her
school. To find out, she administers an anonymous
survey to a random sample of 250 of the school's
students. All 250 respond, and 63 admit to sending
or posting suggestive images. Do the data provide
convincing evidence at the $\alpha = 0.05$ significance
level that the counselor's belief is correct?

4. The coffee challenge People of taste are supposed to
prefer fresh-brewed coffee to the instant variety. Or
maybe many coffee drinkers just want their caffeine
fix. A skeptic claims that only half of all coffee drinkers prefer fresh-brewed coffee. To test this claim, we
ask a random sample of 50 coffee drinkers in a small
city to take part in a study. Each person tastes two
unmarked cups—one containing instant coffee and
one containing fresh-brewed coffee—and says which
he or she prefers. We find that 36 of the 50 choose the
fresh coffee. Do these results give convincing evidence
at the $\alpha = 0.05$ significance level that coffee drinkers
favor fresh-brewed over instant coffee?

*In Exercises 5–8, assume that the conditions for carrying
out the test are met. Calculate the standardized test statistic and find the P-value using Table A or technology.*

5. Two-sided test Suppose that you want to perform a
pg 523 test of $H_0: p = 0.7$ versus $H_a: p \neq 0.7$. An SRS of size
80 from the population of interest yields 59 successes.

6. Two-tailed test Suppose that you want to perform
a test of $H_0: p = 0.65$ versus $H_a: p \neq 0.65$. An SRS
of size 80 from the population of interest yields 41
successes.

7. Another two-sided test Suppose that you want to
perform a test of $H_0: p = 0.25$ versus $H_a: p \neq 0.25$.
An SRS of size 60 from the population of interest
yields 8 successes.

8. Another two-tailed test Suppose that you want to
perform a test of $H_0: p = 0.45$ versus $H_a: p \neq 0.45$.
An SRS of size 60 from the population of interest
yields 37 successes.

9. Drivers passing A state's Department of Motor
pg 524 Vehicles (DMV) claims that 60% of teens pass their
driving test on the first attempt. An investigative
reporter examines an SRS of the DMV records for
125 teens; 86 of them passed the test on their first
try. Is there convincing evidence at the $\alpha = 0.05$
significance level that the DMV's claim is incorrect?

10. Dreams of wealth In a recent year, 73% of first-year
college students identified "being very well-off finan-
cially" as an important personal goal. A state university finds that 132 of an SRS of 200 of its first-year
students say that this goal is important. Is there convincing evidence at the $\alpha = 0.05$ significance level
that the proportion of all first-year students at this
university who think being very well-off is important differs from the national value of 73%?

11. Who the bullies are A media report claims that 75%
of middle school students engage in bullying behavior.
A University of Illinois study on aggressive behavior
surveyed a random sample of 558 middle school students. When asked to describe their behavior in the
last 30 days, 445 students admitted that they had
engaged in physical aggression, social ridicule, teas-
ing, name-calling, and issuing threats—all of which
would be classified as bullying.[12] Do these data provide convincing evidence at the $\alpha = 0.05$ level that the
media report's claim is incorrect?

12. Spinning heads? When a fair coin is flipped, we all
know that the probability the coin lands on heads is
0.50. However, what if a coin is spun? According to
the article "Euro Coin Accused of Unfair Flipping"
in the *New Scientist*, two Polish math professors and
their students spun a Belgian euro coin 250 times. It
landed heads 140 times. One of the professors concluded that the coin was minted asymmetrically. A
representative from the Belgian mint indicated the
result was just chance. Is there convincing evidence at
the $\alpha = 0.05$ level that the coin is unfair?

Applying the Concepts

13. Selling upgrades A software company is trying to
decide whether to produce an upgrade of one of

its programs. Customers would have to pay $100 for the upgrade. For the upgrade to be profitable, the company has to sell it to more than 20% of their customers. You contact a random sample of 60 customers and find that 16 would be willing to pay $100 for the upgrade.

(a) Which would be a more serious mistake in this setting—a Type I error or a Type II error? Justify your answer.

(b) Do the sample data give convincing evidence at the $\alpha = 0.05$ significance level that more than 20% of the company's customers are willing to purchase the upgrade?

14. **Flu vaccine** A drug company has developed a new vaccine for preventing the flu. The company claims that less than 5% of adults who use its vaccine will get the flu. To test the claim, researchers give the vaccine to a random sample of 1000 adults. Of these, 43 get the flu.

(a) Which would be a more serious mistake in this setting—a Type I error or a Type II error? Justify your answer.

(b) Do these data provide convincing evidence at the $\alpha = 0.05$ significance level to support the company's claim?

15. **Teen drivers passing** Refer to Exercise 9.

(a) Construct and interpret a 95% confidence interval for the true proportion of all teens in the state who passed their driving test on the first attempt.

(b) Explain what the interval in part (a) tells you about the DMV's claim.

16. **Rich, rich, rich** Refer to Exercise 10.

(a) Construct and interpret a 95% confidence interval for the true proportion of all first-year students at the university who would identify being well-off as an important personal goal.

(b) Explain what the interval in part (a) tells you about whether the national value holds at this university.

17. **Confidently losing weight** A Gallup Poll found that 59% of the people in its sample said "Yes" when asked, "Would you like to lose weight?" Gallup announced: "For results based on the total sample of national adults, one can say with 95% confidence that the margin of (sampling) error is ±3 percentage points."[13] Does this interval provide convincing evidence that the actual proportion of U.S. adults who would say they want to lose weight differs from 0.60? Justify your answer.

18. **Who tweets?** The Pew Internet and American Life Project asked a random sample of U.S. adults, "Do you ever . . . use Twitter or another service to share updates about yourself or to see updates about others?" According to Pew, the resulting 95% confidence interval is 0.167 to 0.213.[14] Does this interval provide convincing evidence that the actual proportion of U.S. adults who would say they use Twitter differs from 0.17? Justify your answer.

Extending the Concepts

19. **Cranky mower** A company has developed an "easy-start" mower that cranks the engine with the push of a button. The company claims that the probability this model of mower will start on any push of the button is 0.9. A consumer testing agency suspects that this claim is exaggerated. To test the claim, an agency researcher takes a random sample of 20 of these mowers and attempts to start each one by pushing the button. Only 15 of the mowers start.

(a) State appropriate hypotheses for performing a significance test. Be sure to define the parameter of interest.

(b) Show that the Large Counts condition is not met.

When the Large Counts condition is violated, you shouldn't perform a one-sample z test about a proportion. But you can carry out a significance test using the binomial distribution.

(c) Assuming that the null hypothesis from part (a) is true, calculate the probability that 15 or fewer of the 20 randomly selected mowers would start.

(d) Based on your result in part (c), what conclusion would you make? Explain.

Recycle and Review

20. **Cheese nutrition (2.2, 2.3, 2.4)** The scatterplot shows the relationship between x = protein content in grams and y = total fat content in grams for 1 ounce of 25 different types of cheese.[15]

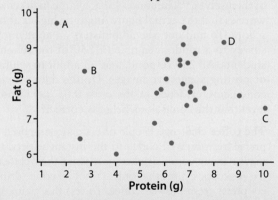

(a) Is the correlation between these two variables close to −1, close to +1, or close to 0? Explain.

(b) What effect do points A and B have on the correlation?

(c) If you want to buy a cheese that has a high protein content relative to its fat content, would you choose the cheese whose point is indicated by A, B, C, or D? Explain.

21. **Photo file sizes (7.6)** Rafiq notices that the file sizes of digital photographs he takes with his camera vary, depending on the image. He selects a random sample of 49 photo files and finds that the mean size is 8.05 megabytes and the standard deviation is 1.96 megabytes. Calculate and interpret a 99% confidence interval for the mean file size of photographs taken with Rafiq's camera.

Lesson 8.5

Testing a Claim about a Mean

You learned how to perform a significance test about a population proportion in Lessons 8.3 and 8.4. Now we'll examine the details of testing a claim about a population mean μ.

Conditions for Testing a Claim about μ

In Lesson 7.5, we introduced conditions that should be met before we construct a confidence interval for a population mean. We called them Random and Normal/Large Sample. These same conditions must be verified before carrying out a significance test.

How to Check Conditions for Performing a Significance Test about μ

- **Random:** The data come from a random sample from the population of interest.
- **Normal/Large Sample:** The data come from a normally distributed population or the sample size is large ($n \geq 30$). When the sample size is small and the shape of the population distribution is unknown, a graph of the sample data shows no strong skewness or outliers.

EXAMPLE

Does this radio station play enough music?

Checking conditions

PROBLEM: A "classic rock" radio station claims to play an average of 50 minutes of music every hour. However, it seems that every time you tune into this station, a commercial is playing. To investigate the station's claim, you randomly select 12 different hours during the next week and record how many minutes of music the station plays in each of those hours. Here are the data:

$$44 \quad 49 \quad 45 \quad 51 \quad 49 \quad 53 \quad 49 \quad 44 \quad 47 \quad 50 \quad 46 \quad 48$$

You would like to perform a test of

$$H_0: \mu = 50$$
$$H_a: \mu < 50$$

where μ = true mean amount of music played (in minutes) during each hour by this station. Check that the conditions for performing the test are met.

SOLUTION:

Random? Random sample of 12 different hours ✓

Normal/Large Sample? The sample size is small, but the boxplot doesn't show any outliers or strong skewness. ✓

> Because the sample size is less than 30, graph the sample data to see if it is plausible that they came from a normal population.

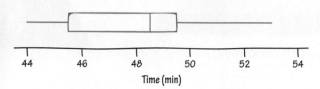

Time (min)

FOR PRACTICE TRY EXERCISE 1.

Calculating the Standardized Test Statistic

In the radio station example, the sample mean amount of music played during 12 randomly selected hours was $\bar{x} = 47.9$ minutes. Because this result is less than 50 minutes, there is *some* evidence against $H_0: \mu = 50$ and in favor of $H_a: \mu < 50$. But do we have *convincing* evidence that the true mean amount of music played during each hour by this station is less than 50 minutes? To answer this question, we want to know if it is likely to get a sample mean of 47.9 minutes or less by chance alone when the null hypothesis is true. As with proportions, we calculate a standardized test statistic and a *P*-value.

Suppose for now that the null hypothesis $H_0: \mu = 50$ is true. Consider the sample mean amount of music played $\bar{x}$ (in minutes) for a random sample of $n = 12$ hours. You learned in Lesson 6.5 that the sampling distribution of $\bar{x}$ will have mean

$$\mu_{\bar{x}} = \mu = 50 \text{ minutes}$$

and standard deviation

$$\sigma_{\bar{x}} = \frac{\sigma}{\sqrt{n}}$$

Because the population standard deviation σ is usually unknown, we use the sample standard deviation s_x in its place. For the radio station data, $s_x = 2.81$. Our resulting estimate of $\sigma_{\bar{x}}$ is the standard error of the mean:

$$SE_{\bar{x}} = \frac{s_x}{\sqrt{n}} = \frac{2.81}{\sqrt{12}} = 0.811$$

To assess how far the sample mean $\bar{x}$ is from the null hypothesis value μ_0, we standardize the statistic:

$$\text{standardized test statistic} = \frac{\text{statistic} - \text{null value}}{\text{standard deviation of statistic}}$$

When we use the sample standard deviation s_x to estimate the unknown population standard deviation σ_x, the standardized test statistic is denoted by t (you will learn why shortly). So the formula becomes

$$t = \frac{\bar{x} - \mu_0}{\frac{s_x}{\sqrt{n}}}$$

The standardized test statistic for the radio station data is

$$t = \frac{47.9 - 50}{\frac{2.81}{\sqrt{12}}} = \frac{47.9 - 50}{0.811} = -2.59$$

EXAMPLE

Anyone for t?

Calculating the standardized test statistic

PROBLEM: Suppose that you want to perform a test at the $\alpha = 0.10$ significance level of

$$H_0: \mu = 7$$
$$H_a: \mu \neq 7$$

A random sample of size $n = 25$ from the population of interest yields $\bar{x} = 7.045$ and $s_x = 0.2$. Calculate the standardized test statistic.

SOLUTION:

$$t = \frac{7.045 - 7}{\frac{0.2}{\sqrt{25}}} = 1.125$$

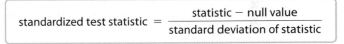

$$\text{standardized test statistic} = \frac{\text{statistic} - \text{null value}}{\text{standard deviation of statistic}}$$

FOR PRACTICE **TRY EXERCISE 5.**

Finding *P*-Values

When the Normal/Large Sample condition is met, the standardized test statistic

$$t = \frac{\bar{x} - \mu_0}{\frac{s_x}{\sqrt{n}}}$$

can be modeled by a **t distribution.** There is a different t distribution for each sample size. As you learned in Lesson 7.5, we specify a particular t distribution by giving its **degrees of freedom (df).** When we perform inference about a population mean μ using a t distribution, the appropriate degrees of freedom are found by subtracting 1 from the sample size n, making df $= n - 1$.

Figure 8.6 (on the next page) compares the density curves of the standard normal distribution and the t distributions with 2 and 9 degrees of freedom. The figure illustrates these facts about the t distributions:

- The t distributions are similar in shape to the standard normal distribution. They are symmetric about 0, single-peaked, and bell-shaped.

- The t distributions have more variability than the standard normal distribution. It is more likely to get an extremely large value of t (say, greater than 3) than an extremely large value of z because the t distributions have more area in the tails of the distribution.

- As the degrees of freedom increase, the t distributions approach the standard normal distribution.

FIGURE 8.6
Density curves for the *t* distributions with 2 and 9 degrees of freedom and the standard normal distribution. All are symmetric with center 0. The *t* distributions have more variability and a slightly different shape from the standard normal distribution.

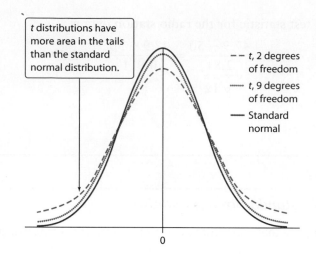

The *t* distribution and the *t* inference procedures were invented by William S. Gosset (1876–1937). Gosset worked for the Guinness brewery, and his goal was to help the company make a better product. He used his new *t* procedures to find the best varieties of barley and hops. Gosset's statistical work helped him become head brewer. Because Gosset published under the pen name "Student," you will often see the *t* distribution called "Student's *t*" in his honor.

> **DEFINITION** *t* Distribution, Degrees of freedom
>
> A **t distribution** is described by a symmetric, single-peaked, bell-shaped density curve. Any *t* distribution is completely specified by its **degrees of freedom (df).** When performing inference about a population mean based on a random sample of size *n*, use df = $n - 1$.

We can use Table B to find a *P*-value from the appropriate *t* distribution when performing a test about a population mean. In the radio station example, we planned to carry out a test of

$$H_0: \mu = 50$$
$$H_a: \mu < 50$$

where μ = the true mean amount of music played (in minutes) during each hour by this station. In $n = 12$ randomly selected hours, the radio station played an average of $\overline{x} = 47.9$ minutes of music. The *P*-value is the probability of getting a result this small or smaller by chance alone when $H_0: \mu = 50$ is true. Earlier, we calculated the standardized test statistic to be $t = -2.59$. So we estimate the *P*-value by finding $P(t \le -2.59)$ in a *t* distribution with df = $12 - 1 = 11$. The shaded area in Figure 8.7 shows this probability.

FIGURE 8.7 The shaded area shows the *P*-value for the radio station example as the area to the left of $t = -2.59$ in a *t* distribution with 11 degrees of freedom.

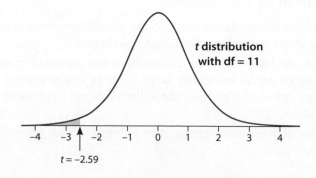

Our *P*-value is $P(t \leq -2.59)$ in a *t* distribution with df = 11. Table B shows only *positive t*-values and areas in the *right* tail of the *t* distributions. But the *t* distributions are symmetric around their center of 0. So $P(t \leq -2.59) = P(t \geq 2.59)$. Go to the df = 11 row. The value *t* = 2.59 falls between the values 2.328 and 2.718. Now look at the top of the corresponding columns in Table B. You see that the "Upper-tail probability *p*" is between 0.02 and 0.01 (see the excerpt from Table B). Therefore, the *P*-value for this test is between 0.01 and 0.02.

	Upper-tail probability *p*		
df	0.02	0.01	0.005
10	2.359	2.764	3.169
11	2.328	2.718	3.106
12	2.303	2.681	3.055

EXAMPLE

How likely is that t?

Finding the P-value

PROBLEM: Suppose that you want to perform a test at the $\alpha = 0.10$ significance level of $H_0: \mu = 7$ versus $H_a: \mu \neq 7$. A random sample of size *n* = 25 from the population of interest yields $\bar{x} = 7.045$ and $s_x = 0.2$. In the previous example, we calculated the standardized test statistic to be *t* = 1.125. Find the *P*-value using Table B. Assume that the conditions for carrying out the test are met.

SOLUTION:

df = 25 − 1 = 24

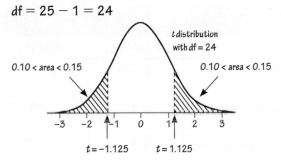

t distribution with df = 24

0.10 < area < 0.15   0.10 < area < 0.15

t = −1.125   *t* = 1.125

	Upper-tail probability *p*		
df	0.15	0.10	0.05
23	1.060	1.319	1.714
24	1.059	1.318	1.711
25	1.058	1.316	1.708

P-value between 2(0.10) = 0.20 and 2(0.15) = 0.30

FOR PRACTICE   TRY EXERCISE 9.

As you can see, Table B gives only an interval of possible *P*-values for a significance test. We can still draw a conclusion from the test in the same way as if we had a single probability. In the radio station example, because the *P*-value of between 0.01 and 0.02 is less than our default $\alpha = 0.05$ significance level, we would reject $H_0: \mu = 50$. We have convincing evidence that the radio station is playing fewer than 50 minutes of music per hour, on average.

What conclusion should we make for the two-sided test in the example above? Because the *P*-value of between 0.20 and 0.30 is greater than the stated significance level of $\alpha = 0.10$, we would fail to reject H_0. We do not have convincing evidence that the population mean differs from 7.

Table B has another limitation for finding *P*-values. It includes probabilities only for *t* distributions with degrees of freedom from 1 to 30 and then skips to df = 40, 50, 60, 80, 100, and 1000. (The bottom row—labeled z^*—gives probabilities for the standard normal distribution.)

If the correct df isn't listed, use the greatest df available that is less than the correct df. It's not fair to "round up" to a larger df. This is like pretending that your sample size is larger than it really is. Doing so would give you a smaller *P*-value than is true and therefore make you more likely to incorrectly reject H_0 when it's true (thereby making a Type I error).

CAUTION

LESSON APP 8.5

Who needs an aspirin?

The makers of Aspaway brand aspirin want to be sure that their tablets contain the right amount of active ingredient (acetylsalicylic acid). So they inspect a random sample of 30 tablets from a batch in production. When the production process is working properly, Aspaway tablets have an average of $\mu = 320$ milligrams (mg) of active ingredient. The amount of active ingredient in the 30 selected tablets has a mean of 319 mg and a standard deviation of 3 mg. We want to perform a test at the $\alpha = 0.05$ significance level of

$$H_0: \mu = 320$$
$$H_a: \mu \neq 320$$

where $\mu =$ the mean amount of active ingredient (in mg) in all Aspaway brand aspirin tablets.

1. Check that the conditions for performing the test are met.

2. Calculate the standardized test statistic.

3. Find the P-value using Table B. Show this result as an area under a t distribution curve.

4. What conclusion would you make?

Lesson 8.5

WHAT DID YOU LEARN?

LEARNING TARGET	EXAMPLES	EXERCISES
Check the Random and Normal/Large Sample conditions for performing a significance test about a population mean.	p. 529	1–4
Calculate the standardized test statistic for a significance test about a population mean.	p. 531	5–8
Find the P-value for a significance test about a population mean using Table B.	p. 533	9–12

Exercises   Lesson 8.5

Mastering Concepts and Skills

1. **The flow of reading** Does the use of fancy type fonts
pg 529 slow down the reading of text on a computer screen? Adults can read four paragraphs of text in the common Times New Roman font in an average time of 22 seconds. Researchers asked a random sample of 24 adults to read these same four paragraphs in the ornate font named Gigi. Here are their times, in seconds.

23.2 21.2 28.9 27.7 29.1 27.3 16.1 22.6 25.6 34.2 23.9 26.8
20.5 34.3 21.4 32.6 26.2 34.1 31.5 24.6 23.0 28.6 24.4 28.1

The researchers want to perform a test at the $\alpha = 0.05$ significance level of $H_0: \mu = 22$ versus $H_a: \mu > 22$, where μ is the true mean time to read the four paragraphs of text in Gigi font in the population of adults. Check that the conditions for performing the test are met.

2. **Tired?** A professor suspects that students at his school are getting fewer than 8 hours of sleep a night, on average. To test his belief, the professor asks a random sample of 28 students, "How much sleep did you get last night?" Here are the data (in hours).

9 6 8 6 8 8 6 6.5 6 7 9 4 3 4
5 6 11 6 3 6 6 10 7 8 4.5 9 7 7

The professor wants to perform a test at the $\alpha = 0.05$ significance level of $H_0: \mu = 8$ versus $H_a: \mu < 8$, where μ is the true mean amount of sleep last night in the population of students at his college. Check that the conditions for performing the test are met.

3. **Hefty pineapples** At the Hawaii Pineapple Company, managers are interested in the size of the pineapples grown in the company's fields. Last year, the mean weight of the pineapples harvested

from one large field was 31 ounces. A new irrigation system was installed in this field after the growing season. Managers wonder if the mean weight of pineapples grown in the field this year will be greater than last year. So they take an SRS of 50 pineapples from this year's crop. The managers want to perform a test at the $\alpha = 0.05$ significance level of H_0: $\mu = 31$ versus H_a: $\mu > 31$, where μ is the mean weight of all pineapples grown in the field this year. Check that the conditions for performing the test are met.

4. **Attitudes toward school** The Survey of Study Habits and Attitudes (SSHA) is a psychological test with scores that range from 0 to 200. The mean score for U.S. college students is about 115. A teacher suspects that older students have better attitudes toward school. She gives the SSHA to an SRS of 45 students at her college who are at least 30 years of age. The teacher wants to perform a test at the $\alpha = 0.05$ significance level of H_0: $\mu = 115$ versus H_a: $\mu > 115$, where μ is the mean SSHA score in the population of students at her college who are at least 30 years old. Check that the conditions for performing the test are met.

5. **Fonts and reading** Refer to Exercise 1. Find the
pg 531   sample mean and standard deviation. Then calculate the standardized test statistic.

6. **More sleepless nights?** Refer to Exercise 2. Find the sample mean and standard deviation. Then calculate the standardized test statistic.

7. **Two-tailed test** Suppose that you want to perform a test at the $\alpha = 0.10$ significance level of
$$H_0: \mu = 5$$
$$H_a: \mu \neq 5$$
A random sample of size $n = 20$ from the population of interest yields $\overline{x} = 5.31$ and $s_x = 0.79$. Calculate the standardized test statistic.

8. **Another two-tailed test** Suppose that you want to perform a test at the $\alpha = 0.10$ significance level of
$$H_0: \mu = 64$$
$$H_a: \mu \neq 64$$
A random sample of size $n = 16$ from the population of interest yields $\overline{x} = 62.8$ and $s_x = 2.85$. Calculate the standardized test statistic.

9. **Fonts and reading: P-value** Refer to Exercises 1 and
pg 533   5. Find the P-value using Table B.

10. **Sleepless, counting P-values** Refer to Exercises 2 and 6. Find the P-value using Table B.

11. **Two-tailed P-value** Refer to Exercise 7. Find the P-value using Table B. Assume that the conditions for carrying out the test are met.

12. **Another two-tailed P-value** Refer to Exercise 8. Find the P-value using Table B. Assume that the conditions for carrying out the test are met.

Applying the Concepts

13. **Gasoline prices** Anne reads that the average price of regular gas in her state is $3.06 per gallon. To see if the average price of gas is different in her city, she selects 10 gas stations at random and records the price per gallon for regular gas at each station. Here are the data.

3.13   3.01   3.09   3.05   2.97   2.99   3.05   2.98   3.09   3.02

Anne wants to perform a test at the $\alpha = 0.05$ significance level of H_0: $\mu = \$3.06$ versus H_a: $\mu \neq \$3.06$, where $\mu = $ the true mean price of regular gasoline at all stations in her city.

(a) Check that the conditions for performing the test are met.

(b) Calculate the standardized test statistic.

(c) Find the P-value using Table B.

(d) What conclusion should Anne make?

14. **Construction zones** Every road has one at some point—construction zones that have much lower speed limits. To see if drivers obey these lower speed limits, a police officer uses a radar gun to measure the speed (in miles per hour, or mph) of a random sample of 10 drivers in a 25 mph construction zone. Here are the data.

27   33   32   21   30   30   29   25   27   34

We want to perform a test at the $\alpha = 0.05$ significance level of H_0: $\mu = 25$ versus H_a: $\mu > 25$, where $\mu = $ the true mean speed of all cars that drive through the construction zone.

(a) Check that the conditions for performing the test are met.

(b) Calculate the standardized test statistic.

(c) Find the P-value using Table B.

(d) What conclusion should we make?

15. **Sampling shoppers** A marketing consultant observes 50 consecutive shoppers at a supermarket, recording how much each shopper spends in the store. Explain why it would not be wise to use these data to carry out a significance test about the mean amount spent by all shoppers at this supermarket.

16. **Paying high prices?** A retailer entered into an exclusive agreement with a supplier who guaranteed to provide all products at competitive prices. To be sure the supplier honored the terms of the agreement, the retailer had an audit performed on a random sample of 25 invoices. The percent of purchases on each invoice for which an alternative supplier offered a lower price than the original supplier was recorded.[16] For example, a data value of 38 means that the price would be lower with a different supplier for 38% of the items on the invoice. A histogram of these data is shown on the next page. Explain why we should not perform a significance test about the mean percent of purchases for which an alternative supplier offered a lower price on all of this company's invoices.

Recycle and Review

17. **Sipping gasoline (5.7)** In its 2015 *Fuel Economy Guide,* the Environmental Protection Agency gives data on 1164 car models that have gasoline engines and are not hybrids. The combined city and highway gas mileage for these vehicles is approximately normal with mean 22.3 miles per gallon (mpg) and standard deviation 5.0 mpg.[17]

(a) The Chevrolet Malibu with a standard four-cylinder engine has a combined gas mileage of 29 mpg. What percent of all vehicles have worse gas mileage than the Malibu?

(b) How high must a vehicle's gas mileage be to fall in the top 10% of all these car models?

18. **Spoofing (2.1, 3.1)** To collect information such as passwords, online criminals use "spoofing" to direct Internet users to fraudulent websites. In one study of Internet fraud, students were warned about spoofing and then asked to log into their university account starting from the university's home page. In some cases, the log-in link led to the genuine dialog box. In others, the box looked genuine but, in fact, was linked to a different site that recorded the ID and password the student entered. The box that appeared for each student was determined at random. An alert student could detect the fraud by looking at the true Internet address displayed in the browser status bar, but most just entered their ID and password. Is this study an experiment? Why or why not? What are the explanatory and response variables?

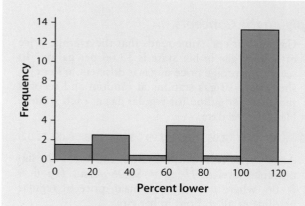

Lesson 8.6

Significance Tests for a Mean

LEARNING TARGETS

- Use the four-step process to perform a significance test about a population mean.
- Use a confidence interval to draw a conclusion about a two-sided test for a population mean.

In Lesson 8.5, you learned how to check conditions and perform calculations for a test about a population mean. This lesson shows you how to use the four-step process to carry out a significance test about μ, and how to use a confidence interval to make a conclusion about a two-sided test for μ.

Putting It All Together: Testing a Claim about a Population Mean

A significance test for a population mean is often referred to as a *one-sample t test for μ.*

EXAMPLE

Are these subs too short?

Performing a significance test about μ

PROBLEM: Abby and Raquel like to eat sub sandwiches. However, they noticed that the lengths of the "6-inch sub" sandwiches they get at their favorite restaurant seemed shorter than the advertised length. To investigate, they randomly selected 24 different times during the next month and ordered a "6-inch" sub. Here are the actual lengths (in inches) of each of the 24 sandwiches.

4.50 4.75 4.75 5.00 5.00 5.00 5.50 5.50 5.50 5.50 5.50 5.50
5.75 5.75 5.75 6.00 6.00 6.00 6.00 6.00 6.50 6.75 6.75 7.00

Do these data provide convincing evidence at the $\alpha = 0.10$ level that the sandwiches at this restaurant are shorter than advertised, on average?

SOLUTION:

STATE: We want to perform a test at the $\alpha = 0.10$ significance level of

$$H_0: \mu = 6$$
$$H_a: \mu < 6$$

where μ = the true mean length of all 6-inch subs from this restaurant.

Follow the four-step process!

STATE: State the hypotheses you want to test and the significance level, and define any parameters you use.

PLAN: One-sample t test for μ

Random? They randomly selected 24 times to buy a sub. ✓

Normal/Large Sample? The sample size is small, but the dotplot doesn't show any outliers or strong skewness. ✓

PLAN: Identify the appropriate inference method and check conditions.

Because the sample size is less than 30, graph the sample data to see if it is plausible that they came from a normal population.

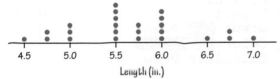

DO:

- $\bar{x} = 5.677$ inches, $s_x = 0.657$ inches

- $t = \dfrac{5.677 - 6}{\dfrac{0.657}{\sqrt{24}}} = -2.41$

- P-value: df $= 24 - 1 = 23$

DO: If the conditions are met, perform calculations: • Identify the sample statistic. • Calculate the standardized test statistic. • Find the P-value.

Use one-variable statistics or an applet to compute the sample mean and standard deviation.

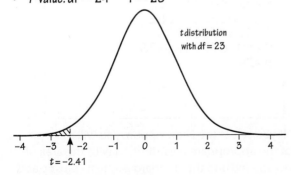

The sample result gives *some* evidence in favor of H_a because $5.677 < 6$.

P-value between 0.01 and 0.02

CONCLUDE: Because the P-value of between 0.01 and 0.02 is less than $\alpha = 0.10$, we reject H_0. There is convincing evidence that the true mean length of all "6-inch" subs at this restaurant is less than 6 inches.

CONCLUDE: Make a conclusion about the hypotheses in the context of the problem.

FOR PRACTICE TRY EXERCISE 1.

Table B allows us to report the *P*-value in the example only as somewhere between 0.01 and 0.02. We can still interpret the *P*-value in context: It is very unlikely to get a random sample of 24 subs with a mean length of 5.677 inches or less if H_0: $\mu = 6$ is true. Given the limitations of Table B, our advice is to use technology to find *P*-values when carrying out a significance test about a population mean.

One-Sample *t*-Test for a Mean

You can use technology to perform the calculations for a significance test about a population mean. We'll illustrate using the preceding example.

Applet

1. Launch the *One Quantitative Variable* applet at highschool.bfwpub.com/spa3e.

2. Enter "Length (inches)" as the variable name. Then input the data. Be sure to separate the data values with spaces or commas as you type them.

One Quantitative Variable

Variable name: Length (inches)
Number of groups: 1 ▼
Input: Raw data ▼

Input data separated by commas or spaces.
Group 1 data: 4.50 4.75 4.75 5.00 5.00 5.00 5.50 5.50 5.50 5.50 5.

[Begin analysis] [Edit inputs] [Reset everything]

3. Press the "Begin analysis" button. Notice that the dotplot shows no strong skewness or outliers, so the Normal/Large Sample condition is met.

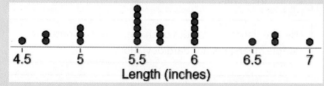

4. Now scroll down to the "Perform inference" section. Choose the 1-sample *t* test for the procedure, $\mu <$ for the alternative hypothesis, and enter 6 for the hypothesized mean. Then press the "Perform inference" button.

Perform Inference

Inference procedure: [1-sample t test ▼] Alternative
hypothesis: [μ < ▼] Hypothesized mean: 6

[Perform inference]

t	P-value	df
-2.407	0.012	23

TI-83/84

1. Enter the data values into list L_1.

2. Press STAT, then choose TESTS and T-Test.

3. When the T-Test screen appears, choose Data as the input method. Then, enter $\mu_0 = 6$, List: L_1, Freq: 1, and choose $\mu < \mu_0$.

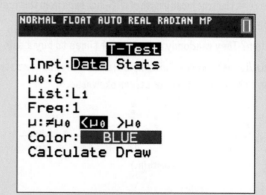

4. Highlight Calculate and press ENTER.

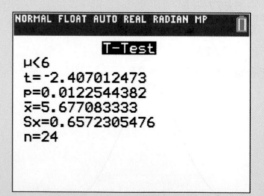

Note: If you select the Draw option, you will get a picture of the appropriate *t* distribution with the area of interest shaded and the standardized test statistic and *P*-value labeled.

TWO-SIDED TESTS AND CONFIDENCE INTERVALS

Who broke the ice?

Two-sided test

PROBLEM: In the children's game Don't Break the Ice, small plastic ice cubes are squeezed into a square frame. Each child takes turns tapping out a cube of "ice" with a plastic hammer, hoping that the remaining cubes don't collapse. For the game to work correctly, the cubes must be big enough so that they hold each other in place in the plastic frame, but not so big that they are too difficult to tap out. The machine that produces the plastic cubes is designed to make cubes that are 25.4 millimeters (mm) wide, but the width varies a little. To ensure that the machine is working well, a supervisor inspects a random sample of 50 cubes every hour and measures their width. The data from a sample taken during 1 hour yield a mean of 25.4154 mm and a standard deviation of 0.09 mm. Do these data give convincing evidence at the $\alpha = 0.05$ significance level that the mean width of cubes produced this hour is different from 25.4 mm?

Jeremy Harrison

SOLUTION:

STATE: We want to perform a test at the $\alpha = 0.05$ significance level of

$H_0: \mu = 25.4$

$H_a: \mu \neq 25.4$

where μ = the true mean width (mm) of all plastic ice cubes produced this hour.

> Follow the four-step process!

> **STATE:** State the hypotheses you want to test and the significance level, and define any parameters you use.

PLAN: One-sample t test for μ

Random? Random sample of 50 ice cubes. ✓

Normal/Large Sample? The sample size is large ($n = 50 \geq 30$). ✓

> **PLAN:** Identify the appropriate inference method and check conditions.

DO:

• $\bar{x} = 25.4154$ mm, $s_x = 0.09$ mm

• $t = \dfrac{25.4154 - 25.4}{\dfrac{0.09}{\sqrt{50}}} = 1.21$

• P-value: df $= 50 - 1 = 49$

> **DO:** If the conditions are met, perform calculations:
> • Identify the sample statistic.
> • Calculate the standardized test statistic.
> • Find the P-value.

> The sample result gives *some* evidence in favor of H_a because $25.4154 \neq 25.4$.

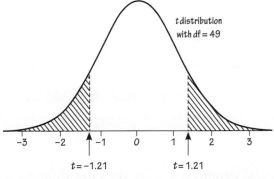

t distribution with df $= 49$

$t = -1.21$ $t = 1.21$

Using Table B: Using df $= 40$, the P-value is between $2(0.10) = 0.20$ and $2(0.15) = 0.30$.

Using technology: Applet / T-test using df $= 49$ gives P-value $= 0.2321$.

CONCLUDE: Because the P-value of 0.2321 is greater than $\alpha = 0.05$, we fail to reject H_0. There is not convincing evidence that the true mean width of all plastic ice cubes produced this hour is different from 25.4 mm.

> **CONCLUDE:** Make a conclusion about the hypotheses in the context of the problem.

FOR PRACTICE   TRY EXERCISE 3.

The significance test in the example led to a simple decision: fail to reject H_0. This result tells us that 25.4 millimeters is a *plausible* value for the true mean width of the plastic ice cubes made during this hour of production. But we saw in Lesson 8.4 that a confidence interval gives more information than a test does—it provides the entire set of plausible values for the parameter based on the data. A 95% confidence interval for the mean width of the plastic ice cubes produced this hour is

$$\bar{x} \pm t^* \frac{s_x}{\sqrt{n}} \rightarrow 25.4154 \pm 2.01 \frac{0.09}{\sqrt{50}} \rightarrow 25.4154 \pm 0.026 \rightarrow 25.3894 \text{ to } 25.4414$$

(We obtained the critical value $t^* = 2.01$ from a t distribution with df $= 50 - 1 = 49$.) Because the interval includes 25.4, we know that it is a plausible value for the true mean width. This is consistent with the conclusion from our two-sided test.

The link between two-sided tests and confidence intervals for a population mean allows us to make a conclusion directly from a confidence interval.

- If a 95% confidence interval for μ does not capture the null value, then we can reject H_0 at the $\alpha = 0.05$ significance level.

- If a 95% confidence interval for μ captures the null value, then we should fail to reject H_0 at the $\alpha = 0.05$ significance level.

The same logic applies for other confidence levels, but *only* for a two-sided test.

EXAMPLE

Do postal workers stick around?

Confidence intervals and two-sided tests

PROBLEM: Twenty years ago, the mean time of employment for the population of postal employees was 7.5 years. Is that value still true today? To find out, researchers take a random sample of 100 current postal employees and record how long each person has worked at the postal service. A 99% confidence interval for the mean time of employment is 6.603 years to 7.397 years. What conclusion would you make for a test at the $\alpha = 0.01$ significance level of H_0: $\mu = 7.5$ versus H_a: $\mu \neq 7.5$, where μ is the true mean time of employment among all current postal employees? Explain.

SOLUTION:

Because the null value of 7.5 is not captured by the 99% confidence interval, we would reject H_0: $\mu = 7.5$ at the $\alpha = 0.01$ significance level. There is convincing evidence that the true mean time of employment among all postal employees is different now than it was 20 years ago.

> The confidence interval suggests that the mean time of employment now is somewhat less than it once was— somewhere between 6.603 years and 7.397 years.

FOR PRACTICE   TRY EXERCISE 7.

CAUTION

Remember the wise saying: Statistical significance is not the same thing as practical importance. The sample mean in the preceding example was $\bar{x} = 7$ years (the value at the center of the confidence interval). This sample result is statistically significant at the $\alpha = 0.01$ level. But the change in the average time of employment for postal employees suggested by the 99% confidence interval may not be practically important, especially if the true mean has only dropped from 7.5 years to 7.397 years!

LESSON APP 8.6

Do our employees have high blood pressure?

The health director of a large company wants to monitor the health of the company's middle-aged male employees. According to the National Center for Health Statistics, the mean systolic blood pressure for males 35 to 44 years of age is 124. The health director takes a random sample of 72 male employees in this age group and records their blood pressures. The data have a mean of 126.93 and a standard deviation of 14.90.

1. Do the data give convincing evidence at the $\alpha = 0.05$ significance level that the mean blood pressure μ for all the company's 35- to 44-year-old male employees differs from the national average?

2. A 95% confidence interval for μ is 123.43 to 130.43. Explain how this interval gives results consistent with the test in Question 1.

Lesson 8.6

WHAT DID YOU LEARN?

LEARNING TARGET	EXAMPLES	EXERCISES
Use the four-step process to perform a significance test about a population mean.	pp. 537, 539	1–6
Use a confidence interval to draw a conclusion about a two-sided test for a population mean.	p. 540	7–10

Exercises   Lesson 8.6

Mastering Concepts and Skills

1. **Healthy waterways** The level of dissolved oxygen
pg 537 (DO) in a stream or river is an important indicator of the water's ability to support aquatic life. A researcher measures the DO level at 15 randomly chosen locations along a stream. Here are the results in milligrams per liter (mg/l):

    4.53 5.04 3.29 5.23 4.13 5.50 4.83 4.40

    5.42 6.38 4.01 4.66 2.87 5.73 5.55

    A mean dissolved oxygen level below 5 mg/l puts aquatic life at risk. Do we have convincing evidence at the $\alpha = 0.10$ significance level that aquatic life in this stream is at risk?

2. **Women need calcium** The recommended daily allowance (RDA) of calcium for women between the ages of 18 and 24 years is 1200 milligrams (mg). Researchers who were involved in a large-scale study of women's bone health suspected that their participants had significantly lower calcium intakes than the RDA. To test this suspicion, the researchers measured the daily calcium intake of a random sample of 36 women from the study who fell in the desired age range. The sample mean was 856.2 mg and the standard deviation was 306.7 mg. Do these data give convincing evidence at the $\alpha = 0.01$ significance level that the researchers' suspicion is correct?

3. **The cost of food** A government report says that
pg 539 the average amount of money spent per U.S. household per week on food is about $158. A random sample of 50 households in a small city is selected, and their weekly spending on food is recorded. The sample data have a mean of $165 and a standard deviation of $20. Is there convincing evidence at the $\alpha = 0.05$ significance level that the mean weekly spending on food in this city differs from the national figure of $158?

4. **Filling cola** Bottles of a popular cola are supposed to contain 300 milliliters (ml) of cola. There is some variation from bottle to bottle because the filling machinery is not perfectly precise. An inspector measures the contents (in ml) of six randomly selected bottles from a single day's production. Here are the data.

299.4   297.7   301.0   298.9   300.2   297.0

Do these data provide convincing evidence at the $\alpha = 0.10$ significance level that the mean amount of cola in all the bottles filled that day differs from the target value of 300 ml?

5. **Improving SAT scores** A national chain of SAT-preparation schools wants to know if using a smartphone app in addition to its regular program will help increase student scores more than using just the regular program. On average, the students in the regular program increase their scores by 128 points during the 3-month class. To investigate using the smartphone app, a random sample of 5000 students uses the app along with the regular program. After 3 months, the average improvement was $\bar{x} = 130$ with a standard deviation of $s_x = 65$. Is there convincing evidence at the $\alpha = 0.05$ significance level that the average score increase for students who use the smartphone app in addition to the regular program is greater than 128 points?

6. **Healing wounds quickly** Suppose researchers are testing a new antibacterial cream, "Formulation NS," on a small cut made on the inner forearm. Previous research indicates that with no medication, the mean healing time (defined as the time for the scab to fall off) is 7.6 days. A random sample of 250 college students who agreed to participate in the study apply Formulation NS to their wounds. The mean healing time for these subjects is $\bar{x} = 7.5$ days and the standard deviation is $s_x = 0.9$ days. Test the claim that Formulation NS speeds healing at the $\alpha = 0.05$ significance level.

7. **Sniffing out radon** Radon is a colorless, odorless gas that is naturally released by rocks and soils and may concentrate in tightly closed houses. It is slightly radioactive, so there is some concern that it may pose a health hazard. Radon detectors are sold to homeowners, but the detectors may be inaccurate. University researchers placed a random sample of 11 detectors in a chamber and exposed them to 105 picocuries per liter of radon over 3 days. A graph of the radon readings from the 11 detectors shows no strong skewness or outliers. The 90% confidence interval for the true mean reading is 99.61 to 110.03. What conclusion would you make for a test at the $\alpha = 0.10$ significance

pg 540

level of H_0: $\mu = 105$ versus H_a: $\mu \neq 105$, where μ is the true mean reading in the population of radon detectors? Explain.

8. **Pressing pills** A drug manufacturer forms tablets by compressing grains of material that contain the active ingredient and fillers. To assure quality, the hardness of a sample from each batch of tablets produced is measured. The target value for the hardness is $\mu = 11.5$. A random sample of 20 tablets is taken, and their hardness is measured. A graph of the data is roughly symmetric with no outliers. The 95% confidence interval for the mean hardness of the tablets is 11.472 to 11.561. What conclusion would you make for a test at the $\alpha = 0.05$ significance level of H_0: $\mu = 11.5$ versus H_a: $\mu \neq 11.5$, where μ is the true mean hardness in the population of tablets? Explain.

9. **How fast?** How long does it take for a chunk of information to travel from one server to another and back on the Internet? The site internettrafficreport.com reports the typical response time as 200 milliseconds (about one-fifth of a second). Researchers collected data on response times of a random sample of 14 servers in Europe. A graph of the data reveals no strong skewness or outliers. The 95% confidence interval for the mean response time is 158.22 to 189.64. Is there convincing evidence at the 5% significance level that the site's claim is incorrect? Justify your answer.

10. **Drink water!** A blogger claims that U.S. adults drink an average of five 8-ounce glasses of water a day. To test this claim, researchers ask a random sample of 24 U.S. adults about their daily water intake. A graph of the data shows a roughly symmetric shape with no outliers. The 90% confidence interval for the mean daily water intake is 3.794 to 4.615. Is there convincing evidence at the 10% significance level that the blogger's claim is incorrect? Justify your answer.

Applying the Concepts

11. **Ending insomnia** An experiment was carried out with a random sample of 10 patients who suffer from insomnia to investigate the effectiveness of a drug designed to increase sleep time. The following data show the number of additional hours of sleep per night gained by each subject after taking the drug.[18] A negative value indicates that the subject got *less* sleep after taking the drug.

1.9   0.8   1.1   0.1   −0.1   4.4   5.5   1.6   4.6   3.4

(a) Is there convincing evidence at the $\alpha = 0.01$ significance level that the drug is effective?

(b) Given your conclusion in part (a), which kind of mistake—a Type I error or a Type II error—could you have made? Explain what this mistake would mean in context.

12. Big ears? A certain variety of sweet corn produces individual ears with a mean weight of 8 ounces. A farmer is testing a new fertilizer designed to produce heavier ears of corn. He finds that 32 randomly selected ears of corn grown with this fertilizer have a mean weight of 8.23 ounces and a standard deviation of 0.8 ounce.

(a) Do the data provide convincing evidence at the $\alpha = 0.05$ level that the fertilized ears of corn weigh more than 8 ounces, on average?

(b) Given your conclusion in part (a), which kind of mistake—a Type I error or a Type II error—could you have made? Explain what this mistake would mean in context.

13. Tests and CIs The P-value for a two-sided test of the null hypothesis H_0: $\mu = 10$ is 0.06.

(a) Does the 95% confidence interval for μ include 10? Why or why not?

(b) Does the 90% confidence interval for μ include 10? Why or why not?

14. More tests and CIs The P-value for a two-sided test of the null hypothesis H_0: $\mu = 15$ is 0.03.

(a) Does the 99% confidence interval for μ include 15? Why or why not?

(b) Does the 95% confidence interval for μ include 15? Why or why not?

15. Improved scores? Refer to Exercise 5. Although the result of the study was statistically significant, it is not practically important. Explain why.

16. Practical healing? Refer to Exercise 6. Although the result of the study was statistically significant, it is not practically important. Explain why.

Extending the Concepts

17. Confidence intervals and one-sided tests There is a connection between a one-sided test and a confidence interval for a population mean. But it's a little complicated. Consider a one-sided test at the $\alpha = 0.05$ significance level of H_0: $\mu = 10$ versus H_a: $\mu > 10$ based on an SRS of $n = 20$ observations.

(a) Complete this statement: We should reject H_0 if the standardized test statistic $t >$ _____ . In other words, we should reject H_0 if the sample mean is more than _____ standard errors greater than $\mu = 10$.

(b) Find the t^* critical value for a 90% confidence level in this setting. The resulting interval is $\bar{x} \pm t^*(\text{SE})$.

(c) Suppose that the sample mean $\bar{x}$ leads us to reject H_0. Use parts (a) and (b) to explain why the 90% confidence interval *cannot* contain $\mu = 10$.

Recycle and Review

18. Who can get ahead? (7.1, 7.2) A recent CBS/*New York Times* poll asked a random sample of Americans, "Which comes closer to your view: In today's economy, everyone has a fair chance to get ahead in the long run, or in today's economy, it's mainly just a few people at the top who have a chance to get ahead?" The 95% confidence interval for the proportion of people who would choose "it's mainly just a few people at the top" is 58% to 64%.[19]

(a) Interpret the confidence interval in context.

(b) Interpret the confidence level in context.

19. Stop doing homework! (2.3, 3.6) Researchers in Spain interviewed 7725 13-year-olds about their homework habits—how much time they spent per night on homework and whether they got help from their parents or not—and then had them take a test with 24 math questions and 24 science questions. They found that students who spent between 90 and 100 minutes on homework did only a little better on the test than those who spent 60 to 70 minutes on homework. Beyond 100 minutes, students who spent more time did worse than those who spent less time. The researchers concluded that 60 to 70 minutes per night is the optimum time for students to spend on homework.[20] Is it appropriate to conclude that students who reduce their homework time from 120 minutes to 70 minutes will likely improve their performance on tests such as those used in this study? Why or why not?

Chapter 8

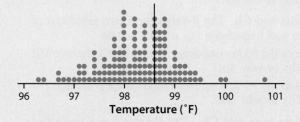

STATS applied!

What is normal body temperature?

At the beginning of the chapter, we described a study investigating whether "normal" human body temperature is really 98.6°F. Here is a dotplot and some summary statistics for the data described in the STATS applied! (page 499) for the random sample of 130 healthy 18- to 40-year-olds.

Temperature (°F)

- The mean temperature was $\bar{x} = 98.25$°F.
- The standard deviation of the temperature readings was $s_x = 0.73$°F.
- 62.3% of the temperature readings were less than 98.6°F.

1. If "normal" body temperature really is 98.6°F, we would expect that half of all healthy 18- to 40-year-olds will have a body temperature lower than 98.6°F. Do the data from this study provide convincing evidence at the $\alpha = 0.05$ level that this is not the case?

2. Based on the conclusion in Question 1, which type of error could have been made: a Type I error or a Type II error? Explain.

3. Do the data provide convincing evidence that the true mean body temperature in all healthy 18- to 40-year-olds is not 98.6°F?

4. A 95% confidence interval for the population mean is 98.123°F to 98.377°F. Explain how the confidence interval is consistent with, but gives more information than, the significance test in Question 3.

Main Points   *Chapter 8*

The Idea of a Significance Test

- A **significance test** is a procedure for using observed data to decide between two competing claims, called hypotheses. The hypotheses are usually statements about a parameter, like the population proportion p or the population mean μ.

- The claim we weigh evidence *against* in a statistical test is called the **null hypothesis** (H_0). The null hypothesis often has the form H_0: parameter = null value.

- The claim about the population that we are trying to find evidence *for* is the **alternative hypothesis** (H_a).
  - A **one-sided** alternative hypothesis has the form H_a: parameter < null value or H_a: parameter > null value.
  - A **two-sided** alternative hypothesis has the form H_a: parameter ≠ null value.

- A **standardized test statistic** measures how far a sample statistic is from what we would expect if the null hypothesis H_0 were true, in standardized units. That is,

$$\text{standardized test statistic} = \frac{\text{statistic} - \text{null value}}{\text{standard deviation of statistic}}$$

- The **P-value** of a test is the probability, computed assuming H_0 is true, that the statistic (such as $\hat{p}$ or $\bar{x}$) would take a value as extreme as or more extreme than the one observed, in the direction specified by H_a.

- Small *P*-values are evidence against the null hypothesis and for the alternative hypothesis because they say that the observed result is unlikely to occur when H_0 is true. To determine if a *P*-value should be considered small, we compare it to the **significance level** α.

- We make a conclusion in a significance test based on the *P*-value.

- If *P*-value < α: Reject H_0 and conclude that there is convincing evidence for H_a (in context).
- If *P*-value ≥ α: Fail to reject H_0 and conclude that there is not convincing evidence for H_a (in context).

- When we make a conclusion in a significance test, there are two kinds of mistakes we can make.
  - A **Type I error** occurs if we find convincing evidence that H_a is true when it really isn't.
  - A **Type II error** occurs if we do not find convincing evidence that H_a is true when it really is.

- The probability of making a Type I error is equal to the significance level α. There is a tradeoff between P(Type I error) and P(Type II error): As one increases, the other decreases. So it is important to consider the possible consequences of each type of error before choosing a significance level.

How Significance Tests Work

- When you perform a significance test, follow the four-step process:

STATE: State the hypotheses you want to test and the significance level, and define any parameters you use.

PLAN: Identify the appropriate inference method and check conditions.

DO: If the conditions are met, perform calculations.
- Give the sample statistic.
- Calculate the standardized test statistic.
- Find the *P*-value.

CONCLUDE: Make a conclusion about the hypotheses in the context of the problem.

Significance Tests for a Proportion

- ■ The conditions for performing a significance test of H_0: $p = p_0$ are:
  - ■ **Random:** The data come from a random sample from the population of interest.
  - ■ **Large Counts:** Both np_0 and $n(1 - p_0)$ are at least 10.
- ■ The standardized test statistic for a *one-sample z test for a proportion* is

$$z = \frac{\hat{p} - p_0}{\sqrt{\dfrac{p_0(1 - p_0)}{n}}}$$

- ■ When the Large Counts condition is met, the standardized test statistic has approximately a standard normal distribution. You can use Table A or technology to find the *P*-value.
- ■ Confidence intervals provide additional information that significance tests do not—namely, a set of plausible values for the population proportion p. A 95% confidence interval for p usually gives consistent results with a two-sided test of H_0: $p = p_0$ at the $\alpha = 0.05$ significance level.

Significance Tests for a Mean

- ■ The conditions for performing a significance test of H_0: $\mu = \mu_0$ are:
  - ■ **Random:** The data come from a random sample from the population of interest.
  - ■ **Normal/Large Sample:** The data come from a normally distributed population or the sample size is large ($n \geq 30$). When the sample size is small and the shape of the population distribution is unknown, a graph of the sample data shows no strong skewness or outliers.
- ■ The standardized test statistic for a *one-sample t test about a mean* is

$$t = \frac{\bar{x} - \mu_0}{\dfrac{s_x}{\sqrt{n}}}$$

- ■ When the Normal/Large Sample condition is met, the standardized test statistic can be modeled by a *t* **distribution** with $n - 1$ **degrees of freedom** (df). You can use Table B or technology to find the *P*-value.
- ■ Confidence intervals provide additional information that significance tests do not—namely, a set of plausible values for the population mean μ. A 95% confidence interval for μ gives consistent results with a two-sided test of H_0: $\mu = \mu_0$ at the $\alpha = 0.05$ significance level.

Chapter 8 Review Exercises

1. **Stating hypotheses (8.1)** For each of the following settings, define the parameter of interest and write the appropriate null and alternative hypotheses for the test that is described.

(a) The average height of 18-year-old American women is 64.2 inches. Researchers wonder if the mean height of this year's female graduates from a large local high school differs from the national average. You measure an SRS of 48 female graduates and find that $\bar{x} = 63.1$ inches.

(b) According to the Humane Society, 47% of U.S. households own at least 1 dog. You suspect that the proportion of households that own at least one dog among students at your school is higher than the national proportion.

2. **Finding P-values (8.3, 8.5)** Use a table or technology to find the P-value in each of the following settings. Assume that the conditions for performing inference are met.

(a) A test of H_0: $p = 0.7$ versus H_a: $p > 0.7$ based on a random sample of size $n = 40$ yields a standardized test statistic of 1.83.

(b) A test of H_0: $\mu = 30$ versus H_a: $\mu \neq 30$ based on a random sample of size $n = 20$ yields a standardized test statistic of -2.35.

3. **Identifying flavors (8.1, 8.2, 8.3)** Are the five flavors of a name-brand fruit snack identifiable? Monica and Shivani conducted a taste test of 50 randomly selected students at their school by giving each student a randomly selected snack and asking them to identify which of the five possible flavors it was. They found that 17 of the 50 subjects correctly identified the flavor of their snack. If the subjects were randomly guessing which of the five flavors they'd just eaten, they would guess correctly 20% of the time. Monica and Shivani want to determine if their data provide convincing evidence that students at their school can identify which flavor of fruit snack they are eating better than random guessing.

(a) State appropriate hypotheses for performing a significance test. Be sure to define the parameter of interest.

(b) Check that the conditions for performing the significance test are met.

(c) The P-value for this test is 0.0067. Interpret the P-value in context.

(d) What conclusion would you make at the $\alpha = 0.01$ significance level?

4. **Underemployment (8.4)** Economists often track employment trends by measuring the proportion of people who are "underemployed," meaning they are either unemployed or would like to work full-time but are only working part-time. In the spring of 2015, 14.7% of Americans were underemployed. The mayor of Carrboro wants to know if the situation in his city is the same as in the rest of the country. A staff member takes a simple random sample of 300 Carrboro residents and finds that 54 of them are underemployed. Do the data give convincing evidence at the $\alpha = 0.05$ significance level that the proportion of underemployed workers in Carrboro is different than elsewhere in the country?

5. **Breaking strength? (8.2)** A company that manufactures classroom chairs for high school students claims that the mean breaking strength of the chairs it makes is 300 pounds. One of the chairs collapsed beneath a 220-pound student last week. You wonder if the manufacturer exaggerates the breaking strength.

(a) State appropriate null and alternative hypotheses for performing a significance test.

(b) Describe a Type I error and a Type II error in this setting.

(c) Give a consequence of each type of error in this setting. Which type of error is more serious?

6. **Glass conducting heat (8.5, 8.6)** How well materials conduct heat matters when designing houses, for example. Conductivity is measured in terms of watts of heat power transmitted per square meter of surface per degree Celsius of temperature difference on the two sides of the material. In these units, glass should have conductivity of about 1.10. Here are measurements of the heat conductivity of 11 randomly selected pieces of a particular type of glass:

1.11 1.07 1.11 1.07 1.12 1.08 1.08 1.18 1.18 1.17 1.12

Is there convincing evidence at the $\alpha = 0.05$ significance level that the mean conductivity of this type of glass is greater than 1.10?

7. **(8.6) How old am I?** A 40-year-old high school teacher asks 75 randomly selected students at his school to guess his age. The 90% confidence interval for the mean age guess is 38.8 ± 0.75 years. Is there convincing evidence at the $\alpha = 0.10$ significance level that students are not able to accurately guess the teacher's age, on average? Justify your answer.

Chapter 8 Practice Test

Section I: Multiple Choice *Select the best answer for each question.*

1. Experiments on learning in animals sometimes measure how long it takes mice to find their way through a maze. The mean time is 18 seconds for one particular maze. A researcher thinks that a loud noise will cause the mice to complete the maze faster. She measures how long each of 10 mice takes with a noise stimulus. The appropriate hypotheses for the significance test are

(a) $H_0: \mu = 18; H_a: \mu \neq 18$.   (c) $H_0: \mu < 18; H_a: \mu = 18$.

(b) $H_0: \mu = 18; H_a: \mu > 18$.   (d) $H_0: \mu = 18; H_a: \mu < 18$.

2. Some people say that more babies are born in September than in any other month. To test this claim, you take a simple random sample of 150 students at your school and find that 21 of them were born in September. You are interested in whether the proportion born in September is greater than 1/12—what you would expect if September was no different from any other month. Thus, the null hypothesis is $H_0: p = \dfrac{1}{12}$. The P-value for your significance test is 0.0056. Which of the following statements best describes what the P-value measures?

(a) The probability that September birthdays are no more common than any other month is 0.0056.

(b) The probability that the proportion of September birthdays in the population is equal to 1/12 is 0.0056.

(c) 0.0056 is the probability of getting a sample with a proportion of September birthdays that is 21/150 or higher, given that the true proportion is 1/12.

(d) 0.0056 is the probability of getting a sample with a proportion of September birthdays that is 21/150 or higher, given that the true proportion is greater than 1/12.

3. You collect test scores from 4 randomly selected members of a population that is approximately normally distributed and test the hypotheses $H_0: \mu = 100$ versus $H_a: \mu > 100$ at the $\alpha = 0.05$ significance level. You obtain a P-value of 0.052. Which of the following statements is true?

(a) You have convincing evidence that H_0 is true.

(b) You have failed to obtain any evidence for H_a.

(c) There is some evidence against H_0, but there is not convincing evidence for H_a.

(d) You can reject H_0.

4. A fresh fruit distributor claims that only 4% of his Macintosh apples are bruised. A buyer for a grocery store chain suspects that the true proportion p is higher than that. She takes a random sample of 30 apples to test the null hypothesis $H_0: p = 0.04$ against the alternative hypothesis $H_a: p > 0.04$. Which of the following statements about conditions for performing a one-sample z test for the population proportion is correct?

(a) The test cannot be performed because the Random condition has not been met.

(b) The test cannot be performed because the Large Counts condition has not been met.

(c) We can't determine if the conditions have been met until we have the sample proportion, $\hat{p}$.

(d) All conditions for performing the test have been met.

5. A random sample of 100 likely voters in a small city produced 59 voters in favor of Candidate A. The observed value of the standardized test statistic for performing a test of $H_0: p = 0.5$ versus $H_a: p > 0.5$ is

(a) $z = \dfrac{0.59 - 0.5}{\sqrt{\dfrac{0.59(0.41)}{100}}}$ (c) $z = \dfrac{0.5 - 0.59}{\sqrt{\dfrac{0.59(0.41)}{100}}}$

(b) $z = \dfrac{0.59 - 0.5}{\sqrt{\dfrac{0.5(0.5)}{100}}}$ (d) $z = \dfrac{0.5 - 0.59}{\sqrt{\dfrac{0.5(0.5)}{100}}}$

6. Your teacher claims to produce random integers from 1 to 5 (inclusive) on her calculator, but you've been keeping track. In the past 100 integers generated, the number 5 has come up only 12 times. You suspect that the calculator is producing fewer 5s than it should. Let p = the actual long-run proportion of 5s produced by the calculator. You decide to perform a test of $H_0: p = 0.2$ versus $H_a: p < 0.2$. The P-value for this test is closest to:

(a) 0.0228   (b) 0.0241   (c) 0.0455   (d) 0.0482

7. You are thinking of conducting a one-sample t test about a population mean μ using a 0.05 significance level. Which of the following statements is correct?

(a) You should not carry out the test if the sample does not have a normal distribution.

(b) You can safely carry out the test if there are no outliers, regardless of the sample size.

(c) You can carry out the test only if the population standard deviation is known.

(d) You can safely carry out the test if your sample size is at least 30.

8. A significance test was performed to test the null hypothesis $H_0: \mu = 2$ versus the alternative $H_a: \mu \neq 2$. A sample of size 28 produced a test statistic of $t = 2.051$. Assuming all conditions for inference were met, which of the following intervals contains the P-value for this test?

(a) $0.01 \leq P < 0.02$ (c) $0.025 \leq P < 0.05$

(b) $0.02 \leq P < 0.025$ (d) $0.05 \leq P < 0.10$

9. You have data on rainwater collected at 16 locations in the Adirondack Mountains of northern New York. One measurement is the acidity of the water, measured by pH on a scale of 0 to 14 (the pH of distilled water is 7.0). Which inference procedure would you use to estimate the average acidity of rainwater in the Adirondacks?

(a) One-sample z interval   (c) One-sample z test

(b) One-sample t interval   (d) One-sample t test

10. A 95% confidence interval for the proportion of viewers of a certain reality television show who are over 30 years old is (0.26, 0.35). Suppose the show's producers want to test the hypotheses $H_0: p = 0.25$ against $H_a: p \neq 0.25$. Which of the following is an appropriate conclusion for them to draw at the $\alpha = 0.05$ significance level?

(a) Reject H_0 and conclude that there is not convincing evidence that the true proportion of viewers who are over 30 years old differs from 0.25.

(b) Reject H_0 and conclude that there is convincing evidence that the true proportion of viewers who are over 30 years old differs from 0.25.

(c) Fail to reject H_0 and conclude that there is not convincing evidence that the true proportion of viewers who are over 30 years old differs from 0.25.

(d) Fail to reject H_0 and conclude that there is convincing evidence that the true proportion of viewers who are over 30 years old differs from 0.25.

Section II: Free Response

11. When a petition is submitted to government officials to put a political candidate's name on a ballot, a certain number of valid voters' signatures is required. Rather than check the validity of all the signatures, officials often randomly select a sample of signatures for verification and perform a significance test to see if the true proportion of signatures is above the required value. Suppose a petition has 30,000 signatures and 18,000 are required for a candidate to be on the ballot, so 60% of the signatures must be valid. The officials take a random sample of 300 signatures and find that 201 are valid.

(a) Does the sample provide convincing evidence at the $\alpha = 0.01$ significance level that the true proportion of valid signatures on the petition is greater than 0.6?

(b) Interpret the P-value of your test from part (a) in context.

12. In the spring of 2015, the U.S. Census Bureau reported that the average price of a single-family home in the United States was $341,500. A random sample of 36 homes sold during that time period in Portland, Oregon, had a mean sale price of $406,700 and a standard deviation of $153,400. Do these data provide convincing evidence at the $\alpha = 0.05$ significance level that the mean price for a home in Portland differs from the national average?

13. As a construction engineer for a city, you are responsible for ensuring that the company that is providing gravel for a new road puts as much gravel in each truckload as they claim to. It has been estimated that 500 truckloads of gravel will be needed to complete this road, so you plan to measure the volume of gravel in an SRS of 25 trucks to make sure that the company isn't delivering less gravel per truckload than they claim. Each truckload is supposed to have 20 cubic meters (m³) of gravel, so you will test the hypotheses $H_0: \mu = 20$ versus $H_a: \mu < 20$ at the $\alpha = 0.05$ level.

(a) Describe a Type I error in this setting and a consequence of making such an error.

(b) Describe a Type II error in this setting and a consequence of making such an error.

9

Comparing Two Populations or Treatments

Lesson 9.1	Estimating a Difference Between Two Proportions	552
Lesson 9.2	Testing a Claim about a Difference Between Two Proportions	561
Lesson 9.3	Estimating a Difference Between Two Means	571
Lesson 9.4	Testing a Claim about a Difference Between Two Means	581
Lesson 9.5	Analyzing Paired Data: Estimating a Mean Difference	591
Lesson 9.6	Testing a Claim about a Mean Difference	601
Chapter 9	Main Points	611
Chapter 9	Review Exercises	613
Chapter 9	Practice Test	615

© Ian Dagnall/Alamy Stock Photo

STATS applied!

Are fast-food drive-thrus fast and accurate?

More than $70 billion is spent each year in the drive-thru lanes of America's fast-food restaurants. Having quick, accurate, and friendly service at a drive-thru window translates directly into revenue for the restaurant. Therefore, industry executives, stockholders, and analysts closely follow the ratings of fast-food drive-thru lanes that appear each year in *QSR*, a publication that reports on the quick-service restaurant industry.

The 2013 *QSR* magazine drive-thru study involved visits to a random sample of restaurants in 7 major fast-food chains in all 50 states. During each visit, the researcher ordered a modified main item (for example, a hamburger with no pickles), a side item, and a drink. If any item was not received as ordered, or if the restaurant failed to give the correct change or supply a straw and a napkin, the order was considered "inaccurate." Service time, which is the time from when the car stopped at the speaker to when the entire order was received, was measured for each visit.[1]

Here are some results from the 2013 *QSR* study:

- Of the 246 visits to Burger King, 82.1% resulted in accurate orders. Of the 282 visits to Wendy's, 86.9% resulted in accurate orders.

- McDonald's average service time for 317 drive-thru visits was 189.49 seconds with a standard deviation of 18.51 seconds. Chick-fil-A's service time for 299 drive-thru visits had a mean of 203.88 seconds and a standard deviation of 19.61 seconds.

> Was there a significant difference in accuracy at Burger King and Wendy's? How much better was the average service time at McDonald's than at Chick-fil-A restaurants in 2013?

We'll revisit STATS applied! at the end of the chapter, so you can use what you have learned to help answer these questions.

Lesson 9.1

Estimating a Difference Between Two Proportions

In Lessons 7.3 and 7.4, you learned how to calculate and interpret a confidence interval for a population proportion p. What if we want to estimate the difference $p_1 - p_2$ between the proportions of successes in Population 1 and Population 2? For instance, maybe we want to estimate the difference in the proportion of all cars and the proportion of all trucks in a large city that have their tires filled with nitrogen. The ideal strategy is to take a separate random sample from each population and to use the difference $\hat{p}_1 - \hat{p}_2$ between the sample proportions as our point estimate. To do inference about $p_1 - p_2$, we have to know about the sampling distribution of $\hat{p}_1 - \hat{p}_2$.

This lesson focuses on constructing confidence intervals for a difference between two proportions. You will learn how to perform significance tests about $p_1 - p_2$ in Lesson 9.2.

The Sampling Distribution of a Difference Between Two Proportions

To explore the sampling distribution of $\hat{p}_1 - \hat{p}_2$, let's start with two populations having a known proportion of successes. Suppose that there are two large high schools in a certain town. At School 1, 70% of students did their homework last night ($p_1 = 0.70$). Only 50% of the students at School 2 did their homework last night ($p_2 = 0.50$). The counselor at School 1 takes an SRS of 100 students and records the proportion $\hat{p}_1$ who did their homework. School 2's counselor takes an SRS of 200 students and records the proportion $\hat{p}_2$ who did their homework. What can we say about the difference $\hat{p}_1 - \hat{p}_2$ in the sample proportions?

Let's do a simulation to find out. We used software to randomly select 100 students from School 1 and 200 students from School 2. Our first set of samples gave $\hat{p}_1 = 0.72$ and $\hat{p}_2 = 0.47$, resulting in a difference of $\hat{p}_1 - \hat{p}_2 = 0.72 - 0.47 = 0.25$. A red dot for this value appears in Figure 9.1. The dotplot shows the results of repeating this process 1000 times.

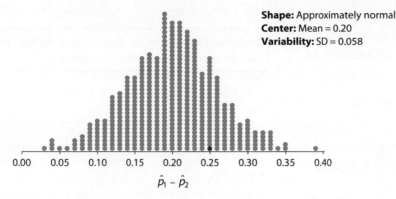

Shape: Approximately normal
Center: Mean = 0.20
Variability: SD = 0.058

FIGURE 9.1 Simulated sampling distribution of the difference in sample proportions $\hat{p}_1 - \hat{p}_2$ in 1000 SRSs of size $n_1 = 100$ from a population with $p_1 = 0.70$ and 1000 SRSs of size $n_2 = 200$ from a population with $p_2 = 0.50$.

The figure suggests that the sampling distribution of $\hat{p}_1 - \hat{p}_2$ has an approximately normal shape. This will be true whenever the Large Counts condition is met for *both* samples. The mean of the sampling distribution is 0.20. This result makes sense because we expect the values of $\hat{p}_1 - \hat{p}_2$ to center on the actual difference in the population proportions, $p_1 - p_2 = 0.70 - 0.50 = 0.20$. The standard deviation of the sampling distribution can be found using the formula

$$\sqrt{\frac{p_1(1 - p_1)}{n_1} + \frac{p_2(1 - p_2)}{n_2}} = \sqrt{\frac{0.7(0.3)}{100} + \frac{0.5(0.5)}{200}} = 0.058$$

That is, in repeated random samples of 100 students from School 1 and 200 students from School 2, the difference in the *sample* proportions of students at the two schools who did their homework last night is typically 0.058 from the true difference in proportions of 0.20.

Describing the Sampling Distribution of $\hat{p}_1 - \hat{p}_2$

Suppose that $\hat{p}_1$ is the proportion of successes in an SRS of size n_1 drawn from a large population with proportion of successes p_1, and $\hat{p}_2$ is the proportion of successes in an SRS of size n_2 drawn from a large population with proportion of successes p_2. Then:

- The **shape** of the sampling distribution of $\hat{p}_1 - \hat{p}_2$ is approximately normal if n_1p_1, $n_1(1 - p_1)$, n_2p_2, and $n_2(1 - p_2)$ are all at least 10.

- The **mean** of the sampling distribution of $\hat{p}_1 - \hat{p}_2$ is $\mu_{\hat{p}_1 - \hat{p}_2} = p_1 - p_2$

- The **standard deviation** of the sampling distribution of $\hat{p}_1 - \hat{p}_2$ is

$$\sigma_{\hat{p}_1 - \hat{p}_2} = \sqrt{\frac{p_1(1 - p_1)}{n_1} + \frac{p_2(1 - p_2)}{n_2}}$$

EXAMPLE

How much do the proportions of U.S.-made cars differ in two states?

Describing the sampling distribution of $\hat{p}_1 - \hat{p}_2$

PROBLEM: Angelica and Kyle both work for the Department of Motor Vehicles (DMV), but they live in different states. In Angelica's state, 84% of the registered cars are made by U.S. manufacturers. In Kyle's state, only 63% of the registered cars are made by U.S. manufacturers. Angelica selects a random sample of 100 cars in her state and Kyle selects a random sample of 70 cars in his state. Let $\hat{p}_1 - \hat{p}_2$ be the difference (Angelica's state − Kyle's state) in the sample proportion of cars made by American manufacturers.

(a) What is the shape of the sampling distribution of $\hat{p}_1 - \hat{p}_2$? Why?

(b) Find the mean of the sampling distribution.

(c) Calculate the standard deviation of the sampling distribution.

SOLUTION:

(a) Approximately normal, because $n_1 p_1 = 100(0.84) = 84$,
$n_1(1 - p_1) = 100(0.16) = 16$,
$n_2 p_2 = 70(0.63) = 44.1$, and
$n_2(1 - p_2) = 70(0.37) = 25.9$ are all ≥ 10.

> Note that these values are the expected numbers of successes and failures in the two samples.

(b) $\mu_{\hat{p}_1 - \hat{p}_2} = 0.84 - 0.63 = 0.21$

> $\mu_{\hat{p}_1 - \hat{p}_2} = p_1 - p_2$

(c) $\sigma_{\hat{p}_1 - \hat{p}_2} = \sqrt{\dfrac{0.84(0.16)}{100} + \dfrac{0.63(0.37)}{70}} = 0.0684$

> $\sigma_{\hat{p}_1 - \hat{p}_2} = \sqrt{\dfrac{p_1(1 - p_1)}{n_1} + \dfrac{p_2(1 - p_2)}{n_2}}$

FOR PRACTICE **TRY EXERCISE 1.**

Conditions for Estimating $p_1 - p_2$

In Lesson 7.3, we introduced two conditions that should be met before we construct a confidence interval for a population proportion: Random and Large Counts. Now that we are comparing two proportions, we have to modify these conditions slightly.

> ### How to Check the Conditions for Constructing a Confidence Interval for $p_1 - p_2$
>
> • **Random:** The data come from separate random samples from the two populations of interest or from two groups in a randomized experiment.
>
> • **Large Counts:** The counts of "successes" and "failures" in each sample or group—$n_1\hat{p}_1$, $n_1(1 - \hat{p}_1)$, $n_2\hat{p}_2$, and $n_2(1 - \hat{p}_2)$—are all at least 10.

Recall from Lesson 3.9 that the Random condition is important for determining the scope of inference. Random sampling allows us to generalize our results to the populations of interest, and random assignment in an experiment permits us to draw cause-and-effect conclusions. We will consider randomized experiments more thoroughly in Lesson 9.2.

The method we use to calculate a confidence interval for $p_1 - p_2$ requires that the sampling distribution of $\hat{p}_1 - \hat{p}_2$ be approximately normal. Earlier in the lesson, we noted that this will be true whenever $n_1 p_1$, $n_1(1 - p_1)$, $n_2 p_2$, and $n_2(1 - p_2)$ are all at least 10. Because we don't know the value of p_1 or p_2 when we are estimating their difference, we use $\hat{p}_1$ and $\hat{p}_2$ when checking the Large Counts condition.

EXAMPLE

Do more teens or adults use Facebook?

Checking conditions

PROBLEM: As part of the Pew Internet and American Life Project, researchers conducted two surveys in 2014. The first survey asked a random sample of 1060 U.S. teens about their use of social media.[2] A second survey posed similar questions to a random sample of 2003 U.S. adults.[3] In these two studies, 71.0% of teens and 58.0% of adults used Facebook. Let p_1 = the true proportion of all U.S. teens who use Facebook and p_2 = the true proportion of all U.S. adults who use Facebook. Check if the conditions for calculating a confidence interval for $p_1 - p_2$ are met.

SOLUTION:

Random? Separate random samples of 1060 U.S. teens and 2003 U.S. adults. ✓

Large Counts?

$n_1\hat{p}_1 = 1060(0.71) = 752.6 \rightarrow 753$

$n_1(1 - \hat{p}_1) = 1060(0.29) = 307.4 \rightarrow 307$

$n_2\hat{p}_2 = 2003(0.58) = 1161.74 \rightarrow 1162$

$n_2(1 - \hat{p}_2) = 2003(0.42) = 841.26 \rightarrow 841$

All counts are at least 10. ✓

> The Random condition requires that the data come from separate random samples from the two populations of interest or from the two groups in a randomized experiment.

> Be sure to check that the counts of "successes" and "failures" in each sample or group—$n_1\hat{p}_1$, $n_1(1 - \hat{p}_1)$, $n_2\hat{p}_2$, and $n_2(1 - \hat{p}_2)$—are *all* at least 10.

> Because these are the observed counts of successes and failures, they should be rounded to the nearest integer.

FOR PRACTICE   TRY EXERCISE 5.

Constructing and Interpreting a Confidence Interval for $p_1 - p_2$

When data come from two separate random samples or two groups in a randomized experiment (the Random condition), the statistic $\hat{p}_1 - \hat{p}_2$ is our point estimate for $p_1 - p_2$. We can calculate a confidence interval for $p_1 - p_2$ using the familiar formula

point estimate ± margin of error

or, in slightly expanded form,

point estimate ± critical value · standard error of statistic

The standard deviation of the sampling distribution of $\hat{p}_1 - \hat{p}_2$ is

$$\sigma_{\hat{p}_1 - \hat{p}_2} = \sqrt{\frac{p_1(1 - p_1)}{n_1} + \frac{p_2(1 - p_2)}{n_2}}$$

Because we don't know the values of the parameters p_1 and p_2, we replace them with $\hat{p}_1$ and $\hat{p}_2$. The result is the *standard error* of $\hat{p}_1 - \hat{p}_2$:

$$SE_{\hat{p}_1 - \hat{p}_2} = \sqrt{\frac{\hat{p}_1(1 - \hat{p}_1)}{n_1} + \frac{\hat{p}_2(1 - \hat{p}_2)}{n_2}}$$

How to Calculate a Confidence Interval for $p_1 - p_2$

When the Random and Large Counts conditions are met, a C% confidence interval for the difference $p_1 - p_2$ between two proportions is

$$(\hat{p}_1 - \hat{p}_2) \pm z^*\sqrt{\frac{\hat{p}_1(1 - \hat{p}_1)}{n_1} + \frac{\hat{p}_2(1 - \hat{p}_2)}{n_2}}$$

where z^* is the critical value for the standard normal distribution with C% of its area between $-z^*$ and z^*.

This interval is often called a *two-sample z interval for a difference between two proportions*.

Let's return to the social media example. We already checked that the Random and Large Counts conditions are met. A 95% confidence interval for $p_1 - p_2$ is

$$(0.71 - 0.58) \pm 1.96\sqrt{\frac{0.71(1 - 0.71)}{1060} + \frac{0.58(1 - 0.58)}{2003}} \rightarrow 0.13 \pm 0.035 \rightarrow 0.095 \text{ to } 0.165$$

Interpretation: We are 95% confident that the interval from 0.095 to 0.165 captures the true difference in the proportion of all U.S. teens and adults who use Facebook. The interval suggests that teen Facebook use is between 9.5 and 16.5 percentage points higher than adult Facebook use in the United States.

Notice that we have added a second sentence to our usual interpretation of the confidence interval. It describes what the interval tells us about the difference between the two population proportions—which is larger and by how much—in plain language. Be sure to include a similar sentence any time you interpret a confidence interval for the difference between two parameters.

EXAMPLE

How much did President Obama's approval rating change?

Confidence interval for $p_1 - p_2$

PROBLEM: Many news organizations conduct polls asking adults in the United States if they approve of the job the president is doing. According to a Gallup poll of 1500 randomly selected U.S. adults in October 2012, 780 approved of President Obama's job performance. A Gallup poll of 1500 randomly selected U.S. adults in October 2014 showed that 615 approved of Obama's job performance. Construct and interpret a 90% confidence interval for the change in Obama's approval rating among all U.S. adults from October 2012 to October 2014.

SOLUTION:

STATE: We want to estimate $p_1 - p_2$ at the 90% confidence level, where $p_1 =$ the true proportion of all U.S. adults who approved of President Obama's job performance in October 2014 and $p_2 =$ the true proportion of all U.S. adults who approved of Obama's job performance in October 2012.

> **STATE:** State the parameters you want to estimate and the confidence level.

PLAN: Two-sample z interval for $p_1 - p_2$

Random? The data came from separate random samples in 2012 and 2014. ✓

> **PLAN:** Identify the appropriate inference method and check conditions.

Large Counts? $1500\left(\dfrac{615}{1500}\right) = 615, 1500\left(\dfrac{885}{1500}\right) = 885,$

$1500\left(\dfrac{780}{1500}\right) = 780, 1500\left(\dfrac{720}{1500}\right) = 720$

are all at least 10. ✓

> These are just the counts of successes and failures in the two samples. So you could write 615, 1500 − 615 = 885, 780, 1500 − 780 = 720 are all at least 10.

DO: $\hat{p}_1 = \dfrac{615}{1500} = 0.41, \hat{p}_2 = \dfrac{780}{1500} = 0.52$

> **DO:** If the conditions are met, perform calculations.

$(0.41 - 0.52) \pm 1.645\sqrt{\dfrac{0.41(1 - 0.41)}{1500} + \dfrac{0.52(1 - 0.52)}{1500}}$

$\rightarrow -0.11 \pm 0.030 \rightarrow -0.140 \text{ to } -0.080$

CONCLUDE: We are 90% confident that the interval from -0.140 to -0.080 captures the true change in the proportion of all U.S. adults who approved of President Obama's job performance from October 2012 to October 2014. The interval suggests that President Obama's approval rating decreased by between 8 and 14 percentage points over this two-year period.

> **CONCLUDE:** Interpret your interval in the context of the problem.

> Don't forget the second sentence when you interpret the interval!

FOR PRACTICE TRY EXERCISE 9.

Note that the confidence interval in the example does not include 0 as a plausible value for $p_1 - p_2$. In fact, all of the values in the interval are negative. Therefore, the interval gives convincing evidence of a decrease in President Obama's approval rating from October 2012 to October 2014.

LESSON APP 9.1

Who likes rap music more?

A study compared 634 randomly chosen black people aged 15 to 25 with 567 randomly selected white people in the same age group. It found that 368 of the blacks and 130 of the whites listened to rap music every day.[4]

1. Construct and interpret a 99% confidence interval for the difference between the proportions of black and white people aged 15 to 25 who listen to rap daily.

2. Based on your interval, is there convincing evidence of a difference in the percentage of black and white people aged 15 to 25 who listen to rap music? Explain.

Chad Springer/Cultura RM
Exclusive/Getty Images

TECH CORNER

Confidence Intervals for a Difference Between Two Proportions

You can use technology to calculate a confidence interval for $p_1 - p_2$. We'll illustrate using the presidential approval example. Of 1500 people surveyed in 2014, $X_1 = 615$ approved of President Obama's job performance. Of 1500 people surveyed in 2012, $X_2 = 780$ approved of Obama's job performance. To calculate a 90% confidence interval for $p_1 - p_2$, use either of the following.

Applet

1. Launch the *One Categorical Variable* applet at **highschool.bfwpub.com/spa3e**.

2. Enter "Presidential approval" as the variable name.

3. For groups, select "Multiple." Choose to input data as "Counts in categories."

4. Type the group names in the boxes at the top of the columns and the category names in the boxes at the left of the rows, as shown.

5. Input the number of successes and failures for each group, as shown.

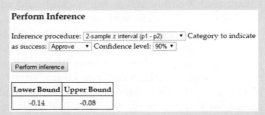

6. Press the "Begin analysis" button. Then, scroll down to the "Perform inference" section. Choose the 2-sample *z*-interval for the procedure, Approve as the category to indicate success, and 90% for the confidence level. Then, press the "Perform inference" button.

TI-83/84

1. Press STAT, then choose TESTS and 2-PropZInt.

2. When the 2-PropZInt screen appears, enter the values shown. Note that the values of x1, n1, x2, and n2 must all be integers!

```
NORMAL FLOAT AUTO REAL RADIAN MP

         2-PropZInt
x1:615
n1:1500
x2:780
n2:1500
C-Level:.9
Calculate
```

3. Highlight "Calculate" and press ENTER.

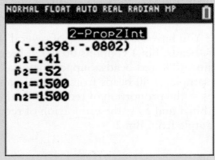

```
NORMAL FLOAT AUTO REAL RADIAN MP

         2-PropZInt
(-.1398,-.0802)
p̂1=.41
p̂2=.52
n1=1500
n2=1500
```

Lesson 9.1

WHAT DID YOU LEARN?

LEARNING TARGET	EXAMPLES	EXERCISES
Describe the shape, center, and variability of the sampling distribution of a difference $\hat{p}_1 - \hat{p}_2$ between two sample proportions.	p. 553	1–4
Check the Random and Large Counts conditions for constructing a confidence interval for a difference between two proportions.	p. 554	5–8
Use the four-step process to construct and interpret a confidence interval for the difference between two proportions.	p. 556	9–12

Exercises

Lesson 9.1

Mastering Concepts and Skills

1. **I want red!** A candy maker offers child- and adult-size bags of jelly beans with different color mixes. The company claims that the child mix has 30% red jelly beans, while the adult mix contains 15% red jelly beans. Assume that the candy maker's claim is true. Suppose we take a random sample of 50 jelly beans from the child mix and a separate random sample of 100 jelly beans from the adult mix. Let $\hat{p}_1$ and $\hat{p}_2$ be the sample proportions of red jelly beans from the child and adult mixes, respectively.
   (a) What is the shape of the sampling distribution of $\hat{p}_1 - \hat{p}_2$? Why?
   (b) Find the mean of the sampling distribution.
   (c) Calculate the standard deviation of the sampling distribution.

2. **Yummy goldfish!** Your teacher brings two large bags of colored goldfish crackers to class. Bag 1 has 25% red crackers and Bag 2 has 35% red crackers. Using a paper cup, your teacher takes an SRS of 50 crackers from Bag 1 and a separate SRS of 40 crackers from Bag 2. Let $\hat{p}_1 - \hat{p}_2$ be the difference in the sample proportions of red crackers.
   (a) What is the shape of the sampling distribution of $\hat{p}_1 - \hat{p}_2$? Why?
   (b) Find the mean of the sampling distribution.
   (c) Calculate the standard deviation of the sampling distribution.

3. **Beads** You have two large bins of several thousand plastic beads. Bin 1 contains 60% red beads; Bin 2 contains 48% red beads. Suppose you take a random sample of 80 beads from each bin and calculate $\hat{p}_1$ = the proportion of red beads in the sample from Bin 1 and $\hat{p}_2$ = the proportion of red beads in the sample from Bin 2.

   (a) What is the shape of the sampling distribution of $\hat{p}_1 - \hat{p}_2$? Why?
   (b) Find the mean of the sampling distribution.
   (c) Calculate the standard deviation of the sampling distribution.

4. **Literacy** A researcher reports that 80% of high school graduates, but only 40% of high school dropouts, would pass a basic literacy test.[5] Assume that the researcher's claim is true. Suppose we give a basic literacy test to a random sample of 60 high school graduates and a separate random sample of 75 high school dropouts. Let $\hat{p}_1$ and $\hat{p}_2$ be the sample proportions of graduates and dropouts, respectively, who pass the test.

   (a) What is the shape of the sampling distribution of $\hat{p}_1 - \hat{p}_2$? Why?
   (b) Find the mean of the sampling distribution.
   (c) Calculate the standard deviation of the sampling distribution.

5. **Unsafe to drink?** The movie *A Civil Action* (1998) tells the story of a major legal battle that took place in the small town of Woburn, Massachusetts. A town well that supplied water to East Woburn residents was contaminated by industrial chemicals. During the period that residents drank water from this well, 16 of the 414 babies born had birth defects. On the west side of Woburn, 3 of the 228 babies born during the same time period had birth defects. Let p_1 = the true proportion of all babies born with birth defects in West Woburn and p_2 = the true proportion of all babies born with birth defects in East Woburn. Check if the conditions for calculating a confidence interval for $p_1 - p_2$ are met.

6. **Ticks!** Lyme disease is spread in the northeastern United States by infected ticks. The ticks are infected

mainly by feeding on mice, so more mice result in more infected ticks. The mouse population, in turn, rises and falls with the abundance of acorns, their favored food. Experimenters studied two similar forest areas in a year when the acorn crop failed. They added hundreds of thousands of acorns to one area to imitate an abundant acorn crop, while leaving the other area untouched. The next spring, 54 of the 72 mice trapped in the first area were in breeding condition, versus 10 of the 17 mice trapped in the second area.[6] Let p_1 = the true proportion of mice in the acorn-supplemented area that are in breeding condition and p_2 = the true proportion of mice in the unsupplemented area that are in breeding condition. Check if the conditions for calculating a confidence interval for $p_1 - p_2$ are met.

7. **Response bias** Does the appearance of the interviewer influence how people respond to a survey question? Ken (white, with blond hair) and Hassan (darker, with Middle-Eastern features) conducted an experiment to address this question. They took turns (in random order) walking up to people on the main street of a small town, identifying themselves as students from a local high school, and asking, "Do you support President Obama's decision to launch airstrikes in Iraq?" Of the 50 people Hassan spoke to, 11 said "Yes," while 21 of the 44 people Ken spoke to said "Yes." If p_1 and p_2 are the proportion of people like these who would say yes when asked by Hassan and Ken, respectively, check if the conditions for calculating a confidence interval for $p_1 - p_2$ are met.

8. **Digital video disks** A company that records and sells rewritable DVDs wants to compare the reliability of DVD-fabricating machines produced by two different manufacturers. Employees randomly select 500 DVDs produced by each fabricator and find that 494 of the disks produced by the first machine are acceptable and 480 of the disks produced by the second machine are acceptable. If p_1 and p_2 are the proportion of acceptable DVDs produced by the first and second machines, respectively, check if the conditions for calculating a confidence interval for $p_1 - p_2$ are met.

9. **Who tweets more?** Do younger people use Twitter more often than older people? In a random sample of 585 adult Internet users aged 18 to 29, 32% used Twitter. In a separate random sample of 595 adult Internet users aged 30 to 49, 13% used Twitter.[7] Construct and interpret a 90% confidence interval for the difference between the true proportions of adult Internet users in these age groups who use Twitter.

pg 556

10. **Summer jobs** A sample survey interviews SRSs of 500 female college students and 550 male college students. Researchers want to determine whether there is a difference in the proportion of male and female college students who worked for pay last summer. In all, 410 of the females and 484 of the males say they worked for pay last summer. Construct and interpret a 99% confidence interval for the difference in the proportion of male and female college students who worked for pay last summer.

11. **Still living at home** A surprising number of young adults (ages 19 to 25) still live in their parents' homes. The National Institutes of Health surveyed random samples of 2253 men and 2629 women in this age group and found that 986 of the men and 923 of the women lived with their parents.[8] Construct and interpret a 99% confidence interval for the difference in the true proportions of men and women aged 19 to 25 who live in their parents' homes.

12. **Who owns iPods?** At a baseball game, 42 of 65 randomly selected people own an iPod. At a rock concert occurring at the same time across town, 34 of 52 randomly selected people own an iPod. Construct and interpret a 99% confidence interval for the difference in the proportion of people at the baseball game and at the rock concert who own an iPod.

Applying the Concepts

13. **Where's Egypt?** In a recent Pew Research poll, 287 out of 522 randomly selected men in the United States were able to identify Egypt when it was highlighted on a map of the Middle East. When 520 randomly selected women were asked, 233 were able to do so.

(a) Construct and interpret a 95% confidence interval for the difference in the proportion of U.S. men and the proportion of U.S. women who can identify Egypt on a map.

(b) Based on your interval, is there convincing evidence of a difference in the true proportions of U.S. men and women who can identify Egypt on a map? Explain.

14. **Quit smoking** Nicotine patches are often used to help smokers quit. Does giving medicine to fight depression also help? A randomized double-blind experiment assigned 244 smokers to receive nicotine patches and another 245 to receive both a patch and an antidepressant drug. A year later, 40 subjects in the nicotine patch group still abstained from smoking, as did 87 in the patch-plus-drug group.

(a) Construct and interpret a 99% confidence interval for the difference in the true proportion of smokers like these who would abstain when using an antidepressant and a nicotine patch and the proportion who would abstain when using only a patch.

(b) Based on your interval, is there convincing evidence that an antidepressant plus a nicotine patch is more effective than the patch alone in helping subjects like these abstain from smoking? Explain.

15. **Artificial trees?** An association of Christmas tree growers in Indiana wants to know if there is a difference in preference for natural trees between urban and rural households. So the association sponsored a survey of Indiana households that had a Christmas tree last year to find out. In a random sample of 160 rural households, 64 had a natural tree. In a separate random sample of 261 urban households, 89 had a natural tree. A 95% confidence interval for the true difference (Rural − Urban) in the proportion of households that had a natural tree is −0.036 to 0.154. Does the confidence interval provide convincing evidence that the two population proportions are equal? Explain.

16. **Ban junk food!** A recent CBS News poll asked 606 randomly selected women and 442 randomly selected men, "Do you think putting a special tax on junk food would encourage more people to lose weight?" 170 of the women and 102 of the men said "Yes."[9] A 99% confidence interval for the true difference (Women − Men) in the proportion of people who said "Yes" is −0.020 to 0.120. Does the confidence interval provide convincing evidence that the two population proportions are equal? Explain.

Extending the Concepts

17. **I want red!** Refer to Exercise 1. Find the probability that the proportion of red jelly beans is higher in the sample from the child-size mix than in the sample from the adult mix.

18. **SE versus ME** In a random sample of 125 workers from Company 1, a total of 35 admitted to using sick leave when they weren't really ill. In a separate random sample of 68 employees from Company 2, a total of 17 workers admitted that they had used sick leave when they weren't ill. The difference in the sample proportions of workers from the two companies who admitted to using sick leave when they weren't ill is $\hat{p}_1 - \hat{p}_2 = 0.28 - 0.25 = 0.03$.

(a) The standard error of $\hat{p}_1 - \hat{p}_2$ is 0.066. Interpret this value in context.

(b) The margin of error for a 95% confidence interval for $p_1 - p_2$ is 0.13. Interpret this value in context.

Recycle and Review

19. **Prayer and conception** (3.6, 3.7) Some women would like to have children but are unable to do so for medical reasons. One option for many of these women is a procedure called in vitro fertilization (IVF), which involves injecting a fertilized egg into the woman's uterus. A large group of women who were about to undergo IVF served as subjects in an experiment. Each subject was randomly assigned to either a treatment group or a control group. Several people (called intercessors), who did not know the women in the treatment group, intentionally prayed for them. (This is called intercessory prayer.) The praying continued for three weeks following IVF. The intercessors did not pray for the women in the control group. Here are the results: 44 of the 88 women in the treatment group got pregnant, compared to 21 out of 81 in the control group.[10]

(a) Explain the purpose of the control group in this experiment.

(b) Describe how researchers could have randomly assigned the women to the two groups.

(c) What is the purpose of random assignment in this experiment?

20. **Does prayer promote pregnancy?** (3.8, 3.9, 8.2) Refer to Exercise 19.

(a) The results of the study were statistically significant. Explain what "statistically significant" means in the context of this study.

(b) Which type of error might have occurred in this study, a Type I error or a Type II error? Explain.

(c) Based on the results of this study, would it be reasonable to say that intercessory prayer caused an increase in the pregnancy rate for women like those in this study? Explain.

Lesson 9.2

Testing a Claim about a Difference Between Two Proportions

LEARNING TARGETS

- State hypotheses and check conditions for performing a significance test about a difference between two proportions.

- Calculate the standardized test statistic and *P*-value for a significance test about a difference between two proportions.

- Use the four-step process to perform a significance test about a difference between two proportions.

An observed difference between two sample proportions can reflect an actual difference in the parameters, or it may just be due to chance variation in random sampling or random assignment. In Lesson 3.8, we used simulation to determine which explanation makes more sense in an experiment about distracted driving. This lesson shows you how to perform a formal significance test for the difference between two proportions.

Stating Hypotheses and Checking Conditions for a Test about $p_1 - p_2$

In a test for comparing two proportions, the null hypothesis has the general form

$$H_0: p_1 - p_2 = \text{hypothesized value}$$

We'll restrict ourselves to situations in which the hypothesized difference is 0. Then the null hypothesis says that there is no difference between the two parameters:

$$H_0: p_1 - p_2 = 0$$

(You will sometimes see the null hypothesis written in the equivalent form $H_0: p_1 = p_2$.) The alternative hypothesis says what kind of difference we expect.

The conditions for performing a significance test about $p_1 - p_2$ are the same as those for constructing a confidence interval that you learned in Lesson 9.1.

> ### How to Check the Conditions for Performing a Significance Test about $p_1 - p_2$
>
> - **Random:** The data come from separate random samples from the two populations of interest or from two groups in a randomized experiment.
>
> - **Large Counts:** The counts of "successes" and "failures" in each sample or group—$n_1\hat{p}_1$, $n_1(1 - \hat{p}_1)$, $n_2\hat{p}_2$, and $n_2(1 - \hat{p}_2)$—are all at least 10.

EXAMPLE

Who eats breakfast?

Stating hypotheses and checking conditions

PROBLEM: Researchers designed a study to compare the proportion of children who come to school having eaten breakfast in two low-income elementary schools. An SRS of 80 students from School 1 found that 61 ate breakfast today. At School 2, an SRS of 150 students included 124 who had breakfast today. Do these data give convincing evidence of a difference in the population proportions at the $\alpha = 0.05$ significance level?

(a) State appropriate hypotheses for performing a significance test. Be sure to define the parameters of interest.

(b) Check that the conditions for performing the test are met.

SOLUTION:

(a) $H_0: p_1 - p_2 = 0$

$H_a: p_1 - p_2 \neq 0$

where p_1 = the true proportion of all students at School 1 who ate breakfast today and p_2 = the true proportion of all students at School 2 who ate breakfast today.

> H_a is two-sided because we're asked if the population proportions differ.

(b) Random? The data came from separate random samples from the two schools. ✓

Large Counts? 61, 80 − 61 = 19, 124, and 150 − 124 = 26 are all at least 10. ✓

FOR PRACTICE TRY EXERCISE 1.

Calculations: Standardized Test Statistic and *P*-Value

If the conditions are met, we can proceed with calculations. To do a test of $H_0: p_1 - p_2 = 0$, standardize $\hat{p}_1 - \hat{p}_2$ to get a z statistic:

$$\text{standardized test statistic} = \frac{\text{statistic} - \text{null value}}{\text{standard deviation of statistic}}$$

$$z = \frac{(\hat{p}_1 - \hat{p}_2) - 0}{\sqrt{\dfrac{p_1(1 - p_1)}{n_1} + \dfrac{p_2(1 - p_2)}{n_2}}}$$

You might be tempted to replace p_1 and p_2 in the denominator with the corresponding sample proportions. Don't do it! Here's why.

A significance test begins by assuming that $H_0: p_1 - p_2 = 0$ is true. In that case, $p_1 = p_2$. We call the common value of these two parameters p_C. To estimate p_C, we combine (or "pool") the data from the two samples as if they came from one larger sample. This *combined sample proportion* is

$$\hat{p}_C = \frac{\text{number of successes in both samples combined}}{\text{number of individuals in both samples combined}} = \frac{X_1 + X_2}{n_1 + n_2}$$

In other words, $\hat{p}_C$ gives the overall proportion of successes in the combined samples. Use $\hat{p}_C$ in place of both p_1 and p_2 in the denominator of the standardized test statistic:

$$z = \frac{(\hat{p}_1 - \hat{p}_2) - 0}{\sqrt{\dfrac{\hat{p}_C(1 - \hat{p}_C)}{n_1} + \dfrac{\hat{p}_C(1 - \hat{p}_C)}{n_2}}}$$

When the Large Counts condition is met, this z statistic will have approximately the standard normal distribution. We can find the appropriate *P*-value using Table A or technology.

EXAMPLE

Which children had breakfast?

Calculating the standardized test statistic and P-value

PROBLEM: Refer to the previous example. The two-way table summarizes the data from the separate random samples of children at School 1 and School 2.

(a) Calculate the standardized test statistic.

(b) Find the *P*-value using Table A or technology.

		School		
		1	2	Total
Breakfast	No	19	26	45
	Yes	61	124	185
	Total	80	150	230

SOLUTION:

(a) Statistics: $\hat{p}_1 = \dfrac{61}{80} = 0.7625$, $\hat{p}_2 = \dfrac{124}{150} = 0.8267$

$$\hat{p}_c = \dfrac{61 + 124}{80 + 150} = \dfrac{185}{230} = 0.804$$

$$z = \dfrac{(0.7625 - 0.8267) - 0}{\sqrt{\dfrac{0.804(1 - 0.804)}{80} + \dfrac{0.804(1 - 0.804)}{150}}} = -1.17$$

> First calculate the sample proportions and the combined sample proportion. Then standardize.
>
> $$z = \dfrac{(\hat{p}_1 - \hat{p}_2) - 0}{\sqrt{\dfrac{\hat{p}_c(1 - \hat{p}_c)}{n_1} + \dfrac{\hat{p}_c(1 - \hat{p}_c)}{n_2}}}$$

(b)

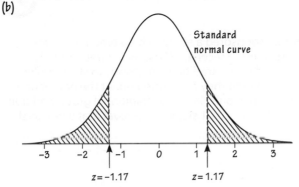

Standard normal curve

$z = -1.17$ $z = 1.17$

> The total column in the table above makes it easy to see that the overall (combined) proportion of successes in the two samples is $\hat{p}_C = 185/230$.

> The observed result gives *some* evidence in favor of H_a because $\hat{p}_1 - \hat{p}_2 = 0.7625 - 0.8267 = -0.0642 \neq 0$.

Using Table A: $P(z \leq -1.17 \text{ or } z \geq 1.17) = 2(0.1210) = 0.2420$

Using technology: Applet/normalcdf(lower: 1.17, upper: 1000, mean: 0, SD: 1)*2 = 0.2420.

FOR PRACTICE   TRY EXERCISE 5.

What does the *P*-value in the example tell us? If there is no difference in the population proportions of students who ate breakfast at the two schools and we repeated the random sampling process many times, we'd get a difference in sample proportions as large as or larger than 0.064 in either direction about 24% of the time. With such a high probability of getting a result like this just by chance when the null hypothesis is true, we don't have enough evidence to reject H_0. That is, we don't have convincing evidence that there is a difference in the proportion of all students who eat breakfast at these two schools.

We can get additional information about the difference between the population proportions who eat breakfast at School 1 and School 2 with a confidence interval. Technology gives the 95% confidence interval for $p_1 - p_2$ as -0.175 to 0.047. That is, we are 95% confident that the interval from -0.175 to 0.047 captures the difference in the true proportions of students who eat breakfast at the two schools. The interval suggests that breakfast-eating frequency at School 1 is between 17.5 percentage points lower and 4.7 percentage points higher than at School 2. This is consistent with our "fail to reject H_0" conclusion, because 0 is included in the interval of plausible values for $p_1 - p_2$.

Performing a Test about $p_1 - p_2$

A significance test about $p_1 - p_2$ is called a *two-sample z test for a difference between two proportions*. As with any test, be sure to follow the four-step process.

All of the examples so far have involved doing inference about $p_1 - p_2$ using data that were produced by random sampling. In such cases, the parameters p_1 and p_2 are the true proportions of successes in the corresponding populations. However, many important statistical results come from randomized experiments. Defining the parameters in experimental settings is more challenging.

Most experiments on people use recruited volunteers as subjects. When subjects are not randomly selected, researchers should not generalize the results of an experiment to some larger populations of interest. But researchers can draw cause-and-effect conclusions that apply to people like those who took part in the experiment. This same logic applies to experiments on animals or things.

EXAMPLE

Do financial incentives help people quit smoking?

Significance test for $p_1 - p_2$

PROBLEM: In an effort to reduce health care costs, General Motors sponsored a study to help employees stop smoking. In the study, half of the subjects (439) were randomly assigned to receive $750 if they agreed to quit smoking for a year, while the other half (439) were simply encouraged to use traditional methods to stop smoking. None of the 878 volunteers knew that there was a financial incentive when they first signed up. At the end of one year, 15% of those in the financial rewards group had quit smoking, while only 5% in the traditional group had quit smoking.[11] Do the results of this study give convincing evidence at the $\alpha = 0.01$ significance level that a financial incentive helps people stop smoking?

SOLUTION:

STATE: We want to perform a test at the $\alpha = 0.01$ significance level of
$$H_0 : p_1 - p_2 = 0$$
$$H_a : p_1 - p_2 > 0$$

where p_1 = the true quitting rate for employees like these who are offered a financial incentive to stop smoking and p_2 = the true quitting rate for employees like these who are encouraged to use traditional methods to stop smoking.

PLAN: Two-sample z test for $p_1 - p_2$

Random? The data came from two groups in a randomized experiment. ✓

Large Counts? The numbers of successes (quitters) and failures in the two groups are
$$n_1\hat{p}_1 = 439(0.15) \approx 66,$$
$$n_1(1 - \hat{p}_1) = 439(0.85) \approx 373,$$
$$n_2\hat{p}_2 = 439(0.05) \approx 22, \text{ and}$$
$$n_2(1 - \hat{p}_2) = 439(0.95) \approx 417.$$
These are all at least 10. ✓

STATE: State the hypotheses you want to test and the significance level, and define any parameters you use.

This experiment used volunteer subjects, so we can only determine whether a financial incentive causes an increase in the proportion of *people like these* who would stop smoking.

PLAN: Identify the appropriate inference method and check conditions.

DO: • $\hat{p}_1 = 0.15$, $\hat{p}_2 = 0.05$, $\hat{p}_c = \dfrac{66 + 22}{439 + 439} = 0.10$

• $z = \dfrac{(0.15 - 0.05) - 0}{\sqrt{\dfrac{0.10(1 - 0.10)}{439} + \dfrac{0.10(1 - 0.10)}{439}}} = 4.94$

• P-value:

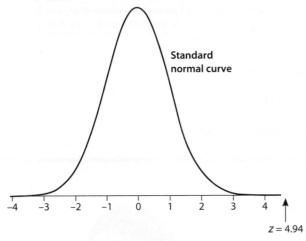

Standard
normal curve

$z = 4.94$

Using Table A: $P(z > 4.94) \approx 0$

Using technology: Applet/normalcdf(lower: 4.94, upper: 1000, mean: 0, SD: 1) ≈ 0

CONCLUDE: *Because the P-value of approximately 0 is less than $\alpha = 0.01$, we reject H_0. We have convincing evidence that employees like these are more likely to quit smoking when given a financial incentive than when they use more traditional methods.*

> **DO:** If the conditions are met, perform calculations:
> • Identify the sample statistics.
> • Calculate the standardized test statistic.
> • Find the P-value.

> The observed result gives *some* evidence in favor of H_a because $\hat{p}_1 - \hat{p}_2 = 0.15 - 0.05 = 0.10 > 0$.

> **CONCLUDE:** Make a conclusion about the hypotheses in the context of the problem.

FOR PRACTICE   TRY EXERCISE 9.

We chose $\alpha = 0.01$ in the example to reduce the chance of making a Type I error—finding convincing evidence that financial incentives help people quit smoking when they really don't. This error could result in a company spending lots of money on a nontraditional method that doesn't help employees like these quit smoking!

THINK ABOUT IT   **Will inference methods for random sampling give accurate P-values for randomized experiments?** Confidence intervals and tests for $p_1 - p_2$ are based on the sampling distribution of $\hat{p}_1 - \hat{p}_2$. But in most experiments, researchers don't select subjects at random from any larger populations. They do randomly assign subjects to treatments. We can think about what would happen if the random assignment were repeated many times under the assumption that H_0: $p_1 - p_2 = 0$ is true. That is, we assume that the specific treatment received doesn't affect an individual subject's response. We can use the simulation approach of Lesson 3.8 to determine if the observed difference in proportions is statistically significant.

We used the *One Categorical Variable* applet to randomly reassign the 878 subjects in the financial incentives study to the two groups, assuming the treatment received doesn't affect whether or not each individual quits smoking. Figure 9.2 on the next page shows the values of $\hat{p}_1 - \hat{p}_2$ in 500 trials of the simulation. The observed difference in the proportions of subjects who quit smoking was $0.15 - 0.05 = 0.10$. How likely is it that a difference this large or larger would happen just by chance when H_0 is true? The figure provides a rough answer: None of the 500 random reassignments yielded a difference in proportions greater than or equal to 0.10. That is, our estimate of the P-value is approximately 0. This is the same result as in the previous example.

FIGURE 9.2 Dotplot of the values of $\hat{p}_1 - \hat{p}_2$ in each of 500 random reassignments of subjects to treatment groups in the stop smoking experiment.

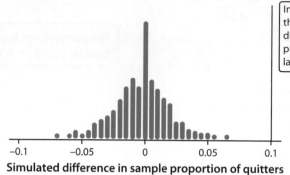

In 500 random reassignments, there were 0 times when the difference in sample proportions was as large as or larger than the observed 0.10.

Simulated difference in sample proportion of quitters

LESSON APP 9.2

Does taking aspirin help prevent heart attacks?

The Physicians' Health Study I was a medical experiment that helped answer this question. The subjects in this experiment were 22,071 male physicians. About half (11,037) of these subjects took an aspirin tablet every other day, and the remaining subjects (11,034) took a dummy pill that looked and tasted like aspirin but contained no active ingredient. After several years, 239 of the control group but only 139 of the aspirin group had suffered heart attacks.

1. Does this study provide convincing evidence that aspirin helps prevent heart attacks for healthy male physicians like those in this study? Justify your answer.

2. Based on your conclusion in Question 1, could you have made a Type I error or a Type II error? Explain.

3. Should you generalize the result in Question 1 to all healthy males? Why or why not?

All kind of people/Shutterstock

Significance Tests for a Difference Between Two Proportions

You can use technology to perform significance tests about $p_1 - p_2$. We'll illustrate using the stop smoking example. In the financial incentives group, there were $n_1 = 439$ subjects and $X_1 = 66$ who quit smoking. In the traditional methods group, there were $n_2 = 439$ subjects and $X_2 = 22$ who quit smoking. To perform a test of $H_0: p_1 - p_2 = 0$ versus $H_a: p_1 - p_2 > 0$, use either of the following.

Applet

1. Launch the *One Categorical Variable* applet at **highschool.bfwpub.com/spa3e**.

2. Enter "Smoking status" as the variable name.

3. For groups, select "Multiple". Choose to input data as "Counts in categories".

4. Type the group names in the boxes at the top of the columns and the category names in the boxes at the left of the rows, as shown.

5. Input the number of successes and failures for each group, as shown.

6 Press the "Begin analysis" button. Then, scroll down to the "Perform inference" section. Choose the 2-sample z test for the procedure, Quit as the category to indicate success, and $p_1 - p_2 > 0$ as the alternative hypothesis. Then press the "Perform inference" button.

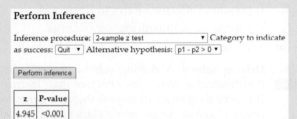

TI-83/84

1. Press $\boxed{\text{STAT}}$, then choose TESTS and 2-PropZTest.

2. When the 2-PropZTest screen appears, enter the values shown. Specify the alternative hypothesis $p_1 > p_2$.

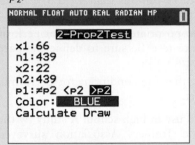

3. Highlight "Calculate" and press $\boxed{\text{ENTER}}$.

Note: If you select the "Draw" option, you will get a picture of the standard normal distribution with the area of interest shaded and the standardized test statistic and P value labeled.

Lesson 9.2

WHAT DID YOU LEARN?

LEARNING TARGET	EXAMPLES	EXERCISES
State hypotheses and check conditions for performing a significance test about a difference between two proportions.	p. 562	1–4
Calculate the standardized test statistic and P-value for a significance test about a difference between two proportions.	p. 563	5–8
Use the four-step process to perform a significance test about a difference between two proportions.	p. 564	9–12

Exercises

Lesson 9.2

Mastering Concepts and Skills

1. Children make choices Many new products introduced into the market are targeted toward children. The choice behavior of children with regard to new products is of particular interest to companies that design marketing strategies for these products. As part of one study, randomly selected children in different age groups were compared on their ability to sort new products into the correct product category (milk or juice).[12] Here are some of the data.

pg **562**

Age group	n	Number who sorted correctly
4- to 5-year-olds	50	10
6- to 7-year-olds	53	28

Researchers want to know if a greater proportion of 6- to 7-year-olds than 4- to 5-year-olds can sort correctly.

(a) State appropriate hypotheses for performing a significance test. Be sure to define the parameters of interest.

(b) Check that the conditions for performing the test are met.

2. **Steroid use in high school** A study by the National Athletic Trainers Association surveyed random samples of 1679 high school freshmen and 1366 high school seniors in Illinois. Results showed that 34 of the freshmen and 24 of the seniors had used anabolic steroids. Steroids, which are dangerous, are sometimes used in an attempt to improve athletic performance.[13] Researchers want to know if there is a difference in the proportion of all Illinois high school freshmen and seniors who have used anabolic steroids.

(a) State appropriate hypotheses for performing a significance test. Be sure to define the parameters of interest.

(b) Check that the conditions for performing the test are met.

3. **Southwestern food** Does the order in which options are presented influence people's responses? Eric randomly assigned 100 people to one of two groups. He asked the first group, "Which southwestern restaurant chain do you like better, Chain C or Chain Q?" He asked the second group, "Which southwestern restaurant chain do you like better, Chain Q or Chain C?" In the "C first" group, 31 of 50 people said they preferred Chain C. In the "Q first" group, 25 of 50 people said they preferred Chain C. Researchers want to know if there is a difference in the proportion of people like these who would say they prefer Chain C for the two versions of the survey question.

(a) State appropriate hypotheses for performing a significance test. Be sure to define the parameters of interest.

(b) Check that the conditions for performing the test are met.

4. **Does preschool help?** To study the long-term effects of preschool programs for poor children, researchers designed an experiment. They recruited 123 children from low-income families in Michigan who had never attended preschool. Researchers randomly assigned 62 of the children to attend preschool (paid for by the study budget) and the other 61 to serve as a control group who would not go to preschool. One response variable of interest was the children's need for social services as adults. Over a 10-year period, 38 children in the preschool group and 49 in the control group have needed social services.[14] Researchers want to know if attending preschool reduces the need for social services as an adult for children like these.

(a) State appropriate hypotheses for performing a significance test. Be sure to define the parameters of interest.

(b) Check that the conditions for performing the test are met.

5. **Children making choices** Refer to Exercise 1.
pg 563
(a) Calculate the standardized test statistic.

(b) Find the P-value using Table A or technology.

6. **Which athletes use steroids?** Refer to Exercise 2.

(a) Calculate the standardized test statistic.

(b) Find the P-value using Table A or technology.

7. **Choosing southwestern food** Refer to Exercise 3.

(a) Calculate the standardized test statistic.

(b) Find the P-value using Table A or technology.

8. **Is preschool effective?** Refer to Exercise 4.

(a) Calculate the standardized test statistic.

(b) Find the P-value using Table A or technology.

9. **Driving school** A driving school owner believes
pg 564
that Instructor A is more effective than Instructor B at preparing students to pass the state exam for a driver's license. An incoming class of 100 students is randomly assigned to two groups, each of size 50. One group is taught by Instructor A; the other is taught by Instructor B. At the end of the course, 30 of Instructor A's students and 22 of Instructor B's students pass the state exam. Do these results give convincing evidence at the $\alpha = 0.05$ level that Instructor A is more effective?

10. **Lower cholesterol?** High levels of cholesterol in the blood are associated with a higher risk of heart attacks. Will using a drug to lower blood cholesterol reduce heart attacks? To investigate this question, the Helsinki Heart Study recruited middle-aged men with high cholesterol but no history of other serious medical problems. The volunteer subjects were assigned at random to one of two treatments: 2051 men took the drug gemfibrozil to reduce their cholesterol levels, and a control group of 2030 men took a placebo. During the next five years, 56 men in the gemfibrozil group and 84 men in the placebo group had heart attacks. Do these data give convincing evidence at the $\alpha = 0.01$ significance level that gemfibrozil reduces heart attack risk in comparison to a placebo?

11. **Are teenagers going deaf?** In a study of 3000 randomly selected teenagers in 1990, 450 showed

some hearing loss. In a similar study of 1800 teenagers reported in 2010, 351 showed some hearing loss.[15] Do these data give convincing evidence that the proportion of all teens with hearing loss has increased at the $\alpha = 0.01$ significance level?

12. **Fix the brakes!** A random sample of 100 of last year's model of a popular car found that 20 had a minor defect in the brakes. The car company made an adjustment in the production process to try to reduce the proportion of cars with the brake problem. A random sample of 350 of this year's model found that 50 had the minor brake defect. Do these data give convincing evidence at the $\alpha = 0.05$ significance level that the company's adjustment was successful?

Applying the Concepts

13. **Controlling cholesterol** Which of two popular drugs—Lipitor or Pravachol—is more effective at helping lower "bad cholesterol"? Researchers designed an experiment, called the PROVE-IT Study, to find out. They used about 4000 people with heart disease as subjects. These volunteers were randomly assigned to one of two treatment groups: Lipitor or Pravachol. At the end of the study, researchers compared the proportion of subjects in each group who died, had a heart attack, or suffered other serious consequences within two years. For the 2063 subjects using Pravachol, the proportion was 0.263. For the 2099 subjects using Lipitor, the proportion was 0.224.[16]

(a) Does this study provide convincing evidence of a difference in the effectiveness of Lipitor and Pravachol?

(b) Based on your conclusion in part (a), which mistake—a Type I error or a Type II error—could you have made? Explain.

(c) Should you generalize the result in part (a) to all people with heart disease? Why or why not?

14. **Peanut allergies** A recent study of peanut allergies—the LEAP trial—explored the relationship between early exposure to peanuts and the subsequent development of an allergy to peanuts. Infants (4 to 11 months old) who had shown evidence of other kinds of allergies were randomly assigned to one of two groups: Group 1 consumed a baby-food form of peanut butter, and Group 2 avoided peanut butter. At 5 years old, 10 of 307 children in the peanut-consumption group were allergic to peanuts, and 55 of 321 children in the peanut-avoidance group were allergic to peanuts.[17]

(a) Does this study provide convincing evidence of a difference in the development of peanut allergies in infants who consume or avoid peanut butter?

(b) Based on your conclusion in part (a), which mistake—a Type I error or a Type II error—could you have made? Explain.

(c) Should you generalize the result in part (a) to all infants? Why or why not?

15. **Reducing cholesterol** Refer to Exercise 13.

(a) Construct and interpret a 95% confidence interval for the difference between the true proportions. If you already defined parameters and checked conditions in Exercise 13, you don't need to do them again here.

(b) Explain how the confidence interval provides more information than the test in Exercise 13.

16. **Exposure to peanuts** Refer to Exercise 14.

(a) Construct and interpret a 95% confidence interval for the difference between the true proportions. If you already defined parameters and checked conditions in Exercise 14, you don't need to do them again here.

(b) Explain how the confidence interval provides more information than the test in Exercise 14.

17. **Distracted driving** Is talking on a cell phone while driving more distracting than talking to a passenger? David Strayer and his colleagues at the University of Utah designed an experiment to help answer this question. They used 48 undergraduate students as subjects. The researchers randomly assigned half of the subjects to drive in a simulator while talking on a cell phone, and the other half to drive in the simulator while talking to a passenger. One response variable was whether or not the driver stopped at a rest area that was specified by researchers before the simulation started. Here are the results.

		Treatment		
		Cell phone	Passenger	Total
Response	Stopped at rest area	12	21	33
	Didn't stop	12	3	15
	Total	24	24	48

In Lesson 3.8, we used simulation to determine if the difference between the proportions who stopped at the rest area in the two groups was statistically significant.

(a) State appropriate hypotheses for performing a significance test. Be sure to define the parameters of interest.

(b) Explain why you should not use the methods of this lesson to calculate the P-value.

(c) We performed 100 trials of a simulation to see what differences in proportions (Passenger – Cell phone) would occur due only to chance variation in the random assignment, assuming that the type of distraction doesn't matter. A dotplot of the results is shown on the next page. What is the estimated P-value?

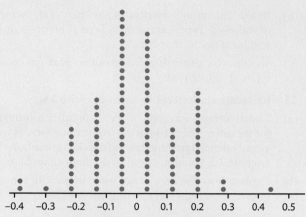

Simulated difference in sample proportion who stop

(d) What conclusion would you draw?

18. **Texting and driving** Does providing additional information affect responses to a survey question? Two statistics students decided to investigate this issue by asking different versions of a question about texting and driving. Fifty mall shoppers were divided into two groups of 25 at random. The first group was asked Version A and the second group was asked Version B. Here are the actual questions:

    • **Version A:** A lot of people text and drive. Are you one of them?

    • **Version B:** About 6000 deaths occur per year due to texting and driving. Knowing the potential consequences, do you text and drive?

    Of the 25 shoppers assigned to Version A, 16 admitted to texting and driving. Of the 25 shoppers assigned to Version B, only 12 admitted to texting and driving.

    (a) State appropriate hypotheses for performing a significance test. Be sure to define the parameters of interest.

    (b) Explain why you should not use the methods of this lesson to calculate the *P*-value.

    (c) We performed 100 trials of a simulation to see what differences in proportions (Version A − Version B) would occur due only to chance variation in the random assignment, assuming that the question asked doesn't matter. A dotplot of the results is shown here. What is the estimated *P*-value?

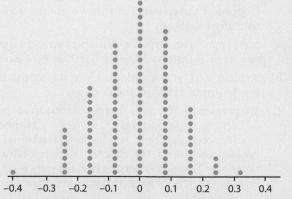

Simulated difference in sample proportion of yes answers

(d) What conclusion would you draw?

Recycle and Review

19. **Car of the Year (1.2)** Every year, *Motor Trend* magazine gives a car manufacturer the coveted "Car of the Year" award.

    (a) The pie chart provides the distribution of winning manufacturer from 1949 through 2014.[18] In about what percentage of years did General Motors win this award? Justify your answer.

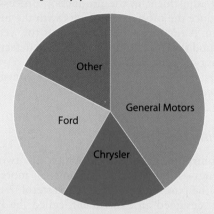

    (b) The pictograph shows the same data about *Motor Trend's* Car of the Year as in part (a). Explain how this graph is misleading.

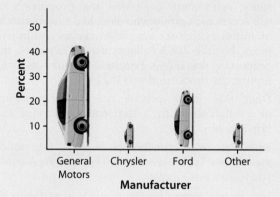

20. **Watching TV (5.2, 6.6)** Choose a young person (aged 19 to 25) at random and ask, "In the past seven days, how many days did you watch television?" Call the response *X* for short. Here is the probability distribution for *X*.

Days	0	1	2	3	4	5	6	7
Probability	0.04	0.03	0.06	0.08	0.09	0.08	0.05	???

    (a) What is the probability that $X = 7$? Justify your answer.

    (b) Calculate the mean and standard deviation of the random variable X.

    (c) Suppose that you asked 100 randomly selected young people (aged 19 to 25) to respond to the question and found that the mean $\bar{x}$ of their responses was 4.96. Would this result surprise you? Explain.

Lesson 9.3

Estimating a Difference Between Two Means

- Describe the shape, center, and variability of the sampling distribution of a difference $\bar{x}_1 - \bar{x}_2$ between two sample means.
- Check the Random and Normal/Large Sample conditions for constructing a confidence interval for a difference between two means.
- Use the four-step process to construct and interpret a confidence interval for the difference between two means.

In Lesson 9.1, you learned how to construct and interpret a confidence interval for a difference between two proportions $p_1 - p_2$. What if we want to estimate the difference $\mu_1 - \mu_2$ between the means of Population 1 and Population 2? For instance, maybe we want to estimate the difference in the mean amount of time spent playing video games last week by male and female students at a large university. The ideal strategy is to take a separate random sample from each population and to use the difference $\bar{x}_1 - \bar{x}_2$ between the sample means as our point estimate. To do inference about $\mu_1 - \mu_2$, we have to know about the sampling distribution of $\bar{x}_1 - \bar{x}_2$.

This lesson focuses on constructing confidence intervals for the difference between two means. You will learn how to perform significance tests about $\mu_1 - \mu_2$ in Lesson 9.4.

The Sampling Distribution of a Difference Between Two Means

To explore the sampling distribution of $\bar{x}_1 - \bar{x}_2$, let's start with two normally distributed populations having known means and standard deviations. The level of cholesterol in the blood for young men follows a normal distribution with mean $\mu_1 = 188$ milligrams per deciliter (mg/dl) and standard deviation $\sigma_1 = 41$ mg/dl. For teenage boys, blood cholesterol levels follow a normal distribution with mean $\mu_2 = 170$ mg/dl and standard deviation $\sigma_2 = 30$ mg/dl. The table summarizes this information.

Population	Shape	Mean	Standard deviation
Young men	Normal	$\mu_1 = 188$ mg/dl	$\sigma_1 = 41$ mg/dl
Teenage boys	Normal	$\mu_2 = 170$ mg/dl	$\sigma_2 = 30$ mg/dl

Suppose we take random samples of $n_1 = 49$ young men and $n_2 = 25$ teenage boys and measure their cholesterol levels. What can we say about the difference $\bar{x}_1 - \bar{x}_2$ in the sample means?

Let's do a simulation to find out. We used software to randomly select 49 young men and 25 teenage boys. Our first set of samples gave $\bar{x}_1 = 199.91$ and $\bar{x}_2 = 173.83$, resulting in a difference of $\bar{x}_1 - \bar{x}_2 = 199.91 - 173.83 = 26.08$ mg/dl. A red dot for this value appears in Figure 9.3 (on the next page). The dotplot shows the results of repeating this process 500 times.

FIGURE 9.3 Simulated sampling distribution of the difference in sample mean cholesterol levels $\bar{x}_1 - \bar{x}_2$ in 500 random samples of $n_1 = 49$ young men and $n_2 = 25$ teenage boys.

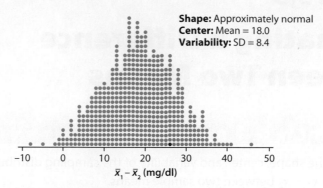

Shape: Approximately normal
Center: Mean = 18.0
Variability: SD = 8.4

The figure suggests that the sampling distribution of $\bar{x}_1 - \bar{x}_2$ has an approximately normal shape. This will be true whenever the Normal/Large Sample condition is met for *both* samples. The mean of the sampling distribution is 18.0. This result makes sense, because we expect the values of $\bar{x}_1 - \bar{x}_2$ to center on the true difference in the population means:

$$\mu_1 - \mu_2 = 188 - 170 = 18 \text{ mg/dl}$$

The standard deviation of the sampling distribution can be found using the formula

$$\sqrt{\frac{\sigma_1^2}{n_1} + \frac{\sigma_2^2}{n_2}} = \sqrt{\frac{41^2}{49} + \frac{30^2}{25}} = 8.38 \text{ mg/dl}$$

That is, the difference in the mean cholesterol level for a sample of 49 young men and the mean cholesterol level for a sample of 25 teenage boys is typically 8.38 mg/dl from the true difference in means of 18 mg/dl.

Describing the Sampling Distribution of $\bar{x}_1 - \bar{x}_2$

Suppose that $\bar{x}_1$ is the sample mean in an SRS of size n_1 drawn from a large population with mean μ_1 and standard deviation σ_1, and that $\bar{x}_2$ is the sample mean in an SRS of size n_2 drawn from a large population with mean μ_2 and standard deviation σ_2. Then:

- The **shape** of the sampling distribution of $\bar{x}_1 - \bar{x}_2$ is approximately normal if (1) both population distributions are approximately normal *or* (2) both sample sizes are large: $n_1 \geq 30$ and $n_2 \geq 30$.
- The **mean** of the sampling distribution of $\bar{x}_1 - \bar{x}_2$ is

$$\mu_{\bar{x}_1 - \bar{x}_2} = \mu_1 - \mu_2$$

- The **standard deviation** of the sampling distribution of $\bar{x}_1 - \bar{x}_2$ is

$$\sigma_{\bar{x}_1 - \bar{x}_2} = \sqrt{\frac{\sigma_1^2}{n_1} + \frac{\sigma_2^2}{n_2}}$$

EXAMPLE

Which supplier's potatoes are heavier?

Describing the sampling distribution of $\bar{x}_1 - \bar{x}_2$

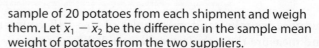

PROBLEM: A potato chip manufacturer buys potatoes from two different suppliers, Camberley Fields and Riderwood Farms. The weights of potatoes from Camberley Fields are approximately normally distributed with a mean of $\mu_1 = 180$ grams and a standard deviation of $\sigma_1 = 30$ grams. The weights of potatoes from Riderwood Farms are approximately normally distributed with a mean of $\mu_2 = 175$ grams and a standard deviation of $\sigma_2 = 25$ grams. When shipments arrive at the factory, inspectors randomly select a sample of 20 potatoes from each shipment and weigh them. Let $\bar{x}_1 - \bar{x}_2$ be the difference in the sample mean weight of potatoes from the two suppliers.

(a) What is the shape of the sampling distribution of $\bar{x}_1 - \bar{x}_2$? Why?

(b) Find the mean of the sampling distribution.

(c) Calculate the standard deviation of the sampling distribution.

SOLUTION:

(a) Approximately normal, because both population distributions are approximately normal.

(b) $\mu_{\bar{x}_1 - \bar{x}_2} = 180 - 175 = 5$ grams

(c) $\sigma_{\bar{x}_1 - \bar{x}_2} = \sqrt{\dfrac{30^2}{20} + \dfrac{25^2}{20}} = 8.73$ grams

$$\mu_{\bar{x}_1 - \bar{x}_2} = \mu_1 - \mu_2$$

$$\sigma_{\bar{x}_1 - \bar{x}_2} = \sqrt{\dfrac{\sigma_1^2}{n_1} + \dfrac{\sigma_2^2}{n_2}}$$

FOR PRACTICE   TRY EXERCISE 1.

The standard deviation in part (c) of the example tells us that the difference in the sample mean weights of potatoes will typically be about 8.73 grams from the true difference of 5 grams in the mean weights of potatoes from Camberley Fields and Riderwood Farms.

Conditions for Estimating $\mu_1 - \mu_2$

In Lesson 7.5, we introduced two conditions that should be met before we construct a confidence interval for a population mean: Random and Normal/Large Sample. Now that we are comparing *two* means, we have to modify these conditions slightly.

How to Check the Conditions for Constructing a Confidence Interval for $\mu_1 - \mu_2$

- **Random:** The data come from separate random samples from the two populations of interest or from two groups in a randomized experiment.

- **Normal/Large Sample:** Both population distributions (or the true distributions of responses to the two treatments) are normal *or* both sample sizes are large ($n_1 \geq 30$ and $n_2 \geq 30$). If the population (treatment) distributions have an unknown shape and the sample sizes are less than 30, a graph of the data from each sample shows no strong skewness or outliers.

EXAMPLE

Which brand of cookies has more chocolate chips?

Checking conditions

PROBLEM: Ashtyn and Olivia wanted to know if generic chocolate-chip cookies have as many chocolate chips as name-brand chocolate-chip cookies, on average. To investigate, they randomly selected 10 bags of Chips Ahoy cookies and 10 bags of Great Value cookies and randomly selected 1 cookie from each bag. Then they carefully broke apart each cookie and counted the number of chocolate chips in each. Here are their results.

Marie C. Fields/Shutterstock

Chips Ahoy	17	19	21	16	17	18	20	21	17	18
Great Value	22	20	14	17	21	22	15	19	26	18

Check if the conditions for calculating a 99% confidence interval for $\mu_1 - \mu_2$ are met.

SOLUTION:

Random? Separate random samples of Chips Ahoy and Great Value cookies. ✓

Normal/Large Sample? The sample sizes are small, but the boxplots don't show any outliers or strong skewness. ✓

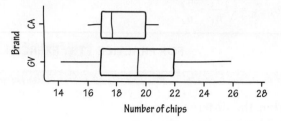

> Because the sample sizes are less than 30, graph the data from each sample to see if it is plausible that they came from normal populations.

FOR PRACTICE   **TRY EXERCISE 5.**

Constructing and Interpreting a Confidence Interval for $\mu_1 - \mu_2$

When data come from two separate random samples or two groups in a randomized experiment, the statistic $\bar{x}_1 - \bar{x}_2$ is our point estimate for $\mu_1 - \mu_2$. We can calculate a confidence interval for $\mu_1 - \mu_2$ using the familiar formula

$$\text{point estimate} \pm \text{critical value} \cdot \text{standard error of statistic}$$

The standard deviation of the sampling distribution of $\bar{x}_1 - \bar{x}_2$ is

$$\sigma_{\bar{x}_1 - \bar{x}_2} = \sqrt{\frac{\sigma_1^2}{n_1} + \frac{\sigma_2^2}{n_2}}$$

Because we usually don't know the values of σ_1 and σ_2, we replace them with the sample standard deviations s_1 and s_2. The result is the *standard error* of $\bar{x}_1 - \bar{x}_2$:

$$SE_{\bar{x}_1 - \bar{x}_2} = \sqrt{\frac{s_1^2}{n_1} + \frac{s_2^2}{n_2}}$$

When the Normal/Large Sample condition is met, we find the critical value t^* for the given confidence level using Table B or technology. We'll discuss the technology approach at the end of the lesson.

How to Calculate a Confidence Interval for $\mu_1 - \mu_2$

When the Random and Normal/Large Sample conditions are met, a C% confidence interval for the difference between two means $\mu_1 - \mu_2$ is

$$(\bar{x}_1 - \bar{x}_2) \pm t^* \sqrt{\frac{s_1^2}{n_1} + \frac{s_2^2}{n_2}}$$

where t^* is the critical value with C% of its area between $-t^*$ and t^* in the t distribution with degrees of freedom equal to the *smaller* of $n_1 - 1$ and $n_2 - 1$ or given by technology.

This interval is often called a *two-sample t interval for a difference between two means*. Let's return to the chocolate-chip cookie example. We already checked that the Random and Normal/Large Sample conditions are met. The table shows summary statistics for the two samples of cookies.

Brand	Sample size	Mean	SD
Chips Ahoy	$n_1 = 10$	$\bar{x}_1 = 18.4$	$s_1 = 1.78$
Great Value	$n_2 = 10$	$\bar{x}_2 = 19.4$	$s_2 = 3.60$

Using df = 9 (the smaller of $n_1 - 1$ and $n_2 - 1$), the critical value for 99% confidence is $t^* = 3.250$.

So a 99% confidence interval for $\mu_1 - \mu_2$ is

$$(18.4 - 19.4) \pm 3.250\sqrt{\frac{1.78^2}{10} + \frac{3.60^2}{10}} \to -1 \pm 4.127 \to -5.127 \text{ to } 3.127$$

	Tail probability p		
df	0.01	0.005	0.0025
8	2.896	3.355	3.833
9	2.821	3.250	3.690
10	2.764	3.169	3.581
z^*	2.326	2.576	2.807
	98%	99%	99.5%
	Confidence level C		

We are 99% confident that the interval from -5.127 to 3.127 captures the true difference in the mean number of chocolate chips in all Chips Ahoy and all Great Value cookies. The interval suggests that Chips Ahoy cookies have between 5.127 fewer and 3.127 more chocolate chips than Great Values cookies, on average.

EXAMPLE

Do helium-filled balloons deflate more quickly?

Confidence interval for $\mu_1 - \mu_2$

PROBLEM: After buying many helium balloons only to see them deflate within a few days, Erin and Jenna decided to test whether helium-filled balloons deflate more quickly than air-filled balloons. To find out, they bought 60 balloons and randomly divided them into two piles of 30, filling the balloons in the first pile with helium and the balloons in the second pile with air. Then they measured the circumference of each balloon immediately after it was filled and again 3 days later. The average decrease in circumference of the helium-filled balloons was 26.5 centimeters (cm) with a standard deviation of 1.92 cm. The average decrease of the air-filled balloons was 2.1 cm with a standard deviation of 2.79 cm. Construct and interpret a 95% confidence interval for the true difference in the mean circumference loss of helium- and air-filled balloons like the ones in this study.

Chris Gramley/iStock/Getty Images Plus

SOLUTION:

STATE: We want to estimate $\mu_1 - \mu_2$ at the 95% confidence level, where $\mu_1 =$ the true mean circumference loss for helium-filled balloons like these and $\mu_2 =$ the true mean circumference loss for air-filled balloons like these.

STATE: State the parameters you want to estimate and the confidence level.

PLAN: Two-sample t interval for $\mu_1 - \mu_2$

Random? The data came from two groups in a randomized experiment. ✓

Normal/Large Sample? Both sample sizes are large ($n_1 = 30 \geq 30$ and $n_2 = 30 \geq 30$). ✓

DO: $df = 30 - 1 = 29, t^* = 2.045$

$$(26.5 - 2.1) \pm 2.045\sqrt{\frac{(1.92)^2}{30} + \frac{(2.79)^2}{30}}$$

$\rightarrow 24.4 \pm 1.26 \rightarrow 23.14 \text{ to } 25.66$

CONCLUDE: We are 95% confident that the interval from 23.14 cm to 25.66 cm captures the true difference in the mean circumference loss for helium-filled and air-filled balloons like these. The data suggest that the helium-filled balloons lose between 23.14 and 25.66 cm more in circumference than the air-filled balloons.

> **PLAN:** Identify the appropriate inference method and check conditions.

> **DO:** If the conditions are met, perform calculations.

> **CONCLUDE:** Interpret your interval in the context of the problem.

> These balloons weren't randomly selected, so the scope of inference is limited to a cause-and-effect conclusion about balloons like these.

FOR PRACTICE   TRY EXERCISE 9.

The confidence interval we calculated in the example using df = 29 is 23.14 to 25.66. Technology uses a complicated formula to compute the degrees of freedom for a two-sample t interval. As the Tech Corner at the end of the lesson shows, technology reports df = 51.44 and the 95% confidence interval for $\mu_1 - \mu_2$ as 23.16 to 25.64. The larger df used by technology always results in a narrower confidence interval (more precise estimate) than the more conservative approach using df = the smaller of $n_1 - 1$ and $n_2 - 1$.

LESSON APP 9.3

Do bigger apartments cost more money?

A college student wants to compare the cost of one- and two-bedroom apartments near campus. She collects the following data on monthly rents (in dollars) for a random sample of 10 apartments of each type.

| 1 bedroom | 500 | 650 | 600 | 505 | 450 | 550 | 515 | 495 | 650 | 395 |
| 2 bedroom | 595 | 500 | 580 | 650 | 675 | 675 | 750 | 500 | 495 | 670 |

1. Construct and interpret a 90% confidence interval for the difference between the mean monthly rents for all one- and two-bedroom apartments near the college campus.

2. Based on your interval, is there convincing evidence of a difference in the population means? Explain.

TECH CORNER

Confidence Intervals for a Difference Between Two Means

You can use technology to calculate a confidence interval for $\mu_1 - \mu_2$. We'll illustrate using the balloon example. For $n_1 = 30$ helium-filled balloons, $\bar{x}_1 = 26.5$ cm and $s_1 = 1.92$ cm. For $n_2 = 30$ air-filled balloons $\bar{x}_2 = 2.1$ cm and $s_2 = 2.79$ cm. To calculate a 95% confidence interval for $\mu_1 - \mu_2$, use either of the following.

Applet

1. Launch the *One Quantitative Variable* applet at highschool.bfwpub.com/spa3e.

2. Enter "Circumference loss (cm)" as the variable name.

3. Choose 2 for the Number of groups and "Mean and standard deviation" for the Input.

4. Enter the name, mean, standard deviation, and sample size for both groups, as shown.

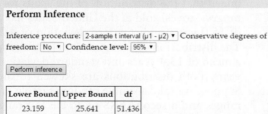

One Quantitative Variable

Variable name: Circumference loss (cm)		
Number of groups: 2 ▾		
Input: Mean and standard deviation ▾		

Group 1 name: Helium	Group 1 mean: 26.5	SD: 1.92	n: 30
Group 2 name: Air	Group 2 mean: 2.1	SD: 2.79	n: 30

Begin analysis | Edit inputs | Reset everything

5. Press the "Begin analysis" button. Then, scroll down to the "Perform inference" section. Choose the 2-sample t-interval ($\mu_1 - \mu_2$) for the Inference procedure, "No" for the Conservative degrees of freedom, and 95% for the Confidence level. Then, press the "Perform inference" button. (*Note:* If you choose "Yes" for the Conservative degrees of freedom, the output from the Applet should match the confidence interval from the example.)

Perform Inference

Inference procedure: 2-sample t interval (μ1 - μ2) ▾ Conservative degrees of freedom: No ▾ Confidence level: 95% ▾

Perform inference

Lower Bound	Upper Bound	df
23.159	25.641	51.436

Note: To calculate a confidence interval for a difference in means from raw data, start by choosing Raw data as the Input. Then, enter the values from each of the two samples and proceed as in Step 4.

TI-83/84

1. Press STAT, then choose TESTS and 2-SampTInt.

2. When the 2-SampTInt screen appears, choose Stats as the input method. Then, enter the summary statistics, as shown. Choose "No" for Pooled.

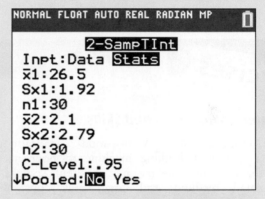

NORMAL FLOAT AUTO REAL RADIAN MP

```
        2-SampTInt
 Inpt:Data Stats
 x̄1:26.5
 Sx1:1.92
 n1:30
 x̄2:2.1
 Sx2:2.79
 n2:30
 C-Level:.95
↓Pooled:No Yes
```

3. Highlight "Calculate" and press ENTER.

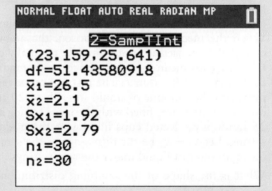

NORMAL FLOAT AUTO REAL RADIAN MP

```
        2-SampTInt
 (23.159,25.641)
 df=51.43580918
 x̄1=26.5
 x̄2=2.1
 Sx1=1.92
 Sx2=2.79
 n1=30
 n2=30
```

Note: To calculate a confidence interval for a difference in means from raw data, start by entering the values into lists L1 and L2. Then, choose the Data option in 2-SampTInt.

Lesson 9.3

WHAT DID YOU LEARN?

LEARNING TARGET	EXAMPLES	EXERCISES
Describe the shape, center, and variability of the sampling distribution of a difference $\bar{x}_1 - \bar{x}_2$ between two sample means.	p. 572	1–4
Check the Random and Normal/Large Sample conditions for constructing a confidence interval for a difference between two means.	p. 573	5–8
Use the four-step process to construct and interpret a confidence interval for the difference between two means.	p. 575	9–12

Exercises *Lesson 9.3*

Mastering Concepts and Skills

1. Automated soda A fast-food restaurant uses
pg 572 an automated filling machine to pour its soft drinks. The machine has different settings for small, medium, and large drink cups. When the large setting is chosen, the amount of liquid dispensed by the machine is approximately normally distributed with mean $\mu_1 = 27$ ounces and standard deviation $\sigma_1 = 0.8$ ounce. When the medium setting is chosen, the amount of liquid dispensed is approximately normally distributed with mean $\mu_2 = 17$ ounces and standard deviation $\sigma_2 = 0.5$ ounce. The restaurant manager measures the amount of liquid in each of 20 randomly selected cups filled with the large setting and 25 randomly selected cups filled with the medium setting. Let $\bar{x}_1 - \bar{x}_2$ be the difference in the sample mean amount of liquid under the two settings.

(a) What is the shape of the sampling distribution of $\bar{x}_1 - \bar{x}_2$? Why?

(b) Find the mean of the sampling distribution.

(c) Calculate the standard deviation of the sampling distribution.

2. Tall kids Based on information from the U.S. National Health and Nutrition Examination Survey (NHANES), the heights of 10-year-old girls are approximately normally distributed with mean $\mu_1 = 56.4$ inches and standard deviation $\sigma_1 = 2.7$ inches. The heights of 10-year-old boys are approximately normally distributed with mean $\mu_2 = 55.7$ inches and standard deviation $\sigma_2 = 3.8$ inches.[19] Suppose we take separate SRSs of 12 girls and 8 boys of this age and measure their heights. Let $\bar{x}_1 - \bar{x}_2$ be the difference in the sample mean heights.

(a) What is the shape of the sampling distribution of $\bar{x}_1 - \bar{x}_2$? Why?

(b) Find the mean of the sampling distribution.

(c) Calculate the standard deviation of the sampling distribution.

3. Dogs and cats Dogs vary in size a great deal more than cats do. The weights of all dogs in a certain city have a mean of 28 pounds and standard deviation of 14 pounds. The weights of all cats in the same city have a mean of 9.5 pounds and standard deviation of 2 pounds. Cat weights are roughly normally distributed, but dog weights are skewed to the right. Suppose you take separate random samples of 10 dogs and 8 cats from this city. Let $\bar{x}_1 - \bar{x}_2$ be the difference (Dogs – Cats) in the sample mean weights.

(a) Is the shape of the sampling distribution of $\bar{x}_1 - \bar{x}_2$ approximately normal? Explain.

(b) Find the mean of the sampling distribution.

(c) Calculate the standard deviation of the sampling distribution.

4. How old is your stove? A consumer group has determined that the distribution of life spans for all gas ranges (stoves) sold in the United States has a mean of 15.0 years and standard deviation of 4.2 years. The distribution of life spans for electric ranges has a mean of 13.4 years and standard deviation of 3.7 years. Both distributions are skewed to the right. Suppose we take a simple random sample of 25 gas ranges and a second SRS of 20 electric ranges. Let $\bar{x}_1 - \bar{x}_2$ be the difference (Gas – Electric) in the sample mean life spans.

(a) Is the shape of the sampling distribution of $\bar{x}_1 - \bar{x}_2$ approximately normal? Explain.

(b) Find the mean of the sampling distribution.

(c) Calculate the standard deviation of the sampling distribution.

Determine if the conditions for calculating a confidence interval for $\mu_1 - \mu_2$ are met in Exercises 5–10.

5. **Do females text more than males?** For their final project, a group of statistics students investigated their belief that females text more than males, on average. They asked separate random samples of 15 males and 16 females from their school to record the number of text messages they sent and received over a 2-day period. Boxplots of their data are shown here.

pg 573

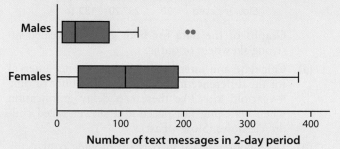

Number of text messages in 2-day period

6. **Comparing household size** How do the numbers of people living in households in the United Kingdom (U.K.) and South Africa compare? To help answer this question, we used an online random data selector to choose separate random samples of 50 students from each country. Here is a dotplot of the household sizes reported by the students in the survey.

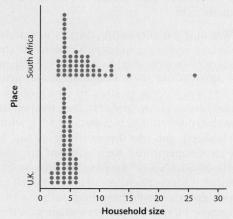

Household size

7. **Big words?** Mary was interested in comparing the mean word length in articles from a medical journal and from an airline's in-flight magazine. She counted the number of letters in the first 400 words of an article in the medical journal and in the first 100 words of an article in the airline magazine. Mary then used statistical software to produce the histograms shown. Note that J stands for journal and M for magazine.

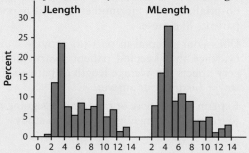

8. **Boys' and girls' shoes** A group of statistics students wanted to estimate the difference in the mean number of pairs of shoes that teenage boys and girls own. They surveyed a random sample of 20 female students and a separate random sample of 20 male students from their school. Then they recorded the number of pairs of shoes that each student reported owning. The back-to-back stemplot displays the data.

Females		Males
	0	4
	0	555677778
333	1	0000124
95	1	
4332	2	2
66	2	
410	3	
8	3	58
	4	
9	4	
100	5	
7	5	

KEY: 2 | 2 represents a student with 22 pairs of shoes.

9. **Different shades of squirrel** In many parts of the northern United States, two color variants of the Eastern Gray Squirrel—gray and black—are found in the same habitats. A scientist studying squirrels in a large forest wonders if there is a difference in the sizes of the two color variants. He collects random samples of 40 squirrels of each color from a large forest and weighs them. The 40 black squirrels have a mean weight of $\bar{x}_1 = 20.3$ ounces and standard deviation of $s_1 = 2.1$ ounces. The 40 gray squirrels have a mean weight of $\bar{x}_2 = 19.2$ ounces and standard deviation of $s_2 = 1.9$ ounces. Construct and interpret a 95% confidence interval for the difference in mean weight of black and gray squirrels in this forest.

pg 575

10. **Beta blockers** In a study of heart surgery, one issue was the effect of drugs called beta blockers on the pulse rate of patients during surgery. The available subjects were randomly split into two groups of 30 patients each. One group received a beta blocker; the other group received a placebo. The pulse rate of each patient at a critical point during the operation was recorded. The treatment group had a mean pulse rate of 65.2 and standard deviation of 7.8. For the control group, the mean pulse rate was 70.3 and the standard deviation was 8.3. Construct and interpret a 99% confidence interval for the difference in mean pulse rates for patients like these who receive a beta blocker or a placebo.

11. **Is red wine healthier than white wine?** Observational studies suggest that moderate use of alcohol by adults reduces heart attacks and that red wine may have special benefits. One reason may be that red wine contains polyphenols, substances that do good things to cholesterol in the blood and so may reduce the risk of heart attacks. In an experiment, healthy men were assigned at random to drink half

a bottle of either red or white wine each day for 2 weeks. The level of polyphenols in their blood was measured before and after the 2-week period. Here are the percent changes in level for the subjects in both groups.[20]

Red wine	3.5	8.1	7.4	4.0	0.7	4.9	8.4	7.0	5.5
White wine	3.1	0.5	−3.8	4.1	−0.6	2.7	1.9	−5.9	0.1

Construct and interpret a 90% confidence interval for the difference in mean percent change in polyphenol levels for subjects like these who are assigned to drink red wine or white wine.

12. **Tropical flowers** Different varieties of the tropical flower *Heliconia* are fertilized by different species of hummingbirds. Researchers believe that over time, the lengths of the flowers and the forms of the hummingbirds' beaks have evolved to match each other. Here are data on the lengths in millimeters for random samples of two color varieties of the same species of flower on the island of Dominica.[21]

H. *caribaea* red

41.90	42.01	41.93	43.09	41.17	41.69	39.78	40.57
39.63	42.18	40.66	37.87	39.16	37.40	38.20	38.07
38.10	37.97	38.79	38.23	38.87	37.78	38.01	

H. *caribaea* yellow

36.78	37.02	36.52	36.11	36.03	35.45	38.13	37.10
35.17	36.82	36.66	35.68	36.03	34.57	34.63	

Construct and interpret a 95% confidence interval for the difference in the mean lengths of these two varieties of flowers on the island of Dominica.

Applying the Concepts

13. **SpongeBob in hot water** Do fast-paced shows affect a child's ability to delay gratification? A study randomly assigned twenty 4-year-olds to view a fast-paced children's show about "an animated sponge that lives under the sea." Another twenty 4-year-olds were randomly assigned to watch a slower-paced PBS children's show about "a typical American preschool-aged boy." In the fast-paced show, there was a scene change every 11 seconds, on average. In the slower-paced show, there was a scene change every 34 seconds, on average.

After watching 9 minutes of the assigned program, children were asked if they preferred mini-marshmallows or goldfish crackers. Once the children had chosen their snack, two plates were prepared: one with 10 pieces and one with 2 pieces. A bell was placed between them and children were told that they could have the plate with 10 pieces if they waited for the researcher to return. Otherwise,

they could ring the bell at any time and get the plate with 2 pieces. The amount of time (in seconds) it took for each child to ring the bell was recorded (330 seconds if the child waited the full time for the researcher to return). The table summarizes the data.[22]

Group	n	Mean	SD
Fast-paced	20	146.15	151.29
Slower-paced	20	257.20	132.16

Graphs of the data for both groups revealed no strong skewness or outliers.

(a) Calculate and interpret a 95% confidence interval for the difference in the true mean amount of time 4-year-old kids like these can delay gratification when watching fast-paced and slower-paced children's shows (Slow − Fast).

(b) Does your interval in part (a) provide convincing evidence that watching fast-paced shows reduces the average amount of time that 4-year-olds like these can delay gratification compared to watching slower-paced shows? Justify your answer.

(c) Explain why it was important for the researchers to randomly determine which children watched each show rather than letting the children choose for themselves.

14. **Warming patients during surgery** Researchers wondered whether maintaining a patient's body temperature close to normal by heating the patient during surgery would affect wound infection rates. Patients were assigned at random to two groups: the normothermic group (patients' core temperatures were maintained at near normal, 36.5°C, with heating blankets) and the hypothermic group (patients' core temperatures were allowed to decrease to about 34.5°C). If keeping patients warm during surgery alters the chance of infection, patients in the two groups should have hospital stays of different lengths. Here are summary statistics on their hospital stays (in number of days) for the two groups.[23]

Group	n	Mean	SD
Normothermic	104	12.1	4.4
Hypothermic	96	14.7	6.5

(a) Construct and interpret a 95% confidence interval for the difference in the true mean length of hospital stays for normothermic and hypothermic patients like these.

(b) Does your interval in part (a) provide convincing evidence that keeping patients warm during surgery affects the average length of patients' hospital stays? Justify your answer.

(c) Explain the purpose of randomly assigning patients to the two treatment groups in this study.

15. **Bird eggs** A researcher wants to see if birds that build larger nests lay larger eggs. She selects two random samples of nests: one of small nests and the other of large nests. Then she weighs one egg (chosen at random if there is more than one egg) from each nest. A 95% confidence interval for the difference (Large – Small) between the mean mass (in grams) of eggs in small and large nests is 1.6 ± 2.0.

(a) Does the interval provide convincing evidence of a difference in the mean egg mass of birds with small nests and birds with large nests? Explain.

(b) Does the interval provide convincing evidence that the mean egg mass of birds with small nests and birds with large nests is the same? Explain.

16. **Reaction times** Catherine and Ana wanted to know if student athletes (students on at least one varsity team) have faster reaction times than non-athletes. They took separate random samples of 33 athletes and 30 non-athletes from their school and tested their reaction times using an online reaction test, which measured the time (in seconds) between when a green light went on and the subject pressed a key on the computer keyboard. A 95% confidence interval for the difference (Non-athlete – Athlete) in the mean reaction times was 0.018 ± 0.034 seconds.

(a) Does the interval provide convincing evidence of a difference in the mean reaction times of athletes and non-athletes? Explain.

(b) Does the interval provide convincing evidence that the mean reaction time of athletes and the mean reaction time of non-athletes is the same? Explain.

Extending the Concepts

17. **Taller girls?** Refer to Exercise 2. Calculate the probability that the mean height for the sample of 12 girls is greater than the mean height for the sample of 8 boys.

Recycle and Review

18. **Who has the flu?** (4.5, 5.3) Central High School is in the midst of a flu epidemic. The probability that a randomly selected student has the flu is 0.35, and the probability that a student who has the flu also has a high fever is 0.90. But there are other illnesses making the rounds, and the probability that a student who doesn't have the flu has a high fever (as a result of some other ailment) is 0.12.

(a) Suppose that a randomly selected Central High School student has a high fever. Find the probability that the student has the flu.

(b) Suppose we randomly select 5 Central High School students. Find the probability that exactly 3 of them have the flu and a high fever.

19. **Mobile phone quality** (7.4) A quality-control inspector is testing mobile phones made during a single day at a factory to determine the proportion of phones with minor defects. She selects an SRS of 200 phones and finds that 12 of them have minor defects. Calculate and interpret a 99% confidence interval for the true proportion of blemished phones produced that day.

Lesson 9.4

Testing a Claim about a Difference Between Two Means

LEARNING TARGETS

● State hypotheses and check conditions for performing a significance test about a difference between two means.

● Calculate the standardized test statistic and *P*-value for a significance test about a difference between two means.

● Use the four-step process to perform a significance test about a difference between two means.

An observed difference between two sample means can reflect an actual difference in the parameters, or it may just be due to chance variation in random sampling or random assignment. In Lesson 3.8, we used simulation to determine which explanation makes more sense in an experiment about whether caffeine affects pulse rate. This lesson shows you how to perform a significance test for the difference between two means.

Stating Hypotheses and Checking Conditions for a Test about $\mu_1 - \mu_2$

In a test for comparing two means, the null hypothesis has the general form

$$H_0: \mu_1 - \mu_2 = \text{hypothesized value}$$

We'll focus on situations in which the hypothesized difference is 0. Then the null hypothesis says that there is no difference between the two parameters:

$$H_0: \mu_1 - \mu_2 = 0$$

You will sometimes see the null hypothesis written in the equivalent form $H_0: \mu_1 = \mu_2$. The alternative hypothesis says what kind of difference we expect.

The conditions for performing a significance test about $\mu_1 - \mu_2$ are the same as those for constructing a confidence interval for $\mu_1 - \mu_2$ that you learned in Lesson 9.3.

Checking the Conditions for Performing a Significance Test about $\mu_1 - \mu_2$

- **Random:** The data come from separate random samples from the two populations of interest or from two groups in a randomized experiment.

- **Normal/Large Sample:** Both population distributions (or the true distributions of responses to the two treatments) are normal or both sample sizes are large ($n_1 \geq 30$ and $n_2 \geq 30$). If the population (treatment) distributions have an unknown shape and the sample sizes are less than 30, a graph of the data from each sample shows no strong skewness or outliers.

EXAMPLE

Where do the big trees grow?

Stating hypotheses and checking conditions

PROBLEM: The Wade Tract Preserve in Georgia is an old-growth forest of longleaf pines that has survived in a relatively undisturbed state for hundreds of years. One question of interest to foresters who study the area is "How do the sizes of longleaf pine trees in the northern and southern halves of the forest compare?" To find out, researchers took random samples of 30 trees from each half and measured the diameter at breast height (DBH) in centimeters. The table shows summary statistics for the two samples of trees.[24]

Forest location	Sample size	Mean	SD
Northern half	$n_1 = 30$	$\bar{x}_1 = 23.70$ cm	$s_1 = 17.50$ cm
Southern half	$n_2 = 30$	$\bar{x}_2 = 34.53$ cm	$s_2 = 14.26$ cm

Do these data give convincing evidence of a difference in the population means at the $\alpha = 0.05$ significance level?

(a) State appropriate hypotheses for performing a significance test. Be sure to define the parameters of interest.

(b) Check that the conditions for performing the test are met.

SOLUTION:

(a) $H_0: \mu_1 - \mu_2 = 0$
$H_a: \mu_1 - \mu_2 \neq 0$, where μ_1 = the true mean DBH of all

> H_a is two-sided because we're asked if the population means differ.

trees in the northern half of the forest and μ_2 = the true mean DBH of all trees in the southern half of the forest.

(b) Random? The data came from separate random samples of trees from the two halves of the forest. ✓

Normal/Large Sample? Both sample sizes are large ($n_1 = 30 \geq 30$ and $n_2 = 30 \geq 30$). ✓

FOR PRACTICE   TRY EXERCISE 1.

Calculations: Standardized Test Statistic and *P*-Value

If the conditions are met, we can proceed with calculations. To do a test of $H_0: \mu_1 - \mu_2 = 0$, start by standardizing $\bar{x}_1 - \bar{x}_2$:

$$\text{standardized test statistic} = \frac{\text{statistic} - \text{null value}}{\text{standard deviation of statistic}}$$

Because we usually don't know the population standard deviations, we use the *standard error* of $\bar{x}_1 - \bar{x}_2$ in the denominator of the standardized test statistic:

$$t = \frac{(\bar{x}_1 - \bar{x}_2) - 0}{\sqrt{\dfrac{s_1^2}{n_1} + \dfrac{s_2^2}{n_2}}}$$

To find the *P*-value, use the *t* distribution with degrees of freedom equal to the *smaller* of $n_1 - 1$ and $n_2 - 1$ or given by technology.

EXAMPLE

Where are the bigger trees?

Calculating the standardized test statistic and P-value

PROBLEM: Refer to the preceding example. The table shows summary statistics for the two samples of trees.

(a) Calculate the standardized test statistic.

(b) Find the *P*-value using Table B.

Forest location	Sample size	Mean	SD
Northern half	$n_1 = 30$	$\bar{x}_1 = 23.70$ cm	$s_1 = 17.50$ cm
Southern half	$n_2 = 30$	$\bar{x}_2 = 34.53$ cm	$s_2 = 14.26$ cm

SOLUTION:

(a) Statistics: $\bar{x}_1 = 23.70, s_1 = 17.50, \bar{x}_2 = 34.53, s_2 = 14.26$

$$t = \frac{(23.70 - 34.53) - 0}{\sqrt{\dfrac{17.50^2}{30} + \dfrac{14.26^2}{30}}} = -2.628$$

$$t = \frac{(\bar{x}_1 - \bar{x}_2) - 0}{\sqrt{\dfrac{s_1^2}{n_1} + \dfrac{s_2^2}{n_2}}}$$

(b) df = 30 - 1 = 29

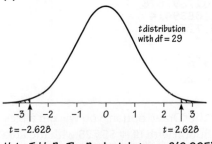

t distribution with df = 29

> The observed result gives *some* evidence in favor of H_a because $\bar{x}_1 - \bar{x}_2 = 23.70 - 34.53 = -10.83 \neq 0$.

-3 -2 -1 0 1 2 3
$t = -2.628$ $t = 2.628$

Using Table B: The P-value is between 2(0.005) = 0.01 and 2(0.01) = 0.02.

Using technology: Applet/T-Test gives P-value = 0.0111.

FOR PRACTICE   TRY EXERCISE 5.

Using the conservative df = 29 from the example, the *P*-value of the test is 0.0136. As the following Tech Corner shows, the actual *P*-value is 0.0111 with df = 55.73. Using technology for a two-sample *t* test results in a larger df and smaller *P*-value than the more conservative method using df = smaller of $n_1 - 1$ and $n_2 - 1$.

What does the *P*-value tell us? If there is no difference in the population mean DBH of the longleaf pine trees in the northern and southern halves of the forest, and we repeated the random sampling process many times, we'd get a difference in sample means of -10.83 or more extreme in either direction about 1.1% of the time. With such a small probability of obtaining a result like this by chance alone when the null hypothesis is true, we would reject H_0. That is, we have convincing evidence that there is a difference in the true mean DBH of the trees in the two halves of the forest.

Significance Tests for a Difference Between Two Means

You can use technology to perform significance tests about $\mu_1 - \mu_2$. We'll illustrate using the big trees example. To perform a test of H_0: $\mu_1 - \mu_2 = 0$ versus H_a: $\mu_1 - \mu_2 \neq 0$, use either of the following.

Applet

1. Launch the *One Quantitative Variable* applet at **highschool.bfwpub.com/spa3e**.

2. Enter "DBH (cm)" as the variable name.

3. Choose 2 for the Number of groups and "Mean and standard deviation" for the Input.

4. Enter the name, mean, standard deviation, and sample size for both groups, as shown.

5. Press the "Begin analysis" button. Then, scroll down to the "Perform inference" section. Choose the 2-sample *t* test for the procedure, "No" for the Conservative degrees of freedom (smaller of $n_1 - 1$ and $n_2 - 1$), and $\mu_1 - \mu_2 \neq 0$ for the alternative hypothesis. Then, press the "Perform inference" button. Compare your results with those from the example.

Note: To perform a significance test for a difference in means from raw data, start by choosing Raw data as the Input. Then enter the values from each of the two samples and proceed as in Step 5.

TI-83/84

1. Press $\boxed{\text{STAT}}$, then choose TESTS and 2-SampTTest.

2. When the 2-SampTTest screen appears, choose Stats as the input method. Then enter $\bar{x}_1 = 23.70$, $s_1 = 17.50$, $n_1 = 30$, $\bar{x}_2 = 34.53$, $s_2 = 14.26$, $n_2 = 30$; choose $\mu_1 \neq \mu_2$ as the alternative hypothesis; and choose "No" for Pooled.

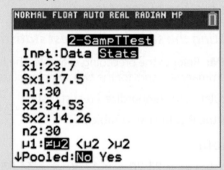

3. Highlight "Calculate" and press $\boxed{\text{ENTER}}$.

Note: If you select the "Draw" option, you will get a picture of the *t* distribution with df = 55.728 with the area of interest shaded and the standardized test statistic and *P*-value labeled.

Performing a Test about $\mu_1 - \mu_2$

A significance test about $\mu_1 - \mu_2$ is called a *two-sample t test for a difference between two means*. As with any test, be sure to follow the four-step process.

EXAMPLE

Does caffeine increase pulse rates?

Significance test for $\mu_1 - \mu_2$

PROBLEM: Mr. Luckow's class performed an experiment to investigate whether drinking a caffeinated beverage would increase pulse rates. Twenty students in the class volunteered to take part in the experiment. All of the students measured their initial pulse rates. Then Mr. Luckow randomly assigned the students into two groups of 10. Each student in the first group drank 12 ounces of cola with caffeine. Each student in the second group drank 12 ounces of caffeine-free cola. All students then measured their pulse rates again. The table displays the change in pulse rate for the students in both groups.

Juanmonino/E+/Getty Images

	Change in pulse rate (Final pulse rate – Initial pulse rate)										Mean change
Caffeine	8	3	5	1	4	0	6	1	4	0	3.2
No caffeine	3	−2	4	−1	5	5	1	2	−1	4	2.0

Do these data give convincing evidence at the $\alpha = 0.01$ significance level that drinking caffeine increases pulse rates, on average?

SOLUTION:

STATE: We want to perform a test at the $\alpha = 0.01$ significance level of

$$H_0: \mu_1 - \mu_2 = 0$$
$$H_a: \mu_1 - \mu_2 > 0$$

where μ_1 = true mean change in pulse rate for students like these who drink 12 ounces of cola with caffeine and μ_2 = true mean change in pulse rate for students like these who drink 12 ounces of caffeine-free cola.

PLAN: Two-sample t test for $\mu_1 - \mu_2$

Random? The data came from two groups in a randomized experiment. ✓

Normal/Large Sample? The sample sizes are small, but the dotplots don't show any outliers or strong skewness. ✓

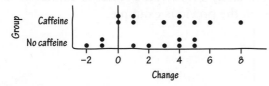

DO:
• $\bar{x}_1 = 3.2, s_1 = 2.70, \bar{x}_2 = 2.0, s_2 = 2.62$

• $t = \dfrac{(3.2 - 2.0) - 0}{\sqrt{\dfrac{2.70^2}{10} + \dfrac{2.62^2}{10}}} = 1.01$

STATE: State the hypotheses you want to test and the significance level, and define any parameters you use.

Because the subjects in this experiment were volunteers, the scope of inference is limited to "students like these."

PLAN: Identify the appropriate inference method and check conditions.

Because the sample sizes are less than 30, graph the data from each sample to see if it is plausible that they came from normal distributions.

DO: If the conditions are met, perform calculations:
• Identify the sample statistics.
• Calculate the standardized test statistic.
• Find the *P*-value.

Use technology to calculate the standard deviations for the two groups.

The observed result gives *some* evidence in favor of H_a because $\bar{x}_1 - \bar{x}_2 = 3.2 - 2.0 = 1.2 > 0$.

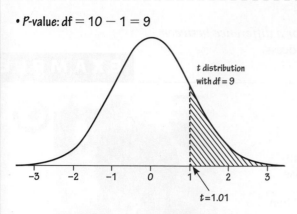

• *P*-value: df $= 10 - 1 = 9$

t distribution with df $= 9$

$t = 1.01$

Using Table B: $P(t > 1.01)$ is between 0.15 and 0.20.

Using technology: $t = 1.01$, *P*-value $= 0.163$ using df $= 17.99$.

CONCLUDE: Because the *P*-value of 0.163 is greater than $\alpha = 0.01$, we fail to reject H_0. We do not have convincing evidence that drinking caffeinated soda increases the pulse rate for students like these, on average.

> **CONCLUDE:** Make a conclusion about the hypotheses in the context of the problem.

FOR PRACTICE   TRY EXERCISE 9.

THINK ABOUT IT **Will inference methods for random sampling give accurate *P*-values for randomized experiments?** Confidence intervals and tests for $\mu_1 - \mu_2$ are based on the sampling distribution of $\bar{x}_1 - \bar{x}_2$. But in most experiments, researchers don't select subjects at random from any larger populations. They do randomly assign subjects to treatments. We can think about what would happen if the random assignment were repeated many times under the assumption that H_0: $\mu_1 - \mu_2 = 0$ is true. That is, we assume that the specific treatment received doesn't affect an individual subject's response. We can use the simulation approach of Lesson 3.8 to determine if the observed difference in means is statistically significant.

We used the *One Quantitative Variable* applet to randomly reassign the 20 subjects in the caffeine study to the two groups many times, assuming the treatment received doesn't affect each individual's change in pulse rate. Figure 9.4 shows the values of $\bar{x}_1 - \bar{x}_2$ in 1000 trials of the simulation. The observed difference in the means was $3.2 - 2.0 = 1.2$. How likely is it that a difference this large or larger would happen just by chance when H_0 is true? The figure provides a rough answer: 177 of the 1000 random reassignments yielded a difference in means greater than or equal to 1.2. That is, our estimate of the *P*-value is approximately 0.177. This result is similar to the 0.163 *P*-value from the two-sample *t* test in the previous example.

FIGURE 9.4 Simulation of the randomization distribution of $\bar{x}_1 - \bar{x}_2$ from 1000 random reassignments of subjects to treatment groups in the caffeine experiment.

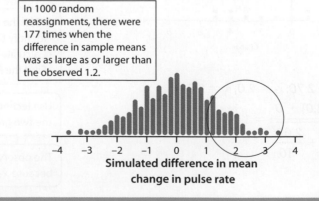

In 1000 random reassignments, there were 177 times when the difference in sample means was as large as or larger than the observed 1.2.

Simulated difference in mean change in pulse rate

LESSON APP 9.4

Is name-brand popcorn better than store-brand popcorn?

bernashafo/Shutterstock

Will using name-brand microwave popcorn result in a greater percentage of popped kernels than store-brand microwave popcorn? To find out, Briana and Maggie randomly selected 10 bags of name-brand microwave popcorn and 10 bags of store-brand microwave popcorn. The chosen bags were arranged in random order. Then each bag was popped for 3.5 minutes, and the percentage of popped kernels was calculated. The data are displayed in the table.[25]

| Name brand | 95 | 88 | 84 | 94 | 81 | 90 | 97 | 93 | 91 | 86 |
| Store brand | 91 | 89 | 82 | 82 | 77 | 78 | 84 | 86 | 86 | 90 |

1. Make parallel boxplots of the data. Write a few sentences comparing the distributions of the percent of popped kernels in the two samples.

2. Do the data provide convincing evidence at the $\alpha = 0.05$ significance level that using name-brand microwave popcorn will result in a greater mean percentage of popped kernels?

Lesson 9.4

WHAT DID YOU LEARN?

LEARNING TARGET	EXAMPLES	EXERCISES
State hypotheses and check conditions for performing a significance test about a difference between two means.	p. 582	1–4
Calculate the standardized test statistic and *P*-value for a significance test about a difference between two means.	p. 583	5–8
Use the four-step process to perform a significance test about a difference between two means.	p. 585	9–12

Exercises   Lesson 9.4

Mastering Concepts and Skills

1. **Got calcium?** Does increasing the amount of calcium in our diet reduce blood pressure? Researchers designed a randomized comparative experiment to find out. The subjects were 21 healthy black men who volunteered to take part in the experiment. They were randomly assigned to two groups: 10 of the men received a calcium supplement for 12 weeks, while the control group of 11 men received a placebo pill that looked identical. The experiment was double-blind. The response variable is the decrease in systolic (top number) blood pressure for a subject after 12 weeks, in millimeters of mercury. An increase appears as a negative number.[26] Here are the data.

pg 582

| Group 1 (calcium) | 7 | −4 | 18 | 17 | −3 | −5 | 1 | 10 | 11 | −2 | |
| Group 2 (placebo) | −1 | 12 | −1 | −3 | 3 | −5 | 5 | 2 | −11 | −1 | −3 |

Researchers want to know if a calcium supplement reduces blood pressure more than a placebo, on average, for healthy black men like these.

(a) State appropriate hypotheses for performing a significance test. Be sure to define the parameters of interest.

(b) Check that the conditions for performing the test are met.

2. **Renting an apartment** Pat wants to compare the cost of one- and two-bedroom apartments in the area of her college campus. She collects data for a

random sample of 10 advertisements of each type. The table shows the rents (in dollars per month) for the selected apartments.

| 1 bedroom | 500 | 650 | 600 | 505 | 450 | 550 | 515 | 495 | 650 | 395 |
| 2 bedroom | 595 | 500 | 580 | 650 | 675 | 675 | 750 | 500 | 495 | 670 |

Pat wonders if two-bedroom apartments rent for significantly more, on average, than one-bedroom apartments.

(a) State appropriate hypotheses for performing a significance test. Be sure to define the parameters of interest.

(b) Check that the conditions for performing the test are met.

3. **Bank on it** As the Hispanic population in the United States continues to grow, businesses want to understand what Hispanics like. One study interviewed separate random samples of Hispanic and Anglo customers leaving a bank. Customers were classified as Hispanic if they preferred to be interviewed in Spanish or as Anglo if they preferred English. Each customer rated the importance of several aspects of bank service on a 10-point scale.[27] Here are summary results for the importance of "reliability" (the accuracy of account records, and so on).

Group	n	$\bar{x}$	s_x
Anglo	95	6.37	0.60
Hispanic	86	5.91	0.93

Researchers want to know if there is a difference in the mean reliability ratings of all Anglo and Hispanic bank customers.

(a) State appropriate hypotheses for performing a significance test. Be sure to define the parameters of interest.

(b) Check that the conditions for performing the test are met.

4. **Does music help or hinder memory?** Many students at Matt's school claim they can think more clearly while listening to their favorite kind of music. Matt believes that music interferes with thinking clearly. To find out which assumption is true, Matt recruits 84 volunteers and randomly assigns them to two groups. The "Music" group listens to their favorite music while playing a "match the animals" memory game. The "No music" group plays the same game in silence. Here are some descriptive statistics for the number of turns it took the subjects in each group to complete the game (fewer turns indicates a better performance).

Group	Sample size	Mean	SD
Music	$n_1 = 42$	$\bar{x}_1 = 15.833$	$s_1 = 3.944$
No music	$n_2 = 42$	$\bar{x}_2 = 13.714$	$s_2 = 3.550$

Matt wants to know if listening to music affects the average number of turns required to finish the memory game for students like these.

(a) State appropriate hypotheses for performing a significance test. Be sure to define the parameters of interest.

(b) Check that the conditions for performing the test are met.

5. **Does calcium affect blood pressure?** Refer to Exercise 1.
   pg 583
   (a) Calculate the standardized test statistic.
   (b) Find the P-value using Table B.

6. **Big apartment** Refer to Exercise 2.
   (a) Calculate the standardized test statistic.
   (b) Find the P-value using Table B.

7. **English or Español?** Refer to Exercise 3.
   (a) Calculate the standardized test statistic.
   (b) Find the P-value using Table B.

8. **Let the music play** Refer to Exercise 4.
   (a) Calculate the standardized test statistic.
   (b) Find the P-value using Table B.

9. **Do you pay less at Payless?** Mary Ann and Abigail were shopping for prom shoes and wondered how the prices at Payless compared to the prices at Famous Footwear. At each store, they randomly selected 30 pairs of shoes and recorded the price of each pair. The table shows summary statistics for the two samples of shoes.[28]
   pg 585

Store	Mean	SD
Famous Footwear	$45.66	$16.54
Payless	$21.39	$7.47

Do these data provide convincing evidence at the $\alpha = 0.05$ significance level that shoes cost less, on average, at Payless than at Famous Footwear?

10. **Do women or men talk more?** Researchers equipped random samples of 56 male and 56 female students from a large university with a small device that secretly records sound for a random 30 seconds during each 12.5-minute period over two days. Then they counted the number of words spoken by each subject during each recording period and, from this, estimated how many words per day each subject speaks. The estimates for female students had a mean of 16,177 words per day with a standard deviation of 7520 words. For the male students, the mean was 16,569 and the standard deviation was 9108.[29] Do these data provide convincing evidence at the $\alpha = 0.05$ significance level of a difference in the average number of words spoken in a day by male and female students at this university?

11. **Music and plant growth** For their final statistics project, two students performed an experiment to determine whether plants grow better if they are exposed to classical music or to metal music. Ten bean seeds

were selected and each was planted in a Styrofoam cup. Half of these cups were randomly assigned to be exposed to metal music each night, while the other half were exposed to classical music each night. The amount of growth, in millimeters, was recorded for each plant after 2 weeks. Here are the data.

Metal	22	36	73	57	3
Classical	87	78	124	121	19

Do these data provide convincing evidence at the $\alpha = 0.05$ significance level of a difference in the mean growth of plants like these that are exposed to classical music or to metal music?

12. **Fish oil** To see if fish oil can help reduce blood pressure, males with high blood pressure were recruited and randomly assigned to different treatments. Seven of the men were randomly assigned to a 4-week diet that included fish oil. Seven other men were assigned to a 4-week diet that included a mixture of oils that approximated the types of fat in a typical diet. At the end of the 4 weeks, each volunteer's blood pressure was measured again and the reduction in diastolic blood pressure was recorded. These differences are shown in the table. Note that a negative value means that the subject's blood pressure increased.[30]

Fish oil	8	12	10	14	2	0	0
Regular oil	−6	0	1	2	−3	−4	2

Do these data provide convincing evidence at the $\alpha = 0.05$ significance level that fish oil helps reduce blood pressure more, on average, than regular oil?

Applying the Concepts

13. **Teaching reading** An educator believes that new reading activities in the classroom will help elementary school pupils improve their reading ability. She recruits 44 third-grade students and randomly assigns them to two groups. One group of 21 students does these new activities for an 8-week period. A control group of 23 third-graders follows the same curriculum without the activities. At the end of the 8 weeks, all students are given the Degree of Reading Power (DRP) test, which measures the aspects of reading ability that the treatment is designed to improve. Parallel boxplots and summary statistics for the data are shown.[31]

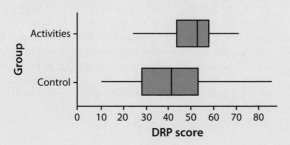

Group	n	Mean	SD
Activities	21	51.48	11.01
Control	23	41.52	17.15

(a) Write a few sentences comparing the DRP scores for the two groups.

(b) Is there convincing evidence at the $\alpha = 0.05$ significance level that the mean DRP score is greater for students like these who do the reading activities?

(c) Can we conclude that the new reading activities caused an increase in the mean DRP score? Justify your answer.

(d) Based on your conclusion in part (b), which type of error—a Type I error or a Type II error—could you have made? Explain.

14. **Native speakers** As a non-native English speaker, Sanda is convinced that people find more grammar and spelling mistakes in essays when they think the writer is a non-native English speaker. To test this, she randomly sorts a group of 40 volunteers into two groups of 20. Both groups are given the same paragraph to read. One group is told that the author of the paragraph is someone whose native language is not English. The other group is told nothing about the author. The subjects are asked to count the number of spelling and grammar mistakes in the paragraph. While the two groups found about the same number of *real* mistakes in the passage, the number of things that were *incorrectly* identified as mistakes was more interesting. A dotplot and some numerical summaries of the data follow.

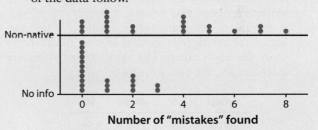

Number of "mistakes" found

Group	n	Mean	SD
Non-native	20	3.15	2.58
No info	20	0.85	1.09

(a) Write a few sentences comparing the distributions of number of "mistakes" found by the two groups.

(b) Is there convincing evidence at the $\alpha = 0.05$ significance level that the mean number of "mistakes" found will be greater when people like these volunteers are told that the author is a non-native speaker?

(c) Can we conclude that telling volunteers the author is a non-native speaker caused an increase in the mean number of "mistakes" found? Justify your answer.

(d) Based on your conclusion in part (b), which type of error—a Type I error or a Type II error—could you have made? Explain.

15. **Activities for reading** Refer to Exercise 13.

(a) Construct and interpret a 95% confidence interval for the difference between the true means. If you already defined parameters and checked conditions in Exercise 13, you don't need to do them again here.

(b) Explain how the confidence interval provides more information than the test in Exercise 13.

16. **Non-native speakers** Refer to Exercise 14.

(a) Construct and interpret a 95% confidence interval for the difference between the true means. If you already defined parameters and checked conditions in Exercise 14, you don't need to do them again here.

(b) Explain how the confidence interval provides more information than the test in Exercise 14.

17. **Enhancing creativity** Do external rewards—things like money, praise, fame, and good grades—promote creativity? Or is creativity motivated by internal rewards, like the joy of sharing one's work with others? Researcher Teresa Amabile recruited 47 experienced creative writers who were college students and divided them at random into two groups. The students in one group were given a list of statements about external reasons (E) for writing, such as public recognition, making money, or pleasing their parents. Students in the other group were given a list of statements about internal reasons (I) for writing, such as expressing yourself and enjoying playing with words. Both groups were then instructed to write a poem about laughter. Each student's poem was rated separately by 12 different poets using a creativity scale.[32] These ratings were averaged to obtain an overall creativity score for each poem. The table shows summary statistics for the two groups.

Group	n	Mean	SD
Internal	24	19.883	4.44
External	23	15.739	5.253

We used the *One Quantitative Variable* applet to randomly reassign the 47 subjects to the two groups 100 times, assuming the treatment received doesn't affect each individual's creativity rating. A dotplot of the simulated difference in mean creativity rating (Internal − External) is shown here.

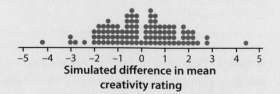

Simulated difference in mean creativity rating

(a) What is the estimated *P*-value?

(b) What conclusion would you make?

18. **Sleep deprivation** Does sleep deprivation linger for more than a day? Researchers designed a study using 21 volunteer subjects between the ages of 18 and 25. All 21 participants took a computer-based visual discrimination test at the start of the study. Then the subjects were randomly assigned into two groups. The 11 subjects in one group were deprived of sleep for an entire night in a laboratory setting. The 10 subjects in the other group were allowed unrestricted sleep for the night. Both groups were allowed as much sleep as they wanted for the next two nights. On Day 4, all the subjects took the same visual discrimination test on the computer. Researchers recorded the improvement in time (measured in milliseconds) from Day 1 to Day 4 on each subject's tests.[33] The table shows summary statistics for the two groups.

Group	n	Mean	SD
Unrestricted sleep	10	19.82	12.17
Sleep deprived	11	3.90	14.73

We used the *One Quantitative Variable* applet to randomly reassign the 21 subjects to the two groups 100 times, assuming the treatment received doesn't affect each individual's time improvement on the test. A dotplot of the simulated difference in mean time improvement (Unrestricted − Sleep deprived) is shown here.

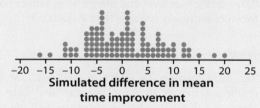

Simulated difference in mean time improvement

(a) What is the estimated *P*-value?

(b) What conclusion would you make?

Extending the Concepts

19. **Down the tubes** A company that makes hotel toilets claims that its new pressure-assisted toilet reduces the average amount of water used by more than 0.5 gallon per flush when compared to its current model. To test this claim, the company randomly selects 30 toilets of each type and measures the amount of water that is used when each toilet is flushed once. For the current-model toilets, the mean amount of water used is 1.64 gallons with a standard deviation of 0.29 gallon. For the new toilets, the mean amount of water used is 1.09 gallons with a standard deviation of 0.18 gallon.

(a) Carry out an appropriate significance test. What conclusion would you draw? (Note that the null hypothesis is *not* $H_0: \mu_1 - \mu_2 = 0$.)

(b) Based on your conclusion in part (a), could you have made a Type I error or a Type II error? Explain.

Recycle and Review

20. **Buy this car (2.5, 2.6)** A random sample of 30 cars selected from a website for buying used cars produced the following scatterplot for the relationship between the age of each car and the mileage showing on its odometer.

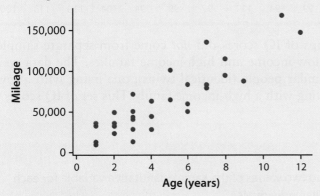

The equation of the least-squares regression line for predicting car mileage from age is

$$\widehat{\text{Mileage}} = 7842.42 + 12085.87 \,(\text{Age})$$

(a) Interpret the slope of the regression line in context.

(b) Predict the mileage for a car that is 15 years old. Comment on the reliability of this prediction.

(c) One car in the sample was 4 years old and had been driven 64,000 miles. Calculate the residual for this observation.

21. **Europe's economies (1.5, 1.6, 1.7)** The histogram shows the distribution of per capita Gross Domestic Product (GDP), expressed in thousands of U.S. dollars, for 30 countries in Europe.

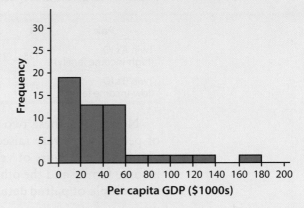

(a) In what percentage of countries is per capita GDP less than $20,000?

(b) Describe the shape of the distribution and identify any possible outliers.

(c) What measures of center and variability would be appropriate for describing this distribution? Explain.

(d) Is it possible to find the exact value of the median GDP from this histogram? Explain.

Lesson 9.5

Analyzing Paired Data: Estimating a Mean Difference

LEARNING TARGETS

- Use a graph to analyze the distribution of differences in a paired data set.
- Calculate the mean and standard deviation of the differences in a paired data set, and interpret the mean difference in context.
- Use the four-step process to construct and interpret a confidence interval for the true mean difference.

Lessons 9.3 and 9.4 showed how to perform inference about the difference between two means when data come from two separate random samples or two groups in a randomized experiment. What if we want to compare means in a setting that involves measuring a quantitative variable twice for the same individual or for two very similar individuals? For instance, a researcher studied a random sample of identical twins who had been separated and adopted at birth. In each case, one twin (Twin A) was adopted by a high-income family and the other (Twin B) by a low-income family. Both twins were given an IQ test as adults. Here are their scores.[34]

Pair	1	2	3	4	5	6	7	8	9	10	11	12
Twin A's IQ (high-income family)	128	104	108	100	116	105	100	100	103	124	114	112
Twin B's IQ (low-income family)	120	99	99	94	111	97	99	94	104	114	113	100

Notice that these two groups of IQ scores did *not* come from separate samples of people who were raised in low-income and high-income families. The data were obtained from *pairs* of very similar people (identical twins), one living with a low-income family and the other living with a high-income family. This set of IQ scores is an example of **paired data.**

|||| DEFINITION Paired data

Paired data result from recording two values of the same quantitative variable for each individual or for each pair of similar individuals.

This lesson focuses on how to analyze paired data and how to estimate a true *mean difference*.

Analyzing Paired Data

The graph in Figure 9.5 shows a parallel dotplot of the IQ scores from the study of identical twins. We can see that the twins raised in high-income households had a larger mean IQ ($\bar{x}_A = 109.5$) than the twins raised in low-income households ($\bar{x}_B = 103.667$). But with so much overlap between the groups, it does not appear that the difference in means is statistically significant.

FIGURE 9.5 Parallel dotplots of the IQ scores for pairs of identical twins raised in high-income (Twin A) and low-income (Twin B) households.

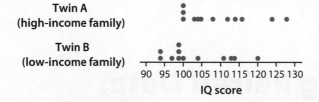

The previous analysis ignores the fact that these are *paired* data. Let's look at the difference in IQ scores for each pair of twins.

Pair	1	2	3	4	5	6	7	8	9	10	11	12
Twin A's IQ (high-income family)	128	104	108	100	116	105	100	100	103	124	114	112
Twin B's IQ (low-income family)	120	99	99	94	111	97	99	94	104	114	113	100
Difference (A − B)	8	5	9	6	5	8	1	6	−1	10	1	12

The dotplot in Figure 9.6 displays these differences. Almost all the differences are positive. For 11 of the 12 pairs, the twin raised in a high-income household had a higher IQ score as an adult than the twin raised in a low-income household. This graph suggests that the mean difference in IQ scores (Twin A − Twin B) is significantly greater than 0.

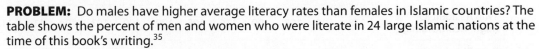

Difference (Twin A – Twin B) in IQ scores

FIGURE 9.6 Dotplot of difference in IQ scores for each pair of twins.

Analyzing Paired Data Graphically

To analyze paired data, start by computing the difference for each pair. Then use a graph to see if the differences are consistently greater than or consistently less than 0.

EXAMPLE

Who knows how to read?

Analyzing paired data with graphs

PROBLEM: Do males have higher average literacy rates than females in Islamic countries? The table shows the percent of men and women who were literate in 24 large Islamic nations at the time of this book's writing.[35]

Country	Male literacy (%)	Female literacy (%)	Country	Male literacy (%)	Female literacy (%)	Country	Male literacy (%)	Female literacy (%)
Afghanistan	43.1	12.6	Jordan	96.6	90.2	Saudi Arabia	90.4	81.3
Algeria	86.0	86.0	Kazakhstan	99.8	99.3	Syria	86.0	73.6
Azerbaijan	99.9	99.7	Kyrgyzstan	99.3	98.1	Tajikistan	99.8	99.6
Bangladesh	62.0	53.4	Lebanon	93.4	86.0	Tunisia	95.1	80.3
Egypt	82.2	65.4	Libya	98.6	90.7	Turkey	99.3	98.2
Indonesia	97.0	89.6	Malaysia	95.4	90.7	Turkmenistan	99.3	98.3
Iran	91.2	82.5	Morocco	76.1	57.6	Uzbekistan	99.6	99.0
Iraq	89.0	73.6	Pakistan	67.0	42.0	Yemen	81.2	46.8

Make a dotplot of the difference in literacy rates (Male – Female) for each country. What does the graph reveal?

SOLUTION:

Country	Difference (Male – Female)	Country	Difference (Male – Female)	Country	Difference (Male – Female)
Afghanistan	30.5	Jordan	6.4	Saudi Arabia	9.1
Algeria	0.0	Kazakhstan	0.5	Syria	12.4
Azerbaijan	0.2	Kyrgyzstan	1.2	Tajikistan	0.2
Bangladesh	8.6	Lebanon	7.4	Tunisia	14.8
Egypt	16.8	Libya	7.9	Turkey	1.1
Indonesia	7.4	Malaysia	4.7	Turkmenistan	1.0
Iran	8.7	Morocco	18.5	Uzbekistan	0.6
Iraq	15.4	Pakistan	25.0	Yemen	34.4

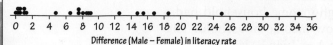

Difference (Male − Female) in literacy rate

> Start by computing the difference for each pair.

None of the difference values is negative, so the male literacy rate equals or exceeds the female literacy rate in all of these Islamic countries. The difference values vary from 0.0% in Algeria to 34.4% in Yemen.

FOR PRACTICE   TRY EXERCISE 1.

THE MEAN AND STANDARD DEVIATION OF THE DIFFERENCES

Now that we have looked at a graph of the differences for paired data, it's time to calculate numerical summaries. Because our goal in this lesson is to compare means, we start by computing the *mean difference*.

Here again are the data and dotplot of IQ scores from the study of identical twins.

Pair	1	2	3	4	5	6	7	8	9	10	11	12
Twin A's adult IQ (high-income family)	128	104	108	100	116	105	100	100	103	124	114	112
Twin B's adult IQ (low-income family)	120	99	99	94	111	97	99	94	104	114	113	100
Difference (A − B)	8	5	9	6	5	8	1	6	−1	10	1	12

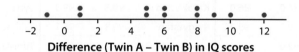

Difference (Twin A − Twin B) in IQ scores

The mean difference for these data is

$$\bar{x}_{\text{diff}} = \bar{x}_{\text{A}-\text{B}} = \frac{8 + 5 + 9 + \cdots + 1 + 12}{12} = \frac{70}{12} = 5.833 \text{ points}$$

This value tells us that the IQ score of the twin in each pair who was raised in a high-income household is 5.83 points higher than that of the twin who was raised in a low-income household, on average. The standard deviation of the differences is $s_{\text{diff}} = 3.93$ points.

Analyzing Paired Data Numerically

When analyzing paired data, use the mean difference $\bar{x}_{\text{diff}}$ and the standard deviation of the differences s_{diff} as summary statistics.

EXAMPLE

Can the women read, too?

Analyzing paired data numerically

PROBLEM: Refer to the preceding example. The table shows the literacy rates for males and females in each of the 24 large Islamic countries, along with the difference in literacy rates.

Country	Male literacy (%)	Female literacy (%)	Difference (Male – Female)	Country	Male literacy (%)	Female literacy (%)	Difference (Male – Female)
Afghanistan	43.1	12.6	30.5	Libya	98.6	90.7	7.9
Algeria	86.0	86.0	0.0	Malaysia	95.4	90.7	4.7
Azerbaijan	99.9	99.7	0.2	Morocco	76.1	57.6	18.5
Bangladesh	62.0	53.4	8.6	Pakistan	67.0	42.0	25.0
Egypt	82.2	65.4	16.8	Saudi Arabia	90.4	81.3	9.1
Indonesia	97.0	89.6	7.4	Syria	86.0	73.6	12.4
Iran	91.2	82.5	8.7	Tajikistan	99.8	99.6	0.2
Iraq	89.0	73.6	15.4	Tunisia	95.1	80.3	14.8
Jordan	96.6	90.2	6.4	Turkey	99.3	98.2	1.1
Kazakhstan	99.8	99.3	0.5	Turkmenistan	99.3	98.3	1.0
Kyrgyzstan	99.3	98.1	1.2	Uzbekistan	99.6	99.0	0.6
Lebanon	93.4	86.0	7.4	Yemen	81.2	46.8	34.4

Calculate the mean and standard deviation of the differences in literacy rates (Male – Female). Interpret the mean difference in context.

SOLUTION:

Mean: $\bar{x}_{\text{diff}} = 9.70\%$, SD: $s_{\text{diff}} = 9.751\%$.

The male literacy rate in these 24 large Islamic countries is 9.7 percentage points higher than the female literacy rate, on average.

> Use the *One Quantitative Variable* applet or your calculator's 1-Var Stats command to compute the mean and standard deviation of the differences.

FOR PRACTICE   TRY EXERCISE 5.

It would not be appropriate to perform inference about the true mean difference in literacy rates (Male − Female) for all Islamic countries because the data did not come from a random sample of countries.

THINK ABOUT IT   **What's the relationship between the mean difference and the difference in the means?** For the study of male and female literacy rates in large Islamic countries, the mean literacy rates were $\bar{x}_M = 88.6375\%$ for the males and $\bar{x}_F = 78.9375\%$ for the females. So the difference in the average literacy rates is

$$\bar{x}_M - \bar{x}_F = 88.6375\% - 78.9375\% = 9.70\%$$

But that is exactly the same value as the mean difference: $\bar{x}_{\text{diff}} = \bar{x}_{M-F} = 9.70\%$. This result holds in general: *The mean difference is equal to the difference in the means.* However, the standard deviation of the differences is *not* equal to the difference in the standard deviations! The standard deviations of the literacy rates for males and females are $s_M = 14.31\%$ and $s_F = 22.43\%$. So $s_M - s_F = 14.31 - 22.43 = -8.12\%$, a negative number! But $s_{\text{diff}} = 9.751\%$.

Constructing and Interpreting a Confidence Interval for μ_{diff}

When paired data come from a random sample or a randomized experiment, the statistic $\bar{x}_{\text{diff}}$ is a point estimate for the true mean difference μ_{diff}. Before constructing a confidence interval for μ_{diff}, we also have to check that the Normal/Large Sample condition is met.

How to Check the Conditions for Constructing a Confidence Interval for μ_{diff}

- **Random:** Paired data come from a random sample from the population of interest or from a randomized experiment.

- **Normal/Large Sample:** The population distribution of differences (or the true distribution of differences in response to the treatments) is normal or the number of differences in the sample is large ($n_{\text{diff}} \geq 30$). If the population (true) distribution of differences has an unknown shape and the number of differences in the sample is less than 30, a graph of the sample differences shows no strong skewness or outliers.

If the conditions are met, we can safely calculate a *one-sample t interval for a mean difference*. The method is the same as in Lesson 7.5.

How to Calculate a Confidence Interval for μ_{diff}

When the Random and Normal/Large Sample conditions are met, a $C\%$ confidence interval for the true mean difference μ_{diff} is

$$\bar{x}_{\text{diff}} \pm t^* \frac{s_{\text{diff}}}{\sqrt{n_{\text{diff}}}}$$

where t^* is the critical value for a t distribution with df $= n_{\text{diff}} - 1$ and $C\%$ of its area between $-t^*$ and t^*.

As with any inference procedure, follow the four-step process.

EXAMPLE

Which twin is smarter?

Constructing a confidence interval for μ_{diff}

PROBLEM: The data from the random sample of identical twins are shown again in the table. Construct and interpret a 95% confidence interval for the true mean difference in IQ scores among twins raised in high-income and low-income households.

Pair	1	2	3	4	5	6	7	8	9	10	11	12
Twin A's IQ (high-income family)	128	104	108	100	116	105	100	100	103	124	114	112
Twin B's IQ (low-income family)	120	99	99	94	111	97	99	94	104	114	113	100
Difference (A – B)	8	5	9	6	5	8	1	6	−1	10	1	12

SOLUTION:

STATE: We want to estimate μ_{diff} = the true mean difference in IQ scores (High − Low) for pairs of identical twins raised in high-income and low-income homes with 95% confidence.

PLAN: One-sample t interval for μ_{diff}

- *Random?* Random sample of 12 pairs of identical twins, one raised in a high-income home and the other in a low-income home. ✓

- *Normal/Large Sample?* The sample size is small, but the dotplot doesn't show any strong skewness or outliers. ✓

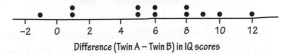

Difference (Twin A − Twin B) in IQ scores

DO: For these data, $\bar{x}_{diff} = 5.833$, $s_{diff} = 3.93$, and $n_{diff} = 12$. With 95% confidence and df = 12 − 1 = 11, $t^* = 2.201$. The confidence interval is

$$5.833 \pm 2.201\frac{3.93}{\sqrt{12}} \rightarrow 5.833 \pm 2.497 \rightarrow 3.336 \text{ to } 8.330$$

CONCLUDE: We are 95% confident that the interval from 3.336 to 8.330 captures the true mean difference in IQ scores (High − Low) among pairs of identical twins raised in high-income and low-income households. The interval suggests that IQs are between 3.336 and 8.330 points higher, on average, for twins raised in high-income households.

STATE: State the parameter you want to estimate and the confidence level.

PLAN: Identify the appropriate inference method and check conditions.

DO: If the conditions are met, perform calculations.

Use the *One Quantitative Variable* applet or your calculator's 1-Var Stats command to compute the mean and standard deviation of the differences.

CONCLUDE: Interpret your interval in the context of the problem.

FOR PRACTICE   TRY EXERCISE 9.

The 95% confidence interval in the example tells us that the plausible values for μ_{diff} are all positive. This gives convincing evidence that twins raised in high-income homes have higher IQs as adults than twins raised in low-income homes, on average. However, we can't conclude that household income level *caused* an increase in average IQ score because this was an observational study, not an experiment.

There are two ways that an experiment with two treatments can yield paired data:

1. Each subject can be given both treatments in a random order.

2. The researcher can form pairs of similar subjects and randomly assign each treatment to exactly one member of every pair.

These are both known as a *matched pairs* design. Notice that the Random condition for an experiment requires random assignment of treatments.

Is caffeine dependence real?

Researchers designed an experiment to study the effects of caffeine withdrawal. They recruited 11 volunteers who were diagnosed as being caffeine dependent to serve as subjects. Each subject was barred from coffee, colas, and other substances with caffeine for the duration of the experiment. During one 2-day period, subjects took capsules containing their normal caffeine intake. During another 2-day period, they took placebo capsules. The order in which subjects took caffeine and the placebo was randomized. At the end of each 2-day period, a test for depression was given to all 11 subjects. Researchers wanted to know whether being deprived of caffeine would lead to an increase in depression.[36]

Eric Simard/iStock/Getty Images

The table contains data on the subjects' scores on the depression test. Higher scores correspond to more symptoms of depression.

Subject	1	2	3	4	5	6	7	8	9	10	11
Depression (caffeine)	5	5	4	3	8	5	0	0	2	11	1
Depression (placebo)	16	23	5	7	14	24	6	3	15	12	0

1. Make a dotplot of the difference (Placebo – Caffeine) in depression test scores for each subject. What does the graph reveal about the difference in depression test scores?

2. Calculate the mean and standard deviation of the difference (Placebo – Caffeine) in depression test scores. Interpret the mean difference in context.

3. Construct and interpret a 90% confidence interval for the true mean difference μ_{diff} in the depression test score (Placebo – Caffeine).

4. Does the interval from Question 3 provide convincing evidence that caffeine deprivation *causes* an increase in the average depression test score for subjects like these? Explain.

Lesson 9.5

WHAT DID YOU LEARN?

LEARNING TARGET	EXAMPLES	EXERCISES
Use a graph to analyze the distribution of differences in a paired data set.	p. 593	1–4
Calculate the mean and standard deviation of the differences in a paired data set, and interpret the mean difference in context.	p. 595	5–8
Use the four-step process to construct and interpret a confidence interval for the true mean difference.	p. 596	9–12

Lesson 9.5

Mastering Concepts and Skills

1. **Fuel efficiency: city versus highway** Here
pg **593** are the EPA estimates of city gas mileage and
highway gas mileage in miles per gallon (mpg)
for a sample of 21 model-year 2014 midsize
cars.[37]

Model	City	Highway	Model	City	Highway
Acura RLX	20	31	Kia Optima	20	31
Audi A8	18	28	Lexus ES 350	21	31
BMW 550I	17	25	Lincoln MKZ	22	31
Buick Lacrosse	18	28	Mazda 6	28	40
Cadillac CTS	18	27	Mercedes-Benz E350	21	30
Chevrolet Malibu	21	30	Nissan Maxima	19	26
Chrysler 200	21	30	Subaru Legacy	24	32
Dodge Avenger	21	30	Toyota Prius	51	48
Ford Fusion	22	31	Volkswagen Passat	24	34
Hyundai Elantra	28	38	Volvo S80	18	25
Jaguar XF	19	30			

Make a dotplot of the difference (Highway – City)
in gas mileage for each car model. What does the
graph reveal?

2. **World champs** How good was the 2015 U.S.
women's soccer team? With players like Abby
Wambach, Megan Rapinoe, and Carli Lloyd, the
team put on an impressive showing en route to
winning the 2015 World Cup in Canada. Here are
data on the number of goals scored by the U.S.
team and its opponent in each game played that
year up to the World Cup victory in early July.[38]

U.S.	Opponent	U.S.	Opponent
0	2	0	0
1	0	3	1
2	1	0	0
3	0	1	0
0	0	2	0
2	0	1	0
4	0	2	0
3	0	5	2
5	1		

Make a dotplot of the difference (U.S. – Opponent)
in goals scored for each game. What does the graph
reveal?

3. **How much faster is the express lane?** Two statistics
students, Libby and Kathryn, decided to investigate which line was faster in the supermarket: the
express lane or the regular lane. They randomly
selected 15 times during a week, went to the same
store, and bought the same item. One of the students used the express lane and the other used the
closest regular lane. To decide which lane each of
them would use, they flipped a coin. They entered
their lanes at the same time, paid in the same way,
and recorded the time in seconds it took them to
complete the transaction. The table displays the
data.[39]

Time in express lane (sec)	Time in regular lane (sec)
337	342
226	472
502	456
408	529
151	181
284	339
150	229
357	263
349	332
257	352
321	341
383	397
565	694
363	324
85	127

(a) Calculate the difference (Regular lane – Express
lane) in transaction time for each visit to the
supermarket.

(b) A dotplot of the differences from part (a) is shown
here. What does the graph reveal about the difference in service times?

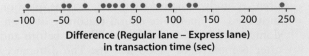

Difference (Regular lane – Express lane)
in transaction time (sec)

4. **Internet speed** Ramon has found that his computer's Internet connection is slower when he is farther
from his wireless modem, located in his living room.
To examine this difference, he randomly selects 14
times during the day and uses an online "speed test"
to determine download speed to his computer in his
bedroom and living room from each location at each
time (choosing which location to test first by flipping
a coin). Here are the data, with download speeds in
megabits per second (Mbps).

Time	1	2	3	4	5	6	7
Bedroom	13.5	15.5	18.4	14.8	14.9	12.1	9.8
Living room	16.6	24.1	25.0	20.4	29.7	12.5	22.2
Time	8	9	10	11	12	13	14
Bedroom	16.0	11.1	14.3	15.6	10.5	15.6	11.3
Living room	17.6	26.7	18.5	28.7	15.7	22.8	27.0

(a) Calculate the difference (Living room − Bedroom) in download speed for each time.

(b) A dotplot of the differences from part (a) is shown here. What does the graph reveal about the difference in service times?

Difference (Living room − Bedroom) in download speed (Mbps)

5. Fill 'er up Refer to Exercise 1. Calculate the mean and standard deviation of the difference (Highway − City) in gas mileages. Interpret the mean difference in context.
pg 595

6. USA! USA! Refer to Exercise 2. Calculate the mean and standard deviation of the difference (U.S. − Opponent) in goals scored. Interpret the mean difference in context.

7. Express lane Refer to Exercise 3. Calculate the mean and standard deviation of the difference (Regular lane − Express lane) in transaction times. Interpret the mean difference in context.

8. Slow modem Refer to Exercise 4. Calculate the mean and standard deviation of the difference (Living room − Bedroom) in download speeds. Interpret the mean difference in context.

9. Life in the fast lane Refer to Exercises 3 and 7. Construct and interpret a 95% confidence interval for the true mean difference in transaction times.
pg 596

10. Fast modem Refer to Exercises 4 and 8. Construct and interpret a 95% confidence interval for the true mean difference in download speeds.

11. Smarter piano players? Do piano lessons improve the spatial–temporal reasoning of preschool children? A study designed to investigate this question measured the spatial–temporal reasoning of a random sample of 34 preschool children before and after 6 months of piano lessons. The differences (After − Before) in the reasoning scores have mean 3.618 and standard deviation 3.055.[40]

(a) Construct and interpret a 90% confidence interval for the mean difference in reasoning scores for all preschool students who take 6 months of piano lessons.

(b) Based on your interval from part (a), can you conclude that taking 6 months of piano lessons would cause an increase in preschool students' average reasoning scores? Why or why not?

12. No annual fee? A bank wonders whether omitting the annual credit card fee for customers who charge at least $2400 in a year will increase the amount they decide to charge on its credit cards. The bank makes this offer to an SRS of 200 of its credit card customers. It then compares these customers' charges for this year with their charges for last year. The mean increase in the sample is $332 and the standard deviation is $108.

(a) Construct and interpret a 99% confidence interval for the true mean increase in the amount spent by this bank's credit card customers with no annual fee.

(b) Based on the interval from part (a), can you conclude that dropping the annual fee would cause an increase in the average amount spent by this bank's credit card customers? Why or why not?

Applying the Concepts

13. Groovy tires Researchers were interested in comparing two methods for estimating tire wear. The first method used the amount of weight lost by a tire. The second method used the amount of wear in the grooves of the tire. A random sample of 16 tires was obtained. Both methods were used to estimate the total distance traveled by each tire. The table provides the two estimates (in thousands of miles) for each tire.[41]

Tire	Weight	Groove	Tire	Weight	Groove
1	45.9	35.7	9	30.4	23.1
2	41.9	39.2	10	27.3	23.7
3	37.5	31.1	11	20.4	20.9
4	33.4	28.1	12	24.5	16.1
5	31.0	24.0	13	20.9	19.9
6	30.5	28.7	14	18.9	15.2
7	30.9	25.9	15	13.7	11.5
8	31.9	23.3	16	11.4	11.2

(a) Make a dotplot of the differences in the estimates of tire wear for the two methods (Weight − Groove). What does the graph reveal?

(b) Construct and interpret a 95% confidence interval for the true mean difference in the estimates from these two methods (Weight − Groove) in the population of tires.

(c) Does your interval in part (b) give convincing evidence of a mean difference in the two methods of estimating tire wear? Justify your answer.

14. Well, well? Trace metals found in wells affect the taste of drinking water, and high concentrations can pose a health risk. Researchers measured the concentration of zinc (in milligrams per liter, or mg/l) near the top and the bottom of 10 randomly selected wells in a large region. The data are provided in the table.[42]

Well	1	2	3	4	5
Bottom	0.430	0.266	0.567	0.531	0.707
Top	0.415	0.238	0.390	0.410	0.605
Well	6	7	8	9	10
Bottom	0.716	0.651	0.589	0.469	0.723
Top	0.609	0.632	0.523	0.411	0.612

(a) Make a dotplot of the differences (Bottom − Top) in the zinc concentrations for these 10 wells. What does the graph reveal?

(b) Construct and interpret a 95% confidence interval for the true mean difference in the zinc concentrations at the top and bottom of the wells in this region.

(c) Does your interval in part (b) give convincing evidence of a mean difference in zinc concentrations at the top and bottom of wells in this region? Justify your answer.

15. **No inference** Refer to Exercise 1. Explain why it would not be appropriate to construct a confidence interval for the true mean difference (Highway − City) in gas mileage for all model-year 2014 mid-size car models.

16. **No CI** Refer to Exercise 2. Explain why it would not be appropriate to construct a confidence interval for the true mean difference (U.S. − Opponent) in goals scored for all women's soccer games played in 2015.

Recycle and Review

17. **Catching food (2.1)** Tempe thinks she's pretty skilled at "mouth-catching" (no hands) bits of snack food that have been thrown to her from across a room. She wonders, though, if she's better at catching some foods than others, so she conducts an experiment. A friend tosses 150 pieces of each of three different kinds of snack food (in random sequence) and then records how often Tempe catches the food in her mouth. Here are the results.

		Snack food			
		Goldfish	Grapes	Gummies	Total
Caught?	Yes	90	100	115	305
	No	60	50	35	145
	Total	150	150	150	450

(a) Identify the explanatory and response variables.

(b) Use the information in the two-way table to make a segmented bar chart to show the relationship between snack food type and Tempe's catching success.

(c) Is there an association between the two variables? Explain your reasoning. If an association does exist, briefly describe it.

18. **Probability and urban legends (4.1)** Describe the misunderstanding about probability in the following scenario. To pass the time during a long drive, you and a friend are keeping track of the makes and models of cars that pass by in the other direction. At one point, you realize that among the last 20 cars, there hasn't been a single Ford. (Currently, about 16% of cars sold in America are Fords.) Your friend says, "By the law of averages, the next car is almost certain to be a Ford."

Lesson 9.6

Testing a Claim about a Mean Difference

LEARNING TARGETS

- Use the four-step process to perform a significance test about a mean difference.
- Determine whether you should use two-sample t procedures for inference about $\mu_1 - \mu_2$ or one-sample t procedures for inference about μ_{diff} in a given setting.

In Lesson 9.5, you learned to analyze paired data by looking at the difference within each pair. You also saw how to construct and interpret a confidence interval for a mean difference μ_{diff}. This lesson shows you how to perform a significance test for μ_{diff}. The second half of the lesson focuses on distinguishing inference about a difference between two means when the data are paired and when the data come from two separate samples or treatment groups.

Performing a Test about μ_{diff}

When paired data come from a random sample or a randomized experiment, we may want to perform a significance test about the true mean difference μ_{diff}. The null hypothesis has the general form

$$H_0: \mu_{diff} = \text{hypothesized value}$$

We'll focus on situations where the hypothesized difference is 0. Then the null hypothesis says that the true mean difference is 0:

$$H_0: \mu_{diff} = 0$$

The alternative hypothesis says what kind of difference we expect.

The conditions for performing a significance test about μ_{diff} are the same as those for constructing a confidence interval for μ_{diff} in Lesson 9.5.

How to Check the Conditions for Performing a Significance Test about μ_{diff}

- **Random:** Paired data come from a random sample from the population of interest or from a randomized experiment.

- **Normal/Large Sample:** The population distribution of differences (or the true distribution of differences in response to the treatments) is normal or the number of differences in the sample is large ($n_{diff} \geq 30$). If the population (true) distribution of differences has an unknown shape and the number of differences in the sample is less than 30, a graph of the sample differences shows no strong skewness or outliers.

When conditions are met, we can carry out a *one-sample t test for a mean difference*. The standardized test statistic is

$$t = \frac{\bar{x}_{diff} - 0}{\dfrac{s_{diff}}{\sqrt{n_{diff}}}}$$

We can use Table B or technology to find the P-value from the t distribution with df $= n - 1$.

Consumers Union designed an experiment to test whether nitrogen-filled tires would maintain pressure better than air-filled tires. They obtained two tires from each of several brands and then randomly assigned one tire in each pair to be filled with air and the other to be filled with nitrogen. All tires were inflated to the same pressure and placed outside for a year. At the end of the year, Consumers Union measured the pressure in each tire. The pressure loss (in pounds per square inch) during the year for the tires of each brand is shown in the table.[43]

Brand	Air	Nitrogen	Brand	Air	Nitrogen
BF Goodrich Traction T/A HR	7.6	7.2	Pirelli P6 Four Seasons	4.4	4.2
Bridgestone HP50 (Sears)	3.8	2.5	Sumitomo HTR H4	1.4	2.1
Bridgestone Potenza G009	3.7	1.6	Yokohama Avid H4S	4.3	3.0
Bridgestone Potenza RE950	4.7	1.5	BF Goodrich Traction T/A V	5.5	3.4
Bridgestone Potenza EL400	2.1	1.0	Bridgestone Potenza RE950	4.1	2.8
Continental Premier Contact H	4.9	3.1	Continental ContiExtreme Contact	5.0	3.4
Cooper Lifeliner Touring SLE	5.2	3.5	Continental ContiPro Contact	4.8	3.3
Dayton Daytona HR	3.4	3.2	Cooper Lifeliner Touring SLE	3.2	2.5
Falken Ziex ZE-512	4.1	3.3	General Exclaim UHP	6.8	2.7
Fuzion Hrl	2.7	2.2	Hankook Ventus V4 H105	3.1	1.4
General Exclaim	3.1	3.4	Michelin Energy MXV4 Plus	2.5	1.5
Goodyear Assurance Tripletred	3.8	3.2	Michelin Pilot Exalto A/S	6.6	2.2
Hankook Optimo H418	3.0	0.9	Michelin Pilot HX MXM4	2.2	2.0
Kumho Solus KH16	6.2	3.4	Pirelli P6 Four Seasons	2.5	2.7
Michelin Energy MXV4 Plus	2.0	1.8	Sumitomo HTR$^+$	4.4	3.7
Michelin Pilot XGT H4	1.1	0.7			

A dotplot of the difference (Air-filled − Nitrogen-filled) in pressure loss for each pair of tires is shown below. We can see that 28 of the 31 differences are positive. That is, the air-filled tire lost more pressure than the nitrogen-filled tire for all but three brands. The mean difference in pressure loss is $\bar{x}_{\text{diff}} = 1.252$ psi, meaning that the air-filled tires lost 1.252 psi more than the nitrogen-filled tires, on average. The standard deviation of the differences is $s_{\text{diff}} = 1.202$ psi.

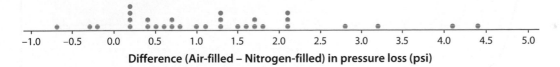

Difference (Air-filled – Nitrogen-filled) in pressure loss (psi)

These data provide *some* evidence that filling tires with nitrogen instead of air reduces pressure loss. But do they give *convincing* evidence that the true mean difference μ_{diff} in pressure loss (Air-filled − Nitrogen-filled) is positive for tire brands like these? To answer that question, we need to perform a significance test.

EXAMPLE

Is filling tires with nitrogen helpful?

Performing a significance test for μ_{diff}

PROBLEM: Refer to the tire experiment carried out by Consumers Union. Do the data give convincing evidence at the $\alpha = 0.05$ significance level that air-filled tires lose more pressure, on average, than nitrogen-filled tires for brands like these?

SOLUTION:

STATE: We want to perform a test at the $\alpha = 0.05$ significance level of

$$H_0: \mu_{\text{diff}} = 0$$
$$H_a: \mu_{\text{diff}} > 0$$

where $\mu_{\text{diff}} =$ the true mean difference (Air-filled – Nitrogen-filled) in pressure loss for tire brands like these.

PLAN: One-sample t test for μ_{diff}

Random? Random assignment of one tire in each pair to be filled with nitrogen and one to be filled with air. ✓

Normal/Large Sample? The number of differences in the sample is large: ($n_{\text{diff}} = 31 \geq 30$). ✓

DO:

• $\bar{x}_{\text{diff}} = 1.252$ psi, $s_{\text{diff}} = 1.202$ psi

• $t = \dfrac{1.252 - 0}{\dfrac{1.202}{\sqrt{31}}} = 5.80$

• P-value: df $= 31 - 1 = 30$

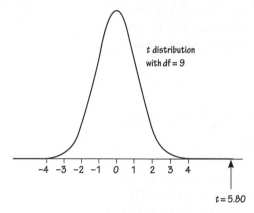

t distribution with df = 9

$t = 5.80$

Using Table B: The P-value is less than 0.0005.

Using technology: Applet/T-test gives P-value ≈ 0.

CONCLUDE: Because the P-value of approximately 0 is less than $\alpha = 0.05$, we reject H_0. There is convincing evidence that air-filled tires lose more pressure, on average, than nitrogen-filled tires for brands like these.

Follow the four-step process!

STATE: State the hypotheses you want to test and the significance level, and define any parameters you use.

PLAN: Identify the appropriate inference method and check conditions.

DO: If the conditions are met, perform calculations:
• Identify the sample statistic.
• Calculate the standardized test statistic.
• Find the P-value.

The sample result gives *some* evidence in favor of H_a because $1.252 > 0$.

CONCLUDE: Make a conclusion about the hypotheses in the context of the problem.

We can draw a cause-and-effect conclusion because this was a randomized experiment.

FOR PRACTICE **TRY EXERCISE 1.**

The significance test in the example led to a simple decision: reject H_0. But we saw in Lesson 8.4 that a confidence interval gives more information than a test—it provides the entire set of plausible values for the parameter based on the data. A 95% confidence interval for μ_{diff} is

$$\bar{x}_{diff} \pm t^* \frac{s_{\text{diff}}}{\sqrt{n_{\text{diff}}}} \rightarrow 1.252 \pm 2.042 \frac{1.202}{\sqrt{31}} \rightarrow 1.252 \pm 0.44 \rightarrow 0.812 \text{ to } 1.692$$

We are 95% confident that the interval from 0.812 to 1.692 pounds per square inch captures the true mean difference (Air-filled − Nitrogen-filled) in pressure loss for all tire brands like these. The interval suggests that air-filled tires lose an average of 0.812 to 1.692 psi more pressure than nitrogen-filled tires in a one-year period.

Paired Data or Two Samples?

In Lessons 9.3 and 9.4, we used two-sample t procedures to perform inference about the difference $\mu_1 - \mu_2$ between two means. These methods require data that come from separate random samples from the two populations of interest or from two groups in a randomized experiment. In Lessons 9.5 and 9.6, we use one-sample t procedures to perform inference about the true mean difference μ_{diff}. These methods require *paired data* that come from a random sample from the population of interest or from a randomized experiment. *The proper inference method depends on how the data were produced.*

EXAMPLE

What do teens do for fun?

Two samples or paired data?

PROBLEM: In each of the following settings, decide whether you should use two-sample t procedures to perform inference about a difference in means or one-sample t procedures to perform inference about a mean difference. Explain your choice.

(a) Which of two brands of sunscreen is more effective at preventing sunburn? To find out, researchers conducted an experiment with a group of volunteer teens at the Jersey shore. The researchers applied the same amount of Brand A sunscreen on one arm and Brand B sunscreen on the other arm of each teen. Which arm got which brand was randomly determined for each volunteer. After the teens sunbathed for an hour, researchers compared the amount of redness on each teen's left and right arms.

(b) Do experienced computer-game players earn higher scores when they play with someone present to cheer them on or when they play alone? Fifty teenagers with experience playing a particular computer game have volunteered for a study. Researchers randomly assign 25 of them to play the game alone and the other 25 to play the game with a supporter present. Each player's score is recorded.

(c) How do young adults look back on adolescent romance? Investigators interviewed a random sample of 40 couples in their mid-20s. The female and male partners were interviewed separately. Each was asked about a romantic relationship that lasted at least two months when they were aged 15 or 16. One response variable was a measure on a numerical scale of how much the attractiveness of the adolescent partner mattered. Researchers want to find out how much men and women differ on this measure.

SOLUTION:

(a) *One-sample t procedures. The data come from a matched pairs experiment with the two treatments (Brand A and Brand B sunscreen) being randomly assigned to the left and right arms of each teen.*

(b) *Two-sample t procedures. The data come from two randomly assigned groups of teenagers in an experiment.*

(c) *One-sample t procedures. The data are paired by couple, and come from a random sample of 40 couples in their mid-20s.*

FOR PRACTICE   TRY EXERCISE 5.

When designing an experiment to compare two treatments, a completely randomized design may not be the best option. A matched pairs design might be a better choice, as the following activity shows.

ACTIVITY

Get your heart beating!

Are standing pulse rates higher, on average, than sitting pulse rates? In this activity, you will perform two experiments to try to answer this question.

Experiment 1: Completely randomized design

1. Your teacher will randomly assign half of the students in your class to stand and the other half to sit. Once the two treatment groups have been formed, students should stand or sit as required. Then they should measure their pulses for 1 minute, and each group should record their data on the board.

2. Analyze the data for the completely randomized design. Make parallel dotplots and calculate the mean pulse rate for each group. Is there *some* evidence that standing pulse rates are higher, on average? Explain.

Experiment 2: Matched pairs design

3. To produce paired data in this setting, each student should receive both treatments in random order. Because you already sat or stood in Step 1, you just need to do the opposite now. As before, everyone should measure their pulses for 1 minute after completing the treatment (that is, once they are standing or sitting). Then each subject should calculate his or her difference (Standing − Sitting) in pulse rate and record this value on the board.

4. Analyze the data for the matched pairs design. Make a dotplot of these differences and calculate their mean. Is there *some* evidence that standing pulse rates are higher, on average? Explain.

5. Which design provides more convincing evidence that standing pulse rates are higher, on average, than sitting pulse rates? Justify your answer.

A statistics class with 24 students performed the Get Your Heart Beating! activity. Figure 9.7 shows a dotplot of the pulse rates for their completely randomized design. The mean pulse rate for the standing group is $\bar{x}_1 = 74.83$; the mean for the sitting group is $\bar{x}_2 = 68.33$. So the average pulse rate is 6.5 beats per minute higher in the standing group. However, the variability in pulse rates for the two groups creates a lot of overlap in the dotplots. A two-sample t test of $H_0: \mu_1 - \mu_2 = 0$ versus $H_a: \mu_1 - \mu_2 > 0$ yields $t = 1.42$ and a P-value of 0.09. These data do not provide convincing evidence that standing pulse rates are higher, on average, than sitting pulse rates for people like the students in this class.

FIGURE 9.7 Parallel dotplots of the pulse rates for the standing and sitting groups in a statistics class's completely randomized design.

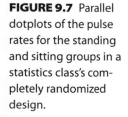

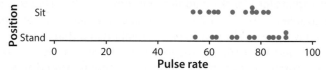

What about the class's matched pairs design? Figure 9.8 shows a dotplot of the difference (Standing − Sitting) in pulse rate for each of the 24 students. We can see that 21 of the 24 students recorded a positive difference, indicating that their standing pulse rate was higher. The mean difference is $\bar{x}_{\text{diff}} = 6.83$ beats per minute. A one-sample t test of $H_0: \mu_{\text{diff}} = 0$ versus $H_a: \mu_{\text{diff}} > 0$ gives $t = 6.483$ and a P-value of approximately 0. These data provide *very* convincing evidence that standing pulse rates are higher, on average, than sitting pulse rates for people like the students in this class.

FIGURE 9.8 Dotplot of the difference in pulse rate (Standing − Sitting) for each student in a statistics class's matched pairs design.

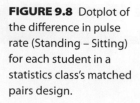

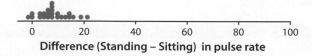

Let's take one more look at the two figures. Notice that we used the same scale for both graphs. The matched pairs design reduced the variability in the response variable by accounting for a big source of variability—the differences between individual students. That made it easier to detect the fact that standing causes an increase in the average pulse rate. With the larger amount of variability in the completely randomized design, we could not draw such a conclusion.

LESSON APP 9.6

Does generic ice cream melt faster?

Few people enjoy melted ice cream. Being from the sunny state of Arizona, Megan and Jenna decided to test if generic vanilla ice cream melts faster than premium-brand vanilla ice cream. At 25 randomly selected times during the day and night, the girls put a single scoop of each type of ice cream in the same location outside and waited for them to melt completely.[44]

1. Why should you use one-sample t procedures to perform inference about a mean difference in this setting?

2. A dotplot of the differences (Premium – Generic) in time to melt completely is shown here. The mean difference is $\bar{x}_{\text{diff}} = 1.32$ minutes and the standard deviation is $s_{\text{diff}} = 1.14$ minutes. Explain why there is *some* evidence that premium-brand vanilla ice cream takes longer to melt than the generic brand.

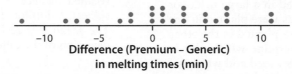

**Difference (Premium – Generic)
in melting times (min)**

3. Do the data provide convincing evidence at the $\alpha = 0.05$ significance level that the premium-brand vanilla ice cream takes longer to melt, on average, than generic vanilla ice cream?

Lesson 9.6

WHAT DID YOU LEARN?

LEARNING TARGET	EXAMPLES	EXERCISES
Use the four-step process to perform a significance test about a mean difference.	p. 603	1–4
Determine whether you should use two-sample t procedures for inference about $\mu_1 - \mu_2$ or one-sample t procedures for inference about μ_{diff} in a given setting.	p. 605	5–10

Exercises   *Lesson 9.6*

Mastering Concepts and Skills

1. **Go in or drive-thru?** For high school students who
pg 603 can leave campus for lunch, this is a very important question. Ben and Maya decided to investigate by visiting a local fast-food restaurant on 10 randomly selected days during lunch. Each time

they went, they flipped a coin to determine which of them would go inside and which would use the drive-thru. Both of them ordered the same item, paid with the same amount of cash, and recorded how long it took (in seconds) to wait in line, pay, and receive their item. Here are their data.[45]

Visit	Drive-thru	Inside
1	325	170
2	608	110
3	90	52
4	519	158
5	216	66
6	263	128
7	559	81
8	154	163
9	449	64
10	512	120

Do these data provide convincing evidence at the $\alpha = 0.05$ level of a mean difference in the service time inside and at the drive-thru for this local fast-food restaurant?

2. **Better barley** Does drying barley seeds in a kiln increase the yield of barley? A famous experiment by William S. Gosset (who discovered the t distributions) investigated this question. Eleven pairs of adjacent plots were marked in a large field. For each pair, regular barley seeds were planted in one plot and kiln-dried seeds were planted in the other. A coin flip was used to determine which plot in each pair got the regular barley seed and which got the kiln-dried seed. The table displays the data on barley yield (in pounds per acre) for each plot.[46]

Plot	Regular	Kiln
1	1903	2009
2	1935	1915
3	1910	2011
4	2496	2463
5	2108	2180
6	1961	1925
7	2060	2122
8	1444	1482
9	1612	1542
10	1316	1443
11	1511	1535

Do these data provide convincing evidence at the $\alpha = 0.05$ level that drying barley seeds in a kiln increases the yield of barley, on average?

3. **Friday the 13th** Does Friday the 13th have an effect on people's behavior? Researchers collected data on the number of shoppers at a random sample of 45 grocery stores on Friday the 6th and Friday the 13th in the same month. A dotplot of the difference in the number of shoppers at each store on these 2 days (Friday the 6th – Friday the 13th) is shown here. The mean difference is −46.5 and the standard deviation of the differences is 178.0.[47]

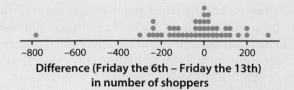

Difference (Friday the 6th – Friday the 13th) in number of shoppers

(a) Do these data provide convincing evidence at the $\alpha = 0.05$ level of a mean difference in the number of shoppers at grocery stores on these 2 days?

(b) Can you conclude that the result in part (a) is due to Friday the 13th affecting people's behavior? Why or why not?

4. **Music and memory** Does listening to music while studying help or hinder students' learning? Two statistics students designed an experiment to find out. They selected a random sample of 30 students from their medium-sized high school to participate. Each subject was given 10 minutes to memorize two different lists of 20 words, once while listening to music and once in silence. The order of the two word lists was determined at random; so was the order of the treatments. The difference in the number of words recalled (Silence − Music) was recorded for each subject. A dotplot of these differences is shown here. The mean difference was 1.57 and the standard deviation of the differences was 2.70.

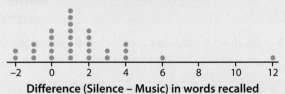

Difference (Silence – Music) in words recalled

(a) Do the data provide convincing evidence at the $\alpha = 0.01$ significance level of a mean difference in the number of words recalled in silence or when listening to music by students at this school?

(b) Can you conclude that the result in part (a) is due to whether students attempt to memorize words in silence or with music playing? Why or why not?

For the settings in Exercises 5–10, decide whether you should use two-sample t procedures to perform inference about a difference in means or one-sample t procedures to perform inference about a mean difference.[48] Explain your choice in each case.

5. **Wearing thin** To compare the wear characteristics of two tire brands, two new Brand A tires are mounted on one side of each of 30 cars of the same model, while two new Brand B tires are mounted on the other side. Which side gets which brand is determined by flipping a coin. After 5000 miles of driving, the amount of wear is measured for each tire.
pg **605**

6. **Gender gap?** To compare funding for men's and women's sports teams, researchers randomly select 100 colleges and record the amount of money spent on the men's basketball program and the women's basketball program at each college.

7. **Work, work, work** Can listening to music while working increase productivity? A total of 20 factory workers agree to take part in a study to investigate this question. Researchers randomly assign 10 workers to do a repetitive task while listening to music and the other 10 workers to do the task in silence. They record the time it takes each worker to complete the task.

8. **Wax on, wax off** Which of two brands of ski wax works better? To find out, researchers randomly assign 15 skiers to do a downhill race with one brand of wax and 15 different skiers to do a downhill race with the other brand of wax. Then researchers compare the average times of the skiers in the two groups.

9. **Shooting basketballs** Is it easier for male basketball players to shoot foul shots with a men's basketball or with a women's basketball, which is slightly smaller? Grayson and Sam asked the 24 members of the varsity and JV basketball teams at their school to shoot 25 foul shots with a men's basketball and 25 foul shots with a women's basketball, resorting to a coin flip to decide which ball they used first. The two students recorded how many shots each player made with each type of ball.

10. **The sound of silence** To test the effect of background music on productivity, researchers recruit 50 factory workers to take part in a study. For one month, each subject works without music. For another month, the subject works while listening to music on an MP3 player. The month in which each subject listens to music is determined by a coin toss. Researchers record the productivity of each worker in both months.

Applying the Concepts

11. **Sleepless nights** An experiment was carried out with 10 randomly selected patients suffering from insomnia to investigate the effectiveness of a drug that was designed to increase sleep time. The data show the number of additional hours of sleep gained by each subject after taking the drug.[49] (A negative value indicates that the subject got less sleep after taking the drug.)

    1.9   0.8   1.1   0.1   −0.1   4.4   5.5   1.6   4.6   3.4

(a) Why should you use one-sample t procedures to perform inference about a mean difference in this setting?

(b) Is there convincing evidence at the $\alpha = 0.05$ significance level that patients like these would get more sleep, on average, when taking the drug?

(c) Can we conclude that the drug is effective at increasing the average amount of sleep time? Justify your answer.

(d) Based on your conclusion in part (b), which type of error—a Type I error or a Type II error—could you have made? Explain.

12. **Running barefoot** Felipe and Armani conducted an experiment to determine if high school sprinters run faster with shoes on or barefoot. They had 20 athletes run a 40-yard dash in their regular track shoes and another 40-yard dash barefoot, randomly determining which they ran first. The data show the difference in 40-yard dash times (in seconds) for each runner (Shoes − Barefoot). A positive number means the athlete took longer with shoes on.

    0.4   −0.3   −0.2   0.3   −0.3   0.4   0.1   0.2   0.2   0

    0.3   0.1   0.1   0.3   −0.1   0.4   0.5   0.1   0.5   0

(a) Why should you use one-sample t procedures to perform inference about a mean difference in this setting?

(b) Is there convincing evidence at the $\alpha = 0.05$ significance level that athletes like the ones in this study run faster, on average, when they are barefoot?

(c) Can we conclude that runners like these will run faster without shoes? Justify your answer.

(d) Based on your conclusion in part (b), which type of error—a Type I error or a Type II error—could you have made? Explain.

13. **Sleep longer** Refer to Exercise 11.

(a) Construct and interpret a 95% confidence interval for the true mean amount of additional sleep that insomnia patients get when taking the drug. If you already defined parameters and checked conditions in Exercise 11, you don't need to do them again here.

(b) Explain how the confidence interval provides more information than the test in Exercise 11.

14. **Shoes off?** Refer to Exercise 12.

(a) Construct and interpret a 95% confidence interval for the true mean difference (Shoes − Barefoot) in 40-yard-dash times for athletes like these. If you already defined parameters and checked conditions in Exercise 12, you don't need to do them again here.

(b) Explain how the confidence interval provides more information than the test in Exercise 12.

Extending the Concepts

Exercises 15 and 16 refer to the following setting. Coaching companies claim that their courses can raise the SAT scores of high school students. Of course, the scores of students who retake the SAT without paying for coaching generally improve. A random sample of students who took the SAT twice found 427 who were coached and 2733 who were uncoached.[50] Starting with their Verbal scores on the first and second tries, here are the summary statistics.

		Try 1		Try 2		Gain	
	n	$\bar{x}$	s_x	$\bar{x}$	s_x	$\bar{x}$	s_x
Coached	427	500	92	529	97	29	59
Uncoached	2733	506	101	527	101	21	52

15. **Coaching and SAT scores** Do the scores of students who are coached increase significantly?

(a) You could use the information on the Coached line to carry out either a two-sample t test comparing Try 1 with Try 2 for coached students or a one-sample t test using Gain. Which is the correct test? Why?

(b) Carry out the proper test. What do you conclude?

16. **Does coaching help?** What we really want to know is whether coached students improve more than uncoached students, and whether any advantage is large enough to be worth paying for. Use the information previously given to answer the following questions.

(a) How much more do coached students gain, on average? Construct and interpret a 99% confidence interval.

(b) Does the interval in part (a) give convincing evidence that coached students gain more, on average, than uncoached students? Explain.

(c) Based on your work, do you think coaching courses are worth paying for?

Recycle and Review

17. **Graduate students (5.4, 6.3)** In the fall of 2014, 29% of the more than 8000 graduate students at the University of North Carolina at Chapel Hill were 25 years old or younger.[51]

(a) Suppose we take a random sample of 200 graduate students. Calculate the mean and standard deviation of the sampling distribution of the sample count of "young" students.

(b) What is the approximate shape of the sampling distribution? Justify your answer.

(c) What is the probability that fewer than 50 of the students in your sample are 25 years old or younger?

Photodisc/Getty Images

Chapter 9 RESOLVED

STATS applied!

Are fast-food drive-thrus fast and accurate?

At the beginning of the chapter, we described a study about drive-thru service at fast-food restaurants. Here is a summary of the details we provided in the STATS applied! (page 551) from the 2013 *QSR* study.

- Of the 246 visits to Burger King, 82.1% resulted in accurate orders. Of the 282 visits to Wendy's, 86.9% resulted in accurate orders.

- McDonald's average service time for 317 drive-thru visits was 189.49 seconds with a standard deviation of 18.51 seconds. Chick-fil-A's service time for 299 drive-thru visits had a mean of 203.88 seconds and standard deviation of 19.61 seconds.

1. Is there convincing evidence at the $\alpha = 0.05$ significance level of a difference in the proportion of accurate drive-thru orders at Burger King and Wendy's restaurants?

2. A 95% confidence interval for the difference in the population proportions of accurate drive-thru orders (Wendy's – Burger King) is −0.014 to 0.110. Explain how the confidence interval is consistent with, but gives more information than, the significance test in Question 1.

3. Should you use one-sample t procedures or two-sample t procedures to compare the average drive-thru service times for all McDonald's and Chick-fil-A restaurants? Justify your answer.

4. Construct and interpret a 99% confidence interval for the difference in the mean drive-thru service times at McDonald's and Chick-fil-A restaurants.

Main Points   *Chapter 9*

Comparing Two Proportions

- Suppose that $\hat{p}_1$ is the proportion of successes in an SRS of size n_1 drawn from a large population with proportion of successes p_1 and that $\hat{p}_2$ is the proportion of successes in an SRS of size n_2 drawn from a large population with proportion of successes p_2. The sampling distribution of $\hat{p}_1 - \hat{p}_2$ has the following properties:
  - **Shape:** Approximately normal if $n_1 p_1$, $n_1(1 - p_1)$, $n_2 p_2$, and $n_2(1 - p_2)$ are all at least 10.
  - **Center:** The mean is $\mu_{\hat{p}_1 - \hat{p}_2} = p_1 - p_2$.
  - **Variability:** The standard deviation is

  $$\sigma_{\hat{p}_1 - \hat{p}_2} = \sqrt{\frac{p_1(1 - p_1)}{n_1} + \frac{p_2(1 - p_2)}{n_2}}$$

- Be sure to follow the four-step process whenever you construct a confidence interval or perform a significance test for comparing two proportions.

- Before estimating or testing a claim about $p_1 - p_2$, check that these conditions are met:
  - **Random:** The data come from separate random samples from the two populations of interest or from two groups in a randomized experiment.
  - **Large Counts:** The counts of "successes" and "failures" in each sample or group—$n_1\hat{p}_1$, $n_1(1 - \hat{p}_1)$, $n_2\hat{p}_2$, and $n_2(1 - \hat{p}_2)$—are all at least 10.

- When conditions are met, a C% confidence interval for $p_1 - p_2$ is

  $$(\hat{p}_1 - \hat{p}_2) \pm z^*\sqrt{\frac{\hat{p}_1(1 - \hat{p}_1)}{n_1} + \frac{\hat{p}_2(1 - \hat{p}_2)}{n_2}}$$

  where z^* is the critical value for the standard normal distribution with C% of its area between $-z^*$ and z^*. This is called a *two-sample z interval for a difference between two proportions*.

- To estimate the standard deviation of the sampling distribution of $\hat{p}_1 - \hat{p}_2$ in a significance test

of H_0: $p_1 - p_2 = 0$, use the *combined (pooled) sample proportion*

$$\hat{p}_C = \frac{\text{number of successes in both samples combined}}{\text{number of individuals in both samples combined}} = \frac{X_1 + X_2}{n_1 + n_2}$$

- The standardized test statistic for a *two-sample z test for a difference between two proportions* is

  $$z = \frac{(\hat{p}_1 - \hat{p}_2) - 0}{\sqrt{\dfrac{\hat{p}_C(1 - \hat{p}_C)}{n_1} + \dfrac{\hat{p}_C(1 - \hat{p}_C)}{n_2}}}$$

  When the Large Counts condition is met, the standardized test statistic has approximately a standard normal distribution. You can use Table A or technology to find the *P*-value.

Comparing Two Means

- Suppose that $\bar{x}_1$ is the sample mean in an SRS of size n_1 drawn from a large population with mean μ_1 and standard deviation σ_1, and that $\bar{x}_2$ is the sample mean in an SRS of size n_2 drawn from a large population with mean μ_2 and standard deviation σ_2. The sampling distribution of $\bar{x}_1 - \bar{x}_2$ has the following properties:
  - **Shape:** Approximately normal if (1) both population distributions are approximately normal *or* (2) both sample sizes are large: $n_1 \geq 30$ and $n_2 \geq 30$.
  - **Center:** The mean is $\mu_{\bar{x}_1 - \bar{x}_2} = \mu_1 - \mu_2$.
  - **Variability:** The standard deviation is

  $$\sigma_{\bar{x}_1 - \bar{x}_2} = \sqrt{\frac{\sigma_1^2}{n_1} + \frac{\sigma_2^2}{n_2}}.$$

- Be sure to follow the four-step process whenever you construct a confidence interval or perform a significance test for comparing two means.

- Before estimating or testing a claim about $\mu_1 - \mu_2$, check that these conditions are met:
  - **Random:** The data come from separate random samples from the two populations of interest or from two groups in a randomized experiment.

- **Normal/Large Sample:** Both population distributions (or the true distributions of responses to the two treatments) are normal *or* both sample sizes are large ($n_1 \geq 30$ and $n_2 \geq 30$). If the population (treatment) distributions have an unknown shape and the sample sizes are less than 30, a graph of the data from each sample shows no strong skewness or outliers.

- When the conditions are met, a C% confidence interval for $\mu_1 - \mu_2$ is

$$(\bar{x}_1 - \bar{x}_2) \pm t^* \sqrt{\frac{s_1^2}{n_1} + \frac{s_2^2}{n_2}}$$

where t^* is the critical value with C% of its area between $-t^*$ and t^* in the t distribution with degrees of freedom equal to the *smaller* of $n_1 - 1$ and $n_2 - 1$ or given by technology. This is called a *two-sample t interval for a difference between two means*.

- A significance test of H_0: $\mu_1 - \mu_2 = 0$ is called a *two-sample t test for a difference between two means*. The standardized test statistic is

$$t = \frac{(\bar{x}_1 - \bar{x}_2) - 0}{\sqrt{\frac{s_1^2}{n_1} + \frac{s_2^2}{n_2}}}$$

When the Normal/Large Sample condition is met, find the P-value using the t distribution with degrees of freedom equal to the *smaller* of $n_1 - 1$ and $n_2 - 1$ or given by technology.

Inference about a Mean Difference: Paired Data

- **Paired data** result from recording two values of the same quantitative variable for each individual or for each pair of similar individuals.

- There are two ways that an experiment with two treatments can yield paired data:
  - Each experimental unit can be given both treatments in random order.
  - The researcher can form pairs of similar experimental units and randomly assign each treatment to exactly one member of every pair.

  These are both known as a *matched pairs* design.

- To analyze paired data, start by calculating the difference for each pair. Then use a graph to see if the differences are consistently greater than or consistently less than 0.

- When analyzing paired data, use the mean difference $\bar{x}_{\text{diff}}$ and the standard deviation of the differences s_{diff} as summary statistics.

- Be sure to follow the four-step process whenever you construct a confidence interval or perform a significance test for a mean difference.

- Before estimating or testing a claim about μ_{diff}, check that these conditions are met:
  - **Random:** Paired data come from a random sample from the population of interest or from a randomized experiment.
  - **Normal/Large Sample:** The population distribution of differences (or the true distribution of differences in response to the treatments) is normal or the number of differences in the sample is large ($n_{\text{diff}} \geq 30$). If the population (true) distribution of differences has an unknown shape and the number of differences in the sample is less than 30, a graph of the sample differences shows no strong skewness or outliers.

- When the conditions are met, a C% confidence interval for the true mean difference μ_{diff} is

$$\bar{x}_{\text{diff}} \pm t^* \frac{s_{\text{diff}}}{\sqrt{n_{\text{diff}}}}$$

where t^* is the critical value for a t distribution with df $= n_{\text{diff}} - 1$ and C% of its area between $-t^*$ and t^*. This is called a *one-sample t interval for a mean difference*.

- A significance test of H_0: $\mu_{\text{diff}} = 0$ is called a *one-sample t test for a mean difference*. The standardized test statistic is

$$t = \frac{\bar{x}_{\text{diff}} - 0}{\frac{s_{\text{diff}}}{\sqrt{n_{\text{diff}}}}}$$

When the Normal/Large Sample condition is met, find the P-value using the t distribution with degrees of freedom equal to $n_{\text{diff}} - 1$.

- The proper inference method depends on how the data were produced. For paired data, use one-sample t procedures for μ_{diff}. For quantitative data that come from separate random samples from two populations of interest or from two groups in a randomized experiment, use two-sample t procedures for $\mu_1 - \mu_2$.

Chapter 9 Review Exercises

1. Multiracial populations (9.1) According to the 2010 U.S. Census, the states with the highest percentage of residents identifying themselves as biracial or multiracial are Hawaii, at 24%, and Alaska, at 7%. Suppose we take separate random samples of 300 residents of Hawaii and 200 residents of Alaska.

(a) What is the shape of the sampling distribution of $\hat{p}_1 - \hat{p}_2$? Why?

(b) Find the mean of the sampling distribution.

(c) Calculate the standard deviation of the sampling distribution.

2. Restless legs (9.1, 9.2) Restless legs syndrome (RLS) causes a powerful urge to move your legs—so much so that it becomes uncomfortable to sit or lie down. Sleep is difficult. A randomized trial of the drug Pramipexole was undertaken to determine its effectiveness in treating this disease. Patients were randomly assigned to one of two groups: One group was treated with Pramipexole, the other with a placebo. Of the 193 subjects in the Pramipexole group, 81.9% reported "much improved" symptoms. In comparison, 54.3% of the 92 subjects in the placebo group reported "much improved" symptoms.[52]

(a) Does this experiment provide convincing evidence at the $\alpha = 0.05$ level of a difference in the proportion of RLS sufferers like the ones in this experiment who will experience much improved symptoms if they take Pramipexole than if they take a placebo?

(b) Construct and interpret a 95% confidence interval for the difference (Pramipexole – Placebo) in the true proportions of RLS patients like these who will experience much improved symptoms. If you already defined parameters and checked conditions in part (a), you don't need to do them again here.

(c) Explain how the confidence interval from part (b) provides more information than the significance test in part (a).

3. Fuel economy (9.3) According to the website fueleconomy.gov, the mean highway fuel efficiency of all 1078 car models tested in 1990 was 22.3 miles per gallon (mpg) and the standard deviation was 5.7 mpg. In 2015, when 1279 car models were tested, the mean fuel efficiency was 28.9 mpg and the standard deviation was 9.9 mpg. The distribution of fuel efficiency in 1990 is approximately normal, but the 2015 distribution has several high outliers. Suppose we take separate random samples of 35 car models from 1990 and 40 from 2015

and calculate $\bar{x}_1 - \bar{x}_2$, the difference in sample mean fuel efficiency in 1990 ($\bar{x}_1$) and 2015 ($\bar{x}_2$).

(a) What is the shape of the sampling distribution of $\bar{x}_1 - \bar{x}_2$? Why?

(b) Find the mean of the sampling distribution.

(c) Calculate the standard deviation of the sampling distribution.

4. Aren't you Glad? (9.4) Do name-brand trash bags hold more weight than store-brand trash bags? Two statistics students, Janelle and Brittany, decided to find out. They randomly selected 15 Glad trash bags and 15 Walmart trash bags and loaded them with oranges until they broke. Why oranges? One of the students had a bumper crop from the orange tree in her yard. Here are the data.[53]

Glad	147	145	143	148	147	156	147	150
Walmart	132	138	140	139	133	133	141	131
Glad	139	148	146	149	151	150	145	
Walmart	135	139	145	132	150	139	135	

Do these data provide convincing evidence at the $\alpha = 0.01$ level that Glad trash bags can hold more weight, on average, than Walmart trash bags?

5. Cell-phone addiction (9.3) In a study of cell-phone usage by college students, Baylor University professor James Roberts found that in a random sample of 84 male college students, the mean number of minutes per day of total cell-phone usage (talking, texting, playing games, and so on) was 458.5 minutes and the standard deviation was 344.2 minutes. For a random sample of 80 female students, the mean was 600.1 minutes and the standard deviation was 392.3 minutes.[54] Construct and interpret a 95% confidence interval for the difference in mean daily cell-phone usage for males and females.

6. Paired or unpaired? (9.6) In each of the following settings, decide whether you should use two-sample t procedures to perform inference about a difference in means or one-sample t procedures to perform inference about a mean difference. Explain your choice in each case.

(a) A city planner is comparing traffic patterns at two different intersections. He randomly selects 12 times between 6:00 A.M. and 10:00 P.M., and he and his assistant count the number of cars passing through each intersection during the 10-minute interval that begins at that time.

(b) How much greater is the percent of senior citizens who attend a play at least once per year than the percent of 20-somethings who do so? Random samples of 100 senior citizens and 100 people in their 20s were surveyed.

(c) Do people smoke less when cigarettes cost more? A random sample of 500 smokers was selected. The number of cigarettes each person smoked per day was recorded over a 1-month period before a 30% cigarette tax was imposed and again for 1 month after the tax was imposed.

7. **Reaction time (9.5, 9.6)** Does exercise improve reaction time? Francie and Sebastian used an online game to measure the reaction time of 15 randomly selected students from their school before and after each student had run two laps on a track. Reaction time was measured in milliseconds, so smaller numbers indicate faster reactions. Here are their results.

Student	Before	After	Student	Before	After
1	275	263	9	254	275
2	329	358	10	317	324
3	297	264	11	292	278
4	476	426	12	314	276
5	314	288	13	299	287
6	366	334	14	295	276
7	317	328	15	259	256
8	313	286			

(a) Make a dotplot of the difference in reaction times (Before – After) for each student. What does the graph reveal?

(b) Calculate the mean and standard deviation of the difference in reaction times, and interpret the mean difference in context.

(c) Do the data provide convincing evidence at the $\alpha = 0.05$ significance level that exercise is associated with a decrease in reaction time, on average, for students at this school?

Chapter 9 Practice Test

Section I: Multiple Choice *Select the best answer for each question.*

1. Mike and Ike are experts at pitching horseshoes. A total of 70% of Mike's tosses are ringers (that is, the horseshoe encircles the stake when it lands), while 67% of Ike's tosses are ringers. Suppose we take separate random samples of 50 horseshoe tosses from Mike and Ike, and record the proportion of ringers each one gets. Which of the following best describes the sampling distribution of $\hat{p}_M - \hat{p}_I$?

(a) Normal with mean 0.03 and standard deviation 0.131

(b) Approximately normal with mean 0.03 and standard deviation 0.131

(c) Normal with mean 0.03 and standard deviation 0.0929

(d) Approximately normal with mean 0.03 and standard deviation 0.0929

2. A total of 35 people from a random sample of 125 workers from Company A admitted to using sick leave when they weren't really ill. And 17 employees from a random sample of 68 workers from Company B admitted that they had used sick leave when they weren't ill. What would be a 95% confidence interval for the difference in the proportions of workers at the two companies who would admit to using sick leave when they weren't ill?

(a) $0.03 \pm 1.96\sqrt{\dfrac{(0.28)(0.72)}{125} + \dfrac{(0.25)(0.75)}{68}}$

(b) $0.03 \pm 1.645\sqrt{\dfrac{(0.28)(0.72)}{125} + \dfrac{(0.25)(0.75)}{68}}$

(c) $0.03 \pm 1.96\sqrt{\dfrac{(0.269)(0.731)}{125} + \dfrac{(0.269)(0.731)}{68}}$

(d) $0.03 \pm 1.645\sqrt{\dfrac{(0.269)(0.731)}{125} + \dfrac{(0.269)(0.731)}{68}}$

3. A sample survey interviews separate SRSs of 500 female college students and 550 male college students. Researchers want to determine whether there is a difference in the proportion of male and female college students who worked for pay last summer. In all, 410 of the females and 484 of the males say they worked for pay last summer. Let p_M and p_F be the proportions of all college males and females who worked last summer. The hypotheses to be tested are:

(a) $H_0: p_M - p_F = 0$ versus $H_a: p_M - p_F \neq 0$.

(b) $H_0: p_M - p_F = 0$ versus $H_a: p_M - p_F > 0$.

(c) $H_0: p_M - p_F = 0$ versus $H_a: p_M - p_F < 0$.

(d) $H_0: p_M - p_F \neq 0$ versus $H_a: p_M - p_F = 0$.

4. Phoebe has a theory that older students at her high school are more likely to bring a bag lunch than are younger students, because they have grown tired of cafeteria food. She takes a simple random sample of 80 sophomores and finds that 52 of them bring a bag lunch. A simple random sample of 104 seniors reveals that 78 of them bring a bag lunch. Letting $p_1 =$ proportion of sophomores who bring a bag lunch, and $p_2 =$ proportion of seniors who bring a bag lunch, Phoebe tests the hypotheses $H_0: p_1 - p_2 = 0$ versus $H_a: p_1 - p_2 < 0$ at the $\alpha = 0.05$ level. What expression should Phoebe use in the denominator of the standardized test statistic?

(a) $\sqrt{\dfrac{(0.65)(0.35)}{80} + \dfrac{(0.75)(0.25)}{104}}$

(b) $\sqrt{\dfrac{(0.707)(0.293)}{80} + \dfrac{(0.707)(0.293)}{104}}$

(c) $\sqrt{\left(\dfrac{(0.707)(0.293)}{80}\right)^2 + \left(\dfrac{(0.707)(0.293)}{104}\right)^2}$

(d) $\sqrt{\left(\dfrac{(0.65)(0.35)}{80}\right)^2 + \left(\dfrac{(0.75)(0.25)}{104}\right)^2}$

5. At a large state university, the heights of male students who are interscholastic athletes is approximately normally distributed with a mean of 74.3 inches and standard deviation of 3.5 inches. The heights of male students who don't play interscholastic sports (we'll call them "non-interscholastics") is approximately normally distributed with a mean of 70.3 inches and standard deviation of 3.2 inches. In an SRS of 10 interscholastic athletes and a separate SRS of 12 non-interscholastics, which of the following best describes the sampling distribution of $\bar{x}_A - \bar{x}_N$, the difference in mean heights of interscholastic athletes and non-interscholastics?

(a) Mean = 4.0 inches, standard deviation = 1.44 inches, shape cannot be determined.

(b) Mean = 4.0 inches, standard deviation = 2.03 inches, shape approximately normal.

(c) Mean = 4.0 inches, standard deviation = 2.03 inches, shape cannot be determined.

(d) Mean = 4.0 inches, standard deviation = 1.44 inches, shape approximately normal.

6. A random sample of 30 words from Jane Austen's *Pride and Prejudice* had a mean length of 4.08 letters with a standard deviation of 2.40 letters. A random sample of 30 words from Henry James's *What Maisie Knew* had a mean length of 3.85 letters with a standard deviation of 2.26 letters. We wish to test the null hypothesis that the mean word length is the same in the two books. Our samples produced a 95% confidence interval for the true difference in mean word length of (−0.97, 1.43). Which of the following gives a correct conclusion for the test?

(a) Because the confidence interval includes 0, we have convincing evidence that the true difference in means is 0.

(b) Because the center of the interval is 0.23, we have convincing evidence that the mean length of words in *Pride and Prejudice* is greater than the mean length of words in *What Maisie Knew*.

(c) Because the confidence interval includes 0, we don't have convincing evidence that the mean word length in the two novels is different.

(d) Because the confidence interval includes more positive than negative values, we have convincing evidence that the mean length of words in *Pride and Prejudice* is greater than the mean length of words in *What Maisie Knew*.

7. An ecologist is studying differences in the populations of a certain species of lizard on two different islands. The ecologist collects separate random samples of lizards from the two islands, weighs the lizards, and then releases them again. The table summarizes the data.

	n	Mean (g)	SD (g)
Sheep Island	24	46.5	5.97
Pig Island	30	44.2	4.24

Which of the following is the correct expression for the standardized test statistic to test the null hypothesis that the mean weights on the two islands are equal?

(a) $t = \dfrac{46.5 - 44.2}{\left(\dfrac{5.97}{\sqrt{24}} + \dfrac{4.24}{\sqrt{30}}\right)^2}$

(b) $t = \dfrac{46.5 - 44.2}{\sqrt{\dfrac{5.97}{24} + \dfrac{4.24}{30}}}$

(c) $t = \dfrac{46.5 - 44.2}{\sqrt{\dfrac{5.97^2}{24} + \dfrac{4.24^2}{30}}}$

(d) $t = \dfrac{46.5 - 44.2}{\sqrt{\dfrac{5.97^2}{24}} + \sqrt{\dfrac{4.24^2}{30}}}$

8. A study of road rage asked separate random samples of 596 men and 523 women about their behavior while driving. Based on their answers, each person was assigned a road rage score on a scale of 0 to 20. The participants were chosen by random digit dialing of phone numbers. The researchers performed a test of the following hypotheses: $H_0: \mu_M = \mu_F$ versus $H_a: \mu_M \neq \mu_F$. Which of the following describes a Type II error in the context of this study?

(a) Finding convincing evidence that the true means are different for males and females, when in reality the true means are the same

(b) Finding convincing evidence that the true means are different for males and females, when in reality the true means are different

(c) Not finding convincing evidence that the true means are different for males and females, when in reality the true means are different

(d) Not finding convincing evidence that the true means are different for males and females, when in reality the true means are the same

9. A study of the impact of caffeine consumption on reaction time was designed to correct for the impact of subjects' prior sleep deprivation by dividing the 24 subjects into 12 pairs on the basis of the average hours of sleep they had gotten the previous 5 nights. That is, the two with the highest average sleep were a pair, then the two with the next highest average sleep, and so on. One randomly assigned member of each pair drank 2 cups of caffeinated coffee, and the other drank 2 cups of decaf. Each subject's performance on a standard reaction-time test was recorded. Which of the following is the correct check of the "Normal/Large Sample" condition for this significance test?

(a) Confirm (graphically) that the scores of the caffeine drinkers and the scores of the decaf drinkers in the study could have come from normal distributions.

(b) Confirm (graphically) that the scores of the caffeine drinkers and the scores of the decaf drinkers in the study, *as well as* the differences in scores within each pair of subjects, could have come from normal distributions.

(c) Confirm that the differences in scores within each pair of subjects could have come from a normal distribution.

(d) Because the central limit theorem applies to this situation, it is not necessary to check the data for normality.

10. The following data are the percents of fat found in 5 randomly selected cartons of each of two brands (A and B) of ice cream:

Brand A	5.7	4.5	6.2	6.3	7.3
Brand B	6.3	5.7	5.9	6.4	5.1

Which of the following procedures is appropriate to test the hypothesis of equal average fat content in the two types of ice cream?

(a) Paired t test

(b) Two-sample t test

(c) Paired z test

(d) Two-sample z test

Section II: Free Response

11. *Daphnia pulicaria* is a water flea—a small crustacean that lives in lakes and is a major food supply for many species of fish. When fish are present in the lake water, they release chemicals called kairomones that induce water fleas to grow long tail spines that make them more difficult for the fish to eat. One study of this phenomenon compared the relative length of tail spines in two populations of *Daphnia pulicaria,* one grown when kairomones were present and one grown when they were not. Below are data on the relative tail spine lengths, measured as a percentage of the entire length of the water flea, for random samples from the two populations.

	Relative tail spine length		
	n	$\bar{x}$	s_x
Fish kairomone present	214	37.26	4.68
Fish kairomone absent	152	30.67	4.19

Do the data provide convincing evidence at the $\alpha = 0.05$ level that the mean relative tail spine length of *Daphnia* is longer in the presence of fish kairomones?

12. **Postcard response rate** In a study of response rates to mailed questionnaires, Morton Brown of the National Center for Health Statistics divided a random sample of 523 physicians into two subsamples at random. The "long" group of 261 physicians was asked to fill out and return a two-page questionnaire that first asked if they had seen patients with a certain disease and then asked for details about four such patients. The "postcard" group of 262 physicians was asked to return a postcard that asked only if they had ever treated patients with the disease (a longer form was then sent to those who said "Yes"). In the long group, 138 physicians returned the questionnaire, while 178 of the physicians in the postcard group returned the card.[55] Construct and interpret a 99% confidence interval for the true difference in response rate for physicians like these who received the postcard versus the long questionnaire.

13. **Pleasant smells** Do pleasant odors change the rate at which people work? To test this idea, 21 subjects worked two different but roughly equivalent paper-and-pencil mazes, one while wearing an unscented mask and the other while wearing a mask with a floral scent. The order in which the masks were worn by each subject was determined at random. Here are the differences (Unscented − Scented) in their average times (in seconds).[56]

−7.37	−3.14	4.10	−4.40	19.47	−10.80	−0.87
8.70	2.94	−17.24	14.30	−24.57	16.17	−7.84
8.60	−10.77	24.97	−4.47	11.90	−6.26	6.67

(a) Calculate and interpret a 95% confidence interval for the mean change in time to solve a maze when wearing a mask with a floral scent for subjects like these.

(b) Based on the confidence interval in part (a), do the data provide convincing evidence of a difference in the mean time it takes to complete the maze with and without the floral scent for subjects like these? Explain.

10

Inference for Distributions and Relationships

Lesson 10.1 Testing the Distribution of a Categorical Variable   620

Lesson 10.2 Chi-Square Tests for Goodness of Fit   627

Lesson 10.3 Testing the Relationship Between
Two Categorical Variables   636

Lesson 10.4 Chi-Square Tests for Association   645

Lesson 10.5 Testing the Relationship Between
Two Quantitative Variables   655

Lesson 10.6 Inference for the Slope of a
Least-Squares Regression Line   667

Chapter 10   Main Points   679
Chapter 10   Review Exercises   680
Chapter 10   Practice Test   682

STATS applied!

Does background music influence what customers buy?

Market researchers suspect that background music may affect the mood and buying behavior of customers. A study compared three randomly assigned treatments in a European-style restaurant: no music, French accordion music, and Italian string music. Under each condition, the researchers recorded the number of customers who ordered French, Italian, and other entrées. Here is a table that summarizes the data.[1]

		Type of background music			
		None	French	Italian	Total
Entrée ordered	French	30	39	30	99
	Italian	11	1	19	31
	Other	43	35	35	113
	Total	84	75	84	243

Do these data provide convincing evidence of an association between type of background music and type of entrée ordered? If so, is it reasonable to conclude that there is a cause-and-effect relationship?

We'll revisit STATS applied! at the end of the chapter, so you can use what you have learned to help answer these questions.

Lesson 10.1

Testing the Distribution of a Categorical Variable

LEARNING TARGETS

- State hypotheses for a test about the distribution of a categorical variable.
- Calculate expected counts for a test about the distribution of a categorical variable.
- Calculate the test statistic for a test about the distribution of a categorical variable.

In Lessons 8.3 and 8.4, you learned how to perform a one-sample z test for a population proportion. This test is appropriate when the data come from a single sample and can be divided into two categories: success and failure. In this lesson, you will again encounter categorical data from a single sample. However, there is no longer a limit to the number of categories for the variable of interest. The following activity gives you a taste of what's ahead.

ACTIVITY

The Color of Candy

Mars, Incorporated, which is headquartered in McLean, Virginia, makes milk chocolate candies. Here's what the company's Consumer Affairs Department says about the color distribution of its M&M'S® Milk Chocolate Candies:

> On average, M&M'S Milk Chocolate Candies will contain 13 percent of each of browns and reds, 14 percent yellows, 16 percent greens, 20 percent oranges, and 24 percent blues.

The purpose of this activity is to determine if we have reason to doubt this claim.

1. Your teacher will take a random sample of 60 M&M'S from a large bag and give one or more pieces of candy to each student. As a class, count the number of M&M'S Milk Chocolate Candies of each color. Make a table on the board that summarizes these *observed counts*.

2. How can you tell if the sample data give convincing evidence to dispute the company's claim? Each team of three or four students should discuss this question and devise a formula for a test statistic that measures the difference between the observed and expected color distributions. The test statistic should yield a single number when

the observed and expected values are plugged in. Also, larger differences between the observed and expected distributions should result in a larger statistic. Here are some questions for your team to consider:

- Should we look at the difference between the observed and expected *proportions* in each color category or between the observed and expected *counts* in each category?

- Should we use the differences themselves, the absolute value of the differences, or the square of the differences?

- Should we divide each difference value by the sample size, expected count, or nothing at all?

3. Each team will share its proposed test statistic with the class. Your teacher will then reveal how the *chi-square test statistic* χ^2 is calculated.

4. Discuss as a class: If your sample is consistent with the company's claimed distribution of M&M'S Milk Chocolate Candies colors, will the value of χ^2 be large or small? If your sample is not consistent with the company's claimed color distribution, will the value of χ^2 be large or small?

5. Compute the value of the chi-square test statistic for the class's data.

We can use simulation to determine if your class's chi-square test statistic is large enough to provide convincing evidence that the distribution of colors in the large bag is different from the company's claim. To conduct the simulation, 100 random samples of size 60 were selected from a population of M&M'S® Milk Chocolate Candies that matches the company's claim. For each random sample, the value of the χ^2 test statistic was calculated and plotted on the dotplot.

6. There is one dot at $\chi^2 = 16$. Explain what this dot represents.

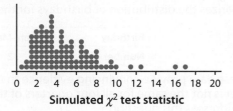

Simulated χ^2 test statistic

7. Where does your class's value of χ^2 fall relative to the other dots on the dotplot? What conclusion can you make about the distribution of colors in the large bag?

Stating Hypotheses

As with any test, we begin by stating hypotheses. The null hypothesis in a test about the distribution of a categorical variable should state a claim about the distribution of the variable in the population of interest. In the case of the Color of Candy Activity, the categorical variable we're measuring is color and the population of interest is the large bag of M&M'S Milk Chocolate Candies. The appropriate null hypothesis is

H_0: The distribution of color in the large bag of M&M'S Milk Chocolate Candies is the same as the claimed distribution.

The alternative hypothesis for this test is that the categorical variable does *not* have the specified distribution. For the M&M'S, our alternative hypothesis is

H_a: The distribution of color in the large bag of M&M'S Milk Chocolate Candies is *not* the same as the claimed distribution.

THINK ABOUT IT **Do we have to use words when stating hypotheses for a chi-square test for goodness of fit?** No. We can also write the hypotheses in symbols as

H_0: $p_{\text{brown}} = 0.13$, $p_{\text{red}} = 0.13$, $p_{\text{yellow}} = 0.14$, $p_{\text{green}} = 0.16$, $p_{\text{orange}} = 0.20$, $p_{\text{blue}} = 0.24$
H_a: At least two of the p_i's are incorrect.

where p_i = the true proportion of M&M'S Milk Chocolate Candies of each color in the large bag. Why don't we write the alternative hypothesis as H_a: At least one of the p_i's is incorrect? If the stated proportion in one category is wrong, then the stated proportion in at least one other category must be wrong because the sum of the p_i's must be 1.

EXAMPLE

Does your birthday matter in hockey?

Stating hypotheses

PROBLEM: In his book *Outliers*, Malcolm Gladwell suggests that a hockey player's birth month has a big influence on his chance to make it to the highest levels of the game. Because January 1 is the cut-off date for youth leagues in Canada [where many National Hockey League (NHL) players come from], players born in January will be competing against players up to 12 months younger. The older players tend to be bigger, stronger, and more coordinated and hence get more playing time, more coaching, and have a better chance of being successful.

To see if birth date is related to whether a player makes it into the NHL, a random sample of 80 NHL players from a recent season was selected and their birthdays were recorded.[2] The table summarizes the distribution of birthdays for these 80 players:

Birthday	Jan–Mar	Apr–Jun	Jul–Sep	Oct–Dec
Number of players	32	20	16	12

Do these data provide convincing evidence that the proportion of all NHL players born in each quarter is not the same for all four quarters of the year? State appropriate hypotheses for a test that addresses this question.

SOLUTION:

H_0: The proportion of all NHL players born in each quarter is the same for all four quarters of the year.

H_a: The proportion of all NHL players born in each quarter is not the same for all four quarters of the year.

> The hypotheses could also be stated symbolically:
> H_0: $p_{\text{jan-mar}} = p_{\text{apr-jun}} = p_{\text{jul-sep}} = p_{\text{oct-dec}} = 0.25$
> H_a: At least 2 of the p_i's are different from 0.25.

FOR PRACTICE **TRY EXERCISE 1.**

Calculating Expected Counts

Andre's class did the Color of Candy Activity. The table summarizes the data from the class's sample of M&M'S® Milk Chocolate Candies.

Color	Brown	Red	Yellow	Green	Orange	Blue	Total
Observed count	6	10	15	12	8	9	60

To begin their analysis, Andre's class compared the observed counts from their sample with the counts that would be expected if the manufacturer's claim is true:

> On average, M&M'S Milk Chocolate Candies will contain 13 percent of each of browns and reds, 14 percent yellows, 16 percent greens, 20 percent oranges, and 24 percent blues.

Assuming that the claimed distribution is true, 24% of all M&M'S Milk Chocolate Candies produced are blue. For random samples of 60 candies, the expected count of blue M&M'S should be $(60)(0.24) = 14.40$.

How to Calculate Expected Counts

The expected count for category i in the distribution of a categorical variable is

$$np_i$$

where p_i is the relative frequency for category i specified by the null hypothesis.

Using the same method, here are the expected counts for the remaining five colors:

Brown: $(60)(0.13) = 7.80$

Red: $(60)(0.13) = 7.80$

Yellow: $(60)(0.14) = 8.40$

Green: $(60)(0.16) = 9.60$

Orange: $(60)(0.20) = 12.00$

Don't worry that most of the expected counts are not integers. Like expected values from Lesson 5.2, the expected counts here represent the *average* number of M&M'S® Milk Chocolate Candies of each color in many, many samples of size 60.

EXAMPLE

Are birthdays uniformly distributed for NHL players?

Calculating expected counts

PROBLEM: To see if birth date is related to whether a player makes it into the NHL, a random sample of 80 NHL players from a recent season was selected and their birthdays were recorded. The table summarizes the data on birthdays for these 80 players.

Birthday	Jan–Mar	Apr–Jun	Jul–Sep	Oct–Dec
Number of players	32	20	16	12

Calculate the expected counts for a test of the null hypothesis that the proportion of all NHL players born in each quarter is the same for all four quarters of the year.

SOLUTION:

Jan–Mar: 80(0.25) = 20

Apr–Jun: 80(0.25) = 20

Jul–Sep: 80(0.25) = 20

Oct–Dec: 80(0.25) = 20

> If the proportion of players born in each of the 4 quarters is the same, then $\dfrac{100\%}{4}$ = 25% or 0.25 of NHL players should be born in each quarter.

FOR PRACTICE   TRY EXERCISE 5.

The Chi-Square Test Statistic

The graph in Figure 10.1 compares the observed and expected counts of M&M'S Milk Chocolate Candies for Andre's class.

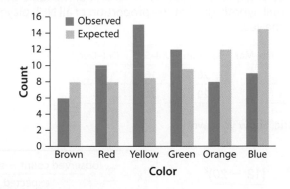

FIGURE 10.1 Bar chart comparing observed and expected counts for Andre's class sample of 60 M&M'S Milk Chocolate Candies.

Because the observed counts are different from the expected counts, there is some evidence that the claimed distribution is not correct. However, it may be that the claimed distribution is correct and the differences found by Andre's class were due to sampling variability.

To assess if these differences are larger than what is likely to happen by chance alone when H_0 is true, we have to calculate a test statistic and *P*-value. For a test about the distribution of a categorical variable, we use the **chi-square test statistic.**

> ▌▌▌ **DEFINITION** Chi-square test statistic
>
> The **chi-square test statistic** is a measure of how different the observed counts are from the expected counts. The formula for the statistic is
>
> $$\chi^2 = \sum \frac{(\text{observed count} - \text{expected count})^2}{\text{expected count}}$$
>
> where the sum is over all categories of the categorical variable.

The table shows the observed and expected counts for Andre's class data.

Color	Observed	Expected
Brown	6	7.8
Red	10	7.8
Yellow	15	8.4
Green	12	9.6
Orange	8	12.0
Blue	9	14.4

For these data, the chi-square test statistic is

$$\chi^2 = \sum \frac{(\text{observed count} - \text{expected count})^2}{\text{expected count}}$$
$$= \frac{(6-7.8)^2}{7.8} + \frac{(10-7.8)^2}{7.8} + \frac{(15-8.4)^2}{8.4} + \frac{(12-9.6)^2}{9.6} + \frac{(8-12.0)^2}{12.0} + \frac{(9-14.4)^2}{14.4}$$
$$= 0.415 + 0.621 + 5.186 + 0.600 + 1.333 + 2.025 = 10.180$$

EXAMPLE

How different is the distribution of NHL birthdays?

The chi-square test statistic

PROBLEM: To see if birth date is related to whether a player makes it into the NHL, a random sample of 80 NHL players from a recent season was selected and their birthdays were recorded. The table shows the observed and expected counts for a test of the null hypothesis that the proportion of all NHL players born in each quarter is the same for all four quarters of the year.

Birthday	Jan–Mar	Apr–Jun	Jul–Sep	Oct–Dec
Observed counts	32	20	16	12
Expected counts	20	20	20	20

Calculate the value of the chi-square test statistic. Show your work.

SOLUTION:

$$\chi^2 = \frac{(32-20)^2}{20} + \frac{(20-20)^2}{20} + \frac{(16-20)^2}{20} + \frac{(12-20)^2}{20}$$
$$\chi^2 = 7.2 + 0 + 0.8 + 3.2 = 11.2$$

> $$\chi^2 = \sum \frac{(\text{observed count} - \text{expected count})^2}{\text{expected count}}$$

FOR PRACTICE TRY EXERCISE 9.

The bigger the value of χ^2, the stronger the evidence against the null hypothesis and for the alternative hypothesis. This is because larger values of χ^2 arise when the differences between the observed counts and expected counts are larger. But how big does the value of χ^2 have to be for the evidence to be convincing? We can use simulation to get an idea.

The dotplot in Figure 10.2 shows the results of a simulation in which 100 random samples of size 80 were selected from a population of NHL players where the proportion of players born in each quarter was the same. For each simulated sample of 80 players, the chi-square test statistic was calculated.

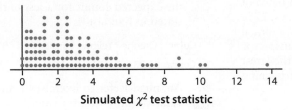

Simulated χ^2 test statistic

FIGURE 10.2 Dotplot showing values of the chi-square test statistic in 100 simulated samples of size $n = 80$ from a population of NHL players where players are equally likely to be born in each of the four quarters.

Because there is only one value greater than or equal to the observed chi-square test statistic of 11.2, it is unusual to get differences between the observed and expected counts this large by chance alone. These data provide convincing evidence that the proportion of NHL players born in each quarter is not the same for all quarters.

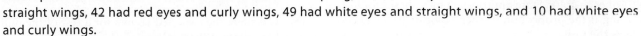

LESSON APP 10.1

Are fruit flies predictable?

Biologists wish to mate pairs of fruit flies having genetic makeup RrCc, indicating that each has one dominant gene (R) and one recessive gene (r) for eye color, along with one dominant (C) and one recessive (c) gene for wing type. Each offspring will receive one gene for each of the two traits from each parent.

The biologists predict a ratio of 9 red-eyed, straight-winged : 3 red-eyed, curly-winged : 3 white-eyed, straight-winged : 1 white-eyed, curly-winged offspring. To test their hypothesis about the distribution of offspring, the biologists randomly select pairs of fruit flies and mate them. Of the 200 offspring, 99 had red eyes and straight wings, 42 had red eyes and curly wings, 49 had white eyes and straight wings, and 10 had white eyes and curly wings.

1. State the hypotheses that the biologists are interested in testing.

2. Calculate the expected count for each possible outcome, assuming that the biologists' prediction is correct.

3. Calculate the value of the chi-square test statistic for the biologist's data.

Andrew Syred/Science Source

Lesson 10.1

WHAT DID YOU LEARN?

LEARNING TARGET	EXAMPLES	EXERCISES
State hypotheses for a test about the distribution of a categorical variable.	p. 621	1–4
Calculate expected counts for a test about the distribution of a categorical variable.	p. 623	5–8
Calculate the test statistic for a test about the distribution of a categorical variable.	p. 624	9–12

Exercises

Lesson 10.1

Mastering Concepts and Skills

1. **Are you nuts?** A company claims that each batch of its deluxe mixed nuts contains 52% cashews, 27% almonds, 13% macadamia nuts, and 8% brazil nuts. To test this claim, a quality-control inspector takes a random sample of 150 nuts from the latest batch. State appropriate hypotheses for performing a test of the company's claim.

pg 621

2. **Spin and win** Casinos are required to verify that their games operate as advertised. American roulette wheels have 38 slots—18 red, 18 black, and 2 green. The managers of a casino record data from a random sample of 200 spins of one of their American roulette wheels. State appropriate hypotheses for performing a test of this roulette wheel.

3. **A uniform rainbow?** Are the colors of Original Skittles® uniformly distributed? That is, are the proportions of green, red, purple, yellow, and orange Skittles the same? To investigate, Jacob and Dotan bought 4 bags of Original Skittles and counted how many candies there were of each color. State appropriate hypotheses for testing if the proportion of each color of Original Skittles is the same.

4. **Uniformly loopy?** Kellogg's Froot Loops® cereal comes in 6 colors: orange, yellow, purple, red, blue, and green. Are these colors uniformly distributed? Charise poured out her morning bowl of cereal and methodically counted the number of cereal pieces of each color. State appropriate hypotheses for testing if the proportion of each color of Froot Loops is the same.

5. **Counting nuts** Refer to Exercise 1. Here are the counts for each type of nut in the sample. Calculate the expected counts for a test of the null hypothesis stated in Exercise 1.

pg 623

Nut	Cashew	Almond	Macadamia	Brazil	Total
Count	83	29	20	18	150

6. **Counting spins** Refer to Exercise 2. Here are the results of the sample of 200 spins. Calculate the expected counts for a test of the null hypothesis stated in Exercise 2.

Color	Red	Black	Green	Total
Count	85	99	16	200

7. **Counting Skittles** Refer to Exercise 3. Here are the counts for each color in the sample. Calculate the expected counts for a test of the null hypothesis stated in Exercise 3.

Color	Green	Red	Purple	Yellow	Orange	Total
Count	39	49	42	53	50	233

8. **Counting Froot Loops** Refer to Exercise 4. Here are the counts for each color in the sample. Calculate the expected counts for a test of the null hypothesis stated in Exercise 4.

Color	Orange	Yellow	Purple	Red	Blue	Green	Total
Count	28	21	16	25	14	16	120

9. **Ah, nuts!** Refer to Exercises 1 and 5. Calculate the value of the χ^2 test statistic.

pg 624

10. **Roulette** Refer to Exercises 2 and 6. Calculate the value of the χ^2 test statistic.

11. **Skittles** Refer to Exercises 3 and 7. Calculate the value of the χ^2 test statistic.

12. **Feeling loopy?** Refer to Exercises 4 and 8. Calculate the value of the χ^2 test statistic.

Applying the Concepts

13. **Bird watching** Researchers want to know if birds prefer particular types of trees when they're searching for seeds and insects. In an Oregon forest, 54% of the trees were Douglas firs, 40% were ponderosa pines, and 6% were other types of trees. At a randomly selected time during the day, the researchers observed 156 red-breasted nuthatches: 70 were seen in Douglas firs, 79 in ponderosa pines, and 7 in other types of trees.[3]

(a) State the hypotheses that the researchers are interested in testing.

(b) Calculate the expected counts for a test of the null hypothesis in part (a).

(c) Calculate the value of the chi-square test statistic for these data.

14. **Watch out for seagulls** Do seagulls show a preference for where they land? To answer this question, biologists conducted a study in an enclosed outdoor space with a piece of shore made up of 56% sand, 29% mud, and 15% rocks. The biologists chose 200 seagulls at random. Each seagull was released into the outdoor space on its own and observed until it landed somewhere on the piece of shore. In all, 128 seagulls landed on the sand, 61 landed in the mud, and 11 landed on the rocks.

(a) State the hypotheses that the biologists are interested in testing.

(b) Calculate the expected counts for a test of the null hypothesis in part (a).

(c) Calculate the value of the chi-square test statistic for these data.

15. **Simulating nuthatches** Refer to Exercise 13. The dotplot shows the results of a simulation in which

100 random samples of size 156 were selected from a population where 54% of red-breasted nuthatches are in Douglas firs, 40% are in ponderosa pines, and 6% are in other types of trees. For each simulated sample of 156 nuthatches, the chi-square test statistic was calculated. Based on the χ^2 statistic from Exercise 13 and the results of the simulation, what conclusion would you draw?

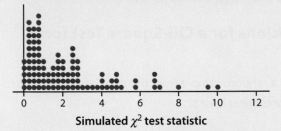

Simulated χ^2 test statistic

16. **Simulating seagulls** Refer to Exercise 14. The dotplot shows the results of a simulation in which 100 random samples of size 200 were selected from a population where 56% of seagulls land on sand, 29% land on mud, and 15% land on rocks. For each simulated sample of 200 seagulls, the chi-square test statistic was calculated. Based on the χ^2 statistic from Exercise 14 and the results of the simulation, what conclusion would you draw?

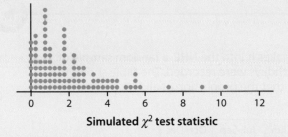

Simulated χ^2 test statistic

Extending the Concepts

17. **Symbolic hypotheses** Refer to Exercise 13. State the hypotheses the researchers are interested in testing, using symbols (for example, p_{sand}). Make sure to define the parameters used in the hypotheses.

Recycle and Review

18. **Seed weights (5.7)** Biological measurements on the same species often follow a normal distribution quite closely. The weights of seeds of a variety of winged bean are approximately normal with a mean of 525 milligrams (mg) and standard deviation of 110 milligrams.

(a) What percent of seeds weigh more than 500 milligrams?

(b) If we discard the lightest 10% of these seeds, what is the smallest weight among the remaining seeds?

19. **More attacks? (3.2)** Recently, the *New York Times Magazine* conducted a poll of its subscribers and reported for several weeks on the responses of "3244 subscribers who chose to participate." One question posed was, "Are you afraid that there will be another terrorist attack in the United States in your lifetime on the order of the September 11 attacks?" A total of 49% of the respondents said "Yes." Give one reason why this poll might produce biased results and describe the direction of bias.[4]

Lesson 10.2

Chi-Square Tests for Goodness of Fit

LEARNING TARGETS

- Check conditions for a test about the distribution of a categorical variable.
- Calculate the *P*-value for a test about the distribution of a categorical variable.
- Use the four-step process to perform a chi-square test for goodness of fit.

In the preceding lesson, you learned how to state hypotheses, calculate expected counts, and calculate the chi-square test statistic for a test about the distribution of a categorical variable. In this lesson, you will learn how to check conditions, calculate a *P*-value, and complete the four-step process for this test, called a *chi-square test for goodness of fit*.

Conditions for a Chi-Square Test for Goodness of Fit

Like the other significance tests you have learned about, there are two conditions that need to be satisfied in a chi-square test for goodness of fit.

> ### How to Check the Conditions for a Chi-Square Test for Goodness of Fit
>
> - **Random:** The data come from a random sample from the population of interest.
> - **Large Counts:** All expected counts are at least 5.

For Andre's class data, given in Lesson 10.1, the conditions for a chi-square test for goodness of fit are met: Andre's teacher selected a random sample of M&M'S® Milk Chocolate Candies from the large bag and all the expected counts were at least 5:

Color	Brown	Red	Yellow	Green	Orange	Blue
Expected count	7.8	7.8	8.4	9.6	12.0	14.4

EXAMPLE

Can we test the distribution of NHL birthdays?

Checking conditions

PROBLEM: To see if birth date is related to whether a player makes it into the NHL, a random sample of 80 NHL players from a recent season was selected and their birthdays were recorded. The table shows the observed and expected counts for a test of the null hypothesis that the proportion of all NHL players born in each quarter is the same for all four quarters of the year.

Birthday	Jan–Mar	Apr–Jun	Jul–Sep	Oct–Dec
Observed counts	32	20	16	12
Expected counts	20	20	20	20

Check if the conditions for performing a chi-square test for goodness of fit are met.

SOLUTION:
- Random? The data come from a random sample of NHL players. ✓
- Large Counts? All expected counts (20, 20, 20, 20) are at least 5. ✓

> Make sure to give the values of the *expected* counts when checking the Large Counts condition.

FOR PRACTICE TRY EXERCISE 1.

Calculating *P*-values

Andre's class did the Color of Candy Activity in Lesson 10.1 and got a chi-square test statistic of $\chi^2 = 10.180$. Is this an unusually large value? Or is this a value that is likely to happen by chance alone when the distribution of colors in the bag is the same as the company claims? To investigate, we used computer software to simulate taking 1000 random samples of size 60 from the population distribution of M&M'S Milk Chocolate Candies given by Mars, Inc. The dotplot in Figure 10.3 shows the values of the chi-square test statistic for these 1000 samples.

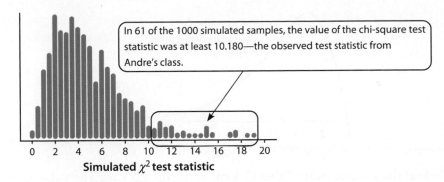

In 61 of the 1000 simulated samples, the value of the chi-square test statistic was at least 10.180—the observed test statistic from Andre's class.

Simulated χ^2 test statistic

FIGURE 10.3 Dotplot showing values of the chi-square test statistic in 1000 simulated samples of size $n = 60$ from the population distribution of M&M'S® Milk Chocolate Candies stated by the company.

In the simulation, 61 of the 1000 samples produced a χ^2 statistic of 10.180 or larger. This means that the *P*-value is about 0.061. That is, there is about a 6.1% chance of getting an observed distribution of colors this different or more different from the expected distribution of colors by chance alone, assuming that the distribution of colors in the bag is the same as the company claims.

As Figure 10.3 suggests, the sampling distribution of the chi-square test statistic is *not* a normal distribution. It is a right-skewed distribution that allows only values greater than or equal to 0 because χ^2 can never be negative. *When the expected counts are all at least 5,* the sampling distribution of the χ^2 statistic is modeled well by a **chi-square distribution** with degrees of freedom (df) equal to the number of categories minus 1. As with the *t* distributions, there is a different chi-square distribution for each possible value of df.

DEFINITION Chi-square distribution

A **chi-square distribution** is described by a density curve that takes only non-negative values and is skewed to the right. A particular chi-square distribution is specified by its degrees of freedom.

Figure 10.4 shows the density curves for three members of the chi-square family of distributions. As the degrees of freedom (df) increase, the density curves become less skewed, and larger values of χ^2 become more plausible.

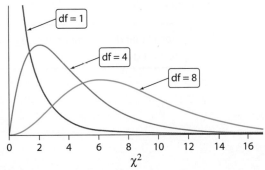

χ^2

FIGURE 10.4 The density curves for three members of the chi-square family of distributions.

To get *P*-values from a chi-square distribution, we can use technology or Table C in the back of the book. For Andre's class data, the conditions were met and $\chi^2 = 10.180$. Because there are 6 different categories for color, df $= 6 - 1 = 5$. The *P*-value is the probability of getting a value of χ^2 as large as or larger than 10.180 when H_0 is true. Figure 10.5 (on the next page) shows this probability as an area under the chi-square density curve with 5 degrees of freedom.

FIGURE 10.5 The *P*-value for a chi-square test for goodness of fit using Andre's M&M'S® Milk Chocolate Candies class data.

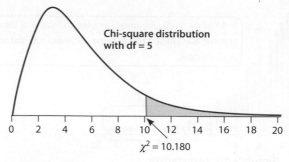

$\chi^2 = 10.180$

Tail probability p		
df	.15	.10	.05
4	6.74	7.78	9.49
5	8.12	9.24	11.07
6	9.45	10.64	12.59

- To find the *P*-value using Table C, look in the df = 5 row. The value $\chi^2 = 10.180$ falls between the critical values 9.24 and 11.07. Looking to the top of the corresponding columns, we find that the right tail area of the chi-square distribution with 5 degrees of freedom is between 0.10 and 0.05. So the *P*-value for a test based on Andre's data is between 0.05 and 0.10.

- On the TI-83/84, the command χ^2cdf(lower:10.180,upper:1000,df:5) gives a *P*-value of 0.070. Find this command in the DISTR menu.

For tests of goodness of fit, we always find the area under the chi-square distribution to the right of the χ^2 test statistic. This is because larger values of χ^2 indicate bigger differences between the observed and expected counts.

EXAMPLE

How unusual are those NHL birthdays?

Calculating P-values

PROBLEM: To see if birth date is related to whether a player makes it into the NHL, a random sample of 80 NHL players from a recent season was selected and their birthdays were recorded. The table shows the observed and expected counts for a test of the null hypothesis that the proportion of all NHL players born in each quarter is the same for all four quarters of the year. The chi-square test statistic for these data is $\chi^2 = 11.2$. Calculate the *P*-value.

Birthday	Jan–Mar	Apr–Jun	Jul–Sep	Oct–Dec
Observed counts	32	20	16	12
Expected counts	20	20	20	20

SOLUTION:

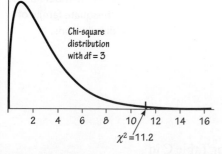

$\chi^2 = 11.2$

> Recall that df = number of categories − 1 for a chi-square test for goodness of fit.

Using Table C: df = 4 − 1 = 3 and the *P*-value is between 0.01 and 0.02.

Using technology: The command χ^2 cdf(lower:11.2,upper: 1000,df:3) gives a *P*-value of 0.011.

Tail probability p		
df	0.025	0.02	0.01
2	7.38	7.82	9.21
3	9.35	9.84	11.34
4	11.14	11.67	13.28

FOR PRACTICE TRY EXERCISE 5.

If the proportion of all NHL players born in each quarter is the same for all four quarters of the year, there is only a 0.011 probability of getting an observed distribution of birthdays this different or more different from the expected distribution by chance alone.

The Chi-Square Test for Goodness of Fit

Now that we know how to calculate *P*-values for a chi-square test for goodness of fit, we can use the four-step process to complete a test from start to finish.

EXAMPLE

Warmer climate, lighter car colors?

Chi-square tests for goodness of fit

PROBLEM: Does the warm, sunny weather in Arizona affect a driver's choice of car color? Cass thinks that Arizona drivers might opt for a lighter color with the hope that it will reflect some of the heat from the sun. To see if the distribution of car colors in Oro Valley, near Tucson, is different from the distribution of car colors across North America, she selected a random sample of 300 cars in Oro Valley. The table shows the distribution of car color for Cass's sample in Oro Valley and the distribution of car color in North America, according to www.ppg.com.[5] Do these data provide convincing evidence that the distribution of car color in Oro Valley differs from the North American distribution?

Color	White	Black	Gray	Silver	Red	Blue	Green	Other	Total
Oro Valley sample	84	38	31	46	27	29	6	39	300
North America	23%	18%	16%	15%	10%	9%	2%	7%	100%

SOLUTION:

STATE: H_0: The distribution of car color in Oro Valley is the same as the distribution of car color in North America.

H_a: The distribution of car color in Oro Valley is *not* the same as the distribution of car color in North America.

Use an $\alpha = 0.05$ significance level.

PLAN: Chi-square test for goodness of fit

Random? Cass randomly selected 300 cars in Oro Valley, AZ. ✓

Large Counts? All expected counts are ≥ 5 (see table). ✓

Color	White	Black	Gray	Silver	Red	Blue	Green	Other	Total
Expected count	69	54	48	45	30	27	6	21	300

DO:

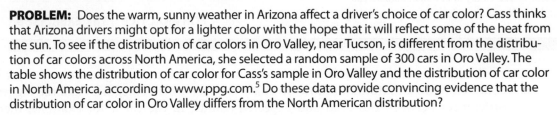

• $\chi^2 = \dfrac{(84 - 69)^2}{69} + \dfrac{(38 - 54)^2}{54}$

$+ \cdots + \dfrac{(39 - 21)^2}{21} = 29.921$

• *P*-value: df $= 8 - 1 = 7$

STATE: State the hypotheses you want to test and the significance level.

PLAN: Identify the appropriate inference method and check conditions.

To find the expected count for white, multiply the sample size by the expected proportion: 300(0.23) = 69. Do the same for the remaining colors.

DO: If the conditions are met, perform calculations:
• Calculate the chi-square test statistic.
• Find the *P*-value.

Because the observed counts differ from the expected counts, the sample gives some evidence that the distribution of car color is different in Oro Valley.

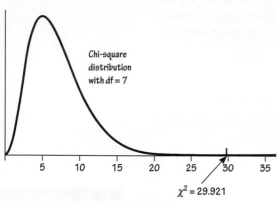

Chi-square
distribution
with df = 7

$\chi^2 = 29.921$

Using Table C: P-value < 0.0005

Using technology: The command χ^2 cdf(lower:29.921, upper:1000, df:7) gives a P-value of 0.0001.

CONCLUDE: Because the P-value of 0.0001 is less than $\alpha = 0.05$, we reject H_0. There is *convincing evidence that the distribution of car color in Oro Valley is not the same as the distribution of car color in North America.*

CONCLUDE: Make a conclusion about the hypotheses in the context of the problem.

FOR PRACTICE   TRY EXERCISE 9.

When the null hypothesis is rejected in a chi-square test for goodness of fit, it is common to do an informal follow-up analysis to see which categories had the biggest differences between observed and expected counts. In the car color example, there were 15 more white cars and 18 more "other" cars than expected. There were also 17 fewer gray cars and 16 fewer black cars than expected. This agrees with Cass's belief that drivers in Arizona prefer lighter-colored cars.

LESSON APP 10.2

Is this die fair?

Josephina made a 6-sided die in her ceramics class and rolled it 60 times to test if each side was equally likely to show up. The table summarizes the outcomes of her 60 rolls. Do these data provide convincing evidence that Josephina's die is unfair?

Outcome of roll	1	2	3	4	5	6	Total
Frequency	13	11	6	12	10	8	60

TECH CORNER

Chi-Square Test for Goodness of Fit

You can use technology to perform the calculations for a chi-square test for goodness of fit. We'll illustrate using the car color example on page 631.

Applet

1. Launch the *One Categorical Variable* applet at highschool.bfwpub.com/spa3e.

2. Enter "Color" as the variable name. Then, input each of the different color choices along with the observed frequency. Press the + button in the lower right to get more rows in the table.

One Categorical Variable

Variable name: Color

Groups: Single ▼

Input data as: Counts in categories ▼

	Category Name	Frequency	
1	White	84	-
2	Black	38	-
3	Grey	31	-
4	Silver	46	-
5	Red	27	-
6	Blue	29	-
7	Green	6	-
8	Other	39	-
			+

Begin analysis   Edit inputs   Reset everything

3. Press the "Begin analysis" button. Then, scroll down to the "Perform inference" section. Choose the Chi-square goodness-of-fit test for the procedure, enter the expected count for each category, and press the "Perform inference" button.

Perform Inference

Inference procedure: Chi-square goodness-of-fit test ▼

Enter expected counts for each category.

Category	Observed Frequency	Expected Frequency
White	84	69
Black	38	54
Grey	31	48
Silver	46	45
Red	27	30
Blue	29	27
Green	6	6
Other	39	21

Perform inference

x^2	P-value	df
29.921	<0.001	7

TI-84

Note: The chi-square test for goodness of fit isn't available on the TI-83 and some older models of TI-84s.

1. Enter the observed counts into list L_1 and the expected counts into list L_2.

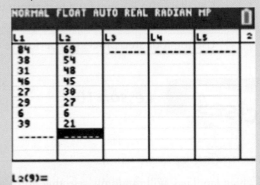

2. Press $\boxed{\text{STAT}}$, then choose TESTS and χ^2 GOF-Test.

3. When the χ^2 GOF-Test screen appears, enter L_1 for Observed, L_2 for Expected, and 7 for df.

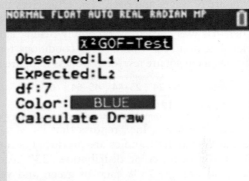

4. Highlight "Calculate" and press $\boxed{\text{ENTER}}$.

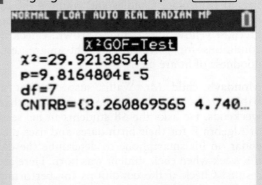

Note: If you select the "Draw" option, you will get a picture of the appropriate χ^2 distribution with the area of interest shaded and the test statistic and *P*-value labeled.

Lesson 10.2

WHAT DID YOU LEARN?

LEARNING TARGET	EXAMPLES	EXERCISES
Check conditions for a test about the distribution of a categorical variable.	p. 628	1–4
Calculate the P-value for a test about the distribution of a categorical variable.	p. 630	5–8
Use the four-step process to perform a chi-square test for goodness of fit.	p. 631	9–12

Exercises · Lesson 10.2

Mastering Concepts and Skills

1. Calling landlines Will randomly selecting households
pg 628 with landlines produce a representative sample? According to the 2010 census, of all U.S. residents aged 18 and older, 13% are 18–24 years old, 35% are 25–44 years old, 35% are 45–64 years old, and 17% are 65 years and over.[6] The table gives the age distribution for a sample of U.S. residents aged 18 and older that was chosen by randomly dialing landline telephone numbers. Check if the conditions for performing a chi-square test for goodness of fit are met.

Age	18–24	25–44	45–64	65+
Count	17	118	161	91

2. Peanut time Mars, Inc., reports that its M&M'S® Peanut Chocolate Candies are produced according to the following color distribution: 23% each of blue and orange, 15% each of green and yellow, and 12% each of red and brown. To test this claim, Eli bought a randomly selected bag of M&M'S Peanut Chocolate Candies and counted the colors of the candies in his sample: 10 blue, 6 orange, 9 green, 4 yellow, 7 red, and 2 brown. Check if the conditions for performing a chi-square test for goodness of fit are met.

3. Monday's child Mr. Wallis suspects that more babies are born in the middle of the week than on weekends. He asks the 68 students in his sections of Algebra II for their birth dates and uses the calendar on his smartphone to determine the day of the week when each student was born. Here are his results. Check if the conditions for performing a chi-square test for goodness of fit are met.

Day	Monday	Tuesday	Wednesday	Thursday
Count	13	10	9	11
Day	Friday	Saturday	Sunday	
Count	10	7	8	

4. Which lane? A statistically minded toll collector wonders if drivers are equally likely to choose each of the 3 lanes at his toll booth. During one of his work shifts, he selects a random sample from all the cars that approach the booth when all 3 lanes are empty so that the driver's choice isn't influenced by the cars already at the booth. Here are his results. Check if the conditions for performing a chi-square test for goodness of fit are met.

Lane	Left	Center	Right
Number of cars	17	27	31

5. Finding P-values Calculate the P-value in each of
pg 630 the following settings.

(a) $\chi^2 = 19.03$, df $= 11$ (b) $\chi^2 = 19.03$, df $= 3$

6. More P-values Calculate the P-value in each of the following settings.

(a) $\chi^2 = 4.49$, df $= 5$ (b) $\chi^2 = 4.49$, df $= 1$

7. More landlines Calculate the χ^2 test statistic and P-value using the data in Exercise 1.

8. Tollbooth lanes Calculate the χ^2 test statistic and P-value using the data in Exercise 4.

9. Wrinkled or smooth? Gregor Mendel (1822–1884),
pg 631 an Austrian monk, is considered the father of genetics. Mendel studied the inheritance of various traits in pea plants. One such trait is if the pea is smooth or wrinkled. Mendel predicted a ratio of 3 smooth peas for every 1 wrinkled pea. In one experiment, he observed 423 smooth and 133 wrinkled peas. Assume that the random condition is met. Carry out an appropriate test of the genetic model that Mendel predicted.

10. Tall tomatoes The paper "Linkage Studies of the Tomato" (*Transactions of the Canadian Institute*, 1931) reported the following data about the characteristics resulting from crossing tall cut-leaf tomatoes with dwarf potato-leaf tomatoes.

Characteristics	Tall cut-leaf	Tall potato-leaf	Dwarf cut-leaf	Dwarf potato-leaf
Frequency	926	288	293	104

We wish to investigate if these frequencies are consistent with genetic laws, which state that the characteristics of height and leaf shape should occur in the ratio 9:3:3:1. Assume that the random condition is met. Carry out an appropriate test of the proposed genetic model.

11. **Personality types?** In some countries, people believe that blood type has a strong impact on personality. For example, type B blood is thought to be associated with passion and creativity. A statistics student at a large U.S. university decides to test this hypothesis. Reasoning that people involved in the arts should be passionate and creative, she takes a random sample of students majoring or minoring in the arts at her university and asks them for their blood type. Here are her results.

Blood type	Type A	Type B	Type AB	Type O
Count	50	23	10	67

The distribution of blood type among all U.S. residents is as follows: type A: 40%; type B: 11%; type AB: 4%; type O: 45%. Do the data provide convincing evidence that the blood type distribution among performing arts majors at this university differs from the national distribution?

12. **Hazardous grading** An ambitious reporter for a university's newspaper suspects that Mr. Hazzard, a new statistics teacher, is grading his introductory statistics students too harshly. From school records, the reporter determines that over the past 2 years, students in *all* sections of introductory statistics (other than sections taught by Mr. Hazzard) received grades of A, B, C, D, or F in the following proportions: A: 0.20; B: 0.30; C: 0.30; D: 0.10; and F: 0.10. The reporter then takes a random sample of 50 students who completed introductory statistics with Mr. Hazzard over the past 2 years and gathers the following information.

Grade	A	B	C	D	F
Number of students	7	16	14	8	5

Is there convincing evidence to conclude that Mr. Hazzard's grade distribution is different from that of other teachers of introductory statistics?

Applying the Concepts

13. **Don't break the law** Faked numbers in tax returns, invoices, or expense account claims often display patterns that aren't present in legitimate records. Some patterns are obvious and easily avoided by a clever crook. Others are more subtle. It is a striking fact that the first digits of numbers in legitimate records often follow a model known as Benford's law.[7] Call the first digit of a randomly chosen record X for short. Benford's law gives this probability model for X (note that a first digit can't be 0).

First digit	1	2	3	4	5
Probability	0.301	0.176	0.125	0.097	0.079
First digit	6	7	8	9	
Probability	0.067	0.058	0.051	0.046	

A forensic accountant who is familiar with Benford's law inspects a random sample of 250 invoices from a company that is accused of committing fraud. The table displays the sample data. Is there convincing evidence that the invoices from this company don't follow Benford's law?

First digit	1	2	3	4	5	6	7	8	9
Count	61	50	43	34	25	16	7	8	6

14. **Housing and ethnicity** According to the U.S. Census Bureau, the distribution by ethnic background of the New York City population in a recent year was as follows: Hispanic: 28%; Black: 24%; White: 35%; Asian: 12%; Others: 1%. The manager of a large housing complex in the city wonders if the distribution by ethnicity of the complex's residents is consistent with the population distribution. To find out, she records data from a random sample of 800 residents.[8] The table displays the sample data. Is there convincing evidence that the ethnic distribution in the housing complex is different from the ethnic distribution in New York City?

Ethnicity	Hispanic	Black	White	Asian	Other
Count	212	202	270	94	22

15. **Benford's law** Refer to Exercise 13.

(a) Describe a Type I error and a Type II error in this setting, and give a possible consequence of each. Which do you think is more serious?

(b) For which first digits were the observed counts much greater than expected? For which first digits were the observed counts much less than expected? Justify your answer.

16. **More on NYC housing** Refer to Exercise 14.

(a) Describe a Type I error and a Type II error in this setting, and give a possible consequence of each. Which do you think is more serious?

(b) For which ethnic backgrounds were the observed counts much greater than expected? For which ethnic backgrounds were the observed counts much less than expected? Justify your answer.

Extending the Concepts

17. **Homework time** A school's principal wants to know if students spend about the same amount of time on homework each night of the week. She asks a random sample of 50 students to keep track of their homework time for a week. The following table displays the average amount of time (in minutes) students reported per night. Explain carefully why it would not be appropriate to perform a chi-square test for goodness of fit using these data.

Night	Sunday	Monday	Tuesday	Wednesday
Average time	130	108	115	104
Night	Thursday	Friday	Saturday	
Average time	99	37	62	

		Soccer experience			
		Elite	Non-elite	Did not play	Total
Arthritis	Yes	10	9	24	43
	No	61	206	548	815
	Total	71	215	572	858

18. Is it really random? Use your calculator's RandInt function to generate 200 digits from 0 to 9 and store them in a list. Based on this sample of random digits, is there convincing evidence that each digit doesn't have an equal chance to be generated?

Recycle and Review

19. Risks of playing soccer (4.3, 4.4) A study in Sweden looked at former elite soccer players, people who had played soccer but not at the elite level, and people of the same age who did not play soccer. Here is a two-way table that classifies these individuals by whether or not they had arthritis of the hip or knee by their mid-50s.[9]

Suppose we choose one of these players at random.

(a) What is the probability that the player has arthritis?

(b) What is the probability that the player has arthritis, given that he or she was classified as an elite soccer player?

20. More risks of playing soccer (2.1) Refer to Exercise 19. We suspect that the more serious soccer players have more arthritis later in life.

(a) Make a graph to show the relationship between soccer experience and arthritis for these players.

(b) Based on your graph, is there an association between soccer experience and arthritis for these players? Explain.

Lesson 10.3

Testing the Relationship Between Two Categorical Variables

LEARNING TARGETS

● State hypotheses for a test about the relationship between two categorical variables.

● Calculate expected counts for a test about the relationship between two categorical variables.

● Calculate the test statistic for a test about the relationship between two categorical variables.

In Lesson 2.1, we considered the association between two categorical variables, such as gender and superpower preference for a random sample of 200 students from the United Kingdom. Because you hadn't learned about inference yet, we could only talk about the association *in that sample*. In this lesson, you will learn how to state hypotheses, calculate expected counts, and calculate the test statistic for a test of association between two categorical variables *in a population*.

Stating Hypotheses

In Chapter 2, we explored relationships between two variables to determine if the variables have an **association.**

DEFINITION Association

There is an **association** between two variables if knowing the value of one variable helps us predict the value of the other.

Is there an association between gender and handedness for Australian high school students? To investigate, we selected a random sample of 100 Australian high school students. The two-way table summarizes the relationship between gender and handedness for these students.

		Gender		
		Male	Female	Total
Dominant hand	Right	39	51	90
	Left	7	3	10
	Total	46	54	100

In Lesson 4.4, we analyzed these data by calculating and displaying conditional probabilities. When selecting a student at random from this sample, P(left-handed | male) $= 7/46 = 0.152$ and P(left-handed | female) $= 3/54 = 0.056$. These probabilities are displayed in Figure 10.6 with the red segments in the segmented bar chart.

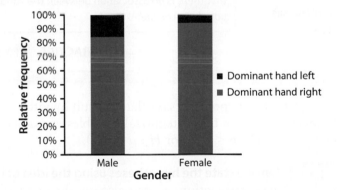

FIGURE 10.6 Segmented bar chart showing the relationship between gender and handedness for a sample of 100 Australian high school students.

Because these probabilities differ, we concluded that the events "male" and "left-handed" are not independent. That is, we concluded that there is an association between gender and handedness *for the members of the sample.*

Does the association in the sample provide convincing evidence that there is an association between gender and handedness for *all* Australian high school students? Or is it plausible that there really is no association between these variables in the population and that the differences observed in the sample were due to sampling variability alone? We use a significance test to find out.

Like other significance tests, we begin a test for the relationship between two categorical variables by stating hypotheses. The null hypothesis for this test states that there is no association between the variables in the population. The alternative hypothesis states that there is an association. Here are the hypotheses for the gender and handedness example:

H_0: There is no association between gender and handedness in the population of Australian high school students.

H_a: There is an association between gender and handedness in the population of Australian high school students.

EXAMPLE

Do angry people have a greater risk of heart disease?

Stating hypotheses

PROBLEM: A study followed a random sample of 8474 people with normal blood pressure for about 4 years. All the individuals were free of heart disease at the beginning of the study. Each person took the Spielberger Trait Anger Scale test, which measures how prone a person is to sudden anger. Researchers also recorded if each individual developed coronary heart disease (CHD). This includes people who had heart attacks and those who needed medical treatment for heart disease. Here is a two-way table that summarizes the data.[10]

© Ryan Jorgensen/Alamy Stock Photo

		Anger level			
		Low	Moderate	High	Total
CHD status	Yes	53	110	27	190
	No	3057	4621	606	8284
	Total	3110	4731	633	8474

State hypotheses for a test about the relationship between anger level and CHD status in the population of people with normal blood pressure.

SOLUTION:

H_0: There is no association between anger level and CHD status in the population of people with normal blood pressure.

H_a: There is an association between anger level and CHD status in the population of people with normal blood pressure.

> The null hypothesis for this type of test always states that there is *no* association between the variables in the *population of interest*.

FOR PRACTICE **TRY EXERCISE 1.**

Notice that the alternative hypothesis says that the null hypothesis is *not* true. For a test of the relationship between two categorical variables, we don't have one-sided alternative hypotheses like H_a: $p < 0.50$ or H_a: $\mu > 100$.

THINK ABOUT IT **Can we state the hypotheses using the idea of independence?** Yes. If two variables have no association, we can also say the two variables are independent. Using the anger and heart disease example, we can state the hypotheses as

H_0: Anger level and CHD status are independent in the population of people with normal blood pressure.

H_a: Anger level and CHD status are not independent in the population of people with normal blood pressure.

Note that the alternative hypothesis says that the null hypothesis is not true, just like when we stated the hypotheses using the idea of association.

Calculating Expected Counts

How do we determine if an association between two variables in a sample provides convincing evidence of an association in the population? We measure how different the observed association is from what we expect if there was no association between the variables in the population. If there is no association between gender and handedness, then the proportion of males who are left-handed should be the same as the proportion of females who are left-handed. Because 10% (10/100) of the students in the sample were left-handed, 10% of the males should be left-handed and 10% of the females should be left-handed.

Gender

		Male	Female	Total
Dominant hand	Right			90
	Left			10
	Total	46	54	100

That is, under the null hypothesis of no association, the expected number of left-handed males is 10% of the 46 males in the sample:

$$\frac{10}{100} \cdot 46 = 4.6$$

Don't worry that the expected count isn't an integer—we aren't saying that a sample can have 4.6 left-handed males! The expected count tells us that if many samples are selected from a population where 10% are left-handed, 46% are male, and there is no association between gender and handedness, we would expect an average of 4.6 left-handed males in samples of size 100.

Notice that the calculation can be rewritten as

$$\frac{10}{100} \cdot 46 = \frac{10 \cdot 46}{100}$$

This is the total number of lefties multiplied by the total number of males, divided by the total number in the sample. This leads to the general rule for computing expected counts in a two-way table.

How to Calculate Expected Counts

For a test of the relationship between two categorical variables, the expected count for a cell in the two-way table is

$$\text{expected count} = \frac{\text{row total} \times \text{column total}}{\text{table total}}$$

You can complete the table of expected counts by using the formula for each cell or by using the formula for some cells and subtracting to find the remaining cells. For example, if the expected count of left-handed males is 4.6, we know the expected count of right-handed males is $46 - 4.6 = 41.4$. Here is the completed table of expected counts.

Gender

		Male	Female	Total
Dominant hand	Right	41.4	48.6	90
	Left	4.6	5.4	10
	Total	46	54	100

EXAMPLE

Can angry people expect heart disease?

Calculating expected counts

PROBLEM: Here are the data from the study of anger and heart disease.

Calculate the expected counts for a test of the null hypothesis that there is no association between anger level and CHD status in the population of people with normal blood pressure.

		Anger level			
		Low	Moderate	High	Total
CHD status	Yes	53	110	27	190
	No	3057	4621	606	8284
	Total	3110	4731	633	8474

SOLUTION:

		Anger level			
		Low	Moderate	High	Total
CHD status	Yes	$\frac{190 \cdot 3110}{8474}$ = 69.73	$\frac{190 \cdot 4731}{8474}$ = 106.08	$190 - 69.73 - 106.08$ = 14.19	190
	No	$3110 - 69.73$ = 3040.27	$4731 - 106.08$ = 4624.92	$633 - 14.19$ = 618.81	8284
	Total	3110	4731	633	8474

> To find the expected count for any cell, use the formula
> $$\text{expected count} = \frac{\text{row total} \times \text{column total}}{\text{table total}}$$

> You can use subtraction to calculate the expected counts for the cells in the last row and last column.

FOR PRACTICE   TRY EXERCISE 5.

The Chi-Square Test Statistic

For a test of the relationship between two categorical variables, we use the familiar chi-square test statistic:

$$\chi^2 = \sum \frac{(\text{observed count} - \text{expected count})^2}{\text{expected count}}$$

This time, the sum is over all cells in the two-way table, not including the totals. Here are the observed and expected counts for the gender and handedness example.

Observed counts

		Gender		
		Male	Female	Total
Dominant hand	Right	39	51	90
	Left	7	3	10
	Total	46	54	100

Expected counts

		Gender		
		Male	Female	Total
Dominant hand	Right	41.4	48.6	90
	Left	4.6	5.4	10
	Total	46	54	100

The chi-square test statistic is

$$\chi^2 = \frac{(39 - 41.4)^2}{41.4} + \frac{(7 - 4.6)^2}{4.6} + \frac{(51 - 48.6)^2}{48.6} + \frac{(3 - 5.4)^2}{5.4} = 2.58$$

EXAMPLE

How strong is the association between anger and heart disease?

Calculating χ^2

PROBLEM: Here are the observed and expected counts from the study of anger and heart disease. Calculate the value of the χ^2 test statistic.

Observed counts

		Low	Moderate	High	Total
CHD status	Yes	53	110	27	190
	No	3057	4621	606	8284
	Total	3110	4731	633	8474

Anger level

Expected counts

		Low	Moderate	High	Total
CHD status	Yes	69.73	106.08	14.19	190
	No	3040.27	4624.92	618.81	8284
	Total	3110	4731	633	8474

Anger level

SOLUTION:

$$\chi^2 = \frac{(53 - 69.73)^2}{69.73} + \frac{(110 - 106.08)^2}{106.08} + \frac{(27 - 14.19)^2}{14.19}$$

$$+ \frac{(3057 - 3040.27)^2}{3040.27} + \frac{(4621 - 4624.92)^2}{4624.92} + \frac{(606 - 618.81)^2}{618.81}$$

$$\chi^2 = 16.08$$

> $$\chi^2 = \sum \frac{(\text{observed count} - \text{expected count})^2}{\text{expected count}}$$

FOR PRACTICE   TRY EXERCISE 9.

How unusual is it to get a χ^2 test statistic as large as or larger than 16.08 just by chance? We will learn how to use a chi-square distribution to estimate the *P*-value in the next lesson. For now, we can estimate the *P*-value with a simulation. The dotplot in Figure 10.7 shows the χ^2 test statistic for 100 simulated random samples of 8474 people with normal blood pressure, assuming 36.7% have low anger, 55.8% have moderate anger, 7.5% have high anger, 2.2% have CHD, and there is no association between anger level and CHD status.

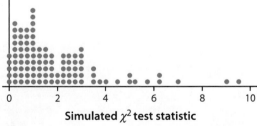

Simulated χ^2 test statistic

FIGURE 10.7 The χ^2 test statistic from each of 100 simulated random samples of size 8474 taken from a specific population where there is no association between anger and CHD status.

Because the observed χ^2 test statistic of 16.08 is greater than any of the simulated χ^2 test statistics, the *P*-value is approximately 0. There is almost no chance of getting an association between anger level and CHD status as strong as or stronger than the one observed in the sample by chance alone when there is no association between these variables in the population. This sample provides convincing evidence of an association between anger level and CHD status in the population of people with normal blood pressure. Of course, this was only an observational study, so it would be incorrect to conclude that anger causes heart disease based on this study. Remember that association doesn't imply causation!

CAUTION

LESSON APP 10.3

Is there an association between gender and superpower preference?

The two-way table shows the gender and superpower preference for a random sample of 200 children ages 9–17 from the United Kingdom.

		Gender		
		Female	Male	Total
Superpower preference	Invisibility	17	13	30
	Super strength	3	17	20
	Telepathy	39	5	44
	Fly	36	18	54
	Freeze time	20	32	52
	Total	115	85	200

1. Make a graph that displays the relationship between gender and superpower preference for the members of the sample. Briefly describe the association.

2. State the hypotheses for a test of the relationship between gender and superpower preference in the population of all children ages 9–17 in the United Kingdom.

3. Calculate the expected counts for a test of the hypotheses stated in Question 2.

4. Calculate the value of the chi-square test statistic.

Jennifer Hardt, Sweethardt Photography/ Moment/Getty Images

Lesson 10.3

WHAT DID YOU LEARN?

LEARNING TARGET	EXAMPLES	EXERCISES
State hypotheses for a test about the relationship between two categorical variables.	p. 638	1–4
Calculate expected counts for a test about the relationship between two categorical variables.	p. 640	5–8
Calculate the test statistic for a test about the relationship between two categorical variables.	p. 641	9–12

Exercises

Lesson 10.3

Mastering Concepts and Skills

1. **Cell phones** The Pew Research Center asked a random sample of 2024 adult cell-phone owners from the United States to identify their age and which type of cell phone they own: iPhone, Android, or other (including non-smartphones). The two-way table displays the results.[11] State the hypotheses for a test of the relationship between age and cell-phone type in the population of U.S. cell-phone owners.

pg 638

		Age			
		18–34	35–54	55+	Total
Type of cell phone	iPhone	169	171	127	467
	Android	214	189	100	503
	Other	134	277	643	1054
	Total	517	637	870	2024

2. **Colorful bears** Courtney and Lexi wondered if the distribution of color was the same for name-brand

gummy bears (Haribo Gold) and store-brand gummy bears (Great Value). To investigate, they randomly selected 6 bags of each type and counted the number of gummy bears of each color. Here are the data.[12] State the hypotheses for a test of the relationship between brand and color for gummy bears.

		Brand		
		Name	Store	Total
Color	Red	137	212	349
	Green	53	104	157
	Yellow	50	85	135
	Orange	81	127	208
	White	52	94	146
	Total	373	622	995

3. **Getting rich** A survey of 4826 randomly selected young adults (aged 19 to 25) asked, "What do you think the chances are you will have much more than a middle-class income at age 30?" The table shows the responses.[13] State the hypotheses for a test of the relationship between gender and opinion about getting rich for young adults.

		Gender		
		Female	Male	Total
Opinion	Almost no chance	96	98	194
	Some chance but probably not	426	286	712
	A 50-50 chance	696	720	1416
	A good chance	663	758	1421
	Almost certain	486	597	1083
	Total	2367	2459	4826

4. **Finger length** Is there a relationship between gender and relative finger length? To investigate, a random sample of 452 U.S. high school students was selected. The two-way table shows the gender of each student and which finger was longer on his or her left hand (index finger or ring finger).[14] State the hypotheses for a test of the relationship between gender and relative finger length for U.S. high school students.

		Gender		
		Female	Male	Total
Longer finger	Index finger	78	45	123
	Ring finger	82	152	234
	Same length	52	43	95
	Total	212	240	452

5. **Expected cell phones** Calculate the expected counts for a test of the hypotheses stated in Exercise 1. pg 640

6. **Expected gummy bears** Calculate the expected counts for a test of the hypotheses stated in Exercise 2.

7. **Expected opinions about getting rich** Calculate the expected counts for a test of the hypotheses stated in Exercise 3.

8. **Expected finger length** Calculate the expected counts for a test of the hypotheses stated in Exercise 4.

9. **Strong cell phones?** Refer to Exercises 1 and 5. Calculate the value of the chi-square test statistic. pg 641

10. **Strong gummy bears?** Refer to Exercises 2 and 6. Calculate the value of the chi-square test statistic.

11. **Strong opinions?** Refer to Exercises 3 and 7. Calculate the value of the chi-square test statistic.

12. **Strong fingers?** Refer to Exercises 4 and 8. Calculate the value of the chi-square test statistic.

Applying the Concepts

13. **Gender and housing** Using data from the 2000 census, a random sample of 348 U.S. residents aged 18 and older was selected. The two-way table summarizes the relationship between gender and housing status for these residents.

		Gender		
		Male	Female	Total
Housing status	Own	132	122	254
	Rent	50	44	94
	Total	182	166	348

(a) Make a graph that displays the relationship between gender and housing status for the members of the sample. Briefly describe the association.

(b) State the hypotheses for a test of the relationship between gender and housing status for U.S. residents.

(c) Calculate the expected counts for a test of the hypotheses stated in part (b).

(d) Calculate the value of the chi-square test statistic.

14. **Marriage and housing** Using data from the 2000 census, a random sample of 348 U.S. residents aged 18 and older was selected. The two-way table summarizes the relationship between marital status and housing status for these residents.

		Marital status		
		Married	Not married	Total
Housing status	Own	172	82	254
	Rent	40	54	94
	Total	212	136	348

(a) Make a graph that displays the relationship between marital status and housing status for the members of the sample. Briefly describe the association.

(b) State the hypotheses for a test of the relationship between marital status and housing status for U.S. residents.

(c) Calculate the expected counts for a test of the hypotheses stated in part (b).

(d) Calculate the value of the chi-square test statistic.

Extending the Concepts

15. **Which is stronger?** Based on your answers to Exercises 13 and 14, which association is stronger: gender and housing status or marital status and housing status? Explain.

16. **No association?** In Exercise 14(a), you constructed a graph to display the relationship between marital status and housing status for the members of the sample. Assuming the overall proportion of homeowners remains the same, construct a new graph that shows no association between marital status and housing status.

17. **Simulating handedness** In this lesson, we analyzed the relationship between gender and handedness for a sample of 100 Australian high school students. The chi-square test statistic for these data was $\chi^2 = 2.58$. The dotplot shows the results of a simulation in which 200 random samples of size 100 were selected from a population of students where there is no association between gender and handedness. For each sample of 100 students, the chi-square test statistic was calculated. Based on the χ^2 test statistic and the results of the simulation, what conclusion would you draw?

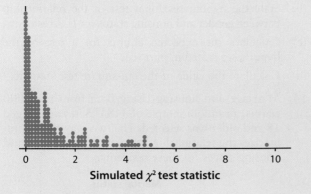

Simulated χ^2 test statistic

Recycle and Review

18. **Reading and grades** (1.8, 8.2, 9.4) Do students who read more books for pleasure tend to earn higher grades in English? The boxplots show data from a simple random sample of 79 students at a large high school. Students were classified as light readers if they read fewer than 3 books for pleasure per year. Otherwise, they were classified as heavy readers. Each student's average English grade for the previous two marking periods was converted to a GPA scale, where A = 4.0, A– = 3.7, B + = 3.3, and so on.

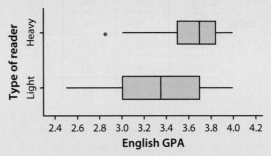

English GPA

(a) Write a few sentences comparing the distributions of English grades for light and heavy readers.

Here are summary statistics for the two groups.

Type of Reader	n	Mean	SD
Heavy	47	3.640	0.324
Light	32	3.356	0.380

(b) Explain why it is acceptable to use two-sample t procedures in this setting.

(c) Do the data provide convincing evidence that the mean English grade is greater for heavy readers than for light readers at this high school? Justify your answer.

(d) Based on your conclusion in part (c), which kind of error might you have made, Type I or Type II? Describe that error in context.

19. **Ravenous caterpillars** (7.2, 7.4) The gypsy moth is a non-native species whose caterpillar stage does considerable damage to hardwoods in the eastern United States. A forester takes a random sample of 200 trees in a certain forest and finds that 124 of them show damage from gypsy moth caterpillars.

(a) Construct and interpret a 99% confidence interval for the proportion of trees in this forest that show damage from gypsy moth caterpillars.

(b) Explain what 99% confidence means in the context of this problem.

Lesson 10.4

Chi-Square Tests for Association

- Check conditions for a test about the relationship between two categorical variables.
- Calculate the *P*-value for a test about the relationship between two categorical variables.
- Use the four-step process to perform a chi-square test for association.

In Lesson 10.3, you learned how to state hypotheses, calculate expected counts, and calculate the chi-square test statistic for a test of the relationship between two categorical variables. In this lesson, you will learn how to check the conditions, calculate the *P*-value, and perform the four-step process for this test, called a *chi-square test for association*.

Conditions for a Chi-Square Test for Association

Like the chi-square test for goodness of fit, there are two conditions to be satisfied in a chi-square test for association.

How to Check the Conditions for a Chi-Square Test for Association

- **Random:** The data come from a random sample from the population of interest, from separate random samples from multiple populations, or from multiple groups in a randomized experiment.
- **Large Counts:** All expected counts are at least 5.

In the anger and heart disease example from the previous lesson, the conditions are satisfied:

- Random: The study used a random sample of 8474 people with normal blood pressure. ✓
- Large Counts: The expected counts shown in the table are all at least 5. ✓

		Anger level			
		Low	Moderate	High	Total
CHD status	Yes	69.73	106.08	14.19	190
	No	3040.27	4624.92	618.81	8284
	Total	3110	4731	633	8474

However, the conditions are *not* satisfied in the gender and handedness example. Even though the data came from a random sample, one of the expected counts shown in the table (4.6) is less than 5.

		Gender		
		Male	Female	Total
Dominant hand	Right	41.4	48.6	90
	Left	4.6	5.4	10
	Total	46	54	100

EXAMPLE

Has technology changed the distribution of birthdays?

Checking conditions

PROBLEM: With many babies being delivered by planned cesarean section, Mrs. McDonald's statistics class hypothesized that there is an association between the age of a person and the day of the week that person was born. After all, why would a doctor plan a C-section for the weekend? To investigate, they selected a random sample of people born before 1980 and a separate random sample of people born after 1993. In addition to year of birth, they also recorded the day of the week on which each person was born. The results are shown in the table.[15] Check the conditions for performing a chi-square test for association.

		Year of birth		
		Before 1980	After 1993	Total
Day of birth	Sunday	12	9	21
	Monday	12	11	23
	Tuesday	14	11	25
	Wednesday	12	10	22
	Thursday	7	17	24
	Friday	9	9	18
	Saturday	11	6	17
	Total	77	73	150

SOLUTION:

- Random: The class selected separate random samples of people born before 1980 and people born after 1993. ✓
- Large Counts: The expected counts are all at least 5 (see table). ✓

To find the expected counts, use the formula
$$\text{expected count} = \frac{\text{row total} \times \text{column total}}{\text{table total}}$$

		Year of birth		
		Before 1980	After 1993	Total
Day of birth	Sunday	10.78	10.22	21
	Monday	11.81	11.19	23
	Tuesday	12.83	12.17	25
	Wednesday	11.29	10.71	22
	Thursday	12.32	11.68	24
	Friday	9.24	8.76	18
	Saturday	8.73	8.27	17
	Total	77	73	150

FOR PRACTICE TRY EXERCISE 1.

Calculating *P*-Values

When the conditions are met, we use a chi-square distribution to calculate the *P*-value for a chi-square test for association. For the chi-square test for goodness of fit, we used df = number of categories − 1. This method for calculating degrees of freedom is correct when there is only one sample and one variable. Because there are two variables in a test for association, we use a different rule for calculating the degrees of freedom.

How to Calculate the Degrees of Freedom in a Test for Association

To calculate the *P*-value in a chi-square test for association, use the chi-square distribution with

$$df = (\text{number of rows} - 1)(\text{number of columns} - 1)$$

where the number of rows and number of columns do not include the totals.

Here is the two-way table for the anger and heart disease example.

<table>
<tr><td colspan="2"></td><td colspan="4" align="center">Anger level</td></tr>
<tr><td colspan="2"></td><td>Low</td><td>Moderate</td><td>High</td><td>Total</td></tr>
<tr><td>CHD</td><td>Yes</td><td>53</td><td>110</td><td>27</td><td>190</td></tr>
<tr><td>status</td><td>No</td><td>3057</td><td>4621</td><td>606</td><td>8284</td></tr>
<tr><td></td><td>Total</td><td>3110</td><td>4731</td><td>633</td><td>8474</td></tr>
</table>

There are 2 rows and 3 columns in this two-way table (not counting the totals). Thus,

$$df = (2 - 1)(3 - 1) = 2$$

To calculate the *P*-value, use Table C or technology as in Lesson 10.2. Recall that $\chi^2 = 16.08$.

df	0.0025	0.001	0.0005
1	9.14	10.83	12.12
2	11.98	13.82	15.20
3	14.32	16.27	17.73

Table heading: **Tail probability p**

- Using Table C, the observed chi-square test statistic of 16.08 is greater than all the critical values in the row for df = 2, so the *P*-value is less than 0.0005.
- Using the TI-83/84, the command χ^2 cdf(lower:16.08,upper:1000,df:2) gives a *P*-value of 0.0003. Find this command in the DISTR menu.

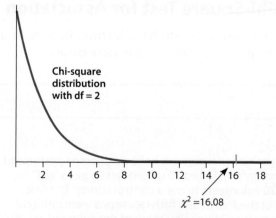

Chi-square distribution with df = 2

$\chi^2 = 16.08$

Because the *P*-value is very small, there is convincing evidence of an association between anger and CHD status in the population of people with normal blood pressure. This matches our conclusion from the end of Lesson 10.3, where we estimated the *P*-value using simulation.

EXAMPLE

Who was born on the weekend?

Calculating P-values

PROBLEM: In the previous example, you checked the conditions for a test of the association between age and day of birth. The chi-square test statistic for this test is $\chi^2 = 6.55$. Calculate the *P*-value.

SOLUTION:
$df = (7 - 1)(2 - 1) = 6$

> $df = $ (number of rows $-$ 1)(number of columns $-$ 1).

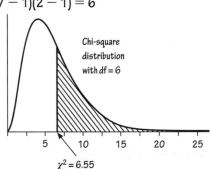

Chi-square distribution with $df = 6$

$\chi^2 = 6.55$

Using Table C: The *P*-value is greater than 0.25.

Using technology: The command χ^2 cdf(lower:6.55,upper:1000, df:6) gives a *P*-value of 0.3645.

	Tail probability *p*		
df	0.25	0.20	0.15
5	6.63	7.29	8.12
6	7.84	8.56	9.45
7	9.04	9.80	10.75

FOR PRACTICE **TRY EXERCISE 5.**

Because the *P*-value is large, Mrs. McDonald's class doesn't have convincing evidence of an association between the age of a person and the day of the week on which that person was born. Of course, it is possible that there was a Type II error—failing to find convincing evidence of an association between age and day of birth when there really is an association in the population.

Performing a Chi-Square Test for Association

As with the other significance tests you have learned, be sure to follow the four-step process when performing a chi-square test for association.

EXAMPLE

Can a survey influence behavior?

Chi-square tests for association

PROBLEM: To investigate subconscious ways to influence a person's behavior, Lorena performed a clever experiment. Using 60 volunteers, she randomly assigned 20 volunteers to get a "red" survey, 20 volunteers to get a "blue" survey, and 20 volunteers to get a control survey. The first three questions on each survey were the same, but the fourth and fifth questions were different. For example, the fourth question on the "red" survey was "When you think of the color red, what do you think about?" On the blue survey, the question replaced the word "red" with "blue." On the control survey, the questions were not about color. As a reward, Lorena let each volunteer choose a chocolate candy in a red wrapper or a chocolate candy in a blue wrapper. Here are the results.

	Survey type			
	Red	Blue	Control	Total
Color of candy Red	13	5	8	26
Blue	7	15	12	34
Total	20	20	20	60

Do these data provide convincing evidence at the $\alpha = 0.05$ level that there is an association between survey type and choice of candy color for subjects like these?

SOLUTION:

STATE: H_0: There is no association between survey type and choice of candy color for subjects like these.

H_a: There is an association between survey type and choice of candy color for subjects like these.

Use an $\alpha = 0.05$ significance level.

PLAN: Chi-square test for association

Random? The three treatments were randomly assigned to subjects. ✓

Large Counts? All expected counts are ≥ 5 (see table). ✓

	Survey type			
	Red	Blue	Control	Total
Color of candy Red	8.67	8.67	8.67	26
Blue	11.33	11.33	11.33	34
Total	20	20	20	60

DO:

- $\chi^2 = \dfrac{(13 - 8.67)^2}{8.67} + \dfrac{(5 - 8.67)^2}{8.67} + \cdots + \dfrac{(12 - 11.33)^2}{11.33} = 6.65$

- P-value: df $= (2 - 1)(3 - 1) = 2$

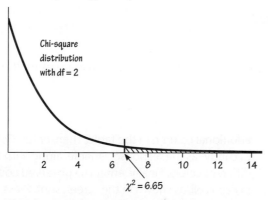

Chi-square distribution with df = 2

$\chi^2 = 6.65$

Using Table C: $0.025 < P\text{-value} < 0.05$
Using technology: See the Tech Corner at the end of the lesson.

CONCLUDE: Because the P-value of between 0.025 and 0.05 is less than $\alpha = 0.05$, we reject H_0. There is convincing evidence of an association between survey type and choice of candy color for subjects like these.

STATE: State the hypotheses you want to test and the significance level.

PLAN: Identify the appropriate inference method and check conditions.

To find the expected counts, use the formula
$$\text{expected count} = \frac{\text{row total} \times \text{column total}}{\text{table total}}$$

DO: If the conditions are met, perform calculations:
- Calculate the chi-square test statistic.
- Find the P-value.

Because the observed counts differ from the expected counts, the sample gives some evidence of an association between survey type and choice of candy color for subjects like these.

CONCLUDE: Make a conclusion about the hypotheses in the context of the problem.

FOR PRACTICE **TRY EXERCISE 9.**

Lorena can only make an inference about people like the ones in the study because she didn't randomly select her subjects. However, because Lorena performed an experiment with randomly assigned treatments, she can infer a cause-and-effect relationship between the survey type and the choice of candy color. Furthermore, the association seen in her study matched her expectations: People who were assigned the red survey chose red candy more often than expected ($13 > 8.67$), and those who were assigned the blue survey chose the blue candy more often than expected ($15 > 11.33$).

LESSON APP 10.4

Should angry people go to the sauna?

Researchers followed a random sample of 2315 middle-aged men from eastern Finland for up to 30 years. They recorded how often each man went to a sauna and whether or not he suffered sudden cardiac death (SCD). The two-way table shows the data from the study.[16]

Sandra Kemppainen/Hemera/
Getty Images

	Weekly sauna frequency			
	1 or fewer	2–3	4 or more	Total
SCD Yes	61	119	10	190
No	540	1394	191	2125
Total	601	1513	201	2315

1. Make a graph that displays the relationship between sauna frequency and SCD for the members of the sample. Briefly describe the association.

2. Perform a test at the $\alpha = 0.05$ level to determine if there is convincing evidence of an association between weekly sauna frequency and sudden cardiac death for middle-aged men from eastern Finland.

3. Based on the results of this study, can you conclude that frequent use of a sauna reduces the risk of sudden cardiac death for middle-aged men from eastern Finland? Explain.

Chi-Square Test for Association

You can use technology to perform the calculations for a chi-square test for association. We'll illustrate using the candy color example on page 649.

Applet

1. Launch the *Two Categorical Variables* applet at highschool.bfwpub.com/spa3e.

2. Enter "Survey type" as the explanatory variable. Then, input each of the different survey types. Press the + button in the upper right to get an

additional column in the table. Then, enter "Color of candy" as the response variable, along with the different colors. Finally, enter the observed counts in each cell, as shown in the screen shot. Press the "Begin analysis" button.

Two Categorical Variables

Input data as: [Counts in categories ▾]

		Explanatory variable: Survey type		
		Red	Blue	Control
Response variable: Color of candy	Red	13	5	8
	Blue	7	15	12

[–] [+]

[Begin analysis] [Edit inputs] [Reset everything]

3. Then, scroll down to the "Perform inference" section and press the button to perform the chi-square test for association. The test statistic, *P*-value, degrees of freedom, and expected counts will be displayed.

Perform Inference

Perform chi-squared 2-way test

χ^2	P-value	df
6.652	0.036	2

Expected counts:

		Survey type		
		Red	Blue	Control
Color of candy	Red	8.667	8.667	8.667
	Blue	11.333	11.333	11.333

TI-83/84

1. Enter the observed counts into matrix [A].

- Press [2nd] [X^{-1}] (MATRIX), arrow to EDIT, and choose [A].

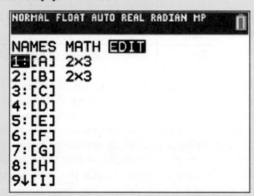

- Enter the dimensions of the matrix: 2×3.

- Enter the observed counts in the same order as the two-way table.

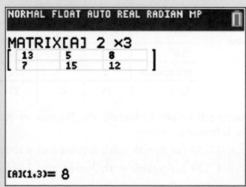

2. Press [STAT], arrow to TESTS, and choose χ^2 Test. Then choose [A] for the observed counts and [B] for the expected counts. Highlight "Calculate" and press [ENTER].

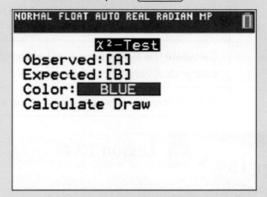

Note: If you select the "Draw" option, you will get a picture of the appropriate χ^2 distribution with the area of interest shaded and the test statistic and *P*-value labeled.

3. To see the expected counts, view matrix [B]. Press [2nd] [X^{-1}] (MATRIX), arrow to [B], and press [ENTER].

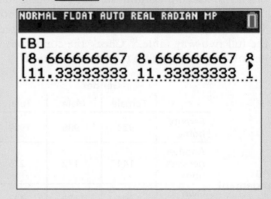

Lesson 10.4

WHAT DID YOU LEARN?

LEARNING TARGET	EXAMPLES	EXERCISES
Check conditions for a test about the relationship between two categorical variables.	p. 646	1–4
Calculate the *P*-value for a test about the relationship between two categorical variables.	p. 648	5–8
Use the four-step process to perform a chi-square test for association.	p. 648	9–12

Exercises

Lesson 10.4

Mastering Concepts and Skills

1. Tuition bills A random sample of U.S. residents was recently asked the following question: "Would you support or oppose major new spending by the federal government that would help undergraduates pay tuition at public colleges without needing loans?" The two-way table shows the responses, grouped by age.[17] Check the conditions for performing a chi-square test for association.

pg 646

		Age				
		18–34	35–49	50–64	65+	Total
Response	Support	91	161	272	332	856
	Oppose	25	74	211	255	565
	Don't know	4	13	20	51	88
	Total	120	248	503	638	1509

2. Moving out? A survey by the National Institutes of Health asked a random sample of young adults (aged 19 to 25 years), "Where do you live now? That is, where do you stay most often?" Here is the full two-way table.[18] Check the conditions for performing a chi-square test for association.

		Gender		
		Female	Male	Total
Living arrangement	Parents' home	923	986	1909
	Another person's home	144	132	276
	Own place	1294	1129	2423
	Group quarters	127	119	246
	Total	2488	2366	4854

3. Surviving the *Titanic* In 1912 the luxury liner *Titanic* struck an iceberg and sank. Some passengers got off the ship in lifeboats, but many died. The two-way table gives information about adult passengers who survived and who died, by class of travel. Check the conditions for performing a chi-square test for association.

		Class of travel			
		First	Second	Third	Total
Survival status	Survived	197	94	151	442
	Died	122	167	476	765
	Total	319	261	627	1207

4. Those good old cartoons Justin wondered if the boys and girls in his school have different preferences for animated shows on a popular television network. He asked a simple random sample of 75 students at his school to choose their favorite show from a list of four shows, recording their gender as well. Here are his data. Check the conditions for performing a chi-square test for association.

		Gender		
		Male	Female	Total
Preferred show	"RR"	8	10	18
	"RP"	10	13	23
	"HA"	5	5	10
	"AB"	7	5	12
	No opinion	5	7	12
	Total	35	40	75

5. Finding *P*-values Calculate the *P*-value in each of the following settings.

pg 648

(a) $\chi^2 = 17.34$ for a table with 5 rows and 4 columns

(b) $\chi^2 = 17.34$ for a table with 3 rows and 2 columns

6. **More *P*-values** Calculate the *P*-value in each of the following settings.
(a) $\chi^2 = 8.03$ for a table with 2 rows and 3 columns
(b) $\chi^2 = 8.03$ for a table with 5 rows and 3 columns

7. **Paying your tuition** Refer to Exercise 1. The chi-square test statistic for these data is $\chi^2 = 39.75$. Calculate the *P*-value.

8. **Moving in?** Refer to Exercise 2. The chi-square test statistic for these data is $\chi^2 = 11.038$. Calculate the *P*-value.

9. **Warm nests** How is the hatching of water python eggs influenced by the temperature of the snake's nest? Researchers randomly assigned newly laid eggs to one of three water temperatures: hot, neutral, or cold. Hot duplicates the extra warmth provided by the mother python, and cold duplicates the absence of the mother. Here are the data on the number of eggs that hatched and didn't hatch.[19] Is there convincing evidence of an association between temperature and hatching for python eggs like these?

		Water temperature			
		Cold	Neutral	Hot	Total
Hatched?	Yes	16	38	75	129
	No	11	18	29	58
	Total	27	56	104	187

10. **Oh, that's cold!** Gastric freezing was once a recommended treatment for ulcers in the upper intestine. One experiment compared 82 subjects randomly assigned to gastric freezing and 78 subjects randomly assigned to receive a placebo. The two-way table shows the results of the experiment.[20] Is there convincing evidence of an association between treatment and outcome for subjects like these?

		Treatment		
		Gastric freezing	Placebo	Total
Outcome	Improved	28	30	58
	Didn't improve	54	48	102
	Total	82	78	160

11. **Online banking** A recent poll conducted by the Pew Research Center asked a random sample of 1846 Internet users if they do any of their banking online. The table summarizes their responses by age.[21] Is there convincing evidence of an association between the age of subjects and whether they do any banking online?

		Age				
		18–29	30–49	50–64	65+	Total
Online banking	Yes	265	352	304	167	1088
	No	130	190	249	189	758
	Total	395	542	553	356	1846

12. **Four seasons** Do male and female students have different favorite seasons? The two-way table shows the favorite season and gender for a simple random sample of 89 high school juniors and seniors in the United States from the Census At School database.[22] Is there convincing evidence of an association between gender and favorite season for students like those who participated in the Census At School survey?

		Gender		
		Female	Male	Total
Favorite season	Fall	22	7	29
	Winter	6	12	18
	Spring	6	5	11
	Summer	12	19	31
	Total	46	43	89

Applying the Concepts

13. **Stopping strokes** Aspirin prevents blood from clotting and helps prevent strokes. The Second European Stroke Prevention Study asked if adding another anticlotting drug named dipyridamole would help. Patients who had already had a stroke were randomly assigned to receive either aspirin only, dipyridamole only, both, or a placebo. Here are the data about strokes during the two years of the study.[23]

		Treatment				
		Placebo	Aspirin	Dipyridamole	Both	Total
Response	Stroke	250	206	211	157	824
	No stroke	1399	1443	1443	1493	5778
	Total	1649	1649	1654	1650	6602

(a) Make a graph that displays the relationship between treatment and response for these patients. Briefly describe the association.

(b) Perform a test at the $\alpha = 0.05$ level to determine if there is convincing evidence of an association between treatment and response for patients like these.

14. **Time to quit** It's hard for smokers to quit. Perhaps prescribing a drug to fight depression will work as well as the usual nicotine patch. Perhaps combining

the patch and the drug will work better than either treatment alone. Here are data from a randomized, double-blind trial that compared four treatments. A "success" means that the subject did not smoke for a year following the beginning of the study.[24]

Response		Treatment				
		Nicotine patch	Drug	Patch plus drug	Placebo	Total
	Success	40	74	87	25	226
	Failure	204	170	158	135	667
	Total	244	244	245	160	893

(a) Make a graph that displays the relationship between treatment and response for these patients. Briefly describe the association.

(b) Perform a test at the $\alpha = 0.05$ level to determine if there is convincing evidence of an association between treatment and response for subjects like these.

15. **More about strokes** Refer to Exercise 13. Based on the observed and expected counts, which treatment(s) appear most effective for preventing additional strokes in patients like these? Justify your answer.

16. **More about quitting** Refer to Exercise 14. Based on the observed and expected counts, which treatment(s) appear most effective for helping subjects like these quit smoking? Justify your answer.

Extending the Concepts

17. **Cracking the mystery** A company sells deluxe and premium nut mixes, both of which contain only cashews, brazil nuts, almonds, and peanuts. The premium nuts are much more expensive than the deluxe nuts. A consumer group suspects that the two nut mixes are really the same. To find out, the group took separate random samples of 20 pounds of each nut mix and recorded the weights of each type of nut in the sample. Here are the data.[25] Explain why we can't use a chi-square test to determine if there is an association between type of nut and type of mix.

Type of nut		Type of mix	
		Deluxe	Premium
	Cashew	6 lb	5 lb
	Brazil nut	3 lb	4 lb
	Almond	5 lb	6 lb
	Peanut	6 lb	5 lb

18. **Gastric freezing alternative** Refer to Exercise 10. Because there are only two treatments, each with two outcomes, you can also analyze the gastric freezing data in Exercise 10 with a two-sample z test.

Let p_g = the proportion of patients like these who improve with gastric freezing and p_c = the proportion of patients like these who improve with a placebo.

(a) For a test of H_0: $p_g - p_c = 0$ versus H_a: $p_g - p_c \neq 0$, calculate the test statistic and P-value.

(b) How does the P-value for this test compare to the P-value from Exercise 10? What is the relationship between z and χ^2 for these data?

Recycle and Review

Exercises 19 and 20 refer to the following setting. Many chess masters and chess advocates believe that chess play develops general intelligence, analytical skill, and the ability to concentrate. To investigate if training in chess increases reading ability, researchers conducted a study. All of the subjects in the study participated in a comprehensive chess program, and their reading performances were measured before and after the program.

Descriptive Statistics: Pretest, Posttest, Post–pre

Variable	N	Mean	Median	StDev	Min	Max	Q_1	Q_3
Pretest	53	57.70	58.00	17.84	23.00	99.00	44.50	70.50
Posttest	53	63.08	64.00	18.70	28.00	99.00	48.00	76.00
Post–pre	53	5.38	3.00	13.02	–19.00	42.00	–3.50	14.00

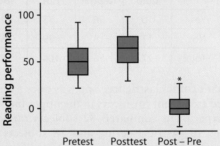

19. **Chess and reading** (3.8, 3.9) The boxplots and numerical summaries provide information on the subjects' pretest scores, posttest scores, and the difference (Post – Pre) between these two scores.

(a) Based only on the graphs and summary statistics, is there evidence that students have greater reading scores after participating in the chess program?

(b) If the study found a statistically significant improvement in reading scores, could you conclude that playing chess causes an increase in reading skills? Justify your answer.

20. **How well does it fit?** (2.7) Here is a scatterplot displaying the relationship between pretest and posttest scores, along with a residual plot based on the least-squares regression line:

$$\widehat{\text{Posttest}} = 0.783 \, (\text{Pretest}) + 17.887$$

For this line, $s = 12.55$ and $r^2 = 55.8\%$. Discuss what s, r^2, and the residual plot reveal about this linear regression model.

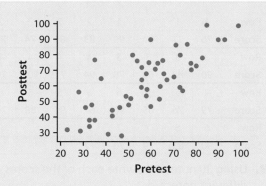
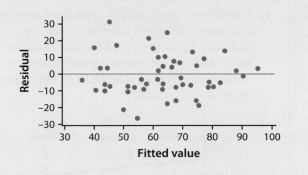

Lesson 10.5

Testing the Relationship Between Two Quantitative Variables

LEARNING TARGETS

- State hypotheses for a test about the relationship between two quantitative variables.
- Check conditions for a test about the relationship between two quantitative variables.
- Calculate the test statistic and *P*-value for a test about the relationship between two quantitative variables given summary statistics.

In Lessons 10.3 and 10.4, you learned how to perform a test about the relationship between two *categorical* variables, such as anger level and heart disease status. In this lesson, you will learn how to state hypotheses, check conditions, and calculate the test statistic and *P*-value for a test about the relationship between two *quantitative* variables. The following activity gives you a preview of what's to come.

ACTIVITY

Should you sit in front?

Many people believe that students learn better if they sit closer to the front of the classroom. Does sitting closer *cause* higher achievement, or do better students simply choose to sit in the front? To investigate, a statistics teacher randomly assigned students to seat locations in his classroom for a particular chapter and recorded the test score for each student at the end of the chapter. In this activity, you will use simulation to determine if these data provide convincing evidence that sitting closer to the front improves test scores.

1. The scatterplot shows the relationship between x = row number (the rows are equally spaced, with row 1 closest to the front and row 7 farthest away) and y = test score. Explain how these data provide some evidence that sitting closer improves test scores.

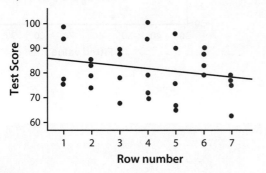

2. The equation of the least-squares regression line shown on the scatterplot is $\hat{y} = 86 - 1.12x$. Interpret the slope, $b = -1.12$.

   Does the negative slope provide *convincing* evidence that sitting closer to the front improves test scores for students like these, or is it plausible that the association is due to the chance variation in the random assignment? Let's investigate. Here are the data.

Row	1	1	1	1	2	2	2	2
Score	76	77	94	99	83	85	74	79
Row	3	3	3	3	4	4	4	4
Score	90	88	68	78	94	72	101	70
Row	4	5	5	5	5	5	6	6
Score	79	76	65	90	67	96	88	79
Row	6	6	7	7	7	7		
Score	90	83	79	76	77	63		

3. Using 30 notecards, write each of the scores on a different notecard.

4. Shuffle the 30 cards and divide them into seven piles (4 cards each for Rows 1, 2, 3, 6, and 7; 5 cards each for Rows 4 and 5). This assigns test scores to rows under the assumption that distance from the front of the classroom has no effect.

5. Using technology, calculate b, the slope of the least-squares regression line using the re-randomized data (x = row number and y = test score). Add your simulated slope to the dotplot provided by your teacher.

6. Is the observed slope of $b = -1.12$ unusual, or is it likely to get a slope this small due to the chance variation in random assignment? What conclusion should you make based on this experiment?

Stating Hypotheses

In Lesson 10.3, you learned how to state hypotheses for a test about the relationship between two *categorical* variables. Stating hypotheses for a test about the relationship between two *quantitative* variables is very similar, except that we can now specify a direction for the relationship in the alternative hypothesis.

In the seating chart activity, we wanted to test if sitting closer to the front improves test scores for students like the ones in the study. That is, we wanted to know if smaller row numbers are associated with larger test scores (a negative association). Here is one way to state the hypotheses for this test:

H_0: There is no association between row number and test score for students like these.

H_a: There is a negative association between row number and test score for students like these.

We can also state the hypotheses symbolically using β, the slope of the true least-squares regression line. If there is no association between these variables, the slope of the true least-squares regression line will be 0. On the other hand, if there is a negative association between these variables, the slope of the true least-squares regression line will be negative. In symbols, the hypotheses are

$$H_0: \beta = 0$$
$$H_a: \beta < 0$$

where β is the slope of the true least-squares regression line relating y = test score to x = row number for students like the ones in this experiment.

As in tests for a population mean and tests for a population proportion, a two-sided alternative hypothesis is possible. For example, if we had no initial belief about the relationship between seat location and test scores, the alternative hypothesis would be H_a: $\beta \neq 0$ (there *is* an association between row number and test score for students like these). As always, the alternative hypothesis is formulated based on the statistical question being investigated—before the data are collected.

EXAMPLE

Can foot length predict height?

Stating hypotheses

PROBLEM: Are students with longer feet typically taller than their classmates with shorter feet? Fifteen high school students will be selected at random and asked to measure their foot length and height, both in centimeters. State hypotheses for a test about the relationship between foot length and height in the population of high school students.

SOLUTION:

$$H_0: \beta = 0$$
$$H_a: \beta > 0$$

where β is the slope of the true least-squares regression line relating y = height (centimeters) to x = foot length (centimeters).

> The alternative hypothesis specifies a positive association because we suspect that students with above-average foot length are also above average in height.

FOR PRACTICE TRY EXERCISE 1.

You could also state the hypotheses for this example in words:

H_0: There is no association between foot length and height for high school students.

H_a: There is a positive association between foot length and height for high school students.

Checking Conditions for a Test about the Slope

Like the chi-square test for association, there are important conditions that need to be satisfied to perform a test about the slope of a least-squares regression line. Figure 10.8 shows the regression model for the seating chart example *when the conditions are met.*

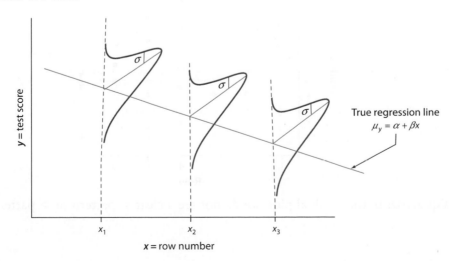

FIGURE 10.8 The regression model for the seating chart example when the conditions for inference are met.

Here are some key observations:

- **Shape:** For each value of x = row number, the distribution of y = test score is normal.
- **Center:** For each value of x = row number, the mean of the distribution of y = test score falls on the true regression line.
- **Variability:** For each value of x = row number, the distribution of y = test score has the same standard deviation.

Before performing a test about the slope of a least-squares regression line, be sure to check the random condition and use data to verify that a regression model like this one is reasonably accurate.

How to Check the Conditions for a Test about the Slope

- **Random:** The data come from a random sample from the population of interest or from a randomized experiment.
- **Normal:** For all values of x, the distribution of y is approximately normal. Check this condition with a dotplot of the residuals. There should be no strong skewness or outliers.
- **Linear:** The form of the association between the two variables is linear. Check this condition with a scatterplot or residual plot. The residual plot should have no leftover curved patterns.
- **Equal SD:** For all values of x, the standard deviation of y is about the same. Check this condition with a residual plot. The residual plot should not have a $<$ pattern (residuals tend to grow in size as x increases) or a $>$ pattern (residuals tend to shrink in size as x increases).

In the seating chart activity, the conditions are met:

- **Random:** The students were randomly assigned to seats for the experiment. ✓
- **Normal:** A dotplot of the residuals does not show strong skewness or outliers. ✓

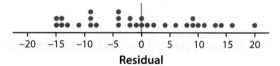

- **Linear:** There is a linear association between row number and test score in the scatterplot shown earlier. Also, there is no leftover curved pattern in the residual plot. ✓

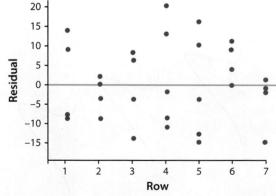

- **Equal SD:** In the residual plot, we do not see a clear $<$ pattern or $>$ pattern. ✓

EXAMPLE

Bigger shoes, taller kids?

Checking conditions

PROBLEM: Here are the data from the 15 randomly selected high school students described in the previous example about foot length and height, along with several graphs.

Foot length (cm)	26	25	26	24	29	26	28	23	23	21	22	30	23	24	22
Height (cm)	164	175	187	156	177	181	179	164	177	169	164	192	168	163	156

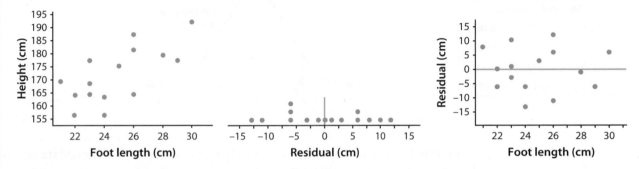

Check the conditions for performing a test about the slope of the least-squares regression line relating y = height to x = foot length.

SOLUTION:
- Random: A random sample of students was selected.✓
- Normal: A dotplot of the residuals does not show strong skewness or outliers.✓
- Linear: There is a linear association between foot length and height in the scatterplot. Also, there is no leftover curved pattern in the residual plot. ✓
- Equal SD: In the residual plot, we do not see a clear $<$ pattern or $>$ pattern.✓

FOR PRACTICE   TRY EXERCISE 5.

Calculating the Test Statistic and *P*-value

The standardized test statistic for a test about the slope of a least-squares regression line has the same form as the standardized test statistic for a test about a mean or a test about a proportion:

$$\text{standardized test statistic} = \frac{\text{statistic} - \text{null value}}{\text{standard deviation of statistic}}$$

The statistic we use is b, the slope of the estimated least-squares regression line. When the null hypothesis is no association between two variables, the null value is $\beta_0 = 0$. We use the standard error of the slope SE_b in the denominator, which is almost always calculated with technology. Here is the formula:

$$t = \frac{b - \beta_0}{SE_b}$$

When the conditions are met for a test about the slope of a least-squares regression line, we use a t distribution with $n - 2$ degrees of freedom to calculate the *P*-value.

In the seating chart example, the estimated slope is $b = -1.12$ and the standard error of the slope is $SE_b = 0.95$. The value of the standardized test statistic is

$$t = \frac{-1.12 - 0}{0.95} = -1.18$$

Because there were 30 students in the experiment, we should use the t distribution with $30 - 2 = 28$ degrees of freedom to calculate the P-value. Using Table B, we want to find $P(t \leq -1.18) = P(t \geq 1.18)$. According to the table, the P-value is between 0.10 and 0.15.

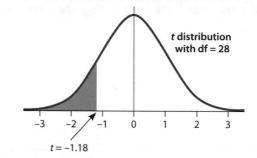

df	Tail probability p		
	0.15	0.10	0.05
27	1.057	1.314	1.703
28	1.056	1.313	1.701
29	1.055	1.311	1.699

t distribution with df = 28

$t = -1.18$

Because this P-value is larger than $\alpha = 0.05$, we fail to reject H_0. These data do not provide convincing evidence that sitting closer to the front of the classroom improves test scores for students like the ones in this experiment.

EXAMPLE

A foot taller?

Calculating test statistics and P-values

PROBLEM: In the example about foot length and height for 15 randomly selected high school students, the equation of the least-squares regression line is $\hat{y} = 102.7 + 2.77x$, where $x =$ foot length (centimeters) and $y =$ height (centimeters). The standard error of the slope is 0.806. Calculate the standardized test statistic and P-value for a test of the hypotheses $H_0: \beta = 0$ versus $H_a: \beta > 0$, where β is the slope of the true least-squares regression line relating $y =$ height to $x =$ foot length.

SOLUTION:

- Test statistic: $t = \dfrac{2.77 - 0}{0.806} = 3.44$

$$t = \frac{b - \beta_0}{SE_b}$$

- P-value: df $= 15 - 2 = 13$.

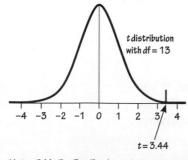

t distribution with df = 13

$t = 3.44$

df	Tail probability p		
	0.0025	0.001	0.0005
12	3.428	3.930	4.318
13	3.372	3.852	4.221
14	3.326	3.787	4.140

Using Table B: The P-value is between 0.001 and 0.0025.

FOR PRACTICE   TRY EXERCISE 9.

Because the *P*-value in this example is less than $\alpha = 0.05$, we reject H_0. These data provide convincing evidence of a positive association between foot length and height in the population of high school students.

THINK ABOUT IT   **What information is provided by the standard error of the slope?** Like other standard errors, the standard error of the slope measures how much the estimated slope typically varies from the true slope in many samples of the same size from the same population. In the previous example about foot length and height, the standard error of the slope was 0.806. This means that if we were to take many samples of 15 high school students and compute the slope of the least-squares regression line relating x = foot length and y = height, the slopes of these lines would typically vary from the slope of the true regression line by about 0.806.

LESSON APP 10.5

Do beavers benefit beetles?

Researchers laid out 23 circular plots, each 4 meters in diameter, at random in an area where beavers were cutting down cottonwood trees. In each plot, they counted the number of stumps from trees cut by beavers and the number of clusters of beetle larvae. Ecologists think that the new sprouts from stumps are more tender than other cottonwood growth, such that beetles prefer them. If so, we would expect more beetle larvae when there are more stumps.[26]

1. State hypotheses for a test about the relationship between x = number of stumps and y = number of beetle larvae.

2. Use the graphs to check the conditions for a test of the hypotheses stated in Question 1.

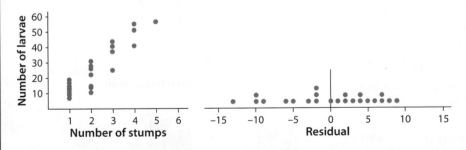

3. The equation of the least-squares regression line relating y = number of beetle larvae to x = number of stumps is $\hat{y} = -1.286 + 11.894x$. The standard error of the slope is 1.136. Calculate the standardized test statistic and *P*-value for a test of the hypotheses stated in Question 1.

4. Based on the *P*-value in Question 3, what conclusion would you make?

Lesson 10.5

WHAT DID YOU LEARN?

LEARNING TARGET	EXAMPLES	EXERCISES
State hypotheses for a test about the relationship between two quantitative variables.	p. 657	1–4
Check conditions for a test about the relationship between two quantitative variables.	p. 659	5–8
Calculate the test statistic and *P*-value for a test about the relationship between two quantitative variables given summary statistics.	p. 660	9–12

Exercises | *Lesson 10.5*

Mastering Concepts and Skills

1. **Dropping helicopters** Mrs. Barrett's class did an
pg 657 experiment in which they dropped paper helicopters from various heights. Each helicopter was assigned at random to a drop height. The class suspects that helicopters dropped from a greater height will take longer to land on the ground. State hypotheses for a test about the relationship between x = drop height (in centimeters) and y = drop time (in seconds) for helicopters like the ones in the experiment.[27]

2. **Picking weeds** Lamb's-quarter is a common weed that interferes with the growth of corn. An agriculture researcher planted corn at the same rate in 16 small plots of ground and then weeded the plots by hand to allow a fixed number of lamb's-quarter plants to grow in each meter of corn row. The decision of how many of these plants to leave in each plot was made at random. No other weeds were allowed to grow. State hypotheses for a test about the relationship between x = number of weeds per meter and y = corn yield (bushels per acre) for plots like the ones in the experiment.[28]

3. **Texting trouble** Mac and Nick suspect that students at their school who spend a lot of time texting are less likely to do well in school. They select a random sample of students and ask them to estimate the typical number of texts they send in a day and their grade-point average for the previous grading period. State hypotheses for a test about the relationship between x = typical texts per day and y = grade-point average for students at this school.[29]

4. **Car statistics** Ross and Eric selected a random sample of teachers from their school and asked them to report the age (in years) of their main vehicles and number of miles driven. The students suspect that older cars will have more miles driven. State hypotheses for a test about the relationship between x = age and y = miles driven for teachers at Ross and Eric's school.

5. **Checking helicopters** Refer to Exercise 1. Use the
pg 659 graphs to check the conditions for performing a test of the hypotheses you stated in Exercise 1.

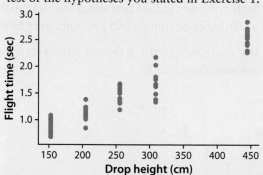

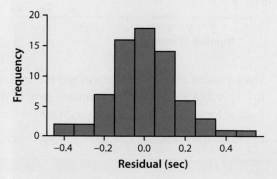

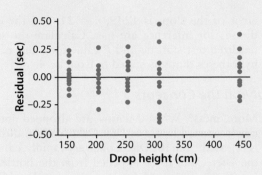

6. **Checking for weeds** Refer to Exercise 2. Use the graphs to check the conditions for performing a test of the hypotheses you stated in Exercise 2.

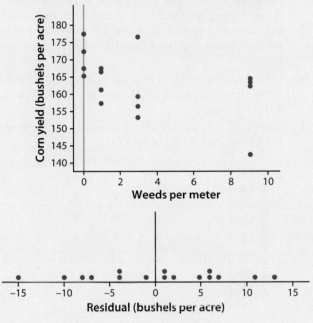

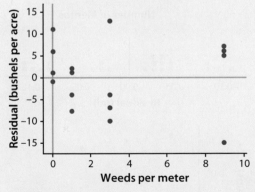

7. **Income and mortality** What does a country's income per person (measured in adjusted gross domestic product per person in dollars) tell us about the mortality rate for children under 5 years of age (per 1000 live births)? A random sample of 14 countries was selected to investigate. Use the graphs to check the conditions for performing a test about the slope of the least-squares regression line relating y = mortality rate to x = income.[30]

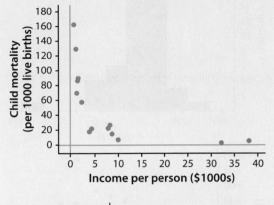

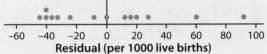

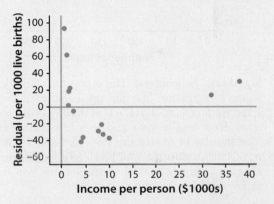

8. **Fouls and points** Is the number of points scored by an NBA player related to the number of fouls he commits? A random sample of 50 NBA players from the 2015 season was selected to investigate. Use the graphs to check the conditions for performing a test about the slope of the least-squares regression line relating y = number of points to x = number of fouls.[31]

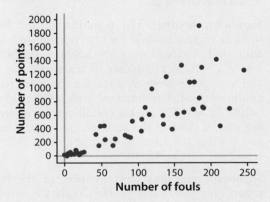

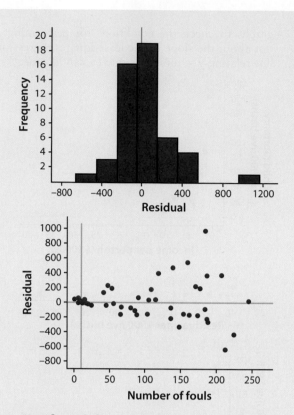

9. Significant helicopters? The equation of the least-squares regression line for the helicopter data in Exercise 5 is $\hat{y} = -0.03761 + 0.0057244x$, where x = drop height (cm) and y = drop time (sec). The standard error of the slope is 0.0002018 and $n = 70$. Calculate the standardized test statistic and P-value for a test of the hypotheses that you stated in Exercise 1.

10. Significant weeds? The equation of the least-squares regression line for the weeds and corn data in Exercise 6 is $\hat{y} = 166.4500 - 1.0808x$, where x = weeds per meter and y = corn yield (bushels per acre). The standard error of the slope is 0.5777 and $n = 16$. Calculate the standardized test statistic and P-value for a test of the hypotheses that you stated in Exercise 2.

11. Significant texting? The equation of the least-squares regression line for the texting and GPA data in Exercise 3 is $\hat{y} = 3.804 - 0.00474x$, where x = typical number of texts sent per day and y = grade-point average for the most recent grading period. The standard error of the slope is 0.000778, $n = 47$, and the conditions for inference are met. Calculate the standardized test statistic and P-value for a test of the hypotheses that you stated in Exercise 3.

12. Significant mileage? The equation of the least-squares regression line for the car data referred to in Exercise 4 is $\hat{y} = 7288.54 + 11630.6x$, where x = age (years) and y = miles driven. The standard

error of the slope is 1249, $n = 21$, and the conditions for inference are met. Calculate the standardized test statistic and P-value for a test of the hypotheses that you stated in Exercise 4.

Applying the Concepts

13. More mess? When Mentos are dropped into a newly opened bottle of Diet Coke, carbon dioxide is released from the Diet Coke very rapidly, causing the Diet Coke to be expelled from the bottle. To see if using more Mentos causes more Diet Coke to be expelled, Brittany and Allie used twenty-four 2-cup bottles of Diet Coke and randomly assigned each bottle to receive either 2, 3, 4, or 5 Mentos. After waiting for the fizzing to stop, they measured the amount expelled (in cups) by subtracting the amount remaining from the original amount in the bottle.[32]

(a) State hypotheses for a test about the relationship between x = number of Mentos and y = amount expelled (cups).

(b) Use the graphs to check the conditions for a test of the hypotheses stated in part (a).

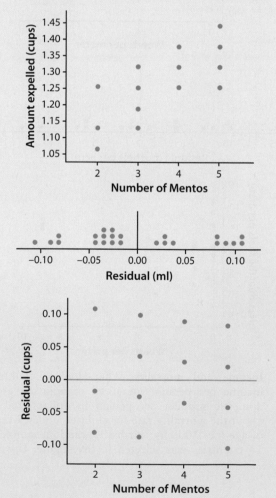

(c) The equation of the least-squares regression line relating y = amount expelled (cups) to x = number of Mentos is $\hat{y} = 1.00208 + 0.07083x$. The standard error of the slope is 0.01228. Calculate the standardized test statistic and P-value for a test of the hypotheses stated in part (a).

(d) Based on the P-value in part (c), what conclusion would you make?

14. **Less mess?** Kerry and Danielle wanted to investigate if tapping on a can of soda would reduce the amount of soda expelled after the can has been shaken. For their experiment, they vigorously shook 40 cans of soda and randomly assigned each can to be tapped for 0 seconds, 4 seconds, 8 seconds, or 12 seconds. After waiting for the fizzing to stop, they measured the amount expelled (in milliliters) by subtracting the amount remaining from the original amount in the can.[33]

(a) State hypotheses for a test about the relationship between x = tapping time (sec) and y = amount expelled (ml).

(b) Use the graphs to check the conditions for a test of the hypotheses stated in part (a).

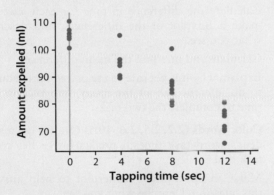

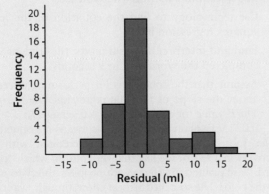

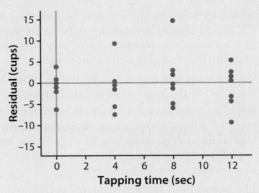

(c) The equation of the least-squares regression line relating y = amount expelled (ml) to x = tapping time is $\hat{y} = 106.36 - 2.635x$. The standard error of the slope is 0.1769. Calculate the standardized test statistic and P-value for a test of the hypotheses stated in part (a).

(d) Based on the P-value in part (c), what conclusion would you make?

Extending the Concepts

15. **Testing the correlation** Refer to the examples about foot length and height beginning on page 657. To test the association between two quantitative variables, it is possible to use the correlation r instead of the slope.

(a) If there is no association between x = foot length (cm) and y = height (cm), what should the value of the correlation equal? Explain.

(b) State the hypotheses for a test to determine if there is a positive association between foot length and height in the population of high school students. Use the Greek letter ρ (rho) to represent the true correlation.

(c) The estimated correlation for these data is $r = 0.690$. A total of 100 trials of a simulation were performed, assuming that there is no association between foot length and height. The value of the simulated correlation was recorded for each trial and is displayed on the dotplot. Based on the dotplot, estimate the P-value.

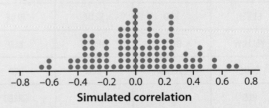

(d) Based on the P-value in part (c), what would you conclude?

16. **New cars** Refer to the car data in Exercises 4 and 12. In addition to performing tests about the slope of a least-squares regression line, it is also possible to perform a test about the y intercept.

    (a) If x = age (years) and y = miles driven, what should the value of the y intercept equal? Explain.

    (b) Based on your answer to part (a), state the hypotheses for a test about the y intercept of the true least-squares regression line. Use the Greek letter α (alpha) to represent the y intercept of the true least-squares regression line.

    (c) The equation of the least-squares regression line is $\hat{y} = 7288.54 + 11630.6x$, where x = age (years) and y = miles driven. The standard error of the y intercept is 6591. Calculate the test statistic and P-value for a test of the hypotheses in part (b), assuming the conditions for inference are met.

    (d) Based on the P-value in part (c), what would you conclude?

17. **Texting slope** Interpret the standard error of the slope provided in Exercise 11.

Recycle and Review

Exercises 18–20 refer to the following setting. Does the color in which words are printed affect your ability to read them? Do the words themselves affect your ability to name the color in which they are printed? Mr. Starnes has his students investigate these questions using the 16 students in his statistics class as subjects. Each student performed two tasks in a random order while a partner timed the activity: (1) Read 32 words aloud as quickly as possible, and (2) say the color in which each of 32 words is printed as quickly as possible. Try both tasks for yourself using the word list below.

YELLOW	RED	BLUE	GREEN
RED	GREEN	YELLOW	YELLOW
GREEN	RED	BLUE	BLUE
YELLOW	BLUE	GREEN	RED
BLUE	YELLOW	RED	RED
RED	BLUE	YELLOW	GREEN
BLUE	GREEN	GREEN	BLUE
GREEN	YELLOW	RED	YELLOW

18. **Color words (3.1, 3.6)** Let's review the design of the study.

    (a) Explain why this was an experiment and not an observational study.

    (b) Explain the purpose of the random assignment in the context of the study.

    Here are the data from Mr. Starnes's experiment. For each subject, the time to perform the two tasks is given to the nearest second.

Subject	Words	Colors	Subject	Words	Colors
1	13	20	9	10	16
2	10	21	10	9	13
3	15	22	11	11	11
4	12	25	12	17	26
5	13	17	13	15	20
6	11	13	14	15	15
7	14	32	15	12	18
8	16	21	16	10	18

19. **Color words (9.5, 9.6)**

    (a) Calculate the difference in time for each student, make a boxplot of the differences, and describe what you see.

    (b) Calculate and interpret the mean difference.

    (c) Explain why it is not safe to use paired t procedures to do inference about the mean difference in the time to complete the two tasks.

20. **Color words (2.2, 2.5, 2.6, 10.5)** Can we use a student's word task time to predict his or her color task time?

    (a) Make an appropriate scatterplot to help answer this question. Describe what you see.

    (b) Use technology to find the equation of the least-squares regression line.

    (c) Find and interpret the residual for the student who completed the word task in 9 seconds.

    (d) Assume that the conditions for performing inference about the slope of the true regression line are met. The P-value for a test of $H_0: \beta = 0$ versus $H_a: \beta > 0$ is 0.0215. Explain what this value means in context.

    Note: John Ridley Stroop is often credited with the discovery in 1935 of the fact that the color in which "color words" are printed interferes with people's ability to identify the color. The so-called Stroop effect, though, was originally discussed by German researchers in a 1929 paper.

Lesson 10.6

Inference for the Slope of a Least-Squares Regression Line

LEARNING TARGETS

- Use technology to calculate the test statistic and *P*-value for a test about the relationship between two quantitative variables.

- Use the four-step process to perform a test for the slope of a least-squares regression line.

- Use the four-step process to calculate and interpret a confidence interval for the slope of a least-squares regression line.

In Lesson 10.5, you learned how to state hypotheses, check conditions, and calculate the test statistic and *P*-value for a test about the relationship between two quantitative variables. In this lesson, you will learn how to use technology to calculate the test statistic and *P*-value when provided with raw data. You will also learn how to use the four-step process to perform a *t test for the slope of a least-squares regression line* and to calculate a *t interval for the slope of a least-squares regression line*.

Calculating the Test Statistic and *P*-Value Using Technology

Because the standard error of the slope is difficult to calculate by hand, we typically use technology to calculate the standardized test statistic and *P*-value for tests about the slope of a least-squares regression line. We can also use technology to check the conditions. The following Tech Corner shows you how.

t Test for the Slope of a Least-Squares Regression Line

You can use technology to check conditions and perform the calculations for a *t* test for the slope of a least-squares regression line. We'll illustrate using the seating chart data from page 656.

Two Quantitative Variables

Variable	Name	Observations (separated by commas or spaces) Keep individuals in the same order.
Explanatory	Row	1,1,1,1,2,2,2,2,3,3,3,3,4,4,4,4,5,5,5,5,6,6,6,6,7,7,7,7
Response	Score	76,77,94,99,83,85,74,79,90,88,68,78,94,72,101,70,79,76,65,90,67

Begin analysis   Edit inputs   Reset everything

Applet

1. Launch the *Two Quantitative Variables* applet at highschool.bfwpub.com/spa3e.

2. Enter "Row number" as the explanatory variable. Then, input the 30 values for this variable. Then, enter "Test score" as the response variable, along with the 30 corresponding values for this variable.

3. Press the "Begin analysis" button. To check the conditions, begin by pressing the "Calculate least-squares regression line" button. This will display the residual plot, allowing you to check the Linear and Equal SD conditions. Then, press the "Show dotplot of residuals" button to check the Normality condition.

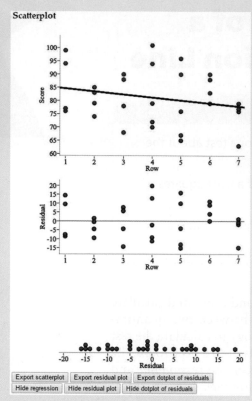

4. To do the calculations, scroll down to the "Perform inference" section and choose *t* test for slope from the drop-down menu. For the alternative hypothesis, choose $\beta < 0$. Press the "Perform inference" button. The test statistic, *P*-value, and degrees of freedom will be displayed.

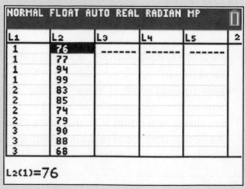

TI-83/84

1. Enter the *x* values (row number) in L1 and the *y* values (test score) in L2.

2. To do the calculations, press STAT, arrow to TESTS, and choose LinRegTTest. Enter L1 for the Xlist, L2 for the YList, and $\beta < 0$ for the alternative hypothesis. Highlight "Calculate" and press ENTER. The test statistic (t), *P*-value (p), and degrees of freedom (df) are reported, along with the *y* intercept (a), slope (b), standard deviation of the residuals (s), r^2, and correlation (r).

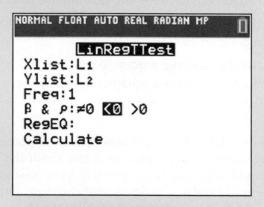

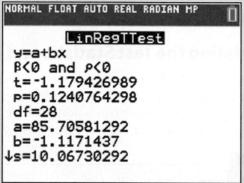

Note: The symbol ρ is the Greek letter rho, which represents the value of the true correlation. The test for slope and the test for correlation give the same standardized test statistic and *P*-value.

3. To check the conditions, begin by making a histogram (or boxplot) of the residuals. The TI-83/84 automatically calculates the residuals when the LinRegTTest is performed (or the least-squares regression line is calculated), so you must run the test or calculate the line *before* plotting the residuals.

• Press 2nd Y= (STAT PLOT).
• Press ENTER or 1 to go into Plot1.
• Adjust the settings as shown.
• Press ZOOM and choose 9:ZoomStat.

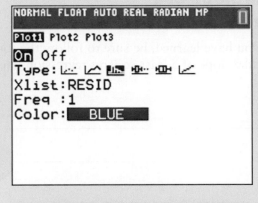

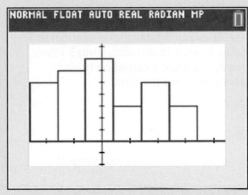

To make the residual plot, adjust the STAT PLOT settings as shown, press ZOOM and choose 9:ZoomStat.

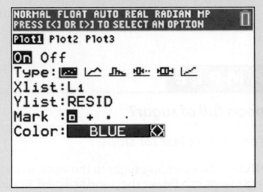

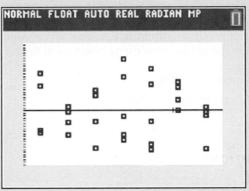

Both the calculator and applet gave a *P*-value of 0.124, which is in the interval from 0.10 to 0.15 that we found using Table B in the previous lesson.

EXAMPLE

Is there a positive association between foot length and height?

Calculating test statistics and P-values using technology

PROBLEM: Here are data from the 15 randomly selected high school students described in the example about foot length and height from Lesson 10.5.

Foot length (cm)	26	25	26	24	29	26	28	23	23	21	22	30	23	24	22
Height (cm)	164	175	187	156	177	181	179	164	177	169	164	192	168	163	156

Use technology to calculate the standardized test statistic and *P*-value for a test of

$$H_0: \beta = 0$$
$$H_a: \beta > 0$$

where β is the slope of the true least-squares regression line relating y = height (cm) to x = foot length (cm).

SOLUTION:

Using the applet, the standardized test statistic is $t = 3.441$, the *P*-value is 0.002, and the df is 13.

Perform Inference for Slope

Inference procedure: [t test for slope ▼] Alternative hypothesis: [β > 0 ▼]

[Perform inference]

t	P-value	df
3.441	0.002	13

FOR PRACTICE   TRY EXERCISE 1.

Performing a *t* Test for the Slope of a Least-Squares Regression Line

As with the other significance tests you have learned, be sure to follow the four-step process when performing a *t* test for the slope of a least-squares regression line.

EXAMPLE

A spoon full of sugar?

Performing a t test for slope

PROBLEM: Does adding sugar to the water in a vase help flowers stay fresh? To find out, two statistics students went to a flower shop and selected 12 carnations. When they got home, the students prepared 12 identical vases with exactly the same amount of water in each vase. They put 1 tablespoon of sugar in 3 vases, 2 tablespoons of sugar in 3 vases, and 3 tablespoons of sugar in 3 vases. In the remaining 3 vases, they put no sugar. After the vases were prepared, the students randomly assigned 1 carnation to each vase and observed how many hours each flower continued to look fresh. Here are the data.

Amount of sugar (tbs)	0	0	0	1	1	1	2	2	2	3	3	3
Freshness time (h)	168	180	192	192	204	204	204	210	210	222	228	234

Do these data provide convincing evidence that adding sugar helps carnations like these last longer?

SOLUTION:

STATE: We want to test

$$H_0: \beta = 0$$
$$H_a: \beta > 0$$

where β = the slope of the true least-squares regression line relating y = freshness time to x = amount of sugar. We'll use a significance level of $\alpha = 0.05$.

PLAN: *t* test for the slope of a least-squares regression line
• Random: Carnations were randomly assigned to different amounts of sugar. ✓
• Normal: A dotplot of the residuals does not show strong skewness or outliers. ✓
• Linear: There is a linear association between sugar and freshness in the scatterplot. Also, there is no leftover curved pattern in the residual plot. ✓
• Equal SD: In the residual plot, we do not see a clear < pattern or > pattern. ✓

> Follow the four-step process!

> **STATE:** State the hypotheses you want to test and the significance level, and define any parameters you use.

> **PLAN:** Identify the appropriate inference method and check conditions.

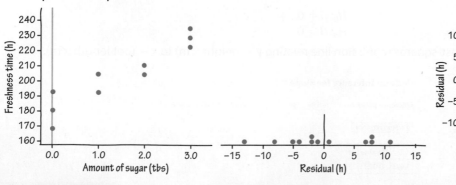

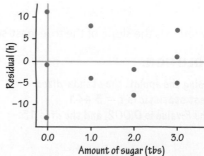

DO: Using the applet
- Estimated slope = b = 15.2
- t = 7.822
- P-value < 0.001 with df = 10

Perform Inference for Slope

Inference procedure: [t test for slope ▾] Alternative hypothesis: [β > 0 ▾]

[Perform inference]

t	P-value	df
7.822	<0.001	10

CONCLUDE: Because the P-value of less than 0.001 is smaller than $\alpha = 0.05$, we reject H_0. There is convincing evidence that adding sugar helps carnations like these last longer.

DO: If the conditions are met, perform calculations:
- Identify the sample statistic.
- Calculate the standardized test statistic.
- Find the P-value.

The sample result gives *some* evidence in favor of H_a because $b = 15.2 > 0$.

CONCLUDE: Make a conclusion about the hypotheses in the context of the problem.

Because carnations were randomly assigned to treatments, a cause-and-effect conclusion is appropriate.

FOR PRACTICE   TRY EXERCISE 5.

Confidence Intervals for the Slope of a Least-Squares Regression Line

Besides testing hypotheses about the slope of a least-squares regression line, it is also possible to estimate the true slope β using a confidence interval. In the sugar and carnations example, there is convincing evidence that the true slope of the least-squares regression line relating y = freshness time to x = amount of sugar is positive. A confidence interval for the slope provides more information—it describes how many extra hours of freshness we can expect for each additional tablespoon of sugar.

When the Random, Normal, Linear, and Equal SD conditions are met, we can use technology to calculate a *t interval for the slope of a least-squares regression line.*

TECH CORNER

t Interval for the Slope of a Least-Squares Regression Line

You can use technology to calculate a *t* interval for the slope of a least-squares regression line. We'll illustrate using the sugar and carnations example on page 670. *Note:* See the previous Tech Corner for instructions about checking conditions.

Applet

1. Launch the *Two Quantitative Variables* applet at highschool.bfwpub.com/spa3e.

2. Enter "Amount of sugar" as the explanatory variable. Then, input the 12 values for this variable. Enter "Freshness time" as the response variable, along with the 12 values for this variable.

Two Quantitative Variables

Variable	Name	Observations (separated by commas or spaces) Keep individuals in the same order.
Explanatory	Amount of sugar	0,0,0,1,1,1,2,2,2,3,3,3
Response	Freshness	168,180,192,192,204,204,204,210,210,222,228,234

[Begin analysis] [Edit inputs] [Reset everything]

3. Press the "Begin analysis" button. Then, scroll down to the "Perform inference" section and choose "*t* interval for slope" from the drop-down menu. For the confidence level, choose 95%. Press the "Perform inference" button. The confidence interval and degrees of freedom will be displayed.

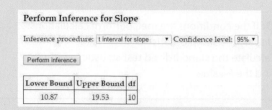

Perform Inference for Slope

Inference procedure: [t interval for slope ▼]   Confidence level: [95% ▼]

[Perform inference]

Lower Bound	Upper Bound	df
10.87	19.53	10

TI-84

Note: The TI-83 and older models of the TI-84 do not have an option for calculating a confidence interval for slope.

1. Enter the *x* values (amount of sugar) in L1 and the *y* values (freshness time) in L2.

NORMAL FLOAT AUTO REAL RADIAN MP

L1	L2	L3	L4	L5	4
0	168	------	------	------	
0	180				
0	192				
1	192				
1	204				
1	204				
2	204				
2	210				
2	210				
3	222				
3	228				

L4(1)=

2. Press STAT, arrow to TESTS, and choose LinRegTInt. Enter L1 for the Xlist, L2 for the YList, and 0.95 for the confidence level.

Highlight "Calculate" and press ENTER. The confidence interval, estimated slope (b), and degrees of freedom (df) are reported, along with the standard deviation of the residuals (s), *y* intercept (a), r^2, and correlation (*r*).

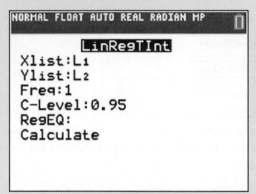

NORMAL FLOAT AUTO REAL RADIAN MP

LinRegTInt
Xlist:L1
Ylist:L2
Freq:1
C-Level:0.95
RegEQ:
Calculate

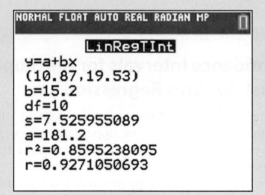

NORMAL FLOAT AUTO REAL RADIAN MP

LinRegTInt
y=a+bx
(10.87,19.53)
b=15.2
df=10
s=7.525955089
a=181.2
r²=0.8595238095
r=0.9271050693

We are 95% confident that the interval from 10.87 to 19.53 captures the slope of the true least-squares regression line relating *y* = freshness time to *x* = amount of sugar. That is, we expect that each additional tablespoon of sugar will add between 10.87 and 19.53 more hours of freshness to carnations like the ones in the experiment.

EXAMPLE

How much is that truck worth?

Calculating a confidence interval for slope

PROBLEM: In Chapter 2, we analyzed the relationship between *y* = asking price and *x* = miles driven for a random sample of Ford F-150 trucks. Here are the data.

Miles driven	70,583	129,484	29,932	29,953	24,495	75,678	8,359	4,447
Price ($)	21,994	9,500	29,875	41,995	41,995	28,986	31,891	37,991
Miles driven	34,077	58,023	44,447	68,474	144,162	140,776	29,397	131,385
Price ($)	34,995	29,988	22,896	33,961	16,883	20,897	27,495	13,997

Calculate and interpret a 95% confidence interval for the slope of the true least-squares regression line relating *y* = asking price to *x* = miles driven.

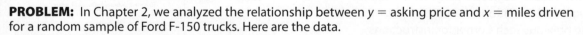

SOLUTION:

STATE: We want to estimate β = the slope of the true least-squares regression line relating y = asking price to x = miles driven with 95% confidence.

> **STATE:** State the parameter you want to estimate and the confidence level.

PLAN: t interval for the slope of a least-squares regression line.
- Random: Random sample of 16 Ford F-150s. ✓
- Normal: A dotplot of the residuals does not show strong skewness or outliers. ✓
- Linear: There is a linear association between number of miles and asking price in the scatterplot. Also, there is no leftover curved pattern in the residual plot. ✓
- Equal SD: In the residual plot, we do not see a clear < pattern or > pattern. ✓

> **PLAN:** Identify the appropriate inference method and check conditions.

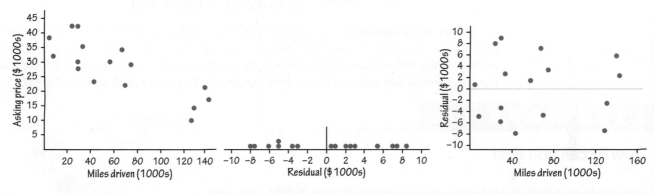

DO: Using the applet, the interval is −0.229 to −0.097 with $df = 14$.

> **DO:** If the conditions are met, perform calculations.

Lower Bound	Upper Bound	df
−0.229	−0.097	14

CONCLUDE: We are 95% confident that the interval from −0.229 to −0.097 captures the slope of the true least-squares regression line relating y = asking price to x = miles driven for Ford F-150s.

> **CONCLUDE:** Interpret your interval in the context of the problem.

FOR PRACTICE   TRY EXERCISE 9.

Because all of the plausible values in the interval are less than 0, these data provide convincing evidence of a negative association between asking price and miles driven for Ford F-150s. For each additional mile driven, we expect the asking price to go down between about 10 and 23 cents.

It is also possible to calculate a confidence interval for the slope by hand when provided with the slope, sample size, and standard error of the slope.

How to Calculate a Confidence Interval for β

When the Random, Normal, Linear, and Equal SD conditions are met, a C% confidence interval for the slope of the true least-squares regression line is

$$b \pm t^* SE_b$$

where t^* is the critical value with C% of its area between $-t^*$ and t^* in the t distribution with degrees of freedom equal to $n - 2$.

In the example about foot length and height for 15 randomly selected high school students, the equation of the least-squares regression line is $\hat{y} = 102.7 + 2.77x$, where x = foot length (cm) and y = height (cm). The standard error of the slope is 0.806.

df	Tail probability p		
	0.05	**0.025**	**0.02**
12	1.782	2.179	2.303
13	1.771	2.160	2.282
14	1.761	2.145	2.264
	90%	**95%**	**96%**
	Confidence level C		

Using $15 - 2 = 13$ degrees of freedom, the t^* critical value for 95% confidence is 2.160. The 95% confidence interval for the slope of the true least-squares regression line relating $y =$ height (cm) to $x =$ foot length (cm) is

$$2.77 \pm 2.160(0.806) \rightarrow 1.03 \text{ to } 4.51$$

We are 95% confident that the interval from 1.03 to 4.51 captures the slope of the true least-squares regression line relating $y =$ height (cm) to $x =$ foot length (cm). Because all of the plausible values in the interval are positive, there is convincing evidence that a positive association exists between foot length and height for high school students.

LESSON APP 10.6

How fit can you get?

Josh wears an activity tracker to record the number of steps he takes each day along with several other health-related measurements. Based on these variables, the device calculates an estimate of the number of calories he burns each day. For 10 randomly selected days, here are the values of $x =$ number of steps and $y =$ calories burned.

Number of steps	7997	6318	6620	7708	12,627	10,961	10,819	7715	10,958	9694
Calories burned	2669	2328	2549	2419	2840	2864	2771	2596	2884	2566

1. Make a scatterplot to display the relationship between number of steps and calories burned. Describe the relationship.

2. Do these data provide convincing evidence of a positive association between number of steps and calories burned for Josh?

3. Calculate and interpret a 95% confidence interval for the slope of the true least-squares regression line relating $y =$ calories burned to $x =$ number of steps for Josh.

Lesson 10.6

WHAT DID YOU LEARN?

LEARNING TARGET	EXAMPLES	EXERCISES
Use technology to calculate the test statistic and P-value for a test about the relationship between two quantitative variables.	p. 669	1–4
Use the four-step process to perform a test for the slope of a least-squares regression line.	p. 670	5–8
Use the four-step process to calculate and interpret a confidence interval for the slope of a least-squares regression line.	p. 672	9–12

© Geoff Goldswain/Alamy Stock Photo

Lesson 10.6

Mastering Concepts and Skills

1. Tipping at a buffet Do customers who stay longer at buffets give larger tips? Charlotte, a statistics student who worked at an Asian buffet, decided to investigate this question for a project. While she was doing her job as a hostess, she obtained a random sample of receipts, which included the length of time (in minutes) a party was at the restaurant and the amount of tip left (in dollars). Here are her data.[34]

pg 669

Time (min)	Tip ($)	Time (min)	Tip ($)
23	5.00	67	9.01
39	2.75	70	5.00
44	7.75	74	7.29
55	5.00	85	7.50
61	7.00	90	6.00
65	8.88	99	6.50

Use technology to calculate the standardized test statistic and P-value for a test of $H_0: \beta = 0$ versus $H_a: \beta > 0$, where β is the slope of the true least-squares regression line relating y = tip (dollars) to x = time (minutes). Assume the conditions for inference are met.

2. Flying from Philly Is there a positive association between air fare and distance traveled? The table gives the distance (in miles) from Philadelphia to a random sample of 6 cities and the cost (in dollars) of the cheapest flight to that city on a popular discount airline (in dollars).[35]

Airport	Distance from Philadelphia (mi)	Cheapest fare ($)
Raleigh-Durham	340	131
Orlando	860	120
St. Louis	810	139
Denver	1550	180
Houston	1330	174
Phoenix	2070	188

Use technology to calculate the standardized test statistic and P-value for a test of $H_0: \beta = 0$ versus $H_a: \beta > 0$, where β is the slope of the true least-squares regression line relating y = fare (dollars) to x = distance (miles). Assume the conditions for inference are met.

3. Turn up the volume? Nicole and Elena wanted to know if listening to music at a louder volume hurts test performance. To investigate, they recruited 30 volunteers and randomly assigned 10 to listen to music at 30 decibels, 10 to listen to music at 60 decibels, and 10 to listen to music at 90 decibels. While listening to the music, each student took a 10-question math test. The table shows the number correct for each volunteer.[36]

Volume (dB)	Score	Volume (dB)	Score	Volume (dB)	Score
30	9	60	10	90	3
30	8	60	8	90	5
30	9	60	8	90	7
30	9	60	7	90	6
30	6	60	9	90	6
30	10	60	6	90	5
30	8	60	7	90	4
30	7	60	8	90	7
30	7	60	5	90	8
30	9	60	7	90	2

Use technology to calculate the standardized test statistic and P-value for a test of $H_0: \beta = 0$ versus $H_a: \beta < 0$, where β is the slope of the true least-squares regression line relating y = number correct to x = volume (decibels). Assume the conditions for inference are met.

4. Hungry fish? Does a larger concentration of fish attract more predators? One study looked at kelp perch and their common predator, the kelp bass. The researcher set up four large circular pens on sandy ocean bottoms off the coast of southern California. He chose young perch at random from a large group and placed 10, 20, 40, and 60 perch in the four pens. Then he dropped the nets protecting the pens, allowing bass to swarm in, and counted the perch left after 2 hours. Here are data on the proportions of perch eaten in four repetitions of this setup.[37]

Number of perch	Proportion killed	Number of perch	Proportion killed
10	0	40	0.075
10	0.1	40	0.3
10	0.3	40	0.6
10	0.3	40	0.725
20	0.2	60	0.517
20	0.3	60	0.55
20	0.3	60	0.7
20	0.6	60	0.817

Use technology to calculate the standardized test statistic and P-value for a test of $H_0: \beta = 0$ versus $H_a: \beta > 0$, where β is the slope of the true least-squares regression line relating y = proportion killed to x = number of perch. Assume the conditions for inference are met.

5. Heartbeats Do people with higher resting pulses have higher body temperatures? Mr. Jordan collected data from a random sample of 20 students at his school on x = body temperature (in degrees Fahrenheit) and y = pulse rate (in beats per minute). The data are given in the table.[38]

Temperature	97.9	97.1	98.6	98.5	98.1	97	97.2	97.8	98.3	97.3
Heart rate	78	64	72	80	60	56	75	70	90	76
Temperature	98.8	98	97.5	98	97.6	97.7	96.6	96.6	96.6	96.8
Heart rate	80	82	84	75	87	61	69	69	71	74

Do these data provide convincing evidence that faster heartbeats are associated with warmer body temperatures for students at Mr. Jordan's school?

6. Literacy and birth rates The table shows x = female literacy rate (for women 15 years of age or older) and y = birth rate (births per 1000 population) for 20 randomly selected countries.[39]

Country	Female literacy (%)	Birth rate
Bhutan	39	23
Poland	99	9.8
Sao Tome and Principe	68	38
Venezuela	94	22
Belize	70	35
Malta	94	9.2
Congo Democratic Republic	49	46
Vietnam	89	17
Zimbabwe	80	35
Madagascar	63	38
Central African Rep.	39	37
Ghana	58	34
Nicaragua	78	24
Egypt	59	25
Ecuador	84	23
Suriname	89	21
Mongolia	98	20
Macedonia	95	12
Gabon	80	32
Mauritania	47	37

Do these data provide convincing evidence that countries with greater female literacy rates have smaller birth rates?

7. Drive for show? In golf, does greater distance off the tee translate to better (lower) scores? We collected data on mean drive distance (in yards) and mean score per round from a random sample of 19 players on the Professional Golfers Association (PGA) Tour in a recent year. The scatterplot shows the relationship between x = mean distance (yards) and y = mean score, along with the least-squares regression line $\hat{y} = 76.90 - 0.02016x$. The standard error of the slope is 0.01319 and the conditions for inference are met. Do these data provide convincing evidence that lower mean scores are associated with greater mean driving distances for PGA golfers?[40]

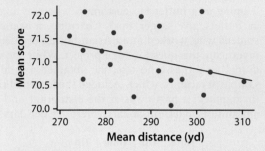

8. Waiting to eat? Is there an association between the length of time a toddler sits at the lunch table and the number of calories consumed? We collected data on a random sample of 20 toddlers observed over several months.

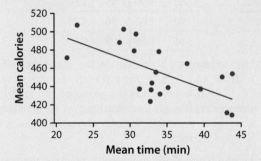

The scatterplot shows the relationship between x = the mean time a child spent at the lunch table (minutes) and y = the mean number of calories consumed, along with the least-squares regression line $\hat{y} = 560.65 - 3.0771x$. The standard error of the slope is 0.8498 and the conditions for inference are met. Do these data provide convincing evidence of an association between time at the table and calories consumed for toddlers?[41]

9. Hot heartbeats Refer to Exercise 5. Calculate and interpret a 95% confidence interval for the slope of the true least-squares regression line relating y = resting pulse rate (in beats per minute) to x = body temperature (in degrees Fahrenheit) for students at Mr. Jordan's school.

10. Literacy and population growth Refer to Exercise 6. Calculate and interpret a 95% confidence interval for the slope of the true least-squares regression line relating y = birth rate (births per 1000 population) to x = female literacy rate (% of women 15 years or older) for all countries.

11. Drive for dough? Refer to Exercise 7. Calculate and interpret a 95% confidence interval for the slope of the true least-squares regression line relating y = mean score to x = mean driving distance (yards) for PGA golfers.

12. **Still waiting?** Refer to Exercise 8. Calculate and interpret a 95% confidence interval for the slope of the true least-squares regression line relating y = mean number of calories consumed to x = mean time at the lunch table for toddlers.

Applying the Concepts

13. **Oh, that smarts!** Infants who cry easily may be more easily stimulated than others. This may be a sign of higher IQ. Using a random sample of 38 infants, child-development researchers explored the relationship between the crying of infants 4 to 10 days old and their IQ test scores at age 3. A snap of a rubber band on the sole of the foot caused the infants to cry. The researchers recorded the crying and measured its intensity by the number of peaks in the most active 20 seconds.[42] Here are the data.

Cry count	IQ	Cry count	IQ	Cry count	IQ	Cry count	IQ
10	87	20	90	17	94	12	94
12	97	16	100	19	103	12	103
9	103	23	103	13	104	14	106
16	106	27	108	18	109	10	109
18	109	15	112	18	112	23	113
15	114	21	114	16	118	9	119
12	119	12	120	19	120	16	124
20	132	15	133	22	135	31	135
16	136	17	141	30	155	22	157
33	159	13	162				

(a) Make a scatterplot to display the relationship between x = cry count and y = IQ at age 3. Describe the relationship.

(b) Do these data provide convincing evidence of a positive association between cry count and IQ at age 3 for infants?

(c) Calculate and interpret a 95% confidence interval for the slope of the true least-squares regression line relating y = IQ at age 3 to x = cry count for infants.

14. **Mass and metabolism** We have data on the lean body mass and resting metabolic rate for a random sample of 12 women who are subjects in a study of dieting. Lean body mass, given in kilograms (kg), is a person's weight leaving out all fat. Metabolic rate is measured in calories burned per 24 hours (cal/24 h). The researchers believe that lean body mass is positively associated with metabolic rate.

Lean body mass (kg)	36.1	54.6	48.5	42.0	50.6	42.0
Metabolic rate (cal/24 h)	995	1425	1396	1418	1502	1256
Lean body mass (kg)	40.3	33.1	42.4	34.5	51.1	41.2
Metabolic rate (cal/24 h)	1189	913	1124	1052	1347	1204

(a) Make a scatterplot to display the relationship between x = lean body mass and y = metabolic rate. Describe the relationship.

(b) Do these data provide convincing evidence of a positive association between lean body mass and metabolic rate for women like these?

(c) Calculate and interpret a 95% confidence interval for the slope of the true least-squares regression line relating y = metabolic rate to x = lean body mass for women like these.

15. **Oh, snap!** Refer to Exercise 13. Explain how the interval from part (c) is consistent with your conclusion in part (b).

16. **More mass, more metabolism?** Refer to Exercise 14. Explain how the interval from part (c) is consistent with your conclusion in part (b).

Extending the Concepts

17. **Don't cry** In Exercise 13, you used the slope to perform a test about the relationship between x = cry count and y = IQ at age 3 for toddlers. It is also possible to classify the values for each variable into categories and use a chi-square test for association. The two-way table summarizes these data.

		Cry count		
		Less than 16	At least 16	Total
IQ	Less than 115	10	12	22
	At least 115	5	11	16
	Total	15	23	38

(a) Do these data provide convincing evidence of an association between cry count and IQ at age 3 for infants?

(b) Describe a disadvantage of using a chi-square test instead of a t test for slope.

Recycle and Review

18. **Shower time (1.8, 5.7, 6.5)** Marcella takes a shower every morning when she gets up. Her time in the shower varies according to a normal distribution with mean 4.5 minutes and standard deviation 0.9 minute.

(a) Find the probability that Marcella's shower lasts between 3 and 6 minutes on a randomly selected day. Show your work.

(b) If Marcella took a 7-minute shower, would it be classified as an outlier by the $1.5 \times IQR$ rule? Justify your answer.

(c) Find the probability that the *mean* length of her shower times on 10 randomly selected days exceeds 5 minutes. Show your work.

19. **Which procedure? (Chapters 7–10)** In each of the following settings, state which inference procedure from Chapters 7–10 you would use. For example, you might say "two-sample z test for the difference between two proportions." You do not have to carry out any procedures.[43]

(a) What is the average voter turnout during an election? A random sample of 38 cities was asked to report the percent of registered voters who actually voted in the most recent election.

(b) Are blondes more likely to have a boyfriend than the rest of the single world? Independent random samples of 300 blondes and 300 non-blondes were asked if they have a boyfriend.

(c) Is there a relationship between attendance at religious services and alcohol consumption? A random sample of 1000 adults was asked if they regularly attend religious services and if they drink alcohol daily.

(d) Separate random samples of 75 college students and 75 high school students were asked how much time, on average, they spend watching television each week. We want to estimate the difference in the average amount of TV watched by high school and college students.

Chapter 10 RESOLVED

STATS applied!

Does background music influence what customers buy?

In the STATS applied! on page 619, we described a study that compared three randomly assigned treatments in a European-style restaurant: no music, French accordion music, and Italian string music. Under each condition, the researchers recorded the number of customers who ordered French, Italian, and other entrées. Here is a table that summarizes the data.[44]

		Type of background music			
		None	French	Italian	Total
	French	30	39	30	99
Entrée ordered	Italian	11	1	19	31
	Other	43	35	35	113
	Total	84	75	84	243

1. Make a graph to display the relationship between type of background music and entrée ordered at this restaurant. Describe the relationship.

2. Do these data provide convincing evidence of an association between type of background music and type of entrée ordered at this restaurant?

3. Based on the answer to Question 2 and the design of the study, is it reasonable to conclude that background music influences what customers order at this restaurant?

Main Points — *Chapter 10*

Tests for the Distribution of One Categorical Variable

■ The test for a distribution of one categorical variable is called a **chi-square test for goodness of fit.**

■ The **null hypothesis** in a test for goodness of fit is that the distribution of a categorical variable in a population is the same as the claimed distribution. The **alternative hypothesis** is that the population distribution is *not* the same as the claimed distribution.

■ There are two conditions for a chi-square test for goodness of fit:

■ **Random:** The data come from a random sample from the population of interest.

■ **Large Counts:** All *expected* counts are at least 5.

■ Calculate the **expected count** for each category by multiplying the sample size by the expected proportion in that category, assuming the null hypothesis is true.

■ The **chi-square test statistic** is a measure of how different the observed counts are from the expected counts. The formula for the statistic is

$$\chi^2 = \sum \frac{(\text{observed count} - \text{expected count})^2}{\text{expected count}}$$

In a chi-square test for goodness of fit, the sum is over all categories of the categorical variable.

■ A **chi-square distribution** is described by a density curve that takes only non-negative values and is skewed to the right. A particular chi-square distribution is specified by its degrees of freedom.

■ If the conditions are met, find the *P*-value for a chi-square test for goodness of fit using Table C or technology and the chi-square distribution with **degrees of freedom** = number of categories − 1.

■ Be sure to follow the four-step process whenever you perform a chi-square test for goodness of fit.

Tests for the Relationship Between Two Categorical Variables

■ Two categorical variables have an **association** if knowing the value of one variable helps us predict the value of the other.

■ The test for a relationship between two *categorical* variables is called the **chi-square test for association.**

■ The **null hypothesis** for this test is that there is no association between the two variables in the population of interest. The **alternative hypothesis** is that there *is* an association between the two variables in the population of interest.

■ There are two **conditions** for a chi-square test for association:

■ **Random:** The data come from a random sample from the population of interest, from separate random samples from multiple populations, or from multiple groups in a randomized experiment.

■ **Large Counts:** All *expected* counts are at least 5.

■ Calculate **expected counts** using the formula

$$\text{expected count} = \frac{\text{row total} \times \text{column total}}{\text{table total}}$$

■ Use the **chi-square test statistic,** where the sum is over all the cells in the table (not including the totals):

$$\chi^2 = \sum \frac{(\text{observed count} - \text{expected count})^2}{\text{expected count}}$$

■ If conditions are met, find the *P*-value using Table C or technology and the chi-square distribution with **degrees of freedom** = (number of rows − 1)(number of columns − 1).

■ Be sure to follow the four-step process whenever you perform a chi-square test for association.

Inference for the Relationship Between Two Quantitative Variables

- There are four **conditions** for performing inference about the relationship between two quantitative variables using the slope of a least-squares regression line:

  - **Random:** The data come from a random sample from the population of interest or from a randomized experiment.

  - **Normal:** For all values of x, the distribution of y is approximately normal.

  - **Linear:** The form of the association between the two variables is linear.

  - **Equal SD:** For all values of x, the standard deviation of y is about the same.

- The test for a relationship between two quantitative variables is called the **t test for the slope of a least-squares regression line.**

- The **null hypothesis** for this test is that there is no association between the two variables in the population of interest. That is, $\beta = 0$, where β is the slope of the true least-squares regression line.

- The **alternative hypothesis** is that there is some kind of association between the two variables in the population of interest. That is, $\beta > 0$ (positive association), $\beta < 0$ (negative association), or $\beta \neq 0$ (an association).

- For the **standardized test statistic**, use

$$t = \frac{b - \beta_0}{SE_b}$$

where β_0 is the hypothesized value of the slope and SE_b is the standard error of the slope (provided by technology).

- If the conditions are met, find the **P-value** using Table B or technology and the t distribution with **degrees of freedom** = $n - 2$.

- If the conditions are met, the **t interval for the slope of a least-squares regression line** β is

$$b \pm t^* SE_b$$

where SE_b is the standard error of the slope (provided by technology) and t^* is the critical value with C% of its area between $-t^*$ and t^* in the t distribution with degrees of freedom equal to $n - 2$.

- Be sure to follow the four-step process whenever you perform a significance test or calculate a confidence interval for β, the slope of the true least-squares regression line.

Chapter 10 Review Exercises

1. **Genetics of color (10.1, 10.2)** Biologists wish to cross pairs of tobacco plants having genetic makeup Gg, indicating that each plant has one dominant gene (G) and one recessive gene (g) for color. Each offspring plant will receive one gene for color from each parent. The Punnett square shows the possible combinations of genes received by the offspring.

		Parent 2 passes on	
		G	**g**
Parent 1	**G**	GG	Gg
passes on	**g**	Gg	gg

The Punnett square suggests that the expected ratio of green (GG) to yellow-green (Gg) to albino (gg) tobacco plants should be 1:2:1. In other words, the biologists predict that 25% of the offspring will be green, 50% will be yellow-green, and 25% will be albino. To test

their hypothesis, the biologists mate 84 randomly selected pairs of yellow-green parent plants. Of 84 offspring, 23 plants were green, 50 were yellow-green, and 11 were albino. Do these data provide convincing evidence at the $\alpha = 0.01$ level that the true distribution of offspring is different from what the biologists predict?

2. **A smash or a hit? (10.3, 10.4)** Two researchers asked 150 subjects to recall the details of a car accident they watched on video. Fifty subjects were selected at random and asked, "About how fast were the cars going when they smashed into each other?" For another 50 randomly selected subjects, the words "smashed into" were replaced with "hit." The remaining 50 subjects—the control group—were not asked to estimate speed. A week later, all subjects were asked if they saw any broken glass at the accident (there wasn't any). The table shows each group's response to the broken glass question.[45]

		Treatment			
		"Smashed into"	"Hit"	Control	Total
Response	Yes	16	7	6	29
	No	34	43	44	121
	Total	50	50	50	150

(a) State the null and alternative hypotheses for a test to determine if there is an association between type of question and response for subjects like these.

(b) Calculate the expected counts for a test of the null hypothesis stated in part (a).

(c) Check that the conditions for this test have been met.

(d) Calculate the test statistic and *P*-value for this test. What conclusion would you make?

3. **Who's cool at school? (10.3, 10.4)** Who were the popular kids at your elementary school? Did they get good grades or have good looks? Were they good at sports? A study was performed to examine factors that determine social status for children in grades 4, 5, and 6. Researchers gave a questionnaire to a random sample of 478 students in these grades. One of the questions they asked was, "What would you most like to do at school: Make good grades, be good at sports, or be popular?" The two-way table summarizes the students' responses.[46]

		Gender		
		Female	Male	Total
Goal	Grades	130	117	247
	Popular	91	50	141
	Sports	30	60	90
	Total	251	227	478

(a) Construct an appropriate graph to compare male and female responses. Briefly describe the relationship between gender and goals for these students.

(b) Do these data provide convincing evidence of an association between gender and goals for elementary school students?

4. **Big chickens (10.5)** Growth hormones are often used to increase the weight gain of chickens. In an experiment using 15 chickens, 3 chickens were randomly assigned to each of 5 different doses of growth hormone (0, 0.2, 0.4, 0.8, and 1.0 milligrams). The subsequent weight gain (in ounces) was recorded for each chicken. The equation of the least-squares regression line is $\hat{y} = 4.5459 + 4.8323x$, where $x =$ dose of growth hormone and $y =$ weight gain. The standard error of the slope is 1.0164. Assume that the conditions for performing inference about the slope of the true least-squares regression line are met.

(a) State hypotheses for a test about the relationship between dose of growth hormone and weight gain.

(b) Calculate the test statistic and *P*-value for a test of the hypothesis stated in part (a).

(c) Based on the *P*-value in part (b), what conclusion would you make?

5. **Short or tall, shoot the ball (10.6)** In basketball, taller players tend to play nearer to the basket than shorter players. Does this mean that taller players make a higher percentage of their shots? The table shows the relationship between $x =$ height (in inches) and $y =$ field goal percentage (shots made divided by shots attempted) for 18 randomly selected players in the National Basketball Association (NBA) during the 2014–2015 season.[47]

Player [Team]	Height (in.)	Field goal percentage
Ellis, Monta [Indiana]	75	44.5
Vonleh, Noah [Portland]	82	39.5
Whittington, Shayne [Indiana]	83	45.2
Olynyk, Kelly [Boston]	84	47.5
Dellavedova, Matthew [Cleveland]	76	36.2
Bogut, Andrew [Golden State]	84	56.3
Afflalo, Arron [New York]	77	42.8
McCallum, Ray [San Antonio]	75	43.8
Pierce, Paul [L.A. Clippers]	79	44.7
James, LeBron [Cleveland]	80	48.8
Allen, Tony [Memphis]	76	49.5
Durant, Kevin [Oklahoma City]	81	48.4
Mirotic, Nikola [Chicago]	82	40.5
Gortat, Marcin [Washington]	83	47.2
Young, Nick [L.A. Lakers]	79	41.7
Rubio, Ricky [Minnesota]	76	35.6
Belinelli, Marco [Sacramento]	77	42.3
Crabbe, Allen [Portland]	78	41.2

(a) Make a scatterplot to display the relationship between $y =$ field goal percentage and $x =$ height. Briefly describe the relationship.

(b) Do these data provide convincing evidence of a positive association between height and field goal percentage for 2014–2015 NBA players?

(c) Calculate and interpret a 90% confidence interval for the slope of the true least-squares regression line relating $y =$ field goal percentage to $x =$ height.

Chapter 10 Practice Test

Section I: Multiple Choice *Select the best answer for each question.*

Questions 1–3 refer to the following setting. A well-known chewing-gum manufacturer wants to find out if any of its four flavors of gum is more popular than the others. A random sample of 80 people who say they chew gum regularly is asked to identify their favorite flavor of gum. Here are the results.

Flavor	Peppermint	Cinnamon	Wintergreen	Spearmint
Frequency	25	19	22	14

1. Which of the following would be an appropriate null hypothesis for the company to test?

(a) $\mu_1 = \mu_2 = \mu_3 = \mu_4$

(b) Flavor preferences for the population are uniformly distributed across the four flavors.

(c) At least one of the four flavor preferences in the population is different from the other three.

(d) The observed counts are equal to the expected counts.

2. Which of the following conditions must be met to use a chi-square distribution to test the null hypothesis from Exercise 1?
  I.   The sample size is at least 30.
  II. All expected cell counts are at least 5.
  III. Respondents come from a random sample of all people who regularly chew gum.

(a) I and II only

(b) II and III only

(c) III only

(d) I, II, and III

3. Which of the following is the correct chi-square test statistic for this test?

(a) $\chi^2 = \dfrac{(25-20)^2}{20} + \dfrac{(19-20)^2}{20} + \dfrac{(22-20)^2}{20} + \dfrac{(14-20)^2}{20}$

(b) $\chi^2 = \dfrac{(25-20)^2}{25} + \dfrac{(19-20)^2}{19} + \dfrac{(22-20)^2}{22} + \dfrac{(14-20)^2}{14}$

(c) $\chi^2 = \dfrac{(25-20)}{20} + \dfrac{(19-20)}{20} + \dfrac{(22-20)}{20} + \dfrac{(14-20)}{20}$

(d) $\chi^2 = \dfrac{(25-20)}{25} + \dfrac{(19-20)}{19} + \dfrac{(22-20)}{22} + \dfrac{(14-20)}{14}$

Questions 4–6 refer to the following setting. Ida wants to know if males and females prefer different brands of ready-made chocolate-chip cookie dough. She bakes 8 dozen cookies of each of 4 brands of cookie dough (A, B, C, and D). She then selects a random sample of 96 students from her school and gives each student one cookie of each brand. For each student, Ida records two variables: gender and cookie brand preference. Here are her results.

		Gender		
		Male	Female	Total
Cookie brand preference	A	4	22	26
	B	6	12	18
	C	13	11	24
	D	15	13	28
	Total	38	58	96

Ida decides to perform a chi-square test to investigate the relationship between gender and cookie brand preference.

4. The alternative hypothesis for Ida's test is:

(a) The observed count in each cell is different from the expected count for that cell.

(b) The proportion who prefer cookie A is the same for both genders at her school.

(c) There is an association between preferred brand of cookie and gender in the sample.

(d) There is an association between preferred brand of cookie and gender at her school.

5. The expected count for the "Males/Brand B" cell is given by which of the following expressions?

(a) $\dfrac{(18)(6)}{38}$

(b) $\left(\dfrac{6}{38}\right)(96)$

(c) $\dfrac{(18)(38)}{96}$

(d) $\dfrac{(18)(6)}{96}$

6. The value of the test statistic is $\chi^2 = 11.09$. Which of the following intervals contains the P-value?

(a) $P\text{-value} < 0.01$

(b) $0.01 < P\text{-value} < 0.05$

(c) $0.05 < P\text{-value} < 0.10$

(d) $P\text{-value} > 0.10$

Questions 7–9 refer to the following setting. An old saying in golf is "You drive for show and you putt for dough." It means that good putting matters more than long driving for achieving better scores and winning more money. To see if this is the case, data from a random sample of 69 of the nearly 1000 players on the PGA Tour's world money list are examined. The average number of putts per hole (x) and the player's total winnings for the previous season (y) are recorded. The equation of the least-squares regression line is $\hat{y} = 7{,}897{,}179 - 4{,}139{,}198x$ and the standard error of the slope is 1,698,371.

7. Let β = the slope of the true least-squares regression line relating y = previous year's winnings to x = putts per hole. Which of the following are appropriate hypotheses for a test to determine if taking fewer putts per hole is associated with greater winnings?

(a) $H_0: \beta = 0$; $H_a: \beta > 0$

(b) $H_0: \beta = 0$; $H_a: \beta < 0$

(c) $H_0: \beta \neq 0$; $H_a: \beta = 0$

(d) $H_0: \beta < 0$; $H_a: \beta > 0$

8. Which of the following is the value of the test statistic for this test?

(a) $t = 1.91$

(b) $t = 2.44$

(c) $t = -1.91$

(d) $t = -2.44$

9. Which of the following is the 95% confidence interval for the slope of the true least-squares regression line relating winnings to putts per hole?

(a) $-4{,}139{,}198 \pm 2.00(1{,}698{,}371)$

(b) $-4{,}139{,}198 \pm 1.96(1{,}698{,}371)$

(c) $7{,}897{,}179 \pm 1.96(1{,}698{,}371)$

(d) $7{,}897{,}179 \pm 1.671(1{,}698{,}371)$

10. Which of the following is *not* one of the conditions that must be satisfied to perform inference about the slope of a least-squares regression line?

(a) For each value of x, the distribution of y-values is normally distributed.

(b) The data come from a random sample or a randomized experiment.

(c) The sample size—that is, the number of paired observations (x, y)—exceeds 30.

(d) The association between x and y is roughly linear.

Section II: Free Response

11. A large distributor of gasoline claims that 60% of all cars that stop at their service stations choose regular unleaded gas. Premium and supreme are each selected 20% of the time. To investigate this claim, researchers collected data from a random sample of drivers who put gas in their vehicles at the distributor's service stations in a large city. Here are the results. Carry out a test of the distributor's claim at the 5% significance level.

Gasoline selected		
Regular	Premium	Supreme
261	51	88

12. Just after President Obama announced plans to re-establish diplomatic relations with Cuba, a CBS poll asked 1002 randomly selected U.S. residents if they favored or opposed the president's action. The table summarizes the relationship between response and political party affiliation for the members of the sample.[48]

		Political party			
		Republican	Democrat	Independent	Total
Opinion	Favor	98	247	239	584
	Oppose	91	45	100	236
	Don't know/No answer	35	51	96	182
	Total	224	343	435	1002

Do these data provide convincing evidence of an association between political party and opinion about the president's policy shift in the population of U.S. residents?

13. A high school guidance counselor wonders if there is an association between y = a student's GPA in his or her senior year and x = the student's GPA in his or her freshman year. The counselor selects a random sample of 15 seniors from the graduating class of 468 and records both values for each student. Here are the data.

Freshman GPA	2.2	3.2	2.7	2.1	3.2	2.9	3.3	2.9
Senior GPA	2.7	3.8	2.7	2.3	3.3	3.6	3.0	3.0
Freshman GPA	2.1	3.3	2.3	3.6	3.6	3.3	3.3	
Senior GPA	2.7	3.6	3.5	3.5	3.2	3.7	3.2	

Do these data provide convincing evidence that senior GPA is associated with freshman GPA for students at this school?

Notes and Data Sources

Note: *The urls listed here were live when we researched this text.*

Chapter 1

1. Data from a student project by Daniel Molina and Kate Gallagher, Canyon del Oro High School, 2012.

2. Adapted from Franklin et al., *Guidelines for Assessment and Instruction in Statistics Education,* The American Statistical Association, 2005.

3. Roller coaster data for 2015 from www.rcdb.com.

4. Data obtained using the Random Sampler tool on the American Statistical Association's Census at School website: www.amstat.org/censusatschool/index.cfm.

5. Data obtained from The Numbers website: www.the-numbers.com/movie/records/All-Time-Worldwide-Box-Office on December 6, 2015.

6. U.S. Census Bureau, Current Population Survey, 2014 Annual Social and Economic Supplement.

7. Data obtained using the Random Data Selector tool on the Census at School International website: www.censusatschool.com.

8. http://pewinternet.org/Reports/2013/Smartphone-Ownership-2013.aspx.

9. *Radio Today 2013*, Executive Summary, downloaded from www.arbitron.com.

10. *Global Automotive 2014 Color Popularity Report* by Axalta Coating Systems, LLC.

11. Found at spam-filter-review.toptenreviews.com, which claims to have compiled data "from a number of different reputable sources."

12. Data for 2010 from the 2012 *Statistical Abstract of the United States* at the Census Bureau website, www.census.gov.

13. *The Asian Population: 2010,* at www.census.gov, based on the data collected from the 2010 U.S. Census.

14. K. Eagan, J. B. Lozano, S. Hurtado, & M. H. Case. *The American Freshman: National Norms, Fall 2013.* (2013) Los Angeles: Higher Education Research Institute, UCLA.

15. Data for 2008 from the 2012 *Statistical Abstract of the United States* Table 1239 at the Census Bureau website, www.census.gov.

16. Data obtained from skyscraperpage.com and Wikipedia.

17. *Model Year 2014 Fuel Economy Guide,* from the website www.fueleconomy.gov.

18. *Consumer Reports* magazine, January 2011.

19. *Model Year 2014 Fuel Economy Guide,* from the website www.fueleconomy.gov.

20. Data on the U.S. Women's National Soccer Team are from the United States Soccer Federation website, www.ussoccer.com.

21. The cereal data came from the Data and Story Library, http://lib.stat.cmu.edu/DASL/.

22. The original paper is T. M. Amabile, "Motivation and creativity: Effects of motivational orientation on creative writers," *Journal of Personality and Social Psychology,* 48, No. 2 (February 1985), pp. 393–399. The data for Exercise 43 came

from Fred L. Ramsey and Daniel W. Schafer, *The Statistical Sleuth,* 3rd ed., Brooks/Cole Cengage Learning, 2013.

23. Data for 2010 from the 2012 *Statistical Abstract of the United States* at the Census Bureau website, www.census.gov. The 1980 data were obtained from www.census.gov/prod/2001pubs/statab/sec01.pdf.

24. From the Electronic Encyclopedia of Statistics Examples and Exercises (EESEE) story "Acorn Size and Oak Tree Range."

25. USDA National Nutrient Database for Standard Reference 26 Software v.1.4

26. *The Older Population: 2010* by Carrie Werner, U.S. Census Bureau.

27. Basketball scores from the *Los Angeles Times.*

28. Fuel efficiency data from www.fueleconomy.gov.

29. Data obtained from www.petersons.com.

30. Pew Hispanic Center tabulations of 2011 American Community Survey.

31. CO_2 emissions data from the World Bank's website: http://data.worldbank.org/indicator/EN.ATM.CO2E.PC.

32. From the American Community Survey, at factfinder2.census.gov.

33. James T. Fleming, "The measurement of children's perception of difficulty in reading materials," *Research in the Teaching of English,* 1 (1967), pp. 136–156.

34. Monthly stock returns from the website of Professor Kenneth French of Dartmouth, http://lib.stat.cmu.edu/DASL/. A fine point: the data are the "excess returns" on stocks, which are actual returns less the small monthly returns on Treasury bills.

35. The cereal data came from the Data and Story Library, www.stat.cmu.edu/StatDat/.

36. Data from the Bureau of Labor Statistics, Annual Demographic Supplement www.bls.gov.

37. Data from the New York Road Runners Club website: www.nyrr.org.

38. From the College Board's AP® Central website: apcentral.collegeboard.com.

39. Runs scored data from www.baseball-reference.com.

40. Temperature data from www.wunderground.com/history.

41. From the American Community Survey, at the Census Bureau website, www.census.gov. The data are a subsample of the individuals in the ACS New York sample who had travel times greater than zero.

42. McDonald's USA Nutrition Facts for Popular Menu Items. This list was generated at www.mcdonalds.com on 07-23-2014.

43. *Model Year 2014 Fuel Economy Guide,* from the website www.fueleconomy.gov.

44. From the American Community Survey, at the Census Bureau website, www.census.gov. The data are a subsample of the 13,194 individuals in the ACS North Carolina sample who had travel times greater than zero.

45. The graph is based on an episode in the original Annenberg/Corporation for Public Broadcasting telecourse *Against All Odds: Inside Statistics.*

46. Tom Lloyd et al., "Fruit consumption, fitness, and cardiovascular health in female adolescents: the Penn State Young Women's Health Study," *American Journal of Clinical Nutrition,* 67 (1998), pp. 624–630.

47. C. B. Williams, *Style and Vocabulary: Numerological Studies,* Griffin, 1970.

48. Home price data from www.realtor.org.

49. See Note 41.

50. Data from http://pirate.shu.edu/~wachsmut/Teaching/MATH1101/Descriptives/variability.html.

51. Pets data set from *Guidelines for Assessment and Instruction in Statistics Education: Pre-K–12 Report,* American Statistical Association, 2007.

52. See Note 18.

53. See Note 50.

54. Sidney Crosby data from www.hockey-reference.com.

55. Hurricane data from www.wunderground.com.

56. *Nutrition Action,* April 2014.

57. These are some of the data from the EESEE story "Stress among Pets and Friends." The study results appear in K. Allen, J. Blascovich, J. Tomaka, and R. M. Kelsey, "Presence of human friends and pet dogs as moderators of autonomic responses to stress in women," *Journal of Personality and Social Psychology,* 83 (1988), pp. 582–589.

58. See Note 18.

59. Data from the Florida Fish and Wildlife Conservation Commission website, myfwc.com.

60. Data from Wikipedia, https://en.wikipedia.org/wiki/List_of_Best_in_Show_winners_of_the_Westminster_Kennel_Club_Dog_Show.

61. https://www.census.gov/topics/income-poverty/income.html.

62. Sheldon Ross, *Introduction to Probability and Statistics for Engineers and Scientists,* 3rd ed., Academic Press, 2004.

63. Information on bone density in the reference populations was found at www.courses.washington.edu/bonephys/opbmd.html.

64. This exercise is based on information from www.minerandcostudio.com.

Chapter 2

1. *Nutrition Action,* "Weighing the options: Do extra pounds mean extra years?" (March 2013).

2. Data from CensusAtSchool, www.amstat.org/censusatschool.

3. Data from http://www.pewinternet.org/2013/06/05/smartphone-ownership-2013/.

4. R. Shine et al., "The influence of nest temperatures and maternal brooding on hatchling phenotypes in water pythons," *Ecology,* 78 (1997), pp. 1713–1721.

5. The National Longitudinal Study of Adolescent Health interviewed a stratified random sample of 27,000 adolescents, then reinterviewed many of the subjects six years later, when most were aged 19 to 25. These data are from the Wave III reinterviews in 2000 and 2001, found at the website of the Carolina Population Center, www.cpc.unc.edu.

6. Pew Research Center for the People and the Press, "The cellphone challenge to survey research," news release for May 15, 2006, at www.people-press.org.

7. Data from: www.toyota-global.com/company/profile/figures/vehicle_production_sales_and_exports_by_region.html.

8. Data from CensusAtSchool, www.amstat.org/censusatschool.

9. Wikipedia, https://en.wikipedia.org/wiki/113th_United_States_Congress.

10. www.lumeradiamonds.com/diamonds/results?price=1082-1038045&carat=0.30-16.03&shapes=B&cut=EX&clarity=FL,IF&color=D,E,F#.

11. *Nutrition Action,* December 2009.

12. Based on T. N. Lam, "Estimating fuel consumption from engine size," *Journal of Transportation Engineering,* 111 (1985), pp. 339–357. The data for 10 to 50 km/h are measured; those for 60 and higher are calculated from a model given in the paper and are therefore smoothed.

13. www.rcdb.com.

14. www.weatherbase.com.

15. www.gapminder.org.

16. www.nutrition411.com/pdf/Bean%20Comparison%20Chart.pdf.

17. www.pewsocialtrends.org/2013/05/29/appendix-1-additional-charts/.

18. Data from the most recent Annual Demographic Supplement can be found at www.census.gov/cps.

19. http://myfwc.com/research/manatee/rescue-mortality-response/mortality-statistics/ and http://myfwc.com/boating/safety-education/accidents/.

20. http://www.tylervigen.com/.

21. http://www.nejm.org/doi/full/10.1056/NEJMon1211064.

22. www.basketball-reference.com.

23. Samuel Karelitz et al., "Relation of crying activity in early infancy to speech and intellectual development at age three years," *Child Development,* 35 (1964), pp. 769–777.

24. www.weatherbase.com.

25. From a graph in Bernt-Erik Saether, Steiner Engen, and Erik Mattysen, "Demographic characteristics and population dynamical patterns of solitary birds," *Science,* 295 (2002), pp. 2070–2073.

26. Data from Jordynn Watson, Canyon del Oro High School.

27. Data from Charlotte Lee, Glen A. Wilson High School.

28. www.gapminder.com.

29. www.basketball-reference.com/wnba/years/2013.html.

30. *Consumer Reports,* February 2010.

31. M. A. Houck et al., "Allometric scaling in the earliest fossil bird, *Archaeopteryx lithographica,*" *Science,* 247 (1990), pp. 195–198. The authors conclude from a variety of evidence that all specimens represent the same species.

32. www.basketball-reference.com.

33. http://www.ope.ed.gov/security/ 2012 data randomly selected from California universities.

34. Cheapest "wanna-get-away" fare on Southwest Airlines as of August 8, 2014.

35. Data from George W. Pierce, *The Songs of Insects,* Cambridge, MA, Harvard University Press, 1949, pp. 12–21.

36. Laurel Robertson, Carol Flinders, and Bronwen Godfrey, *Laurel's kitchen: A handbook for vegetarian cookery and nutrition.* Nilgiri Press, Petaluma, CA, 1976.

37. http://fivethirtyeight.com/features/should-travelers-avoid-flying-airlines-that-have-had-crashes-in-the-past/.

38. http://fivethirtyeight.com/features/should-travelers-avoid-flying-airlines-that-have-had-crashes-in-the-past/.

39. *Consumer Reports,* November 2005.

40. G. A. Sacher and E. F. Staffelt, "Relation of gestation time to brain weight for placental mammals: Implications for the theory of vertebrate growth," *American Naturalist,* 108 (1974), pp. 593–613. We found the data in Fred L. Ramsey and Daniel W. Schafer, *The Statistical Sleuth: A Course in Methods of Data Analysis,* Duxbury, 1997, p. 228.

41. Data on used car prices from autotrader.com, September 8, 2012. We searched for F-150 4 × 4's on sale within 50 miles of College Station, Texas.

42. Data from Kerry Lane and Danielle Neal, Canyon del Oro High School.

43. *Consumer Reports,* June 1986, pp. 366–367.

44. N. R. Draper and J. A. John, "Influential observations and outliers in regression," *Technometrics,* 23 (1981), pp. 21–26.

45. www.pro-football-reference.com.

46. *Nutrition Action,* December 2009.

47. Prices for LG LED televisions as of 8/21/14, http://www.lg.com/us/led-tvs.

48. From a graph in G. D. Martinsen, E. M. Driebe, and T. G. Whitham, "Indirect interactions mediated by changing plant chemistry: Beaver browsing benefits beetles," *Ecology,* 79 (1998), pp. 192–200.

49. http://www.soccerstats.com/widetable.asp?league= england.

50. C. J. Krebs, *Ecology: The experimental analysis of distribution and abundance,* 2nd edition. Harper and Row, 1978, p. 281.

51. www.edmunds.com.

52. *The World Almanac and Book of Facts* (2009).

53. Debora L. Arsenau, "Comparison of diet management instruction for patients with non–insulin dependent diabetes mellitus: Learning activity package vs. group instruction," M.S. thesis, Purdue University, 1993.

54. http://www.nutrition411.com/pdf/Bean%20Comparison%20Chart.pdf.

55. Frank J. Anscombe, "Graphs in statistical analysis," *The American Statistician,* 27 (1973), pp 17–21.

56. B. H. West, R. N. McGill, J. W. Hodgson, S. S. Sluder, and D. E. Smith, Development and verification of light-duty modal emissions and fuel consumption values for traffic models, FHWA-RD-99-068, U.S. Department of Transportation, Federal Highway Administration, Washington, D.C., March 1999.

57. David M. Fergusson and L. John Horwood, "Cannabis use and traffic accidents in a birth cohort of young adults," *Accident Analysis and Prevention,* 33 (2001), pp. 703–711.

58. Table 1 of E. Thomassot et al., "Methane-related diamond crystallization in the earth's mantle: Stable isotopes evidence from a single diamond-bearing xenolith," *Earth and Planetary Science Letters,* 257 (2007), pp. 362–371.

59. http://www.huffingtonpost.com/business/the-news/2013/02/02/.

60. Consumer Reports, January 2013.

61. College Board, https://research.collegeboard.org/programs/ap/data.

62. www.lpga.com.

63. www.amstat.org/censusatschool.

64. Data from Brittany Foley and Allie Dutson, Canyon del Oro High School.

65. From a graph in Christer G. Wiklund, "Food as a mechanism of density-dependent regulation of breeding numbers in the merlin *Falco columbarius,*" *Ecology,* 82 (2001), pp. 860–867.

66. http://www.pro-football-reference.com/years/2013/.

67. http://nutrition.mcdonalds.com/getnutrition/nutritionfacts.pdf.

68. www.pro-football-reference.com.

69. http://en.wikipedia.org/wiki/Transistor_count.

70. www.lpga.com.

71. Sample Pennsylvania female rates provided by Life Quotes, Inc., in *USA Today,* December 20, 2004.

72. http://www.lumeradiamonds.com/diamonds/results?price=1082-1038045&carat=0.30-16.03&shapes=B&cut=EX&clarity=FL,IF&color=D,E,F#.

73. Data estimated from a graph in Kyle G. Ashton, Russell L. Burke, and James N. Layne, "Geographic variation in body and clutch size of gopher tortoises," *Copeia,* 2007, No. 2 (2007), pp. 355–363.

74. www.baseball-reference.com.

75. http://en.wikipedia.org/wiki/World_population.

76. http://en.wikipedia.org/wiki/World_population.

77. S. Chatterjee and B. Price, *Regression Analysis by Example,* Wiley, 1977.

78. Data was estimated from a graph in M. H. Bornstein, "The pace of life: Revisited," *International Journal of Psychology,* 14 (1979), pp. 83–90.

79. Major League Baseball Players Association.

80. College Board, https://research.collegeboard.org/programs/ap/data.

81. Planetary data from http://hyperphysics.phy-astr.gsu.edu/hbase/solar/soldata2.html.

82. http://www.1stock1.com/1stock1_112.htm.

83. S. M. Stigler, "Do robust estimators work with real data?" *Annals of Statistics,* 5 (1977), pp. 1055–1078.

84. From the EESEE story "What makes a pre-teen popular?"

85. From the EESEE story "Is it tough to crawl in March?"

86. www.weatherbase.com.

87. http://www.thocp.net/companies/microsoft/microsoft_company.htm.

88. Daniel I. Bolnick and Thomas J. Near, "Tempo of hybrid inviability in centrarchid fishes (Teleostei: *centrarchidae*)." *Evolution,* 59, No. 8, (2005), pp. 1754–1767.

89. www.gapminder.com.

90. http://www.baseball-reference.com/players/j/jeterde01.shtml.

91. www.gapminder.com.

92. http://www.basketball-reference.com/teams/BRK/2014.html.

Chapter 3

1. http://www.plosmedicine.org/article/info%3Adoi%2F10.1371%2Fjournal.pmed.1001595.

2. "Family dinner linked to better grades for teens: Survey finds regular meal time yields additional benefits," written by John Mackenzie for ABC News's *World News Tonight,* September 13, 2005.

3. *Early Human Development,* 76, No. 2 (February 2004), pp. 139–145.

4. National Institute of Child Health and Human Development, Study of Early Child Care and Youth Development. The article appears in the July 2003 issue of *Child Development.* The quotation is from the summary on the NICHD website, www.nichd.nih.gov.

5. www.pediatrics.org/cgi/doi/10.1542/peds.2010-1919.

6. http://bigstory.ap.org/content/coffee-buzz-study-finds-java-drinkers-live-longer.

7. *Arizona Daily Star,* February 11, 2009.

8. From a graph in Bernt-Erik Saether, Steiner Engen, and Erik Mattysen, "Demographic characteristics and population dynamical patterns of solitary birds," *Science,* 295 (2002), pp. 2070–2073.

9. Frederick Mosteller and David L. Wallace, *Inference and disputed authorship: The Federalist.* Addison-Wesley, Reading, MA, 1964. Other information obtained from http://en.wikipedia.org/wiki/ and http://www.constitution.org/fed/federa51.htm.

10. Data obtained on September 3, 2014 from http://espn.go.com/tennis/rankings and http://www.wtatennis.com/singles-rankings.

11. Ed O'Brien and Phoebe C. Ellsworth, "Saving the last for best: A positivity bias for end experiences," *Psychological Science,* 23, No. 2 (2012), pp. 163–165.

12. Table 1 in Robert C. Parker and Patrick A. Glass, "Preliminary results of double-sample forest inventory of pine and mixed stands with high- and low-density LiDAR," in Kristina F. Connoe (Ed.), *Proceedings of the 12th Biennial Southern Silvicultural Research Conference,* U.S. Department of Agriculture, Forest Service, Southern Research Station, 2004. The researchers actually sampled every tenth plot. This is a systematic sample.

13. Gary S. Foster and Craig M. Eckert, "Up from the grave: A socio-historical reconstruction of an African American community from cemetery data in the rural Midwest," *Journal of Black Studies,* 33 (2003), pp. 468–489.

14. http://www.washingtonpost.com/blogs/monkey-cage/wp/2014/04/07/the-less-americans-know-about-ukraines-location-the-more-they-want-u-s-to-intervene/?tid=pm_pop.

15. According to the article "Euro Coin Accused of Unfair Flipping" in the *New Scientist* (January 4, 2002).

16. *San Gabriel Valley Tribune* (February 13, 2003).

17. Michele L. Head, "Examining college students' ethical values," Consumer Science and Retailing honors project, Purdue University, 2003.

18. This and similar results from the Pew Internet and American Life Project can be found at www.pewinternet.org.

19. These data are from "Results report on the vitamin C pilot program," prepared by SUSTAIN (Sharing United States Technology to Aid in the Improvement of Nutrition) for the U.S. Agency for International Development. The report was used by the Committee on International Nutrition of the National Academy of Sciences/Institute of Medicine (NAS/IOM) to make recommendations on whether or not the vitamin C content of food commodities used in U.S. food aid programs should be increased. The program was directed by Peter Ranum and Françoise Chomé.

20. Data from Melissa Silva and Madeline Dunlap, Canyon del Oro High School.

21. Data from Carly Myers and Maysem Ahmad, Canyon del Oro High School.

22. Cory Heid, "Tootsie Pops: How many licks to the chocolate?" *Significance Magazine,* October 2013, p. 47.

23. Bryan E. Porter and Thomas D. Berry, "A nationwide survey of self-reported red light running: Measuring prevalence, predictors, and perceived consequences," *Accident Analysis and Prevention,* 33 (2001), pp. 735–741.

24. Data from Christina Lesnewski and Rachel Polsky, Canyon del Oro High School.

25. Pew Research Center tabulation of Current Population Survey data.

26. Wins from http://www.basketball-reference.com/leagues/NBA_2015.html, attendance from http://espn.go.com/nba/attendance.

27. http://fivethirtyeight.com/features/is-the-polling-industry-in-stasis-or-in-crisis/.

28. Data from Hailey Kiernan, Canyon del Oro High School.

29. www.nytimes.com/1994/07/08/us/poll-on-doubt-of-holocaust-is-corrected.html.

30. http://www.cleaninginstitute.org/assets/1/AssetManager/2010%20Hand%20Washing%20Findings.pdf.

31. http://en.wikipedia.org/wiki/The_Literary_Digest.

32. See Note 23.

33. Mario A. Parada et al., "The validity of self-reported seatbelt use: Hispanic and non-Hispanic drivers in El Paso," *Accident Analysis and Prevention,* 33 (2001), pp. 139–143.

34. Data from Emma Merry, Canyon del Oro High School.

35. Data from Marcos Chavez-Martinez, Canyon del Oro High School.

36. Cynthia Crossen, "Margin of error: Studies galore support products and positions, but are they reliable?" *Wall Street Journal,* November 14, 1991.

37. See Note 23.

38. mlb.mlb.com.

39. http://www.nbcwashington.com/news/health/ADHD_Linked_To_Lead_and_Mom_s_Smoking.html.

40. The placebo effect examples are from Sandra Blakeslee, "Placebos prove so powerful even experts are surprised," *New York Times,* October 13, 1998.

41. The "three-quarters" estimate is cited by Martin Enserink, "Can the placebo be the cure?" *Science,* 284 (1999), pp. 238–240. An extended treatment is Anne Harrington (ed.), *The Placebo Effect: An Interdisciplinary Exploration,* Harvard University Press, 1997.

42. Carlos Vallbona et al., "Response of pain to static magnetic fields in postpolio patients, a double blind pilot study," *Archives of Physical Medicine and Rehabilitation,* 78 (1997), pp.1200–1203.

43. Steering Committee of the Physicians' Health Study Research Group, "Final report on the aspirin component of the ongoing Physicians' Health Study," *New England Journal of Medicine,* 321 (1989), pp. 129–135.

44. *Nutrition Action,* July/August 2008.

45. *Nutrition Action,* October 2013.

46. *Wall Street Journal,* August 15, 2011 http://tinyurl.com/l5hnxl3.

47. *Time,* December 9, 2010. http://tinyurl.com/2dt8hjj.

48. K. B. Suttle, Meredith A. Thomsen, and Mary E. Power, "Species interactions reverse grassland responses to changing climate," *Science,* 315 (2007), pp. 640–642.

49. Esther Duflo, Rema Hanna, and Stephan Ryan, "Monitoring works: Getting teachers to come to school," November 21, 2007, at http://web.stanford.edu/group/SITE/archive/SITE_2008/segment_5/papers/ryan_paper_and_tables.pdf.

50. Information found online at http://ssw.unc.edu/about/news/careerstart_1-13-09.

51. *New York Times,* November 11, 2014; http://www.nytimes.com/2014/11/11/science/dead-jellyfish-are-more-nutrition-than-nuisance.html?_r=0.

52. *Nutrition Action,* October 2013 (*Archives of Physical Medicine and Rehabilitation,* 93, p. 1269).

53. Marielle H. Emmelot-Vonk et al., "Effect of testosterone supplementation on functional mobility, cognition, and other parameters in older men," *Journal of the American Medical Association,* 299 (2008), pp. 39–52.

54. From the Electronic Encyclopedia of Statistical Examples and Exercises (EESEE), "Anecdotes of Placebos."

55. http://media.jamanetwork.com/news-item/flu-vaccine-associated-lower-risk-cardiovascular-events/.

56. Naomi D. L. Fisher, Meghan Hughes, Marie Gerhard-Herman, and Norman K. Hollenberg, "Flavonol-rich cocoa induces nitricoxide-dependent vasodilation in healthy humans," *Journal of Hypertension,* 21, No. 12 (2003), pp. 2281–2286.

57. http://www.nytimes.com/2008/03/05/health/research/05placebo.html?_r=0.

58. From the Electronic Encyclopedia of Statistical Examples and Exercises (EESEE), "Anecdotes of Placebos."

59. S. M. Stigler, "Do robust estimators work with real data?" *Annals of Statistics,* 5 (1977), pp. 1055–1078.

60. http://archinte.jamanetwork.com/article.aspx?articleid=1899554.

61. http://www.sciencedirect.com/science/article/pii/S0360131512002254.

62. http://foodpsychology.cornell.edu/OP/buffet_pricing.

63. http://www.sciencedaily.com/releases/2014/11/141124081040.htm.

64. Joel Brockner et al., "Layoffs, equity theory, and work performance: Further evidence of the impact of survivor guilt," *Academy of Management Journal,* 29 (1986), pp. 373–384.

65. http://cg4tx.org/wp-content/uploads/2014/11/kalla-broockman-donor-access-to-lege.pdf.

66. *Nutrition Action,* July/August 2014 (*Journal of the American Medical Association,* 311, p. 2083).

67. *Nutrition Action,* March 2009.

68. Details of the Carolina Abecedarian Project, including references to published work, can be found online at abc.fpg.unc.edu.

69. Christopher Anderson, "Measuring what works in health care," *Science,* 263 (1994), pp. 1080–1082.

70. www.azstarnet.com/news/science/health-med-fit/article_28f1f958-0193-5a3e-949a-d8bd21ac57b4.html.

71. *New York Times,* November 19, 2014; http://tinyurl.com/k8xm6rj.

72. http://www.nbc.com/the-voice/vote/rules.

73. Karpinski and A. Duberstein, "A description of Facebook use and academic performance among undergraduate and graduate students," (Ch. 1, ref 50 A), paper presented at the American Educational Research Association annual meeting, April 2009. Thanks to Aryn Karpinski for providing us with some original data from the study.

74. David L. Strayer, Frank A. Drews, and William A. Johnston, "Cell phone–induced failures of visual attention during simulated driving," *Journal of Experimental Psychology: Applied,* 9 (2003), pp. 23–32.

75. *New England Journal of Medicine,* 320 (1989), pp. 1037–1043; cited in Fred Ramsey and Daniel Schafer, *The Statistical Sleuth,* Pacific Grove, CA: Duxbury Press, 2002, p. 23.

76. *Scientific American,* September 13, 2011. http://www.scientificamerican.com/article/how-to-improve-your-life-with-story-editing/.

77. http://annals.org/article.aspx?articleid=746567.

78. W. E. Paulus et al., "Influence of acupuncture on the pregnancy rate in patients who undergo assisted reproductive therapy," *Fertility and Sterility,* 77, No. 4 (2002), pp. 721–724.

79. http://www.nejm.org/doi/full/10.1056/NEJMoa1107911.

80. "Antibiotics no better than placebo for sinus infections" https://www.minnpost.com/second-opinion/2012/02/antibiotics-no-better-placebo-treating-sinus-infections-study-finds.

81. Data from Erin Fung and Jenna Parker, Canyon del Oro High School.

82. Data from Michael Khawam, Canyon del Oro High School.

83. http://www.nejm.org/doi/full/10.1056/NEJMoa0905471.

84. Data from Jantzen Hale, Canyon del Oro High School.

85. http://www.plosone.org/article/info%3Adoi%2F10.1371%2Fjournal.pone.0016782.

86. Data from Maddi Edwards and Natasha Kunzler, Canyon del Oro High School.

87. Niels Juel-Nielsen, *Individual and Environment: Monozygotic Twins Reared Apart,* International Universities Press, 1980.

88. Scott DeCarlo, with Michael Schubach and Vladimir Naumovski, "A decade of new issues," *Forbes,* March 5, 2001, www.forbes.com.

89. The sleep deprivation study is described in R. Stickgold, L. James, and J. Hobson, "Visual discrimination learning requires post-training sleep," *Nature Neuroscience,* 2000, pp. 1237–1238. We obtained the data from Allan Rossman, who got it courtesy of the authors.

90. *Arizona Daily Star,* February 16, 2015.

91. See the details on the website of the Office for Human Research Protections of the Department of Health and Human Services, hhs.gov/ohrp.

92. Charles A. Nelson III et al., "Cognitive recovery in socially deprived young children: The Bucharest Early Intervention Project," *Science,* 318 (2007), pp. 1937–1940.

93. *Nutrition Action,* December 2014 (*American Journal of Clinical Nutrition,* 100, p. 1182).

94. Marilyn Ellis, "Attending church found factor in longer life," *USA Today,* August 9, 1999.

95. *Nutrition Action,* December 2014 (*Acta Psychologica,* 153, p. 13).

96. *Nutrition Action,* March 2013 (*Circulation* 127, p. 188), http://www.ncbi.nlm.nih.gov/pubmed/23319811.

97. The article describing this study is Nikhil Swaminathan, "Gender jabber: Do women talk more than men?" *Scientific American,* July 6, 2007.

98. http://www.pnas.org/content/111/24/8788.full.

99. *Nutrition Action,* March 2014 (JAMA: Pediatr. 2013, doi:10.1001/jamapediatrics.2013.4139), http://archpedi.jamanetwork.com/article.aspx?articleid=1793699.

100. *Nutrition Action,* March 2014 (Psychosom. Med. 68: 809, 2006), http://www.ncbi.nlm.nih.gov/pubmed/17101814.

101. Data from the report "Is our tuna family-safe?" prepared by Defenders of Wildlife, 2006.

102. Cyrus Vance Jr.'s 'Moneyball' Approach to Crime. *New York Times Magazine*, December 7, 2014. http://www.nytimes.com/2014/12/07/magazine/cyrus-vance-jrs-moneyball-approach-to-crime.html?smid=fb-share&_r=0.

103. http://www.gallup.com/poll/180260/americans-rate-nurses-highest-honesty-ethical-standards.aspx.

104. L. E. Moses and F. Mosteller, "Safety of anesthetics," in J. M. Tanur et al. (eds.), *Statistics: A Guide to the Unknown*, 3rd ed., Wadsworth, 1989, pp. 15–24.

105. http://well.blogs.nytimes.com/2014/12/22/e-books-may-interfere-with-sleep/?module=Search&mabReward=relbias%3As%2C%7B%221%22%3A%22RI%3A11%22%7D.

106. Based on a news item "Bee off with you," *Economist*, November 2, 2002, p. 78.

107. http://en.wikipedia.org/wiki/Milgram_experiment.

Chapter 4

1. Poll results from www.yuricareport.com/Miscellaneous/AARP_Rig%20Its%20Own%20Poll.htm.

2. Data for 2006 from the website of Statistics Canada, www.statcan.gc.ca.

3. 2009 National Household Travel Survey: http://nhts.ornl.gov/2009/pub/stt.pdf.

4. These figures are cited by Dr. William Dement at "The Sleep Well" website, www.stanford.edu/~dement/.

5. Lenhart, Amanda, "Teen, Social Media and Technology Overview 2015," Pew Research Center, April 2015.

6. Gail Burrill, "Two-way tables: Introducing probability using real data," paper presented at the Mathematics Education into the Twenty-first Century Project, Czech Republic, September 2003. Burrill cites as her source H. Kranendonk, P. Hopfensperger, and R. Scheaffer, *Exploring Probability*, Dale Seymour Publications, 1999.

7. Stephen J. Blumberg, Ph.D., and Julian V. Luke, "Wireless substitution: Early release of estimates from the National Health Interview Survey," July–December 2012, National Center for Health Statistics.

8. Maeve Duggan and Aaron Smith, "Social Media Update 2013," Pew Research Center, January 2014.

9. Data for 2015 obtained from Wikipedia.

10. From the EESEE story "What Makes a Pre-teen Popular?"

11. From the website of the Carolina Population Center, www.cpc.unc.edu.

12. The table closely follows the grade distributions for these three schools at the University of New Hampshire in the fall of 2000, found in a self-study document at www.unh.edu/academic-affairs/neasc/. The counts of grades mirror the proportions of UNH students in these schools. The table is simplified to refer to a university with only these three schools.

13. The National Longitudinal Study of Adolescent Health interviewed a stratified random sample of 27,000 adolescents, then reinterviewed many of the subjects six years later, when most were aged 19 to 25. These data are from the Wave III reinterviews in 2000 and 2001, found at the website of the Carolina Population Center, www.cpc.unc.edu.

14. See Note 7.

15. Data on David Ortiz from www.baseball-reference.com.

16. http://www.washingtonpost.com/wp-dyn/content/article/2009/09/23/AR2009092301947.html, http://www.collegesports scholarships.com/percentage-high-school-athletes-ncaa-college.htm.

17. Information about Internet users comes from sample surveys carried out by the Pew Internet and American Life Project, at www.pewinternet.org.

18. Victoria J. Rideout, M.A., Ulla G. Foehr, Ph.D., and Donald F. Roberts, Ph.D., "GENERATION M: Media in the Lives of 8- to 18-year-olds," Kaiser Family Foundation, January 2010.

19. From the National Institutes of Health's National Digestive Diseases Information Clearinghouse, found at http://digestive.niddk.nih.gov/.

20. Information about Internet users comes from sample surveys carried out by the Pew Internet and American Life Project, found online at www.pewinternet.org. The music-downloading data were collected in 2003.

21. We got these data from the Energy Information Administration on their website at http://www.eia.gov/dnav/pet/pet_sum_mkt_dcu_sct_m.htm.

22. Thanks to Michael Legacy for suggesting this exercise.

23. Data on Roger Federer's serve percentages from www.atpworldtour.com.

24. Probabilities from trials with 2897 people known to be free of HIV antibodies and 673 people known to be infected are reported in J. Richard George, "Alternative specimen sources: Methods for confirming positives," 1998 Conference on the Laboratory Science of HIV, found online at the Centers for Disease Control and Prevention, www.cdc.gov.

25. This exercise was inspired by the report at www.cbsnews.com/stories/2010/06/01/health/webmd/main6537635.shtml.

26. The probabilities given are realistic, according to the fundraising firm SCM Associates, at scmassoc.com.

27. *Arizona Daily Star*, November 13, 2014, http://tucson.com/news/science/environment/tucson-country-club-area-uses-most-water/article_fd2acb0d-6663-52ae-aa81-9e00bc4c7a76.html.

28. See Chapter 1, Note 17.

29. This is one of several tests discussed in Bernard M. Branson, "Rapid HIV testing: 2005 update," a presentation by the Centers for Disease Control and Prevention, at www.cdc.gov. The Malawi clinic result is reported by Bernard M. Branson, "Point-of-care rapid tests for HIV antibody," *Journal of Laboratory Medicine*, 27 (2003), pp. 288–295.

30. Information about the First Trimester Screen from www.americanpregnancy.org/prenataltesting/firstscreen.html.

31. The probabilities in this exercise are taken from Tommy Bennett's article "Expanded Horizons: Perfection," June 8, 2010 at www.baseballprospectus.com.

32. See Note 7.

33. Robert P. Dellavalle et al., "Going, going, gone: Lost Internet references," *Science*, 302 (2003), pp. 787–788.

34. Thanks to Corey Andreasen for suggesting the idea for this exercise.

35. http://www.dailymail.co.uk/sciencetech/article-2410532/Why-toast-falls-butter-Scientists-finally-uncover-reason-height-table.html.

36. Nutritional information about bagels from www.einsteinbros.com.

37. Data on workplace injuries from the Bureau of Labor Statistics website: www.bls.gov/news.release/osh2.t13.htm.

38. Home run data from www.baseball-reference.com.

39. The table in this exercise was constructed using the search function at the GSS archive, sda.berkeley.edu/archive.htm.

40. We found the article online at www.theguardian.com/news/datablog/2011/aug/04/live-to-100-likely.

41. The probability distribution was based on data found at http://online.wsj.com/mdc/public/page/2_3022-autosales.html.

Chapter 5

1. The Apgar score data came from National Center for Health Statistics, *Monthly Vital Statistics Reports*, Vol. 30, No. 1, Supplement, May 6, 1981.

2. Grade distribution obtained from the Office of the Registrar, Indiana University Bloomington. Data accessed online at http://gradedistribution.registrar.indiana.edu/.

3. You can find a mathematical explanation of Benford's law in Ted Hill, "The first-digit phenomenon," *American Scientist*, 86 (1996), pp. 358–363; and Ted Hill, "The difficulty of faking data," *Chance*, 12, No. 3 (1999), pp. 27–31. Applications in fraud detection are discussed in the second paper by Hill and in Mark A. Nigrini, "I've got your number," *Journal of Accountancy*, May 1999, available online at www.journalofaccountancy.com/issues/1999/may/nigrini.

4. The National Longitudinal Study of Adolescent Health interviewed a stratified random sample of 27,000 adolescents, then reinterviewed many of the subjects six years later, when most were aged 19 to 25. These data are from the Wave III reinterviews in 2000 and 2001, found at the website of the Carolina Population Center, www.cpc.unc.edu.

5. Probability distribution based on a sample of students from the U.S. Census at School database, www.amstat.org/censusatschool/.

6. Aaron Smith, "The Best and Worst of Mobile Connectivity," Pew Research Center, November 30, 2012. Accessed online at www.pewinternet.org.

7. See Note 2.

8. See Note 3.

9. Data from the Census Bureau's American Housing Survey.

10. Emily Oster, "It's hard to know where gluten sensitivity stops and the placebo effect begins," published February 11, 2015 at www.fivethirtyeight.com.

11. We obtained the Super Bowl television viewing data from http://tvbythenumbers.zap2it.com/2013/01/30/will-super-bowl-xlvii-tv-viewership-set-another-record-poll-ratings-history/167078/.

12. We got the 9% figure from The Pew Research Center's "Assessing the Representativeness of Public Opinion Surveys," published on May 15, 2012.

13. Office of Technology Assessment, *Scientific Validity of Polygraph Testing: A Research Review and Evaluation*, Government Printing Office, 1983.

14. Ed O'Brien and Phoebe C. Ellsworth, "Saving the last for best: A positivity bias for end experiences," *Psychological Science*, 23, No. 2 (September 2011), pp. 163–165.

15. Data from Gary Community School Corporation, courtesy of Celeste Foster, Department of Education, Purdue University.

16. Detailed data appear in P. S. Levy et al., *Total Serum Cholesterol Values for Youths 12–17 Years*, Vital and Health Statistics, Series 11, No. 155, National Center for Health Statistics, 1976.

17. Kevin Quealy, Amanda Cox, and Josh Katz, "At Chipotle, how many calories do people really eat?" *New York Times*, February 17, 2015.

18. We found the information on birth weights of Norwegian children on the National Institute of Environmental Health Sciences website. The relevant article can be accessed here: www.ncbi.nlm.nih.gov/pubmed/1536353.

19. We obtained the data on number of televisions per household from www.statista.com.

20. John C. Kern. *Journal of Statistics Education* Volume 14, Number 3 (2006), www.amstat.org/publications/jse/v14n3/datasets.kern.html.

Chapter 6

1. This activity is based on a similar activity suggested in Richard L. Schaeffer, Ann Watkins, Mrudulla Gnanadesikan, and Jeffrey A. Witmer, *Activity-Based Statistics*, Springer, 1996.

2. http://www.census.gov/hhes/www/cpstables/032014/perinc/pinc01_000.htm.

3. http://www.pewinternet.org/files/old-media/Files/Reports/2009/PIP%20Teens%20and%20Mobile%20Phones%20Data%20Memo.pdf.

4. Libertystreeteconomics.newyorkfed.org/2012/03/grading-student-loans.html.

5. The survey question is reported in Trish Hall, "Shop? Many say 'Only if I must,'" *New York Times*, November 28, 1990. In fact, 66% (1650 of 2500) in the sample said "Agree."

6. http://www.offa.org/stats_hip.html.

7. http://www.pewsocialtrends.org/2014/09/24/record-share-of-americans-have-never-married/.

8. Data from Zenon Kane, Canyon del Oro High School.

9. http://www.nber.org/papers/w20689, referenced in http://fivethirtyeight.com/datalab/summer-reading-government-jobs-and-charitable-donations/.

10. Karl Pearson and A. Lee, "On the laws of inheritance in man," *Biometrika*, 2 (1902), p. 357. These data also appear in D. J. Hand et al., *A Handbook of Small Data Sets*, Chapman & Hall, 1994. This book offers more than 500 data sets that can be used in statistical exercises.

11. www.gallup.com.

12. www.bls.gov/news.release/pdf/famee.pdf.

13. https://www.kickstarter.com/help/stats.

14. David J. LeBlanc, Michael Sivak, and Scott Bogard, "Using naturalistic driving data to assess variations in fuel efficiency among individual drivers," The University of Michigan Transportation Research Institute, http://deepblue.lib.umich.edu/bitstream/handle/2027.42/78449/102705.pdf.

15. Based on a figure in Peter R. Grant. *Ecology and Evolution of Darwin's Finches*, Princeton University Press, 1986.

16. A. J. Wilcox and I. T. Russell, "Birthweight and perinatal mortality: I. On the frequency distribution of birthweight," *International Journal of Epidemiology*, 12 (1983), pp. 314–318.

17. https://sleepfoundation.org/sites/default/files/2014-NSF-Sleep-in-America-poll-summary-of-findings---FINAL-Updated-3-26-14-.pdf.

18. http://media.wix.com/ugd/80ea24_edd136e3b72b07c93775906aee3dfa35.pdf.

Chapter 7

1. Thanks to Floyd Bullard for sharing the idea for this activity.

2. http://www.gallup.com/poll/181217/americans-healthcare-low-wages-top-financial-problems.aspx.

3. http://www.gallup.com/poll/181283/obama-approval-hits-first-time-2013.aspx.

4. http://www.census.gov/content/dam/Census/library/publications/2014/demo/p60-249.pdf.

5. http://www.smithsonianmag.com/ideas-innovations/How-Much-Do-Americans-Know-About-Science.html.

6. See Note 5.

7. Amanda Lenhart and Mary Madden, "Teens, privacy and online social networks," Pew Internet and American Life Project, 2007, at www.pewinternet.org.

8. www.basketball-reference.com.

9. This and similar results from the Pew Internet and American Life Project can be found at www.pewinternet.org.

10. Data from Ellery Page, Canyon del Oro High School.

11. http://www.pewresearch.org/files/2014/01/Survey-Questions_Facebook.pdf.

12. This and similar results of Gallup polls are from the Gallup Organization website, www.gallup.com.

13. Data from 2013 Current Population Survey, found at http://www.eeps.com/zoo/acs/source/index.php.

14. See Note 13.

15. Data from 2014 Current Population Survey, found at http://www.census.gov/hhes/socdemo/education/. Does not include institutionalized residents.

16. http://www.nap.edu/openbook.php?record_id=1195&page=537.

17. http://cdn.annenbergpublicpolicycenter.org/wp-content/uploads/Civics-survey-press-release-09-17-2014-for-PR-Newswire.pdf.

18. Eric Sanford et al., "Local selection and latitudinal variation in a marine predator–prey interaction," *Science,* 300 (2003), pp. 1135–1137.

19. Data from Jordan Tabor.

20. Michele L. Head, "Examining college students' ethical values," Consumer Science and Retailing honors project, Purdue University, 2003.

21. http://www.usatoday.com/picture-gallery/news/2015/04/07/usa-today-snapshots/6340793/.

22. "Poll: Men, women at odds on sexual equality," Associated Press dispatch appearing in the *Lafayette (Ind.) Journal and Courier,* October 20, 1997.

23. http://www.indexmundi.com/facts/united-states/quick-facts/all-states/percent-of-population-under-18#table.

24. http://sleepfoundation.org/sleep-polls-data/2015-sleep-and-pain.

25. https://kaiserfamilyfoundation.files.wordpress.com/2013/01/8010.pdf.

26. Pew Research Center's Teens Relationships Survey, September 25–October 9, 2014, and February 10–March 16, 2015.

27. http://www.gallup.com/poll/163238/americans-reject-size-limit-soft-drinks-restaurants.aspx?utm_source=soft%20drinks&utm_medium=search&utm_campaign=tiles.

28. http://www.quinnipiac.edu/news-and-events/quinnipiac-university-poll/connecticut/release-detail?ReleaseID=2176.

29. Based on information in "NCAA 2003 national study of collegiate sports wagering and associated health risks," which can be found on the NCAA website, www.ncaa.org.

30. Data from Tori Heimink and Ann Perry, Canyon del Oro High School.

31. Cory Heid, "Tootsie pops: How many licks to the chocolate?" *Significance Magazine,* October 2013, page 47.

32. Alan S. Banks et al., "Juvenile hallux abducto valgus association with metatarsus adductus," *Journal of the American Podiatric Medical Association,* 84 (1994), pp. 219–224.

33. TUDA results for 2003 from the National Center for Education Statistics, at nces.ed.gov/nationsreportcard.

34. M. Ann Laskey et al., "Bone changes after 3 mo of lactation: Influence of calcium intake, breast-milk output, and vitamin D–receptor genotype," *American Journal of Clinical Nutrition,* 67 (1998), pp. 685–692.

35. http://www.gallup.com/poll/182261/four-americans-look-forward-checking-mail.aspx?utm_source=Politics&utm_medium=newsfeed&utm_campaign=tiles.

36. Data from Christina Lesnewski and Rachel Polsky.

37. Harry B. Meyers, "Investigations of the life history of the velvetleaf seed beetle, *Althaeus folkertsi* Kingsolver," M.S. thesis, Purdue University, 1996.

38. These data are from "Results report on the vitamin C pilot program," prepared by SUSTAIN (Sharing United States Technology to Aid in the Improvement of Nutrition) for the U.S. Agency for International Development. The report was used by the Committee on International Nutrition of the National Academy of Sciences/Institute of Medicine (NAS/IOM) to make recommendations on whether or not the vitamin C content of food commodities used in U.S. food aid programs should be increased. The program was directed by Peter Ranum and Françoise Chomé.

39. Data provided by Drina Iglesia, Purdue University. The data are part of a larger study reported in D. D. S. Iglesia, E. J. Cragoe, Jr., and J. W. Vanable, "Electric field strength and epithelization in the newt (*Notophthalmus viridescens*)," *Journal of Experimental Zoology,* 274 (1996), pp. 56–62.

40. http://www.crosswordtournament.com/2015/index.htm.

41. Data from Melissa Silva and Madeline Dunlap, Canyon del Oro High School.

42. Data from Carly Myers and Maysem Ahmad, Canyon del Oro High School.

43. Data from Ben Garcia and Maya Kraft, Canyon del Oro High School.

44. These data are a partial subset of the dataset in Plank, Martin E. 1981. Estimating Value and Volume of Ponderosa Pine Trees by Equations. U.S. Forest Service Research Paper PNW-283, http://www.fs.fed.us/pnw/pubs/pnw_rp283.pdf.

45. Population base from the 2006 Labor Force Survey, at www.statistics.gov.uk. Smoking data from Action on Smoking and Health, *Smoking and Health Inequality,* at www.ash.org.uk.

46. http://www.washingtonpost.com/page/2010-2019/WashingtonPost/2014/03/23/National-Politics/Polling/question_13300.xml?uuid=07HdFrI_EeO4s0Sx0c1MHw.

47. http://www.cdc.gov/healthyyouth/data/yrbs/results.htm.

48. James T. Fleming, "The measurement of children's perception of difficulty in reading materials," *Research in the Teaching of English,* 1 (1967), pp. 136–156.

49. http://www.gallup.com/poll/159881/americans-call-term-limits-end-electoral-college.aspx.

Chapter 8

1. P. A. Mackowiak, S. S. Wasserman, and M. M. Levine, "A critical appraisal of 98.6 degrees F, the upper limit of the normal body temperature, and other legacies of Carl Reinhold

August Wunderlich," *Journal of the American Medical Association,* 268 (1992), pp. 1578–1580.

2. Results of this study were reported in the *San Gabriel Valley Tribune* on February 13, 2003.

3. Julie Ray, "Few teens clash with friends," May 3, 2005, on the Gallup Organization website, www.gallup.com.

4. Kevin M. Simmons, Daniel Sutter, and Roger Pielke, "Normalized tornado damage in the United States: 1950–2011," *Environmental Hazards,* 12, No. 2 (2013), pp. 132–147, http://sciencepolicy.colorado.edu/admin/publication_files/2012.31.pdf.

5. Thanks to Josh Tabor for suggesting the idea for this example.

6. Julie Ray, "Few teens clash with friends," May 3, 2005, on the Gallup Organization website, www.gallup.com.

7. The idea for this exercise was provided by Michael Legacy and Susan McGann.

8. See Note 6.

9. Laura Kann, Ph.D., Steve Kinchen, Shari L. Shanklin, et al., Youth Risk Behavior Surveillance—United States, 2013.

10. National Institute for Occupational Safety and Health, *Stress at Work,* 2000, available online at http://www.cdc.gov/niosh/docs/99-101/. Results of this survey were reported in *Restaurant Business,* September 15, 1999, pp. 45–49.

11. From the report "Sex and tech: Results from a study of teens and young adults," published by the National Campaign to Prevent Teen and Unplanned Pregnancy, www.thenationalcampaign.org/sextech.

12. Dorothy Espelage et al., "Factors associated with bullying behavior in middle school students," *Journal of Early Adolescence,* 19, No. 3 (August 1999), pp. 341–362.

13. This and similar results of Gallup polls are from the Gallup Organization website, www.gallup.com.

14. Aaron Smith and Joanna Brenner, "Twitter Use 2012," published by the Pew Internet and American Life Project, May 31, 2012.

15. USDA Agricultural research service, http://ndb.nal.usda.gov/ndb/foods?format=&count=&max=35&sort=&fgcd=Dairy+and+Egg+Products&manu=&lfacet=&qlookup=cheese&offset=70&order=desc.

16. This exercise is based on events that are real. The data and details have been altered to protect the privacy of the individuals involved.

17. *Model Year 2014 Fuel Economy Guide,* from the website www.fueleconomy.gov.

18. W. S. Gosset, "The probable error of a mean," *Biometrika,* 6 (1908), pp. 1–25. We obtained the sleep data from the Data and Story Library (DASL) website, http://lib.stat.cmu.edu/DASL/. They cite as a reference R. A. Fisher, *The Design of Experiments,* 3rd ed., Oliver and Boyd, 1942, p. 27.

19. www.cbsnews.com/news/poll-who-can-get-ahead-in-the-u-s/.

20. http://blogs.edweek.org/edweek/curriculum/2015/03/homework_math_science_study.html from this study: http://www.apa.org/pubs/journals/releases/edu-0000032.pdf.

Chapter 9

1. This case study is based on the story "Drive-Thru Competition" in the Electronic Encyclopedia of Statistical Examples and Exercises (EESEE). Updated data were obtained from the *QSR* magazine website: www.qsrmagazine.com/drive-thru.

2. Amanda Lenhart, "Teens, social media & technology overview 2015," Pew Research Center, April 9, 2015.

3. Maeve Duggan, Nicole B. Ellison, Cliff Lampe, Amanda Lenhart, and Mary Madden, "Social Media Update 2014," Pew Research Center, January 9, 2015.

4. From the website of the Black Youth Project, blackyouthproject.com.

5. The idea for this exercise was inspired by an example in David M. Lane's *Hyperstat Online* textbook at http://davidmlane.com/hyperstat.

6. Clive G. Jones et al., "Chain reactions linking acorns to gypsy moth outbreaks and Lyme disease risk," *Science,* 279 (1998), pp. 1023–1026.

7. Maeve Duggan, "Mobile messaging and social media 2015," Pew Research Center, August 19, 2015.

8. The National Longitudinal Study of Adolescent Health interviewed a stratified random sample of 27,000 adolescents, then reinterviewed many of the subjects six years later, when most were aged 19 to 25. These data are from the Wave III reinterviews in 2000 and 2001, found at the website of the Carolina Population Center, www.cpc.unc.edu.

9. www.cbsnews.com/htdocs/pdf/poll_whereamericastands_obesity_010710.pdf.

10. Kwang Y. Cha, Daniel P. Wirth, and Rogerio A. Lobo, "Does prayer influence the success of *in vitro* fertilization–embryo transfer?" *Journal of Reproductive Medicine,* 46 (2001), pp. 781–787.

11. These data are reported in the *Arizona Daily Star,* February 11, 2009.

12. Based on Deborah Roedder John and Ramnath Lakshmi-Ratan, "Age differences in children's choice behavior: The impact of available alternatives," *Journal of Marketing Research,* 29 (1992), pp. 216 226.

13. National Athletic Trainers Association, press release, dated September 30, 1994.

14. The study is reported in William Celis III, "Study suggests Head Start helps beyond school," *New York Times,* April 20, 1993. See www.highscope.org.

15. *Arizona Daily Star,* August 18, 2010.

16. C. P. Cannon et al., "Intensive versus moderate lipid lowering with statins after acute coronary syndromes," *New England Journal of Medicine,* 350 (2004), pp. 1495–1504.

17. George Du Toit, M.B., B.Ch., et al., "Randomized Trial of Peanut Consumption in Infants at Risk for Peanut Allergy," *New England Journal of Medicine,* 372 (February 2015), pp. 803–813.

18. Data for this exercise were obtained from Wikipedia, https://en.wikipedia.org/wiki/Car_of_the_Year.

19. We obtained the National Health and Nutrition Examination Survey data from the Centers for Disease Control and Prevention website at www.cdc.gov/nchs/nhanes.htm.

20. Shailija V. Nigdikar et al., "Consumption of red wine polyphenols reduces the susceptibility of low-density lipoproteins to oxidation in vivo," *American Journal of Clinical Nutrition,* 68 (1998), pp. 258–265.

21. Ethan J. Temeles and W. John Kress, "Adaptation in a plant–hummingbird association," *Science,* 300 (2003), pp. 630–633. We thank Ethan J. Temeles for providing the data.

22. http://pediatrics.aappublications.org/content/early/2011/09/08/peds.2010-1919.

23. From the EESEE story "Surgery in a Blanket."

24. Data for this example from Noel Cressie, *Statistics for Spatial Data,* Wiley, 1993.

25. Data from a student project by Briana Dohogne and Maggie McCord, Canyon del Oro High School.

26. This study is reported in Roseann M. Lyle et al., "Blood pressure and metabolic effects of calcium supplementation in normotensive white and black men," *Journal of the American Medical Association,* 257 (1987), pp. 1772–1776. The data were provided by Dr. Lyle.

27. Gabriela S. Castellani, "The effect of cultural values on Hispanics' expectations about service quality," M.S. thesis, Purdue University, 2000.

28. Data from a student project by Mary Ann McRae and Abigail O'Conner, Canyon del Oro High School.

29. Matthias R. Mehl, Simine Vazire, Nairán Ramírez-Esparza, Richard B. Slatcher, and James W. Pennebaker, "Are women really more talkative than men?" *Science,* 317 (July 2007), www.sciencemag.org.

30. *New England Journal of Medicine,* 320 (1989), pp. 1037–1043; cited in Fred Ramsey and Daniel Schafer, *The Statistical Sleuth,* Pacific Grove, CA: Duxbury Press, 2002, p. 23.

31. Adapted from Maribeth Cassidy Schmitt, "The effects of an elaborated directed reading activity on the metacomprehension skills of third graders," Ph.D. dissertation, Purdue University.

32. The original paper is T. M. Amabile, "Motivation and creativity: Effects of motivational orientation on creative writers," *Journal of Personality and Social Psychology,* 48, No. 2 (February 1985), pp. 393–399. The data for Exercise 43 came from Fred L. Ramsey and Daniel W. Schafer, *The Statistical Sleuth,* 3rd ed., Brooks/Cole Cengage Learning, 2013.

33. The data for this exercise came from Rossman, Cobb, Chance, and Holcomb's National Science Foundation project shared at the Joint Mathematics Meeting (JMM) 2008 in San Diego. Their original source was Robert Stickgold, LaTanya James, and J. Allan Hobson, "Visual discrimination learning requires sleep after training," *Nature Neuroscience,* 3 (2000), pp. 1237–1238.

34. Niels Juel-Nielsen, *Individual and Environment: Monozygotic Twins Reared Apart,* International Universities Press, 1980.

35. United Nations data on literacy were found at http://en.openei.org/wiki/WRI-Earth_Trends_Data.

36. E. C. Strain et al., "Caffeine dependence syndrome: Evidence from case histories and experimental evaluation," *Journal of the American Medical Association,* 272 (1994), pp. 1604–1607.

37. *Model Year 2014 Fuel Economy Guide,* from the website www.fueleconomy.gov.

38. http://www.ussoccer.com/womens-national-team.

39. Data from a student project by Libby Foulk and Kathryn Hilton, Canyon del Oro High School.

40. F. H. Rauscher et al., "Music training causes long-term enhancement of preschool children's spatial-temporal reasoning," *Neurological Research,* 19 (1997), pp. 2–8.

41. R. D. Stichler, G. G. Richey, and J. Mandel, "Measurement of treadware of commercial tires," *Rubber Age,* 73, No. 2 (May 1953).

42. Data from Pennsylvania State University Stat 500 Applied Statistics online course, https://onlinecourses.science.psu.edu/stat500/.

43. We obtained the tire pressure loss data from the Consumer Reports website: http://news.consumerreports.org/cars/2007/10/nitrogen-tires-.html.

44. Data from a student project by Megan Zeeb and Jenna Wilson, Canyon del Oro High School.

45. Data from a student project by Ben Garcia and Maya Kraft, Canyon del Oro High School.

46. W. S. Gosset, "The probable error of a mean," *Biometrika,* 6 (1908), 1–25.

47. From the story "Friday the 13th," at the Data and Story Library, lib.stat.cmu.edu/DASL.

48. The idea for this example was provided by Robert Hayden.

49. W. S. Gosset, "The probable error of a mean," *Biometrika,* 6 (1908), pp. 1–25. We obtained the sleep data from the Data and Story Library (DASL) website, http://lib.stat.cmu.edu/DASL/. They cite as a reference R. A. Fisher, *The Design of Experiments,* 3rd ed., Oliver and Boyd, 1942, p. 27.

50. Wayne J. Camera and Donald Powers, "Coaching and the SAT I," *TIP* (online journal at www.siop.org/tip), July 1999.

51. UNC-Chapel Hill Office of Institutional Research and Assessment.

52. http://trials.boehringer-ingelheim.com/content/dam/internet/opu/clinicaltrial/com_EN/results/248/248.630_U09-3886-01-DS_CO.pdf.

53. Data from a student project by Janelle Christman and Brittany Marimow, Canyon del Oro High School.

54. Roberts, J. A. et al., "The invisible addiction: Cell-phone activities and addiction among male and female college students," *Journal of Behavioral Addictions,* 3, No. 4 (2014), pp. 254–265, http://www.akademiai.com/doi/pdf/10.1556/JBA.3.2014.015.

55. Morton R. Brown, "Use of a postcard query in mail surveys," *Public Opinion Quarterly,* 29, No. 4 (Winter 1965–1966), http://www.jstor.org/stable/2747040.

56. A. R. Hirsch and L. H. Johnston, "Odors and learning," *Journal of Neurological and Orthopedic Medicine and Surgery,* 17 (1996), pp. 119–126. We found the data in the EESEE case study "Floral Scents and Learning."

Chapter 10

1. The context of this example was inspired by C. M. Ryan et al., "The effect of in-store music on consumer choice of wine," *Proceedings of the Nutrition Society,* 57 (1998), p. 1069A.

2. www.hockey-reference.com.

3. R. W. Mannan and E. C. Meslow, "Bird populations and vegetation characteristics in managed and old-growth forests, northwestern Oregon," *Journal of Wildlife Management,* 48 (1984), pp. 1219–1238.

4. *New York Times Magazine* August 2, 2015, page 6.

5. http://newsroom.ppg.com/getmedia/ef974015-f89b-4211-97ab-18d5ff3e7303/2014-NA-PPG.jpg.aspx.

6. http://www.census.gov/prod/cen2010/briefs/c2010br-03.pdf.

7. You can find a mathematical explanation of Benford's law in Ted Hill, "The first-digit phenomenon," *American Scientist,* 86 (1996), pp. 358–363; and Ted Hill, "The difficulty of faking data," *Chance,* 12, No. 3 (1999), pp. 27–31. Applications in fraud detection are discussed in the second paper by Hill and in Mark A. Nigrini, "I've got your number," *Journal of Accountancy,* May 1999, available online at http://www.journalofaccountancy.com/issues/1999/may/nigrini.

8. The idea for this exercise came from a post to the AP® Statistics electronic discussion group by Joshua Zucker.

9. H. Lindberg, H. Roos, and P. Gardsell, "Prevalence of coxarthritis in former soccer players," *Acta Orthopedica Scandinavica*, 64 (1993), pp. 165–167.

10. Janice E. Williams et al., "Anger proneness predicts coronary heart disease risk," *Circulation*, 101 (2000), pp. 63–95.

11. See Chapter 1, Note 8.

12. Data from Lexi Epperson and Courtney Johnson, Canyon del Oro High School.

13. The National Longitudinal Study of Adolescent Health interviewed a stratified random sample of 27,000 adolescents, then reinterviewed many of the subjects six years later, when most were aged 19 to 25. These data are from the Wave III reinterviews in 2000 and 2001, found at the website of the Carolina Population Center, www.cpc.unc.edu.

14. See Chapter 2, Note 2.

15. Data from DeAnna McDonald, University High School.

16. http://archinte.jamanetwork.com/article.aspx?articleid=2130724#tab1.

17. http://www.quinnipiac.edu/news-and-events/quinnipiac-university-poll/national/release-detail?ReleaseID=2275.

18. See Note 13.

19. R. Shine et al., "The influence of nest temperatures and maternal brooding on hatchling phenotypes in water pythons," *Ecology*, 78 (1997), pp. 1713–1721.

20. Lillian Lin Miao, "Gastric freezing: an example of the evaluation of medical therapy by randomized clinical trials," in John P. Bunker, Benjamin A. Barnes, and Frederick Mosteller (Eds.), *Costs, Risks, and Benefits of Surgery*, Oxford University Press, 1977, pp. 198–211.

21. http://www.pewinternet.org/files/old-media//Files/Reports/2013/PIP_OnlineBanking.pdf.

22. http://www.amstat.org/censusatschool/RandomSampleForm.cfm.

23. Martin Enserink, "Fraud and ethics charges hit stroke drug trial," *Science*, 274 (1996), pp. 2004–2005.

24. Douglas E. Jorenby et al., "A controlled trial of sustained release bupropion, a nicotine patch, or both for smoking cessation," *New England Journal of Medicine*, 340 (1990), pp. 685–691.

25. The idea for this exercise came from Bob Hayden.

26. Based on a plot in G. D. Martinsen, E. M. Driebe, and T. G. Whitham, "Indirect interactions mediated by changing plant chemistry: Beaver browsing benefits beetles," *Ecology*, 79 (1998), pp. 192–200.

27. The idea for this activity came from Gloria Barrett, Floyd Bullard, and Dan Teague at the North Carolina School of Science and Math.

28. Data from Samuel Phillips, Purdue University.

29. Data from Mac Cordrey and Nick Ogden, Lawrenceville School.

30. www.gapminder.org.

31. www.basketball-reference.com.

32. Data from Brittany Foley and Allie Dutson, Canyon del Oro High School.

33. Data from Kerry Lane and Danielle Neal, Canyon del Oro High School.

34. Data from Charlotte Lee, Glen A. Wilson High School.

35. See Chapter 2, Note 34.

36. Data from Nicole Enos and Elena Tesluk, Canyon del Oro High School.

37. Todd W. Anderson, "Predator responses, prey refuges, and density-dependent mortality of a marine fish," *Ecology*, 81 (2001), pp. 245–257.

38. Data from James R. Jordan, Lawrenceville School.

39. www.gapminder.org.

40. Data obtained from the PGA Tour website, www.pgatour.com, by Kevin Stevick, The Lawrenceville School.

41. Based on Marion E. Dunshee, "A study of factors affecting the amount and kind of food eaten by nursery school children," *Child Development*, 2 (1931), pp. 163–183. This article gives the means, standard deviations, and correlation for 37 children but does not give the actual data.

42. Samuel Karelitz et al., "Relation of crying activity in early infancy to speech and intellectual development at age three years," *Child Development*, 35 (1964), pp. 769–777.

43. Thanks to Larry Green, Lake Tahoe Community College, for giving us permission to use several of the contexts from his website at www.ltcconline.net/greenl/java/Statistics/catStatProb/categorizingStatProblems12.html.

44. See Note 1.

45. Elizabeth F. Loftus and John C. Palmer, "Reconstruction of automobile destruction: An example of the interaction between language and memory," *Journal of Verbal Learning and Verbal Behavior*, 13 (1974), pp. 585–589, https://www.researchgate.net/publication/222307973_Reconstruction_of_Automobile_Destruction_An_Example_of_the_Interaction_Between_Language_and_Memory.

46. Based on the the EESEE story "What Makes Pre-teens Popular?"

47. http://stats.nba.com.

48. http://www.cbsnews.com/news/cbs-news-poll-resuming-relations-with-cuba/.

Solutions

CHAPTER 1

Lesson 1.1 Exercises

1. *Individuals:* Top 10 movies. *Categorical variables:* Year, rating, genre. *Quantitative variables:* Time, box office sales.

3. *Individuals:* 10 U.S. residents. *Categorical variables:* State, gender, marital status. *Quantitative variables:* Number of family members, age, income, travel time to work.

5.

Frequency table	
Superpower	**Frequency**
Fly	14
Freeze time	8
Telepathy	12
Super strength	3
Invisibility	3
Total	40

Relative frequency table	
Superpower	**Relative frequency**
Fly	14/40 = 0.350 or 35.0%
Freeze time	8/40 = 0.200 or 20.0%
Telepathy	12/40 = 0.300 or 30.0%
Super strength	3/40 = 0.075 or 7.5%
Invisibility	3/40 = 0.075 or 7.5%
Total	40/40 = 1.000 or 100%

7.

Frequency table	
Hours of sleep	**Frequency**
4	1
4.5	0
5	2
5.5	0
6	6
6.5	1
7	13
7.5	1
8	11
8.5	1
9	4
Total	40

Relative frequency table	
Hours of sleep	**Relative frequency**
4	1/40 = 0.025 or 2.5%
4.5	0/40 = 0.000 or 0%
5	2/40 = 0.050 or 5.0%
5.5	0/40 = 0.000 or 0%
6	6/40 = 0.150 or 15.0%
6.5	1/40 = 0.025 or 2.5%
7	13/40 = 0.325 or 32.5%
7.5	1/40 = 0.025 or 2.5%
8	11/40 = 0.275 or 27.5%
8.5	1/40 = 0.025 or 2.5%
9	4/40 = 0.100 or 10.0%
Total	40/40 = 1.000 or 100%

9. *Individuals:* 100 spring break trips. *Categorical variables:* State or country visited, mode of transportation. *Quantitative variables:* Number of nights, distance from home, average cost per night.

11. Student answers will vary. Two possible answers are area code and football jersey number.

13. Student answers will vary. Two categorical variables are transmission type (automatic or manual) and wheel drive (front-wheel drive, rear-wheel drive, or all-wheel drive). Two quantitative variables are horsepower and gas mileage.

Lesson 1.2 Exercises

1.

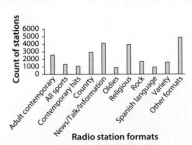

The most frequent programs are Other formats, News/Talk/Information, and Religious, while the least frequent programs are Oldies, Spanish language, and Contemporary hits.

3. (a) $100 - (19 + 6 + 5 + 12 + 1 + 9 + 14 + 29 + 3) = 2$ percent

(b)

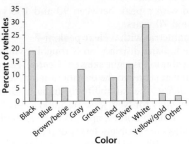

White and black were the most popular colors, while green was the least popular color. (c) Yes, we have the percents from all the categories that make up the whole 100%.

5. $2893/25{,}452 = 0.114$ or 11.4%

7. For business, about 14%; for education, about 8%

9. The balls are different sizes and the key for each ball is different.

11. By starting the vertical scale at 53 instead of 0.

13.

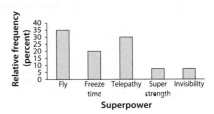

The most popular choices for superpower are to fly or have telepathy, while the least popular choices are super strength and invisibility.

15. Students who attend a private college are more likely to travel a farther distance to attend college.

17.

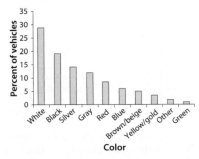

It is very easy to see the colors that are most preferred and the relative size of each category.

19. *Individuals:* 10 tallest buildings in the world. *Categorical variables:* Country, use, year completed. *Quantitative variables:* Height, number of floors.

Lesson 1.3 Exercises

1. (a)

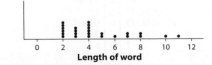

(b) The length of one of the first 25 words on a randomly selected page. (c) $9/25 = 0.36$ or 36%

3.

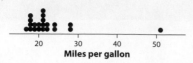

Miles per gallon

5. (a) The Toyota Prius has a mileage rating of 3 mpg lower on the highway than in the city.
(b) 8/21 = 0.38 or about 38%.

7. Left-skewed with a peak between 90 and 100 years; there is a small gap around 70 years.

9. Roughly symmetric with a clear peak at 7

11. *Shape:* The Indiana distribution is roughly symmetric, while the New Jersey distribution is right-skewed. *Center:* The distribution for Indiana has a similar center to the distribution for New Jersey (around $45,000). *Variability:* The variability for Indiana ($0 to $125,000) is less than the variability for New Jersey ($5,000 to $165,000). *Outliers:* $160,000 in the Indiana dotplot appears to be an outlier; there aren't any obvious outliers for the New Jersey dotplot.

13. (a) Yes, the middle shelf has the cereals with the most sugar, putting them at "eye level" for kids.
(b) The variability of the sugar content of the cereals appears to be similar for all three shelves.

15. The scale on the *x* axis is missing 4, 7, 8, 10, and 11.

17.

Frequency table	
Camera brand	**Frequency**
Canon	23
Sony	6
Nikon	11
Fujifilm	3
Olympus	2
Total	45

Relative frequency table	
Camera brand	**Relative frequency**
Canon	23/45 = 0.511 or 51.1%
Sony	6/45 = 0.133 or 13.3%
Nikon	11/45 = 0.244 or 24.4%
Fujifilm	3/45 = 0.067 or 6.7%
Olympus	2/45 = 0.044 or 4.4%
Total	45/45 = 1.000 or 100%

Lesson 1.4 Exercises

1.

```
4 | 8
5 | 5
6 | 012345668
7 | 0011244567789
8 | 16
```
KEY: 6|1 is a ninth-grade biology student with a resting heart rate of 61.

3.

```
62 | 24
62 | 566
63 | 122234
63 | 667899
64 | 234
64 | 588
65 | 23
```
KEY: 64|2 is a salmon fillet with a pH of 6.42.

5. (a) 11/26 = 0.423 or 42.3%. **(b)** Fairly symmetric with a single peak on the 70 − 74 stem and a gap from 48 to 55. **(c)** 48 appears to be an outlier. Key: 4|8 is a ninth-grade biology student with a resting heart rate of 48.

7. (a) 4/51 = 0.078 or 7.8%. **(b)** Fairly symmetric with a single peak on the 13.0 − 13.9 stem and a gap from 7.7 to 9.0. **(c)** 7.7% appears to be an outlier. This state is Alaska.

9. (a)

```
     Division I        Division V
           8 |  3 | 677
        7761 |  4 | 4456
     8654321 |  5 | 48
 87554443322 |  6 | 0266779
      876411 |  7 | 24
           7 |  8 | 6
           1 |  9 | 23688
           6 | 10 |
```
KEY: 3|6 represents a game in which 36 points were scored.

(b) Teams in Division V tended to score more points than teams in Division I as the center is slightly higher. **(c)** Different because Division I is fairly symmetric and single peaked while Division V is more uniform and has several peaks.

11. The distribution for women is fairly symmetric and single peaked, while the distribution for males is slightly right-skewed. Females tend to have higher study times than men. Both distributions have similar variability. The distribution for women has a possible outlier at 360 minutes, and the distribution for men has a possible outlier at 300 minutes.

13. (a)

```
0 | 399
1 | 1345677889
2 | 000123455668888
3 | 25699
4 | 1345579
5 | 0359
6 | 1
7 | 0
8 | 366
9 | 3
```
KEY: 4|3 is a shopper who spent $43 (rounded to the nearest dollar).

(b)

```
0 | 3
0 | 99
1 | 134
1 | 5677889
2 | 0001234
2 | 55668888
3 | 2
3 | 5699
4 | 134
4 | 5579
5 | 03
5 | 59
6 | 1
6 |
7 | 0
7 |
8 | 3
8 | 66
9 | 3
```
KEY: 4|3 is a shopper who spent $43 (rounded to the nearest dollar).

The split stem graph shows the shape better. **(c)** The shape of the distribution is right-skewed with a single peak. The center of the distribution is around $28, and the data vary from $3 to $93.

15.

```
4 | 5699
5 | 001122333344444
5 | 555556667777788889
```
KEY: 4|9 represents an orange with a weight of 4.9 ounces.

Too few stems. It is very difficult to see overall shape, peaks, and gaps (if any).

17. By starting the vertical scale at 20 instead of 0.

Lesson 1.5 Exercises

1.

[Histogram: vertical axis "Frequency of CO_2" from 2 to 18; horizontal axis "CO_2" from 0 to 20. Bars decreasing from about 17 near 0–2 with a small rise near 16–18.]

3. **(a)** Answers may vary depending on interval width.

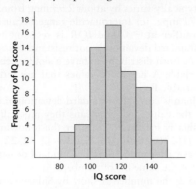

(b) Yes, the distribution is single peaked, symmetric, and is close to a bell shape.

5. **(a)** $(1 + 1 + 1 + 1 + 2)/48 = 6/48$ or 12.5%. **(b)** Skewed to the right with a single peak in the 0 to 1 interval; there are gaps in the 10 to <11, 13 to <14, and 15 to <16 intervals. **(c)** Australia, Saudi Arabia, and the United States

7. **(a)** $(2 + 1 + 4 + 5 + 10 + 28 + 50)/273 = 100/273 = 37\%$. **(b)** Slightly skewed to the left with a single peak in the 0 to <2.5% interval; there are gaps in the −22.5 to <−20% interval and the −15 to <−12.5% interval. **(c)** The −25 to <−22.5% interval and the −17.5 to <−15% interval

9. *Shape:* Both distributions are skewed to the right and single-peaked with low-income more strongly skewed. *Center:* The center of the distribution is higher for high income. *Variability:* Similar for low-income and high-income households. *Outliers:* No apparent outliers in either distribution.

11. **(a)** Because the groups have such different sizes. **(b)** The center of the histogram for Australia is higher than for Vietnam. Also, the histogram for Australia is lower and more uniform in shape than the histogram for Vietnam.

13.

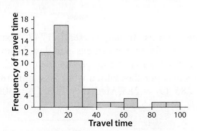

The histogram of travel times is skewed to the right with a center in the 10 to <20 interval and a gap in the 70 to <80 interval. The times vary from 1 minute to 90 minutes. The travel times of 80 and 90 minutes are possible outliers.

15. The graph shown is a bar chart and not a histogram. A bar chart is used to display categorical data. It doesn't make sense to describe a distribution of categorical data as being skewed to the right.

17. **(a)** Use a relative frequency histogram to show percent.

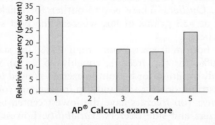

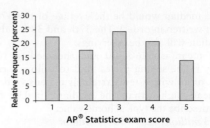

(b) The AP® Calculus scores have a peak at 1 and another, slightly lower, peak at 5 whereas the scores for AP® Statistics are more uniform, with scores of 1 and 3 being the most frequent and scores of 5 being the least frequent. The center of both distributions is 3. There are more scores close to the center on the AP® Statistics exam and more scores at the extremes on the AP® Calculus exam.

19. **(a)**

Stem	Leaf
0	246778
1	459
2	4
3	8
4	1366
5	7
6	08
7	
8	4

KEY: 1|9 represents a player who scored 19 runs.

(b) Right-skewed with two peaks—on the 0−9 stem and the 40−49 stem—and a gap from 68 to 84 runs.

Lesson 1.6 Exercises

1. 83; about half of Joey's quiz scores are higher than 83, and about half are less.

3. $\dfrac{166 + 167}{2} = 166.5$ grams; about half of the orders of fries sampled weighed more than 166.5 grams, and about half weighed less.

5. **(a)** $\bar{x} = \dfrac{73 + 77 + 79 + \cdots + 93 + 95}{13} = \dfrac{1092}{13} \approx 84$

(b) $\bar{x} = \dfrac{73 + 77 + 79 + \cdots + 93 + 95 + 0}{14} = \dfrac{1092}{14} \approx 78$

With the 14th quiz score of zero, the mean decreases a fair amount from 84 to 78, and the median only slightly decreases from 83 to 82.5. This is because the mean is not resistant to outliers, whereas the median is resistant to outliers.

7. **(a)** $\bar{x} = \dfrac{152 + 159 + 160 + \cdots + 171 + 173}{14}$

$= \dfrac{2318}{14} \approx 165.57$ grams

(b) The order that weighed 152 grams brought the mean down; without this order, the mean increases from 165.57 grams to 166.62 grams.

9. **(a)** The mean will be less than the median because the mean is pulled toward the long tail in this left-skewed distribution. **(b)** The median better summarizes the center of the distribution because the birth rate distribution is skewed to the left.

11. The mean, because the distribution is roughly symmetric with no outliers.

13. $\bar{x} =$

$\dfrac{15 \cdot 0 + 11 \cdot 1 + 15 \cdot 2 + 11 \cdot 3 + 8 \cdot 4 + 5 \cdot 5 + 3 \cdot 6 + 3 \cdot 7 + 3 \cdot 8}{74}$

$= \dfrac{194}{74} = 2.62$

servings. The median would be the average of the 37th and 38th values (when written in order). The 37th and 38th values are both 2, so the median will also be 2 servings.

15. (a) The mean is $260,700 and the median is $213,600. Single-family home selling prices tend to follow a right-skewed distribution, often with a few extremely expensive homes, which pulls the mean higher than the median. **(b)** Use the mean of $260,700 to describe the typical home value.

17. (a) ($2.3 *million*) $\cdot$ 25 = $57.5 *million*
(b) No. The median value is calculated using the relative positions of all the values in a data set and does not depend on the actual values of the individuals in the set.

19. (a) $\bar{x} =$

$$\frac{4+5+5+5+6+7+7+7+7+8+10+10+10+10+11+12+14+22+35+38}{20}$$

$$= \frac{233}{20} = 11.65$$

(b) 10% of 20 values = 2 values, so we trim 2 values on the low end (4,5) and the high end (35,38):

$$\bar{x} = \frac{5+5+6+7+7+7+7+8+10+10+10+10+11+12+14+22}{16} = \frac{151}{16} = 9.44$$

(c) Because it is more resistant to outliers than is the mean.

21. *Shape:* The distribution of points for Kevin Durant and Lebron James are both roughly symmetric. *Center:* Higher for Kevin Durant. *Variability:* More variable for Lebron James. *Outliers:* Lebron James has a possible outlier in the 60 to <65 interval, while the distribution for Kevin Durant doesn't appear to have any outliers.

Lesson 1.7 Exercises

1. Range = 38 − 4 = 34 pairs of shoes
3. Range = 321 − 285 = 36 pounds
5. $IQR = Q_3 − Q_1 = 11.5 − 6.5 = 5$ pairs of shoes. The range of the middle half of number of pairs of shoes for a random sample of 20 male students is 5 pairs of shoes.
7. $IQR = Q_3 − Q_1 = 300 − 285 = 15$ pounds. The range of the middle half of weights of Dallas Cowboys defensive linemen is 15 pounds.
9. $\bar{x} = \dfrac{7+7+8+10}{4} = \dfrac{32}{4} = 8$ hours

$$s_x = \sqrt{\frac{\sum(x_i - \bar{x})^2}{n-1}}$$

$$= \sqrt{\frac{(7-8)^2 + (7-8)^2 + (8-8)^2 + (10-8)^2}{4-1}}$$

$$= \sqrt{\frac{6}{3}} = 1.41 \text{ hours}$$

The number of hours of sleep last night for the first four students to arrive in a first period statistics class typically varies by about 1.41 hours from the mean of 8 hours.

11. $\bar{x} = 5.4$ mg/deciliter

$$s_x = \sqrt{\frac{\sum(x_i - \bar{x})^2}{n-1}}$$

$$= \sqrt{\frac{0.04 + 0.04 + 0.64 + 0.25 + 0.09 + 1}{6-1}}$$

$$= \sqrt{\frac{2.06}{5}} = 0.64 \text{ mg/deciliter}$$

The level of phosphate in the blood of a patient on 6 consecutive visits to a clinic typically varies by about 0.64 mg/deciliter from the mean of 5.4 mg/deciliter.

13. (a) $IQR = Q_3 − Q_1 = 10 − 8 = 2$ mpg. The range of the middle half of differences in mpg (Highway − City) for 21 model year 2014 midsize cars is 2 mpg. **(b)** $\bar{x} = 8.71$ mpg;

$$s_x = \sqrt{\frac{\sum(x_i - \bar{x})^2}{n-1}} = \sqrt{\frac{178.29}{21-1}} = 2.99 \text{ mpg}$$

The difference in mpg (Highway − City) for 21 model year 2014 midsize cars typically varies by about 2.99 mpg from the mean difference of 8.71 mpg. **(c)** Interquartile range because of the presence of the outlier at −3. The IQR is resistant to the outlier, whereas the standard deviation is not resistant.

15. Variable A. Both distributions have a similar center (around 4 or 5), but variable A has more values that are farther from the center than variable B.

17. (a) Yes, Juan is correct. A standard deviation of 0 indicates no variability in the values, which means they must all be the same. **(b)** No, Letishia is incorrect. Consider the following two sets of data: Set 1 = {10, 10, 10, 20, 30}. Set 2 = {2, 12, 22, 22, 22}. Both sets of data have a mean of 16 and a standard deviation of 8.944, but they are not the same list of numbers.

19. Closest to 2; the number of shots by Sidney Crosby per game typically varies by about 2 shots from the mean of 3 or 4 shots.

21. (a) Skewed to the right, with a single peak in the 3 to <4 hurricanes interval and a gap from 13 to <15 hurricanes. **(b)** The median because the distribution of number of hurricanes per year is skewed to the right and has a possible outlier in the 15 to <16 hurricanes interval.

Lesson 1.8 Exercises

1. $Q_1 − 1.5 \times IQR = 285 − 1.5 \times 15 = 262.5$; $Q_3 + 1.5 \times IQR = 300 + 1.5 \times 15 = 322.5$. Because there are no data values less than 262.5 or greater than 322.5, this distribution has no outliers.

3. $Q_1 − 1.5 \times IQR = 315 − 1.5 \times 35 = 262.5$; $Q_3 + 1.5 \times IQR = 350 + 1.5 \times 35 = 402.5$. Because 260 is less than 262.5, it is considered an outlier.

5. (a) Min = 0, $Q_1 = 3$, Med = 9, $Q_3 = 43$, Max = 118. 118 is an outlier.

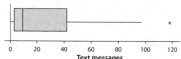

(b) The article claims teenagers send 1742 text messages a month, or roughly 58 text messages per day. From the boxplot, $Q_3 = 43$, so at least 75% of the teenagers in the class sent 43 or fewer text messages per day, which is far less than 58.

7. (a) Min = 285, $Q_1 = 285$, Median = 293, $Q_3 = 300$, Max = 321

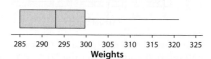

(b) The IQR is more appropriate because the distribution is skewed to the right.

9. *Shape:* Both distributions of fat content are roughly symmetric. *Center:* Chicken or fish sandwiches at McDonald's tend to have less fat than the beef sandwiches. *Variability:* There is more variation for beef sandwiches than for chicken or fish sandwiches. *Outliers:* There is one outlier for chicken or fish sandwiches at ≈33 grams of fat, whereas the beef sandwiches have no outliers.

11. (a) For bottom freezers, more than 75%; for side freezers, roughly 100%; for top freezers, roughly 25%. **(b)** *Shape:* The energy cost distribution for bottom freezers is skewed to the right, while the distributions for side and top freezers are roughly symmetric. *Center:* Top freezers have the lowest center, followed by bottom freezers and side freezers. *Variability:* Lowest variability for the top freezers, followed by the side freezers and then the bottom freezers. *Outliers:* Two outliers for bottom freezers at ≈ $141 and $149. There are no outliers for the other two distributions.

13. Yes. The Q_1 of the differences is at zero, indicating that 75% of the students in Mr. Williams's class were sending more text messages than making phone calls.

15. (a) *Women:* Min = 101, Q_1 = 126, Median = 138.5, Q_3 = 154, Max = 200, outlier = 200. *Men:* Min = 70, Q_1 = 98, Median = 114.5, Q_3 = 143, Max = 187, no outliers.

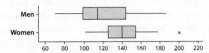

(b) Women tend to have higher scores on the SSHA (median = 138.5) than men (median = 114.5), while scores for men tend to have more variability (IQR = 45) than they do for women (IQR = 28).

17. (a) $\dfrac{\text{maximum} - \text{median}}{\text{median} - \text{minimum}} = \dfrac{55 - 8}{8 - 3} = \dfrac{47}{5} = 9.4$

The top of the ratio indicates the range of the upper 50% of the data, while the bottom of the ratio indicates the range of the lower 50% of the data. If the range is larger in the upper 50% of the data (right-skewed), then the ratio of these values will be greater than 1.
(b) Answers may vary. One possible solution would be:

$$\dfrac{\text{maximum} - Q_3}{Q_1 - \text{minimum}} = \dfrac{55 - 12}{4 - 3} = \dfrac{43}{1} = 43$$

The top of the ratio indicates the range of the top 25% of the data, while the bottom of the ratio indicates the range of the bottom 25% of the data. If the range is larger in the top 25% of the data (right-skewed), then the ratio of these values will be greater than 1.

Lesson 1.9 Exercises

1. (a) 5/30 = 0.17 = 17th percentile. (b) The New York Yankees won more games than 60% of the 30 teams (0.60 × 30 − 18 teams) in the 2014 season. The Yankees won 85 games in 2014.
3. (a) 17/20 = 0.85 = 85th percentile. (b) 0.25 × 20 = 5, so Luis has more pairs of shoes than 5 of the 20 male students. Luis has 7 pairs of shoes.
5. (a) $19.50 ($x$ axis) is at about the 25th percentile (y axis). (b) The 80th percentile (y axis) is about 50 (x axis). About 80% of shoppers spent less than $50.
7. (a) 30 minutes (x axis) is at about the 65th percentile (y axis). About 100% − 65% = 35% of phone calls that are 30 minutes or more. (b) The 25th percentile (y axis) is about 13 (x axis) and the 75th percentile (y axis) is about 32 (x axis). So $IQR = Q_3 - Q_1$ = 32 − 13 = 19 minutes.
9. $z = \dfrac{96 - 81}{9.6} = 1.56$. The Washington Nationals' number of wins in 2014 is 1.56 standard deviations above the mean of 81 wins.

11. (a) $z = \dfrac{9.7 - 13.26}{1.67} = -2.13$. Colorado has a percent of residents aged 65 or older that is 2.13 standard deviations below the mean of 13.26%. (b) $2.60 = \dfrac{x - 13.26}{1.67}$ gives x = 17.60. Florida has 17.6% of residents aged 65 or older.
13. The speed limit is set at the value that will have 85% of vehicles on that road traveling at a speed less than the speed limit.
15. 48% of boys the same age as Mrs. Munson's son have weights that are less than that of her son, while 76% of boys the same age as Mrs. Munson's son have heights that are less than that of her son.

17. SAT: $z = \dfrac{680 - 514}{117} = 1.42$ ACT: $z = \dfrac{27 - 21}{5.3} = 1.13$

Courtney scored better on the SAT relative to her peers because her z-score was higher.

19. (a) Judy's hip bone density is 1.45 standard deviations below the mean hip bone density (956 g/cm²) of all 25-year-old women. This means that Judy's hip bone density is lower than most women her age. (b) $-1.45 = \dfrac{948 - 956}{X}$ gives X = 5.52. This means that the standard deviation of hip bone density in the population of 25-year-old women is 5.52 g/cm².

Chapter 1 Review Exercises

1. *Individuals:* The first ten U.S. Presidents. *Categorical variables:* Political party, state of birth. *Quantitative variables:* age at inauguration, age at death.
2. (a)

Frequency table	
Breed group	**Frequency**
Whippet	3
Mixed breed	14
Other purebred	2
Labrador retriever	3
Border collie	9
Australian shepherd	6
Total	37

(b)

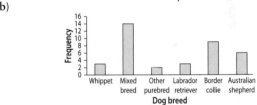

Mixed breed dogs won the most often (14), followed by border collies (9) and Australian shepherds (6). Whippets (3), Labrador retrievers (3), and other purebreds (2) were the least represented in the group.
3. In 2012, the most preferred music type was pop at around 29%, but in 2013 the top preference switched to hip hop at about 34%. In both years, the music type that was least preferred was jazz. From 2012 to 2013, hip hop had the largest increase in percent, while rock and jazz saw the largest decreases in percent.
4. By starting the vertical scale at 12 instead of 0, relative sizes of the groups are misrepresented by the size of the bars.
5. *Shape:* All three income distributions are skewed to the right, with the Connecticut distribution of total family income being more strongly skewed than the distributions for Indiana and Maine. *Center:* The center is largest for Connecticut, followed by Indiana; the center is lowest for Maine. *Variability:* Connecticut family incomes vary the most, with Indiana and Maine having similar variability. *Outliers:* The two large values of $280,000 and $350,000 in the Connecticut dotplot and the one large value of $210,000 in the Indiana dotplot appear to be outliers; there aren't any obvious outliers for the Maine dotplot.
6. (a) The data vary from 7 to 25. We chose intervals of width 2, beginning at 6.

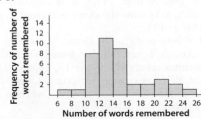

(b) The median is 13 and the mean is 14.3. Because the shape of the distribution is skewed to the right, the mean will be higher than the median.

7. **(a)** Range = Maximum − Minimum = 25 − 7 = 18 words. **(b)** $IQR = Q_3 − Q_1 = 16 − 11.5 = 4.5$ words. **(c)** Outliers $< Q_1 − 1.5 × IQR = 11.5 − 1.5 × 4.5 = 4.75$ or outliers $> Q_3 + 1.5 × IQR = 16 + 1.5 × 4.5 = 22.75$. Because 23 and 25 are more than 22.75, they are considered outliers. **(d)** The number of words that students remember typically varies by about 4.05 words from the mean of 14.3 words.

8. The data do provide evidence for Catherine and Ana's suspicion. The median of the distribution of reaction time for athletes (around 260 milliseconds) is less than the median for other students (around 295 milliseconds). This suggests that athletes typically have a faster reaction time than other students.

9. **(a)** 26/40 = 0.65, so the house indicated by the triangle is at the 65th percentile of the 40 houses in Ames, Iowa sold during a recent month. **(b)** $z = \dfrac{\text{value} − \text{mean}}{\text{standard deviation}} = \dfrac{234,000 − 203,388}{87,609} = 0.35$

The house that sold for $234,000 is 0.35 standard deviation above the mean selling price of $203,388.

10. **(a)** Ace, who took 120 minutes to finish the exam (x axis) is at about the 90th percentile (y axis). **(b)** The 50th percentile (y axis) is at around 90 minutes (x axis). The 25th percentile (y axis) is about 80 minutes (x axis), and the 75th percentile (y axis) is about 110 minutes (x axis). So estimated $IQR = Q_3 − Q_1 = 110 − 80 = 30$ minutes.

Chapter 1 Practice Test

1. c
2. d
3. c
4. b
5. b
6. c
7. b
8. a
9. d
10. d
11. c
12. **(a)** Answers may vary due to choice of intervals.

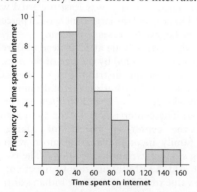

(b) $IQR = Q_3 − Q_1 = 77 − 30 = 47$ minutes
Outliers $< Q_1 − 1.5 × IQR = 30 − 1.5 × 47 = −40.5$
Outliers $> Q_3 + 1.5 × IQR = 77 + 1.5 × 47 = 147.5$
Because 151 is greater than 147.5, it is considered an outlier. **(c)** Median and IQR because the shape is skewed to the right and there is an outlier at 151.
13. **(a)** 13/30 = 0.43 or 43%. **(b)** Brand X batteries tend to have lifetimes of at least 30 hours, whereas Brand Y has the possibility of very low lifetimes (as shown by the 17, 22, and 26 in the dotplot). **(c)** Brand Y batteries tend to have a longer lifetime than Brand X. Brand Y has a median of 43 hours, whereas Brand X has a median of 39 hours.

14.

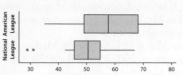

Shape: Both distributions are roughly symmetric. *Center*: American League teams tend to have more home runs than National League teams. *Variability*: The variability for American League teams is higher than the variability for National League teams. *Outliers*: The American League has no outliers. There are two outliers in the National League home run distribution—at 29 and 31 home runs.
15. **(a)** 7 hours (x axis) is at about the 60th percentile (y axis).
(b) The 25th percentile (y axis) is about 2.5 hours (x axis) and the 75th percentile (y axis) is about 11 hours (x axis). So estimated $IQR = Q_3 − Q_1 = 11 − 2.5 = 8.5$ hours.

CHAPTER 2

Lesson 2.1 Exercises

1. The explanatory variable is the year in college, because the year in college might help predict whether or not a person lives on- or off-campus.
3. The explanatory variable is the gender, because the gender might help predict a person's favorite sport.
5.

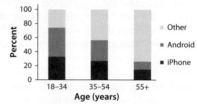

7.

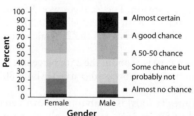

9. There is an association between type of phone use and political affiliation because the distribution of political affiliation is not the same for each phone status. In other words, knowing a person's type of phone use helps us predict the person's political affiliation. People who use cell only are more likely to be Democrat or Republican than people who use landlines, and people who use landlines are more likely to have no political leaning than people who use cell only.
11. There is an association between color preference and where you live because the distribution of color preference is not the same for the U.S. and Europe. In other words, knowing where a person lives helps us predict the person's favorite vehicle color. People from the U.S. prefer white/pearl and red much more often than people from Europe, who prefer black, silver, and gray more often than people from the U.S.
13. **(a)**

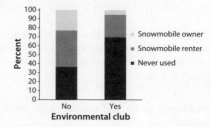

(b) There is an association between environmental club affiliation and snowmobile use because the distribution of snowmobile use is not the same for people who are in an environmental club and people who are not. In other words, knowing whether a person belongs to an environmental club helps us predict the person's snowmobile use. People who are not in an environmental club are more likely to have owned or rented a snowmobile than are people who do belong to an environmental club.

15. Student answers will vary. One possible answer is shown below.

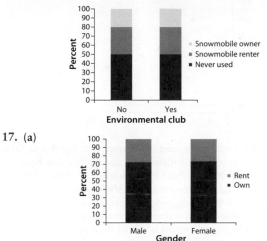

17. (a)

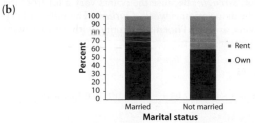

There is an association between gender and housing status, because the distribution of housing status is not the same for males and females. Because the percentages in the distributions are so close (72.5% vs. 73.5% for home owners and 27.5% vs. 26.5% for renters), we would say that the association is very weak.

(b)

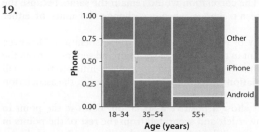

There is an association between marital status and housing status, because the distribution of housing status is not the same for married and not married adults. People who are married are more likely to own than people who are not married. Because the percentages in the distributions are much farther apart (81% vs. 60% for home owners and 19% vs. 40% for renters) than they were for the different genders, we would say that the association between marital status and housing status is a stronger association than the one between gender and housing status. Knowing the person's marital status would better help us predict the person's housing status than would knowing the person's gender.

19.

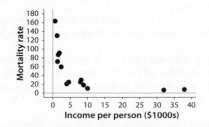

21. (a) $Q_1 = 25,000$; $Q_3 = 65,000$; $IQR = Q_3 - Q_1 = 65,000 - 25,000 = \$40,000$

Interpretation: The range of the middle half of family incomes in the Indiana sample is \$40,000. **(b)** The family income of individuals in Indiana typically varies by about \$29,400 from the mean of \$47,400.

Lesson 2.2 Exercises

1. The explanatory variable is the amount of time studying for a statistics exam, because the amount of time studying helps explain the grade on the exam.

3. The explanatory variable could be either the arm span or the height, as either variable could be used to explain or predict the other.

5.

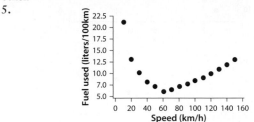

7.

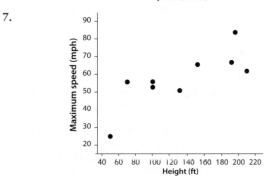

9. *Direction:* There is a negative association between percent taking SAT and mean math score. States with a higher percentage of graduates taking the SAT tended to have lower mean math scores, and vice versa. *Form:* There is a curved (nonlinear) pattern in the scatterplot. *Strength:* Because the points vary somewhat from the curved pattern, the association has moderate strength. *Outliers:* There are two states that depart from the overall curved pattern, (20, 500) and (87, 460).

11. *Direction:* There is a positive association between height and maximum speed. Taller roller coasters tend to have a higher maximum speed and shorter roller coasters tend to have a lower maximum speed. *Form:* There is a linear pattern in the scatterplot. That is, the overall pattern follows a straight line. *Strength:* Because the points vary somewhat from the linear pattern, the association has moderate strength. *Outliers:* There don't appear to be any individuals that depart from the linear pattern.

13.

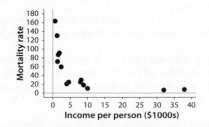

Direction: There is a negative association between income per person and mortality rate for countries. Countries with higher income per person tend to have a lower mortality rate, and vice versa. *Form:* There is a curved (nonlinear) pattern in the scatterplot. *Strength:* Because the points do not vary much from the curved pattern, the association is strong. *Outliers:* There don't appear to be any individuals that depart from the curved pattern.

15. Student answers will vary. One possible answer is shown below. Both scatterplots have strong positive association, but the upper one has a linear form and the lower one has a curved form.

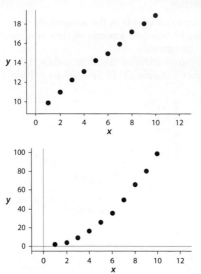

17. Higher outside temperatures tend to correspond to lower amounts of fuel oil usage, and lower outside temperatures tend to correspond to higher amounts of fuel oil usage.

19. (a) Southern states typically have lower mean SAT math scores than other states with a similar percent of students taking the SAT. **(b)** It has a much lower mean SAT Math score than the other states that have a similar percent of students taking the exam.

21. There is an association between year and employment arrangement because the distribution of employment arrangement is not the same for each year. In other words, knowing the year helps us predict the couple's employment arrangement. Couples in 1960 are more likely to have only the father employed than couples in 2011, and couples in 2011 are more likely to have only the mother employed or dual income than are couples in 1960.

Lesson 2.3 Exercises

1. Because the relationship is positive, $r > 0$. Also, r is closer to 0 than 1 because the relationship is weak—there is a lot of scatter from the linear pattern.

3. Because the relationship is negative, $r < 0$. Also, r is closer to -1 than 0 because the relationship is strong—there isn't much scatter from the linear pattern.

5. The linear relationship between number of turnovers and number of points scored for players in the NBA in the 2013−2014 season is strong and positive.

7. The linear relationship between latitude and average July temperature for the 12 cities with the largest populations is moderate and negative.

9. Probably not. Although there is a weak, negative correlation, a decrease in the number of pencils carried is not likely to cause a higher GPA. It is likely that both of these variables are changing due to several other variables, such as work ethic or organizational skills.

11. Probably not. Although there is a strong, positive correlation, an increase in turnovers is not likely to cause an increase in points for NBA players. It is likely that both of these variables are changing due to several other variables, such as time played.

13. (a) The linear relationship between the length of time a party spends at the restaurant and the amount of tip they leave is weak

and positive. **(b)** Probably not. Although there is a weak, positive correlation, an increase in the amount of time a party spends at the restaurant is not likely to cause an increase in the tip. It is likely that both of these variables are changing due to several other variables, such as the amount of food consumed.

15. Answers may vary. Here is one possibility. This is the graph of $y = 1/x$ (nonlinear), for x values between 3 and 10; $r = -0.94$.

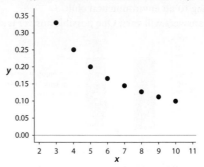

17. The largest correlation is height of women at age 4 and their height as women at age 18, followed by the correlation between heights of male parents and their adult children, followed by correlation between heights of husbands and their wives.

19. The author incorrectly suggests that a correlation of zero indicates a negative association between research productivity and teaching rating. The paper's report is wrong because the correlation of $r \approx 0$ means that there is no linear association between research productivity and teaching rating.

21. *Direction:* There is a negative association between average points per game scored by opposing team and total wins. WNBA teams with higher average points per game scored by opposing teams tend to have less total wins. *Form:* There is a linear pattern in the scatterplot. *Strength:* Because the points vary a lot from the linear pattern, the association is weak. *Outliers:* The Indiana Fury, with 16 total wins, and the Phoenix Mercury, with 19 total wins depart from the linear pattern.

Lesson 2.4 Exercises

1. $r = \dfrac{\sum z_x z_y}{n - 1} = \dfrac{1.96 + -0.06 + 2.03 + -0.02 + 0.28 + 0.33 + 0.51}{7 - 1} = 0.84$

3.

$r = \dfrac{\sum z_x z_y}{n - 1} = \dfrac{-0.31 + 0.00 + -1.31 + -1.08 + -1.13 + -0.26 + -0.55 + -2.24}{8 - 1} = -0.98$

5. (a) The correlation would still be $r = 0.88$, because the correlation makes no distinction between explanatory and response variables. **(b)** The correlation would still be $r = 0.88$, because the correlation doesn't change when we change the units of either variable. **(c)** No. The correlation doesn't have units, so including "cal/kg" is incorrect.

7. (a) The correlation would remain the same, because the correlation makes no distinction between explanatory and response variables. **(b)** The correlation would remain the same, because the correlation doesn't change when we change the units of either variable.

9. The majority of points shows a positive association. However, because the point in red is in the lower right and separated from the rest of the points in the x-direction, it will be influential and will make the correlation closer to 0. Without this point, the association is much stronger.

11. The points show a positive association. Because the point in red is in the lower left and separated from the rest of the points in the x-direction, it will be influential and will make the correlation closer to 1. With this point, the correlation is more strongly positive.

13. **(a)** Gender is a categorical variable, and correlation r is for two quantitative variables. **(b)** The stated correlation is greater than 1, but the largest possible value of the correlation is $r = 1$.
15. The correlation would still be $r = -0.07$. On the scatterplot, the reduction of \$50 for each netbook would shift all of the points left 50 units. The pattern of all the points would be exactly the same as before, so the strength of the linear relationship would not change.
17. **(a)** A scatterplot of mileage versus speed is shown here.

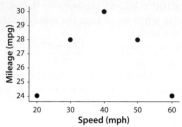

(b) The correlation is $r = 0$. **(c)** The correlation measures the strength of a *linear* association between two quantitative variables. This plot shows a *nonlinear* relationship between speed and mileage.
19. **(a)** The correlation of -0.71 indicates that the linear relationship between the number of car washes per year and annual repair costs is moderately strong and negative. Cars that had more car washes per year tended to have lower annual repair costs. **(b)** No. We know that there is a negative correlation between number of car washes per year and annual repair costs, but we do not know if increasing the number of car washes per year will cause the repair costs to go down. It might be that people who are more likely to wash their cars have newer cars that are less likely to need repair than older cars.

Lesson 2.5 Exercises

1. $\hat{y} = 33.35 + 13.29(4) = 86.51$ minutes
3. **(a)** $\hat{y} = 415 - 45.7(6) = 140.8$ inches **(b)** No, this would be an extrapolation because a dash time of 3 seconds is far outside the interval of x values used to obtain the regression line. This prediction may not be accurate.
5. **(a)** $\hat{y} = 33.35 + 13.29(2) = 59.93$ minutes $residual = y - \hat{y} = 62 - 59.93 = 2.07$ minutes. **(b)** This time between eruptions lasted 2.07 more minutes than expected, based on the regression line using x = duration of previous eruption.
7. **(a)** $\hat{y} = 415 - 45.7(7.17) = 87.33$ inches $residual = y - \hat{y} = 65 - 87.331 = -22.331$ inches. **(b)** This student had a long-jump distance that was 22.331 inches less than expected, based on the regression line using x = dash time.
9. **(a)** The predicted time between eruptions increases by 13.29 minutes for each additional minute of time for the previous eruption. **(b)** No, because it does not make sense to have an eruption of duration 0 minutes.
11. **(a)** The predicted long-jump distance decreases by 45.7 inches for each additional second of time for the 40-yard dash. **(b)** No, because it is not possible for a student to have a 40-yard dash time of 0 seconds.
13. **(a)** $\hat{y} = 1425 - 19.87(49.4) = 443.422$ cubic feet per day $residual = y - \hat{y} = 520 - 443.422 = 76.578$ cubic feet per day. This month had 76.578 more cubic feet per day of gas consumption than expected, based on the regression line using x = average temperature. **(b)** The predicted gas consumption decreases by 19.87 cubic feet per day for each additional degree Fahrenheit in the average monthly temperature. **(c)** Yes. If a month has an average temperature of 0 degrees Fahrenheit, the predicted gas consumption is 1425 cubic feet per day. However, because this prediction is an extrapolation, it might not be accurate.
15. $(19.87)(10) = 198.7$ cubic feet per day

17. **(a)** $\hat{y} = 293.56 - 31.05(5.41) + 42.02(1) = 167.5995$ inches. **(b)** $residual = y - \hat{y} = 171 - 167.5995 = 3.4005$ inches. This student had a long-jump distance of 3.4005 more inches than expected, based on the regression line using x_1 = dash time and x_2 = gender. **(c)** 42.02 inches. Because x_2 can only take on the values 0 and 1, going from 0 (female) to 1 (male) will increase the predicted value by $42.02(1) = 42.02$ inches.
19. **(a)** The linear relationship between calories and salt content for these hot dogs is moderately strong and positive. **(b)** The points show a positive association. Because the hot dog with the lowest calorie count is in the lower left and separated from the rest of the points in the x-direction, it will be influential and will make the correlation closer to 1. With this point, the correlation is more strongly positive.

Lesson 2.6 Exercises

1. $\hat{y} = 300.04 + 2.83x$
3. $\hat{y} = -1.29 + 11.89x$
5. slope $= b = r\dfrac{s_y}{s_x} = 0.5\left(\dfrac{2.7}{2.5}\right) = 0.54$; y intercept $= a = \bar{y} - b\bar{x} = 68.5 - 0.54(64.5) = 33.67$; $\hat{y} = 33.67 + 0.54x$
7. slope $= b = r\dfrac{s_y}{s_x} = 0.596\left(\dfrac{15.35}{5.36}\right) = 1.71$; y intercept $= a = \bar{y} - b\bar{x} = 9.07 - 1.71(1.75) = 6.08$; $\hat{y} = 6.08 + 1.71x$
9. It is making the regression line less steep (slope less negative) and increasing the y intercept. (Note that the y intercept is to the right of the plot because all of the x values are negative.)
11. **(a)** It is making the regression line less steep (slope less positive) and increasing the y intercept. **(b)** It pulls the line up a little, which increases the y intercept but doesn't change the slope much.
13. **(a)**

(b) $\hat{y} = 66.4 + 10.4x$. **(c)** It is making the regression line less steep (less positive) and increasing the y intercept. With Subject 18 removed from the data set, $\hat{y} = 52.3 + 12.1x$. As expected, without the point, the slope is steeper and the line has a smaller y intercept.
15. **(a)** $\hat{y} = 4.77 + 0.14x$. **(b)** No. In Exercise 1, we had a slope of 2.83 and a y intercept of 300.04, and now the slope is 0.14 and the y intercept is 4.77.
17. The best model for a set of data is the one that minimizes the sum of the squares of the residuals (SSR). $k = 31$ gives the lowest SSR and should be used. This gives a best model of $\hat{y} = 31x$.
19. **(a)**

![Stacked bar chart of Percent versus Frequency of marijuana use per year, with legend: Accident caused, No accidents caused](attachment)

There is an association between frequency of marijuana use per year and accidents caused because the distribution of accidents caused is not the same for each frequency of marijuana use group.

(b) Even if there is a strong association between two variables, we should not conclude that changes in one variable necessarily cause changes in the other variable.

Lesson 2.7 Exercises

1. Because there is a leftover pattern (a U-shaped curve) in the residual plot, the least-squares regression line is not an appropriate model for relating the number of AP® Statistics exams to the year.
3. Because there is no leftover pattern in the residual plot, the least-squares regression line is an appropriate model for relating a student's grade point average to the number of text messages a student sent on a previous day.
5. The actual height is typically about 8.61 centimeters away from its predicted height using the least-squares regression line with x = age (in years).
7. The actual average gas consumption is typically about 46.4 cubic feet per day away from its predicted average gas consumption using the least-squares regression line with x = average temperature (in degrees Fahrenheit).
9. 27.4% of the variability in height is accounted for by the least-squares regression line with x = age (in years).
11. 96.6% of the variability in average gas consumption is accounted for by the least-squares regression line with x = average temperature (in degrees Fahrenheit).
13. (a) Because there is no leftover pattern in the residual plot, the least-squares regression line is an appropriate model for relating the percent of the grass area burned to the number of wildebeests. **(b)** The actual percent of the grass area burned is typically about 15.99% away from its predicted percent of the grass area burned using the least-squares regression line with x = wildebeest abundance (in thousands of animals). **(c)** 64.6% of the variability in the percent of the grass area burned is accounted for by the least-squares regression line with x = wildebeest abundance.
15.

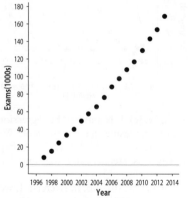

17. (a)

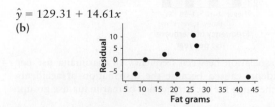

$\hat{y} = 129.31 + 14.61x$
(b)

Because there is no leftover pattern in the residual plot, the least-squares regression line is an appropriate model for relating the number of calories to the amount of fat. **(c)** $s = 7.667$; the actual number of calories is typically about 7.667 away from its predicted number of calories using the least-squares regression line with x = fat (grams). **(d)** $r^2 = 0.998$; 99.8% of the variability in the number of calories is accounted for by the least-squares regression line with x = fat (grams).
19. $s = 8.61$ cm/100 = 0.0861 meters; $r^2 = 0.274$. The correlation r remains the same when we change the units of either variable (so r^2 also remains the same).
21. (a) $z = \dfrac{\text{value} - \text{mean}}{\text{standard deviation}} = \dfrac{179 - 170}{7.5} = 1.2$. Paul has a height that is 1.2 standard deviations above the mean of 170 centimeters. **(b)** Roughly 85% of 15-year-old males are shorter than Paul.

Lesson 2.8 Exercises

1. (a) $\hat{y} = 34472.709x^2 - 30697.592x + 13150.905$
(b)

(c) $\hat{y} = 34472.709(3.61)^2 - 30697.592(3.61) + 13150.905 = \$351,584.39$. The residual is $y - \hat{y} = 365,700 - 351,584.39 = \$14,115.61$. The price for the 3.61-carat diamond is \$14,115.61 greater than expected, based on the quadratic model with x = carat weight.
3. (a) $\hat{y} = 0.0089x^2 - 0.5583x + 9.8111$
(b)

(c) $\hat{y} = 0.0089(32)^2 - 0.5583(32) + 9.8111 = 1.059$. The residual is $y - \hat{y} = 0.853 - 1.059 = -0.206$. The WHIP for the 1968 season is 0.206 less than expected, based on the quadratic model with x = age (years).
5. (a) $\hat{y} = 120.12151(1.00592)^x$
(b)

(c) $\hat{y} = 120.12151(1.00592)^{312} = 758$; the residual is $y - \hat{y} = 740 - 758 = -18$. The population for 2012 was 18 million less than expected, based on the exponential model with x = years since 1700.

7. (a) $\hat{y} = 392.74489(0.80378)^x$

(b)

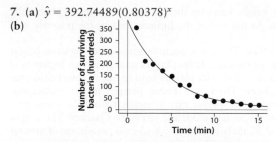

(c) $\hat{y} = 392.74489(0.80378)^8 = 68.4$ hundred; the residual is $y - \hat{y} = 60 - 68.4 = -8.4$. The number of surviving bacteria for the time $t = 8$ was 8.4 hundred (840) less than expected, based on the exponential model with x = time exposed.

9. Because the residual plot for the exponential model seems more randomly scattered, the exponential model is more appropriate for these data than the quadratic model.

11. (a) The quadratic model is $\hat{y} = 195.1686x^2 + 6426.3002x + 3059.3382$ where y = number of AP® Statistics exams and x = year. **(b)** The exponential model is $\hat{y} = 15179.08488(1.16762)^x$ where y = number of AP® Statistics exams and x = year. **(c)** Quadratic residual plot

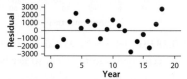

Exponential residual plot

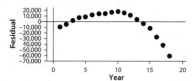

Because the residual plot for the quadratic model seems more randomly scattered, the quadratic model is more appropriate for these data than the exponential model. **(d)** Using the quadratic model, $\hat{y} = 195.1686(17)^2 + 6426.3002(17) + 3059.3382 = 168,710$. The residual is $y - \hat{y} = 169,508 - 168,710 = 798$. The number of exams is 798 greater than expected, based on the quadratic model with x = year.

13. $x = \dfrac{-b}{2a} = \dfrac{-835.5}{2(-13.65)} = 30.6$ According to our quadratic model, NFL quarterbacks peak at age 30.6.

15. The exponential model for Europe is $\hat{y} = 120.12151(1.00592)^x$ and for North America is $\hat{y} = 1.43398(1.01825)^x$. Because $1.01825 > 1.00592$, North America has a population with the higher growth rate.

17. (a) The predicted percent change in 2013 increases by 0.396% for each additional percent increase in 2012. **(b)** If a stock has a 2012 percent increase of 0, the predicted percent increase for 2013 is 24.9%. **(c)** Because the point for Boeing Company is near $\bar{x}$ but above the rest of the points, it pulls the line up a little, which increases the y intercept but doesn't change the slope very much.

Chapter 2 Review Exercises

1. (a) The explanatory variable is the grade level of the student, and the response variable is the student's most important goal. Grade level may help predict what goal was most important to them.

(b)

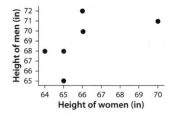

(c) There is a weak association between grade level and most important goal because the distribution of most important goal is not exactly the same for each grade level. In other words, knowing the grade level of a student helps us predict the student's most important goal. Students in the fourth grade more often choose good grades as their most important goal than do fifth- and sixth-graders, whereas fifth-graders are more likely to choose athletic ability than are fourth- or sixth-graders. The percent of students in each grade level who chose being popular as their main goal is about the same for all grade levels.

2. (a) The explanatory variable is the average temperature (in degrees Fahrenheit) for the month that is six months after the birth month, and the response variable is the average crawling age (in weeks). Average temperature for the month that is six months after the birth month might help predict the average crawling age.

(b)

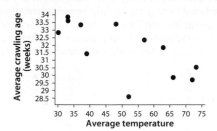

(c) *Direction:* There is a negative association between average temperature and average crawling age. Crawling months with a higher average temperature tended to have lower average crawling age, and vice versa. *Form:* There is a linear pattern in the scatterplot. *Strength:* Because the points vary somewhat from the linear pattern, the association has moderate strength. *Outliers:* The birth month of May (or crawling month November) at point (52, 28.58) departs from the overall linear pattern. **(d)** Because the relationship is negative, $r < 0$. Also, r is closer to -1 than 0 because the relationship is moderately strong—there isn't much scatter from the linear pattern.

3. Although there is a positive correlation, an increase in the number of calculators is not likely to cause an increase in math achievement. It is likely that both of these variables are changing due to several other variables, such as family income and how much the student values education.

4. (a)

(b) $r = \dfrac{0 + 0.37680 + 0 + 0.18840 + 1.50717 + 0.75358}{6 - 1} = 0.57$

The correlation of 0.57 indicates that the linear relationship between the height of women and the height of men that they date is moderate and positive. **(c)** The majority of points shows a positive association. However, because the point $(70, 71)$ is in the lower right and separated from the rest of the points in the x-direction, it will be influential and will make the correlation weaker (closer to 0). **(d)** The correlation would remain the same, because the correlation doesn't change when we change the units of either variable.

5. (a) $\hat{y} = 106.5 - 0.782(65) = 55.67°F$; we can't be confident in this prediction because this would be an extrapolation, as a latitude of $65°$ is far outside the interval of x values used to obtain the regression line. This prediction could be inaccurate.
(b) $\hat{y} = 106.5 - 0.782(34) = 79.9°F$; $residual = y - \hat{y} = 74 - 79.9 = -5.9°F$ **(c)** The predicted mean July temperature decreases by $0.782°F$ for each additional degree of latitude. **(d)** At the equator, the predicted mean July temperature is $106.5°F$. However, this predicted value is an extrapolation and could be inaccurate.

6. (a) $\hat{y} = 35.6781 - 0.0777x$. **(b)** The point representing May does not follow the linear trend of the rest of the data points and could be considered an outlier. It has a much lower average crawling age than the linear trend of the rest of the data points would predict. **(c)** Because the point for May is near $\bar{x}$ but below the rest of the points, it pulls the line down a little, which decreases the y intercept but doesn't change the slope much.

7. slope $= b = r\dfrac{s_y}{s_x} = 0.95\left(\dfrac{12.3}{14}\right) = 0.835$; y intercept $= a = \bar{y} - b\bar{x} = 55.1 - 0.835(61.7) = 3.581$. The equation of the least-squares regression line is $\hat{y} = 3.581 + 0.835x$.

8. (a) Because there is no leftover pattern in the residual plot, the least-squares regression line is an appropriate model for relating mean July temperature (°F) to the latitude of the city. **(b)** The actual mean July temperature of a city is typically about 6.4 degrees away from its predicted actual mean July temperature using the least-squares regression line with x = latitude (in degrees). **(c)** 27.7% of the variability in the mean July temperatures is accounted for by the least-squares regression line with x = latitude (in degrees).

9. (a)

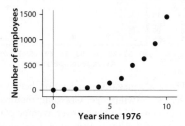

(b) $\hat{y} = 23.2914x^2 - 108.9138x + 82.1888$ where y = number of employees and x = years since 1976. **(c)** $\hat{y} = 5.44026(1.78947)^x$ where y = number of employees and x = years since 1976.
(d) Quadratic: $\hat{y} = 23.2914(5)^2 - 108.9138(5) + 82.1888 = 120$ employees. The residual is $y - \hat{y} = 128 - 120 = 8$. The number of employees for the year 1981 is 8 greater than expected, based on the quadratic model with x = years since 1976.
Exponential: $\hat{y} = 5.44026(1.78947)^5 = 100$ employees. The residual is $y - \hat{y} = 128 - 100 = 28$. The number of employees for the year 1981 is 28 greater than expected, based on the exponential model with x = years since 1976. **(e)** To decide which model is better, use the model with the most randomly scattered residual plot.

Chapter 2 Practice Test

1. a
2. b
3. c
4. a
5. c
6. c
7. b
8. c
9. c
10. d
11. (a) The explanatory variable is the size of the business, because the size of the business may help predict the response rate.
(b)

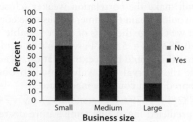

(c) There is an association between business size and response rate because the distribution of response is not the same for each business

size. In other words, knowing the business size helps us predict the response rate. As the size of the business grows, the response rate decreases.

12. (a) *Direction:* There is a positive association between height and field goal percentage. Taller players tend to have a higher field goal percentage, and shorter players tend to have a lower field goal percentage. *Form:* There is a linear pattern in the scatterplot. *Strength:* Because the points vary somewhat from the linear pattern, the association has moderate strength. *Outliers:* The player who is around 82 inches tall with a field-goal percentage of around 65 percent departs from the linear pattern.

(b) slope $= b = r\dfrac{s_y}{s_x} = 0.59\left(\dfrac{7.1}{3.29}\right) = 1.27$; y intercept $= a = \bar{y} - b\bar{x} = 45.3 - 1.27(78.9) = -54.9$. The equation of the least-squares regression line is $\hat{y} = -54.9 + 1.27x$

13. (a)

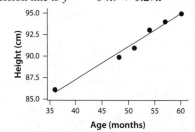

(b) $\hat{y} = 71.95 + 0.3833x$. **(c)** $\hat{y} = 71.95 + 0.3833(480) = 255.934$ cm; there are 2.54 centimeters to the inch, so her height in inches would be $\dfrac{255.943}{2.54} = 100.76$ inches. **(d)** Extrapolation. The linear trend will not continue until she is 40 years old. Our data were based on only the first 5 years of life, and predictions should be made for only ages 0–5.

CHAPTER 3

Lesson 3.1 Exercises

1. (a) Valid; the answer to this question is based on data that vary—the amount of money that each student carries. **(b)** Not valid; this question can be answered with a single value that doesn't vary.

3. (a) Not valid; this question can be answered with a single value that doesn't vary. **(b)** Valid; the answer to this question is based on data that vary—some U.S. residents think the president is doing a good job and some do not.

5. Population: A large batch of wood. Sample: The 5 pieces chosen to be tested.

7. Population: All of the artifacts collected on the archaeological dig. Sample: The 2% of artifacts that are chosen to be checked.

9. Population: The 45,000 people that made credit card purchases. Sample: The 137 people who returned the survey form.

11. Experiment; treatments were imposed on the high school juniors. Some students were assigned to using the computer and others were assigned to using the text.

13. Experiment; treatments were imposed on the laptop computers. Some computers were assigned to adjust screens to the brightest setting and others were assigned to moderate brightness.

15. Observational study; there were no treatments imposed on the statistics students. In other words, students weren't told to sleep a certain amount of hours.

17. (a) Is there an association between coffee drinking and long life? **(b)** Population: All of the older people in eight states. Sample: The 400,000 older people who participated in the survey. **(c)** Observational study; there were no treatments imposed on the older people. In other words, the older people weren't told to drink a certain amount of coffee.

19. Answers will vary. Possible questions include: (1) How many calories do college students consume per day? (2) Are college students more likely to eat at the dormitory or fast food? (3) Is there an association between the number of classes a college student is taking and calorie consumption?

21. (a) 40 hours; 8% (less than 40 hours) + 42% (40 hours) = 50% of the workers. **(b)** Skewed right; with a mean of 47, which is much higher than the median of 40, we expect the distribution to be skewed right with possible large outliers.

Lesson 3.2 Exercises

1. Convenience sample; the sample is likely to overestimate the unemployment rate because people without jobs have more time to be at the mall than those who are employed.

3. Convenience sample; it is likely that the first 100 students to arrive at school are getting less sleep than other students because they have to wake up earlier to get to school first. As a result, the average of the sample is likely to be less than the true average amount of sleep last night for all students at the school.

5. Voluntary response sample; it is likely that those customers who volunteered to leave reviews feel strongly about the hotel, often due to a negative experience. Customers who had an average or high-quality experience are less likely to leave a review. As a result, the 26% from the sample is likely greater than the true percentage of all the hotel's customers who would give the hotel 1 star.

7. Voluntary response sample; it is likely that the newspaper readers who volunteered to write letters feel strongly about the mayor's decision, often in opposition. Readers who don't have a strong opinion are less likely to write a letter. As a result, the 90% from the sample is likely greater than the true percentage of all the readers who oppose the mayor's decision.

9. Instead of sampling at the mall, the interviewer could obtain a list of all people in the city and randomly select a sample of them to be surveyed. Because no personal choice is involved, this sample should be more representative of the people in the city.

11. Instead of looking at Yelp reviews, the interviewer could obtain a list of all the recent customers of the hotel and randomly select a sample of them to be surveyed.

13. (a) Convenience sample. **(b)** It is likely that the students in Sammy's statistics class are older than most high school students, as juniors and seniors more often take a statistics class. Students who are older are more likely to have a driver's license. As a result, the 68% from the sample is likely greater than the true percentage of all students at her high school who have a driver's license. **(c)** Instead of using her statistics class as the sample, Sammy could obtain a list of all the students in her high school and randomly select a sample of them to be surveyed.

15. (a) Voluntary response. **(b)** It is likely that only those people who feel strongly about the proposal will volunteer to respond, often in opposition. Listeners who don't have a strong opinion are less likely to call in. As a result, the 78% from the sample is likely greater than the percent of all Springdale residents who oppose the proposal. **(c)** Instead of asking for volunteers, the radio station could obtain a list of all of the residents of Springdale and randomly select a sample of them to be surveyed.

17. Answers will vary. Here is one example. There was a recent online poll in Taiwan that asked, "Do you support establishing a Same-Sex Partnership Act?" The poll concluded that 75% of people support the idea. This poll suffered from voluntary response. It is likely that only those people who feel strongly about the proposal will volunteer to respond, often in favor. People who don't have a strong opinion are less likely to vote online. As a result, the 75% from the sample is likely greater than the percent of all Taiwan residents who support the proposal.

19. (a)

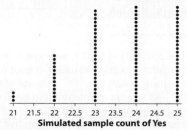

(b) Because there is a leftover pattern (a U-shaped curve) in the residual plot, the least-squares regression line is not an appropriate model. **(c)** Because the residual plot for the quadratic model seems more randomly scattered (and generally has smaller residuals), the quadratic model is more appropriate for these data than the exponential model.

Lesson 3.3 Exercises

1. (a) Give each hotel a distinct label from 1 to 28. Number 28 identically sized slips of paper from 1 to 28. Put these slips of paper into a hat and mix well. Select 4 slips of paper and call the 4 hotels that are labeled with the numbers selected. **(b)** Give each hotel a distinct label from 1 to 28. Using a random number generator, generate 4 integers from 1 to 28, ignoring repeated integers. Call the 4 hotels that are labeled with the generated integers.

3. (a) This method would require making and labeling 1410 slips of paper to put into a hat and would be very time consuming. **(b)** Give each circular plot a distinct label from 1 to 1410. Using a random number generator, generate 141 integers from 1 to 1410, ignoring repeated integers. Visit the 141 circular plots that are labeled with the generated integers.

5. (a) No; because different random samples will produce different correlations, it is unlikely that this sample provides a correlation that is *exactly* correct. **(b)** Because the sample size has increased, we could expect our estimate to be closer to the true correlation between arm span and height for all students at Washington High School.

7. (a) No; because different random samples will produce different means, it is unlikely that this sample provides a mean that is *exactly* correct. **(b)** Because the sample size has decreased, we could expect our estimate to be farther from the true mean time that all callers to the customer service line were placed on hold.

9. (a) If we repeatedly take random samples of size 250 from a fair coin, the number of heads in a sample varies from about 105 to 142. **(b)** Yes; in the Polish math professors' classroom, there were 140 heads in their sample of 250 spins. This is greater than what we would expect to happen by chance alone. Of the 100 trials of a simulation, 140 or more heads happened only 3 times.

11. (a) If we repeatedly take random samples of size 50 from a population of students in which 10% are left-handed, the number of left-handed students in a sample varies from about 1 to 11. **(b)** No; in Simon's sample of 50 students, 7 of them were left-handed. Of the 100 trials of a simulation, 7 or more left-handed students happened 29 times. This suggests that Simon's results may have happened purely by chance.

13. (a) Give each student a distinct label from 1 to the total number of students, N. Using a random number generator, generate 100 integers from 1 to N, ignoring repeated integers. Survey the 100 students who are labeled with the generated integers. **(b)** No; because different random samples will produce different means, it is unlikely that this sample provides a mean that is *exactly* correct. **(c)** An SRS of 100 students; this is because increasing sample size reduces sampling variability.

15. (a) Divide a spinner into 20 equal sectors. Label 19 (95%) of the 20 sectors as "on time" and the remaining sector as "late." **(b)**

(c) No; in the sample, 22 of 25 planes were on time. Of the above 100 trials of a simulation, a count of 22 or lower happened 17 times. This suggests that our results may have happened purely by chance.

17. (a) Moving alphabetically, give each apartment complex a label from 01 to 33 and go through the random-digit table looking at sets of 2 digits until we find 3 distinct numbers from 01 to 33. Survey the complexes labeled with the selected integers. **(b)** 38<u>16</u>7 98<u>532</u> 62<u>183</u> 70632 23417 26185 41448 75532. The sample is 16 Fairington, 32 Waterford Court, 18 Fowler.

19. (a) $b = r\frac{s_y}{s_x} = 0.6\left(\frac{8}{30}\right) = 0.16$. $a = \bar{y} - b\bar{x} = 75 - 0.16(280) = 30.2$. So $\hat{y} = 30.2 + 0.16x$, where y = final exam score and x = total score before the final examination. **(b)** $\hat{y} = 30.2 + 0.16(300) = 78.2$. **(c)** The least-squares regression line is the line that has the smallest sum of squared residuals.

Lesson 3.4 Exercises

1. (a) margin of error $\approx 2(0.014) = 0.028$. **(b)** We expect the true proportion of adults who are satisfied with the way things are going in the United States to be at most 0.028 from the estimate of 0.25.

3. (a) margin of error $\approx 2(0.009) = 0.018$. **(b)** We expect the true proportion of U.S. adults who use Twitter or another service to share updates about themselves to be at most 0.018 from the estimate of 0.19.

5. (a) margin of error $\approx 2(2.405) = 4.81$ mg/100 g. **(b)** We expect the true mean amount of vitamin C in the blend to be at most 4.81 mg/100 g from the estimate of 22.5 mg/100 g.

7. (a) margin of error $\approx 2(2.127) = 4.254$ pepperonis. **(b)** We expect the true mean number of pepperonis on a large pizza from this restaurant to be at most 4.254 pepperonis from the estimate of 37.4 pepperonis.

9. (a) We expect the true proportion of adults aged 18 and older who say football is their favorite sport to watch on television to be at most 0.031 from the estimate of 0.37. **(b)** Yes; according to the sample, the true proportion could be as low as $0.37 - 0.031 = 0.339$ or as high as $0.37 + 0.031 = 0.401$. Therefore, any proportion between 0.339 and 0.401 is plausible. $p = 0.50$ is not plausible. **(c)** Increase the number of adults in the sample.

11. (a) We expect the true mean travel time for employed adults in New York State to be at most 4.47 minutes from the estimate of 31.25 minutes. **(b)** No; according to the sample, the true mean could be as low as $31.25 - 4.47 = 26.78$ minutes or as high as $31.25 + 4.47 = 35.72$ minutes. Therefore, any mean between 26.78 minutes and 35.72 minutes is plausible (which includes values less than 30 minutes). **(c)** Increase the number of employed adults in the sample.

13. The margin of error for each of the estimates will be larger. If the total sample size of males and females is 1000, the individual sample size of males and the individual sample size of females will both be less than 1000. Reducing the sample size will increase the sampling variability, thus increasing the margin of error.

15. (a) $171/880 = 0.194$. **(b)** Answers may vary due to sampling variability.

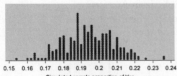

Distribution of simulated proportion		
# samples	mean	SD
300	0.195	0.014

Margin of error $\approx 2(0.014) = 0.028$; we expect the true proportion of drivers who admit to going through at least one red light of the last 10 to be at most 0.028 from the estimate of 0.194.

17. $(0.58)(600) = 348$ seniors are planning to go to the prom. The margin of error for this estimate is $(0.14)(600) = 84$ seniors.

19. The percentage of 18- to 31-year-olds who live with a spouse was much lower in 2012 (around 22%) than it was in 1968

(around 55%). Also, the percentage of 18- to 31-year-olds who have "other" living conditions was much higher in 2012 (around 28%) than it was in 1968 (around 6%).

Lesson 3.5 Exercises

1. The fans in the club seats and box seats can't be part of the sample. These fans would likely spend more money at the game than fans in the cheaper seats, meaning your estimate for the average amount of money spent will likely be too low.

3. This study displays undercoverage because the U.S. residents who are not registered voters can't be part of the sample. Residents who are not registered voters are more likely to support a "pathway to citizenship" because this group includes some people who came to the U.S. illegally. This means the news organization's estimate for the percent of U.S. residents who support a "pathway to citizenship" will likely be too low.

5. People who didn't improve their golf game might be frustrated and choose not to reply to the survey. On the other hand, people who improved their golf game are probably happy to share their success with the website. This means that the estimate from the sample is likely to be larger than the mean reduction in golf score for all golfers who paid for the secret technique.

7. This survey will yield a biased result because the people who drive more miles per day are less likely to be at home to respond to the survey. This will likely produce an estimate for the average driving miles per day that is too low, as the people who are at home to answer the survey probably drive less than those people who are not at home.

9. People likely claim to wear their seat belts because they know they should; they are embarrassed or ashamed to say that they do not always wear seat belts.

11. The boys who were asked directly might claim that they don't cry because they are embarrassed or ashamed to admit it. Boys who were given an anonymous survey are more likely to be honest about their experiences.

13. By making it sound as if they are not a problem in the landfill, this question will result in fewer people suggesting that we should ban disposable diapers. The proportion who would say "Yes" to this survey question is likely to be smaller than the proportion who would say "Yes" to a more fairly worded question.

15. (a) People who do not have a residential telephone are not included in the sample. People with residential telephones tend to be much older, and older people tend to be much more cautious drivers; thus, they would be less likely to drive through red lights. **(b)** People who went through red lights might be embarrassed or ashamed of their actions or worried what people might think of them, so they lie by saying they didn't go through red lights. **(c)** This would not reduce the bias due to undercoverage because people without a telephone are still going to be left out. Also, it would not reduce the response bias, as people would continue to lie about their driving habits (there would just be more people lying).

17. Answers will vary. Here is one possible response. *Neutral:* The city is proposing a ban on plastic grocery bags. Do you agree or disagree with the proposal? *Slanted:* One in three leatherback sea turtles has plastic in its stomach, most often a plastic bag, based on a study of over 370 autopsies. Do you agree or disagree with the city's proposal to ban plastic grocery bags?

19. (a) July 1 to August 31—when families are likely to be vacationing or spending time outdoors. **(b)** Answers will vary. Here is one possible response. How many hours did you spend outside this week?

21. (a) *Direction:* There is a negative association between the year and winning time for the men's Olympic 100-meter dash from 1900 through 2012. *Form:* Linear. *Strength:* Because the points don't vary much from the linear pattern, the association is strong. *Outliers:* There are no obvious outliers. **(b)** The predicted winning time for the men's Olympic 100-meter dash decreases by 0.01153 second for each additional year. **(c)** $\hat{y} = 32.84 - 0.01153(2100) = 8.627$ seconds. This prediction may not be realistic because it is an extrapolation.

Lesson 3.6 Exercises

1. No; although eating seafood may decrease the risk of colon cancer, it is possible that the physicians who ate seafood were also more likely to exercise. Perhaps it was the exercise that caused the decrease in colon cancer risk, not eating seafood.

3. No; although acting less agreeable may increase the chance of a higher salary, it is possible that people who act less agreeable were also more likely to work more hours. Perhaps it was the extra number of hours working that caused the increase in salary, not acting less agreeable.

5. The purpose of the control group is to provide a baseline for comparing the effects of the other treatments. Otherwise, we wouldn't be able to tell if the two levels of added water caused an increase in total plant biomass.

7. The purpose of the control group is to provide a baseline for comparing the effect of the treatment. Otherwise, we wouldn't be able to tell if the CareerStart program caused an improvement in students' attendance, behavior, standardized test scores, level of engagement, and graduation rate.

9. This experiment could be double-blind if the treatment (ASU or placebo) assigned to a subject was unknown to both the subject and those responsible for assessing the effectiveness of that treatment. It is important for the subjects to be blind because even though the placebo possesses no medical properties, some subjects may show improvement or benefits because they have positive expectations. It is important for the experimenters to be blind so that they will be unbiased in how they interact and assess the subjects.

11. The experiment was not blind because the subjects and the experimenter knew which treatment each person was receiving. The subjects who received the meditation training may have assumed that the meditation would reduce their anxiety, thus influencing the results. Furthermore, the experimenter might have had high hopes that the meditation would reduce anxiety, and that could have influenced the experimenter's judgment of the anxiety level ratings at the end of the experiment.

13. **(a)** An experiment is needed to show a cause-and-effect relationship between ultrasounds and baby birth weight. Simply asking pregnant women whether or not they had an ultrasound is an observational study and might suffer from confounding. For example, women who have ultrasounds might have access to better medical care. We wouldn't know if it was the ultrasound or the medical care that influenced birth weight. **(b)** The purpose of the control group is to provide a baseline for comparing the effects of the treatment. Otherwise, we wouldn't be able to tell if the ultrasound caused an increase in birth weight. **(c)** This experiment could be blind if there was a placebo ultrasound in which the machine was not turned on. To prevent mothers from figuring out the machine was turned off, the ultrasound screen would have to be turned away from all the mothers. Blinding the mothers is important because it's possible that seeing the image of their unborn child encouraged the mothers who had an ultrasound to eat a better diet, resulting in healthier babies.

15. There was no control group for comparison purposes. We don't know if this was a placebo effect or if the flavonols affected the blood flow.

17. In this experiment, both treatment groups experienced pain relief, even though neither group was receiving medication. Furthermore, because people generally expect that more expensive products work better, a greater proportion of people in the "expensive" group experienced pain relief.

19. Answers will vary. Here is one possible response. Randomly assign the piglets into two groups. For one of the groups, feed them 6 cups of the new piglet food per day for 30 days, and then measure their weight gain. For the other group, feed them 4 cups of the old piglet food per day for 30 days, and then measure their weight gain. In the end, we may find that the group with the new piglet food has greater weight gains, but we will not know if it is because of the new piglet food, or simply the increase in the amount of daily food (confounding).

21. **(a)**

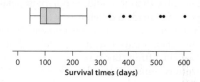

The survival times are right skewed, as expected.

(b) Outliers $> Q_3 + 1.5 \times IQR = 151.5 + 1.5 \times 69 = 255$. Because 329, 380, 403, 511, 522, 598 are all more than 255, they are considered outliers.

(c) Because the distribution is strongly skewed with high outliers, it would be better to use the median and IQR to summarize the distribution.

Lesson 3.7 Exercises

1. **(a)** Use 139 slips of paper. Write the letter "A" on 70 slips of paper and write the letter "B" on the remaining 69 slips of paper. Shuffle the slips of paper, and hand out one slip of paper to each volunteer. Customers who get an "A" slip will pay $4 for the buffet and customers who get a "B" slip will pay $8 for the buffet. **(b)** Label each customer with a different integer from 1 to 139. Then randomly generate 70 *different* integers from 1 to 139. The customers with these labels will pay $4 for the buffet. The remaining 69 customers will pay $8 for the buffet.

3. **(a)** Use 120 slips of paper. Write the number "1" on 40 slips of paper, the number "2" on 40 slips of paper, and the number "3" on 40 slips of paper. Shuffle the slips of paper and hand out one slip of paper to each of the subjects. Subjects who get a "1" slip will be given treatment 1, those with a "2" slip will be given treatment 2, and those with a "3" slip will be given treatment 3. **(b)** Label each subject with a different integer from 1 to 120. Then randomly generate 40 *different* integers from 1 to 120. The subjects with these labels will be given treatment 1. Next, randomly generate 40 more *different* integers from 1 to 120 (being careful not to choose subjects who have already been selected for treatment 1). The subjects with these labels will be given treatment 2. The remaining 40 subjects will be given treatment 3.

5. The purpose of random assignment is to create roughly equivalent groups at the beginning of the experiment. If there is a difference between the groups in pizza rating, it is likely due to the different prices paid.

7. The purpose of random assignment is to create roughly equivalent groups at the beginning of the experiment. If there is a difference between the groups in work performance, it is likely due to the different treatments.

9. **(a)** The details of the phone call to the member of Congress. **(b)** If one of the treatment groups was allowed to discuss different details during the phone call, we wouldn't know if the details of the phone call or the caller's designation (campaign donor or constituent) was the cause of a difference in success rate in obtaining meetings.

11. (a) The amount of time that subjects were given between the treatment and the bowl of pasta (15 minutes). **(b)** If the researchers let subjects within each group wait for as little or as long as they wanted before eating the pasta, the number of pasta calories consumed would be much more variable than it would be otherwise, making it harder to tell if the treatments made a difference.

13. (a) Label each subject with a different integer from 1 to 132. Then randomly generate 66 *different* integers from 1 to 132. The subjects with these labels will be assigned the low-carbohydrate diet. The remaining 66 subjects will be assigned the low-fat diet. **(b)** The purpose of random assignment is to create roughly equivalent groups at the beginning of the experiment. If there is a difference between the groups in weight loss, it is likely due to the difference in diet. **(c)** Length of time for treatment. If one of the treatment groups was allowed to be on their diet for a longer period of time, we wouldn't know if the amount of time or the type of diet was the cause of a difference in weight loss. Also, if the researchers let subjects within each group diet for as little or as long as they wanted, the weight loss would be much more variable than it would be otherwise, making it harder to tell if the type of diet made a difference.

15. The coach did not use random assignment. When he allowed players to choose their treatment, it is possible that the strongest players chose weight lifting. Because the strongest players will be able to do more push-ups, we wouldn't know if any difference in numbers of push-ups was due to initial strength of the players or the strength program.

17. Recruit 20 volunteers from the football team. Label each player with a different integer from 1 to 20. Then randomly generate 10 *different* integers from 1 to 20. The players with these labels will be assigned milk. The remaining 10 players will be assigned sports drinks. Each player will do the same 60-minute workout, wait 15 minutes, and then drink 12 ounces of either milk or sports drink. One hour later, each player will take a fitness test, which measures their muscle recovery.

19. The purpose of random assignment is to create roughly equivalent groups at the beginning of the experiment. Many variables may have an effect on the crash rate of 16- to 18-year-old drivers, including driver preparation course, driving and texting, number of hours of practice driving, and others. Only with random assignment can we be comfortable that all of these variables have been equally distributed into the two groups, so that we may be convinced that the difference in crash rates is due to the school starting time.

21.

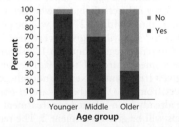

There is an association between age and Facebook use because the Facebook use percentages are not the same for each age group. In other words, knowing a person's age group helps us predict whether or not they use Facebook. Young people are much more likely to use Facebook.

Lesson 3.8 Exercises

1.

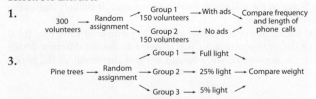

3.

5. The difference in the mean lengths of the cold in the two groups was small enough that it could be due to chance variation in the random assignment to treatments.

7. The difference in the balance test scores in the tai chi group and the other two groups was too large to be due only to chance variation in the random assignment to treatments.

9. (a) $26.5 - 2.1 = 24.4$ cm. **(b)** Because a difference of means of 24.4 cm or higher never occurred in the simulation, the difference is statistically significant. It is extremely unlikely to get a difference this big simply due to chance variation in the random assignment.

11. (a) $16/25 - 12/25 = 0.64 - 0.48 = 0.16$. **(b)** Because a difference of proportions of 0.16 or higher occurred 26 out of 100 times in the simulation, the difference is not statistically significant. It is quite plausible to get a difference this big simply due to chance variation in the random assignment.

13. (a)

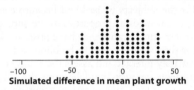

(b) $90.6 - 86.1 = 4.5\%$. **(c)** Because a difference of means of 4.5 or higher rarely occurred in the simulation ($3/100 = 3\%$), the difference is statistically significant. It is unlikely to get a difference this big simply due to chance variation in the random assignment.

15. (a) $85.8 - 38.2 = 47.6$ mm. **(b)** Answers for parts **(b)** and **(c)** may vary.

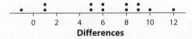

(c) Because a difference of means of 47.6 mm or higher never occurred in the simulation ($0/100 = 0\%$), the difference is statistically significant. It is extremely unlikely to get a difference this big simply due to chance variation in the random assignment.

17. (a)

There is a reasonably strong linear relationship between the IQs of the twins with a correlation of $r = 0.91$. So the IQ of one twin will do a good job of predicting the IQ of the other twin. **(b)** If we subtract IQs (Twin B − Twin A), we get the following dotplot of the differences.

Because all but one of the differences are positive, this suggests that Twin B (the one living in the higher income homes) tends to have a higher IQ than Twin A.

Lesson 3.9 Exercises

1. (a) Yes, because the study used a random sample of batteries from the population of all batteries in the warehouse. **(b)** Yes, because batteries were randomly assigned to be stored in the freezer or assigned to be stored at room temperature.

3. Because this study involved a random sample of adults, we can make an inference about the population of adults. It appears that adults who attend religious services regularly have a lower risk of dying. However, because subjects were not randomly assigned to attend religious services (or not), we cannot infer cause and effect.

We do not know that attending religious services is the cause of the lower risk.

5. Because this study does not involve random assignment to the treatments, we cannot infer that the difference in blueberry and strawberry intake caused the difference in heart attack risk. Also, we should not generalize this result to any larger population because these 93,600 women were not randomly selected from a larger population.

7. In this case, the subjects were not able to give informed consent. They did not know what was happening to them and they were not old enough to understand the ramifications in any event.

9. Many would consider this to be an appropriate use of collecting data without participants' knowledge because the data are, in effect, anonymous and confidential.

11. (a) This is an observational study and not an experiment. Perhaps mothers who ate more nuts exercised more and it was the exercise that caused the decrease in allergies. (b) This could be an ethical experiment if an institutional review board reviewed it, subjects gave their informed consent, and individual data were kept confidential.

13. Answers will vary. Possible answers include the following. (a) A non-scientist might be more likely to consider the subjects as people and not be blinded by the scientific results that might be discovered. (b) One may consider at least two outside members. A religious leader may be chosen because we would expect him or her to help the committee in ethical and moral discussions. You may also choose a patient advocate to speak for the subjects involved.

15. (a) Students do not have to feel guilty about answering "yes" to the question because it is possible that their coin landed on heads and they are simply following the protocol. (b) If 100 students each flipped a coin, we would expect 50 of them to get heads and answer "yes" according to the protocol. We hope that the other 50 students are answering the question honestly. This means that $63 - 50 = 13$ of the 50 admit to cheating. We estimate that the proportion of students that have cheated is $13/50 = 0.26$.

17. Several other variables could have contributed to the decrease in crime rate, including economic conditions, police force decisions, and quality of eduction. It is hard to know if the decrease in crime was due to Lance's leadership or these other variables.

Chapter 3 Review Exercises

1. (a) Not valid; this question can be answered with a single value that doesn't vary. (b) Valid; the answer to this question is based on data that vary—the number of people who visit Acadia National Park on each weekday in August. (c) Valid; the answer to this question is based on data that vary—some soda bottles contain less soda than the label indicates and some contain more.

2. (a) Sample: The 805 adult U.S. residents who were randomly sampled. Population: All U.S. residents. (b) No; because different random samples will produce different proportions, it is unlikely that this sample provides a proportion that is *exactly* correct. (c) Margin of error $\approx 2(0.012) = 0.024$. We expect the true proportion of U.S. residents who rate the honesty and ethical standards of nurses as high or very high to be at most 0.024 from the estimate of 0.85. (d) Increasing the sample size would reduce the margin of error. With a larger sample size, we can expect that our estimate will be closer to the true proportion of U.S. residents who rate the honesty and ethical standards of nurses as high or very high.

3. Answers will vary. Here is one possible answer. (a) Announce in daily bulletin that there is a survey concerning student parking available in the main office for students who want to respond. Because only those who feel strongly about the issue generally respond to voluntary response surveys, this method will likely be biased. (b) Personally interview a group of students as they come in from the parking lot. Because convenience samples are not generally representative of the population, this method will likely be biased. (c) Give each student a distinct label from 1 to 1800. Using a random number generator, generate 50 integers from 1 to 1800, ignoring repeated integers. Interview the 50 students who are labeled with the generated integers. (d) An SRS is more likely to accurately represent the whole population, whereas a voluntary response or convenience sample often results in a sample that does not have the same characteristics as the whole population.

4. (a) The random samples of 40 flying disk flips are producing different values for the number that land right side up, some as small as 11 and some as large as 27. (b) No; 16 right-side-up flips or fewer occurred 12 out of 100 times in the simulation, so the result is not statistically significant. It is quite plausible to get 16 right-side-up flips out of 40 simply due to chance variation in the random sample.

5. (a) When the interviewer provides the additional information that "Box-office revenues are at an all-time high," listeners may believe that they contributed to this fact and be more likely to overestimate the number of movies they've seen in the past 12 months. A change that could fix this problem is simply to eliminate that sentence. (b) This sample is likely to underrepresent younger adults who use only cell phones. If younger adults go to movies more often than older adults, the estimated mean will be too small. (c) People who do not go to the movie theater very often might be more likely to respond to the poll because they are at home. Because the frequent moviegoers will not be at home to respond, the estimated mean will be too small.

6. (a) Hospital records were used to "observe" the death rates, rather than to impose different anesthetics on the subjects. (b) One variable that might be confounded with choice of anesthetic is type of surgery. If anesthesia C is used more often with a type of surgery that has a higher death rate, we wouldn't know if the death rate was higher because of the anesthesia or the type of surgery.

7. (a) The purpose of the control group is to provide a baseline for comparing the effects of the treatment. Otherwise, we wouldn't be able to tell if the student-made outline or something else (e.g., upcoming parent–teacher conferences) caused an increase in test scores. (b) No; while it may be possible to set up an experiment in which the teacher doesn't know the assignment of treatments, there is no way to set up the experiment so that students don't know which treatment they are getting. (c) Answers will vary. One possible answer is that Mr. Chen graded both sets of tests himself to avoid confounding. If a different teacher graded the tests for each group, we wouldn't know if the difference in test scores was the result of the student-made outlines or of the teacher grading the test. Mr. Chen also gave the same 50-minute test to both groups to reduce variability in the test scores. If Mr. Chen let students work for as long as they wanted on the test, some students would quit early and some would work a very long time. This would increase the variability in the test scores and make it harder to tell if the treatment made a difference. (d) Use 28 slips of paper. Write the letter "A" on 14 of them and write the letter "B" on the remaining 14. Shuffle the slips of paper, and hand out one slip of paper to each student. Students who get an "A" slip will do the list of exercises from the textbook, and students who get a "B" slip will do those same exercises, as well as make an outline of ideas that the test covers. The purpose of random assignment is to create roughly equivalent groups at the beginning of the experiment. If there is a difference between the groups in scores on the test, it is likely due to the difference in review assignments.

8. (a) $87.9 - 80.4 = 7.5$. (b) Because a difference of means of 7.5 or higher only occurred 4 out of 200 times in the simulation, the difference is statistically significant. It is unlikely to get a difference this big simply due to chance variation in the random assignment.

9. (a)

(b) Because this experiment involved random assignment to the treatments (ebook or printed book), we can infer that the difference in the type of book caused the difference in the responses. However, we should not generalize this result to any larger population because these 50 students were not randomly selected from a larger population.

10. (a) This does not meet the requirements of informed consent because the subjects did not know the nature of the experiment before they agreed to participate. (b) All individual data should be kept confidential and the experiment should go before an institutional review board before being carried out.

Chapter 3 Practice Test

1. c
2. d
3. c
4. d
5. c
6. c
7. c
8. c
9. b
10. d
11. c
12. (a) Experiment; treatments were imposed on the students. Some students were given the "Markup" version of the question and others were given the "Convenience" version of the question. (b) $18/30 - 7/30 = 0.60 - 0.23 = 0.37$. (c) Because a difference of proportions of 0.37 or higher never occurred in the simulation, the difference is statistically significant. It is extremely unlikely to get a difference this big simply due to chance variation in the random assignment. (d) Because this experiment involved random assignment to the treatments, we can infer that the difference between the versions of the question caused the difference in response. Also, we can generalize this result to the larger population of all students at the school because the 60 students were randomly selected.
13. (a) Give each resident of the dormitory a distinct label from 1 to 816. Using a random number generator, generate 50 integers from 1 to 816, ignoring repeated integers. Survey the 50 residents who are labeled with the generated integers. (b) Not necessarily; because different random samples will produce different percents, it is unlikely that another sample of 50 would provide a percent that is *exactly* the same as the first. (c) Because these interviews were done in person, the students may lie to avoid offending the director, making the estimate of the proportion of residents who answered "yes" too high. Increasing the sample size would not eliminate this bias, because the in-person interview would continue to elicit similar responses (just more of them).
14. (a) The experimental units are the acacia trees. The treatments are placing either active beehives, empty beehives, or nothing in the trees. The response variable is damage to trees caused by elephants.

(b)

Assign the trees numbers from 1 to 72, and use a random number generator to generate 24 different numbers in this range. Those trees will get the active beehives. The trees associated with the next 24 different numbers generated in the range will get the empty beehives, and the remaining 24 trees will remain empty.
15. It is unlikely that an institutional review board reviewed the Milgram experiment in advance. The review board would likely have rejected his experiment because participants were subjected to a stressful situation that could have caused them psychological harm.

CHAPTER 4
Lesson 4.1 Exercises

1. (a) If you take a very large random sample of children whose parents both carry the gene for cystic fibrosis but don't have the disease themselves, about 25% of the children will develop cystic fibrosis. (b) No; probability describes what happens in many, many repetitions (way more than 4) of a chance process. We would expect to get *about* 1 child who develops cystic fibrosis in a random sample of 4 children, but this result is not guaranteed.
3. If you take a very large random sample of times during the day and record the color of the traffic light, about 55% of the times the light will be red.
5. The commentator's claim is based on the erroneous "law of averages." Even after the player failed to hit safely in six straight at-bats, he will continue to have the same 35% chance of getting a hit on the next at-bat.
7. The two sequences are equally likely. Any specific sequence of 6 coin tosses has a 1/64 probability of occurring.
9. STATE: What is the probability that at least one of four randomly selected adult males is red-green color-blind (assuming the probability of having some form of red-green color blindness is 0.07)?
PLAN: Let $1-7$ = adult male has red-green color blindness and $8-100$ = adult male does not have red-green color-blindness. Use randInt(1,100,4) to simulate taking a random sample of 4 adult males. Record the number of adult males in the sample of 4 who have red-green color-blindness.
DO: Rep 1: 58 61 38 82   0 color-blind
Rep 2: 61 92 45 15   0 color-blind
Rep 3: 19 69 29 29   0 color-blind
Rep 4: 86 47 92 98   0 color-blind
Rep 5: 19 <u>8</u> 73 24   1 color-blind
CONCLUDE: In 50 repetitions, there was at least 1 color-blind adult male 12 times. Assuming the probability of having some form of red-green color blindness is 0.07, the estimated probability that at least one of four randomly selected adult males is red-green color-blind is approximately 12/50 = 24%.
11. (a) Let 1 = made shot and 2 = missed shot. Use randInt(1,2,30) to simulate taking a random sample of 30 shots. Record the longest streak of made shots in the sample of 30 shots. (b) Assuming the player makes 50% of the shots and that the results of a shot don't depend on previous shots, the estimated probability that the player would have a streak of 10 or more made shots is 1/50 = 0.02. This is convincing evidence that the player is streaky.
13. In the short run, the player's percentage of made 3-point shots is extremely variable. In the long run, the percentage of made shots approaches a specific value of around 30%.
15. (a) 43/200 = 0.215; it is quite plausible that there is not a majority of students that recycle, and the statistics class got their results purely by chance. (b) 1/200 = 0.005; because this result is unlikely to happen purely by chance, we have convincing evidence that a majority of the school's students recycle.
17. STATE: What is the probability of getting all vowels when selecting a random sample of 7 tiles from a Scrabble bag (which contains 42 vowels, 56 consonants, and 2 blank tiles)?
PLAN: Let $1-42$ = vowels, $43-98$ = consonants, and $99-100$ blank. Use randInt(1,100) to simulate taking a random tile. Hit Enter as many times as needed to get seven unique integers (no repeats). Record the number of vowels in the sample.
DO: Rep 1: 61 82 <u>35</u> <u>28</u> 72 64 <u>15</u>   3 vowels
Rep 2: <u>7</u> <u>17</u> 98 83 61 54 62   2 vowels
Rep 3: 61 60 <u>4</u> <u>37</u> <u>21</u> <u>18</u> 80   4 vowels
Rep 4: <u>1</u> 64 <u>12</u> 95 85 83 <u>28</u>   3 vowels
Rep 5: 83 <u>33</u> <u>7</u> 56 85 59 98   2 vowels
CONCLUDE: In 50 repetitions, there were 7 vowels 0 times. The estimated probability of getting all 7 vowels in Scrabble is approximately 0/50 = 0%. Cait should be surprised at her result.

19. (a)

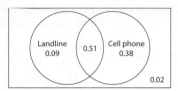

(b) *Shape:* The distribution of quiz scores for the 10 A.M. class is skewed left, while the distribution is fairly symmetric for the 8 A.M. group. *Center:* The center is higher for the 10 A.M. group (median = 24) than it is for the 8 A.M. group (median = 22). *Variability:* The 10 A.M. group has more variability ($IQR = 8$) than the 8 A.M. group ($IQR = 1.5$). *Outliers:* The 10 A.M. group has no outliers, while the 8 A.M. group has outliers at 17 and 18. **(c)** Not necessarily. It is possible that the difference in student performance for the two classes is caused by some other variable besides the time the class met.

Lesson 4.2 Exercises

1. (a) $(1, 1), (1, 2), (1, 3), (1, 4), (2, 1), (2, 2), (2, 3), (2, 4), (3, 1), (3, 2),$ $(3, 3), (3, 4), (4, 1), (4, 2), (4, 3), (4, 4)$. Because both dice are fair, each of these 16 outcomes will be equally likely and have probability 1/16.

(b) $(1, 4), (2, 3), (3, 2), (4, 1)$; $P(\text{sum of 5}) = \dfrac{4}{16} = 0.25$.

3. (a) rock/rock, rock/paper, rock/scissors, paper/rock, paper/paper, paper/scissors, scissors/rock, scissors/paper, and scissors/scissors. Because each player is equally likely to choose any of the three, each of these 9 outcomes will be equally likely and have probability 1/9. **(b)** rock/scissors, paper/rock, and scissors/paper. $P(\text{Player 1 wins}) = \dfrac{3}{9} = 0.33$.

5. (a) $1 - (0.49 + 0.27 + 0.20) = 1 - 0.96 = 0.04$. **(b)** $1 - 0.04 = 0.96$

7. (a) $1 - (0.25 + 0.32 + 0.07 + 0.03 + 0.01) = 1 - 0.68 = 0.32$. Each probability must be $0.32/2 = 0.16$. **(b)** $0.32 + 0.16 + 0.16 + 0.07 + 0.03 + 0.01 = 0.75$. Alternate solution: $1 - 0.25 = 0.75$.

9. (a) $0.49 + 0.20 = 0.69$ **(b)** $1 - 0.69 = 0.31$

11. $0.07 + 0.16 + 0.16 + 0.32 + 0.25 = 0.96$. Alternate solution: $1 - (0.03 + 0.01) = 1 - 0.04 = 0.96$.

13. (a) $1 - (0.13 + 0.29 + 0.30) = 1 - 0.72 = 0.28$ **(b)** $1 - 0.13 = 0.87$. **(c)** $0.28 + 0.30 = 0.58$

15. Using R = rock, P = paper, S = scissors, L = lizard, Sp = Spock. The sample space is R/R, R/P, R/S, R/L, R/Sp, P/R, P/P, P/S, P/L, P/Sp, S/R, S/P, S/S, S/L, S/Sp, L/R, L/P, L/S, L/L, L/Sp, Sp/R, Sp/P, Sp/S, Sp/L, Sp/Sp. Because each player is equally likely to choose any of the five, each of these 25 outcomes will be equally likely and have probability 1/25. Player 2 wins the following plays: P/S, R/P, L/R, Sp/L, S/Sp, L/S, P/L, Sp/P, R/Sp, S/R.

$P(\text{winning}) = \dfrac{\text{number of outcomes in event "Winning"}}{\text{total number of outcomes in sample space}} = \dfrac{10}{25} = 0.40$

17. (a) The predicted number of hours of sleep debt increases by 3.17 hours for each additional day. **(b)** predicted sleep debt = $2.23 + 3.17(5) = 18.08$ hours. The students' results are similar to the researcher's report.

Lesson 4.3 Exercises

1. (a) $295/595 = 0.496$; there is about a 50% chance that a randomly selected student from this school does not eat breakfast regularly.

(b) $165/595 = 0.277$ **(c)** $\dfrac{110 + 130 + 165}{595} = \dfrac{405}{595} = 0.681$

3. (a) $\dfrac{94 + 167 + 151 + 476}{1207} = \dfrac{888}{1207} = 0.736$

(b) $\dfrac{94 + 151}{1207} = \dfrac{245}{1207} = 0.203$

(c) $\dfrac{94 + 167 + 151 + 476 + 197}{1207} = \dfrac{1085}{1207} = 0.899$

5. $P(\text{landline or cell phone}) = P(\text{landline}) + P(\text{cell phone}) - P(\text{landline and cell phone}) = 0.60 + 0.89 - 0.51 = 0.98$

7. $P(\text{graduate student or Mac user}) = P(\text{graduate student}) + P(\text{Mac user}) - P(\text{graduate student and Mac user}) = 0.33 + 0.60 - 0.23 = 0.70$

9. (a)

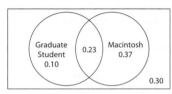

(b) $P(\text{cell phone only}) = 0.38$

11. (a)

(b) $P(G^C \cap M^C) = P(\text{not a graduate student and not a Mac user}) = 0.30$. There is a 30% chance that a randomly selected student from this major university is an undergraduate student who uses a PC.

13. (a) $P(\text{sixth grader or good grades}) = P(\text{sixth grader}) + P(\text{good grades}) - P(\text{sixth grader and good grades}) = 135/335 + 168/335 - 69/335 = 234/335 = 0.699$ **(b)** $\dfrac{24 + 36 + 19 + 22}{335} = \dfrac{101}{335} = 0.301$

15. (a)

	Number 8	Not number 8	Total
Red	1	0	9
Blue	1	8	9
Green	1	8	9
Yellow	1	8	9
Total	4	32	36

(b) $P(B) = P(\text{blue}) = 9/36 = 0.25$; $P(E) = P(\text{number 8}) = 4/36 = 0.111$. **(c)** "Blue eight" is the event (B and E); $P(B \text{ and } E) = 1/36 = 0.028$. **(d)** The events "Disk is blue" and "Disk is the number eight" are not mutually exclusive. $P(B \cup E) = P(B) + P(E) - P(B \text{ and } E) = 9/36 + 4/36 - 1/36 = 12/36 = 0.333$.

17. (a) True; an outcome cannot simultaneously be in event A and A^C at the same time. Therefore, these two events must be mutually exclusive. **(b)** False; two mutually exclusive events do not necessarily make up the entire sample space. Complementary events are mutually exclusive events that make up the entire sample space.

19. There is an association between gender and responses to this question because the percentages are not the same for each gender. In other words, knowing a person's gender helps us predict which answer he or she has to the question.

Lesson 4.4 Exercises

1. (a) $82/212 = 0.387$; given that the person chosen is a female, there is a 38.7% chance that her ring finger is longer. **(b)** $P(F^C | R^C) = \dfrac{45 + 43}{123 + 95} = \dfrac{88}{218} = 0.404$

3. (a) $\dfrac{197}{197 + 122} = \dfrac{197}{319} = 0.618$

(b) $\dfrac{197 + 94}{197 + 94 + 151} = \dfrac{291}{442} = 0.658$

5. $P(C|L) = \dfrac{P(C \cap L)}{P(L)} = \dfrac{0.51}{0.09 + 0.51} = \dfrac{0.51}{0.60} = 0.85$

7. $P(\text{Twitter} \mid \text{Facebook}) = P(T|FB) = \dfrac{P(T \cap FB)}{P(FB)} = \dfrac{0.15}{0.71} = 0.211$

9. $P(\text{right-handed} \mid \text{female}) = 18/21 = 0.857$; $P(\text{right-handed} \mid \text{male}) = 6/7 = 0.857$. Because the probabilities are equal, the events "Female" and "Right-handed" are independent. Knowing that a student is female does not change the probability that she is right-handed.

11. $P(\text{survived} \mid \text{first class}) = \dfrac{197}{197 + 122} = \dfrac{197}{319} = 0.618$

$P(\text{survived} \mid \text{not first class}) = \dfrac{94 + 151}{94 + 167 + 151 + 476} = \dfrac{245}{888} = 0.276$

Because the probabilities are not equal, the events "Survived" and "First class" are not independent. Knowing that a randomly selected person comes from first class increases the probability that the person survived.

13. $P(\text{Spanish} \mid \text{some other language besides English}) =$

$P(S|L) = \dfrac{P(S \cap L)}{P(L)} = \dfrac{0.26}{0.26 + 0.09 + 0.03 + 0.03} = \dfrac{0.26}{0.41} = 0.634$

15. $P(B) < P(B|T) < P(T) < P(T|B)$; there are very few professional basketball players, so $P(b)$ should be the smallest probability. If you are a professional basketball player, it is quite likely that you are tall, so $P(T|B)$ should be the largest probability. Finally, it's much more likely to be over 6 feet tall than it is to be a professional basketball player if you're over 6 feet tall.

17. (blue die, red die). The sums of 7 are bolded.
$(1, 1), (1, 2), (1, 3), (1, 4), (1, 5), \mathbf{(1, 6)}$
$(2, 1), (2, 2), (2, 3), (2, 4), \mathbf{(2, 5)}, (2, 6)$
$(3, 1), (3, 2), (3, 3), \mathbf{(3, 4)}, (3, 5), (3, 6)$
$(4, 1), (4, 2), \mathbf{(4, 3)}, (4, 4), (4, 5), (4, 6)$
$(5, 1), \mathbf{(5, 2)}, (5, 3), (5, 4), (5, 5), (5, 6)$
$\mathbf{(6, 1)}, (6, 2), (6, 3), (6, 4), (6, 5), (6, 6)$
$P(\text{sum is 7} \mid \text{blue die is a 4}) = 1/6$; $P(\text{sum is 7} \mid \text{blue die is not a 4}) = 5/30 = 1/6$. Because the probabilities are equal, the events "Blue die shows a 4" and "Sum is 7" are independent. Knowing that a roll has a blue die showing a 4 does not change the probability that the sum is 7.

19. **(a)** The second segment from the top in each graph shows $P(\text{fly} \mid \text{female})$ on the left and $P(\text{fly} \mid \text{male})$ on the right. Because the probabilities are not equal, the events "Male" and "Fly" are not independent. Knowing that the person selected is a male decreases the probability that he chose fly as the superpower. **(b)** The bars for invisibility look roughly the same size, indicating that $P(\text{invisibility} \mid \text{female})$ is the same as $P(\text{invisibility} \mid \text{male})$. Because the probabilities are equal, the events "Male" and "Invisibility" are independent. Knowing that the person selected is a male does not change the probability that he chose invisibility as the superpower. **(c)** There is an association between gender and superpower preference because the probabilities for "fly" are not the same for each gender [as shown in part **(a)**]. In other words, knowing a person's gender helps us predict which superpower the person prefers.

21. Set 1 has the largest standard deviation. Set 1 has the highest range, and the values in Set 1 are at the extremes (very low and very high) with none in the middle. Set 4 has the smallest standard deviation. Set 4 has the lowest range, and the values in Set 4 are closer to the middle.

Lesson 4.5 Exercises

1. $(0.29)(0.67) = 0.194$

3. $(14/20)(13/19) = 0.479$

5. (a)

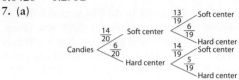

(b) $(0.88)(0.28) + (0.02)(0.34) + (0.10)(0.42) = 0.2464 + 0.0068 + 0.0420 = 0.2952$

7. (a)

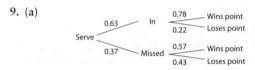

(b) $\left(\dfrac{14}{20}\right)\left(\dfrac{6}{19}\right) + \left(\dfrac{6}{20}\right)\left(\dfrac{14}{19}\right) = 0.22 + 0.22 = 0.44$

9. (a)

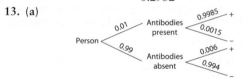

(b) $\dfrac{(0.37)(0.57)}{(0.63)(0.78) + (0.37)(0.57)} = \dfrac{0.2109}{0.7023} = 0.300$

11. See Exercise 5 for the tree diagram and calculation to get $P(\text{credit card}) = 0.2952$. $\dfrac{(0.10)(0.42)}{0.2952} = 0.142$

13. (a)

Person
0.01 Antibodies present
0.9985 +
0.0015 −
0.99 Antibodies absent
0.006 +
0.994 −

(b) $(0.01)(0.9985) + (0.99)(0.006) = 0.009985 + 0.00594 = 0.015925$. **(c)** $P(\text{antibody} \mid \text{positive}) = \dfrac{P(\text{antibody and positive})}{P(\text{positive})}$

$= \dfrac{(0.01)(0.9985)}{0.015925} = \dfrac{0.009985}{0.015925} = 0.6270.$

15. (a) $(0.5)(0.4)(0.8) + (0.3)(0.3)(0.6) + (0.2)(0.1)(0.5) = 0.16 + 0.054 + 0.01 = 0.224$ **(b)** $P(\text{recent donor} \mid \text{contributed}) = \dfrac{P(\text{recent donor and contributed})}{P(\text{contributed})} = \dfrac{0.16}{0.224} = 0.714$

17. $P(\text{has HIV antibody} \mid 2+\text{ tests})$

$= \dfrac{P(\text{has HIV antibody and } 2+\text{ tests})}{P(2+\text{ tests})}$

$= \dfrac{(0.01)(0.9985)(0.9985)}{(0.01)(0.9985)(0.9985) + (0.99)(0.006)(0.006)}$

$= \dfrac{0.009970}{0.010006} = 0.9964$

19. *Shape*: Fairly symmetric. *Center*: The typical miles per gallon for "Mini hwy" (median ≈ 25 mpg) is greater than for "Two hwy" (median ≈ 22 mpg), which is greater than for "Mini city" (median ≈ 18 mpg), with "Two city" having the lowest miles per gallon (median ≈ 16 mpg). *Variability*: There is more variation in miles per gallon for "Two city" and "Two hwy" ($IQR \approx 9$ mpg) than for "Mini city" and "Mini hwy" ($IQR \approx 6$ mpg). *Outliers*: There are no outliers for any of the distributions.

Lesson 4.6 Exercises

1. $(0.98)(0.98) \ldots (0.98) = (0.98)^{20} = 0.668$

3. $(0.90)(0.90)(0.90)(0.90)(0.90) = (0.90)^5 = 0.5905$

5. P(none are universal donors) $= (1 - 0.072) (1 - 0.072) \ldots$
$(1 - 0.072) = (0.928)^{10} = 0.474$; P(at least 1 is universal donor) $=$
$1 - P$(none are universal donors) $= 1 - 0.474 = 0.526$
7. P(none are six) $= (5/6) (5/6) \ldots (5/6) = (5/6)^{10} = 0.1615$;
P(at least one six) $= 1 - P$(none are six) $= 1 - 0.1615 = 0.8385$
9. No, because the 4 consecutive flights being on time are not independent events. Knowing that the first flight is on time makes it much more likely that the next flight is on time.

11. P(*USA Today* and *New York Times*) $\overset{?}{=} P$(*USA Today*) $\cdot$
P(*New York Times*)

$$(0.05) \overset{?}{=} (0.40)(0.25)$$
$$0.05 \neq 0.10$$

Because (*USA Today* and *New York Times*) does not equal P(*USA Today*) $\cdot P$(*New York Times*), the two events are not independent.
13. (a) P(all 4 calls are for medical help) $= (0.81)(0.81)(0.81)(0.81) =$
$(0.81)^4 = 0.4305$. (b) P(at least 1 not for medical) $= 1 - P$(all 4 calls are for medical help) $= 1 - 0.4305 = 0.5695$. (c) The calculation in part (a) might not be valid because the 4 consecutive calls being medical are not independent events. Knowing that the first call is medical might make it more likely that the next call is medical (for example, several people might call for the same medical emergency).
15. (a) Answers may vary.

	Male	Female	Total
Blue	0	10	10
Brown	20	20	40
Total	20	30	50

The events "Student is male" and "Student has blue eyes" are mutually exclusive because they don't occur at the same time.
(b)

	Male	Female	Total
Blue	4	6	10
Brown	16	24	40

If the event "Student is male" and the event "Student has blue eyes" are independent, then P(male and blue) $= P$(male) $\cdot P$(blue)
$(4/50) \overset{?}{=} (20/50)(10/50)$; $0.08 = 0.08$. Because P(male and blue) does equal P(male) $\cdot P$(blue), the two events are independent.
17. (a) If you take a very large random sample of pieces of buttered toast and dropped them from a table 2.5 feet high, about 81% of them will land butter side down. (b) No; probability describes what happens in many, many repetitions (way more than 100) of a chance process. We would expect to get *about* 81 pieces of toast that land butter side down, but this result is not guaranteed. (c) Let $1-81 =$ butter side down and $82-100 =$ butter side up. Use randInt(1,100) to simulate dropping one piece of toast. Press Enter 10 times to simulate dropping 10 pieces of toast. Record the number of pieces of toast out of 10 that land butter side down. (d) Assuming the probability that each piece of toast lands butter side down is 0.81, the estimate of the probability that dropping 10 pieces of toast yields 4 or fewer butter side down is $1/50 = 0.02$. There is convincing evidence that the 0.81 claim is false. Maria does have reason to be surprised, as these results are unlikely to have occurred purely by chance.

Lesson 4.7 Exercises

1. $\underline{10} \cdot \underline{10} \cdot \underline{10} \cdot \underline{20} \cdot \underline{23} \cdot \underline{23} = 10,580,00$ possible license plates
3. (a) $\underline{1} \cdot \underline{26} \cdot \underline{26} = 676$ possible three-letter radio call signs
(b) $676 + 676 = 1352$
5. $\underline{6} \cdot \underline{5} \cdot \underline{4} \cdot \underline{3} \cdot \underline{2} \cdot \underline{1} = 6!$ or 720 possible orders to complete all of the assignments.
7. $15 \cdot 14 \cdot 13 \cdot 12 \cdot 11 \cdot 10 \cdot 9 \cdot 8 \cdot 7 \cdot 6 \cdot 5 \cdot 4 \cdot 3 \cdot 2 \cdot 1 = 15! =$
$1,307,674,368,000$ possible ways to arrange the students in a

single-file line. With over 1 trillion possibilities, the class should not take the offer, because they will never have time to do it.
9. $\underline{100} \cdot \underline{99} \cdot \underline{98} \cdot \underline{97} \cdot \underline{96} \cdot \underline{95} \cdot \underline{94} \cdot \underline{93} = {}_{100}P_8$ or 7.5×10^{15} possible lists of songs
11. $\underline{11} \cdot \underline{10} \cdot \underline{9} \cdot \underline{8} \cdot \underline{7} = {}_{11}P_5$ or 55,440 possible ways the team can take penalty kicks
13. (a) $\underline{10} \cdot \underline{10} \cdot \underline{10} \cdot \underline{10} = 10,000$ possible four-digit password
(b) $\underline{9} \cdot \underline{9} \cdot \underline{9} \cdot \underline{9} = 6561$ possible four-digit password (no 3s)
15. (a) $\underline{28} \cdot \underline{28} \cdot \underline{28} \cdot \underline{28} \cdot \underline{28} = 17,210,368$ possible ways to select the students. (b) $\underline{28} \cdot \underline{27} \cdot \underline{26} \cdot \underline{25} \cdot \underline{24} = {}_{28}P_5$ or 11,793,600 possible ways to select five different students.
17. In Exercise 15, we found that there are 17,210,368 possible ways to select the five students, of which 11,793,600 will select five different students. Thus, define event A as "All five students different."
$$P(A) = \frac{\text{number of outcomes in event A}}{\text{total number of outcomes in sample space}}$$
$$= \frac{11,793,600}{17,210,368} = 0.685$$
19. (a) The linear relationship between calories and carbohydrates for these 19 varieties of bagels is strong and positive. (b) The correlation would still be $r = 0.915$, because the correlation doesn't change when we change the units of either variable. (c) Because the three points are in the lower left and separated from the rest of the points in the x-direction, they will be influential and make the correlation stronger and closer to 1. With these points removed, the correlation is less strongly positive and will get closer to 0.

Lesson 4.8 Exercises

1. ${}_{25}C_2 = \dfrac{{}_{25}P_2}{2!} = \dfrac{25 \cdot 24}{2 \cdot 1} = 300$

3. ${}_{30}C_4 = \dfrac{{}_{30}P_4}{4!} = \dfrac{30 \cdot 29 \cdot 28 \cdot 27}{4 \cdot 3 \cdot 2 \cdot 1} = 27,405$

5. P(all 4 batteries good) $= \dfrac{\text{no. of ways to pick 4 good batteries}}{\text{total no. of ways to pick 4 batteries}}$
$= \dfrac{{}_6C_4}{{}_8C_4} = \dfrac{15}{70} = 0.214$

7. (a) P(all 3 from Mr. Wilder's class) $= \dfrac{{}_{28}C_3}{{}_{95}C_3} = \dfrac{3276}{138,415} = 0.024$
(b) No; there is less than a 5% chance that 3 randomly selected students would all be from Mr. Wilder's class. It is unlikely that this result happened purely by chance.
9. (a)
$${}_6C_6 \cdot {}_{12}C_3 = \left(\frac{6 \cdot 5 \cdot 4 \cdot 3 \cdot 2 \cdot 1}{6 \cdot 5 \cdot 4 \cdot 3 \cdot 2 \cdot 1}\right)\left(\frac{12 \cdot 11 \cdot 10}{3 \cdot 2 \cdot 1}\right) = 1 \cdot 220 = 220$$
(b) P(all 6 statistics majors selected) $= \dfrac{{}_6C_6 \cdot {}_{12}C_3}{{}_{18}C_9} = \dfrac{220}{48,620} = 0.0045$
(c) Yes; there is less than a 1% chance that a random selection of 9 players would include all 6 statistics majors. It is unlikely that this happened purely by chance.
11. P(6 men, 4 women selected) $= \dfrac{{}_8C_6 \cdot {}_{12}C_4}{{}_{20}C_{10}} = \dfrac{13,860}{184,756} = 0.0750.$

The researchers should not be surprised. It is quite plausible that this happened purely by chance.
13. P(none of 6 are winners) $= P$(all 6 are losers)
$$= \frac{\text{no. of ways to pick 6 losers}}{\text{total no. ways to pick 6 numbers}} = \frac{{}_{43}C_6}{{}_{49}C_6} = \frac{6,096,454}{13,983,816} = 0.436$$
15. (a) ${}_{12}C_5 = \dfrac{{}_{12}P_5}{5!} = \dfrac{12 \cdot 11 \cdot 10 \cdot 9 \cdot 8}{5 \cdot 4 \cdot 3 \cdot 2 \cdot 1} = 792$

(b) ${}_5C_2 \cdot {}_7C_3 = \left(\dfrac{5 \cdot 4}{2 \cdot 1}\right)\left(\dfrac{7 \cdot 6 \cdot 5}{3 \cdot 2 \cdot 1}\right) = 10 \cdot 35 = 350$ (c) P(ideal lineup) $= P$ (2 guards, 3 forwards/centers) $= \dfrac{{}_5C_2 \cdot {}_7C_3}{{}_{12}C_5} = \dfrac{350}{792} = 0.442$

17. $_{25}C_8 + _{25}C_7 + _{25}C_6 + _{25}C_5 + _{25}C_4 + _{25}C_3 + _{25}C_2 + _{25}C_1 + _{25}C_0$
$= 1,081,575 + 480,700 + 177,100 + 53,130 + 12,650 + 2300$
$+ 300 + 25 + 1 = 1,807,781$ different pizzas
19. Min $= 95$, $Q_1 = 123$, Med $= 135$, $Q_3 = 155$, Max $= 211$;
$IQR = Q_3 - Q_1 = 155 - 123 = 32$ home runs. Outliers $> Q_3 +$
$1.5 \times IQR = 155 + 1.5 \times 32 = 203$. Because 211 is more than
203, it is considered an outlier.

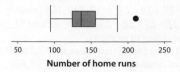

Number of home runs

Chapter 4 Review Exercises

1. (a) No; probability describes what happens in many, many repetitions (way more than 200) of a chance process. We would expect
to get *about* 54 women who live to at least 100, but exactly 54 is
not guaranteed. **(b)** P(none live to 100 years) $= (1 - 0.27)(1 -$
$0.27)(1 - 0.27)(1 - 0.27)(1 - 0.27) = (0.73)^5 = 0.207$; P(at
least 1 lives to 100 years) $= 1 - P$(none live to 100 years) $= 1 -$
$0.207 = 0.793$
2. (a) $1 - (0.46 + 0.15 + 0.10 + 0.05) = 1 - 0.76 = 0.24$
(b) $\dfrac{0.15}{0.15 + 0.10 + 0.24 + 0.05} = \dfrac{0.15}{0.54} = 0.278$
(c) $0.15 + 0.10 + 0.05 = 0.30$
(d) Let $1-46 = $ passenger car, $47-61 = $ pickup truck, $62-71 =$
SUV, $72-95 = $ crossover, and $96-100 = $ minivan. Use rand-
Int(1,100) to simulate randomly selecting one vehicle. Press Enter
3 times to simulate randomly selecting 3 vehicles. Record whether
or not the three vehicles are of three different types. Perform many
repetitions of the simulation, and calculate the proportion of times
when the three vehicles are of three different types.
3. (a) $\dfrac{13 + 11}{4 + 6 + 13 + 13 + 22 + 11 + 11 + 14} = \dfrac{24}{94} = 0.255$
(b) P(brand C or female) $= P$(brand C) $+ P$(female) $- P$(brand C
and female) $= 24/94 + 36/94 - 13/94 = 47/94 = 0.50$
(c) $\dfrac{13}{4 + 6 + 13 + 13} = \dfrac{13}{36} = 0.361$
4. (a)

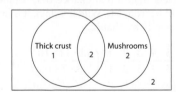

(b) P(thick crust | mushrooms) $= 2/4 = 0.50$; P(thick crust | no
mushrooms) $= 1/3 = 0.33$. Because the probabilities are not equal,
the events "Thick crust" and "Mushrooms" are not independent.
(c) P(thick crust | mushrooms) $= 2/4 = 0.50$; P(thick crust | no
mushrooms) $= 2/4 = 0.50$. Because the probabilities are equal, the
events "Thick crust" and "Mushrooms" are independent.

5. (a) $= \left(\dfrac{8}{18}\right)\left(\dfrac{7}{17}\right) = \dfrac{56}{306} = 0.183$. **(b)** P(both same) $= P$(both

blue or both brown or both gray) $= \dfrac{_8C_2}{_{18}C_2} + \dfrac{_6C_2}{_{18}C_2} + \dfrac{_4C_2}{_{18}C_2} = 0.183 +$
$0.098 + 0.039 = 0.32$. Alternative solution using a tree diagram:
P(same color) $= P$(both blue) $+ P$(both brown) $+ P$(both gray)
$= \left(\dfrac{8}{18}\right)\left(\dfrac{7}{17}\right) + \left(\dfrac{6}{18}\right)\left(\dfrac{5}{17}\right) + \left(\dfrac{4}{18}\right)\left(\dfrac{3}{17}\right) = 0.183 + 0.098 +$
$0.039 = 0.32$
6. (a)

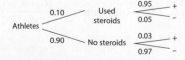

(b) P(positive) $= (0.10)(0.95) + (0.90)(0.03) = 0.095 + 0.027 =$
0.122
(c) P(used steroids | positive) $= \dfrac{P(\text{used steroids and positive})}{P(\text{positive})}$

$= \dfrac{(0.10)(0.95)}{0.122} = \dfrac{0.095}{0.122} = 0.779$

7. (a) $\underline{16} \cdot \underline{16} \cdot \underline{16} \cdot \underline{16} \cdot \underline{16} = (16)^5 = 1,048,576$
(b) $\underline{16} \cdot \underline{15} \cdot \underline{14} \cdot \underline{13} \cdot \underline{12} = _{16}P_5 = 524,160$. **(c)** P(all 5 rolls are

different) $= \dfrac{524,160}{1,048,576} = 0.4999$; P(at least 2 rolls are the same)

$= 1 - P$(all 5 rolls are different) $= 1 - 0.4999 = 0.5001$

8. (a) $_{24}C_4 = \dfrac{_{24}P_4}{4!} = \dfrac{24 \cdot 23 \cdot 22 \cdot 21}{4 \cdot 3 \cdot 2 \cdot 1} = 10,626$

(b) $\dfrac{_3C_2 \cdot _{21}C_2}{10,626} = \dfrac{630}{10,626} = 0.059$. **(c)** No; there is a 5.9% chance that
a random selection of 4 students would result in 2 strong students. It
is quite plausible that this happened purely by chance.

Chapter 4 Practice Test

1. d
2. c
3. d
4. a
5. b
6. c
7. d
8. c
9. b
10. b
11. (a)

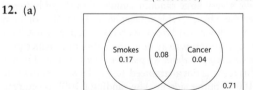

(b) $(0.60)(0.10) + (0.30)(0.30) + (0.10)(0.40) = 0.06 + 0.09 + 0.04 =$
0.19 **(c)** P(C | defective) $= \dfrac{P(\text{C and defective})}{P(\text{defective})} = \dfrac{0.04}{0.19} = 0.211$

12. (a)

	Cancer	No cancer	Total
Smokes	0.08	0.17	0.25
Does not smoke	0.04	0.71	0.75
Total	0.12	0.88	1

(b) P(cancer | smoker) $= \dfrac{P(\text{cancer and smoker})}{P(\text{smoker})} = \dfrac{0.08}{0.25} = 0.32$ **(c)**
P(smokes or cancer) $= P$(smokes) $+ P$(cancer) $- P$(smokes and
cancer) $= 0.25 + 0.12 - 0.08 = 0.29$ **(d)** P(neither get cancer) $=$
$(0.88)(0.88) = 0.774$; P(at least 1 gets cancer) $= 1 - P$(neither gets
cancer) $= 1 - 0.774 = 0.226$
13. (a) $_{14}C_{10} = \dfrac{_{14}P_{10}}{10!} = \dfrac{14 \cdot 13 \cdot 12 \cdot 11 \cdot 10 \cdot 9 \cdot 8 \cdot 7 \cdot 6 \cdot 5}{10 \cdot 9 \cdot 8 \cdot 7 \cdot 6 \cdot 5 \cdot 4 \cdot 3 \cdot 2 \cdot 1} = 1001$

(b) P(all 6 electric acts selected) $= \dfrac{_6C_6 \cdot _8C_4}{_{14}C_{10}} = \dfrac{70}{1001} = 0.0699$

(c) P(5 electric, 5 acoustic) $= \dfrac{_6C_5 \cdot _8C_5}{_{14}C_{10}} = \dfrac{336}{1001} = 0.336$

CHAPTER 5

Lesson 5.1 Exercises

1. The probabilities are all between 0 and 1. Also, the sum of the probabilities is $0.301 + 0.176 + 0.125 + 0.097 + 0.079 + 0.067 + 0.058 + 0.051 + 0.046 = 1$.

3. Because the probabilities must add to 1, $P(X = 5) = 1 - (0.03 + 0.16 + 0.30 + 0.23 + 0.17) = 1 - 0.89 = 0.11$.

5. $X \geq 6$ is the event that the first digit of a randomly chosen legitimate record is greater than or equal to 6. $P(X \geq 6) = 0.067 + 0.058 + 0.051 + 0.046 = 0.222$.

7. (a) "Plays with at most two toys" is equivalent to $X \leq 2$. $P(X \leq 2) = 0.03 + 0.16 + 0.30 = 0.49$.
(b) $P(X < 2) = 0.03 + 0.16 = 0.19$; the outcome $X = 2$ is included in "less than or equal to" but is not included in "less than."

9. (a) Continuous **(b)** Discrete **(c)** Continuous

11. (a) Discrete **(b)** Continuous

13. (a) Discrete; X could take any of the integer values $0, 1, 2, \ldots, 5$ but not values like 2.359543. **(b)** The probabilities are all between 0 and 1. Also, the sum of the probabilities is $0.630 + 0.295 + 0.065 + 0.008 + 0.002 = 1$. **(c)** $P(X \geq 3)$ is the probability that a randomly selected U.S. high school student speaks at least 3 languages. $P(X \geq 3) = 0.065 + 0.008 + 0.002 = 0.075$.

15. (a) The possible outcomes are TTTT, TTTH, TTHT, THTT, HTTT, TTHH, THTH, THHT, HTTH, HTHT, HHTT, THHH, HTHH, HHTH, HHHT, HHHH.

Number of heads	0	1	2	3	4
Probability	1/16	4/16	6/16	4/16	1/16

(b) $P(X \leq 3) = \frac{1}{16} + \frac{4}{16} + \frac{6}{16} + \frac{4}{16} = \frac{15}{16} = 0.9375$; there is about a 94% chance of getting less than or equal to 3 heads when tossing a fair coin 4 times.

17. (a) $P(\text{all 5 check phones}) = (0.67)(0.67)(0.67)(0.67)(0.67) = (0.67)^5 = 0.135$. **(b)** $P(\text{at least 1 doesn't check}) = 1 - P(\text{all 5 check phones}) = 1 - 0.135 = 0.865$

Lesson 5.2 Exercises

1.

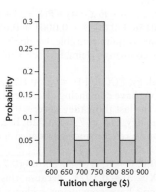

The graph is fairly symmetric with three peaks at $600, $750, and $900.

3.

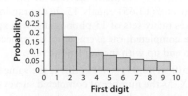

The graph is skewed to the right and single peaked at 1.

5. $\mu_T = E(T) = 600(0.25) + 650(0.10) + 700(0.05) + \cdots + 900 (0.15) = 732.5$. If many, many students from El Dorado Community College are randomly selected, their average tuition charge will be about $732.50.

7. $\mu_X = E(X) = 1(0.301) + 2(0.176) + 3(0.125) + \cdots + 9(0.046) = 3.441$. If many, many legitimate reports are randomly selected, the average of the first digits in the reports will be about 3.441.

9.
$$\sigma_T = \sqrt{(600 - 732.5)^2(0.25) + (650 - 732.5)^2(0.10) + (700 - 732.5)^2 (0.05) + \cdots + (900 - 732.5)^2 (0.15)}$$ $\sigma_T = \sqrt{10{,}568.66} = 102.80$. A randomly selected student from El Dorado Community College has a tuition that typically varies from the mean ($732.50) by about $102.80.

11.
$$\sigma_X = \sqrt{(1 - 3.441)^2(0.301) + (2 - 3.441)^2(0.176) + (3 - 3.441)^2 (0.125) + \cdots + (9 - 3.441)^2 (0.046)}$$
$\sigma_X = \sqrt{6.061} = 2.462$. A randomly selected legitimate record has a first digit that typically varies from the mean (3.441) by about 2.462.

13. (a)

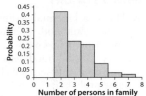

The distribution of number of persons in a family appears to have a larger center and less variability than the distribution of number of persons in a household. **(b)** $\mu_H = E(H) = 1(0.25) + 2(0.32) + 3(0.17) + \cdots + 7(0.01) = 2.6$. $\mu_F = E(F) = 1(0) + 2(0.42) + 3(0.23) + \cdots + 7(0.02) = 3.14$. There are no families with only 1 person, which will bring the family average higher than for the households (which can have only 1 person). **(c)** The possible values for families go from 2 to 7, while the possible values for households are more spread out from 1 to 7.

15. (a) If a client dies at age 25, the client paid the company $250 for 5 years ($1250 total), but then the company had to pay $100,000 because the client died. The company's overall profit is $1250 - $100,000 = -$98,750 or a $98,750 loss. **(b)** $\mu_Y = E(Y) = -99,750 (0.00183) + -99,500(0.00186) + \cdots + 1250(0.99058) = 303.35$. If many, many 21-year-old males bought term life insurance policies, the average profit for the company will be about $303.35 per client. **(c)**
$$\sigma_Y = \sqrt{(-99,750 - 303.35)^2(0.00183) + (-99,500 - 303.35)^2 (0.00186) + \cdots + (1250 - 303.35)^2 (0.99058)};$$
$\sigma_Y = 9707.57$. The profit from a randomly selected client typically varies from the mean ($303.35) by about $9707.57. This large amount of variability is dangerous for a company because it is very possible that they could have huge losses.

17. (a) If all values from 1 to 9 are equally likely, the center of the distribution will be at the exact center between 1 and 9, which is 5. **(b)** $\mu_X = E(X) = 1(0.301) + 2(0.176) + 3(0.125) + \ldots + 9(0.046) = 3.441$. If there was a company suspected of creating false reports, we could find the average value of the first digit on all of their reports. If the average is far above the expected 3.441, we would have evidence that this company is being dishonest. **(c)**
$$\sigma_Y = \sqrt{(1 - 5)^2\left(\frac{1}{9}\right) + (2 - 5)^2\left(\frac{1}{9}\right) + (3 - 5)^2\left(\frac{1}{9}\right) + \ldots + (9 - 5)^2\left(\frac{1}{9}\right)}$$
$= \sqrt{6.667} = 2.582$. **(d)** No, because the standard deviation of the distribution of first digits for legitimate records (2.462) is so close to the standard deviation when assuming all digits $1 - 9$ are equally likely (2.582) to be the first digit. The mean would be a much better way to detect fraud.

19. (a) $P(\text{drives and late}) = P(\text{drives}) \cdot P(\text{late} \mid \text{drives})$
$0.05 = (0.20) \cdot P(\text{late} \mid \text{drives})$

$$P(\text{late} \mid \text{drives}) = \frac{0.05}{0.20} = 0.25$$

(b) Because $P(\text{late} \mid \text{drives}) = 0.25$, we know $P(\text{on time} \mid \text{drives}) = 1 - 0.25 = 0.75$. Because $P(\text{late} \mid \text{rides bike}) = 0.30$, we know $P(\text{on time} \mid \text{rides bike}) = 1 - 0.30 = 0.70$.

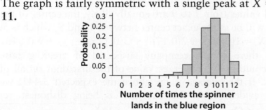

(c) $P(\text{biked} \mid \text{late}) = \dfrac{P(\text{biked and late})}{P(\text{late})} = \dfrac{(0.80)(0.30)}{(0.20)(0.25) + (0.80)(0.30)}$

$\qquad = \dfrac{0.24}{0.05 + 0.24} = \dfrac{0.24}{0.29} = 0.828$

Lesson 5.3 Exercises

1. This is a binomial setting. **Binary?** "Success" = baby elk survives to adulthood. "Failure" = baby elk does not survive. **Independent?** Knowing whether or not one randomly selected elk survives to adulthood tells you nothing about whether or not another randomly selected elk survives to adulthood. **Number?** $n = 7$. **Same probability?** $p = 0.44$.

3. This is not a binomial setting because the **independent** condition is not met. Selecting four names from a hat is the same as sampling without replacement. If the first person selected has a last name with more than six letters, the next person chosen is less likely to have a last name with more than six letters because you're not replacing the first person's name back into the hat.

5. $P(X = 4) = {}_7C_4(0.44)^4(1 - 0.44)^3 = 35(0.44)^4(0.56)^3 = 0.2304$

7. (a) This is a binomial setting. **Binary?** "Success" = spinner lands in the blue region. "Failure" = spinner does not land in the blue region. **Independent?** Knowing whether or not one spin lands in the blue region tells you nothing about whether or not another spin lands in the blue region. **Number?** $n = 12$. **Same probability?** $p = 0.80$.

(b) $P(X = 8) = {}_{12}C_8(0.80)^8(1 - 0.80)^4 = 495(0.80)^8(0.20)^4 = 0.1329$. If we were to spin the spinner 12 times, there is about a 13% chance that the spinner will land in the blue region exactly 8 times.

9.

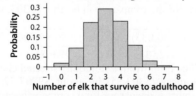

The graph is fairly symmetric with a single peak at $X = 3$.

11.

The graph is left-skewed with a single peak at $X = 10$.

13. (a) Binary? "Success" = the light is red. "Failure" = the light is not red. **Independent?** Knowing whether or not one passenger has a red light tells you nothing about whether or not another passenger gets a red light. **Number?** $n = 20$. **Same probability?** $p = 0.30$.

(b)

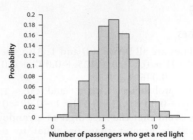

The graph is fairly symmetric with a single peak at $R = 6$. **(c)** $P(R = 6) = {}_{20}C_6(0.30)^6(1 - 0.30)^{14} = 38{,}760(0.30)^6(0.70)^{14} = 0.1916$

15. This is not a binomial setting because the **independent** condition is not met. Selecting eight names from a hat is the same as sampling without replacement. If the first pilot selected is a female, the next pilot chosen is less likely to be a female ($9/24 = 0.375$) because you're not replacing the first pilot's name into the hat.

17. (a) It is possible that those people who eat a gluten-free diet are feeling benefits simply because they have an expectation that the diet will work.

(b)

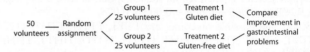

Lesson 5.4 Exercises

1. (a) $\mu_X = 15(0.09) = 1.35$; if many, many sets of 15 phone calls are made by the random dialing machine, we expect about 1.35 surveys will be completed, on average.

(b) $\sigma_X = \sqrt{15(0.09)(0.91)} = 1.11$; the number of completed surveys would typically vary by about 1.11 from the mean of 1.35.

3. (a) $\mu_X = 12(0.8) = 9.6$; if many, many sets of 12 spins are done with the spinner, we expect about 9.6 of them will land in the blue region, on average. **(b)** $\sigma_X = \sqrt{12(0.8)(0.2)} = 1.39$; the number of spins that land in the blue region would typically vary by about 1.39 from the mean of 9.6.

5. $P(X < 3) = P(X = 0) + P(X = 1) + P(X = 2) = 0.0173 + 0.0950 + 0.2239 = 0.3362$

7. $P(X \geq 3) = P(X = 3) + P(X = 4) + P(X = 5) + P(X = 6) + P(X = 7) = 0.0781 + 0.0156 + 0.0019 + 0.0001 + 0.0000 = 0.0957$; because this outcome is quite plausible, we don't have convincing evidence against the company's claim. It is believable that the group just got lucky.

9. Using technology, $P(X \geq 14) = 0.00001$. Because this outcome is very unlikely, we have convincing evidence that participants have a preference for the last thing they taste.

11. Using technology, $P(X \leq 7) = 0.0726$. Because this outcome is plausible, we don't have convincing evidence against the company's claim.

13. (a) $\mu_X = 20(0.3) = 6$; $\sigma_X = \sqrt{20(0.3)(0.7)} = 2.05$. **(b)** Using technology, $P(X \leq 3) = 0.1071$. Because this outcome is quite plausible, we don't have convincing evidence against the custom agent's claim.

15. (a) $\mu_Y = 15(0.91) = 13.65$; the sum of the mean of $X(1.35)$ and the mean of $Y(13.65)$ equals 15. If the random digit dialing machine makes many sets of 15 phone calls, we expect about 1.35 surveys to be completed, on average, so we expect an average of 13.65 calls to end up with incomplete surveys.

(b) $\sigma_Y = \sqrt{15(0.91)(0.09)} = 1.11$; this is the same as the standard deviation of X. As the number of completed surveys varies from its mean, so does the number of incompleted surveys vary from its mean, because the sum of X and Y must be 15.

17. (a)

$$\frac{5}{\text{Bread}} \cdot \frac{6}{\text{Meat}} \cdot \frac{4}{\text{Cheese}} \cdot \frac{2}{\substack{\text{Lettuce}\\(\text{or no})}} \cdot \frac{2}{\substack{\text{Tomato}\\(\text{or no})}} \cdot \frac{2}{\substack{\text{Hot}\\\text{pepper}\\(\text{or no})}} \cdot \frac{2}{\substack{\text{Mayo}\\(\text{or no})}} \cdot \frac{2}{\substack{\text{Mustard}\\(\text{or no})}} = 3840$$

(b)

$$\frac{5}{\text{Bread}} \cdot \frac{_6C_2}{\text{Meat}} \cdot \frac{_4C_2}{\text{Cheese}} \cdot \frac{2}{\substack{\text{Lettuce}\\(\text{or no})}} \cdot \frac{2}{\substack{\text{Tomato}\\(\text{or no})}} \cdot \frac{2}{\substack{\text{Hot}\\\text{pepper}\\(\text{or no})}} \cdot \frac{2}{\substack{\text{Mayo}\\(\text{or no})}} \cdot \frac{2}{\substack{\text{Mustard}\\(\text{or no})}} = 14,400$$

Lesson 5.5 Exercises

1. (a) The density curve is entirely above the horizontal axis, and the area under the density curve is length × width = $3 \times (1/3) = 1$.

(b) Area $= 0.3 \times \frac{1}{3} = 0.1$; the probability that an accident occurs between the 0.8 and 1.1 mile mark is $P(0.8 \le Y \le 1.1) = 0.1$.

3. (a) Any valid density curve has an area of exactly 1 underneath it, thus the height must be $\frac{1}{3}$ so that the area is 1; area $= 3 \cdot \frac{1}{3} = 1$.

(b) Area $= 1.5 \cdot \frac{1}{3} = 0.5$; there is a 50% chance that the light turns on between 2.5 and 4 seconds after the subject clicks "start."

5. Median = B, mean = C; B is the equal areas point of the distribution. The mean will be greater than the median due to its right-skewed shape.

7. Median = B, mean = B; B is the equal areas point of the distribution. The mean will be the same as the median due to its symmetric shape.

9.

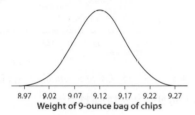

8.97 9.02 9.07 9.12 9.17 9.22 9.27
Weight of 9-ounce bag of chips

11. The mean is at the center of 115 miles per hour. The standard deviation is $121 - 115 = 6$ miles per hour.

13. (a)

Height = 1/4
−2   0   2
Amount of time (min)

(b) 10 seconds $= 10/60 = 1/6$ minute, so we want to find $P\left(-\frac{1}{6} \le X \le \frac{1}{6}\right)$. Area $= \frac{1}{3} \cdot \frac{1}{4} = \frac{1}{12} = 0.083$; there is about an 8% chance that the alarm goes off within 10 seconds of the time it is set for on a randomly selected day.

15. The center appears to be at 10, so 10 is the estimate for the mean. The curve seems to change curvature at 8 and 12, which are the points that are 1 standard deviation away from the mean, so 2 is the estimate for the standard deviation.

17. (a) The density curve is entirely above the horizontal axis and the area under the density curve is = 1 (see below).
Area (rectangle) $= lw = (0.8)(1) = 0.8$
Area (triangle) $= \frac{1}{2}bh = \frac{1}{2}(0.4)(1) = 0.2$
Total Area = Area(rectangle) + Area(triangle) $= 0.8 + 0.2 = 1$

(b) $P(0 \le X \le 0.2) =$ Area(trapezoid) $= \frac{1}{2}h(b_1 + b_2) = \frac{1}{2}(0.2)$ $(2 + 1.5) = 0.35$

(c) $P(0 \le X \le 0.4) =$ Area(trapezoid) $= \frac{1}{2}h(b_1 + b_2) = \frac{1}{2}(0.4)(2 + 1)$ $= 0.60$; if $P(0 \le X \le 0.2) = 0.35$

from part **(b)**, and $P(0 \le X \le 0.4) = 0.60$, then we know somewhere between $X = 0.2$ and $X = 0.4$ is the equal-areas point with an area of 0.50 to the left, which is the definition of the median.

(d) The mean will be greater than the median due to its right-skewed shape.

19. (a) (Gains, Gains) Value $= 1000(1.3)(1.3) = \$1690$; P(Gains, Gains) $= (0.5)(0.5) = 0.25$. (Gains, Loses) Value $= 1000(1.3)(0.75)$ $= \$975$; P(Gains, Loses) $= (0.5)(0.5) = 0.25$. (Loses, Gains) Value $= 1000(0.75)(1.3) = \$975$; P(Loses, Gains) $= (0.5)(0.5) = 0.25$. (Loses, Loses) Value $= 1000(0.75)(0.75) = \$562.50$; P(Loses, Loses) $= (0.5)(0.5) = 0.25$

Possible values (X)	1690	975	562.5
Probability	0.25	0.5	0.25

(b) The only possibility where the stock is worth more after two days is $1690 with a probability of 0.25 (suggesting one shouldn't invest). **(c)** $\mu_X = E(X) = 1690(0.25) + 975(0.5) + 562.50(0.25)$ $= \$1050.63$ (suggesting that one should invest).

Lesson 5.6 Exercises

1. (a) $P(X < 9.02) = \dfrac{1 - 0.95}{2} = \dfrac{0.05}{2} = 0.025$

(b) $P(8.97 \le X \le 9.17) = 0.68 + \dfrac{0.997 - 0.68}{2} = 0.68 + 0.1585$ $= 0.8385$

3. (a) $P(X > 121) = \dfrac{1 - 0.68}{2} = \dfrac{0.32}{2} = 0.16$

(b) $P(109 \le X \le 133) = 0.68 + \dfrac{0.997 - 0.68}{2} = 0.68 + 0.1585$ $= 0.8385$

5. (a) Using Table A, $P(Z > -2.15) = 1 - 0.0158 = 0.9842$.
(b) Using Table A, $P(-0.56 \le Z \le 1.87) = 0.9693 - 0.2877$ $= 0.6816$.

7. (a) Using Table A, $P(Z < 2.46) = 0.9931$. **(b)** Using Table A, $P(0.50 \le Z \le 1.79) = 0.9633 - 0.6915 = 0.2718$.

9. From Table A, the closest area to $1 - 0.34 = 0.66$ is 0.6591, which corresponds to a z-score of $z = 0.41$.

11. From Table A, the closest area to 0.25 is 0.2514, which corresponds to a z-score of $z = -0.67$. So $Q_1 = -0.67$. By the symmetry of the normal curve, $Q_3 = 0.67$.

13. With 16% of women having heights less than 62 inches, we know 62 is about 1 standard deviation below the mean. With 97.5% of women having heights less than 70 inches, we know 70 is about 2 standard deviations above the mean. There is a difference of 3 standard deviations between 62 and 70, so the estimate of the standard deviation is $\dfrac{(70 - 62)}{3} = 2.67$ inches. Add this standard deviation to 62 to get our estimate of the mean: $62 + 2.67 = 64.67$ inches.

15. From Exercise 11, $Q_1 = -0.67$ and $Q_3 = 0.67$, so $IQR = Q_3 - Q_1 = 0.67 - (-0.67) = 1.34$. Outlier $< Q_1 - 1.5$ $IQR = -0.67 - 1.5(1.34) = -2.68$. $P(Z < -2.68) = 0.0037$. Outlier $> Q_3 + 1.5IQR = 0.67 + 1.5(1.34) = 2.68$. $P(Z > 2.68)$ $= 1 - 0.9963 = 0.0037$. About $0.37\% + 0.37\% = 0.74\%$ of the total values in a normal distribution are considered outliers.

17. (a) Yes; there is no leftover pattern in the residual plot, so the least-squares regression line is appropriate.
(b) $\hat{y} = 1.4146 + 0.4399(2.01) = 2.3$ offspring; the residual is about -1.2 from the residual plot; $residual = y - \hat{y}$, so $y = residual + \hat{y} = -1.2 + 2.3 = 0.9$.

Lesson 5.7 Exercises

1.

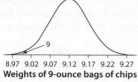

Weights of 9-ounce bags of chips

$z = \dfrac{9 - 9.12}{0.05} = -2.40$; $P(X < 9) = P(Z < -2.40) = 0.0082$. *Using technology*: Applet/normalcdf(lower:−1000, upper:9, mean: 9.12, SD:0.05) = 0.0082. There's about a 1% chance that a randomly selected bag of chips will have a weight less than 9 ounces.

3.

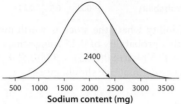

Sodium content (mg)

$z = \dfrac{2400 - 2000}{500} = 0.80$; $P(X > 2400) = P(Z > 0.80) = 1 - 0.7881$ $= 0.2119 = 21.19\%$. *Using technology*: Applet/normalcdf(lower: 2400, upper:10000, mean:2000, SD:500) = 0.2119 = 21.19%.

5.

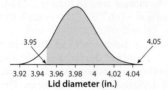

Sodium content (mg)

$z = \dfrac{1200 - 2000}{500} = -1.60$; $z = \dfrac{1800 - 2000}{500} = -0.40$

$P(1200 \le X \le 1800) = P(-1.60 \le Z \le -0.40) = 0.3446$ $- 0.0548 = 0.2898 = 28.98\%$. *Using technology*: Applet/normalcdf (lower:1200, upper:1800, mean:2000, SD:500) = 0.2898 = 28.98%.

7.

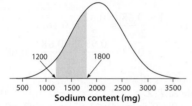

Lid diameter (in.)

$z = \dfrac{3.95 - 3.98}{0.02} = -1.50$; $z = \dfrac{4.05 - 3.98}{0.02} = 3.50$

$P(3.95 \le X \le 4.05) = P(-1.50 \le Z \le 3.50) = 1 - 0.0668$ $= 0.9332$. *Using technology*: Applet/normalcdf(lower:3.95,upper: 4.05, mean:3.98, SD:0.02) = 0.9330. There's about a 93% chance that a randomly selected lid will fit a large cup by having a diameter between 3.95 and 4.05 inches.

9.

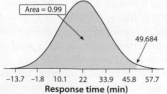

Response time (min)

$2.33 = \dfrac{x - 22}{11.9} \Rightarrow x = 49.727$ minutes. *Using technology*: Applet/ invNorm(area:0.99, mean:22, SD:11.9) = 49.684 minutes.

11.

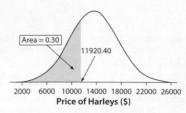

Price of Harleys ($)

$-0.52 = \dfrac{x - 14,000}{4000} \Rightarrow x = \$11,920$. *Using technology*: Applet/ invNorm(area:0.30, mean:14,000, SD:4000) = $11,902.40.

13. (a)

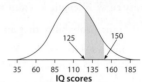

IQ scores

$z = \dfrac{125 - 110}{25} = 0.60$; $z = \dfrac{150 - 110}{25} = 1.60$. $P(125 \le X \le 150)$ $= P(0.60 \le Z \le 1.60) = 0.9452 - 0.7257 = 0.2195 = 21.95\%$ *Using technology*: Applet/normalcdf(lower:125, upper:150, mean: 110, SD:25) = 0.2195 = 21.95%.

(b)

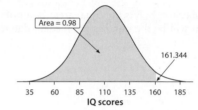

IQ scores

$2.05 = \dfrac{x - 110}{25} \Rightarrow x = 161.25$. *Using technology*: Applet/invNorm (area:0.98, mean:110, SD:25) = 161.344.

15. (a) $z = \dfrac{x - \mu}{\sigma}$; $-2.33 = \dfrac{3.95 - \mu}{0.02}$; $\mu = 3.9966$ inches

(b) $z = \dfrac{x - \mu}{\sigma}$; $-2.33 = \dfrac{3.95 - 3.98}{\sigma}$; $\sigma = 0.012876$ inches

17. (a) No, because the **independent** condition is not met. Selecting seven cards from a deck is the same as sampling without replacement. **(b)** No, because the **same probability** condition is not met. The probability of selecting a female in each class will not be the same for all 10 classrooms. **(c)** No, because the **number** condition is not met. The number of cars is not fixed in advance, because we are counting the number of cars *until* we get one with more than two occupants.

Chapter 5 Review Exercises

1. (a) The probabilities are all between 0 and 1. Also, the sum of the probabilities is $0.01 + 0.21 + 0.33 + 0.23 + 0.13 + 0.09 = 1$. **(b)** Discrete; the random variable X takes a fixed set of possible values (0, 1, 2, 3, 4 or 5) with gaps between (there are no households with 2.5943 televisions). **(c)** $P(X \ge 2) = 0.33 + 0.23 + 0.13 + 0.09 = 0.78$

2. (a)

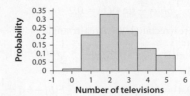

Number of televisions

(b) $\mu_X = E(X) = 0(0.01) + 1(0.21) + 2(0.33) + 3(0.23) + 4(0.13) + 5(0.09) = 2.53$; if many, many U.S. households are randomly selected, their average number of televisions will be about 2.53.

(c)

$$\sigma_X = \sqrt{(0 - 2.53)^2(0.01) + (1 - 2.53)^2(0.21) + (2 - 2.53)^2(0.33)}$$
$$+ (3 - 2.53)^2 (0.23) + (4 - 2.53)^2 (0.13) + (5 - 2.53)^2 (0.09)$$

$\sigma_X = \sqrt{1.529} = 1.237$; the number of televisions in a randomly selected U.S. household typically varies from the mean (2.53) by about 1.237.

3. (a) Binary? "Success" = correctly identifies the card. "Failure" = incorrectly identifies the card. **Independent?** Knowing whether or not one card is correctly identified tells you nothing about whether or not another card is correctly identified. **Number?** $n = 20$. **Same probability?** $p = \frac{1}{4}$. **(b)** $\mu_X = 20\left(\frac{1}{4}\right) = 5$; if many, many different sets of 20 cards are selected, we expect about 5 of them will be correctly identified, on average. **(c)** $\sigma_X = \sqrt{20\left(\frac{1}{4}\right)\left(1 - \frac{1}{4}\right)}$

$= \sqrt{20\left(\frac{1}{4}\right)\left(\frac{3}{4}\right)} = 1.94$; the number of correctly identified cards would typically vary by about 1.94 from the mean of 5.

4. (a) $P(X = 5) = {}_{20}C_5\left(\frac{1}{4}\right)^5\left(1 - \frac{1}{4}\right)^{15} = 15{,}504\left(\frac{1}{4}\right)^5\left(\frac{3}{4}\right)^{15} = 0.202$;

if someone were asked to guess the symbol on a randomly chosen card 20 times, there is about a 20% chance that the person would correctly identify exactly 5 of them. **(b)** Using technology, $P(X \geq 8) = 0.1018$. Assuming the person does not have ESP and is just guessing on each card, there is about a 10% chance that the person will correctly identify 8 or more cards out of 20. Because this outcome is quite plausible, we don't have convincing evidence that Alec has ESP.

5. (a)

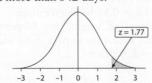

(b) The rectangle has a length of $13 - 8 = 5$ and a height of $\frac{1}{7}$, so the probability = area = $5 \cdot \frac{1}{7} = 0.714$. **(c)** Because the distribution is symmetric, the mean (or balance point) would be directly in the middle at $(8 + 15)/2 = 11.5$ seconds.

6. (a)

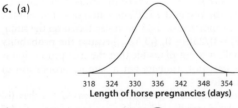

(b)

68% of horse pregnancies last between 330 and 342 days. The other 32% fall outside this range. Because normal distributions are symmetric, half of these horse pregnancies last fewer than 330 days and half last more than 342 days. So about 32%/2 = 16% of horse pregnancies last more than 342 days.

7. (a)

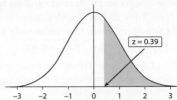

Using Table A, $P(Z > 1.77) = 1 - 0.9616 = 0.0384$.

(b)

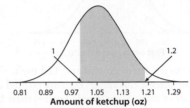

For an area of 0.35 to the right of z, we will be looking at Table A for an area to the left of z of $1 - 0.35 = 0.65$. The closest area to 0.65 is 0.6517, which corresponds to a z-score of $z = 0.39$.

8. (a)

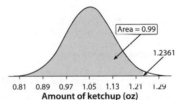

$z = \dfrac{1 - 1.05}{0.08} = -0.63$; $z = \dfrac{1.2 - 1.05}{0.08} = 1.88$. $P(1 \leq X \leq 1.2)$

$= P(-0.63 \leq Z \leq 1.88) = 0.9699 - 0.2643 = 0.7056 = 70.56\%$. *Using technology*: Applet/normalcdf(lower:1, upper:1.2, mean: 1.05, SD:0.08) = 0.7036 = 70.36%.

(b)

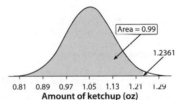

$2.33 = \dfrac{x - 1.05}{0.08} \Rightarrow x = 1.2364$ ounces. *Using technology*: Applet/invNorm(area:0.99, mean:1.05, SD:0.08) = 1.2361 ounces.

Chapter 5 Practice Test

1. c
2. a
3. d
4. c
5. d
6. b
7. c
8. c
9. b
10. c
11. (a) $P(Y \geq 2) = 0.08 + 0.03 + 0.01 = 0.12$. **(b)** $\mu_Y = E(Y) = 0(0.77) + 1(0.11) + 2(0.08) + 3(0.03) + 4(0.01) = 0.4$; if many, many cartons of "store brand" eggs were randomly selected, the average number of broken eggs will be about 0.4.
(c)

$$\sigma_Y = \sqrt{(0 - 0.4)^2(0.77) + (1 - 0.4)^2(0.11) + (2 - 0.4)^2(0.08)}$$
$$+ (3 - 0.4)^2 (0.03) + (4 - 0.4)^2 (0.01). \ \sigma_Y = \sqrt{0.7} = 0.837.$$ The number of broken eggs from a randomly selected carton of "store brand" eggs varies from the mean (0.4) by about 0.837.

12. (a) Binary? "Success" = flight arrives on time. "Failure" = flight arrives late. **Independent?** Knowing whether or not the flight is on time for one randomly selected day tells you nothing about whether or not the flight is on time for another randomly selected day. **Number?** $n = 20$. **Same probability?** $p = 0.85$.

(b) $P(X = 19) = {}_{20}C_{19}(0.85)^{19}(1 - 0.85)^1 = 20(0.85)^{19}(0.15)^1$
$= 0.1368$; if we were to choose 20 days at random to observe the 7 A.M. flight from New York to Los Angeles, there is about a 14% chance that the flight would arrive on time for exactly 19 of the days.
(c) $\mu_X = 20(0.85) = 17$; $\sigma_X = \sqrt{20(0.85)(1 - 0.85)}$
$= \sqrt{20(0.85)(0.15)} = 1.597$. **(d)** Using technology, $P(X \leq 14)$ $= 0.0673$. No, we don't have convincing evidence against the airline's claim because this outcome is plausible.

13. (a)

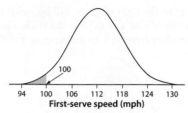

95% of serve speeds are between 100 and 124 mph. The other 5% of serve speeds are outside this range. Because normal distributions are symmetric, half of these serve speeds are less than 100 mph and half are greater than 124 mph. So about 5%/2 = 2.5% of serve speeds are less than 100 mph.

(b)

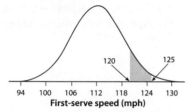

$z = \dfrac{120 - 112}{6} = 1.33$; $z = \dfrac{125 - 112}{6} = 2.17$. $P(120 \leq Y \leq 125)$
$= P(1.33 \leq Z \leq 2.17) = 0.9850 - 0.9082 = 0.0768$. *Using technology:* Applet/normalcdf(lower:120, upper:125, mean:112, SD:6) = 0.0761. There's about an 8% chance that a randomly selected first serve from Novak Djokovic will be between 120 and 125 mph.

(c)

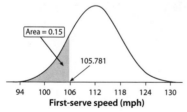

$-1.04 = \dfrac{y - 112}{6} \Rightarrow y = 105.76$ miles per hour. *Using technology:* Applet/invNorm(area:0.15, mean:112, SD:6) = 105.781 miles per hour.

CHAPTER 6

Lesson 6.1 Exercises

1. (a) *Population:* All people who signed a card saying that they intend to quit smoking. *Parameter:* p = the true proportion of the population who quit smoking. *Sample:* A random sample of 1000 people who signed the cards. *Statistic:* The proportion of the sample who quit smoking; $\hat{p} = 0.21$. **(b)** *Population:* All 10-year-old boys. *Parameter:* The true 75th percentile of all 10-year-old boys. *Sample:* Sample of 50 patients. *Statistic:* The 75th percentile of the sample, 56 inches.

3. (a) *Population:* All bottles of iced tea filled in a plant on Tuesday. *Parameter:* μ = the true mean amount of tea in the population. *Sample:* A random sample of 50 bottles. *Statistic:* The mean amount of tea in the sample; $\bar{x}$ = 19.6 ounces. **(b)** *Population:* All passengers in the airport. *Parameter:* p = the true proportion of the population who are chosen for random screening. *Sample:* The 125 passengers on a New York-to-Denver flight. *Statistic:* The proportion of the sample selected for security screening; $\hat{p} = 0.08$.

5.

Sample #1: Abigail (10), Bobby (5) $\bar{x} = 7.5$
Sample #2: Abigail (10), Carlos (10) $\bar{x} = 10$
Sample #3: Abigail (10), DeAnna (7) $\bar{x} = 8.5$
Sample #4: Abigail (10), Emily (9) $\bar{x} = 9.5$
Sample #5: Bobby (5), Carlos (10) $\bar{x} = 7.5$
Sample #6: Bobby (5), DeAnna (7) $\bar{x} = 6$
Sample #7: Bobby (5), Emily (9) $\bar{x} = 7$
Sample #8: Carlos (10), DeAnna (7) $\bar{x} = 8.5$
Sample #9: Carlos (10), Emily (9) $\bar{x} = 9.5$
Sample #10: DeAnna (7), Emily (9) $\bar{x} = 8$

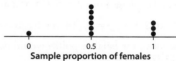

7.

Sample #1: Abigail, Bobby $\hat{p} = 0.50$
Sample #2: Abigail, Carlos $\hat{p} = 0.50$
Sample #3: Abigail, DeAnna $\hat{p} = 1$
Sample #4: Abigail, Emily $\hat{p} = 1$
Sample #5: Bobby, Carlos $\hat{p} = 0$
Sample #6: Bobby, DeAnna $\hat{p} = 0.50$
Sample #7: Bobby, Emily $\hat{p} = 0.50$
Sample #8: Carlos, DeAnna $\hat{p} = 0.50$
Sample #9: Carlos, Emily $\hat{p} = 0.50$
Sample #10: DeAnna, Emily $\hat{p} = 1$

9. (a) In one SRS of size $n = 100$, 73% of the students did all their assigned homework. **(b)** Yes; in the 250 simulated samples, 0 of the SRSs had a sample proportion of 0.45 or lower. Based on the simulation, $P(\hat{p} \leq 0.45) = 0/250 = 0$. **(c)** Yes; because the probability from part **(b)** is small, it is not plausible that the true proportion is $p = 0.60$ and the statistics students got a sample proportion of $\hat{p} = 0.45$ by chance alone.

11. (a) In one SRS of size $n = 20$, the mean height was 62.5 inches. **(b)** No; in the 200 simulated samples, 23 of the SRSs had a mean of 64.7 or more. Based on the simulation, $P(\bar{x} \geq 64.7) = 23/200 = 0.115$. **(c)** No; because the probability from part **(b)** isn't small, it is plausible that the true mean is $\mu = 64$ and the class got a sample mean of $\bar{x} = 64.7$ by chance alone.

13. (a) *Population:* All game pieces. *Parameter:* p = the true proportion of the population that are instant winners. *Sample:* The 20 game pieces collected by the frequent diner. *Statistic:* The proportion of the sample that are instant winners; $\hat{p} = 3/20 = 0.15$. **(b)** No; in the 100 simulated samples, 18 of the SRSs had a sample proportion of 0.15 or lower. Based on the simulation, $P(\hat{p} \leq 0.15) = 18/100 = 0.18$. **(c)** No; because the probability from part **(b)** isn't small, it is plausible that the true proportion is $p = 0.25$ and the frequent diner got a sample proportion of $\hat{p} = 0.15$ by chance alone.

15. (a) The distribution of heights for 16-year-old females is approximately normal with a mean of $\mu = 64$ inches and a standard deviation of $\sigma = 2.5$ inches.

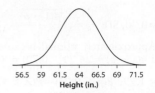

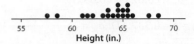

(b) Answers will vary. This is the distribution of one possible sample.

17. (a) In ten cases of taking a random sample of size $n = 50$ from each high school, the difference in proportions of students with Internet access at home is 0%. This means the proportion of students with Internet access was the same for each high school in ten pairs of simulated samples. **(b)** Yes; in the 100 pairs of simulated samples, 0 of the pairs had a difference in proportions of 0.20 or higher. Based on the simulation, $P(\hat{p}_N - \hat{p}_S \geq 0.20) = 0/100 = 0$. **(c)** Yes; because the probability from part **(b)** is small, it is not plausible that the true difference in proportions is $p_N - p_S = 0$ and the superintendent got a sample difference of proportion of $\hat{p}_N - \hat{p}_S = 0.20$ by chance alone.

19. (a) $z = \dfrac{x - \mu}{\sigma}$; $0.67 = \dfrac{28{,}000 - 23{,}300}{\sigma}$; $\sigma = 7014.93$

(b) $z = \dfrac{54{,}000 - 23{,}300}{7014.93} \approx 4.38$; $P(X \geq 54{,}000) \approx P(Z \geq 4.38) \approx 0$

Using technology: Applet/normalcdf(lower:54000, upper:100000, mean:23300, SD:7014.93) = 0.000006. **(c)** If the distribution of loan balances is approximately normal, then we would expect almost no one to have a balance that large. Because 10% of borrowers owe more than $54,000, we can conclude that the distribution of loan balances isn't normal and is right skewed. **(d)** Yes; because the mean ($23,300) is so much larger than the median ($12,800), we can conclude that the distribution of loan balances is skewed to the right.

Lesson 6.2 Exercises

1. Yes; the mean of the sampling distribution is very close to 22.96, the value of the population median.

3. No; the mean of the sampling distribution is clearly less than 153.53, the value of the population range. Population range = max − min = 153.73 − 0.20 = 153.53.

5. (a) It will be less variable because the sample size is larger. **(b)** The estimated median amount spent will typically be closer to the true median amount spent. In other words, the estimate will be more precise.

7. (a) It will be more variable because the sample size is smaller. **(b)** The estimated range amount spent will typically be farther from the true range amount spent. In other words, the estimate will be less precise.

9. If we chose many SRSs and calculated the sample mean $\bar{x}$ for each sample, the distribution of $\bar{x}$ would be centered at the value of μ. In other words, when we use $\bar{x}$ to estimate μ, we will not consistently underestimate μ or consistently overestimate μ.

11. A larger random sample will provide more information about the population and, therefore, more precise results. The variability of the distribution of $\bar{x}$ decreases as the sample size increases.

13. $n = 10$; the sampling distribution of the sample median will be more variable with $n = 10$ than with $n = 100$. Because the distribution is more variable, it is more likely to get a sample median ($250,000) that is far away from the true median ($200,000).

15. (a) Statistics (ii) and (iii) both appear to be unbiased, because the mean of each sampling distribution is very close to the value of the population parameter. **(b)** Statistic (ii); while both statistics (ii) and (iii) are unbiased, statistic (ii) has lower variability.

17. (a) $p = \dfrac{3}{5} = 0.6$

(b)
Sample #1: Abigail, Bobby	$\hat{p} = 0.5$
Sample #2: Abigail, Carlos	$\hat{p} = 0.5$
Sample #3: Abigail, DeAnna	$\hat{p} = 1$
Sample #4: Abigail, Emily	$\hat{p} = 1$
Sample #5: Bobby, Carlos	$\hat{p} = 0$
Sample #6: Bobby, DeAnna	$\hat{p} = 0.5$
Sample #7: Bobby, Emily	$\hat{p} = 0.5$
Sample #8: Carlos, DeAnna	$\hat{p} = 0.5$
Sample #9: Carlos, Emily	$\hat{p} = 0.5$
Sample #10: DeAnna, Emily	$\hat{p} = 1$

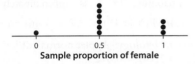

(c) $\mu_{\hat{p}} = \dfrac{0.5 + 0.5 + 1 + 1 + 0 + 0.5 + 0.5 + 0.5 + 0.5 + 1}{10}$

$= \dfrac{6}{10} = 0.6$. Yes, the sample proportion is an unbiased estimator of the population proportion. The mean of the sampling distribution is equal to 0.6, which is the value of the population proportion.

19. (a) Yes. **Binary?** "Success" = female is selected. "Failure" = male is selected. **Independent?** Knowing whether or not one randomly selected student is a female tells you nothing about whether or not another randomly selected student is a female. **Number?** $n = 40$. **Same probability?** $p = \dfrac{55}{104} = 0.529$. **(b)** $\mu_X = np = 40(0.529) = 21.16$

$\sigma_X = \sqrt{np(1-p)} = \sqrt{40(0.529)(1-0.529)} = \sqrt{40(0.529)(0.471)}$

$= \sqrt{9.97} = 3.16$. **(c)** $P(X = 0) = {}_{10}C_0(0.529)^0(1 - 0.529)^{10}$

$= 1(0.529)^0(0.471)^{10} = 0.0005$. $P(X = 1) = {}_{10}C_1(0.529)^1(1 - 0.529)^9$

$= 10(0.529)^1(0.471)^9 = 0.006$. $P(X \leq 1) = P(X = 0) + P(X = 1)$

$= 0.0005 + 0.006 = 0.0065$. If the professor were to randomly choose students for the first 10 meetings, there is less than a 1% chance that he would select 1 female or fewer purely by chance. Because this is unlikely, we have convincing evidence that his selection process is not really random.

Lesson 6.3 Exercises

1. (a) $\mu_X = np = 100(0.11) = 11$; $\sigma_X = \sqrt{np(1-p)}$

$= \sqrt{100(0.11)(1 - 0.11)} = 3.13$. **(b)** If many samples of size 100 were taken, the number of students who are left-handed would typically vary by about 3.13 from the mean of 11.

3. (a) $\mu_X = np = 12\left(\dfrac{1}{5}\right) = 2.4$; $\sigma_X = \sqrt{np(1-p)} = \sqrt{12\left(\dfrac{1}{5}\right)\left(1 - \dfrac{1}{5}\right)}$

$= 1.39$. **(b)** If many samples of size 12 were taken, the number of Kyle Busch cards would typically vary by about 1.39 from the mean of 2.4.

5. Yes; because $np = 100(0.11) = 11 \geq 10$ and $n(1 - p) = 100(1 - 0.11) = 89 \geq 10$, the sampling distribution of X is approximately normal.

7. No; because $np = 12\left(\dfrac{1}{5}\right) = 2.4 < 10$, the sampling distribution of X is not approximately normal.

9. From Exercises 1 and 5, X is approximately normal with a mean of 11 and a standard deviation of 3.13. $z = \dfrac{15 - 11}{3.13} \approx 1.28$;

$P(X \geq 15) \approx P(Z \geq 1.28) = 1 - 0.8997 = 0.1003$. *Using technology*: Applet/normalcdf(lower:15, upper:1000, mean:11, SD:3.13) = 0.1006.

11. X = the number of residents who use public transportation at least once per week. **Mean:**$\mu_X = np = 200(0.34) = 68$. **SD:** $\sigma_X = \sqrt{np(1-p)} = \sqrt{200(0.34)(1-0.34)} = 6.70$. **Shape:** Approximately normal because $np = 200(0.34) = 68 \geq 10$ and $n(1-p) = 200(1-0.34) = 132 \geq 10$. $z = \dfrac{60-68}{6.70} \approx -1.19$; $P(X \leq 60) \approx P(Z \leq -1.19) = 0.1170$. *Using technology*: Applet/normalcdf(lower:−1000, upper:60, mean:68, SD:6.70) = 0.1162.

13. (a) $\mu_X = np = 30(0.5) = 15$ students; $\sigma_X = \sqrt{np(1-p)}$ $= \sqrt{30(0.5)(1-0.5)} = 2.74$ students. If many samples of size 30 were taken, the number of students who prefer name-brand chips would typically vary by about 2.74 from the mean of 15. **(b)** Because $np = 30(0.5) = 15 \geq 10$ and $n(1-p) = 30(1-0.5) = 15 \geq 10$, the sampling distribution of X is approximately normal. **(c)** $z = \dfrac{19-15}{2.74} \approx 1.46$; $P(X \geq 19) \approx P(Z \geq 1.46) = 1 - 0.9279$ $= 0.0721$. *Using technology*: Applet/normalcdf(lower:19, upper: 1000, mean:15, SD:2.74) = 0.0722.

15. No; assuming that 50% of students prefer-name brand chips, there's about a 7% chance that the number of students who prefer name-brand chips (out of an SRS of 30) is 19 or more. The results from Zenon's study could have happened purely by chance, so we do not have convincing evidence that more than half of the students prefer name-brand chips.

17. X = the number of residents who use public transportation at least once per week. Using the applet (Probability, Binomial distribution, $n = 200$, $p = 0.34$, "at most 60 successes") or binomcdf ($n = 200$, $p = 0.34$, $X = 60$) gives $P(X \leq 60) = 0.131$. (Compare with the answer of 0.1162 from Exercise 11.)

19. (a)

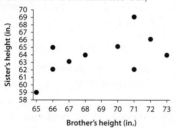

Direction: Positive. *Form*: Linear. *Strength*: Moderate. *Outliers*: No obvious outliers. **(b)** $\hat{y} = 27.635 + 0.527x$, where x = brother's height and y = sister's height. The slope of the line is 0.527, which tells us the predicted sister's height increases by 0.527 inch for each additional increase of 1 inch in the brother's height. **(c)** $\hat{y} = 27.635 + 0.527(70) = 64.525$ inches **(d)** The actual height of a sister is typically about 2.247 inches away from her predicted height using the least-squares regression line.

Lesson 6.4 Exercises

1. (a) $\mu_{\hat{p}} = p = 0.20$; $\sigma_{\hat{p}} = \sqrt{\dfrac{p(1-p)}{n}} = \sqrt{\dfrac{0.20(1-0.20)}{20}} = 0.089$

(b) In SRSs of size $n = 20$, the sample proportion of orange Skittles® will typically vary by about 0.089 from the true proportion of $p = 0.20$.

3. (a) $\mu_{\hat{p}} = p = 0.90$; $\sigma_{\hat{p}} = \sqrt{\dfrac{p(1-p)}{n}} = \sqrt{\dfrac{0.90(1-0.90)}{100}} = 0.03$

(b) In SRSs of size $n = 100$, the sample proportion of orders shipped within three working days will typically vary by about 0.03 from the true proportion of $p = 0.90$.

5. No; because $np = (10)\left(\dfrac{16}{120}\right) = 1.33 < 10$, the sampling distribution of $\hat{p}$ is not approximately normal.

7. Yes; because $np = (100)(0.90) = 90 \geq 10$ and $n(1-p) = (100)(1-0.90) = 10 \geq 10$, the sampling distribution of $\hat{p}$ is approximately normal.

9. Mean: $\mu_{\hat{p}} = p = 0.70$. **SD:** $\sigma_{\hat{p}} = \sqrt{\dfrac{p(1-p)}{n}} = \sqrt{\dfrac{0.70(1-0.70)}{267}}$ $= 0.028$. **Shape:** Approximately normal because $np = 267(0.70) = 186.9 \geq 10$ and $n(1-p) = 267(1-0.70) = 80.1 \geq 10$. $z = \dfrac{0.75-0.70}{0.028} = 1.79$; $P(\hat{p} \geq 0.75) = P(Z \geq 1.79) = 1 - 0.9633$ $= 0.0367$. *Using technology*: Applet/normalcdf(lower:0.75, upper: 1000, mean:0.70, SD:0.028) = 0.0371.

11. Mean: $\mu_{\hat{p}} = p = 0.387$. **SD:** $\sigma_{\hat{p}} = \sqrt{\dfrac{p(1-p)}{n}}$ $= \sqrt{\dfrac{0.387(1-0.387)}{50}} = 0.069$. **Shape:** Approximately normal because $np = (50)(0.387) = 19.35 \geq 10$ and $n(1-p) = 50(1-0.387)$ $= 30.65 \geq 10$. $z = \dfrac{0.30-0.387}{0.069} = -1.26$; $P(\hat{p} < 0.30)$ $= P(Z < -1.26) = 0.1038$. *Using technology*: Applet/normalcdf (lower:−1000, upper:0.30, mean:0.387, SD:0.069) = 0.1037.

13. (a) $\mu_{\hat{p}} = p = 0.70$; $\sigma_{\hat{p}} = \sqrt{\dfrac{p(1-p)}{n}} = \sqrt{\dfrac{0.70(1-0.70)}{1012}} = 0.014$ In SRSs of size $n = 1012$, the sample proportion of people who drink the cereal milk will typically vary by about 0.014 from the true proportion of $p = 0.70$. **(b)** Because $np = (1012)(0.70) = 708.4 \geq 10$ and $n(1-p) = (1012)(1-0.70) = 303.6 \geq 10$, the sampling distribution of $\hat{p}$ is approximately normal. **(c)** $z = \dfrac{0.67-0.70}{0.014} = -2.14$; $P(\hat{p} \leq 0.67) = P(Z \leq -2.14)$ $= 0.0162$. *Using technology*: Applet/normalcdf(lower:−1000, upper: 0.67, mean:0.70, SD:0.014) = 0.0161.

15. Yes; assuming the true proportion of people who drink the cereal milk is 0.70, there is only about a 2% chance of getting a sample proportion of 0.67 or lower purely by chance. Because this result is unlikely (less than 5%), we have convincing evidence that less than 70% of all U.S. adults drink the cereal milk.

17. Because the standard deviation is found by dividing by $\sqrt{n}$, using $4n$ for the sample size halves the standard deviation $\left(\sqrt{4n} = 2\sqrt{n}\right)$; we would have to sample $1012(4) = 4048$ adults.

19. (a) *Shape*: Both distributions are skewed to the right. *Center*: Drivers generally take longer to leave when someone is waiting for the space. *Spread*: There is more variability for the drivers with someone waiting. *Outliers*: There were no outliers for those with someone waiting, but there were two high outliers for those with no one waiting. **(b)** Not necessarily; the researchers merely observed what was happening, and they did not randomly assign the treatments of either having a person waiting or not to the drivers of the cars leaving the lot.

Lesson 6.5 Exercises

1. (a) $\mu_{\bar{x}} = \mu = 225$ seconds; $\sigma_{\bar{x}} = \dfrac{\sigma}{\sqrt{n}} = \dfrac{60}{\sqrt{10}} = 18.97$ seconds.

(b) In SRSs of size $n = 10$, the sample mean play time of songs will typically vary by about 18.97 seconds from the true mean of 225 seconds.

3. (a) $\mu_{\bar{x}} = \mu = 250$ grams; $\sigma_{\bar{x}} = \dfrac{\sigma}{\sqrt{n}} = \dfrac{42}{\sqrt{20}} = 9.39$ grams. **(b)** In SRSs of size $n = 20$, the sample mean weight of tomatoes will typically vary by about 9.39 grams from the true mean of 250 grams.

5. Mean: $\mu_{\bar{x}} = \mu = 9.70$. **SD:** $\sigma_{\bar{x}} = \dfrac{\sigma}{\sqrt{n}} = \dfrac{0.03}{\sqrt{5}} = 0.013$. **Shape:** Normal because the population distribution is normal. $z = \dfrac{9.65-9.70}{0.013} = -3.85$; $P(\bar{x} \leq 9.65) = P(Z \leq -3.85) \approx 0$ *Using technology*: Applet/normalcdf(lower:−1000, upper:9.65, mean:9.70, SD:0.013) = 0.0001.

7. Mean: $\mu_{\bar{x}} = \mu = 188$. **SD:** $\sigma_{\bar{x}} = \dfrac{\sigma}{\sqrt{n}} = \dfrac{41}{\sqrt{100}} = 4.1$.

Shape: Normal because the population distribution is normal.

$z = \dfrac{185 - 188}{4.1} = -0.73$; $z = \dfrac{191 - 188}{4.1} = 0.73$

$P(185 \le \bar{x} \le 191) = P(-0.73 \le Z \le 0.73) = 0.7673 - 0.2327$

$= 0.5346$. *Using technology:* Applet/normalcdf(lower:185, upper:191, mean:188, SD:4.1) = 0.5357.

9. (a) $\mu_{\bar{x}} = \mu = 48$ months; $\sigma_{\bar{x}} = \dfrac{\sigma}{\sqrt{n}} = \dfrac{8.2}{\sqrt{8}} = 2.90$ months. **(b)** In SRSs of size $n = 8$, the sample mean battery life will typically vary by about 2.90 months from the true mean of 48 months. **(c)** The sampling distribution is normal because the population distribution is normal.

$z = \dfrac{42.2 - 48}{2.90} = -2.00$; $P(\bar{x} < 42.2) = P(Z < -2.00) = 0.0228$

Using technology: Applet/normalcdf(lower:–1000, upper:42.2, mean:48, SD:2.90) = 0.0228.

11. Yes; assuming the true mean battery life is 48 months, there is only about a 2% chance of getting a sample mean of 42.2 or lower purely by chance. Because this result is unlikely (less than 5%), we have convincing evidence that the population mean battery life is less than 48 months.

13. (a) Less likely. Individual values vary more than averages, so getting an individual value close to the true mean is less likely.

(b) $z = \dfrac{185 - 188}{41} = -0.07$; $z = \dfrac{191 - 188}{41} = 0.07$

$P(185 \le X \le 191) = P(-0.07 \le Z \le 0.07) = 0.5279 - 0.4721$

$= 0.0558$. *Using technology:* Applet/normalcdf(lower:185, upper:191, mean:188, SD:41) = 0.0583. This probability of 0.0583 is much less than the probability calculated in Exercise 7 (0.5357).

15. $\sigma_{\bar{x}} = \dfrac{\sigma}{\sqrt{n}} \rightarrow 30 = \dfrac{60}{\sqrt{n}} \rightarrow n = 4$; we would have to sample 4 songs.

17. A bag of 20 oranges that weighs 65 ounces would give an average orange weight of $\bar{x} = 3.25$ ounces. So we want to find $P(\bar{x} > 3.25)$.

Mean: $\mu_{\bar{x}} = \mu = 3$. **SD:** $\sigma_{\bar{x}} = \dfrac{\sigma}{\sqrt{n}} = \dfrac{0.5}{\sqrt{20}} = 0.11$. **Shape:** Normal

because the population distribution is normal. $z = \dfrac{3.25 - 3}{0.11} = 2.27$.

$P(\text{total weight} > 65) = P(\bar{x} > 3.25) = P(Z > 2.27) = 1 - 0.9884$

$= 0.0116$. *Using technology:* Applet/normalcdf(lower:3.25, upper:1000, mean:3, SD:0.11) = 0.0115.

19. (a) The probabilities are all between 0 and 1. Also, the sum of the probabilities is $0.28 + 0.27 + 0.18 + 0.16 + 0.07 + 0.04 = 1$.
(b) Using the complement rule, $P(X \ge 1) = 1 - P(X = 0) = 1 - 0.28 = 0.72$.
(c) $E(X) = \mu_X = 0(0.28) + 1(0.27) + 2(0.18) + 3(0.16) + 4(0.07) + 5(0.04) = 1.59$ devices; $\sigma_X = 1.422$ devices

Lesson 6.6 Exercises

1. (a) Because $n = 5 < 30$, the sampling distribution of $\bar{x}$ will also be skewed to the right but not quite as strongly as the population. **(b)** Because $n = 100 \ge 30$, the sampling distribution of $\bar{x}$ is approximately normal by the central limit theorem.
3. (a) No; a count only takes on whole-number values, so it cannot be normally distributed. **(b)** Because $n = 100 \ge 30$, the sampling distribution of $\bar{x}$ is approximately normal by the central limit theorem.
5. Mean: $\mu_{\bar{x}} = \mu = 225$ seconds. **SD:** $\sigma_{\bar{x}} = \dfrac{\sigma}{\sqrt{n}} = \dfrac{60}{\sqrt{100}} = 6$ seconds.

Shape: Because $n = 100 \ge 30$, the sampling distribution of $\bar{x}$ is

approximately normal by the central limit theorem. $z = \dfrac{240 - 225}{6}$

$= 2.5$; $P(\bar{x} < 240) = P(Z < 2.5) = 0.9938$. *Using technology:* Applet/normalcdf(lower:–1000, upper:240, mean:225, SD:6) = 0.9938.

7. Mean: $\mu_{\bar{x}} = \mu = 1.5$ people. **SD:** $\sigma_{\bar{x}} = \dfrac{\sigma}{\sqrt{n}} = \dfrac{0.75}{\sqrt{100}} = 0.075$ people.

Shape: Because $n = 100 \ge 30$, the sampling distribution of $\bar{x}$ is approx-

imately normal by the central limit theorem. $z = \dfrac{1.7 - 1.5}{0.075} = 2.67$;

$P(\bar{x} > 1.7) = P(Z > 2.67) = 1 - 0.9962 = 0.0038$. *Using technology:* Applet/normalcdf(lower:1.7, upper:1000, mean:1.5, SD:0.075) = 0.0038.

9. (a) Because $n = 50 \ge 30$, the sampling distribution of $\bar{x}$ is approximately normal by the central limit theorem. **(b) Mean:** $\mu_{\bar{x}} = \mu = 6$

lightning strikes. **SD:** $\sigma_{\bar{x}} = \dfrac{\sigma}{\sqrt{n}} = \dfrac{2.4}{\sqrt{50}} = 0.339$ lightning strikes.

$z = \dfrac{5 - 6}{0.339} = -2.95$; $P(\bar{x} < 5) = P(Z < -2.95) = 0.0016$. *Using technology:* Applet/normalcdf(lower:–1000, upper:5, mean:6, SD:0.339) = 0.0016.

11. (a) Because $n = 10 < 30$, we can't be sure that the sampling distribution of $\bar{x}$ will be approximately normal. **(b)** Greater than. The variability of the sampling distribution of $\bar{x}$ will be greater with the smaller sample size of 10, making it more likely to get a sample mean that is far away from the true mean of 6 lightning strikes (such as $\bar{x} = 5$ or lower).

13. From Exercise 7, we know that when the true mean number of people in the car is 1.5, there is almost a 0% chance that the mean number of people in the car will be at least 1.7 in a random sample of 100 cars. Because the observed result is unlikely to happen purely by chance (less than 5%), the researcher has good evidence to conclude that people are more likely to drive with other people in the car on Sundays.

15. No; the graph of the sample will resemble the shape of the population distribution, regardless of the sample size. The student should say that the graph of the *sampling distribution of the sample mean* ($\bar{x}$) looks more and more normal as you take larger and larger samples from a population.

17. A total of $45,000 for 50 students yields an average cost per student of $900. We want to find $P(\bar{x} > 900)$.

Mean: $\mu_{\bar{x}} = \mu = \$836$. **SD:** $\sigma_{\bar{x}} = \dfrac{\sigma}{\sqrt{n}} = \dfrac{388}{\sqrt{50}} = \$54.87$. **Shape:**

Because $n = 50 \ge 30$, the sampling distribution of $\bar{x}$ is approxi-

mately normal by the central limit theorem. $z = \dfrac{900 - 836}{54.87} = 1.17$;

$P(\text{total cost} > 45,000) = P(\bar{x} > 900) = P(Z > 1.17) = 1 - 0.8790$

$= 0.1210$. *Using technology:* Applet/normalcdf(lower:900, upper:10000, mean:836, SD:54.87) = 0.1217.

19. $P(\text{yielded to pedestrians} \mid \text{expensive car}) = \dfrac{32}{32 + 26} = \dfrac{32}{58} = 0.55$

$P(\text{yielded to pedestrians} \mid \text{inexpensive car}) = \dfrac{67}{67 + 27} = \dfrac{67}{94} = 0.71$

Because the probabilities are not equal, the events "Yielded to pedestrians" and "Expensive car" are not independent. Knowing that a randomly selected car is expensive decreases the probability that the car yielded to pedestrians.

Chapter 6 Review Exercises

1. *Population:* All eggs shipped in one day. *Sample:* The 200 eggs examined. *Parameter:* The proportion p of eggs shipped that day that had salmonella. *Statistic:* The proportion of eggs in the sample that had salmonella, $\hat{p} = \dfrac{9}{200} = 0.045$.

2. (a)

Sample #1: 64, 66, 71	Median = 66
Sample #2: 64, 66, 73	Median = 66
Sample #3: 64, 66, 76	Median = 66
Sample #4: 64, 71, 73	Median = 71
Sample #5: 64, 71, 76	Median = 71
Sample #6: 64, 73, 76	Median = 73
Sample #7: 66, 71, 73	Median = 71
Sample #8: 66, 71, 76	Median = 71
Sample #9: 66, 73, 76	Median = 73
Sample #10: 71, 73, 76	Median = 73

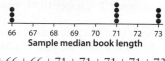

(b) $\mu_{\text{median}} = \dfrac{66+66+66+71+71+71+71+73+73+73}{10} = \dfrac{701}{10}$

$= 70.1$. The sample median is a biased estimator of the population median. The mean of the sampling distribution is equal to 70.1, which is less than the value of the population median of 71. **(c)** The sampling distribution of the sample median will be less variable because the sample size is larger. The estimated median book length will typically be closer to the true median book length. In other words, the estimate will be more precise.

3. (a) $\mu_X = np = 500(0.24) = 120$ people; $\sigma_X = \sqrt{np(1-p)}$ $= \sqrt{500(0.24)(1-0.24)} = 9.55$ people. **(b)** If many samples of size 500 were taken, the number of people who are under 18 years old would typically vary by about 9.55 from the mean of 120. **(c)** The sampling distribution of X is approximately normal because $np = 500(0.24) = 120 \geq 10$ and $n(1-p) = 500$ $(1-0.24) = 380 \geq 10$. **(d)** $z = \dfrac{100-120}{9.55} = -2.09$; $z = \dfrac{110-120}{9.55}$ $= -1.05$; $P(100 \leq X \leq 110) \approx P(-2.09 \leq Z \leq -1.05)$ $= 0.1469 - 0.0183 = 0.1286$. *Using technology*: Applet/normalcdf (lower:100, upper:110, mean:120, SD:9.55) = 0.1294.

4. (a) $\mu_{\hat{p}} = p = 0.20$; $\sigma_{\hat{p}} = \sqrt{\dfrac{p(1-p)}{n}} = \sqrt{\dfrac{0.20(1-0.20)}{80}} = 0.045$
(b) In SRSs of size $n = 80$, the sample proportion of people who subscribe to the five-second rule will typically vary by about 0.045 from the true proportion of $p = 0.20$. **(c)** Because $np = (80)(0.2) = 16 \geq 10$ and $n(1-p) = (80)(1-0.20) = 64 \geq 10$, the sampling distribution of $\hat{p}$ is approximately normal. **(d)** $z = \dfrac{0.10 - 0.20}{0.045} = -2.22$; $P(\hat{p} \leq 0.10) = P(Z \leq -2.22) = 0.0132$. *Using technology*: Applet/normalcdf(lower:−1000, upper:0.10, mean:0.20, SD:0.045) = 0.0131. Assuming the true proportion of people who subscribe to the 5-second rule is 0.20, there is only about a 1% chance of getting a sample proportion of 0.10 or less purely by chance. Because this result is unlikely (less than 5%), we have convincing evidence that the proportion of all U.S. adults who subscribe to the 5-second rule is less than 0.20.

5. (a) $\mu_{\bar{x}} = \mu = 100$; $\sigma_{\bar{x}} = \dfrac{\sigma}{\sqrt{n}} = \dfrac{15}{\sqrt{10}} = 4.74$. **(b)** In SRSs of size $n = 10$, the sample mean WAIS score will typically vary by about 4.74 from the true mean of 100. **(c)** The sampling distribution of $\bar{x}$ is approximately normal, because the population distribution is approximately normal. $z = \dfrac{105 - 100}{4.74} = 1.05$; $P(\bar{x} \geq 105) = P(Z \geq 1.05)$ $= 1 - 0.8531 = 0.1469$. *Using technology*: Applet/normalcdf (lower:105, upper:1000, mean:100, SD:4.74) = 0.1457. **(d)** The answer to parts **(a)** and **(b)** would be the same because the mean and standard deviation do not depend on the shape of the population distribution. We could not answer part **(c)**, because we could not be sure that the sampling distribution is approximately normal.

6. (a) Because $n = 50 \geq 30$, the sampling distribution of $\bar{x}$ is approximately normal by the central limit theorem. **(b) Mean:** $\mu_{\bar{x}} = \mu = 0.5$ moths. **SD:** $\sigma_{\bar{x}} = \dfrac{\sigma}{\sqrt{n}} = \dfrac{0.7}{\sqrt{50}} = 0.099$ moths. $z = \dfrac{0.6 - 0.5}{0.099} = 1.01$; $P(\bar{x} \geq 0.6) = P(Z \geq 1.01) = 1 - 0.8438 = 0.1562$. *Using technology*: Applet/normalcdf(lower:0.6, upper:1000, mean:0.5, SD:0.099) = 0.1562. **(c)** Assuming the true mean number of moths is 0.5, there is about a 16% chance that the mean number of moths will be at least 0.6 in a sample of 50 traps. Because this result is plausible (more than 5%), we do not have convincing evidence that the moth population is getting larger.

Chapter 6 Practice Test

1. b
2. c
3. c
4. b
5. a
6. a
7. b
8. b
9. a
10. d
11. Statistic A. Both statistics A and B appear to be unbiased, with the center of their sampling distributions equal to the value of the parameter, but statistic A has less variability than statistic B.

12. (a) $\mu_{\bar{x}} = \mu = \$48$; $\sigma_{\bar{x}} = \dfrac{\sigma}{\sqrt{n}} = \dfrac{20}{\sqrt{500}} = \$0.89$. In SRSs of size $n = 500$, the sample mean amount paid for Internet will typically vary by about \$0.89 from the true mean of \$48. **(b)** Because $n = 500 \geq 30$, the sampling distribution of $\bar{x}$ is approximately normal by the central limit theorem. **(c)** $z = \dfrac{50 - 48}{0.89} = 2.25$; $P(\bar{x} > 50) = P(Z > 2.25) = 1 - 0.9878 = 0.0122$. *Using technology*: Applet/normalcdf(lower:50, upper:1000, mean:48, SD:0.89) = 0.0123.

13. (a) The sampling distribution of X is approximately normal because $np = 300(0.22) = 66 \geq 10$ and $n(1-p) = 300(1-0.22)$ $= 234 \geq 10$. **(b)** 20% of 300 is 60, so we want to find $P(X > 60)$. $\mu_X = np = 300(0.22) = 66$; $\sigma_X = \sqrt{np(1-p)} = \sqrt{300(0.22)(1-0.22)}$ $= 7.17$. $z = \dfrac{60 - 66}{7.17} = -0.84$; $P(X > 60) = P(Z > -0.84) = 0.7995$. *Using technology*: Applet/normalcdf(lower:60, upper:1000, mean:66, SD:7.17) = 0.7987.

Chapter 7

Lesson 7.1 Exercises

1. We are 90% confident that the interval from \$51,484 to \$52,394 captures the true median income for all U.S. households in 2013.
3. We are 95% confident that the interval from 0.175 to 0.225 captures the true proportion of all U.S. adults who would correctly identify nitrogen as the answer.
5. point estimate = $\dfrac{51,484 + 52,394}{2} = \$51,939$; margin of error $= 52,394 - 51,939 = \$455$ or $51,939 - 51,484 = \$455$.
7. point estimate $= \dfrac{0.175 + 0.225}{2} = 0.20$; margin of error $= 0.225 - 0.20 = 0.025$ or $0.20 - 0.175 = 0.025$.
9. No; \$51,759 is a plausible value for the median household income for all U.S. adults in 2013 because it is captured in the confidence interval. There is not convincing evidence that the median household income increased in 2013.
11. Yes; all the values of the confidence interval are less than 0.25. There is convincing evidence that less than 25% of all U.S. adults would answer this question correctly.

13. (a) We are 95% confident that the interval from 26.8 to 41.0 minutes captures the true mean delivery time to my home for all deliveries from the local pizza place. **(b)** point estimate = $\dfrac{26.8 + 41.0}{2}$ = 33.9 minutes; margin of error = 41.0 − 33.9 = 7.1 minutes or 33.9 − 26.8 = 7.1 minutes. **(c)** Yes; there are many values less than 30 minutes that are captured in the confidence interval (29 minutes, for example). It is plausible that the true average delivery time is less than 30 minutes.

15. Not confident; while we can say that we are 95% confident that the *interval* from 26.8 to 41.0 minutes captures the true mean delivery time, it is very unlikely that the true mean delivery time is *exactly* 33.9 minutes.

17. (a) We are 95% confident that the interval from 0.120 to 0.297 captures the true difference in the proportion of all younger teens who include false information in their profiles and the proportion of all older teens who do the same (Younger teens − Older teens). **(b)** Yes; because all the values of the confidence interval are greater than 0, there is convincing evidence that the younger teens are more likely than older teens to include false information in their profiles.

19. (a)

Sample #1: Violet (9), Patty (5), Lucy (4)	$\bar{x}$ = 6
Sample #2: Violet (9), Patty (5), Sally (4)	$\bar{x}$ = 6
Sample #3: Violet (9), Patty (5), Marcie (8)	$\bar{x}$ = 7.33
Sample #4: Violet (9), Lucy (4), Sally (4)	$\bar{x}$ = 5.67
Sample #5: Violet (9), Lucy (4), Marcie (8)	$\bar{x}$ = 7
Sample #6: Violet (9), Sally (4), Marcie (8)	$\bar{x}$ = 7
Sample #7: Patty (5), Lucy (4), Sally (4)	$\bar{x}$ = 4.33
Sample #8: Patty (5), Lucy (4), Marcie (8)	$\bar{x}$ = 5.67
Sample #9: Patty (5), Sally (4), Marcie (8)	$\bar{x}$ = 5.67
Sample #10: Lucy (4), Sally (4), Marcie (8)	$\bar{x}$ = 5.33

Sample mean litter size

(b) Population mean: $\mu = \dfrac{9 + 5 + 4 + 4 + 8}{5} = 6$. Mean of sample means:

$\mu_{\bar{x}} = \dfrac{6 + 6 + 7.33 + 5.67 + 7 + 7 + 4.33 + 5.67 + 5.67 + 5.33}{10}$

$= \dfrac{60}{10} = 6$. The sample mean is an unbiased estimator of the population mean because the mean of the sampling distribution is equal to 6, the value of the population mean.

Lesson 7.2 Exercises

1. If the principal took many random samples of students and constructed a 90% confidence interval for each sample, about 90% of these intervals would capture the true mean number of hours doing homework per week for all students of the school.

3. If the *New York Times*/CBS News poll took many random samples of U.S. adults and constructed a 95% confidence interval for each sample, about 95% of these intervals would capture the true proportion of all U.S. adults who favor the amendment.

5. (a) Wider because increasing the confidence level increases the margin of error. **(b)** Narrower because increasing the sample size decreases the margin of error.

7. (a) Narrower because decreasing the confidence level decreases the margin of error. **(b)** Wider because decreasing the sample size increases the margin of error.

9. Many students may lie about their homework habits when having to respond to the principal without anonymity. The mean number of hours spent on homework for all students may be even less than 2.3 hours.

11. Those people selected for the sample may not be available to respond or may refuse to answer (nonresponse). Also, people may answer "Yes" because they think they should, even if they don't support the amendment (response bias).

13. (a) If we took many random samples of employed California adults and constructed a 90% confidence interval for each sample,

about 90% of these intervals would capture the true mean travel time to work for all employed California adults. **(b)** We could decrease the confidence level. The drawback is that we can't be as confident that our interval will capture the true proportion. We could also increase the sample size. The drawback is that larger samples cost more time and money to obtain. **(c)** People who have longer travel times to work may have less time to respond to a survey. This would cause our estimate from the sample to be less than the true mean travel time to work. The bias due to nonresponse is not accounted for by the margin of error, because the margin of error accounts only for variability we expect from random sampling.

15. The figure shows that 4 of the 25 confidence intervals (16%) did not capture the true parameter. Therefore, 84% of the intervals capture the true parameter, which suggests that these were 80% or 90% confidence intervals.

17. (a) Answers may vary but should be close to 95%. **(b)** Answers may vary but should be close to 95%. **(c)** Answers may vary but should be close to 95%. **(d)** No; a 95% confidence level will have a success rate of about 95%, regardless of the sample size. As the sample size increases, the confidence intervals become narrower, but the success rate (determined by the confidence level) stays about the same.

19. (a) Shape: Approximately normal because $np = 120(0.88) = 105.6 \geq 10$ and $n(1 − p) = 120(1 − 0.88) = 14.4 \geq 10$. **Center:** $\mu_{\hat{p}} = p = 0.88$. **Spread:** $\sigma_{\hat{p}} = \sqrt{\dfrac{p(1-p)}{n}} = \sqrt{\dfrac{0.88(1-0.88)}{120}} = 0.0297$.

(b) $z = \dfrac{0.90 - 0.88}{0.0297} = 0.67$; $P(\hat{p} \geq 0.90) = P(Z \geq 0.67) = 1 − 0.7486 = 0.2514$. *Using technology:* Applet/normalcdf(lower:0.90, upper:1000, mean:0.88, SD:0.0297) = 0.2503. **(c)** Yes; let X = the number of residents who have earned a high school diploma. The random variable X has a binomial distribution with $n = 10$ and $p = 0.88$ (BINS all satisfied). Using the applet (Probability, Binomial distribution, $n = 10$, $p = 0.88$, "at least 9 sucesses") or $1 −$ binomcdf ($n = 10$, $p = 0.88$, $X = 8$) gives $P(X \geq 9) = 0.6583$.

Lesson 7.3 Exercises

1. Random? Latoya interviews an SRS of students. ✓ **Large counts?** $n\hat{p} = 50\left(\dfrac{14}{50}\right) = 14 > 10$ and $n(1-\hat{p}) = 50\left(1 - \dfrac{14}{50}\right) = 36 \geq 10$. ✓

3. Random? The 200 messages that the representative received did not come from a random sample. We can't generalize to the larger population of "all registered voters." **Large counts?** $n\hat{p} = 200\left(\dfrac{2}{200}\right) = 2 < 10$, so the sampling distribution of $\hat{p}$ is not approximately normal.

5. Because $\dfrac{1 - 0.98}{2} = 0.01$, z^* for a 98% confidence interval can be found by looking in Table A for a left-tail area of 0.01. The closest area is 0.0099, corresponding to a critical value of $z^* = 2.33$. *Using technology:* Applet/invNorm(left-tail area:0.01, mean:0, SD:1) gives $z^* = 2.326$.

7. Because $\dfrac{1 - 0.75}{2} = 0.125$, z^* for a 75% confidence interval can be found by looking in Table A for a left-tail area of 0.125. The closest area is 0.1251, corresponding to a critical value of $z^* = 1.15$. *Using technology:* Applet/invNorm(left-tail area:0.125, mean:0, SD:1) gives $z^* = 1.150$.

9. Because $\dfrac{1 - 0.90}{2} = 0.05$, z^* for a 90% confidence interval can be found by looking in Table A for a left-tail area of 0.05. The closest area is 0.0495, corresponding to a critical value of $z^* = 1.65$. *Using technology:* Applet/invNorm(left-tail area:0.05, mean:0, SD:1) gives $z^* = 1.645$.

$\hat{p} \pm z^*\sqrt{\dfrac{\hat{p}(1-\hat{p})}{n}} \rightarrow 0.28 \pm 1.645\sqrt{\dfrac{0.28(1-0.28)}{50}} \rightarrow 0.28 \pm 0.104 \rightarrow$ 0.176 to 0.384.

11. Because $\dfrac{1-0.95}{2} = 0.025$, z^* for a 95% confidence interval can be found by looking in Table A for a left-tail area of 0.025. The closest area is 0.025, corresponding to a critical value of $z^* = 1.96$. *Using technology*: Applet/invNorm(left-tail area:0.025, mean:0, SD:1) gives $z^* = 1.960$.

$\hat{p} \pm z^*\sqrt{\dfrac{\hat{p}(1-\hat{p})}{n}} \rightarrow 0.66 \pm 1.960\sqrt{\dfrac{0.66(1-0.66)}{50}} \rightarrow 0.66 \pm 0.131$
$\rightarrow 0.529$ to 0.791.

13. **(a) Random?** The student project put the question to an SRS of undergraduate students. ✓ **Large counts?** $n\hat{p} = 172\left(\dfrac{19}{172}\right) = 19 \geq 10$
and $n(1-\hat{p}) = 172\left(1 - \dfrac{19}{172}\right) = 153 \geq 10$. ✓
(b) Because $\dfrac{1-0.99}{2} = 0.005$, z^* for a 99% confidence interval can be found by looking in Table A for a left-tail area of 0.005. The closest area is 0.0049, corresponding to a critical value of $z^* = 2.58$. *Using technology*: Applet/invNorm(left-tail area:0.005, mean:0, SD:1) gives $z^* = 2.576$.

$\hat{p} \pm z^*\sqrt{\dfrac{\hat{p}(1-\hat{p})}{n}} \rightarrow 0.11 \pm 2.576\sqrt{\dfrac{0.11(1-0.11)}{172}} \rightarrow 0.11 \pm 0.061$
$\rightarrow 0.049$ to 0.171. **(c)** We are 95% confident that the interval from 0.049 to 0.171 captures the true proportion of all undergraduate students at this university who would answer "Yes."

15. $SE_{\hat{p}} = \sqrt{\dfrac{\hat{p}(1-\hat{p})}{n}} = \sqrt{\dfrac{0.11(1-0.11)}{172}} = 0.0239$. In SRSs of size $n = 172$, the sample proportion of students at this university who would answer "Yes" will typically vary by about 0.0239 from the true proportion.

17. (a) We do not know the sample sizes for the men and for the women. **(b)** The margin of error for women alone would be greater than 0.03 because the sample size for women alone is smaller than 1019.

19. (a) $\left(\dfrac{1}{6}\right)\left(\dfrac{1}{6}\right)\left(\dfrac{1}{6}\right) = \dfrac{1}{216}$. **(b)** The friend's claim is based on the erroneous "law of averages." Even after rolling doubles three turns in a row, there is still the same 1/6 chance of getting doubles on the next roll. **(c)** Because each roll is independent of the next, the probability of rolling doubles on the fourth roll is 1/6.

Lesson 7.4 Exercises

1. **STATE:** We want to estimate p = the true proportion of all seniors at Leticia's school who plan to go to the prom at a 90% confidence level. **PLAN:** One-sample z interval for p.
Random? Leticia interviews an SRS of 50 seniors. ✓

Large counts? $50\left(\dfrac{36}{50}\right) = 36 \geq 10$ and $50\left(1 - \dfrac{36}{50}\right) = 14 \geq 10$. ✓

DO: $\hat{p} \pm z^*\sqrt{\dfrac{\hat{p}(1-\hat{p})}{n}} \rightarrow 0.72 \pm 1.645\sqrt{\dfrac{0.72(1-0.72)}{50}} \rightarrow 0.72$
$\pm 0.104 \rightarrow 0.616$ to 0.824
CONCLUDE: We are 90% confident that the interval from 0.616 to 0.824 captures the true proportion of all seniors at Leticia's school who plan to go to the prom.

3. STATE: We want to estimate p = the true proportion of all teens with cell phones who use messaging apps at a 95% confidence level. **PLAN:** One-sample z interval for p.
Random? Pew surveyed 929 randomly selected teens with cell phones. ✓
Large counts? $929(0.33) = 307 \geq 10$ and $929(1-0.33) = 622 \geq 10$. ✓

DO: $\hat{p} \pm z^*\sqrt{\dfrac{\hat{p}(1-\hat{p})}{n}} \rightarrow 0.33 \pm 1.960\sqrt{\dfrac{0.33(1-0.33)}{929}} \rightarrow 0.33 \pm$
$0.030 \rightarrow 0.300$ to 0.360
CONCLUDE: We are 95% confident that the interval from 0.300 to 0.360 captures the true proportion of all teens with cell phones who use messaging apps.

5. $1.645\sqrt{\dfrac{0.75(1-0.75)}{n}} \leq 0.04 \rightarrow n \geq 317.11$. We have to survey at least 318 Americans of Italian descent.

7. $1.960\sqrt{\dfrac{0.5(1-0.5)}{n}} \leq 0.03 \rightarrow n \geq 1067.11$. We have to survey at least 1068 registered voters.

9. (a) STATE: We want to estimate p = the true proportion of all baseball fans in Connecticut who would say their favorite team is the Yankees at a 95% confidence level.
PLAN: One-sample z interval for p.
Random? The poll included 803 randomly selected baseball fans in Connecticut. ✓
Large counts? $803(0.44) = 353 \geq 10$ and $803(1-0.44) = 450 \geq 10$. ✓
DO: $\hat{p} \pm z^*\sqrt{\dfrac{\hat{p}(1-\hat{p})}{n}} \rightarrow 0.44 \pm 1.960\sqrt{\dfrac{0.44(1-0.44)}{803}} \rightarrow 0.44$
$\pm 0.034 \rightarrow 0.406$ to 0.474.
CONCLUDE: We are 95% confident that the interval from 0.406 to 0.474 captures the true proportion of all baseball fans in Connecticut who would say their favorite team is the Yankees.
(b) No; all the values of the confidence interval are less than 0.50, so it is not plausible that more than 50% (a majority) of all baseball fans in Connecticut would say their favorite team is the Yankees. **(c)** $1.960\sqrt{\dfrac{0.44(1-0.44)}{n}} \leq 0.02 \rightarrow n \geq 2366.43$.
We have to survey at least 2367 baseball fans in Connecticut. This would require that an *additional* $2367 - 803 = 1564$ baseball fans in Connecticut be randomly selected.

11. The sample size would have to be multiplied by 4 to cut the
margin of error in half: $ME_{new} = z^*\sqrt{\dfrac{\hat{p}(1-\hat{p})}{4n}} = \sqrt{\dfrac{1}{4}}z^*\sqrt{\dfrac{\hat{p}(1-\hat{p})}{n}}$
$= \dfrac{1}{2}z^*\sqrt{\dfrac{\hat{p}(1-\hat{p})}{n}} = \dfrac{1}{2}ME_{old}$. A sample of $4(929) = 3716$ is needed.

13. Multiply each proportion by the total number of seniors in Leticia's school (750). We are 90% confident that the interval from $750(0.616) = 462$ to $750(0.824) = 618$ captures the true *total* number of seniors at Leticia's school who plan to go to the prom.

15. (a) $ME = z^*\sqrt{\dfrac{\hat{p}(1-\hat{p})}{n}} \rightarrow 0.01 = z^*\sqrt{\dfrac{0.6341(0.3659)}{5594}}$, so
$z^* = 1.55$. The area between $z = -1.55$ and $z = 1.55$ under the standard normal curve is $0.9394 - 0.0606 = 0.8788$. The confidence level is about 88%. **(b)** We do not know if those who *did* respond can reliably represent those who did not. Also, the athletes may not have been truthful in their responses.

17. (a) Mean: $\mu_{\bar{x}} = \mu = \$290{,}000$. **SD:** $\sigma_{\bar{x}} = \dfrac{\sigma}{\sqrt{n}} = \dfrac{145{,}000}{\sqrt{100}} = \$14{,}500$.
Shape: Because $n = 100 \geq 30$, the sampling distribution of $\bar{x}$ is approximately normal by the central limit theorem.
$z = \dfrac{325{,}000 - 290{,}000}{14{,}500} = 2.41$. $P(\bar{x} > 325{,}000) = P(Z > 2.41)$
$= 1 - 0.9920 = 0.0082$. *Using technology*: Applet/normalcdf(lower: 325,000, upper:1,000,000, mean:290,000, SD:14,500) $= 0.0079$.
(b) Because $n = 5 < 30$, the sampling distribution of $\bar{x}$ may not be approximately normal.

Lesson 7.5 Exercises

1. Random? Random sample of 32 students from the school. ✓ **Normal/Large Sample?** We don't know the shape of the population distribution, but the sample size is large: $n = 32 \geq 30$. ✓

3. Random? This is not a random sample, so this condition is not met. Inference is not needed here because we have the age at death for the entire population. **Normal/Large Sample?** We don't know the shape of the population distribution, but the sample size is large: $n = 38 \geq 30$. ✓

5. (a) 95% confidence and df $= 10 - 1 = 9$ give $t^* = 2.262$. **(b)** 90% confidence and df $= 77 - 1 = 76$ (use df $= 60$) give $t^* = 1.671$.

7. (a) 99% confidence and df $= 20 - 1 = 19$ give $t^* = 2.861$. **(b)** 95% confidence and df $= 85 - 1 = 84$ (use df $= 80$) give $t^* = 1.990$.

9. 95% confidence and df $= 92 - 1 = 91$ (use df $= 80$) give $t^* = 1.990$.

$$\bar{x} \pm t^* \frac{s_x}{\sqrt{n}} \to 356.1 \pm 1.990 \frac{185.7}{\sqrt{92}} \to 356.1 \pm 38.5 \to 317.6 \text{ to } 394.6 \text{ licks}$$

11. 90% confidence and df $= 15 - 1 = 14$ give $t^* = 1.761$.

$$\bar{x} \pm t^* \frac{s_x}{\sqrt{n}} \to 183,100 \pm 1.761 \frac{29,200}{\sqrt{15}} \to 183,100 \pm 13,277 \to$$

$169,823$ to $196,377$

13. (a) Random? Random sample of 1470 eighth-graders. ✓ **Normal/Large Sample?** We don't know the shape of the population distribution, but the sample size is large: $n = 1470 \geq 30$. ✓ **(b)** 99% confidence and df $= 1470 - 1 = 1469$ (use df $= 1000$) give $t^* = 2.581$.

$$\bar{x} \pm t^* \frac{s_x}{\sqrt{n}} \to 240 \pm 2.581 \frac{42.17}{\sqrt{1470}} \to 240 \pm 2.84 \to 237.16 \text{ to }$$

242.84. **(c)** Yes; all the values of the confidence interval are less than 243, so there is convincing evidence that the mean score for all Atlanta eighth-graders is less than the basic level.

15. $SE_{\bar{x}} = \frac{s_x}{\sqrt{n}} = \frac{9.3}{\sqrt{27}} = 1.79$; if we take many samples of size 27, the sample mean systolic blood pressure will typically vary from the true mean by about 1.79.

17. Using the endpoints of the interval from Exercise 11 ($169,823 to $196,377), Tax Revenue $- 169,823 \times (.01) \times 300 = \$509,469$ and $196,377 \times (.01) \times 300 = 589,131$. We are 90% confident that the interval from $509,469 to $589,131 captures the true total amount of tax revenue generated by the new houses.

19. (a) If a social scientist repeatedly takes random samples of size 50 from a population of people in their 20s, the number of people in a sample who look forward to opening their mailbox each day varies from about 13 to 29. **(b)** Yes; of the 100 simulated samples, there were only 4 samples where 14 or fewer people in their 20s responded that they look forward to opening their mail.

Lesson 7.6 Exercises

1. Yes; we don't know the shape of the population distribution, but the sample size is large: $n = 100 \geq 30$. ✓

3. No; the sample size is small ($16 < 30$), and the histogram shows possible outliers.

5. STATE: We want to estimate μ = the true mean amount of vitamin C in the CSB from this production run at a 95% confidence level.

PLAN: One-sample t interval for μ.

Random? Random sample of 8 measurements. ✓

Normal/Large Sample? The sample size is small, but the dotplot doesn't show any outliers or strong skewness. ✓

Amount of vitamin C (mg/100 g)

DO: For these data, $\bar{x} = 22.5$, $s_x = 7.191$, and $n = 8$. With 95% confidence and df $= 8 - 1 = 7$, $t^* = 2.365$.

$$22.5 \pm 2.365 \frac{7.191}{\sqrt{8}} \to 22.5 \pm 6.01 \to 16.49 \text{ to } 28.51$$

CONCLUDE: We are 95% confident that the interval from 16.49 mg/100 g to 28.51 mg/100 g captures the true mean amount of vitamin C in the CSB from this production run.

7. STATE: We want to estimate μ = the true mean wrist circumference for Canadian high school students at a 90% confidence level.

PLAN: One-sample t interval for μ.

Random? Random sample of 19 Canadian high school students. ✓

Normal/Large Sample? The sample size is small, but the dotplot doesn't show any outliers or strong skewness. ✓

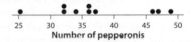

Wrist circumference (mm)

DO: For these data, $\bar{x} = 155.526$, $s_x = 19.993$, and $n = 19$. With 90% confidence and df $= 19 - 1 = 18$, $t^* = 1.734$.

$$155.526 \pm 1.734 \frac{19.993}{\sqrt{19}} \to 155.526 \pm 7.953 \to 147.573 \text{ to } 163.479$$

CONCLUDE: We are 90% confident that the interval from 147.573 mm to 163.479 mm captures the true mean wrist circumference for Canadian high school students.

9. STATE: We want to estimate μ = the true mean number of pepperonis on a large pizza at this restaurant at a 95% confidence level.

PLAN: One-sample t interval for μ.

Random? Random sample of 10 pizzas. ✓

Normal/Large Sample? The sample size is small, but the dotplot doesn't show any outliers or strong skewness. ✓

Number of pepperonis

DO: For these data, $\bar{x} = 37.4$, $s_x = 7.662$, and $n = 10$. With 95% confidence and df $= 10 - 1 = 9$, $t^* = 2.262$.

$$37.4 \pm 2.262 \frac{7.662}{\sqrt{10}} \to 37.4 \pm 5.481 \to 31.919 \text{ to } 42.881$$

CONCLUDE: We are 95% confident that the interval from 31.919 pepperonis to 42.881 pepperonis captures the true mean number of pepperonis on a large pizza at this restaurant.

11. No; the 40 pepperonis observation is captured within the confidence interval, so this is a plausible value for the average number of pepperonis. We don't have convincing evidence that the average number of pepperonis is less than 40.

13. $z^* \frac{\sigma}{\sqrt{n}} \leq ME \to 2.576 \frac{7.5}{\sqrt{n}} \leq 1$, so $n \geq 373.26$. Take an SRS of 374 women.

15. Four of the 8 values (only 50%) are within the 95% confidence interval 16.49 to 28.51 mg/100 g. This is because the confidence interval estimates the *mean* vitamin C level and does not tell us anything about *individual* values. Because means are less variable than individual values, we would expect that less than 95% of individual values would lie inside the 95% confidence interval for the mean.

17. (a) Because there is no leftover pattern in the residual plot, the least-squares regression line is an appropriate model. **(b)** The predicted tree height increases by 1.950 feet for each additional inch of diameter at breast height. **(c)** $\hat{y} = 65.34 + 1.950(23) = 110.19$ feet; *residual* $= y - \hat{y} = 120 - 110.19 = 9.81$ feet. **(d)** No, this would be an extrapolation, because a diameter of 15 inches is far outside the interval of x values used to obtain the regression line. This prediction may not be accurate.

Chapter 7 Review Exercises

1. (a) We are 95% confident that the interval from 0.295 to 0.365 captures the true proportion of all U.S. residents who support salaries for college athletes. **(b)** If the pollsters took many random samples of U.S. residents and constructed a 95% confidence interval for each sample, about 95% of these intervals would capture the true proportion of all U.S. residents who support salaries for college athletes. **(c)** point estimate $= \dfrac{0.295 + 0.365}{2} = 0.33$; margin of error $= 0.365 - 0.33 = 0.035$ or $0.33 - 0.295 = 0.035$. **(d)** The pollsters could decrease the confidence level. The drawback is that they can't be as confident that their interval will capture the true proportion. The pollsters could also increase the sample size. The drawback is that larger samples take more time and money to obtain.

2. (a) Because $\dfrac{1 - 0.94}{2} = 0.03$, z^* for a 94% confidence interval can be found by looking for a left-tail area of 0.03. The closest area in Table A is 0.0301, corresponding to a critical value of $z^* = 1.88$. *Using technology*: Applet/invNorm(left-tail area:0.03, mean:0, SD:1) gives $z^* = 1.881$. **(b)** 99% confidence and df $= 8 - 1 = 7$ give $t^* = 3.499$. **(c)** 90% confidence and df $= 95 - 1 = 94$ (use df $= 80$) give $t^* = 1.664$.

3. (a) Random? The principal asked a random sample of 180 students. ✓ **Large counts?** $n\hat{p} = 180\left(\dfrac{21}{180}\right) = 21 \geq 10$ and

$n(1 - \hat{p}) = 180\left(1 - \dfrac{21}{180}\right) = 159 \geq 10$. ✓

(b) $\hat{p} \pm z^*\sqrt{\dfrac{\hat{p}(1-\hat{p})}{n}} \rightarrow 0.117 \pm 1.960\sqrt{\dfrac{0.117(1-0.117)}{180}} \rightarrow 0.117$

$\pm\ 0.047 \rightarrow 0.070$ to 0.164. **(c)** No; 0.157 is a plausible value for the true proportion of all students at this school who have smoked in the last 30 days because it is captured in the confidence interval. There is not convincing evidence that the percentage of students at this school who have smoked in the last 30 days is less than 15.7%.

4. Response bias; for example, students might lie and answer "No" because they are being surveyed by the principal, and they don't want to get in trouble.

5. $2.576\sqrt{\dfrac{0.5(1 - 0.5)}{n}} \leq 0.05 \rightarrow n \geq 663.58$. We have to sample at least 664 students.

6. (a) Random? SRS of 60 students. ✓ **Normal/Large Sample?** We don't know the shape of the population distribution, but the sample size is large: $n = 60 \geq 30$. ✓ **(b)** 90% confidence and df $= 60 - 1 = 59$ (use df $= 50$) give $t^* = 1.676$.

$\bar{x} \pm t^*\dfrac{s_x}{\sqrt{n}} \rightarrow 114.98 \pm 1.676\dfrac{14.80}{\sqrt{60}} \rightarrow 114.98 \pm 3.20 \rightarrow 111.78$ to 118.18

(c) We are 90% confident that the interval from 111.78 to 118.18 captures the true mean IQ score of all students at the school.

7. STATE: We want to estimate $\mu =$ the true mean fuel efficiency for this model of pick-up truck at a 95% confidence level.
PLAN: One-sample t interval for μ.
Random? Randomly selected 20 owners. ✓
Normal/Large Sample? The sample size is small, but the dotplot doesn't show any outliers or strong skewness. ✓

Miles per gallon

DO: For these data, $\bar{x} = 18.48$, $s_x = 3.116$, and $n = 20$. With 95% confidence and df $= 20 - 1 = 19$, $t^* = 2.093$.

$18.48 \pm 2.093\dfrac{3.116}{\sqrt{20}} \rightarrow 18.48 \pm 1.458 \rightarrow 17.022$ to 19.938

CONCLUDE: We are 95% confident that the interval from 17.022 mpg to 19.938 mpg captures the true mean fuel efficiency for this model of pick-up truck.

Chapter 7 Practice Test

1. a
2. d
3. d
4. b
5. c
6. c
7. c
8. c
9. b
10. d

11. (a) STATE: We want to estimate $\mu =$ the true mean delivery time to deliver a pizza to the college dorms at a 90% confidence level.
PLAN: One-sample t interval for μ.
Random? Random sample of 15 dormitory residents. ✓
Normal/Large Sample? The sample size is small, but the dotplot doesn't show any outliers or strong skewness. ✓

Delivery times (min)

DO: For these data, $\bar{x} = 33.8$, $s_x = 7.72$, and $n = 15$. With 90% confidence and df $= 15 - 1 = 14$, $t^* = 1.761$.

$33.8 \pm 1.761\dfrac{7.72}{\sqrt{15}} \rightarrow 33.8 \pm 3.51 \rightarrow 30.29$ to 37.31

CONCLUDE: We are 90% confident that the interval from 30.29 minutes to 37.31 minutes captures the true mean delivery time to deliver a pizza to the college dorms. **(b)** Yes; all the values in the confidence interval are greater than 30 minutes, giving us convincing evidence that the true mean delivery time to deliver a pizza to the college dorms is greater than 30 minutes.

12. STATE: We want to estimate $p =$ the true proportion of all Americans who are not likely at all to make a New Year's resolution at a 95% confidence level.
PLAN: One-sample z interval for p.
Random? 1104 randomly selected Americans. ✓
Large counts? $1104\left(\dfrac{618}{1104}\right) = 618 \geq 10$ and $1104\left(1 - \dfrac{618}{1104}\right) = 486 \geq 10$. ✓

DO: $\hat{p} \pm z^*\sqrt{\dfrac{\hat{p}(1 - \hat{p})}{n}} = 0.56 \pm 1.960\sqrt{\dfrac{0.56(1 - 0.56)}{1104}} \rightarrow$

$0.56 \pm 0.029 \rightarrow 0.531$ to 0.589

CONCLUDE: We are 95% confident that the interval from 0.531 to 0.589 captures the true proportion of all Americans who are not likely at all to make a New Year's resolution. **(b)** If Marist College took many random samples of Americans and constructed a 95% confidence interval for each sample, about 95% of these intervals would capture the true proportion of all Americans who are not likely at all to make a New Year's resolution.

Chapter 8

Lesson 8.1 Exercises

1. $H_0: \mu = 31$
$H_a: \mu > 31$, where $\mu =$ the true mean weight (in ounces) of pineapples grown in the field this year.

3. $H_0: p = 0.72$
$H_a: p \neq 0.72$, where $p =$ the true proportion of all students at Yvonne's school who would say they seldom or never argue with friends.

5. Assuming that the true mean weight of pineapples grown in the field this year is 31 ounces, there is a 0.1317 probability of getting a sample mean of 31.4 ounces or greater just by chance in a random sample of 50 pineapples.
7. Assuming that the true proportion of all boards that would crack when dried with the new method is 0.16, there is a 0.027 probability of getting a sample proportion of 0.11 or less just by chance in a random sample of 200 boards.
9. The *P*-value of 0.1317 is large, so we fail to reject H_0. We don't have convincing evidence that the mean weight of the pineapples grown this year is greater than last year.
11. The *P*-value of 0.027 is small, so we reject H_0. We have convincing evidence that the new method of drying wood has a lower proportion of boards that crack.
13. (a) $H_0: \mu = 1$
$H_a: \mu < 1$, where μ = the true mean weight (in pounds) of bread loaves made by employees at the bakery. **(b)** Assuming that the true mean weight of bread loaves made by the bakery employees is 1 pound, there is a 0.0806 probability of getting a sample mean of 0.975 pound or less just by chance in a random sample of 50 bread loaves. **(c)** The *P*-value of 0.0806 is large, so we fail to reject H_0. We don't have convincing evidence that the mean weight of the bread loaves made by the bakery employees is less than 1 pound.
15. The student can't conclude that the null hypothesis is true just because he didn't find convincing evidence for the alternative hypothesis. Never "accept H_0" or conclude that H_0 is true.
17. (a) $H_0: \mu = 6.9$
$H_a: \mu > 6.9$, where μ = the true mean number of new words for all sonnets in the new manuscript. **(b)** Estimated *P*-value = 13/200 = 0.065. Assuming that the true mean number of new words for all sonnets in the new manuscript is 6.9, there is an estimated 0.065 probability of getting a sample mean of 9.2 or greater just by chance in a random sample of 5 sonnets. **(c)** The *P*-value of 0.065 is large, so we fail to reject H_0. We don't have convincing evidence that the new manuscript came from an author other than William Shakespeare.
19. (a) Skewed to the right, with a single peak in the 0 to <500 tornadoes interval and a gap from 4000 to 7500 tornadoes. The obvious outlier is Texas with 7990 tornadoes. **(b)** The median and *IQR* better summarize this distribution because it is skewed to the right and has a possible outlier, and these measures are resistant.

Lesson 8.2 Exercises

1. (a) Because the *P*-value of 0.0098 < $\alpha = 0.05$, we reject H_0. We have convincing evidence that the true mean amount of liquid in the company's bottles is less than 180 ml. **(b)** Because the *P*-value of 0.0098 < $\alpha = 0.01$, we reject H_0. We have convincing evidence that the true mean amount of liquid in the company's bottles is less than 180 ml.
3. (a) Because the *P*-value of 0.0291 < $\alpha = 0.05$, we reject H_0. We have convincing evidence that the true proportion of teens in Yvonne's school who would say they rarely or never argue with their friends differs from 0.72. **(b)** Because the *P*-value of 0.0291 $\geq$ $\alpha = 0.01$, we fail to reject H_0. We do not have convincing evidence that the true proportion of teens in Yvonne's school who would say they rarely or never argue with their friends differs from 0.72.
5. Type I error: You find convincing evidence that the true mean income level in this area is greater than $85,000, when it is really equal to $85,000. **Type II error:** You don't find convincing evidence that the true mean income level in this area is greater than $85,000, when it really is greater than $85,000.
7. Type I error: The city manager finds convincing evidence that the true proportion of calls involving life-threatening injuries for which emergency personnel took more than 8 minutes to arrive

during this 6-month period is less than 0.22, when it is really equal to 0.22. **Type II error:** The city manager doesn't find convincing evidence that the true proportion of calls involving life-threatening injuries for which emergency personnel took more than 8 minutes to arrive during this 6-month period is less than 0.22, when it really is less than 0.22.
9. Type I error: You decide to open a restaurant in this area, but there is not enough money in the area and the restaurant fails. **Type II error:** You decide not to open a restaurant in this area when it would have been successful.
11. (a) Type I error: The city manager concludes that the new guidelines were successful in saving more lives, when they really were not. **Type II error:** The city manager concludes that the new guidelines were not successful in saving more lives, when they really were. **(b) Type I error:** The risk for life-threatened patients has not improved; therefore, more lives are at risk.
13. (a) Because the *P*-value of 0.04 < $\alpha = 0.05$, we reject H_0. We have convincing evidence that the true mean lifetime of all the bulbs is less than 3 hours. **(b) Type I error:** We found convincing evidence that the true mean lifetime of all the bulbs is less than 3 hours (rejected H_0), when it could be equal to 3 hours. **(c)** The manufacturer decides that this batch of light bulbs does not meet its standard and destroys them, when in fact the bulbs did meet the requirements.
15. $\alpha = 0.01$ would be most appropriate, because it would decrease the probability of the more serious Type I error.
17. (a) For an observed result to be statistically significant at the 1% level, it must yield a *P*-value that is less than 1%. If the *P*-value is less than 1%, it must also be less than 5%; therefore, the result is also significant at the 5% level. **(b)** For a result to be statistically significant at the 5% level, it must yield a *P*-value that is less than 5%. This *P*-value may or may not be less than 1%; therefore, the result may or may not be significant at the 1% level.
19. (a) If the true proportion of patients experiencing nausea is 0.08, there is a 0.29 probability that this test will find convincing evidence that the true proportion is less than 0.10. **(b)** *P*(Type II error) = 1 − power = 1 − 0.29 = 0.71 **(c)** Increase the sample size (*n*) or increase the significance level (α).
21. (a) $(0.65)^3 = 0.275$ **(b)** $(0.65)^3 + (0.35)^3 = 0.275 + 0.043 = 0.318$ **(c)** $1 - (0.65)^4 = 1 - 0.179 = 0.821$

Lesson 8.3 Exercises

1. Random? Jason chooses an SRS of 60 students. ✓ **Large counts?** $np_0 = 60(0.80) = 48 \geq 10$ and $n(1-p_0) = 60(1-0.80) = 12 \geq 10$. ✓
3. Random? The data came from a voluntary response sample. This condition is not satisfied. **Large counts?** $np_0 = 200\left(\dfrac{2}{3}\right) = 133.33 \geq 10$ and $n(1-p_0) = 200\left(1-\dfrac{2}{3}\right) = 66.67 \geq 10$. ✓
5. Statistic: $\hat{p} = \dfrac{41}{60} = 0.6833$; $z = \dfrac{0.6833 - 0.80}{\sqrt{\dfrac{0.80(1-0.80)}{60}}} = -2.26$
7. Statistic: $\hat{p} = \dfrac{96}{100} = 0.96$; $z = \dfrac{0.96 - 0.90}{\sqrt{\dfrac{0.90(1-0.90)}{100}}} = 2.00$
9. *Using Table A:* $P(z \leq -2.26) = 0.0119$. *Using technology:* Applet/normalcdf(lower: −1000, upper: −2.26, mean:0, SD:1) = 0.0119.
11. *Using Table A:* $P(z \geq 2.00) = 1 - 0.9772 = 0.0228$. *Using technology:* Applet/normalcdf(lower:2.00, upper:1000, mean:0, SD:1) = 0.0228.
13. (a) Random? Zenon randomly selected 50 students. ✓ **Large counts?** $np_0 = 50(0.5) = 25 \geq 10$ and $n(1-p_0) = 50(1=0.50) = 25 \geq 10$. ✓

(b) Statistic: $\hat{p} = \dfrac{32}{50} = 0.64$; $z = \dfrac{0.64 - 0.50}{\sqrt{\dfrac{0.50(1 - 0.50)}{50}}} = 1.98$. (c)

Using Table A: $P(z \geq 1.98) = 1 - 0.9761 = 0.0239$. *Using technology:* Applet/normalcdf(lower:1.98, upper:1000, mean:0, SD:1) $= 0.0239$. (d) Because the P-value of $0.0239 < \alpha = 0.05$, we reject H_0. We have convincing evidence that a majority of students at Zenon's school prefer the name-brand chips.

15. (a) Because the P-value of $0.0116 < \alpha = 0.05$, we reject H_0. We have convincing evidence that more than 80% of students at Jason's high school have a computer at home. (b) Type I error; Jason found convincing evidence that the true proportion of students at his school who have a computer at home is greater than 0.80 (rejected H_0), but it is possible that the true proportion is really equal to 80% (H_0 is true).

17. This sample result doesn't give *any* evidence to support the alternative hypothesis: $H_a: p > 0.9$ because $\hat{p} = \dfrac{80}{100} = 0.80$ is *less than* 0.90.

19. (a) $\mu_{\bar{x}} = \mu = 14{,}000$ miles. (b) $\sigma_{\bar{x}} = \dfrac{\sigma}{\sqrt{n}} = \dfrac{4{,}000}{\sqrt{12}} = 1154.7$

miles. In SRSs of size $n = 12$, the sample mean number of miles driven per year will typically vary by about 1154.7 miles from the true mean of 14,000 miles. (c) Because $n = 12 < 30$, the sampling distribution of $\bar{x}$ will also be skewed to the right, but not quite as strongly as the population.

Lesson 8.4 Exercises

1. STATE: $H_0: p = 0.37$ vs. $H_a: p > 0.37$, where $p =$ the proportion of all the school's students who approve of the parking that is provided. PLAN: One-sample z test for p. Random? SRS. Large counts? $200(0.37) = 74 \geq 10$ and $200(1 - 0.37) = 126 \geq 10$. DO: $z = 1.32$, P-value $= 0.0934$. CONCLUDE: Because the P-value of $0.0934 \geq \alpha = 0.05$, we fail to reject H_0. We do not have convincing evidence that the proportion of all the school's students who would say they approve of the parking provided is $>37\%$.

3. STATE: $H_0: p = 0.20$ vs. $H_a: p > 0.20$, where $p =$ the true proportion of the school's students who have sent or posted suggestive images. PLAN: One-sample z test for p. Random? Random sample. Large counts? $250(0.20) = 50 \geq 10$ and $250(1 - 0.20) = 200 \geq 10$. DO: $z = 2.06$, P-value $= 0.0197$. CONCLUDE: Because the P-value of $0.0197 < \alpha = 0.05$, we reject H_0. We have convincing evidence that the true proportion of the school's students who have sent or posted suggestive images is $>20\%$.

5. $z = \dfrac{0.7375 - 0.7}{\sqrt{\dfrac{0.70(1 - 0.70)}{80}}} = 0.73$. P-value: *Using Table A:*

$P(z \geq 0.73 \text{ or } z \leq -0.73) = 2(0.2327) = 0.4654$. *Using technology:* Applet/normalcdf(lower:0.73, upper:1000, mean:0, SD:1) $\times 2$ $= 0.4654$.

7. $z = \dfrac{0.1333 - 0.25}{\sqrt{\dfrac{0.25(1 - 0.25)}{60}}} = -2.09$. P-value: *Using Table A:*

$P(z \leq -2.09 \text{ or } z \geq 2.09) = 2(0.0183) = 0.0366$. *Using technology:* Applet/normalcdf(lower: -1000, upper: -2.09, mean:0, SD:1) $\times 2 = 0.0366$.

9. STATE: $H_0: p = 0.60$ vs. $H_a: p \neq 0.60$, where $p =$ the proportion of all teens who pass their driving test on the first attempt. PLAN: One-sample z test for p. Random? SRS. Large counts? $125(0.60) = 75 \geq 10$ and $125(1 - 0.60) = 50 \geq 10$. DO: $z = 2.01$, P-value $= 0.0444$. CONCLUDE: Because the P-value of $0.0444 < \alpha = 0.05$, we reject H_0. We have convincing evidence

that the DMV's claim of a 60% pass rate for teens on their first driving test is incorrect.

11. STATE: $H_0: p = 0.75$ vs. $H_a: p \neq 0.75$, where $p =$ the true proportion of middle school students who engage in bullying behavior. PLAN: One-sample z test for p. Random? Random sample. Large counts? $558(0.75) = 418.5 \geq 10$ and $558(1 - 0.75) = 139.5 \geq 10$. DO: $z = 2.59$, P-value $= 0.0096$. CONCLUDE: Because the P-value of $0.0096 < \alpha = 0.05$, we reject H_0. We have convincing evidence that the media report's claim that 75% of middle school students engage in bullying behavior is incorrect.

13. (a) Type I error: We find convincing evidence that more than 20% of the company's customers are willing to buy the upgrade, when the true proportion is only 0.20. The company produces the upgrade and it is not profitable. Type II error: We don't find convincing evidence that more than 20% of the company's customers are willing to purchase the upgrade, when the true proportion really is greater than 0.20. The company doesn't produce the upgrade and misses out on the potential profit. The Type I error is more serious because the company is losing money. (b) STATE: $H_0: p = 0.20$ vs. $H_a: p > 0.20$, where $p =$ the proportion of all the company's customers that would be willing to buy the upgrade. PLAN: One-sample z test for p. Random? Random sample. Large counts? $60(0.20) = 12 \geq 10$ and $60(1 - 0.20) = 48 \geq 10$. DO: $z = 1.29$, P-value $= 0.0985$. CONCLUDE: Because the P-value of $0.0985 \geq \alpha = 0.05$, we fail to reject H_0. We do not have convincing evidence that $>20\%$ of the company's customers are willing to buy the upgrade.

15. STATE: We want to estimate $p =$ the true proportion of all teens in the state who passed their driving test on the first attempt at a 95% confidence level. PLAN: One-sample z interval for p. Random? SRS.

Large counts? $125\left(\dfrac{86}{125}\right) = 86 \geq 10$ and $125\left(1 - \dfrac{86}{125}\right) = 39$

≥ 10. DO: $0.688 \pm 1.960\sqrt{\dfrac{0.688(1 - 0.688)}{125}} \rightarrow 0.688$

$\pm 0.081 \rightarrow 0.607$ to 0.769. CONCLUDE: We are 95% confident that the interval from 0.607 to 0.769 captures the true proportion of all teens in the state who passed their driving test on the first attempt. (b) All the values in the confidence interval are more than 0.60, so it is not plausible that 60% of teens pass the driving test on the first attempt. This gives us convincing evidence against the DMV's claim.

17. No; (0.56, 0.62). Because 0.60 is contained in the interval, it is a plausible value and we do not have convincing evidence that the proportion of U.S. adults who would say they want to lose weight differs from 0.60.

19. (a) $H_0: p = 0.90$ vs. $H_a: p < 0.90$, where $p =$ the true proportion of times this model of mower will start on many pushes of the button. (b) Large counts? $20(0.9) = 18 \geq 10$ and $20(1 - 0.90)$ $= 2 < 10$. This condition is not met. (c) Using technology, $P(X \leq 15) = 0.043$. (d) Because the P-value of $0.043 < \alpha = 0.05$, we reject H_0. We have convincing evidence that the probability that this mower will start on any push of the button is less than the company's claim of 0.90.

21. STATE: We want to estimate $\mu =$ the true mean file size of photographs taken with Rafiq's camera at a 99% confidence level. PLAN: One-sample t interval for μ. Random? Random sample. Normal/Large Sample? Sample size is large: $n = 49 \geq 30$.

DO: $8.05 \pm 2.704 \dfrac{1.96}{\sqrt{49}} \rightarrow 8.05 \pm 0.757 \rightarrow 7.293$ to 8.807.

CONCLUDE: We are 99% confident that the interval from 7.293 megabytes to 8.807 megabytes captures the true mean file size of photographs taken with Rafiq's camera.

Lesson 8.5 Exercises

1. Random? Random sample of 24 adults. ✓ **Normal/Large Sample?** The sample size is small ($n < 30$), but the dotplot doesn't show any outliers or strong skewness. ✓

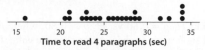

Time to read 4 paragraphs (sec)

3. Random? Random sample of 50 pineapples. ✓ **Normal/Large Sample?** We don't know the shape of the population distribution, but the sample size is large: $n = 50 \geq 30$. ✓

5. $t = \dfrac{26.496 - 22}{\dfrac{4.728}{\sqrt{24}}} = 4.659$

7. $t = \dfrac{5.31 - 5}{\dfrac{0.79}{\sqrt{20}}} = 1.755$

9. df $= 24 - 1 = 23$; $t = 4.659$; P-value less than 0.0005.
11. df $= 20 - 1 = 19$; $t = 1.755$; P-value between $2(0.025) = 0.05$ and $2(0.05) = 0.10$.
13. (a) Random? Random sample of 10 gas stations in the city. ✓ **Normal/Large Sample?** The sample size is small, but the dotplot doesn't show any outliers or strong skewness. ✓

Gas prices ($)

(b) $t = \dfrac{3.038 - 3.06}{\dfrac{0.053}{\sqrt{10}}} = -1.313$. **(c)** df $= 10 - 1 = 9$; $t = -1.313$;

P-value between $2(0.10) = 0.20$ and $2(0.15) = 0.30$. **(d)** Because the P-value of between 0.20 and $0.30 \geq \alpha = 0.05$, we fail to reject H_0. We do not have convincing evidence that the true mean price of gas in Anne's city is different from that in the rest of her state.
15. The Random condition is not satisfied because the marketing consultant has used a convenience sample.

17. (a) $z = \dfrac{29 - 22.3}{5} = 1.34$; $P(X < 29) = P(Z < 1.34) = 0.9099$

(b) $1.28 = \dfrac{x - 22.3}{5} \rightarrow x = 28.7$ mpg

Lesson 8.6 Exercises

1. STATE: $H_0: \mu = 5$ vs. $H_a: \mu < 5$, where μ = the true mean dissolved oxygen level (mg/l) in this stream. **PLAN:** One-sample t test for a mean. **Random?** Random sample. **Normal/Large Sample?** The sample size is small, but the dotplot doesn't show any outliers or strong skewness.

Dissolved oxygen level (mg/l)

DO: $t = -0.943$, P-value $= 0.181$. **CONCLUDE:** Because the P-value of $0.181 \geq \alpha = 0.10$, we fail to reject H_0. There is not convincing evidence that aquatic life in this stream is at risk.
3. STATE: $H_0: \mu = 158$ vs. $H_a: \mu \neq 158$, where μ = the true mean amount of money spent ($) per U.S. household per week on food in a small city. **PLAN:** One-sample t test for a mean. **Random?** Random sample. **Normal/Large Sample?** We don't know the shape of the population distribution, but the sample size is large: $n = 50 \geq 30$. **DO:** $t = 2.475$, P-value $= 0.017$. **CONCLUDE:** Because the P-value of $0.017 < \alpha = 0.05$, we reject H_0. There is convincing evidence that the mean weekly spending on food in this city differs from the national figure of $158.

5. STATE: $H_0: \mu = 128$ vs. $H_a: \mu > 128$, where μ = the mean score increase (points) for all students who would use the smartphone app in addition to the regular program. **PLAN:** One-sample t test for a mean. **Random?** Random sample. **Normal/Large Sample?** We don't know the shape of the population distribution, but the sample size is large: $n = 5000 \geq 30$. **DO:** $t = 2.178$, P-value $= 0.015$. **CONCLUDE:** Because the P-value of $0.015 < \alpha = 0.05$, we reject H_0. There is convincing evidence that the true average score increase for students who use the smartphone app in addition to the regular program is >128 points.
7. Because the null value of 105 is captured by the 90% confidence interval, we would fail to reject $H_0: \mu = 105$ at the $\alpha = 0.10$ level. There is not convincing evidence that the true mean reading in the population of radon detectors is different from the actual value of 105 picocuries per liter.
9. Because the null value of 200 is not captured by the 95% confidence interval, we would reject $H_0: \mu = 200$ at the $\alpha = 0.05$ level. There is convincing evidence that the true mean server response time is different from the site's claim of 200 milliseconds.
11. (a) STATE: $H_0: \mu = 0$ vs. $H_a: \mu > 0$, where μ = the mean number of additional hours of sleep per night gained by using the drug for all people who would take it. **PLAN:** One-sample t test for a mean. **Random?** Random sample. **Normal/Large Sample?** The sample size is small, but the dotplot doesn't show any outliers or strong skewness.

Number of additional hours of sleep

DO: $t = 3.680$, P-value $= 0.00254$. **CONCLUDE:** Because the P-value of $0.00254 < \alpha = 0.05$, we reject H_0. There is convincing evidence that the drug is effective at increasing the average sleep time for patients who suffer from insomnia. **(b) Type I error:** The test finds convincing evidence that the drug is effective, when it really is not effective.
13. (a) Yes; because the P-value of $0.06 \geq \alpha = 0.05$, we fail to reject H_0. This means that the null value is plausible and, therefore, must be contained in the 95% confidence interval. **(b)** No; because the P-value of $0.06 < \alpha = 0.10$, we reject H_0. This means that the null value is not plausible and, therefore, must not be contained in the 90% confidence interval.
15. The sample mean from the data was $\bar{x} = 130$, which is only 2 points higher than that for students who didn't use the smartphone app. Because the increase in scores is so small, it is not practically important.
17. (a) 1.729, 1.729. **(b)** At df $= 20 - 1 = 19$ and an upper tail probability of 0.05, $t^* = 1.729$. **(c)** If we reject H_0, we know from part **(a)** that the sample mean x is more than 1.729 standard errors greater than $\mu = 10$. If x is more than 1.729 standard errors away from the mean, we know from part **(b)** that the 90% confidence interval cannot contain $\mu = 10$.
19. Probably not; an experiment is needed to show cause and effect.

Chapter 8 Review Exercises

1. (a) $\begin{aligned} H_0: \mu &= 64.2 \\ H_a: \mu &\neq 64.2 \end{aligned}$, where μ = the true mean height (in inches) of this year's female graduates from a large local high school.
(b) $\begin{aligned} H_0: p &= 0.47 \\ H_a: p &> 0.47 \end{aligned}$, where p = the true proportion of households of students at your school that own at least one dog.
2. (a) $z = 1.83$. *Using Table A:* $P(z \geq 1.83) = 1 - 0.9664 = 0.0336$. *Using technology:* Applet/normalcdf(lower:1.83, upper:1000, mean:0, SD:1) $= 0.0336$. **(b)** $t = -2.35$. *Using Table B:* Using df $= 20 - 1 = 19$, the P-value is between $2(0.01) = 0.02$ and $2(0.02) = 0.04$.

3. (a) $H_0{:}p = 0.20$
$H_a{:}p > 0.20$, where p = the true proportion of all students

at Monica and Shivani's school who can correctly identify which flavor of fruit snack they are eating. (b) **Random?** 50 randomly selected students at their school. ✓ **Large counts?** $np_0 = 50(0.20) = 10 \geq 10$ and $n(1 - p_0) = 50(1 - 0.20) = 40 \geq 10.$ ✓ (c) Assuming that the true proportion of all students at Monica and Shivani's school who can correctly identify which flavor of fruit snack they are eating is 0.20, there is a 0.0067 probability of getting a sample proportion of $\frac{17}{50} = 0.34$ or greater who make correct identifications just by chance in a random sample of 50 students. (d) Because the P-value of $0.0067 < \alpha = 0.01$, we reject H_0. We have convincing evidence that students at Monica and Shivani's school can identify which flavor of fruit snack they are eating better than by random guessing.

4. STATE: $H_0{:}p = 0.147$ vs. $H_a{:}p \neq 0.147$, where p = the true proportion of Carrboro residents who are underemployed. **PLAN:** One-sample z test for p. **Random?** Random sample. **Large counts?** $300(0.147) = 44.1 \geq 10$ and $300(1 - 0.147) = 255.9 \geq 10$. **DO:** $z = 1.61$, P-value = 0.1065. **CONCLUDE:** Because the P-value of $0.1065 \geq \alpha = 0.05$, we fail to reject H_0. We do not have convincing evidence that the proportion of Carrboro residents who are underemployed is different from the national proportion of 0.147.

5. (a) $H_0{:}\mu = 300$
$H_a{:}\mu < 300$, where μ = the true mean breaking strength (in

pounds) of the manufacturer's classroom chairs for high school students. (b) **Type I error:** The test finds convincing evidence that the true mean breaking strength (in pounds) of the manufacturer's classroom chairs is less than 300, when it is really equal to 300. **Type II error:** The test doesn't find convincing evidence that the true mean breaking strength (in pounds) of the manufacturer's classroom chairs is less than 300, when it really is less than 300. (c) **Type I error:** You send all of the chairs back to the manufacturer, even though the chairs are as advertised. This decision would likely upset the manufacturer. **Type II error:** You keep the chairs, even though the chairs are not as advertised. More students have chairs that collapse beneath them. The Type II error is more serious because we are risking injuries to students from falling due to broken chairs.

6. STATE: $H_0{:}\mu = 1.10$ vs. $H_a{:}\mu > 1.10$, where μ = the true mean conductivity $\left(\dfrac{\text{watts}}{\text{m}^2 \cdot {}^{\circ}\text{C}}\right)$ of this type of glass. **PLAN:** One-sample t test for a mean. **Random?** Random sample. **Normal/Large Sample?** The sample size is small, but the dotplot doesn't show any outliers or strong skewness. ✓

Conductivity

DO: $t = 1.35$, P-value = 0.103. **CONCLUDE:** Because the P-value of $0.103 \geq \alpha = 0.05$, we fail to reject H_0. There is not convincing evidence that the true mean conductivity of this type of glass is greater than $1.10 \dfrac{\text{watts}}{\text{m}^2 \cdot {}^{\circ}\text{C}}$.

7. $38.8 \pm 0.75 \rightarrow (38.05, 39.55)$; because the null value of 40 is not captured by the 90% confidence interval, we have convincing evidence that the students are not able to accurately guess the teacher's age.

Chapter 8 Practice Test

1. d
2. c
3. c
4. b

5. b
6. a
7. d
8. d
9. b
10. b

11. (a) **STATE:** $H_0{:}p = 0.60$ vs. $H_a{:}p > 0.60$, where p = the proportion of all signatures on the petition that are valid. **PLAN:** One-sample z test for p. **Random?** Random sample. **Large counts?** $300(0.60) = 180 \geq 10$ and $300(1 - 0.60) = 120 \geq 10$. **DO:** $z = 2.47$, P-value = 0.0068. **CONCLUDE:** Because the P-value of $0.0068 < \alpha = 0.01$, we reject H_0. We have convincing evidence that the proportion of all signatures on the petition that are valid is greater than 60%. (b) Assuming that the true proportion of signatures on the petition that are valid is 0.60, there is a 0.0068 probability of getting a sample proportion of 0.67 or greater just by chance in a random sample of 300 signatures.

12. STATE: $H_0{:}\mu = 341,500$ vs. $H_a{:}\mu \neq 341,500$, where μ = the true mean price (\$) for a home in Portland, Oregon, in the spring of 2015. **PLAN:** One-sample t test for a mean. **Random?** Random sample. **Normal/Large Sample?** We don't know the shape of the population distribution, but the sample size is large: $n = 36 \geq 30$. **DO:** $t = 2.550$, P-value = 0.0153. **CONCLUDE:** Because the P-value of $0.0153 < \alpha = 0.05$, we reject H_0. There is convincing evidence that the true mean price (\$) for a home in Portland, Oregon, in the spring of 2015 differs from the national average of \$341,500.

13. (a) **Type I error:** You find convincing evidence that the true mean amount of gravel per truckload is less than 20 m³, when it is really equal to 20 m³. You tell the company they lied about their claim, even though they delivered the appropriate amount of gravel. This decision would likely upset the gravel company. (b) **Type II error:** You don't find convincing evidence that the true mean amount of gravel per truckload is less than 20 m³, when it is really less than 20 m³. You don't say anything to the company, even though they did not deliver the appropriate amount of gravel. We won't have enough gravel to complete the road.

Chapter 9

Lesson 9.1 Exercises

1. (a) Approximately normal, because $n_1p_1 = 50(0.30) = 15$, $n_1(1 - p_1) = 50(0.70) = 35$, $n_2p_2 = 100(0.15) = 15$, $n_2(1 - p_2) = 100(0.85) = 85$ all ≥ 10. (b) $\mu_{\hat{p}_1 - \hat{p}_2} = 0.30 - 0.15 = 0.15$. (c) $\sigma_{\hat{p}_1 - \hat{p}_2} = \sqrt{\dfrac{0.30(0.70)}{50} + \dfrac{(0.15)(0.85)}{100}} = 0.0740$

3. (a) Approximately normal, because $n_1p_1 = 80(0.60) = 48$, $n_1(1 - p_1) = 80(0.40) = 32$, $n_2p_2 = 80(0.48) = 38.4$, $n_2(1 - p_2) = 80(0.52) = 41.6$ all ≥ 10. (b) $\mu_{\hat{p}_1 - \hat{p}_2} = 0.60 - 0.48 = 0.12$. (c) $\sigma_{\hat{p}_1 - \hat{p}_2} = \sqrt{\dfrac{0.60(0.40)}{80} + \dfrac{(0.48)(0.52)}{80}} = 0.0782$

5. Random? The data represent the whole population of babies born on the east and west side and are not a random sample, so this condition is not met. **Large Counts?** $414\left(\dfrac{16}{414}\right) = 16$, $414\left(\dfrac{398}{414}\right) = 398$, $228\left(\dfrac{3}{228}\right) = 3$, $228\left(\dfrac{225}{228}\right) = 225$ are not all ≥ 10, so this condition is not met.

7. Random? The data come from two groups in a randomized experiment. ✓ **Large Counts?** $50\left(\dfrac{11}{50}\right) = 11$, $50\left(\dfrac{39}{50}\right) = 39$, $44\left(\dfrac{21}{44}\right) = 21$, $44\left(\dfrac{23}{44}\right) = 23$ all ≥ 10. ✓

9. STATE: We want to estimate $p_1 - p_2$ at the 90% confidence level, where p_1 = the true proportion of Internet users aged 18–29 who use Twitter and p_2 = the true proportion of Internet users aged 30–49 who use Twitter. **PLAN:** Two-sample z interval for $p_1 - p_2$. **Random?** The data come from separate random samples of younger and older Internet users. ✓ **Large Counts?** $316(0.32) = 101.12$, $316(0.74) = 234$, $532(0.13) = 69.16$, $532(0.86) = 458$ all ≥ 10. ✓ **DO:**

$$(0.32 - 0.13) \pm 1.645 \sqrt{\frac{0.32(1 - 0.32)}{316} + \frac{0.13(1 - 0.13)}{532}} \to 0.141$$

to 0.239. **CONCLUDE:** We are 90% confident that the interval from 0.141 to 0.239 captures the difference between the true proportions of adult Internet users in these two age groups who use Twitter. The interval suggests that the percent of Internet users aged 18–29 who use Twitter is between 14.1 and 23.9 percentage points higher than the percent for people aged 30–49.

11. STATE: We want to estimate $p_1 - p_2$ at the 99% confidence level, where p_1 = the true proportion of male young adults aged 19–25 who live in their parents' home and p_2 = the true proportion of female young adults aged 19–25 who live in their parents' home. **PLAN:** Two-sample z interval for $p_1 - p_2$. **Random?** The data come from separate random samples of male and female young adults.

✓ **Large Counts?** $2253\left(\frac{986}{2253}\right) = 986$, $2253\left(\frac{1267}{2253}\right) = 1267$, $2629\left(\frac{923}{2629}\right) = 923$, $2629\left(\frac{1706}{2629}\right) = 1706$ all ≥ 10. ✓ **DO:**

$$(0.438 - 0.351) \pm 2.576 \sqrt{\frac{0.438(1 - 0.438)}{2253} + \frac{0.351(1 - 0.351)}{2629}}$$

$\to$ 0.051 to 0.123. **CONCLUDE:** We are 99% confident that the interval from 0.051 to 0.123 captures the difference in the true proportions of men and women aged 19–25 who live in their parents' homes. The interval suggests that the percent of male young adults who live in their parents' home is between 5.1 and 12.3 percentage points higher than the percent for female young adults.

13. (a) STATE: We want to estimate $p_1 - p_2$ at the 95% confidence level, where p_1 = the true proportion of men in the United States who can identify Egypt on a map and p_2 = the true proportion of women in the United States who can identify Egypt on a map. **PLAN:** Two-sample z interval for $p_1 - p_2$. **Random?** The data come from separate random samples of males and females in the United States. ✓ **Large Counts?** $522\left(\frac{287}{522}\right) = 287$, $522\left(\frac{235}{522}\right) = 235$, $520\left(\frac{233}{520}\right) = 233$, $520\left(\frac{287}{520}\right) = 287$ all ≥ 10. ✓ **DO:** $(0.550 - 0.448) \pm 1.960$

$$\sqrt{\frac{0.550(1 - 0.550)}{522} + \frac{0.448(1 - 0.448)}{520}} \to \quad 0.042 \quad \text{to} \quad 0.162.$$

CONCLUDE: We are 95% confident that the interval from 0.042 to 0.162 captures the difference in the proportion of U.S. men and the proportion of U.S. women who can identify Egypt on a map. The interval suggests that the percent of U.S. men who can identify Egypt on a map is between 4.2 and 16.2 percentage points higher than the percent for U.S. women. **(b)** Yes, because our interval does not capture a difference of 0 (no difference) as a plausible value for $p_1 - p_2$.

15. No; because 0 is captured in the interval, we think it is *plausible* that the proportions are equal, but this does not provide *convincing evidence* that the two proportions are equal.

17. From Exercise 1, we know that the sampling distribution of $\hat{p}_1 - \hat{p}_2$ is approximately normal with $\mu_{\hat{p}_1 - \hat{p}_2} = 0.15$ and $\sigma_{\hat{p}_1 - \hat{p}_2} = 0.0740$; $z = \dfrac{0 - 0.15}{0.074} = -2.03$. $P(\hat{p}_1 - \hat{p}_2 > 0) = P(Z > -2.03) = 1 - 0.0212 = 0.9788$. *Using technology:* Applet/normalcdf(lower:0, upper:1000, mean:0.15, SD:0.074) = 0.9787.

19. (a) To provide a baseline for comparing the effects of the treatment. **(b)** Label each subject with a different integer from 1 to 169. Then randomly generate 88 *different* integers from 1 to 169.

The subjects with these labels will be in the intercessory prayer group. The remaining 81 subjects will be in the control group. **(c)** To create roughly equivalent groups at the beginning of the experiment. This way, if there is a large difference in the results between the two groups, we have good evidence that the treatment caused the difference.

Lesson 9.2 Exercises

1. (a) $\begin{aligned}H_0{:}p_1 - p_2 &= 0 \\ H_a{:}p_1 - p_2 &< 0\end{aligned}$, where p_1 = true proportion of 4- to 5-year-olds who can sort correctly and p_2 = true proportion of 6- to 7-year-olds who can sort correctly. **(b) Random?** The data come from two separate random samples from the population of 4- to 5-year-olds and 6- to 7-year-olds. ✓ **Large Counts?** 10, $50 - 10 = 40$, 28, and $53 - 28 = 25$ all ≥ 10. ✓

3. (a) $\begin{aligned}H_0{:}p_1 - p_2 &= 0 \\ H_a{:}p_1 - p_2 &\neq 0\end{aligned}$, where p_1 = true proportion of people like those in the study who would say they prefer Chain C if C was mentioned first and p_2 = true proportion of people like those in the study who would say they prefer Chain C if Q was mentioned first. **(b) Random?** The data come from two groups in a randomized experiment. ✓ **Large Counts?** 31, $50 - 31 = 19$, 25, and $50 - 25 = 25$ all ≥ 10. ✓

5. (a) $z = \dfrac{(0.2000 - 0.5283) - 0}{\sqrt{\dfrac{0.3689(1 - 0.3689)}{50} + \dfrac{0.3689(1 - 0.3689)}{53}}} = -3.45$

(b) *P-value:* Using Table A: $P(z < -3.45) = 0.0003$. *Using technology:* Applet/normalcdf(lower: -1000, upper: -3.45, mean: 0, SD: 1) = 0.0003.

7. (a) $z = \dfrac{(0.62 - 0.50) - 0}{\sqrt{\dfrac{0.56(1 - 0.56)}{50} + \dfrac{0.56(1 - 0.56)}{50}}} = 1.21$. **(b)** *P-value:*

Using Table A: $P(z \leq -1.21)$ or $P(z \geq 1.21) = 2(0.1131) = 0.2262$. *Using technology:* Applet/normalcdf(lower: 1.21, upper: 1000, mean: 0, SD: 1)*2 = 0.2262.

9. STATE: $H_0{:}p_1 - p_2 = 0$ vs. $H_a{:}p_1 - p_2 > 0$, where p_1 = true proportion of students like these who would pass the state exam if they took the course with Instructor A and p_2 = true proportion of students like these who would pass the state exam if they took the course with Instructor B. **PLAN:** Two-sample z test for $p_1 - p_2$. **Random?** Two groups in a randomized experiment. ✓ **Large Counts?** 30, $50 - 30 = 20$, 22, and $50 - 22 = 28$ all ≥ 10. ✓ **DO:** $z = 1.60$, P-value = 0.0548. **CONCLUDE:** Because the P-value of 0.0548 is greater than $\alpha = 0.05$, we fail to reject H_0. We do not have convincing evidence that Instructor A is more effective.

11. STATE: $H_0{:}p_1 - p_2 = 0$ vs. $H_a{:}p_1 - p_2 < 0$, where p_1 = true proportion of teenagers in 1990 who showed some hearing loss and p_2 = true proportion of teenagers in 2010 who showed some hearing loss. **PLAN:** Two-sample z test for $p_1 - p_2$. **Random?** Two separate random samples. ✓ **Large Counts?** 450, $3000 - 450 = 2550$, 351, and $1800 - 351 = 1449$ all ≥ 10. ✓ **DO:** $z = -4.05$, P-value ≈ 0. **CONCLUDE:** Because the P-value of approximately 0 is less than $\alpha = 0.01$, we reject H_0. We have convincing evidence that the proportion of all teens with hearing loss has increased from 1990 to 2010.

13. (a) STATE: $H_0{:}p_1 - p_2 = 0$ vs. $H_a{:}p_1 - p_2 \neq 0$, where p_1 = true proportion of subjects like these who would die, have a heart attack, or suffer other serious consequences within two years if they took Pravachol and p_2 = true proportion of subjects like these who would die, have a heart attack, or suffer other serious consequences within two years if they took Lipitor. **PLAN:** Two-sample z test for $p_1 - p_2$. **Random?** Two groups in a randomized experiment. ✓ **Large Counts?** $2063(0.263) = 543$, $2063 - 543 = 1520$, $2099(0.224) = 470$, and $2099 - 470 = 1629$ all ≥ 10. ✓ **DO:** $z = 2.95$, P-value = 0.0032.

CONCLUDE: Because the P-value of 0.0032 is less than $\alpha = 0.05$, we reject H_0. We have convincing evidence that there is a difference in the effectiveness of Pravachol and Lipitor for people like those in the study. **(b) Type I error:** The experiment finds convincing evidence that there is a difference in the effectiveness of Pravachol and Lipitor when they are equally effective. **(c)** No; this experiment used recruited volunteers as subjects. When subjects are not randomly selected, we should not generalize the results of an experiment to some larger population of interest.

15. (a) See Exercise 13 for the definition of parameters and condition check. **STATE:** We want to estimate $p_1 - p_2$ at the 95% confidence level. **PLAN:** Two-sample z interval for $p_1 - p_2$. **DO:**
$$(0.263 - 0.224) \pm 1.960\sqrt{\frac{0.263(1 - 0.263)}{2063} + \frac{0.224(1 - 0.224)}{2099}}$$
$\rightarrow$ 0.013 to 0.065. **CONCLUDE:** We are 95% confident that the interval from 0.013 to 0.065 captures the true difference between the proportions of subjects like these who would die, have a heart attack, or suffer other serious consequences within two years if they took Pravachol versus if they took Lipitor. The interval suggests that the percent of those who use Pravachol who would die, have a heart attack, or suffer other serious consequences is between 1.3 and 6.5 percentage points higher than the percent for those who use Lipitor. **(b)** The confidence interval provides more information than the significance test—namely, a set of plausible values for the true difference in proportions, $p_1 - p_2$.

17. (a) $H_0 : p_1 - p_2 = 0$ vs. $H_a : p_1 - p_2 \neq 0$, where $p_1 =$ true proportion of subjects like these who would stop at the rest area when talking on a cell phone and $p_2 =$ true proportion of subjects like these who would stop at the rest area when talking to a passenger. **(b)** The Large Counts condition is not satisfied. The number of subjects who didn't stop when talking to a passenger (3) is less than 10. **(c)** Estimated P-value = 3/100 = 0.03. **(d)** Because the estimated P-value of 0.03 is less than $\alpha = 0.05$, we reject H_0. We have convincing evidence that there is a difference between the proportions of subjects like these who would stop at the rest area when talking on a cell phone versus when talking to a passenger.

19. (a) About 40% of the years (more than ¼ and less than ½ of the pie, and looks as though it is closer to ½ than it is to ¼). **(b)** The pictograph makes it seem that the number of times General Motors has won the award is much larger than for the other companies, which isn't the case.

Lesson 9.3 Exercises

1. (a) Approximately normal, because both population distributions are approximately normal. **(b)** $\mu_{\bar{x}_1 - \bar{x}_2} = 27 - 17 = 10$ ounces.

(c) $\sigma_{\bar{x}_1 - \bar{x}_2} = \sqrt{\dfrac{0.8^2}{20} + \dfrac{0.5^2}{25}} = 0.205$ ounces

3. (a) No; the sample size for dogs ($n_1 = 10$) is not large enough for the central limit theorem to apply, and the population distribution of dog weights is skewed. **(b)** $\mu_{\bar{x}_1 - \bar{x}_2} = 28 - 9.5 = 18.5$ pounds.

(c) $\sigma_{\bar{x}_1 - \bar{x}_2} = \sqrt{\dfrac{14^2}{10} + \dfrac{2^2}{8}} = 4.48$ pounds.

5. Random? Separate random samples of males and females at the school. ✓ **Normal/Large Sample?** The sample sizes are small and the boxplot for males shows outliers, so this condition is not satisfied.

7. Random? The sample of words was not a random sample, so this condition is not satisfied. **Normal/Large Sample?** Both sample sizes are large ($n_1 = 400 \geq 30$ and $n_2 = 100 \geq 30$). ✓

9. STATE: We want to estimate $\mu_1 - \mu_2$ at the 95% confidence level, where $\mu_1 =$ the true mean weight of black squirrels (ounces) in the large forest and $\mu_2 =$ the true mean weight of gray squirrels (ounces) in the large forest. **PLAN:** Two-sample t interval for

$\mu_1 - \mu_2$. **Random?** Separate random samples of black and gray squirrels. ✓ **Normal/Large Sample?** Both sample sizes are large; $n_1 = 40 \geq 30$ and $n_2 = 40 \geq 30$. ✓ **DO:** $(20.3 - 19.2) \pm 2.042$
$\sqrt{\dfrac{(2.1)^2}{40} + \dfrac{(1.9)^2}{40}} \rightarrow$ 0.19 to 2.01 ounces with df = 39 (use df = 30).
Using technology: 0.21 to 1.99 with df = 77.23. **CONCLUDE:** We are 95% confident that the interval from 0.19 to 2.01 ounces captures the true difference in mean weight of black squirrels and gray squirrels in this forest. The interval suggests that the mean weight for black squirrels is between 0.19 ounces and 2.01 ounces heavier than the mean weight for the gray squirrels.

11. STATE: We want to estimate $\mu_1 - \mu_2$ at the 90% confidence level, where $\mu_1 =$ the true mean percent change in polyphenol levels for healthy men like these who drink red wine and $\mu_2 =$ the true mean percent change in polyphenol levels for healthy men like these who drink white wine. **PLAN:** Two-sample t interval for $\mu_1 - \mu_2$. **Random?** The data come from two groups in a randomized experiment. ✓ **Normal/Large Sample?** The sample sizes are small, but the boxplots don't show any outliers or strong skewness. ✓

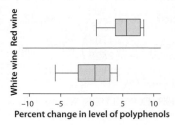

DO: $(5.50 - 0.23) \pm 1.860\sqrt{\dfrac{(2.52)^2}{9} + \dfrac{(3.29)^2}{9}} \rightarrow$ 2.70 to 7.84 with df = 8. *Using technology:* 2.85 to 7.69 with df = 14.97. **CONCLUDE:** We are 90% confident that the interval from 2.70 to 7.84 captures the true difference in mean percent change in polyphenol levels for healthy men like these who drink red wine versus white wine. The interval suggests that the mean percent change for healthy men like these who drink red wine is between 2.70% and 7.84% higher than the mean percent change for subjects like these who drink white wine.

13. (a) STATE: We want to estimate $\mu_1 - \mu_2$ at the 95% confidence level, where $\mu_1 =$ the true mean amount of time 4-year-old kids like these can delay gratification after watching slower-paced shows and $\mu_2 =$ the true mean amount of time 4-year-old kids like these can delay gratification after watching fast-paced shows. **PLAN:** Two-sample t interval for $\mu_1 - \mu_2$. **Random?** The data come from two groups in a randomized experiment. ✓ **Normal/Large Sample?** The sample sizes are small, but graphs of the data for both groups revealed no strong skewness or outliers. ✓ **DO:**
$$(257.20 - 146.15) \pm 2.093\sqrt{\frac{(132.16)^2}{20} + \frac{(151.29)^2}{20}} \rightarrow 17.03 \text{ to}$$
205.07 seconds with df = 19. *Using technology:* 20.06 to 202.04 with df = 37.33. **CONCLUDE:** We are 95% confident that the interval from 17.03 seconds to 205.07 seconds captures the difference in the true mean amount of time 4-year-old kids like these can delay gratification after watching slower-paced versus fast-paced shows. The interval suggests that the mean amount of time 4-year-old kids like these can delay gratification after watching slower-paced shows is between 17.03 and 205.07 seconds longer than the mean amount of time 4-year-old kids like these can delay gratification after watching fast-paced shows. **(b)** Yes, because our interval does not include a difference of 0 (no difference) as a plausible value. **(c)** Random assignment to treatments allows the researchers to make conclusions about cause and effect.

15. (a) No, because our interval includes a difference of 0 (no difference) as a plausible value. **(b)** No, we do not have convincing evidence that the mean egg masses of birds with small nests and birds with large nests are the same. Instead, we *don't have convincing evidence* that the mean egg masses *are different*.

17. From Exercise 1, we know $\mu_{\bar{x}_1-\bar{x}_2} = 0.7$ inch and $\sigma_{\bar{x}_1-\bar{x}_2} = 1.553$ inches. $z = \dfrac{0 - 0.7}{1.553} = -0.45;$ $P(\bar{x}_1 - \bar{x}_2 > 0) = P(Z > -0.45)$ $= 1 - 0.3264 = 0.6736.$

19. STATE: We want to estimate p = the true proportion of all phones produced in the factory on this day that have minor blemishes at a 99% confidence level. **PLAN:** One-sample z interval for p. **Random?** 200 randomly selected phones from the factory on this day. ✓ **Large counts?** $12 \geq 10$ and $200 - 12 = 188 \geq 10$. ✓ **DO:**

$$0.06 \pm 2.576\sqrt{\dfrac{0.06(1-0.06)}{200}} \rightarrow 0.06 \pm 0.043 \rightarrow 0.017 \text{ to } 0.103.$$

CONCLUDE: We are 99% confident that the interval from 0.017 to 0.103 captures the true proportion of all phones produced in the factory on this day that have minor blemishes.

Lesson 9.4 Exercises

1. (a) $\begin{aligned} &H_0{:}\mu_1 - \mu_2 = 0 \\ &H_a{:}\mu_1 - \mu_2 > 0 \end{aligned}$, where μ_1 = true mean reduction in blood pressure for healthy black men like these who take calcium and μ_2 = true mean reduction in blood pressure for healthy black men like these who take a placebo. **(b) Random?** The data come from two groups in a randomized experiment. ✓ **Normal/Large Sample?** The sample sizes are small, but the boxplots don't show any outliers or strong skewness. ✓

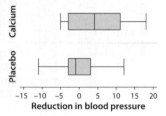

3. (a) $\begin{aligned} &H_0{:}\mu_1 - \mu_2 = 0 \\ &H_a{:}\mu_1 - \mu_2 \neq 0 \end{aligned}$, where μ_1 = the true mean reliability ratings of all Anglo bank customers and μ_2 = the true mean reliability ratings of all Hispanic bank customers.
(b) Random? Separate random samples of Anglo and Hispanic bank customers. ✓ **Normal/Large Sample?** Both sample sizes are large. $n_1 = 95 \geq 30$ and $n_2 = 86 \geq 30$. ✓

5. (a) $t = \dfrac{(5.00 - (-0.27)) - 0}{\sqrt{\dfrac{8.74^2}{10} + \dfrac{(5.90)^2}{11}}} = 1.60.$ **(b)** df $= 10 - 1 = 9$. The P-value is between 0.05 and 0.10.

7. (a) $t = \dfrac{(6.37 - 5.91) - 0}{\sqrt{\dfrac{0.60^2}{95} + \dfrac{0.93^2}{86}}} = 3.91.$ **(b)** df $= 86 - 1 = 85$ (use df $= 80$). The P-value is less than $2(0.005) = 0.01$.

9. STATE: $H_0{:}\mu_1 - \mu_2 = 0$ vs. $H_a{:}\mu_1 - \mu_2 > 0$, where μ_1 = true mean cost of shoes at Famous Footwear and μ_2 = true mean cost of shoes at Payless. **PLAN:** Two-sample t test for $\mu_1 - \mu_2$. **Random?** Two separate random samples. ✓ **Normal/Large Sample?** Both sample sizes are large. $n_1 = 30 \geq 30$ and $n_2 = 30 \geq 30$. ✓ **DO:** $t = 7.32$, P-value ≈ 0. **CONCLUDE:** Because the P-value ≈ 0 is less than $\alpha = 0.05$, we reject H_0. We have convincing evidence that shoes cost less, on average, at Payless than at Famous Footwear.

11. STATE: $H_0{:}\mu_1 - \mu_2 = 0$ vs. $H_a{:}\mu_1 - \mu_2 \neq 0$, where μ_1 = true mean growth (mm) of plants like these that are exposed to heavy metal music and μ_2 = true mean growth (mm) of plants like these that are exposed to classical music. **PLAN:** Two-sample t test for $\mu_1 - \mu_2$. **Random?** Two groups in a randomized experiment. ✓ **Normal/Large Sample?** The sample sizes are small, but the dotplots don't show any outliers or strong skewness. ✓

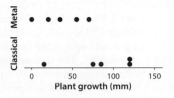

DO: $t = -2.10$, P-value $= 0.075$. **CONCLUDE:** Because the P-value $= 0.075$ is greater than $\alpha = 0.05$, we fail to reject H_0. We do not have convincing evidence that there is a difference in the mean growth of plants like these that are exposed to classical music versus heavy metal music.

13. (a) The shape of the distribution of DRP scores is slightly skewed to the left for the activities group and slightly skewed to the right for the control group. The center of the distribution is higher for the activities group and the variability is higher for the control group. There are no outliers. **(b) STATE:** $H_0{:}\mu_1 - \mu_2 = 0$ vs. $H_a{:}\mu_1 - \mu_2 > 0$, where μ_1 = true mean DRP score for students like these who do the reading activities and μ_2 = true mean DRP score for students like these who do not do the reading activities. **PLAN:** Two-sample t test for $\mu_1 - \mu_2$. **Random?** Two groups in a randomized experiment. ✓ **Normal/Large Sample?** The sample sizes are small, but the boxplots don't show any outliers or strong skewness. ✓ **DO:** $t = 2.31$, P-value $= 0.013$. **CONCLUDE:** Because the P-value $= 0.013$ is less than $\alpha = 0.05$, we reject H_0. We have convincing evidence that the mean DRP score is higher for students like these who do reading activities. **(c)** Yes; the data come from two groups in a randomized experiment, and the results are statistically significant. **(d)** Type I error; we found convincing evidence that the true mean DRP score was higher for students like these who do reading activities (rejected H_0) when the mean DRP score is the same for both treatments.

15. (a) See Exercise 13 for the definition of parameters and check of conditions. **STATE:** We want to estimate $\mu_1 - \mu_2$ at the 95% confidence level. **PLAN:** Two-sample t interval for $\mu_1 - \mu_2$. **DO:**

$$(51.48 - 41.52) \pm 2.086\sqrt{\dfrac{(11.01)^2}{21} + \dfrac{(17.15)^2}{23}} \rightarrow 0.97 \text{ to } 18.95.$$

CONCLUDE: We are 95% confident that the interval from 0.97 to 18.95 points captures the true difference in mean DRP score for students like these who do the reading activities and those who don't. The interval suggests that the mean DRP score of students like these who do the reading activities is between 0.97 points and 18.95 points higher than the mean DRP score for students like these who do not do the reading activities. **(b)** The confidence interval provides a set of plausible values for the difference in population means, $\mu_1 - \mu_2$.

17. (a) $1/100 = 0.01$. **(b)** Because the estimated P-value $= 0.01$ is less than $\alpha = 0.05$, we have convincing evidence that intrinsic rewards promote creativity for students like these.

19. (a) STATE: $H_0{:}\mu_1 - \mu_2 = 0.5$ vs. $H_a{:}\mu_1 - \mu_2 > 0.5$, where μ_1 = true mean amount of water per flush for the current model toilets and μ_2 = true mean amount of water per flush for the new pressure-assisted toilets. **PLAN:** Two-sample t test for $\mu_1 - \mu_2$. **Random?** Two separate random samples. ✓ **Normal/Large Sample?** Both sample sizes are large. $n_1 = n_2 = 30 \geq 30$. ✓ **DO:** $t = 0.80$, P-value $= 0.215$. **CONCLUDE:** Because the P-value is greater than $\alpha = 0.05$, we fail to reject H_0. We do not have convincing evidence that the new pressure-assisted toilet reduces the average amount of water used by more than 0.5 gallon per flush. **(b)** Type II error;

we didn't find convincing evidence that the new pressure-assisted toilet reduces the average amount of water used by more than 0.5 gallon per flush (failed to reject H_0) when the new pressure-assisted toilet does reduce the average amount of water used by more than 0.5 gallon per flush.

21. (a) $19/30 = 0.633 = 63.3\%$. **(b)** Right-skewed with a possible outlier between $160,000 and $180,000. **(c)** Because the distribution is skewed and has a possible outlier, the median should be used for center and IQR for variability. The median and IQR are resistant to outliers. **(d)** No; a histogram gives the frequency of data values within certain intervals but does not provide information about the exact values in the data set.

Lesson 9.5 Exercises

1.

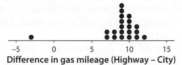

Difference in gas mileage (Highway – City)

Only 1 of the difference values is negative, so the highway gas mileage exceeds the city gas mileage for 20 out of the 21 vehicles. The difference values vary from −3 mpg (Toyota Prius) to 12 mpg (Mazda 6).

3. (a) Differences: 5, 246, −46, 121, 30, 55, 79, −94, −17, 95, 20, 14, 129, −39, 42. **(b)** Four of the difference values are negative, so the transaction time for the regular lane exceeds the transaction time for the express lane in 11 out of the 15 visits to the supermarket. The difference values vary from −94 to 246 seconds.

5. $\bar{x}_{\text{diff}} = 8.81$, $s_{\text{diff}} = 2.994$; the highway gas mileage for these 21 vehicles is 8.81 mpg higher than the city gas mileage, on average.

7. $\bar{x}_{\text{diff}} = 42.667$, $s_{\text{diff}} = 84.019$; the transaction time for the regular lane in these 15 visits to the supermarket is 42.667 seconds longer than the transaction time for the express lane, on average.

9. STATE: We want to estimate μ_{diff} = the true mean difference in transaction time (Regular lane – Express lane) at a 95% confidence level. **PLAN:** One-sample t interval for μ_{diff}. **Random?** Random sample of times during the week. ✓ **Normal/Large Sample?** The sample size is small, but the dotplot in Exercise 3 doesn't show any outliers or strong skewness. ✓ **DO:** $42.667 \pm 2.145\dfrac{84.019}{\sqrt{15}} \rightarrow$ −3.866 to 89.2. **CONCLUDE:** We are 95% confident that the interval from −3.866 seconds to 89.2 seconds captures the true mean difference in transaction time (Regular lane – Express lane). This interval suggests that the transaction times are between 3.866 seconds shorter and 89.2 seconds longer, on average, when using the regular lane.

11. (a) STATE: We want to estimate μ_{diff} = the true mean difference in reasoning scores for all preschool students who take six months of piano lessons (After – Before) at a 90% confidence level. **PLAN:** One-sample t interval for μ_{diff}. **Random?** Random sample of 34 preschool children. ✓ **Normal/Large Sample?** We don't know the shape of the population distribution, but the sample size is large: $n = 34 \geq 30$. ✓ **DO:** $3.618 \pm 1.697\dfrac{3.055}{\sqrt{34}} \rightarrow 2.729$ to 4.507. **CONCLUDE:** We are 90% confident that the interval from 2.729 to 4.507 captures the true mean difference in reasoning scores for all preschool students who take six months of piano lessons (After – Before). This interval suggests that the reasoning scores are between 2.729 and 4.507 higher, on average, after six months of piano lessons. **(b)** No; this was an observational study rather than an experiment.

13. (a)

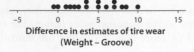

Difference in estimates of tire wear (Weight – Groove)

Only 1 of the difference values is negative, so the estimates of tire wear using the weight method equal or exceed the estimates of tire wear using the groove method for 15 out of the 16 tires. The difference values vary from −0.5 to 10.2 thousand miles. **(b) STATE:** We want to estimate μ_{diff} = the true mean difference in the estimates of tire wear from these two methods (Weight – Groove) at a 95% confidence level. **PLAN:** One-sample t interval for μ_{diff}. **Random?** Random sample of 16 tires. ✓ **Normal/Large Sample?** The sample size is small, but the dotplot doesn't show any outliers or strong skewness. ✓ **DO:** $4.556 \pm 2.131\dfrac{3.226}{\sqrt{16}} \rightarrow 2.837$ to 6.275. **CONCLUDE:** We are 95% confident that the interval from 2.837 to 6.275 thousand miles captures the true mean difference in the estimates of tire wear from these two methods (Weight – Groove). This interval suggests that the estimates of tire wear are between 2837 miles and 6275 miles higher, on average, when using the weight method. **(c)** Yes; zero is not included within the confidence interval and is, therefore, not a plausible value for the true mean difference in estimates (Weight – Groove).

15. The Random condition would not be satisfied, as there was no random selection of car models. Also, the Normal/Large Sample condition would not be satisfied because the dotplot of differences has an outlier.

17. (a) Explanatory: type of snack food. Response: whether or not she catches the food in her mouth.

(b)

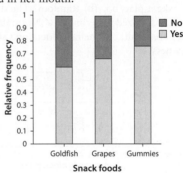

(c) Yes; the percentages of "Yes" are not the same for each type of food.

Lesson 9.6 Exercises

1. STATE: $H_0: \mu_{\text{diff}} = 0$ vs. $H_a: \mu_{\text{diff}} \neq 0$, where μ_{diff} = the true mean difference in service times (Drive-thru – Inside) for this local fast-food restaurant. **PLAN:** One-sample t test for μ_{diff}. **Random?** Randomly selected days. ✓ **Normal/Large Sample?** The sample size is small, but the dotplot doesn't show any outliers or strong skewness. ✓

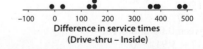

Difference in service times (Drive-thru – Inside)

DO: $t = 4.42$, P-value $= 0.0017$. **CONCLUDE:** Because the P-value of 0.0017 is less than $\alpha = 0.05$, we reject H_0. There is convincing evidence of a difference in the average service time inside and at the drive-thru for this local fast-food restaurant.

3. (a) STATE: $H_0: \mu_{\text{diff}} = 0$ vs. $H_a: \mu_{\text{diff}} \neq 0$, where μ_{diff} = the true mean difference in number of shoppers (Friday the 6th – Friday the 13th) for all grocery stores. **PLAN:** One-sample t test for μ_{diff}. **Random?** Random sample. ✓ **Normal/Large Sample?** The number of differences in the sample is large ($n_{\text{diff}} = 45 \geq 30$). ✓ **DO:** $t = -1.75$, P-value $= 0.087$. **CONCLUDE:** Because the P-value of 0.087 is greater than $\alpha = 0.05$, we fail to reject H_0. There is not convincing evidence of a difference in the mean number of shoppers at grocery stores on these two days. **(b)** No; this was an observational study, not a randomized experiment, so we can't conclude causation.

5. One-sample t procedures; the data come from a matched pairs experiment with the two treatments (tire Brand A and tire Brand B) being randomly assigned to the left and right side of each car.

7. Two-sample t procedures; the data come from two randomly assigned groups of workers in an experiment.

9. One-sample t procedures; the data come from a matched pairs experiment with the order of the two treatments (men's basketball and women's basketball) being randomly assigned.

11. (a) The data are paired by subject, and they come from a random sample of 10 subjects. (b) **STATE:** $H_0{:}\mu_{\text{diff}} = 0$ vs. $H_a{:}\mu_{\text{diff}} > 0$, where $\mu_{\text{diff}} =$ the true mean difference in sleep time in hours (After drug – Before drug) for insomnia patients. **PLAN:** One-sample t test for μ_{diff}. **Random?** Random sample. ✓ **Normal/Large Sample?** The sample size is small, but the dotplot doesn't show any outliers or strong skewness. ✓

```
        ••  •• ••   •    ••    •
    ┼───┼───┼───┼───┼───┼───┼───┼
   -1   0   1   2   3   4   5   6
         Difference in sleep hours
            (Drug – No drug)
```

DO: $t = 3.68$, P-value $= 0.003$. **CONCLUDE:** Because the P-value of 0.003 is less than $\alpha = 0.05$, we reject H_0. There is convincing evidence that patients like these would get more sleep, on average, when taking the drug. (c) No; there is no control group to compare results. It may be that the increase in sleep hours is due to the placebo effect. (d) Type I error; we found convincing evidence that patients like these would get more sleep, on average, when taking the drug (rejected H_0) when they wouldn't get more sleep with the drug.

13. (a) See Exercise 11 for the definition of parameters and check of conditions. **STATE:** We want to estimate μ_{diff} at a 95% confidence level. **PLAN:** One-sample t interval for μ_{diff}. **DO:** $2.33 \pm 2.262\dfrac{2.002}{\sqrt{10}} \rightarrow 0.90$ to 3.76. **CONCLUDE:** We are 95% confident that the interval from 0.90 to 3.76 hours captures the true mean difference in sleep time in hours (After drug – Before drug). This interval suggests that the sleep times are between 0.90 and 3.76 hours longer, on average, when insomnia patients are given the sleep drug. (b) The confidence interval provides a set of plausible values for the true mean difference, μ_{diff}.

15. (a) A one-sample t test using Gain; the data are paired by student (Try 1 and Try 2), and they come from a random sample of 427 students who were coached. (b) **STATE:** $H_0{:}\mu_{\text{diff}} = 0$ vs. $H_a{:}\mu_{\text{diff}} > 0$, where $\mu_{\text{diff}} =$ the true mean difference in SAT score (Try 2 – Try 1) for all students who are coached. **PLAN:** One-sample t test for μ_{diff}. **Random?** Random sample. ✓ **Normal/Large Sample?** The number of differences in the sample is large $n_{\text{diff}} = 427 \geq 30$. ✓ **DO:** $t = 10.16$, P-value ≈ 0. **CONCLUDE:** Because the P-value of approximately 0 is less than $\alpha = 0.05$, we reject H_0. There is convincing evidence that students who are coached increase their SAT scores, on average.

17. (a) $\mu_X = np = 200(0.29) = 58$, $\sigma_X = \sqrt{np(1-p)} = \sqrt{200(0.29)(1-0.29)} = 6.42$. (b) Approximately normal because $np = 200(0.29) = 58 \geq 10$ and $n(1-p) = 200(1-0.29) = 142 \geq 10$. (c) $z = \dfrac{50-58}{6.42} = -1.25$; $P(X < 50) = P(Z < -1.25) = 0.1056$.

Chapter 9 Review Exercises

1. (a) Approximately normal, because $n_1p_1 = 72$, $n_1(1-p_1) = 228$, $n_2p_2 = 14$, $n_2(1-p_2) = 186$ are all ≥ 10. (b) $\mu_{\hat{p}_1-\hat{p}_2} = 0.24 - 0.07 = 0.17$. (c) $\sigma_{\hat{p}_1-\hat{p}_2} = \sqrt{\dfrac{0.24(0.76)}{300} + \dfrac{(0.07)(0.93)}{200}} = 0.031$.

2. (a) **STATE:** $H_0{:}p_1 - p_2 = 0$ vs. $H_a{:}p_1 - p_2 > 0$, where p_1 is the true proportion of RLS sufferers like the ones in this experiment who will experience much improved symptoms if they take Pramipexole and p_2 is the true proportion of RLS sufferers like the ones in this

experiment who will experience much improved symptoms if they take a placebo. **PLAN:** Two-sample z test for $p_1 - p_2$. **Random?** Two groups in a randomized experiment. ✓ **Large Counts?** $n_1\hat{p}_1 = 158$, $n_1(1-\hat{p}_1) = 35$, $n_2\hat{p}_2 = 50$, $n_2(1-\hat{p}_2) = 42$ are all ≥ 10. ✓ **DO:** $z = 4.89$, P-value ≈ 0. **CONCLUDE:** Because the P-value of approximately 0 is less than $\alpha = 0.05$, we reject H_0. We have convincing evidence that a greater proportion of RLS sufferers like the ones in this experiment will experience much improved symptoms if they take Pramipexole than if they take a placebo. (b) See part (a) for the definition of parameters and check of conditions. **STATE:** We want to estimate $p_1 - p_2$ at the 95% confidence level. **PLAN:** Two-sample z interval for $p_1 - p_2$. **DO:** $(0.819 - 0.543) \pm 1.960\sqrt{\dfrac{0.819(1-0.819)}{193} + \dfrac{0.543(1-0.543)}{92}}$ $\rightarrow 0.161$ to 0.391. **CONCLUDE:** We are 95% confident that the interval from 0.161 to 0.391 captures the difference in the true proportions of RLS patients like these who will experience much improved symptoms (Pramipexole – Placebo). The interval suggests that the percent of RLS patients like these who would experience much improved symptoms when taking Pramipexole is between 16.1 and 39.1 percentage points higher than the percent who would experience much improved symptoms when taking a placebo. (c) Confidence intervals provide a set of plausible values for the difference in population proportions, $p_1 - p_2$.

3. (a) Approximately normal because both sample sizes are large ($n_1 = 35 \geq 30$ and $n_2 = 40 \geq 30$). (b) $\mu_{\bar{x}_1-\bar{x}_2} = 22.3 - 28.9 = -6.6$ miles per gallon. (c) $\sigma_{\bar{x}_1-\bar{x}_2} = \sqrt{\dfrac{5.7^2}{35} + \dfrac{9.9^2}{40}} = 1.84$ miles per gallon.

4. **STATE:** $\mu_1 - \mu_2 = 0$ vs. $\mu_1 - \mu_2 > 0$, where $\mu_1 =$ true mean number of oranges that Glad trash bags will hold until breaking and $\mu_2 =$ true mean number of oranges that Walmart trash bags will hold until breaking. **PLAN:** Two-sample t test for $\mu_1 - \mu_2$. **Random?** Two separate random samples. ✓ **Normal/Large Sample?** The sample sizes are small, but the boxplots don't show any outliers or strong skewness. ✓

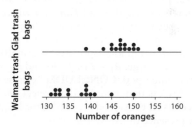

DO: $t = 5.87$, P-value ≈ 0. **CONCLUDE:** Because the P-value of approximately 0 is less than $\alpha = 0.01$, we reject H_0. We have convincing evidence that Glad trash bags can hold more weight, on average, than Walmart trash bags.

5. **STATE:** We want to estimate $\mu_1 - \mu_2$ at the 95% confidence level, where $\mu_1 =$ the true mean daily cell-phone usage for all male college students and $\mu_2 =$ the true mean daily cell-phone usage for all female college students. **PLAN:** Two-sample t interval for $\mu_1 - \mu_2$. **Random?** Two separate random samples. ✓ **Normal/Large Sample?** The sample sizes are large ($n_1 = 84 \geq 30$ and $n_2 = 80 \geq 30$). ✓ **DO:** $(458.5 - 600.1) \pm 2.000\sqrt{\dfrac{(344.2)^2}{84} + \dfrac{(392.3)^2}{80}} \rightarrow -257.1$ to -26.1 minutes with df $= 60$. *Using technology:* Applet/2-SampTInt gives $(-255.7$ to $-27.55)$ with df $= 157$. **CONCLUDE:** We are 95% confident that the interval from -257.1 minutes to -26.1 minutes captures the true difference in mean daily cell-phone usage for male and female college students. The interval suggests that the mean daily cell-phone usage for male college students is between 26.1 and 257.1 minutes less than the mean daily cell-phone usage for female college students.

6. (a) One-sample t procedures; the data are paired by the time of day, and they come from a random sample of 12 times. **(b)** Two-sample t procedures; the data come from two separate random samples. **(c)** One-sample t procedures; the data are paired for each smoker, and they come from a random sample of 500 smokers.

7. (a)

-30 -20 -10 0 10 20 30 40 50
**Difference in reaction time in seconds
(Before – After)**

Of the difference values, 4 are negative, so the reaction time before exceeds the reaction time after for 11 of the 15 subjects. The difference values vary from -29 (Subject 2) to 50 (Subject 4). **(b)** $\bar{x}_{diff} = 13.2$ seconds, $s_{diff} = 22.606$ seconds. The reaction time before running two laps for these 15 subjects is 13.2 seconds higher than their reaction time after running two laps, on average. **(c) STATE:** $H_0{:}\mu_{diff} = 0$ vs. $H_a{:}\mu_{diff} > 0$, where $\mu_{diff} =$ the true mean difference in reaction time (Before – After) for all students at this school. **PLAN:** One-sample t test for μ_{diff}. **Random?** Random sample. ✓ **Normal/Large Sample?** The sample size is small, but the dotplot doesn't show any outliers or strong skewness. ✓ **DO:** $t = 2.26$, P-value $= 0.0201$. **CONCLUDE:** Because the P-value of 0.0201 is less than $\alpha = 0.05$, we reject H_0. There is convincing evidence that exercise is associated with a decrease in reaction time, on average, for subjects like these.

Chapter 9 Practice Test

1. d
2. a
3. a
4. b
5. d
6. c
7. c
8. c
9. c
10. b
11. STATE: $H_0{:}\mu_1 - \mu_2 = 0$ vs. $H_a{:}\mu_1 - \mu_2 > 0$, where $\mu_1 =$ true mean relative tail spine length of all *Daphnia* where kairomone is present and $\mu_2 =$ true mean relative tail spine length of all *Daphnia* where kairomone is absent. **PLAN:** Two-sample t test for $\mu_1 - \mu_2$. **Random?** Two separate random samples. ✓ **Normal/Large Sample?** Both sample sizes are large. $n_1 = 214 \geq 30$ and $n_2 = 152 \geq 30$. ✓ **DO:** $t = 14.12$, P-value ≈ 0. **CONCLUDE:** Because the P-value of approximately 0 is less than $\alpha = 0.05$, we reject H_0. We have convincing evidence that the mean relative tail spine length of *Daphnia* is longer in the presence of fish kairomones.
12. STATE: We want to estimate $p_1 - p_2$ at the 99% confidence level, where $p_1 =$ the true proportion of physicians who would return the postcard and $p_2 =$ the true proportion of physicians who would return the long questionnaire. **PLAN:** Two-sample z interval for $p_1 - p_2$. **Random?** The data come from a random sample of 523 physicians who were randomly assigned to the two treatment groups in an experiment. ✓ **Large Counts?** $262\left(\dfrac{178}{262}\right) = 178$, $262\left(\dfrac{84}{262}\right)$
$= 84$, $261\left(\dfrac{138}{261}\right) = 138$, $261\left(\dfrac{123}{261}\right) = 123$ are all at least 10. ✓ **DO:**
$(0.679 - 0.529) \pm 2.576\sqrt{\dfrac{0.679(1-0.679)}{262} + \dfrac{0.529(1-0.529)}{261}} \rightarrow$
0.041 to 0.259. **CONCLUDE:** We are 99% confident that the interval from 0.041 to 0.259 captures the true difference in response rate for all physicians who receive the postcard versus the long questionnaire. The interval suggests that the percent of physicians who receive the postcard and return it is between 4.1 and 25.9 percentage points higher than the percent for physicians who receive the long questionnaire.

13. (a) STATE: We want to estimate $\mu_{diff} =$ the true mean difference in time to solve a maze (Unscented – Scented) at a 95% confidence level. **PLAN:** One-sample t interval for μ_{diff}. **Random?** Paired data come from a randomized experiment. ✓ **Normal/Large Sample?** The sample size is small, but the dotplot doesn't show any outliers or strong skewness. ✓

-30 -20 -10 0 10 20 30
**Difference in time to solve a maze in seconds
(Unscented – Scented)**

DO: $0.957 \pm 2.086\dfrac{12.548}{\sqrt{21}} \rightarrow -4.755$ to 6.669 seconds. **CONCLUDE:** We are 95% confident that the interval from -4.755 seconds to 6.669 seconds captures the true mean change in time to solve a maze (Unscented – Scented). This interval suggests that the times to solve a maze are between 4.755 seconds lower and 6.669 seconds higher, on average, when subjects are given the unscented mask. **(b)** No; zero is included in the confidence interval and is, therefore, a plausible value for the true mean difference in time to finish the maze (Unscented – Scented).

Chapter 10

Lesson 10.1 Exercises

1. H_0: The distribution of type of nuts in this batch is the same as the claimed distribution. H_a: The distribution of type of nuts in this batch is not the same as the claimed distribution.
3. H_0: The colors of Original Skittles are uniformly distributed. H_a: The colors of Original Skittles are not uniformly distributed.
5. Cashew: $150(0.52) = 78$; Almond: $150(0.27) = 40.5$; Macadamia: $150(0.13) = 19.5$; Brazil: $150(0.08) = 12$
7. The expected count for each color is $233\left(\dfrac{1}{5}\right) = 46.6$.

9. $\chi^2 = \dfrac{(83-78)^2}{78} + \dfrac{(29-40.5)^2}{40.5} + \dfrac{(20-19.5)^2}{19.5} + \dfrac{(18-12)^2}{12} = 6.60$

11. $\chi^2 = \dfrac{(39-46.6)^2}{46.6} + \dfrac{(49-46.6)^2}{46.6} + \dfrac{(42-46.6)^2}{46.6} + \dfrac{(53-46.6)^2}{46.6}$
$+\dfrac{(50-46.6)^2}{46.6} = 2.94$

13. (a) H_0: Nuthatches do not prefer particular types of trees when searching for seeds and insects. H_a: Nuthatches do prefer particular types of trees when searching for seeds and insects. **(b)** Douglas firs: $156(0.54) = 84.24$; Ponderosa pines: $156(0.40) = 62.4$; Other types: $156(0.06) = 9.36$. **(c)** $\chi^2 = \dfrac{(70-84.24)^2}{84.24} + \dfrac{(79-62.4)^2}{62.4}$
$+\dfrac{(7-9.36)^2}{9.36} = 7.43$

15. Because there are only three simulated chi-square values greater than or equal to the observed chi-square test statistic of 7.43, it is unusual to get differences between the observed and expected counts this large by chance alone. These data provide convincing evidence that nuthatches do prefer particular types of trees when searching for seeds and insects.
17. H_0: $p_{douglas\ fir} = 0.54$, $p_{ponderosa\ pine} = 0.40$, $p_{other} = 0.06$. H_a: At least two of the proportions are incorrect. $p_{tree} =$ the true proportion of nuthatches that would land on that type of tree.
19. Voluntary response; subscribers who are fearful of terrorism are probably more likely to volunteer to respond. As a result, the 49% from the sample is likely greater than the percent of all subscribers who are afraid there will be another terrorist attack.

Lesson 10.2

1. Random? This sample was chosen by randomly dialing landline telephone numbers. ✓ **Large Counts?** All expected counts are ≥ 5.

387(0.13) = 50.31, 387(0.35) = 135.45, 387(0.35) = 135.45, 387(0.17) = 65.79. ✓

3. Random? Mr. Wallis did not take a random sample. The sample is 68 students from his Algebra II classes. **Large Counts?** All expected counts are ≥ 5. The expected count for each day is the same, $68\left(\dfrac{1}{7}\right) = 9.71$. ✓

5. (a) *Using Table C:* df = 11 and the P-value is between 0.05 and 0.10. *Using technology:* χ^2cdf(lower:19.03, upper:1000, df:11) gives a P-value of 0.061. **(b)** *Using Table C:* df = 3 and the P-value is less than 0.0005. *Using technology:* χ^2cdf(lower:19.03, upper:1000, df:3) gives a P-value of 0.0003.

7. $\chi^2 = \dfrac{(17-50.31)^2}{50.31} + \dfrac{(118-135.45)^2}{135.45} + \dfrac{(161-135.45)^2}{135.45}$
$+ \dfrac{(91-65.79)^2}{65.79} = 38.78$. *Using Table C:* df = 3 and the P-value is less than 0.0005. *Using technology:* χ^2cdf(lower:38.78, upper:1000, df:3) gives a P-value of approximately 0.

9. STATE: H_0: Mendel's predicted model is correct. H_a: Mendel's predicted model is incorrect. **PLAN:** Chi-square test for goodness of fit. **Random?** Assume the random condition is met. ✓ **Large Counts?** All expected counts are ≥ 5. $556(0.75) = 417$ and $556(0.25) = 139$. ✓ **DO:** $\chi^2 = 0.35$, P-value = 0.557. **CONCLUDE:** Because the P-value of 0.557 is greater than $\alpha = 0.05$, we fail to reject H_0. There is not convincing evidence that Mendel's predicted model is incorrect.

11. STATE: H_0: The blood type distribution among performing arts majors at this university is the same as the national distribution. H_a: The blood type distribution among performing arts majors at this university differs from the national distribution. **PLAN:** Chi-square test for goodness of fit. **Random?** Random sample. ✓ **Large Counts?** All expected counts (60, 16.5, 6, 67.5) are ≥ 5. ✓ **DO:** $\chi^2 = 6.90$, P-value = 0.075. **CONCLUDE:** Because the P-value of 0.075 is greater than $\alpha = 0.05$, we fail to reject H_0. There is not convincing evidence that the blood type distribution among performing arts majors at this university differs from the national distribution.

13. STATE: H_0: The company's invoices follow Benford's law. H_a: The company's invoices do not follow Benford's law. **PLAN:** Chi-square test for goodness of fit. **Random?** Random sample. ✓ **Large Counts?** All expected counts are ≥ 5 (75.25, 44, 31.25, 24.25, 19.75, 16.75, 14.5, 12.75, 11.5). ✓ **DO:** $\chi^2 = 21.56$, P-value = 0.006. **CONCLUDE:** Because the P-value of 0.006 is less than $\alpha = 0.05$, we reject H_0. There is convincing evidence that the company's invoices do not follow Benford's law.

15. (a) Type I: The accountant concludes that the company is faking invoices, even though the company is being honest. This decision would upset the company and could cost the accountant his job. **Type II:** The accountant doesn't conclude that the company is faking invoices, even though the company is faking invoices. This means the company gets away with being unethical. The Type II error is more serious because we want businesses to have high ethical standards. **(b)** Much greater than expected for a first digit of 3 $(43 > 31.25)$ and 4 $(34 > 24.25)$, and much less than expected for a first digit of 1 $(61 < 75.25)$ and 7 $(7 < 14.5)$.

17. Time spent doing homework is quantitative. Chi-square tests for goodness of fit should only be used for distributions of categorical data.

19. (a) $\dfrac{43}{858} = 0.050$. **(b)** $\dfrac{10}{71} = 0.141$

Lesson 10. 3 Exercises

1. H_0: There is no association between age and cell-phone type in the population of U.S. cell-phone owners. H_a: There is an association between age and cell-phone type in the population of U.S. cell-phone owners.

3. H_0: There is no association between gender and opinion about getting rich in the population of young adults. H_a: There is an association between gender and opinion about getting rich in the population of young adults.

5.

	18−34	35−54	55+
iPhone	119.29	146.98	200.74
Android	128.48	158.31	216.21
Other	269.23	331.72	453.05

7.

	Female	Male
Almost no chance	95.15	98.85
Some chance but probably not	349.21	362.79
A 50-50 chance	694.5	721.5
A good chance	696.96	724.04
Almost certain	531.18	551.82

9. $\chi^2 = \dfrac{(169-119.29)^2}{119.29} + \dfrac{(171-146.98)^2}{146.98} + \cdots + \dfrac{(643-453.05)^2}{453.05}$
$= 333.65$

11. $\chi^2 = \dfrac{(96-95.15)^2}{95.15} + \dfrac{(98-98.85)^2}{98.85} + \cdots + \dfrac{(597-551.82)^2}{551.82} = 43.95$

13. (a)

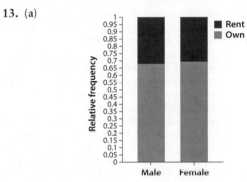

There is a weak association between gender and housing status, because females are slightly more likely to own property than males. **(b)** H_0: There is no association between gender and housing status for U.S. residents. H_a: There is an association between gender and housing status for U.S. residents.

(c)

	Male	Female
Own	132.84	121.16
Rent	49.16	44.84

(d) $\chi^2 = \dfrac{(132-132.84)^2}{132.84} + \dfrac{(122-121.16)^2}{121.16} + \dfrac{(50-49.16)^2}{49.16}$
$+ \dfrac{(44-44.84)^2}{44.84} = 0.041$

15. Marital status and housing status, because the difference between the observed and expected values is much larger in Exercise 14 than in Exercise 13. This leads to a much greater chi-square test statistic $(18.25 > 0.041)$.

17. Estimated P-value = $27/200 = 0.135$. These data do not provide convincing evidence that gender and handedness are associated in the population of Australian high school students.

19. (a) STATE: We want to estimate p = the true proportion of trees in this forest that show damage from gypsy moth caterpillars at a 99% confidence level. **PLAN:** One-sample z interval for p. **Random?** Random sample. ✓ **Large counts?** $200(0.62) = 124 \geq 10$ and $200(1 - 0.62) = 76 \geq 10$. ✓ **DO:**

$0.62 \pm 2.576 \sqrt{\dfrac{0.62(1-0.62)}{200}} \rightarrow 0.62 \pm 0.088 \rightarrow 0.532$ to 0.708.

CONCLUDE: We are 99% confident that the interval from 0.532 to 0.708 captures the true proportion of trees in this forest that show damage from gypsy moth caterpillars. **(b)** If the forester took many random samples of trees and constructed a 99% confidence interval for each sample, about 99% of these intervals would capture the true proportion of trees in this forest that show damage from gypsy moth caterpillars.

Lesson 10.4 Exercises

1. Random? This was a random sample of U.S. residents. ✓ **Large Counts?** All expected counts (68.07, 140.68, 285.33, 361.91, 44.93, 92.86, 188.33, 238.88, 7.00, 14.46, 29.33, 37.21) are ≥ 5. ✓
3. Random? This was not a random sample. **Large Counts?** All expected counts (116.82, 95.58, 229.61, 202.18, 165.42, 397.39) are ≥ 5. ✓
5. (a) P-value: df = $(5 - 1)(4 - 1) = 12$. *Using Table C:* P-value is between 0.10 and 0.15. *Using technology:* χ^2cdf(lower:17.34, upper:1000, df:12) gives a P-value of 0.137. **(b)** P-value: df = $(3 - 1)(2 - 1) = 2$. *Using Table C:* P-value is less than 0.0005. *Using technology:* χ^2cdf(lower:17.34, upper:1000, df:2) gives a P-value of 0.0002.
7. P-value: df = $(3 - 1)(4 - 1) = 6$. *Using Table C:* P-value is less than 0.0005. *Using technology:* χ^2cdf(lower:39.75, upper:1000, df:6) gives a P-value ≈ 0.
9. STATE: H_0: There is no association between temperature and hatching for python eggs like these. H_a: There is an association between temperature and hatching for python eggs like these. **PLAN:** Chi-square test for association. **Random?** Random assignment. ✓ **Large Counts?** All expected counts (18.63, 38.63, 71.74, 8.37, 17.37, 32.26) are ≥ 5. ✓ **DO:** $\chi^2 = 1.703$, P-value $= 0.427$. **CONCLUDE:** Because the P-value of 0.427 is greater than $\alpha = 0.05$, we fail to reject H_0. There is not convincing evidence of an association between temperature and hatching for python eggs like these.
11. STATE: H_0: There is no association between age of subjects and whether they do any banking online. H_a: There is an association between age of subjects and whether they do any banking online. **PLAN:** Chi-square test for association. **Random?** Random sample. ✓ **Large Counts?** All expected counts (232.81, 319.45, 325.93, 209.82, 162.19, 222.55, 227.07, 146.18) are ≥ 5. ✓ **DO:** $\chi^2 = 43.797$, P-value ≈ 0. **CONCLUDE:** Because the P-value of ≈ 0 is less than $\alpha = 0.05$, we reject H_0. There is convincing evidence of an association between age of subjects and whether they do any banking online.
13. (a)

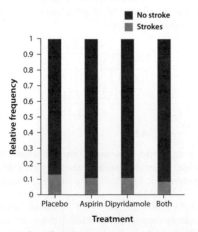

The graph shows that for the patients in the sample, using both drugs led to the lowest stroke rate, followed by aspirin only and dipyridamole only. The placebo group had the highest stroke rate. **(b) STATE:** H_0: There is no association between stroke treatment and response for patients like these. H_a: There is an association between stroke treatment and response for patients like these. **PLAN:** Chi-square test for association. **Random?** Random assignment. ✓ **Large Counts?** All expected counts (205.81, 205.81,

206.44, 205.94, 1443.19, 1443.19, 1447.56, 1444.06) are ≥ 5. ✓ **DO:** $\chi^2 = 24.243$, P-value ≈ 0. **CONCLUDE:** Because the P-value of ≈ 0 is less than $\alpha = 0.05$, we reject H_0. There is convincing evidence of an association between stroke treatment and response for patients like these.
15. Both aspirin and dipyridamole; the observed number of strokes in this group (157) was much less than the expected count (205.938).
17. These data are quantitative. A chi-square test for association requires categorical data.
19. (a) Yes; it appears that a majority of the values for the differences (Post − Pre) in the boxplot are greater than 0, and the mean of the differences (5.38) is greater than 0. **(b)** No; there was no control group to compare to, so we don't know if the scores increased due to the chess training or to some other confounding variable, such as what the students were learning in school.

Lesson 10.5 Exercises

1. H_0:$\beta = 0$ vs. H_a:$\beta > 0$, where β is the slope of the true least-squares regression line relating y = drop time (seconds) to x = drop height (cm).
3. H_0:$\beta = 0$ vs. H_a:$\beta < 0$, where β is the slope of the true least-squares regression line relating y = grade-point average to x = typical texts per day.
5. Random? Each helicopter was assigned at random to a drop height. ✓ **Normal?** A histogram of the residuals does not show strong skewness or outliers. ✓ **Linear?** There is a linear association between drop height and drop time in the scatterplot. Also, there is no leftover curved pattern in the residual plot. ✓ **Equal SD?** In the residual plot, we do not see a clear < pattern or > pattern. ✓
7. Random? Random sample. ✓ **Normal?** A dotplot of the residuals shows a strong right skew. This condition is not met. **Linear?** The association between the income and mortality rate in the scatterplot does not appear to be linear. Also, there is a leftover curved pattern in the residual plot. This condition is not met. **Equal SD?** In the residual plot, we do not see a clear < pattern or > pattern. ✓
9. $t = \dfrac{0.0057244 - 0}{0.0002018} = 28.37$. With df = $70 - 2 = 68$, the P-value is less than 0.0005.
11. $t = \dfrac{-0.00474 - 0}{0.000778} = -6.09$. With df = $47 - 2 = 45$, the P-value is less than 0.0005.
13. (a) H_0:$\beta = 0$ vs. H_a:$\beta > 0$, where β is the slope of the true least-squares regression line relating y = amount expelled (cups) to x = number of Mentos. **(b) Random?** Random assignment. ✓ **Normal?** A dotplot of the residuals does not show strong skewness or outliers. ✓ **Linear?** There is a linear association between number of Mentos and amount expelled. Also, there is no leftover curved pattern in the residual plot. ✓ **Equal SD?** In the residual plot, we do not see a clear < pattern or > pattern. ✓. **(c)** $t = \dfrac{0.07083 - 0}{0.01228} = 5.77$. With df = $24 - 2 = 22$, the P-value is less than 0.0005. **(d)** Because the P-value is less than $\alpha = 0.05$, we reject H_0. There is convincing evidence of a positive association between number of Mentos and amount expelled.
15. (a) $r = 0$ indicates that there is no linear relationship (no association) between two variables. **(b)** H_0:$\rho = 0$ vs. H_a:$\rho > 0$, where ρ is the true correlation between y = height (cm) to x = foot length (cm). **(c)** $1/100 = 0.01$. **(d)** Because the estimated P-value is less than $\alpha = 0.05$, we reject H_0. There is convincing evidence of a positive association between foot length and height.
17. If Mac and Nick were to take many samples of high school students and compute the least-squares regression line relating x = typical number of texts and y = GPA for each sample, the slopes of these lines would typically vary from the slope of the true regression line by about 0.000778.

19. (a) The differences (Colors − Words): 7, 11, 7, 13, 4, 2, 18, 5, 6, 4, 0, 9, 5, 0, 6, 8.

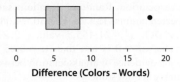

Difference (Colors – Words)

The shape of the distribution of differences is slightly skewed right, with a center at the median of 6 and an IQR of $8.5 − 4 = 4.5$. There is one outlier at 18. **(b)** $\bar{x}_{diff}$

$$= \frac{7+11+7+13+4+2+18+5+6+4+0+9+5+0+6+8}{16} = 6.563.$$

On average, each subject took 6.563 more seconds to identify all the colors than to identify all the words. **(c)** The Normal/Large Sample condition is not met because the sample size is small, and the boxplot shows an outlier at 18.

Lesson 10.6 Exercises

1. $t = 1.231$, the P-value is 0.123, and the df is 10.

3. $t = −4.163$, the P-value is less than 0.001, and the df is 28.

5. STATE: $H_0:\beta = 0$ vs. $H_a:\beta > 0$, where β is the slope of the true least-squares regression line relating y = pulse rate (beats per minute) to x = body temperature (in degrees Fahrenheit). **PLAN:** t test for the slope of a least-squares regression line. **Random?** Random sample. ✓ **Normal?** A dotplot of the residuals does not show strong skewness or outliers. ✓ **Linear?** There is a linear association between body temperature (in degrees Fahrenheit) and pulse rate (beats per minute) in the scatterplot. Also, there is no leftover curved pattern in the residual plot. ✓ **Equal SD?** In the residual plot, we do not see a clear < pattern or > pattern. ✓

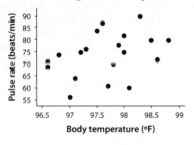

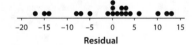

Residual

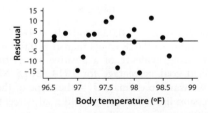

DO: $t = 1.722$, P-value $= 0.051$ with df $= 18$. **CONCLUDE:** Because the P-value of 0.051 is greater than $\alpha = 0.05$, we fail to reject H_0. There is not convincing evidence that faster heartbeats are associated with warmer body temperatures for students at Mr. Jordan's school.

7. STATE: $H_0:\beta = 0$ vs. $H_a:\beta < 0$, where β is the slope of the true least-squares regression line relating y = mean score to x = mean distance (yards). **PLAN:** t test for the slope of a least-squares regression line. Conditions for inference are met. **DO:**

$$t = \frac{-0.02016 - 0}{0.01319} = -1.53,$$ with df $= 19 − 2 = 17$, the P-value is between 0.05 and 0.10. **CONCLUDE:** Because the P-value of

between 0.05 and 0.10 is greater than $\alpha = 0.05$, we fail to reject H_0. There is not convincing evidence that lower mean scores are associated with greater mean driving distances for PGA golfers.

9. STATE: We want to estimate β = the slope of the true least-squares regression line relating y = pulse rate (beats per minute) to x = body temperature (degrees Fahrenheit) for students in Mr. Jordan's school with 95% confidence. **PLAN:** t interval for the slope of a least-squares regression line. Conditions were checked in Exercise 5. **DO:** Using the applet, the interval is −1.088 to 10.976 with df = 18. **CONCLUDE:** We are 95% confident that the interval from −1.088 to 10.976 captures the slope of the true least-squares regression line relating y = pulse rate (beats per minute) to x = body temperature (degrees Fahrenheit) for students at Mr. Jordan's school.

11. STATE: We want to estimate β = the slope of the true least-squares regression line relating y = mean score to x = mean distance (yards) for PGA golfers with 95% confidence. **PLAN:** t interval for the slope of a least-squares regression line. Conditions for inference are met. **DO:** $−0.02016 \pm 2.110(0.01319) \rightarrow −0.0480$ to 0.008. **CONCLUDE:** We are 95% confident that the interval from −0.0480 to 0.008 captures the slope of the true least-squares regression line relating y = mean score to x = mean distance (yards) for PGA golfers.

13 (a)

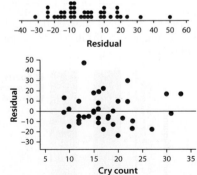

There is a weak, positive, linear relationship between cry count and IQ at age 3. **(b) STATE:** $H_0:\beta = 0$ vs. $H_a:\beta > 0$, where β is the slope of the true least-squares regression line relating y = IQ at age 3 to x = cry count for infants. **PLAN:** t test for the slope of a least-squares regression line. **Random?** Random sample. ✓ **Normal?** A dotplot of the residuals does not show strong skewness or outliers. ✓ **Linear?** There is a linear association between cry count and IQ at age 3 in the scatterplot. Also, there is no leftover curved pattern in the residual plot. ✓ **Equal SD?** In the residual plot, we do not see a clear < pattern or > pattern. ✓

DO: $t = 3.065$, P-value $= 0.002$ with df $= 36$. **CONCLUDE:** Because the P-value of 0.002 is less than $\alpha = 0.05$, we reject H_0. There is convincing evidence of a positive association between cry count and IQ at age 3 for infants. **(c) STATE:** We want to estimate β = the slope of the true least-squares regression line relating y = IQ at age 3 to x = cry count for infants with 95% confidence. **PLAN:** t interval for the slope of a least-squares regression line. Conditions were checked in part **(b)**. **DO:** Using the applet, the interval is 0.505 to 2.481 with df = 36. **CONCLUDE:** We are 95% confident that the interval from

0.505 to 2.481 captures the slope of the true least-squares regression line relating $y =$ IQ at age 3 to $x =$ cry count for infants.

15. Because all of the plausible values in the interval $(0.505, 2.481)$ are greater than 0, these data provide convincing evidence of a positive association between cry count and IQ at age 3.

17. STATE: H_0: There is no association between cry count and IQ at age 3 for infants. H_a: There is an association between cry count and IQ at age 3 for infants. **PLAN:** Chi-square test for association. **Random?** Random sample. ✓ **Large Counts?** All expected counts $(8.68, 13.32, 6.32, 9.68)$ are ≥ 5. ✓ **DO:** $\chi^2 = 0.782$, P-value $= 0.38$. **CONCLUDE:** Because the P-value of 0.38 is greater than $\alpha = 0.05$, we fail to reject H_0. There is not convincing evidence of an association between cry count and IQ at age 3 for infants. **(b)** The magnitude of the values for IQ and cry count is lost when turning the quantitative variables into categorical variables needed for the chi-square test. As a result, the P-value for the chi-square test (0.38) is much larger than the P-value for the test for slope (0.002) and the results wouldn't be considered statistically significant.

19. (a) one-sample t interval for μ. **(b)** two-sample z test for $p_1 - p_2$. **(c)** chi-square test for association. **(d)** two-sample t interval for $\mu_1 - \mu_2$.

Chapter 10 Review Exercises

1. (a) STATE: H_0: The biologists' predicted ratio of 1:2:1 for the distribution of offspring is correct. H_a: The biologists' predicted ratio of 1:2:1 for the distribution of offspring is not correct. **PLAN:** Chi-square test for goodness of fit. **Random?** Random sample. ✓ **Large Counts?** All expected counts $(21, 42, 21)$ are ≥ 5. ✓ **DO:** $\chi^2 = 6.48$, P-value $= 0.039$. **CONCLUDE:** Because the P-value of 0.039 is greater than $\alpha = 0.01$, we fail to reject H_0. There is not convincing evidence that the true distribution of offspring is different from what the biologists predict.

2. (a) H_0: There is no association between type of question and response for subjects like these. H_a: There is an association between type of question and response for subjects like these.

(b)

	"Smashed into"	"Hit"	Control
Yes	9.67	9.67	9.67
No	40.33	40.33	40.33

(c) Random? The subjects were randomly assigned to the three treatment groups. ✓ **Large Counts?** All expected counts are ≥ 5 (see part b).

(d) $\chi^2 = \dfrac{(16 - 9.67)^2}{9.67} + \dfrac{(7 - 9.67)^2}{9.67} + \cdots + \dfrac{(44 - 40.33)^2}{40.33} = 7.78$.

P-value: *Using Table C:* P-value is between 0.01 and 0.025. *Using technology:* The command χ^2cdf(lower:7.78, upper:1000, df:2) gives a P-value of 0.020.

3. (a)

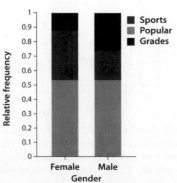

The graph shows that for the students in the study, a higher proportion of females than males chose "be popular," whereas a higher proportion of males than females chose "sports." The proportion who chose "grades" was about the same for both genders.

(b) STATE: H_0: There is no association between gender and goals

for elementary school students. H_a: There is an association between gender and goals for elementary school students. **PLAN:** Chi-square test for association. **Random?** Random sample. ✓ **Large Counts?** All expected counts $(129.70, 117.30, 74.04, 66.96, 47.26, 42.74)$ are ≥ 5. ✓ **DO:** $\chi^2 = 21.455$, P-value ≈ 0. ✓ **CONCLUDE:** Because the P-value of ≈ 0 is less than $\alpha = 0.05$, we reject H_0. There is convincing evidence of an association between gender and goals for elementary school students.

4. (a) $H_0: \beta = 0$ vs. $H_a: \beta > 0$ where β is the slope of the true least-squares regression line relating $y =$ weight gain (ounces) to $x =$ dose of growth hormone (milligrams). **(b)** $t = \dfrac{4.8323 - 0}{1.0164} = 4.75$.

P-value: Using df $= 15 - 2 = 13$, the P-value is less than 0.0005. **(c)** Because the P-value is less than $\alpha = 0.05$, we reject H_0. There is convincing evidence of a positive association between growth hormone and weight gain.

5. (a)

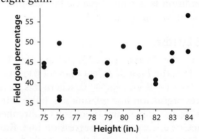

There is a moderate, positive, linear relationship between height and field goal percentage. **(b) STATE:** $H_0: \beta = 0$ vs. $H_a: \beta > 0$, where β is the slope of the true least-squares regression line relating $y =$ field goal percentage to $x =$ height. **PLAN:** t test for the slope of a least-squares regression line. **Random?** Random sample. ✓ **Normal?** A dotplot of the residuals does not show strong skewness or outliers. ✓ **Linear?** There is a linear association between height and field goal percentage in the scatterplot. Also, there is no leftover curved pattern in the residual plot. ✓ **Equal SD?** In the residual plot, we do not see a clear $<$ pattern or $>$ pattern. ✓

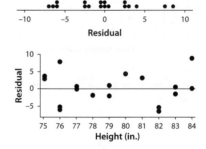

DO: $t = 2.069$, P-value $= 0.028$ with df $= 16$. **CONCLUDE:** Because the P-value of 0.028 is less than $\alpha = 0.05$, we reject H_0. There is convincing evidence of a positive association between height and field goal percentage for 2014–2015 NBA players. **(c) STATE:** We want to estimate $\beta =$ the slope of the true least-squares regression line relating $y =$ field goal percentage to $x =$ height with 90% confidence. **PLAN:** t interval for the slope of a least-squares regression line. Conditions were checked in part **(b)**. **DO:** Using the applet, the interval is 0.133 to 1.340 with df $= 16$. **CONCLUDE:** We are 90% confident that the interval from 0.133 to 1.340 captures the slope of the true least-squares regression line relating $y =$ field goal percentage to $x =$ height.

Chapter 10 Practice Test

1. b
2. b
3. a
4. d

5. c

6. b

7. b

8. d

9. a

10. c

11. STATE: H_0: The distributor's claimed distribution of gas selection is correct for customers in this large city. H_a: The distributor's claimed distribution of gas selection is not correct for customers in this large city. **PLAN:** Chi-square test for goodness of fit. **Random?** Random sample. ✓ **Large Counts?** All expected counts are ≥ 5. $400(0.6) = 240$, $400(0.20) = 80$ and $400(0.20) = 80$. ✓ **DO:** $\chi^2 = 13.15$, P-value $= 0.0014$. **CONCLUDE:** Because the P-value of 0.0014 is less than $\alpha = 0.05$, we reject H_0. There is convincing evidence that the distributor's claimed distribution of gas selection is not correct for customers in this large city.

12. STATE: H_0: There is no association between political party and opinion about the president's policy shift in the population of U.S. residents. H_a: There is an association between political party and opinion about the president's policy shift in the population of U.S. residents. **PLAN:** Chi-square test for association. **Random?** Random sample. ✓ **Large Counts?** All expected counts (130.56, 199.91, 253.53, 52.76, 80.79, 102.46, 40.69, 62.30, 79.01) are ≥ 5. ✓ **DO:** $\chi^2 = 70.17$, P-value. ≈ 0. ✓ **CONCLUDE:** Because the P-value of ≈ 0 is less than $\alpha = 0.05$, we reject H_0. There is convincing evidence of an association between political party and opinion about the president's policy shift in the population of U.S. residents.

13. STATE: H_0:$\beta = 0$ vs. H_a:$\beta \neq 0$, where β is the slope of the true least-squares regression line relating $y =$ senior year GPA to $x =$ freshman year GPA. **PLAN:** t test for the slope of a least-squares regression line. **Random?** Random sample. ✓ **Normal?** A dotplot of the residuals does not show strong skewness or outliers.

✓ **Linear?** There is a linear association between freshman year GPA and senior year GPA in the scatterplot. Also, there is no clear left-over curved pattern in the residual plot. ✓ **Equal SD?** In the residual plot, we do not see a clear $<$ pattern or $>$ pattern. ✓

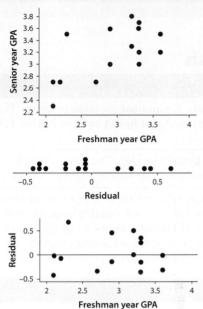

DO: $t = 2.964$, P-value $= 0.011$ with df $= 13$. **CONCLUDE:** Because the P-value of 0.011 is less than $\alpha = 0.05$, we reject H_0. There is convincing evidence that senior GPA is associated with freshman GPA for students at this school.

Glossary/Glosario

English	Español
A	
addition rule for mutually exclusive events For mutually exclusive events A and B, $P(\text{A or B}) = P(\text{A}) + P(\text{B})$. (p. 273)	**regla de la suma para sucesos mutuamente excluyentes** Para los sucesos mutuamente excluyentes A y B, $P(\text{A o B}) = P(\text{A}) + P(\text{B})$. (pág. 273)
alternative hypothesis (H_a) The claim about the population for which we are trying to find evidence. (p. 501)	**hipótesis alternativa** (H_a) Aseveración sobre la población para la que intentamos hallar evidencia. (pág. 501)
association A relationship between two variables in which knowing the value of one variable helps us predict the value of the other. If knowing the value of one variable does not help us predict the value of the other, then there is no association between the variables. (pp. 99, 637)	**asociación** Relación entre dos variables en la que conocer el valor de una de ellas nos ayuda a predecir el valor de la otra. Si conocer el valor de una variable no nos permite predecir el valor de la otra, entonces no hay asociación entre ellas. (págs. 99, 637)
B	
bar chart A chart that shows each category as a bar. The heights of the bars show the category frequencies or relative frequencies. (p. 11)	**gráfica de barras** Gráfica que muestra cada categoría como una barra. La altura de las barras muestra las frecuencias de las categorías o las frecuencias relativas. (pág. 11)
bias The design of a statistical study shows bias if it consistently underestimates or consistently overestimates the value you want to know when the study is repeated many times. (p. 189)	**sesgo** El diseño de un estudio estadístico muestra sesgo si desestima consistentemente o sobreestima consistentemente el valor que se quiere conocer cuando se repite el estudio muchas veces. (pág.189)
binomial distribution In a binomial setting where X = the number of successes, the probability distribution of X is a binomial distribution with parameters n and p, where n is the number of trials of the chance process and p is the probability of a success on any one trial. (p. 351)	**distribución binomial** En una configuración binomial donde X = el número de éxitos, la distribución de probabilidades de X es una distribución binomial con parámetros n y p, donde n es el número de pruebas del proceso aleatorio p es la probabilidad de un éxito en cualquier prueba. (pág. 351)
binomial random variable The count of successes X in a binomial setting. The possible values of X are 0, 1, 2, . . . , n. (p. 349)	**variable aleatoria binomial** Conteo de éxitos X en una configuración binomial. Los valores posibles de X son 0, 1, 2, . . . , n. (pág. 349)
binomial setting Arises when we perform n independent trials of the same chance process and count the number of times that a particular outcome (called a "success") occurs. The four conditions for a binomial setting are: • Binary? The possible outcomes of each trial can be classified as "success" or "failure". • Independent? Trials must be independent. That is, knowing the result of one trial must not tell us anything about the result of any other trial. • Number? The number of trials n of the chance process must be fixed in advance. • Same probability? There is the same probability of success p on each trial. (p. 348)	**configuración binomial** Surge cuando realizamos n pruebas independientes del mismo proceso aleatorio y contamos el número de veces que se obtiene un resultado específico (llamado "éxito"). Las cuatro condiciones para una configuración binomial son: • ¿Binaria? Los resultados posibles de cada prueba se clasifican como "éxito" o "fracaso". • ¿Independiente? Las pruebas deben ser independientes. Es decir, conocer el resultado de una prueba no debe decirnos nada sobre el resultado de ninguna otra. • ¿Número? El número de pruebas n del proceso aleatorio debe ser fijo de antemano. • ¿Misma probabilidad? Existe la misma probabilidad de éxito p en cada prueba. (pág. 348)
boxplot A visual representation of the five-number summary for a distribution of quantitative data. (p. 69)	**gráfica de caja** Representación visual del resumen de cinco números para una distribución de valores cuantitativos. (pág. 69)

C

categorical variable A variable that assigns labels placing individuals into particular groups. (p. 5)

variable categórica Variable que asigna rótulos que colocan a los individuos en grupos específicos. (pág. 5)

---

census A method of collecting data from every individual in the population. (p. 183)

censo Método de recolección de datos sobre cada individuo de la población. (pág. 183)

---

central limit theorem (CLT) Draw an SRS of size n from any population with mean μ and finite standard deviation σ. The central limit theorem (CLT) says that when n is large, the sampling distribution of the sample mean $\bar{x}$ is approximately normal. (p. 440)

teorema del límite central (TLC) Dibujo de una MAS de tamaño n de cualquier población con la media μ y una desviación estándar finita σ. El teorema del límite central (TLC) dice que cuando n es grande, la distribución muestral de la muestra significa que $\bar{x}$ es aproximadamente normal. (pág. 440)

---

chi-square distribution A distribution described by a density curve that takes only non-negative values and is skewed to the right. A particular chi-square distribution is specified by its degrees of freedom. (p. 629)

distribución de chi-cuadrado Una distribución se describe con una curva de densidad que toma solamente valores negativos y está sesgada hacia la derecha. Una distribución particular de chi-cuadrado está especificada por sus grados de libertad. (pág. 629)

---

chi-square test statistic A measure of how different the observed counts are from the expected counts. The formula for the statistic is

$$\chi^2 = \frac{\sum(\text{observed count} - \text{expected count})^2}{\text{expected count}}$$

where the sum is over all categories of the categorical variable or all cells in a two-way table. (p. 624)

estadístico de la prueba de chi-cuadrado Medida de la diferencia que existe entre los datos observados y los datos esperados. La fórmula del estadístico es

$$\chi^2 = \frac{\sum(\text{Datos observado} - \text{Dato esperado})^2}{\text{Dato esperado}}$$

donde la suma es sobre todas las categorías de la variable categórica o todas las celdas de una tabla de doble entrada. (pág. 624)

---

coefficient of determination r^2 A measure of the percent reduction in the sum of squared residuals when using the least-squares regression line to make predictions, rather than the mean value of y. In other words, r^2 measures the percent of the variability in the response variable that is accounted for by the least-squares regression line. (p. 152)

coeficiente de determinación r^2 Medida de la reducción porcentual en la suma de cuadrados de los residuos cuando se usa la recta de regresión por mínimos cuadrados para hacer predicciones, en lugar de la media del valor de y. En otras palabras, r^2 mide el porcentaje de la variabilidad en la variable de respuesta que se explica con la recta de regresión por mínimos cuadrados. (pág. 152)

---

combination A selection of individuals from some group in which the order of selection doesn't matter. (p. 318)

combinación Selección de individuos de un grupo en el que no importa el orden de selección. (pág. 318)

---

complement The complement of event A, written A^C, is the event that A does not happen. (p. 272)

complemento El complemento del suceso A, escrito A^C, es el suceso en que A no ocurre. (pág. 272)

---

complement rule $P(A^C) = 1 - P(A)$, where A^C is the complement of event A. (p. 272)

regla del complemento $P(A^C) = 1 - P(A)$, donde A^C es el complemento del suceso A. (pág. 272)

---

completely randomized design Design in which the experimental units are assigned to the treatments completely by chance. (p. 236)

diseño completamente aleatorizado Diseño en el que las unidades experimentales se asignan a los tratamientos de manera completamente aleatoria. (pág. 236)

---

conditional probability The probability that one event happens given that another event is known to have happened. The conditional probability that event B happens given that event A has happened is denoted by $P(B \mid A)$. (p. 287)

probabilidad condicional Probabilidad de que un suceso ocurra dado que se conoce que otro suceso ha ocurrido. La probabilidad condicional de que el suceso B ocurra dado que el suceso A ha ocurrido se indica con $P(B \mid A)$. (pág. 287)

---

confidence interval An interval of plausible values for a parameter. (p. 453)

intervalo de confianza Intervalo de valores posibles para un parámetro. (pág. 453)

---

confidence level C Gives the long-run success rate of confidence intervals calculated with $C\%$ confidence. That is, in $C\%$ of all possible samples, the interval computed from the sample data will capture the true parameter value. (p. 453)

nivel de confianza C Da la tasa de éxito a largo plazo de los intervalos de confianza calculado con una confianza de $C\%$. Es decir, en $C\%$ de todas las muestras posibles, el intervalo calculado a partir de los datos de muestra captará el verdadero valor de parámetro. (pág. 453)

---

confounding Occurs when two variables are associated in such a way that their effects on a response variable cannot be distinguished from each other. (p. 221)

confusión Ocurre cuando se asocian dos variables de manera tal que no pueden distinguirse sus efectos sobre una variable de respuesta. (pág. 221)

continuous random variable A variable X that can take any value in an interval on the number line. (p. 335)	**variable aleatoria continua** Una variable X que puede tomar cualquier valor en un intervalo sobre la recta numérica. (pág. 335)
control group A group in an experiment used to provide a baseline for comparing the effects of other treatments. Depending on the purpose of the experiment, a control group may be given an inactive treatment or an active treatment. (p. 222)	**grupo de control** Grupo de un experimento que se usa para proporcionar una base que permita comparar los efectos de otros tratamientos. Según cuál sea el propósito del experimento, se puede dar a un grupo de control un tratamiento inactivo o activo. (pág. 222)
convenience sample Choosing individuals from the population who are easy to reach. (p. 188)	**muestra de conveniencia** Elegir individuos de una población a los que es fácil acceder. (pág. 188)
correlation r A measure of the strength and direction of a linear relationship between two quantitative variables. (p. 114)	**correlación r** Medida de la fuerza y la dirección de una relación lineal entre dos variables cuantitativas. (pág. 114)
critical value A multiplier that makes a confidence interval wide enough to have the stated capture rate. (p. 466)	**valor crítico** Multiplicador que hace que un intervalo de confianza sea suficientemente amplio para tener la tasa de captura enunciada. (pág. 466)
cumulative relative frequency graph A graph that plots a point corresponding to the cumulative relative frequency in each interval at the smallest value of the next interval, starting with a point at a height of 0% at the smallest value of the first interval. Consecutive points are then connected with a line segment to form the graph. (p. 79)	**gráfica de frecuencia relativa acumulada** Gráfica donde se marca un punto que se corresponde con la frecuencia relativa acumulada de cada intervalo en el menor valor del siguiente intervalo, que comienza con un punto a la altura de 0% en el menor valor del primer intervalo. Los puntos consecutivos luego se conectan con un segmento de recta para formar la gráfica. (pág. 79)

D

density curve A curve that describes the probability distribution of a continuous random variable; this curve is always on or above the horizontal axis, and has an area of exactly 1 underneath it. (p. 366)	**curva de densidad** Curva que describe la probabilidad de distribución de una variable aleatoria continua; la curva está siempre sobre o por encima del eje horizontal y tiene un área de exactamente 1 debajo de ella. (pág. 366)
discrete random variable A variable X that takes a fixed set of possible values with gaps between. (p. 332)	**variable aleatoria discreta** Variable X que toma un conjunto fijo de valores posibles con brechas entre ellos. (pág. 332)
distribution of a variable Indicates what values a variable takes and how often it takes these values. (p. 7)	**distribución** De una variable nos dice qué valores toma la variable y con qué frecuencia los toma. (pág. 7)
dotplot A graph that displays the distribution of a quantitative variable by plotting each data value as a dot above its location on a number line. (p. 21)	**diagrama de puntos** Gráfica que muestra la distribución de una variable cuantitativa mediante la representación de cada valor de datos como un punto sobre su ubicación en una recta numérica. (pág. 21)
double-blind An experiment in which neither the subjects nor those who interact with them and measure the response variable know which treatment a subject received. (p. 224)	**doble ciego** Experimento en el que ni los sujetos ni quienes interactúan con ellos y miden la variable de respuesta saben qué tratamiento recibió un sujeto. (pág. 224)

E

event Any collection of outcomes from some chance process. (p. 271)	**suceso** Conjunto de resultados de un proceso aleatorio. (pág. 271)
experiment A study that deliberately imposes some treatment on individuals to measure their responses. (p. 184)	**experimento** Estudio que impone deliberadamente algún tratamiento sobre individuos para medir sus respuestas. (pág. 184)
experimental units The smallest collection of individuals to which treatments are randomly assigned. (p. 222)	**unidades experimental** Conjunto más pequeño de individuos a los que se asignan tratamientos de manera aleatoria. (pág. 222)
explanatory variable A variable that may help predict or explain changes in a response variable. (p. 96)	**variable explicativa** Variable que permite predecir o explicar cambios en una variable de respuesta. (pág. 96)
exponential model A model in the form $\hat{y} = a(b)^x$, where $b > 0$. If $b > 1$, the graph will show exponential growth. If $0 < b < 1$, the graph will show exponential decay. (p. 164)	**modelo exponencial** Modelo en forma $\hat{y} = a(b)^x$, donde $b > 0$. Si $b > 1$, la gráfica mostrará crecimiento exponencial. Si $0 < b < 1$, la gráfica mostrará disminución exponencial. (pág. 164)

extrapolation The use of a regression line for prediction far outside the interval of x values used to obtain the line. Such predictions are often not accurate. (p. 131)

extrapolación Uso de una recta de regresión para la predicción muy por fuera de los valores de x que se usan para obtener la recta. Las predicciones no suelen ser precisas. (pág. 131)

F

factorial For any positive integer n, we define $n!$ (n factorial) as $n! = n(n - 1)(n - 2) \ldots \cdot 3 \cdot 2 \cdot 1$. (p. 311)

factorial Por cada entero positivo n, definimos $n!$ (n factorial) como $n! = n(n - 1)(n - 2) \ldots \cdot 3 \cdot 2 \cdot 1$. (pág. 311)

first quartile Q_1 The median of the data values that are to the left of the median in an ordered list. (p. 60)

primer cuartil Q_1 Mediana de los valores de datos que están a la izquierda de la mediana en una lista ordenada. (pág. 60)

five-number summary A summary of a distribution of quantitative data that consists of the minimum, the first quartile Q_1, the median, the third quartile Q_3, and the maximum. (p. 69)

resumen de cinco números Resumen de una distribución de datos cuantitativos que consiste en el mínimo, el primer cuartil Q_1, la mediana, el tercer cuartil Q_3 y el máximo. (pág. 69)

frequency table A table that shows the number of individuals having each data value. (p. 7)

tabla de frecuencia Tabla que muestra el número de individuos que tienen valores de datos. (pág. 7)

G

general addition rule for two events If A and B are any two events resulting from some chance process, $P(\text{A or B}) = P(\text{A}) + P(\text{B}) - P(\text{A and B})$. (p. 279)

regla general de la suma de dos sucesos Si A y B son dos sucesos cualesquiera que resultan de un proceso aleatorio, $P(\text{A o B}) = P(\text{A}) + P(\text{B}) - P(\text{A y B})$. (pág. 279)

general multiplication rule For any chance process, the probability that events A and B both occur can be found using the general multiplication rule: $P(\text{A and B}) = P(\text{A}) \cdot P(\text{B} \mid \text{A})$. (p. 295)

regla general de la multiplicación Para cualquier proceso aleatorio, la probabilidad de que ocurran los sucesos A y B puede hallarse mediante la regla general de la multiplicación: $P(\text{A y B}) = P(\text{A}) \cdot P(\text{B} \mid \text{A})$. (pág. 295)

H

histogram A graph that displays the distribution of a quantitative variable where each interval is represented by a bar. The heights of the bars show the frequencies or relative frequencies of values in each interval. (p. 38)

histograma Gráfica que muestra la distribución de una variable cuantitativa en la que cada intervalo está representado por una barra. La altura de las barras muestra las frecuencias o las frecuencias relativas de los valores de cada intervalo. (pág. 38)

I

independent events A and B are independent events if knowing whether or not one event has occurred does not change the probability that the other event will happen. Events A and B are independent if $P(\text{A} \mid \text{B}) = P(\text{A} \mid \text{B}^C) = P(\text{A})$ and $P(\text{B} \mid \text{A}) = P(\text{B} \mid \text{A}^C) = P(\text{B})$. (p. 289)

sucesos independientes A y B son sucesos independientes si saber si un suceso ha ocurrido no cambia la probabilidad de que el otro ocurra. Los sucesos A y B son independientes si $P(\text{A} \mid \text{B}) = P(\text{A} \mid \text{B}^C) = P(\text{A})$ y $P(\text{B} \mid \text{A}) = P(\text{B} \mid \text{A}^C) = P(\text{B})$. (pág. 289)

individual A person, animal, or thing described in a set of data. (p. 5)

individuo Persona, animal o cosa descrita en un conjunto de datos. (pág. 5)

interquartile range (IQR) The distance between the first and third quartiles of a distribution: $IQR = Q_3 - Q_1$. (p. 61)

rango intercuartílico (RIC) Distancia entre el primer y el tercer cuartil de una distribución: $RIC = Q_3 - Q_1$. (pág. 61)

intersection The event "A and B" is called the intersection of events A and B. It consists of all outcomes that are common to both events, and is denoted $\text{A} \cap \text{B}$. (p. 281)

intersección El suceso "A y B" se llama intersección de los sucesos A y B. Consiste en todos los resultados que son comunes a ambos sucesos, y se representa con $\text{A} \cap \text{B}$. (pág. 281)

L

Large Counts condition Suppose X is the number of successes and $\hat{p}$ is the proportion of successes in a random sample of size n from a population with proportion of successes p. The Large Counts condition says that the distribution of X and the distribution of $\hat{p}$ will be approximately normal when $np \geq 10$ and $n(1 - p) \geq 10$. For a chi-square test, the Large Counts condition says that the expected counts must all be at least 5. (pp. 420, 428, 628)

condición de datos grandes Suponga que X es el número de éxitos y $\hat{p}$ es la proporción de éxitos en una muestra aleatoria de tamaño n de una población con una proporción de éxitos p. La condición de datos grandes dice que la distribución de X y la distribución de $\hat{p}$ será aproximadamente normal cuando $np \geq 10$ y $n(1 - p) \geq 10$. Para una prueba de chi-cuadrado, la condición de datos grandes dice que todos los datos esperados deben ser al menos de 5. (págs. 420, 428, 628)

law of large numbers If we observe more and more repetitions of any chance process, the proportion of times that a specific outcome occurs approaches its probability. (p. 264)	**ley de números grandes** Si observamos cada vez más repeticiones de cualquier proceso aleatorio, la proporción de veces que ocurre un resultado específico se acerca a su probabilidad. (pág. 264)
least-squares regression line The line that makes the sum of the squared residuals as small as possible. (p. 138)	**recta de regresión por mínimos cuadrados** Recta que hace que la suma del cuadrado de los residuos sea tan pequeña como sea posible. (pág. 138)

M

margin of error Describes how far, at most, we expect an estimate to vary from the true population value. That is, in a $C\%$ confidence interval, the distance between the point estimate and the true parameter value will be less than the margin of error in $C\%$ of all samples. (pp. 204, 454)	**margen de error** Describe qué tan lejos, como máximo, esperamos que varíe una estimación con respecto al verdadero valor poblacional. Es decir, en un intervalo de confianza $C\%$, la distancia entre la estimación puntual y el verdadero valor de parámetro será menor que el margen de error en $C\%$ de todas las muestras. (págs. 204, 454)
mean The mean of a quantitative data set is the average of all n data values. To find the mean, add all of the values and divide by n. (p. 51)	**media** La media de un conjunto de datos cuantitativos es el promedio de todos los valores de datos n. Para hallar la media, sume todos los valores y divida por n. (pág. 51)
mean of a continuous random variable The point at which the probability distribution of a continuous random variable would balance if made of solid material. (p. 368)	**media de una variable aleatoria continua** Punto en el que la distribución de las probabilidades de una variable aleatoria continua se equilibrará si está formada por material sólido. (pág. 368)
mean (expected value) of a discrete random variable A discrete random variable's long-run average value over many, many repetitions of the same chance process. (p. 340)	**media (valor esperado) de una variable aleatoria discreta** Valor promedio a largo plazo de una variable aleatoria discreta durante muchísimas repeticiones del mismo proceso aleatorio. (pág. 340)
median The midpoint of a distribution, which is the number such that about half the observations are smaller and about half are larger. (p. 49)	**mediana** Punto medio de una distribución, que es el número tal que aproximadamente la mitad de las observaciones son más pequeñas y aproximadamente la mitad son más grandes. (pág. 49)
median of a continuous random variable The equal-areas point, the point that divides the area under the probability distribution of a continuous random variable in half. (p. 368)	**mediana de una variable aleatoria continua** Punto de áreas iguales, punto que divide el área bajo la distribución de las probabilidades de una variable aleatoria continua por la mitad. (pág. 368)
multiplication counting principle The result of a process involving multiple (k) steps; suppose that there are n_1 ways to do Step 1, n_2 ways to do Step 2, . . . , and n_k ways to do Step k. The total number of different ways to complete the process is $n_1 \cdot n_2 \cdot \ldots \cdot n_k$. (p. 310)	**principio de conteo de la multiplicación** Resultado de un proceso que implica varios pasos (k); imagine que hay n_1 maneras de hacer el paso 1, n_2 maneras de hacer el paso 2, . . . , y n_k maneras de hacer el paso k. El número total de maneras diferentes de completar el proceso es $n_1 \cdot n_2 \cdot \ldots \cdot n_k$. (pág. 310)
multiplication rule for independent events If A and B are independent events, the probability that A and B both occur is $P(A \text{ and } B) = P(A) \cdot P(B)$. (p. 303)	**regla de la multiplicación para sucesos independientes** Si A y B son sucesos independientes, la probabilidad de que tanto A como B ocurran es $P(A \text{ y } B) = P(A) \cdot P(B)$. (pág. 303)
mutually exclusive events Two events A and B that have no outcomes in common and so can never occur together—that is, $P(A \text{ and } B) = 0$. (p. 273)	**sucesos mutuamente excluyentes** Dos sucesos A y B que no tienen resultados en común, y por lo tanto, no pueden ocurrir juntos—es decir, $P(A \text{ y } B) = 0$. (pág. 273)

N

$_nC_k$ If there are n individuals, then the notation $_nC_k$ represents the number of different combinations of k individuals chosen from the entire group of n. (p. 318)	$_nC_k$ Si hay n individuos, entonces la notación $_nC_k$ representa el número de combinaciones diferente de k individuos elegidos del grupo completo de n. (pág. 318)
$_nP_k$ If there are n individuals, then the notation $_nP_k$ represents the number of different permutations of k individuals selected from the entire group of n. (p. 312)	$_nP_k$ Si hay n individuos, entonces la notación $_nP_k$ representa el número de permutaciones diferente de k individuos seleccionados del grupo completo de n. (pág. 312)

nonresponse Occurs when an individual chosen for the sample can't be contacted or refuses to participate. (p. 215)

no respuesta Ocurre cuando no se puede contactar a un individuo elegido para la muestra o este se niega a participar. (pág. 215)

normal distribution A distribution described by a symmetric, single-peaked, bell-shaped density curve. Any normal distribution is completely specified by two numbers: its mean μ and standard deviation σ. (p. 370)

distribución normal Distribución que se describe con una curva de densidad simétrica, de un solo pico y con forma de campana. Cualquier distribución normal se especifica completamente con dos números: su media μ y la desviación estándar σ. (pág. 370)

Normal/Large Sample condition A condition for performing inference about a mean, which requires that the data come from a normally distributed population or that the sample size is large ($n \geq 30$). When the sample size is small and the shape of the population distribution is unknown, a graph of the sample data shows no strong skewness or outliers. When performing inference about a difference between two means, check that this condition is met for both groups. (pp. 440, 529, 573)

Condición de muestra normal/grande Condición para realizar una inferencia sobre una media, que requiere que los datos provengan de una población distribuida normalmente o que el tamaño de la muestra sea grande ($n \geq 30$). Cuando el tamaño de la muestra es pequeño y la forma de la distribución de la población no se conoce, una gráfica de los datos de la muestra no presenta un sesgo marcado o valores atípicos. Cuando realice una inferencia sobre una diferencia entre dos medias, verifique que ambos grupos cumplan con esta condición. (págs. 440, 529, 573)

null hypothesis (H_0) The claim about the population that we weigh evidence against in a statistical test. (p. 501)

hipótesis nula (H_0) Aseveración sobre la población contra la que evaluamos la evidencia en una prueba estadística. (pág. 501)

O

observational study A study that observes individuals and measures variables of interest, but does not attempt to influence the responses. (p. 184)

estudio observable Estudio que observa a los individuos y mide las variables de interés, pero no pretende influir en las respuestas. (pág. 184)

one-sided alternative hypothesis An alternative hypothesis is one-sided if it states that a parameter is *greater than* the null value or if it states that the parameter is *less than* the null value. (p. 502)

hipótesis alternativa unilateral Una hipótesis alternativa es unilateral si establece que un parámetro es *mayor que* el valor nulo o si establece que el parámetro es *menor que* el valor nulo. (pág. 502)

outlier Individual value that falls outside the overall pattern of a distribution. Call an observation an outlier if it falls more than $1.5 \times IQR$ above the third quartile or more than $1.5 \times IQR$ below the first quartile. (pp. 25, 68)

valor atípico Valor individual que está fuera del patrón general de una distribución. Llamamos valor atípico a una observación si cae más de $1.5 \times RIC$ por encima del tercer cuartil o más de $1.5 \times RIC$ por debajo del primer cuartil. (págs. 25, 68)

P

paired data The result of recording two values of the same quantitative variable for each individual or for each pair of similar individuals. (p. 592)

datos pareados Resultado de registrar dos valores de la misma variable cuantitativa para cada individuo o para cada par de individuos similares. (pág. 592)

parameter A number that describes some characteristic of the population. (p. 401)

parámetro Número que describe algunas características de la población. (pág. 401)

percentile The percent of values in a distribution that are less than the individual's data value. (p. 78)

percentil Porcentaje de valores en una distribución que son menores que el valor de los datos del individuo. (pág. 78)

permutation A distinct arrangement of some group of individuals. (p. 311)

permutación Conjunto distintivo de un grupo de individuos. (pág. 311)

pie chart A graph that shows each category as a slice of the "pie." The areas of the slices are proportional to the category frequencies or relative frequencies. (p. 11)

gráfica circular Gráfica que muestra cada categoría como porción de una "torta". Las áreas de las porciones son proporcionales a las frecuencias de las categorías o las frecuencias relativas. (pág. 11)

placebo A treatment that has no active ingredient, but is otherwise like other treatments. (p. 223)

placebo Tratamiento que no tiene ingredientes activos, pero que por lo demás es como otros tratamientos. (pág. 223)

placebo effect The fact that some subjects in an experiment will respond favorably to any treatment, even an inactive treatment. (p. 223)

efecto placebo Hecho de que algunos sujetos de un experimento respondan favorablemente a cualquier tratamiento, aunque sea un tratamiento inactivo. (pág. 223)

point estimate A single-value estimate of a population parameter. (p. 454)	**estimación puntual** Estimación de un solo valor de un parámetro de población. (pág. 454)
population The entire group of individuals we want information about in a statistical study. (p. 183)	**población** Grupo completo de individuos de quienes queremos obtener información en un estudio estadístico. (pág. 183)
probability A number between 0 and 1 that describes the proportion of times an outcome of a chance process would occur in a very large number of repetitions. (p. 263)	**probabilidad** Número entre 0 y 1 que describe la proporción de veces que un resultado de un proceso aleatorio ocurriría en un número muy grande de repeticiones. (pág. 263)
probability distribution Gives the possible values of a random variable and their probabilities. (p. 332)	**distribución de probabilidad** Da sus posibles valores a una variable aleatoria y sus probabilidades. (pág. 332)
probability model A description of some chance process that consists of two parts: a list of all possible outcomes and the probability for each outcome. (p. 270)	**modelo probabilístico** Descripción de un proceso aleatorio que consiste en dos partes: una lista de todos los resultados posibles y la probabilidad de cada resultado. (pág. 270)
P-value The probability of getting evidence for the alternative hypothesis H_a as strong as or stronger than the observed evidence when the null hypothesis H_0 is true. (p. 503)	**valor P** Probabilidad de obtener evidencia de la hipótesis alternativa H_a igual o mayor que la evidencia observada cuando la hipótesis nula H_0 es verdadera. (pág. 503)

Q

quadratic model A model in the form $\hat{y} = ax^2 + bx + c$. The graph of a quadratic model is a parabola. (p. 160)	**modelo cuadrático** Modelo en forma $\hat{y} = ax^2 + bx + c$. La gráfica de un modelo cuadrático es una parábola. (pág. 160)
quantitative variable A variable that takes number values for which it makes sense to find an average. (p. 5)	**variable cuantitativa** Variable que toma valores numéricos para los que tiene sentido hallar un promedio. (pág. 5)
quartiles The quartiles of a distribution divide the ordered data set into four groups having roughly the same number of values. (p. 60)	**cuartiles** Los cuartiles de una distribución dividen el conjunto de datos ordenados en cuatro grupos que tienen el mismo número de valores. (pág. 60)

R

random assignment In an experiment, when the experimental units are assigned to treatments using a chance process. (p. 229)	**asignación aleatoria** En un experimento, cuando las unidades experimentales se asignan a tratamientos mediante un proceso aleatorio. (pág. 229)
random condition A condition for performing inference, which requires that the data come from a random sample from the population of interest or from a randomized experiment. When comparing two populations or treatments, check that the data come from separate random samples from the two populations of interest or from two groups in a randomized experiment. (pp. 465, 573)	**condición aleatoria** Condición para realizar una inferencia, que requiere que los datos provengan de una muestra aleatoria de la población de interés o de un experimento aleatorizado. Cuando compare dos poblaciones o tratamientos, verifique que los datos provengan de muestras aleatorias distintas de las dos poblaciones de interés o de dos grupos de un experimento aleatorizado. (págs. 465, 573)
random sampling A method of sampling that involves using a chance process to determine which members of a population are included in the sample. (p. 190)	**muestreo aleatorio** Método de muestreo que implica usar un proceso aleatorio para determinar qué miembros de una población están incluidos en la muestra. (pág. 190)
random variable A variable that takes numerical values describing the outcomes of a chance process. (p. 332)	**variable aleatoria** Variable que toma valores numéricos que describen los resultados de un proceso aleatorio. (pág. 332)
range The distance between the minimum value and the maximum value of a distribution. That is, range = maximum − minimum. (p. 59)	**rango** Distancia entre el valor mínimo y el valor máximo de una distribución. Es decir, rango = máximo − mínimo. (pág. 59)
regression line A line that describes how a response variable y changes as an explanatory variable x changes. Regression lines are expressed in the form $\hat{y} = a + bx$, where $\hat{y}$ (pronounced "y hat") is the predicted value of y for a given value of x. (p. 131)	**recta de regresión** Recta que describe cómo una variable de respuesta y cambia al mismo tiempo que cambia una variable explicativa x. Las rectas de regresión se expresan en forma $\hat{y} = a + bx$, donde $\hat{y}$ es el valor predicho de y para un valor dado de x. (pág. 131)

relative frequency table A table that shows the proportion or percent of individuals having each data value. (p. 7)	**tabla de frecuencia relativa** Tabla que muestra la proporción, o porcentaje, de individuos que tienen cada valor de datos. (pág. 7)
residual The difference between an actual value of y and the value of y predicted by the regression line. That is, residual = actual y − predicted $y = y - \hat{y}$. (p. 132)	**residuo** Diferencia entre un valor real de y y el valor de y predicho por la recta de regresión. Es decir, residuo = real y real − y predicha = $y - \hat{y}$. (pág. 132)
residual plot A scatterplot that plots the residuals on the vertical axis and the explanatory variable on the horizontal axis. (p. 149)	**diagrama de residuos** Diagrama de puntos donde se marcan los residuos en el eje vertical y la variable explicativa en el eje horizontal. (pág. 149)
resistant A measure of center (or variability) is resistant if it isn't influenced by unusually large or unusually small values in a distribution. (p. 52)	**resistente** Una medida de tendencia central (o variabilidad) es resistente si los valores inusualmente grandes o inusualmente pequeños de una distribución no influyen en ella. (pág. 52)
response bias Occurs when there is a consistent pattern of inaccurate responses to a survey question. (p. 216)	**sesgo de respuesta** Ocurre cuando hay un patrón consistente de respuestas imprecisas a una pregunta de encuesta. (pág. 216)
response variable A variable that measures an outcome of a study. (p. 96)	**variable de respuesta** Variable que mide un resultado de un estudio. (pág. 96)

S

sample A subset of individuals in the population from which we collect data. (p. 183)	**muestra** Subconjunto de individuos en la población de los que se reúnen datos. (pág. 183)
sample space The list of all possible outcomes of some chance process. (p. 270)	**espacio muestral** Lista de todos los resultados posibles de un proceso aleatorio. (pág. 270)
sampling distribution The distribution of values taken by a statistic in all possible samples of the same size from the same population. (p. 402)	**distribución muestral** Distribución de valores que toma una estadística en todas las muestras posibles del mismo tamaño de la misma población. (pág. 402)
sampling distribution of the sample count X The distribution of values taken by the sample count X in all possible samples of the same size from the same population. (p. 417)	**distribución muestral del dato X** Distribución de valores que toma el dato de muestra X en todas las muestras posibles del mismo tamaño de la misma población. (pág. 417)
sampling distribution of the sample mean $\bar{x}$ The distribution of values taken by the sample mean $\bar{x}$ in all possible samples of the same size from the same population. (p. 433)	**distribución muestral de la media $\bar{x}$** Distribución de valores que toma la media de la muesta $\bar{x}$ en todas las muestras posibles del mismo tamaño de la misma población. (pág. 433)
sampling distribution of the sample proportion $\hat{p}$ The distribution of values taken by the sample proportion $\hat{p}$ in all possible samples of the same size from the same population. (p. 425)	**distribución muestral de la proporción $\hat{p}$** Distribución de valores que toma la proporción de la muestra $\hat{p}$ en todas las muestras posibles del mismo tamaño de la misma población. (pág. 425)
sampling variability The fact that different random samples of the same size from the same population produce different estimates. (p. 196)	**variabilidad muestral** Hecho de que diferentes muestras aleatorias del mismo tamaño de la misma población produzcan diferentes estimaciones. (pág. 196)
scatterplot A graph that shows the relationship between two quantitative variables measured on the same individuals. The values of one variable appear on the horizontal axis, and the values of the other variable appear on the vertical axis. Each individual in the data set appears as a point in the graph. (p. 105)	**diagrama de dispersión** Gráfica que muestra la relación entre dos variables cuantitativas medidas en el mismo individuo. Los valores de una variable aparecen en el eje horizontal y los valores de la otra aparecen en el eje vertical. Cada individuo del conjunto de datos aparece como un punto en la gráfica. (pág. 105)
segmented bar chart A graph that displays the possible values of a categorical variable as segments of a rectangle, with the area of each segment proportional to the percent of individuals in the corresponding category. (p. 97)	**gráfica de barras segmentadas** Gráfica que muestra los valores posibles de una variable categórica como segmentos de un rectángulo, con el área de cada segmento proporcional al porcentaje de individuos en la categoría correspondiente. (pág. 97)
significance level α The value that we use as a cutoff to decide if an observed result is unlikely to happen by chance alone when the null hypothesis is true. (p. 508)	**nivel de significancia α** Valor que se usa como atajo para decidir si es probable que ocurra solo por casualidad un resultado observado cuando la hipótesis nula es verdadera. (pág. 508)

significance test A formal procedure for using observed data to decide between two competing claims (called *hypotheses*). The claims are usually statements about a parameter, like the population proportion p or the population mean μ. (p. 500)	**prueba de significancia** Procedimiento formal para usar datos observados y decidir entre dos afirmaciones opuestas (llamadas *hipótesis*). Las afirmaciones son generalmente enunciados sobre un parámetro, como la proporción poblacional p o la media poblacional μ. (pág. 500)
simple random sample (SRS) A sample of size n chosen in such a way that every group of n individuals in the population has an equal chance of being selected as the sample. (p. 194)	**muestra aleatoria simple (MAS)** Muestra de tamaño n elegida de manera tal que cada grupo de n individuos de la población tenga iguales posibilidades de ser elegido como muestra. (pág. 194)
simulation The imitation of chance behavior, based on a model that accurately reflects the situation. (p. 265)	**simulación** Imitación de comportamiento aleatorio, basada en un modelo que refleja la situación con precisión. (pág. 265)
single-blind An experiment where either the subjects don't know which treatment they are receiving or the people who interact with them and measure the response variable don't know which subjects are receiving which treatment. (p. 224)	**simple ciego** Experimento en el que los individuos no saben qué tratamiento reciben o quiénes interactúan con ellos y miden la variable de respuesta y no saben qué sujetos reciben cada tratamiento. (pág. 224)
68–95–99.7 rule In any normal distribution with mean μ and standard deviation σ, about 68% of the values fall within σ of the mean μ, about 95% of the values fall within 2σ of the mean μ, and about 99.7% of the values fall within 3σ of the mean μ. (p. 376)	**regla del 68–95–99.7** En cualquier distribución normal con la media μ y la desviación estándar σ, aproximadamente el 68% de los valores caen dentro de σ de la media μ, aproximadamente el 95% de los valores caen dentro de 2σ de la media μ, y aproximadamente 99.7% de los valores caen dentro de 3σ de la media μ. (pág. 376)
skewed distribution A distribution is *skewed to the right* if the right side of the graph is much longer than the left side. A distribution is *skewed to the left* if the left side of the graph is much longer than the right side. (p. 23)	**distribución sesgada** Una distribución es *sesgada a la derecha* si el lado derecho de la gráfica es mucho más largo que el lado izquierdo. Una distribución es *sesgada a la izquierda* si el lado izquierdo de la gráfica es mucho más largo que el lado derecho. (pág. 23)
slope the slope b of a regression line describes the predicted change in the y variable for each 1-unit increase in the x variable. (p. 133)	**pendiente** La pendiente b de una recta de regresión describe el cambio predicho en la variable y por cada aumento de 1 unidad en la variable x. (pág. 133)
standard deviation Measures the typical distance of the values in a distribution from the mean. To find the standard deviation s_x of a quantitative data set with n values: (1) Find the mean of the distribution. (2) Calculate the deviation of each value from the mean: deviation = value − mean. (3) Square each of the deviations. (4) Add all the squared deviations, divide $n - 1$, and take the square root. (p. 61)	**desviación estándar** Mide la distancia típica de los valores en una distribución desde la media. Para hallar la desviación estándar s_x de un conjunto de datos cuantitativos con n valores: (1) Halle la media de la distribución. (2) Calcule la desviación de cada valor desde la media: desviación = valor − media. (3) Calcule la raíz cuadrada de cada una de las desviaciones. (4) Sume el cuadrado de todas las desviaciones, divida $n - 1$, y calcule la raíz cuadrada. (pág. 61)
standard deviation of a discrete random variable Measures how much the values of the variable typically differ from the mean. (p. 342)	**desviación estándar de una variable aleatoria discreta** Mide cuánto difieren generalmente los valores de la variable respecto de la media. (pág. 342)
standard deviation of the residuals s A measure of the size of a typical residual. That is, s measures the typical distance between the actual y values and the predicted y values. (p. 151)	**desviación estándar de los residuos s** Medida del tamaño de un residuo típico. Es decir, s mide la distancia típica entre los valores reales de y y los valores predichos de y. (pág. 151)
standard error of $\hat{p}$ An estimate of the standard deviation of the sampling distribution of $\hat{p}$: $$SE_{\hat{p}} = \sqrt{\frac{\hat{p}(1-\hat{p})}{n}}$$ The standard error of $\hat{p}$ estimates how much $\hat{p}$ typically varies from p. (p. 467)	**error estándar de $\hat{p}$** Estimación de la desviación estándar de la distribución muestral de $\hat{p}$: $$SE_{\hat{p}} = \sqrt{\frac{\hat{p}(1-\hat{p})}{n}}$$ El error estándar de $\hat{p}$ estima cuánto varía generalmente $\hat{p}$ respecto de p. (pág. 467)

standard error of $\bar{x}$ An estimate of the standard deviation of the sampling distribution of $\bar{x}$:

$$SE_{\bar{x}} = \frac{s_x}{\sqrt{n}}$$

The standard error of $\bar{x}$ estimates how much $\bar{x}$ typically varies from μ. (p. 480)

error estándar de $\bar{x}$ Estimación de la desviación estándar de la distribución muestral de $\bar{x}$:

$$SE_{\bar{x}} = \frac{s_x}{\sqrt{n}}$$

El error estándar de $\bar{x}$ estima cuánto varía generalmente $\bar{x}$ respecto de μ. (pág. 480)

---

standard normal distribution The normal distribution with mean 0 and standard deviation 1. (p. 378)

distribución normal estándar Distribución normal con media 0 y desviación estándar 1. (pág. 378)

---

standardized score (z-score) For an individual value in a distribution, indicates how many standard deviations from the mean the value falls, and in what direction. To find the standardized score (z-score), compute $z = \dfrac{\text{value} - \text{mean}}{\text{standard deviation}}$. (p. 81)

puntuación estandarizada (puntuación z) Para un valor individual de una distribución, indica a cuántas desviaciones estándar respecto de la media cae el valor, y en qué dirección. Para hallar el puntaje estandarizado (puntaje z), halle $z = \dfrac{\text{valor} - \text{desviación media}}{\text{estándar}}$. (pág. 81)

---

standardized test statistic A value that measures how far a sample statistic is from what we would expect if the null hypothesis H_0 were true, in standard deviation units. That is, standardized test statistic $= \dfrac{\text{statistic} - \text{null value}}{\text{standard deviation of statistic}}$. (p. 517)

estadístico de prueba estandarizada Valor que mide a qué distancia está el estadístico de una muestra de lo que esperaríamos si la hipótesis nula H_0 fuera verdadera, en unidades de desviación estándar. Es decir, e de prueba estandarizada $= \dfrac{\text{estadístico} - \text{valor nulo}}{\text{desviación estándar del estadístico}}$ (pág. 517)

---

statistic A number that describes some characteristic of a sample. (p. 401)

estadístico Número que describe alguna característica de una muestra. (pág. 401)

---

statistical problem-solving process The steps involved in solving statistics problems, including (1) ask questions, (2) collect data, (3) analyze data, and (4) interpret results. (p. 5)

proceso estadístico de resolución de problemas Pasos que intervienen en la resolución de problemas estadísticos. Incluyen: (1) hacer preguntas, (2) reunir datos, (3) analizar datos e (4) interpretar resultados. (pág. 5)

---

statistically significant When an observed difference in responses between the groups in an experiment is too large to be explained by chance variation in the random assignment. (p. 237)

estadísticamente significativo Cuando una diferencia observada en respuestas entre los grupos de un experimento es demasiado grande para explicarse con una variación aleatoria en la asignación aleatoria. (pág. 237)

---

statistics The science and art of collecting, analyzing, and drawing conclusions from data. (p. 4)

estadística Ciencia y arte de reunir y analizar datos y luego sacar conclusiones a partir de ellos. (pág. 4)

---

stemplot A graph that shows each data value separated into two parts: a *stem*, which consists of all but the final digit, and a *leaf*, the final digit. The stems are ordered from least to greatest and arranged in a vertical column. The leaves are arranged in increasing order out from the appropriate stems. (p. 30)

diagrama de tallo y hoja Gráfica que muestra cada valor de datos separado en dos partes: un *tallo*, que consiste en todos los dígitos menos el último, y una *hoja*, el último dígito. Los tallos están ordenados de menor a mayor en una columna vertical. Las hojas están en orden creciente desde los tallos correspondientes. (pág. 30)

---

subjects Experimental units that are human beings. (p. 222)

sujetos Unidades experimentales que son seres humanos. (pág. 222)

---

symmetric distribution A distribution is roughly symmetric if the right side of the graph (containing the half of the observations with larger values) is approximately a mirror image of the left side. (p. 23)

distribución simétrica Una distribución es aproximadamente simétrica si el lado derecho de la gráfica (que contiene la mitad de las observaciones con valores más grandes) es aproximadamente una imagen especular del lado izquierdo. (pág. 23)

T

t distribution A distribution described by a symmetric, single-peaked, bell-shaped density curve. Any t distribution is completely described by its *degrees of freedom*. (p. 532)

distribución t Distribución descrita por una curva de densidad simétrica de un solo pico con forma de campana. Cualquier distribución t se describe con sus *grados de libertad*. (pág. 532)

---

third quartile Q_3 The median of the data values that are to the right of the median in an ordered list. (p. 60)

tercer cuartil Q_3 Mediana de los valores de datos que están a la derecha de la mediana en una lista ordenada. (pág. 60)

treatment A specific condition applied to the individuals in an experiment. (p. 222)	**tratamiento** Condición específica que se aplica a los individuos en un experimento. (pág. 222)
tree diagram A diagram that shows the sample space of a chance process involving multiple stages. The probability of each outcome is shown on the corresponding branch of the tree. All probabilities after the first stage are conditional probabilities. (p. 296)	**diagrama de árbol** Diagrama que presenta el espacio muestral de un proceso aleatorio. La probabilidad de cada resultado se muestra en la rama correspondiente del árbol. Todas las probabilidades después de la primera etapa son probabilidades condicionales. (pág. 296)
two-sided alternative hypothesis An alternative hypothesis is two-sided if it states that the parameter is *different from* the null value (it could be either greater than or less than). (p. 502)	**hipótesis alternativa bilateral** Una hipótesis alternativa es bilateral si establece que el parámetro es *diferente del* valor nulo (podría ser mayor o menor). (pág. 502)
Type I error An error that occurs if a test rejects H_0 when H_0 is true. That is, the test finds convincing evidence that H_a is true when it really isn't. (p. 509)	**Error de tipo I** Error que ocurre si una prueba rechaza H_0 cuando H_0 es verdadero. Es decir, las pruebas hallan evidencia convincente de que H_a es verdadero cuando en realidad no lo es. (pág. 509)
Type II error An error that occurs if a test fails to reject H_0 when H_a is true. That is, the test does not find convincing evidence that H_a is true when it really is. (p. 509)	**Error de tipo II** Error que ocurre si una prueba no rechaza H_0 cuando H_a es verdadero. Es decir, la prueba no halla evidencia convincente de que H_a es verdadero cuando realmente lo es. (pág. 509)

U

unbiased estimator A statistic used to estimate a parameter where the mean of its sampling distribution is equal to the value of the parameter being estimated. (p. 410)	**estimador imparcial** Estadística que se usa para estimar un parámetro en el que la media de su distribución muestral equivale al valor del parámetro que se estima. (pág. 410)
undercoverage Occurs when some members of the population are less likely to be chosen or cannot be chosen for the sample. (p. 214)	**error de cobertura** Se presenta cuando algunos miembros de la población tienen menos probabilidades de ser seleccionados o no pueden ser seleccionados para ser parte de la muestra. (pág. 214)
union The event "A or B" is called the union of events A and B. It consists of all outcomes that are in event A or event B, or both, and is denoted $A \cup B$. (p. 281)	**unión** El suceso "A o B" se llama unión de los sucesos A y B. Consiste en todos los resultados que están en el suceso A o el suceso B, o ambos, y se representa $A \cup B$. (pág. 281)

V

variable Any attribute that can take different values for different individuals. (p. 5)	**variable** Atributo que puede tomar diferentes valores para cada individuo. (pág. 5)
Venn diagram A diagram that consists of one or more circles surrounded by a rectangle. Each circle represents an event. The region inside the rectangle represents the sample space of the chance process. (p. 279)	**diagrama de Venn** Diagrama que consiste en uno o más círculos rodeados por un rectángulo. Cada círculo representa un suceso. La región interior del rectángulo representa el espacio muestral del proceso aleatorio. (pág. 279)
voluntary response sample A sample that consists of people who choose to be in the sample by responding to a general invitation. Voluntary response samples are sometimes called *self-selected samples*. (p. 189)	**muestra de respuesta voluntaria** Muestra que consiste en personas que eligen estar en la muestra al responder a una invitación general. Las muestras de respuesta voluntaria a veces se llaman *muestras auto seleccionadas*. (pág. 189)

Y

y intercept The predicted value of y when $x = 0$. (p. 133)	**intercepto y** valor predicho y cuando $x = 0$. (pág. 133)

Index

A page number in **boldface** indicates a definition; a figure is followed by f; and a table is followed by t

Aaron, Hank, 68
AARP, 269
abandoned children, 249
accuracy, 413
acetylsalicylic acid, 534
ACT scores, 85
activity trackers, 674
acupuncture, 241
addition rule
  general, 277–279, **279**, 294
  for mutually exclusive
    events, 272, **273**
ADHD, 221
age-related macular
    degeneration (AMD),
    225
AIDS, 301, 304
airliner crashes, 128–129
alarm clocks, 373
alligator attacks, 48, *76*
alternative hypothesis (H_a),
    501
  one-sided, **502**
  two-sided, **502**
Alzheimer's disease, 514, 527
Amabile, Teresa, 29
AMD. *See* age-related
    macular degeneration
American Crossword Puzzle
    Tournament, 489
American Heart Association,
    274
American Statistical
    Association, 6
Android phones, 642
anger, heart disease and, 96
animal testing, attitudes
    towards, 507
anonymity, 248
antibacterial creams, 542
anticlotting drugs, 186
anxiety, 226
AP® Calculus AB exam, grade
    distributions, 48
AP® Statistics exam, 156,
    159, 171, 273
  grade distributions, 48
Apgar, Virginia, 333
Apgar scores, 333, 334
  mean of, 341

standard deviation of, 343
Arbitron, 17
Archaeopteryx, 126
Arizona Daily Star, 234, 302
Aspaway, 534
aspirin, 534, 566, 653
association, **99**, 99f, 100,
    293, **637**
  causation and, 641
  chi-square tests for,
    645–651
  correlation and, 116
  degrees of freedom in test
    for, 647
  direction of, 108
  hypotheses for, 637
  negative, 107
  nonlinear, 160, 175
  positive, 107
asthma, 233
atmospheric gases, 456
Australia, 47, 103
authorship, 507
Avatar (film), 9
axes
  bar chart, 12
  boxplots, 69
  dotplots, 21
  histograms, 39, 40
  scatterplots, 106
  segmented bar chart, 97

Babe the Blue Ox, 514
background music, 619, 678
backpacking, 111
back-to-back stemplots, 33
Bangor Daily News, 514
bar chart, **11**, 12–13, 12f,
    673f
  distribution comparison
    with, 14
  making with applet, 16–17
  segmented, 97f
  side-by-side, 97f
barefoot running, 609
barley, 608
baseball fans, 475
basketball, 681
bathroom scales, 407
batteries

lifetime of, 412, 445
  quality control of, 399
  storing in freezer, 249
  types, 322
batting averages, 371,
    381–382, 391
beads, 403–404, 464, 465
beans, 112, 147, 244
beavers, 144–145, 661
beetles, 144–145, 661
Belgian Euro coin, 201, 527
Benford's law, 336, 345, 347,
    635
Bentley, Dierks, 45
Berkshire Hathaway, 347
Bernstein, M. H., 171
berries, 250
beta blockers, 579
bias, **189**, 412, 413f
  margin of error and, 461
  mean influenced by, 190
  minimizing in surveys, 217
  response, **216**, 217, 559
  sampling methods and,
    410
  in surveys, 214–217
  in surveys, checking for, 422
Big Bang Theory (television
    show), 276
biking accidents, 372
binomial distribution, **351**
  graphing, 352
  mean of, 356–358, 418
  normal approximation to,
    419
  shape and, 351–352
  shape of, 419–420
  standard deviation of,
    356–358, 418
  with technology, 353
binomial probabilities,
    350–351
  cumulative, 358–360
binomial random variable,
    349, 394
binomial setting, **348**, 349
bird beaks, 437
bird watching, 626
birth rates
  in Africa, 56, 85

literacy and, 676
birth weights, 438
blood pressure, 390, 391,
    482, 513, 541
  calcium and, 587
  fish oil and, 240
  memory and, 184
blood types, 275, 307,
    349–351, 356–358, 423
  personality and, 635
BMC. *See* bone mineral
    content
BMD. *See* bone mineral
    density
body mass index (BMI), 490
body temperature, 499, 542
  pulse rate and, 676
Bonds, Barry, 68, 69–70
bone density testing, 85, 476
bone health, 541
bone mineral content (BMC),
    482
bone mineral density (BMD),
    476
The Book of Odds, 267
Boston Red Sox, 83, 475
bottle filling, 542
boundary values, 508
box-and-whisker plot. *See*
    boxplot
boxplot, **69**, 86
  distribution comparison
    with, 72
  making and interpreting,
    69–71
  making with technology,
    73–74
braking distance, 163
breakfast, 283
breast cancer, 298–299
Brizendine, Louann, 250
Brooklyn half-marathon, 48
Bucharest Early Intervention
    Project, 249
Buffett, Warren, 347
bunions, 482
Bureau of Labor Statistics,
    430
Burger King, 55, 70, 551, 610
Busch, Kyle, 422

caffeine dependence, 598
caffeine experiment, 231,
    231f, 236f, 239f,
    585–586
  completely randomized
    design for, 235–236
  design for, 222–223,
    228–231
  determining statistical
    significance, 238–240
calcium, 503–504, 541
  blood pressure and, 587
California
  employment in, 463
  license plates, 310
  speed limits on highways,
    84
  traffic in, 462–463
  unemployment in, 463
call-in polls, 189
calls on hold, 451, 491
Canada
  communication methods
    in, 48
  languages in, 275, 431
  Lotto 6/49 game in, 307
  prostate treatment research
    in, 234
  wrist circumference in, 489
candy calories, 144, 158
car accidents, 680
car colors, 17–18, 20, 103,
    212, 363
  climate and, 631–632
car gas mileage, 27, 37
car tires, 600, 608
  nitrogen-filled, 602–605,
    603t
car value, 128, 201
carbon dioxide emissions, 45
  measuring, 44
career education, 226
Carolina Abecedarian Project,
    234
cases, 5
categorical data, 6
  displaying, 11–17
categorical variables, 5, 6
  analyzing with technology,
    101
  relationships between,
    96–101
  testing distribution of,
    620–625, 679
  testing relationship between,
    636–642, 679
causation
  association and, 641
  correlation and, 116
cause and effect, inference
    about, 245

Cavendish, Henry, 172
CBS News, 462, 560
CDC. See Centers for Disease
    Control and Prevention
cell phones, 284, 292,
    337–338. See also
    smartphones
  drivers using, 191
  teen ownership of, 401
  types of, 102, 285, 642
census, 183
Census At School, 25, 653
center, 25
  choosing measure of, 54, 63
  measuring, 49–55
  resistant measure of, 52
  of sampling distribution,
    409–413
Centers for Disease Control
    and Prevention (CDC),
    524
central limit theorem (CLT),
    439–442, 440, 444
Challenger (space shuttle),
    304
chance behavior, 263
charts, 16–17
  bar, 11
  pie, 11, 12–14
CHD. See coronary heart
    disease
cheating, reporting, 210
Chebyshev's inequality, 383
cherry trees, 95, 173
chess, 654
Chick-fil-A, 610
Chipotle, 390
chi-square distribution, 629,
    629f
chi-square test for
    association, 645–651
  conditions for, 645, 646
  performing, 648–650
  P-values, 646–648
chi-square test for goodness
    of fit, 631–633
  conditions for, 628
  performing, 631–632
  P-values, 628–631, 630f
chi-square test statistic,
    620–621, 624, 624–625
  for relationship between
    categorical variables,
    640–641
chocolate, 117, 186,
    198–199
cholesterol, 274, 373, 382,
    389, 415, 568, 569
  estimating, 437
church attendance, 431
circle graph. See pie chart

city gas mileage, 27
A Civil Action (film), 558
clementines, 456
Cleveland Indians, 218
climate change, 226
CLT. See central limit theorem
CNN, 475
cocoa, 227
coefficient of determination
    (r^2), 152, 153–154
  s and, 154
coin tosses, 262–263, 527
college costs, 345, 652
college debt, 408
college sports
  gambling and, 475
  gender and funding of, 608
colon cancer, 225
color words, 666
combinations ($_nC_k$), 317, 318
  calculating with technology,
    321
  computing, 318
  probability and, 320–321
combined sample proportion,
    562
common cold, 241
commute travel times, 44–45,
    46, 49, 212
  in North Carolina, 65
comparison in experiments,
    222–223
complement, 272, 280, 281,
    285
complement rule, 272
completely randomized
    design, 235, 236, 236f
conclusions
  experimental, 238–240
  making in significance test,
    504–505, 508, 509
conditional probability,
    286–291, 287, 325
  calculating, 287
  independence and, 289–290
  tree diagrams and, 296–299
confidence intervals,
    452–493
  building, 454
  critical values and, 466, 467
  decision making with, 455
  for difference between
    proportions, 552–557
  conditions, 554–555
  for difference between
    means, 574–576
  conditions, 573–574
  four-step process for, 471
  interpreting, 453–454
  for mean, 476–481,
    484–488, 493

conditions, 477, 484–485
  for mean difference,
    596–597
  conditions, 596
  for proportions, 464–468,
    470–474, 492–493
  conditions, 464–466
  for slope of least-squares
    regression line, 671–674
  conditions, 657–659
  two-sided tests and, 540
confidence level, 453, 458
  interpreting, 459
confidentiality, 248
confounding, 221, 222, 231
construction zones, 535
Consumer Reports
    (magazine), 75
Consumers Union, 602
continuous random variable,
    335, 365–371, 368f
  density curves and, 367
  finding probabilities for,
    366–367
  mean of, 368, 368f, 369
  median of, 368
control group, 222, 223,
    456
convenience sample, 188,
    188–189
cookie brands, 197–198
copyright, 300
coronary heart disease
    (CHD), 96, 638, 641
correlation (r), 113–117, 114,
    114f, 116f, 174
  association and, 116
  calculating, 121–123
  calculating with applet,
    126
  causation and, 116
  estimating and interpreting,
    114–116
  outlier and, 124–126
  properties of, 123–124
counting, 326
  probability and, 319–320,
    326
CPS. See Current Population
    Survey
creativity, 29, 590
cricket chirps, 128, 135
critical values, 466, 467, 479,
    482
crossbills, 145
cumulative binomial
    probabilities, 358–360
cumulative frequency, 79
cumulative relative frequency
    graph, 79, 79–81,
    80f, 82

Current Population Survey (CPS), 11, 245, 400, 405
curved relationships
  choosing models for, 167–168
  fitting models for, 160–169
customs inspection, 355, 363–364
cystic fibrosis, 267

Dallas Cowboys, 55, 74
data, 4
  categorical, 6, 11–17
  classifying, 5–6
  collecting, 182–185
  organizing, 86
  paired, 592–595
  quantitative, 6, 21–27, 30–35, 38–44, 67–74, 87
  summarizing, 7–8
  two-variable, 174
  unrepresentative, 189
data ethics, 247–248
deciles, 383
decision making
  with confidence intervals, 455
  significance test and, 508–512
Degree of Reading Power (DRP), 589
degrees of freedom (df), 531–533, **532**
  for two-sample t procedures, 574, 576–577
  in test for association, 647
  in test for goodness of fit, 629
density curve, **366**, 367
  chi-square distributions, 629, 629f
  normal distribution, 369–371
  symmetric, 368
  for t distribution, 532, 532f
Denver Broncos, 143–144
Department of Agriculture, 391
Department of Motor Vehicles (DMV), 527, 553–554
departures from patterns, 107
DEXA. See dual-energy X-ray absorptiometry
df. See degrees of freedom
diabetes, glucose measurement and, 146–147
diamonds, weights of, 148, 169
dice

fairness of, 632
possible outcomes, 270–271, 270f
Diet Coke, 157, 664
dipyridamole, 186
direction of association, 108
discrete probability distributions, 338–339
discrete random variable, **332–333**, 335
  analyzing, 338–344
  analyzing with technology, 344
  mean of, 339–341, **340**
  median of, 341
  probabilities for, 334–335
  standard deviation of, 341–343, **342**
dissolved oxygen (DO), 208–209, 489, 541
distractions, 232
  in driving, 236, 239–240, 569
distributions, **7**. See also specific distributions
  bar charts for comparing, 14
  boxplots for comparing, 72
  boxplots summarizing, 71
  Chebyshev's inequality and, 383
  chi-square, **629**
  describing and comparing, 24–26
  describing location in, 77–82, 87
  dotplots for comparing, 25
  histograms for comparing, 42
  median of, **49**, 49–51
  normal, 369–371, **370**, 384–390, 394
  probability, **332**, 333
  sampling, 400–405, **402**, 403–404
  shape of, 23–24
  skewed, **23**
  standard normal, 375–382, **378**, 378–380
  stemplots for comparing, 33
  symmetric, **23**
  t, 531–533, **532**, 532f
  types of, 403
  uniform, 367
DJIA. See Dow Jones Industrial Average
DMV. See Department of Motor Vehicles
DO. See dissolved oxygen
Dobbs, Lou, 192
Don't Break the Ice, 539

dotplot, **21**, 21–27, 21f, 22, 24f, 38f, 53f, 77, 402f
  distribution comparison with, 25
  making with applet, 26
double-blind, **224**, 225
double-counting problem, 278
Dow Jones Industrial Average (DJIA), 84, 172
drink cup lids, 391
drinking water, 512–513, 558
  trace metals in, 600–601
drive-thrus, 551, 607, 610
driving
  average distance, 218
  cellphones and, 191
  distractions in, 236, 239–240, 569
  texting and, 242, 570
driving school, 568
DRP. See Degree of Reading Power
drug tests, 261, 301, 324
dual-energy X-ray absorptiometry (DEXA), 85
Durant, Kevin, 58
DVDs, 559

earned run average (ERA), 37
Earnhardt, Dale, Jr., 265–266, 422
Earth, density of, 172
echinacea, 241
education
  attitudes toward school, 535
  career, 226
  home ownership and, 292
  homework habits, 542
  income and, 113
  level attained, 463
  reading and grades, 644
  teaching reading, 589
  vouchers for, 475
  among young adults, 275
eggs
  tortoise, 169
  water python, 102, 284, 292, 653
  weights of, 391
Egypt, 559
EIA. See enzyme immunoassay
El Dorado Community College, 345
election predictions, 415
Electoral College, 56, 74
election predictions, 415
electricity, conserving, 223
electronic communication, 483

elk, 354, 362
e-mail spam, 18
emergency response times, 513
emotional contagion study, 250
empirical rule, 377
Environmental Protection Agency (EPA), 22, 512–513
  car mileage ratings, 22, 27, 37, 66, 536, 599
enzyme immunoassay (EIA), 301
EPA. See Environmental Protection Agency
equally likely outcomes, 271
ERA. See earned run average
error. See also margin of error
  significance test types of, 509–511, 510f
  standard, 467, 480, 583, 661
ethics, data, 247–248
Euro coin, 527
Europe, population growth in, 170, 172
event, 270, **271**
  independent, **289**
  mutually exclusive, 273
  probability of, 271–272
exercise, memory and, 250
expected counts
  calculating, 622–623
  for categorical variable relationships, 638–640
  chi-square distribution and, 629
expected value, **340**
experimental units, **222**
experiments, **184**, 220–225, 254
  comparison, 222–223
  data ethics for, 247–248
  design of, 222
  inference for, 235–240
  other sources of variability, 231
  placebo effect, **223**, 223–225
  random assignment, 229–230
explanatory variable, **96**
  correlation and, 121, 123
  identifying, 105, 106
exponential models, **164**
  calculating, 165–167
  choosing, 167–168
  using, 166–167
express lanes, 599
extrapolation, **131**

Facebook, 219, 277, 279, 461
  age of users, 235
  emotional contagion study, 250
  number of friends on, 285
  usage of, 284, 292
factorials, **311**
  calculating, 314
false negatives, 261, 301
false positives, 261, 299, 301
fast-food drive-thrus, 551
Federal Reserve Bank of New York, 408
Federalist Papers, 188, 190, 196, 410
Federer, Roger, 300
The Female Brain (Brizendine), 250
fertility, acupuncture and, 241
finch beaks, 437
finger length, 291, 643
first quartile (Q_1), **60**
  in boxplot, 69
  percentile of, 78
First Trimester Screen, 306
first-serve percentage, 406
fish oil, 240, 589
five-number summary, **69**
flavonols, 227
flight capacity, 482
Florida Fish and Wildlife Conservation Commission, 76
flowers, 670
flu vaccines, 227, 528
fluoride varnish, 338
fonts, 534
food costs, 541
foot length, 121–124, 122f, 657, 659, 660
foster care, 249
free throws, 500, 501
french fries, 55, 56
frequency distribution, 7
frequency histogram, 41f
frequency table, **7**, 8, 79
Friday the 13th, 608
Froot Loops, 626
fruit flies, 625
fuel consumption, speed and, 110–111, 130, 148
fuel economy ratings, 22, 27, 37, 66, 536
fuel efficiency, 437
  in city driving, 146
  city *versus* highway, 599
  speed and, 148, 170

Gallup, 185, 206, 210, 424, 431, 453, 462, 475, 506, 513, 528, 556

gambling, college sports and, 475
gasoline prices, 36, 405, 535
gastric freezing, 653
gender
  college sports funding and, 608
  finger length and, 291, 643
  handedness and, 637, 640
  political party and, 284
  superpower preference and, 97, 97f, 99, 293, 642
  talking and, 588
  texting differences by, 579
general addition rule, 277–279, **279**, 294
General Motors, 186, 564
general multiplication rule, 294, **295**
  tree diagrams and, 296–297
General Social Survey, 251, 324, 507
genetic testing, 267
germination rate, 520
Gesell Adaptive Score, 142, 143f
gestational period, 146
Gibson, Bob, 170
Gladwell, Malcolm, 621
glucose measurement, 146–147
gluten-free diet, 355
GMAT. *See* Graduate Management Admission Test
Golden State Warriors, 457
golf drive distance, 385–386, 676
Gordon, Jeff, 265–266
GPA, 85, 119, 477
  IQ and, 111
  texting and, 157
Graduate Management Admission Test (GMAT), 276
grapefruit juice, 512
graphs: good and bad, 14–15
growth hormones, 681
GrubHub, 390
Guinness, 532
gummy bears, 642–643

H_0. *See* null hypothesis
H_a. *See* alternative hypothesis
Hamilton, Alexander, 188
hand sanitizer, 3, 85–86
hand washing, 217
Harley Davidson, 391, 430–431
head lice, 243

headaches, 233, 234
health care costs, 453–454, 459
health insurance, 219
hearing loss, 568–569
heart attacks, 566
heart disease
  anger and, 96, 641
  coronary, 96, 638, 641
heart rates, 35, 77, 482, 606, 676
  caffeine and, 585–586
height
  distribution for women of, 406–407
  foot length and, 121–124, 122f, 123f, 124f, 657, 659, 660
  smoking and, 405
helium, 242, 575–576
helmet sizes, 391
hepatitis, 250
Herd, Pamela, 226
hip dysplasia, 422, 423
Hispanic population, 588
histogram, **38**, 38f
  of binomial distribution, 352f
  distributions compared with, 42
  frequency, 41f
  making and interpreting, 38–41
  making with applet, 43–44
  of probability distribution, 338–339, 339f, 352f, 419f
HIV, 301
HIV testing, 304–305
hockey, 621–622, 623, 624, 628
Holocaust, 217
home computers, 519, 520
home insurance, 346
home ownership, 104, 282–283
  education and, 292
homework habits, 542, 635
household size, 275, 346
housing prices, 415, 576
hypotheses
  alternative, **501**
  for associations, 637
  for difference between means, 582–583
  for difference between proportions, 561, 562
  independence and, 638
  null, **501**, 502, 505
  one-sided alternative, **502**

for relationship between categorical variables, 637–638
for relationship between quantitative variables, 656–657
  stating, 501–503
  two-sided alternative, **502**

ice cream melting speed, 607
illegal immigration, 192
illegal music downloading, 300
immigration reform, 218
in vitro fertilization (IVF), 560
income
  education and, 113
  mortality and, 111–112
independence, 293, 305, 325–326
  checking for, 290
  conditional probability and, 289–290
  hypotheses using, 638
  mutually exclusive and, 306
independent events, **289**
  multiplication rule for, **303**, 303–306
India, teachers in, 226
Indiana University Bloomington, 336, 343
individual, **5**
inference
  about cause and effect, 245
  for experiments, 235–240
  about populations, 245
  for sampling, 197–198
  scope of, 245–247
informed consent, 248
insomnia, 542, 609
Instagram, usage of, 284
institutional review board, 248
insurance claims, 443
integrated circuit chips, 164
Intel Corporation, 164, 164f
Internet
  access to, 408
  broken links on, 308
  connection speeds, 599–600
  file transfer on, 371
  spoofing on, 536
Internet Movie Database, 9
Internet telephone calls, 241
interquartile range (IQR), 59, 60, **61**
  in boxplots, 69–71
  outlier identification with, 68
intersection, 280, **281**

Iowa Test of Basic Skills (ITBS), 365, 365f, 366, 370
distribution of scores, 375, 375f, 384, 384f, 387, 387f
interpreting vocabulary scores, 377
iPhones, 452, 642
iPod play list, 321
iPods, 559
IQ scores, 45, 56
distribution of, 391
GPA and, 111
reading and, 228
of twins, 592–594, 592f
IQR. See interquartile range
ITBS. See Iowa Test of Basic Skills
ivermectin, 243
IVF. See in vitro fertilization

James, Lebron, 58
Jay, John, 188
job stress, 525
Johnson, Jimmie, 265–266, 422

Kahne, Kasey, 265–266, 422
Kaiser Family Foundation, 297
Kellogg's Froot Loops, 626
Kenya, 181
keys, 30
Kickstarter, 431
kissing, 201, 505

lactose intolerance, 299
Ladies Professional Golf Association (LPGA), 156, 166–167
lamb's-quarter, 662
Landers, Ann, 189
landline phone service, 102, 284, 292, 634
Landon, Alf, 217
Lane, David, 434
Large Counts condition, 420, 428, 628
law of averages, 264
law of large numbers, 263, 264
layoffs, 233, 323
leafy greens, 225
lean body mass, 112, 127–128, 147–148
metabolic rate and, 677
LEAP trial, 569
least-squares regression line, 138, 143f, 657
calculating equation of, 138–140

calculating with summary statistics, 141
calculating with technology, 139–140
confidence intervals for slope of, 671–674
inference for slope of, 667–674
outlier and, 142–143
sum of squared residuals and, 153f
t interval for slope of, 671–672
t test for slope of, 667–669, 670–671
leaves, 30
left-handedness, 201–202, 422, 423, 506, 513, 639
left-skewed, 23, 369
"Letting Our Fingers Do the Talking" (New York Times), 76
lice, 243
license plates
California, 310
New Jersey, 315
life insurance, 168–169, 346
light bulbs, 513–514
lightning, 443–444
time of day of, 56
linear associations, 174–175
linear relationships, 113
Lipitor, 569
literacy, 558, 593, 595
birth rates and, 676
Literary Digest, 217
Lloyd, Carli, 27, 599
long run, 264
longleaf pine trees, 582
Los Angeles Times, 84
lotto games, 318, 319
low birth weight, 391
LPGA. See Ladies Professional Golf Association
lutein, 225
Lyme disease, 558–559

macular degeneration, 225
Madison, James, 188
magnets, 224, 225, 237, 237f
mail, 483
Major League Baseball, 83, 84, 323
batting averages in, 371, 381–382, 391
mean yearly salary of teams in, 57
player salaries, 171
malaria, 181
malathion, 243

mammograms, 298–299
manatees, 115
Mandarin oranges, 438
margin of error, 204, 454, 455
bias and, 461
determining sample size for estimating a mean, 486
determining sample size for estimating a proportion, 472–473
estimating, 204–209
factors affecting, 459–461
factors not accounted for by, 461
marijuana use, traffic accidents and, 148
Mars, Inc., 620, 628
matched pairs design, 597
McDonalds, 159, 551, 610
McIlroy, Rory, 385–386
mean, 49, 51, 51–53. See also population mean; sample mean
as balance point, 53
of binomial distribution, 356–358, 418
conditions for estimating, 477
conditions for estimating difference in, 573–574
conditions for test about difference in, 582–583
conditions for testing claim about, 529–530
confidence intervals for, 484–488, 493
of continuous random variable, 368, 368f, 369
in correlation calculation, 121, 126
of discrete random variable, 339–341, 340
estimating, 476–481
estimating difference between, 571–577
finding P-values for testing claim about, 531–533
in least-squares regression line calculation, 141
margin of error estimation for, 206–209
median compared with, 53–54
outliers and, 126
not resistant, 52
sampling distribution of, 432–436, 433
sampling distribution of difference between, 571–573

significance tests for, 536–540, 546
standard deviation and, 63
standardized test statistic for testing claim about, 530–531
testing claim about, 529–533
testing claim about difference between, 581–587
trimmed, 57
unbiased estimators and, 411
unknown σ problem in estimating, 477
z-scores and, 81
mean difference, 592, 594, 595
conditions for confidence interval or significance test about, 596, 602
confidence interval for, 596–597
inference about, 612
significance test for, 602–605
testing claim about, 601–607
median, 49, 49–51
of continuous random variable, 368
of discrete random variable, 341
with even number of values, 50–51
mean compared with, 53–54
with odd number of values, 50
percentile of, 78
resistant, 54
medical jargon, 488
Medicare, 269
meditation, 226
memory
blood pressure and, 184
exercise and, 250
music and, 588, 608
reaction times and, 146
Mendel, Gregor, 634
Mentos, 157, 664
mercury, 251
metabolic rate, 66, 112, 127–128, 147–148
lean body mass and, 677
microprocessors, 164, 164f
mixed nuts, 626, 654
M&M'S®, 272, 620, 621, 622, 628, 630f
models
exponential, 164, 165–168

fitting to curved
    relationships, 160–169
power, 172
probability, **270**, 271, 325,
    336
quadratic, **160**, 161–163,
    167–168
regression, assessing,
    149–156
Moore, Gordon, 164
Moore's law, 164, 165
mortality, income and,
    111–112
mosaic plots, 104
Motor Trend (magazine), 570
motorcycle braking distance,
    163
movie box-office receipts, 9
MSNBC, 470
multiple distractions, 232
multiple regression, 137
multiplication counting
    principle, 309, **310**
multiplication rules, 325–326
    general, 294–297, **295**
    for independent events, **303**,
    303–306
multitasking, 232
music
    background, 619, 678
    downloading, 300
    memory and, 588, 608
    productivity and, 609
    test performance and, 675
mussels, 469
mutually exclusive events,
    272, **273**, 277, 285
    independence and, 306

Nadal, Rafael, 373, 382
NASCAR, 265–266, 422,
    423
National Athletic Trainers
    Association, 568
National Basketball
    Association (NBA), 58,
    127, 457, 663, 681
National Campaign to
    Prevent Teen and
    Unplanned Pregnancy,
    527
National Center for Health
    Statistics, 159, 284, 292,
    308, 406
National Collegiate Athletic
    Association (NCAA),
    294–295, 475
National Football League
    quarterback ages, 160–161,
    160f, 161f, 172
    quarterback ratings, 158

National Health and
    Nutrition Examination
    Survey (NHANES), 249,
    578
National Hockey League
    (NHL), 621–622, 623,
    624, 628
National Household Travel
    Survey, 275
National Institutes of Health
    (NIH), 503, 559, 652
national insurance numbers,
    315
National Lightning Detection
    Network (NLDN), 443
National Opinion Research
    Center, 251
National Sleep Foundation,
    438, 471
native English speakers, 589
natural gas consumption,
    127, 136
NBA. *See* National Basketball
    Association
NCAA. *See* National
    Collegiate Athletic
    Association
$_nC_k$. *See* combinations
negative association, 107
netbooks, 125
*New England Journal of
    Medicine*, 117, 243, 306
New Jersey
    license plates, 315
    Pick Six lotto game in, 323
New Jersey Transit, 266, 352
New York City, 635
New York Times (newspaper),
    76, 282, 307, 462
New York Yankees, 83, 475
New Zealand, 103, 148
NHANES. *See* National
    Health and Nutrition
    Examination Survey
NHL. *See* National Hockey
    League
nicotine patches, 559, 653
Nielsen Mobile, 74
Nielsen ratings, 355, 522
NIH. *See* National Institutes
    of Health
nitrates, 26
NLDN. *See* National
    Lightning Detection
    Network
nonlinear associations, 160,
    175
non-native English speakers,
    589
non-normal population,
    sampling from, 439, 440

nonresponse, **215**, 216
normal body temperature,
    499, 542
normal distribution, 369–371,
    370, 370f, 394, 423,
    436f
    approximation with, 375
    calculations with
    technology, 389, 390
    finding values from
    probabilities, 387, 388
    probability calculation with,
    384–386
    68–95–99.7 rule, 375–377,
    378
    standard, 375–382, **378**,
    378–380, 379f
normal population, sampling
    from, 434–435
Normal/Large Sample
    condition, **440**, 484, 485,
    582, 596, 602
North America, population
    growth in, 170, 172
null hypothesis (H_0), **501**,
    502, 505
numerical summaries, 87
    computing with technology,
    64–65
nursing, 482
nuthatches, 626–627

Obama, Barack, 185, 206,
    455, 556
observational study, **184**,
    220–225, 254
observational units, 5
observed counts, 620
Okeeheelee County Park, 169
Old Faithful geyser, 134, 136
"1 in 6 wins" game, 4
$1.5 \times IQR$ rule, 68
O-negative blood, 307
one-sample t interval for a
    mean, 486
one-sample t interval for
    mean difference, 596
one-sample t test for a mean,
    536, 538
one-sample t test for mean
    difference, 602
one-sample z interval for p,
    471
one-sample z test for a
    proportion, 526
one-sided alternative
    hypothesis, **502**
online banking, 653
oranges, 438
Oreo cookies, 481
O-rings, 304

orphanages, 249
Ortiz, David, 293
osteoporosis, 476
outlier, 25
    boxplots and, 69
    correlation and, 124–126
    identifying, 68–69
    least-squares regression line
    and, 142–143
Outliers (Gladwell), 621
overall patterns, 107

pace of life, 171
pain, magnets and, 224, 225,
    237, 237f
paired data, **592**, 592–598,
    602–607, 612
    analyzing numerically,
    594–595
    analyzing with graphs,
    593–594
paper helicopters, 662
parabola, 160–161
Parade (magazine), 191
parallel dotplot, 592f, 596f
parameter, **401**
    plausible values, 453, 454
    statistic choice for, 411
Pareto, Vilfredo, 20
Pareto charts, 20
parking, 527
parking tickets, 322
Parkinson's disease, 241–242
Patrick, Danica, 265–266,
    422
Paul Bunyan, 514
peanuts, 435
    allergies to, 569
percentiles, **78**
    finding and interpreting, 78
    in normal distributions,
    380
perfect game (baseball), 307
permutations, **311**
    computing, 313–314
    denoting, 312
Pew Internet and American
    Life Project, 210, 295,
    459, 528, 554–555
Pew Research Center, 19,
    102, 284, 285, 401, 456,
    461, 475, 559
Philadelphia Phillies, 506
Phoenix, Arizona, high
    temperatures, 35
phone surveys, landlines and,
    634
phosphate, 66
Physicians' Health Study I,
    225, 238, 566
piano lessons, 600

Pick Six lotto game, 318, 319, 323
pictograph, 15
pie chart, **11**, 12–14
  making with applet, 16–17
Pierce, George Washington, 128
pierced ears, 286–287, 289
Piff, Paul, 444
pineapples, 505, 506, 534–535
placebo, **223**, 224
placebo effect, **223**, 223–225
point estimate, **454**, 455
political contributions, 233
political party, gender and, 284
popcorn, 587
Popular Science (magazine), 47
population, **183**
  inference about, 245
  standard deviation of, 62
population mean, 52
  conditions for estimating, 477, 484
  confidence intervals for, 487–488
  estimating, 206–209
  testing claim about, 536–540
population parameter, 401
population proportion
  conditions for estimating, 464–466
  confidence interval for, 471
  estimating, 204–206
  testing a claim about, 521–525
positive association, 107
postal workers, 540
potato chips, 423, 429, 520, 572
power model, 172
Pravachol, 569
precision, 413
predictions, using regression line for, 131
Premier League soccer, 145
preschool, 234, 568
president job performance, 455, 556
presidential elections, 56, 74
probability, 262, **263**, 325
  as area, 366, 367, 376, 377, 385, 387
  basic rules of, 270–274
  binomial, 350–351, 358–360
  calculating with normal distribution, 378–380, 384–386

central limit theorem and, 441–442
combinations and, 320–321
conditional, 286–291, 325
for continuous random variables, 366, 367
counting and, 319–320, 326
for discrete random variables, 334–335
of event, 271–272
finding values from, 387–390
general multiplication rule, 295
interpreting, 264
multiplication rule for independent events, 303–306
standard normal, 378
two-way tables and, 277–278
urban legends and, 601
Venn diagrams and, 282
z-scores and, 380–381
probability distribution, **332**, 333
  continuous, 366
  discrete, 338–339
  histogram and shape of, 338–339, 339f, 352f, 368f, 419f
probability involving sample count, 417–421
probability involving sample mean, 435–436, 441–442
probability involving sample proportion, 428–429
probability model, **270**, 325
  Benford's law, 336
  equally likely outcomes, 271
probability rules, 325
proportions. *See also* population proportion, sample proportion
  combined sample, 562
  comparing, 611
  conditions for estimating, 464–466
  conditions for inference about difference between, 554–555
  conditions for tests about, 515
  confidence intervals for, 464–468, 470–474, 492–493
  confidence intervals for difference between, 552–557
  estimating, 464–468

estimating difference between, 552–557
estimating margin of error for, 204–206
sampling distribution of difference between, 552–554
significance test for, 522, 546
testing claims about, 515–519
testing claims about difference between, 561–567
prostate disease treatment, 234
PROVE-IT Study, 569
Prudential Insurance Company, 28
PTC, 475
public transportation, 423
pulse rates
  body temperature and, 676
  caffeine and, 222, 228, 231, 231f, 235–236, 236f, 238–239, 239f, 585–586
  standing and sitting, 606
P-values, **503**, 504, 516–518, 537
  for chi-square test for association, 646–648
  for chi-square test for goodness of fit, 628–631, 630f
  for difference between means, 583–584
  for difference between proportions, 562–564
  inference and, 564, 565, 586
  for mean, 531f
  for mean difference, 602f
  for proportion, 516f
  for slope, 659, 667
python eggs, 102, 284, 292, 653

quadratic model, **160**, 161
  calculating, 161–162, 163
  choosing, 167–168
  using, 163
quantitative data, 6
  back-to-back stemplot for comparing, 33
  distribution of, 8, 24, 25
  dotplots of, 21–27
  histograms of, 38–44
  numerical summaries, 87
  shape of, 23–24
  stemplots of, 30–35
  summarizing, 67–74

quantitative variables, 5, 6.
  See also quantitative data
nonlinear associations, 160
  relationships between, 105–110
  testing relationship between, 656–661
quartiles, **60**

r. See correlation
*r*². *See* coefficient of determination
radio frequencies, 17
radio stations, 17, 18
  call signs, 315
radon, 542
random assignment, **229**
  methods, 229
  purpose, 230
random digit dialing, 362
random number generators, 194, 265
random sampling, **190**
random variable, **332**, 393
  binomial, **349**, 394
  continuous, **335**, 365–371, 368f, 378
  discrete, **332–333**, 335, 338–344
  normal, 369
"random walk" theory, 514
randomness, 325
  myths about, 264–265
range, **59**, 60
  bias in, 414
rap music, 557
Rapinoe, Megan, 27, 599
RDA. *See* recommended daily allowance
reaction times, 372, 581
  memory and, 146
reading
  chess and, 654
  fonts and, 534
  grades and, 644
  IQ scores and, 228
  teaching, 589
recommended daily allowance (RDA), 541
regression lines, 130–134, **131**, 132f, 138f. *See also* least-squares regression line
  extrapolation with, 131
  interpreting, 133
  residuals and, 132–133
regression models
  assessing, 149–156
  multiple regression, 137
  in test for slope, 657–659

relative frequency distribution, 7
relative frequency graph, cumulative, 79–81, 80f, 82
relative frequency table, 7, 8, 11
replication, 230
residual plots, **149,** 149f
    calculating, 154–156
    interpreting, 150–151
    patterns in, 150
residuals, **132,** 132f, 133, 138f
    least-squares regression line and, 138
    regression lines and, 132–133
    squared, 152, 153
    standard deviation of, **151**
    sum of squared, 152–153, 153f
resistant, **52**
response bias, **216,** 217, 559
response rates, 215
response variable, **96**
    distinguishing, 106
restaurant employees, 525
resting heart rates, 35
returns on stocks, 46
right-skewed, 23
rock-paper-scissors, 274, 276
roller coasters, 111, 136
Roosevelt, Franklin D., 217
Roper (polling organization), 217

s. *See* standard deviation of the residuals
sales tax, 514
salt, 390–391
sample, **183**
    convenience, **189,** 190
    mean of, 52
    random, **190**
    standard deviation of, 62
    voluntary response, **189,** 190
sample count, 417, 446
    center and variability, 418
    probabilities involving, 421
    sampling distribution of, 417–421
    shape of, 419
sample maximum, unbiased estimators and, 410
sample mean, 196f, 446
    center and variability, 433–434
    central limit theorem and, 441–442
    probabilities involving, 435–436, 441–442

sampling distribution of, 432–436
    shape of, 434–435
    standard error of, **480**
sample median, unbiased estimators and, 410
sample proportion, 446
    center and variability, 425–427
    combined, 562
    probabilities involving, 428
    sampling distribution of, 424–429, **425**
    shape of, 427–428
    standard error of, **467**
sample size, for desired margin of error, 472–473
sample space, **270**
sample statistic, 401
sample surveys, 421
sampling, 187–191, 253
    inference for, 197–198
    surveys and, 214–217
    variability in, 411–412
sampling distribution, 400–405, **402,** 402f, 410f–412f, 425f, 433f, 445–446, 516
    of difference between means, 571–573
    of difference between proportions, 552–554, 553f
    evaluating claims with, 403–404
    of sample count X. *See* sample count
    of sample mean. *See* sample mean
    of sample proportion. *See* sample proportion
sampling frame, 215
sampling variability, **196,** 197, 411–412
    decreasing, 412
San Antonio Spurs, 127
sandwich calories, 159
SAT prep, 218
SAT scores, 85, 105, 111, 112, 609, 610
    improving, 542
saunas, 650
scales
    bar charts, 12
    dotplots, 21
    histograms, 39, 40
    pictographs, 15
    scatterplots, 106
    segmented bar chart, 97
scatterplot, **105,** 105f

correlations of, 114, 114f, 116f
    curved relationships in, 160–161
    describing, 107–108
    making, 106–107
    making with technology, 109–110
    outliers and correlation in, 125, 126
    regression line and, 132f, 138f
SCD. *See* sudden cardiac death
school, attitudes toward, 535
Scientific American (magazine), 250
Scrabble, 269, 322, 430
SD. *See* standard deviation
seafood, 225
seagulls, 626, 627
seat-belt use, 218–219
Second European Stroke Prevention Study, 186, 653
security screening, 405
segmented bar charts, 96, **97,** 97f
    making, 97, 98
Serengeti National Park, 158
Shakespeare, William, 507
shape, 25
    binomial distributions and, 351–352, 419–420
    discrete probability distributions, 338–339
    sample mean, 434–435
    sample proportion, 427–428
shipping, 430
shoe sizes, 579
shopping, 421
side effects, 227, 527
side-by-side bar chart, 14, 97f
The Signal and the Noise (Silver), 392
significance level, **508,** 509
significance test, **500,** 500–505, 545
    for association between categorical variables, 636, 645
    decision making and, 508–512
    for difference between means, 584
    for difference between proportions, 561
    for distribution of a categorical variable, 620, 627

four-step process for, 521–523
    making conclusions in, 504–505, 508, 509
    for mean, 536–540, 546
    for mean difference, 602–605
    for proportion, 522, 546
    for slope, 655, 667
    two-sided, 523–525
Silver, Nate, 128–129, 215, 392
simple random sample (SRS), 193–199, **194,** 196f
simulation, **265,** 325
    in difference between means, 586f
    performing, 265
    statistical significance determination with, 238–240
    with technology, 265–266
singing competitions, 519
single-blind, **224**
sinus infections, 242
68–95–99.7 rule, 375–377, **378**
Skee Ball, 345
skewed distributions, **23**
skewed to the left, **23**
skewed to the right, **23**
skewness, 23, 76
Skittles, 430, 626
Skype, 241
skyscrapers, 20
sleep, 206–208, 207f, 534, 535, 542, 609
Sleep Awareness Week, 471
sleep deprivation, 590
sleep habits, 471
slope, **133**
    confidence intervals for, 671–674
    of least-squares regression line, 141, 667–669
    standard error of, 661
    t interval for, 671–672
    t test for, 667–669, 670–671
smartphones, 426, 642
    battery life, 452
    SAT study apps, 542
Smithsonian (magazine), 456
smoking, 186, 221
    height and, 405
    by high school students, 524–525
    quitting, 559, 564–565, 653–654
    in United Kingdom, 490
snowmobiles, 103

social media, 483, 554–555
social-networking sites, 295
sodium, 390–391
song lengths, 437, 443
sources of error, 462
Space Shuttle, 304
sparrowhawks, 119, 187
spatial-temporal reasoning, 600
speech patterns, 232–233
speed, braking distance and, 163
speed limits, 84
spell-checking software, 337
splitting stems, 32
spoofing, 536
squared residuals, 152, 153
squirrels, 383, 579
SRS. *See* simple random sample
SSHA. *See* Survey of Study Habits and Attitudes
St. Louis Cardinals, 170
stacked bar charts, 97
stair climber machine, 159
standard deviation (SD), 59, **61**, 62–63
   of binomial distribution, 356–358, 418
   calculating and interpreting, 62
   comparing, 66
   in correlation calculation, 121, 126
   of discrete random variable, 341–343, **342**
   estimating, 67
   in least-squares regression line calculation, 141
   resistance of, 63
   of sampling distribution, 418, 425, 433
   68–95–99.7 rule, 375–377, **378**
   variation from mean, 63
standard deviation of the residuals (*s*), **151**
   calculating, 154–156
   r^2 and, 154
standard error
   of difference between means, 583
   of difference between proportions, 555
   of sample mean, **480**
   of sample proportion, **467**
   of slope, 661
standard normal distribution, 375–382, **378**, 379f
   converting probabilities to z-scores in, 380–381

finding probabilities in, 378–380
standardized score (z-score), **81**, 82
   in correlation calculation, 121
   probabilities and, 380–381, 381f
standardized test statistic, 516–518, **517**
   for difference between means, 583–584
   for difference between proportions, 562–564
   for mean, 530–531
   for proportion, 516
   for slope, 659
statistic, **401**
   choice for parameters, 411
   sampling distribution of, 402
statistical inference, 197, 245
statistical problem-solving process, **5**, 253
   asking questions, 182–183
statistical questions, 182–183
statistical significance, **237**, 238, 504
   determining, 508–509
   determining with simulation, 238–240
   practical importance and, 540
statistics, **4**
stem-and-leaf plot. *See* stemplots
stemplots, **30**, 30f, 32f
   back-to-back, 33
   distributions compared with, 33
   interpreting, 32
   making, 30–32, 31
   making with applet, 34
stems, 30
   splitting, 32
Stern, Linda, 234
steroids, 568
stock market
   in January, 145
   random walk theory and, 514
stock returns, 46
strength training, 234, 336
strokes, 186, 653
Stroop, John Ridley, 666
Stroop effect, 666
student housing, 416
study design, bias from bad, 189
subjects, **222**

sudden cardiac death (SCD), 650
sugar, 670
sum of squared residuals, 152–153, 153f
summary statistics, 64–65, 81
   in least-squares regression line calculation, 141
   outliers and, 69
sunscreen, 605
Super Bowl
   ads, 149, 149f, 168
   Nielsen ratings, 355
superpowers, 10, 14, 19
   gender and preference of, 97, 97f, 99, 293, 642
Survey of Study Habits and Attitudes (SSHA), 76, 506, 535
surveys, 253
   bias in, 214–217
   checking for bias, 422
   minimizing bias in, 217
   sampling and, 214–217
Survivor (television show), 522, 526
Suttle, Kenwyn, 226
Sweetman, Andrew, 226
symmetric density curve, 368
symmetric distributions, **23**
symmetric probability distribution, 368, 368f

*t** critical values, 479–480, 482
t distribution, 531–533, **532**
   density curve for, 532, 532f
t interval, for slope of least-squares regression line, 671–672
t test, for slope of least-squares regression line, 667–669, 670–671
tai chi, 241–242
Tanzania, 158
tax reform, 96
tax returns, 635
tea, 405
Teen Relationship Survey, 475
telephone surveys, 102
television prices, 144
television ratings, 513, 522
test statistic
   for chi-square tests, 623, 640. *See also* standardized test statistic
testosterone supplements, 226
texting, 76, 662
   driving and, 242, 570
   gender differences in, 579
   GPA and, 157

third quartile (Q_3), **60**
   in boxplot, 69
   percentile of, 78
ticks, 558–559
tipping, 675
Titanic (ship), 98, 100, 104, 283, 292, 652
tobacco plants, 680
tomatoes, 437, 634–635
tortoise eggs, 169
trace metals, 600–601
traffic accidents, marijuana use and, 148
traffic lights, 303
transistors, 164, 164f, 165
Transportation Security Administration (TSA), 269, 322, 405, 430
travel times to work, 44–45, 46, 49, 212
   in North Carolina, 65
treatment, **222**
tree diagrams, **296**, 296–299, 296f, 297f
trimmed mean, 57
tropical flowers, 580
truck price, 131, 133, 150, 152–153
TSA. *See* Transportation Security Administration
tuition, 652
turkey, cooking temperature, 401
Twitter, 199, 277, 279, 459, 460, 528, 559
   usage of, 292
two-income couples, 113
two-sample *t* interval, 575
two-sample *t* test, 585, 605
two-sample *z* interval, 555
two-sample *z* test, 564
two-sided alternative hypothesis, **502**
two-sided tests, 523–525, 533, 539
   confidence intervals and, 540
two-variable data, 174
two-way tables, 98f, 277–279, 278f, 280, 637f
type B blood, 423
type O blood, 349, 350, 351, 356–357, 358
Type I error, **509**, 509–511
Type II error, **509**, 509–511

unbiased estimators, 409–411, **410**
undercoverage, **214**, 215
unemployment, 405
   in California, 463

uniform distribution, 24, 367
union, **281**
United Kingdom
  national insurance numbers
    in, 315
  smoking in, 490
United States
  commute travel time in,
    44–45
  high school student home
    computer access in, 519,
    520
  Hispanic population in, 588
  household size in, 275, 346
  median income in, 456
  men's heights in, 382
  presidential elections in,
    56, 74
University of Helsinki, 186
upper-tail probability p, 533
urban legends, 601
U.S. Agency for International
  Development, 211
U.S. Bureau of Labor
  Statistics, 430
U.S. Census Bureau, 11, 18,
  47, 245, 463, 635
  commute travel time data
    from, 65
  median income data from,
    456
U.S. Constitution, 188
U.S. Forest Service, 200, 490

U.S. Mint, 423
U.S. women's soccer team,
  27, 599
USA Today (newspaper), 282,
  307, 431

Vance, Cyrus, Jr., 251
variability, 25, 58f, 86, 413,
  413f
  choosing measures of, 63
  continuous random,
    365–371
  measuring, 58–65
  other sources in
    experiments, 231
  of sample mean, 433–434
  of sample proportion,
    425–427
  sampling, 196, 197,
    411–412
  sampling distribution,
    409–413, 418
variables, 5. *See also*
    categorical variables;
    quantitative variables
  binomial random, **349**,
    394
  continuous random, **335**,
    365–371, 368f, 378
  discrete random, **332–333**,
    334–335, 338–344
  distribution of, 7
  explanatory, **96**, 105, 106

random, **332**, 393
  response, **96**, 105, 106
variance, 357
  unbiased estimators and,
    410
vehicle colors, 17–18, 20,
    103, 212, 363
  climate and, 631–632
velvetleaf, 488
vending machines, 331,
    392–393
Venn diagrams, **279**, 279–
    282, 279f, 280f, 281f
video screen tension, 487
Vietnam, 47
vitamin C, 211, 489, 490
vitamin D, 233
The Voice (television show),
    235
voluntary response samples,
    189, 190
  nonresponse and, 216

Wade Tract Preserve, 582
Waggoner, Aaron, 163
Wambach, Abby, 27, 599
Washington Nationals, 84
water python eggs, 102, 284,
    292, 653
waterways, 541
The Weather Channel, 392
Web servers, 371
web-based polls, 189

Wechsler Adult Intelligence
  Scale, 391
weekend birthdays, 202
Wendy's, 610
whelks, 469
WHIP values, 170
whiskers, 69, 71
WHO. *See* World Health
  Organization
wildebeest, 158
Wilson, Timothy, 241
WNBA, 120
women's heights, 383, 436
  distribution of,
    406–407
Woods, Tiger, 69
World Health Organization
  (WHO), 476
World War II, 413
write-in polls, 189
writing style, 47

y intercept, **133**
  of least-squares regression
    line, 141
Yellowstone National Park,
    103, 134, 287

$z*$ critical value, 467, 479
zeaxanthin, 225
zip codes, 5
z-score. *See* standardized
    score

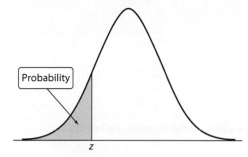

Probability

Table entry for *z* is the area under the standard normal curve to the left of *z*.

Table A	**Standard normal probabilities**									
z	.00	.01	.02	.03	.04	.05	.06	.07	.08	.09
−3.4	.0003	.0003	.0003	.0003	.0003	.0003	.0003	.0003	.0003	.0002
−3.3	.0005	.0005	.0005	.0004	.0004	.0004	.0004	.0004	.0004	.0003
−3.2	.0007	.0007	.0006	.0006	.0006	.0006	.0006	.0005	.0005	.0005
−3.1	.0010	.0009	.0009	.0009	.0008	.0008	.0008	.0008	.0007	.0007
−3.0	.0013	.0013	.0013	.0012	.0012	.0011	.0011	.0011	.0010	.0010
−2.9	.0019	.0018	.0018	.0017	.0016	.0016	.0015	.0015	.0014	.0014
−2.8	.0026	.0025	.0024	.0023	.0023	.0022	.0021	.0021	.0020	.0019
−2.7	.0035	.0034	.0033	.0032	.0031	.0030	.0029	.0028	.0027	.0026
−2.6	.0047	.0045	.0044	.0043	.0041	.0040	.0039	.0038	.0037	.0036
−2.5	.0062	.0060	.0059	.0057	.0055	.0054	.0052	.0051	.0049	.0048
−2.4	.0082	.0080	.0078	.0075	.0073	.0071	.0069	.0068	.0066	.0064
−2.3	.0107	.0104	.0102	.0099	.0096	.0094	.0091	.0089	.0087	.0084
−2.2	.0139	.0136	.0132	.0129	.0125	.0122	.0119	.0116	.0113	.0110
−2.1	.0179	.0174	.0170	.0166	.0162	.0158	.0154	.0150	.0146	.0143
−2.0	.0228	.0222	.0217	.0212	.0207	.0202	.0197	.0192	.0188	.0183
−1.9	.0287	.0281	.0274	.0268	.0262	.0256	.0250	.0244	.0239	.0233
−1.8	.0359	.0351	.0344	.0336	.0329	.0322	.0314	.0307	.0301	.0294
−1.7	.0446	.0436	.0427	.0418	.0409	.0401	.0392	.0384	.0375	.0367
−1.6	.0548	.0537	.0526	.0516	.0505	.0495	.0485	.0475	.0465	.0455
−1.5	.0668	.0655	.0643	.0630	.0618	.0606	.0594	.0582	.0571	.0559
−1.4	.0808	.0793	.0778	.0764	.0749	.0735	.0721	.0708	.0694	.0681
−1.3	.0968	.0951	.0934	.0918	.0901	.0885	.0869	.0853	.0838	.0823
−1.2	.1151	.1131	.1112	.1093	.1075	.1056	.1038	.1020	.1003	.0985
−1.1	.1357	.1335	.1314	.1292	.1271	.1251	.1230	.1210	.1190	.1170
−1.0	.1587	.1562	.1539	.1515	.1492	.1469	.1446	.1423	.1401	.1379
−0.9	.1841	.1814	.1788	.1762	.1736	.1711	.1685	.1660	.1635	.1611
−0.8	.2119	.2090	.2061	.2033	.2005	.1977	.1949	.1922	.1894	.1867
−0.7	.2420	.2389	.2358	.2327	.2296	.2266	.2236	.2206	.2177	.2148
−0.6	.2743	.2709	.2676	.2643	.2611	.2578	.2546	.2514	.2483	.2451
−0.5	.3085	.3050	.3015	.2981	.2946	.2912	.2877	.2843	.2810	.2776
−0.4	.3446	.3409	.3372	.3336	.3300	.3264	.3228	.3192	.3156	.3121
−0.3	.3821	.3783	.3745	.3707	.3669	.3632	.3594	.3557	.3520	.3483
−0.2	.4207	.4168	.4129	.4090	.4052	.4013	.3974	.3936	.3897	.3859
−0.1	.4602	.4562	.4522	.4483	.4443	.4404	.4364	.4325	.4286	.4247
−0.0	.5000	.4960	.4920	.4880	.4840	.4801	.4761	.4721	.4681	.4641

(Continued)

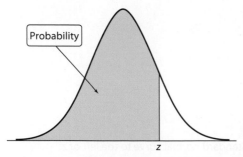

Table entry for *z* is the area under the standard normal curve to the left of *z*.

z	.00	.01	.02	.03	.04	.05	.06	.07	.08	.09
				Table A	**Standard normal probabilities (continued)**					
0.0	.5000	.5040	.5080	.5120	.5160	.5199	.5239	.5279	.5319	.5359
0.1	.5398	.5438	.5478	.5517	.5557	.5596	.5636	.5675	.5714	.5753
0.2	.5793	.5832	.5871	.5910	.5948	.5987	.6026	.6064	.6103	.6141
0.3	.6179	.6217	.6255	.6293	.6331	.6368	.6406	.6443	.6480	.6517
0.4	.6554	.6591	.6628	.6664	.6700	.6736	.6772	.6808	.6844	.6879
0.5	.6915	.6950	.6985	.7019	.7054	.7088	.7123	.7157	.7190	.7224
0.6	.7257	.7291	.7324	.7357	.7389	.7422	.7454	.7486	.7517	.7549
0.7	.7580	.7611	.7642	.7673	.7704	.7734	.7764	.7794	.7823	.7852
0.8	.7881	.7910	.7939	.7967	.7995	.8023	.8051	.8078	.8106	.8133
0.9	.8159	.8186	.8212	.8238	.8264	.8289	.8315	.8340	.8365	.8389
1.0	.8413	.8438	.8461	.8485	.8508	.8531	.8554	.8577	.8599	.8621
1.1	.8643	.8665	.8686	.8708	.8729	.8749	.8770	.8790	.8810	.8830
1.2	.8849	.8869	.8888	.8907	.8925	.8944	.8962	.8980	.8997	.9015
1.3	.9032	.9049	.9066	.9082	.9099	.9115	.9131	.9147	.9162	.9177
1.4	.9192	.9207	.9222	.9236	.9251	.9265	.9279	.9292	.9306	.9319
1.5	.9332	.9345	.9357	.9370	.9382	.9394	.9406	.9418	.9429	.9441
1.6	.9452	.9463	.9474	.9484	.9495	.9505	.9515	.9525	.9535	.9545
1.7	.9554	.9564	.9573	.9582	.9591	.9599	.9608	.9616	.9625	.9633
1.8	.9641	.9649	.9656	.9664	.9671	.9678	.9686	.9693	.9699	.9706
1.9	.9713	.9719	.9726	.9732	.9738	.9744	.9750	.9756	.9761	.9767
2.0	.9772	.9778	.9783	.9788	.9793	.9798	.9803	.9808	.9812	.9817
2.1	.9821	.9826	.9830	.9834	.9838	.9842	.9846	.9850	.9854	.9857
2.2	.9861	.9864	.9868	.9871	.9875	.9878	.9881	.9884	.9887	.9890
2.3	.9893	.9896	.9898	.9901	.9904	.9906	.9909	.9911	.9913	.9916
2.4	.9918	.9920	.9922	.9925	.9927	.9929	.9931	.9932	.9934	.9936
2.5	.9938	.9940	.9941	.9943	.9945	.9946	.9948	.9949	.9951	.9952
2.6	.9953	.9955	.9956	.9957	.9959	.9960	.9961	.9962	.9963	.9964
2.7	.9965	.9966	.9967	.9968	.9969	.9970	.9971	.9972	.9973	.9974
2.8	.9974	.9975	.9976	.9977	.9977	.9978	.9979	.9979	.9980	.9981
2.9	.9981	.9982	.9982	.9983	.9984	.9984	.9985	.9985	.9986	.9986
3.0	.9987	.9987	.9987	.9988	.9988	.9989	.9989	.9989	.9990	.9990
3.1	.9990	.9991	.9991	.9991	.9992	.9992	.9992	.9992	.9993	.9993
3.2	.9993	.9993	.9994	.9994	.9994	.9994	.9994	.9995	.9995	.9995
3.3	.9995	.9995	.9995	.9996	.9996	.9996	.9996	.9996	.9996	.9997
3.4	.9997	.9997	.9997	.9997	.9997	.9997	.9997	.9997	.9997	.9998

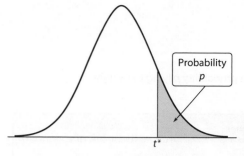

Table entry for p and C is the point t^* with probability p lying to its right and probability C lying between $-t^*$ and t^*.

						Tail probability p						
df	.25	.20	.15	.10	.05	.025	.02	.01	.005	.0025	.001	.0005
1	1.000	1.376	1.963	3.078	6.314	12.71	15.89	31.82	63.66	127.3	318.3	636.6
2	0.816	1.061	1.386	1.886	2.920	4.303	4.849	6.965	9.925	14.09	22.33	31.60
3	0.765	0.978	1.250	1.638	2.353	3.182	3.482	4.541	5.841	7.453	10.21	12.92
4	0.741	0.941	1.190	1.533	2.132	2.776	2.999	3.747	4.604	5.598	7.173	8.610
5	0.727	0.920	1.156	1.476	2.015	2.571	2.757	3.365	4.032	4.773	5.893	6.869
6	0.718	0.906	1.134	1.440	1.943	2.447	2.612	3.143	3.707	4.317	5.208	5.959
7	0.711	0.896	1.119	1.415	1.895	2.365	2.517	2.998	3.499	4.029	4.785	5.408
8	0.706	0.889	1.108	1.397	1.860	2.306	2.449	2.896	3.355	3.833	4.501	5.041
9	0.703	0.883	1.100	1.383	1.833	2.262	2.398	2.821	3.250	3.690	4.297	4.781
10	0.700	0.879	1.093	1.372	1.812	2.228	2.359	2.764	3.169	3.581	4.144	4.587
11	0.697	0.876	1.088	1.363	1.796	2.201	2.328	2.718	3.106	3.497	4.025	4.437
12	0.695	0.873	1.083	1.356	1.782	2.179	2.303	2.681	3.055	3.428	3.930	4.318
13	0.694	0.870	1.079	1.350	1.771	2.160	2.282	2.650	3.012	3.372	3.852	4.221
14	0.692	0.868	1.076	1.345	1.761	2.145	2.264	2.624	2.977	3.326	3.787	4.140
15	0.691	0.866	1.074	1.341	1.753	2.131	2.249	2.602	2.947	3.286	3.733	4.073
16	0.690	0.865	1.071	1.337	1.746	2.120	2.235	2.583	2.921	3.252	3.686	4.015
17	0.689	0.863	1.069	1.333	1.740	2.110	2.224	2.567	2.898	3.222	3.646	3.965
18	0.688	0.862	1.067	1.330	1.734	2.101	2.214	2.552	2.878	3.197	3.611	3.922
19	0.688	0.861	1.066	1.328	1.729	2.093	2.205	2.539	2.861	3.174	3.579	3.883
20	0.687	0.860	1.064	1.325	1.725	2.086	2.197	2.528	2.845	3.153	3.552	3.850
21	0.686	0.859	1.063	1.323	1.721	2.080	2.189	2.518	2.831	3.135	3.527	3.819
22	0.686	0.858	1.061	1.321	1.717	2.074	2.183	2.508	2.819	3.119	3.505	3.792
23	0.685	0.858	1.060	1.319	1.714	2.069	2.177	2.500	2.807	3.104	3.485	3.768
24	0.685	0.857	1.059	1.318	1.711	2.064	2.172	2.492	2.797	3.091	3.467	3.745
25	0.684	0.856	1.058	1.316	1.708	2.060	2.167	2.485	2.787	3.078	3.450	3.725
26	0.684	0.856	1.058	1.315	1.706	2.056	2.162	2.479	2.779	3.067	3.435	3.707
27	0.684	0.855	1.057	1.314	1.703	2.052	2.158	2.473	2.771	3.057	3.421	3.690
28	0.683	0.855	1.056	1.313	1.701	2.048	2.154	2.467	2.763	3.047	3.408	3.674
29	0.683	0.854	1.055	1.311	1.699	2.045	2.150	2.462	2.756	3.038	3.396	3.659
30	0.683	0.854	1.055	1.310	1.697	2.042	2.147	2.457	2.750	3.030	3.385	3.646
40	0.681	0.851	1.050	1.303	1.684	2.021	2.123	2.423	2.704	2.971	3.307	3.551
50	0.679	0.849	1.047	1.299	1.676	2.009	2.109	2.403	2.678	2.937	3.261	3.496
60	0.679	0.848	1.045	1.296	1.671	2.000	2.099	2.390	2.660	2.915	3.232	3.460
80	0.678	0.846	1.043	1.292	1.664	1.990	2.088	2.374	2.639	2.887	3.195	3.416
100	0.677	0.845	1.042	1.290	1.660	1.984	2.081	2.364	2.626	2.871	3.174	3.390
1000	0.675	0.842	1.037	1.282	1.646	1.962	2.056	2.330	2.581	2.813	3.098	3.300
z^*	0.674	0.841	1.036	1.282	1.645	1.960	2.054	2.326	2.576	2.807	3.091	3.291
	50%	60%	70%	80%	90%	95%	96%	98%	99%	99.5%	99.8%	99.9%

Table B *t* distribution critical values

Confidence level C

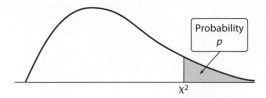

Table entry for p is the point χ^2 with probability p lying to its right.

					Table C Chi−square distribution critical values						
					Tail probability p						
df	.25	.20	.15	.10	.05	.025	.02	.01	.005	.0025	.001	.0005
1	1.32	1.64	2.07	2.71	3.84	5.02	5.41	6.63	7.88	9.14	10.83	12.12
2	2.77	3.22	3.79	4.61	5.99	7.38	7.82	9.21	10.60	11.98	13.82	15.20
3	4.11	4.64	5.32	6.25	7.81	9.35	9.84	11.34	12.84	14.32	16.27	17.73
4	5.39	5.99	6.74	7.78	9.49	11.14	11.67	13.28	14.86	16.42	18.47	20.00
5	6.63	7.29	8.12	9.24	11.07	12.83	13.39	15.09	16.75	18.39	20.51	22.11
6	7.84	8.56	9.45	10.64	12.59	14.45	15.03	16.81	18.55	20.25	22.46	24.10
7	9.04	9.80	10.75	12.02	14.07	16.01	16.62	18.48	20.28	22.04	24.32	26.02
8	10.22	11.03	12.03	13.36	15.51	17.53	18.17	20.09	21.95	23.77	26.12	27.87
9	11.39	12.24	13.29	14.68	16.92	19.02	19.68	21.67	23.59	25.46	27.88	29.67
10	12.55	13.44	14.53	15.99	18.31	20.48	21.16	23.21	25.19	27.11	29.59	31.42
11	13.70	14.63	15.77	17.28	19.68	21.92	22.62	24.72	26.76	28.73	31.26	33.14
12	14.85	15.81	16.99	18.55	21.03	23.34	24.05	26.22	28.30	30.32	32.91	34.82
13	15.98	16.98	18.20	19.81	22.36	24.74	25.47	27.69	29.82	31.88	34.53	36.48
14	17.12	18.15	19.41	21.06	23.68	26.12	26.87	29.14	31.32	33.43	36.12	38.11
15	18.25	19.31	20.60	22.31	25.00	27.49	28.26	30.58	32.80	34.95	37.70	39.72
16	19.37	20.47	21.79	23.54	26.30	28.85	29.63	32.00	34.27	36.46	39.25	41.31
17	20.49	21.61	22.98	24.77	27.59	30.19	31.00	33.41	35.72	37.95	40.79	42.88
18	21.60	22.76	24.16	25.99	28.87	31.53	32.35	34.81	37.16	39.42	42.31	44.43
19	22.72	23.90	25.33	27.20	30.14	32.85	33.69	36.19	38.58	40.88	43.82	45.97
20	23.83	25.04	26.50	28.41	31.41	34.17	35.02	37.57	40.00	42.34	45.31	47.50
21	24.93	26.17	27.66	29.62	32.67	35.48	36.34	38.93	41.40	43.78	46.80	49.01
22	26.04	27.30	28.82	30.81	33.92	36.78	37.66	40.29	42.80	45.20	48.27	50.51
23	27.14	28.43	29.98	32.01	35.17	38.08	38.97	41.64	44.18	46.62	49.73	52.00
24	28.24	29.55	31.13	33.20	36.42	39.36	40.27	42.98	45.56	48.03	51.18	53.48
25	29.34	30.68	32.28	34.38	37.65	40.65	41.57	44.31	46.93	49.44	52.62	54.95
26	30.43	31.79	33.43	35.56	38.89	41.92	42.86	45.64	48.29	50.83	54.05	56.41
27	31.53	32.91	34.57	36.74	40.11	43.19	44.14	46.96	49.64	52.22	55.48	57.86
28	32.62	34.03	35.71	37.92	41.34	44.46	45.42	48.28	50.99	53.59	56.89	59.30
29	33.71	35.14	36.85	39.09	42.56	45.72	46.69	49.59	52.34	54.97	58.30	60.73
30	34.80	36.25	37.99	40.26	43.77	46.98	47.96	50.89	53.67	56.33	59.70	62.16
40	45.62	47.27	49.24	51.81	55.76	59.34	60.44	63.69	66.77	69.70	73.40	76.09
50	56.33	58.16	60.35	63.17	67.50	71.42	72.61	76.15	79.49	82.66	86.66	89.56
60	66.98	68.97	71.34	74.40	79.08	83.30	84.58	88.38	91.95	95.34	99.61	102.7
80	88.13	90.41	93.11	96.58	101.9	106.6	108.1	112.3	116.3	120.1	124.8	128.3
100	109.1	111.7	114.7	118.5	124.3	129.6	131.1	135.8	140.2	144.3	149.4	153.2

Table D Random digits

Line								
101	19223	95034	05756	28713	96409	12531	42544	82853
102	73676	47150	99400	01927	27754	42648	82425	36290
103	45467	71709	77558	00095	32863	29485	82226	90056
104	52711	38889	93074	60227	40011	85848	48767	52573
105	95592	94007	69971	91481	60779	53791	17297	59335
106	68417	35013	15529	72765	85089	57067	50211	47487
107	82739	57890	20807	47511	81676	55300	94383	14893
108	60940	72024	17868	24943	61790	90656	87964	18883
109	36009	19365	15412	39638	85453	46816	83485	41979
110	38448	48789	18338	24697	39364	42006	76688	08708
111	81486	69487	60513	09297	00412	71238	27649	39950
112	59636	88804	04634	71197	19352	73089	84898	45785
113	62568	70206	40325	03699	71080	22553	11486	11776
114	45149	32992	75730	66280	03819	56202	02938	70915
115	61041	77684	94322	24709	73698	14526	31893	32592
116	14459	26056	31424	80371	65103	62253	50490	61181
117	38167	98532	62183	70632	23417	26185	41448	75532
118	73190	32533	04470	29669	84407	90785	65956	86382
119	95857	07118	87664	92099	58806	66979	98624	84826
120	35476	55972	39421	65850	04266	35435	43742	11937
121	71487	09984	29077	14863	61683	47052	62224	51025
122	13873	81598	95052	90908	73592	75186	87136	95761
123	54580	81507	27102	56027	55892	33063	41842	81868
124	71035	09001	43367	49497	72719	96758	27611	91596
125	96746	12149	37823	71868	18442	35119	62103	39244
126	96927	19931	36809	74192	77567	88741	48409	41903
127	43909	99477	25330	64359	40085	16925	85117	36071
128	15689	14227	06565	14374	13352	49367	81982	87209
129	36759	58984	68288	22913	18638	54303	00795	08727
130	69051	64817	87174	09517	84534	06489	87201	97245
131	05007	16632	81194	14873	04197	85576	45195	96565
132	68732	55259	84292	08796	43165	93739	31685	97150
133	45740	41807	65561	33302	07051	93623	18132	09547
134	27816	78416	18329	21337	35213	37741	04312	68508
135	66925	55658	39100	78458	11206	19876	87151	31260
136	08421	44753	77377	28744	75592	08563	79140	92454
137	53645	66812	61421	47836	12609	15373	98481	14592
138	66831	68908	40772	21558	47781	33586	79177	06928
139	55588	99404	70708	41098	43563	56934	48394	51719
140	12975	13258	13048	45144	72321	81940	00360	02428
141	96767	35964	23822	96012	94591	65194	50842	53372
142	72829	50232	97892	63408	77919	44575	24870	04178
143	88565	42628	17797	49376	61762	16953	88604	12724
144	62964	88145	83083	69453	46109	59505	69680	00900
145	19687	12633	57857	95806	09931	02150	43163	58636
146	37609	59057	66967	83401	60705	02384	90597	93600
147	54973	86278	88737	74351	47500	84552	19909	67181
148	00694	05977	19664	65441	20903	62371	22725	53340
149	71546	05233	53946	68743	72460	27601	45403	88692
150	07511	88915	41267	16853	84569	79367	32337	03316